Subjects:

Morning:
1) Computer programming
2) Dynamics
3) Mathematics
4) Mech. of Materials
5) Statics
6) Math. Modeling of Engr. Systems
7) Materials Science
8) Structure of Matter
? 9) Chemistry
? 10) Electrical circuits
? 11) Thermo.
? 12) Fluid mechanics
? 13) Engr. economics

Ser #	Appl. #	Seat
14	000-17-0226	F3

IMPORTANT NAMES, DATES AND ADDRESSES

This book belongs to

____

____

____

phone ______________________________

examination date: ______________ hours: __________

examination location: ______________________________

____

____

____

tape your cancelled check here

phone number of your registration board: ______________________________

address of your registration board: ______________________________

____

____

names of contacts at your registration board: ______________________________

____

____

tape your proof of mailing here

date you sent your application: ______________________________

registered/certified mail receipt number: ______________________________

date confirmation was received: ______________________________

names of examination proctors: ______________________________

____

____

tape dime here

booklet number: ______________ (A.M.) ______________ (P.M.)

problems you disagreed with on the examination

problem no.	reason

ENGINEER IN TRAINING REVIEW MANUAL

Sixth Edition

A complete review and reference for the E-I-T examination

Michael R. Lindeburg, P.E.
Director, Professional Engineering Institute

PROFESSIONAL PUBLICATIONS
San Carlos, CA 94070

In the *ENGINEERING REVIEW MANUAL SERIES*

Engineer-In-Training Review Manual
Quick Reference Cards for the E-I-T Exam
Mini-Exams for the E-I-T Exam
Civil Engineering Review Manual
Seismic Design for the Civil P.E. Exam
Timber Design for the Civil P.E. Exam
Structural Engineering Practice Problem Manual
Mechanical Engineering Review Manual
Electrical Engineering Review Manual
Chemical Engineering Review Manual
Chemical Engineering Practice Exam Set
Land Surveyor Reference Manual
Expanded Interest Tables
Engineering Law, Design Liability, and Professional Ethics

Distributed by: Professional Publications, Inc
Post Office Box 199
Department 77
San Carlos, CA 94070
(415) 593-9119

ENGINEER-IN-TRAINING REVIEW MANUAL
Sixth Edition

Printed in the United States of America

Library of Congress catalog number: 81-84850

ISBN: 0-932276-31-8

Professional Engineering Registration Program
Post Office Box 911, San Carlos, CA 94070

Current printing of this edition (last number) 12 11 10 9 8 7 6

Table of Contents

2 ENGINEERING ECONOMIC ANALYSIS

3 SYSTEMS OF UNITS

4 FLUID STATICS AND DYNAMICS

5 OPEN CHANNEL FLOW

6 THERMODYNAMICS

7 VAPOR, COMBUSTION, REFRIGERATION AND COMPRESSION CYCLES

8 CHEMISTRY

9 STATICS

10 MATERIALS SCIENCE

11 MECHANICS OF MATERIALS

12 DYNAMICS

13 DIRECT CURRENT ELECTRICITY

14 ALTERNATING CURRENT ELECTRICITY

15 PERIPHERAL SCIENCES

16 MODELING OF ENGINEERING SYSTEMS

17 GENERAL SYSTEMS MODELING

18 COMPUTER SCIENCE

Preface to the Sixth Edition

With this book, you will be able to prepare effectively for the national Engineer-In-Training (also known as the E-I-T, Fundamentals of Engineering, and Intern Engineer) certification examination. The contents, examples, and problems in this book have been extensively tested in actual classroom courses. The material in this book will prove to be highly valuable during your preparation.

An important requirement for passing the examination is adequate exposure to the types of questions typically found on the exam. Since this book is based on research covering more than 15 years of examinations, almost all types of problems likely to be encountered will be found in the practice problems at the end of each chapter.

This sixth edition is consistent with the new examination structure in use since November 1980. The introductory chapter fully describes this new structure, as well as the new scoring method introduced in November 1981.

This edition differs from previous editions in several ways: (1) Solutions to all practice problems are included. (2) A representative 4-hour examination is included for self-evaluation. (3) Extensive reorganization and revisions have been made to conform to the changed examination emphasis.

Michael R. Lindeburg, P.E.
San Carlos, CA
May 1982

Acknowledgments

page 4-17
(Figure 4.13, "Moody Friction Factor Chart") reprinted with permission from *Fluid Mechanics and Hydraulics,* 2nd ed., Ranald V. Giles, McGraw-Hill Book Company, 1962

page 4-23
(Figure 4.23, "Flow Coefficients For Orifice Plates") reprinted from *Elementary Fluid Mechanics,* 2nd ed., John K. Vennard, John Wiley & Sons, Inc., 1947

page 4-32
(Appendix A, "Properties of Water at Atmospheric Pressure") reprinted with permission from *Fluid Mechanics,* 6th ed., Victor L. Streeter and E. Benjamin Wylie, McGraw-Hill Book Company, 1975

page 4-32
(Appendix B, "Properties of Air at Atmospheric Pressure") reprinted with permission from *Fluid Mechanics and Hydraulics,* 2nd ed., Ranald V. Giles, McGraw-Hill Book Company, 1962

page 4-33
(Appendix C, "Properties of Liquids") reprinted from *Cameron Hydraulic Data,* 15th ed., C.R., Westaway and A.W. Loomis, Ingersoll-Rand, 1977

page 6-16
(Figures 6.11 and 6.12, "Compressibility Factors for Low Pressures" and "Compressibility Factors for High Pressures") reprinted with permission from *ASME Transactions,* 76, 1057, Edward F. Obert and L.C. Nelson, American Society of Mechanical Engineers, 1954

page 6-19
(Figure 6.14, "Psychrometric Chart") reproduced by permission of Carrier Corporation, copyright 1982 Carrier Corporation

page 6-30
(Appendix A, "Saturated Steam: Temperatures") reprinted from *Thermodynamic Properties of Steam,* Joseph H. Keenan and Frederick G. Keyes, John Wiley & Sons, Inc., 1937

page 6-31
(Appendix B, "Saturated Steam: Pressures") reprinted from *Thermodynamic Properties of Steam,* Joseph H. Keenan and Frederick G. Keyes, John Wiley & Sons, Inc., 1937

pages 6-32, 33
(Appendix C, "Superheated Steam") reprinted from *Thermodynamic Properties of Steam,* Joseph H. Keenan and Frederick G. Keyes, John Wiley & Sons, Inc., 1937

page 6-34
(Appendix D, "Compressed Water") reprinted from *Thermodynamic Properties of Steam,* Joseph H. Keenan and Frederick G. Keyes, John Wiley & Sons, Inc., 1937

page 6-35
(Appendix E, "Mollier Diagram") reprinted with permission from *Steam: Its Generation & Use,* Babcock & Wilcox, 1963

page 6-36
(Appendix F, "Air Table") reprinted from *Gas Tables: Thermodynamic Properties of Air,* Joseph H. Keenan and Joseph Kaye, John Wiley & Sons, Inc.

page 6-40
(Appendix I, "Saturated Freon-12") reprinted with permission from du Pont Bulletin T-12, copyright by du Pont 1956

page 6-41
(Appendix J, "Superheated Freon-12") reprinted with permission from du Pont Bulletin T-12, copyright by du Pont 1956

page 8-14
(Table 8.9, "Approximate Heats of Combustion") reprinted from *Steam: Its Generation & Use,* Babcock & Wilcox, 1963

page 10-16
(Table 10.12, "Approximate Recrystallization Temperatures") reprinted with permission, Guy/*Elements of Physical Metallurgy,* copyright 1960, Addison-Wesley, Reading, MA, p. 428

pages 11-23, 24
(Appendix B, "Mechanical Properties of Representative Metals") M.F. Spotts, *Design of Machine Elements,* 2nd ed., copyright 1953, renewed 1981, pp 478, 483. Reprinted by permission of Prentice-Hall, Inc., Englewood Cliffs, NJ

page 13-2
(Table 13.1, "Approximate Dielectric Constants") reprinted with permission from Angus/*Electrical Engineering Fundamentals,* copyright 1961, Addison-Wesley, Reading, MA, p. 471

page 13-9
(Table 13.5, "Approximate Temperature Coefficients of Resistance") reprinted with permission of Angus/*Electrical Engineering Fundamentals,* copyright 1961, Addison-Wesley, Reading, MA, p. 473

page 13-9
(Table 13.6, "Approximate Percent Conductivities") reprinted with permission from Angus/*Electrical Engineering Fundamentals,* copyright 1961, Addison-Wesley, Reading, MA, p. 472

pages 18-22, 23
(Appendix D, "Hexadecimal-Decimal Conversion Table") reprinted from IBM

page 19-7
(Table 19.6, "The Major Elementary Particles") reprinted from *Physics* (combined edition), 2nd printing, David Halliday and Robert Resnick, John Wiley & Sons, Inc., 1967

Introduction

Purpose of Registration

As an engineer, you may have to obtain your Professional Engineering license through procedures which have been established by the state in which you reside. These procedures are designed to protect the public by preventing unqualified individuals from legally practicing as engineers.

There are many reasons for wanting to become a Professional Engineer. Among them are the following:

- You may wish to become an independent consultant. By law, consulting engineers must be registered.

- Your company may require a professional engineering license as a requirement for employment or advancement.

- Your state may require registration as a Professional Engineer if you use the title *engineer*.

The Registration Procedure

The registration procedure is similar in most states. You will probably take two 8-hour written examinations. The first examination is the *Engineering-In-Training examination,* also known as the *Intern Engineer exam* and the *Fundamentals of Engineering exam.* The initials E-I-T, I.E., and F.E. are also used. The second examination is the *Professional Engineering (P.E.) exam* which is different from the E-I-T exam in format and content.

If you have significant experience in engineering, you may be allowed to skip the E-I-T examination. However, actual details of registration, experience requirements, minimum education levels, fees, and examination schedules vary from state to state. You should contact your state's Board of Registration for Professional Engineers.

Reciprocity Among States

All states except Illinois use the NCEE[1] E-I-T examination. If you take and pass the E-I-T examination in one state, your certificate will be honored by the other states using the NCEE examination. It will not be necessary to retake the E-I-T examination.

$1\frac{3}{4}$ min/q. $= \dfrac{240 \text{ min.}}{140 \text{ quest.}}$

The Examination Format

Since November 1980, the NCEE Engineer-In-Training examination has had the following format:

- There are two four-hour sessions separated by a one-hour lunch.
- The morning session has 140 multiple choice questions with five possible answers (lettered *A* to *E*) each. The subjects represented and the approximate distribution of questions are given in the following table:

Morning E-I-T Subjects $35\text{q}/\text{hr.}$

subject	number of questions
chemistry	10
computer programming	8
dynamics	13
engineering economics	6
electrical circuits	18
fluid mechanics	14
materials science	6
mathematics	12
mathematical modeling of engineering systems	8
mechanics of materials	13
statics	13
structure of matter	5
thermodynamics	14

$72 \times \frac{1}{2} = 36$

Questions in each subject are grouped together.

---

[1]The National Council of Engineering Examiners (NCEE) in Seneca, South Carolina, produces, distributes, and grades the national E-I-T examination used by all states except Illinois. It does not distribute applications to take the E-I-T examination.

$$\frac{240 \text{ min.}}{70 \text{ q.}} = 3\tfrac{1}{2} \text{ min/q.}$$

All questions requiring numerical calculations are stated twice—once in English units and once in *SI* units. Each statement will have its own selection of five answers, but the correct choice (e.g., answer *E*) will be the same for both statements. Either the English or the *SI* question may be answered.

Phone Numbers of State Boards of Registration

Alabama	(205) 261-5568
Alaska	(907) 465-2540
Arizona	(602) 255-4053
Arkansas	(501) 371-2517
California	(916) 445-5544
Colorado	(303) 866-2396
Connecticut	(203) 566-3386
Delaware	(302) 656-7311
District of Columbia	(202) 727-7454
Florida	(904) 488-9912
Georgia	(404) 656-3926
Guam	(671) 646-8643
Hawaii	(808) 548-4100
Idaho	(208) 334-3860
Illinois	(217) 785-0872
Indiana	(317) 232-1840
Iowa	(515) 281-6566
Kansas	(913) 296-3053
Kentucky	(502) 564-2680
Louisiana	(504) 568-8450
Maine	(207) 289-3236
Maryland	(301) 659-6322
Massachusetts	(617) 727-3088
Michigan	(517) 373-3880
Minnesota	(612) 296-2388
Mississippi	(601) 354-7241
Missouri	(314) 751-2334
Montana	(406) 444-3737
Nebraska	(402) 471-2021
Nevada	(702) 329-1955
New Hampshire	(603) 271-2219
New Jersey	(201) 648-2660
New Mexico	(505) 827-9940
New York	(518) 474-3833
North Carolina	(919) 781-9499
North Dakota	(701) 258-0786
Ohio	(614) 466-8948
Oklahoma	(405) 521-2874
Oregon	(503) 378-4180
Pennsylvania	(717) 783-7049
Puerto Rico	(809) 722-2121
Rhode Island	(401) 277-2565
South Carolina	(803) 758-2855
South Dakota	(605) 394-2510
Tennessee	(615) 741-3221
Texas	(512) 475-3141
Utah	(801) 530-6632
Vermont	(802) 828-2363
Virginia	(804) 257-8512
Virgin Islands	(809) 774-1301
Washington	(206) 753-6966
West Virginia	(304) 348-3554
Wisconsin	(608) 266-1397
Wyoming	(307) 777-6156

- Each problem in the morning session is worth one-half point. The total score possible in the morning is 70 points. No points are subtracted for incorrect answers.

- The afternoon session consists of 100 questions. However, you may work only 70 of these questions. Questions from the areas of mechanics, mathematics, electricity, and economics must be part of those 70. The questions are divided into nine subjects. Questions within each subject may be grouped into related problem sets containing between two and ten questions each.

Afternoon E-I-T Subjects

required subjects	number of questions
engineering mechanics	15 ✓
mathematics	15 ✓
electrical circuits	10
engineering economics	10 ✓

additional subjects (choose two)	
computer programming	10 ✓
electronics & electrical machinery	10
fluid mechanics	10
mechanics of materials	10 ✓
thermodynamics/heat transfer	10

60

Each afternoon question consists of a problem statement followed by multiple-choice questions. Five answer choices (lettered *A* to *E*) are given from which you must choose the best answer. Dual dimensioning is used to incorporate *SI* units.

- Each question in the afternoon is worth one point. The total score possible in the afternoon is 70 points. No points are subtracted for incorrect answers.

- The scores from both sessions are added together to determine your total score. Both sessions are given equal weight. It is not necessary to achieve any minimum score on either the morning or afternoon sessions. For the November 1980 examination, the mean score for examinees from all states was 72.4.

- The minimum passing score is established by each individual State Board of Registration for Professional Engineers, although most boards use the cut-off score recommended by NCEE. A modification of the *Angoff procedure* is used to develop the cut-off score. With this method, experts estimate the fraction of minimally-qualified engineers who will be able to answer each question correctly. The summation over all test questions of these estimated fractions becomes the cut-off score. Grading on a curve and the concept of a 70% passing rate are not employed.

For the November 1980 examination, the modified Angoff grading procedure resulted in the following passing rates.

November 1980 Passing Rates

category of examinee	per cent passing
ABET[2] accredited, 4-year engineering degrees	72
ABET accredited, 4-year technology degrees	42
non-accredited, 4-year engineering degrees	44
non-accredited, 4-year technology degrees	33
non-graduates	36
total, all states	64

- All grading is done by computer optical sensing. Choices to the multiple-choice questions must be recorded on score sheets with number 2 pencils. No credit is given for answers recorded in ink.
- The E-I-T examination is open-book. Most states do not limit the number and types of books you can bring into the exam. Loose-leaf papers and writing tablets are usually forbidden, although you may be able to bring in loose reference pages in a three-ring binder. References used in the afternoon session need not be the same as for the morning session.
- Any battery-powered, silent calculator may be used. There are no restrictions on programmable or preprogrammed calculators. Printers must not be used unless they are totally silent.
- You will not be permitted to share books, calculators, or any other items with other examinees.
- You will receive the results of your examination by mail. Allow 12–14 weeks for notification. Your score may or may not be revealed to you depending on your state's procedure.

Examination Dates

The national E-I-T examination is administered on the same weekend in all states. Each state decides independently whether to offer the E-I-T examination on Thursday, Friday, or Saturday of the examination period. The upcoming examination dates are given in the accompanying table.

[2]The Accreditation Board for Engineering and Technology (ABET) was known as the Engineers Council for Professional Development (ECPD) prior to 1980. ABET is the only organization which reviews and approves engineering degree programs in the United States. No engineering degree programs offered by universities outside of the United States are accredited by ABET.

National E-I-T Examination Dates

year	Spring exam	Fall exam
1982	April 15–17	October 28–30
1983	April 14–16	October 27–29
1984	April 12–14	October 25–27
1985	April 18–20	October 24–26
1986	April 10–12	October 23–25
1987	April 9–11	October 29–31
1988	April 14–16	October 27–29
1989	April 13–15	October 26–28

Preparing for the Examination

Since all 140 morning problems must be worked to get full credit, it will be necessary for you to prepare in all of the subjects. Do not make the mistake of studying only a few subjects in hopes of finding *enough* questions to pass with it. The number of questions and their relative proportions by subject are known in advance. You must work in all subjects to pass the E-I-T exam.

More important than strategy are fast recall and stamina. You must be able to quickly recall solution procedures, formulas, and important data — and, this sharpness must be maintained for eight hours.

In order to develop this recall and stamina, you should work the sample problems at the end of each chapter and compare your answers to the solutions provided. This will enable you to become familiar with problem types and solution methods. You will not have time in the exam to derive solution methods — you must know them instinctively.

It is imperative that you develop and adhere to a review outline and schedule. If you are not taking a classroom review course where the order of your preparation is determined by the lectures, you should use the accompanying *Outline of Subjects for Self-Study* to schedule your preparation.

It is unnecessary to bring a large quantity of books to the examination. This book, a dictionary, and one or two other references of your choice should be sufficient. The examination is very fast-paced. You will not have time to look up solution procedures, data, or equations with which you are not familiar. Although the examination is open-book, there is insufficient time to use books with which you are not thoroughly familiar.

To minimize time spent searching for often-used formulas and data, you should prepare a one-page summary of all important formulas and information in each subject area. You can then use these summaries during the examination instead of searching for the correct page in your book. If you do not want to prepare your own summary pages, you can obtain QUICK REFERENCE SUMMARY CARDS from Professional Publications, Inc. See the back cover for ordering information.

Outline of Subjects for Self-Study
(Subjects do not have to be studied in the order listed.)

subject	chapter	recommended number of weeks	date to be started	date to be completed	check when complete
mathematics	1	2			☐
engineering economy	2	1			☐
systems of units	3	0			☐
fluid statics and dynamics	4	2			☐
open channel flow	5	1			☐
thermodynamics	6	1			☐
power and refrigeration cycles	7	1			☐
chemistry	8	1			☐
statics	9	1			☐
materials science	10	1			☐
mechanics of materials	11	1			☐
dynamics	12	1			☐
DC electricity	13	1			☐
AC electricity	14	1			☐
peripheral sciences	15	1			☐
modeling of engineering systems	16	1			☐
general systems modeling	17	1			☐
computer science	18	1			☐
atomic and nuclear theory	19	1			☐
postscripts	20	0			☐
final exam	22	1			☐

What to Do Before the Exam

Here are some suggestions for making your examination experience comfortable and successful.

- Keep a copy of your examination application. Send the original application by certified mail and request a receipt of delivery. Tape your delivery receipt in the space indicated on the first page of this book.
- Visit the exam site the day before your examination. This is especially important if you are not familiar with the area. Find the examination room, the parking area, and the rest rooms.
- Plan on arriving at least 30 minutes before the examination starts. This will assure you a convenient parking place and adequate time for site, room, and seating changes.
- If you live a considerable distance from the examination site, consider getting a hotel room in which to spend the night before.
- Take off the day before the examination to relax. Don't cram the last night. Rather, get a good night's sleep.
- Be prepared to find that the examination room is not ready at the designated time. Bring an interesting novel or magazine to read in the interim and at lunch.
- If you make arrangements for babysitters or transportation, allow for a delayed completion.
- Prepare your examination kit the day before. Here is a checklist of items to bring with you to the examination.

  ☐ this and your other reference books
  ☐ course notes in a binder
  ☐ calculator and a spare
  ☐ spare calculator batteries or battery pack
  ☐ battery charger and 20′ extension cord
  ☐ chair cushions. A large, thick bath mat works well.
  ☐ earplugs
  ☐ desk expander. If you are taking the exam in theater chairs with tiny, fold-up writing surfaces, you should bring a long, wide board to place across the arm rests.
  ☐ a cardboard box cut to fit your references
  ☐ twist-to-advance pencils
  ☐ extra leads
  ☐ machine-scoring pencils
  ☐ snacks such as raisins, nuts, or trail mix
  ☐ thermos filled with hot chocolate
  ☐ a light lunch
  ☐ a collection of graph paper

- ☐ scissors, stapler, and staple puller
- ☐ construction paper for stopping drafts and sunlight
- ☐ scotch and masking tape
- ☐ sunglasses
- ☐ extra prescription glasses if you wear them
- ☐ aspirin
- ☐ travel pack of Kleenex
- ☐ Webster's dictionary
- ☐ dictionary of scientific terms
- ☐ $2 in change
- ☐ a light comfortable sweater
- ☐ comfortable shoes or slippers for the exam room
- ☐ raincoat, boots, gloves, hat, and umbrella
- ☐ local street maps
- ☐ photographic identification
- ☐ letter admitting you to the examination
- ☐ your copy of the original application and delivery receipt
- ☐ note to the parking patrol for your windshield
- ☐ pad of scratch paper with holes for 3-ring binder
- ☐ straight-edge, ruler, compass, and protractor
- ☐ battery-powered desk lamp
- ☐ watch
- ☐ wire coat hanger

What to Do in the Exam

- Do not spend more than 2 minutes per morning question. If you have not finished a question by then, make a note of it and continue on.
- Read through all of the optional afternoon questions before starting. You will have sufficient time to do so. It is very likely you will find an easy problem in the areas that you did not intend to work.
- Stop five minutes before the end of each session and guess at all remaining unsolved problems. Do not work up to the end. You will be lucky with about 20% of your guesses. These points will more than make up for the few points you would have earned by working during the last five minutes.
- Record the details of any problem for which you cannot find a correct response. Your being able to point out an error may later give you the margin needed to pass.
- Make sure all of your responses on the answer sheet are dark and that they completely fill the 'bubbles.'

A Personal Note to Instructors Using This Book

This book started as a series of handouts for an E-I-T review course taught at a local community college. It was originally intended as a reference for all of the long formulas, illustrations, and tables of data that I did not have time to put on the chalkboard. Starting with the 4th edition, however, the chapters were rewritten to more closely parallel the organization and contents of my lectures.

The written words in this 6th edition so closely parallel my verbal presentations that I must surely now be accused of giving lectures by reading from the book. However, this embarrassment for me can work to your advantage as an instructor.

If you are unfamiliar with the E-I-T examination, you can use the chapters in this book as guides to preparing your lectures. You should emphasize the subjects in each chapter and avoid subjects omitted. You can feel confident that subjects omitted from this book are relatively absent from the E-I-T exam.

It has always been my goal to over-prepare my students. For that reason, the examples and practice problems are more difficult than actual examination problems. Also, it is more efficient to cover several procedural steps in one problem than to ask simple *one-liners* having multiple choice answers. You might want to mention this fact when your students have trouble solving the practice[3] problems in 2 minutes each.

The practice examination, however, is completely representative of the actual examination. The types of problems, format, emphasis, and degree of difficulty all match that which your students will experience in the actual E-I-T examination.

There are between 30 and 50 practice problems for each major examination subject. Every problem is assigned in my review courses. This requires between five and ten hours of preparation on the part of the students each week.

If you assign ten hours of practice problems and a student can only put in eight hours of preparation, then that student worked to capacity. *Capacity assignment* is the goal of my courses. After the actual E-I-T examination, your students will honestly say they could not have prepared any more than they did.

Assignments are not individually graded. The students have the solutions to all practice problems in the back of this book. However, each student must turn in the completed set of problems for credit each week. Any special problems or questions written on the assignment are answered at that time.

I have found that a 14-week format works well for an E-I-T review course. Each week contains one three-hour lecture with a short intermediate break. The course format is shown in the accompanying table.

[3] I prefer the word *practice* to *homework*.

E-I-T Review Course Outline

meeting	subject covered
1	introduction to the examination
2	mathematics
3	mathematics
4	engineering economy
5	thermodynamics
6	power and refrigeration cycles
7	fluids and hydraulic machines
8	chemistry
9	DC electricity
10	AC electricity
11	materials science
12	mechanics of materials
13	statics
14	dynamics

I have tried to put the more difficult subjects toward the beginning of the course. This gives students time to learn the subjects which are conceptually more difficult.

Some examination subjects are not covered at all by lecture. These omissions are intentional— they are not the result of scheduling limitations. For example, I have found that very few people try to learn systems modeling. Unless you have two quarters in which to teach your E-I-T review course, your students' time can be better spent covering other subjects.

Computer science is another subject absent from the review schedule. I have found it difficult to teach FORTRAN to non-programming students. Those that know programming do not need the lecture. Those that do not know programming or FORTRAN cannot learn it in one lecture. Few seem interested in analog computers.

Other chapters cover subjects (e.g., atomic theory and peripheral sciences) which contribute only a small number of problems to the examination.

All of the skipped chapters are assigned as floating assignments to be made up on the students' free time.

I do not use the practice examination as an in-class exercise. Since the review course usually ends only a few days prior to the real E-I-T examination, I cannot see making students sit for four hours in the late evening. The practice examination is assigned as a take-home assignment.

If the practice examination is to be used as an indication of preparedness, your students should not even look at it prior to taking it. Looking at the examination or otherwise using it to direct their study will produce unwarranted specialization in subjects contained in the practice examination.

There are many ways to organize an E-I-T review course depending on the amount of time available for the instructor, students, and course. However, all good organizations have the same result: the students complain about the work load during the course . . . and then they breeze through the examination after the course.

MATH

1 Mathematics and Related Subjects

1. Introduction

The majority of engineering problems on the Engineer-In-Training examination requires only algebra and simple trigonometry. Due to the limited time, difficult and time-consuming mathematical techniques are seldom required during the solution of engineering problems.

However, the examination does have sections containing problems from the various fields of pure mathematics. These problems are limitless in possibility, but they are almost always oriented to the higher-level subjects such as linear algebra, calculus, differential equations, probability theory, etc.

This chapter is designed as a reference for the most frequently required mathematical techniques. Since the majority of engineering problems requires only algebra and trigonometry, most of this chapter will be useful only for the pure mathematical problems.

2. Symbols Used in This Book

Many symbols, letters, and Greek characters are used to represent variables in the formulas used throughout this book. These symbols and characters are defined in the nomenclature section of each chapter. However, some of the symbols which are used in this book as operators are listed below.

Symbol	Name	Use	Example
Σ	sigma	series addition	$\sum_{i=1}^{3} x_i = x_1 + x_2 + x_3$
Π	pi	series multiplication	$\prod_{i=1}^{3} x_i = x_1x_2x_3$
Δ	delta	change in quantity	$\Delta h = h_2 - h_1$
$\overline{\ }$	over bar	average value	$\overline{x}$
$\cdot$	over dot	per unit time	$\dot{Q}$ = quantity flowing per second
!	factorial		$x! = x(x-1)(x-2) \cdots (2)(1)$
$\lvert\ \rvert$	absolute value		$\lvert -3 \rvert = +3$
$\approx$	approximately equal to		$x \approx 1.5$
$\propto$	proportional to		$x \propto y$
∞	infinity		$x \to \infty$
log	base 10 logarithm		log(5.74)
ln	natural logarithm		ln(5.74)
EE	scientific notation		EE − 4
exp	exponential power		$\exp(x) = e^x$

3. *Mensuration*

Nomenclature

A	total surface area
d	distance
h	height
p	perimeter
r	radius
s	side (edge) length, arc length
V	volume
θ	vertex angle, in radians
ø	central angle, in radians

Circle

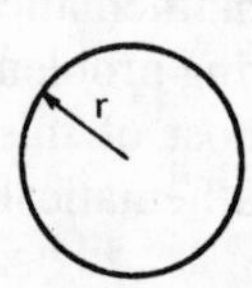

$$p = 2\pi r \qquad 1.1$$

$$A = \pi r^2 = \frac{p^2}{4\pi} \qquad 1.2$$

Circular Segment

$$A \approx \tfrac{1}{2} r^2 (\phi - \sin\phi) \qquad 1.3$$

$$\phi = s/r = 2\left(\arccos\frac{r-d}{r}\right) \qquad 1.4$$

Triangle

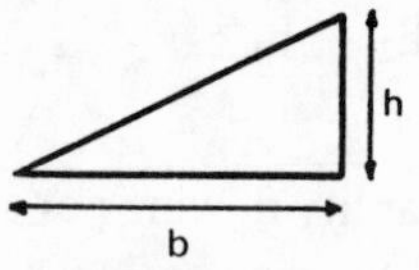

$$A = \tfrac{1}{2} bh \qquad 1.5$$

Parabola

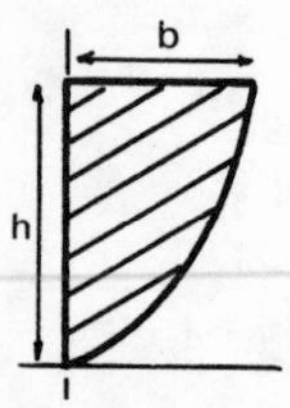

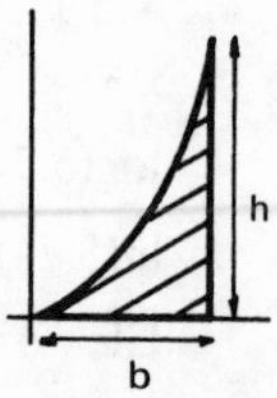

$$A = \frac{2bh}{3} \qquad 1.6$$

$$A = \tfrac{1}{3} bh \qquad 1.7$$

Circular Sector

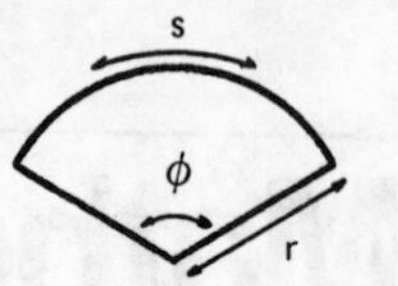

$$A = \tfrac{1}{2}\phi r^2 = \tfrac{1}{2} sr \qquad 1.8$$

$$\phi = s/r \qquad 1.9$$

Ellipse

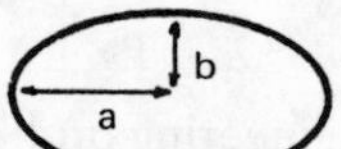

$$A = \pi ab \qquad 1.10$$

$$p = 2\pi\sqrt{\tfrac{1}{2}(a^2+b^2)} \qquad 1.11$$

Trapezoid

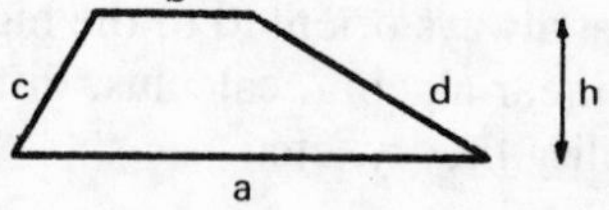

$$p = a + b + c + d \qquad 1.12$$

$$A = \tfrac{1}{2} h(a+b) \qquad 1.13$$

The trapezoid is isosceles if c = d

Parallelogram

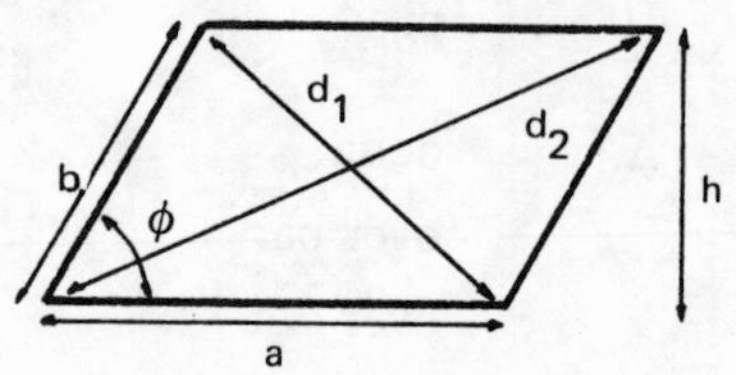

$$p = 2(a+b) \qquad 1.14$$

$$d_1 = \sqrt{a^2 + b^2 - 2ab(\cos\phi)} \qquad 1.15$$

$$d_2 = \sqrt{a^2 + b^2 + 2ab(\cos\phi)} \qquad 1.16$$

$$d_1^2 + d_2^2 = 2(a^2 + b^2) \qquad 1.17$$

$$A = ah = ab(\sin\phi) \qquad 1.18$$

If a = b, the parallelogram is a rhombus.

Regular Polygon
(n equal sides)

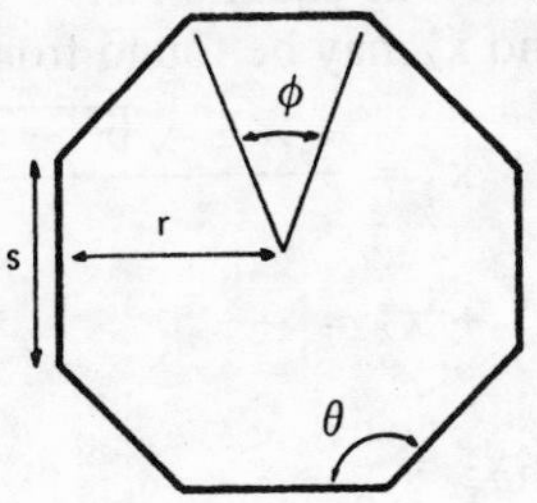

$$\phi = 2\pi/n \qquad 1.19$$

$$\theta = \frac{\pi(n-2)}{n} \qquad 1.20$$

$$p = ns \qquad 1.21$$

$$s = 2r(\tan(\tfrac{1}{2}\phi)) \qquad 1.22$$

$$A = \tfrac{1}{2}nsr \qquad 1.23$$

Table 1.1 Polygons

Number of Sides	Name of Polygon
3	triangle
4	rectangle
5	pentagon
6	hexagon
7	heptagon
8	octagon
9	nonagon
10	decagon

Sphere

$$V = \frac{4\pi r^3}{3} \qquad 1.24$$

$$A = 4\pi r^2 \qquad 1.25$$

Right Circular Cone

$$V = \frac{\pi r^2 h}{3} \qquad 1.26$$

$$A = \pi r\sqrt{r^2 + h^2} \qquad 1.27$$

(does not include base area)

Right Circular Cylinder

$$V = \pi h r^2 \qquad 1.28$$

$$A = 2\pi r h \qquad 1.29$$

(does not include end area)

Paraboloid of Revolution

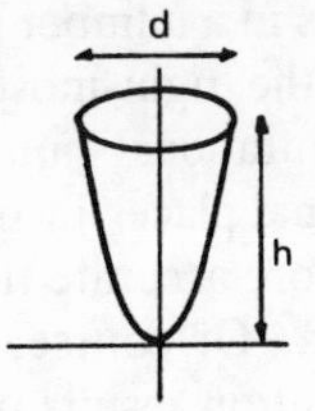

$$V = \frac{\pi h d^2}{8} \qquad 1.30$$

Regular Polyhedron

$$r = \frac{3V}{A} \qquad 1.31$$

Table 1.2 Polyhedrons

Number of Faces	Form of Faces	Total Surface Area	Volume
4	equilateral triangle	1.7321 s^2	0.1179 s^3
6	square	6.0000 s^2	1.0000 s^3
8	equilateral triangle	3.4641 s^2	0.4714 s^3
12	regular pentagon	20.6457 s^2	7.6631 s^3
20	equilateral triangle	8.6603 s^2	2.1817 s^3

Example 1.1

What is the hydraulic radius of a 6″ pipe filled to a depth of 2″?

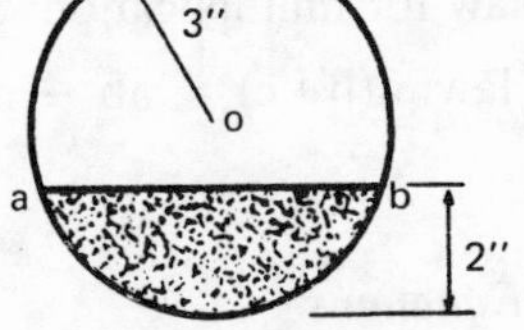

The hydraulic radius is defined as

$$r_h = \frac{\text{area in flow}}{\text{length of wetted perimeter}} = \frac{A}{s}$$

Points o, a, and b may be used to find the central angle of the circular segment.

$$\tfrac{1}{2}(\text{angle aob}) = \arccos(1/3) = 70.53°$$

$$\phi = 141.06° = 2.46 \text{ radians}$$

Then,

$$A \approx \tfrac{1}{2}(3)^2(2.46 - .63) = 8.235$$

$$s = (3)(2.46) = 7.38$$

$$r_h = \frac{8.235}{7.38} = 1.12$$

MATH

4. Significant Digits

The significant digits in a number include the left-most, non-zero digits to the right-most digit written. Final answers from computations should be rounded off to the number of decimal places justified by the data. The answer can be no more accurate than the least-accurate number in the data. Of course, rounding should be done on final calculation results only. It should not be done on interim results.

number as written	number of significant digits	implied range
341	3	340.5 to 341.5
34.1	3	34.05 to 34.15
.00341	3	.003405 to .003415
3410.	4	3409.5 to 3410.5
341 EE7	3	340.5 EE7 to 341.5 EE7
3.41 EE−2	3	3.405 EE−2 to 3.415 EE−2

5. Algebra

Algebra provides the rules which allow complex mathematical relationships to be expanded or condensed. Algebraic laws may be applied to complex numbers, variables, and numbers. The general rules for changing the form of a mathematical relationship are given here:

Commutative law for addition: $a + b = b + a$

Commutative law for multiplication: $ab = ba$

Associative law for addition: $a+(b+c) = (a+b)+c$

Associative law for multiplication: $a(bc) = (ab)c$

Distributive law: $a(b+c) = ab + ac$

A. Polynomial Equations

1. Standard Forms

$$(a+b)(a-b) = a^2 - b^2 \qquad 1.32$$

$$(a\pm b)^2 = a^2 \pm 2ab + b^2 \qquad 1.33$$

$$(a\pm b)^3 = a^3 \pm 3a^2b + 3ab^2 \pm b^3 \qquad 1.34$$

$$(a^3\pm b^3) = (a\pm b)(a^2 \mp ab + b^2) \qquad 1.35$$

$$(a^n+b^n) = (a+b)(a^{n-1} - a^{n-2}b + \cdots + b^{n-1}) \text{ (for n odd)} \qquad 1.36$$

$$(a^n-b^n) = (a-b)(a^{n-1} + a^{n-2}b + \cdots + b^{n-1}) \qquad 1.37$$

2. Quadratic Equations

Given a quadratic equation $ax^2 + bx + c = 0$, the roots x_1^* and x_2^* may be found from

$$x_1^*, x_2^* = \frac{-b \pm \sqrt{b^2 - 4ac}}{2a} \qquad 1.38$$

$$x_1^* + x_2^* = -\frac{b}{a} \qquad 1.39$$

$$x_1^* x_2^* = \frac{c}{a} \qquad 1.40$$

3. Cubic Equations

Cubic and higher order equations occur infrequently in most engineering problems. However, they usually are difficult to factor when they do occur. Trial and error solutions are usually unsatisfactory except for finding the general region in which a root occurs. Graphical means can only be used to obtain a fair approximation to the root.

Numerical analysis techniques must be used if extreme accuracy is needed. The more efficient numerical analysis techniques are too complicated to present here. However, the bisection method illustrated in example 1.2 can usually provide the required accuracy with only a few simple iterations.

The bisection method starts out with two values of the independent variable, L_o and R_o, which straddle a root. Since the function has a value of zero at a root, $f(L_o)$ and $f(R_o)$ will have opposite signs. The following algorithm describes the remainder of the bisection method.

Let n be the iteration number. Then, for n = 0, 1, 2, . . . perform the following steps until sufficient accuracy is attained.

Set $m = ½(L_n+R_n)$

Calculate $f(m)$

If $f(L_n)f(m) \leq 0$, set $L_{n+1} = L_n$ and $R_{n+1} = m$

Otherwise, set $L_{n+1} = m$ and $R_{n+1} = R_n$

$f(x)$ has at least one root in the interval (L_{n+1}, R_{n+1})

The estimated value of that root, x^*, is

$$x^* \approx ½(L_{n+1}+R_{n+1}) \qquad 1.41$$

The maximum error is $½(R_{n+1}-L_{n+1})$. The iterations continue until the maximum error is reasonable for the accuracy of the problem.

Example 1.2

Use the bisection method to find the roots of

$$f(x) = x^3 - 2x - 7.$$

The first step is to find L_o and R_o, which are the values of x which straddle a root and have opposite signs. A table can be made and values of f(x) calculated for random values of x:

x	−2	−1	0	+1	+2	+3
f(x)	−11	−6	−7	−8	−3	+14

Since f(x) changes sign between x = 2 and x = 3,

$$L_o = 2 \text{ and } R_o = 3$$

Iteration 0: $m = \frac{1}{2}(2+3) = 2.5$

$f(2.5) = (2.5)^3 - 2(2.5) - 7 = 3.625$

Since f(2.5) is positive, a root must exist in the interval (2, 2.5) . Therefore,

$L_1 = 2$ and $R_1 = 2.5$

At this point, the best estimate of the root is

$x^* \approx \frac{1}{2}(2+2.5) = 2.25$

The maximum error is $\frac{1}{2}(2.5-2) = .25$

Iteration 1: $m = \frac{1}{2}(2+2.5) = 2.25$

$f(2.25) = -.1094$

Since f(m) is negative, a root must exist in the interval (2.25, 2.5). Therefore,

$L_2 = 2.25$ and $R_2 = 2.5$

The best estimate of the root is

$x^* \approx \frac{1}{2}(2.25+2.5) = 2.375$

The maximum error is $\frac{1}{2}(2.5-2.25) = .125$

This procedure continues until the maximum error is acceptable. Of course, this method does not automatically find any other roots that may exist on the real number line.

4. Finding Roots to General Expressions

There is no specific technique that will work with all general expressions for which roots are needed. If graphical means are not used, some combination of factoring and algebraic simplification must be used. However, multiplying each side of an equation by a power of a variable may introduce extraneous roots. Such an extraneous root will not satisfy the original equation, even though it was derived correctly according to the rules of algebra.

Although it is always a good idea to check your work, this step is particularly necessary whenever you have squared an expression or multiplied it by a variable.

Example 1.3

Find the value of x which will satisfy the following expression:

$$\sqrt{x-2} = \sqrt{x} + 2$$

First, square both sides.

$$x - 2 = x + 4\sqrt{x} + 4$$

Next, subtract x from both sides and combine constants.

$$4\sqrt{x} = -6$$

Solving for x yields $x^* = 9/4$. However, 9/4 does not satisfy the original expression since it is an extraneous root.

B. Simultaneous Linear Equations

Given n independent equations and n unknowns, the n values which simultaneously solve all n equations can be found by the methods illustrated below.

1. By Substitution (shown by example)

Example 1.4

Solve $2x + 3y = 12$ (a)
$3x + 4y = 8$ (b)

step 1: From equation (a), solve for $x = 6 - 1.5y$

step 2: Substitute $(6-1.5y)$ into equation (b) wherever x appears. $3(6-1.5y) + 4y = 8$ or $y^* = 20$

step 3: Solve for x^* from either equation: $x^* = 6 - 1.5(20) = -24$

step 4: Check that (−24, 20) solves both original equations.

2. By Reduction (same example)

step 1: Multiply each equation by a number chosen to make the coefficient of one of the variables the same in each equation.

3 times equation (a): $6x + 9y = 36$ (c)

2 times equation (b): $6x + 8y = 16$ (d)

step 2: Subtract one equation from the other. Solve for one of the variables.

(c) − (d): $y^* = 20$

step 3: Solve for the remaining variable.

step 4: Check that the calculated values of (x^*,y^*) solve both original equations.

3. By Cramer's Rule

This method is best for 3 or more simultaneous equations. (The calculation of determinants is covered later in this chapter.)

To find x^* and y^* which satisfy

$$a_1x + b_1y = c_1$$
$$a_2x + b_2y = c_2$$

calculate the determinants

$$D_1 = \begin{vmatrix} a_1 & b_1 \\ a_2 & b_2 \end{vmatrix} \qquad 1.42$$

$$D_2 = \begin{vmatrix} c_1 & b_1 \\ c_2 & b_2 \end{vmatrix} \qquad 1.43$$

$$D_3 = \begin{vmatrix} a_1 & c_1 \\ a_2 & c_2 \end{vmatrix} \qquad 1.44$$

Then, if $\mathbf{D_1} \neq 0$, the unique numbers satisfying the two simultaneous equations are:

$$x^* = \mathbf{D_2/D_1} \qquad 1.45$$

$$y^* = \mathbf{D_3/D_1} \qquad 1.46$$

If $\mathbf{D_1}$ (the determinant of the coefficients matrix) is zero, the system of simultaneous equations may still have a solution. However, Cramer's rule cannot be used to find that solution. If the system is homogeneous (i.e., has the general form $\mathbf{Ax} = \mathbf{0}$), then a non-zero solution exists if and only if $\mathbf{D_1}$ is zero.

Example 1.5

Solve the following system of simultaneous equations:

$$2x + 3y - 4z = 1$$
$$3x - y - 2z = 4$$
$$4x - 7y - 6z = -7$$

Calculate the determinants:

$$\mathbf{D_1} = \begin{vmatrix} 2 & 3 & -4 \\ 3 & -1 & -2 \\ 4 & -7 & -6 \end{vmatrix} = 82$$

$$\mathbf{D_2} = \begin{vmatrix} 1 & 3 & -4 \\ 4 & -1 & -2 \\ -7 & -7 & -6 \end{vmatrix} = 246$$

$$\mathbf{D_3} = \begin{vmatrix} 2 & 1 & -4 \\ 3 & 4 & -2 \\ 4 & -7 & -6 \end{vmatrix} = 82$$

$$\mathbf{D_4} = \begin{vmatrix} 2 & 3 & 1 \\ 3 & -1 & 4 \\ 4 & -7 & -7 \end{vmatrix} = 164$$

Then, $x^* = \mathbf{D_2/D_1} = 3$

$y^* = \mathbf{D_3/D_1} = 1$

$z^* = \mathbf{D_4/D_1} = 2$

C. Simultaneous Quadratic Equations

Although simultaneous non-linear equations are best solved graphically, a specialized method exists for simultaneous quadratic equations. This method is known as 'Eliminating the Constant Term.'

step 1: Isolate the constant terms of both equations on the right-hand side of the equalities.

step 2: Multiply both sides of one equation by a number chosen to make the constant terms of both equations the same.

step 3: Subtract one equation from the other to obtain a difference equation.

step 4: Factor the difference equation into terms.

step 5: Solve for one of the variables from one of the factor terms.

step 6: Substitute the formula for the variable into one of the original equations and complete the solution.

step 7: Check the solution.

Example 1.6

Solve for the simultaneous values of x and y:

$$2x^2 - 3xy + y^2 = 15$$
$$x^2 - 2xy + y^2 = 9$$

steps 2&3:

$$\begin{array}{r} 6x^2 - 9xy + 3y^2 = 45 \\ -(5x^2 - 10xy + 5y^2) = 45 \\ \hline x^2 + xy - 2y^2 = 0 \end{array}$$

steps 4&5: $x^2 + xy - 2y^2$ factors into $(x+2y)(x-y)$ from which we obtain $x = -2y$.

step 6: Substituting $x = -2y$ into $(2x^2 - 3xy + y^2 = 15)$ gives $y^* = \pm 1$, from which $x^* = \pm 2$ can be derived by further substitution.

D. Exponentiation

(x is any variable or constant)

$$x^m x^n = x^{(n+m)} \qquad 1.47$$

$$x^m/x^n = x^{(m-n)} \qquad 1.48$$

$$(x^n)^m = x^{(mn)} \qquad 1.49$$

$$a^{m/n} = \sqrt[n]{a^m} \qquad 1.50$$

$$(a/b)^n = a^n/b^n \qquad 1.51$$

$$\sqrt[n]{x} = (x)^{1/n} \qquad 1.52$$

$$x^{-n} = 1/(x^n) \qquad 1.53$$

$$x^0 = 1 \qquad 1.54$$

E. Logarithms

Logarithms are exponents. That is, the exponent x in the expression $b^x = n$ is the logarithm of n to the base b. Therefore, $(\log_b n) = x$ is equivalent to $(b^x = n)$.

The base for common logs is 10. Usually, 'log' will be written when common logs are desired, although '$\log_{10}$' appears occasionally. The base for natural (naperien) logs is 2.718 · · ·, a number which is given the symbol 'e'. When natural logs are desired, usually 'ln' will be written, although '$\log_e$' is also used.

Most logarithms will contain an integer part (the characteristic) and a fractional part (the mantissa). The logarithm of any number less than one is negative. If the number is greater than one, its logarithm is positive. Although the logarithm may be negative, the mantissa is always positive.

If the number is greater than one, the characteristic will be positive and equal to one less than the number of digits in front of the decimal. If the number is less than one, the characteristic will be negative and equal to one more than the number of zeros immediately following the decimal point.

Example 1.7

What is $\log_{10}(.05)$?

Since the number is less than one and there is one leading zero, the characteristic is -2. From the logarithm tables, the mantissa of 5.0 is .699. Two ways of combining the mantissa and characteristic are possible:

Method 1: $\bar{2}.699$

Method 2: $8.699 - 10$

If the logarithm is to be used in a calculation, it must be converted to operational form: $-2 + .699 = -1.301$. Notice that -1.301 is not the same as $\bar{1}.301$.

F. Logarithm Identities

$$x^a = \text{antilog}[a \log(x)] \quad 1.55$$

$$\log(x^a) = a \log(x) \quad 1.56$$

$$\log(xy) = \log(x) + \log(y) \quad 1.57$$

$$\log(x/y) = \log(x) - \log(y) \quad 1.58$$

$$\ln(x) = (\log_{10}x)/(\log_{10}e) \approx 2.3(\log_{10}x) \quad 1.59$$

$$\log_b(b) = 1 \quad 1.60$$

$$\log(1) = 0 \quad 1.61$$

$$\log_b(b^n) = n \quad 1.62$$

Example 1.8

The surviving fraction, x, of a radioactive isotope is given by

$$x = e^{-.005t}$$

For what value of t will the surviving fraction be 7%?

$$.07 = e^{-.005t}$$

Taking the natural log of both sides,

$$\ln(.07) = \ln(e^{-.005t})$$
$$-2.66 = -.005t$$
$$t = 532$$

G. Partial Fractions

Given some rational fraction $H(x) = P(x)/Q(x)$ where $P(x)$ and $Q(x)$ are polynomials, the polynomials and constants A_i and $Y_i(x)$ are needed such that

$$H(x) = \sum_i \frac{A_i}{Y_i(x)} \quad 1.63$$

case 1: Q(x) factors into n different linear terms. That is,

$$Q(x) = (x-a_1)(x-a_2)\cdots(x-a_n) \quad 1.64$$

Then, set

$$H(x) = \sum_{i=1}^{n} \frac{A_i}{(x-a_i)} \quad 1.65$$

case 2: Q(x) factors into n identical linear terms. That is,

$$Q(x) = (x-a)(x-a)\cdots(x-a) \quad 1.66$$

Set

$$H(x) = \sum_{i=1}^{n} \frac{A_i}{(x-a)^i} \quad 1.67$$

case 3: Q(x) factors into n different quadratic terms, $(x^2+p_ix+q_i)$

Set

$$H(x) = \sum_{i=1}^{n} \frac{A_ix + B_i}{x^2 + p_ix + q_i} \quad 1.68$$

case 4: Q(x) factors into n identical quadratic terms, (x^2+px+q)

Set

$$H(x) = \sum_{i=1}^{n} \frac{A_ix + B_i}{(x^2+px+q)^i} \quad 1.69$$

case 5: Q(x) factors into any combination of the above. The solution is illustrated by example 1.9.

Example 1.9

Resolve $H(x) = \dfrac{x^2+2x+3}{x^4+x^3+2x^2}$ into partial fractions.

Here, $Q(x) = x^4+x^3+2x^2$ which factors into $x^2(x^2+x+2)$. This is a combination of cases 2 and 3. We set

$$H(x) = \frac{A_1}{x} + \frac{A_2}{x^2} + \frac{A_3 + A_4x}{x^2+x+2}$$

Cross multiplying to obtain a common denominator yields

$$\frac{(A_1+A_4)x^3 + (A_1+A_2+A_3)x^2 + (2A_1+A_2)x + 2A_2}{x^4+x^3+2x^2}$$

Since the original numerator is known, the following simultaneous equations result:

$$A_1 + A_4 = 0$$
$$A_1 + A_2 + A_3 = 1$$
$$2A_1 + A_2 = 2$$
$$2A_2 = 3$$

The solutions are: $A_1^* = .25$; $A_2^* = 1.5$; $A_3^* = -.75$; $A_4^* = -.25$

So that

$$H(x) = \frac{1}{4x} + \frac{3}{2x^2} - \frac{x+3}{4(x^2+x+2)}$$

H. Series Theory

1. Common Series: All series having a finite number of terms converge. That is, $\lim\limits_{i \to n} B_n$ exists.

Some series with an infinite number of terms also converge. In the material below, the following nomenclature is used:

B_n is the sum of n terms
b_i is the ith term
b_n is the last term

Geometric Series:

$$B_n = \sum_{i=0}^{n-1} aR^i = a(1+R+R^2+\cdots+R^{n-1}) \qquad 1.70$$

$$b_n = aR^{n-1} \qquad 1.71$$

$$B_n = \frac{a(1-R^n)}{1-R} \text{ (finite series)} \qquad 1.72$$

$$B_n = \frac{a}{(1-R)} \text{ (infinite series)} \qquad 1.73$$

The infinite geometric series converges for $-1<R<1$ and diverges otherwise.

Arithmetic Series:

$$B_n = \sum_{i=1}^{n} (a+(i-1)d) = {}^1\!/\!_2 n(2a+(n-1)d) \qquad 1.74$$

$$b_n = (a+(n-1)d) \qquad 1.75$$

The infinite series always diverges.

Harmonic Series:

$$B_n = \sum_{i=1}^{n} \frac{1}{a+(i-1)d} \qquad 1.76$$

$$b_n = \frac{1}{a+(n-1)d} \qquad 1.77$$

$$B_n = \frac{2}{n(2a+(n-1)d)} \text{ (finite series)} \qquad 1.78$$

The infinite series always diverges.

p Series:

$$B_n = \sum_{i=1}^{n} (1/i^p) = 1 + (1/2^p) + (1/3^p) + \cdots \qquad 1.79$$

The infinite series converges if $p > 1$.
The infinite series diverges if $p \leqslant 1$.

2. Rth Order Differences: If several elements of a series or sequence are known and the next element is needed, the following method may be used if the general term is not obvious.

step 1: Find the first-order differences by subtracting each element from the following element.

step 2: Find the 2nd-order, 3rd-order, etc. differences until the Rth-order differences are all equal. By working backwards, the unknown element can be found.

Example 1.10

Find the 6th term in the series (7,16,29,46,67,?)

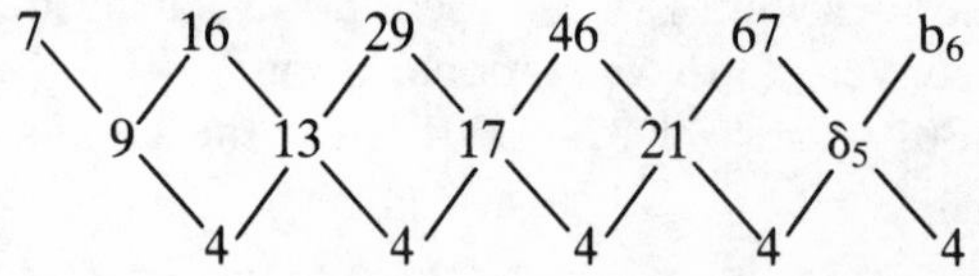

Since $\delta_5 - 21 = 4$, $\delta_5 = 25$. Since $b_6 - 67 = 25$, $b_6 = 92$.

3. Tests for Convergence of Infinite Series

Theorem: a) If $\lim\limits_{i \to \infty} B_i$ exists, then the series converges.

b) If $\lim\limits_{i \to \infty} b_i \neq 0$, then the series diverges.

c) If $\lim\limits_{i \to \infty} b_i = 0$, then further tests are required.

Ratio Test: Calculate $\lim\limits_{i \to \infty} \left(\frac{b_{i+1}}{b_i}\right) = k \qquad 1.80$

The series converges for k less than 1.
The series diverges for k greater than 1.
The test is inconclusive if k equals 1.

Comparison Test: If A_n and B_n are both series such that $a_i<b_i$ for all i, then
(1) If A_n diverges, then B_n diverges.
(2) If B_n converges, then A_n converges.
Generally, it is convenient to compare the series to a geometric or 'p' series.

MATH

4. Series of Alternating Sign

The ratio test will determine convergence if the absolute value of the ratio is used. The same criteria apply. If a series containing all positive terms converges, then the same series with some negative terms also converges. Therefore, make all the signs positive and determine if this 'new' series converges. If it does, then the original series will also converge.

B_n converges if $|b_{i+1}| < |b_i|$ for all i and $\lim_{i\to\infty}(b_n) = 0$.

Example 1.11

Does the series below converge?

$$B_n = 2/1 + 3/4 + 4/9 + 5/16 + \cdots$$

The general term is $b_i = (i+1)/i^2$

Using partial fractions, the general term is $b_i = \frac{1}{i} + \frac{1}{i^2}$

However, $\frac{1}{i}$ is the harmonic series with d=1. Since the harmonic series is divergent and since this series exceeds the harmonic series, this series also diverges.

Example 1.12

Does the series below converge?

$$B_n = 3 + 9/2 + 27/6 + 81/24 + \cdots$$

The general term is $b_i = 3^i/i!$ and $b_{i+1} = 3^{(i+1)}/(i+1)!$

By using the ratio test,

$$\lim_{i\to\infty}\left(\frac{b_{i+1}}{b_i}\right) = \lim_{i\to\infty}\frac{3}{(i+1)} = 0$$

so B_n converges.

I. Boolean Algebra

Boolean algebra is a system wherein the variables are constrained to two values, usually 0 and 1. The basic operations and corresponding gate symbols are shown in the following figure.

Gates may have more than two inputs, except for the 'not' gate which may have only one input.

The basic algebraic laws governing Boolean variables are listed below.

Commutative: $A+B = B+A$

$A \cdot B = B \cdot A$

Associative: $A+(B+C) = (A+B)+C$

$A \cdot (B \cdot C) = (A \cdot B) \cdot C$

Distributive: $A \cdot (B+C) = (A \cdot B)+(A \cdot C)$

$A+(B \cdot C) = (A+B) \cdot (A+C)$

Absorptive: $A+(A \cdot B) = A$

$A \cdot (A+B) = A$

$0+0 = 0$

$0+1 = 1$

$1+0 = 1$

$1+1 = 1$

$0 \cdot 0 = 0$

$0 \cdot 1 = 0$

$1 \cdot 0 = 0$

$1 \cdot 1 = 1$

$A+0 = A$

$A+1 = 1$

$A+A = A$

$A+(-A) = 1$

$A \cdot 0 = 0$

$A \cdot 1 = A$

$A \cdot A = A$

$A \cdot (-A) = 0$

$-0 = 1$

$-1 = 0$

$-(-A) = A$

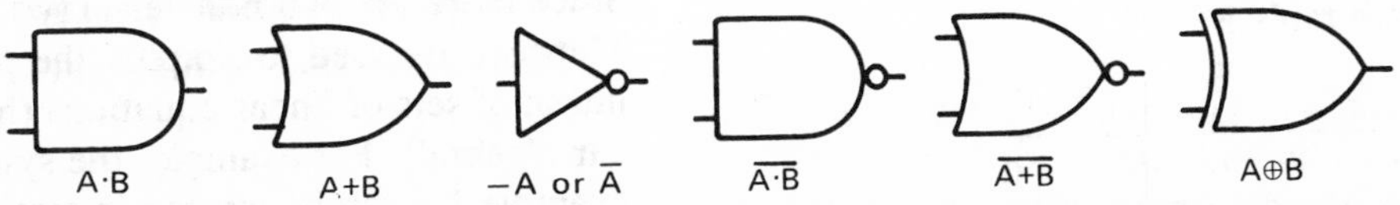

Figure 1.1 Logic Gates

For a (2 × 2) matrix, $\begin{pmatrix} + & - \\ - & + \end{pmatrix}$

For a (3 × 3) matrix, $\begin{pmatrix} + & - & + \\ - & + & - \\ + & - & + \end{pmatrix}$

Example 1.18

What is the cofactor of the (−3) in the following matrix?

$$\begin{pmatrix} 2 & 9 & 1 \\ -3 & 4 & 0 \\ 7 & 5 & 9 \end{pmatrix}$$

The resulting matrix is $\begin{pmatrix} 9 & 1 \\ 5 & 9 \end{pmatrix}$ with determinant 76. The cofactor is −76.

4. The *classical adjoint* is a matrix formed from the transposed cofactor matrix with the conventional sign arrangement. The resulting matrix is represented as $\mathbf{A}_{adj}$.

Example 1.19

What is the classical adjoint of $\begin{pmatrix} 2 & 3 & -4 \\ 0 & -4 & 2 \\ 1 & -1 & 5 \end{pmatrix}$?

The matrix of cofactors (considering the sign convention) is

$$\begin{pmatrix} -18 & 2 & 4 \\ -11 & 14 & 5 \\ -10 & -4 & -8 \end{pmatrix}$$

The transposed cofactor matrix is

$$\mathbf{A}_{adj} = \begin{pmatrix} -18 & -11 & -10 \\ 2 & 14 & -4 \\ 4 & 5 & -8 \end{pmatrix}$$

5. The *inverse,* $\mathbf{A}^{-1}$, of **A** is a matrix such that $(\mathbf{A})(\mathbf{A}^{-1}) = \mathbf{I}$. (**I** is a square matrix with ones along the left-to-right diagonal and zeros elsewhere.)

For a (2 × 2) matrix, $\begin{pmatrix} a & b \\ c & d \end{pmatrix}$, the inverse is

$$\frac{1}{\mathbf{D}}\begin{pmatrix} d & -b \\ -c & a \end{pmatrix} \qquad 1.85$$

For larger matrices, the inverse is best calculated by dividing every entry in the classical adjoint by the determinant of the original matrix.

Example 1.20

What is the inverse of $\begin{pmatrix} 4 & 5 \\ 2 & 3 \end{pmatrix}$?

The determinant is 2. The inverse is

$$\tfrac{1}{2}\begin{pmatrix} 3 & -5 \\ -2 & 4 \end{pmatrix} = \begin{pmatrix} 3/2 & -5/2 \\ -1 & 2 \end{pmatrix}$$

6. Trigonometry

A. Degrees and Radians

360 degrees = one complete circle = 2π radians

90 degrees = right angle = $\tfrac{1}{2}\pi$ radians

one radian = 57.3 degrees

one degree = .0175 radians

multiply degrees by $(\pi/180)$ to obtain radians

multiply radians by $(180/\pi)$ to obtain degrees

B. Right Triangles

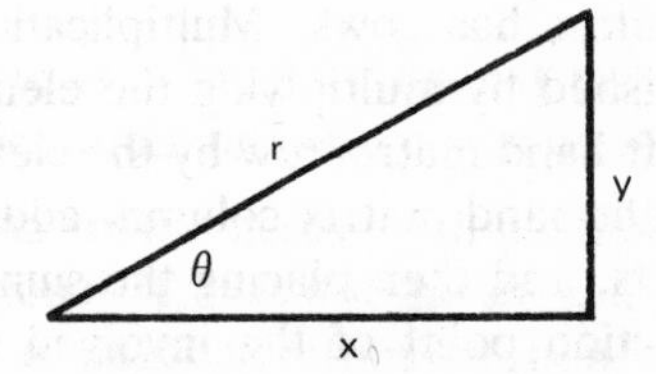

Figure 1.2 A Right Triangle

1. Pythagorean Theorem

$$x^2 + y^2 = r^2 \qquad 1.86$$

2. Trigonometric Functions

$$\sin\theta = y/r \qquad 1.87$$

$$\cos\theta = x/r \qquad 1.88$$

$$\tan\theta = y/x \qquad 1.89$$

$$\csc\theta = r/y \qquad 1.90$$

$$\sec\theta = r/x \qquad 1.91$$

3. Relationship of the Trigonometric Functions to the Unit Circle

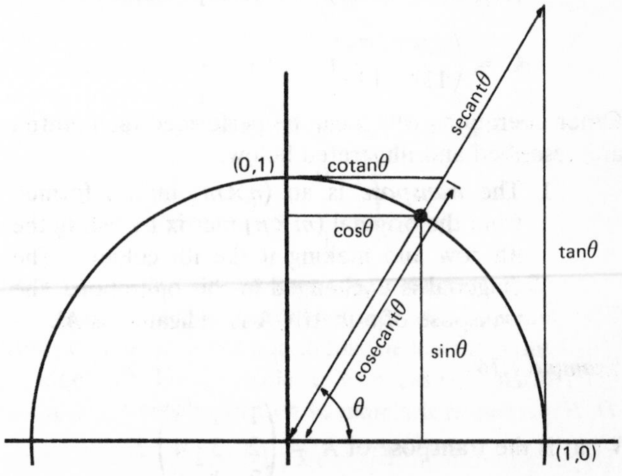

Figure 1.3 The Unit Circle

4. *Signs of the Trigonometric Functions*

quadrants		quadrant	I	II	III	IV
II	I	sin	+	+	−	−
		cos	+	−	−	+
III	IV	tan	+	−	+	−

5. *Functions of the Related Angles*

	−θ	90−θ	90+θ	180−θ	180+θ
sin	−sinθ	cosθ	cosθ	sinθ	−sinθ
cos	cosθ	sinθ	−sinθ	−cosθ	−cosθ
tan	−tanθ	cotθ	−cotθ	−tanθ	tanθ

6. *Trigonometric Identities*

$$\sin^2\theta + \cos^2\theta = 1 \qquad 1.92$$

$$1 + \tan^2\theta = \sec^2\theta \qquad 1.93$$

$$1 + \cot^2\theta = \csc^2\theta \qquad 1.94$$

$$\sin 2\theta = 2(\sin\theta)(\cos\theta) \qquad 1.95$$

$$\cos 2\theta = \cos^2\theta - \sin^2\theta$$

$$= 1 - 2\sin^2\theta \qquad 1.96$$

$$\sin\theta = 2[\sin(\tfrac{1}{2}\theta)\cos(\tfrac{1}{2}\theta)] \qquad 1.97$$

$$\sin(\tfrac{1}{2}\theta) = \pm\sqrt{\tfrac{1}{2}(1-\cos\theta)} \qquad 1.98$$

7. *Two-Angle Formulas*

$$\sin(\theta + ø) = [\sin\theta][\cos ø] + [\cos\theta][\sin ø] \qquad 1.99$$

$$\sin(\theta - ø) = [\sin\theta][\cos ø] - [\cos\theta][\sin ø] \qquad 1.100$$

$$\cos(\theta + ø) = [\cos\theta][\cos ø] - [\sin\theta][\sin ø] \qquad 1.101$$

$$\cos(\theta - ø) = [\cos\theta][\cos ø] + [\sin\theta][\sin ø] \qquad 1.102$$

C. *General Triangles*

Law of Sines: $\frac{\sin A}{a} = \frac{\sin B}{b} = \frac{\sin C}{c}$ 1.103

Law of Cosines: $a^2 = b^2 + c^2 - 2bc(\cos A)$ 1.104

Area = $\tfrac{1}{2}ab(\sin C)$ 1.105

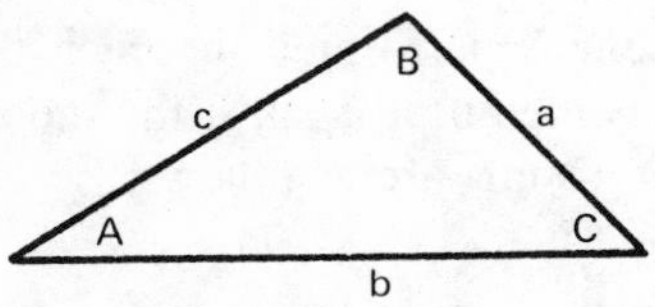

Figure 1.4 A General Triangle

D. *Hyperbolic Functions*

Hyperbolic functions are specific equations containing the terms e^x and e^{-x}. These combinations of e^x and e^{-x} appear regularly in certain types of problems. In order to simplify the mathematical equations they appear in, these hyperbolic functions are given special names and symbols.

$$\sinh x = \frac{e^x - e^{-x}}{2} \qquad 1.106$$

$$\cosh x = \frac{e^x + e^{-x}}{2} \qquad 1.107$$

$$\tanh x = \frac{e^x - e^{-x}}{e^x + e^{-x}} = \frac{\sinh x}{\cosh x} \qquad 1.108$$

$$\coth x = \frac{e^x + e^{-x}}{e^x - e^{-x}} = \frac{\cosh x}{\sinh x} \qquad 1.109$$

$$\text{sech}\, x = \frac{2}{e^x + e^{-x}} = \frac{1}{\cosh x} \qquad 1.110$$

$$\text{csch}\, x = \frac{2}{e^x - e^{-x}} = \frac{1}{\sinh x} \qquad 1.111$$

The hyperbolic identities are somewhat different than the standard trigonometric identities. Several of the most common identities are presented below.

$$\cosh^2 x - \sinh^2 x = 1 \qquad 1.112$$

$$1 - \tanh^2 x = \text{sech}^2 x \qquad 1.113$$

$$1 - \coth^2 x = \text{csch}^2 x \qquad 1.114$$

$$\cosh x + \sinh x = e^x \qquad 1.115$$

$$\cosh x - \sinh x = e^{-x} \qquad 1.116$$

$$\sinh(x+y) = [\sinh x][\cosh y] + [\cosh x][\sinh y] \qquad 1.117$$

$$\cosh(x+y) = [\cosh x][\cosh y] + [\sinh x][\sinh y] \qquad 1.118$$

7. *Straight Line Analytic Geometry*

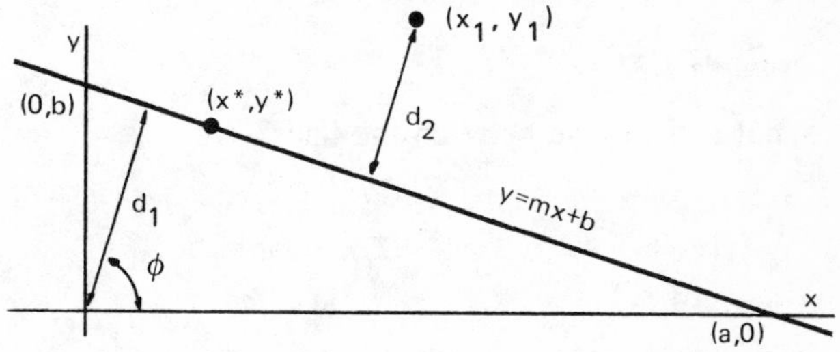

Figure 1.5 A Straight Line

A. *Equations of a Straight Line*

General Form: $Ax + By + C = 0$ 1.119

Slope Form: $y = mx + b$ 1.120

Point-Slope Form: $(y - y^*) = m(x - x^*)$ 1.121

(x^*, y^*) is any point on the line

Intercept Form: $\frac{x}{a} + \frac{y}{b} = 1$ 1.122

Two-Point Form: $\frac{y - y_1^*}{x - x_1^*} = \frac{y_2^* - y_1^*}{x_2^* - x_1^*}$ 1.123

MATH

Normal Form: $x(\cos\phi) + y(\sin\phi) - d_1 = 0$ 1.124

Polar Form: $r(\sin\phi) = d_1$ 1.125

B. Points, Lines, and Distances

The distance d_2 between a point and a line is:

$$d_2 = \frac{|Ax_1 + By_1 + C|}{\sqrt{A^2 + B^2}} \qquad 1.126$$

The distance between two points is:

$$d = \sqrt{(x_2-x_1)^2 + (y_2-y_1)^2} \qquad 1.127$$

Parallel lines:

$$A_1/A_2 = B_1/B_2 \qquad 1.128$$

$$m_1 = m_2 \qquad 1.129$$

Perpendicular lines:

$$A_1A_2 = -B_1B_2 \qquad 1.130$$

$$m_1 = \frac{-1}{m_2} \qquad 1.131$$

Point of intersection of two lines:

$$x_1 = \frac{B_2C_1 - B_1C_2}{A_2B_1 - A_1B_2} \qquad 1.132$$

$$y_1 = \frac{A_1C_2 - A_2C_1}{A_2B_1 - A_1B_2} \qquad 1.133$$

Smaller angle between two intersecting lines:

$$\tan\phi = \frac{A_1B_2 - A_2B_1}{A_1A_2 + B_1B_2} = \frac{m_2 - m_1}{1 + m_1m_2} \qquad 1.134$$

$$\phi = |\arctan(m_1) - \arctan(m_2)| \qquad 1.135$$

Example 1.21

What is the angle between the lines?

$$y_1 = -.577x + 2$$

$$y_2 = +.577x - 5$$

method 1:

$$\arctan\left[\frac{m_2 - m_1}{1 + m_1m_2}\right] =$$

$$\arctan\left[\frac{.577 - (-.577)}{1 + (.577)(-.577)}\right] = 60°$$

method 2: Write both equations in general form:

$$-.577x - y_1 + 2 = 0$$

$$.577x - y_2 - 5 = 0$$

$$\arctan\left[\frac{A_1B_2 - A_2B_1}{A_1A_2 + B_1B_2}\right] =$$

$$\arctan\left[\frac{(-.577)(-1) - (.577)(-1)}{(-.577)(.577) + (-1)(-1)}\right] = 60°$$

method 3:

$$\phi = |\arctan(-.577) - \arctan(.577)| = |-30° - 30°| = 60°$$

C. Linear Regression

If it is necessary to draw a straight line through n data points (x_1,y_1), (x_2,y_2), . . . , (x_n,y_n), the following method based on the theory of least squares may be used:

step 1: Calculate the following quantities:

Σx_i $\quad \Sigma x_i^2$ $\quad (\Sigma x_i)^2$ $\quad \bar{x} = (\Sigma x_i/n)$ $\quad \Sigma x_iy_i$

Σy_i $\quad \Sigma y_i^2$ $\quad (\Sigma y_i)^2$ $\quad \bar{y} = (\Sigma y_i/n)$

step 2: Calculate the slope of the line, $y = mx + b$.

$$m = \frac{n\Sigma(x_iy_i) - (\Sigma x_i)(\Sigma y_i)}{n\Sigma x_i^2 - (\Sigma x_i)^2} \qquad 1.136$$

step 3: Calculate the y intercept.

$$b = \bar{y} - m\bar{x} \qquad 1.137$$

step 4: To determine the goodness of fit, calculate the correlation coefficient.

$$r = \frac{n\Sigma(x_iy_i) - (\Sigma x_i)(\Sigma y_i)}{\sqrt{[n\Sigma x_i^2 - (\Sigma x_i)^2][n\Sigma y_i^2 - (\Sigma y_i)^2]}} \qquad 1.138$$

If m is positive, r will be positive. If m is negative, r will be negative. As a general rule, if the absolute value of r exceeds .85, the fit is good. Otherwise, the fit is poor. r equals one if the fit is a perfect straight line.

A low value of r does not eliminate the possibility of a non-linear relationship existing between x and y. It is possible that the data describe a parabolic, logarithmic, or other non-linear relationship. (Usually this will be apparent if the data are graphed.) It may be necessary to convert one or both variables to new variables by taking squares, square roots, cubes, or logs to name a few of the possibilities.

Example 1.22

An experiment is performed in which the dependent variable (y) is measured against the independent variable (x). The results are as follows:

x	y
1.2	.602
4.7	5.107
8.3	6.984
20.9	10.031

What is the least squares straight line equation which represents this data?

step 1: $\Sigma x_i = 35.1$
$\Sigma y_i = 22.72$
$\Sigma x_i^2 = 529.23$
$\Sigma y_i^2 = 175.84$
$(\Sigma x_i)^2 = 1232.01$
$(\Sigma y_i)^2 = 516.19$
$\overline{x} = 8.775$
$\overline{y} = 5.681$
$\Sigma x_i y_i = 292.34$
$n = 4$

step 2: $m = \dfrac{(4)(292.34) - (35.1)(22.72)}{(4)(529.23) - (35.1)^2} = .42$

step 3: $b = 5.681 - (.42)(8.775) = 2.0$

D. Vector Operations

A vector is a directed straight line of a given magnitude. Two directed straight lines with the same magnitudes and directions are said to be equivalent. Thus, the actual end-points of a vector are often irrelevant as long as the direction and magnitude are known.

A vector defined by its end-points and direction is designated as

$$\mathbf{V} = \overrightarrow{p_1 p_2}$$

Usually, p_1 will be the origin, in which case **V** will be designated by its end-point, $p_2 = (x,y)$. Such a zero-based vector is equivalent to all other vectors of the same magnitude and direction. Any vector p_1p_2 can be transformed into a zero-based vector by subtracting (x_1,y_1) from all points along the vector line.

A vector may also be specified in the terms of the unit vectors (**i**,**j**,**k**). Thus,

$$\mathbf{V} = (x,y) = (x\mathbf{i} + y\mathbf{j})$$

Or,

$$\mathbf{V} = (x,y,z) = (x\mathbf{i} + y\mathbf{j} + z\mathbf{k})$$

Important operations on vectors based at the origin are:

$c\mathbf{V} = (cx, cy)$ (vector multiplication by a scalar)

$$\mathbf{V}_1 + \mathbf{V}_2 = (x_1 + x_2, y_1 + y_2) \qquad 1.139$$

$$|\mathbf{V}| = \sqrt{x^2 + y^2} \quad \text{(vector magnitude)} \qquad 1.140$$

α = angle between vector **V** and x axis

$$= \arccos(x/|\mathbf{V}|) = \arcsin(y/|\mathbf{V}|) \qquad 1.141$$

$$m = \text{slope of vector} = y/x \qquad 1.142$$

$$\theta = \left\{\begin{matrix}\text{angle between}\\ \text{two vectors}\end{matrix}\right\} = \arccos\frac{(x_1x_2 + y_1y_2)}{|\mathbf{V}_1|\,|\mathbf{V}_2|} \qquad 1.143$$

$\mathbf{V}_1 \cdot \mathbf{V}_2$ = dot product

$$= |\mathbf{V}_1|\,|\mathbf{V}_2|\cos\theta = x_1x_2 + y_1y_2 + z_1z_2 \qquad 1.144$$

$\mathbf{V}_1 \times \mathbf{V}_2$ = cross product

$$= \begin{vmatrix} \mathbf{i} & x_1 & x_2 \\ \mathbf{j} & y_1 & y_2 \\ \mathbf{k} & z_1 & z_2 \end{vmatrix} \qquad 1.145$$

$$\mathbf{V}_1 \times \mathbf{V}_2 = -\mathbf{V}_2 \times \mathbf{V}_1 \qquad 1.146$$

$$|\mathbf{V}_1 \times \mathbf{V}_2| = |\mathbf{V}_1|\,|\mathbf{V}_2|\sin\theta \qquad 1.147$$

Example 1.23

What is the angle between the vectors $\mathbf{V}_1 = (-\sqrt{3},1)$ and $\mathbf{V}_2 = (2\sqrt{3},2)$?

$$\cos\theta = \frac{\mathbf{V}_1 \cdot \mathbf{V}_2}{|\mathbf{V}_1|\,|\mathbf{V}_2|} = \frac{(-\sqrt{3})(2\sqrt{3}) + (1)(2)}{\sqrt{3+1}\ \sqrt{12+4}} = -1/2$$

$$\theta = 120°$$

(Graph and compare this result to example 1.21 in which the lines were not directed.)

Example 1.24

Find a unit vector orthogonal to $\mathbf{V}_1 = \mathbf{i} - \mathbf{j} + 2\mathbf{k}$ and $\mathbf{V}_2 = 3\mathbf{j} - \mathbf{k}$.

The cross product is orthogonal to $\mathbf{V}_1$ and $\mathbf{V}_2$, although its length may not be equal to one.

$$\mathbf{V}_1 \times \mathbf{V}_2 = \begin{vmatrix} \mathbf{i} & 1 & 0 \\ \mathbf{j} & -1 & 3 \\ \mathbf{k} & 2 & -1 \end{vmatrix} = -5\mathbf{i} + \mathbf{j} + 3\mathbf{k}$$

Since the length of $|\mathbf{V}_1 \times \mathbf{V}_2|$ is $\sqrt{35}$, it is necessary to divide $|\mathbf{V}_1 \times \mathbf{V}_2|$ by this amount to obtain a unit vector. Thus,

$$\mathbf{V}_3 = \frac{-5\mathbf{i} + \mathbf{j} + 3\mathbf{k}}{\sqrt{35}}$$

The orthogonality can be proved from

$$\mathbf{V}_1 \cdot \mathbf{V}_3 = 0 \quad \text{and} \quad \mathbf{V}_2 \cdot \mathbf{V}_3 = 0$$

That $\mathbf{V}_3$ is a unit vector can be proved from

$$\mathbf{V}_3 \cdot \mathbf{V}_3 = +1$$

E. Direction Numbers, Direction Angles, and Direction Cosines

Given a directed line from (x_1,y_1,z_1) to (x_2,y_2,z_2), the direction numbers are:

$$L = x_2 - x_1 \qquad 1.148$$

$$M = y_2 - y_1 \qquad 1.149$$

$$N = z_2 - z_1 \qquad 1.150$$

The distance between the two points is:

$$d = \sqrt{L^2 + M^2 + N^2} \qquad 1.151$$

The direction cosines are:

$$\cos\alpha = L/d \qquad 1.152$$

$$\cos\beta = M/d \qquad 1.153$$

$$\cos\gamma = N/d \qquad 1.154$$

Note that $\cos^2\alpha + \cos^2\beta + \cos^2\gamma = 1$ 1.155

The direction angles are the angles between the axes and the lines. They are found from the inverse functions of the direction cosines. That is

$$\alpha = \arccos(L/d) \qquad 1.156$$

$$\beta = \arccos(M/d) \qquad 1.157$$

$$\gamma = \arccos(N/d) \qquad 1.158$$

Once the direction cosines have been found, they can be used to write the equation of the straight line in terms of the unit vectors. The line **R** would be defined as

$$\mathbf{R} = \mathbf{i}\cos\alpha + \mathbf{j}\cos\beta + \mathbf{k}\cos\gamma \qquad 1.159$$

Similarly, the line may be written in terms of its direction numbers,

$$\mathbf{R} = L\mathbf{i} + M\mathbf{j} + N\mathbf{k} \qquad 1.160$$

Given two directed lines $\mathbf{R}_1$ and $\mathbf{R}_2$, the angle between $\mathbf{R}_1$ and $\mathbf{R}_2$ is defined as the angle between the two arrow heads.

$$\cos\phi = \cos\alpha_1\cos\alpha_2 + \cos\beta_1\cos\beta_2 + \cos\gamma_1\cos\gamma_2$$

$$= \frac{L_1L_2 + M_1M_2 + N_1N_2}{d_1d_2} \qquad 1.161$$

If $\mathbf{R}_1$ and $\mathbf{R}_2$ are parallel and in the same direction, then

$$\alpha_1 = \alpha_2; \quad \beta_1 = \beta_2; \quad \gamma_1 = \gamma_2$$

If $\mathbf{R}_1$ and $\mathbf{R}_2$ are parallel but in opposite directions, then

$$\alpha_1 + \alpha_2 = 180 \text{ (etc.)}$$

If $\mathbf{R}_1$ and $\mathbf{R}_2$ are normal to each other, then

$$\phi = 90° \text{ and } \cos\phi = 0$$

Example 1.25

A line passes through the points (4,7,9) and (0,1,6). Write the equation of the line in terms of its direction cosines and direction numbers.

$$L = 4 - 0 = 4$$

$$M = 7 - 1 = 6$$

$$N = 9 - 6 = 3$$

The line now may be written in terms of its direction numbers.

$$\mathbf{R} = 4\mathbf{i} + 6\mathbf{j} + 3\mathbf{k}$$

The distance between the two points is

$$d = \sqrt{(4)^2 + (6)^2 + (3)^2} = 7.81$$

The line may now be written in terms of its direction cosines.

$$\mathbf{R} = \frac{4\mathbf{i} + 6\mathbf{j} + 3\mathbf{k}}{7.81} = .512\mathbf{i} + .768\mathbf{j} + .384\mathbf{k}$$

8. Planes

A plane **P** is uniquely determined by one of three combinations of parameters:

1. three non-collinear points in space
2. two non-parallel vectors ($\mathbf{V}_1$ and $\mathbf{V}_2$) and their intersection point p_o
3. a point p_o and a normal vector **N**

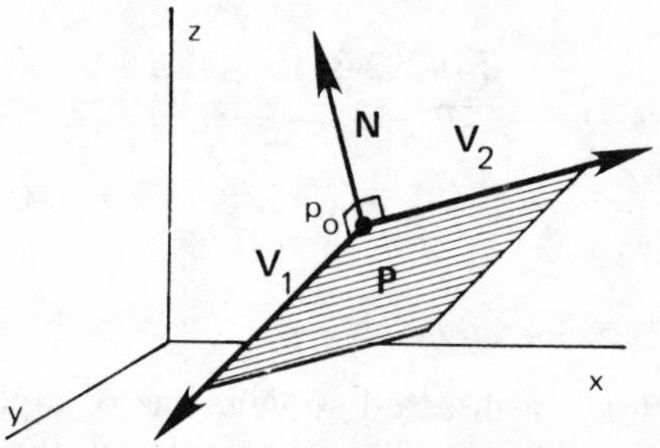

Figure 1.6 A Plane in 3-Space

The plane consists of all points such that the coordinates can be written as a linear combination of $\mathbf{V}_1$ and $\mathbf{V}_2$. That is, points in the plane can be written as

$$(x,y,z) = s\mathbf{V}_1 + t\mathbf{V}_2 \qquad 1.162$$

where s and t are constants and

$$\mathbf{V}_1 = a_1\mathbf{i} + b_1\mathbf{j} + c_1\mathbf{k} \qquad 1.163$$

$$\mathbf{V}_2 = a_2\mathbf{i} + b_2\mathbf{j} + c_2\mathbf{k} \qquad 1.164$$

If the intersection point $p_o = (x_o,y_o,z_o)$ is known, then points in the plane can be represented by the parametric equations given below. Notice the similarity to the slope form of an equation for a straight line.

$$x = sa_1 + ta_2 + x_o \qquad 1.165$$

$$y = sb_1 + tb_2 + y_o \qquad 1.166$$

$$z = sc_1 + tc_2 + z_o \qquad 1.167$$

The plane is also defined by its rectangular equations:

$$A(x - x_o) + B(y - y_o) + C(z - z_o) = 0 \qquad 1.168$$

or

$$Ax + By + Cz + D = 0 \qquad 1.169$$

where

$$D = -(Ax_o + By_o + Cz_o) \qquad 1.170$$

Constants A, B, and C are found from the cross product giving the normal vector **N**.

$$\mathbf{N} = \mathbf{V}_1 \times \mathbf{V}_2 = A\mathbf{i} + B\mathbf{j} + C\mathbf{k} \qquad 1.171$$

Example 1.26

A plane is defined by a point (2,1,−4) and two vectors:

$$\mathbf{V}_1 = (2\mathbf{i} - 3\mathbf{j} + \mathbf{k}) \qquad \mathbf{V}_2 = (2\mathbf{j} - 4\mathbf{k})$$

Find the parametric and rectangular plane equations.

The parametric equations (for any values of s and t) are:

$$x = 2 + 2s$$
$$y = 1 - 3s + 2t$$
$$z = -4 + s - 4t$$

The normal vector is found by evaluating the determinant

$$\mathbf{N} = \begin{vmatrix} \mathbf{i} & 2 & 0 \\ \mathbf{j} & -3 & 2 \\ \mathbf{k} & 1 & -4 \end{vmatrix}$$
$$= \mathbf{i}(12-2) - 2(-4\mathbf{j}-2\mathbf{k}) = 10\mathbf{i}+8\mathbf{j}+4\mathbf{k}$$

One form of the rectangular equation is

$$10(x-2) + 8(y-1) + 4(z+4) = 0$$

Another form can be derived from equations 1.169 and 1.170.

$$D = -[(10)(2) + (8)(1) + (4)(-4)] = -12$$
$$\mathbf{P} = 10x + 8y + 4z - 12 = 0$$

Three noncollinear points may be used to describe a plane using the following procedure:

step 1: Form vectors $\mathbf{V}_1$ and $\mathbf{V}_2$ from two pairs of the points.

step 2: Find the normal vector $\mathbf{N} = \mathbf{V}_1 \times \mathbf{V}_2$.

step 3: Write the rectangular form of the plane using A, B, and C from the normal vector and any one of the three points.

If the rectangular form of the plane is known, it may be used to write parametric equations. In this case, 2 of the three variables (x, y, z) replace the parameters s and t.

Example 1.27

Find the rectangular and parametric equations of a plane containing the following points: (2,1,−4); (4,−2,−3); (2,3,−8).

Use the first two points to find $\mathbf{V}_1$:

$$\mathbf{V}_1 = (4-2)\mathbf{i}+(-2-1)\mathbf{j}+(-3-(-4))\mathbf{k}$$
$$= 2\mathbf{i} - 3\mathbf{j} + \mathbf{k}$$

Similarly,

$$\mathbf{V}_2 = (2-2)\mathbf{i}+(3-1)\mathbf{j}+(-8-(-4))\mathbf{k}$$
$$= 2\mathbf{j}-4\mathbf{k}$$

From the previous example,

$$\mathbf{N} = 10\mathbf{i} + 8\mathbf{j} + 4\mathbf{k}$$
$$\mathbf{P} = 10x + 8y + 4z - 12 = 0$$

Dividing the rectangular form by 4 gives

$$2.5x + 2y + z - 3 = 0$$

or

$$z = 3 - 2y - 2.5x$$

Using x and y as the parameters, the parametric equations are

$$x = x$$
$$y = y$$
$$z = 3 - 2y - 2.5x$$

The angle between two planes is the same as the angle between their normal vectors, as calculated from the following equation:

$$\cos\phi = \frac{|\mathbf{N}_1 \cdot \mathbf{N}_2|}{|\mathbf{N}_1|\;|\mathbf{N}_2|}$$
$$= \frac{|A_1A_2+B_1B_2+C_1C_2|}{\sqrt{A_1^2+B_1^2+C_1^2}\,\sqrt{A_2^2+B_2^2+C_2^2}} \qquad 1.172$$

A vector equation of the line formed by the intersection of two planes is given by the cross product ($\mathbf{N}_1 \times \mathbf{N}_2$). The distance from a point (x′,y′,z′) to a plane is given by

$$d = \frac{Ax'+By'+Cz'+D}{\sqrt{A^2+B^2+C^2}} \qquad 1.173$$

9. Conic Sections

A. Circle: The center-radius form of a circle with radius r and center at (h,k) is

$$(x-h)^2 + (y-k)^2 = r^2 \qquad 1.174$$

The x-intercept is found by letting y = 0 and solving for x. The y-intercept is found similarly.

The general form is

$$x^2 + y^2 + Dx + Ey + F = 0 \qquad 1.175$$

This can be converted to the center-radius form.

$$(x+\tfrac{1}{2}D)^2 + (y+\tfrac{1}{2}E)^2 = \tfrac{1}{4}(D^2+E^2-4F) \qquad 1.176$$

If the right-hand side is greater than zero, the equation is that of a circle with center at $(-\tfrac{1}{2}D, -\tfrac{1}{2}E)$ and radius given by the square root of the right-hand side. If the right-hand side is zero, the equation is that of a point. If the right-hand side is negative, the plot is imaginary.

B. Parabola: A parabola is formed by a locus of points equidistant from point F and the directrix.

$$(y-k)^2 = 4p(x-h) \qquad 1.177$$

The above equation represents a parabola with vertex at (h,k), focus at (p+h,k), and directrix equation $x = h-p$. The parabola points to the left if $p>0$ and points to the right if $p<0$.

$$(x-h)^2 = 4p(y-k) \qquad 1.178$$

The above equation represents a parabola with vertex at (h,k), focus at $(h,p+k)$, and directrix equation $y = k-p$. The parabola points down if $p>0$ and points up if $p<0$.

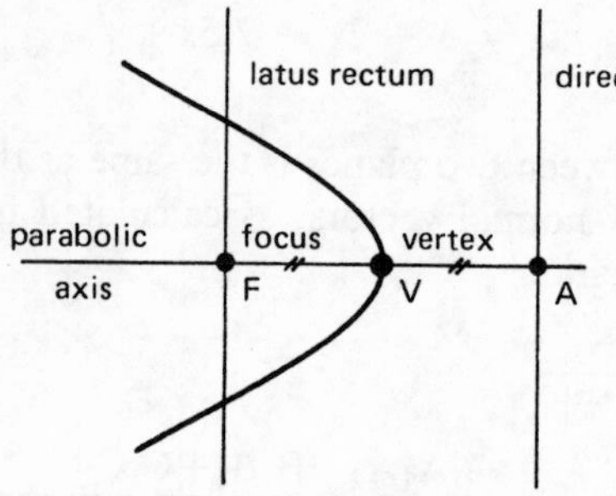

Figure 1.7 A Parabola

An alternate form of the vertically-oriented parabola is

$$y = Ax^2 + Bx + C \qquad 1.179$$

This parabola has a vertex at $\left(\frac{-B}{2A}, C-\frac{B^2}{4A}\right)$ and points down if $A>0$ and points up if $A<0$.

C. Ellipse: An ellipse is formed from a locus of points such that the sum of distances from the two foci is constant. The distance between the two foci is 2c. The sum of those distances is

$$F_1P + PF_2 = 2a \qquad 1.180$$

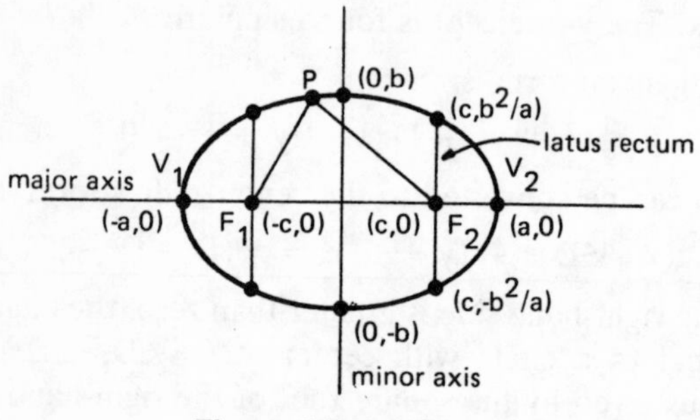

Figure 1.8 An Ellipse

The eccentricity of an ellipse is less than 1, and is equal to

$$e = \frac{\sqrt{a^2 - b^2}}{a} \qquad 1.181$$

For an ellipse centered at the origin,

$$(x/a)^2 + (y/b)^2 = 1 \qquad 1.182$$

$$b^2 = a^2 - c^2 \qquad 1.183$$

If $a>b$, the ellipse is wider than it is tall. If $a<b$, it is taller than it is wide.

For an ellipse centered at (h,k),

$$\frac{(x-h)^2}{a^2} + \frac{(y-k)^2}{b^2} = 1 \qquad 1.184$$

The general form of an ellipse is

$$Ax^2 + Cy^2 + Dx + Ey + F = 0 \qquad 1.185$$

If $A \neq C$ and both have the same sign, the general form can be written as

$$A\left(x + \frac{D}{2A}\right)^2 + C\left(y + \frac{E}{2C}\right)^2 = M \qquad 1.186$$

$$M = D^2/4A + E^2/4C - F \qquad 1.187$$

If $M = 0$, the graph is a single point at

$$(-D/2A,\ -E/2C).$$

If $M < 0$, the graph is the null set.

If $M > 0$, then the ellipse is centered at

$$(-D/2A,\ -E/2C)$$

and the equation can be rewritten

$$\frac{(x+D/2A)^2}{M/A} + \frac{(y+E/2C)^2}{M/C} = 1 \qquad 1.188$$

D. Hyperbola: A hyperbola is a locus of points such that $F_1P - PF_2 = 2a$. The distance between the foci is 2c, and $a<c$.

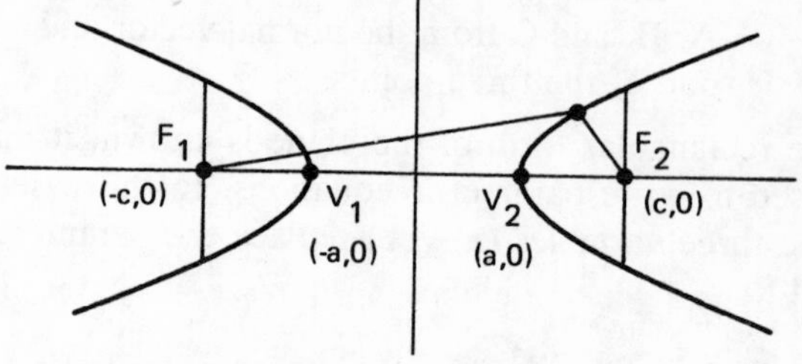

Figure 1.9 A Hyperbola

For a hyperbola centered at the origin with foci on the x-axis,

$$(x/a)^2 - (y/b)^2 = 1 \quad \text{with } b^2 = c^2 - a^2 \qquad 1.189$$

If the foci are on the y-axis,

$$(y/a)^2 - (x/b)^2 = 1 \qquad 1.190$$

The coordinates and length of the latus recta are the same as for the ellipse. The hyperbola is asymptotic to the lines

$$y = \pm\left(\frac{b}{a}\right)x \qquad 1.191$$

The asymptotes need not be perpendicular, but if they are, the hyperbola is known as a rectangular hyperbola. If the asymptotes are the x and y axes, the equation of the hyperbola is

$$xy = \pm a^2 \qquad 1.192$$

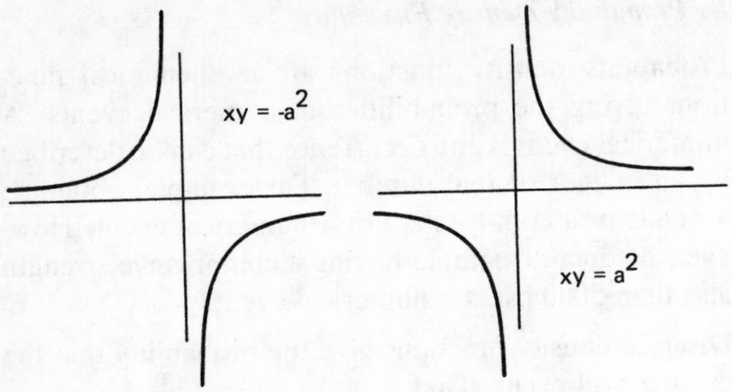

Figure 1.10 Rectangular Hyperbolas

In general, for a hyperbola with transverse axis parallel to the x-axis and center at (h,k),

$$\frac{(x-h)^2}{a^2} - \frac{(y-k)^2}{b^2} = 1 \qquad 1.193$$

The general form of the hyperbolic equation is

$$Ax^2 + Cy^2 + Dx + Ey + F = 0 \qquad 1.194$$

If AC<0, the equation can be rewritten as

$$A(x+\frac{D}{2A})^2 + C(y+\frac{E}{2C})^2 = M \qquad 1.195$$

$$\text{where } M = \frac{D^2}{4A} + \frac{E^2}{4C} - F \qquad 1.196$$

If $M = 0$, the graph is two intersecting lines.

If $M \neq 0$, the graph is a hyperbola with center at

$$(-D/2A, \ -E/2C)$$

The transverse axis is horizontal if (M/A) is positive. It is vertical if (M/C) is positive.

10. Spheres

The equation of a sphere whose center is at the point (h,k,l) and whose radius is r is

$$(x-h)^2 + (y-k)^2 + (z-l)^2 = r^2 \qquad 1.197$$

If the sphere is centered at the origin, its equation is

$$x^2 + y^2 + z^2 = r^2 \qquad 1.198$$

11. Permutations and Combinations

Suppose you have *n* objects, and you wish to work with a subset of *r* of them. An order-conscious arrangement of *n* objects taken *r* at a time is known as a *permutation.* The permutation is said to be order-conscious because the arrangement of two objects (say A and B) as AB is different than the arrangement BA. There are a number of ways of taking *n* objects *r* at a time. The total number of possible permutations is

$$P(n,r) = \frac{n!}{(n-r)!} \qquad 1.199$$

Example 1.28

A shelf has room for only 3 vases. If 4 different vases are available, how many ways can the shelf be arranged?

$$P(4,3) = \frac{4!}{(4-3)!} = \frac{(4)(3)(2)(1)}{(1)} = 24$$

The special cases of *n* objects taken *n* at a time are illustrated by the following examples.

Example 1.29

How many ways can 7 resistors be connected end-to-end into a single unit?

$$P(7,7) = \frac{7!}{(7-7)!} = \frac{7!}{0!} = 7! = 5040$$

Example 1.30

5 people are to sit at a round table with 5 chairs. How many ways can these 5 people be arranged so that they all have different companions?

This is known as a *ring permutation.* Since the starting point of the arrangement around the circle does not affect the number of permutations, the answer is $(5-1)! = 4! = 24$.

An arrangement of *n* objects taken *r* at a time is known as a *combination* if the arrangement is not order-conscious. The total number of possible combinations is

$$C(n,r) = \frac{n!}{(n-r)!r!} \qquad 1.200$$

Example 1.31

How many possible ways can 6 people fit into a 4-seat boat?

$$C(6,4) = \frac{6!}{(6-4)!4!} = \frac{(6)(5)(4)(3)(2)(1)}{(2)(1)(4)(3)(2)(1)} = 15$$

12. Probability and Statistics

A. Probability Rules:

The following rules are applied to sample spaces **A** and **B**:

$\mathbf{A} = [A_1, A_2, A_3, \ldots, A_n]$ and $\mathbf{B} = [B_1, B_2, B_3, \ldots, B_n]$ where the A_i and B_i are independent.

Rule 1:

$p\{\emptyset\} =$ probability of an impossible event $= 0$ 1.201

Example 1.32

An urn contains 5 white balls, 2 red balls, and 3 green balls. What is the probability of drawing a blue ball from the urn?

$$p\{\text{blue ball}\} = p\{\emptyset\} = 0$$

Rule 2:

$$p\{A_1 \text{ or } A_2 \text{ or} \ldots \text{ or } A_n\} = p\{A_1\} + p\{A_2\} + \cdots + p\{A_n\} \qquad 1.202$$

Example 1.33

Returning to the urn described in example 1.32, what is the probability of getting either a white ball or a red ball in one draw from the urn?

$$p\{\text{red or white}\} = p\{\text{red}\} + p\{\text{white}\} = .5 + .2 = .7$$

Rule 3:

$$p\{A_i \text{ and } B_i \text{ and} \ldots Z_i\} = p\{A_i\}p\{B_i\} \ldots p\{Z_i\} \qquad 1.203$$

Example 1.34

Given two identical urns (as described in example 1.32), what is the probability of getting a red ball from the first urn and a green ball from the second urn, given one draw from each urn?

$$p\{\text{red and green}\} = p\{\text{red}\}p\{\text{green}\} = (.2)(.3) = .06$$

Rule 4:

$$p\{\text{not } A\} = \text{probability of event A not occurring} = 1 - p\{A\} \qquad 1.204$$

Example 1.35

Given the urn of example 1.32, what is the probability of not getting a red ball from the urn in one draw?

$$p\{\text{not red}\} = 1 - p\{\text{red}\} = 1 - .2 = .8$$

Rule 5:

$$p\{A_i \text{ or } B_i\} = p\{A_i\} + p\{B_i\} - p\{A_i\}p\{B_i\} \qquad 1.205$$

Example 1.36

Given one urn as described in example 1.32 and a second urn containing 8 red balls and 2 black balls, what is the probability of drawing either a white ball from the first urn or a red ball from the second urn, given one draw from each?

$$\begin{aligned} p\{\text{white or red}\} &= p\{\text{white}\} + p\{\text{red}\} \\ &\quad - p\{\text{white}\}p\{\text{red}\} \\ &= .5 + .8 - (.5)(.8) = .9 \end{aligned}$$

Rule 6:

$p\{A|B\}$ = probability that A will occur given that B has already occurred, where the two events are dependent.

$$= \frac{p\{A \text{ and } B\}}{p\{B\}} \qquad 1.206$$

The above equation is known as Bayes theorem.

B. Probability Density Functions

Probability density functions are mathematical functions giving the probabilities of numerical events. A numerical event is any occurrence that can be described by an integer or real number. For example, obtaining a heads in a coin-toss is not a numerical event. However, a concrete sample having a compressive strength less than 5000 psi is a numerical event.

Discrete density functions give the probability that the event x will occur. That is,

$$f\{x\} = \text{probability of a process having a value of x}$$

Important discrete functions are the binomial and poisson distributions.

1. Binomial

- n is the number of trials
- x is the number of successes
- p is the probability of a success in a single trial
- q is the probability of failure, $1 - p$
- $\binom{n}{x}$ is the binomial coefficient $= \dfrac{n!}{(n-x)!x!}$
- $x! = x(x-1)(x-2) \cdots (2)(1)$

Then, the probability of obtaining x successes in n trials is

$$f\{x\} = \binom{n}{x} p^x q^{(n-x)} \qquad 1.207$$

The mean of the binomial distribution is np. The variance of the distribution is npq.

Example 1.37

In a large quantity of items, 5% are defective. If 7 items are sampled, what is the probability that exactly three will be defective?

$$f\{3\} = \binom{7}{3}(.05)^3(.95)^4 = .0036$$

2. Poisson

Suppose an event occurs, on the average, λ times per period. The probability that the event will occur x times per period is

$$f\{x\} = \frac{e^{-\lambda}\lambda^x}{x!} \qquad 1.208$$

λ is both the distribution mean and variance. λ must be a number greater than zero.

Example 1.38

The number of customers arriving in some period is distributed as poisson with a mean of 8. What is the probability that 6 customers will arrive in any given period?

$$f\{6\} = \frac{e^{-8}8^6}{6!} = .122$$

Continuous probability density functions are used to find the cumulative distribution functions, F{x}. Cumulative distribution functions give the probability of event x or less occurring.

x = any value, not necessarily an integer

$$f\{x\} = \frac{dF\{x\}}{dx} \qquad 1.209$$

F{x} = probability of x or less occurring

3. Exponential

$$f\{x\} = u(e^{-ux}) \qquad 1.210$$

$$F\{x\} = 1 - e^{-ux} \qquad 1.211$$

The mean of the exponential distribution is 1/u. The variance is $(1/u)^2$.

Example 1.39

The reliability of a unit is exponentially distributed with mean time to failure (MTBF) of 1000 hours. What is the probability that the unit will be operational at t = 1200 hours?

The reliability of an item is (1 − probability of failing before time t). Therefore,

$$R\{t\} = 1 - F\{t\} = 1 - (1 - e^{-ux}) = e^{-ux}$$

$$u = 1/\text{MTBF} = 1/1000 = .001$$

$$R\{1200\} = e^{-(.001)(1200)} = .3$$

4. Normal

Although f{x} may be expressed mathematically for the normal distribution, tables are used to evaluate F{x} since f{x} cannot be easily integrated. Since the x axis of the normal distribution will seldom correspond to actual sample variables, the sample values are converted into standard values. Given the mean, u, and the standard deviation, σ, the standard normal variable is

$$z = \frac{(\text{sample value} - u)}{\sigma} \qquad 1.212$$

Then, the probability of a sample exceeding the given sample value is equal to the area in the tail past point z.

Example 1.40

Given a population that is normally distributed with mean of 66 and standard deviation of 5, what percent of the population exceeds 72?

$$z = \frac{(72-66)}{5} = 1.2$$

Then, from table 1.4,

$$p\{\text{exceeding } 72\} = .5 - .3849 = .1151 \text{ or } 11.5\%$$

C. Statistical Analysis of Experimental Data

Experiments can take on many forms. An experiment might consist of measuring the weight of one cubic foot of concrete. Or, an experiment might consist of measuring the speed of a car on a roadway. Generally, such experiments are performed more than once to increase the precision and accuracy of the results.

Of course, the intrinsic variability of the process being measured will cause the observations to vary, and we would not expect the experiment to yield the same result each time it was performed. Eventually, a collection of experimental outcomes (observations) will be available for analysis.

One fundamental technique for organizing random observations is the frequency distribution. The frequency distribution is a systematic method for ordering the observations from small to large, according to some convenient numerical characteristic.

Example 1.41

The number of cars that travel through an intersection between 12 noon and 1 p.m. is measured for 30 consecutive working days. The results of the 30 observations are:

79, 66, 72, 70, 68, 66, 68, 76, 73, 71, 74, 70, 71, 69, 67, 74, 70, 68, 69, 64, 75, 70, 68, 69, 64, 69, 62, 63, 63, 61

What is the frequency distribution using an interval of 2 cars per hour?

cars per hour	frequency of occurrence
60–61	1
62–63	3
64–65	2
66–67	3
68–69	8
70–71	6
72–73	2
74–75	3
76–77	1
78–79	1

In example 1.41, 2 cars per hour is known as the step interval. The step interval should be chosen so that the data is presented in a meaningful manner. If there are too many intervals, many of them will have zero fre-

Table 1.4 Areas Under The Standard Normal Curve
(0 to z)

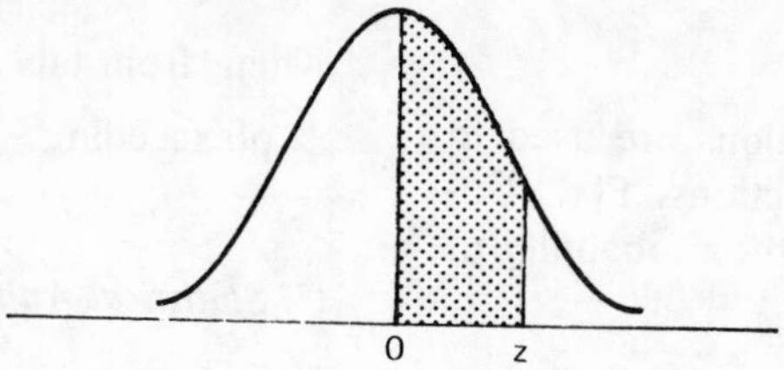

z	0	1	2	3	4	5	6	7	8	9
0.0	.0000	.0040	.0080	.0120	.0160	.0199	.0239	.0279	.0319	.0359
0.1	.0398	.0438	.0478	.0517	.0557	.0596	.0636	.0675	.0714	.0754
0.2	.0793	.0832	.0871	.0910	.0948	.0987	.1026	.1064	.1103	.1141
0.3	.1179	.1217	.1255	.1293	.1331	.1368	.1406	.1443	.1480	.1517
0.4	.1554	.1591	.1628	.1664	.1700	.1736	.1772	.1808	.1844	.1879
0.5	.1915	.1950	.1985	.2019	.2054	.2088	.2123	.2157	.2190	.2224
0.6	.2258	.2291	.2324	.2357	.2389	.2422	.2454	.2486	.2518	.2549
0.7	.2580	.2612	.2642	.2673	.2704	.2734	.2764	.2794	.2823	.2852
0.8	.2881	.2910	.2939	.2967	.2996	.3023	.3051	.3078	.3106	.3133
0.9	.3159	.3186	.3212	.3238	.3264	.3289	.3315	.3340	.3365	.3389
1.0	.3413	.3438	.3461	.3485	.3508	.3531	.3554	.3577	.3599	.3621
1.1	.3643	.3665	.3686	.3708	.3729	.3749	.3770	.3790	.3810	.3830
1.2	.3849	.3869	.3888	.3907	.3925	.3944	.3962	.3980	.3997	.4015
1.3	.4032	.4049	.4066	.4082	.4099	.4115	.4131	.4147	.4162	.4177
1.4	.4192	.4207	.4222	.4236	.4251	.4265	.4279	.4292	.4306	.4319
1.5	.4332	.4345	.4357	.4370	.4382	.4394	.4406	.4418	.4429	.4441
1.6	.4452	.4463	.4474	.4484	.4495	.4505	.4515	.4525	.4535	.4545
1.7	.4554	.4564	.4573	.4582	.4591	.4599	.4608	.4616	.4625	.4633
1.8	.4641	.4649	.4656	.4664	.4671	.4678	.4686	.4693	.4699	.4706
1.9	.4713	.4719	.4726	.4732	.4738	.4744	.4750	.4756	.4761	.4767
2.0	.4772	.4778	.4783	.4788	.4793	.4798	.4803	.4808	.4812	.4817
2.1	.4821	.4826	.4830	.4834	.4838	.4842	.4846	.4850	.4854	.4857
2.2	.4861	.4864	.4868	.4871	.4875	.4878	.4881	.4884	.4887	.4890
2.3	.4893	.4896	.4898	.4901	.4904	.4906	.4909	.4911	.4913	.4916
2.4	.4918	.4920	.4922	.4925	.4927	.4929	.4931	.4932	.4934	.4936
2.5	.4938	.4940	.4941	.4943	.4945	.4946	.4948	.4949	.4951	.4952
2.6	.4953	.4955	.4956	.4957	.4959	.4960	.4961	.4962	.4963	.4964
2.7	.4965	.4966	.4967	.4968	.4969	.4970	.4971	.4972	.4973	.4974
2.8	.4974	.4975	.4976	.4977	.4977	.4978	.4979	.4979	.4980	.4981
2.9	.4981	.4982	.4982	.4983	.4984	.4984	.4985	.4985	.4986	.4986
3.0	.4987	.4987	.4987	.4988	.4988	.4989	.4989	.4989	.4990	.4990
3.1	.4990	.4991	.4991	.4991	.4992	.4992	.4992	.4992	.4993	.4993
3.2	.4993	.4993	.4994	.4994	.4994	.4994	.4994	.4995	.4995	.4995
3.3	.4995	.4995	.4995	.4996	.4996	.4996	.4996	.4996	.4996	.4997
3.4	.4997	.4997	.4997	.4997	.4997	.4997	.4997	.4997	.4997	.4998
3.5	.4998	.4998	.4998	.4998	.4998	.4998	.4998	.4998	.4998	.4998
3.6	.4998	.4998	.4999	.4999	.4999	.4999	.4999	.4999	.4999	.4999
3.7	.4999	.4999	.4999	.4999	.4999	.4999	.4999	.4999	.4999	.4999
3.8	.4999	.4999	.4999	.4999	.4999	.4999	.4999	.4999	.4999	.4999
3.9	.5000	.5000	.5000	.5000	.5000	.5000	.5000	.5000	.5000	.5000

quencies. If there are too few intervals, the frequency distribution will have little value. Generally, 10 to 15 intervals are used.

Once the frequency distribution is complete, it may be represented graphically as a histogram. The procedure in drawing a histogram is to mark off the interval limits on a number line and then draw bars with lengths that are proportional to the frequencies in the intervals. If it is necessary to show the continuous nature of the data, a frequency polygon can be drawn.

Example 1.42

Draw the frequency histogram and frequency polygon for the data given in example 1.41.

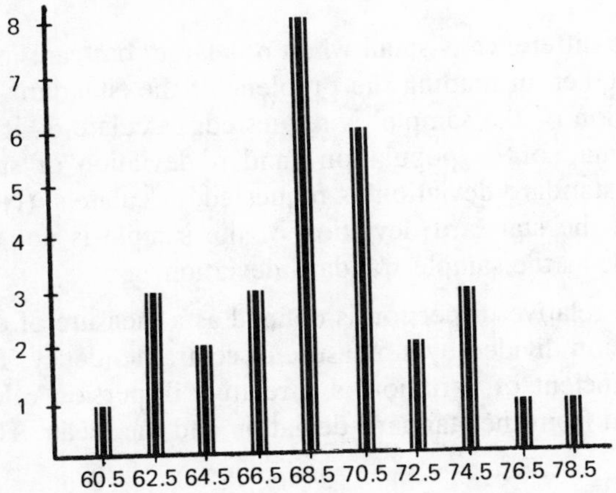

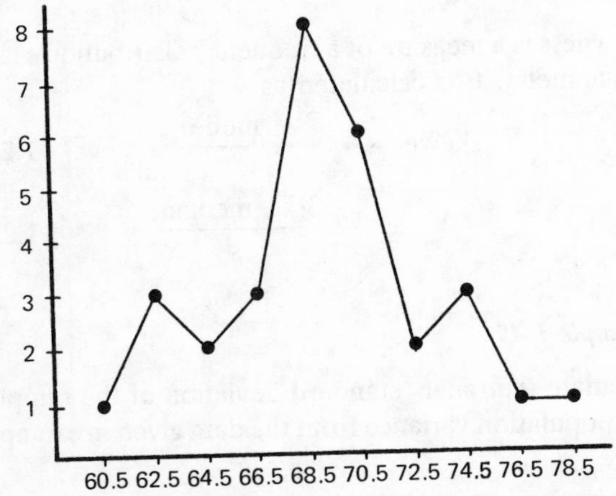

If it is necessary to know the number or percentage of observations that occur up to and including some value, the cumulative frequency table can be formed. This procedure is illustrated in the following example.

Example 1.43

Form the cumulative frequency distribution and graph for the data given in example 1.41.

cars per hour	frequency	cumulative frequency	cumulative per cent
60–61	1	1	3
62–63	3	4	13
64–65	2	6	20
66–67	3	9	30
68–69	8	17	57
70–71	6	23	77
72–73	2	25	83
74–75	3	28	93
76–77	1	29	97
78–79	1	30	100

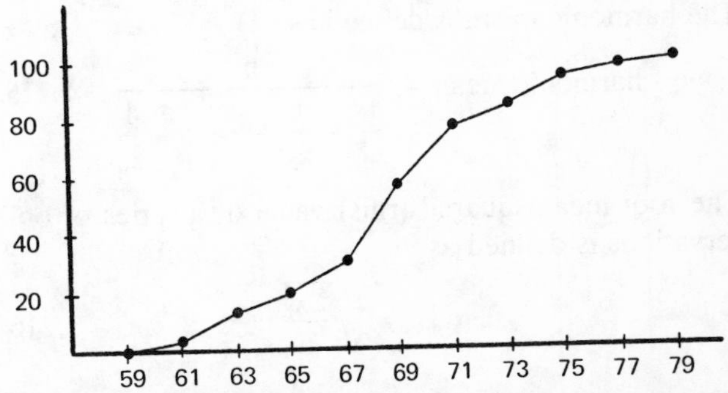

It is often unnecessary to present the experimental data in its entirety, either in tabular or graphical form. In such cases, the data and distribution can be represented by various parameters. One type of parameter is a measure of central tendency. The mode, median, and mean are measures of central tendency. The other type of parameter is a measure of dispersion. Standard deviation and variance are measures of dispersion.

The mode is the observed value which occurs most frequently. The mode may vary greatly between series of observations. Therefore, its main use is as a quick measure of the central value since no computation is required to find it. Beyond this, the usefulness of the mode is limited.

The median is the point in the distribution which divides the total observations into two parts containing equal numbers of observations. It is not influenced by the extremity of scores on either side of the distribution. The median is found by counting up (from either end of the frequency distribution) until half of the observations have been accounted for. The procedure is more difficult if the median falls within an interval, as is illustrated in example 1.44.

Similar in concept to the median are percentile ranks, quartiles, and deciles. The median could also have been called the 50th percentile observation. Similarly, the 80th percentile would be the number of cars per hour for which the cumulative frequency was 80%. The quartile and decile points on the distribution divide the observations or distribution into segments of 25% and 10% respectively.

The arithmetic mean is the arithmetic average of the observations. The mean may be found without ordering the data (which was necessary to find the mode and median). The mean can be found from the following formula:

$$\bar{x} = \left(\frac{1}{n}\right)(x_1 + x_2 + \cdots + x_n) = \frac{\Sigma x_i}{n} \qquad 1.213$$

The geometric mean is occasionally used when it is necessary to average ratios. The geometric mean is calculated as

$$\text{geometric mean} = \sqrt[n]{x_1 x_2 x_3 \cdots x_n} \qquad 1.214$$

The harmonic mean is defined as

$$\text{harmonic mean} = \frac{n}{\frac{1}{x_1} + \frac{1}{x_2} + \cdots + \frac{1}{x_n}} \qquad 1.215$$

The root-mean-squared (rms) value of a series of observations is defined as

$$x_{rms} = \sqrt{\frac{\Sigma x_i^2}{n}} \qquad 1.216$$

Example 1.44

Find the mode, median, and arithmetic mean of the distribution represented by the data given in example 1.41.

The mode is the interval 68–69, since this interval has the highest frequency. If 68.5 is taken as the interval center, then 68.5 would be the mode.

Since there are 30 observations, the median is the value which separates the observations into 2 groups of 15. From example 1.43, the median occurs some place within the 68–69 interval. Up through interval 66–67, there are 9 observations, so 6 more are needed to make 15. Interval 68–69 has 8 observations, so the mean is found to be (6/8) or (3/4) of the way through the interval. Since the real limits of the interval are 67.5 and 69.5, the median is located at

$$67.5 + \tfrac{3}{4}(69.5 - 67.5) = 69$$

The mean can be found from the raw data or from the grouped data using the interval center as the assumed observation value. Using the raw data,

$$\bar{x} = \frac{\Sigma x}{n} = \frac{2069}{30} = 68.97$$

The simplest statistical parameter which describes the variation in observed data is the range. The range is found by subtracting the smallest value from the largest. Since the range is influenced by extreme (low probability) observations, its use is limited as a measure of variability.

The standard deviation is a better estimate of variability because it considers every observation. The standard deviation can be found from:

$$\sigma = \sqrt{\frac{\Sigma(x_i - \bar{x})^2}{n}} = \sqrt{\frac{\Sigma x_i^2}{n} - (\bar{x})^2} \qquad 1.217$$

The above formula assumes that n is a large number, such as above 50. Theoretically, n is the size of the entire population. If a small sample (less than 50) is used to calculate the standard deviation of the distribution, the formulas are changed. The *sample* standard deviation is

$$s = \sqrt{\frac{\Sigma(x_i - \bar{x})^2}{n-1}} = \sqrt{\frac{\Sigma x_i^2 - (\Sigma x_i)^2/n}{n-1}} \qquad 1.218$$

The difference is small when n is large, but care must be taken in reading the problem. If the 'standard deviation of the sample' is requested, calculate σ. If an estimate of the 'population standard deviation' or 'sample standard deviation' is requested, calculate s. (Note that the standard deviation of the sample is not the same as the sample standard deviation.)

The relative dispersion is defined as a measure of dispersion divided by a measure of central tendency. The coefficient of variation is a relative dispersion calculated from the standard deviation and the mean. That is,

$$\text{coefficient of variation} = \frac{s}{\bar{x}} \qquad 1.219$$

Skewness is a measure of a frequency distribution's lack of symmetry. It is calculated as

$$\text{skewness} = \frac{\bar{x} - \text{mode}}{s} \qquad 1.220$$

$$\approx \frac{3(\bar{x} - \text{median})}{s} \qquad 1.221$$

Example 1.45

Calculate the range, standard deviation of the sample, and population variance from the data given in example 1.41.

$\Sigma x = 2069 \quad (\Sigma x)^2 = 4280761 \quad \Sigma x^2 = 143225$

$n = 30 \quad \bar{x} = 68.97$

$$\sigma = \sqrt{\frac{143225}{30} - (68.97)^2} = 4.16$$

$$s = \sqrt{\frac{143225 - (4280761)/30}{29}} = 4.29$$

$$s^2 = 18.4$$

$$R = 79 - 61 = 18$$

Referring again to example 1.41, suppose that the hourly through-put for 15 similar intersections is measured over a 30 day period. At the end of the 30 day period, there will be 15 ranges, 15 medians, 15 means, 15 standard deviations, and so on. These parameters themselves constitute distributions.

The mean of the sample means is an excellent estimator of the average hourly through-put of an intersection, μ.

$$\mu = (1/15)\Sigma\bar{x}$$

The standard deviation of the sample means is known as the standard error of the mean to distinguish it from the standard deviation of the raw data. The standard error is written as $\sigma_{\bar{x}}$.

The standard error is not a good estimator of the population standard deviation, σ'.

In general, if k sets of n observations each are used to estimate the population mean (μ) and the population standard deviation (σ'), then

$$\mu \approx \left(\frac{1}{k}\right)\Sigma\bar{x} \qquad 1.222$$

$$\sigma' \approx \sqrt{k}\,\sigma_{\bar{x}} \qquad 1.223$$

13. Complex Number Arithmetic

Despite the fact that no negative number has a square root in the real number system, it is possible to develop an algebraic expression which contains such a square root. The imaginary operator, i[1], is used to indicate the square root of -1. That is,

$$i = \sqrt{-1} \qquad 1.224$$

The square root of all other negative numbers can be written in terms of i. For example, $\sqrt{-4} = i\sqrt{4} = 2i$.

A complex number consists of a real part and an imaginary part, e.g., $(a + ib)$. The complex number can be plotted in the real-imaginary coordinate system, as shown in figure 1.11.

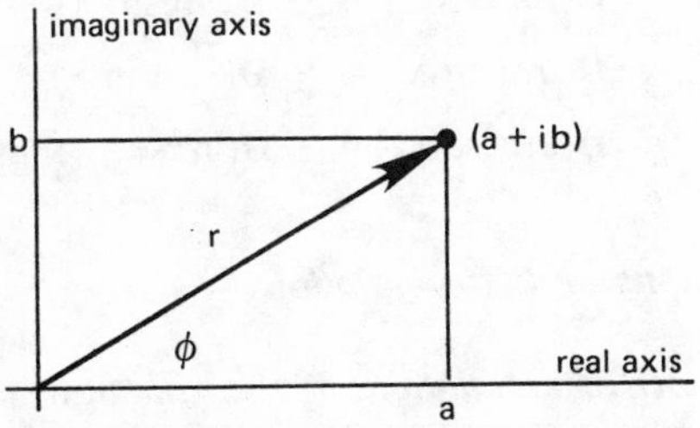

Figure 1.11 A Complex Number

[1]The symbol j is commonly used in electrical engineering to indicate the imaginary operator.

Most standard algebraic expressions work with complex numbers. Notable exceptions are the inequality operators. The concept of one complex number being less than or greater than another complex number is meaningless.

The following examples illustrate the straight-forward method of adding and multiplying complex numbers.

Example 1.46

Add the complex numbers $(3+4i)$ and $(2+i)$.

$$\begin{aligned}(3+4i) + (2+i) &= ((3+2) + (4+1)i)\\ &= (5+5i)\end{aligned}$$

Example 1.47

Multiply the complex numbers $(7+2i)$ and $(5-3i)$.

$$\begin{aligned}(7+2i)(5-3i) &= ((7)(5)-(7)(3i)+(2i)(5)-(2i)(3i))\\ &= (35-21i+10i-6i^2)\\ &= (35-21i+10i-6(-1))\\ &= (41-11i)\end{aligned}$$

Division is not as easy to perform on complex numbers in the rectangular form. If necessary, the complex conjugate of a complex number can be used, as shown in the following example. The complex conjugate of a complex number $(a+bi)$ is $(a-bi)$.

Example 1.48

Divide $(2+3i)$ by $(4-5i)$.

$$\begin{aligned}\frac{(2+3i)}{(4-5i)} &= \frac{(2+3i)(4+5i)}{(4-5i)(4+5i)}\\ &= \frac{(-7+22i)}{(4)^2 + (5)^2}\\ &= (-7/41) + (22/41)i\end{aligned}$$

The complex number $(a+ib)$ can also be expressed in phasor (polar) form by using the equations 1.225 and 1.226.

$$r = \sqrt{a^2 + b^2} \qquad 1.225$$

$$\phi = \arctan(b/a) \qquad 1.226$$

The phasor form of $(a+ib)$ is, then, $r\angle\phi$. Since

$$a = r(\cos\phi) \qquad 1.227$$

$$b = r(\sin\phi) \qquad 1.228$$

the rectangular form of the imaginary number can be written as

$$a + ib = r(\cos\phi) + r(i\sin\phi) = r(\cos\phi + i\sin\phi) \qquad 1.229$$

$$= r(\text{cis}\phi) \qquad 1.230$$

From Euler's theorem,

$$e^{i\phi} = \text{cis}\phi \qquad 1.231$$

MATH

So, the imaginary number $(a + ib)$ can also be expressed as an exponential.

$$a + ib = r(e^{i\phi}) \qquad 1.232$$

Table 1.5 Important Exponential Relationships

$$e^{i\phi} = \cos\phi + i\sin\phi$$
$$e^{-i\phi} = \cos\phi - i\sin\phi$$
$$\cos\phi = \tfrac{1}{2}(e^{i\phi} + e^{-i\phi})$$
$$\sin\phi = \tfrac{1}{2}{}_i(e^{i\phi} - e^{-i\phi})$$

Example 1.49

What is the exponential form of the vector $\mathbf{A} = (3 + i4)$?

$$r = \sqrt{3^2 + 4^2} = 5$$
$$\phi = \arctan(4/3) = 53.1°$$
$$\mathbf{A} = 5\, e^{i(53.1°)}$$

The addition and subtraction of complex numbers is best done in rectangular form.

$$(a_1 + ib_1) + (a_2 + ib_2) = (a_1 + a_2) + i(b_1 + b_2) \qquad 1.233$$
$$(a_1 + ib_1) - (a_2 + ib_2) = (a_1 - a_2) + i(b_1 - b_2) \qquad 1.234$$

The multiplication and division of complex numbers is best done in phasor or exponential form.

$$(r_1 \angle\phi_1)(r_2 \angle\phi_2) = r_1 r_2 \angle\phi_1 + \phi_2 \qquad 1.235$$
$$(r_1 e^{i\phi_1})(r_2 e^{i\phi_2}) = r_1 r_2 e^{i(\phi_1 + \phi_2)} \qquad 1.236$$
$$\frac{r_1 \angle\phi_1}{r_2 \angle\phi_2} = \left(\frac{r_1}{r_2}\right) \angle\phi_1 - \phi_2 \qquad 1.237$$
$$\frac{r_1 e^{i\phi_1}}{r_2 e^{i\phi_2}} = \left(\frac{r_1}{r_2}\right) e^{i(\phi_1 - \phi_2)} \qquad 1.238$$

Taking powers and roots of complex numbers requires the use of DeMoivre's theorem. That theorem is

$$(r\, cis\phi)^n = r^n\, cis(n\phi) \qquad 1.239$$

Since trigonometric functions of angles ϕ and $(\phi + 360°)$ are the same, DeMoivre's theorem may be rewritten in the case of a root of a complex number. The n values (corresponding to $k = 0,1,2,\cdots,n-1$) of the nth root of a complex number are

$$\sqrt[n]{r cis\phi} = \sqrt[n]{r}\; cis\left(\frac{\phi + k(360)}{n}\right) \qquad 1.240$$

14. Differential Calculus

A. Terminology: Given y, a function of x, the first derivative with respect to x may be written as $\mathbf{D}y$, y', or (dy/dx). The first derivative corresponds to the slope of the line described by the function y. The second derivative may be written as $\mathbf{D}^2y$, (d^2y/dx^2) or y''.

B. Basic Operations: In the formulas below, f and g are functions of x. **D** is the derivative operator.

$$\mathbf{D}(a) = 0 \qquad 1.241$$
$$\mathbf{D}(af) = a\mathbf{D}(f) \qquad 1.242$$
$$\mathbf{D}(f + g) = \mathbf{D}(f) + \mathbf{D}(g) \qquad 1.243$$
$$\mathbf{D}(f - g) = \mathbf{D}(f) - \mathbf{D}(g) \qquad 1.244$$
$$\mathbf{D}(f \cdot g) = f\mathbf{D}(g) + g\mathbf{D}(f) \qquad 1.245$$
$$\mathbf{D}(f/g) = (g\mathbf{D}(f) - f\mathbf{D}(g))/g^2 \qquad 1.246$$
$$\mathbf{D}(x^n) = nx^{n-1} \qquad 1.247$$
$$\mathbf{D}(f^n) = nf^{n-1}\mathbf{D}(f) \qquad 1.248$$
$$\mathbf{D}(f(g)) = \mathbf{D}(f(g))\mathbf{D}(g) \qquad 1.249$$
$$\mathbf{D}(\ln x) = 1/x \qquad 1.250$$
$$\mathbf{D}(e^{ax}) = ae^{ax} \qquad 1.251$$

Example 1.50

A function is given as $f(x) = x^3 - 2x$. What is the slope of the line at $x = 3$?

$$y' = 3x^2 - 2$$
$$y'(3) = 27 - 2 = 25$$

C. Transcendental Functions:

$$\mathbf{D}(\sin x) = \cos x \qquad 1.252$$
$$\mathbf{D}(\cos x) = -\sin x \qquad 1.253$$
$$\mathbf{D}(\tan x) = \sec^2 x \qquad 1.254$$
$$\mathbf{D}(\cot x) = -\csc^2 x \qquad 1.255$$
$$\mathbf{D}(\sec x) = (\sec x)(\tan x) \qquad 1.256$$
$$\mathbf{D}(\csc x) = (-\csc x)(\cot x) \qquad 1.257$$
$$\mathbf{D}(\arcsin x) = 1/\sqrt{1 - x^2} \qquad 1.258$$
$$\mathbf{D}(\arctan x) = 1/(1 + x^2) \qquad 1.259$$
$$\mathbf{D}(\text{arcsec } x) = 1/x\sqrt{x^2 - 1} \qquad 1.260$$
$$\mathbf{D}(\arccos x) = -\mathbf{D}(\arcsin x) \qquad 1.261$$
$$\mathbf{D}(\text{arccot } x) = -\mathbf{D}(\arctan x) \qquad 1.262$$
$$\mathbf{D}(\text{arccsc } x) = -\mathbf{D}(\text{arcsec } x) \qquad 1.263$$

D. Variations on Differentiation

1. Partial Differentiation: If the function has two or more independent variables, a partial derivative is found by considering all extraneous variables as constants. The geometric interpretation of the partial derivative $(\partial z/\partial x)$ is the slope of a line tangent to the 3-dimensional surface in a plane of constant y, and parallel to the x axis. Similarly, the interpretation of $(\partial z/\partial y)$ is the slope of a line tangent to the surface in a plane of constant x, and parallel to the y axis.

Example 1.51

A surface has the equation $x^2 + y^2 + z^2 = 9$. What is the slope of a line tangent to (1,2,2) and parallel to the x axis?

$$z = (9 - x^2 - y^2)^{1/2}$$

$$\frac{\partial z}{\partial x} = \frac{-x}{\sqrt{9 - x^2 - y^2}}$$

At the point (1,2,2), $\frac{\partial z}{\partial x} = -\frac{1}{2}$

2. Implicit Differentiation: If a relationship between n variables cannot be manipulated to yield an explicit function of $(n-1)$ independent variables, the relationship implicitly defines the nth remaining variable. The derivative of the implicit variable taken with respect to any other variable is found by a process known as implicit differentiation.

If $f(x,y) = 0$ is a function, the implicit derivative is

$$\frac{dy}{dx} = -\frac{\partial f}{\partial x}\bigg/\frac{\partial f}{\partial y} \qquad 1.264$$

If $f(x,y,z) = 0$ is a function, the implicit derivatives are

$$\frac{\partial z}{\partial x} = -\frac{\partial f}{\partial x}\bigg/\frac{\partial f}{\partial z} \qquad 1.265$$

$$\frac{\partial z}{\partial y} = -\frac{\partial f}{\partial y}\bigg/\frac{\partial f}{\partial z} \qquad 1.266$$

Example 1.52

If $f = x^2 + xy + y^3$, what is $\frac{dy}{dx}$?

Since this function cannot be written as an explicit function of x, implicit differentiation is required.

$$\frac{\partial f}{\partial x} = 2x + y \qquad \frac{\partial f}{\partial y} = x + 3y^2$$

$$\frac{dy}{dx} = \frac{-(2x+y)}{(x+3y^2)}$$

Example 1.53

Solve example 1.51 using implicit differentiation.

$$f = x^2 + y^2 + z^2 - 9$$

$$\frac{\partial f}{\partial x} = 2x \qquad \frac{\partial f}{\partial z} = 2z$$

$$\frac{\partial z}{\partial x} = \frac{-2x}{2z} = -\frac{x}{z}$$

and at (1,2,2), $\frac{\partial z}{\partial x} = -\frac{1}{2}$

3. The Gradient Vector

The slope of a function is defined as the change in one variable with respect to a distance in another direction. Usually, this direction is parallel to an axis. However, the maximum slope at a point on a 3-dimensional object may not be in a direction parallel to one of the coordinate axes.

The gradient vector function $\nabla f(x,y,z)$ (pronounced del f) gives the maximum rate of change of the function $f(x,y,z)$. The gradient vector function is defined as

$$\nabla f(x,y,z) = \frac{\partial f(x,y,z)}{\partial x}\mathbf{i} + \frac{\partial f(x,y,z)}{\partial y}\mathbf{j} + \frac{\partial f(x,y,z)}{\partial z}\mathbf{k} \qquad 1.267$$

Example 1.54

Find the maximum slope of $f(x,y) = 2x^2 - y^2 + 3x - y$ at the point $(1,-2)$. What is the equation of the maximum-slope tangent?

This is a 2-dimensional problem.

$$\frac{\partial f(x,y)}{\partial x} = 4x + 3$$

$$\frac{\partial f(x,y)}{\partial y} = -2y - 1$$

$$\nabla f(x,y) = (4x+3)\mathbf{i} + (-2y-1)\mathbf{j}$$

The equation of the maximum-slope tangent is

$$\nabla f(1,-2) = 7\mathbf{i} + 3\mathbf{j}$$

The magnitude of the slope is

$$\sqrt{(7)^2 + (3)^2} = \sqrt{58}$$

4. The Directional Derivative

The rate of change of a function in the direction of some given vector **U** can be found from the directional derivative function, $\nabla_u f(x,y,z)$. This directional derivative function depends on the gradient vector and the direction cosines of the vector **U**.

$$\nabla_u f(x,y,z) = \frac{\partial f(x,y,z)}{\partial x}\cos\alpha + \frac{\partial f(x,y,z)}{\partial y}\cos\beta + \frac{\partial f(x,y,z)}{\partial z}\cos\gamma \qquad 1.268$$

Example 1.55

What is the rate of change of $f(x,y) = 3x^2 + xy - 2y^2$ at the point $(1,-2)$ in the direction $4\mathbf{i} + 3\mathbf{j}$?

$$\cos\alpha = \frac{4}{\sqrt{(4)^2 + (3)^2}} = 4/5$$

$$\cos\beta = 3/5$$

$$\frac{\partial f(x,y)}{\partial x} = 6x + y$$

$$\frac{\partial f(x,y)}{\partial y} = x - 4y$$

$$\nabla_u f(x,y) = (4/5)(6x+y) + (3/5)(x-4y)$$

$$\nabla_u f(1,-2) = (4/5)[(6)(1)-2] + (3/5)[1-(4)(-2)] = 8.6$$

5. Tangent Plane Function

Partial derivatives may be used to find the tangent plane to a 3-dimensional surface at some point p_o. If the surface is defined by the function $f(x,y,z) = 0$, the equation of the tangent plane is

$$\mathbf{T}(x_o,y_o,z_o) = (x-x_o)\left.\frac{\partial f(x,y,z)}{\partial x}\right|_{p_o} + (y-y_0)\left.\frac{\partial f(x,y,z)}{\partial y}\right|_{p_o} + (z-z_0)\left.\frac{\partial f(x,y,z)}{\partial z}\right|_{p_o} \quad 1.269$$

Example 1.56

What is the equation of the plane tangent to $f(x,y,z) = 4x^2 + y^2 - 16z$ at the point (2,4,2)?

$$\left.\frac{\partial f(x,y,z)}{\partial x}\right|_{p_o} = \left.8x\right|_{(2,4,2)} = (8)(2) = 16$$

$$\left.\frac{\partial f(x,y,z)}{\partial y}\right|_{p_o} = \left.2y\right|_{(2,4,2)} = (2)(4) = 8$$

$$\left.\frac{\partial f(x,y,z)}{\partial z}\right|_{p_o} = \left.-16\right|_{(2,4,2)} = -16$$

Therefore

$$\mathbf{T}(2,4,2) = 16(x-2) + 8(y-4) - 16(z-2) = 2x + y - 2z - 4$$

6. Normal Line Function

Partial derivatives may be used to find the equation of a straight line normal to a 3-dimensional surface at some point p_o. If the surface is defined by the function $f(x,y,z) = 0$, the equation of the normal line is

$$\mathbf{N} = A\mathbf{i} + B\mathbf{j} + C\mathbf{k} \quad 1.270$$

where

$$A = \left.\frac{\partial f(x,y,z)}{\partial x}\right|_{p_o} \quad 1.271$$

$$B = \left.\frac{\partial f(x,y,z)}{\partial y}\right|_{p_o} \quad 1.272$$

$$C = \left.\frac{\partial f(x,y,z)}{\partial z}\right|_{p_o} \quad 1.273$$

E. Extrema and Optimization

Derivatives may be used to locate local maxima, minima, and points of inflection. No distinction is made between local and global extrema. The end-points of the interval should always be checked against the local extrema located by the method below. The following rules define the extreme points.

$f'(x) = 0$ at any extrema

$f''(x) = 0$ at an inflection point

$f''(x)$ is negative at a maximum

$f''(x)$ is positive at a minimum

There is always an inflection point between a maximum and a minimum.

Example 1.57

Find the global extreme points of the function $f(x) = x^3+x^2-x+1$ on the interval $[-2,+2]$.

$$f'(x) = 3x^2 + 2x - 1$$

$$f'(x) = 0 \text{ at } x = 1/3 \text{ and } x = -1$$

$$f''(x) = 6x + 2$$

$f(-1) = 2$, $f''(-1) = -4$, so $x = -1$ is a maximum

$f(1/3) = 22/27$, $f''(1/3) = +4$, so $x = 1/3$ is a minimum

Checking the endpoints,

$$f(-2) = -1 \qquad f(+2) = +11$$

Therefore, the absolute extreme points are the endpoints.

F. Limit Theory

a and b are constants in the following paragraphs.

1. Limit Theorems

$$\lim_{x \to a} x = a \quad 1.274$$

$$\lim_{x \to a}(mx+b) = ma+b \quad 1.275$$

$$\lim_{x \to b} b = b \quad 1.276$$

$$\lim_{x \to a}(kF(x)) = k\lim_{x \to a}F(x) \quad 1.277$$

$$\lim_{x \to a}(F_1(x)\#F_2(x)) = \lim_{x \to a}(F_1(x)) \# \lim_{x \to a}(F_2(x)) \quad 1.278$$

where # is an operation such as addition, subtraction, multiplication, or division.

2. *L'Hopital's Rule*

If F(a) and G(a) are both zero, the ratio F(a)/G(a) will be undefined. L'Hopital's rule may be used in this case. F^k and G^k are the kth derivatives of the functions F and G respectively.

$$\lim_{x \to a}\left(\frac{F(x)}{G(x)}\right) = \lim_{x \to a}\left(\frac{F^k(x)}{G^k(x)}\right) \qquad 1.279$$

3. *Miscellaneous Rules*

$$\lim_{x \to 0} \sin x = 0 \qquad 1.280$$

$$\lim_{x \to 0} \frac{\sin x}{x} = 1 \qquad 1.281$$

$$\lim_{x \to 0} \cos x = 1 \qquad 1.282$$

If the limit is taken to infinity, divide through by the largest power of x in a polynomial. Try to cancel terms first.

Example 1.58

Evaluate the following limits:

a) $\lim_{x \to 3} \frac{x^3 - 27}{x^2 - 9}$

b) $\lim_{x \to \infty} \frac{3x - 2}{4x + 3}$

c) $\lim_{x \to 2} \frac{x^2 + x - 6}{x^2 - 3x + 2}$

d) $\lim_{x \to 1} \frac{x^2 + x - 2}{(x-1)^2}$

a) Factor the numerator and denominator.

$$\lim_{x \to 3} \frac{x^3 - 27}{x^2 - 9} = \lim_{x \to 3} \frac{(x-3)(x^2+3x+9)}{(x-3)(x+3)} = \frac{9+9+9}{6} = \frac{9}{2}$$

b) Divide through by the largest power of x.

$$\lim_{x \to \infty} \frac{3x - 2}{4x + 3} = \lim_{x \to \infty} \frac{3 - \frac{2}{x}}{4 + \frac{3}{x}} = \frac{3}{4}$$

c) Use L'Hopital's rule.

$$\lim_{x \to 2} \frac{x^2 + x - 6}{x^2 - 3x + 2} = \lim_{x \to 2} \frac{2x + 1}{2x - 3} = \frac{5}{1} = 5$$

d) Factor the numerator.

$$\lim_{x \to 1} \frac{x^2 + x - 2}{(x-1)^2} = \lim_{x \to 1} \frac{(x-1)(x+2)}{(x-1)(x-1)} = \frac{3}{0} = \infty$$

15. *Integral Calculus*

A. *Fundamental Theorem*

The Fundamental Theorem of Calculus is

$$\int_{x_1}^{x_2} f'(x) = f(x_2) - f(x_1) \qquad 1.283$$

B. *Integration By Parts*

If f and g are functions, then

$$\int f\,dg = fg - \int g\,df \qquad 1.284$$

Example 1.59

Evaluate the following integral: $\int xe^x\,dx$

Use integration by parts.

Let f = x. Then df = 1 = dx.

Let dg = $e^x dx$. Then g = $\int e^x dx = e^x$.

Therefore, $\int xe^x dx = xe^x - \int e^x dx + c$

$= xe^x - e^x + c$

C. *Indefinite Integrals* (. . . + C omitted)

$$\int dx = x \qquad 1.285$$

$$\int au\,dx = a\int u\,dx \qquad 1.286$$

$$\int (u+v)dx = \int u\,dx + \int v\,dx \qquad 1.287$$

$$\int x^m\,dx = \frac{x^{(m+1)}}{m+1} \quad m \neq -1 \qquad 1.288$$

$$\int \frac{dx}{x} = \ln|x| \qquad 1.289$$

$$\int e^{ax}dx = \frac{1}{a}e^{ax} \qquad 1.290$$

$$\int xe^{ax}dx = \frac{1}{a^2}e^{ax}(ax-1) \qquad 1.291$$

$$\int \cosh x\,dx = \sinh x \qquad 1.292$$

$$\int \sinh x\,dx = \cosh x \qquad 1.293$$

$$\int \sin x\,dx = -\cos x \qquad 1.294$$

$$\int \cos x\,dx = \sin x \qquad 1.295$$

$$\int \tan x\,dx = \ln|\sec x| \qquad 1.296$$

$$\int \cot x\,dx = \ln|\sin x| \qquad 1.297$$

$$\int \sec x\,dx = \ln|\sec x + \tan x| \qquad 1.298$$

$$\int \csc x\,dx = \ln|\csc x - \cot x| \qquad 1.299$$

$$\int dx/(1+x^2) = \arctan x \qquad 1.300$$

$$\int dx/\sqrt{1 - x^2}) = \arcsin x \qquad 1.301$$

$$\int dx/x\sqrt{x^2 - 1} = \text{arcsec}\ x \qquad 1.302$$

D. Uses of Integrals

1. Finding Areas: The area bounded by $x = a$, $x = b$, $f_1(x)$ above, and $f_2(x)$ below is given by

$$A = \int_a^b [f_1(x) - f_2(x)]\, dx \qquad 1.303$$

2. Surfaces of Revolution: The surface area obtained by rotating $f(x)$ about the y axis is

$$A_s = 2\pi \int_a^b f(x)\sqrt{1 + [f'(x)]^2}\, dx \qquad 1.304$$

3. Rotation of a Function: The volume of a function rotated about the x axis is

$$V = \pi \int_a^b (f(x))^2\, dx \qquad 1.305$$

The volume of a function rotated about the y axis is

$$V = 2\pi \int_a^b x\, f(x)\, dx \qquad 1.306$$

4. Length of a Curve: The length of a curve given by $f(x)$ is

$$L = \int_a^b \sqrt{1 + (f'(x))^2}\, dx \qquad 1.307$$

Example 1.60

For the shaded area shown below, find (a) the area, and (b) the volume enclosed by the curve rotated about the x axis.

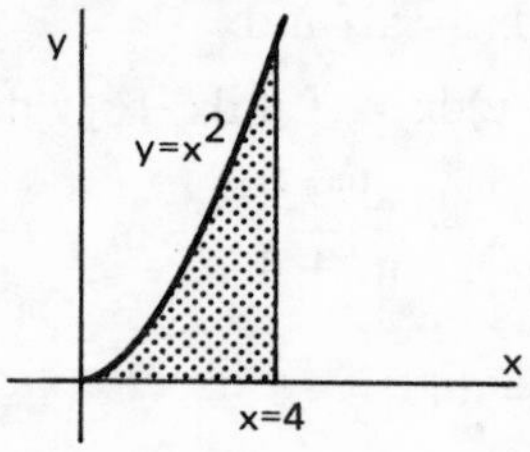

(a) $f_2(x) = 0;\ f_1(x) = x^2$

$$A = \int_0^4 x^2\, dx = \left[\frac{x^3}{3}\right]_0^4 = 21.33$$

(b) $$V = \pi \int_0^4 (x^2)^2\, dx = \pi\left[\frac{x^5}{5}\right]_0^4 = 204.8\pi$$

16. Differential Equations

A differential equation is a mathematical expression containing a dependent variable and one or more of that variable's derivatives. First order differential equations contain only the first derivative of the dependent variable. Second order equations contain the second derivative.

The differential equation is said to be *linear* if all terms containing the dependent variable are multiplied only by real scalars. The equation is said to be *homogeneous* if there are no terms which do not contain the dependent variable or one of its derivatives.

Most differential equations are difficult to solve. However, there are several forms which are fairly simple. These are presented below.

A. First Order Linear

The first order linear differential equation has the general form given below. $p(t)$ and $g(t)$ may be constants or any function of t.

$$x' + p(t)x = g(t) \qquad 1.308$$

The solution depends on an integrating factor defined as

$$u = \exp[\int p(t)\, dt] \qquad 1.309$$

The solution to the first order linear differential equation is

$$x = \frac{1}{u}[\int ug(t)\, dt + c] \qquad 1.310$$

Example 1.61

Find a solution to the differential equation

$$x' - x = 2te^{2t} \qquad x(0) = 1$$

This meets the definition of a first-order linear equation with

$$p(t) = -1 \text{ and } g(t) = 2te^{2t}$$

The integrating constant is

$$u = \exp[\int -1\, dt] = e^{-t}$$

Then, x is

$$x = (1/e^{-t})[\int e^{-t}2te^{2t}dt + c] = e^t[\int 2te^t dt + c]$$
$$= e^t[2te^t - 2e^t + c]$$

But, $x(0) = 1$, so $c = +3$, and

$$x = e^t[2e^t(t-1) + 3]$$

B. Second Order Homogeneous with Constant Coefficients

This type of differential equation has the following general form:

$$c_1x'' + c_2x' + c_3x = 0 \qquad 1.311$$

The solution can be found by first solving the characteristic quadratic equation for its roots k_1^* and k_2^*. This characteristic equation is derived directly from the differential equation:

$$c_1k^2 + c_2k + c_3 \qquad 1.312$$

The form of the solution depends on the values of k_1^* and k_2^*. If $k_1^* \neq k_2^*$ and both are real, then

$$x = a_1(e^{k_1^*t}) + a_2(e^{k_2^*t}) \qquad 1.313$$

If $k_1^* = k_2^*$, then

$$x = a_1(e^{k_1^* t}) + a_2 t(e^{k_2^* t}) \quad 1.314$$

If $k^* = (r \pm iu)$, then

$$x = a_1(e^{rt})\cos(ut) + a_2(e^{rt})\sin(ut) \quad 1.315$$

In all of the above cases, a_1 and a_2 must be found from the given initial conditions.

Example 1.62

Solve the following differential equation for x.

$$x'' + 6x' + 9x = 0 \qquad x(0) = 0,\ x'(0) = 1$$

The characteristic equation is

$$k^2 + 6k + 9 = 0$$

This has roots of $k_1^* = k_2^* = -3$, therefore, the solution has the form

$$x(t) = a_1 e^{-3t} + a_2 t e^{-3t}$$

But, $x(0) = 0$,

$$0 = a_1(e^0) + a_2(0)(e^0)$$

$$0 = a_1(1) + 0$$

$$0 = a_1$$

Also, $x'(0) = 1$. The derivative of x(t) is

$$x'(t) = -3a_2 t e^{-3t} + a_2 e^{-3t}$$

$$1 = -3a_2(0)(e^0) + a_2 e^0$$

$$1 = 0 + a_2$$

$$1 = a_2$$

The final solution is

$$x = te^{-3t}$$

17. Laplace Transforms

Traditional methods of solving non-homogeneous differential equations are very difficult. The Laplace transformation can be used to reduce the solution of many complex differential equations to simple algebra.

Every mathematical function can be converted into a Laplace function by use of the following transformation definition.

$$\mathcal{L}[f(t)] = \int_0^\infty e^{-st} f(t)dt \quad 1.316$$

The variable s will be shown in later chapters to be equivalent to the derivative operator. For the time being, however, it may be thought of as a simple variable.

Example 1.63

Let f(t) be the unit step. That is, $f(t) = 0$ for $t<0$ and $f(t) = 1$ for $t \geq 0$.

Then, the Laplace transform of $f(t) = 1$ is

$$\mathcal{L}[f(t)] = \int_0^\infty e^{-st}(1)\, dt = -\frac{e^{-st}}{s}\Big]_0^\infty$$

$$= 0 - \left(-\frac{1}{s}\right) = \frac{1}{s}$$

Example 1.64

What is the Laplace transform of $f(t) = e^{at}$?

$$\mathcal{L}[e^{at}] = \int_0^\infty e^{-st} e^{at} dt$$

$$= \int_0^\infty e^{-(s-a)t}\, dt$$

$$= -\frac{e^{-(s-a)t}}{(s-a)}\Big]_0^\infty$$

$$= \frac{1}{s-a}$$

It is generally unnecessary to actually obtain a function's Laplace transform by use of equation 1.316. Tables of these transforms are readily available. A small collection of the most frequently required transforms is given at the end of this chapter.

The Laplace transform method may be used with any linear differential equation with constant coefficients. Assuming the dependent variable is x, the basic procedure is as follows:

step 1: Put the differential equation in standard form.

step 2: Use superposition and take the Laplace transform of each term.

step 3: Use the following relationships to expand terms.

$$\mathcal{L}(x'') = s^2\mathcal{L}(x) - sx_0 - x_0' \quad 1.317$$

$$\mathcal{L}(x') = s\,\mathcal{L}(x) - x_0 \quad 1.318$$

step 4: Solve for $\mathcal{L}(x)$. Simplify the resulting expression using partial fractions.

step 5: Find x by applying the inverse transform.

This method reduces the solutions of differential equations to simple algebra. However, a complete set of transforms is required.

Working with Laplace transforms is simplified by the following two theorems:

Linearity Theorm: If c is a constant, then

$$\mathcal{L}[cf(t)] = c\mathcal{L}[f(t)] \quad 1.319$$

Superposition Theorem: If f(t) and g(t) are different functions, then

$$\mathcal{L}[f(t) \pm g(t)] = \mathcal{L}[f(t)] \pm \mathcal{L}[g(t)] \quad 1.320$$

Example 1.65

Suppose the following differential equation results from the analysis of a mechanical system:

$$x'' + 2x' + 2x = \cos(t) \qquad x_0 = 1,\ x_0' = 0$$

x is the dependent variable. We start by taking the Laplace transform of both sides:

$$\mathcal{L}(x'') + 2\mathcal{L}(x') + 2\mathcal{L}(x) = \mathcal{L}(\cos(t))$$

$$s^2\mathcal{L}(x) - sx_0 - x_0' + 2s\mathcal{L}(x) - 2x_0 + 2\mathcal{L}(x) = \mathcal{L}\cos(t)$$

But, $x_0 = 1$ and $x_0' = 0$. Also, the Laplace transform of cos(t) can be found from the appendix of this chapter.

$$s^2\mathcal{L}(x) - s + 2s\mathcal{L}(x) - 2 + 2\mathcal{L}(x) = \frac{s}{s^2+1}$$

$$\mathcal{L}(x)[s^2+2s+2] - s - 2 = \frac{s}{s^2+1}$$

$$\mathcal{L}(x) = \frac{s^3+2s^2+2s+2}{(s^2+1)(s^2+2s+2)}$$

This is now expanded by partial fractions:

$$\frac{s^3+2s^2+2s+2}{(s^2+1)(s^2+2s+2)} = \frac{A_1s+B_1}{s^2+1} + \frac{A_2s+B_2}{s^2+2s+2}$$

$$= \frac{s^3(A_1+A_2)+s^2(2A_1+B_1+B_2)+s(2A_1+2B_1+A_2)+2B_1+B_2}{(s^2+1)(s^2+2s+2)}$$

The following simultaneous equations result:

$$A_1 + A_2 = 1$$

$$2A_1 + B_1 + B_2 = 2$$

$$2A_1 + A_2 + 2B_1 = 2$$

$$2B_1 + B_2 = 2$$

These equations have the solutions

$$A_1^* = 1/5 \quad A_2^* = 4/5 \quad B_1^* = 2/5 \quad B_2^* = 6/5$$

Therefore, x can be found by taking the following inverse transform:

$$x = \mathcal{L}^{-1}\left[\frac{1/5s + 2/5}{s^2 + 1} + \frac{4/5s + 6/5}{s^2+2s+2}\right]$$

The solution is

$$x = 1/5\cos(t) + 2/5\sin(t) + 4/5e^{-t}\cos(t) + 2/5e^{-t}\sin(t)$$

18. Applications of Differential Equations

A. Fluid Mixture Problems

The typical fluid mixing problem involves a tank containing some liquid. There may be an initial solute in the liquid, or the liquid may be pure. Liquid and solute are added at known rates. A drain usually removes some of the liquid which is assumed to be thoroughly mixed. The problem is to find the weight or concentration of solute in the tank at some time t. The following symbols are used.

- C(t) concentration of solute in tank at time t
- I(t) liquid inflow rate at time t from all sources
- k a constant
- Φ(t) liquid outflow rate at time t due to all drains
- $S_1(t)$ solute inflow rate at time t (this may have to be calculated from the incoming concentration and I(t))
- $S_2(t)$ solute outflow rate at time t
- V_o original volume of tank at time = 0
- V(t) volume of tank at time = t (equal to $V_o + \int I(t)dt - \int \Phi(t)dt$)
- W_o initial weight of solute in tank at t = 0
- W(t) weight of solute in tank at time = t

case 1: Constant Volume

$$W'(t) = S_1(t) - S_2(t)$$

$$= S_1(t) - \frac{\Phi(t)W(t)}{V_o} \qquad 1.321$$

The differential equation is

$$W'(t) + \frac{\Phi W(t)}{V_o} = S_1 \qquad 1.322$$

This is a first order linear equation because Φ, V_o, and S_1 are constants.

case 2: Changing Volume

$$W'(t) = S_1(t) - S_2(t) \qquad 1.323$$

$$= S_1(t) - \frac{\Phi(t)W(t)}{V(t)} \qquad 1.324$$

The differential equation is

$$W'(t) + \frac{\Phi(t)W(t)}{V(t)} = S_1(t) \qquad 1.325$$

B. Decay Problems

A given quantity is known to decrease at a rate proportional to the amount present. The original amount is known, and the amount at some time t is desired.

- k a negative proportionality constant
- Q_o original amount present
- Q(t) amount present at time t
- t time
- $t_{1/2}$ half-life

The differential equation is

$$Q'(t) = kQ(t) \qquad 1.326$$

The solution is

$$Q(t) = Q_oe^{kt} \qquad 1.327$$

If Q* is known for some time t*, k can be found from

$$k = \left(\frac{1}{t^*}\right)\ln(Q^*/Q_o) \qquad 1.328$$

k can also be found from the half-life:

$$k = \frac{-.693}{t_{1/2}} \qquad 1.329$$

C. Epidemics

It is assumed that members of the p and q groups move about freely, so that the number of contacts is βxy.

n	community size (p+q)
p	number of affected members
q	number of susceptible members
t	time
x	density of affected members (p/n)
x_o	original density of affected members
y	density of susceptible members (q/n)
β	a proportionality constant

$$\frac{dx}{dt} = \beta xy = \beta x(1-x) \qquad 1.330$$

This is a separable differential equation with solution

$$x(t) = \frac{x_o}{x_o + (1-x_o)e^{-\beta t}} \qquad 1.331$$

This solution assumes no quarantine and that contagion does not limit members' activities.

D. Surface Temperature

k	a constant
t	time
T	absolute temperature of the surface
T_o	ambient temperature

Assuming that the surface temperature changes at a rate proportional to the difference in surface and ambient temperatures, the differential equation is

$$\frac{dT}{dt} = k(T - T_o) \qquad 1.332$$

The above equation is known as Newton's Law of Cooling.

E. Surface Evaporation

A	exposed surface area
k	proportionality constant
r	radius
s	side length
t	time
V	object volume

The equation is

$$\frac{dV}{dt} = -kA \qquad 1.333$$

For a spherical drop, this reduces to

$$\frac{dr}{dt} = -k \qquad 1.334$$

For a cube, this reduces to

$$\frac{ds}{dt} = -2k \qquad 1.335$$

19. Fourier Analysis

Any periodic waveform may be written as the sum of an infinite number of sinusoidal terms. In practice, it is possible to obtain a close approximation to the original waveform with a limited number of sinusoidal terms since most series converge rapidly.

Fourier's theorem is given below. The object of Fourier analysis is to determine the coefficients a_n and b_n.

$$f(t) = a_o + a_1\cos\omega_o t + a_2\cos 2\omega_o t + \cdots$$
$$+ b_1\sin\omega_o t + b_2\sin 2\omega_o t + \cdots \qquad 1.336$$

ω_o is known as the fundamental frequency of the waveform. It depends on the actual waveform period.

$$\omega_o = \frac{2\pi}{T} \qquad 1.337$$

To simplify the analysis, the time domain may be normalized to the radian scale. The normalized scale is obtained by dividing all frequencies by ω_o. Then, the Fourier series becomes

$$f(t) = a_o + a_1\cos t + a_2\cos 2t + \cdots$$
$$+ b_1\sin t + b_2\sin 2t + \cdots \qquad 1.338$$

The coefficients a_n and b_n can be found from the following relationships:

$$a_o = \frac{1}{2\pi}\int_o^{2\pi} f(t)\,dt \qquad 1.339$$

$$a_n = \frac{1}{\pi}\int_o^{2\pi} f(t)\cos nt\,dt \qquad 1.340$$

$$b_n = \frac{1}{\pi}\int_o^{2\pi} f(t)\sin nt\,dt \qquad 1.341$$

Notice that a_o is the average value of the function. Usually, this average value can be determined by observation without having to go through the integration process. The equation for a_n cannot be used to find a_o.

Example 1.66

Find the Fourier series for $f(t) = \begin{Bmatrix} 1 & 0<t<\pi \\ 0 & \pi<t<2\pi \end{Bmatrix}$

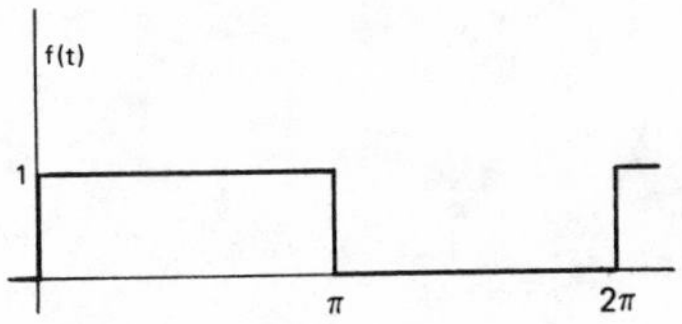

$$a_o = \frac{1}{2\pi}\int_o^{\pi}(1)dt + \frac{1}{2\pi}\int_{\pi}^{2\pi}(0)dt = 1/2$$

This value of ½ corresponds to the average value of f(t). It could have been found by observation.

$$a_1 = \frac{1}{\pi}\int_o^{\pi}(1)\cos t\,dt + \frac{1}{\pi}\int_{\pi}^{2\pi}(0)\cos t\,dt$$

$$= \frac{1}{\pi}\Big[\ \text{sint}\ \Big]_o^{\pi} + 0 = 0$$

$$\text{In general, } a_n = \frac{1}{\pi}\left[\frac{\text{sin nt}}{n}\right]_o^{\pi} = 0$$

$$b_1 = \frac{1}{\pi}\int_o^{\pi}(1)\ \text{sint dt} + \frac{1}{\pi}\int_{\pi}^{2\pi}(0)\ \text{sint dt}$$

$$= \frac{1}{\pi}\Big[-\text{cost}\Big]_o^{\pi} = \frac{2}{\pi}$$

$$\text{In general, } b_n = \frac{1}{\pi}\left[\frac{-\text{cos nt}}{n}\right]_o^{\pi} = \begin{cases} 0 \text{ for n even} \\ \dfrac{2}{\pi n} \text{ for n odd} \end{cases}$$

The series is

$$f(t) = ½ + \frac{2}{\pi}[\text{sint} + ⅓ \sin 3t + ⅕ \sin 5t + \cdots]$$

The sum of the first few terms is illustrated below.

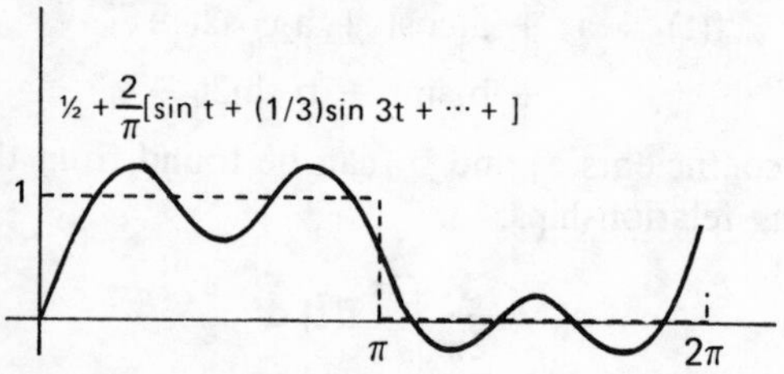

It may be possible to eliminate some of the a_n or b_n coefficients if the function f(t) is symmetrical. There are four types of symmetry.

A function is said to have *even symmetry* if $f(t) = f(-t)$. The cosine is an example of this type of waveform. Even symmetry may be detected from the graph of the function. The function to the left of $x=0$ is a reflection of the function to the right of $x=0$. With even symmetry, all b_n terms are zero.

A function is said to have *odd symmetry* if $f(t) = -f(-t)$. The sine is an example of this type of waveform. With odd symmetry, all a_n terms are zero (but not necessarily a_o).

A function is said to have *rotational symmetry* or *half-wave symmetry* if $f(t) = -f(t+\pi)$. Functions of this type are identical on alternate ½-cycles, except for a sign reversal. All a_n and b_n are zero for even values of n.

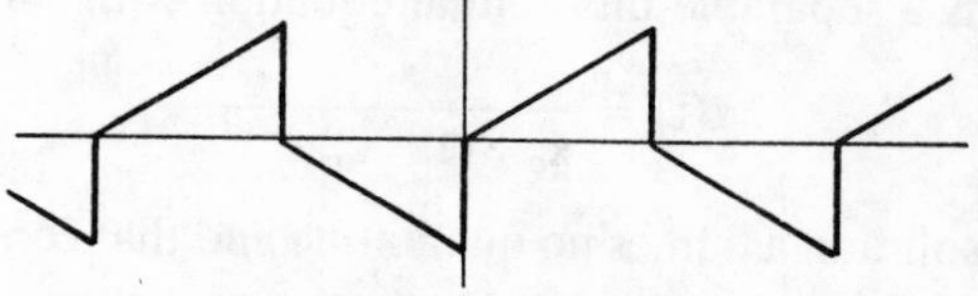

Figure 1.12 Rotational Symmetry

These types of symmetry are not mutually exclusive. For example, it is possible for a function with rotational symmetry to have either odd or even symmetry also. Such a case is known as *quarter-wave symmetry*.

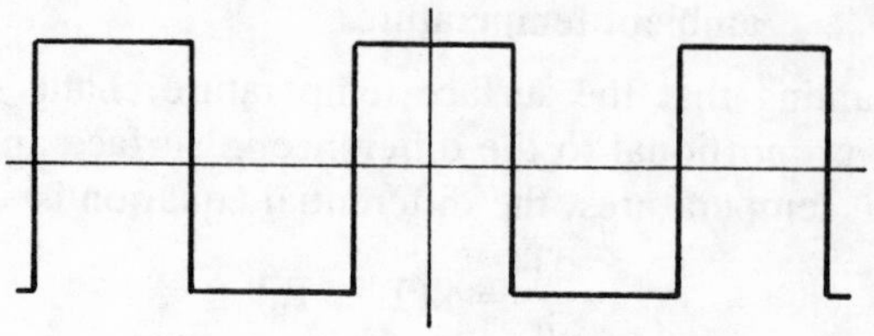

Figure 1.13 Quarter-Wave Symmetry

Appendix A
Laplace Transforms

f(t)	$\mathscr{L}[f(t)]$
Unit impulse at t = 0	1
Unit impulse at t = c	e^{-cs}
Unit step at t = 0	$(1/s)$
Unit step at t = c	$\frac{e^{-cs}}{s}$
t	$\frac{1}{s^2}$
$\frac{t^{n-1}}{(n-1)!}$	$\frac{1}{s^n}$
$\sin At$	$\frac{A}{s^2+A^2}$
$At - \sin At$	$\frac{A^3}{s^2(s^2+A^2)}$
$\sinh(At)$	$\frac{A}{s^2-A^2}$
$t \sin At$	$\frac{2As}{(s^2+A^2)^2}$
$\cos At$	$\frac{s}{s^2+A^2}$
$1 - \cos At$	$\frac{A^2}{s(s^2+A^2)}$
$\cosh(At)$	$\frac{s}{s^2-A^2}$
$t \cos At$	$\frac{s^2-A^2}{(s^2+A^2)^2}$
t^n (n is a positive integer)	$\frac{n!}{s^{(n+1)}}$
e^{At}	$\frac{1}{s-A}$
$e^{At}\sin Bt$	$\frac{B}{(s-A)^2+B^2}$
$e^{At}\cos Bt$	$\frac{s-A}{(s-A)^2+B^2}$
$e^{At}t^n$ (n is positive integer)	$\frac{n!}{(s-A)^{n+1}}$
$1 - e^{-At}$	$\frac{A}{s(s+A)}$
$e^{-At} + At - 1$	$\frac{A^2}{s^2(s+A)}$
$\frac{e^{-At} - e^{-Bt}}{B-A}$	$\frac{1}{(s+A)(s+B)}$
$\frac{(C-A)e^{-At} - (C-B)e^{-Bt}}{B-A}$	$\frac{s + C}{(s+A)(s+B)}$
$\frac{1}{AB} + \frac{Be^{-At} - Ae^{-Bt}}{AB(A-B)}$	$\frac{1}{s(s+A)(s+B)}$

MATH

Appendix B
Conversion Factors

To Convert	Into	Multiply by
Acres	hectares	0.4047
Acres	square feet	43,560.0
Acres	square miles	1.562 EE−3
Ampere hours	coulombs	3,600.0
Angstrom units	inches	3.937 EE−9
Angstrom units	microns	1 EE−4
Astronomical units	kilometers	1.495 EE8
Atmospheres	cms of mercury	76.0
BTU's	horsepower-hrs	3.931 EE−4
BTU's	kilowatt-hrs	2.928 EE−4
BTU/hr	watts	0.2931
Bushels	cubic inches	2,150.4
Calories, gram (mean)	BTU (mean)	3.9685 EE−3
Centares	square meters	1.0
Centimeters	kilometers	1 EE−5
Centimeters	meters	1 EE−2
Centimeters	millimeters	10.0
Centimeters	feet	3.281 EE−2
Centimeters	inches	0.3937
Chains	inches	792.0
Coulombs	faradays	1.036 EE−5
Cubic centimeters	cubic inches	0.06102
Cubic centimeters	pints (U.S. liq.)	2.113 EE−3
Cubic feet	cubic meters	0.02832
Cubic feet/min.	pounds water/min.	62.43
Cubic feet/sec.	gallons/min.	448.831
Cubits	inches	18.0
Days	seconds	86,400.0
Degrees (angle)	radians	1.745 EE−2
Degrees/sec.	revolutions/min.	0.1667
Dynes	grams	1.020 EE−3
Dynes	joules/meter (newtons)	1 EE−5
Ells	inches	45.0
Ergs	BTU's	9.480 EE−11
Ergs	foot-pounds	7.3670 EE−8
Ergs	kilowatt-hours	2.778 EE−14
Faradays/sec.	amperes (absolute)	96,500
Fathoms	feet	6.0
Feet	centimeters	30.48
Feet	meters	0.3048
Feet	miles (nautical)	1.645 EE−4
Feet	miles (statute)	1.894 EE−4
Feet/min.	centimeters/sec.	0.5080
Feet/sec.	knots	0.5921
Feet/sec.	miles/hour	0.6818
Foot-pounds	BTU's	1.286 EE−3
Foot-pounds	kilowatt-hours	3.766 EE−7
Furlongs	miles (U.S.)	0.125
Furlongs	feet	660.0
Gallons	liters	3.785
Gallons of water	pounds of water	8.3453
Gallons/min.	cubic feet/hour	8.0208
Grams	ounces (avoirdupois)	3.527 EE−2
Grams	ounces (troy)	3.215 EE−2
Grams	pounds	2.205 EE−3
Hectares	acres	2.471
Hectares	square feet	1.076 EE5
Horsepower	BTU/min.	42.44
Horsepower	kilowatts	0.7457
Horsepower	watts	745.7
Hours	days	4.167 EE−2
Hours	weeks	5.952 EE−3
Inches	centimeters	2.540
Inches	miles	1.578 EE−5
Joules	BTU's	9.480 EE−4
Joules	ergs	1 EE7
Kilograms	pounds	2.205
Kilometers	feet	3,281.0
Kilometers	meters	1,000.0
Kilometers	miles	0.6214
Kilometers/hr.	knots	0.5396
Kilowatts	horsepower	1.341
Kilowatt-hours	BTU'S	3,413.0
Knots	feet/hour	6,080.0
Knots	nautical miles/hr.	1.0
Knots	statute miles/hr.	1.151
Light years	miles	5.9 EE12
Links (surveyor's)	inches	7.92
Liters	cubic centimeters	1,000.0
Liters	cubic inches	61.02
Liters	gallons (U.S. liq.)	0.2642
Liters	milliliters	1,000.0
Liters	pints (U.S. liq.)	2.113
Meters	centimeters	100.0
Meters	feet	3.281
Meters	kilometers	1 EE−3
Meters	miles (nautical)	5.396 EE−4
Meters	miles (statute)	6.214 EE−4
Meters	millimeters	1,000.0
Microns	meters	1 EE−6
Miles (nautical)	feet	6,080.27
Miles (statute)	feet	5,280.0
Miles (nautical)	kilometers	1.853
Miles (statute)	kilometers	1.609
Miles (nautical)	miles (statute)	1.1516
Miles (statute)	miles (nautical)	0.8684
Miles/hour	feet/min.	88.0
Milligrams/liter	parts/million	1.0
Milliliters	liters	1 EE−3
Millimeters	inches	3.937 EE−2
Newtons	dynes	1 EE5
Ohms (international)	ohms (absolute)	1.0005
Ounces	grams	28.349527
Ounces	pounds	6.25 EE−2
Ounces (troy)	ounces (avoirdupois)	1.09714
Parsecs	miles	19 EE12
Parsecs	kilometers	3.084 EE13
Pints (liq.)	cubic centimeters	473.2
Pints (liq.)	cubic inches	28.87
Pints (liq.)	gallons	0.125
Pints (liq.)	quarts (liq.)	0.5
Pounds	kilograms	0.4536
Pounds	ounces	16.0
Pounds	ounces (troy)	14.5833
Pounds	pounds (troy)	1.21528
Quarts (dry)	cubic inches	67.20

MATH

To Convert	Into	Multiply by
Quarts (liq.)	cubic inches	57.75
Quarts (liq.)	gallons	0.25
Quarts (liq.)	liters	0.9463
Radians	degrees	57.30
Radians	minutes	3,438.0
Revolutions	degrees	360.0
Revolutions/min.	degrees/sec.	6.0
Rods	meters	5.029
Rods	feet	16.5
Rods (surveyor's measure)	yards	5.5
Seconds	minutes	1.667 EE−2
Slugs	pounds	32.17
Tons (long)	kilograms	1,016.0
Tons (short)	kilograms	907.1848
Tons (long)	pounds	2,240.0
Tons (short)	pounds	2,000.0
Tons (long)	tons (short)	1.120
Tons (short)	tons (long)	0.89287
Volt (absolute)	statvolts	3.336 EE−3
Watts	BTU/hour	3.4129
Watts	horsepower	1.341 EE−3
Yards	meters	0.9144
Yards	miles (nautical)	4.934 EE−4
Yards	miles (statute)	5.682 EE−4

MATH

Appendix C
Computational Values of Fundamental Constants

Constant	SI	English
charge on electron	−1.602 EE−19 C	
charge on proton	+1.602 EE−19 C	
atomic mass unit	1.66 EE−27 kg	
electron rest mass	9.11 EE−31 kg	
proton rest mass	1.673 EE−27 kg	
neutron rest mass	1.675 EE−27 kg	
earth weight		1.32 EE25 lb
earth mass	6.00 EE24 kg	4.11 EE23 slug
mean earth radius	6.37 EE3 km	2.09 EE7 ft
mean earth density	5.52 EE3 kg/m³	3.45 lbm/ft³
earth escape velocity	1.12 EE4 m/s	3.67 EE4 ft/sec
distance from sun	1.49 EE11 m	4.89 EE11 ft
Boltzmann constant	1.381 EE−23 J/°K	5.65 EE−24 $\frac{\text{ft-lbf}}{°\text{R}}$
permeability of a vacuum	1.257 EE−6 H/m	
permittivity of a vacuum	8.854 EE−12 F/m	
Planck constant	6.626 EE−34 J·s	
Avogadro's number	6.022 EE23 molecules/gmole	2.73 EE26 $\frac{\text{molecules}}{\text{pmole}}$
Faraday's constant	9.648 EE4 C/gmole	
Stefan-Boltzmann constant	5.670 EE−8 W/m²−K⁴	1.71 EE−9 $\frac{\text{BTU}}{\text{ft}^2\text{-hr-}°\text{R}^4}$
gravitational constant (G)	6.672 EE−11 m³/s²−kg	3.44 EE−8 $\frac{\text{ft}^4}{\text{lbf-sec}^4}$
universal gas constant	8.314 J/°K−gmole	1545 $\frac{\text{ft-lbf}}{°\text{R-pmole}}$
speed of light	3.00 EE8 m/s	9.84 EE8 ft/sec
speed of sound, air, STP	3.31 EE2 m/s	1.09 EE3 ft/sec
speed of sound, air, 70°F, one atmosphere	3.44 EE2 m/s	1.13 EE3 ft/sec
standard atmosphere	1.013 EE7 N/m²	14.7 psia
standard temperature	0°C	32°F
molar ideal gas volume (STP)	22.4138 EE−3 m³/gmole	359 ft³/pmole
standard water density	1 EE3 kg/m³	62.4 lbm/ft³
air density, STP	1.29 kg/m³	8.05 EE−2 lbm/ft³
air density, 70°F, 1 atm	1.20 kg/m³	7.49 EE−2 lbm/ft³
mercury density	1.360 EE4 kg/m³	8.49 EE2 lbm/ft³
gravity on moon	1.67 m/s²	5.47 ft/sec²
gravity on earth	9.81 m/s²	32.17 ft/sec²

MENSURATION

1. An 8″ diameter (I.D.) pipe is filled to a depth equal to (1/3) of its diameter. What is the area in flow?

2. What is the total surface area, including the base, of a 4′ high right circular cone with a 3′ base?

3. What is the largest diameter circle that will fit inside a regular pentagon with 2″ sides?

4. What is the area of an ellipse with an aspect ratio of two and total perimeter of 18″?

SIGNIFICANT DIGITS

5. How many significat digits do the following numbers have?
   (a) 10.097
   (b) 20,540.00
   (c) .001964
   (d) 7.93 EE-2

6. What are the most significant answers to the following operations?
   (a) 12.72 + 9.8
   (b) 18.91 x .006
   (c) 7/2.005
   (d) .001 - .00329

FACTORING

7. Factor the following equations as completely as possible.
   (a) $x^2 + 6x + 8$
   (b) $3x^3 - 3x^2 - 18x$
   (c) $x^4 + 7x^2 + 12$
   (d) $16 - 10x + x^2$

8. Factor the following equations as completely as possible.
   (a) $x^2 - 9$
   (b) $3x^2 - 12$
   (c) $1 - x^8$
   (d) $x^4 - y^4$

9. Factor the following equations as completely as possible.
   (a) $x^2 + 8x + 16$
   (b) $x^2 - 4x + 4$
   (c) $1 + 4y + 4y^2$
   (d) $25x^2 + 60xy + 36y^2$

10. Factor the following equations as completely as possible.
   (a) $x^3 + 8$
   (b) $x^6 - y^6$
   (c) $(x-2)^3 - 8y^3$
   (d) $6x^2 - 4ax - 9bx + 6ab$
   (e) $64 + y^3$
   (f) $z^5 + 32$

QUADRATIC EQUATION

11. Solve for all values of x which satisfy the equations.
   (a) $4x^2 = 12x - 7$
   (b) $2x^2 - 400 = 0$
   (c) $x^2 + 36 = 9 - 2x^2$
   (d) $6x^2 - 7x - 5 = 0$

LINEAR SIMULTANEOUS EQUATIONS

12. Solve for the simultaneous variables that satisfy all equations in each set.
   (a) $4x + 2y = 5$
   $5x - 3y = -2$
   (b) $3x - y = 6$
   $9x - y = 12$
   (c) $2x - 2y + 3z = 1$
   $x - 3y - 2z = -9$
   $x + y + z = 6$
   (d) $x + y = -4$
   $x + z - 1 = 0$
   $2z - y + 3x = 4$

2–VARIABLE SIMULTANEOUS EQUATIONS

13. Solve for the simultaneous variables that satisfy all equations in each set.
   (a) $2x^2 - y^2 - 14 = 0$
   $y = x - 1$
   (b) $y - 3x + 4 = 0$
   $y + (x^2/y) = (24/y)$
   (c) $x^2 + y^2 = 25$
   $x + 2y = 10$
   (d) $2x^2 - 3y^2 = 6$
   $3x^2 + 2y^2 = 35$

LOGARITHMS

14. Solve for x: $38.5^x = 6.5^{(x-2)}$

15. What are the characteristic and mantissa of the common logarithms of the following numbers?
   (a) 40.60
   (b) 9.28
   (c) .0071
   (d) 4070.9

16. Solve for x: $1.4 = (.0613/x)^{1.32}$

SERIES

17. What are the general terms for the following sequences?
   (a) 1, 1/3, 1/5, 1/7
   (b) 4/3, 5/8, 6/15, 7/24
   (c) 1/5, 3/7, 5/9, 7/11
   (d) 2, 5, 8, 11, 14

18. Investigate each of the following series for convergence:
   (a) $(1/3)+(1/6)+(1/11)+\cdots+(1/(2^n+n))$
   (b) $(1/2)+(1/4)+(1/6)+\cdots+(1/2n)$
   (c) $(2/1!)+(2^2/2!)+(2^3/3!)+\cdots+(2^n/n!)$
   (d) $(1/4)-(1/10)+(1/28)-(1/82)+\cdots$
   (e) $(1/4) - (3/6) + (5/8) - (7/10) + \cdots$

PARTIAL FRACTIONS

19. Resolve the following into partial fractions
   (a) $(x+2)/(x^2-7x+12)$
   (b) $(5x+4)/(x^2+2x)$
   (c) $(3x^2-8x+9)/(x-2)^3$
   (d) $(3x)/(x^3-1)$

LINEAR ALGEBRA

20. Perform the following operations:
   (a) $\begin{bmatrix}1 & 7\\2 & 6\end{bmatrix}+\begin{bmatrix}3 & 5\\1 & 1\end{bmatrix}$
   (b) $\begin{bmatrix}7 & 6 & 6\\5 & 4 & 1\\9 & 8 & 2\end{bmatrix}-\begin{bmatrix}1 & 0 & 0\\0 & 1 & 0\\0 & 0 & 1\end{bmatrix}$
   (c) $\begin{bmatrix}2 & 1\\0 & 0\end{bmatrix}\times\begin{bmatrix}3 & 2\\1 & 1\end{bmatrix}$
   (d) $\begin{bmatrix}1 & 9\\7 & 2\end{bmatrix}\times\begin{bmatrix}2 & 1 & 3\\4 & -1 & -7\end{bmatrix}$

21. Find the determinants of the matrixes:
   (a) $\begin{bmatrix}3 & 1\\2 & -4\end{bmatrix}$
   (b) $\begin{bmatrix}4 & -6\\1 & 1\end{bmatrix}$
   (c) $\begin{bmatrix}1 & 2 & 3\\0 & 0 & -1\\4 & -2 & 0\end{bmatrix}$
   (d) $\begin{bmatrix}2 & -1 & 3\\0 & 1 & 2\\3 & -2 & 1\end{bmatrix}$

22. Find the inverses of the following matrixes:
   (a) $\begin{bmatrix}6 & 1\\2 & 4\end{bmatrix}$
   (b) $\begin{bmatrix}2 & 1 & 1\\1 & 3 & 1\\-1 & 4 & 0\end{bmatrix}$

TRIGONOMETRY

23. What are the indicated angles?

(a)

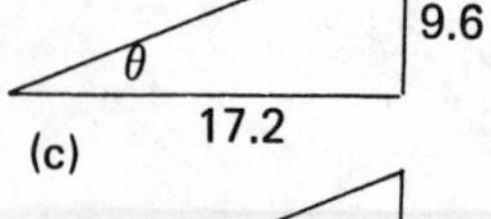

(b)

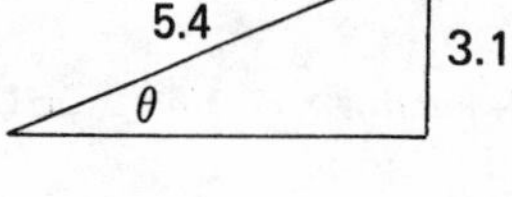

(c)

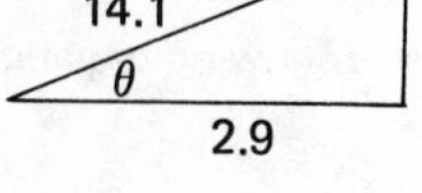

24. Simplify the following relations:
   (a) $\sec X - (\sec X)(\sin^2 X)$
   (b) $\sin^2 X(1+\cot^2 X)$
   (c) $(\tan^2 X)(\cos^2 X) + (\cot^2 X)(\sin^2 X)$

25. Find all sides and angles:

(a)

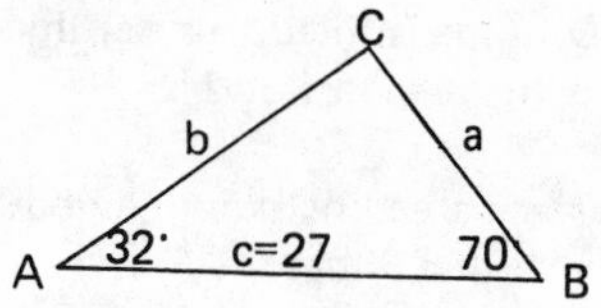

(b)

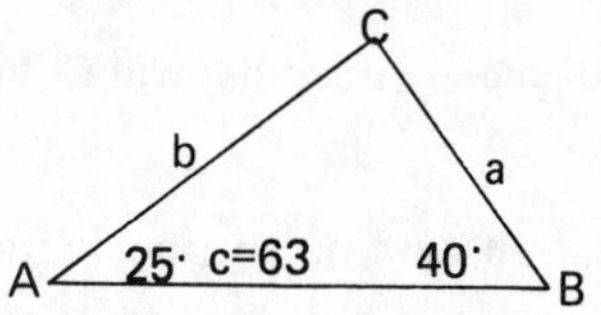

STRAIGHT–LINE GEOMETRY

26. Find the slope form of the line passing through the points.
   (a) (1.7,3.4) and (8.3,9.5)
   (b) (-6,-3.8) and (5.1,-7.2)

27. Put the equations from problem 26 into the following forms:
   (a) general form
   (b) point-slope form
   (c) intercept form

28. What is the angle between the x-axis and the line $y=.4x$?

29. What is the distance between the two points (1,-3,9) and (-4,5,-7)?

30. Find the equation of the straight line passing through (3,1) which is perpendicular to the line passing through (3,-2) and (-6,5).

31. What is the equation of the straight line with an x intercept of 5 and with minimum distance of 3 from the origin?

32. What is the equation of the straight line whose nearest point to the origin is (-6,8)?

VECTOR OPERATIONS

33. Calculate the following dot products:
   (a) $(2,3)\cdot(5,-2)$
   (b) $(2i+3j)\cdot(5i-2j)$
   (c) $(2,-3,6)\cdot(8,2,-3)$

34. What is the angle between the vectors of parts (a) and (b) for problem 33?

35. Calculate the cross products:
   (a) $(1,4)\times(9,-3)$
   (b) $(6,2,3)\times(1,0,1)$
   (c) $(7i-3j)\times(3i+4j)$

36. Give the direction numbers and direction cosines for the following vectors:
   (a) (9,7)
   (b) (3,2,-6)
   (c) (-2i+j-3k)

PLANE GEOMETRY

37. What is the equation of the plane containing the point (3,1,2) which is normal to the vector (1,2,-3)?

38. What is the equation of the plane containing all of the following points? (0,0,2), (2,4,1), and (-2,3,3)?

39. What is the cosine of the angle between the planes of 2x-y-2z-5=0 and 6x-2y+3z+8 = 0?

40. What is the distance from the plane 2x+2y-z-6=0 to the point (2,2,-4)?

CONIC SECTIONS

41. Find the equation of a parabola with
   (a) vertex at (0,0) and focus at (4,0)
   (b) vertex at (2,3) and directrix at x=5
   (c) focus at (-1,2) and directrix at y=-2

42. Which conic section is described by the equation

$$9x^2+16y^2+54x-64y = -1$$

43. What is the equation of an ellipse with
   (a) foci at (±3,0) and major axis of length 8
   (b) eccentricity $\sqrt{33/49}$ and ends of minor axis at (0,±4)
   (c) vertices at (0,±3), passing through the point $(\sqrt{2}, 1.5\sqrt{2})$

COUNTING

44. How many different 3-digit numbers can be formed from the digits 1,2,3,7,8,9 without reusing the digits?

45. How many different ways can a party of 7 politicians be seated
   (a) in a row?
   (b) at a round table?

46. How many different sequences of 6 signal flags can be constructed from 4 red flags and 2 green flags?

47. A student must answer 6 out of 8 questions on an exam.
   (a) How many different ways can he do the exam?
   (b) If the first 2 questions are mandatory, how many ways can he do the exam?

PROBABILITY

48. A coin is weighted so that heads is twice as likely to appear as tails. What is the probability of
   (a) a heads on the first toss?
   (b) 3 heads in a row?
   (c) 2 heads in 4 tosses?

49. Two cards are drawn from an ordinary deck of 52. What is the probability of
   (a) 2 spades?
   (b) a spade and a diamond?
   (c) 2 spades or 2 diamonds?
   (d) 1 spade and (1 heart or 1 diamond)?

PROBABILITY DENSITY FUNCTIONS

50. It is known that 25% of all college students drop out. Your family has 7 children. If they all go to college, what is the probability that
   (a) nobody will drop out?
   (b) 1 will drop out?
   (c) 2 will drop out?

51. The average number of customers serviced per hour in a bank is 38. What is the probability that exactly 34 customers will be serviced during the next hour?

52. 10% of a mixed batch of rubber erasers are red, and the rest are blue. Out of a sample of 5, what is the probability of getting 2 red?

53. 15% of a mixed batch of gum balls are green. If you randomly sample 20, what is the probability of getting 2 green?

54. What is the probability that X will be between 15 and 18 if you sample from a normal distribution with mean 12 and standard deviation 3?

55. An automatic screw machine turns out 200 washers per hour. The size of the hole in the washer is normally distributed with a mean of .502" and standard deviation of .005". Washers are defective if the hole diameter is less than .497" or more than .507".
   (a) What is the probability that a washer chosen at random will be defective?
   (b) What is the probability that 2 out of 3 random samples will be defective?
   (c) How many washers per day will be defective if each work day has 8 hours?

STATISTICS

56. Calculate the mean, standard deviation, and variance of the following set of numbers: 3,4,4,5,8,8,8,10,11,15,18,20.

57. What is the standard deviation of the following 50 data points?

data value	frequency
1.5	3
2.5	8
3.5	18
4.5	12
5.5	9

DERIVATIVES

58. Find the derivatives of the following functions.
(a) $2t^2+8t+9$
(b) $(x+5)/(x^2-1)$
(c) $2(x^2-1)^{-1}$
(d) $\sqrt{2-3x^2}$
(e) $\sin^2(x^2+3x)$

59. Find maximum and minimum values of the functions below on the intervals indicated.
(a) x^3-5x-4 (-3,-1)
(b) x^3+3x^2-9x (-4,4)
(c) $(x+5)/(x-3)$ (-5,2)
(d) x^3+7x^2-5x (-2,2)

60. Find the partial derivitives with respect to x of the following functions:
(a) $4x^2-3xy$
(b) xy^2-5y+6
(c) $x^2+4y^2+9z^2$

LIMITS

61. Find the limits indicated:

(a) $(1-\cos x)/x^2$ $x \to 0$
(b) $(x^2-4)/(x-2)$ $x \to 2$
(c) $(x^3-27)/(x-3)$ $x \to 3$
(d) $(2x+1)/(5x-2)$ $x \to \infty$

INTEGRATION

62. Find the integrals given below:
(a) $\int x^2 e^x \, dx$
(b) $\int \sqrt{1-x} \, dx$
(c) $\int [x/(x^2+1)] \, dx$
(d) $\int [x^2/(x^2+x-6)] \, dx$

63. Calculate the definite integrals below:
(a) $\int_1^3 (x^2+4x) \, dx$
(b) $\int_{-2}^{2} (x^3+1) dx$
(c) $\int_1^2 (4x^3-3x^2) dx$

64. Find the area bounded by x=1, x=3, y+x+1=0, and $y=6x-x^2$.

DIFFERENTIAL EQUATIONS

65. Solve the differential equations below.
(a) $y' = x^2 - 2x - 4$ $y(3) = -6$
(b) $y'' + 4y' + 4y = 0$ $y(0) = 1$ $y'(0) = 1$
(c) $y'' + 3y' + 2y = 0$ $y(0) = 1$ $y'(0) = 0$
(d) $y'' + 2y' + 2y = e^{-t}$ $y(0) = 0$ $y'(0) = 1$

66. A tank contains 100 gallons of brine made by dissolving 60 pounds of salt in water. Salt water containing one pound of salt per gallon runs in at the rate of 2 gallons per minute. A well-stirred mixture runs out at the rate of 3 gallons per minute. Find the amount of salt in the tank at the end of one hour.

2 Engineering Economic Analysis

Nomenclature

A	annual amount or annuity	\$
B	present worth of all benefits	\$
BV_j	book value at the end of the jth year	\$
C	initial cost	\$
d	declining balance depreciation rate	decimal
D_j	depreciation in year j	\$
D.R.	present worth of after-tax depreciation recovery	\$
e	natural logarithm base (2.718)	—
EUAC	equivalent uniform annual cost	\$
f	federal income tax rate	decimal
F	future amount or future worth	\$
G	uniform gradient amount	\$
i	effective rate per period (usually per year)	decimal
k	number of compounding periods per year	—
n	number of compounding periods, or life of asset	—
P	present worth or present value	\$
P_t	present worth after taxes	\$
ROR	rate of return	decimal
ROI	return on investment	\$
r	nominal rate per year (rate per annum)	decimal
s	state income tax rate	decimal
S_n	expected salvage value in year n	\$
t	composite tax rate	decimal
z	a factor equal to $\frac{(1+i)}{(1-d)}$	decimal
ø	nominal rate per period	decimal

1. Equivalence

Unlike most individuals involved with personal financial affairs, industrial decision makers using engineering economics are not as much concerned with the timing of a project's cash flows as with the total profitability of that project. In this situation, a method is required to compare projects involving receipts and disbursements occurring at different times.

By way of illustration, consider \$100 placed in a bank account which pays 5% effective annual interest at the end of each year. After the first year, the account will have grown to \$105. After the second year, the account will have grown to \$110.25.

Assume that you will have no need for money during the next two years and that any money received would immediately go into your 5% bank account. Then, which of the following options would be more desirable?

option a: \$100 now

option b: \$105 to be delivered in one year

option c: \$110.25 to be delivered in two years

In light of the previous illustration, none of the options is superior under the assumptions given. If the first option is chosen, you will immediately place \$100 into a 5% account, and in two years the account will have grown to \$110.25. In fact, the account will contain \$110.25 at the end of two years regardless of the option chosen. Therefore, these alternatives are said to be equivalent.

This is the way that many industrial economic decisions are made. Economically stable companies with many income sources evaluate alternative investments, not on the basis of when income is received, but rather on the basis of which alternative has the largest eventual profitability.

2. Cash Flow Diagrams

Although they are not always necessary in simple problems (and they are often unwieldly in very complex problems), cash flow diagrams may be drawn to help visualize and simplify problems having diverse receipts and disbursements.

The conventions below are used to standardize cash flow diagrams.

a. The horizontal (time) axis is marked off in equal increments, one per period, up to the duration or horizon of the project.

b. All disbursements and receipts (cash flows) are assumed to take place at the end of the year in which they occur. This is known as the year-end convention. The exception to the year-end convention is any initial cost (purchase cost) which occurs at $t = 0$.

c. Disbursements are represented by downward arrows. Receipts are represented by upward arrows. Arrow lengths are approximately proportional to the magnitude of the cash flow.

d. Two or more transfers in the same year are placed end-to-end, and these may be combined.

e. Expenses incurred before $t = 0$ are called sunk costs. Sunk costs are not relevant to the problem.

Example 2.1

A mechanical device will cost $20,000 when purchased. Maintenance will cost $1,000 each year. The device will generate revenues of $5,000 each year for 5 years after which the salvage value is expected to be $7,000. Draw and simplify the cash flow diagram.

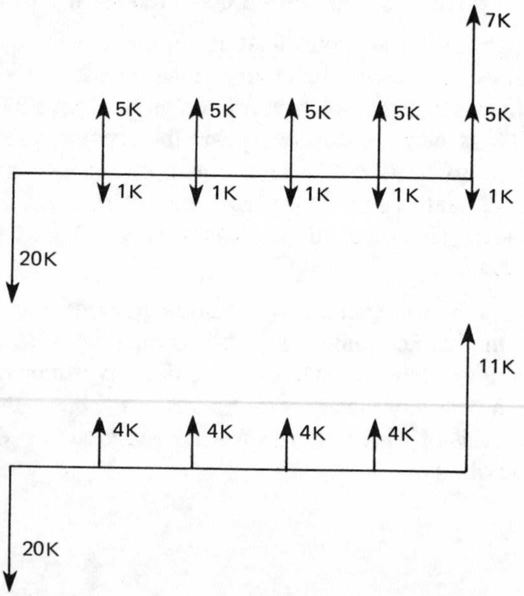

3. Typical Problem Format

With the exception of some investment and rate of return problems, the typical problem involving engineering economics will have the following characteristics:

a. An interest rate will be given.

b. Two or more alternatives will be competing for funding.

c. Each alternative will have its own cash flows.

d. It is necessary to select the best alternative.

Example 2.2

Investment A costs $10,000 today and pays back $11,500 two years from now. Investment B costs $8,000 today and pays back $4,500 each year for two years. If an interest rate of 5% is used, which alternative is superior?

The solution to this example is not difficult, but it will be postponed until methods of calculating equivalence have been covered.

4. Calculating Equivalence

It was previously illustrated that $100 now is equivalent at 5% to $105 in one year. The equivalence of any present amount, P, at $t = 0$ to any future amount, F, at $t = n$ is called the future worth and can be calculated from equation 2.1.

$$F = P(1 + i)^n \qquad 2.1$$

The factor $(1 + i)^n$ is known as the compound amount factor and has been tabulated at the end of this chapter for various combinations of i and n. Rather than actually write the formula for the compound amount factor, the convention is to use the standard functional notation (F/P,i%,n). Thus,

$$F = P(F/P,i\%,n) \qquad 2.2$$

Similarly, the equivalence of any future amount to any present amount is called the present worth and can be calculated from

$$P = F(1 + i)^{-n} = F(P/F,i\%,n) \qquad 2.3$$

The factor $(1 + i)^{-n}$ is known as the present worth factor, with functional notation (P/F,i%,n). Tabulated values are also given for this factor at the end of this chapter.

Example 2.3

How much should you put into a 10% savings account in order to have $10,000 in 5 years?

This problem could also be stated: What is the equivalent present worth of $10,000 5 years from now if money is worth 10%?

$$P = F(1 + i)^{-n} = 10{,}000(1 + .10)^{-5} = 6{,}209$$

The factor .6209 would usually be obtained from the tables.

A cash flow which repeats regularly each year is known as an annual amount. Although the equivalent value for each of the n annual amounts could be calculated and then summed, it is much easier to use one of the uniform series factors, as illustrated in example 2.4 below.

Example 2.4

Maintenance costs for a machine are $250 each year. What is the present worth of these maintenance costs over a 12 year period if the interest rate is 8%?

$$\text{Notice that } (P/A,8\%,12) = (P/F,8\%,1) + (P/F,8\%,2) + \cdots + (P/F,8\%,12)$$

$$P = A(P/A,i\%,n) = -250(7.536) = -1{,}884$$

A common complication involves a uniformly increasing cash flow. Such an increasing cash flow should be handled with the uniform gradient factor, (P/G,i%,n). The uniform gradient factor finds the present worth of a uniformly increasing cash flow which starts in year 2 (not year 1) as shown in example 2.5.

Example 2.5

Maintenance on an old machine is $100 this year, but is expected to increase by $25 each year thereafter. What is the present worth of 5 years of maintenance? Use an interest rate of 10%.

In this problem, the cash flow must be broken down into parts. Notice that the 5-year gradient factor is used even though there are only 4 non-zero gradient cash flows.

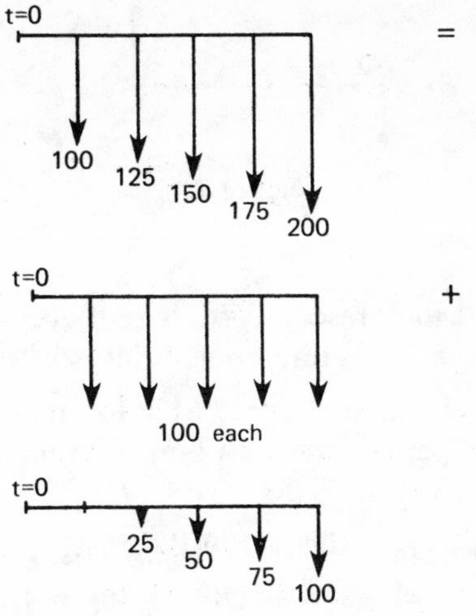

$$P = A(P/A,10\%,5) + G(P/G,10\%,5) = -100(3.791) - 25(6.8618) = -551$$

Various combinations of the compounding and discounting factors are possible. For instance, the annual cash flow that would be equivalent to a uniform gradient may be found from

$$A = G(P/G,i\%,n)(A/P,i\%,n) \qquad 2.4$$

Formulas for all of the compounding and discounting factors are contained in table 2.1. Normally, it will not be necessary to calculate factors from the formulas. The tables at the end of this chapter are adequate for solving most problems.

5. The Meaning of 'Present Worth' and 'i'

It is clear that $100 invested in a 5% bank account will allow you to remove $105 one year from now. If this investment is made, you will clearly receive a return on your investment (ROI) of $5. The cash flow diagram and the present worth of the two transactions are:

$$P = -100 + 105(P/F,5\%,1) = -100 + 105(.95238) = 0$$

TABLE 2.1 Discount Factors for Discrete Compounding

Factor Name	*Converts*	*Symbol*	*Formula*
Single Payment Compound Amount	P to F	(F/P,i%,n)	$(1 + i)^n$
Present Worth	F to P	(P/F,i%,n)	$(1 + i)^{-n}$
Uniform Series Sinking Fund	F to A	(A/F,i%,n)	$\frac{i}{(1 + i)^n - 1}$
Capital Recovery	P to A	(A/P,i%,n)	$\frac{i(1 + i)^n}{(1 + i)^n - 1}$
Compound Amount	A to F	(F/A,i%,n)	$\frac{(1 + i)^n - 1}{i}$
Equal Series Present Worth	A to P	(P/A,i%,n)	$\frac{(1 + i)^n - 1}{i(1 + i)^n}$
Uniform Gradient	G to P	(P/G,i%,n)	$\frac{(1 + i)^n - 1}{i^2(1 + i)^n} - \frac{n}{i(1 + i)^n}$

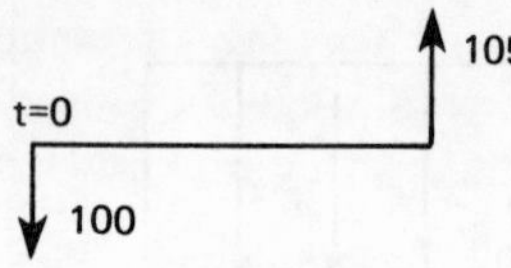

Figure 2.1

Notice that the present worth is zero even though you did receive a 5% return on your investment.

However, if you are offered \$120 for use of \$100 over a one-year period, the cash flow diagram and present worth (at 5%) would be

$$P = -100 + 120(P/F,5\%,1)$$
$$= -100 + 120(.95238) = 14.29$$

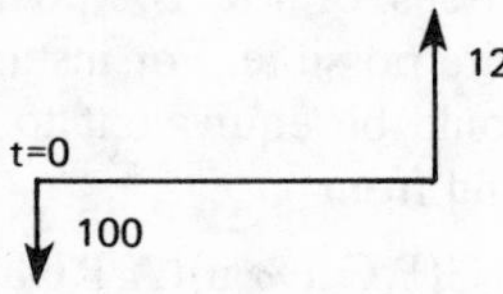

Figure 2.2

Therefore, it appears that the present worth of an alternative is equal to the equivalent value at t = 0 of the increase in return above that which you would be able to earn in an investment offering i% per period. In the above case, \$14.29 is the present worth of (\$20–\$5), the difference in the two ROI's.

Alternatively, the actual earned interest rate, called rate of return (ROR), can be defined as the rate which makes the present worth of the alternative zero.

The present worth is also the amount that you would have to be given to dissuade you from making an investment, since placing the initial investment amount along with the present worth into a bank account earning i% will yield the same eventual ROI. Relating this to the previous paragraphs, you could be dissuaded against investing \$100 in an alternative which would return \$120 in one year by a t = 0 payment of \$14.29. Clearly, (\$100 + \$14.29) invested at t = 0 will also yield \$120 in one year at 5%.

The selection of the interest rate is difficult in engineering economics problems. Usually it is taken as the average rate of return that an individual or business organization has realized in past investments. Fortunately, choosing an interest rate is seldom required as it is usually given.

It should be obvious that alternatives with negative present worths are undesirable, and alternatives with positive present worths are desirable since they increase the average earning power of invested capital.

6. Choice Between Alternatives

A variety of methods exists for selecting a superior alternative from among a group of proposals. Each method has its own merits and applications.

The Present Worth Method has already been implied. When two or more alternatives are capable of performing the same functions, the superior alternative will have the largest present worth. This method is suitable for ranking the desirability of alternatives. The present worth method is restricted to evaluating alternatives that are mutually exclusive and which have the same lives.

Returning to example 2.2, the present worth of each alternative should be found in order to determine which alternative is superior.

Example 2.2, continued

$$P(A) = -10{,}000 + 11{,}500(P/F,5\%,2) = 431$$
$$P(B) = -8{,}000 + 4{,}500(P/A,5\%,2) = 367$$

Alternative A is superior and should be chosen.

The present worth of a project with an infinite life is known as the capitalized cost. Capitalized cost is the amount of money at t = 0 needed to perpetually support the project on the earned interest only. Capitalized cost is a positive number when expenses exceed income.

$$\text{Capitalized Cost} = \text{Initial Cost} + \frac{\text{Annual Costs}}{i} \qquad 2.5$$

If disbursements occur irregularly instead of annually, the capitalized cost is

$$\text{Capitalized Cost} = \text{Initial Cost} + \frac{\text{EUAC}}{i} \qquad 2.6$$

In comparing two alternatives, each of which is infinitely lived, the superior alternative will have the lowest capitalized cost.

Alternatives which accomplish the same purpose but which have unequal lives must be compared by the Annual Cost Method. The annual cost method assumes that each alternative will be replaced by an identical twin at the end of its useful life (infinite renewal). This method, which may also be used to rank alternatives according to their desirability, is also called the Annual Return Method and Capital Recovery Method. Restrictions are that the alternatives must be mutually exclusive and infinitely renewed up to the duration of the longest-lived alternative. The calculated annual cost is known as the equivalent uniform annual cost, EUAC. Cost is a positive number when expenses exceed income.

Example 2.6

Which of the following alternatives is superior over a 30 year period if the interest rate is 7%?

	A	*B*
type	brick	wood
life	30 years	10 years
cost	$1800	$450
maintenance	$5/year	$20/year

$$\text{EUAC(A)} = 1800(\text{A/P},7\%,30) + 5 = 150$$

$$\text{EUAC(B)} = 450(\text{A/P},7\%,10) + 20 = 84$$

Alternative B is superior since its annual cost of operation is the lowest. It is assumed that three wood facilities, each with a life of 10 years and a cost of $450, will be built to span the 30 year period.

The Benefit-Cost Ratio Method is often used in municipal project evaluations where benefits and costs accrue to different segments of the community. With this method, the present worth of all benefits (regardless of the beneficiary) is divided by the present worth of all costs. The project is considered acceptable if the ratio exceeds one.

When the benefit-cost ratio method is used, disbursements by the initiators or sponsors are costs. Disbursements by the users of the project are known as disbenefits. It is often difficult to determine whether a cash flow is a cost or a disbenefit (whether to place it in the numerator or denominator of the benefit-cost ratio calculation).

However, regardless of where the cash flow is placed, an acceptable project will always have a benefit-cost ratio greater than one, although the actual numerical result will depend on the placement. For this reason, the benefit-cost ratio method should not be used to rank competing projects.

The benefit-cost ratio method may be used to rank alternative proposals only if an incremental analysis is used. First determine that the ratio is greater than one for each alternative. Then, calculate the ratio $(B_2\text{-}B_1)/(C_2\text{-}C_1)$ for each possible pair of alternatives. If the ratio exceeds one, alternative 2 is superior to alternative 1. Otherwise, alternative 1 is superior.

Perhaps no method of analysis is less understood than the Rate of Return (ROR) Method. As was stated previously, the ROR is the interest rate that would yield identical profits if all money was invested at that rate. The present worth of any such investment is zero.

The ROR is defined as the interest rate that will discount all cash flows to a total present worth equal to the initial required investment. This definition is used to determine the ROR of an alternative. The ROR should not be used to rank or compare alternatives unless an incremental analysis is used. The advantage of the ROR method is that no knowledge of an interest rate is required.

To find the ROR of an alternative, proceed as follows:

1. Set up the problem as if to calculate the present worth.
2. Arbitrarily select a reasonable value for i. Calculate the present worth.
3. Choose another value of i (not too close to the original value) and again solve for the present worth.
4. Interpolate or extrapolate the value of i which gives a zero present worth.
5. For increased accuracy, repeat steps (2) and (3) with two more values that straddle the value found in step (4).

A common, although incorrect, method of calculating the ROR involves dividing the annual receipts or returns by the initial investment. However, this technique ignores such items as salvage, depreciation, taxes, and the time value of money. This technique also fails when the annual returns vary.

Example 2.7

What is the return on invested capital if $1000 is invested now with $500 being returned in year 4 and $1000 being returned in year 8?

First, set up the problem as a present worth calculation.

$$P = -1000 + 500(\text{P/F},i\%,4) + 1000(\text{P/F},i\%,8)$$

Arbitrarily select i = 5%. The present worth is then found to be $88.50. Next take a higher value of i to reduce the present worth. If i = 10%, the present worth is −$192. The ROR is found from simple interpolation to be approximately 6.6%.

7. *Treatment of Salvage Value in Replacement Studies*

An investigation into the retirement of an existing process or piece of equipment is known as a replacement study. Replacement studies are similar in most respects to other alternative comparison problems: an interest rate is given, two alternatives exist, and one of the previously mentioned methods of comparing alternatives is used to choose the superior alternative.

In replacement studies, the existing process or piece of equipment is known as the defender. The new process or piece of equipment being considered for purchase is known as the challenger.

Because most defenders still have some market value when they are retired, the problem of what to do with

ECON

the salvage value arises. It seems logical to use the salvage value of the defender to reduce the initial purchase cost of the challenger. This is consistent with what would actually happen if the defender were to be retired.

By convention, however, the salvage value is subtracted from the defender's present value. This does not seem logical but is done to keep all costs and benefits related to the defender with the defender. In this case, the salvage value is treated as an opportunity cost which would be incurred if the defender is not retired.

If the defender and the challenger have the same lives and a present worth study is used to choose the superior alternative, the placement of the salvage value will have no effect on the net difference between present worths for the challenger and defender. Although the values of the two present worths will be different depending on the placement, the difference in present worths will be the same.

If the defender and challenger have different lives, an annual cost comparison must be made. Since the salvage value would be 'spread over' a different number of years depending on its placement, it is important to abide by the conventions listed in this section.

There are a number of ways to handle salvage value. The best way is to think of the EUAC of the defender as the cost of keeping the defender from now until next year. In addition to the usual operating and maintenance costs, that cost would include an opportunity interest cost incurred by not selling the defender and also a drop in the salvage value if the defender is kept for one additional year. Specifically,

$$\begin{aligned}\text{EUAC (defender)} &= \text{maintenance costs}\\ &\quad + \text{i(current salvage value)}\\ &\quad + \text{(current salvage} - \text{next year's salvage)}\end{aligned} \qquad 2.7$$

It is important in retirement studies not to double count the salvage value. That is, it would be incorrect to add the salvage value to the defender and at the same time subtract it from the challenger.

8. Basic Income Tax Considerations

Assume that an organization pays f% of it profits to the federal government as income taxes. If the organization also pays a state income tax of s%, and if state taxes paid are recognized by the federal government as expenses, then the composite tax rate is

$$t = s + f - sf \qquad 2.8$$

The basic principles used to incorporate taxation into economic analyses are listed below.

a. Initial purchase cost is unaffected by income taxes and investment credit.

b. Salvage value is unaffected by income taxes.

c. Deductible expenses, such as operating costs, maintenance costs, and interest payments, are reduced by t% (e.g., multiplied by the quantity (1 - t)).

d. Revenues are reduced by t% (e.g., multiplied by the quantity (1 - t)).

e. Depreciation is multiplied by the quantity (t) and added to the appropriate year's cash flow, increasing that year's present worth.

Income taxes and depreciation have no bearing on municipal or governmental projects since municipalities, states, and the U.S. Government pay no taxes.

Example 2.8

A corporation which pays 53% of its revenues in income taxes invests \$10,000 in a project which will result in \$3000 annual revenues for 8 years. If the annual expenses are \$700, salvage after 8 years is \$500, and 9% interest is used, what is the after-tax present worth? Disregard depreciation.

$$\begin{aligned}P_t &= -10{,}000 + 3000(P/A,9\%,8)(1-.53)\\ &\qquad - 700(P/A,9\%,8)(1-.53)\\ &\qquad + 500(P/F,9\%,8)\\ &= -3766\end{aligned}$$

It is interesting that the alternative evaluated in example 2.8 is undesirable if income taxes are considered but is desirable if income taxes are omitted.

9. Depreciation (Also see page 20-2)

Although depreciation calculations may be considered independently in examination questions, it is important to recognize that depreciation has no effect on engineering economic calculations unless income taxes are also considered.

Generally, tax regulations do not allow the cost of equipment[1] to be treated as a deductible expense in the year of purchase. Rather, portions of the cost may be allocated to each of the years of the item's economic life (which may be different from the actual useful life). Each year, the book value (which is initially equal to the purchase price) is reduced by the depreciation in that year. Theoretically, the book value of an item will equal the market value at any time within the economic life of that item.

[1]The IRS tax regulations allow depreciation on almost all forms of *property* except land. The following types of property are distinguished: real (e.g., buildings used for business), residential (e.g., buildings used as rental property), and personal (e.g., equipment used for business). Personal property does not include items for personal use, despite its name. Tangible personal property is distinguished from intangible personal property (e.g., goodwill, copyrights, patents, trademarks, franchises, and agreements not to compete).

Since tax regulations allow the depreciation in any year to be handled as if it were an actual operating expense, and since operating expenses are deductible from the income base prior to taxation, the after-tax profits will be increased. If D is the depreciation, the net result to the after-tax cash flow will be the addition of (t)(D).

The present worth of all depreciation over the economic life of the item is called the depreciation recovery. Although originally established to do so, depreciation recovery can never fully replace an item at the end of its life.

Depreciation is often confused with amortization and depletion. While depreciation spreads the cost of a fixed asset over a number of years, amortization spreads the cost of an intangible asset (e.g., a patent) over some basis such as time or expected units of production.

Depletion is another artificial deductible operating expense designed to compensate mining organizations for decreasing mineral reserves. Since original and remaining quantities of minerals are seldom known accurately, the depletion allowance is calculated as a fixed percentage of the organization's gross income. These percentages are usually in the 10%-20% range and apply to such mineral deposits as oil, natural gas, coal, uranium, and most metal ores.

There are four common methods of calculating depreciation. The book value of an asset depreciated with the Straight Line (SL) Method (also known as the Fixed Percentage Method) decreases linearly from the initial purchase at $t = 0$ to the estimated salvage at $t = n$. The depreciated amount is the same each year. The quantity $(C - S_n)$ in equation 2.9 is known as the depreciation base.

$$D_j = \frac{(C - S_n)}{n} \qquad 2.9$$

Double Declining Balance[2] (DDB) depreciation is independent of salvage value. Furthermore, the book value never stops decreasing, although the depreciation decreases in magnitude. Usually, any remaining book value is written off in the last year of the asset's estimated life. Unlike any of the other depreciation methods, DDB depends on accumulated depreciation.

$$D_j = \frac{2\left(C - \sum_{i=1}^{j-1} D_i\right)}{n} \qquad 2.10$$

In Sum-of-the-Year's-Digits (SOYD) depreciation, the digits from 1 to n inclusive are summed. The total, T, can also be calculated from

$$T = \tfrac{1}{2}n(n + 1) \qquad 2.11$$

The depreciation can be found from

$$D_j = \frac{(C - S_n)(n - j + 1)}{T} \qquad 2.12$$

The Sinking Fund Method is seldom used in industry because the initial depreciation is low. The formula for sinking fund depreciation is

$$D_j = (C - S_n)(A/F,i\%,n) \qquad 2.13$$

The depreciation allowed by the Internal Revenue Service in the first year is usually less than the value D_1 as calculated from formulas 2.9, 2.10, 2.12, and 2.13. This reduction occurs whenever an asset is purchased during the fiscal year (as compared to a purchase made at the beginning of the fiscal year). In such an instance, D_1 must be prorated in proportion to the amount of time remaining until the end of the fiscal year. For example, only $(11/12)D_1$ would be allowed if an asset was purchased on February 1.

The above discussion gives the impression that any form of depreciation may be chosen regardless of the nature and circumstances of the purchase. In reality, the IRS tax regulations place restrictions on the higher-rate ("accelerated") methods such as DDB and SOYD. For example, these two methods cannot be used with used property. There are other restrictions on economic life, purchase date, and property type. Refer to the IRS tax regulations for more information on selecting a depreciation method.

Three other depreciation methods should be mentioned, not because they are currently accepted or in widespread use, but because they have seen recent use on the licensing examinations.

The 'sinking-fund plus interest on first cost' depreciation method, like the following two methods, is an attempt to include the opportunity interest cost on the purchase price with the depreciation. That is, the purchasing company not only incurs an annual loss due to the drop in book value, but it also loses the interest on the purchase price. The formula for this method is

$$D_j = (C - S_n)(A/F,i\%,n) + (C)(i) \qquad 2.14$$

The 'straight-line plus interest on first cost' method is similar. Its formula is

$$D_j = \left(\frac{1}{n}\right)(C - S_n) + (C)(i) \qquad 2.15$$

The 'straight-line plus average interest' assumes that the opportunity interest cost should be based on the book value only, not on the full purchase price. Since

[2]Double declining balance depreciation is a particular form of *declining balance depreciation,* as defined by the IRS tax regulations. Declining balance depreciation also includes 125% declining balance and 150% declining balance depreciations which can be calculated by substituting 1.25 and 1.50 respectively for the 2 in equation 2.10.

ECON

the book value changes each year, an average value is used. The depreciation formula is

$$D_j = \left(\frac{1}{n}\right)(C - S_n) + \tfrac{1}{2}(i)(C - S_n)\left(\frac{n+1}{n}\right) \quad 2.16$$

These three depreciation methods are not to be used in the usual manner, e.g., in conjunction with the income tax rate. These methods are attempts to calculate a more accurate annual cost of an alternative. Sometimes they work, and sometimes they give misleading answers. Their use cannot be recommended. They are included in this chapter only for the sake of completeness.

ECON

Example 2.9

An asset is purchased for $9000. Its estimated economic life is 10 years, after which it will be sold for $1000. Find the depreciation in the first three years using SL, DDB, SOYD, and Sinking Fund at 6%.

SL: $D = (9000 - 1000)/10 = 800$ each year

DDB: $D_1 = 2(9000)/10 = 1800$ in year 1

$D_2 = 2(9000 - 1800)/10 = 1440$ in year 2

$D_3 = 2(9000 - 3240)/10 = 1152$ in year 3

SOYD: $T = \tfrac{1}{2}(10)(11) = 55$

$D_1 = (10/55)(9000 - 1000) = 1454$ in year 1

$D_2 = (9/55)(8000) = 1309$ in year 2

$D_3 = (8/55)(8000) = 1164$ in year 3

Sinking Fund: $D = (9000 - 1000)(A/F,6\%,10) = 607$ each year

Example 2.10

For the asset described in example 2.9, calculate the book value during the first three years if SOYD depreciation is used.

The book value at the beginning of year 1 is $9000. Then,

$BV_1 = 9000 - 1454 = 7546$

$BV_2 = 7546 - 1309 = 6327$

$BV_3 = 6327 - 1164 = 5073$

Example 2.11

For the asset described in example 2.9, calculate the after-tax depreciation recovery with SL and SOYD depreciation methods. Use 6% interest with 48% income taxes.

SL: $D.R. = .48(800)(P/A,6\%,10) = 2826$

SOYD: The depreciation series can be thought of as a constant 1454 term with a negative 145 gradient.

$$D.R. = .48(1454)(P/A,6\%,10) - .48(145)(P/G,6\%,10) = 3076$$

Finding book values, depreciation, and depreciation recovery is particularly difficult with DDB depreciation, since all previous years' quantities seem to be required. It appears that the depreciation in the 6th year cannot be calculated unless the values of depreciation for the first five years are calculated first. Questions asking for depreciation or book value in the middle or at the end of an asset's economic life may be solved from the following equations:

$$d = 2/n \quad 2.17$$

$$z = \frac{(1+i)}{(1-d)} \quad 2.18$$

$$(P/EG) = \frac{z^n - 1}{z^n(z-1)} \quad 2.19$$

Then, assuming that the remaining book value at t = n is written off in one lump sum, the present worth of the depreciation recovery is

$$D.R. = t\left[\frac{(d)(C)}{(1-d)}(P/EG) + (1-d)^n(C)(P/F,i\%,n)\right] \quad 2.20$$

$$D_j = (d)(C)(1-d)^{j-1} \quad 2.21$$

$$BV_j = C(1-d)^j \quad 2.22$$

Example 2.12

What is the after-tax present worth of the asset described in example 2.8 if SL, SOYD, and DDB depreciation methods are used?

The after-tax present worth, neglecting depreciation, was previously found to be −3766.

Using SL, the depreciation recovery is

$$D.R. = (.53)\frac{(10{,}000 - 500)}{8}(P/A,9\%,8) = 3483$$

Using SOYD, the depreciation recovery is calculated as follows:

$$T = \tfrac{1}{2}(8)(9) = 36$$

Depreciation base = (10,000 − 500) = 9500

$D_1 = 8/36(9500) = 2111$

G = gradient = $1/36(9500) = 264$

$$D.R. = (.53)[2111(P/A,9\%,8) - 264(P/G,9\%,8)] = 3829$$

Using DDB, the depreciation recovery is calculated as follows:

$$d = 2/8 = .25$$

$$z = \frac{(1.09)}{.75} = 1.453$$

$$(P/EG) = \frac{(1.453)^8 - 1}{(1.453)^8(.453)} = 2.096$$

$$D.R. = .53\left[\frac{(.25)(10{,}000)}{.75}(2.096) + (.75)^8(10{,}000)(P/F,9\%,8)\right] = 3969$$

The after-tax present worths including depreciation recovery are:

SL: $P_t = -3766 + 3483 = -283$
SOYD: $P_t = -3766 + 3829 = 63$
DDB: $P_t = -3766 + 3969 = 203$

10. Advanced Income Tax Considerations

There are a number of specialized techniques that are infrequently needed. These techniques are related more to the accounting profession than the engineering profession. Nevertheless, it is occasionally necessary to use these techniques. See page 20-1 for recent changes in the tax laws.

A. Additional First-Year Depreciation

The Internal Revenue Service permits the purchaser of business-related equipment to take an additional depreciation allowance in the first year. This additional allowance is known as 'additional first-year depreciation.' Additional first-year depreciation is completely separate from the normal depreciation calculated in the first year (as determined by formulas 2.9, 2.10, 2.12, and 2.13).

The additional first-year depreciation is calculated as 20% of the actual cost of qualifying property. The following special rules apply:

1. Only tangible property qualifies for additional first-year depreciation.
2. The property must have a life of 6 years or greater.
3. Salvage value is not subtracted from the cost of acquisition to find the depreciation base for additional first-year depreciation.
4. The actual cost used as the depreciation base is the actual amount paid or signed for. Therefore, the list price must be reduced by the allowed value of any trade-in equipment.
5. The maximum additional first-year depreciation allowed is $2000.
6. Property acquired during the year (any time before year-end) qualifies for the entire 20% allowance.
7. The additional first-year depreciation must be subtracted from the actual purchase price when calculating the basis to be used with normal depreciation.

Example 2.13

A corporation whose fiscal year ends on December 31 purchases a pile-driver for $14,500 on July 1. The estimated useful life is 10 years. The estimated salvage after 10 years is $500. What is the total depreciation in the first and second years? Use the straight-line method to calculate the normal depreciation.

20% of the actual cost is (.2)(14,500) = 2900. But, additional first-year depreciation is limited to 2000.

The normal depreciation is

$$(1/10)(14{,}500 - 2000 - 500)(1/2 \text{ year}) = 600.$$

The total first-year depreciation is (2000 + 600) = 2600.

The second year depreciation is

$$(1/10)(14{,}500 - 2000 - 500) = 1200.$$

B. Salvage Value Reduction

If an asset is to be kept longer than 3 years, the IRS will allow a reduction in salvage value. Specifically, the salvage value may be reduced by any amount up to 10% of the purchase price.

Example 2.14

A new in-place sewer pipe forming unit is purchased for $85,000. The unit has an expected life of 10 years and a salvage of $12,500. What will be the normal straight-line depreciation?

10% of the purchase price is 8,500. Therefore, the salvage can be taken as (12,500 − 8,500) = 4,000. The normal straight-line depreciation is (1/10)(85,000 − 4,000) = 8,100.

C. Investment Tax Credit

Equipment which is purchased and which will be kept for at least 3 years is eligible for a special investment tax credit. This credit is used to directly reduce the income tax paid. It is not a deductible expense as is depreciation.

The credit allowed is

1. 10% of the purchase price for items kept 7 or more years

ECON

2. 6.67% of the purchase price for items kept 5 to 7 years
3. 3.33% of the purchase price for items kept 3 to 5 years
4. 0% of the purchase price for items kept less than 3 years

ECON

D. Gain or Loss on the Sale of a Depreciated Asset

Each year a non-capital[3] asset is depreciated, its book value declines. Theoretically, this book value corresponds to the actual market value of the asset. If the asset is sold at any time after purchase (even after the asset has been fully depreciated) for an amount different from the book value, an income tax adjustment is required.

If the amount received from the sale exceeds the book value, the difference must be taken as regular taxable income in that year. If the sale price was less than the book value, the difference may be taken as the depreciation in that year.

E. Capital Gains and Losses

A gain is defined as the difference in selling and purchase prices of a capital asset. The gain is called a regular gain if the item sold has been kept less than one year. The gain is called a capital gain if the item sold has been kept for longer than one year. Capital gains are taxed at the taxpayer's usual rate, but 60% of the gain is excluded from taxation. That is, only 40% of the gain is taxed.

Regular (as defined above) losses are fully deductible in the year of their occurrence. The IRS tax regulations should be consulted to determine the treatment of capital losses.

11. Rate and Period Changes

All of the foregoing calculations were based on compounding once a year at an effective interest rate, i. However, some problems specify compounding more frequently than annually. In such cases, a nominal interest rate, r, will be given. The nominal rate does not include the effect of compounding and is not the same as the effective rate, i. A nominal rate may be used to calculate the effective rate by using equation 2.23 or 2.24.

$$i = \left(1 + \frac{r}{k}\right)^k - 1 \qquad 2.23$$

$$= (1 + \phi)^k - 1 \qquad 2.24$$

[3]Non-capital assets are used as a normal part of a business operation. If an item is depreciated, it is usually classified as a non-capital asset. Capital assets are purchased for appreciation (i.e., as investments.)

A problem may also specify an effective rate per period, ø, (e.g., per month). However, that will be a simple problem since compounding for n periods at an effective rate per period is not affected by the definition or length of the period.

The following rules may be used to determine which interest rate is given in a problem:

- Unless specifically qualified in the problem, the interest rate given is an annual rate.
- If the compounding is annually, the rate given is the effective rate. If compounding is other than annually, the rate given is the nominal rate.
- If the type of compounding is not specified, assume annual compounding.

In the case of continuous compounding, the appropriate discount factors may be calculated from the formulas in the following table:

Table 2.2 Discount Factors for Continuous Compounding

(F/P)	e^{rn}
(P/F)	e^{-rn}
(A/F)	$(e^r - 1)/(e^{rn} - 1)$
(F/A)	$(e^{rn} - 1)/(e^r - 1)$
(A/P)	$(e^r - 1)/(1 - e^{-rn})$
(P/A)	$(1 - e^{-rn})/(e^r - 1)$

Example 2.15

A savings and loan offers 5¼% compounded daily. What is the annual effective rate?

method 1: $r = .0525$, $k = 365$

$$i = \left(1 + \frac{.0525}{365}\right)^{365} - 1 = .0539$$

method 2: Assume daily compounding is the same as continuous compounding.

$$i = (F/P) - 1$$
$$= e^{.0525} - 1 = .0539$$

12. Probabilistic Problems

Thus far, all of the cash flows included in the examples have been known exactly. If the cash flows are not known exactly but are given by some implicit or explicit probability distribution, the problem is probabilistic.

Probabilistic problems typically possess the following characteristics:

a. There is a chance of extreme loss that must be minimized.

b. There are multiple alternatives that must be chosen from. Each alternative gives a different degree of protection against the loss or failure.

c. The outcome is independent of the alternative chosen. Thus, as illustrated in example 2.16, the size of the dam that is chosen for construction will not alter the rainfall in successive years. However, it will alter the effects on the downstream watershed areas.

Probabilistic problems are typically solved using annual costs and expected values. An expected value is similar to an 'average value' since it is calculated as the mean of the given probability distribution. If cost 1 has a probability of occurrence of p_1, cost 2 has a probability of occurrence of p_2, and so on, the expected value is

$$E(\text{cost}) = p_1(\text{cost } 1) + p_2(\text{cost } 2) + \cdots \qquad 2.25$$

Example 2.16

Flood damage in any year is given according to the table below. What is the present worth of flood damage for a 10-year period? Use 6%.

Damage	Probability
0	.75
$10,000	.20
$20,000	.04
$30,000	.01

The expected value of flood damage is

$$\begin{aligned} E(\text{damage}) &= (0)(.75) + (10{,}000)(.20) + (20{,}000)(.04) \\ &\quad + (30{,}000)(.01) \\ &= \$3{,}100 \end{aligned}$$

$$\text{present worth} = 3{,}100(P/A,6\%,10) = 22{,}816$$

Probabilities in probabilistic problems may be given to you in the problem (as in the example above) or you may have to obtain them from some named probability distribution. In either case, the probabilities are known explicitly and such problems are known as explicit probability problems.

Example 2.17

A dam is being considered on a river which periodically overflows and causes $600,000 damage The damage is essentially the same each time the river causes flooding. The project horizon is 40 years. A 10% interest rate is being used.

Three different designs are available, each with different costs and storage capacities.

design alternative	cost	maximum capacity
A	500,000	1 unit
B	625,000	1.5 units
C	900,000	2.0 units

The U.S. Weather Service has provided a statistical analysis of annual rainfall in the area draining into the river.

units annual rainfall	probability
0	.10
.1 - .5	.60
.6 - 1.0	.15
1.1 - 1.5	.10
1.6 - 2.0	.04
2.0 or more	.01

Which design alternative would you choose assuming the dam is essentially empty at the start of each rainfall season?

The sum of the construction cost and the expected damage needs to be minimized. If alternative A is chosen, it will have a capacity of 1 unit. Its capacity will be exceeded (causing $600,000 damage) when the annual rainfall exceeds 1 unit. Therefore, the annual cost of A is

$$\begin{aligned} \text{EUAC(A)} &= 500{,}000(A/P,10\%,40) \\ &\quad + 600{,}000(.10 + .04 + .01) = 141{,}150 \end{aligned}$$

Similarly,

$$\begin{aligned} \text{EUAC(B)} &= 625{,}000(A/P,10\%,40) \\ &\quad + 600{,}000(.04 + .01) = 93{,}940 \end{aligned}$$

$$\begin{aligned} \text{EUAC(C)} &= 900{,}000(A/P,10\%,40) + 600{,}000(.01) \\ &= 98{,}070 \end{aligned}$$

Alternative B should be chosen.

In other problems, a probability distribution will not be given even though some parameter (such as the life of an alternative) is not known with certainty. Such problems are known as implicit probability problems since they require a reasonable assumption about the probability distribution.

Implicit probability problems typically involve items whose average lives are known. (Another term for average life is 'mean time before failure,' MTBF.) The key to such problems is in recognizing that an average life is not the same as a fixed life. An item may belong to a class which has an average life of 10 years, but it may fail after only 1 year of service. Conversely, it may not fail until after 25 or more years.

Obviously, it is not possible to 'read the mind' of a grader or question writer. However, if reasonable assumptions are made in writing at the start of the solution, you should receive full credit. One such reasonable assumption is that of a rectangular distribution. A rectangular distribution is one which is assumed to give an equal probability of failure in each year. Such an assumption is illustrated in example 2.18.

Example 2.18

A bridge is needed for 20 years. Failure of the bridge at any time will require a 50% reinvestment. Evaluate the two design alternatives below using 6% interest.

design alternative	initial cost	MTBF	annual costs	salvage at t = 20
A	15,000	9 years	1,200	0
B	22,000	43 years	1,000	0

It will be assumed that each of the alternatives has an annual probability of failure that is inversely proportional to its average life. Then, for alternative A, the probability of failure in any year is (1/9). Similarly, the annual failure probability for alternative B is (1/43).

$$\text{EUAC(A)} = 15{,}000(A/P,6\%,20) + 15{,}000(.5)(1/9) + 1{,}200 = 3{,}341$$

$$\text{EUAC(B)} = 22{,}000(A/P,6\%,20) + 22{,}000(.5)(1/43) + 1{,}000 = 3{,}174$$

Alternative B should be chosen.

13. Estimating Economic Life

As assets grow older, their operating and maintenance costs typically increase each year. Eventually, the cost to keep an asset in operation becomes prohibitive, and the asset is retired or replaced. However, it is not always obvious when an asset should be retired or replaced.

As the asset's maintenance is increasing each year, the amortized cost of its initial purchase is decreasing. It is the sum of these two costs that should be evaluated to determine the point at which the asset should be retired or replaced. Since an asset's initial purchase price is likely to be high, the amortized cost will be the controlling factor in those years when the maintenance costs are low. Therefore, the EUAC of the asset will decrease in the initial part of its life.

However, as the asset grows older, the change in its amortized cost decreases while maintenance increases. Eventually the sum of the two costs reaches a minimum and then starts to increase. The age of the asset at the minimum cost point is known as the economic life of the asset.

The determination of an asset's economic life is illustrated by example 2.19.

Example 2.19

A bus in a municipal transit system has the characteristics listed below. When should the city replace its buses if money can be borrowed at 8%?

Initial cost: $120,000

year	maintenance cost	salvage value
1	35,000	60,000
2	38,000	55,000
3	43,000	45,000
4	50,000	25,000
5	65,000	15,000

If the bus is kept for 1 year and then sold, the annual cost will be

$$\begin{aligned}\text{EUAC}(1) &= 120{,}000(A/P,8\%,1) + 35{,}000(A/F,8\%,1) \\ &\quad - 60{,}000(A/F,8\%,1) \\ &= 104{,}600\end{aligned}$$

If the bus is kept for 2 years and then sold, the annual cost will be

$$\begin{aligned}\text{EUAC}(2) &= [120{,}000+35{,}000(P/F,8\%,1)](A/P,8\%,2) \\ &\quad + (38{,}000-55{,}000)(A/F,8\%,2) \\ &= 77{,}300\end{aligned}$$

If the bus is kept for 3 years and then sold, the annual cost will be

$$\begin{aligned}\text{EUAC}(3) &= [120{,}000 + 35{,}000(P/F,8\%,1) \\ &\quad + 38{,}000(P/F,8\%,2)](A/P,8\%,3) \\ &\quad + (43{,}000-45{,}000)(A/F,8\%,3) \\ &= 71{,}200\end{aligned}$$

This process is continued until EUAC begins to increase. In this example, EUAC(4) is 71,700. Therefore, the buses should be retired after 3 years.

14. Basic Cost Accounting

Cost accounting is the system which determines the cost of manufactured products. Cost accounting is called job cost accounting if costs are accumulated by part number or contract. It is called process cost accounting if costs are accumulated by departments or manufacturing processes.

Three types of costs make up the total manufacturing cost of a product. Those are: direct material, direct labor, and all indirect costs.

Direct material costs are the costs of all materials that go into the product, priced at the original purchase cost.

Direct labor costs are the costs of all labor required to assemble or shape the product.

Indirect material and labor costs are generally limited to costs incurred in the factory, excluding costs incurred in the office area. Examples of indirect materials are cleaning fluids, assembly lubricants, and temporary routing tags. Examples of indirect labor are stock-picking, inspection, expediting, and supervision labor.

Here are some important points concerning basic cost accounting:

a. The sum of direct material and direct labor costs is known as the prime cost.

b. Indirect costs may be called indirect manufacturing expenses (IME).

c. Indirect costs may also include the overhead sector of the company (i.e., secretaries, engineers, and corporate administration). In this case, the indirect cost is usually called burden or overhead. Burden may also include the EUAC of non-regular costs which must be spread evenly over several years.

d. The cost of a product is usually known in advance from previous manufacturing runs or by estimation. Any deviation from this known cost is called a variance. Variance may be broken down into labor variance and material variance.

e. Indirect cost per item is not easily measured. The method of allocating indirect costs to a product is as follows:

   step 1: Estimate the total expected indirect (and overhead) costs for the upcoming year.

   step 2: Decide on some convenient vehicle for allocating the overhead to production. Usually, this vehicle is either the number of units expected to be produced or the number of direct hours expected to be worked in the upcoming year.

   step 3: Estimate the quantity or size of the overhead vehicle.

   step 4: Divide the expected overhead costs by the expected overhead vehicle to obtain the unit overhead.

   step 5: Regardless of the true size of the overhead vehicle during the upcoming year, one unit of overhead cost is allocated per product.

f. Although estimates of production for the next year are always somewhat inaccurate, the cost of the product is assumed to be independent of forecasting errors. Any difference between true cost and calculated cost goes into a variance account.

g. Burden (overhead) variance will be caused by errors in forecasting both the actual overhead for the upcoming year and the vehicle size. In the former case, the variance is called burden budget variance; in the latter, it is called burden capacity variance.

Example 2.20

A small company expects to produce 8,000 items in the upcoming year. The current material cost is $4.54 each. 16 minutes of direct labor are required per unit. Workers are paid $7.50 per hour. 2,133 direct labor hours are forecasted for the product. Miscellaneous overhead costs are estimated at $45,000.

Find the expected direct material cost, the direct labor cost, the prime cost, the burden as a function of production and direct labor, and the total cost.

- The direct material cost was given as $4.54.
- The direct labor cost is $(16/60)(\$7.50) = \$2.00$
- The prime cost is $\$4.54 + \$2.00 = \$6.54$.
- If the burden vehicle is production, the burden rate is $(\$45{,}000/8{,}000) = \$5.63$ per item, making the total cost $\$4.54 + \$2.00 + \$5.63 = \$12.17$.
- If the burden vehicle is direct labor hours, the burden rate is $(45{,}000/2{,}133) = \$21.10$ per hour, making the total cost $\$4.54 + \$2.00 + {}^{16}/_{60}(\$21.10) = \$12.17$.

Example 2.21

The actual performance of the company in example 2.20 is given by the following figures:

actual production: 7,560

actual overhead costs: $47,000

What are the burden budget variance and the burden capacity variance?

The burden capacity variance is

$$\$45{,}000 - 7{,}560(\$5.63) = \$2{,}437$$

The burden budget variance is

$$\$47{,}000 - \$45{,}000 = \$2{,}000$$

The overall burden variance is

$$\$47{,}000 - 7{,}560(\$5.63) = \$4{,}437$$

15. Break-Even Analysis

Break-even analysis is a method of determining when costs exactly equal revenue. If the manufactured quantity is less than the break-even quantity, a loss is incurred. If the manufactured quantity is greater than the break-even quantity a profit is incurred.

Consider the following special variables:

f	a fixed cost which does not vary with production
a	an incremental cost, which is the cost to produce one additional item. It may also be called the marginal cost or differential cost.
Q	the quantity sold
p	the incremental revenue
R	the total revenue
C	the total cost

Assuming no change in the inventory, the break-even point can be found from C = R, where

$$C = f + aQ \quad 2.26$$

$$R = pQ \quad 2.27$$

An alternate form of the break-even problem is to find the number of units per period for which two alternatives have the same total costs. Fixed costs are to be spread over a period longer than one year. One of the alternatives will have a lower cost if production is less than the break-even point. The other will have a lower cost for production greater than the break-even point.

Example 2.22

Two plans are available for a company to obtain automobiles for its salesmen. How many miles must the cars be driven each year for the two plans to have the same costs? Use an interest rate of 10%.

Plan A: Lease the cars and pay $.15 per mile

Plan B: Purchase the cars for $5,000. Each car has an economic life of 3 years, after which it can be sold for $1,200. Gas and oil cost $.04 per mile. Insurance is $500 per year.

Let x be the number of miles driven per year. Then, the EUAC for both alternatives are:

$$\text{EUAC(A)} = .15x$$

$$\begin{aligned}\text{EUAC(B)} &= .04x + 500 + 5{,}000(A/P,10\%,3) \\ &\quad - 1{,}200(A/F,10\%,3) \\ &= .04x + 2{,}148\end{aligned}$$

Setting EUAC(A) and EUAC(B) equal and solving for x yields 19,527 miles per year as the break-even point.

16. Handling Inflation

It is important to perform economic studies in terms of 'constant value dollars.' One method of converting all cash flows to constant value dollars is to divide the flows by some annual economic indicator or price index. Such indicators would normally be given to you as part of a problem.

If indicators are not available, this method can still be used by assuming that inflation is relatively constant at a decimal rate (e) per year. Then, all cash flows can be converted to 't = 0' dollars by dividing by $(1 + e)^n$ where n is the year of the cash flow.

Example 2.23

What is the uninflated present worth of $2,000 in 2 years if the average inflation rate is 6% and i is 10%?

$$P = \frac{\$2{,}000}{(1.10)^2(1.06)^2} = \$1{,}471.07$$

An alternative is to replace i with a value corrected for inflation. This corrected value, i′, is

$$i' = i + e + ie \quad 2.28$$

This method has the advantage of simplifying the calculations. However, pre-calculated factors may not be available for the non-integer values of i′. Therefore, table 2.1 will have to be used to calculate the factors.

Example 2.24

Repeat example 2.23 using i′.

$$i' = .10 + .06 + (.10)(.06) = .166$$

$$P = \frac{\$2{,}000}{(1.166)^2} = \$1{,}471.07$$

17. Learning Curves

The more products that are made, the more efficient the operation becomes due to experience gained. Therefore, direct labor costs decrease. Usually, a learning curve is specified by the decrease in cost each time the quantity produced doubles. If there is a 20% decrease per doubling, the curve is said to be an 80% learning curve.

Consider the following special variables:

- T_1 time or cost for the first item
- T_n the time or cost for the nth item
- n the total number of items produced
- b the learning curve constant

learning curve	b
75%	.415
80	.322
85	.234
90	.152
95	.074

Then, the time to produce the nth item is given by

$$T_n = T_1(n)^{-b} \quad 2.29$$

The total time to produce units from quantity n_1 to n_2 inclusive is

$$\int_{n_1}^{n_2} T_n dn \approx \frac{T_1}{(1-b)}\left[(n_2 + \tfrac{1}{2})^{1-b} - (n_1 - \tfrac{1}{2})^{1-b}\right] \quad 2.30$$

The average time per unit over the production from n_1 to n_2 is the above total time from equation 2.30 divided by the quantity produced, $(n_2 - n_1 + 1)$.

It is important to remember that learning curve reductions apply only to direct labor costs. They are not applied to indirect labor or direct material costs.

ECON

Example 2.25

A 70% learning curve is used with an item whose first production time was 1.47 hours. How long will it take to produce the 11th item? How long will it take to produce the 11th through 27th items?

First, find b.

$$\frac{T_2}{T_1} = .7 = (2)^{-b} \quad \text{or } b = .515$$

Then, $T_{11} = 1.47(11)^{-.515} = .428$ hours

The time to produce the 11th item through 27th item is approximately

$$T = \frac{1.47}{1-.515}[(27.5)^{1-.515} - (10.5)^{1-.515}]$$
$$= 5.643 \text{ hours}$$

18. *Economic Order Quantity*

The Economic Order Quantity (EOQ) is defined as the order quantity which minimizes the inventory costs per unit time. Although there are many different EOQ models, the most simple is based on the following assumptions:

a. Reordering is instantaneous. The time between order placement and receipt is zero.
b. Shortages are not allowed.
c. Demand for the inventory item is deterministic (i.e., is not a random variable).
d. Demand is constant with respect to time.
e. An order is placed when the on-hand quantity is zero.

The following special variables are used:

- a — the constant depletion rate (items/unit time)
- h — the inventory storage cost ($/item-unit time)
- H — the total inventory storage cost between orders ($)
- K — the fixed cost of placing an order ($)
- Q_o — the order quantity

If the original quantity on hand is Q_o, the stock will be depleted at $t^* = (Q_o/a)$. The total inventory storage cost between t_o and t^* is

$$H = \tfrac{1}{2}hQ_o^2/a \qquad 2.31$$

The total inventory and ordering cost per unit time is

$$C_t = \frac{aK}{Q_o} + \tfrac{1}{2}hQ_o \qquad 2.32$$

C_t can be minimized with repect to Q_o. The EOQ and time between orders are:

$$Q_o^* = \sqrt{2aK/h} \qquad 2.33$$

$$t^* = Q_o^*/a \qquad 2.34$$

19. *Consumer Loans*

Consumer loans, typical of home mortgages, cannot be handled by the equivalence formulas presented up to this point. Many different arrangements can be made between lender and borrower. Four of the most common consumer loan arrangements are presented below. Refer to a real estate or investment analysis book for more complex loans.

A. Simple Interest

Interest due does not compound with a simple interest loan. The interest due is merely proportional to the length of time the principal is outstanding. Because of this, simple interest loans are seldom made for long periods (e.g., longer than one year).

Example 2.26

A $12,000 simple interest loan is taken out at 16% per annum. The loan matures in one year with no intermediate payments. How much will be due at the end of the year?

$$\text{Amount due} = (1 + .16)(12{,}000) = 13{,}920$$

For loans less than one year, it is commonly assumed that a year consists of 12 months of 30 days each.

Example 2.27

$4000 is borrowed for 75 days at 16% per annum simple interest. How much will be due at the end of 75 days?

$$\text{Amount due} = 4000 + (.16)(75/360)(4000) = 4{,}133$$

B. Loans with Constant Amount Paid Towards Principal

With this loan type, the payment is not the same each period. The amount paid towards the principal is constant, but the interest varies from period to period. The following special symbols are used.

- BAL_j — balance after the jth payment
- LV — total value loaned (cost minus down payment)
- j — payment or period number
- N — total number of payments to pay off the loan
- PI_j — jth interest payment
- PP_j — jth principal payment
- PT_j — jth total payment
- ø — effective rate per period (r/k)

The equations which govern this type of loan are:

$$BAL_j = LV - (j)(PP) \qquad 2.35$$

ECON

$$PI_j = \phi(BAL_j) \quad 2.36$$

$$PT_j = PP + PI_j \quad 2.37$$

C. Direct Reduction Loans

This is the typical "interest paid on unpaid balance" loan. The amount of the periodic payment is constant, but the amounts paid towards the principal and interest both vary.

The same symbols are used with this type of loan as are listed above.

$$N = -\frac{\ln\left[\frac{-\phi(LV)}{PT} + 1\right]}{\ln(1 + \phi)} \quad 2.38$$

$$BAL_{j-1} = PT\left[\frac{1-(1+\phi)^{j-1-N}}{\phi}\right] \quad 2.39$$

$$PI_j = \phi\,(BAL_{j-1}) \quad 2.40$$

$$PP_j = PT - PI_j \quad 2.41$$

$$BAL_j = BAL_{j-1} - PP_j \quad 2.42$$

Example 2.28

A $45,000 loan is financed at 9.25% per annum. The monthly payment is $385. What are the amounts paid towards interest and principal in the 14th period? What is the remaining principal balance after the 14th payment has been made?

The effective rate per month is

$$\phi = \frac{r}{k} = \frac{.0925}{12} = .007708$$

$$N = -\frac{\ln\left[\frac{(.007708)(45{,}000)}{385} + 1\right]}{\ln(1 + .007708)} = 301$$

$$BAL_{13} = 385\left[\frac{1-(1+.007708)^{14-1-301}}{.007708}\right]$$

$$= 44{,}476.39$$

$$PI_{14} = (.007708)(44{,}476.39) = 342.82$$

$$PP_{14} = 385 - 342.82 = 42.18$$

$$BAL_{14} = 44{,}476.39 - 42.18 = 44{,}434.20$$

D. Direct Reduction Loan with Balloon Payment

This type of loan has a constant periodic payment, but the duration of the loan is insufficient to completely pay back the principal. Therefore, all remaining unpaid principal must be paid back in a lump sum when the loan matures. This large payment is known as a balloon payment.

Equations 2.38 through 2.42 can also be used with this type of loan. The remaining balance after the last payment is the balloon payment. This balloon payment must be repaid along with the last regular payment calculated.

ECON

STANDARD CASH FLOW FACTORS

MULTIPLY		*BY*	*TO OBTAIN*	
	F	(P/F,i%,n)	P	
	P	(F/P,i%,n)	F	
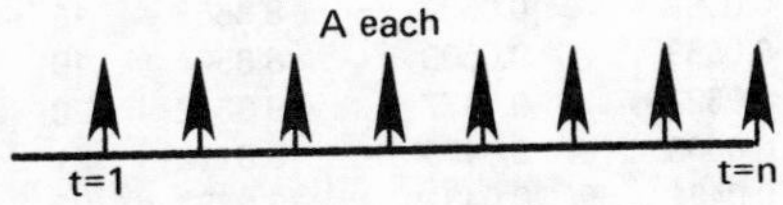	A	(P/A,i%,n)	P	
	P	(A/P,i%,n)	A	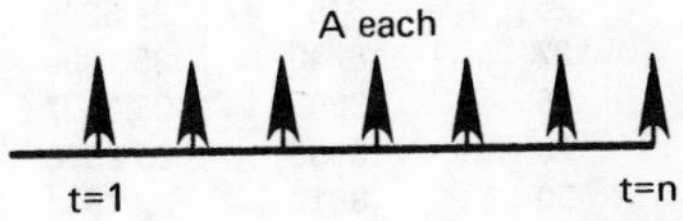
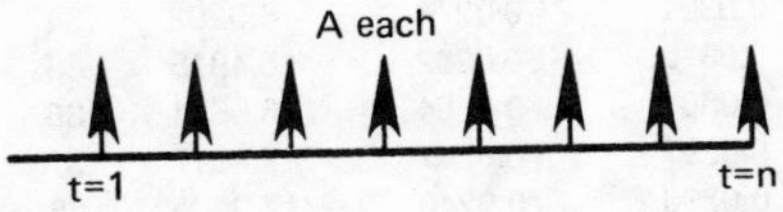	A	(F/A,i%,n)	F	 
	F	(A/F,i%,n)	A	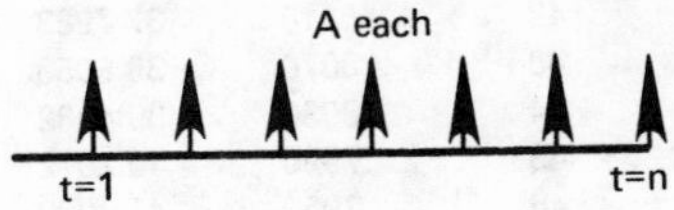
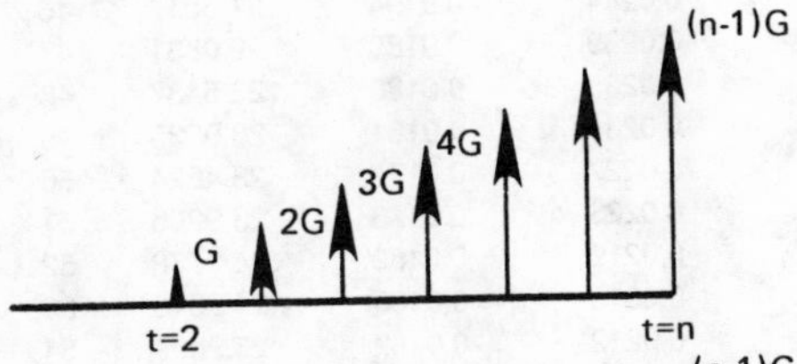	G	(P/G,i%,n)	P	
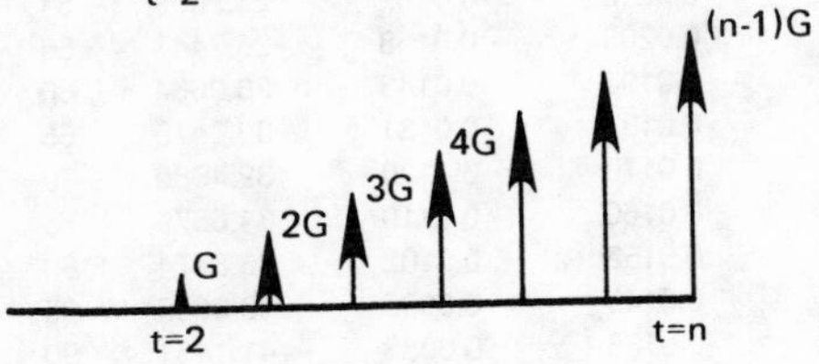	G	(A/G,i%,n)	A	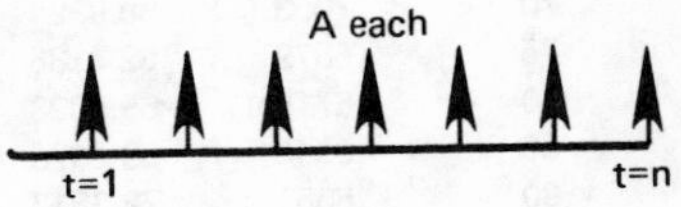

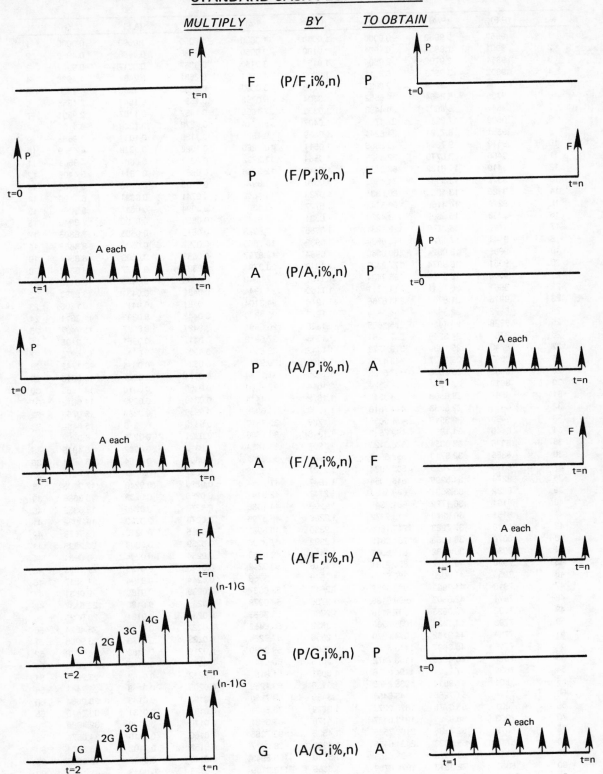

I = 0.50 %

N	(P/F)	(P/A)	(P/G)	(F/P)	(F/A)	(A/P)	(A/F)	(A/G)	N
1	.9950	0.9950	−0.0000	1.0050	1.0000	1.0050	1.0000	−0.0000	1
2	.9901	1.9851	0.9901	1.0100	2.0050	0.5038	0.4988	0.4988	2
3	.9851	2.9702	2.9604	1.0151	3.0150	0.3367	0.3317	0.9967	3
4	.9802	3.9505	5.9011	1.0202	4.0301	0.2531	0.2481	1.4938	4
5	.9754	4.9259	9.8026	1.0253	5.0503	0.2030	0.1980	1.9900	5
6	.9705	5.8964	14.6552	1.0304	6.0755	0.1696	0.1646	2.4855	6
7	.9657	6.8621	20.4493	1.0355	7.1059	0.1457	0.1407	2.9801	7
8	.9609	7.8230	27.1755	1.0407	8.1414	0.1278	0.1228	3.4738	8
9	.9561	8.7791	34.8244	1.0459	9.1821	0.1139	0.1089	3.9668	9
10	.9513	9.7304	43.3865	1.0511	10.2280	0.1028	0.0978	4.4589	10
11	.9466	10.6770	52.8526	1.0564	11.2792	0.0937	0.0887	4.9501	11
12	.9419	11.6189	63.2136	1.0617	12.3356	0.0861	0.0811	5.4406	12
13	.9372	12.5562	74.4602	1.0670	13.3972	0.0796	0.0746	5.9302	13
14	.9326	13.4887	86.5835	1.0723	14.4642	0.0741	0.0691	6.4190	14
15	.9279	14.4166	99.5743	1.0777	15.5365	0.0694	0.0644	6.9069	15
16	.9233	15.3399	113.4238	1.0831	16.6142	0.0652	0.0602	7.3940	16
17	.9187	16.2586	128.1231	1.0885	17.6973	0.0615	0.0565	7.8803	17
18	.9141	17.1728	143.6634	1.0939	18.7858	0.0582	0.0532	8.3658	18
19	.9096	18.0824	160.0360	1.0994	19.8797	0.0553	0.0503	8.8504	19
20	.9051	18.9874	177.2322	1.1049	20.9791	0.0527	0.0477	9.3342	20
21	.9006	19.8880	195.2434	1.1104	22.0840	0.0503	0.0453	9.8172	21
22	.8961	20.7841	214.0611	1.1160	23.1944	0.0481	0.0431	10.2993	22
23	.8916	21.6757	233.6768	1.1216	24.3104	0.0461	0.0411	10.7806	23
24	.8872	22.5629	254.0820	1.1272	25.4320	0.0443	0.0393	11.2611	24
25	.8828	23.4456	275.2686	1.1328	26.5591	0.0427	0.0377	11.7407	25
26	.8784	24.3240	297.2281	1.1385	27.6919	0.0411	0.0361	12.2195	26
27	.8740	25.1980	319.9523	1.1442	28.8304	0.0397	0.0347	12.6975	27
28	.8697	26.0677	343.4332	1.1499	29.9745	0.0384	0.0334	13.1747	28
29	.8653	26.9330	367.6625	1.1556	31.1244	0.0371	0.0321	13.6510	29
30	.8610	27.7941	392.6324	1.1614	32.2800	0.0360	0.0310	14.1265	30
31	.8567	28.6508	418.3348	1.1672	33.4414	0.0349	0.0299	14.6012	31
32	.8525	29.5033	444.7618	1.1730	34.6086	0.0339	0.0289	15.0750	32
33	.8482	30.3515	471.9055	1.1789	35.7817	0.0329	0.0279	15.5480	33
34	.8440	31.1955	499.7583	1.1848	36.9606	0.0321	0.0271	16.0202	34
35	.8398	32.0354	528.3123	1.1907	38.1454	0.0312	0.0262	16.4915	35
36	.8356	32.8710	557.5598	1.1967	39.3361	0.0304	0.0254	16.9621	36
37	.8315	33.7025	587.4934	1.2027	40.5328	0.0297	0.0247	17.4317	37
38	.8274	34.5299	618.1054	1.2087	41.7354	0.0290	0.0240	17.9006	38
39	.8232	35.3531	649.3883	1.2147	42.9441	0.0283	0.0233	18.3686	39
40	.8191	36.1722	681.3347	1.2208	44.1588	0.0276	0.0226	18.8359	40
41	.8151	36.9873	713.9372	1.2269	45.3796	0.0270	0.0220	19.3022	41
42	.8110	37.7983	747.1886	1.2330	46.6065	0.0265	0.0215	19.7678	42
43	.8070	38.6053	781.0815	1.2392	47.8396	0.0259	0.0209	20.2325	43
44	.8030	39.4082	815.6087	1.2454	49.0788	0.0254	0.0204	20.6964	44
45	.7990	40.2072	850.7631	1.2516	50.3242	0.0249	0.0199	21.1595	45
46	.7950	41.0022	886.5376	1.2579	51.5758	0.0244	0.0194	21.6217	46
47	.7910	41.7932	922.9252	1.2642	52.8337	0.0239	0.0189	22.0831	47
48	.7871	42.5803	959.9188	1.2705	54.0978	0.0235	0.0185	22.5437	48
49	.7832	43.3635	997.5116	1.2768	55.3683	0.0231	0.0181	23.0035	49
50	.7793	44.1428	1035.6966	1.2832	56.6452	0.0227	0.0177	23.4624	50
51	.7754	44.9182	1074.4670	1.2896	57.9284	0.0223	0.0173	23.9205	51
52	.7716	45.6897	1113.8162	1.2961	59.2180	0.0219	0.0169	24.3778	52
53	.7677	46.4575	1153.7372	1.3026	60.5141	0.0215	0.0165	24.8343	53
54	.7639	47.2214	1194.2236	1.3091	61.8167	0.0212	0.0162	25.2899	54
55	.7601	47.9814	1235.2686	1.3156	63.1258	0.0208	0.0158	25.7447	55
60	.7414	51.7256	1448.6458	1.3489	69.7700	0.0193	0.0143	28.0064	60
65	.7231	55.3775	1675.0272	1.3829	76.5821	0.0181	0.0131	30.2475	65
70	.7053	58.9394	1913.6427	1.4178	83.5661	0.0170	0.0120	32.4680	70
75	.6879	62.4136	2163.7525	1.4536	90.7265	0.0160	0.0110	34.6679	75
80	.6710	65.8023	2424.6455	1.4903	98.0677	0.0152	0.0102	36.8474	80
85	.6545	69.1075	2695.6389	1.5280	105.5943	0.0145	0.0095	39.0065	85
90	.6383	72.3313	2976.0769	1.5666	113.3109	0.0138	0.0088	41.1451	90
95	.6226	75.4757	3265.3298	1.6061	121.2224	0.0132	0.0082	43.2633	95
100	.6073	78.5426	3562.7934	1.6467	129.3337	0.0127	0.0077	45.3613	100

I = 0.75 %

N	(P/F)	(P/A)	(P/G)	(F/P)	(F/A)	(A/P)	(A/F)	(A/G)	N
1	.9926	0.9926	-0.0000	1.0075	1.0000	1.0075	1.0000	-0.0000	1
2	.9852	1.9777	0.9852	1.0151	2.0075	0.5056	0.4981	0.4981	2
3	.9778	2.9556	2.9408	1.0227	3.0226	0.3383	0.3308	0.9950	3
4	.9706	3.9261	5.8525	1.0303	4.0452	0.2547	0.2472	1.4907	4
5	.9633	4.8894	9.7058	1.0381	5.0756	0.2045	0.1970	1.9851	5
6	.9562	5.8456	14.4866	1.0459	6.1136	0.1711	0.1636	2.4782	6
7	.9490	6.7946	20.1808	1.0537	7.1595	0.1472	0.1397	2.9701	7
8	.9420	7.7366	26.7747	1.0616	8.2132	0.1293	0.1218	3.4608	8
9	.9350	8.6716	34.2544	1.0696	9.2748	0.1153	0.1078	3.9502	9
10	.9280	9.5996	42.6064	1.0776	10.3443	0.1042	0.0967	4.4384	10
11	.9211	10.5207	51.8174	1.0857	11.4219	0.0951	0.0876	4.9253	11
12	.9142	11.4349	61.8740	1.0938	12.5076	0.0875	0.0800	5.4110	12
13	.9074	12.3423	72.7632	1.1020	13.6014	0.0810	0.0735	5.8954	13
14	.9007	13.2430	84.4720	1.1103	14.7034	0.0755	0.0680	6.3786	14
15	.8940	14.1370	96.9876	1.1186	15.8137	0.0707	0.0632	6.8606	15
16	.8873	15.0243	110.2973	1.1270	16.9323	0.0666	0.0591	7.3413	16
17	.8807	15.9050	124.3887	1.1354	18.0593	0.0629	0.0554	7.8207	17
18	.8742	16.7792	139.2494	1.1440	19.1947	0.0596	0.0521	8.2989	18
19	.8676	17.6468	154.8671	1.1525	20.3387	0.0567	0.0492	8.7759	19
20	.8612	18.5080	171.2297	1.1612	21.4912	0.0540	0.0465	9.2516	20
21	.8548	19.3628	188.3253	1.1699	22.6524	0.0516	0.0441	9.7261	21
22	.8484	20.2112	206.1420	1.1787	23.8223	0.0495	0.0420	10.1994	22
23	.8421	21.0533	224.6682	1.1875	25.0010	0.0475	0.0400	10.6714	23
24	.8358	21.8891	243.8923	1.1964	26.1885	0.0457	0.0382	11.1422	24
25	.8296	22.7188	263.8029	1.2054	27.3849	0.0440	0.0365	11.6117	25
26	.8234	23.5422	284.3888	1.2144	28.5903	0.0425	0.0350	12.0800	26
27	.8173	24.3595	305.6387	1.2235	29.8047	0.0411	0.0336	12.5470	27
28	.8112	25.1707	327.5416	1.2327	31.0282	0.0397	0.0322	13.0128	28
29	.8052	25.9759	350.0867	1.2420	32.2609	0.0385	0.0310	13.4774	29
30	.7992	26.7751	373.2631	1.2513	33.5029	0.0373	0.0298	13.9407	30
31	.7932	27.5683	397.0602	1.2607	34.7542	0.0363	0.0288	14.4028	31
32	.7873	28.3557	421.4675	1.2701	36.0148	0.0353	0.0278	14.8636	32
33	.7815	29.1371	446.4746	1.2796	37.2849	0.0343	0.0268	15.3232	33
34	.7757	29.9128	472.0712	1.2892	38.5646	0.0334	0.0259	15.7816	34
35	.7699	30.6827	498.2471	1.2989	39.8538	0.0326	0.0251	16.2387	35
36	.7641	31.4468	524.9924	1.3086	41.1527	0.0318	0.0243	16.6946	36
37	.7585	32.2053	552.2969	1.3185	42.4614	0.0311	0.0236	17.1493	37
38	.7528	32.9581	580.1511	1.3283	43.7798	0.0303	0.0228	17.6027	38
39	.7472	33.7053	608.5451	1.3383	45.1082	0.0297	0.0222	18.0549	39
40	.7416	34.4469	637.4693	1.3483	46.4465	0.0290	0.0215	18.5058	40
41	.7361	35.1831	666.9144	1.3585	47.7948	0.0284	0.0209	18.9556	41
42	.7306	35.9137	696.8709	1.3686	49.1533	0.0278	0.0203	19.4040	42
43	.7252	36.6389	727.3297	1.3789	50.5219	0.0273	0.0198	19.8513	43
44	.7198	37.3587	758.2815	1.3893	51.9009	0.0268	0.0193	20.2973	44
45	.7145	38.0732	789.7173	1.3997	53.2901	0.0263	0.0188	20.7421	45
46	.7091	38.7823	821.6283	1.4102	54.6898	0.0258	0.0183	21.1856	46
47	.7039	39.4862	854.0056	1.4207	56.1000	0.0253	0.0178	21.6280	47
48	.6986	40.1848	886.8404	1.4314	57.5207	0.0249	0.0174	22.0691	48
49	.6934	40.8782	920.1243	1.4421	58.9521	0.0245	0.0170	22.5089	49
50	.6883	41.5664	953.8486	1.4530	60.3943	0.0241	0.0166	22.9476	50
51	.6831	42.2496	988.0050	1.4639	61.8472	0.0237	0.0162	23.3850	51
52	.6780	42.9276	1022.5852	1.4748	63.3111	0.0233	0.0158	23.8211	52
53	.6730	43.6006	1057.5810	1.4859	64.7859	0.0229	0.0154	24.2561	53
54	.6680	44.2686	1092.9842	1.4970	66.2718	0.0226	0.0151	24.6898	54
55	.6630	44.9316	1128.7869	1.5083	67.7688	0.0223	0.0148	25.1223	55
60	.6387	48.1734	1313.5189	1.5657	75.4241	0.0208	0.0133	27.2665	60
65	.6153	51.2963	1507.0910	1.6253	83.3709	0.0195	0.0120	29.3801	65
70	.5927	54.3046	1708.6065	1.6872	91.6201	0.0184	0.0109	31.4634	70
75	.5710	57.2027	1917.2225	1.7514	100.1833	0.0175	0.0100	33.5163	75
80	.5500	59.9944	2132.1472	1.8180	109.0725	0.0167	0.0092	35.5391	80
85	.5299	62.6838	2352.6375	1.8873	118.3001	0.0160	0.0085	37.5318	85
90	.5104	65.2746	2577.9961	1.9591	127.8790	0.0153	0.0078	39.4946	90
95	.4917	67.7704	2807.5694	2.0337	137.8225	0.0148	0.0073	41.4277	95
100	.4737	70.1746	3040.7453	2.1111	148.1445	0.0143	0.0068	43.3311	100

I = 1.00 %

N	(P/F)	(P/A)	(P/G)	(F/P)	(F/A)	(A/P)	(A/F)	(A/G)	N
1	.9901	0.9901	-0.0000	1.0100	1.0000	1.0100	1.0000	-0.0000	1
2	.9803	1.9704	0.9803	1.0201	2.0100	0.5075	0.4975	0.4975	2
3	.9706	2.9410	2.9215	1.0303	3.0301	0.3400	0.3300	0.9934	3
4	.9610	3.9020	5.8044	1.0406	4.0604	0.2563	0.2463	1.4876	4
5	.9515	4.8534	9.6103	1.0510	5.1010	0.2060	0.1960	1.9801	5
6	.9420	5.7955	14.3205	1.0615	6.1520	0.1725	0.1625	2.4710	6
7	.9327	6.7282	19.9168	1.0721	7.2135	0.1486	0.1386	2.9602	7
8	.9235	7.6517	26.3812	1.0829	8.2857	0.1307	0.1207	3.4478	8
9	.9143	8.5660	33.6959	1.0937	9.3685	0.1167	0.1067	3.9337	9
10	.9053	9.4713	41.8435	1.1046	10.4622	0.1056	0.0956	4.4179	10
11	.8963	10.3676	50.8067	1.1157	11.5668	0.0965	0.0865	4.9005	11
12	.8874	11.2551	60.5687	1.1268	12.6825	0.0888	0.0788	5.3815	12
13	.8787	12.1337	71.1126	1.1381	13.8093	0.0824	0.0724	5.8607	13
14	.8700	13.0037	82.4221	1.1495	14.9474	0.0769	0.0669	6.3384	14
15	.8613	13.8651	94.4810	1.1610	16.0969	0.0721	0.0621	6.8143	15
16	.8528	14.7179	107.2734	1.1726	17.2579	0.0679	0.0579	7.2886	16
17	.8444	15.5623	120.7834	1.1843	18.4304	0.0643	0.0543	7.7613	17
18	.8360	16.3983	134.9957	1.1961	19.6147	0.0610	0.0510	8.2323	18
19	.8277	17.2260	149.8950	1.2081	20.8109	0.0581	0.0481	8.7017	19
20	.8195	18.0456	165.4664	1.2202	22.0190	0.0554	0.0454	9.1694	20
21	.8114	18.8570	181.6950	1.2324	23.2392	0.0530	0.0430	9.6354	21
22	.8034	19.6604	198.5663	1.2447	24.4716	0.0509	0.0409	10.0998	22
23	.7954	20.4558	216.0660	1.2572	25.7163	0.0489	0.0389	10.5626	23
24	.7876	21.2434	234.1800	1.2697	26.9735	0.0471	0.0371	11.0237	24
25	.7798	22.0232	252.8945	1.2824	28.2432	0.0454	0.0354	11.4831	25
26	.7720	22.7952	272.1957	1.2953	29.5256	0.0439	0.0339	11.9409	26
27	.7644	23.5596	292.0702	1.3082	30.8209	0.0424	0.0324	12.3971	27
28	.7568	24.3164	312.5047	1.3213	32.1291	0.0411	0.0311	12.8516	28
29	.7493	25.0658	333.4863	1.3345	33.4504	0.0399	0.0299	13.3044	29
30	.7419	25.8077	355.0021	1.3478	34.7849	0.0387	0.0287	13.7557	30
31	.7346	26.5423	377.0394	1.3613	36.1327	0.0377	0.0277	14.2052	31
32	.7273	27.2696	399.5858	1.3749	37.4941	0.0367	0.0267	14.6532	32
33	.7201	27.9897	422.6291	1.3887	38.8690	0.0357	0.0257	15.0995	33
34	.7130	28.7027	446.1572	1.4026	40.2577	0.0348	0.0248	15.5441	34
35	.7059	29.4086	470.1583	1.4166	41.6603	0.0340	0.0240	15.9871	35
36	.6989	30.1075	494.6207	1.4308	43.0769	0.0332	0.0232	16.4285	36
37	.6920	30.7995	519.5329	1.4451	44.5076	0.0325	0.0225	16.8682	37
38	.6852	31.4847	544.8835	1.4595	45.9527	0.0318	0.0218	17.3063	38
39	.6784	32.1630	570.6616	1.4741	47.4123	0.0311	0.0211	17.7428	39
40	.6717	32.8347	596.8561	1.4889	48.8864	0.0305	0.0205	18.1776	40
41	.6650	33.4997	623.4562	1.5038	50.3752	0.0299	0.0199	18.6108	41
42	.6584	34.1581	650.4514	1.5188	51.8790	0.0293	0.0193	19.0424	42
43	.6519	34.8100	677.8312	1.5340	53.3978	0.0287	0.0187	19.4723	43
44	.6454	35.4555	705.5853	1.5493	54.9318	0.0282	0.0182	19.9006	44
45	.6391	36.0945	733.7037	1.5648	56.4811	0.0277	0.0177	20.3273	45
46	.6327	36.7272	762.1765	1.5805	58.0459	0.0272	0.0172	20.7524	46
47	.6265	37.3537	790.9938	1.5963	59.6263	0.0268	0.0168	21.1758	47
48	.6203	37.9740	820.1460	1.6122	61.2226	0.0263	0.0163	21.5976	48
49	.6141	38.5881	849.6237	1.6283	62.8348	0.0259	0.0159	22.0178	49
50	.6080	39.1961	879.4176	1.6446	64.4632	0.0255	0.0155	22.4363	50
51	.6020	39.7981	909.5186	1.6611	66.1078	0.0251	0.0151	22.8533	51
52	.5961	40.3942	939.9175	1.6777	67.7689	0.0248	0.0148	23.2686	52
53	.5902	40.9844	970.6057	1.6945	69.4466	0.0244	0.0144	23.6823	53
54	.5843	41.5687	1001.5743	1.7114	71.1410	0.0241	0.0141	24.0945	54
55	.5785	42.1472	1032.8148	1.7285	72.8525	0.0237	0.0137	24.5049	55
60	.5504	44.9550	1192.8061	1.8167	81.6697	0.0222	0.0122	26.5333	60
65	.5237	47.6266	1358.3903	1.9094	90.9366	0.0210	0.0110	28.5217	65
70	.4983	50.1685	1528.6474	2.0068	100.6763	0.0199	0.0099	30.4703	70
75	.4741	52.5871	1702.7340	2.1091	110.9128	0.0190	0.0090	32.3793	75
80	.4511	54.8882	1879.8771	2.2167	121.6715	0.0182	0.0082	34.2492	80
85	.4292	57.0777	2059.3701	2.3298	132.9790	0.0175	0.0075	36.0801	85
90	.4084	59.1609	2240.5675	2.4486	144.8633	0.0169	0.0069	37.8724	90
95	.3886	61.1430	2422.8811	2.5735	157.3538	0.0164	0.0064	39.6265	95
100	.3697	63.0289	2605.7758	2.7048	170.4814	0.0159	0.0059	41.3426	100

I = 1.50 %

N	(P/F)	(P/A)	(P/G)	(F/P)	(F/A)	(A/P)	(A/F)	(A/G)	N
1	.9852	0.9852	-0.0000	1.0150	1.0000	1.0150	1.0000	-0.0000	1
2	.9707	1.9559	0.9707	1.0302	2.0150	0.5113	0.4963	0.4963	2
3	.9563	2.9122	2.8833	1.0457	3.0452	0.3434	0.3284	0.9901	3
4	.9422	3.8544	5.7098	1.0614	4.0909	0.2594	0.2444	1.4814	4
5	.9283	4.7826	9.4229	1.0773	5.1523	0.2091	0.1941	1.9702	5
6	.9145	5.6972	13.9956	1.0934	6.2296	0.1755	0.1605	2.4566	6
7	.9010	6.5982	19.4018	1.1098	7.3230	0.1516	0.1366	2.9405	7
8	.8877	7.4859	25.6157	1.1265	8.4328	0.1336	0.1186	3.4219	8
9	.8746	8.3605	32.6125	1.1434	9.5593	0.1196	0.1046	3.9008	9
10	.8617	9.2222	40.3675	1.1605	10.7027	0.1084	0.0934	4.3772	10
11	.8489	10.0711	48.8568	1.1779	11.8633	0.0993	0.0843	4.8512	11
12	.8364	10.9075	58.0571	1.1956	13.0412	0.0917	0.0767	5.3227	12
13	.8240	11.7315	67.9454	1.2136	14.2368	0.0852	0.0702	5.7917	13
14	.8118	12.5434	78.4994	1.2318	15.4504	0.0797	0.0647	6.2582	14
15	.7999	13.3432	89.6974	1.2502	16.6821	0.0749	0.0599	6.7223	15
16	.7880	14.1313	101.5178	1.2690	17.9324	0.0708	0.0558	7.1839	16
17	.7764	14.9076	113.9400	1.2880	19.2014	0.0671	0.0521	7.6431	17
18	.7649	15.6726	126.9435	1.3073	20.4894	0.0638	0.0488	8.0997	18
19	.7536	16.4262	140.5084	1.3270	21.7967	0.0609	0.0459	8.5539	19
20	.7425	17.1686	154.6154	1.3469	23.1237	0.0582	0.0432	9.0057	20
21	.7315	17.9001	169.2453	1.3671	24.4705	0.0559	0.0409	9.4550	21
22	.7207	18.6208	184.3798	1.3876	25.8376	0.0537	0.0387	9.9018	22
23	.7100	19.3309	200.0006	1.4084	27.2251	0.0517	0.0367	10.3462	23
24	.6995	20.0304	216.0901	1.4295	28.6335	0.0499	0.0349	10.7881	24
25	.6892	20.7196	232.6310	1.4509	30.0630	0.0483	0.0333	11.2276	25
26	.6790	21.3986	249.6065	1.4727	31.5140	0.0467	0.0317	11.6646	26
27	.6690	22.0676	267.0002	1.4948	32.9867	0.0453	0.0303	12.0992	27
28	.6591	22.7267	284.7958	1.5172	34.4815	0.0440	0.0290	12.5313	28
29	.6494	23.3761	302.9779	1.5400	35.9987	0.0428	0.0278	12.9610	29
30	.6398	24.0158	321.5310	1.5631	37.5387	0.0416	0.0266	13.3883	30
31	.6303	24.6461	340.4402	1.5865	39.1018	0.0406	0.0256	13.8131	31
32	.6210	25.2671	359.6910	1.6103	40.6883	0.0396	0.0246	14.2355	32
33	.6118	25.8790	379.2691	1.6345	42.2986	0.0386	0.0236	14.6555	33
34	.6028	26.4817	399.1607	1.6590	43.9331	0.0378	0.0228	15.0731	34
35	.5939	27.0756	419.3521	1.6839	45.5921	0.0369	0.0219	15.4882	35
36	.5851	27.6607	439.8303	1.7091	47.2760	0.0362	0.0212	15.9009	36
37	.5764	28.2371	460.5822	1.7348	48.9851	0.0354	0.0204	16.3112	37
38	.5679	28.8051	481.5954	1.7608	50.7199	0.0347	0.0197	16.7191	38
39	.5595	29.3646	502.8576	1.7872	52.4807	0.0341	0.0191	17.1246	39
40	.5513	29.9158	524.3568	1.8140	54.2679	0.0334	0.0184	17.5277	40
41	.5431	30.4590	546.0814	1.8412	56.0819	0.0328	0.0178	17.9284	41
42	.5351	30.9941	568.0201	1.8688	57.9231	0.0323	0.0173	18.3267	42
43	.5272	31.5212	590.1617	1.8969	59.7920	0.0317	0.0167	18.7227	43
44	.5194	32.0406	612.4955	1.9253	61.6889	0.0312	0.0162	19.1162	44
45	.5117	32.5523	635.0110	1.9542	63.6142	0.0307	0.0157	19.5074	45
46	.5042	33.0565	657.6979	1.9835	65.5684	0.0303	0.0153	19.8962	46
47	.4967	33.5532	680.5462	2.0133	67.5519	0.0298	0.0148	20.2826	47
48	.4894	34.0426	703.5462	2.0435	69.5652	0.0294	0.0144	20.6667	48
49	.4821	34.5247	726.6884	2.0741	71.6087	0.0290	0.0140	21.0484	49
50	.4750	34.9997	749.9636	2.1052	73.6828	0.0286	0.0136	21.4277	50
51	.4680	35.4677	773.3629	2.1368	75.7881	0.0282	0.0132	21.8047	51
52	.4611	35.9287	796.8774	2.1689	77.9249	0.0278	0.0128	22.1794	52
53	.4543	36.3830	820.4986	2.2014	80.0938	0.0275	0.0125	22.5517	53
54	.4475	36.8305	844.2184	2.2344	82.2952	0.0272	0.0122	22.9217	54
55	.4409	37.2715	868.0285	2.2679	84.5296	0.0268	0.0118	23.2894	55
60	.4093	39.3803	988.1674	2.4432	96.2147	0.0254	0.0104	25.0930	60
65	.3799	41.3378	1109.4752	2.6320	108.8028	0.0242	0.0092	26.8393	65
70	.3527	43.1549	1231.1658	2.8355	122.3638	0.0232	0.0082	28.5290	70
75	.3274	44.8416	1352.5600	3.0546	136.9728	0.0223	0.0073	30.1631	75
80	.3039	46.4073	1473.0741	3.2907	152.7109	0.0215	0.0065	31.7423	80
85	.2821	47.8607	1592.2095	3.5450	169.6652	0.0209	0.0059	33.2676	85
90	.2619	49.2099	1709.5439	3.8189	187.9299	0.0203	0.0053	34.7399	90
95	.2431	50.4622	1824.7224	4.1141	207.6061	0.0198	0.0048	36.1602	95
100	.2256	51.6247	1937.4506	4.4320	228.8030	0.0194	0.0044	37.5295	100

I = 2.00 %

N	(P/F)	(P/A)	(P/G)	(F/P)	(F/A)	(A/P)	(A/F)	(A/G)	N
1	.9804	0.9804	-0.0000	1.0200	1.0000	1.0200	1.0000	-0.0000	1
2	.9612	1.9416	0.9612	1.0404	2.0200	0.5150	0.4950	0.4950	2
3	.9423	2.8839	2.8458	1.0612	3.0604	0.3468	0.3268	0.9868	3
4	.9238	3.8077	5.6173	1.0824	4.1216	0.2626	0.2426	1.4752	4
5	.9057	4.7135	9.2403	1.1041	5.2040	0.2122	0.1922	1.9604	5
6	.8880	5.6014	13.6801	1.1262	6.3081	0.1785	0.1585	2.4423	6
7	.8706	6.4720	18.9035	1.1487	7.4343	0.1545	0.1345	2.9208	7
8	.8535	7.3255	24.8779	1.1717	8.5830	0.1365	0.1165	3.3961	8
9	.8368	8.1622	31.5720	1.1951	9.7546	0.1225	0.1025	3.8681	9
10	.8203	8.9826	38.9551	1.2190	10.9497	0.1113	0.0913	4.3367	10
11	.8043	9.7868	46.9977	1.2434	12.1687	0.1022	0.0822	4.8021	11
12	.7885	10.5753	55.6712	1.2682	13.4121	0.0946	0.0746	5.2642	12
13	.7730	11.3484	64.9475	1.2936	14.6803	0.0881	0.0681	5.7231	13
14	.7579	12.1062	74.7999	1.3195	15.9739	0.0826	0.0626	6.1786	14
15	.7430	12.8493	85.2021	1.3459	17.2934	0.0778	0.0578	6.6309	15
16	.7284	13.5777	96.1288	1.3728	18.6393	0.0737	0.0537	7.0799	16
17	.7142	14.2919	107.5554	1.4002	20.0121	0.0700	0.0500	7.5256	17
18	.7002	14.9920	119.4581	1.4282	21.4123	0.0667	0.0467	7.9681	18
19	.6864	15.6785	131.8139	1.4568	22.8406	0.0638	0.0438	8.4073	19
20	.6730	16.3514	144.6003	1.4859	24.2974	0.0612	0.0412	8.8433	20
21	.6598	17.0112	157.7959	1.5157	25.7833	0.0588	0.0388	9.2760	21
22	.6468	17.6580	171.3795	1.5460	27.2990	0.0566	0.0366	9.7055	22
23	.6342	18.2922	185.3309	1.5769	28.8450	0.0547	0.0347	10.1317	23
24	.6217	18.9139	199.6305	1.6084	30.4219	0.0529	0.0329	10.5547	24
25	.6095	19.5235	214.2592	1.6406	32.0303	0.0512	0.0312	10.9745	25
26	.5976	20.1210	229.1987	1.6734	33.6709	0.0497	0.0297	11.3910	26
27	.5859	20.7069	244.4311	1.7069	35.3443	0.0483	0.0283	11.8043	27
28	.5744	21.2813	259.9392	1.7410	37.0512	0.0470	0.0270	12.2145	28
29	.5631	21.8444	275.7064	1.7758	38.7922	0.0458	0.0258	12.6214	29
30	.5521	22.3965	291.7164	1.8114	40.5681	0.0446	0.0246	13.0251	30
31	.5412	22.9377	307.9538	1.8476	42.3794	0.0436	0.0236	13.4257	31
32	.5306	23.4683	324.4035	1.8845	44.2270	0.0426	0.0226	13.8230	32
33	.5202	23.9886	341.0508	1.9222	46.1116	0.0417	0.0217	14.2172	33
34	.5100	24.4986	357.8817	1.9607	48.0338	0.0408	0.0208	14.6083	34
35	.5000	24.9986	374.8826	1.9999	49.9945	0.0400	0.0200	14.9961	35
36	.4902	25.4888	392.0405	2.0399	51.9944	0.0392	0.0192	15.3809	36
37	.4806	25.9695	409.3424	2.0807	54.0343	0.0385	0.0185	15.7625	37
38	.4712	26.4406	426.7764	2.1223	56.1149	0.0378	0.0178	16.1409	38
39	.4619	26.9026	444.3304	2.1647	58.2372	0.0372	0.0172	16.5163	39
40	.4529	27.3555	461.9931	2.2080	60.4020	0.0366	0.0166	16.8885	40
41	.4440	27.7995	479.7535	2.2522	62.6100	0.0360	0.0160	17.2576	41
42	.4353	28.2348	497.6010	2.2972	64.8622	0.0354	0.0154	17.6237	42
43	.4268	28.6616	515.5253	2.3432	67.1595	0.0349	0.0149	17.9866	43
44	.4184	29.0800	533.5165	2.3901	69.5027	0.0344	0.0144	18.3465	44
45	.4102	29.4902	551.5652	2.4379	71.8927	0.0339	0.0139	18.7034	45
46	.4022	29.8923	569.6621	2.4866	74.3306	0.0335	0.0135	19.0571	46
47	.3943	30.2866	587.7985	2.5363	76.8172	0.0330	0.0130	19.4079	47
48	.3865	30.6731	605.9657	2.5871	79.3535	0.0326	0.0126	19.7556	48
49	.3790	31.0521	624.1557	2.6388	81.9406	0.0322	0.0122	20.1003	49
50	.3715	31.4236	642.3606	2.6916	84.5794	0.0318	0.0118	20.4420	50
51	.3642	31.7878	660.5727	2.7454	87.2710	0.0315	0.0115	20.7807	51
52	.3571	32.1449	678.7849	2.8003	90.0164	0.0311	0.0111	21.1164	52
53	.3501	32.4950	696.9900	2.8563	92.8167	0.0308	0.0108	21.4491	53
54	.3432	32.8383	715.1815	2.9135	95.6731	0.0305	0.0105	21.7789	54
55	.3365	33.1748	733.3527	2.9717	98.5865	0.0301	0.0101	22.1057	55
60	.3048	34.7609	823.6975	3.2810	114.0515	0.0288	0.0088	23.6961	60
65	.2761	36.1975	912.7085	3.6225	131.1262	0.0276	0.0076	25.2147	65
70	.2500	37.4986	999.8343	3.9996	149.9779	0.0267	0.0067	26.6632	70
75	.2265	38.6771	1084.6393	4.4158	170.7918	0.0259	0.0059	28.0434	75
80	.2051	39.7445	1166.7868	4.8754	193.7720	0.0252	0.0052	29.3572	80
85	.1858	40.7113	1246.0241	5.3829	219.1439	0.0246	0.0046	30.6064	85
90	.1683	41.5869	1322.1701	5.9431	247.1567	0.0240	0.0040	31.7929	90
95	.1524	42.3800	1395.1033	6.5617	278.0850	0.0236	0.0036	32.9189	95
100	.1380	43.0984	1464.7527	7.2446	312.2323	0.0232	0.0032	33.9863	100

I = 3.00 %

N	(P/F)	(P/A)	(P/G)	(F/P)	(F/A)	(A/P)	(A/F)	(A/G)	N
1	.9709	0.9709	-0.0000	1.0300	1.0000	1.0300	1.0000	-0.0000	1
2	.9426	1.9135	0.9426	1.0609	2.0300	0.5226	0.4926	0.4926	2
3	.9151	2.8286	2.7729	1.0927	3.0909	0.3535	0.3235	0.9803	3
4	.8885	3.7171	5.4383	1.1255	4.1836	0.2690	0.2390	1.4631	4
5	.8626	4.5797	8.8888	1.1593	5.3091	0.2184	0.1884	1.9409	5
6	.8375	5.4172	13.0762	1.1941	6.4684	0.1846	0.1546	2.4138	6
7	.8131	6.2303	17.9547	1.2299	7.6625	0.1605	0.1305	2.8819	7
8	.7894	7.0197	23.4806	1.2668	8.8923	0.1425	0.1125	3.3450	8
9	.7664	7.7861	29.6119	1.3048	10.1591	0.1284	0.0984	3.8032	9
10	.7441	8.5302	36.3088	1.3439	11.4639	0.1172	0.0872	4.2565	10
11	.7224	9.2526	43.5330	1.3842	12.8078	0.1081	0.0781	4.7049	11
12	.7014	9.9540	51.2482	1.4258	14.1920	0.1005	0.0705	5.1485	12
13	.6810	10.6350	59.4196	1.4685	15.6178	0.0940	0.0640	5.5872	13
14	.6611	11.2961	68.0141	1.5126	17.0863	0.0885	0.0585	6.0210	14
15	.6419	11.9379	77.0002	1.5580	18.5989	0.0838	0.0538	6.4500	15
16	.6232	12.5611	86.3477	1.6047	20.1569	0.0796	0.0496	6.8742	16
17	.6050	13.1661	96.0280	1.6528	21.7616	0.0760	0.0460	7.2936	17
18	.5874	13.7535	106.0137	1.7024	23.4144	0.0727	0.0427	7.7081	18
19	.5703	14.3238	116.2788	1.7535	25.1169	0.0698	0.0398	8.1179	19
20	.5537	14.8775	126.7987	1.8061	26.8704	0.0672	0.0372	8.5229	20
21	.5375	15.4150	137.5496	1.8603	28.6765	0.0649	0.0349	8.9231	21
22	.5219	15.9369	148.5094	1.9161	30.5368	0.0627	0.0327	9.3186	22
23	.5067	16.4436	159.6566	1.9736	32.4529	0.0608	0.0308	9.7093	23
24	.4919	16.9355	170.9711	2.0328	34.4265	0.0590	0.0290	10.0954	24
25	.4776	17.4131	182.4336	2.0938	36.4593	0.0574	0.0274	10.4768	25
26	.4637	17.8768	194.0260	2.1566	38.5530	0.0559	0.0259	10.8535	26
27	.4502	18.3270	205.7309	2.2213	40.7096	0.0546	0.0246	11.2255	27
28	.4371	18.7641	217.5320	2.2879	42.9309	0.0533	0.0233	11.5930	28
29	.4243	19.1885	229.4137	2.3566	45.2189	0.0521	0.0221	11.9558	29
30	.4120	19.6004	241.3613	2.4273	47.5754	0.0510	0.0210	12.3141	30
31	.4000	20.0004	253.3609	2.5001	50.0027	0.0500	0.0200	12.6678	31
32	.3883	20.3888	265.3993	2.5751	52.5028	0.0490	0.0190	13.0169	32
33	.3770	20.7658	277.4642	2.6523	55.0778	0.0482	0.0182	13.3616	33
34	.3660	21.1318	289.5437	2.7319	57.7302	0.0473	0.0173	13.7018	34
35	.3554	21.4872	301.6267	2.8139	60.4621	0.0465	0.0165	14.0375	35
36	.3450	21.8323	313.7028	2.8983	63.2759	0.0458	0.0158	14.3688	36
37	.3350	22.1672	325.7622	2.9852	66.1742	0.0451	0.0151	14.6957	37
38	.3252	22.4925	337.7956	3.0748	69.1594	0.0445	0.0145	15.0182	38
39	.3158	22.8082	349.7942	3.1670	72.2342	0.0438	0.0138	15.3363	39
40	.3066	23.1148	361.7499	3.2620	75.4013	0.0433	0.0133	15.6502	40
41	.2976	23.4124	373.6551	3.3599	78.6633	0.0427	0.0127	15.9597	41
42	.2890	23.7014	385.5024	3.4607	82.0232	0.0422	0.0122	16.2650	42
43	.2805	23.9819	397.2852	3.5645	85.4839	0.0417	0.0117	16.5660	43
44	.2724	24.2543	408.9972	3.6715	89.0484	0.0412	0.0112	16.8629	44
45	.2644	24.5187	420.6325	3.7816	92.7199	0.0408	0.0108	17.1556	45
46	.2567	24.7754	432.1856	3.8950	96.5015	0.0404	0.0104	17.4441	46
47	.2493	25.0247	443.6515	4.0119	100.3965	0.0400	0.0100	17.7285	47
48	.2420	25.2667	455.0255	4.1323	104.4084	0.0396	0.0096	18.0089	48
49	.2350	25.5017	466.3031	4.2562	108.5406	0.0392	0.0092	18.2852	49
50	.2281	25.7298	477.4803	4.3839	112.7969	0.0389	0.0089	18.5575	50
51	.2215	25.9512	488.5535	4.5154	117.1808	0.0385	0.0085	18.8258	51
52	.2150	26.1662	499.5191	4.6509	121.6962	0.0382	0.0082	19.0902	52
53	.2088	26.3750	510.3742	4.7904	126.3471	0.0379	0.0079	19.3507	53
54	.2027	26.5777	521.1157	4.9341	131.1375	0.0376	0.0076	19.6073	54
55	.1968	26.7744	531.7411	5.0821	136.0716	0.0373	0.0073	19.8600	55
60	.1697	27.6756	583.0526	5.8916	163.0534	0.0361	0.0061	21.0674	60
65	.1464	28.4529	631.2010	6.8300	194.3328	0.0351	0.0051	22.1841	65
70	.1263	29.1234	676.0869	7.9178	230.5941	0.0343	0.0043	23.2145	70
75	.1089	29.7018	717.6978	9.1789	272.6309	0.0337	0.0037	24.1634	75
80	.0940	30.2008	756.0865	10.6409	321.3630	0.0331	0.0031	25.0353	80
85	.0811	30.6312	791.3529	12.3357	377.8570	0.0326	0.0026	25.8349	85
90	.0699	31.0024	823.6302	14.3005	443.3489	0.0323	0.0023	26.5667	90
95	.0603	31.3227	853.0742	16.5782	519.2720	0.0319	0.0019	27.2351	95
100	.0520	31.5989	879.8540	19.2186	607.2877	0.0316	0.0016	27.8444	100

I = 4.00 %

N	(P/F)	(P/A)	(P/G)	(F/P)	(F/A)	(A/P)	(A/F)	(A/G)	N
1	.9615	0.9615	-0.0000	1.0400	1.0000	1.0400	1.0000	-0.0000	1
2	.9246	1.8861	0.9246	1.0816	2.0400	0.5302	0.4902	0.4902	2
3	.8890	2.7751	2.7025	1.1249	3.1216	0.3603	0.3203	0.9739	3
4	.8548	3.6299	5.2670	1.1699	4.2465	0.2755	0.2355	1.4510	4
5	.8219	4.4518	8.5547	1.2167	5.4163	0.2246	0.1846	1.9216	5
6	.7903	5.2421	12.5062	1.2653	6.6330	0.1908	0.1508	2.3857	6
7	.7599	6.0021	17.0657	1.3159	7.8983	0.1666	0.1266	2.8433	7
8	.7307	6.7327	22.1806	1.3686	9.2142	0.1485	0.1085	3.2944	8
9	.7026	7.4353	27.8013	1.4233	10.5828	0.1345	0.0945	3.7391	9
10	.6756	8.1109	33.8814	1.4802	12.0061	0.1233	0.0833	4.1773	10
11	.6496	8.7605	40.3772	1.5395	13.4864	0.1141	0.0741	4.6090	11
12	.6246	9.3851	47.2477	1.6010	15.0258	0.1066	0.0666	5.0343	12
13	.6006	9.9856	54.4546	1.6651	16.6268	0.1001	0.0601	5.4533	13
14	.5775	10.5631	61.9618	1.7317	18.2919	0.0947	0.0547	5.8659	14
15	.5553	11.1184	69.7355	1.8009	20.0236	0.0899	0.0499	6.2721	15
16	.5339	11.6523	77.7441	1.8730	21.8245	0.0858	0.0458	6.6720	16
17	.5134	12.1657	85.9581	1.9479	23.6975	0.0822	0.0422	7.0656	17
18	.4936	12.6593	94.3498	2.0258	25.6454	0.0790	0.0390	7.4530	18
19	.4746	13.1339	102.8933	2.1068	27.6712	0.0761	0.0361	7.8342	19
20	.4564	13.5903	111.5647	2.1911	29.7781	0.0736	0.0336	8.2091	20
21	.4388	14.0292	120.3414	2.2788	31.9692	0.0713	0.0313	8.5779	21
22	.4220	14.4511	129.2024	2.3699	34.2480	0.0692	0.0292	8.9407	22
23	.4057	14.8568	138.1284	2.4647	36.6179	0.0673	0.0273	9.2973	23
24	.3901	15.2470	147.1012	2.5633	39.0826	0.0656	0.0256	9.6479	24
25	.3751	15.6221	156.1040	2.6658	41.6459	0.0640	0.0240	9.9925	25
26	.3607	15.9828	165.1212	2.7725	44.3117	0.0626	0.0226	10.3312	26
27	.3468	16.3296	174.1385	2.8834	47.0842	0.0612	0.0212	10.6640	27
28	.3335	16.6631	183.1424	2.9987	49.9676	0.0600	0.0200	10.9909	28
29	.3207	16.9837	192.1206	3.1187	52.9663	0.0589	0.0189	11.3120	29
30	.3083	17.2920	201.0618	3.2434	56.0849	0.0578	0.0178	11.6274	30
31	.2965	17.5885	209.9556	3.3731	59.3283	0.0569	0.0169	11.9371	31
32	.2851	17.8736	218.7924	3.5081	62.7015	0.0559	0.0159	12.2411	32
33	.2741	18.1476	227.5634	3.6484	66.2095	0.0551	0.0151	12.5396	33
34	.2636	18.4112	236.2607	3.7943	69.8579	0.0543	0.0143	12.8324	34
35	.2534	18.6646	244.8768	3.9461	73.6522	0.0536	0.0136	13.1198	35
36	.2437	18.9083	253.4052	4.1039	77.5983	0.0529	0.0129	13.4018	36
37	.2343	19.1426	261.8399	4.2681	81.7022	0.0522	0.0122	13.6784	37
38	.2253	19.3679	270.1754	4.4388	85.9703	0.0516	0.0116	13.9497	38
39	.2166	19.5845	278.4070	4.6164	90.4091	0.0511	0.0111	14.2157	39
40	.2083	19.7928	286.5303	4.8010	95.0255	0.0505	0.0105	14.4765	40
41	.2003	19.9931	294.5414	4.9931	99.8265	0.0500	0.0100	14.7322	41
42	.1926	20.1856	302.4370	5.1928	104.8196	0.0495	0.0095	14.9828	42
43	.1852	20.3708	310.2141	5.4005	110.0124	0.0491	0.0091	15.2284	43
44	.1780	20.5488	317.8700	5.6165	115.4129	0.0487	0.0087	15.4690	44
45	.1712	20.7200	325.4028	5.8412	121.0294	0.0483	0.0083	15.7047	45
46	.1646	20.8847	332.8104	6.0748	126.8706	0.0479	0.0079	15.9356	46
47	.1583	21.0429	340.0914	6.3178	132.9454	0.0475	0.0075	16.1618	47
48	.1522	21.1951	347.2446	6.5705	139.2632	0.0472	0.0072	16.3832	48
49	.1463	21.3415	354.2689	6.8333	145.8337	0.0469	0.0069	16.6000	49
50	.1407	21.4822	361.1638	7.1067	152.6671	0.0466	0.0066	16.8122	50
51	.1353	21.6175	367.9289	7.3910	159.7738	0.0463	0.0063	17.0200	51
52	.1301	21.7476	374.5638	7.6866	167.1647	0.0460	0.0060	17.2232	52
53	.1251	21.8727	381.0686	7.9941	174.8513	0.0457	0.0057	17.4221	53
54	.1203	21.9930	387.4436	8.3138	182.8454	0.0455	0.0055	17.6167	54
55	.1157	22.1086	393.6890	8.6464	191.1592	0.0452	0.0052	17.8070	55
60	.0951	22.6235	422.9966	10.5196	237.9907	0.0442	0.0042	18.6972	60
65	.0781	23.0467	449.2014	12.7987	294.9684	0.0434	0.0034	19.4909	65
70	.0642	23.3945	472.4789	15.5716	364.2905	0.0427	0.0027	20.1961	70
75	.0528	23.6804	493.0408	18.9453	448.6314	0.0422	0.0022	20.8206	75
80	.0434	23.9154	511.1161	23.0498	551.2450	0.0418	0.0018	21.3718	80
85	.0357	24.1085	526.9384	28.0436	676.0901	0.0415	0.0015	21.8569	85
90	.0293	24.2673	540.7369	34.1193	827.9833	0.0412	0.0012	22.2826	90
95	.0241	24.3978	552.7307	41.5114	1012.7846	0.0410	0.0010	22.6550	95
100	.0198	24.5050	563.1249	50.5049	1237.6237	0.0408	0.0008	22.9800	100

ECON

I = 5.00 %

N	(P/F)	(P/A)	(P/G)	(F/P)	(F/A)	(A/P)	(A/F)	(A/G)	N
1	.9524	0.9524	-0.0000	1.0500	1.0000	1.0500	1.0000	-0.0000	1
2	.9070	1.8594	0.9070	1.1025	2.0500	0.5378	0.4878	0.4878	2
3	.8638	2.7232	2.6347	1.1576	3.1525	0.3672	0.3172	0.9675	3
4	.8227	3.5460	5.1028	1.2155	4.3101	0.2820	0.2320	1.4391	4
5	.7835	4.3295	8.2369	1.2763	5.5256	0.2310	0.1810	1.9025	5
6	.7462	5.0757	11.9680	1.3401	6.8019	0.1970	0.1470	2.3579	6
7	.7107	5.7864	16.2321	1.4071	8.1420	0.1728	0.1228	2.8052	7
8	.6768	6.4632	20.9700	1.4775	9.5491	0.1547	0.1047	3.2445	8
9	.6446	7.1078	26.1268	1.5513	11.0266	0.1407	0.0907	3.6758	9
10	.6139	7.7217	31.6520	1.6289	12.5779	0.1295	0.0795	4.0991	10
11	.5847	8.3064	37.4988	1.7103	14.2068	0.1204	0.0704	4.5144	11
12	.5568	8.8633	43.6241	1.7959	15.9171	0.1128	0.0628	4.9219	12
13	.5303	9.3936	49.9879	1.8856	17.7130	0.1065	0.0565	5.3215	13
14	.5051	9.8986	56.5538	1.9799	19.5986	0.1010	0.0510	5.7133	14
15	.4810	10.3797	63.2880	2.0789	21.5786	0.0963	0.0463	6.0973	15
16	.4581	10.8378	70.1597	2.1829	23.6575	0.0923	0.0423	6.4736	16
17	.4363	11.2741	77.1405	2.2920	25.8404	0.0887	0.0387	6.8423	17
18	.4155	11.6896	84.2043	2.4066	28.1324	0.0855	0.0355	7.2034	18
19	.3957	12.0853	91.3275	2.5270	30.5390	0.0827	0.0327	7.5569	19
20	.3769	12.4622	98.4884	2.6533	33.0660	0.0802	0.0302	7.9030	20
21	.3589	12.8212	105.6673	2.7860	35.7193	0.0780	0.0280	8.2416	21
22	.3418	13.1630	112.8461	2.9253	38.5052	0.0760	0.0260	8.5730	22
23	.3256	13.4886	120.0087	3.0715	41.4305	0.0741	0.0241	8.8971	23
24	.3101	13.7986	127.1402	3.2251	44.5020	0.0725	0.0225	9.2140	24
25	.2953	14.0939	134.2275	3.3864	47.7271	0.0710	0.0210	9.5238	25
26	.2812	14.3752	141.2585	3.5557	51.1135	0.0696	0.0196	9.8266	26
27	.2678	14.6430	148.2226	3.7335	54.6691	0.0683	0.0183	10.1224	27
28	.2551	14.8981	155.1101	3.9201	58.4026	0.0671	0.0171	10.4114	28
29	.2429	15.1411	161.9126	4.1161	62.3227	0.0660	0.0160	10.6936	29
30	.2314	15.3725	168.6226	4.3219	66.4388	0.0651	0.0151	10.9691	30
31	.2204	15.5928	175.2333	4.5380	70.7608	0.0641	0.0141	11.2381	31
32	.2099	15.8027	181.7392	4.7649	75.2988	0.0633	0.0133	11.5005	32
33	.1999	16.0025	188.1351	5.0032	80.0638	0.0625	0.0125	11.7566	33
34	.1904	16.1929	194.4168	5.2533	85.0670	0.0618	0.0118	12.0063	34
35	.1813	16.3742	200.5807	5.5160	90.3203	0.0611	0.0111	12.2498	35
36	.1727	16.5469	206.6237	5.7918	95.8363	0.0604	0.0104	12.4872	36
37	.1644	16.7113	212.5434	6.0814	101.6281	0.0598	0.0098	12.7186	37
38	.1566	16.8679	218.3378	6.3855	107.7095	0.0593	0.0093	12.9440	38
39	.1491	17.0170	224.0054	6.7048	114.0950	0.0588	0.0088	13.1636	39
40	.1420	17.1591	229.5452	7.0400	120.7998	0.0583	0.0083	13.3775	40
41	.1353	17.2944	234.9564	7.3920	127.8398	0.0578	0.0078	13.5857	41
42	.1288	17.4232	240.2389	7.7616	135.2318	0.0574	0.0074	13.7884	42
43	.1227	17.5459	245.3925	8.1497	142.9933	0.0570	0.0070	13.9857	43
44	.1169	17.6628	250.4175	8.5572	151.1430	0.0566	0.0066	14.1777	44
45	.1113	17.7741	255.3145	8.9850	159.7002	0.0563	0.0063	14.3644	45
46	.1060	17.8801	260.0844	9.4343	168.6852	0.0559	0.0059	14.5461	46
47	.1009	17.9810	264.7281	9.9060	178.1194	0.0556	0.0056	14.7226	47
48	.0961	18.0772	269.2467	10.4013	188.0254	0.0553	0.0053	14.8943	48
49	.0916	18.1687	273.6418	10.9213	198.4267	0.0550	0.0050	15.0611	49
50	.0872	18.2559	277.9148	11.4674	209.3480	0.0548	0.0048	15.2233	50
51	.0831	18.3390	282.0673	12.0408	220.8154	0.0545	0.0045	15.3808	51
52	.0791	18.4181	286.1013	12.6428	232.8562	0.0543	0.0043	15.5337	52
53	.0753	18.4934	290.0184	13.2749	245.4990	0.0541	0.0041	15.6823	53
54	.0717	18.5651	293.8208	13.9387	258.7739	0.0539	0.0039	15.8265	54
55	.0683	18.6335	297.5104	14.6356	272.7126	0.0537	0.0037	15.9664	55
60	.0535	18.9293	314.3432	18.6792	353.5837	0.0528	0.0028	16.6062	60
65	.0419	19.1611	328.6910	23.8399	456.7980	0.0522	0.0022	17.1541	65
70	.0329	19.3427	340.8409	30.4264	588.5285	0.0517	0.0017	17.6212	70
75	.0258	19.4850	351.0721	38.8327	756.6537	0.0513	0.0013	18.0176	75
80	.0202	19.5965	359.6460	49.5614	971.2288	0.0510	0.0010	18.3526	80
85	.0158	19.6838	366.8007	63.2544	1245.0871	0.0508	0.0008	18.6346	85
90	.0124	19.7523	372.7488	80.7304	1594.6073	0.0506	0.0006	18.8712	90
95	.0097	19.8059	377.6774	103.0347	2040.6935	0.0505	0.0005	19.0689	95
100	.0076	19.8479	381.7492	131.5013	2610.0252	0.0504	0.0004	19.2337	100

I = 6.00 %

N	(P/F)	(P/A)	(P/G)	(F/P)	(F/A)	(A/P)	(A/F)	(A/G)	N
1	.9434	0.9434	-0.0000	1.0600	1.0000	1.0600	1.0000	-0.0000	1
2	.8900	1.8334	0.8900	1.1236	2.0600	0.5454	0.4854	0.4854	2
3	.8396	2.6730	2.5692	1.1910	3.1836	0.3741	0.3141	0.9612	3
4	.7921	3.4651	4.9455	1.2625	4.3746	0.2886	0.2286	1.4272	4
5	.7473	4.2124	7.9345	1.3382	5.6371	0.2374	0.1774	1.8836	5
6	.7050	4.9173	11.4594	1.4185	6.9753	0.2034	0.1434	2.3304	6
7	.6651	5.5824	15.4497	1.5036	8.3938	0.1791	0.1191	2.7676	7
8	.6274	6.2098	19.8416	1.5938	9.8975	0.1610	0.1010	3.1952	8
9	.5919	6.8017	24.5768	1.6895	11.4913	0.1470	0.0870	3.6133	9
10	.5584	7.3601	29.6023	1.7908	13.1808	0.1359	0.0759	4.0220	10
11	.5268	7.8869	34.8702	1.8983	14.9716	0.1268	0.0668	4.4213	11
12	.4970	8.3838	40.3369	2.0122	16.8699	0.1193	0.0593	4.8113	12
13	.4688	8.8527	45.9629	2.1329	18.8821	0.1130	0.0530	5.1920	13
14	.4423	9.2950	51.7128	2.2609	21.0151	0.1076	0.0476	5.5635	14
15	.4173	9.7122	57.5546	2.3966	23.2760	0.1030	0.0430	5.9260	15
16	.3936	10.1059	63.4592	2.5404	25.6725	0.0990	0.0390	6.2794	16
17	.3714	10.4773	69.4011	2.6928	28.2129	0.0954	0.0354	6.6240	17
18	.3503	10.8276	75.3569	2.8543	30.9057	0.0924	0.0324	6.9597	18
19	.3305	11.1581	81.3062	3.0256	33.7600	0.0896	0.0296	7.2867	19
20	.3118	11.4699	87.2304	3.2071	36.7856	0.0872	0.0272	7.6051	20
21	.2942	11.7641	93.1136	3.3996	39.9927	0.0850	0.0250	7.9151	21
22	.2775	12.0416	98.9412	3.6035	43.3923	0.0830	0.0230	8.2166	22
23	.2618	12.3034	104.7007	3.8197	46.9958	0.0813	0.0213	8.5099	23
24	.2470	12.5504	110.3812	4.0489	50.8156	0.0797	0.0197	8.7951	24
25	.2330	12.7834	115.9732	4.2919	54.8645	0.0782	0.0182	9.0722	25
26	.2198	13.0032	121.4684	4.5494	59.1564	0.0769	0.0169	9.3414	26
27	.2074	13.2105	126.8600	4.8223	63.7058	0.0757	0.0157	9.6029	27
28	.1956	13.4062	132.1420	5.1117	68.5281	0.0746	0.0146	9.8568	28
29	.1846	13.5907	137.3096	5.4184	73.6398	0.0736	0.0136	10.1032	29
30	.1741	13.7648	142.3588	5.7435	79.0582	0.0726	0.0126	10.3422	30
31	.1643	13.9291	147.2864	6.0881	84.8017	0.0718	0.0118	10.5740	31
32	.1550	14.0840	152.0901	6.4534	90.8898	0.0710	0.0110	10.7988	32
33	.1462	14.2302	156.7681	6.8406	97.3432	0.0703	0.0103	11.0166	33
34	.1379	14.3681	161.3192	7.2510	104.1838	0.0696	0.0096	11.2276	34
35	.1301	14.4982	165.7427	7.6861	111.4348	0.0690	0.0090	11.4319	35
36	.1227	14.6210	170.0387	8.1473	119.1209	0.0684	0.0084	11.6298	36
37	.1158	14.7368	174.2072	8.6361	127.2681	0.0679	0.0079	11.8213	37
38	.1092	14.8460	178.2490	9.1543	135.9042	0.0674	0.0074	12.0065	38
39	.1031	14.9491	182.1652	9.7035	145.0585	0.0669	0.0069	12.1857	39
40	.0972	15.0463	185.9568	10.2857	154.7620	0.0665	0.0065	12.3590	40
41	.0917	15.1380	189.6256	10.9029	165.0477	0.0661	0.0061	12.5264	41
42	.0865	15.2245	193.1732	11.5570	175.9505	0.0657	0.0057	12.6883	42
43	.0816	15.3062	196.6017	12.2505	187.5076	0.0653	0.0053	12.8446	43
44	.0770	15.3832	199.9130	12.9855	199.7580	0.0650	0.0050	12.9956	44
45	.0727	15.4558	203.1096	13.7646	212.7435	0.0647	0.0047	13.1413	45
46	.0685	15.5244	206.1938	14.5905	226.5081	0.0644	0.0044	13.2819	46
47	.0647	15.5890	209.1681	15.4659	241.0986	0.0641	0.0041	13.4177	47
48	.0610	15.6500	212.0351	16.3939	256.5645	0.0639	0.0039	13.5485	48
49	.0575	15.7076	214.7972	17.3775	272.9584	0.0637	0.0037	13.6748	49
50	.0543	15.7619	217.4574	18.4202	290.3359	0.0634	0.0034	13.7964	50
51	.0512	15.8131	220.0181	19.5254	308.7561	0.0632	0.0032	13.9137	51
52	.0483	15.8614	222.4823	20.6969	328.2814	0.0630	0.0030	14.0267	52
53	.0456	15.9070	224.8525	21.9387	348.9783	0.0629	0.0029	14.1355	53
54	.0430	15.9500	227.1316	23.2550	370.9170	0.0627	0.0027	14.2402	54
55	.0406	15.9905	229.3222	24.6503	394.1720	0.0625	0.0025	14.3411	55
60	.0303	16.1614	239.0428	32.9877	533.1282	0.0619	0.0019	14.7909	60
65	.0227	16.2891	246.9450	44.1450	719.0829	0.0614	0.0014	15.1601	65
70	.0169	16.3845	253.3271	59.0759	967.9322	0.0610	0.0010	15.4613	70
75	.0126	16.4558	258.4527	79.0569	1300.9487	0.0608	0.0008	15.7058	75
80	.0095	16.5091	262.5493	105.7960	1746.5999	0.0606	0.0006	15.9033	80
85	.0071	16.5489	265.8096	141.5789	2342.9817	0.0604	0.0004	16.0620	85
90	.0053	16.5787	268.3946	189.4645	3141.0752	0.0603	0.0003	16.1891	90
95	.0039	16.6009	270.4375	253.5463	4209.1042	0.0602	0.0002	16.2905	95
100	.0029	16.6175	272.0471	339.3021	5638.3681	0.0602	0.0002	16.3711	100

ECON

I = 7.00 %

N	(P/F)	(P/A)	(P/G)	(F/P)	(F/A)	(A/P)	(A/F)	(A/G)	N
1	.9346	0.9346	-0.0000	1.0700	1.0000	1.0700	1.0000	-0.0000	1
2	.8734	1.8080	0.8734	1.1449	2.0700	0.5531	0.4831	0.4831	2
3	.8163	2.6243	2.5060	1.2250	3.2149	0.3811	0.3111	0.9549	3
4	.7629	3.3872	4.7947	1.3108	4.4399	0.2952	0.2252	1.4155	4
5	.7130	4.1002	7.6467	1.4026	5.7507	0.2439	0.1739	1.8650	5
6	.6663	4.7665	10.9784	1.5007	7.1533	0.2098	0.1398	2.3032	6
7	.6227	5.3893	14.7149	1.6058	8.6540	0.1856	0.1156	2.7304	7
8	.5820	5.9713	18.7889	1.7182	10.2598	0.1675	0.0975	3.1465	8
9	.5439	6.5152	23.1404	1.8385	11.9780	0.1535	0.0835	3.5517	9
10	.5083	7.0236	27.7156	1.9672	13.8164	0.1424	0.0724	3.9461	10
11	.4751	7.4987	32.4665	2.1049	15.7836	0.1334	0.0634	4.3296	11
12	.4440	7.9427	37.3506	2.2522	17.8885	0.1259	0.0559	4.7025	12
13	.4150	8.3577	42.3302	2.4098	20.1406	0.1197	0.0497	5.0648	13
14	.3878	8.7455	47.3718	2.5785	22.5505	0.1143	0.0443	5.4167	14
15	.3624	9.1079	52.4461	2.7590	25.1290	0.1098	0.0398	5.7583	15
16	.3387	9.4466	57.5271	2.9522	27.8881	0.1059	0.0359	6.0897	16
17	.3166	9.7632	62.5923	3.1588	30.8402	0.1024	0.0324	6.4110	17
18	.2959	10.0591	67.6219	3.3799	33.9990	0.0994	0.0294	6.7225	18
19	.2765	10.3356	72.5991	3.6165	37.3790	0.0968	0.0268	7.0242	19
20	.2584	10.5940	77.5091	3.8697	40.9955	0.0944	0.0244	7.3163	20
21	.2415	10.8355	82.3393	4.1406	44.8652	0.0923	0.0223	7.5990	21
22	.2257	11.0612	87.0793	4.4304	49.0057	0.0904	0.0204	7.8725	22
23	.2109	11.2722	91.7201	4.7405	53.4361	0.0887	0.0187	8.1369	23
24	.1971	11.4693	96.2545	5.0724	58.1767	0.0872	0.0172	8.3923	24
25	.1842	11.6536	100.6765	5.4274	63.2490	0.0858	0.0158	8.6391	25
26	.1722	11.8258	104.9814	5.8074	68.6765	0.0846	0.0146	8.8773	26
27	.1609	11.9867	109.1656	6.2139	74.4838	0.0834	0.0134	9.1072	27
28	.1504	12.1371	113.2264	6.6488	80.6977	0.0824	0.0124	9.3289	28
29	.1406	12.2777	117.1622	7.1143	87.3465	0.0814	0.0114	9.5427	29
30	.1314	12.4090	120.9718	7.6123	94.4608	0.0806	0.0106	9.7487	30
31	.1228	12.5318	124.6550	8.1451	102.0730	0.0798	0.0098	9.9471	31
32	.1147	12.6466	128.2120	8.7153	110.2182	0.0791	0.0091	10.1381	32
33	.1072	12.7538	131.6435	9.3253	118.9334	0.0784	0.0084	10.3219	33
34	.1002	12.8540	134.9507	9.9781	128.2588	0.0778	0.0078	10.4987	34
35	.0937	12.9477	138.1353	10.6766	138.2369	0.0772	0.0072	10.6687	35
36	.0875	13.0352	141.1990	11.4239	148.9135	0.0767	0.0067	10.8321	36
37	.0818	13.1170	144.1441	12.2236	160.3374	0.0762	0.0062	10.9891	37
38	.0765	13.1935	146.9730	13.0793	172.5610	0.0758	0.0058	11.1398	38
39	.0715	13.2649	149.6883	13.9948	185.6403	0.0754	0.0054	11.2845	39
40	.0668	13.3317	152.2928	14.9745	199.6351	0.0750	0.0050	11.4233	40
41	.0624	13.3941	154.7892	16.0227	214.6096	0.0747	0.0047	11.5565	41
42	.0583	13.4524	157.1807	17.1443	230.6322	0.0743	0.0043	11.6842	42
43	.0545	13.5070	159.4702	18.3444	247.7765	0.0740	0.0040	11.8065	43
44	.0509	13.5579	161.6609	19.6285	266.1209	0.0738	0.0038	11.9237	44
45	.0476	13.6055	163.7559	21.0025	285.7493	0.0735	0.0035	12.0360	45
46	.0445	13.6500	165.7584	22.4726	306.7518	0.0733	0.0033	12.1435	46
47	.0416	13.6916	167.6714	24.0457	329.2244	0.0730	0.0030	12.2463	47
48	.0389	13.7305	169.4981	25.7289	353.2701	0.0728	0.0028	12.3447	48
49	.0363	13.7668	171.2417	27.5299	378.9990	0.0726	0.0026	12.4387	49
50	.0339	13.8007	172.9051	29.4570	406.5289	0.0725	0.0025	12.5287	50
51	.0317	13.8325	174.4915	31.5190	435.9860	0.0723	0.0023	12.6146	51
52	.0297	13.8621	176.0037	33.7253	467.5050	0.0721	0.0021	12.6967	52
53	.0277	13.8898	177.4447	36.0861	501.2303	0.0720	0.0020	12.7751	53
54	.0259	13.9157	178.8173	38.6122	537.3164	0.0719	0.0019	12.8500	54
55	.0242	13.9399	180.1243	41.3150	575.9286	0.0717	0.0017	12.9215	55
60	.0173	14.0392	185.7677	57.9464	813.5204	0.0712	0.0012	13.2321	60
65	.0123	14.1099	190.1452	81.2729	1146.7552	0.0709	0.0009	13.4760	65
70	.0088	14.1604	193.5185	113.9894	1614.1342	0.0706	0.0006	13.6662	70
75	.0063	14.1964	196.1035	159.8760	2269.6574	0.0704	0.0004	13.8136	75
80	.0045	14.2220	198.0748	224.2344	3189.0627	0.0703	0.0003	13.9273	80
85	.0032	14.2403	199.5717	314.5003	4478.5761	0.0702	0.0002	14.0146	85
90	.0023	14.2533	200.7042	441.1030	6287.1854	0.0702	0.0002	14.0812	90
95	.0016	14.2626	201.5581	618.6697	8823.8535	0.0701	0.0001	14.1319	95
100	.0012	14.2693	202.2001	867.7163	12381.6618	0.0701	0.0001	14.1703	100

I = 8.00 %

N	(P/F)	(P/A)	(P/G)	(F/P)	(F/A)	(A/P)	(A/F)	(A/G)	N
1	.9259	0.9259	-0.0000	1.0800	1.0000	1.0800	1.0000	-0.0000	1
2	.8573	1.7833	0.8573	1.1664	2.0800	0.5608	0.4808	0.4808	2
3	.7938	2.5771	2.4450	1.2597	3.2464	0.3880	0.3080	0.9487	3
4	.7350	3.3121	4.6501	1.3605	4.5061	0.3019	0.2219	1.4040	4
5	.6806	3.9927	7.3724	1.4693	5.8666	0.2505	0.1705	1.8465	5
6	.6302	4.6229	10.5233	1.5869	7.3359	0.2163	0.1363	2.2763	6
7	.5835	5.2064	14.0242	1.7138	8.9228	0.1921	0.1121	2.6937	7
8	.5403	5.7466	17.8061	1.8509	10.6366	0.1740	0.0940	3.0985	8
9	.5002	6.2469	21.8081	1.9990	12.4876	0.1601	0.0801	3.4910	9
10	.4632	6.7101	25.9768	2.1589	14.4866	0.1490	0.0690	3.8713	10
11	.4289	7.1390	30.2657	2.3316	16.6455	0.1401	0.0601	4.2395	11
12	.3971	7.5361	34.6339	2.5182	18.9771	0.1327	0.0527	4.5957	12
13	.3677	7.9038	39.0463	2.7196	21.4953	0.1265	0.0465	4.9402	13
14	.3405	8.2442	43.4723	2.9372	24.2149	0.1213	0.0413	5.2731	14
15	.3152	8.5595	47.8857	3.1722	27.1521	0.1168	0.0368	5.5945	15
16	.2919	8.8514	52.2640	3.4259	30.3243	0.1130	0.0330	5.9046	16
17	.2703	9.1216	56.5883	3.7000	33.7502	0.1096	0.0296	6.2037	17
18	.2502	9.3719	60.8426	3.9960	37.4502	0.1067	0.0267	6.4920	18
19	.2317	9.6036	65.0134	4.3157	41.4463	0.1041	0.0241	6.7697	19
20	.2145	9.8181	69.0898	4.6610	45.7620	0.1019	0.0219	7.0369	20
21	.1987	10.0168	73.0629	5.0338	50.4229	0.0998	0.0198	7.2940	21
22	.1839	10.2007	76.9257	5.4365	55.4568	0.0980	0.0180	7.5412	22
23	.1703	10.3711	80.6726	5.8715	60.8933	0.0964	0.0164	7.7786	23
24	.1577	10.5288	84.2997	6.3412	66.7648	0.0950	0.0150	8.0066	24
25	.1460	10.6748	87.8041	6.8485	73.1059	0.0937	0.0137	8.2254	25
26	.1352	10.8100	91.1842	7.3964	79.9544	0.0925	0.0125	8.4352	26
27	.1252	10.9352	94.4390	7.9881	87.3508	0.0914	0.0114	8.6363	27
28	.1159	11.0511	97.5687	8.6271	95.3388	0.0905	0.0105	8.8289	28
29	.1073	11.1584	100.5738	9.3173	103.9659	0.0896	0.0096	9.0133	29
30	.0994	11.2578	103.4558	10.0627	113.2832	0.0888	0.0088	9.1897	30
31	.0920	11.3498	106.2163	10.8677	123.3459	0.0881	0.0081	9.3584	31
32	.0852	11.4350	108.8575	11.7371	134.2135	0.0875	0.0075	9.5197	32
33	.0789	11.5139	111.3819	12.6760	145.9506	0.0869	0.0069	9.6737	33
34	.0730	11.5869	113.7924	13.6901	158.6267	0.0863	0.0063	9.8208	34
35	.0676	11.6546	116.0920	14.7853	172.3168	0.0858	0.0058	9.9611	35
36	.0626	11.7172	118.2839	15.9682	187.1021	0.0853	0.0053	10.0949	36
37	.0580	11.7752	120.3713	17.2456	203.0703	0.0849	0.0049	10.2225	37
38	.0537	11.8289	122.3579	18.6253	220.3159	0.0845	0.0045	10.3440	38
39	.0497	11.8786	124.2470	20.1153	238.9412	0.0842	0.0042	10.4597	39
40	.0460	11.9246	126.0422	21.7245	259.0565	0.0839	0.0039	10.5699	40
41	.0426	11.9672	127.7470	23.4625	280.7810	0.0836	0.0036	10.6747	41
42	.0395	12.0067	129.3651	25.3395	304.2435	0.0833	0.0033	10.7744	42
43	.0365	12.0432	130.8998	27.3666	329.5830	0.0830	0.0030	10.8692	43
44	.0338	12.0771	132.3547	29.5560	356.9496	0.0828	0.0028	10.9592	44
45	.0313	12.1084	133.7331	31.9204	386.5056	0.0826	0.0026	11.0447	45
46	.0290	12.1374	135.0384	34.4741	418.4261	0.0824	0.0024	11.1258	46
47	.0269	12.1643	136.2739	37.2320	452.9002	0.0822	0.0022	11.2028	47
48	.0249	12.1891	137.4428	40.2106	490.1322	0.0820	0.0020	11.2758	48
49	.0230	12.2122	138.5480	43.4274	530.3427	0.0819	0.0019	11.3451	49
50	.0213	12.2335	139.5928	46.9016	573.7702	0.0817	0.0017	11.4107	50
51	.0197	12.2532	140.5799	50.6537	620.6718	0.0816	0.0016	11.4729	51
52	.0183	12.2715	141.5121	54.7060	671.3255	0.0815	0.0015	11.5318	52
53	.0169	12.2884	142.3923	59.0825	726.0316	0.0814	0.0014	11.5875	53
54	.0157	12.3041	143.2229	63.8091	785.1141	0.0813	0.0013	11.6403	54
55	.0145	12.3186	144.0065	68.9139	848.9232	0.0812	0.0012	11.6902	55
60	.0099	12.3766	147.3000	101.2571	1253.2133	0.0808	0.0008	11.9015	60
65	.0067	12.4160	149.7387	148.7798	1847.2481	0.0805	0.0005	12.0602	65
70	.0046	12.4428	151.5326	218.6064	2720.0801	0.0804	0.0004	12.1783	70
75	.0031	12.4611	152.8448	321.2045	4002.5566	0.0802	0.0002	12.2658	75
80	.0021	12.4735	153.8001	471.9548	5886.9354	0.0802	0.0002	12.3301	80
85	.0014	12.4820	154.4925	693.4565	8655.7061	0.0801	0.0001	12.3772	85
90	.0010	12.4877	154.9925	1018.9151	12723.9386	0.0801	0.0001	12.4116	90
95	.0007	12.4917	155.3524	1497.1205	18701.5069	0.0801	0.0001	12.4365	95
100	.0005	12.4943	155.6107	2199.7613	27484.5157	0.0800	0.0000	12.4545	100

I = 9.00 %

N	(P/F)	(P/A)	(P/G)	(F/P)	(F/A)	(A/P)	(A/F)	(A/G)	N
1	.9174	0.9174	-0.0000	1.0900	1.0000	1.0900	1.0000	-0.0000	1
2	.8417	1.7591	0.8417	1.1881	2.0900	0.5685	0.4785	0.4785	2
3	.7722	2.5313	2.3860	1.2950	3.2781	0.3951	0.3051	0.9426	3
4	.7084	3.2397	4.5113	1.4116	4.5731	0.3087	0.2187	1.3925	4
5	.6499	3.8897	7.1110	1.5386	5.9847	0.2571	0.1671	1.8282	5
6	.5963	4.4859	10.0924	1.6771	7.5233	0.2229	0.1329	2.2498	6
7	.5470	5.0330	13.3746	1.8280	9.2004	0.1987	0.1087	2.6574	7
8	.5019	5.5348	16.8877	1.9926	11.0285	0.1807	0.0907	3.0512	8
9	.4604	5.9952	20.5711	2.1719	13.0210	0.1668	0.0768	3.4312	9
10	.4224	6.4177	24.3728	2.3674	15.1929	0.1558	0.0658	3.7978	10
11	.3875	6.8052	28.2481	2.5804	17.5603	0.1469	0.0569	4.1510	11
12	.3555	7.1607	32.1590	2.8127	20.1407	0.1397	0.0497	4.4910	12
13	.3262	7.4869	36.0731	3.0658	22.9534	0.1336	0.0436	4.8182	13
14	.2992	7.7862	39.9633	3.3417	26.0192	0.1284	0.0384	5.1326	14
15	.2745	8.0607	43.8069	3.6425	29.3609	0.1241	0.0341	5.4346	15
16	.2519	8.3126	47.5849	3.9703	33.0034	0.1203	0.0303	5.7245	16
17	.2311	8.5436	51.2821	4.3276	36.9737	0.1170	0.0270	6.0024	17
18	.2120	8.7556	54.8860	4.7171	41.3013	0.1142	0.0242	6.2687	18
19	.1945	8.9501	58.3868	5.1417	46.0185	0.1117	0.0217	6.5236	19
20	.1784	9.1285	61.7770	5.6044	51.1601	0.1095	0.0195	6.7674	20
21	.1637	9.2922	65.0509	6.1088	56.7645	0.1076	0.0176	7.0006	21
22	.1502	9.4424	68.2048	6.6586	62.8733	0.1059	0.0159	7.2232	22
23	.1378	9.5802	71.2359	7.2579	69.5319	0.1044	0.0144	7.4357	23
24	.1264	9.7066	74.1433	7.9111	76.7898	0.1030	0.0130	7.6384	24
25	.1160	9.8226	76.9265	8.6231	84.7009	0.1018	0.0118	7.8316	25
26	.1064	9.9290	79.5863	9.3992	93.3240	0.1007	0.0107	8.0156	26
27	.0976	10.0266	82.1241	10.2451	102.7231	0.0997	0.0097	8.1906	27
28	.0895	10.1161	84.5419	11.1671	112.9682	0.0989	0.0089	8.3571	28
29	.0822	10.1983	86.8422	12.1722	124.1354	0.0981	0.0081	8.5154	29
30	.0754	10.2737	89.0280	13.2677	136.3075	0.0973	0.0073	8.6657	30
31	.0691	10.3428	91.1024	14.4618	149.5752	0.0967	0.0067	8.8083	31
32	.0634	10.4062	93.0690	15.7633	164.0370	0.0961	0.0061	8.9436	32
33	.0582	10.4644	94.9314	17.1820	179.8003	0.0956	0.0056	9.0718	33
34	.0534	10.5178	96.6935	18.7284	196.9823	0.0951	0.0051	9.1933	34
35	.0490	10.5668	98.3590	20.4140	215.7108	0.0946	0.0046	9.3083	35
36	.0449	10.6118	99.9319	22.2512	236.1247	0.0942	0.0042	9.4171	36
37	.0412	10.6530	101.4162	24.2538	258.3759	0.0939	0.0039	9.5200	37
38	.0378	10.6908	102.8158	26.4367	282.6298	0.0935	0.0035	9.6172	38
39	.0347	10.7255	104.1345	28.8160	309.0665	0.0932	0.0032	9.7090	39
40	.0318	10.7574	105.3762	31.4094	337.8824	0.0930	0.0030	9.7957	40
41	.0292	10.7866	106.5445	34.2363	369.2919	0.0927	0.0027	9.8775	41
42	.0268	10.8134	107.6432	37.3175	403.5281	0.0925	0.0025	9.9546	42
43	.0246	10.8380	108.6758	40.6761	440.8457	0.0923	0.0023	10.0273	43
44	.0226	10.8605	109.6456	44.3370	481.5218	0.0921	0.0021	10.0958	44
45	.0207	10.8812	110.5561	48.3273	525.8587	0.0919	0.0019	10.1603	45
46	.0190	10.9002	111.4103	52.6767	574.1860	0.0917	0.0017	10.2210	46
47	.0174	10.9176	112.2115	57.4176	626.8628	0.0916	0.0016	10.2780	47
48	.0160	10.9336	112.9625	62.5852	684.2804	0.0915	0.0015	10.3317	48
49	.0147	10.9482	113.6661	68.2179	746.8656	0.0913	0.0013	10.3821	49
50	.0134	10.9617	114.3251	74.3575	815.0836	0.0912	0.0012	10.4295	50
51	.0123	10.9740	114.9420	81.0497	889.4411	0.0911	0.0011	10.4740	51
52	.0113	10.9853	115.5193	88.3442	970.4908	0.0910	0.0010	10.5158	52
53	.0104	10.9957	116.0593	96.2951	1058.8349	0.0909	0.0009	10.5549	53
54	.0095	11.0053	116.5642	104.9617	1155.1301	0.0909	0.0009	10.5917	54
55	.0087	11.0140	117.0362	114.4083	1260.0918	0.0908	0.0008	10.6261	55
60	.0057	11.0480	118.9683	176.0313	1944.7921	0.0905	0.0005	10.7683	60
65	.0037	11.0701	120.3344	270.8460	2998.2885	0.0903	0.0003	10.8702	65
70	.0024	11.0844	121.2942	416.7301	4619.2232	0.0902	0.0002	10.9427	70
75	.0016	11.0938	121.9646	641.1909	7113.2321	0.0901	0.0001	10.9940	75
80	.0010	11.0998	122.4306	986.5517	10950.5741	0.0901	0.0001	11.0299	80
85	.0007	11.1038	122.7533	1517.9320	16854.8003	0.0901	0.0001	11.0551	85
90	.0004	11.1064	122.9758	2335.5266	25939.1842	0.0900	0.0000	11.0726	90
95	.0003	11.1080	123.1287	3593.4971	39916.6350	0.0900	0.0000	11.0847	95
100	.0002	11.1091	123.2335	5529.0408	61422.6755	0.0900	0.0000	11.0930	100

ECON

ECON

I = 10.00 %

N	(P/F)	(P/A)	(P/G)	(F/P)	(F/A)	(A/P)	(A/F)	(A/G)	N
1	.9091	0.9091	−0.0000	1.1000	1.0000	1.1000	1.0000	−0.0000	1
2	.8264	1.7355	0.8264	1.2100	2.1000	0.5762	0.4762	0.4762	2
3	.7513	2.4869	2.3291	1.3310	3.3100	0.4021	0.3021	0.9366	3
4	.6830	3.1699	4.3781	1.4641	4.6410	0.3155	0.2155	1.3812	4
5	.6209	3.7908	6.8618	1.6105	6.1051	0.2638	0.1638	1.8101	5
6	.5645	4.3553	9.6842	1.7716	7.7156	0.2296	0.1296	2.2236	6
7	.5132	4.8684	12.7631	1.9487	9.4872	0.2054	0.1054	2.6216	7
8	.4665	5.3349	16.0287	2.1436	11.4359	0.1874	0.0874	3.0045	8
9	.4241	5.7590	19.4215	2.3579	13.5795	0.1736	0.0736	3.3724	9
10	.3855	6.1446	22.8913	2.5937	15.9374	0.1627	0.0627	3.7255	10
11	.3505	6.4951	26.3963	2.8531	18.5312	0.1540	0.0540	4.0641	11
12	.3186	6.8137	29.9012	3.1384	21.3843	0.1468	0.0468	4.3884	12
13	.2897	7.1034	33.3772	3.4523	24.5227	0.1408	0.0408	4.6988	13
14	.2633	7.3667	36.8005	3.7975	27.9750	0.1357	0.0357	4.9955	14
15	.2394	7.6061	40.1520	4.1772	31.7725	0.1315	0.0315	5.2789	15
16	.2176	7.8237	43.4164	4.5950	35.9497	0.1278	0.0278	5.5493	16
17	.1978	8.0216	46.5819	5.0545	40.5447	0.1247	0.0247	5.8071	17
18	.1799	8.2014	49.6395	5.5599	45.5992	0.1219	0.0219	6.0526	18
19	.1635	8.3649	52.5827	6.1159	51.1591	0.1195	0.0195	6.2861	19
20	.1486	8.5136	55.4069	6.7275	57.2750	0.1175	0.0175	6.5081	20
21	.1351	8.6487	58.1095	7.4002	64.0025	0.1156	0.0156	6.7189	21
22	.1228	8.7715	60.6893	8.1403	71.4027	0.1140	0.0140	6.9189	22
23	.1117	8.8832	63.1462	8.9543	79.5430	0.1126	0.0126	7.1085	23
24	.1015	8.9847	65.4813	9.8497	88.4973	0.1113	0.0113	7.2881	24
25	.0923	9.0770	67.6964	10.8347	98.3471	0.1102	0.0102	7.4580	25
26	.0839	9.1609	69.7940	11.9182	109.1818	0.1092	0.0092	7.6186	26
27	.0763	9.2372	71.7773	13.1100	121.0999	0.1083	0.0083	7.7704	27
28	.0693	9.3066	73.6495	14.4210	134.2099	0.1075	0.0075	7.9137	28
29	.0630	9.3696	75.4146	15.8631	148.6309	0.1067	0.0067	8.0489	29
30	.0573	9.4269	77.0766	17.4494	164.4940	0.1061	0.0061	8.1762	30
31	.0521	9.4790	78.6395	19.1943	181.9434	0.1055	0.0055	8.2962	31
32	.0474	9.5264	80.1078	21.1138	201.1378	0.1050	0.0050	8.4091	32
33	.0431	9.5694	81.4856	23.2252	222.2515	0.1045	0.0045	8.5152	33
34	.0391	9.6086	82.7773	25.5477	245.4767	0.1041	0.0041	8.6149	34
35	.0356	9.6442	83.9872	28.1024	271.0244	0.1037	0.0037	8.7086	35
36	.0323	9.6765	85.1194	30.9127	299.1268	0.1033	0.0033	8.7965	36
37	.0294	9.7059	86.1781	34.0039	330.0395	0.1030	0.0030	8.8789	37
38	.0267	9.7327	87.1673	37.4043	364.0434	0.1027	0.0027	8.9562	38
39	.0243	9.7570	88.0908	41.1448	401.4478	0.1025	0.0025	9.0285	39
40	.0221	9.7791	88.9525	45.2593	442.5926	0.1023	0.0023	9.0962	40
41	.0201	9.7991	89.7560	49.7852	487.8518	0.1020	0.0020	9.1596	41
42	.0183	9.8174	90.5047	54.7637	537.6370	0.1019	0.0019	9.2188	42
43	.0166	9.8340	91.2019	60.2401	592.4007	0.1017	0.0017	9.2741	43
44	.0151	9.8491	91.8508	66.2641	652.6408	0.1015	0.0015	9.3258	44
45	.0137	9.8628	92.4544	72.8905	718.9048	0.1014	0.0014	9.3740	45
46	.0125	9.8753	93.0157	80.1795	791.7953	0.1013	0.0013	9.4190	46
47	.0113	9.8866	93.5372	88.1975	871.9749	0.1011	0.0011	9.4610	47
48	.0103	9.8969	94.0217	97.0172	960.1723	0.1010	0.0010	9.5001	48
49	.0094	9.9063	94.4715	106.7190	1057.1896	0.1009	0.0009	9.5365	49
50	.0085	9.9148	94.8889	117.3909	1163.9085	0.1009	0.0009	9.5704	50
51	.0077	9.9226	95.2761	129.1299	1281.2994	0.1008	0.0008	9.6020	51
52	.0070	9.9296	95.6351	142.0429	1410.4293	0.1007	0.0007	9.6313	52
53	.0064	9.9360	95.9679	156.2472	1552.4723	0.1006	0.0006	9.6586	53
54	.0058	9.9418	96.2763	171.8719	1708.7195	0.1006	0.0006	9.6840	54
55	.0053	9.9471	96.5619	189.0591	1880.5914	0.1005	0.0005	9.7075	55
60	.0033	9.9672	97.7010	304.4816	3034.8164	0.1003	0.0003	9.8023	60
65	.0020	9.9796	98.4705	490.3707	4893.7073	0.1002	0.0002	9.8672	65
70	.0013	9.9873	98.9870	789.7470	7887.4696	0.1001	0.0001	9.9113	70
75	.0008	9.9921	99.3317	1271.8954	12708.9537	0.1001	0.0001	9.9410	75
80	.0005	9.9951	99.5606	2048.4002	20474.0021	0.1000	0.0000	9.9609	80
85	.0003	9.9970	99.7120	3298.9690	32979.6903	0.1000	0.0000	9.9742	85
90	.0002	9.9981	99.8118	5313.0226	53120.2261	0.1000	0.0000	9.9831	90
95	.0001	9.9988	99.8773	8556.6760	85556.7605	0.1000	0.0000	9.9889	95
100	.0001	9.9993	99.9202	13780.6123	137796.1234	0.1000	0.0000	9.9927	100

I = 12.00 %

N	(P/F)	(P/A)	(P/G)	(F/P)	(F/A)	(A/P)	(A/F)	(A/G)	N
1	.8929	0.8929	-0.0000	1.1200	1.0000	1.1200	1.0000	-0.0000	1
2	.7972	1.6901	0.7972	1.2544	2.1200	0.5917	0.4717	0.4717	2
3	.7118	2.4018	2.2208	1.4049	3.3744	0.4163	0.2963	0.9246	3
4	.6355	3.0373	4.1273	1.5735	4.7793	0.3292	0.2092	1.3589	4
5	.5674	3.6048	6.3970	1.7623	6.3528	0.2774	0.1574	1.7746	5
6	.5066	4.1114	8.9302	1.9738	8.1152	0.2432	0.1232	2.1720	6
7	.4523	4.5638	11.6443	2.2107	10.0890	0.2191	0.0991	2.5515	7
8	.4039	4.9676	14.4714	2.4760	12.2997	0.2013	0.0813	2.9131	8
9	.3606	5.3282	17.3563	2.7731	14.7757	0.1877	0.0677	3.2574	9
10	.3220	5.6502	20.2541	3.1058	17.5487	0.1770	0.0570	3.5847	10
11	.2875	5.9377	23.1288	3.4785	20.6546	0.1684	0.0484	3.8953	11
12	.2567	6.1944	25.9523	3.8960	24.1331	0.1614	0.0414	4.1897	12
13	.2292	6.4235	28.7024	4.3635	28.0291	0.1557	0.0357	4.4683	13
14	.2046	6.6282	31.3624	4.8871	32.3926	0.1509	0.0309	4.7317	14
15	.1827	6.8109	33.9202	5.4736	37.2797	0.1468	0.0268	4.9803	15
16	.1631	6.9740	36.3670	6.1304	42.7533	0.1434	0.0234	5.2147	16
17	.1456	7.1196	38.6973	6.8660	48.8837	0.1405	0.0205	5.4353	17
18	.1300	7.2497	40.9080	7.6900	55.7497	0.1379	0.0179	5.6427	18
19	.1161	7.3658	42.9979	8.6128	63.4397	0.1358	0.0158	5.8375	19
20	.1037	7.4694	44.9676	9.6463	72.0524	0.1339	0.0139	6.0202	20
21	.0926	7.5620	46.8188	10.8038	81.6987	0.1322	0.0122	6.1913	21
22	.0826	7.6446	48.5543	12.1003	92.5026	0.1308	0.0108	6.3514	22
23	.0738	7.7184	50.1776	13.5523	104.6029	0.1296	0.0096	6.5010	23
24	.0659	7.7843	51.6929	15.1786	118.1552	0.1285	0.0085	6.6406	24
25	.0588	7.8431	53.1046	17.0001	133.3339	0.1275	0.0075	6.7708	25
26	.0525	7.8957	54.4177	19.0401	150.3339	0.1267	0.0067	6.8921	26
27	.0469	7.9426	55.6369	21.3249	169.3740	0.1259	0.0059	7.0049	27
28	.0419	7.9844	56.7674	23.8839	190.6989	0.1252	0.0052	7.1098	28
29	.0374	8.0218	57.8141	26.7499	214.5828	0.1247	0.0047	7.2071	29
30	.0334	8.0552	58.7821	29.9599	241.3327	0.1241	0.0041	7.2974	30
31	.0298	8.0850	59.6761	33.5551	271.2926	0.1237	0.0037	7.3811	31
32	.0266	8.1116	60.5010	37.5817	304.8477	0.1233	0.0033	7.4586	32
33	.0238	8.1354	61.2612	42.0915	342.4294	0.1229	0.0029	7.5302	33
34	.0212	8.1566	61.9612	47.1425	384.5210	0.1226	0.0026	7.5965	34
35	.0189	8.1755	62.6052	52.7996	431.6635	0.1223	0.0023	7.6577	35
36	.0169	8.1924	63.1970	59.1356	484.4631	0.1221	0.0021	7.7141	36
37	.0151	8.2075	63.7406	66.2318	543.5987	0.1218	0.0018	7.7661	37
38	.0135	8.2210	64.2394	74.1797	609.8305	0.1216	0.0016	7.8141	38
39	.0120	8.2330	64.6967	83.0812	684.0102	0.1215	0.0015	7.8582	39
40	.0107	8.2438	65.1159	93.0510	767.0914	0.1213	0.0013	7.8988	40
41	.0096	8.2534	65.4997	104.2171	860.1424	0.1212	0.0012	7.9361	41
42	.0086	8.2619	65.8509	116.7231	964.3595	0.1210	0.0010	7.9704	42
43	.0076	8.2696	66.1722	130.7299	1081.0826	0.1209	0.0009	8.0019	43
44	.0068	8.2764	66.4659	146.4175	1211.8125	0.1208	0.0008	8.0308	44
45	.0061	8.2825	66.7342	163.9876	1358.2300	0.1207	0.0007	8.0572	45
46	.0054	8.2880	66.9792	183.6661	1522.2176	0.1207	0.0007	8.0815	46
47	.0049	8.2928	67.2028	205.7061	1705.8838	0.1206	0.0006	8.1037	47
48	.0043	8.2972	67.4068	230.3908	1911.5898	0.1205	0.0005	8.1241	48
49	.0039	8.3010	67.5929	258.0377	2141.9806	0.1205	0.0005	8.1427	49
50	.0035	8.3045	67.7624	289.0022	2400.0182	0.1204	0.0004	8.1597	50
51	.0031	8.3076	67.9169	323.6825	2689.0204	0.1204	0.0004	8.1753	51
52	.0028	8.3103	68.0576	362.5243	3012.7029	0.1203	0.0003	8.1895	52
53	.0025	8.3128	68.1856	406.0273	3375.2272	0.1203	0.0003	8.2025	53
54	.0022	8.3150	68.3022	454.7505	3781.2545	0.1203	0.0003	8.2143	54
55	.0020	8.3170	68.4082	509.3206	4236.0050	0.1202	0.0002	8.2251	55
60	.0011	8.3240	68.8100	897.5969	7471.6411	0.1201	0.0001	8.2664	60
65	.0006	8.3281	69.0581	1581.8725	13173.9374	0.1201	0.0001	8.2922	65
70	.0004	8.3303	69.2103	2787.7998	23223.3319	0.1200	0.0000	8.3082	70
75	.0002	8.3316	69.3031	4913.0558	40933.7987	0.1200	0.0000	8.3181	75
80	.0001	8.3324	69.3594	8658.4831	72145.6925	0.1200	0.0000	8.3241	80
85	.0001	8.3328	69.3935	15259.2057	127151.7140	0.1200	0.0000	8.3278	85
90	.0000	8.3330	69.4140	26891.9342	224091.1185	0.1200	0.0000	8.3300	90
95	.0000	8.3332	69.4263	47392.7766	394931.4719	0.1200	0.0000	8.3313	95
100	.0000	8.3332	69.4336	83522.2657	696010.5477	0.1200	0.0000	8.3321	100

I = 15.00 %

N	(P/F)	(P/A)	(P/G)	(F/P)	(F/A)	(A/P)	(A/F)	(A/G)	N
1	.8696	0.8696	-0.0000	1.1500	1.0000	1.1500	1.0000	-0.0000	1
2	.7561	1.6257	0.7561	1.3225	2.1500	0.6151	0.4651	0.4651	2
3	.6575	2.2832	2.0712	1.5209	3.4725	0.4380	0.2880	0.9071	3
4	.5718	2.8550	3.7864	1.7490	4.9934	0.3503	0.2003	1.3263	4
5	.4972	3.3522	5.7751	2.0114	6.7424	0.2983	0.1483	1.7228	5
6	.4323	3.7845	7.9368	2.3131	8.7537	0.2642	0.1142	2.0972	6
7	.3759	4.1604	10.1924	2.6600	11.0668	0.2404	0.0904	2.4498	7
8	.3269	4.4873	12.4807	3.0590	13.7268	0.2229	0.0729	2.7813	8
9	.2843	4.7716	14.7548	3.5179	16.7858	0.2096	0.0596	3.0922	9
10	.2472	5.0188	16.9795	4.0456	20.3037	0.1993	0.0493	3.3832	10
11	.2149	5.2337	19.1289	4.6524	24.3493	0.1911	0.0411	3.6549	11
12	.1869	5.4206	21.1849	5.3503	29.0017	0.1845	0.0345	3.9082	12
13	.1625	5.5831	23.1352	6.1528	34.3519	0.1791	0.0291	4.1438	13
14	.1413	5.7245	24.9725	7.0757	40.5047	0.1747	0.0247	4.3624	14
15	.1229	5.8474	26.6930	8.1371	47.5804	0.1710	0.0210	4.5650	15
16	.1069	5.9542	28.2960	9.3576	55.7175	0.1679	0.0179	4.7522	16
17	.0929	6.0472	29.7828	10.7613	65.0751	0.1654	0.0154	4.9251	17
18	.0808	6.1280	31.1565	12.3755	75.8364	0.1632	0.0132	5.0843	18
19	.0703	6.1982	32.4213	14.2318	88.2118	0.1613	0.0113	5.2307	19
20	.0611	6.2593	33.5822	16.3665	102.4436	0.1598	0.0098	5.3651	20
21	.0531	6.3125	34.6448	18.8215	118.8101	0.1584	0.0084	5.4883	21
22	.0462	6.3587	35.6150	21.6447	137.6316	0.1573	0.0073	5.6010	22
23	.0402	6.3988	36.4988	24.8915	159.2764	0.1563	0.0063	5.7040	23
24	.0349	6.4338	37.3023	28.6252	184.1678	0.1554	0.0054	5.7979	24
25	.0304	6.4641	38.0314	32.9190	212.7930	0.1547	0.0047	5.8834	25
26	.0264	6.4906	38.6918	37.8568	245.7120	0.1541	0.0041	5.9612	26
27	.0230	6.5135	39.2890	43.5353	283.5688	0.1535	0.0035	6.0319	27
28	.0200	6.5335	39.8283	50.0656	327.1041	0.1531	0.0031	6.0960	28
29	.0174	6.5509	40.3146	57.5755	377.1697	0.1527	0.0027	6.1541	29
30	.0151	6.5660	40.7526	66.2118	434.7451	0.1523	0.0023	6.2066	30
31	.0131	6.5791	41.1466	76.1435	500.9569	0.1520	0.0020	6.2541	31
32	.0114	6.5905	41.5006	87.5651	577.1005	0.1517	0.0017	6.2970	32
33	.0099	6.6005	41.8184	100.6998	664.6655	0.1515	0.0015	6.3357	33
34	.0086	6.6091	42.1033	115.8048	765.3654	0.1513	0.0013	6.3705	34
35	.0075	6.6166	42.3586	133.1755	881:1702	0.1511	0.0011	6.4019	35
36	.0065	6.6231	42.5872	153.1519	1014.3457	0.1510	0.0010	6.4301	36
37	.0057	6.6288	42.7916	176.1246	1167.4975	0.1509	0.0009	6.4554	37
38	.0049	6.6338	42.9743	202.5433	1343.6222	0.1507	0.0007	6.4781	38
39	.0043	6.6380	43.1374	232.9248	1546.1655	0.1506	0.0006	6.4985	39
40	.0037	6.6418	43.2830	267.8635	1779.0903	0.1506	0.0006	6.5168	40
41	.0032	6.6450	43.4128	308.0431	2046.9539	0.1505	0.0005	6.5331	41
42	.0028	6.6478	43.5286	354.2495	2354.9969	0.1504	0.0004	6.5478	42
43	.0025	6.6503	43.6317	407.3870	2709.2465	0.1504	0.0004	6.5609	43
44	.0021	6.6524	43.7235	468.4950	3116.6334	0.1503	0.0003	6.5725	44
45	.0019	6.6543	43.8051	538.7693	3585.1285	0.1503	0.0003	6.5830	45
46	.0016	6.6559	43.8778	619.5847	4123.8977	0.1502	0.0002	6.5923	46
47	.0014	6.6573	43.9423	712.5224	4743.4824	0.1502	0.0002	6.6006	47
48	.0012	6.6585	43.9997	819.4007	5456.0047	0.1502	0.0002	6.6080	48
49	.0011	6.6596	44.0506	942.3108	6275.4055	0.1502	0.0002	6.6146	49
50	.0009	6.6605	44.0958	1083.6574	7217.7163	0.1501	0.0001	6.6205	50
51	.0008	6.6613	44.1360	1246.2061	8301.3737	0.1501	0.0001	6.6257	51
52	.0007	6.6620	44.1715	1433.1370	9547.5798	0.1501	0.0001	6.6304	52
53	.0006	6.6626	44.2031	1648.1075	10980.7167	0.1501	0.0001	6.6345	53
54	.0005	6.6631	44.2311	1895.3236	12628.8243	0.1501	0.0001	6.6382	54
55	.0005	6.6636	44.2558	2179.6222	14524.1479	0.1501	0.0001	6.6414	55
60	.0002	6.6651	44.3431	4383.9987	29219.9916	0.1500	0.0000	6.6530	60
65	.0001	6.6659	44.3903	8817.7874	58778.5826	0.1500	0.0000	6.6593	65
70	.0001	6.6663	44.4156	17735.7200	118231.4669	0.1500	0.0000	6.6627	70
75	.0000	6.6665	44.4292	35672.8680	237812.4532	0.1500	0.0000	6.6646	75
80	.0000	6.6666	44.4364	71750.8794	478332.5293	0.1500	0.0000	6.6656	80
85	.0000	6.6666	44.4402	144316.6470	962104.3133	0.1500	0.0000	6.6661	85
90	.0000	6.6666	44.4422	290272.3252	1935142.1680	0.1500	0.0000	6.6664	90
95	.0000	6.6667	44.4433	583841.3276	3892268.8509	0.1500	0.0000	6.6665	95
100	.0000	6.6667	44.4438	1174313.4507	7828749.6713	0.1500	0.0000	6.6666	100

I = 20.00 %

N	(P/F)	(P/A)	(P/G)	(F/P)	(F/A)	(A/P)	(A/F)	(A/G)	N
1	.8333	0.8333	-0.0000	1.2000	1.0000	1.2000	1.0000	-0.0000	1
2	.6944	1.5278	0.6944	1.4400	2.2000	0.6545	0.4545	0.4545	2
3	.5787	2.1065	1.8519	1.7280	3.6400	0.4747	0.2747	0.8791	3
4	.4823	2.5887	3.2986	2.0736	5.3680	0.3863	0.1863	1.2742	4
5	.4019	2.9906	4.9061	2.4883	7.4416	0.3344	0.1344	1.6405	5
6	.3349	3.3255	6.5806	2.9860	9.9299	0.3007	0.1007	1.9788	6
7	.2791	3.6046	8.2551	3.5832	12.9159	0.2774	0.0774	2.2902	7
8	.2326	3.8372	9.8831	4.2998	16.4991	0.2606	0.0606	2.5756	8
9	.1938	4.0310	11.4335	5.1598	20.7989	0.2481	0.0481	2.8364	9
10	.1615	4.1925	12.8871	6.1917	25.9587	0.2385	0.0385	3.0739	10
11	.1346	4.3271	14.2330	7.4301	32.1504	0.2311	0.0311	3.2893	11
12	.1122	4.4392	15.4667	8.9161	39.5805	0.2253	0.0253	3.4841	12
13	.0935	4.5327	16.5883	10.6993	48.4966	0.2206	0.0206	3.6597	13
14	.0779	4.6106	17.6008	12.8392	59.1959	0.2169	0.0169	3.8175	14
15	.0649	4.6755	18.5095	15.4070	72.0351	0.2139	0.0139	3.9588	15
16	.0541	4.7296	19.3208	18.4884	87.4421	0.2114	0.0114	4.0851	16
17	.0451	4.7746	20.0419	22.1861	105.9306	0.2094	0.0094	4.1976	17
18	.0376	4.8122	20.6805	26.6233	128.1167	0.2078	0.0078	4.2975	18
19	.0313	4.8435	21.2439	31.9480	154.7400	0.2065	0.0065	4.3861	19
20	.0261	4.8696	21.7395	38.3376	186.6880	0.2054	0.0054	4.4643	20
21	.0217	4.8913	22.1742	46.0051	225.0256	0.2044	0.0044	4.5334	21
22	.0181	4.9094	22.5546	55.2061	271.0307	0.2037	0.0037	4.5941	22
23	.0151	4.9245	22.8867	66.2474	326.2369	0.2031	0.0031	4.6475	23
24	.0126	4.9371	23.1760	79.4968	392.4842	0.2025	0.0025	4.6943	24
25	.0105	4.9476	23.4276	95.3962	471.9811	0.2021	0.0021	4.7352	25
26	.0087	4.9563	23.6460	114.4755	567.3773	0.2018	0.0018	4.7709	26
27	.0073	4.9636	23.8353	137.3706	681.8528	0.2015	0.0015	4.8020	27
28	.0061	4.9697	23.9991	164.8447	819.2233	0.2012	0.0012	4.8291	28
29	.0051	4.9747	24.1406	197.8136	984.0680	0.2010	0.0010	4.8527	29
30	.0042	4.9789	24.2628	237.3763	1181.8816	0.2008	0.0008	4.8731	30
31	.0035	4.9824	24.3681	284.8516	1419.2579	0.2007	0.0007	4.8908	31
32	.0029	4.9854	24.4588	341.8219	1704.1095	0.2006	0.0006	4.9061	32
33	.0024	4.9878	24.5368	410.1863	2045.9314	0.2005	0.0005	4.9194	33
34	.0020	4.9898	24.6038	492.2235	2456.1176	0.2004	0.0004	4.9308	34
35	.0017	4.9915	24.6614	590.6682	2948.3411	0.2003	0.0003	4.9406	35
36	.0014	4.9929	24.7108	708.8019	3539.0094	0.2003	0.0003	4.9491	36
37	.0012	4.9941	24.7531	850.5622	4247.8112	0.2002	0.0002	4.9564	37
38	.0010	4.9951	24.7894	1020.6747	5098.3735	0.2002	0.0002	4.9627	38
39	.0008	4.9959	24.8204	1224.8096	6119.0482	0.2002	0.0002	4.9681	39
40	.0007	4.9966	24.8469	1469.7716	7343.8578	0.2001	0.0001	4.9728	40
41	.0006	4.9972	24.8696	1763.7259	8813.6294	0.2001	0.0001	4.9767	41
42	.0005	4.9976	24.8890	2116.4711	10577.3553	0.2001	0.0001	4.9801	42
43	.0004	4.9980	24.9055	2539.7653	12693.8263	0.2001	0.0001	4.9831	43
44	.0003	4.9984	24.9196	3047.7183	15233.5916	0.2001	0.0001	4.9856	44
45	.0003	4.9986	24.9316	3657.2620	18281.3099	0.2001	0.0001	4.9877	45
46	.0002	4.9989	24.9419	4388.7144	21938.5719	0.2000	0.0000	4.9895	46
47	.0002	4.9991	24.9506	5266.4573	26327.2863	0.2000	0.0000	4.9911	47
48	.0002	4.9992	24.9581	6319.7487	31593.7436	0.2000	0.0000	4.9924	48
49	.0001	4.9993	24.9644	7583.6985	37913.4923	0.2000	0.0000	4.9935	49
50	.0001	4.9995	24.9698	9100.4382	45497.1908	0.2000	0.0000	4.9945	50
51	.0001	4.9995	24.9744	10920.5258	54597.6289	0.2000	0.0000	4.9953	51
52	.0001	4.9996	24.9783	13104.6309	65518.1547	0.2000	0.0000	4.9960	52
53	.0001	4.9997	24.9816	15725.5571	78622.7856	0.2000	0.0000	4.9966	53
54	.0001	4.9997	24.9844	18870.6685	94348.3427	0.2000	0.0000	4.9971	54
55	.0000	4.9998	24.9868	22644.8023	113219.0113	0.2000	0.0000	4.9976	55
60	.0000	4.9999	24.9942	56347.5144	281732.5718	0.2000	0.0000	4.9989	60
65	.0000	5.0000	24.9975	140210.6469	701048.2346	0.2000	0.0000	4.9995	65
70	.0000	5.0000	24.9989	348888.9569	1744439.7847	0.2000	0.0000	4.9998	70
75	.0000	5.0000	24.9995	868147.3693	4340731.8466	0.2000	0.0000	4.9999	75

ECON

I = 25.00 %

N	(P/F)	(P/A)	(P/G)	(F/P)	(F/A)	(A/P)	(A/F)	(A/G)	N
1	.8000	0.8000	0.0	1.2500	1.0000	1.2500	1.0000	0.0	1
2	.6400	1.4400	0.6400	1.5625	2.2500	0.6944	0.4444	0.4444	2
3	.5120	1.9520	1.6640	1.9531	3.8125	0.5123	0.2623	0.8525	3
4	.4096	2.3616	2.8928	2.4414	5.7656	0.4234	0.1734	1.2249	4
5	.3277	2.6893	4.2035	3.0518	8.2070	0.3718	0.1218	1.5631	5
6	.2621	2.9514	5.5142	3.8147	11.2588	0.3388	0.0888	1.8683	6
7	.2097	3.1611	6.7725	4.7684	15.0735	0.3163	0.0663	2.1424	7
8	.1678	3.3289	7.9469	5.9605	19.8419	0.3004	0.0504	2.3872	8
9	.1342	3.4631	9.0207	7.4506	25.8023	0.2888	0.0388	2.6048	9
10	.1074	3.5705	9.9870	9.3132	33.2529	0.2801	0.0301	2.7971	10
11	.0859	3.6564	10.8460	11.6415	42.5661	0.2735	0.0235	2.9663	11
12	.0687	3.7251	11.6020	14.5519	54.2077	0.2684	0.0184	3.1145	12
13	.0550	3.7801	12.2617	18.1899	68.7596	0.2645	0.0145	3.2437	13
14	.0440	3.8241	12.8334	22.7374	86.9495	0.2615	0.0115	3.3559	14
15	.0352	3.8593	13.3260	28.4217	109.6868	0.2591	0.0091	3.4530	15
16	.0281	3.8874	13.7482	35.5271	138.1085	0.2572	0.0072	3.5366	16
17	.0225	3.9099	14.1085	44.4089	173.6357	0.2558	0.0058	3.6084	17
18	.0180	3.9279	14.4147	55.5112	218.0446	0.2546	0.0046	3.6698	18
19	.0144	3.9424	14.6741	69.3889	273.5558	0.2537	0.0037	3.7222	19
20	.0115	3.9539	14.8932	86.7362	342.9447	0.2529	0.0029	3.7667	20
21	.0092	3.9631	15.0777	108.4202	429.6809	0.2523	0.0023	3.8045	21
22	.0074	3.9705	15.2326	135.5253	538.1011	0.2519	0.0019	3.8365	22
23	.0059	3.9764	15.3625	169.4066	673.6264	0.2515	0.0015	3.8634	23
24	.0047	3.9811	15.4711	211.7582	843.0329	0.2512	0.0012	3.8861	24
25	.0038	3.9849	15.5618	264.6978	1054.7912	0.2509	0.0009	3.9052	25
26	.0030	3.9879	15.6373	330.8722	1319.4890	0.2508	0.0008	3.9212	26
27	.0024	3.9903	15.7002	413.5903	1650.3612	0.2506	0.0006	3.9346	27
28	.0019	3.9923	15.7524	516.9879	2063.9515	0.2505	0.0005	3.9457	28
29	.0015	3.9938	15.7957	646.2349	2580.9394	0.2504	0.0004	3.9551	29
30	.0012	3.9950	15.8316	807.7936	3227.1743	0.2503	0.0003	3.9628	30
31	.0010	3.9960	15.8614	1009.7420	4034.9678	0.2502	0.0002	3.9693	31
32	.0008	3.9968	15.8859	1262.1774	5044.7098	0.2502	0.0002	3.9746	32
33	.0006	3.9975	15.9062	1577.7218	6306.8872	0.2502	0.0002	3.9791	33
34	.0005	3.9980	15.9229	1972.1523	7884.6091	0.2501	0.0001	3.9828	34
35	.0004	3.9984	15.9367	2465.1903	9856.7613	0.2501	0.0001	3.9858	35
36	.0003	3.9987	15.9481	3081.4879	12321.9516	0.2501	0.0001	3.9883	36
37	.0003	3.9990	15.9574	3851.8599	15403.4396	0.2501	0.0001	3.9904	37
38	.0002	3.9992	15.9651	4814.8249	19255.2994	0.2501	0.0001	3.9921	38
39	.0002	3.9993	15.9714	6018.5311	24070.1243	0.2500	0.0000	3.9935	39
40	.0001	3.9995	15.9766	7523.1638	30088.6554	0.2500	0.0000	3.9947	40
41	.0001	3.9996	15.9809	9403.9548	37611.8192	0.2500	0.0000	3.9956	41
42	.0001	3.9997	15.9843	11754.9435	47015.7740	0.2500	0.0000	3.9964	42
43	.0001	3.9997	15.9872	14693.6794	58770.7175	0.2500	0.0000	3.9971	43
44	.0001	3.9998	15.9895	18367.0992	73464.3969	0.2500	0.0000	3.9976	44
45	.0000	3.9998	15.9915	22958.8740	91831.4962	0.2500	0.0000	3.9980	45
46	.0000	3.9999	15.9930	28698.5925	114790.3702	0.2500	0.0000	3.9984	46
47	.0000	3.9999	15.9943	35873.2407	143488.9627	0.2500	0.0000	3.9987	47
48	.0000	3.9999	15.9954	44841.5509	179362.2034	0.2500	0.0000	3.9989	48
49	.0000	3.9999	15.9962	56051.9386	224203.7543	0.2500	0.0000	3.9991	49
50	.0000	3.9999	15.9969	70064.9232	280255.6929	0.2500	0.0000	3.9993	50
51	.0000	4.0000	15.9975	87581.1540	350320.6161	0.2500	0.0000	3.9994	51
52	.0000	4.0000	15.9980	109476.4425	437901.7701	0.2500	0.0000	3.9995	52
53	.0000	4.0000	15.9983	136845.5532	547378.2126	0.2500	0.0000	3.9996	53
54	.0000	4.0000	15.9986	171056.9414	684223.7658	0.2500	0.0000	3.9997	54
55	.0000	4.0000	15.9989	213821.1768	855280.7072	0.2500	0.0000	3.9997	55
60	.0000	4.0000	15.9996	652530.4468	2610117.7872	0.2500	0.0000	3.9999	60

ECON

BASIC EQUIVALENCE

1. If $250 is invested at 6% on January 1, 1971, how much will be accumulated by January 1, 1981?

2. How much would you have had to invest on January 1, 1975 in order to accumulate $2000 on January 1, 1981 at 6%?

3. What is the present worth on January 1, 1973 of $2000 available at January 1, 1980 if interest is at 6%?

4. If $50 was invested at 6% on January 1, 1968, what equal year-end withdrawals could be made each year for 10 years, leaving nothing in the fund after the tenth withdrawal?

5. How much could be accumulated in a fund earning 6% at the end of 10 years if $20,000 is deposited at the end of each year for 10 years?

6. How much must be deposited at 6% each year for 7 years beginning on January 1, 1977 in order to accumulate $5000 on the date of the last deposit, January 1, 1983?

7. How much would you need to deposit at 6% in order to draw out $400 at the end of each year for 7 years, leaving nothing in the fund at the end?

8. If $500 is invested now, $700 two years from now, and $900 four years from now (all at 4%), what will be the total amount in 10 years?

9. What is the compounded amount of $550 for 8 years with interest at 6% compounded semi-annually?

10. A savings certificate costing $50 now will pay $75 in five years. What is the interest rate?

11. How much must be invested at the end of each month for 30 years in a sinking fund which is to amount to $50,000 at the end of 30 years if interest is 4%?

12. How much would be accumulated in the sinking fund for problem 11 above at the end of 18 years?

13. In 18 years you will need $20,000 for your child's college expenses. How much should you deposit each year starting on his day of birth so that you will meet this goal? Assume that the last payment is on his 18th birthday and that 5% interest is paid.

14. Starting on January 1, 1960 $50 is deposited in an account paying 6% annually. Each January, up to and including January 1, 1970, another $50 will be deposited. Starting January 1, 1975 (the date of the first withdrawal) five uniform annual withdrawals are made. The last withdrawal will exhaust the fund. How much will be withdrawn each year?

PRACTICAL LOANS

15. A $2000 loan is taken out at a bank. Monthly payments are $400 plus interest (10% annual rate) on the unpaid balance. What will be the payments for the loan duration? What principal remains to be paid off after the third payment? What is the interest on the fourth payment.?

PRESENT WORTH

16. Equipment is purchased for $12,000 which is expected to be sold after 10 years for $2,000. The estimated maintenance is $1,000 the first year, but is expected to increase by $200 each year thereafter. Using 10%, find the present worth of the project.

17. A fast-acting brake on a fast-turning lathe is estimated to save 7 seconds per piece produced since the operator (paid at the rate of $3.90 per hour) doesn't have to wait as long for the lathe to stop. 40,000 pieces are produced annually. Assuming a 3 year life, no salvage value, and an 8% interest rate. what should be the maximum purchase price of the brake?

CAPITALIZED COST

18. A $100,000 item is purchased. Annual costs are $18,000. Using 8%, what is the capitalized cost of perpetual service?

19. The heat loss from bare steam pipes costs a motel manager $400 annually. Two brands of insulation are available. Brand A will cost $120 and will reduce losses by 93%; Brand B will cost $70 and will reduce losses by 87%. What are the capitalized costs of perpetual service for both brands? Which insulation is superior? Use 10%.

ANNUAL COST

20. A new machine will cost $17,000 and will have an estimated salvage value of $14,000 in five years. Special tools for the new machine will cost $5000 and will have a resale value of $2,500 at the end of five years. Maintenance costs are estimated at $200 per year. What will be the average annual cost of ownership during the next five years if interest is 6%?

21. A small building can be reroofed for $6,000 (aluminum) or $3,500 (shingles). The estimated lives are 50 years and 15 years respectively. If the interest rate is 10%, which alternative is superior?

22. Designers need to decide how to condition the air in a new building with 40 year life. Two alternatives are available, both of which have 20 year lives. Costs are given below. Use 12% interest to find the best alternative.

	A	B
First cost	90,000	60,000
Salvage in 20 years	10,000	6,000
Annual fuel costs	3,000	5,000
Annual maintenance costs	2,200	3,000
Extra annual income tax	400	0

ECON

RATE OF RETURN

23. A $14,000 plot of land can be paid for with $4,000 down and $1,200 per year for 12 years. What is the annual interest rate being charged?

24. A speculator in land and property pays $14,000 for a house that he expects to hold for 10 years. $1,000 is spent in renovation and a monthly rent of $75 is collected from the tenants who live in the house. (Assume all rent is paid at year-end.) Taxes are $150 per year and maintenance costs are $250 per year. What must the sales price be in 10 years to realize a 10% return on investment?

25. An investor wishes to invest $40,000. Venture A, requiring $40,000, will return 8%. Venture B, requiring $10,000, will return 15%. What return on the remaining $30,000 is required to equal the overall profitability of venture A?

BENEFIT COST RATIO

26. A large sewer system will cost $175,000 annually. There will be favorable consequences to the general public of $500,000 annually, and adverse consequences to a small segment of the public of $50,000 annually. What is the excess of benefits over costs? What is the benefit/cost ratio?

27. A public works project has an initial cost of $1,000,000, benefits with a present worth of $1,500,000, and disbenefits with a present worth of $300,000. What is the benefit/cost ratio? What is the excess of benefits over costs?

USEFUL LIFE

28. An asset costing $10,000 has the salvage values and operating costs shown below. Using 8%, when should the asset be replaced with an identical asset?

End of year	Salvage	Average Annual Cost
1	6000	2300
2	4000	2500
3	3200	3300
4	2500	4800
5	2000	6800
6	1800	9500
7	1700	12000

29. A car costs $6000. Annual costs are $400 the first year, $500 the second year, $600 the third year, and so on, increasing $100 each year. Use 10% to find the year in which the car should be replaced.

DEPRECIATION

30. An asset is purchased for $500,000. Salvage in 25 years is $100,000. What are the depreciations in the first 3 years using SL, DDB, and SOYD?

31. Determine the depreciation charge in the fifth year using SOYD, DDB, and SL methods if an asset is purchased new for $12,000 and has a six-year expected life with $2,000 salvage value.

32. Find the book value at the beginning and end of each year by SL, DDB, SOYD, and Sinking Fund at 6% methods for an asset with initial cost of $2,500 and with salvage of $1,100 in 6 years.

INCOME TAXES (Optional)

33. If the corporate tax rate is 53%, what is the present worth of the 6-year after-tax depreciation recovery for the asset in problem 32?

34. An agricultural corporation paying 53% in income taxes wants to build a grain elevator designed to last 25 years at a cost of $80,000, with no salvage value. Annual income generated will be $22,500 and annual expenditures will be $12,000. Using SL depreciation and a 10% interest rate, what is the 25-year after-tax present worth of the project?

35. Repeat problem 34 using SOYD depreciation.

36. A small company pays 30% in income taxes. A $2,000 machine with a 4-year life and zero salvage value will increase revenues $1,200 annually. Using 8%, compare the present worth of the after-tax profits using SL and SOYD depreciation.

RATE AND PERIOD CHANGES

37. What is the true effective annual rate of a credit card plan that charges 1.5% per month on an unpaid balance?

38. A loan company advertises that $100 borrowed for one year may be repaid by 12 monthly installments of $9.46. Assuming the difference between the amount repaid and the amount borrowed in interest only, what is the effective annual interest rate being charged?

39. What is the effective interest rate for a payment plan of 30 equal payments of $89.30 per month, when a lump sum of $2,000 would have been an outright purchase?

COST ACCOUNTING

40. The total costs of producing 240 and 360 units are $3,400 and $4,000 respectively. What is the average unit cost over the first 240 units? What is the incremental cost? What is the fixed cost? What is the total profit or loss from the sale of units 240 through 249 at $10.47 each?

41. A machine with 8-year life and $5000 salvage is purchased for $40,000. Annual costs of operation at $800. The machine operator is paid $4.90 per hour. Power is consumed by the machine at the rate of $1.15 per hour. 2000 units are produced each year on the machine, which requires 48 minutes per unit. Use a 10% interest rate to find the unit cost.

42. A factory with capacity of 700,000 units per year is operating at 62% of capacity. The annual income is $430,000, annual fixed costs are $190,000, and the variable costs are $.348 per unit. What is the current profit/loss? What is the break-even point?

BREAK-EVEN ANALYSIS

43. A fixture costing $700 will save $.06 per item produced. Maintenance will be $40 annually. 3500 units are produced annually. What is the pay-back period at 10%?

44. A hand tool costs $200 and requires $1.21 labor cost per unit. A machine tool costs $3600 and reduces labor to $.75. What is the break-even point (in years) at 5% for an annual production of 4000 units?

3 Systems of Units

1. Consistent Systems of Units

A set of units used in a problem is said to be *consistent*[1] if no conversion factors are needed. For example, a moment with units of foot-pounds cannot be obtained directly from a moment arm with units of inches. In this illustration, a conversion factor of (1/12) feet/inch is needed, and the set of units used is said to be *inconsistent.*

On a larger scale, a system of units is said to be consistent if Newton's second law of motion can be written without conversion factors. Newton's law states simply that the force required to accelerate an object is proportional to the amount of matter in the item.

$$F = ma \tag{3.1}$$

The definitions of the symbols *F, m,* and *a* are familiar to every engineer. However, the use of Newton's second law is complicated by the multiplicity of available unit systems. For example, *m* may be in kilograms, pounds, or slugs. All three of these are units of mass. However, as figure 3.1 illustrates, these three units do not represent the same amount of mass.

[1]The terms *homogeneous* and *coherent* are also used to describe a consistent set of units.

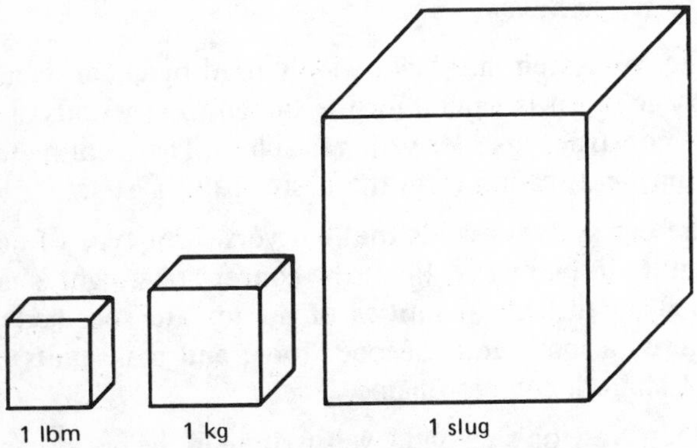

Figure 3.1 Common Units of Mass

It should be mentioned that the decision to work with a consistent set of units is arbitrary and unnecessary. Problems in fluid flow and thermodynamics are commonly solved with inconsistent units. This causes no more of a problem than working with inches and feet in the calculation of a moment.

2. The Absolute English System

Engineers are accustomed to using pounds as a unit of mass. For example, density is typically given in pounds per cubic foot. The abbreviation *pcf* tends to obsure the fact that the true units are pounds of *mass* per cubic foot.

If pounds are the units for mass, and feet per second squared are the units of acceleration, the units of force for a consistent system can be found from Newton's second law.

$$\begin{aligned} \text{units of } F &= (\text{units of } m)(\text{units of } a) \\ &= (\text{lbm})(\text{ft/sec}^2) = \frac{\text{lbm-ft}}{\text{sec}^2} \end{aligned} \tag{3.2}$$

The units for F cannot be simplified any more than they are in equation 3.2. This particular combination of units is known as a *poundal.*[2]

The absolute English system, which requires the poundal as a unit of force, is seldom used. However, the absolute English system does exist. This existence is a direct outgrowth of the requirement to have a consistent system of units.

3. The English Gravitational System

Force is frequently measured in pounds. When the thrust on an accelerating rocket is given as so many pounds, it is understood that the pound is being used as a unit of force.

If acceleration is given in feet per second squared, the units of mass for a consistent system of units can be determined from Newton's second law.

[2]A poundal is equal to .03108 pounds force.

$$\text{units of } m = \frac{\text{units of } F}{\text{units of } a} = \frac{\text{lbf}}{\text{ft/sec}^2} = \frac{\text{lbf-sec}^2}{\text{ft}} \qquad 3.3$$

The combination of units in equation 3.3 is known as a *slug*[3]. Slugs and pounds-mass are not the same, as illustrated in figure 3.1. However, units of mass can be converted using equation 3.4.

$$\#\ \text{slugs} = \frac{\#\ \text{lbm}}{g_c} \qquad 3.4$$

g_c[4] is a dimensional conversion factor having the following value:

$$g_c = 32.1740 \frac{\text{lbm-ft}}{\text{lbf-sec}^2} \qquad 3.5$$

32.1740 is commonly rounded to 32.2 when six significant digits are unjustified. That practice is followed in this book.

Notice that the number of slugs cannot be determined from the number of pounds-mass by dividing by the local gravity. g_c is used regardless of the local gravity. However, the local gravity can be used to find the weight of an object. Weight is defined as the force exerted on a mass by the local gravitational field.

$$\text{weight in lbf} = (\text{m in slugs})(\text{g in ft/sec}^2) \qquad 3.6$$

If the effects of large land and water masses are neglected, the following formula may be used to estimate the local acceleration of gravity in ft/sec² at the earth's surface. ϕ is the latitude in degrees.

$$g_{\text{surface}} = 32.088[1 + (5.305\ \text{EE}-3)\sin^2\phi - (5.9\ \text{EE}-6)\sin^2 2\phi] \qquad 3.7$$

If the effects of the earth's rotation are neglected, the gravitational acceleration at an altitude *h* in miles is given by equation 3.8. R is the earth's radius—approximately 3960 miles.

$$g_h = g_{\text{surface}}\left[\frac{R}{R+h}\right]^2 \qquad 3.8$$

4. The English Engineering System

Many thermodynamics and fluid flow problems freely combine variables containing pound-mass and pound-force terms. For example, the steady-flow energy equation (SFEE) used in chapter 6 mixes enthalpy terms in BTU/lbm with pressure terms in lbf/ft². This requires the use of g_c as a mass conversion factor.

Newton's second law then becomes

$$\text{F in lbf} = \frac{(\text{m in lbm})(\text{a in ft/sec}^2)}{\left(g_c \text{ in } \frac{\text{lbm-ft}}{\text{lbf-sec}^2}\right)} \qquad 3.9$$

Since g_c is required, the English Engineering System is inconsistent. However, that is not particularly troublesome, and the use of g_c does not overly complicate the solution procedure.

Example 3.1

Calculate the weight of a 1.0 lbm object in a gravitational field of 27.5 ft/sec².

Since weight is commonly given in pounds-force, the mass of the object must be converted from pounds-mass to slugs.

$$F = \frac{ma}{g_c} = \frac{(1)\ \text{lbm}\ (27.5)\ \text{ft/sec}^2}{(32.2)\frac{\text{lbm-ft}}{\text{lbf-sec}^2}} = .854\ \text{lbf}$$

Example 3.2

A rocket with a mass of 4000 lbm is traveling at 27,000 ft/sec. What is its kinetic energy in ft-lbf?

The usual kinetic energy equation is $E_k = \frac{1}{2}mv^2$. However, this assumes consistent units. Since energy is wanted in foot-pounds-force, g_c is needed to convert m to units of slugs.

$$E_k = \frac{mv^2}{2g_c} = \frac{(4000)\ \text{lbm}\ (27{,}000)^2\ \text{ft}^2/\text{sec}^2}{(2)(32.2)\frac{\text{lbm-ft}}{\text{lbf-sec}^2}} = 4.53\ \text{EE10 ft-lbf}$$

In the English Engineering System, work and energy are typically measured in ft-lbf (mechanical systems) or in British Thermal Units, BTU (thermal and fluid systems). One BTU is equal to 778.26 ft-lbf.

5. The cgs System

The cgs system has been widely used by chemists and physicists. It is named for the three primary units used to construct its derived variables. The *c*entimeter, *g*ram, and *s*econd form the basis of this system.

The cgs system avoids the lbm versus lbf type of ambiguity in two ways. First, the concept of weight is not used at all. All quantities of matter are specified in grams, a mass unit. Second, force and mass units do not share a common name.

When Newton's second law is written in the cgs system, the following combination of units results.

[3]A slug is equal to 32.1740 lbm.

[4]Three different meanings of the symbol *g* are commonly used. g_c is the dimensional conversion factor given in equation 3.5. g_o is the standard acceleration due to gravity with a value of 32.1740 ft/sec². *g* is the local acceleration due to gravity in ft/sec².

$$\text{units of force} = (\text{m in g})(\text{a in } \frac{\text{cm}}{\text{sec}^2})$$
$$= \frac{\text{g-cm}}{\text{sec}^2} \qquad 3.10$$

This combination of units for force is known as a *dyne*. Energy variables in the cgs system have units of dyne-cm, or equivalently, of $\frac{\text{g-cm}^2}{\text{sec}^2}$. These combinations are known as an *erg*. There is no uniformly accepted unit of power in the cgs system, although calories per second are frequently used. Ergs may be converted to calories by multiplying by 2.389 EE−8.

The fundamental volume unit in the cgs system is the cubic centimeter (cc). Since this is the same volume as one thousandth of a liter, units of milliliters (ml) are freely used in this system.

6. The mks System

The mks system is appropriate when variables take on values larger than can be easily accomodated by the cgs system. This system uses the *m*eter, *k*ilogram, and *s*econd as its primary units. The mks system avoids the lbm versus lbf ambiguity in the same ways as does the cgs system.

The units of force can be derived from Newton's second law.

$$\text{units of force} = (\text{m in kg})(\text{a in } \frac{\text{m}}{\text{sec}^2})$$
$$= \frac{\text{kg-m}}{\text{sec}^2} \qquad 3.11$$

This combination of units for force is known as a *newton*.

Energy variables in the mks system have units of N-m, or equivalently, $\frac{\text{kg-m}^2}{\text{sec}^2}$. Both of these combinations are known as a *joule*. The units of power are joules per second, equivalent to a *watt*. The common volume unit is the liter, equivalent to one-thousandth of a cubic meter.

Example 3.3

A 10 kg block hangs from a cable. What is the tension in the cable?

$$F = ma = (10)\text{ kg } (9.8)\ \frac{\text{m}}{\text{sec}^2}$$
$$= 98\ \frac{\text{kg-m}}{\text{sec}^2} = 98\text{ N}$$

Example 3.4

A 10 kg block is raised vertically 3 meters. What is the change in potential energy?

$$\Delta E_p = mg\Delta h = (10)\text{ kg } (9.8)\ \frac{\text{m}}{\text{sec}^2}\ (3)\text{ m}$$
$$= 294\ \frac{\text{kg-m}^2}{\text{sec}^2} = 294\text{ J}$$

7. The SI System

Strictly speaking, both the cgs and mks systems are *metric* systems. Although the metric units simplify solutions to problems, the multiplicity of possible units for each variable is sometimes confusing.

The SI system ('International System of Units') was established in 1960 by the General Conference of Weights and Measures, an international treaty organization. The SI system is derived from the earlier metric systems, but it is intended to supersede them all.

The SI system has the following features:

(a) There is only one recognized unit for each variable.

(b) The system is fully consistent.

(c) Scaling of units is done in multiples of 1000.

(d) Prefixes, abbreviations, and symbol-syntax are rigidly defined.

Table 3.1 SI Prefixes

Prefix	Symbol	Value
exa	E	EE18
peta	P	EE15
tera	T	EE12
giga	G	EE9
mega	M	EE6
kilo	k	EE3
hecto	h	EE2
deca	da	EE1
deci	d	EE−1
centi	c	EE−2
milli	m	EE−3
micro	μ	EE−6
nano	n	EE−9
pico	p	EE−12
femto	f	EE−15
atto	a	EE−18

Three types of units are used: base units, supplementary units, and derived units. The base units (table 3.2) are dependent only on accepted standards or reproducible phenomena. The supplementary units (table 3.3) have not yet been classified as being base units or derived units. The derived units (tables 3.4 and 3.5) are made up of combinations of base and supplementary units.

UNITS

The expressions for the derived units in symbolic form are obtained by using the mathematical signs of multiplication and division. For example, units of velocity are m/s. Units of torque are N·m (not N-m or Nm).

Table 3.2 SI Base Units

Quantity	Name	Symbol
length	meter	m
mass	kilogram	kg
time	second	s
electric current	ampere	A
temperature	kelvin	K
amount of substance	mole	mol
luminous intensity	candela	cd

Table 3.3 SI Supplementary Units

Quantity	Name	Symbol
plane angle	radian	rad
solid angle	steradian	sr

In addition, there is a set of non-SI units which may be used. This temporary concession is due primarily to the significance and widespread acceptance of these units. Use of the non-SI units listed in table 3.6 will usually create an inconsistent expression requiring conversion factors.

In addition to having standardized units, the SI system also specifies syntax rules for writing the units and combinations of units. Each unit is abbreviated with a specific *symbol*. The following rules for writing these symbols should be adhered to.

(a) The symbols are always printed in roman type, irrespective of the type used in the rest of the text. The only exception to this is in the use of the symbol for liter, where the use of the lower case l (ell) may be confused with the number 1 (one). In this case, "liter" should be written out in full, or the script *l* used. There is no problem with such symbols as cl (centiliter) or ml (milliliter).

(b) Symbols are never pluralized: 1 kg, 45 kg (not 45 kgs).

(c) A period after a symbol is not used, except when the symbol occurs at the end of a sentence.

(d) When symbols consist of letters, there is always a full space between the quantity and the symbols: e.g. 45 kg (not 45kg). However, when the first character of a symbol is not a letter, no space is left: e.g. 32°C (not 32° C or 32 °C); or 42° 12′ 45″ (not 42 ° 12 ′ 45 ″).

(e) All symbols are written in lower case, except when the unit is derived from a proper name. Examples: m for meter; s for second; but A for ampere, Wb for weber, N for newton, W for watt. Prefixes are printed roman type without spacing between the prefix and the unit symbol: e.g. km is the symbol for kilometer.

(f) In text, symbols should be used when associated with a number, however, when no number is involved, the unit should be spelled out. Examples: The area of the carpet is 16 m², not 16 square meters. Carpet is sold by the square meter, not by the m².

(g) A practice in some countries is to use a comma as a decimal marker, while the practice in North America, the United Kingdom and some other countries is to use a period (or dot) as the decimal marker. Further, in some countries using the decimal comma, a dot is frequently used to divide long numbers into groups of three. Be-

Table 3.4 Some SI Derived Units with Special Names

Quantity	Name	Symbol	Expressed in Terms of Other Units
frequency	hertz	Hz	
force	newton	N	
pressure, stress	pascal	Pa	N/m²
energy, work, quantity of heat	joule	J	N·m
power, radiant flux	watt	W	J/s
quantity of electricity, electric charge	coulomb	C	
electric potential, potential difference, electromotive force	volt	V	W/A
electric capacitance	farad	F	C/V
electric resistance	ohm	Ω	V/A
electric conductance	siemen	S	A/V
magnetic flux	weber	Wb	V·s
magnetic flux density	tesla	T	Wb/m²
inductance	henry	H	Wb/A
luminous flux	lumen	lm	
illuminance	lux	lx	lm/m²

Table 3.5 Some SI Derived Units

Quantity	Description	Expressed in Terms of Other Units
area	square meter	m^2
volume	cubic meter	m^3
speed — linear	meter per second	m/s
— angular	radian per second	rad/s
acceleration — linear	meter per second squared	m/s^2
— angular	radian per second squared	rad/s^2
density, mass density	kilogram per cubic meter	kg/m^3
concentration (of amount of substance)	mole per cubic meter	mol/m^3
specific volume	cubic meter per kilogram	m^3/kg
luminance	candela per square meter	cd/m^2
dynamic viscosity	pascal second	Pa·s
moment of force	newton meter	N·m
surface tension	newton per meter	N/m
heat flux density, irradiance	watt per square meter	W/m^2
heat capacity, entropy	joule per kelvin	J/K
specific heat capacity, specific entropy	joule per kilogram kelvin	J/(kg·K)
specific energy	joule per kilogram	J/kg
thermal conductivity	watt per meter kelvin	W/(m·K)
energy density	joule per cubic meter	J/m^3
electric field strength	volt per meter	V/m
electric charge density	coulomb per cubic meter	C/m^3
surface density of charge, flux density	coulomb per square meter	C/m^2
permittivity	farad per meter	F/m
current density	ampere per square meter	A/m^2
magnetic field strength	ampere per meter	A/m
permeability	henry per meter	H/m
molar energy	joule per mole	J/mol
molar entropy, molar heat capacity	joule per mole kelvin	J/(mol·K)
radiant intensity	watt per steradian	W/sr

Table 3.6 Acceptable Non-SI Units

Quantity	Unit Name	Symbol	Relationship to SI Unit
Area	hectare	ha	1 ha = 10 000 m^2
Energy	kilowatt-hour	kWh	1 kWh = 3.6 MJ
Mass	metric ton[5]	t	1 t = 1000 kg
Plane Angle	degree (of arc)	°	1° = 0.017 453 rad
Speed of Rotation	revolution per minute	r/min	1 r/min = $\frac{2\pi}{60}$ rad/s
Temperature Interval	degree Celsius	°C	1 °C = 1 K
Time	minute	min	1 min = 60 s
	hour	h	1 h = 3600 s
	day (mean solar)	d	1 d = 86 400 s
	year (calendar)	a	1 a = 31 536 000 s
Velocity	kilometer per hour	km/h	1 km/h = 0.278 m/s
Volume	liter[6]	*l*	1 *l* = 0.001 m^3

[5]The international name for metric ton is "tonne". The metric ton is equal to the "megagram" (Mg).

[6]The international symbol for liter is the lowercase "l", which can be easily confused with the numeral "1". Several English speaking countries have adopted the script "*l*" as symbol for liter in order to avoid any misinterpretation.

UNITS

cause of these differing practices, spaces must be used intead of commas to separate long lines of digits into easily-readable blocks of three digits with respect to the decimal marker: e.g. 32 453.246 072 5. A space (half space preferred) is optional with a four-digit number: e.g. 1 234 or 1 234.

(h) Where a decimal fraction of a unit is used, a zero should always be placed before the decimal marker, e.g. 0.45 kg (not .45 kg). This practice draws attention to the decimal marker, and helps avoid errors of scale.

(i) Some confusion may arise with the work "tonne" (1 000 kg). When this word occurs in French text of Canadian origin, the meaning may be a ton of 2 000 pounds.

8. Use of the SI System on the E-I-T Exam

In order to make the E-I-T exam fair for engineers who use the SI system, examination problems may be worked in either the SI or English system.

In the morning section of the E-I-T exam, where problems are relatively short, two problems of equal complexity will be presented. One problem will be written using SI units. The other will be written using customary English units. You will be instructed to work either problem. For example, problem 121 might be:

121. (SI VERSION)

The mass of 1.00 cubic meter of air at a pressure of 760 000 Pa absolute and a temperature of 540 K is most nearly . . .

121. (ENGLISH VERSION)

The mass of 1.00 cubic foot of air at a pressure of 110 pounds per square inch absolute and a temperature of 32°F is most nearly . . .

Both problems would have five answer choices. The answer choices for the SI problem would have units of kilograms. The answer choices for the English problem would have units of pounds-mass. The answer key to both problems would be the same.

Afternoon problems are usually more complex and are given in sets of related questions. Some form of dual dimensioning will be used with these problems. For example, the questions will be stated in English units with equivalent SI values and units in parentheses. The five answer choices will also be given in English and SI units.

In future E-I-T examinations, some experimentation with editorial styles will be used to find the most satisfactory method of presenting the more complex afternoon problems. The ideal style will keep the problems equally fair for all examinees.

Appendix A
Selected Conversion Factors to SI Units

	SI Symbol	Multiplier to Convert From Existing Unit to SI Unit	Multiplier to Convert From SI Unit to Existing Unit
Area			
Circular Mil	μm²	506.7	0.001 974
Foot Squared	m²	0.092 9	10.764
Mile Squared	km²	2.590	0.386 1
Yard Squared	m²	0.836 1	1.196
Energy			
Btu (International)	kJ	1.055 1	0.947 8
Erg	μJ	0.1	10.0
Foot Pound-Force	J	1.355 8	0.737 6
Horsepower Hour	MJ	2.684 5	0.372 5
Kilowatt Hour	MJ	3.6	0.277 8
Meter Kilogram-Force	J	9.806 7	0.101 97
Therm	MJ	105.506	0.009 478
Kilogram Calorie (International)	kJ	4.186 8	0.238 8
Force			
Dyne	μN	10.	0.1
Kilogram-Force	N	9.806 7	0.101 97
Ounce-Force	N	0.278 0	3.597
Pound-Force	N	4.448 2	0.224 8
KIP	N	4 448.2	0.000 224 8
Heat			
Btu Per Hour	W	0.293 1	3.412 1
Btu Per (Square Foot Hour)	W/m²	3.154 6	0.317 0
Btu Per (Square Foot Hour °F)	W/(m²·°C)	5.678 3	0.176 1
Btu Inch Per (Square Foot Hour °F)	W/(m·°C)	0.144 2	6.933
Btu Per (Cubic Foot °F)	MJ/(m³·°C)	0.067 1	14.911
Btu Per (Pound °F)	J/(kg·°C)	4 186.8	0.000 238 8
Btu Per Cubic Foot	MJ/m³	0.037 3	26.839
Btu Per Pound	J/kg	2 326.	0.000 430
Length			
Angstrom	nm	0.1	10.0
Foot	m	0.304 8	3.280 8
Inch	mm	25.4	0.039 4
Mil	mm	0.025 4	39.370
Mile	km	1.609 3	0.621 4
Mile (International Nautical)	km	1.852	0.540
Micron	μm	1.0	1.0
Yard	m	0.914 4	1.093 6
Mass (weight)			
Grain	mg	64.799	0.015 4
Ounce (Avoirdupois)	g	28.350	0.035 3
Ounce (Troy)	g	31.103 5	0.032 15
Ton (short 2000 lb.)	kg	907.185	0.001 102
Ton (long 2240 lb.)	kg	1 016.047	0.000 984 2
Slug	kg	14.593 9	0.068 522
Pressure			
Bar	kPa	100.0	0.01
Inch of Water Column (20°C)	kPa	0.248 6	4.021 9
Inch of Mercury (20°C)	kPa	3.374 1	0.296 4
Kilogram-force per Centimeter Squared	kPa	98.067	0.010 2
Millimeters of Mercury (mm·Hg) (20°C)	kPa	0.132 84	7.528
Pounds Per Square Inch (P.S.I.)	kPa	6.894 8	0.145 0
Standard Atmosphere (760 torr)	kPa	101.325	0.009 869
Torr	kPa	0.133 32	7.500 6

UNITS

Appendix A (continued)
Selected Conversion Factors to SI Units

	SI Symbol	Multiplier to Convert From Existing Unit to SI Unit	Multiplier to Convert From SI Unit to Existing Unit
Power			
Btu (International) Per Hour	W	0.293 1	3.412 2
Foot Pound-Force Per Second	W	1.355 8	0.737 6
Horsepower	kW	0.745 7	1.341
Meter Kilogram-Force Per Second	W	9.806 7	0.101 97
Tons of Refrigeration	kW	3.517	0.284 3
Torque			
Kilogram-Force Meter (kg·m)	N·m	9.806 7	0.101 97
Pound-Force Foot	N·m	1.355 8	0.737 6
Pound-Force Inch	N·m	0.113 0	8.849 5
Gram-Force Centimeter	mN·m	0.098 067	10.197
Temperature			
Fahrenheit	°C	$\frac{5}{9}$ (°F − 32)	($\frac{9}{5}$°C) + 32
Rankine	K	(°F + 459.67)$\frac{5}{9}$	(°C + 273.16)$\frac{9}{5}$
Velocity			
Foot Per Second	m/s	0.304 8	3.280 8
Mile Per Hour	m/s	0.447 04	2.236 9
	or	*or*	*or*
	km/h	1.609 34	0.621 4
Viscosity			
Centipoise	mPa·s	1.0	1.0
Centistoke	μm²/s	1.0	1.0
Volume (Capacity)			
Cubic Foot	*l* (dm³)	28.316 8	0.035 31
Cubic Inch	cm³	16.387 1	0.061 02
Cubic Yard	m³	0.764 6	1.308
Gallon (U.S.)	*l*	3.785	0.264 2
Ounce (U.S. Fluid)	ml	29.574	0.033 8
Pint (U.S. Fluid)	*l*	0.473 2	2.113
Quart (U.S. Fluid)	*l*	0.946 4	1.056 7
Volume Flow (Gas-Air)			
Standard Cubic Foot Per Minute	m³/s	0.000 471 9	2119.
	or	*or*	*or*
	l/s	0.471 9	2.119
	or	*or*	*or*
	ml/s	471.947	0.002 119
Standard Cubic Foot Per Hour	ml/s	7.865 8	0.127 133
	or	*or*	*or*
	μ*l*/s	7 866.	0.000 127
Volume Liquid Flow			
Gallons Per Hour (U.S.)	*l*/s	0.001 052	951.02
Gallons Per Minute (U.S.)	*l*/s	0.063 09	15.850

Appendix B
Consistent Electric/Magnetic Units

Numbers in parentheses are the number of ESU or EMU units per single SI unit, except for the permittivity and permeability of free space, where actual values of ε_o and μ_o are given.

Variable	Symbol	SI System	ESU System	EMU System
length	l	meter	cm (100)	cm (100)
mass	m	kg	g (1000)	g (1000)
time	t	sec	sec	sec
force	F	newton	dyne (EE5)	dyne (EE5)
work, energy	W, E	joule	erg (EE7)	erg (EE7)
power	P	watt	erg/sec (EE7)	erg/sec (EE7)
charge	q	coulomb	statcoulomb (3 EE9)	abcoulomb (EE − 1)
current	i	ampere	statcoulomb/sec (3 EE9)	abcoulomb/sec (EE − 1)
electric flux	ϕ	coulomb	statcoulomb (3 EE9)	abcoulomb (EE − 1)
electric flux density	D	coulomb/m^2	--- (12π EE5)	--- (4π EE − 5)
electric field intensity	E	volt/meter = newtons/coulomb	statvolt/cm (3.33 EE − 5)	abvolt/cm (EE6)
electric potential	V	volt	statvolt (3.33 EE − 3)	abvolt (EE8)
resistance	R	ohm	statohm (1.11 EE − 12)	abohm (EE9)
capacitance	C	farad	statfarad (9 EE11)	abfarad (EE − 9)
magnetic pole strength	m	weber	--- ($\frac{1}{12\pi}$ EE2)	--- $\left(\frac{\text{EE8}}{4\pi}\right)$
magnetic flux	ϕ	weber	statweber (3.33 EE − 3)	maxwell (EE8)
magnetic flux density	B	tesla = wb/m^2	--- (3.33 EE − 7)	gauss (EE4)
magnetic field intensity	H	amp-turn/meter = N/wb	--- (12π EE7)	oersted (4π EE − 3)
magnetomotive force	MMF	amp-turn	--- (12π EE9)	gilbert (4π EE − 1)
inductance	L	henry = wb/amp	abhenry (1.11 EE − 12)	stathenry (EE9)
reluctance	$\mathcal{R}$	amp-turn/weber	--- (36π EE11)	gilbert/maxwell (4π EE − 9)
permittivity of free space	ε_o	$\frac{1}{(36\pi\ \text{EE9})}\ \frac{\text{farad*}}{\text{meter}}$	$1\ \frac{\text{statfarad}}{\text{cm}}$	$\frac{1}{(9\ \text{EE20})}\ \frac{\text{abfarad}}{\text{cm}}$
permeability of free space	μ_o	$4\pi\ \text{EE}-7\ \frac{\text{henry**}}{\text{meter}}$	$\frac{1}{9\ \text{EE20}}\ \frac{\text{stathenry***}}{\text{cm}}$	$1\ \frac{\text{abhenry}}{\text{cm}}$

* same as coul^2/n-m^2

** same as (unit-poles)^2/n-m^2 and webers/amp-turn-meter

*** same as (sec/cm)^2

UNITS

There are no practice problems for this chapter.

4 Fluid Statics and Dynamics

FLUIDS

Nomenclature

a	acceleration	ft/sec^2
bhp	brake horsepower	hp
A	area	ft^2
c	speed of sound in fluid	ft/sec
C	compressibility, Hazen-Williams constant, or coefficient	ft^2/lbf, –, –
d	depth, diameter	ft
D	diameter	ft
ehp	electrical horsepower	hp
E	bulk modulus, energy	lbf/ft^2, ft-lbf
f	Darcy friction factor	–
fhp	friction horsepower	hp
F	force	lbf
F_{va}	velocity of approach factor	–
g	local gravitational acceleration	ft/sec^2
g_c	gravitational constant (32.2)	lbm-ft/lbf-sec^2
G	mass flow rate per unit area	lbm/sec-ft^2
h	fluid height, head, depth	ft
H	total head	ft
I	moment of inertia	ft^4
k	ratio of specific heats	–
K	minor loss coefficient	–
L	length of pipe, lift	ft, lbf
m	mass	lbm
$\dot{m}$	mass flow rate	lbm/sec
n	rotational speed	rpm
n_s	specific speed	–
N_{Fr}	Froude number	–
N_{Re}	Reynolds number	–
N_W	Weber number	–
p	pressure	lbf/ft^2
P	power	ft-lbf/sec
Q	flow rate	gpm
r	radius	ft
r_h	hydraulic radius	ft
R	specific gas constant	ft-lbf/lbm-°R
s	length	ft
S.G.	specific gravity	–
t	time	sec
T	absolute temperature	°R
v	velocity	ft/sec
V	volume	ft^3
$\dot{V}$	volumetric flow rate	ft^3/sec
w	weight	lbf
whp	water horsepower	hp
x	x-coordinate	ft
y	distance, y-coordinate	ft
z	height above datum	ft

Symbols

β	contact angle, beta ratio	°
γ	specific weight	lbf/ft^3
ϵ	specific roughness	ft
η	efficiency	–
θ	angle	°
μ	absolute viscosity	lbf-sec/ft^2
ν	kinematic viscosity	ft^2/sec
ρ	density	lbm/ft^3
τ	shear stress	lbf/ft^2
T	surface tension	lbf/ft
υ	specific volume	ft^3/lbm
ø	angle, deflection angle	°
ω	rotational speed	rad/sec

Subscripts

a	atmospheric
A	added
b	blade
c	centroid, contraction
d	discharge
D	drag
e	equivalent, entrance
E	English, extracted
f	friction, flow
i	inside, inlet
j	jet
k	kinetic
L	lift

m	manometer fluid, metacentric, model, motor
M	metric
n	nozzle
o	outside, outlet
p	static pressure, pump
r	ratio
R	resultant
s	stagnation
STP	standard temperature and pressure
t	total, tank, true
v	velocity
vp	vapor pressure

FLUIDS

PART 1: Fluid Properties

Fluids are generally divided into two categories: ideal and real. Ideal fluids are those which have zero viscosity and shearing forces, are incompressible, and have uniform velocity distributions when flowing.

Real fluids are divided into Newtonian and non-Newtonian fluids. Newtonian fluids are typified by gases, thin liquids, and most fluids having simple chemical formulas. Non-Newtonian fluids are typified by gels, emulsions, and suspensions. Both Newtonian and non-Newtonian fluids exhibit finite viscosities and non-uniform velocity distributions. However, Newtonian fluids exhibit viscosities which are independent of the rate of change of shear stress while non-Newtonian fluids exhibit viscosities dependent on the rate of change of shear stress.

Most fluid problems assume Newtonian fluid characteristics.

1. Fluid Density

Most fluid flow calculations are based on an inconsistent system of units which measures density in pounds-mass per cubic foot. That convention is followed in this chapter when the symbol ρ is employed.

$$\rho = \text{fluid density in lbm/ft}^3 \qquad 4.1$$

Hydrostatic pressure and energy conservation equations given in this chapter require a standard local gravity of 32.2 ft/sec^2. However, a short discussion of situations with non-standard gravity is given at the ends of parts 2 and 4 in this chapter.

The density of a fluid in liquid form is usually given, known in advance, or easily obtained from a table (see end of this chapter). The density of a gas may be found from the following formula, which has been derived from the ideal gas law:

$$\rho = \frac{p}{RT} \qquad 4.2$$

2. Specific Volume

Specific volume is the volume occupied by a pound of fluid. It is the reciprocal of the density.

$$\upsilon = 1/\rho \qquad 4.3$$

3. Specific Gravity

Specific gravity is the ratio of a fluid's density to some specified reference density. For liquids, the reference density is the density of pure water. There is some confusion about this reference since the density of water varies with temperature, and various reference temperatures have been used (e.g., 39°F, 60°F, 70°F, etc.).

Strictly speaking, specific gravity of a liquid cannot be given without specifying the reference temperature at which the water's density was evaluated. However, the reference temperature is often omitted since water's density is fairly constant over the normal ambient temperature range. Using three significant digits, this reference density is 62.4 lbm/ft^3.

$$\text{S.G.}_{\text{liquid}} = \frac{\rho}{62.4} \qquad 4.4$$

Specific gravities of petroleum products and aqueous acid solutions may be found from hydrometer readings. There are two basic hydrometer scales. The Baumé scale has been widely used in the past. However, the API (American Petroleum Institute) scale is now recommended for use with all liquids.

For liquids lighter than water, specific gravity may be found from the Baumé hydrometer reading:

$$\text{S.G.} = \frac{140.0}{130.0 + {}^\circ\text{Baumé}} \qquad 4.5$$

For liquids heavier than water, specific gravity may be found from the Baumé hydrometer reading:

$$\text{S.G.} = \frac{145.0}{145.0 - {}^\circ\text{Baumé}} \qquad 4.6$$

The modern API scale may be used with all liquids:

$$\text{S.G.} = \frac{141.5}{131.5 + {}^\circ\text{API}} \qquad 4.7$$

Specific gravities may also be given for gases. The reference density is the density of air at specified conditions of pressure and temperature. The density of air evaluated at STP is approximately .075 lbm/ft³. Therefore,

$$S.G._{gas} = \frac{\rho_{STP}}{.075} \qquad 4.8$$

If the gas and air densities are both evaluated at the same temperature and pressure, the specific gravity is the inverse ratio of specific gas constants.

$$S.G._{gas} = \frac{R_{air}}{R_{gas}} = \frac{53.3}{R_{gas}} \qquad 4.9$$

Example 4.1

Determine the specific gravity of carbon dioxide (150°F, 20 psia) using STP air as a reference.

The specific gas constant for carbon dioxide is approximately 35.1 ft-lbf/lbm-°R. Using equation 4.2, the density is

$$\rho = \frac{(20)(144)}{(35.1)(150+460)} = .135 \text{ lbm/ft}^3$$

From equation 4.8,

$$S.G. = \frac{.135}{.075} = 1.8$$

4. Viscosity

Viscosity of a fluid is a measure of its resistance to flow. Consider two plates separated by a viscous fluid with thickness equal to y. The bottom plate is fixed. The top plate is kept in motion at a constant velocity v by a constant force F.

Experiments with Newtonian fluids have shown that the force required to maintain the velocity is proportional to the velocity and inversely proportional to the separation of the plates. That is,

$$\frac{F}{A} \propto \frac{dv}{dy} \qquad 4.10$$

The constant of proportionality is known as the *absolute*[1] viscosity. Recognizing that the quantity (F/A) is the fluid shear stress allows the following equation to be written.

$$\tau = \mu\frac{dv}{dy} \qquad 4.11$$

Another quantity using the name *viscosity* is the combination of units given by equation 4.12. This combination of units, known as the *kinematic* viscosity, appears sufficiently often in fluids problems to warrant its own symbol and name.

$$\nu = \frac{\mu g}{\rho} \qquad 4.12$$

There are a number of different units used to measure viscosity. Table 4.1 lists the most common units used in the English and SI systems.

Table 4.1 Typical Viscosity Units

	Absolute	Kinematic
English	lbf-sec/ft² (slug/ft-sec)	ft²/sec
Conventional Metric	dyne-sec/cm² (poise)	cm²/sec (stoke)
SI	Pascal-second (N·s/m²)	m²/s

Conversions between the two types of viscosities and between the English and various metric systems may be accomplished with table 4.2.

[1]Another name for *absolute* viscosity is *dynamic* viscosity. The term *absolute* is preferred.

Table 4.2 Viscosity Conversions

To Obtain	Multiply	By	and Divide By
ft²/sec	lb-sec/ft²	32.2	density
ft²/sec	stokes	1.076 EE−3	1
lb-sec/ft²	ft²/sec	density	32.2
lb-sec/ft	poise	1	478.8
m²/s	centistokes	1 EE−6	1
m²/s	stokes	1 EE−4	1
m²/s	ft²/sec	9.29 EE−2	1
pascal-sec	centipoise	1 EE−3	1
pascal-sec	lbm/ft-sec	1.488	1
pascal-sec	lbf-sec/ft²	47.88	1
pascal-sec	poise	.1	1
pascal-sec	slug/ft-sec	47.88	1
poise	lb-sec/ft²	478.8	1
poise	stokes	specific gravity	1
stokes	ft²/sec	929	1
stokes	poise	1	specific gravity

FLUIDS

Example 4.2

Water at 60°F has a specific gravity of 999 and a kinematic viscosity of 1.12 centistokes. What is the absolute viscosity in lbf-sec/ft^2?

$$\nu_M = 1.12/100 = .0112 \text{ stokes}$$

$$\mu_M = (.0112)(.999) = .01119 \text{ poise}$$

$$\mu_E = .01119/478.8 = 2.34 \text{ EE}-5 \text{ lbf-sec/ft}^2$$

Viscosity may also be measured by a viscometer. A viscometer is essentially a container which allows the fluid to leak out through a small hole. The more viscous the fluid, the more time will be required to leak out a given quantity. Viscosity measured in this indirect manner has the units of seconds. Seconds Saybolt Universal (SSU) and Seconds Saybolt Furol (SSF) are two systems of indirect viscosity measurement.

In liquids, molecular cohesion is the dominating cause of viscosity. As the temperature of a liquid increases, these cohesive forces decrease, resulting in an absolute viscosity decrease.

In gases, the dominating cause of viscosity is random collisions between gas molecules. This molecular agitation increases with increases in temperature. Therefore, viscosity in gases increases with temperature.

The absolute viscosity of both gases and liquids is essentially independent of changes in pressure. Of course, kinematic viscosity is greatly dependent on both temperature and pressure since these variables affect density.

5. Vapor Pressure

Molecular activity in a liquid will tend to free some surface molecules. This tendency toward vaporization is dependent on temperature. The partial pressure exerted by these free molecules at the surface is known as the vapor pressure. Boiling occurs when the vapor pressure is increased (by increasing the fluid temperature) to the local ambient pressure. Thus, a liquid's boiling point depends on both the temperature and the external pressure. Liquids with low vapor pressures are used in accurate barometers.

Vapor pressure is a function of temperature only. Typical values are given in table 4.3.

Table 4.3 Typical Vapor Pressures at 68°F

Fluid	Vapor Pressure
Ethyl alcohol	122.4 psf
Turpentine	1.115
Water	48.9
Ether	1231.
Mercury	.00362

6. Surface Tension

The skin which seems to form on the free surface of a fluid is due to the intermolecular cohesive and adhesive forces known as surface tension. Surface tension is the amount of work required to form a new unit of surface area. The units are, therefore, ft-lbf/ft^2 or just lbf/ft.

Surface tension can be measured as the tension between two points a foot in separation on the surface. It decreases with increases in temperature and depends on the gas in contact with the free surface. Surface tension values are usually quoted for air contact.

Table 4.4 Typical Surface Tensions (68°F, air contact)

Fluid	T
ethyl alcohol	.001527 lbf/ft
turpentine	.001857
water	.004985
mercury	.03562
n-octane	.00144
acetone	.00192
benzene	.00192
carbon tetrachloride	.00180

The relationship between surface tension and the pressure in a bubble surrounded by gas is given by equation 4.13. r is the radius of the bubble.

$$T = \tfrac{1}{4}r(p_{inside} - p_{outside}) \qquad 4.13$$

The surface tension in a full spherical droplet or in a bubble in a liquid is given by equation 4.14

$$T = \tfrac{1}{2}r(p_{inside} - p_{outside}) \qquad 4.14$$

7. Capillarity

Surface tension is the cause of capillarity which occurs whenever a liquid comes into contact with a vertical solid surface. In water, adhesive forces dominate. They cause water to readily attach itself to a vertical surface, to climb the wall. In a thin-bore tube, water will rise above the general level as it tries to wet the interior surface.

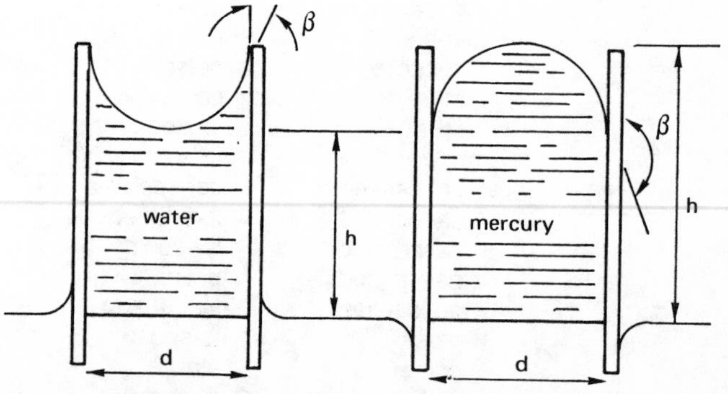

Figure 4.1 Capillarity in Thin-Wall Tubes

FLUIDS

On the other hand, cohesive forces dominate in mercury, since mercury molecules have a great affinity for one another. The curved surface called the *meniscus* formed inside a thin-bore tube inserted into a container of mercury will be below the general level.

Whether adhesive or cohesive forces dominate can be determined by the angle of contact, β, as shown in figure 4.1. For a contact angle less than 90°, adhesive forces dominate. Typical values of β are given in table 4.5.

If the tube diameter is less than .1″, the surface tension inside a circular capillary tube can be approximated by equation 4.15. The meniscus is assumed spherical with radius r. Equation 4.15 may also be used to estimate the capillary rise in a capillary tube, as illustrated in figure 4.1.

$$T = \frac{h\rho d}{4\cos\beta} \qquad 4.15$$

$$r = \frac{d}{2\cos\beta} \qquad 4.16$$

Table 4.5 may be used to determine T and β for various combinations of contacting liquids.

8. Compressibility

Usually fluids are considered to be incompressible. However, fluids are actually somewhat compressible. Compressibility is the percentage change in a unit volume per unit change in pressure.

$$C = \frac{\Delta V/V}{\Delta p} \qquad 4.17$$

9. Bulk Modulus

The bulk modulus of a liquid is the reciprocal of the compressibility.

$$E = \frac{1}{C} \qquad 4.18$$

The bulk modulus of an ideal gas is given by equation 4.19, where k is the ratio of specific heats. k is equal to 1.4 for air.

$$E = kp \qquad 4.19$$

10. Speed of Sound

The speed of sound in a pure liquid or gas is given by equations 4.20 and 4.21.

$$c_{liquid} = \sqrt{Eg_c/\rho} = \sqrt{g_c/C\rho} \qquad 4.20$$

$$c_{gas} = \sqrt{kg_cRT} = \sqrt{kpg_c/\rho} \qquad 4.21$$

The temperature term in equation 4.21 must be in degrees absolute.

Example 4.3

What is the velocity of sound in 150°F water?

From the tables at the end of this chapter, the density of water at 150°F is 61.2 lbm/ft^3. Similarly, the bulk modulus is 328 EE3 psi. From equation 4.20,

$$c = \sqrt{\frac{(328\text{ EE3})(144)(32.2)}{61.2}} = 4985 \text{ ft/sec}$$

Example 4.4

What is the velocity of sound in 150°F air at atmospheric pressure?

The specific gas constant for air is 53.3 ft-lbf/lbm-°R. Using equation 4.21,

$$c = \sqrt{(1.4)(32.2)(53.3)(150+460)}$$
$$= 1210.7 \text{ ft/sec}$$

Table 4.5 Capillary Constants

Combination	Surface Tension, T	Contact Angle, β
Mercury-vacuum-glass	3.29 EE−2 lbf/ft	140°
Mercury-air-glass	2.02 EE−2	140°
Mercury-water-glass	2.60 EE−2	140°
Water-air-glass	5.00 EE−3	0°

FLUIDS

PART 2: Fluid Statics

1. Measuring Pressures

The value of pressure, regardless of the device used to measure it, is dependent on the reference point chosen. Two such reference points exist: zero absolute pressure and standard atmospheric pressure.

If standard atmospheric pressure (approximately 14.7 psia) is chosen as the reference, pressures are known as *gage* pressures. Positive gage pressures are always pressures above atmospheric pressure. Vacuum (negative gage pressure) is the pressure below atmospheric. Maximum vacuum, according to this convention, is −14.7 psig. The term *gage* is somewhat misleading, since a mechanical gauge may not be used to measure gage pressures.

If zero absolute pressure is chosen as the reference, the pressures are known as *absolute* pressures. The barometer is a common device for measuring the absolute pressure of the atmosphere. It is constructed by filling a long, hollow tube open at one end with mercury, and inverting it such that the open end is below the level of a mercury filled container. If the vapor pressure is neglected, the mercury will be supported only by the atmospheric pressure transmitted through the container fluid at the lower, open end. The equation balancing the weight of the fluid against the atmospheric force is:

$$p_a = .491(h)(144) \qquad 4.22$$

h is the height of the mercury column in inches and .491 is the density of mercury in pounds per cubic inch.

Any fluid may be used to measure atmospheric pressure, although vapor pressure may be significant. For any fluid used in a barometer,

$$p_a = [(.0361)(\text{S.G.})(h) + p_v](144) \qquad 4.23$$

.0361 is the density of water in pounds per cubic inch. p_v should be given in psi in equation 4.23.

Example 4.5

A vacuum pump is used to drain a flooded mine shaft of 68°F water. The pump is incapable of lifting the water beyond 400 inches. What is the atmospheric pressure?

From table 4.3, the vapor pressure of 68°F water is $\dfrac{48.9}{144} = .34$ psi.

Then, the atmospheric pressure is

$$p_a = [(.0361)(1)(400) + .34](144) = 2128.3 \text{ psf}$$

This is 14.78 psia.

2. Manometers

Manometers are frequently used to measure pressure differentials. Figure 4.2 shows a simple U-tube manometer whose ends are connected to two pressure vessels. Often, one end will be open to the atmosphere, which then determines that end's pressure.

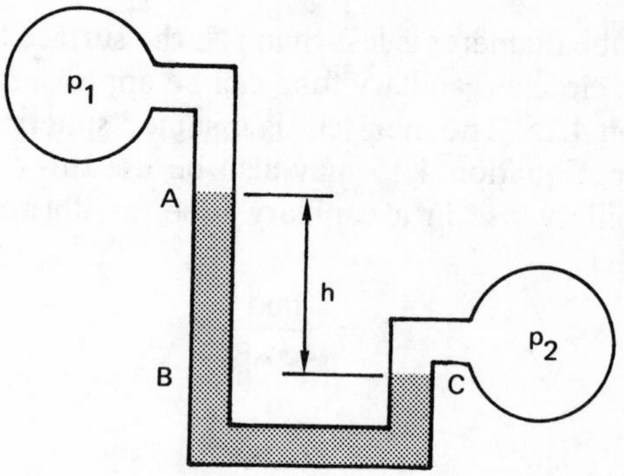

Figure 4.2 A Simple Manometer

Since the pressure at point B is the same as at point C, the pressure differential produces the fluid column of height h.

$$\Delta p = p_2 - p_1 = \rho_m h \qquad 4.24$$

Equation 4.24 assumes that the manometer is small and that only low density gases fill the tubes above the measuring fluid. If a high density fluid (such as water) is present above the measuring fluid, or if the gas columns h_1 or h_2 are very long, corrections must be made:

$$\Delta p = \rho_m h + \rho_1 h_1 - \rho_2 h_2 \qquad 4.25$$

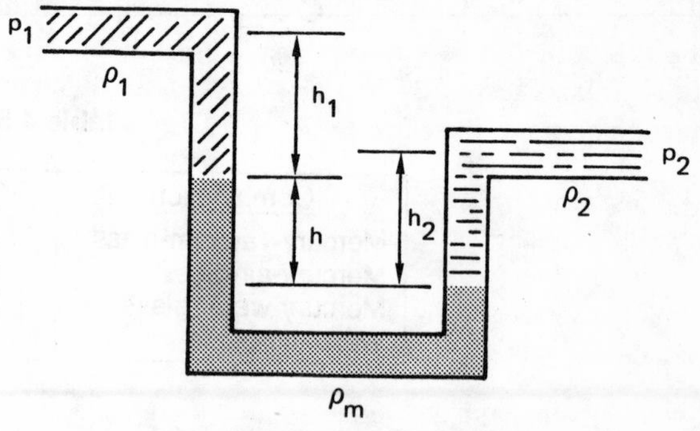

Figure 4.3 A Manometer Requiring Corrections

Corrections for capillarity are seldom needed since manometer tubes are generally large in diameter.

Example 4.6

What is the pressure at the bottom of the water tank?

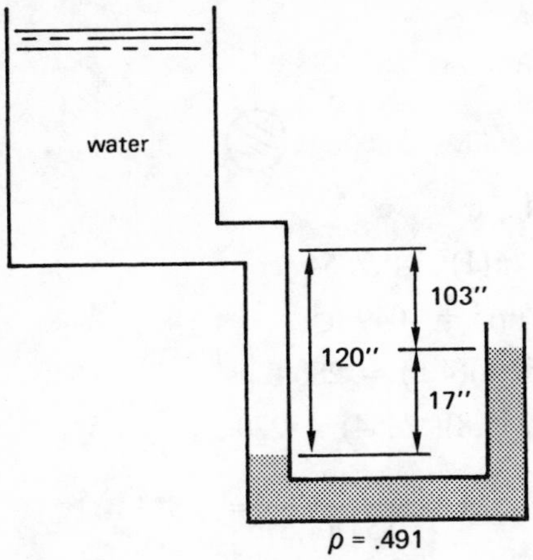

Using equation 4.25, the pressure differential is

$$\Delta p = p_{\text{tank bottom}} - p_a = (.491)(17) - (.0361)(120)$$
$$= 8.347 - 4.332$$
$$= 4.015 \text{ psig}$$

The third term in equation 4.25 was omitted because the density of air is much smaller than that of water or mercury.

3. Hydrostatic Pressure Due to Incompressible Fluids

Hydrostatic pressure is the pressure which a fluid exerts on an object or container walls. It always acts through the center of pressure and is normal to the exposed surface, regardless of the object's orientation or shape. It varies linearly with depth and is a function of depth and density only.

A. Force on a Horizontal Plane Surface

In the case of a horizontal surface, such as the bottom of a container, the pressure is uniform and the center of pressure corresponds to the centroid of the plane surface. The gage pressure is

$$p = \rho h \qquad 4.26$$

The total vertical force on the horizontal plane is

$$R = pA \qquad 4.27$$

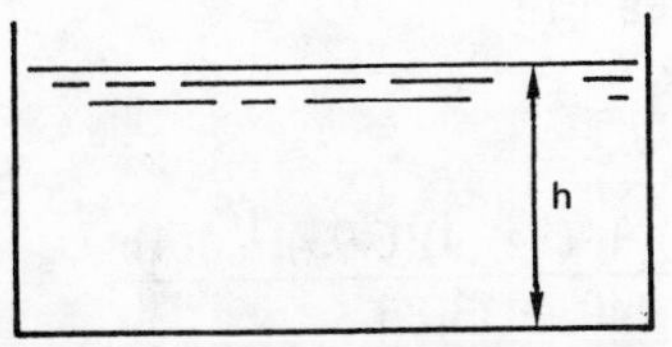

Figure 4.4 Horizontal Plane Surface

B. Vertical and Inclined Rectangular Plane Surfaces

If a rectangular plate is vertical or inclined within a fluid body, the linear variation in pressure with depth is maintained. The pressures at the top and bottom of the plate are

$$p_1 = \rho h_1 \sin\theta \qquad 4.28$$

$$p_2 = \rho h_2 \sin\theta \qquad 4.29$$

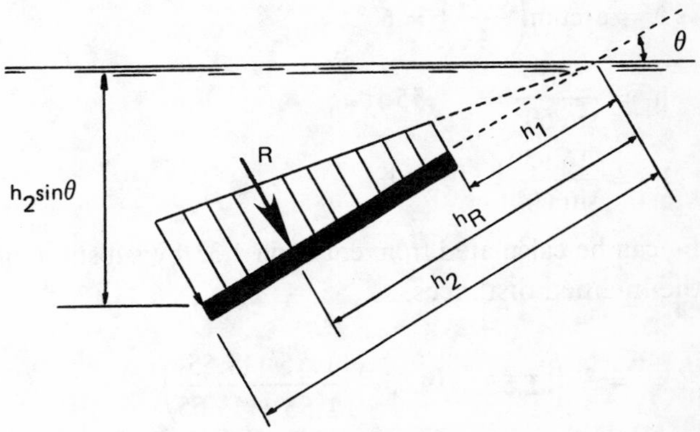

Figure 4.5 Inclined Rectangular Plane

The average pressure occurs at the average depth $(\frac{1}{2})(h_1+h_2)\sin\theta$. The average pressure over the entire vertical or inclined surface is

$$\bar{p} = \tfrac{1}{2}\rho(h_1+h_2)\sin\theta \qquad 4.30$$

The total resultant force on the inclined plane is

$$R = \bar{p}A \qquad 4.31$$

The center of pressure is not located at the average depth, but is located at the centroid of the triangular or trapezoidal pressure distribution. That depth is

$$h_R = \tfrac{2}{3}\left[h_1 + h_2 - \frac{h_1 h_2}{(h_1+h_2)}\right] \qquad 4.32$$

If the object is inclined, h_R must be measured parallel to the object's surface (e.g., an inclined length).

Example 4.7

The tank shown below is filled with water. What is the force on a one-foot length of the inclined portion of the wall? Where is the resultant located on the inclined section?

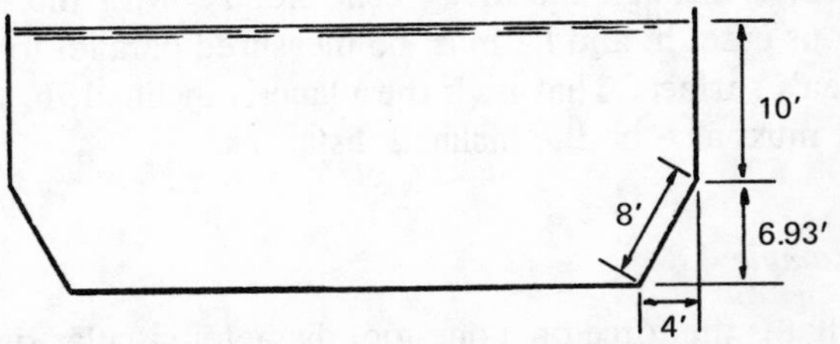

The sinθ terms in equations 4.28, 4.29, and 4.30 convert the inclined distances to vertical distances. Therefore, the sinθ terms may be omitted if the vertical distances are known. The average pressure on the inclined section is

$$\bar{p} = \tfrac{1}{2}(62.4)(10 + 16.93) = 840.2 \text{ psf}$$

The total force is $R = (840.2)(8)(1) = 6721.6$ lbf

In order to determine h_R, θ must be known in order to calculate h_1 and h_2.

$$\theta = \arctan\left(\frac{6.93}{4}\right) = 60°$$

$$h_1 = \frac{10}{\sin 60°} = 11.55 \text{ ft}$$

$$h_2 = \frac{16.93}{\sin 60°} = 19.55 \text{ ft}$$

h_R can be calculated from equation 4.32 by substituting the inclined distances.

$$h_R = \tfrac{2}{3}\left(11.55 + 19.55 - \frac{(11.55)(19.55)}{11.55 + 19.55}\right)$$

$$= 15.89 \text{ ft} \quad \text{inclined}$$

C. General Plane Surface

For any non-rectangular plane surface, the average pressure depends on the location of the surface's centroid, h_c.

$$\bar{p} = \rho h_c \qquad 4.33$$

$$R = \bar{p}A \qquad 4.34$$

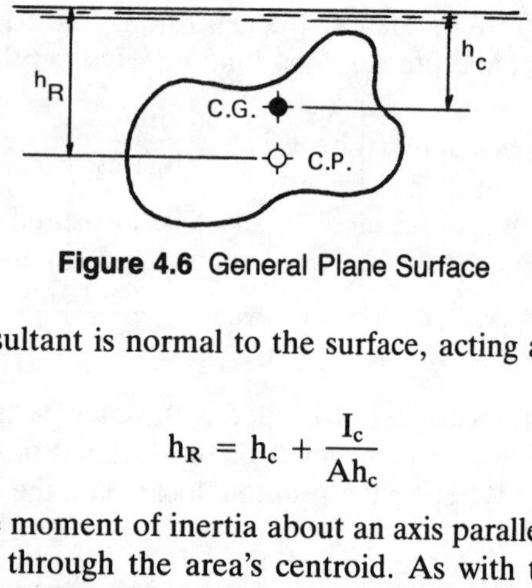

Figure 4.6 General Plane Surface

The resultant is normal to the surface, acting at depth h_R.

$$h_R = h_c + \frac{I_c}{Ah_c} \qquad 4.35$$

I_c is the moment of inertia about an axis parallel to the surface through the area's centroid. As with the previous case, h_c and h_R must be measured parallel to the area's surface. That is, if the plane is inclined, h_c and h_R must also be the inclined distances.

Example 4.8

What is the force on a one-foot diameter circular sighthole whose top edge is located 4′ below the water surface? Where does the resultant act?

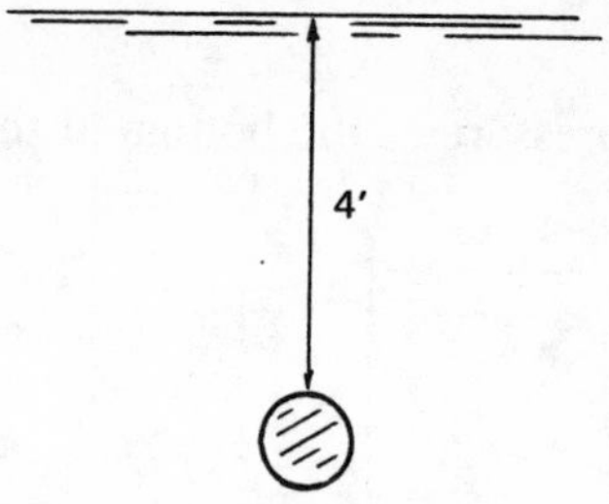

$$h_c = 4.5 \text{ ft}$$

$$A = \tfrac{1}{4}\pi(1)^2 = .7854 \text{ ft}^2$$

$$I_c = \tfrac{1}{4}\pi r^4 = .049 \text{ ft}^4$$

$$\bar{p} = (62.4)(4.5) = 280.8 \text{ psf}$$

$$R = (280.8)(.7854) = 220.5 \text{ lb}$$

$$h_R = 4.5 + \frac{.049}{(.7854)(4.5)} = 4.514 \text{ ft}$$

D. Curved Surfaces

The horizontal component of the resultant force acting on a curved surface can be found by the same method used for a vertical plane surface. The vertical component of force on an area will usually equal the weight of the liquid above it. In figure 4.7, the vertical force on length AB is the weight of area ABCD, with a line of action passing through the centroid of the area ABCD.

The resultant magnitude and direction may be found from conventional component composition.

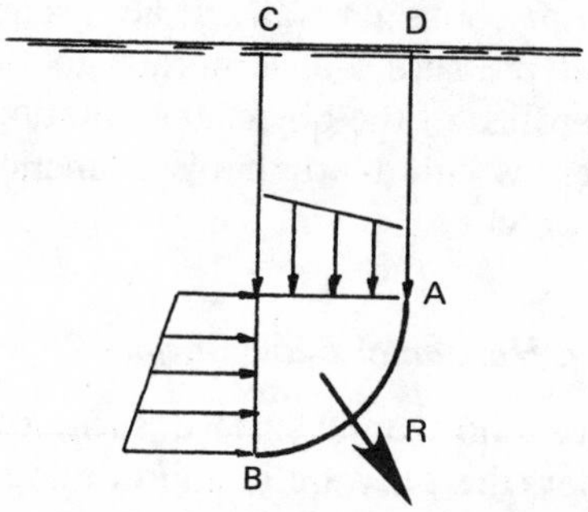

Figure 4.7 Forces on Curved Surface

Example 4.9

What is the total force on a one-foot section of the wall described in example 4.7?

The centroid of section ABCD (with point B serving as the reference) is

$$\bar{x} = \frac{\Sigma A_i \bar{x}_i}{\Sigma A_i} =$$

$$\frac{(4)(10)(2) + (\tfrac{1}{2})(4)(6.93)(\tfrac{1}{3})(4)}{40 + 13.86}$$

$$= 1.83$$

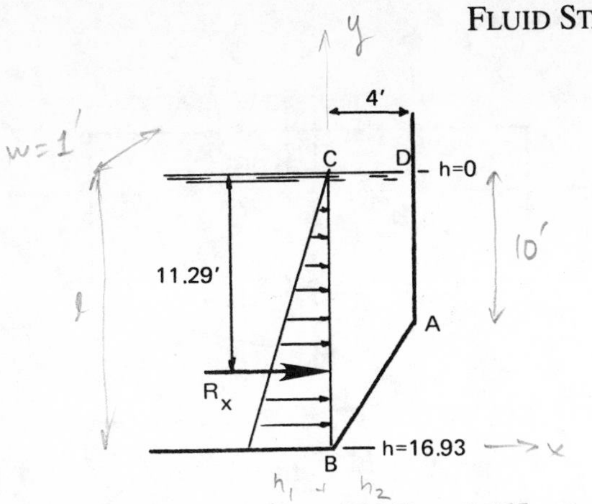

The average depth is ½(0 + 16.93) = 8.465

Using equations 4.30 and 4.31, the average pressure and horizontal component of the resultant are

$$\bar{p} = 62.4(8.465) = 528.2 \text{ psf}$$

$$R_x = (16.93)(1)(528.2) = 8942.4 \text{ lb}$$

From equation 4.32, the horizontal component acts (⅔)(16.93) = 11.29 ft from the top.

The volume of a one foot section of area ABCD is

$$(1)\left[(4)(10) + \tfrac{1}{2}(4)(6.93)\right] = 53.86 \text{ ft}^3$$

Therefore, the vertical component is

$$R_y = (62.4)(53.86) = 3360.9 \text{ lb}$$

The resultant of R_x and R_y is

$$R = \sqrt{(8942.4)^2 + (3360.9)^2} = 9553.1 \text{ lb}$$

$$\phi = \arctan\left(\frac{3360.9}{8942.4}\right) = 20.6$$

In general, it is not correct to calculate the vertical component of force on a submerged surface as being the weight of the fluid above it, as was done in example 4.9. This procedure is valid only when there is no change in the cross section of the tank area.

The *hydrostatic paradox* is illustrated by figure 4.8. The pressure anywhere on the bottom of either container is the same. This pressure is dependent only on the maximum height of the fluid, not the volume.

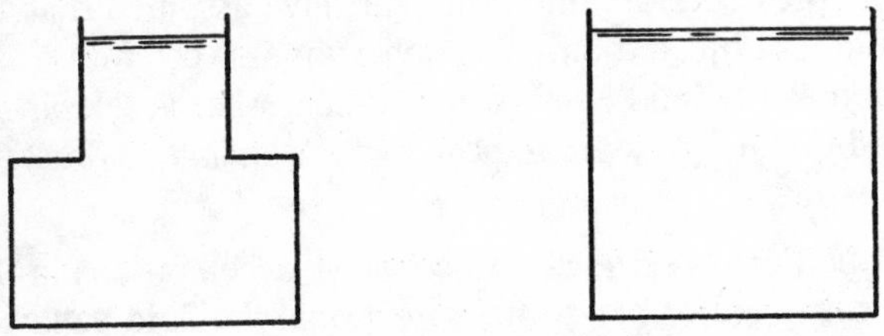

Figure 4.8 Hydrostatic Paradox

4. Hydrostatic Pressure Due to Compressible Fluids

Equation 4.26 is a special case of a more general equation known as the *Fundamental Equation of Fluid Statics*. This equation is presented below. As defined in the nomenclature, h is a variable representing height, and it is assumed that h_2 is greater than h_1. The minus sign in equation 4.36 indicates that pressure decreases when height increases.

$$\int_1^2 \frac{dp}{\rho} = -(h_2 - h_1) \qquad 4.36$$

If the fluid is a compressible layer of perfect gas, and if compression is assumed to be isothermal, equation 4.36 becomes

$$h_2 - h_1 = RT \ln\left(\frac{p_1}{p_2}\right) \qquad 4.37$$

The pressure at height h_2 in a layer of gas which has been isothermally compressed is

$$p_2 = p_1\left[e^{\frac{h_1 - h_2}{RT}}\right] \qquad 4.38$$

The following relationships assume a polytropic compression of the gas layer. These three relationships may be used for adiabatic compression by substituting *k* for *n*.

$$h_2 - h_1 = \frac{n}{n-1} RT_1\left[1 - (p_2/p_1)^{\frac{n-1}{n}}\right] \qquad 4.39$$

$$p_2 = p_1\left[1 - \left(\frac{n-1}{n}\right)\left(\frac{h_2 - h_1}{RT_1}\right)\right]^{\frac{n}{n-1}} \qquad 4.40$$

$$T_2 = T_1\left[1 - \left(\frac{n-1}{n}\right)\left(\frac{h_2 - h_1}{RT_1}\right)\right] \qquad 4.41$$

Example 4.10

The pressure at sea level is 14.7 psia. Assume 70°F isothermal compression and calculate the pressure at 5000 feet altitude.

R = 53.3 ft-lbf/lbm-°R for air. T = (70 + 460) = 530°R.

From equation 4.38,

$$p_{5000} = 14.7\left[e^{\frac{0-5000}{(53.3)(530)}}\right]$$

$$= 12.32 \text{ psia}$$

5. Buoyancy

The buoyancy theorem, also known as *Archimedes' principle,* states that the upward force on an immersed object is equal to the weight of the displaced fluid. A buoyant force due to displaced air is also relevant in the case of partially-submerged objects. For lighter-than-air crafts, the buoyant force results entirely from displaced air.

$$F_{\text{buoyant}} = \begin{pmatrix}\text{displaced}\\ \text{volume}\end{pmatrix}\begin{pmatrix}\text{density of}\\ \text{displaced fluid}\end{pmatrix} \qquad 4.42$$

In the case of floating or submerged objects not moving vertically, the buoyant force and weight are equal. If the forces are not in equilibrium, the object will rise or fall until some equilibrium is reached. The object will sink until it is supported by the bottom or until the density of the supporting fluid increases sufficiently. It will rise until the weight of the displaced fluid is reduced, either by a decrease in the fluid density or by breaking the surface.

Example 4.11

An empty polyethylene telemetry balloon with payload has a mass of 500 pounds. It is charged with helium when the atmospheric conditions are 60°F and 14.8 psia. What volume of helium is required for liftoff from a sea level platform? The specific gas constant of helium is 386.3 ft-lbf/lbm-°R.

The gas densities are

$$\rho_{air} = \frac{p}{RT} = \frac{(14.8)(144)}{(53.3)(60+460)} = .07689 \text{ lbm/ft}^3$$

$$\rho_{helium} = \frac{(14.8)(144)}{(386.3)(60+460)} = .01061 \text{ lbm/ft}^3$$

The total mass of the balloon, payload, and helium is

$$m = 500 + (.01061)\begin{pmatrix}\text{helium}\\ \text{volume}\end{pmatrix}$$

The buoyant force is the weight of the displaced air.

$$F = (.07689)\begin{pmatrix}\text{helium}\\ \text{volume}\end{pmatrix}$$

Equating F and m results in a helium volume of 7544 ft^3.

6. Stability of Floating Objects

The buoyant force on a floating object acts upward through the center of gravity of the displaced *volume,* known as the *center of buoyancy.* The weight acts downward through the center of gravity of the *object.* For totally submerged objects such as balloons and submarines, the center of buoyancy must be above the center of gravity for stability. For partially submerged vessels, the metacenter must be above the center of gravity. Stability exists because a righting moment is created if the vessel heels over, since the center of buoyancy moves outboard of the center of gravity.

Refer to figure 4.9. If the floating body heels through an angle ϕ, the location of the center of gravity does not change. However, the center of buoyancy will shift to the center of gravity of the new submerged section *123.* The centers of buoyancy and gravity are no longer in line. The couple thus formed tends to resist further overturning.

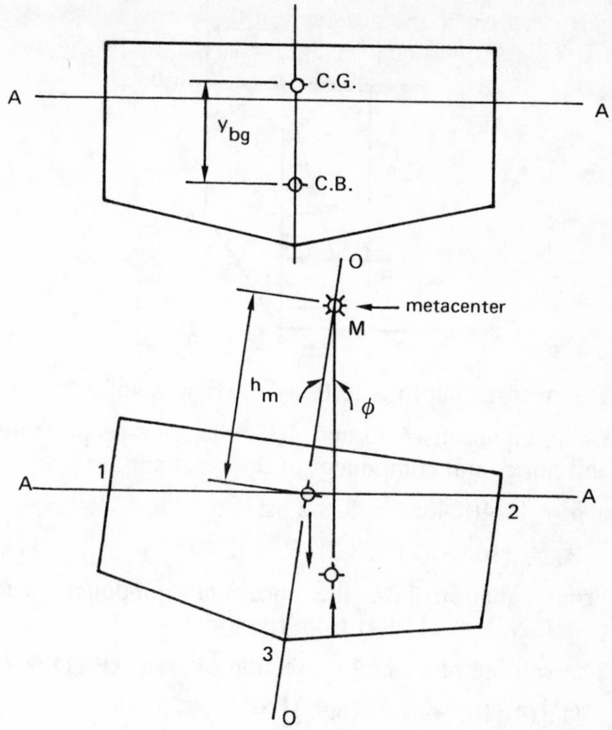

Figure 4.9 Locating the Metacenter

This righting couple exists when the extension of the buoyant force F intersects line O-O above the center of gravity at M, the *metacenter.* If M lies below the center of gravity, an overturning couple will exist. The distance between the center of gravity and the metacenter is called the *metacentric height,* and it is reasonably constant for heel angles less than 10 degrees.

The metacentric height, h_m, can be found from equation 4.43 where I is the moment of inertia of the submerged portion about line A-A, and V is the displaced volume.

$$h_m = \frac{I}{V} \pm y_{bg} \qquad 4.43$$

7. Fluid Masses Under Acceleration

The pressures obtained thus far have assumed that the fluid is subjected only to gravitational acceleration. As soon as the fluid is subjected to any other acceleration, additional forces are imposed which change hydrostatic pressures.

If the fluid is subjected to constant accelerations in the vertical and/or horizontal directions, the fluid behavior will be given by equations 4.44 and 4.45.

$$\theta = \arctan\left[\frac{a_x}{a_y + g}\right] \qquad 4.44$$

$$p_h = \rho h\left(1 + \frac{a_y}{g}\right) \qquad 4.45$$

a_y is negative if the acceleration is downward. Notice that a plane of equal pressure is also inclined if the fluid mass experiences a horizontal acceleration.

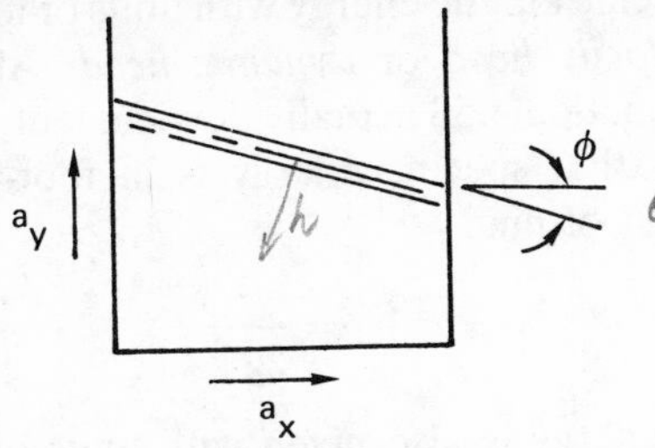

Figure 4.10 Constant Linear Acceleration

If the fluid mass is rotated about a vertical axis, a parabolic fluid surface will result. The distance *h* is measured from the lower-most part of the fluid during rotation. *h* is not measured from the original level of the stationary fluid. *h* is the height of the fluid at a distance *r* from the center of rotation.

$$\theta = \arctan\left(\frac{\omega^2 r}{g}\right) \qquad 4.46$$

$$h = \frac{(\omega r)^2}{2g} = \frac{v^2}{2g} \qquad 4.47$$

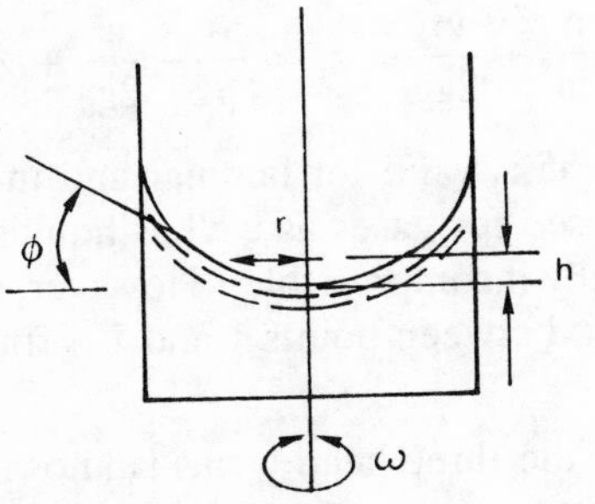

Figure 4.11 Constant Rotational Acceleration

8. Modifications for Other Gravities

All of the equations in Part 2 of this chapter may be used in a standard gravitational field of 32.2 ft/sec^2. In such a standard field, the terms *lbm* and *lbf* may be freely cancelled. However, modifications to the formulas are needed if the local gravity deviates from standard.

The modification that must be made is that of converting the mass density in lbm/ft^3 to a *specific weight* in lbf/ft^3. This can be done through the use of equation 4.48.

$$\gamma = \frac{\rho g_{local}}{g_c} \qquad 4.48$$

g_{local} is the actual gravitational acceleration in ft/sec^2. g_c is the dimensional conversion factor presented in chapter 3. g_c is approximately equal to 32.2 lbm-ft/lbf-sec^2.

The various hydrostatic pressure formulas may be used in non-standard gravitational fields by substituting γ for ρ.

Example 4.12

What is the maximum height that a vacuum pump can lift 60°F water if the atmospheric pressure is 14.6 psia and the local gravity is 28 ft/sec^2?

At 60°F, the vapor pressure of water is .2563 psia. The effective pressure which can be used to lift water is (14.6 − .2563) = 14.3437 psi.

$$p = \left(\frac{g}{g_c}\right)\rho h$$

$$h = \frac{(14.3437)(144)\left(\frac{32.2}{28.0}\right)}{62.4} = 38.07 \text{ ft}$$

FLUIDS

PART 3: Fluid Flow Parameters

1. Introduction

Pressure is commonly measured in pounds per square inch (psi) or pounds per square foot (psf). However, pressure may be changed into a new variable called *head* by dividing by the density of the fluid. This operation does more than just scale the pressure down by a factor equal to the reciprocal of the density. Since density itself possesses dimensional units, the units of head are not the same as the units of pressure.

$$(h \text{ in ft}) = \frac{(p \text{ in lbf/ft}^2)}{(\rho \text{ in lbm/ft}^3)} \qquad 4.49$$

Equation 4.49 is, of course, the same as equation 4.26. As long as the fluid density and local gravitational acceleration remain constant, there is complete interchangeability between the variables of pressure and head.

When Bernoulli's equation is introduced in this chapter, head will also be used as a measure of energy. Actually, head is used as a measure of *specific* energy. This is commonly justified as follows:

$$(h \text{ in ft}) = \frac{(E \text{ in ft-lbf})}{(m \text{ in lbm})} \qquad 4.50$$

A certain amount of care is required in the use of equations 4.49 and 4.50 since lb*f* is being cancelled completely by lb*m*. The actual operation being performed is given by equation 4.51.

$$(h \text{ in ft}) = \frac{(g_c \text{ in } \frac{\text{lbm-ft}}{\text{lbf-sec}^2})(p \text{ in lbf/ft}^2)}{(g \text{ in } \frac{\text{ft}}{\text{sec}^2})(\rho \text{ in } \frac{\text{lbm}}{\text{ft}^3})} \qquad 4.51$$

Since g_c is always equal to 32.2, it can be seen from equation 4.51 that equations 4.49 and 4.50 will give the correct numerical value for head as long as the local gravitational acceleration is 32.2 ft/sec².

2. Fluid Energy

A fluid can possess energy in three[2] forms — as pressure, kinetic, or potential energies. The energy (work) that must be put into a fluid to raise its pressure is known as the *pressure energy* or *static energy.* (This form of energy is also known as *flow work* or *flow energy.*) In keeping with the common convention to put fluid energy terms into units of feet, the *pressure head* or *static head* is defined by equation 4.52.

$$h_p = \frac{p}{\rho} \qquad 4.52$$

Energy is also required to accelerate a fluid to velocity v. The specific kinetic energy with units of feet is known as the *velocity head* or *dynamic head.* Although the units in equation 4.53 actually do yield feet, you should remember that specific energy is in foot-pounds per pound mass of fluid.

$$h_v = \frac{v^2}{2g_c} \qquad 4.53$$

Potential energy is also given with units of feet. This results in a very simple expression for *potential head* or *gravitational head.* z in equation 4.54 is the height of the fluid above some arbitrary reference point. Cancelling the weight (lbf) of the fluid by its mass (lbm) is acceptable under the constraints previously given.

$$z = \frac{wz}{m} \qquad 4.54$$

3. Bernoulli Equation

The Bernoulli equation is essentially an energy conservation equation. It states that the total energy of a fluid flowing without losses in a pipe cannot change. The total energy possessed by a fluid is the sum of its pressure, kinetic, and potential energies.

$$\frac{p_1}{\rho} + \frac{v_1^2}{2g_c} + z_1 = \frac{p_2}{\rho} + \frac{v_2^2}{2g_c} + z_2 \qquad 4.55$$

Equation 4.55 is valid for laminar and turbulent flow. It may be used for gases as well as liquids if the gases are essentially incompressible.[3] However, it is assumed that the flow between points 1 and 2 is frictionless and adiabatic.

The sum of the three head terms is known as the total head, H. The total pressure may be calculated from the total head.

$$H = \frac{p}{\rho} + \frac{v^2}{2g_c} + z \qquad 4.56$$

$$p_t = \rho H \qquad 4.57$$

The total energy of the fluid stream has two definitions. The total specific energy is the same as total head, H. The total energy of all of the fluid flowing can be calculated from equation 4.58.

$$E_t = mH \qquad 4.58$$

A graph of the total specific energy versus distance along a pipe is known as a total energy line. In a frictionless pipe without pumps or turbines, the total specific energy will remain constant. Total specific energy will decrease if fluid friction is present.

[2]Another important energy form is *thermal* energy. Thermal energy terms (internal energy and enthalpy) are not included in this analysis since it is assumed that the temperature of the fluid remains constant.

[3]A gas may be considered to be incompressible as long as its pressure does not change more than 10% between points 1 and 2, and its velocity is less than Mach .3 everywhere.

FLUIDS

Example 4.13

A pipe takes water from the reservoir as shown and discharges it freely 100 feet below. The flow is frictionless. (a) What is the total specific energy at point B? (b) What is the velocity at point C?

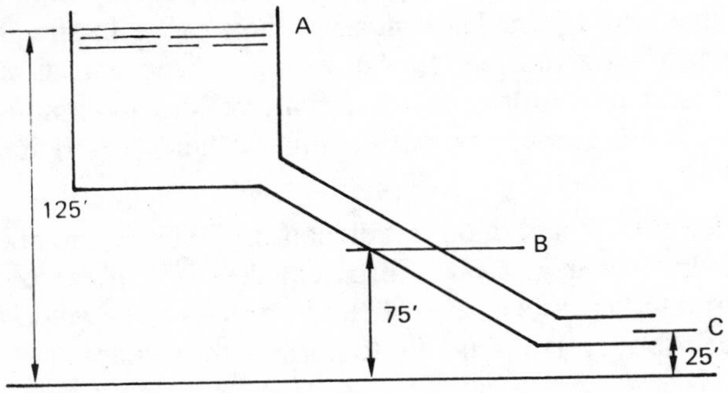

(a) At point A, the velocity and gage pressure are both zero. So, the total specific energy with respect to the reference line is

$$H_A = 0 + 0 + 125 = 125 \text{ ft}$$

At point B, the fluid is moving and possesses kinetic energy. The fluid is also under hydrostatic pressure. However, the flow is frictionless, and the total specific energy is constant (see equation 4.56). The velocity and static heads have increased at the expense of the potential head. Therefore, $H_B = H_A = 125$ ft.

(b) At point C, the pressure head is again zero since the discharge is at atmospheric pressure. The potential head with respect to the energy reference line is 25 ft. From equation 4.56,

$$125 = 0 + \frac{v^2}{2g_c} + 25$$

$$v^2 = (2)(32.2)(100)$$

$$v = 80.2 \text{ ft/sec}$$

Example 4.14

Water is pumped at a rate of 3 cfs through the piping system illustrated. If the pump has a discharge pressure of 150 psig, to what elevation can the tank be raised? Assume the head loss due to friction is 10 feet.

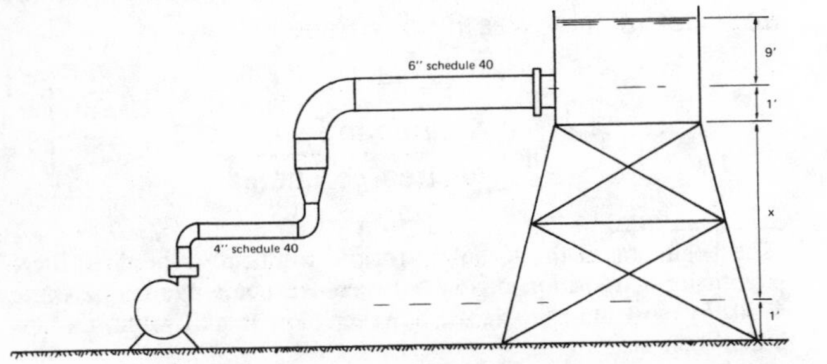

$$h_{p,1} = \frac{(150)(144)}{62.4} = 346.15 \text{ ft}$$

$$v_1 = 3/.0884 = 33.94 \text{ fps}$$

$$h_{v,1} = \frac{(33.94)^2}{(2)(32.2)} = 17.89 \text{ ft}$$

$$z_1 = 0$$

$$h_{p,2} = 0 \text{ (at free surface)}$$

$$v_2 = 0 \text{ (at free surface)}$$

From equation 4.55,

$$346.15 + 17.89 = z_2 + 10$$

$$z_2 = 354.0 \text{ ft}$$

The tank bottom may be raised to $(354.0 - 10 + 1) = 345$ feet above the ground.

4. Impact Energy

Impact energy (also known as stagnation or total[4] energy) is the sum of the kinetic and pressure energies. The impact head is

$$h_s = \frac{p}{\rho} + \frac{v^2}{2g_c} = h_p + h_v \qquad 4.59$$

Impact head represents the effective head in a fluid which has been brought to rest (stagnated) in an adiabatic and reversible manner. Equation 4.59 can be used with a gas as long as the velocity is low — less than 400 ft/sec.

Impact head can be measured directly by use of a pitot tube. This is illustrated in figure 4.12. Equation 4.59 may be used with a pitot tube.

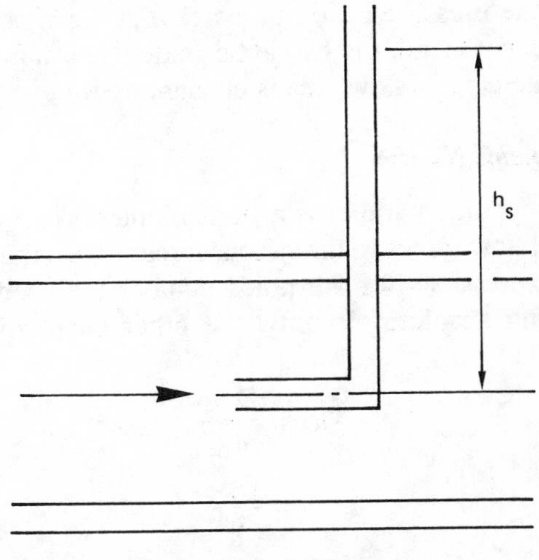

Figure 4.12 A Pitot Tube

[4]There is confusion about *total head* as defined by equations 4.56 and 4.59. The effective pressure in a fluid which has been brought to rest adiabatically does not depend on the potential energy term, z. The application will determine which definition of total head is intended.

FLUIDS

A mercury manometer must be used if the stagnation properties of a gas or high-pressure liquid are being measured. Measurement of stagnation properties is covered in greater detail in part 4 of this chapter.

Example 4.15

The static pressure of air (ρ = .075 lbm/ft³) flowing in a pipe is measured by a precision gage to be 10.0 psig. A pitot tube manometer indicates 20.6 inches of mercury. What is the velocity of the air in the pipe?

The pitot tube measures stagnation pressure. From equation 4.24,

$$p_s = (20.6)(.491) = 10.11 \text{ psig}$$

Since stagnation pressure is the sum of static and velocity pressures, the velocity pressure is

$$p_v = p_s - p_p = 10.11 - 10.0$$
$$= .11 \text{ psig}$$

The velocity head is

$$h_v = p_v/\rho = (.11)(144)/.075$$
$$= 211.2 \text{ ft}$$

From equation 4.53,

$$v = \sqrt{(2)(32.2)(211.2)}$$
$$= 116.6 \text{ ft/sec}$$

5. Hydraulic Grade Line

The hydraulic grade line is a graphical representation of the sum of the static and potential heads versus position along the pipeline.

$$\text{hydraulic grade} = z + h_p \qquad 4.60$$

Since the pressure head can increase at the expense of the velocity head, the hydraulic grade line can increase if an increase in flow area is encountered.

6. Reynolds Number

The Reynolds number is a dimensionless ratio of the inertial flow forces to the viscous forces within the fluid. Two expressions for Reynolds number are used, one requiring absolute viscosity, the other kinematic viscosity:

$$N_{Re} = \frac{D_e v \rho}{\mu g_c} \qquad 4.61$$

$$= \frac{D_e v}{\nu} \qquad 4.62$$

The Reynolds number can also be calculated from the mass flow rate per unit area, G. G must have the units of lbm/sec-ft².

$$N_{Re} = \frac{D_e G}{\mu g_c} \qquad 4.63$$

The Reynolds number is an important indicator in many types of problems. In addition to being used quantitatively in many equations, the Reynolds number is also used to determine whether fluid flow is laminar or turbulent.

A Reynolds number of 2000 or less indicates laminar flow. Fluid particles in laminar flow move in straight paths, parallel to the flow direction. Viscous effects are dominant, resulting in a parabolic velocity distribution with a maximum velocity along the fluid flow centerline.

The fluid is said to be turbulent if the Reynolds number is greater than 2000.[5] Turbulent flow is characterized by random movement of fluid particles. The velocity distribution is essentially uniform with turbulent flow.

7. Equivalent Diameter

The equivalent diameter, D_e, used in equations 4.61 and 4.62 is equal to the inside diameter of a circular pipe. The equivalent diameters of other cross sections in flow are given by table 4.6.

Example 4.16

Determine the equivalent diameter of the open trapezoidal channel shown below:

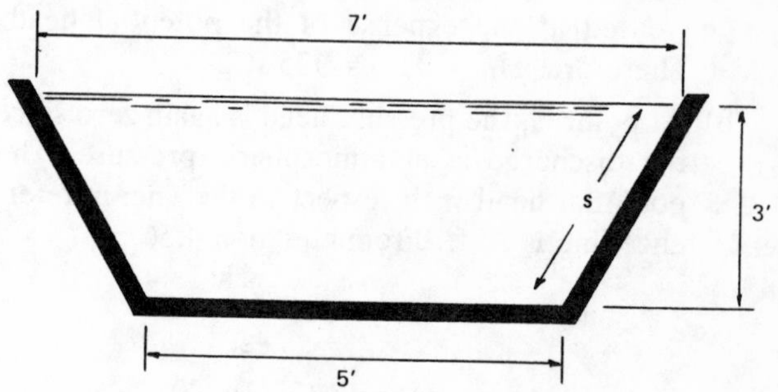

$$s = \sqrt{3^2 + 1^2} = 3.16 \text{ feet}$$

$$D_e = \frac{2(3)(7+5)}{5 + 2(3.16)} = 6.36 \text{ feet}$$

8. Hydraulic Radius

The equivalent diameter can also be found from the hydraulic radius, which is defined as the area in flow divided by the wetted perimeter. The wetted perimeter does not include free fluid surface.

$$D_e = 4r_h \qquad 4.64$$

$$r_h = \frac{\text{area in flow}}{\text{wetted perimeter}} \qquad 4.65$$

[5]The beginning of the turbulent region is difficult to predict. There is actually a transition region between Reynolds numbers 2000 to 4000. In most fluid problems, however, flow is well within the turbulent region.

Table 4.6 Equivalent Diameters

Conduit Cross Section	D_e
Flowing Full	
annulus	$D_o - D_i$
square	L
rectangle	$\dfrac{2L_1L_2}{L_1+L_2}$
Flowing Partially Full	
half-filled circle	D
rectangle (h deep, L wide)	$\dfrac{4hL}{L+2h}$
wide, shallow stream (h deep)	$4h$
triangle (h deep, L broad, s side)	$\dfrac{hL}{s}$
trapezoid (h deep, a wide at top, b wide at bottom, s side)	$\dfrac{2h(a+b)}{b+2s}$

FLUIDS

Consider a circular pipe flowing full. The area in flow is πr^2. The wetted perimeter is the entire circumference, $2\pi r$. The hydraulic radius is $(\pi r^2/2\pi r) = \frac{1}{2}r$. Therefore, the hydraulic radius and the pipe radius are not the same. (The hydraulic radius of a pipe flowing half full is also $\frac{1}{2}r$ since the flow area and wetted perimeter are both halved.)

Example 4.17

What is the hydraulic radius of the trapezoidal channel described in example 4.16?

From equation 4.65

$$r_H = \frac{(5)(3) + (3)(1)}{3.16 + 5 + 3.16} = 1.59 \text{ feet}$$

Using the results of the previous example and equation 4.64,

$$r_H = 6.36/4 = 1.59 \text{ feet}$$

The hydraulic radius of a pipe flowing less than full can be found from table 4.7.

Table 4.7
Hydraulic Radius of Partially Filled Circular Pipes

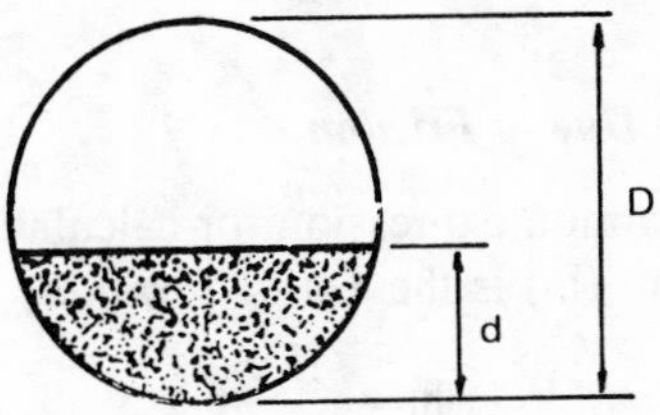

$\frac{d}{D}$	$\frac{\text{hyd. rad.}}{D}$	$\frac{d}{D}$	$\frac{\text{hyd. rad.}}{D}$
0.05	0.0326	0.55	0.2649
0.10	0.0635	0.60	0.2776
0.15	0.0929	0.65	0.2881
0.20	0.1206	0.70	0.2962
0.25	0.1466	0.75	0.3017
0.30	0.1709	0.80	0.3042
0.35	0.1935	0.85	0.3033
0.40	0.2142	0.90	0.2980
0.45	0.2331	0.95	0.2864
0.50	0.2500	1.00	0.2500

PART 4: Fluid Dynamics

1. Fluid Conservation Laws

Many fluid flow problems can be solved using the principles of conservation of mass and energy.

When applied to fluid flow, the principle of mass conservation is known as the *continuity equation:*

$$\rho_1 A_1 v_1 = \rho_2 A_2 v_2 \qquad 4.66$$

$$\dot{m}_1 = \dot{m}_2 \qquad 4.67$$

If the fluid is incompressible, $\rho_1 = \rho_2$, so

$$A_1 v_1 = A_2 v_2 \qquad 4.68$$

$$\dot{V}_1 = \dot{V}_2 \qquad 4.69$$

The energy conservation principle is based on the Bernoulli equation. However, terms for friction loss and hydraulic machines must be included.

$$\left(\frac{p_1}{\rho} + \frac{v_1^2}{2g_c} + z_1\right) + h_A$$

$$= \left(\frac{p_2}{\rho} + \frac{v_2^2}{2g_c} + z_2\right) + h_E + h_f \qquad 4.70$$

2. Head Loss Due to Friction

The most common expression for calculating head loss due to friction (h_f) is the *Darcy formula:*

$$h_f = \frac{fLv^2}{2Dg_c} \qquad 4.71$$

The *Moody friction factor chart* (figure 4.13) is probably the most convenient method of determining the friction factor, f.

The basic parameter required to use the Moody friction factor chart is the Reynolds number. If the Reynolds number is less than 2000, the friction factor is given by equation 4.72.

$$f = \frac{64}{N_{Re}} \qquad 4.72$$

For turbulent flow ($N_{Re} > 2000$), the friction factor depends on the relative roughness of the pipe. This roughness is expressed by the ratio ε/D, where ε is the specific surface roughness and D the inside diameter. Values of ε for various types of pipe are found in Table 4.8.

Another method for finding the friction head loss is the *Hazen-Williams formula.* The Hazen-Williams formula gives good results for liquids which have kinematic viscosities around 1.2 EE−5 ft²/sec (which corresponds to 60°F water). At extremely high and low temperatures, the Hazen-Williams formula can be as much as 20% in error for water. The Hazen-Williams formula should only be used for turbulent flow.

The Hazen-Williams head loss is

$$h_f = \frac{(3.012)(v)^{1.85}L}{(C)^{1.85}(D)^{1.165}} \qquad 4.73$$

Or, in terms of other units,

$$h_f = (10.44)(L)\frac{(\text{gpm})^{1.85}}{(C)^{1.85}(d_{\text{inches}})^{4.8655}} \qquad 4.74$$

Use of these formulas requires a knowledge of the Hazen-Williams coefficient, C, which is assumed independent of the Reynolds number. Table 4.8 gives values of C for various types of pipe.

Example 4.18

50°F water is pumped through 4″ schedule 40 welded steel pipe ($\varepsilon = .0002$) at the rate of 300 gpm. What is the friction head loss calculated by the Darcy formula for 1000 feet of pipe?

First, it is necessary to collect data on the pipe and water. The fluid viscosity and pipe dimensions can be found from the tables at the end of the chapter.

kinematic viscosity = 1.41 EE−5 ft²/sec

inside diameter = .3355 ft

flow area = .0884 ft²

The flow quantity is

$$(300)(.002228) = .6684 \text{ cfs}$$

The velocity is

$$v = \frac{\dot{V}}{A} = \frac{.6684}{.0884} = 7.56 \text{ fps}$$

The Reynolds number is

$$N_{Re} = \frac{(.3355)(7.56)}{1.41\ \text{EE}-5} = 1.8\ \text{EE5}$$

The relative roughness is

$$\varepsilon/D = .0002/.3355 = .0006$$

From the Moody friction chart, f = .0195

From equation 4.71,

$$h_f = \frac{(.0195)(1000)(7.56)^2}{(2)(.3355)(32.2)} = 51.6 \text{ ft}$$

Example 4.19

Repeat example 4.18 using the Hazen-Williams formula. Assume C = 100.

Using equation 4.73,

$$h_f = \frac{(3.012)(7.56)^{1.85}(1000)}{(100)^{1.85}(.3355)^{1.165}} = 90.5 \text{ ft}$$

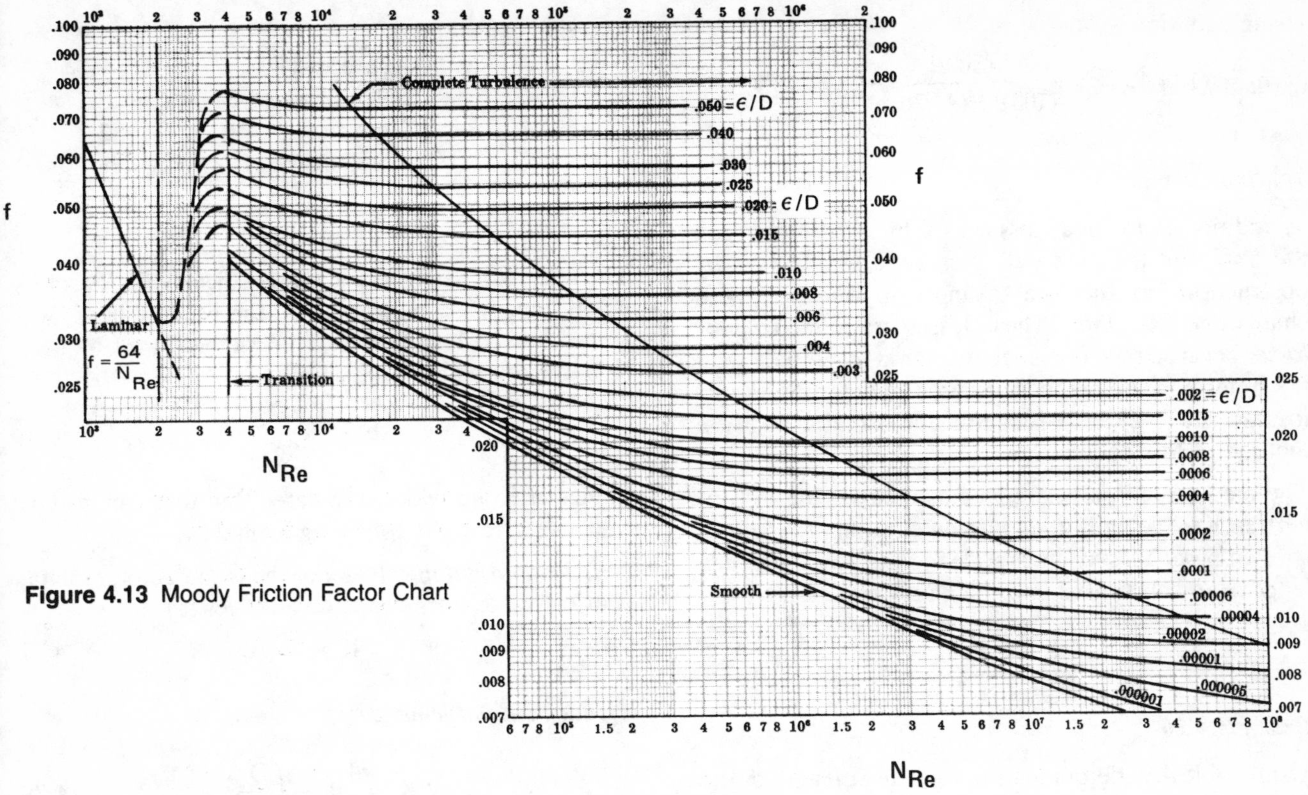

Figure 4.13 Moody Friction Factor Chart

Table 4.8
Specific Roughness and Hazen-Williams Constants for Various Pipe Materials

Type of pipe or surface	ε(ft)		C		
	Range	Design	Range	Clean	Design
STEEL					
welded and seamless	.0001–.0003	.0002	150–80	140	100
interior riveted, no projecting rivets				139	100
projecting girth rivets				130	100
projecting girth and horizontal rivets				115	100
vitrified, spiral-riveted, flow with lap				110	100
vitrified, spiral-riveted, flow against lap				100	90
corrugated				60	60
MINERAL					
concrete	.001–.01	.004	152–85	120	100
cement-asbestos			160–140	150	140
vitrified clays					110
brick sewer					100
IRON					
cast, plain	.0004–.002	.0008	150–80	130	100
cast, tar (asphalt) coated	.0002–.0006	.0004	145–50	130	100
cast, cement lined	.000008	.000008		150	140
cast, bituminous lined	.000008	.000008	160–130	148	140
cast, centrifugally spun	.00001	.00001			
galvanized, plain	.0002–.0008	.0005			
wrought, plain	.0001–.0003	.0002	150–80	130	100
MISCELLANEOUS					
fiber				150	140
copper and brass	.000005	.000005	150–120	140	130
wood stave	.0006–.003	.002	145–110	120	110
transite	.000008	.000008			
lead, tin, glass		.000005	150–120	140	130
plastic		.000005	150–120	140	130

Using equation 4.74,

$$h_f = (10.44)(1000)\frac{(300)^{1.85}}{(100)^{1.85}(4.026)^{4.8655}} = 90.9 \text{ ft}$$

3. *Minor Losses*

In addition to the head loss caused by friction between the fluid and the pipe wall, losses are also caused by obstructions in the line, changes in direction, and changes in flow area. These losses are named *minor losses* because they are much smaller in magnitude than the h_f term. Two methods are used to determine these losses: the method of equivalent lengths and the method of loss coefficients.

The method of equivalent lengths uses a table to convert each valve and fitting into an equivalent length of straight pipe. This length is added to the actual pipeline length and substituted into the Darcy equation for L_e.

$$h_f = \frac{fL_e v^2}{2Dg_c} \qquad 4.75$$

Example 4.20

Using table 4.9, determine the equivalent length of the piping network shown below.

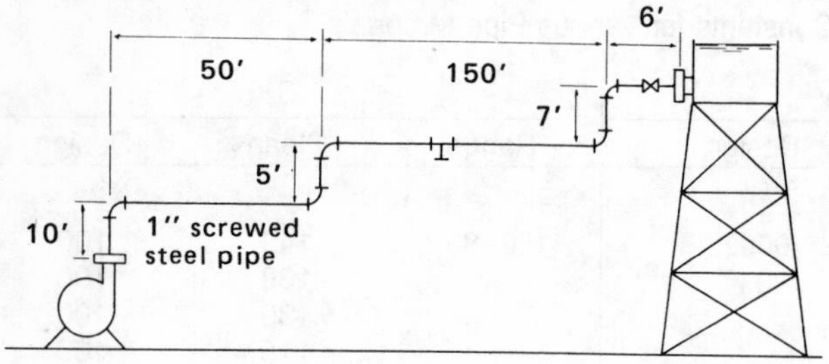

The line consists of:

1 gate valve	.84
5 90° standard elbows	5.2 × 5
1 tee run	3.2
straight pipe	228
L_e =	258 feet

The alternative is to use a loss coefficient, K. This loss coefficient, when multiplied by the velocity head, will give the head loss in feet. This method must be used to find exit and entrance losses.

$$h_f = K\frac{v^2}{2g_c} \qquad 4.76$$

Values of K are widely tabulated, but they can also be calculated from the following formulas:

Valves and Fittings: Refer to the manufacturer's data, or calculate from the equivalent length.

$$K = fL_e/D \qquad 4.77$$

Sudden Enlargements:

$$K = \left[1 - \left(\frac{D_1}{D_2}\right)^2\right]^2 \qquad 4.78$$

Sudden Contractions:

$$K = \tfrac{1}{2}\left[1 - \left(\frac{D_1}{D_2}\right)^2\right] \qquad 4.79$$

Pipe Exit: (projecting exit, sharp edged, and rounded)

$$K = 1.0$$

Table 4.9
Typical Equivalent Lengths of Schedule 40 Straight Pipe For Screwed Steel Fittings and Valves

(For any fluid in turbulent flow)

	Equivalent Length, ft		
	Pipe Size		
Fitting Type	1"	2"	4"
Regular 90° Elbow	5.2	8.5	13.0
Long Radius 90° Elbow	2.7	3.6	4.6
Regular 45° Elbow	1.3	2.7	5.5
Tee, flow through line (run)	3.2	7.7	17.0
Tee, flow through stem	6.6	12.0	21.0
180° Return Bend	5.2	8.5	13.0
Globe Valve	29.0	54.0	110.0
Gate Valve	.84	1.5	2.5
Angle Valve	17.0	18.0	18.0
Swing Check Valve	11.0	19.0	38.0
Coupling or Union	.29	.45	.65

Pipe Entrance:

Reentrant: K = .78
Sharp edged: K = .5
Rounded:

r/D	K
.02	.28
.04	.24
.06	.15
.10	.09
.15	.04

Tapered Diameter Changes:

$$\beta = \frac{\text{small diameter}}{\text{large diameter}}$$

ϕ = wall-to-horizontal angle

	Gradual, $\phi<22°$	Sudden, $\phi>22°$	
Enlargement	$2.6(\sin\phi)(1-\beta^2)^2$	$(1-\beta^2)^2$	4.80
Contraction	$.8(\sin\phi)(1-\beta^2)$	$\frac{1}{2}(1-\beta^2)\sqrt{\sin\phi}$	4.81

4. Head Additions/Extractions

A pump adds head (energy) to the fluid stream. A turbine extracts head from the fluid stream. The amount of head added or extracted can be found by evaluating Bernoulli's equation (equation 4.55) on both sides of the device.

$$h_A = (H_2 - H_1) \quad \text{(pumps)} \qquad 4.82$$

$$h_E = (H_1 - H_2) \quad \text{(turbines)} \qquad 4.83$$

The head increase from a pump is given by equation 4.84.

$$h_A = \frac{(550)(\text{pump input horsepower})\eta_{pump}}{\dot{m}} \qquad 4.84$$

Bernoulli's equation can also be used to calculate the power available to a turbine in a fluid stream by multiplying the total energy by the mass flow rate. This is typically called the 'water horsepower.'

$$P = \dot{m}H = \dot{m}\left(\frac{p}{\rho} + \frac{v^2}{2g_c} + z\right) \qquad 4.85$$

$$\dot{m} = \rho A v \qquad 4.86$$

$$\text{whp} = \frac{P}{550} \qquad 4.87$$

Pumps and turbines are covered in greater detail in part 5 of this chapter.

5. Discharge from Tanks[6]

Flow from a tank discharging liquid to the atmosphere through an opening in the tank wall (figure 4.14) is affected by both the area and shape of the opening. At the orifice, the total head of the fluid is converted into kinetic energy according to equation 4.88.

$$v_o = C_v\sqrt{2gh} \qquad 4.88$$

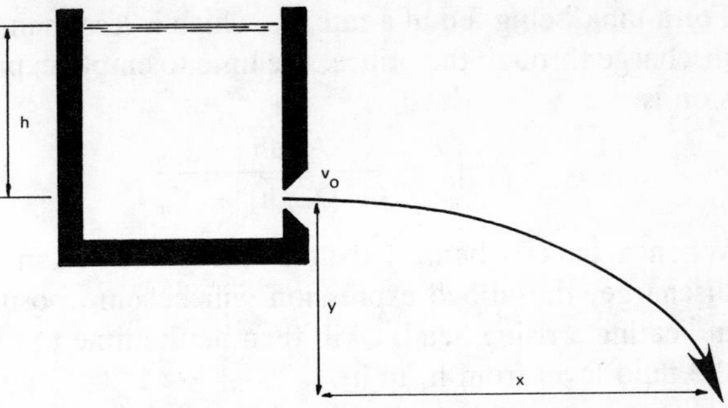

Figure 4.14 Discharge From a Tank

C_v is the coefficient of velocity which can be calculated from the coefficients of discharge and contraction. Typical values of C_v, C_d, and C_c are given in table 4.10.

$$C_v = C_d/C_c \qquad 4.89$$

The discharge from the orifice is

$$\mathring{V} = (C_cA_o)v_o = C_cA_oC_v\sqrt{2gh} = C_dA_o\sqrt{2gh} \qquad 4.90$$

The head loss due to turbulence at the orifice is

$$h_f = \left(\frac{1}{C_v^2} - 1\right)\frac{v_o^2}{2g_c} \qquad 4.91$$

The discharge stream coordinates (see figure 4.14) are

$$x = v_ot = v_o\sqrt{2y/g} = 2C_v\sqrt{hy} \qquad 4.92$$

$$y = \frac{gt^2}{2} \qquad 4.93$$

The fluid velocity at a point downstream of the orifice is

$$v_x = v_o \qquad 4.94$$

$$v_y = gt \qquad 4.95$$

If the liquid in a tank is not constantly being replenished, the static head forcing discharge through the orifice will decrease. For a tank with a constant cross-sectional area, the time required to lower the fluid level from level h_1 to h_2 is calculated from equation 4.96.

$$t = \frac{2A_t(\sqrt{h_1} - \sqrt{h_2})}{C_dA_o\sqrt{2g}} \qquad 4.96$$

If the tank has a varying cross-section, the following basic relationship holds:

$$\mathring{V}\,dt = -A_t\,dh \qquad 4.97$$

An expression for the tank area, A_t, as a function of h must be determined. Then, the time to empty the tank from height h_1 and h_2 is

$$t = \int_{h_1}^{h_2}\frac{A_t\,dh}{C_dA_o\sqrt{2gh}} \qquad 4.98$$

[6]Although the term g_c appears in the equation for velocity head (equation 4.53), it is the local gravity, g, which appears in equation 4.88. An analysis of equation 4.178 will show why this is so.

FLUIDS

For a tank being fed at a rate $\dot{V}_{in}$, which is less than the discharge through the orifice, the time to empty expression is

$$t = \int_{h_1}^{h_2} \frac{A_t\,dh}{(C_d A_o \sqrt{2gh}) - \dot{V}_{in}} \qquad 4.99$$

When a tank is being fed at a rate greater than the discharge, the above expression will become positive indicating a rising head. t will then be the time to raise the fluid level from h_1 to h_2.

The preceding discussion has assumed that the tank has been open or vented to the atmosphere. If the fluid is discharging from a pressurized tank, the total head will be increased by the gage pressure converted to head of fluid by means of equation 4.52.

Example 4.21

A 15′ diameter tank discharges 150°F water through a sharp edged 1″ diameter orifice. If the original water depth is 12′ and the tank is continually pressurized to 50 psig, find the time to empty the tank.

At 150°F, $\rho = 61.20\ \text{lbm/ft}^3$

For the orifice,

$$A_o = .00545\ \text{ft}^2,\ C_d = .62$$

$$h_1 = 12 + \frac{(50)(144)}{61.2} = 129.65\ \text{ft}$$

$$h_2 = \frac{(50)(144)}{61.20} = 117.65\ \text{ft}$$

From equation 4.96,

$$t = \frac{2[\pi(7.5)^2](\sqrt{129.65} - \sqrt{117.65})}{(.62)(.00545)\sqrt{(2)(32.2)}} = 7035\ \text{seconds}$$

6. Culverts and Siphons

A culvert is a water path used to drain runoff from an obstructing geographical feature. Most culvert designs are empirical. However, if the entrance and exit of the culvert are both submerged, the discharge will be independent of the barrel slope. In that case, equation 4.90 can be used to evaluate the discharge. h is the difference in surface levels of the headwater and tailwater.

$$\dot{V} = C_d A \sqrt{2gh} \qquad 4.100$$

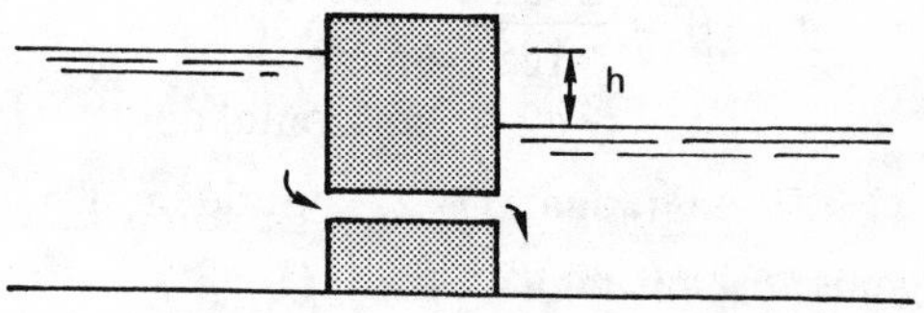

Figure 4.15 A Simple Pipe Culvert

If the culvert length is greater than 50 feet or if the entrance is not smooth, the available energy will be divided between friction and velocity heads. The effective head to be used in equation 4.100 is:

$$h' = h - h_{entrance} - h_f \qquad 4.101$$

The entrance head loss is calculated using loss coefficients:

$$h_{entrance} = K_e\left(\frac{v^2}{2g_c}\right) \qquad 4.102$$

Typical values of K_e are:

.08 for a smooth and tapered entrance
.10 for a flush concrete groove or bell design
.15 for a projecting concrete groove or bell design
.50 for a flush square-edged entrance
.90 for a projecting square-edged entrance

Table 4.10 Orifice Coefficients for Water

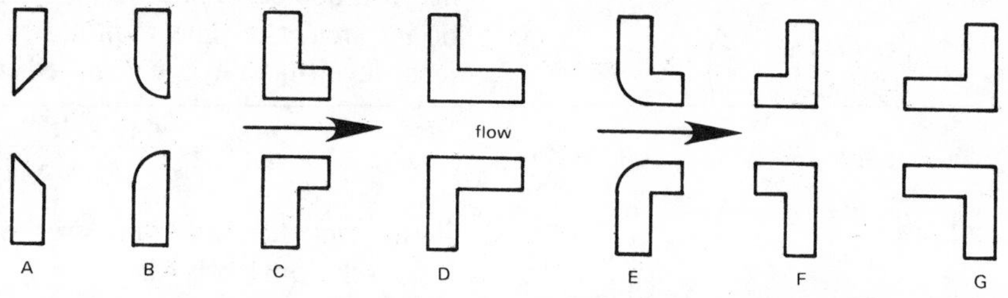

illustration	description	C_d	C_c	C_v
A	sharp-edged	.62	.63	.98
B	round-edged	.98	1.00	.98
C	short tube (fluid separates from walls)	.61	1.00	.61
D	short tube (no separation)	.82	1.00	.82
E	short tube with rounded entrance	.97	.99	.98
F	reentrant tube, length less than one-half of pipe diameter	.54	.55	.99
G	reentrant tube, length 2 to 3 pipe diameters	.72	1.00	.72
not shown	smooth, well-tapered nozzle	.98	.99	.99

The friction loss, h_f, can be found in the usual manner, either from the Darcy equation and Moody friction factor chart or from the Hazen-Williams equation. A trial and error solution may be necessary since v is not known but is needed to find the friction factor.

7. Flow Measuring Devices

The total energy in a fluid flow is the sum of pressure head, velocity head, and gravitational head.

$$H = \frac{p}{\rho} + \frac{v^2}{2g_c} + z \qquad 4.103$$

Change in gravitational head within a flow measuring instrument is negligible. Therefore, if two of the three remaining variables (H, p, or v) are known, then the third can be found from subtraction. The flow measuring devices discussed in this section are capable of measuring total head (H) or pressure head (p).

A. Velocity Measurement

Velocity of a fluid stream is determined by measuring the difference between the static and stagnation pressures and then solving for the velocity head.

A piezometer tap may be used to measure the pressure head directly in feet of fluid.

$$h_p = p/\rho \qquad 4.104$$

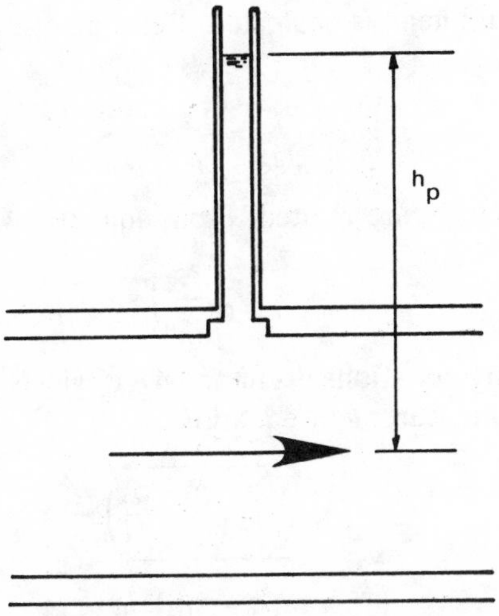

Figure 4.16 Piezometer Tap

For liquids with pressures higher than the capability of the direct reading tap, a manometer may be used with a piezometer tap or with a static probe as shown in figure 4.17.

For either configuration of figure 4.17, the static pressure is

$$p = \rho_m \Delta h_m - \rho y \qquad 4.105$$

$$h_p = \frac{\rho_m \Delta h_m}{\rho} - y \qquad 4.106$$

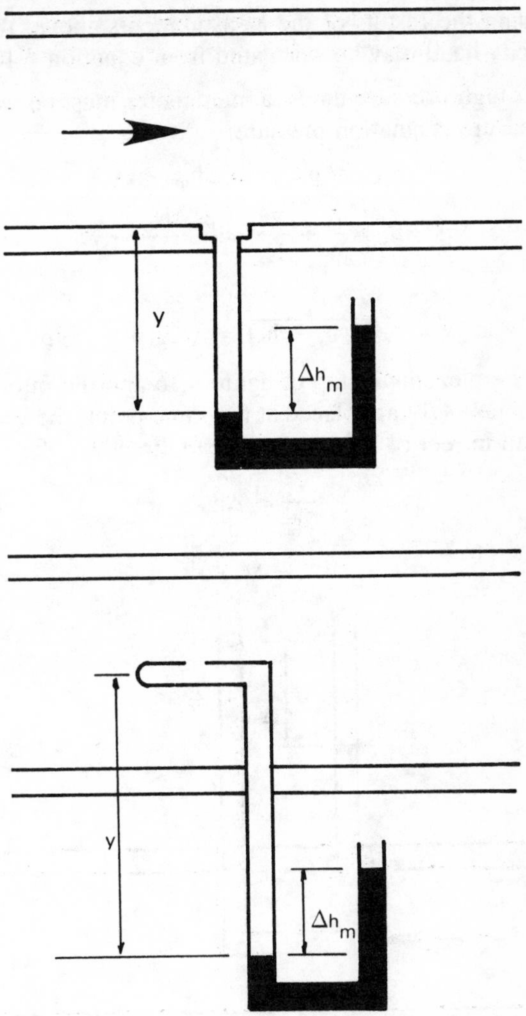

Figure 4.17 Use of Manometers to Measure Static Pressure

Stagnation pressure, also known as total pressure or impact pressure, may be measured directly in feet of fluid by using a pitot tube as shown in figure 4.12.

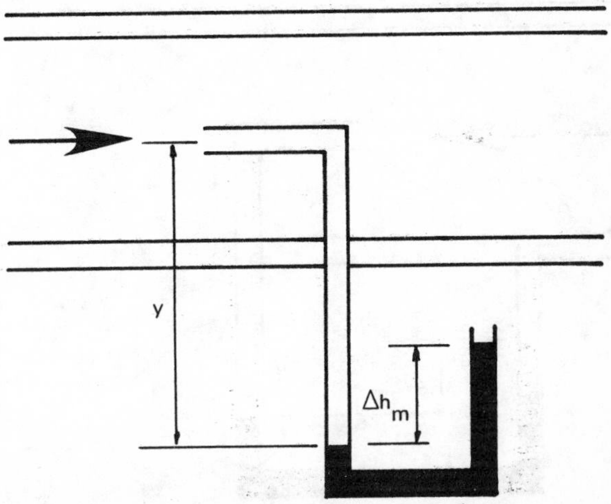

Figure 4.18 Use of Manometer to Measure Total Pressure

Using the results of the measurements above, the velocity head may be calculated from equation 4.109.

For high-pressure fluids, a manometer must be used to measure stagnation pressure:

$$p_s = \rho h_s = \rho_m \Delta h_m - \rho y \qquad 4.107$$

$$h_s = \frac{p}{\rho} + \frac{v^2}{2g_c} = \frac{\rho_m \Delta h_m}{\rho} - y \qquad 4.108$$

$$v = \sqrt{2g(h_s - h_p)} = \sqrt{2g(p_s - p)/\rho} \qquad 4.109$$

If the piezometer tap of figure 4.16 and the pitot tube of figure 4.12 are placed at the same point, the velocity head in feet of fluid may be read directly.

$$\frac{v^2}{2g_c} = \Delta h \qquad 4.110$$

$$v = \sqrt{2g\Delta h} \qquad 4.111$$

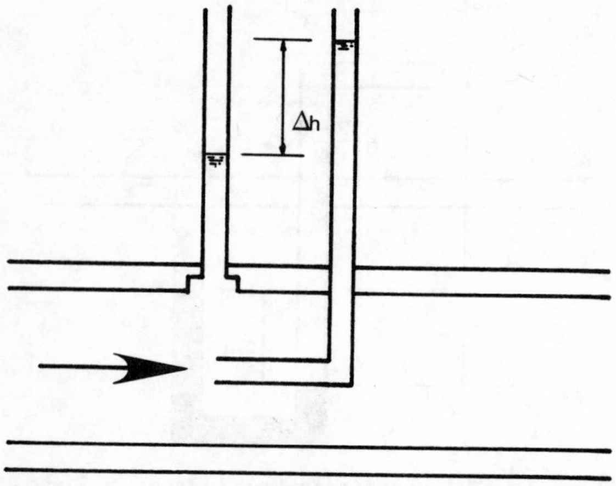

Figure 4.19 Comparative Velocity Head Measurement

The instrumentation arrangement of figures 4.17 and 4.18 may be combined into a single instrument to provide a measurement of velocity head as shown in figure 4.20.

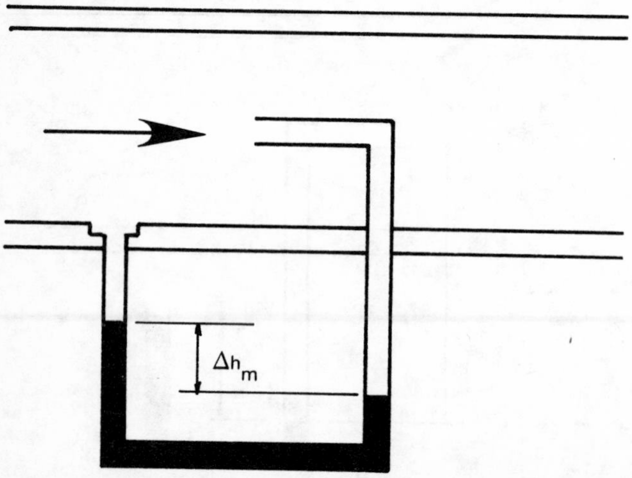

Figure 4.20 Velocity Head Measurement

$$\frac{v^2}{2g_c} = \frac{\Delta h_m(\rho_m - \rho)}{\rho} \qquad 4.112$$

$$v = \sqrt{\frac{2g(\rho_m - \rho)\Delta h_m}{\rho}} \qquad 4.113$$

Example 4.22

50°F water is flowing through a pipe. A pitot-static gage registers a 3" deflection of mercury. What is the velocity within the pipe? (The density of mercury is 848.6 pcf.)

Using equation 4.113,

$$v = \sqrt{\frac{2(32.2)(848.6 - 62.4)(3/12)}{62.4}} = 14.24 \text{ fps}$$

B. Flow Measurement

Using the same techniques described in the preceding section, the flow rate in a line may be determined by measuring the pressure drop across a restriction. Once the geometry of the restriction is known, the Bernoulli equation, along with empirically determined correction coefficients, may be applied to obtain an expression directly relating flow rate with pressure drop.

If potential head is neglected, Bernoulli's equation becomes

$$\frac{p_1}{\rho} + \frac{v_1^2}{2g_c} = \frac{p_2}{\rho} + \frac{v_2^2}{2g_c} \qquad 4.114$$

But, v_1 and v_2 are related. From equation 4.68,

$$v_1 = v_2\left(\frac{A_2}{A_1}\right) \qquad 4.115$$

Combining equations 4.114 and 4.115 yields the standard flow measurement equation.

$$v_2 = \frac{\sqrt{2g\left(\frac{p_1 - p_2}{\rho}\right)}}{\sqrt{1 - \left(\frac{A_2}{A_1}\right)^2}} \qquad 4.116$$

The flow quantity can be found from

$$\dot{V} = v_2 A_2 \qquad 4.117$$

The reciprocal of the denominator of equation 4.116 is known as the *velocity of approach factor*, F_{va}. The *beta ratio* may be incorporated into the formula for F_{va}.

$$F_{va} = \frac{1}{\sqrt{1 - \beta^4}} \qquad 4.118$$

$$\beta = \frac{D_2}{D_1} \qquad 4.119$$

The simplest fluid flow measuring device is the orifice plate. This consists of a thin plate or diaphragm with a central hole through which the fluid flows.

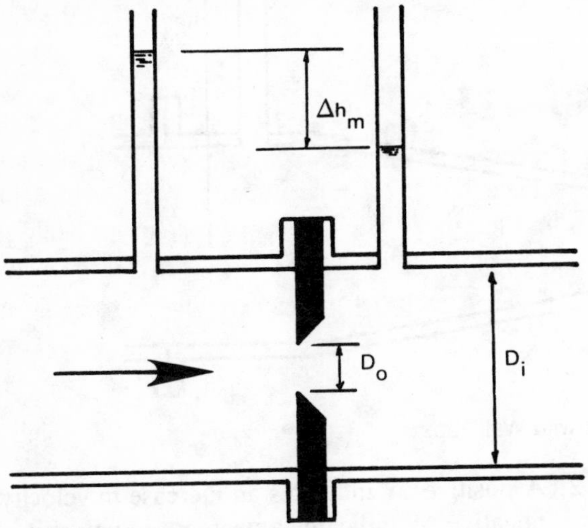

Figure 4.21 Comparative Reading Orifice Plate

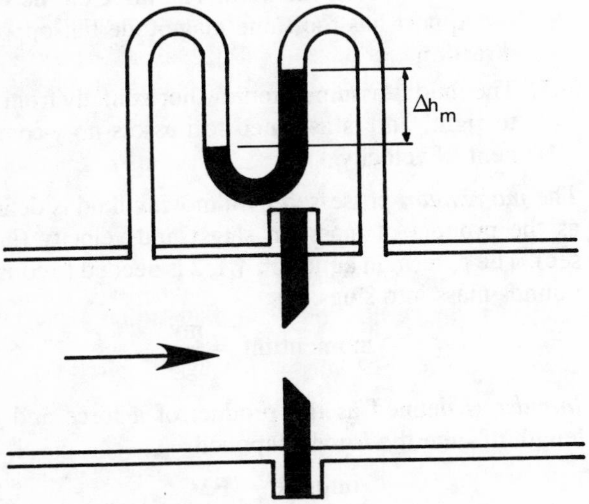

Figure 4.22 Direct Reading Orifice Plate

The governing orifice plate equations for liquid flow are given below.

$$\dot{V} = F_{va}C_dA_o\sqrt{\frac{2g(p_1 - p_2)}{\rho}} \qquad 4.120$$

$$= F_{va}C_dA_o\sqrt{\frac{2g(\rho_m - \rho)\Delta h_m}{\rho}} \qquad 4.121$$

The definition of the velocity of approach factor is modified slightly for the orifice plate.

$$F_{va} = \frac{1}{\sqrt{1 - \left(\frac{C_cA_o}{A_i}\right)^2}} \qquad 4.122$$

The flow coefficient depends on the velocity of approach factor and the discharge coefficient. It may also be obtained from figure 4.23.

$$C_f = F_{va}C_d \qquad 4.123$$

$$C_d = C_vC_c \qquad 4.124$$

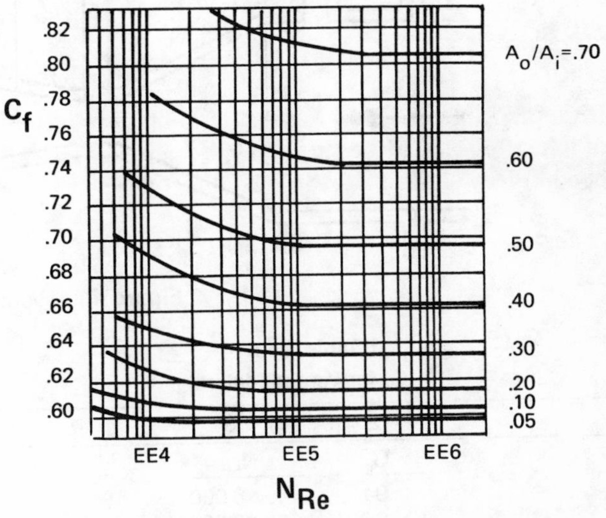

Figure 4.23 Flow Coefficients For Orifice Plates

The flow coefficients can be used to rewrite equations 4.120 and 4.121.

$$\dot{V} = C_fA_o\sqrt{\frac{2g(p_1 - p_2)}{\rho}} \qquad 4.125$$

$$= C_fA_o\sqrt{\frac{2g(\rho_m - \rho)\Delta h_m}{\rho}} \qquad 4.126$$

Operating on the same principles as the orifice plate, the venturi meter induces a smaller pressure drop. However, it is mechanically more complex, as shown by figure 4.24.

The governing equations are similar to those for orifice plates. C_c is usually 1.0 for venturi meters.

$$v_2 = F_{va}\sqrt{\frac{2g(p_1 - p_2)}{\rho}} \qquad 4.127$$

$$\dot{V} = F_{va}C_dA_2\sqrt{\frac{2g(p_1 - p_2)}{\rho}}$$

$$= C_fA_2\sqrt{\frac{2g(p_1 - p_2)}{\rho}} \qquad 4.128$$

$$F_{va} = \frac{1}{\sqrt{1 - (A_2/A_1)^2}} \qquad 4.129$$

$$C_d = C_vC_c \qquad 4.130$$

$$C_f = F_{va}C_d \qquad 4.131$$

FLUIDS

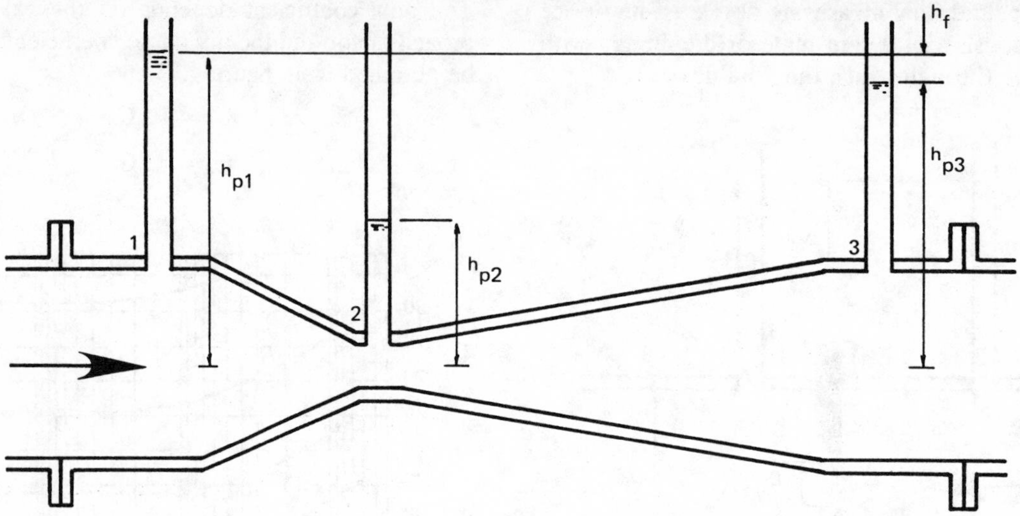

Figure 4.24 Venturi Meter with Wall Taps

Table 4.11
C_d for Venturi Meters

$2 < (A_1/A_2) < 3$	
C_d	N_{Re}
.94	6,000
.95	10,000
.96	20,000
.97	50,000
.98	200,000
.99	2,000,000

Example 4.23

150°F water is flowing in an 8″ schedule 40 steel pipe at 2.23 cfs. If a 7 inch sharp edge orifice plate is bolted across the line, what manometer deflection in inches of Hg would be expected? (Mercury has a density of 848.6 pcf.)

	7″ orifice	8″ schedule 40
flow area	.267 ft^2	.3474 ft^2
diameter	.583 ft	.6651 ft

From table 4.10 for the orifice: $C_c = .63$, $C_d = .62$

Using equation 4.122, $F_{va} = 1.13$

From equation 4.120,

$$\Delta p = \left[\frac{2.23}{(1.13)(.62)(.267)}\right]^2\left[\frac{61.2}{2(32.2)}\right] = 135.06 \text{ psf}$$

$$h_m = 135.06/848.6 = .159 \text{ ft}$$

8. The Impulse/Momentum Principle

A force is required to cause a direction or velocity change in a flowing fluid. Conventions necessary to determine such a force are given here:

1. $\Delta v = v_2 - v_1$
2. A positive Δv indicates an increase in velocity. A negative Δv indicates a decrease in velocity.
3. F and x are positive to the right. F and y are positive upward.
4. F is the force on the fluid. The force on the walls or support has the same magnitude but opposite direction.
5. The fluid is assumed to flow horizontally from left to right, and is assumed to possess no y-component of velocity.

The *momentum* possessed by a moving fluid is defined as the product of mass (in slugs) and velocity (in ft/sec). The g_c term in equation 4.132 is needed to convert pounds-mass into slugs.

$$\text{momentum} = \frac{mv}{g_c} \qquad 4.132$$

Impulse is defined as the product of a force and the length of time the force is applied.

$$\text{impulse} = F\Delta t \qquad 4.133$$

The *impulse-momentum principle* states that the impulse applied to a moving body is equal to the change in momentum. This is expressed by equation 4.134.

$$F\Delta t = \frac{m\Delta v}{g_c} \qquad 4.134$$

Solving for F and combining m and Δt yields equation 4.135.

$$F = \frac{m\Delta v}{g_c\Delta t} = \frac{\dot{m}\Delta v}{g_c} \qquad 4.135$$

Since F is a vector, it may be broken into its components.

$$F_x = \frac{\dot{m}\Delta v_x}{g_c} \qquad 4.136$$

$$F_y = \frac{\dot{m}\Delta v_y}{g_c} \qquad 4.137$$

If the fluid flow is directed through an angle ø, then

$$\Delta v_x = v(\cos ø - 1) \qquad 4.138$$

$$\Delta v_y = v \sin ø \qquad 4.139$$

There are several important fluid applications of the impulse-momentum principle.

A. Jet Propulsion

$$\dot{m}_2 = \dot{m}_1 + \dot{m}_{fuel} = \dot{V}_1\rho_1 + \dot{V}_{fuel}\rho_{fuel} \qquad 4.140$$

$$F_x = \frac{\dot{V}_2\rho_2 v_{2x} - \dot{V}_1\rho_1 v_{1x}}{g_c} \qquad 4.141$$

$$F_y = \frac{\dot{V}_2\rho_2 v_{2y} - \dot{V}_1\rho_1 v_{1y}}{g_c} \qquad 4.142$$

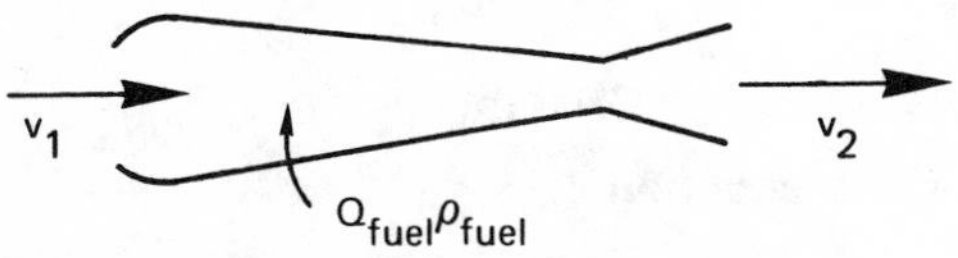

Figure 4.25 Jet Propulsion

B. Open Jet on Vertical Flat Plate

$$\Delta v_y = 0 \qquad 4.143$$

$$\Delta v_x = -v \qquad 4.144$$

$$F_x = \frac{-\dot{m}v}{g_c} = \frac{-\dot{V}\rho v}{g_c} \qquad 4.145$$

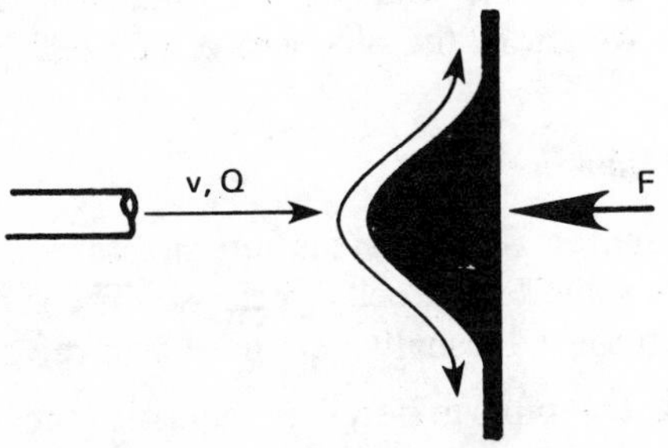

Figure 4.26 Open Jet on Vertical Plate

C. Open Jet on Horizontal Flat Plate

As the jet travels upwards, its velocity decreases since gravity is working against it. By the time the liquid has reached the plate, the velocity has become

$$v_y = \sqrt{v_o^2 - 2gh} \qquad 4.146$$

$$\Delta v_x = 0 \qquad 4.147$$

$$\Delta v_y = -\sqrt{v_o^2 - 2gh} \qquad 4.148$$

$$F = \left(\frac{-\dot{m}}{g_c}\right)\sqrt{v_o^2 - 2gh} = \left(\frac{\dot{V}\rho}{g_c}\right)\sqrt{v_o^2 - 2gh} \qquad 4.149$$

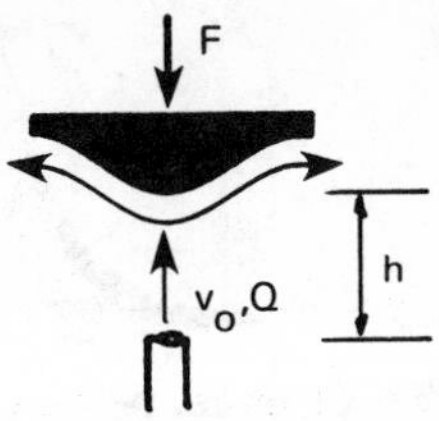

Figure 4.27 Open Jet on Horizontal Plate

D. Open Jet on Single Stationary Blade

v_2 may not be the same as v_1 if friction is present. If no information is given, assume that $v_2 = v_1$.

$$\Delta v_x = v_2\cos ø - v_1 \qquad 4.150$$

$$\Delta v_y = v_2\sin ø \qquad 4.151$$

$$F_x = \left(\frac{\dot{m}}{g_c}\right)\left(v_2\cos ø - v_1\right) = \left(\frac{\dot{V}\rho}{g_c}\right)(v_2\cos ø - v_1) \qquad 4.152$$

$$F_y = \left(\frac{\dot{m}}{g_c}\right)(v_2\sin ø) = \left(\frac{\dot{V}\rho}{g_c}\right)(v_2\sin ø) \qquad 4.153$$

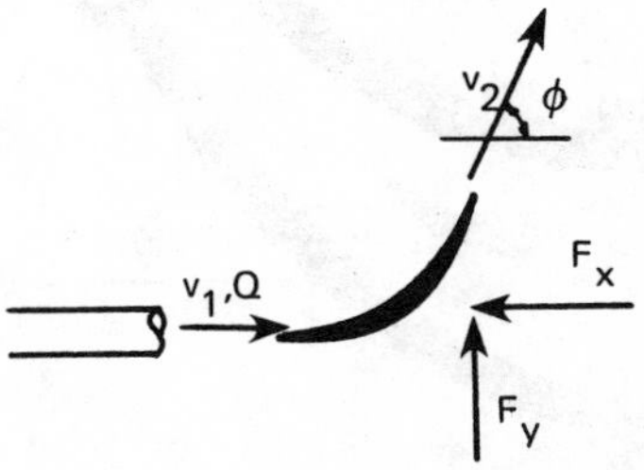

Figure 4.28 Open Jet on Stationary Blade

E. Open Jet on Single Moving Blade

v_b is the blade velocity. Friction is neglected for simplicity. The discharge overtaking the moving blade is $\dot{V}'$.

$$\dot{V}' = \left(\frac{v - v_b}{v}\right)\dot{V} \qquad 4.154$$

$$\Delta v_x = (v - v_b)(\cos ø - 1) \qquad 4.155$$

$$\Delta v_y = (v - v_b)(\sin ø) \qquad 4.156$$

$$F_x = \frac{\dot{m}'\Delta v_x}{g_c} = \left(\frac{\dot{V}'\rho}{g_c}\right)\Delta v_x \qquad 4.157$$

$$F_y = \frac{\dot{m}'\Delta v_y}{g_c} = \left(\frac{\dot{V}'\rho}{g_c}\right)\Delta v_y \qquad 4.158$$

The power transferred to the blade is given by equation 4.159. Power is maximized when ø = 180° and $v_b = \frac{1}{2}v$.

$$P = F_x v_b \qquad 4.159$$

Equations 4.155 through 4.159 can be used with a *multiple*-bladed wheel by using the full $\dot{V}$ instead of $\dot{V}'$.

FLUIDS

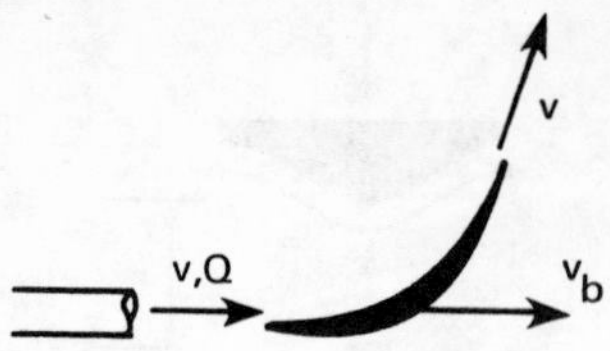

Figure 4.29 Open Jet on Moving Blade

F. Confined Streams in Pipe Bends

Since the fluid is confined, the forces due to static pressure must be included along with the force from momentum changes. Using gage pressures and neglecting the fluid weight,

$$F_x = p_1A_1 - p_2A_2\cos\phi - \left(\frac{\dot{V}\rho}{g_c}\right)(v_2\cos\phi - v_1) \quad 4.160$$

$$F_y = \left[p_2A_2 + \frac{\dot{V}\rho v_2}{g}\right]\sin\phi \quad 4.161$$

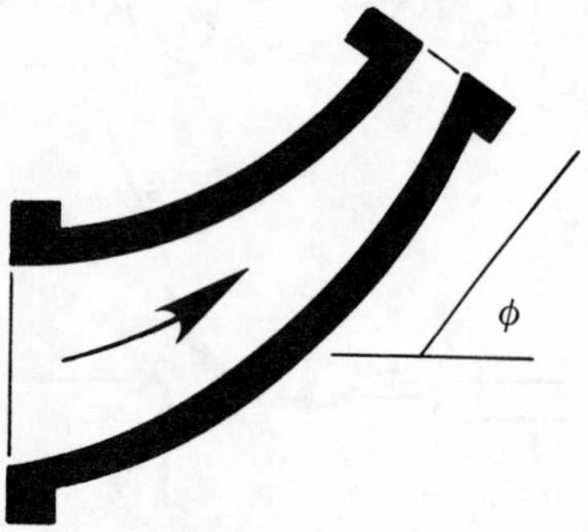

Figure 4.30 A Pipe Bend

Example 4.24

60° water at 40 psig flowing at 8 ft/sec enters a 12″ × 8″ reducing elbow as shown and is turned 30°. (a) What is the resultant force on the water? (b) What other forces should be considered in the design of supports for the fitting?

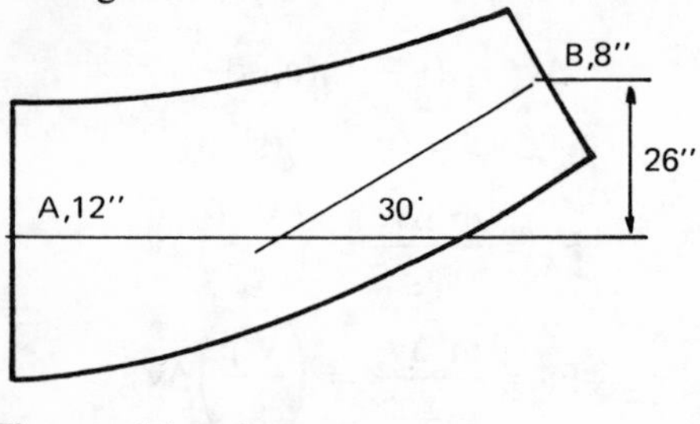

(a) The total head at point A is

$$\frac{(40)(144)}{(62.4)} + \frac{(8)^2}{(2)(32.2)} + 0 = 93.3 \text{ ft}$$

At point B, the velocity is

$$(8)\left(\frac{12}{8}\right)^2 = 18 \text{ ft/sec}$$

The pressure at B can be found from Bernoulli's equation:

$$93.3 = \frac{p_B(144)}{62.4} + \frac{(18)^2}{(2)(32.2)} + \frac{26}{12}$$

So, p_B = 37.3 psig

$$\dot{V} = vA = (8)\left(\frac{1}{4}\right)\pi\left(\frac{12}{12}\right)^2 = 6.28 \text{ cfs}$$

From equation 4.160,

$$F_x = (40)(144)\left(\frac{1}{4}\right)\pi\left(\frac{12}{12}\right)^2 - (37.3)(144)\left(\frac{1}{4}\right)\pi\left(\frac{8}{12}\right)^2\cos30° - \left(\frac{(6.28)(62.4)}{32.2}\right)[(18)(\cos30°) - 8]$$

$$= 2808 \text{ lbf}$$

From equation 4.161,

$$F_y = \left[(37.3)(144)\left(\frac{1}{4}\right)\pi\left(\frac{8}{12}\right)^2 + \left(\frac{(6.28)(62.4)(18)}{32.2}\right)\right]\sin30°$$

$$= 1047 \text{ lbf}$$

The resultant force on the water is

$$R = \sqrt{(2808)^2 + (1047)^2} = 2997 \text{ lbf}$$

(b) The support should be designed to also carry the weight of the water in the pipe and bend, and the weight of the pipe and bend itself.

G. Water Hammer

Water hammer is an increase in pressure in a pipe caused by a sudden velocity decrease. The sudden velocity decrease will usually be caused by a valve closing.

Assuming the pipe material is inelastic, the time required for the water hammer shock wave to travel from a valve to the end of a pipe and back is given by

$$t = 2L/c \quad 4.162$$

The fluid pressure increase resulting from this shock wave is

$$\Delta p = \frac{\rho c \Delta v}{g_c} \quad 4.163$$

Example 4.25

40°F water is flowing at 10 ft/sec through a 4″ schedule 40 welded steel pipe. A check valve is suddenly closed. What increase in fluid pressure will occur?

Assume that the closing valve completely stops the flow. Therefore, Δv is 10 ft/sec.

At 40°F, E = 294 EE3 psi and ρ = 62.43 lbm/ft³. From equation 4.20, the speed of sound in the water is

$$c = \sqrt{\frac{(294\ \text{EE3})(144)(32.2)}{62.43}} = 4673\ \text{fps}$$

From equation 4.163,

$$\Delta p = \frac{(62.43)(4673)(10)}{32.2} = 90{,}600\ \text{psf}$$

9. Lift and Drag

Lift and drag are both forces that are exerted on an object as it passes through a fluid. For example, lift on the wing of an airplane forces the airplane upward, and drag tries to slow it down. Lift and drag are the components of the resultant force on an object, as shown in figure 4.31.

The amounts of lift and drag on an object depend on the shape of the object. Coefficients of lift and drag are used to measure the effectiveness of the object in producing lift and drag. Lift and drag may be calculated from equations 4.164 and 4.165.

$$L = \frac{C_L A \rho v^2}{2g_c} \qquad 4.164$$

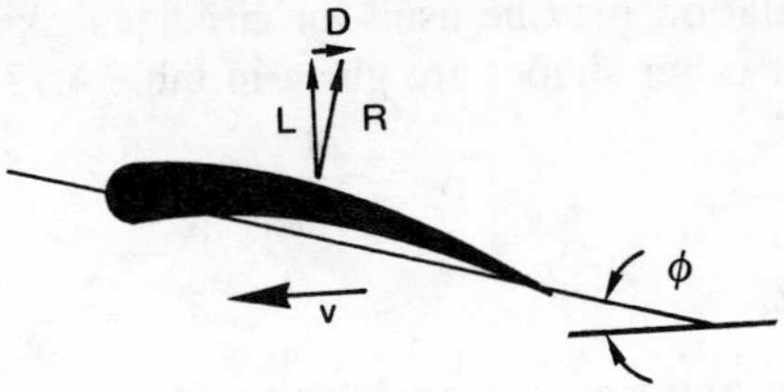

Figure 4.31 Lift and Drag on an Airfoil

$$D = \frac{C_D A \rho v^2}{2g_c} \qquad 4.165$$

A in equation 4.164 is the object's area projected onto a plane parallel to the direction of motion. In equation 4.165, A is the area projected onto a plane normal to the direction of motion.

Values of C_L for various airfoil sections have been correlated with N_{Re}. No simple relationship may be given for airfoils in general. However, the theoretical relationship for a thin flat plate inclined at an angle ø is

$$C_L = 2\pi \sin ø \qquad 4.166$$

The drag coefficient for a sphere moving with N_{Re} less than .4 is predicted by *Stokes' law*, equation 4.167. The

Table 4.12 Approximate Drag Coefficients
(Do not interpolate between N_{Re} = EE5 and N_{Re} = EE6)

Body Shape, and (Characteristic Dimension)	Reynolds Number, N_{Re} EE0	EE1	EE2	EE3	EE4	EE5	EE6	fully turbulent EE6–EE7
sphere (diameter)	(a)	4	1.0	.45	.40	.55	.25	.2
flat disk (diameter)	(a)	4	1.5	1.9(b)	1.1	1.1	1.1	1.1
flat plate, normal to flow, (short side)								
length/breadth = 1				(b)	1.16	1.16	1.16	1.16
4				(b)	1.17	1.17	1.17	1.17
8				(b)	1.23	1.23	1.23	1.23
12.5				(b)	1.34	1.34	1.34	1.34
20				(b)	1.50	1.50	1.50	1.50
25				(b)	1.57	1.57	1.57	1.57
50				(b)	1.76	1.76	1.76	1.76
∞				(b)	2.0	2.0	2.0	2.0
circular cylinder, axis normal to flow (diameter)								
length/diameter = 1				.6	.6	.6		.35
5				.7	.9	.9		
20				.9	.9	.9		
∞	10	2.5	1.3	.9	1.1	1.4	.37	.33
circular cylinder, axis parallel to flow (diameter)								
length/diameter = 1				.91	.91	.91		
2				.85	.85	.85		
4				.87	.87	.87		
7				.99	.99	.99		

Note a: Use Stokes' law, equation 4.167

Note b: Becomes fully turbulent at N_{Re} = 3 EE3

FLUIDS

same equation may be used for circular disks. Values of C_D for other shapes are given in table 4.12

$$C_D = \frac{24}{N_{Re}} \qquad 4.167$$

10. Similarity

Similarity between a model (subscript *m*) and a full-size object (subscript *t*) implies that the model can be used to predict the performance of the full-size object. Such a model is said to be mechanically similar to the full-size object.

Complete mechanical similarity requires geometric and dynamic similarity.[7] *Geometric similarity* means that the model is true to scale in length, area, and volume. *Dynamic similarity* means that the ratios of all types of forces are equal. These forces result from inertia, gravity, viscosity, elasticity (fluid compressibility), surface tension, and pressure.

The model scale or length ratio is

$$L_r = \frac{\text{size of model}}{\text{full size}} \qquad 4.168$$

The area and volume ratios are based on the length ratio.

$$\frac{A_m}{A_t} = (L_r)^2 \qquad 4.169$$

$$\frac{V_m}{V_t} = (L_r)^3 \qquad 4.170$$

The number of possible ratios of forces is large. Fortunately, some force ratios may be neglected because the forces are negligible or self-canceling. Three important cases where the analysis can be simplified are listed below.

A. Viscous and Inertial Forces Dominate

Consider the testing of a completely submerged object such as a submarine. Surface tension effects are negligible. The fluid is assumed incompressible for low velocities. And gravity does not change the path of the fluid particles significantly during the time the submarine is near.

Only inertial, viscous, and pressure forces are significant. Being the only ones that are significant, these three forces are in equilibrium. Since they are in equilibrium, knowing any two will define the third completely. Since it is not an independent force, pressure is omitted from the similarity analysis.

The ratio of the inertial forces to the viscous forces is the Reynolds number. Setting the model and full-size Reynolds numbers equal will ensure similarity. That is,

$$(N_{Re})_m = (N_{Re})_t \qquad 4.171$$

[7]Complete mechanical similarity also requires kinematic and thermal similarity which are not discussed in this book.

This approach works for problems involving fans, pumps, turbines, drainage through holes in tanks, closed-pipe flow with no free surfaces (in the turbulent region with the same relative roughness), and for completely submerged objects such as torpedoes, airfoils, and submarines. It is assumed that the drag coefficients are the same for the model and full-size object.

B. Inertial and Gravitational Forces Dominate

Elasticity and surface tension may be neglected in the analysis of large surface vessels. This leaves pressure, inertia, viscosity, and gravity. Pressure is again omitted as being dependent.

There are only two possible combinations of the remaining three forces. The ratio of inertial and viscous forces is again recognized as the Reynolds number. The ratio of the inertial forces to the gravitational forces is known as the Froude number.

$$N_{Fr} = \frac{v}{\sqrt{Lg}} \qquad 4.172$$

Similarity is ensured when equations 4.173 and 4.174 are satisfied.

$$(N_{Re})_m = (N_{Re})_t \qquad 4.173$$

$$(N_{Fr})_m = (N_{Fr})_t \qquad 4.174$$

As an alternative, equations 4.172 and 4.62 may be solved simultaneously. This results in the following requirement for complete similarity.

$$\frac{\nu_m}{\nu_t} = \left(\frac{L_m}{L_t}\right)^{3/2} = (L_r)^{3/2} \qquad 4.175$$

Sometimes it is not possible to satisfy equation 4.174 or 4.175. This occurs when a model viscosity is called for that is not available. If only equation 4.173 is satisfied, the model is said to be partially similar.

This analysis is valid for surface ships, seaplane hulls, and open channels with varying surface levels such as weirs and spillways.

C. Surface Tension Dominates

Problems involving waves, droplets, bubbles, and air entrainment can be solved by setting the Weber numbers equal.

$$N_W = \frac{v^2 L\rho}{T} \qquad 4.176$$

$$(N_W)_m = (N_W)_t \qquad 4.177$$

11. Effects of Non-Standard Gravity

Most of the equations in part 4 are based on Bernoulli's equation. This equation can be modified to allow for non-standard gravities. Assuming an incompressible fluid, Bernoulli's equation becomes

$$\frac{p_1}{\rho} + \frac{v_1^2}{2g_c} + \frac{gz_1}{g_c} = \frac{p_2}{\rho} + \frac{v_2^2}{2g_c} + \frac{gz_2}{g_c} \qquad 4.178$$

PART 5: Hydraulic Machines

1. Centrifugal Pump Operation

The primary purpose of a pump is to add head. Liquid flowing into the suction side of a centrifugal pump is captured by an impeller and thrown to the outside of the pump casing. Within the casing, the velocity imparted to the fluid by the impeller is converted into pressure head.

A *multiple stage pump* consists of two or more impellers within a single casing, such that the discharge of the first stage becomes the input of the second stage. In this manner, higher heads are achieved than would be possible with a single impeller. A *double suction pump* allows liquid to enter the casing from both sides of the impeller.

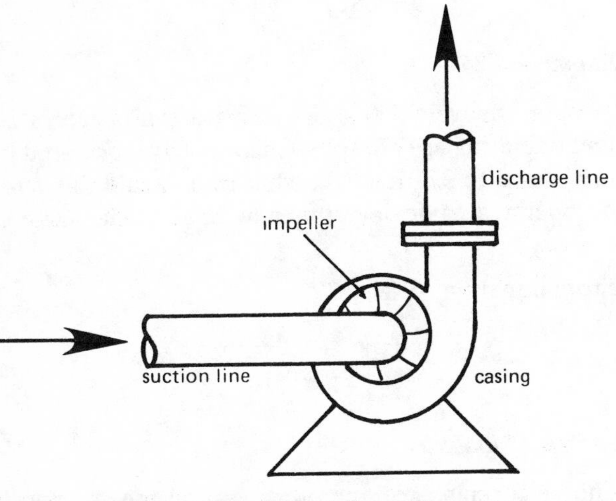

Figure 4.32 A Centrifugal Pump

The head added by a pump can be determined by evaluating the fluid properties at the pump's inlet and discharge. Bernoulli's equation may then be written for these two points.

$$h_A = \frac{p_d}{\rho} - \frac{p_i}{\rho} + \frac{v_d^2}{2g_c} - \frac{v_i^2}{2g_c} + z_d - z_i \qquad 4.179$$

In most applications, the increase in velocity and potential heads is zero or small in comparison to the increase in pressure head. Equation 4.179 then reduces to

$$h_A = \frac{p_d - p_i}{\rho} \qquad 4.180$$

The amount of energy (head) added per pound of fluid and the mass flow rate through the pump can be used to determine the water horsepower delivered to the fluid stream. Equation 4.84 gives the water horsepower equation in its simplest form. Table 4.13 may be used to find the hydraulic horsepower directly from the added head and flow rate.

Table 4.13 Hydraulic Horsepower Equations

	Q in gpm	$\dot{m}$ in lbm/sec	$\dot{V}$ in cfs
h_A is added head in feet	$\frac{h_A Q (S.G.)}{3960}$	$\frac{h_A \dot{m}}{550}$	$\frac{h_A \dot{V} (S.G.)}{8.814}$
Δp is added head in psf	$\frac{\Delta p Q}{2.468 \text{ EE5}}$	$\frac{\Delta p \dot{m}}{(34{,}320)(S.G)}$	$\frac{\Delta p \dot{V}}{550}$

The *brake horsepower* delivered by the motor to the pump is

$$\text{bhp} = \frac{\text{whp}}{\eta_p} \qquad 4.181$$

The difference between hydraulic horsepower and brake horsepower is the power lost within the pump due to mechanical and hydraulic friction. This is referred to as *friction horsepower* and is determined from equation 4.182.

$$\text{fhp} = \text{bhp} - \text{whp} \qquad 4.182$$

Electrical horsepower to the motor is

$$\text{ehp} = \frac{\text{bhp}}{\eta_m} \qquad 4.183$$

Overall efficiency is the pump efficiency multiplied by the motor efficiency.

$$\eta = (\eta_p)(\eta_m) = \frac{\text{whp}}{\text{ehp}} \qquad 4.184$$

2. Pump Suction Head

Liquid is not sucked into a pump. A positive head (normally atmospheric pressure) must push the liquid into the impeller. *Net Positive Suction Head Required* (NPSHR) is the minimum fluid energy required at the inlet by the pump for satisfactory operation. NPSHR is usually specified by the pump manufacturer. *Net Positive Suction Head Available* (NPSHA) is the actual fluid energy at the inlet.

NPSHA can be found by evaluating the total atmospheric and potential heads at the surface of the fluid to be pumped. Friction losses and vapor pressure head are subtracted to yield the net remaining head available to force liquid into the pump.

$$\text{NPSHA} = h_a + z_{\text{liquid level}} - z_{\text{pump inlet}} - h_{f,\ \text{suction line}} - h_{vp} \qquad 4.185$$

NPSHA can also be evaluated if the pressure and velocity at the pump inlet are known.

$$\text{NPSHA} = h_{p,\ \text{pump inlet}} + h_{v,\ \text{pump inlet}} - h_{vp} \qquad 4.186$$

If NPSHA is less than the NPSHR, the fluid will cavitate. Cavitation is the vaporization of fluid within the casing or suction line. If the fluid pressure is less than the vapor pressure, pockets of vapor will form. As vapor pockets reach the surface of the impeller, the local high fluid pressure will collapse them, causing noise, vibration, and possible structural damage.

Cavitation may be caused by any of the following conditions:

a. Discharge heads far below the pump's calibrated head at peak efficiency.
b. Suction lift higher or suction head lower than the manufacturer's recommendation.
c. Speeds higher than the manufacturer's recommendation.
d. Liquid temperatures (thus vapor pressures) higher than that for which the system was designed.

Cavitation can be eliminated by increasing NPSHA or decreasing NPSHR. NPSHA can be increased by the following actions.

a. Increase the height of the free fluid level of the supply tank.
b. Reduce the distance and minor losses between the supply tank and the pump.
c. Reduce the temperature of the fluid.
d. Pressurize the supply tank.

NPSHR may be reduced by the following actions.

a. Place a throttling valve in the discharge line.
b. Use a different pump.

3. Pump Specific Speed

Specific speed is a function of a pump's capacity, head, and rotational speed at peak efficiency. It is used as a guide to selecting the most efficient pump type. For a given pump and impeller configuration, the specific speed remains essentially constant over a range of flow rates and heads. Theoretically, specific speed is the rpm at which a *homologous*[8] pump would have to turn in order to put out 1 gpm at 1 foot total head. (For double suction pumps, Q in equation 4.187 is divided by 2.)

$$n_s = \frac{n\sqrt{Q}}{(h_A)^{.75}} \qquad 4.187$$

4. Pump Affinity Laws

Most parameters (impeller diameter, speed, and flow rate) determining a pump's performance can be varied. If the impeller diameter is held constant and the speed varied, the following ratios are valid.

$$\frac{Q_2}{Q_1} = \frac{n_2}{n_1} \qquad 4.188$$

$$\frac{H_2}{H_1} = \left(\frac{n_2}{n_1}\right)^2 = \left(\frac{Q_2}{Q_1}\right)^2 \qquad 4.189$$

$$\frac{bhp_2}{bhp_1} = \left(\frac{n_2}{n_1}\right)^3 = \left(\frac{Q_2}{Q_1}\right)^3 \qquad 4.190$$

If the speed is held constant and the impeller size varied, the following ratios apply.

$$\frac{Q_2}{Q_1} = \frac{d_2}{d_1} \qquad 4.191$$

$$\frac{H_2}{H_1} = \left(\frac{d_2}{d_1}\right)^2 \qquad 4.192$$

$$\frac{bhp_2}{bhp_1} = \left(\frac{d_2}{d_1}\right)^3 \qquad 4.193$$

Example 4.26

A pump delivers 500 gpm against a total head of 200 feet operating at 1770 rpm. Changes have increased the total head to 375 feet. At what rpm should this pump be operated to achieve this new head at the same efficiency?

From equation 4.189,

$$n_2 = 1700\sqrt{\frac{375}{200}} = 2424 \text{ rpm}$$

5. Pump Similarity

The performance of one pump can be used to predict the performance of a dynamically similar (homologous) pump.

$$\frac{n_1 d_1}{\sqrt{H_1}} = \frac{n_2 d_2}{\sqrt{H_2}} \qquad 4.194$$

$$\frac{Q_1}{d_1^2\sqrt{H_1}} = \frac{Q_2}{d_2^2\sqrt{H_2}} \qquad 4.195$$

$$\frac{bhp_1}{\rho_1 d_1^2 H_1^{1.5}} = \frac{bhp_2}{\rho_2 d_2^2 H_2^{1.5}} \qquad 4.196$$

$$\frac{Q_1}{n_1 d_1^3} = \frac{Q_2}{n_2 d_2^3} \qquad 4.197$$

$$\frac{bhp_1}{\rho_1 n_1^3 d_1^5} = \frac{bhp_2}{\rho_2 n_2^3 d_2^5} \qquad 4.198$$

$$\frac{n_1\sqrt{Q_1}}{H_1^{.75}} = \frac{n_2\sqrt{Q_2}}{H_2^{.75}} \qquad 4.199$$

Example 4.27

A 6″ pump operating at 1770 rpm discharges 1500 gpm of cold water against an 80 foot head at 80% efficiency.

[8]*Homologous* pumps are geometrically similar. This means that each is a scaled up (or down) version of the others. Such pumps are said to belong to a *family*.

A homologous 8″ pump operating at 1170 rpm is being considered as a replacement. What capacity and total head can be expected from the new pump? What would be the new power requirement?

From equation 4.194,

$$H_2 = \left[\frac{(8)(1170)}{(6)(1770)}\right]^2 (80) = 62.14 \text{ ft}$$

From equation 4.197,

$$Q_2 = \left[\frac{(1170)(8)^3}{(1770)(6)^3}\right](1500) = 2350 \text{ gpm}$$

$$\text{whp}_2 = \frac{(2350.3)(62.14)(1.0)}{3960} = 36.88 \text{ hp}$$

$$\text{bhp}_2 = 36.88/.8 = 46.1 \text{ hp}$$

6. Impulse Turbines

As shown in Figure 4.33, an impulse turbine converts the energy of a fluid stream into kinetic energy by use of a nozzle which directs the stream jet against the turbine blades. Impulse turbines are generally employed where the available head exceeds 800 feet.

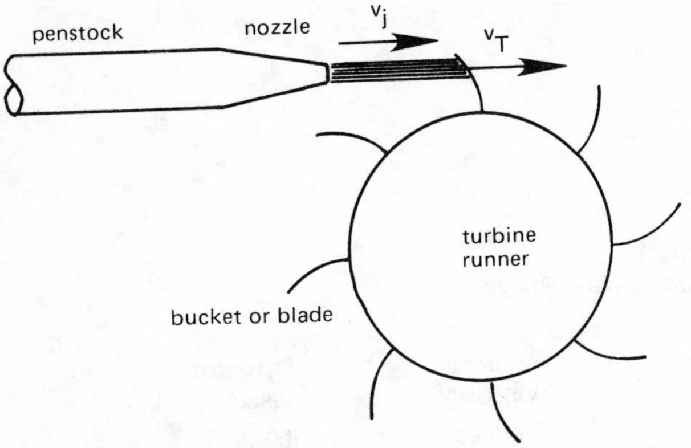

Figure 4.33 A Simple Impulse Turbine

The total head available to an impulse turbine is given by equations 4.200 and 4.201. p is the pressure of the fluid at the nozzle entrance.

$$H = \frac{p}{\rho} + \frac{v^2}{2g_c} - h_n \qquad 4.200$$

$$= h_p - \frac{fL_e v^2}{2Dg_c} - h_n \qquad 4.201$$

The nozzle loss is

$$h_n = \left(\frac{p}{\rho} + \frac{v^2}{2g_c}\right)(1 - C_v^2) \qquad 4.202$$

The velocity of the fluid jet is

$$v_j = \sqrt{2gH} \qquad 4.203$$

The energy transmitted by each pound-mass of fluid to the turbine runner's blades is given by equation 4.204.

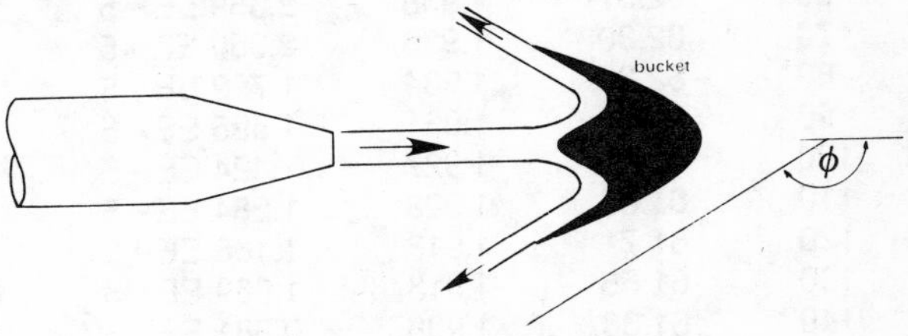

Figure 4.34 Turbine Blade Geometry

$$E = \frac{v_b(v_j - v_b)(1 - \cos\phi)}{g_c} \qquad 4.204$$

The theoretical brake horsepower of the turbine can be derived from the energy per pound and the fluid flow rate. (Compare equation 4.205 with equations 4.157 and 4.159.)

$$\text{bhp} = \frac{\dot{m} v_b(v_j - v_b)(1 - \cos\phi)}{(550)g_c} \qquad 4.205$$

The actual output will be less than the theoretical input. Typical efficiencies range from 80% to 90%.

7. Reaction Turbines

Reaction turbines are essentially centrifugal pumps in reverse. They are used when the total head is small, typically below 800 feet. However, their energy conversion efficiency is higher than for impulse turbines, typically 90% – 95%.

Since reaction turbines are centrifugal pumps in reverse, all of the affinity and similarity relationships (equations 4.188 through 4.199) may be used when comparing homologous turbines.

By convention, the equation for turbine specific speed is not the same as for pumps.

$$n_s = \frac{n\sqrt{\text{bhp}}}{(H)^{1.25}} \qquad 4.206$$

Appendix A
Properties of Water at Atmospheric Pressure

Temp. °F	Density lbm/ft³	Density slug/ft³	Viscosity lbf-sec/ft²	Kinematic Viscosity ft²/sec	Surface Tension lbf/ft	Vapor Pressure Head ft	Bulk Modulus lbf/in²
32	62.42	1.940	3.746 EE−5	1.931 EE−5	0.518 EE−2	0.20	293 EE3
40	62.43	1.940	3.229 EE−5	1.664 EE−5	0.514 EE−2	0.28	294 EE3
50	62.41	1.940	2.735 EE−5	1.410 EE−5	0.509 EE−2	0.41	305 EE3
60	62.37	1.938	2.359 EE−5	1.217 EE−5	0.504 EE−2	0.59	311 EE3
70	62.30	1.936	2.050 EE−5	1.059 EE−5	0.500 EE−2	0.84	320 EE3
80	62.22	1.934	1.799 EE−5	0.930 EE−5	0.492 EE−2	1.17	322 EE3
90	62.11	1.931	1.595 EE−5	0.826 EE−5	0.486 EE−2	1.61	323 EE3
100	62.00	1.927	1.424 EE−5	0.739 EE−5	0.480 EE−2	2.19	327 EE3
110	61.86	1.923	1.284 EE−5	0.667 EE−5	0.473 EE−2	2.95	331 EE3
120	61.71	1.918	1.168 EE−5	0.609 EE−5	0.465 EE−2	3.91	333 EE3
130	61.55	1.913	1.069 EE−5	0.558 EE−5	0.460 EE−2	5.13	334 EE3
140	61.38	1.908	0.981 EE−5	0.514 EE−5	0.454 EE−2	6.67	330 EE3
150	61.20	1.902	0.905 EE−5	0.476 EE−5	0.447 EE−2	8.58	328 EE3
160	61.00	1.896	0.838 EE−5	0.442 EE−5	0.441 EE−2	10.95	326 EE3
170	60.80	1.890	0.780 EE−5	0.413 EE−5	0.433 EE−2	13.83	322 EE3
180	60.58	1.883	0.726 EE−5	0.385 EE−5	0.426 EE−2	17.33	313 EE3
190	60.36	1.876	0.678 EE−5	0.362 EE−5	0.419 EE−2	21.55	313 EE3
200	60.12	1.868	0.637 EE−5	0.341 EE−5	0.412 EE−2	26.59	308 EE3
212	59.83	1.860	0.593 EE−5	0.319 EE−5	0.404 EE−2	33.90	300 EE3

FLUIDS

Appendix B
Properties of Air at Atmospheric Pressure

Temp. °F	Density slug/ft³	Density lbm/ft³	Kinematic Viscosity ft²/sec	Dynamic Viscosity lbf-sec/ft²
0	0.00268	0.0862	12.6 EE−5	3.28 EE−7
20	0.00257	0.0827	13.6 EE−5	3.50 EE−7
40	0.00247	0.0794	14.6 EE−5	3.62 EE−7
60	0.00237	0.0763	15.8 EE−5	3.74 EE−7
68	0.00233	0.0752	16.0 EE−5	3.75 EE−7
80	0.00228	0.0735	16.9 EE−5	3.85 EE−7
100	0.00220	0.0709	18.0 EE−5	3.96 EE−7
120	0.00215	0.0684	18.9 EE−5	4.07 EE−7
250	0.00174	0.0559	27.3 EE−5	4.74 EE−7

Appendix C
Properties of Liquids

Liquid	Temp, °F	Specific Gravity*	Viscosity centistokes	SSU	ft²/sec
Acetone	68	.792	.41		
Alcohol, ethyl	68	.789	1.52	31.7	1.65 EE − 5
(C_2H_5OH)	104	.772	1.2	31.5	
Alcohol, methyl	68	.79			
(CH_3OH)	59		.74		
Ammonia	0	.662	.30		
Butane	−50		.52		
	30		.35		
	60	.584			
Castor Oil	68	.96			1110 EE − 5
	104	.95	259–325	1200–1500	
	130		98–130	450–600	
Ethylene glycol	60	1.125			
	70		17.8	88.4	
Freon-11	70	1.49	21.1	.21	
Freon-12	70	1.33	21.1	.27	
Fuel oils, #1 to #6	60	.82–.95			
Fuel oil #1	70		2.39–4.28	34–40	
	100		−2.69	32–35	
Fuel oil #2	70		3.0–7.4	36–50	
	100		2.11–4.28	33–40	
Fuel oil #3	70		2.69–5.84	35–45	
	100		2.06–3.97	32.8–39	
Fuel oil #5A	70		7.4–26.4	50–125	
	100		4.91–13.7	42–72	
Fuel oil #5B	70		26.4–	125–	
	100		13.6–67.1	72–310	
Fuel oil #6	122		97.4–660	450–3000	
	160		37.5–172	175–780	
Gasoline (regular)	60	.728			.73 EE − 5
	80	.719			.66 EE − 5
	100	.710			.60 EE − 5
Kerosene	60	.78–.82			
	68		2.17	35	
Jet Fuel	−30		7.9	52	
	60	.82			
Mercury	70	13.6	21.1	.118	
	100	13.6	37.8	.11	
Oils, SAE 5 to 150	60	.88–.94			
SAE-5W	0		1295 max	6000 max	
SAE-10W	0		1295–2590	6000–12000	
SAE-20W	0		2590–10350	12000–48000	
SAE-20	210		5.7–9.6	45–58	
SAE-30	210		9.6–12.9	58–70	
SAE-40	210		12.9–16.8	70–85	
SAE-50	210		16.8–22.7	85–110	
Salt Water (5%)	39	1.037			
	68		1.097	31.1	
Salt Water (25%)	39	1.196			
	60	1.19	2.4	34	

*Measured with respect to 60°F water

Appendix D
Internal Dimensions of Schedule 40 Steel Pipe

Nominal Diameter Inches	Internal Diameter Inches	Internal Area Square Inches	Internal Diameter Feet	Internal Area Square Feet
1/8	0.269	0.0568	0.0224	0.00039
1/4	0.364	0.1041	0.0303	0.00072
3/8	0.493	0.1909	0.0411	0.00133
1/2	0.622	0.3039	0.0518	0.00211
3/4	0.824	0.5333	0.0687	0.00370
1	1.049	0.8643	0.0874	0.00600
1 1/4	1.380	1.496	0.1150	0.01039
1 1/2	1.610	2.036	0.1342	0.01414
2	2.067	3.356	0.1723	0.02330
2 1/2	2.469	4.788	0.2058	0.03325
3	3.068	7.393	0.2557	0.05134
3 1/2	3.548	9.887	0.2957	0.06866
4	4.026	12.73	0.3355	0.08841
5	5.047	20.01	0.4206	0.1389
6	6.065	28.89	0.5054	0.2006
8	7.981	50.03	0.6651	0.3474
10	10.020	78.85	0.8350	0.5476
12	11.938	111.93	0.9948	0.7773
14	13.124	135.28	1.0937	0.9394
16	15.000	176.72	1.2500	1.2272
18	16.876	223.68	1.4063	1.5533
20	18.812	277.95	1.5677	1.9802
24	22.624	402.00	1.8853	2.7917

Appendix E
Important Fluid Conversions

Multiply	By	To Obtain
cubic feet	7.4805	gallons
cfs	448.83	gpm
cfs	.64632	MGD
gallons	.1337	cubic feet
gpm	.002228	cfs
inches of mercury	.491	psi
inches of mercury	70.7	psf
inches of mercury	13.60	inches of water
inches of water	5.199	psf
inches of water	.0361	psi
inches of water	.0735	inches of mercury
psi	144	psf
psi	2.308	feet of water
psi	27.7	inches of water
psi	2.037	inches of mercury
psf	.006944	psi

FLUIDS

GENERAL QUESTIONS

1. The barometer reads 29.0 inches of mercury. What is the absolute pressure if a vacuum gage reads 9.5 psi?

2. Find the speed of sound in 32·F water if the coefficient of compressibility is 3.4 EE-6 (1/psi).

3. Two flanges are held together by three bolts. The pipe section is 4" inside diameter. If the pipe is pressurized to 900 psig, what is the force exerted by the bolts?

4. 4.9 cubic feet of water are compressed to 5000 psig. What is the volume decrease if the temperature is 60·F?

CAPILLARITY

5. To what height will pure 68·F water rise in a .1" diameter glass tube open to the atmosphere at the other end?

VAPOR PRESSURES

6. What is the vapor pressure of
   (a) mercury at 68·F?
   (b) ethyl alcohol at 68·F?
   (c) water at 68·F?
   (d) water at 100·F?
   (e) water at 212·F?

7. To what height will a barometer column rise if the atmospheric conditions are 13.9 psia and 68·F, and the barometer fluid is
   (a) mercury?
   (b) water?
   (c) ethyl alcohol?

HYDROSTATIC PRESSURE

8. What is the pressure 8000 feet below the surface of the ocean? Neglect compressibility.

9. What is the pressure at point A?

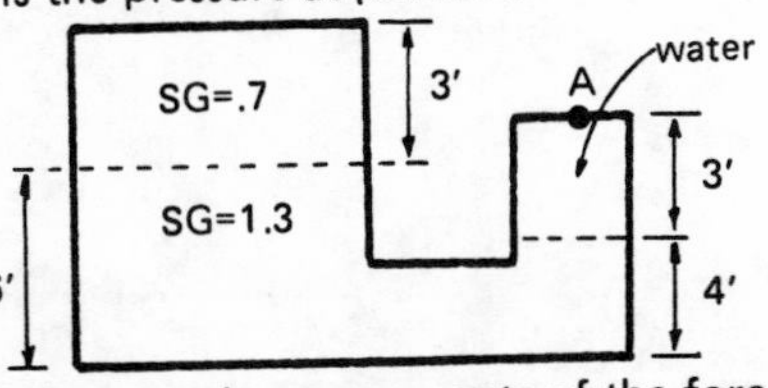

10. Find the x- and y-components of the force on the inclined surface.

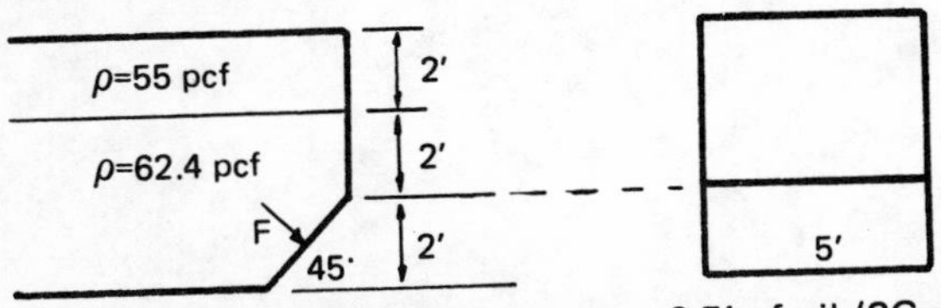

11. A container supports 9.3' of water, 6.5' of oil (SG=.71), and 5.0' of air at 26 psig and 70·F. What is the absolute pressure at the bottom of the container?

12. A water flume is dammed by a rectangular gate 10' wide and 6' high. The water height is 5'. There is no water on the down side of the flume. What is the total force on the gate? How far below the upper edge of the gate does the resultant act?

13. What is the force at B if the tank contains oil with SG = .7?

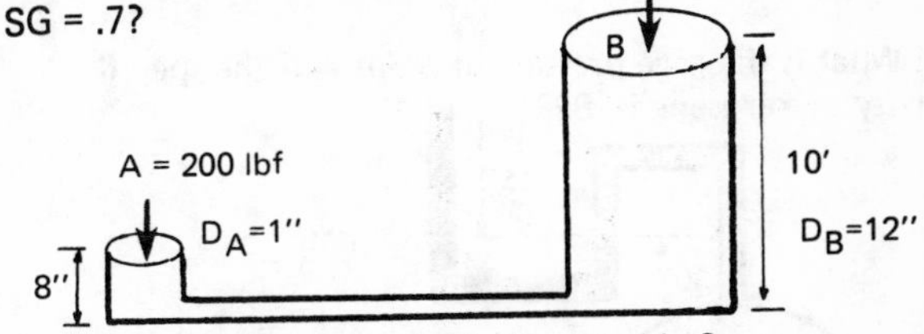

14. What is the pressure in the inverted jar?

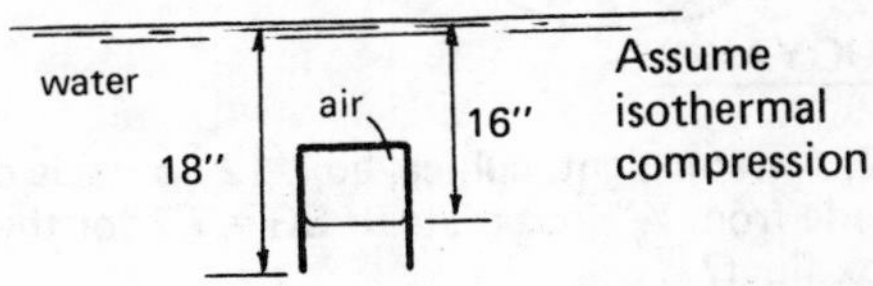

15. What is the normal force acting on the 4' x 8' plate? Where does it act?

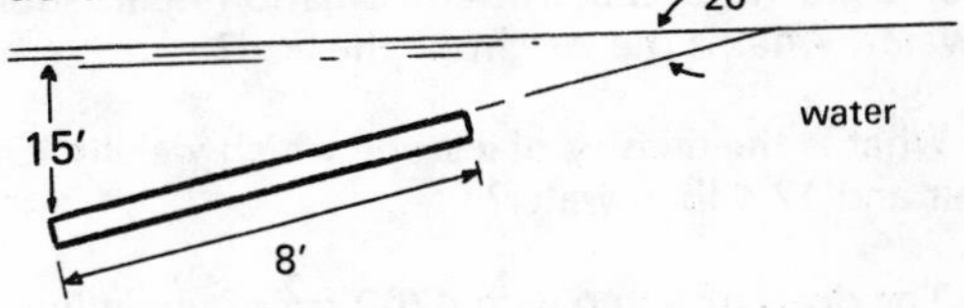

16. Given the water tank shown below, find (for a 1 ft. width)
   (a) the horizontal wall force component over A-C
   (b) the vertical wall force component over A-C
   (c) the magnitude and direction of the water force

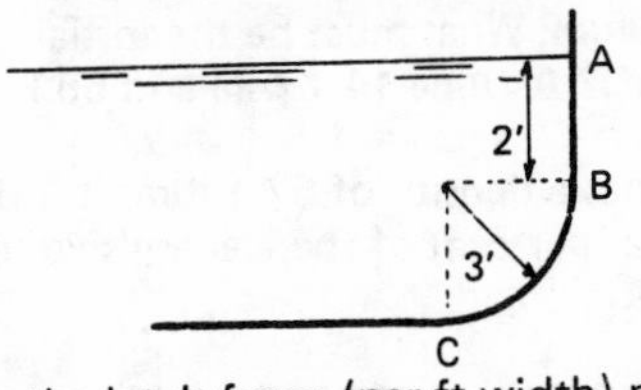

17. What is the latch force (per ft width) necessary at point B?

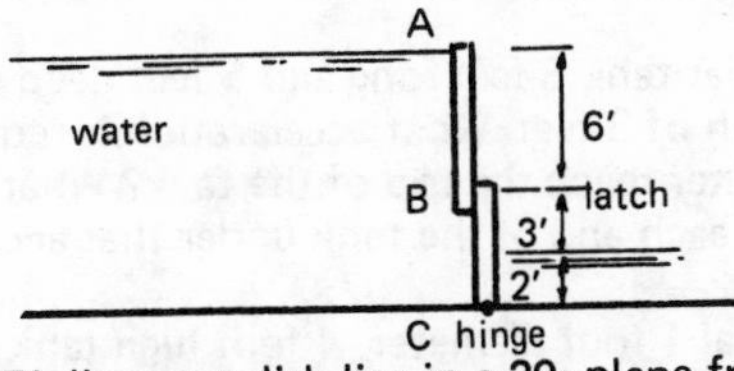

18. A 7' diameter disk lies in a 30· plane from the horizontal and is covered with water to a depth of 12' from its center. Find the magnitude, direction, and location of the resultant force on the disk.

MANOMETRY

19. What is the water pressure in the vessel?

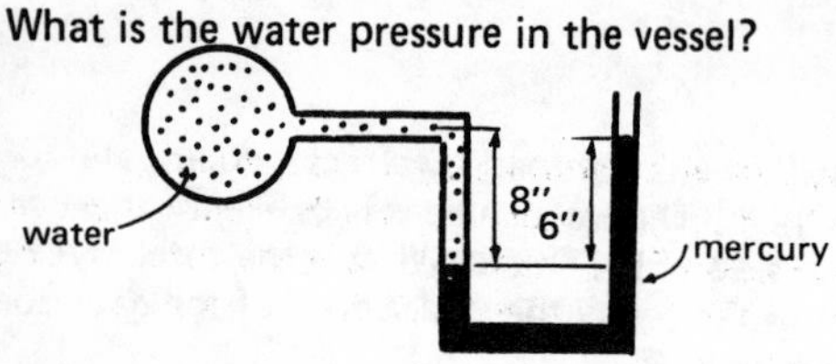

20. A mercury manometer across an orifice in a water line reads 18" Hg. What is the pressure drop across the orifice?

21. What is the gage pressure at point A if the specific gravity of kerosene is .82?

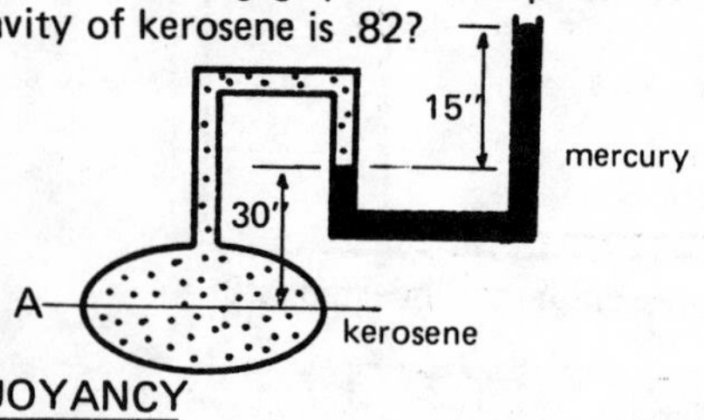

BUOYANCY

22. A water-tight, cubical box, 12" outside dimensions, is made from ¼" iron plate. If SG = 7.7 for the iron, will the box float?

23. A piece of lead is tied to 8 cubic inches of cork with a specific gravity of .25. The combination floats submerged in water. What is the weight of the lead?

24. What is the density of a stone which weighs 19.9 lb in air and 12.4 lb in water?

25. The draft of a ship with 4700 ft^2 of water line cross section is 12 feet. What cargo load is required to increase the draft 3 inches? Assume that sea water has a density of 64.0 lbm/ft^3.

26. A 16,000 ft^3 hydrogen balloon's cargo and structure weigh in at 650 lbm. What must be the initial sand weight to prevent lift off? Assume 14.7 psia and 55·F.

27. An iceberg has a density of 57.1 lbm/ft^3. If it floats in fresh water, what per cent of the iceberg's volume will be visible?

ACCELERATING FLUID MASSES

28. A rectangular tank 8 feet long and 5 feet deep contains water to a depth of 3 feet. What acceleration is required to make the water reach the top of the tank? What is the water force on each end of the tank under that acceleration?

29. A cylindrical 1 foot diameter, 4 foot high tank contains 3 feet of water. What rotational speed (rps) is required to spin water out of the top?

COMPRESSIBLE FLUIDS (Optional)

30. Assuming atmospheric conditions at sea level are 14.7 psia and 59·F, what is the expected pressure at 35,000 feet altitude? Assume the temperature decreases 3.5·F per 1000 feet altitude.

31. If atmospheric air is 14.7 psia and 60·F at sea level, what is the pressure at 12,000 feet altitude if the air is
(a) incompressible?
(b) compressed isothermally?
(c) compressed adiabatically?

Note: Assume all pipe dimensions given are exact, not nominal dimensions. Unless given data to the contrary, assume the density of water to be 62.4 pounds per cubic foot.

VELOCITY HEAD

1. What is the velocity head in feet of
   a) 70°F water flowing in a 2" pipe at 15 fps?
   b) 70°F air flowing in a 2" pipe at 15 fps?

REYNOLDS NUMBER

2. What is the Reynolds number of air at 250°F and 14.7 psia flowing through a 2" diameter pipe at 120 fps?

3. What is the Reynolds number of 100°F water flowing at 20 fps through a 4" diameter pipe?

4. An oil with kinematic viscosity of 0.005 ft^2/sec flows at 10 fps through a 3" pipe. Is the flow laminar or turbulent?

HYDRAULIC RADIUS

5. What is the hydraulic radius of the flows illustrated below?

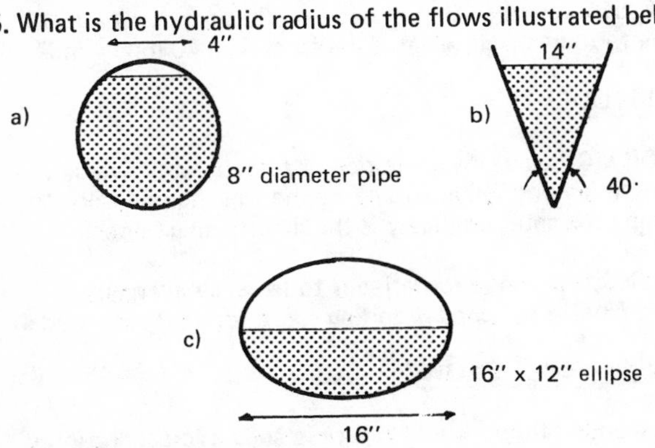

CONSERVATION OF MASS

6. What is the velocity at point B if water is flowing?

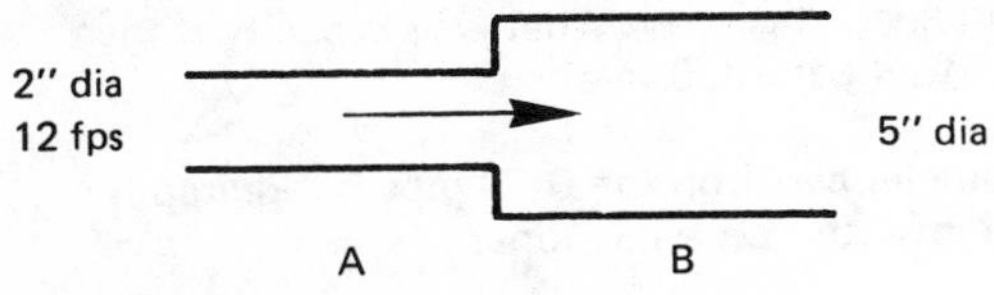

7. What is the density at point B?

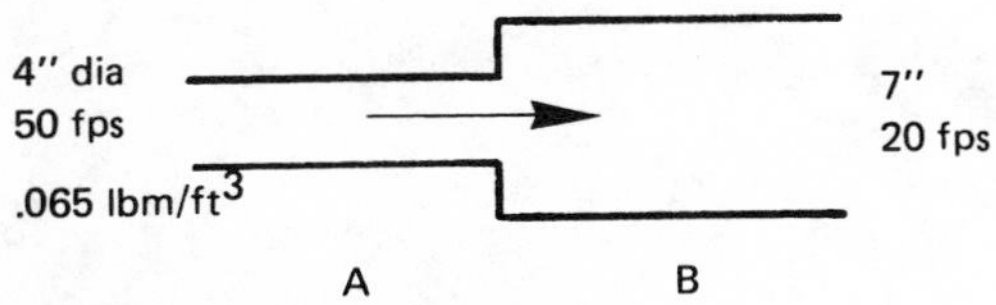

CONSERVATION OF ENERGY

8. Water flows from a source to a turbine, exiting 625 feet lower. The head loss is 58 feet due to friction, the flow rate is 1000 cfs, and the turbine efficiency is 89%. What is the output in kilowatts?

9. What is the water velocity at section B? The pipe area at point A is 12.5 square inches. The area at B is not known. The flow is 2.5 cubic feet per second.

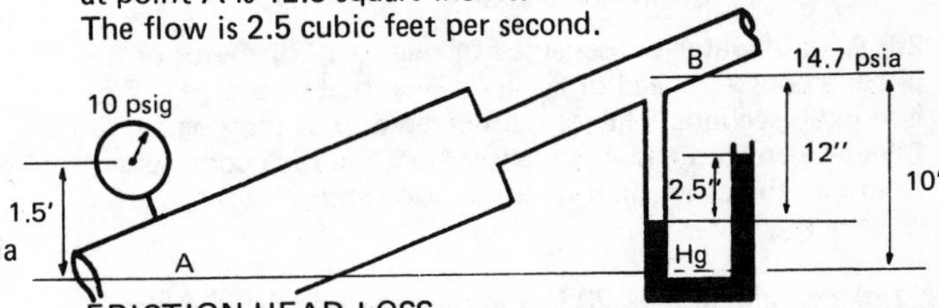

FRICTION HEAD LOSS

10. What is the friction loss per 100 feet of 12" diameter concrete duct if air with an average temperature of 100°F flows at 14 fps?

11. What is the head loss for 1000 feet of 2" galvanized iron pipe if the velocity is 4 fps and the fluid is 60°F water?

12. Points A and B are 3000 feet apart along a new 6" steel pipe. B is 60 feet above A. 750 gpm of fuel oil with a specific gravity of .9 and a dynamic viscosity of .0015 lb-sec/ft^2 flow in the pipe. The flow direction is from A to B. The friction factor is .03. What pressure must be maintained at A if the pressure at B is to be 50 psig?

MINOR LOSSES

13. What is the head loss for a 90° 3" diameter screwed elbow (K = .80) with water flowing at 15 fps?

14. A smooth 12" steel pipe 300 feet long has a flush entrance (K = .5) and a submerged discharge. 70°F water flows at 10 fps. What is the total head loss?

15. 70°F water flows at 12 fps in a 150 foot length of 4" steel pipe through an open globe valve and one regular elbow. What is the friction loss?

PUMPING POWER

16. A 3" diameter pipe 2000 feet long with friction factor of .020 carries water from a reservoir and discharges freely at a point 100 feet below the reservoir's surface level. Find the pump horsepower required to double the gravity flow.

17. Brine (SG=1.2) flows through a 2000 gpm pump. The pump outlet is 12" diameter and is 4 feet above the 12" inlet. The inlet vacuum is 6" mercury. The outlet pressure is 20 psig. What power does the pump add to the fluid?

18. In a centrifugal pump test, the discharge gage reads 100 psig and the suction gage reads 5 psig. Both read pressure above atmospheric. The gage centers are at the same level. The suction diameter is 3" and the discharge diameter is 2". Oil (SG=.85) flows at 100 gpm. With no losses, what is the pump horsepower?

FLUIDS

TANK DISCHARGE

19. Water flows through a perfect nozzle in the side of a water tank. The water level is constant at 20 feet. What height should the nozzle be located to make the stream flow a maximum horizontal distance before the water strikes the ground?

20. A jet of water is discharged through a 1" diameter orifice under a constant head of 2.1 feet. The total discharge is 228 lbm in 90 seconds. The jet is observed to pass through a point 2 feet downward and 4 feet away from the vena contracta. Compute the coefficient of contraction and the coefficient of velocity.

21. A cylindrical tank 20 feet in diameter and 40 feet high has a 4" hole in the bottom with C_d = .98. How long will it take for the water level to drop from 40' to 20'?

VENTURI METERS

22. A venturi meter with an 8" diameter throat is installed in a 12" diameter water line. Assume the venturi coefficient is equal to unity. What is the flow in cfs when a mercury manometer registers a 4" differential?

23. A 3"x1½" venturi (C_d=.98) is used in a pipe flowing air. Barometric pressure is 30.05" Hg, air temperature is 114°F, and the static pressure in the pipe is .287" Hg. How many pounds of air are flowing per second if the pressure drop across the venturi is .838" Hg?

ORIFICE PLATES

24. Oil (density of 55 lbm/ft^3) flows at 1 fps through a 1" diameter pipe. A .2" diameter orifice (C_f=.6) is installed. What is the indicated pressure drop?

25. What will be the measured pressure drop across a .2 foot diameter orifice installed in a 1 foot diameter pipe with water flowing at 2 fps?

OTHER MEASURING DEVICES

26. How high will the liquid rise in the tube shown below? Fluid density is 56 lb/ft^3. The pipe diameter is 12" and the velocity is 15 fps. The centerline pressure is 28 psig.

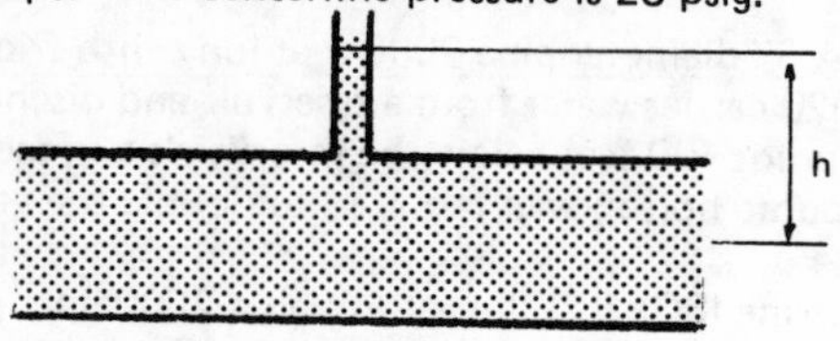

27. The difference between stagnation and static pressures in a pitot/static tube is 2.0" Hg. What is the relative velocity in air at 70°F and 14.7 psia?

IMPULSE-MOMENTUM

28. A pipe necks down from 24" at point A to 12" at point B. The discharge is 8 cfs in the direction of A to B. The pressure head at A is 20 feet. Assume no friction. Find the resultant force and its direction on the fluid if water is flowing.

29. Air flows in a pipe of .75 in^2 area at 250 fps. The density of the air is .075 lbm/ft^3. If the flow splits and both halves leave parallel to the plate, what is the force required to hold the plate?

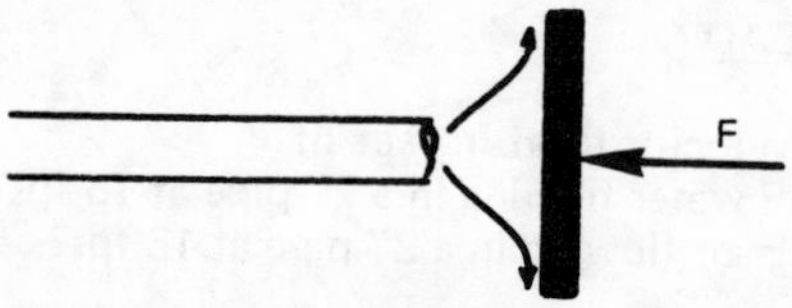

30. A uniform area jet travels at 600 fps and 100 cfs. What horizontal force acts on the water jet if it undergoes a

a) 90° turn?
b) 180° turn?

AERODYNAMICS

31. A sphere (C_D = .2) 1 foot in diameter travels 100 fps in standard atmospheric air. What is the drag force?

32. What area airfoil (C_L= .5) is required to obtain a 5000 lbf lift at 90 mph?

33. If the force of a 20 mph wind is 800 pounds on a 1000 square foot area wall, what is the force of a 40 mph wind?

SIMILITUDE

34. A (1/10) scale airplane model is to be used to evaluate a 60 mph landing. What should be the wind tunnel velocity assuming dynamic similarity is the determining factor?

35. A 600 foot long submarine is to travel submerged at 10 fps. What is the corresponding speed for a 10 foot model?

OPEN CHANNEL FLOW

36. A wooden flume (n=.012) of rectangular cross section is 2 feet wide. The flume carries 3 cfs of water on a 1% slope. What is the depth?

37. An 18" concrete storm drain (n=.013) flows half full under the influence of gravity. What is its capacity if the drain slopes 1 foot per 1000 feet?

38. What diameter cast iron (n=.017) pipe is required to flow 100 gpm of water on a 2% slope?

5 Open Channel Flow

Nomenclature

A	area	ft^2
b	weir width	ft
C	coefficient	–
d	depth	ft
d_H	hydraulic diameter	ft
D	pipe diameter	ft
E	specific energy	ft-lbf/lbm
f	Darcy friction factor	–
g	acceleration due to gravity[1] (32.2)	ft/sec^2
h	head	ft
H	head	ft
K	minor loss coefficient	–
L	channel length	ft
m	Bazin coefficent	–
n	Manning roughness coefficient	–
N	number of end contractions	–
p	pressure	lbf/ft^2
P	wetted perimeter	ft
Q	flow quantity	ft^3/sec
r_H	hydraulic radius	ft
S	slope of energy line (energy gradient)	–
v	velocity	ft/sec
w	channel width	ft
Y	weir height	ft
z	height above datum	ft

Symbols

ρ	density	lbm/ft^3

Subscripts

c	critical
f	friction
H	hydraulic
t	total

[1]Open channels are always constructed close to the surface of the earth. Therefore, this chapter does not distinguish between g and g_c.

1. Introduction

An open channel is a fluid passageway which exposes part of the fluid to the atmosphere. This type of channel includes natural waterways, canals, culverts, flumes, and pipes flowing under the influence of gravity (as opposed to pressure conduits which always flow full).

There are many difficulties in evaluating open channel flow. The unlimited geometric cross sections and variations in roughness have contributed to a small number of scientific observations upon which to estimate the required coefficients and exponents. Therefore, the analysis of open channel flow is more empirical and less exact than that of pressure conduit flow. This lack of precision, however, is more than offset by the percentage error in runoff calculations that generally precede the channel calculations.

Flow in open channels is almost always turbulent. However, within that category are many somewhat confusing categories of flow. Flow can be a function of time and location. If the flow quantity is invariant, it is said to be *steady*. If the flow cross section does not depend on the location along the channel, the flow is said to be *uniform*. Steady flow can be *non-uniform*, as in the case of a river with a varying cross section or on a steep slope. Furthermore, uniform channel construction does not ensure uniform flow, as in the case of a hydraulic jump.

Other types of flow are defined in the chapter glossary.

Due to the adhesion between the wetted surface of the channel and the water, the velocity will not be uniform across the area in flow. The velocity term used in this chapter is the mean velocity. The mean velocity, when multiplied by the flow area, gives the flow quantity.

$$Q = Av \qquad 5.1$$

2. Definitions

Accelerated flow: A form of varied flow in which the velocity is increasing and the depth is decreasing.

Backwater: Water upstream from a dam or other ob-

FLOW

struction which is deeper than it would normally be without the obstruction.

Backwater curve: A plot of depth versus location along the channel containing backwater.

Colloidal state: A mixture of water and extremely fine sediment which will not easily settle out.

Conjugate depth: The depth on either side of a hydraulic jump.

Contraction: A decrease in the width or depth of flow caused by the geometry of a weir, orifice, or obstruction.

Critical flow: Flow at the critical depth and velocity. Critical flow minimizes the specific energy and maximizes discharge.

Critical depth: The depth which minimizes the specific energy of flow.

Critical slope: The slope which produces critical flow.

Critical velocity: The velocity which minimizes specific energy. When water is moving at its critical velocity, a disturbance wave cannot move upstream since it moves at the critical velocity.

Energy gradient: The slope of the specific energy line (i.e., the sum of the potential and velocity heads.)

Flume: An open channel constructed above the earth's surface, usually supported on a trestle or piers.

Freeboard: The height of the channel side above the water level.

Gradient: See 'Slope'.

Hydraulic gradient: Slope of the potential head relative to the channel bottom. Since the potential head is equal to the depth of the channel, the hydraulic gradient and channel bottom are parallel. (Static pressure is omitted from the hydraulic gradient since the pressure is atmospheric at all points on the surface.)

Hydraulic jump: A spontaneous increase in flow depth from a velocity higher than critical to a velocity lower than critical.

Hydraulic mean depth: Same as 'Hydraulic radius'.

Limit slope: The smallest critical slope for a channel with a given shape and roughness.

Normal depth: The depth of uniform flow. This is a unique depth of flow for any combination of channel conditions. Normal depth is found from the Chezy-Manning equation.

Rapid flow: See 'Supercritical flow'.

Retarded flow: A form of varied flow in which the velocity is decreasing and the depth increasing.

Shooting flow: See 'Supercritical flow'.

Slope: The head loss per foot. For almost-level channels in uniform flow, the slope is equal to the tangent of the angle made by the channel bottom.

Standing wave: A stationary wave caused by an obstruction in a water-course. The wave cannot move (propagate) because the water is flowing at its critical speed.

Steady flow: Flow which does not vary with time.

Subcritical flow: Flow at greater than the critical depth (less than the critical velocity).

Supercritical flow: Flow at less than the critical depth (greater than the critical velocity).

Tranquil flow: See 'Subcritical flow'.

Uniform flow: Flow which has a constant depth, volume, and shape along its course.

Varied flow: Flow that has a changing depth along the water course. The variation is with respect to location, not time.

Wetted perimeter: The length of the channel which has water contact. The air-water interface is not included in the wetted perimeter.

3. Parameters

The *hydraulic radius* is the ratio of area in flow to wetted perimeter.

$$r_H = \frac{A}{P} \qquad 5.2$$

For a circular channel flowing either full or half-full, the hydraulic radius is (D/4). Hydraulic radii of other channel shapes is easily calculated from the basic definition. The *hydraulic depth* is the ratio of area in flow to the width of the channel at the fluid surface:

$$d_H = \frac{A}{w} \qquad 5.3$$

The *slope*, S, in open channel equations is the slope of the energy line. If the flow is uniform, the slope of the energy line will parallel the water surface and channel bottom. In general, the slope can be calculated as the energy loss per unit length of channel.

$$S = \frac{h_f}{L} \qquad 5.4$$

4. Governing Equations for Uniform Flow

Although it is of limited value, the incompressibility of the water allows the use of the continuity equation.

$$A_1 v_1 = A_2 v_2 \qquad 5.5$$

The most common equation used to calculate the flow velocity in open channels is the *Chezy equation.*

$$v = C\sqrt{r_H S} \qquad 5.6$$

Various equations for evaluating the coefficient C have been proposed. If the channel is small and very smooth, Chezy's formula may be used. f in the following equation is dependent on the Reynolds number which can be found in the usual manner from the Moody diagram.

$$C = \sqrt{8g/f} \qquad 5.7$$

Table 5.1
Manning's and Kutter's n

Kind of pipe	Variation From	Variation To	Design Use From	Design Use To
clean, uncoated cast iron	.011	.015	.013	.015
clean, coated cast iron	.010	.014	.012	.014
dirty, tuberculated cast iron	.015	.035		
riveted steel	.013	.017	.015	.017
welded steel	.010	.013	.012	.013
galvanized iron	.012	.017	.015	.017
brass and glass	.009	.013		
wood stave	.010	.014		
small diameter			.011	.012
large diameter			.012	.013
concrete	.010	.017		
with rough joints			.016	.017
dry mix, rough forms			.015	.016
wet mix, rough forms			.012	.014
very smooth, finished			.011	.012
vitrified sewer	.010	.017	.013	.015
common-clay drainage tile	.011	.017	.012	.014
asbestos			.011	
planed timber			.011	
canvas			.012	
unplaned timber			.014	
brick			.016	
rubble masonry			.017	
smooth earth			.018	
firm gravel			.023	
corrugated metal pipe			.022	
natural channels, good condition			.025	
natural channels with stones and weeds			.035	
very poor natural channels			.060	

If it is assumed that the channel is large, then the friction loss will not depend so much on the Reynolds number as on the channel roughness. The *Manning formula* is frequently used to evaluate the constant C. Notice that the value of C depends only on the channel roughness and geometry.

$$C = \frac{1.49}{n}(r_H)^{1/6} \qquad 5.8$$

n is the *Manning roughness constant,* and it is found in table 5.1. Putting equations 5.6 and 5.8 together produces the *Chezy-Manning equation,* applicable when the slope is less than .10.

$$v = \frac{1.49}{n}(r_H)^{2/3}\sqrt{S} \qquad 5.9$$

The *Kutter coefficient* has also seen widespread use, although it is more cumbersome than the Manning coefficient. In the following equation, n is the same as for the Manning equation.

$$C = \frac{41.65 + \dfrac{.00281}{S} + \dfrac{1.811}{n}}{1 + \left(41.65 + \dfrac{.00281}{S}\right)\dfrac{n}{\sqrt{r_H}}} \qquad 5.10$$

The Kutter formula has essentially been replaced by the Manning formula because of the former's complexity. There is also evidence that the Kutter equation is in error when S is very small (much smaller, however, than is usually encountered in normal design work). Other than these drawbacks, the two give similar results.

The *Bazin formula* has been used extensively in France. It is given by equation 5.11.

$$C = \frac{157.6}{1 + m/\sqrt{r_H}} \qquad 5.11$$

Values of m are given in table 5.2.

Table 5.2
Typical Bazin Coefficients

type of surface	m
smooth cement	.109
planed wood	.109
brickwork	.290
rough planks	.290
rubble masonry	.833
smooth earth channels	1.540
ordinary earth channels	2.360
rough channels	3.170

Example 5.1

A rectangular channel on a .002 slope is constructed of finished concrete and is 8 feet wide. What is the uniform flow if water is at a depth of 5 feet? Evaluate C with both the Manning and Kutter equations.

The hydraulic radius is:

$$r_h = \frac{(8)(5)}{5+8+5} = 2.22 \text{ ft}$$

From table 5.1, the roughness coefficient for finished concrete is .012. The Manning coefficient is

$$C = \frac{1.49}{.012}(2.22)^{1/6} = 141.8$$

The discharge from equations 5.1 and 5.6 is

$$Q = (141.8)(8)(5)\sqrt{(2.22)(.002)} = 377.9 \text{ cfs}$$

The Kutter coefficient, as calculated from equation 5.10, is 144.0. This results in a flow of 383.8 cfs.

5. *Energy and Friction Relationships*

The Bernoulli equation can be written for two points along the bottom of an open channel experiencing uniform flow.

$$\frac{p_1}{\rho} + \frac{v_1^2}{2g} + z_1 = \frac{p_2}{\rho} + \frac{v_2^2}{2g} + z_2 + h_f \qquad 5.12$$

However, $(p/\rho) = d$, the flow depth. And since $d_1 = d_2$ and $v_1 = v_2$,

$$h_f = z_1 - z_2 \qquad 5.13$$

If the slope is small, then the horizontal run and the channel length are almost identical. So, the hydraulic slope is

$$S = \frac{z_1 - z_2}{L} = \frac{h_f}{L} \qquad 5.14$$

Therefore, the friction loss in a length of channel is

$$h_f = LS \qquad 5.15$$

The friction loss can also be calculated from the *Darcy equation* using $D = 4r_H$, equation 5.7, and equation 5.8 to find f.

$$h_f = \frac{Ln^2v^2}{2.21(r_H)^{4/3}} \qquad 5.16$$

Minor losses from obstructions, curves, and changes in velocity are calculated with loss coefficients as they are in a pressure-conduit flow.

$$h_{minor} = \frac{Kv^2}{2g} \qquad 5.17$$

Example 5.2

In example 5.1, an open channel in normal flow had the following characteristics: $S = .002$, $n = .012$, $v = 9.447$ ft/sec, $r_H = 2.22$ ft. What is the energy lost per 1000 feet of channel?

From equation 5.15,

$$h_f = (1000)(.002) = 2 \text{ ft}$$

From equation 5.16,

$$h_f = \frac{1000(.012)^2(9.447)^2}{2.21(2.22)^{4/3}} = 2 \text{ ft}$$

6. *Most Efficient Cross Section*

The most efficient cross section (from an open channel standpoint) is the one which has maximum discharge for a given slope, area, and roughness. Wetted perimeter will be at a minimum (to minimize friction) when the flow is maximum.

Semicircular cross sections have the smallest wetted perimeter, and therefore the cross section with the highest efficiency is the semi-circle. Although such a shape can be constructed with concrete, it cannot be used with earth channels.

Rectangular channels are frequently used with wooden flumes. The most efficient rectangle is one which has a depth equal to one-half of the width.

Example 5.3

A rubble masonry open channel is being designed to carry 500 cfs of water on a .0001 slope. Using $n = .017$, find the most efficient dimensions for a rectangular channel.

Let the depth and width be d and w respectively. For an efficient rectangle, $d = \frac{1}{2}w$. Therefore,

$$A = dw = \tfrac{1}{2}w^2$$

$$P = d + w + d = 2w$$

$$r_H = \tfrac{1}{2}w^2/2w = \tfrac{1}{4}w$$

From equation 5.1, $Q = Av$. Combining this with equation 5.9,

$$500 = (\tfrac{1}{2}w^2)\left(\frac{1.49}{.017}\right)(\tfrac{1}{4}w)^{2/3}(.0001)^{1/2}$$

$$500 = (.1739)w^{8/3}$$

$$w = 19.82 \text{ ft}$$

$$d = \tfrac{1}{2}w = 9.91 \text{ ft}$$

7. *Circular Sections*

Combining equations 5.1 and 5.9 gives

$$Q = vA = \frac{1.49}{n}(A)(r_H)^{2/3}\sqrt{S} \qquad 5.18$$

The area in flow also depends on the hydraulic radius. If the flow is known, the diameter of a round pipe flowing full is

$$D = 1.33\left(\frac{Qn}{\sqrt{S}}\right)^{3/8} \qquad 5.19$$

If the round pipe is flowing half full, replace the 1.33 with 1.73.

FLOW

8. Flow Measurement with Weirs

A *weir* is an obstruction in an open channel over which flow occurs. Most weirs are designed for flow measurement. These weirs consist of a vertical flat plate with sharpened edges. Because of their construction, they are called sharp-crested weirs.

Sharp-crested weirs are most frequently rectangular, consisting of a straight, horizontal crest. However, weirs may also have trapezoidal and triangular openings.

If a rectangular weir is constructed with an opening width less than the channel width, the overfalling liquid sheet (called the *nappe*) decreases in width as it falls. This *contraction* of the nappe causes these weirs to be called *contracted weirs,* although it is the nappe that is actually contracted. If the opening of the weir extends the full channel width, the weir is called a *suppressed weir,* since the contractions are suppressed.

The derivation of the basic weir equation is not particularly difficult, but it is dependent on many simplifying assumptions. If it is assumed that the contractions are suppressed, upstream velocity is uniform, flow is laminar over the crest, nappe pressure is zero, the nappe is fully ventilated, and viscosity, turbulence, and surface tension effects are negligible, then the following equation may be derived from the Bernoulli equation:

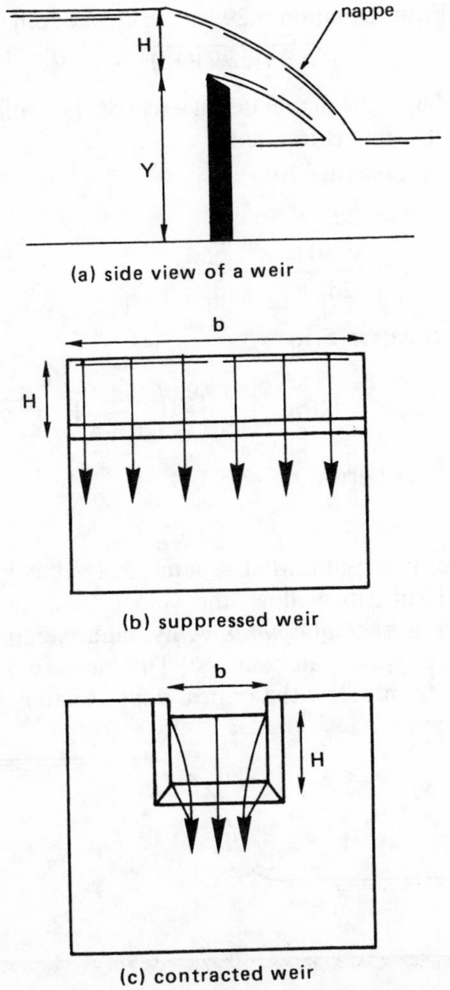

Figure 5.1 Contracted and Suppressed Weirs

$$Q = \tfrac{2}{3} b\sqrt{2g}\left[\left(H + \frac{v_1^2}{2g}\right)^{3/2} - \left(\frac{v_1^2}{2g}\right)^{3/2}\right] \qquad 5.20$$

If v_1 is negligible, then

$$Q = \tfrac{2}{3} b\sqrt{2g}(H)^{3/2} \qquad 5.21$$

Equation 5.21 must be corrected for all of the assumptions made. This is done by introducing a coefficient C_1.

$$Q = \tfrac{2}{3}(C_1) b\sqrt{2g}(H)^{3/2} \qquad 5.22$$

A number of investigations have been done to evaluate C_1. Perhaps the most widely used is the coefficient formula developed by Rehbock:

$$C_1 = \left[.6035 + .0813\left(\frac{H}{Y}\right) + \frac{.000295}{Y}\right]\left[1 + \frac{.00361}{H}\right]^{3/2} \qquad 5.23$$

If the contractions are not suppressed (i.e., one or both sides do not extend to the channel sides) then the actual width, b, should be replaced with the effective width.

$$b_{effective} = b_{actual} - (.1)(N)(H) \qquad 5.24$$

N is one if one side is contracted, and N is two if there are two end contractions.

Example 5.4

A sharp-crested, rectangular weir with two contractions is 2½ feet high and 4 feet long. A 4″ head exists upstream from the weir. What is the velocity of approach?

$$H = 4/12 = .333 \text{ ft}$$

From equation 5.24, N = 2 and the effective width is

$$b_{effective} = 4 - (.1)(2)(.333) = 3.93 \text{ ft}$$

The Rehbock coefficient (from equation 5.23) is

$$C_1 = \left[.6035 + .0813\left(\frac{.333}{2.5}\right) + \frac{.000295}{2.5}\right]\left[1 + \frac{.00361}{.333}\right]^{3/2}$$

$$= .624$$

From equation 5.22, the flow is

$$Q = \tfrac{2}{3}(.624)(3.93)\sqrt{(2)(32.2)}\,(.333)^{3/2}$$

$$= 2.52 \text{ cfs}$$

$$v = \frac{Q}{A} = \frac{2.52}{(4)(2.5 + .333)}$$

$$= .222 \text{ ft/sec}$$

9. Non-Uniform Flow

If water is introduced down a path with a steep slope (as down a spillway), the effect of gravity will be to cause an increasing velocity. This velocity will be op-

FLOW

posed by friction. Since the gravitational force is constant but friction varies as the square of velocity, these two forces eventually become equal. When they become equal, the velocity stops increasing, the depth stops decreasing, and the flow becomes uniform. Until they become equal, however, the flow is non-uniform (varied).

The total head is given by the Bernoulli equation:

$$H_t = z + \frac{p}{\rho} + \frac{v^2}{2g} \qquad 5.25$$

Specific energy is the total head with respect to the channel bottom. In this case, $z=0$ and $(p/\rho) = d$.

$$E = d + \frac{v^2}{2g} \qquad 5.26$$

In uniform flow, total head decreases due to the frictional effects, but specific energy is constant. In non-uniform flow, total head decreases, but specific energy may increase or decrease.

Since $v = Q/A$, equation 5.26 can be written

$$E = d + \frac{Q^2}{2gA^2} \qquad 5.27$$

Since the area depends on the depth, fixing the channel shape and slope and assuming depth will determine Q. This also will determine the specific energy, as illustrated in the specific energy diagram.

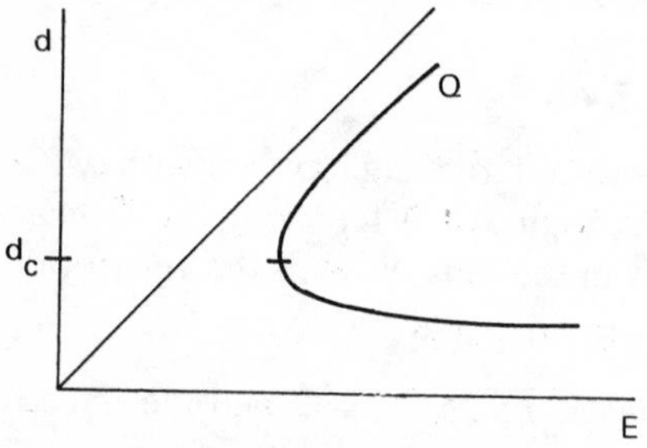

Figure 5.2 Specific Energy Diagram

For a given flow rate, there are two different depths of flow that have the same energy—a high velocity with low depth and a low velocity with high depth. The former is called *rapid (supercritical)* flow; the latter is called *tranquil (subcritical)* flow.

There is one depth, the *critical depth,* which minimizes the energy of flow. Critical depth for a given flow depends on the shape of the channel.

If the channel is rectangular, the critical depth is two-thirds of the critical specific energy.

$$d_c = \tfrac{2}{3}E_c \qquad 5.28$$

Equation 5.29 can be used to calculate the critical depth.

$$d_c = \sqrt[3]{Q^2/g(w)^2} \qquad 5.29$$

Once the critical depth is known, the corresponding velocity and discharge are given by the following two equations:

$$v_c = \sqrt{gd_c} \qquad 5.30$$

$$Q_c = v_c w d_c = w\sqrt{g}(d_c)^{3/2} \qquad 5.31$$

For non-rectangular shapes, the critical depth can be found by trial and error from the following equation in which b is the surface width.

$$\frac{Q^2}{g} = \frac{A^3}{b} \qquad 5.32$$

For any given discharge and cross section, there is a unique slope that will produce and maintain flow at critical depth. Once d_c is known, this critical slope can be found from the Chezy-Manning equation. Generally, the slope will not be critical, and flow will be supercritical (faster) or subcritical (slower).

Example 5.5

At a particular point in an open rectangular channel ($n = .013$, $S = .002$, $w = 10$ feet) the flow is 250 cfs and the depth is 4.2 feet. (a) Is the flow tranquil, normal, critical, or rapid? (b) What is the normal depth?

a) From equation 5.29, the critical depth is

$$d_c = \sqrt[3]{(250)^2/(32.2)(10)^2} = 2.69 \text{ ft}$$

Since the actual depth exceeds the critical depth, the flow is tranquil.

b) $A = (d_n)(10)$

$P = 2d_n + 10$

$$r_H = \frac{10d_n}{2d_n + 10} = \frac{5d_n}{d_n + 5}$$

From equation 5.18,

$$250 = (10)(d_n)\left(\frac{1.49}{.013}\right)\left(\frac{5d_n}{d_n+5}\right)^{2/3}\sqrt{.002}$$

By trial and error, $d_n = 3.1$ ft

If water is introduced at supercritical velocity to a section of subcritical flow, the velocity will be reduced through a *hydraulic jump.* A hydraulic jump is an abrupt rise in the water surface. The increase in depth is always from below the critical depth to above the critical depth.

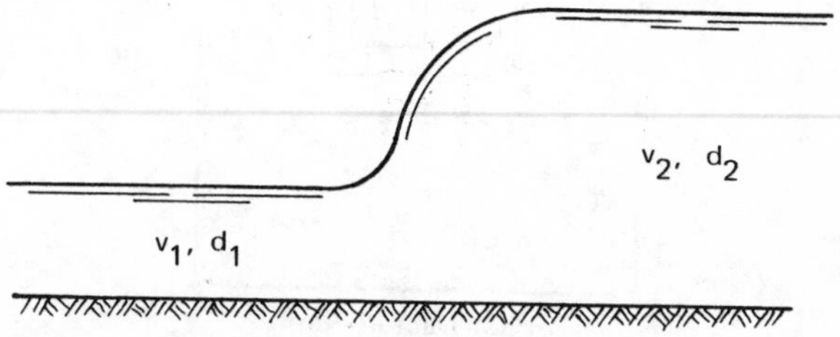

Figure 5.3 A Hydraulic Jump

If the depths d_1 and d_2 are known, then the velocity v_1 can be found from equation 5.33.

$$v_1^2 = \frac{gd_2}{2d_1}(d_1 + d_2) \qquad 5.33$$

d_1 and d_2 are known as conjugate depths because they occur on either side of the jump.

$$d_1 = -\tfrac{1}{2}d_2 + \sqrt{\frac{2v_2^2 d_2}{g} + \frac{d_2^2}{4}} \qquad 5.34$$

$$d_2 = -\tfrac{1}{2}d_1 + \sqrt{\frac{2v_1^2 d_1}{g} + \frac{d_1^2}{4}} \qquad 5.35$$

The specific energy dissipated in the jump is the energy lost per pound of water flowing.

$$\Delta E = \left(d_1 + \frac{v_1^2}{2g}\right) - \left(d_2 + \frac{v_2^2}{2g}\right) \qquad 5.36$$

Example 5.6

A hydraulic jump is produced at a point in a 10 foot wide channel where the depth is 1 foot and the flow is 200 cfs. (a) What is the depth after the jump? (b) What is the total power dissipated?

a) $v_1 = Q/A = 200/(10)(1) = 20$ ft/sec

From equation 5.35,

$$d_2 = -\tfrac{1}{2}(1) + \sqrt{\frac{(2)(20)^2(1)}{32.2} + \frac{(1)^2}{4}} = 4.51$$

b) The mass flow is

$(200)(62.4) = 12{,}480$ lbm/sec

The velocity after the jump is

$v_2 = 200/(10)(4.51) = 4.43$ ft/sec

From equation 5.36, the change in specific energy is

$$\left(1 + \frac{(20)^2}{2(32.2)}\right) - \left(4.51 + \frac{(4.43)^2}{2(32.2)}\right) = 2.4 \text{ ft}$$

The total power dissipated is

$(12480)(2.4) = 29{,}952$ ft-lbf/sec

FLOW

Practice problems for *Open Channel Flow* have been combined in this edition with problems from chapter 4, *Fluid Statics and Dynamics*.

6 Thermodynamics

PART 1: Properties of a Substance

Nomenclature

a	van der Waals' correction factor	atm-ft^6/pmole
A	area	ft^2
b	van der Waals' correction factor	ft^3/pmole
B	volumetric fraction	-
c	specific heat	BTU/lbm-°R
C	specific heat	BTU/pmole-°R
E	energy	BTU/lbm
g	local gravitational acceleration	ft/sec^2
g_c	gravitational constant (32.2)	lbm-ft/lbf-sec^2
G	gravimetric fraction	-
h	enthalpy	BTU/lbm
H	enthalpy	BTU/pmole
J	Joule's constant (778)	ft-lbf/BTU
k	ratio of specific heats, Boltzmann constant	-, J/°K
m	mass	lbm
$\dot{m}$	mass flow rate	lbm/sec
M	molecular weight, Mach number	lbm/pmole, -
n	number of moles, polytropic exponent	-
N	number of molecules	-
N_o	Avogadro's number (6.023 EE23)	molecules/gmole
p	pressure	lbf/ft^2
P	power	BTU/sec
Q	heat	BTU or BTU/lbm
R	specific heat constant	ft-lbf/lbm-°R
R*	universal gas constant (1545.33)	ft-lbf/pmole-°R
s	entropy	BTU/lbm-°R
S	entropy	BTU/pmole-°R
T	temperature	°R
u	internal energy	BTU/lbm
U	internal energy	BTU/pmole
v	velocity	ft/sec
V	volume	ft^3
W	work	BTU or BTU/lbm
x	quality, or mole fraction	-
z	height above datum	ft
Z	compressibility factor	-

Symbols

η	efficiency	-
ρ	density	lbm/ft^3
υ	specific volume	ft^3/lbm
ω	humidity ratio	-
μ	Joule-Thompson coefficient	°R-ft^2/lbf
ϕ	relative humidity	-
Φ	availability function	BTU/lbm

Subscripts

*	at sonic velocity
a	dry air
c	critical
f	saturated liquid
fg	vaporization
g	saturated vapor
k	kinetic
l	latent
m	mean
o	environment
p	potential, constant pressure, probable
r	ratio, reduced
rms	root-mean-squared
s	isentropic, sensible
sat	saturated
th	thermal
T	total
v	constant volume
w	water

1. Phases of a Pure Substance

Thermodynamics is the study of a substance's energy-related properties. This study can be theoretical and based on derivations, or it can be results-oriented. This chapter is practical in its approach, and it provides appropriate background for the useful applications introduced in chapter 7.

The properties of a substance and the procedures used to determine those properties depend on the phase of the substance. It is convenient to distinguish between more than just the usual solid, liquid, and gas phases. Because they behave according to different rules, it is necessary to distinguish between the following[1] phases and sub-phases.

solid — A solid does not take on the shape or volume of its container.

subcooled liquid — If a liquid is not saturated (i.e., the liquid is not at its boiling point), it is said to be subcooled. 60°F water at standard atmospheric pressure is subcooled since the addition of a small amount of heat will not cause vaporization.

saturated liquid — A saturated liquid has absorbed as much heat energy as it can without vaporizing. Liquid water at standard atmospheric pressure and 212°F is an example of a saturated liquid.

liquid-vapor mixture — A liquid and a vapor can coexist at the same temperature and pressure. This is called a two-phase, liquid-vapor mixture.

saturated vapor — A vapor (e.g., steam at standard atmospheric pressure and 212°F) which is on the verge of condensing is said to be saturated.

superheated vapor — A superheated vapor is one which has absorbed more heat than is needed to merely vaporize it. A superheated vapor will not condense when small amounts of heat are removed.

ideal gas — A gas is a highly superheated vapor. If the gas behaves according to the ideal gas laws, it is called an ideal gas.

real gas — A real gas does not behave according to the ideal gas laws.

gas mixtures — Most gases freely mix together. Two or more pure gases together constitute a gas mixture.

vapor/gas mixtures — Atmospheric air is an example of a mixture of several gases and water vapor.

These phases and sub-phases can be illustrated with a pure substance in the piston/cylinder arrangement shown in figure 6.1. The pressure in this system is determined by the weight of the piston, which moves freely to permit volume changes.

[1]Plasma and solids near absolute zero are not discussed in this chapter.

In illustration (a), the volume is minimum. This is the solid phase. The temperature will rise as heat, Q, is added to the solid. This increase in temperature is accompanied by a small increase in volume. The temperature increases until the melting point is reached.

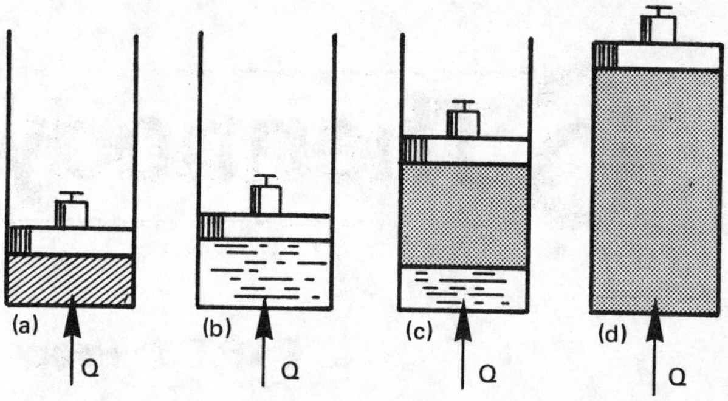

Figure 6.1 Phase Changes at Constant Pressure

The solid will begin to melt as heat is added to it at the melting point. The temperature will not increase until all of the solid has been turned into liquid. The liquid phase, with its small increase in volume, is illustrated by (b).

If the subcooled liquid continues to receive heat, its temperature will rise. This temperature increase continues until evaporation is imminent. The liquid at this point is said to be *saturated.* Any increase in heat energy will cause a portion of the liquid to vaporize. This is shown by (c) in which a liquid/vapor mixture exists.

When a substance exists as part liquid and part vapor at the saturation temperature, its *quality, x,* is defined as the fraction of the total mass which is vapor.

$$x = \frac{m_{vapor}}{m_{vapor} + m_{liquid}} \qquad 6.1$$

As with melting, evaporation occurs at constant temperature and pressure, but with a very large increase in volume. The temperature cannot increase until the last drop of liquid has been evaporated, illustration (d), at which point the vapor is said to be saturated.

Additional heat will result in a high-temperature *superheated vapor.* This vapor may or may not behave according to the ideal gas laws, depending on the temperature.

2. Determining Phase

It is important to know which phase[2] a substance is in since most equipment is incompatible with some phases. For example, you cannot put ice through a turbine. Nor can you put gas through a centrifugal pump.

[2]The word *phase* is always used instead of *state*, which has a different meaning in thermodynamics. The state of a substance will change any time a property changes, even though the phase remains the same.

THERMO

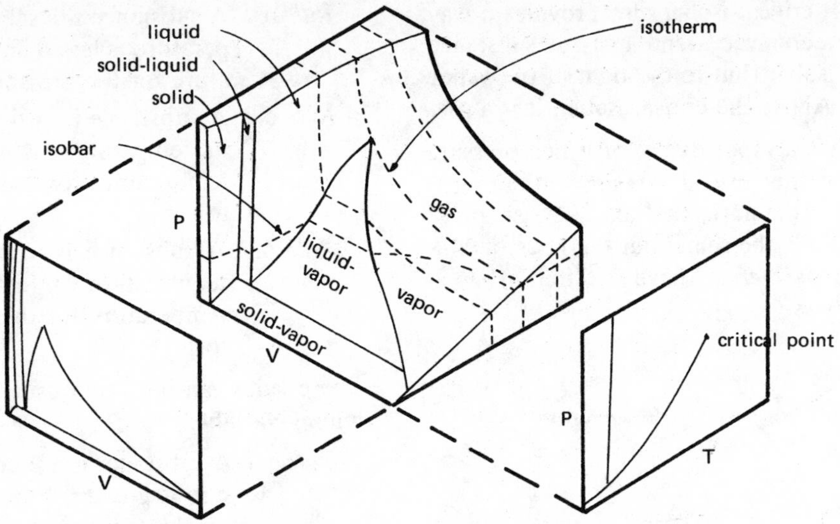

Figure 6.2 An Equilibrium Solid

It is theoretically possible to develop a three-dimensional surface which predicts the substance's phase based on the properties of pressure, temperature, and specific volume. Such an *equilibrium solid* is illustrated by figure 6.2. Equilibrium solids are not of much value in quantitative problems.

If one property is held constant through a process, a two-dimensional projection of the equilibrium solid can be used. This projection is known as an *equilibrium diagram* or *phase diagram*.

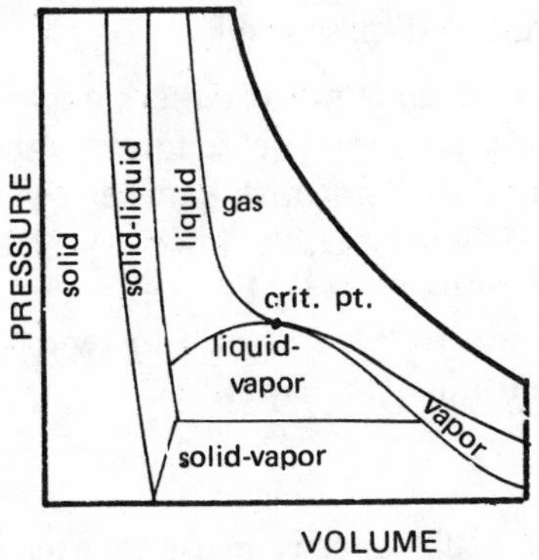

Figure 6.3 Phase Diagram

The important part of a phase diagram is limited to the liquid and vapor regions. A general phase diagram showing this region and the bell-shaped dividing line (known as the *vapor dome*) is given in figure 6.4.

The vapor dome region can be drawn with many variables for the axes. For example, either temperature or pressure can be used for the vertical axis. Energy, volume, or entropy can be chosen for the horizontal axis. However, the principles presented here apply to all combinations.

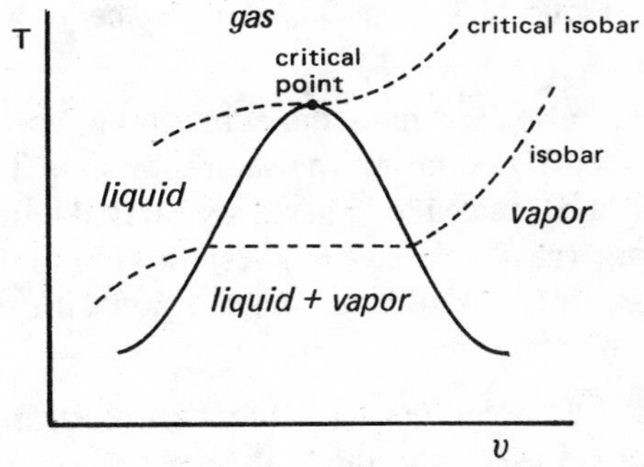

Figure 6.4 The Vapor Dome with Isobars

The left-hand part of the vapor dome line separates the liquid phase from the liquid/vapor phase. This part of the line is known as the *saturated liquid line*. Similarly, the right-hand part of the line separates the liquid/vapor phase from the vapor phase. This line is called the *saturated vapor line*.

Lines of constant pressure (*isobars*) can be superimposed on the vapor dome. Each isobar is horizontal as it passes through the two-phase region, verifying that both temperature and pressure remain unchanged as a liquid vaporizes.

Notice that there is no dividing line between liquid and vapor at the top of the vapor dome. Far above the vapor dome, there is no distinction made between liquids and gases since the properties of each are identical. The phase is assumed to be a gas.

The implied dividing line between liquid and gas is the isobar which passes through the top-most part of the vapor dome. This is known as the *critical isobar*, and the top-most point on the vapor dome is known as the

THERMO

critical point.[3] This critical isobar also provides a way to distinguish between a vapor and a gas. A substance below the critical isobar (but to the right of the vapor dome) is a vapor. Above the critical isobar, it is a gas.

Figure 6.5 illustrates a vapor dome for which pressure has been chosen as the vertical axis and enthalpy (h) has been chosen for the horizontal axis. The shape of the dome is essentially the same, but the lines of constant temperature (*isotherms*) have a different slope direction than isobars.

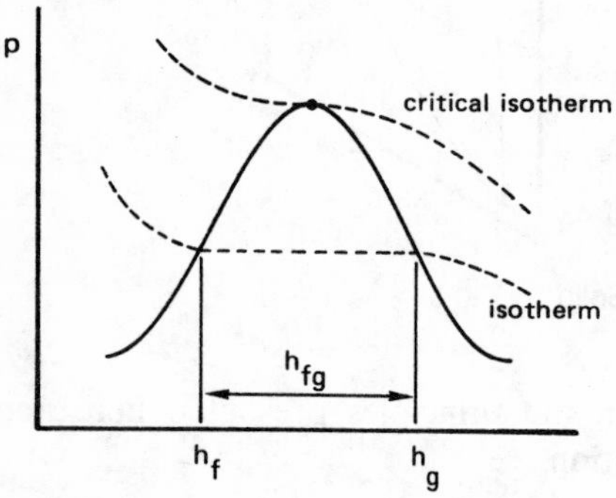

Figure 6.5 Vapor Dome with Isotherms

Figure 6.5 also illustrates the subscripting convention used to identify points on the saturation line. The subscript $_f$ (fluid) is used to indicate a saturated liquid. The subscript $_g$ (gas)[4] is used to indicate a saturated vapor. The subscript $_{fg}$ is used to indicate the difference in saturation properties.

The vapor dome makes a good tool for illustration, but it cannot be used to actually determine a substance's phase. Such a determination must be made based on the substance's pressure and temperature.

For example, consider water in a container surrounded by 14.7 psia air. Water's boiling temperature (the *saturation temperature*) at 14.7 psia is 212°F. If water has a temperature lower than 212°F, say 85°F, we know that the water is a liquid. On the other hand, if the water's properties are 14.7 psia and 270°F, we know that it is in vapor form.

This illustration is valid only for 14.7 psia water. Water at other pressures will have other boiling temperatures. (The lower the pressure, the lower the boiling temperature.) However, the rules given here follow directly from the above example. The rules will become more meaningful as you progress through this chapter.

[3]The critical properties for water are 1165°R and 218.2 atmospheres. Critical properties for other substances are given in table 6.7.

[4]Although this book makes it a rule never to call a vapor a gas, this convention is not adhered to in the field of thermodynamics. The subscript $_g$ is standard for a saturated vapor.

Rule 6.1 A substance is a subcooled liquid if its temperature is less than the saturation temperature for the pressure it is exposed to.

Rule 6.2 A substance is in the liquid-vapor region if its temperature is equal to the saturation temperature for the pressure it is exposed to.

Rule 6.3 A substance is a superheated vapor if its temperature is greater than the saturation temperature for the pressure it is exposed to.

These rules can be stated using pressure as the determining variable.

Rule 6.4 A substance is a subcooled liquid if its pressure is greater than the saturation pressure for the temperature it is exposed to.

Rule 6.5 A substance is in the liquid-vapor region if its pressure is equal to the saturation pressure for the temperature it is exposed to.

Rule 6.6 A substance is a superheated vapor if its pressure is less than the saturation pressure for the temperature it is exposed to.

3. *Properties Possessed by a Substance*

The thermodynamic state or condition of a substance is determined by its properties. *Intensive properties* are independent of the amount of substance present. Temperature, pressure, and stress are examples of intensive properties. *Extensive properties* are dependent on the amount of substance present. Examples are volume, strain, charge, and mass itself.

In this chapter, and in most thermodynamics books, both lower case and upper case forms of the same characters are used to distinguish between the basis for the properties. For example, the lower case letter *h* is used to represent enthalpy in BTU's[5] per pound (BTU/lbm). The upper case letter *H* is used to represent enthalpy in BTU's per mole (BTU/pmole).

A. *Mass: m*

The mass of a substance is measured exclusively in English units in this chapter. Mass can be expressed in either pounds-mass (lbm) or pound-moles (pmole). A pound-mole is an amount of substance which has a mass in pounds equal to the molecular weight. For example, 18 lbm would be one pmole of water.

B. *Temperature: T*

Temperature is a thermodynamic property of a substance which depends on energy content. Heat energy entering a substance will increase that substance's tem-

[5]The *British Thermal Unit* is a measure of heat energy. It is approximately the energy given off by burning one wooden match.

THERMO

perature. Heat energy will normally flow only from a hot object to a cold object. If two objects are at the same temperature (are in thermal equilibrium), no heat will flow.

If two systems are in thermal equilibrium with one another, they must be at the same temperature. If both of these are in equilibrium with a third, then all three are at the same temperature. This concept is known as the *Zeroth Law of Thermodynamics*.

The most commonly used scales for measuring temperature are the Fahrenheit and Celsius[6] scales. The relationship between these two scales is:

$$T_{°F} = 32 + (9/5)T_{°C} \tag{6.2}$$

An absolute temperature scale defines temperature independently of the properties of any particular substance. This is unlike the Celsius and Fahrenheit scales which are based on the freezing point of water. Absolute temperature scales should be used for all calculations unless a temperature difference is needed.

The absolute scale in the English system is the *Rankine scale*.

$$T_{°R} = T_{°F} + 460° \tag{6.3}$$

$$\Delta T_{°R} = \Delta T_{°F} \tag{6.4}$$

The absolute temperature scale in the SI system is the *Kelvin scale*.

$$T_{°K} = T_{°C} + 273° \tag{6.5}$$

$$\Delta T_{°K} = \Delta T_{°C} \tag{6.6}$$

These four temperature scales are compared in figure 6.6

C. Pressure: p

Pressure in thermodynamics problems can be given in psi, psf, or atmospheres. One *standard*[7] *atmosphere* is approximately 14.7 psia. Other pressure units include inches of water (407.1 inches of water equal one atmosphere), millimeters of mercury (760 millimeters of mercury equal one atmosphere), torr (760 torr equal one atmosphere), and bars (one bar equals one atmosphere). Torr and millimeters of mercury are essentially identical.

D. Density: ρ

Density has been covered in chapter 4. As in that chapter, this chapter gives density in lbm/ft³. Density is the reciprocal of specific volume.

$$\rho = \frac{1}{\upsilon} \tag{6.7}$$

E. Specific Volume: υ

Specific volume is the reciprocal of density. It is the volume that is occupied by one pound-mass of the substance. As such, its units are ft³/lbm.

$$\upsilon = \frac{1}{\rho} \tag{6.8}$$

F. Internal Energy: u and U

Internal energy encompasses all of the potential and kinetic energies of the atoms or molecules in a substance. Energies in the translational, rotational, and vibrational modes are included. Since this movement increases as the temperature increases, internal energy is a function of temperature. Internal energy does not depend on the process or path taken to reach a particular temperature.

Internal energy can be represented by a lower case *u* with units of BTU/lbm, or it can be represented by an upper case *U* with units of BTU/pmole. Of course, the relationship between *u* and *U* depends on the molecular weight of the substance.

$$U = Mu \tag{6.9}$$

Calculations in this book use *u* (the *specific internal energy*) exclusively. However, many tabulations of thermodynamic properties in other books are given in BTU/pmole (the *molar internal energy*). A conversion from a molar basis to a specific basis can be performed with equation 6.9.

[6]The name *Centigrade* is no longer correct.

[7]The term *STP (Standard Temperature and Pressure)* has several meanings. The standard pressure is always one atmosphere. The most common standard temperature is 32°F (0°C). However, 60°F, 68°F, and 70°F are also used in specific industrial situations.

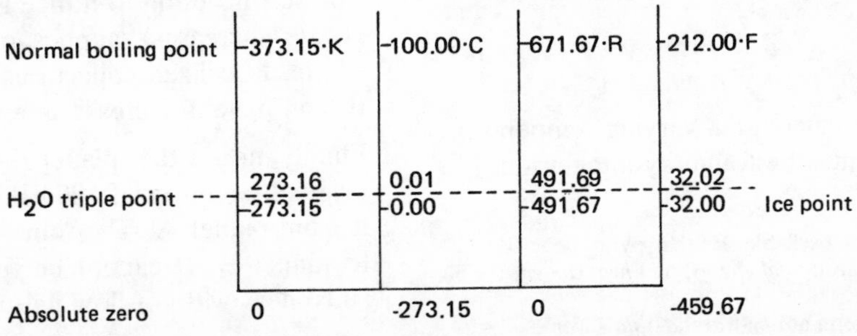

Figure 6.6 The Common Temperature Scales

G. Enthalpy: h and H

Enthalpy[8] is a property which represents the total[9] useful energy in the substance. This useful energy consists of two parts—the internal energy and *flow energy.*[10]

$$h = u + \frac{p\upsilon}{J} \quad \text{(BTU/lbm)} \qquad 6.10$$

$$H = U + \frac{pV}{J} \quad \text{(BTU/pmole)} \qquad 6.11$$

$$H = Mh \qquad 6.12$$

Enthalpy is defined as useful energy because, if the ambient conditions are proper, all of it can be used to perform useful tasks. It takes energy to increase the temperature of a substance. If that internal energy is recovered, it can be used to heat something else (e.g., to vaporize water in a boiler). Also, it takes energy to increase pressure and volume (as in blowing up a balloon). If pressure and volume are decreased, useful energy is given up.

The *J* term in equations 6.10 and 6.11 is known as *Joule's constant.* It has a value of 778 ft-lbf/BTU and is a conversion factor between ft-lbf and BTU. It is needed because internal energy (u) has the units of BTU, but the product of pressure and volume results in units of ft-lbf.

H. Entropy: s and S

Entropy[11] is a measure of energy which is no longer available to perform useful work within the current environment. Other definitions are frequently used (e.g., disorder of the system, randomness, etc.) but these alternate definitions are difficult to use in the framework of a numerical calculation.

The total unavailable energy in a system is equal to the summation of all unavailable energy inputs over the life of the system. That is,

$$s = \Sigma \Delta s \qquad 6.13$$

For an isothermal process (i.e., a process which takes place at a constant temperature), occurring at temperature T_o, the change in entropy is a function of the energy transfer. If *Q* is the energy transfer per pound-mass of substance, then the entropy change is given by equation 6.14.

$$\Delta s = \frac{Q}{T_o} \qquad 6.14$$

For processes that occur over a varying temperature, the entropy change must be found by integration.

$$\Delta s = \int ds = \int \frac{dQ}{T} \qquad 6.15$$

From the above equations, it can be seen that the units of entropy are BTU/lbm-°R. Entropy can also be given with units of BTU/pmole-°R, in which case a capital *S* would be used. Of course, *s* and *S* are related by the substance's molecular weight.

$$S = Ms \qquad 6.16$$

The concept of entropy and how it relates to unavailable energy can be illustrated by the three identical planets in figure 6.7. The average temperatures of planets A, B, and C are 530°R, 520°R, and 510°R respectively. All three planets are large, so small energy transfers to and from them will not change the average temperature. Therefore, such energy transfers can be considered isothermal.

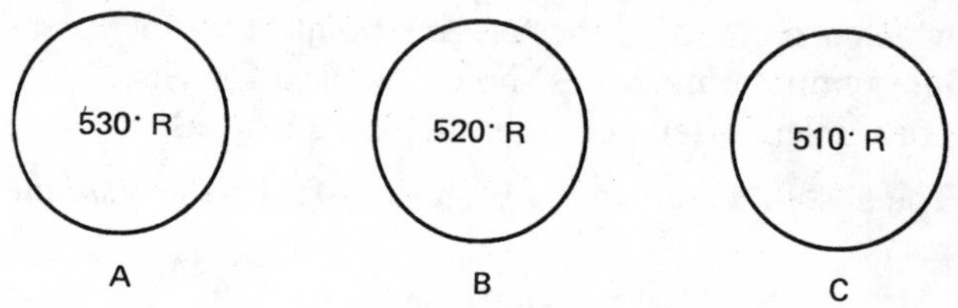

Figure 6.7 Three Planets in Space

Now, suppose that heat radiation transfers 1000 BTU/lbm from planet B to planet C. This transfer will occur naturally because planet B is hotter than planet C. The energy is gone from planet B, and it cannot be used. Furthermore, it cannot be recovered through a natural process, because heat will not flow by itself from a cold object to a hot object.[12]

From equation 6.14, the entropy changes in planets B and C are:

$$\Delta s_B = \frac{(-1000)\ \text{BTU/lbm}}{520°\text{R}} = -1.92\ \text{BTU/lbm-°R}$$

$$\Delta s_C = \frac{(1000)\ \text{BTU/lbm}}{510°\text{R}} = 1.96\ \text{BTU/lbm-°R}$$

Notice several things about this transfer. First, the entropy change is not the same for the two planets. Entropy is not conserved in an energy transfer process. In fact, entropy always increases when the total universe is considered. (The sum of Δs_B and Δs_C is non-zero.)

The second thing to notice is that the phrase "entropy always increases" applies only to the universe as a whole. Localized collections of matter, such as planet B, can have a decrease in entropy if they lose energy.

Finally, notice that planet B can be brought back to its original condition if 1000 BTU/lbm are transferred to it from planet A. The transfer must be from planet A to planet B. It cannot be from planet C to planet B since heat will not flow naturally in that direction.

[8]The accent is on the second syllable: En-thal′-py.
[9]The older terms of *total heat, total energy,* and *heat content* are no longer recommended.
[10]Other names for the pυ term are *pυ-work* and *flow work.*
[11]The accent is on the first syllable: En′-tro-py.

[12]This is one way of stating the *Second Law of Thermodynamics.*

An analogous situation where entropy is continually increasing and decreasing is the water in a closed boiler/turbine installation. The entropy increases when the water is vaporized in the boiler. The entropy decreases when the steam is expanded through the turbine. When the water returns to the boiler, its entropy is increased to its original value.

The Second Law of Thermodynamics may be related to entropy: *A natural process that starts in one equilibrium state and ends in another will go in the direction that causes th entropy of the system and environment to increase.* In this form, the Second Law applies only to *irreversible processes*—those processes having a 'natural direction'.

Although all real-world processes result in an overall increase in entropy, it is possible to conceptualize processes that have a zero entropy change. Such processes are said to be *reversible* or *isentropic*. For a reversible process,

$$\Delta s = 0 \tag{6.17}$$

Processes which contain friction are never reversible. Other processes which are not as obviously irreversible are listed here.

- stirring a viscous liquid
- moving fluid coming to rest
- magnetic hysteresis
- ideal gas expanding into a vacuum
- throttling
- releasing a stretched spring
- chemical reactions
- diffusion of gases
- freezing a supercooled liquid
- condensation of vapor
- heat conduction

I. Specific Heat: c, C

Heat energy is needed to cause a rise in temperature. Substances differ in the quantity of heat needed to produce a given temperature increase. The ratio of heat to the temperature change is called the *specific heat* of the substance, *c*.

$$c = \frac{Q}{m\Delta T} \tag{6.18}$$

Equation 6.18 can be rearranged to give the heat required to change the temperature of an object with mass *m*.

$$Q = mc\Delta T \tag{6.19}$$

The lower case *c* implies that the units are BTU/lbm-°R. The *molar specific heat* may be given, and its units are BTU/pmole-°R. A capital *C* is used to represent the molar specific heat.

$$C = Mc \tag{6.20}$$

Values of specific heat are listed in table 6.1.

Table 6.1
Approximate Specific Heats of
Some Liquids and Solids
(BTU/lbm-°R)

Substance	T, °F	c
Aluminum, pure	100	.225
alloy (2024-T4)	−200-200	.151-.217
Ammonia	−50-100	1.00-1.16
Asbestos	70	.195
Brass, red	70	.093
Bronze	70	.082
Concrete	70	.21
Copper, pure	100	.094
Freon-12	−20-100	.214-.240
Gasoline	0-100	.465-.526
Glass	70	.18
Gold, pure	100	.031
Ice	32	.49
Iron, pure	70	.11
cast (4% C)	70	.10
Lead, pure	100	.031
Magnesium, pure	100	.24
Mercury	0-600	.033-.032
Oil, light	100-300	.46-.54
Silver, pure	100	.06
Steel, (1010)	70	.102
stainless (301)	70	.109
Tin, pure	200	.055
Titanium, pure	100	.13
Tungsten, pure	100	.032
Water	32	1.007
	100	.998
Wood, typical	70	.6
Zinc, pure	100	.088

For gases, the specific heat depends on the conditions under which the heat exchange occurs. The most common conditions are those of constant volume and constant pressure, designated by subscripts $_v$ and $_p$ respectively.

$$Q = mc_v\Delta T \quad \text{(constant volume)} \tag{6.21}$$

$$Q = mc_p\Delta T \quad \text{(constant pressure)} \tag{6.22}$$

Values of c_p and c_v for common gases are given in table 6.4. Specific heats of solids and liquids vary only slightly with temperature. The mean specific heat is generally used for processes covering a large temperature range. c_p and c_V for solids and liquids are essentially the same.

J. Ratio of Specific Heats: k

For gases, the ratio of specific heats is defined by equation 6.23.

$$k = \frac{c_p}{c_v} \tag{6.23}$$

Values of k are given in table 6.4.

THERMO

K. Latent Heats

Energy which changes the phase of a substance is distinguished from the energy which produces only a change in temperature. Energy entering a substance, known as the *total heat*, may be divided into *latent* and *sensible heats*. Latent heat is that energy which produces a change in phase without causing a temperature change. Examples of latent effects are melting, vaporization, sublimation, and changes in crystalline form. Some typical values of latent energy for water are listed in table 6.2.

Table 6.2
Latent Heats for 14.7 psia Water

	BTU/lbm	cal/g	kcal/mole
Fusion (ice to water)	143.4	79.71	1.434
Vaporization (water to vapor)	970.3	539.55	9.703
Sublimation (ice to vapor)	1293.5	719.26	12.935

The latent heat of vaporization is so important in thermodynamic calculations that it has its own symbol—h_{fg}. h_{fg} may be found tabulated in most listings of thermodynamic properties of liquids.

Sensible heat is the heat that changes temperature. The amount of heat required is dependent on the temperature and specific heat. The total heat and sensible heat are equal for substances which undergo a temperature change without a phase change.

Example 6.1

How much energy is required to vaporize 1 pound-mass of water which is originally at 75°F and standard atmospheric pressure?

The sensible heat required to raise the temperature of the water from 75°F to 212°F is

$$Q_s = mc(T_2 - T_1)$$
$$= (1)\text{lbm}(1)\frac{\text{BTU}}{\text{lbm-°R}}(212 - 75)\text{°F}$$
$$= 137 \text{ BTU}$$

The latent heat required to vaporize the water is found in table 6.2 to be 970.3 BTU/lbm.

Therefore, the total heat required is

$$Q = Q_s + Q_l = 137 + 970.3$$
$$= 1107.3 \text{ BTU}$$

4. Using Tables to Find Properties

A. Mollier Diagram

The *Mollier diagram* is a graph of enthalpy versus entropy for steam. It is particularly suitable for determining property changes between the superheated vapor and the liquid-vapor regions. For this reason, the Mollier diagram covers only a limited region, as indicated by the dashed lines in figure 6.8.

The Mollier diagram plots the enthalpy for a pound of steam as the ordinate and entropy as the abscissa. Lines of constant pressure (isobars) slope upward from left to right. (In the low-pressure region at the right-hand side, dotted lines represent absolute pressure in inches of mercury and are convenient for exhaust steam calculations.) Below the saturation line, curves of constant moisture content slope down from left to right. Above the saturation line are lines of constant temperature and lines of constant superheat.

A complete Mollier diagram is given at the end of this chapter.

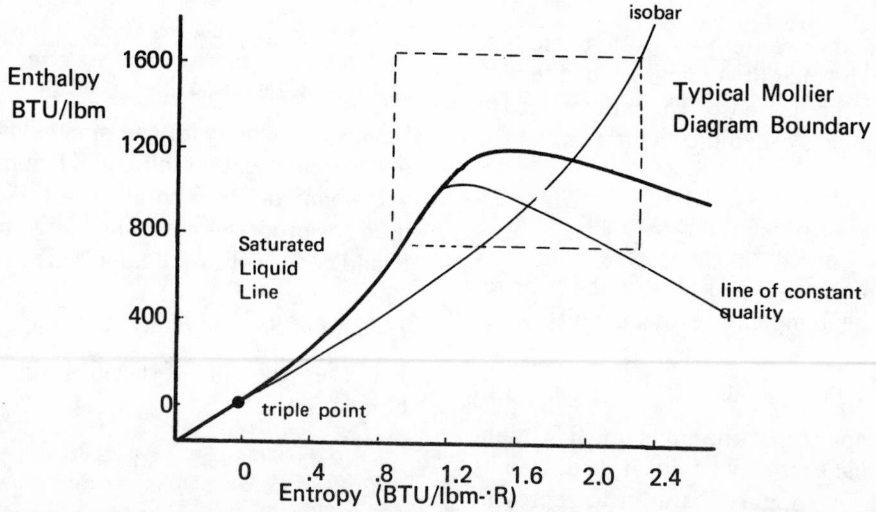

Figure 6.8 General Mollier Diagram

Example 6.2

Find the following properties using the Mollier diagram: (a) enthalpy and entropy of steam at 700 psia and 1000°F, (b) enthalpy of steam at 1 in. Hg absolute and 80% quality, and (c) final temperature of steam throttled from 700 psia and 1000°F to 450 psia.

(a) Reading directly from the Mollier diagram at the end of this chapter: h = 1510 BTU/lbm, s = 1.698 BTU/lbm-°R.

(b) 80% quality is the same as 20% moisture. Reading at the intersection of 20% moisture and 1 in. Hg: h = 885 BTU/lbm.

(c) By definition, a throttling process does not change the enthalpy. This process is represented by a horizontal line to the right on the Mollier diagram. Starting at the intersection of 700 psia and 1000°F and moving horizontally to the right until 450 psia is reached defines the end point of the process. The final temperature is interpolated as 988°F.

B. Steam Tables

The information presented graphically by the Mollier diagram can be obtained with greater accuracy from *steam tables.*[13] The tables represent extensive tabulations of data for liquid and vapor phases of water.

The steam tables contain data about specific volume (υ), enthalpy (h), internal energy (u), and entropy (s). These properties are functions of temperature, and appendix A is organized in this manner.

However, as shown in figure 6.4, there is a unique pressure associated with each temperature (i.e., there is only one horizontal isobar for each temperature). Since the pressure does not vary in even increments when temperature is changed, the second column of appendix A varies irregularly. A second property table, appendix B, is set up so that the pressure increments evenly.

In appendix A, the first column after temperature gives the corresponding saturation pressure. The next three columns give specific volume. The first of these gives the specific volume of the saturated liquid, υ_f; the second column gives the increase in specific volume when the state changes from saturated liquid to saturated vapor, υ_{fg}; the third column gives the specific volume of the saturated vapor, υ_g.

The relationship between υ_f, υ_{fg}, and υ_g is given by equation 6.24.

$$\upsilon_g = \upsilon_f + \upsilon_{fg} \tag{6.24}$$

The subsequent columns list the same data for enthalpy and entropy.

$$h_g = h_f + h_{fg} \tag{6.25}$$

$$s_g = s_f + s_{fg} \tag{6.26}$$

In appendix B, the first column after the pressure gives the corresponding saturation temperature. The next two columns give specific volume in a manner similar to appendix A, except that υ_{fg} is not listed. When necessary, υ_{fg} can be found by subtracting υ_f from υ_g. Appendix B also has a tabulation of internal energy.

C. Superheat Tables

In the superheated region, pressure and temperature are independent properties. Therefore, for each pressure, a large number of temperatures are possible. Appendix C is a superheated steam table which gives the properties of specific volume, enthalpy, and entropy for various combinations of temperature and pressure.

D. Compressed Liquid Tables

Water is only slightly compressible. For most thermodynamic problems, changes in properties for a liquid are negligible. In problems where the exact values are needed, appendix D should be used. This table gives the properties at the saturation state and the corrections to those properties for various pressures.

E. Gas Tables

Gas tables are essentially superheat tables for which an assumption has been made about the pressure. For example, appendix F is a gas table for air at low pressures. 'Low pressure' means less than several hundred psi pressure. However, reasonably good results can be expected even if pressures are higher.

Gas tables are indexed by temperature. That is, implicit in their use is the assumption that properties are functions of temperature only. Gas tables, however, are not arranged in the same way as other property tables.

The *volume ratio* (υ_r) and *pressure ratio* (p_r) columns can be used when gases take part in an isentropic process. These two columns are not the specific volume or pressure. They are ratios with arbitrary references that make analysis of isentropic processes easier. Their use is illustrated by example 6.3 and is based on equations 6.27 and 6.28.

$$\frac{\upsilon_{r,1}}{\upsilon_{r,2}} = \frac{V_1}{V_2} \tag{6.27}$$

$$\frac{p_{r,1}}{p_{r,2}} = \frac{p_1}{p_2} \tag{6.28}$$

Entropy is not listed at all in appendix F. Instead a column of *entropy functions* (ϕ) with the same units as entropy is given. The entropy function is not the same as specific entropy. However, it can be used to calculate

[13] A more general name is *property tables*. Property tables are available in essentially the same format for all common working fluids.

THERMO

the change in entropy as the gas goes through a process. This entropy change is calculated with equation 6.29.

$$s_2 - s_1 = \phi_2 - \phi_1 - \left(\frac{R}{J}\right)\ln\left(\frac{p_2}{p_1}\right) \quad 6.29$$

Example 6.3

Air is originally at 60°F and 14.7 psia. It is compressed isentropically to 86.5 psia. What is its new temperature and enthalpy?

From appendix F for 520°R, $p_{r,1} = 1.2147$. Using equation 6.28,

$$p_{r,2} = \frac{(1.2147)(86.5)}{14.7} = 7.148$$

Searching the p_r column of appendix F results in T = 860°R and h = 206.46 BTU/lbm.

5. Determining Properties

A. Properties of Solids

There are few mathematical relationships that predict the thermodynamic properties of solids. Properties such as temperature, specific heat, and density are usually known or stated in a problem. If the properties are not known or given, they must be found from tables. For example, table 6.1 gives specific heats of some solids.

THERMO

The reference point for properties of solids is usually absolute zero temperature. That is, properties like enthalpy and entropy[14] are defined to be zero at 0°R. This is an arbitrary convention. The choice in reference points does not affect the *change* in properties between two temperatures.

B. Properties of Sub-Cooled and Compressed Liquids

A sub-cooled liquid has a temperature less than the saturation temperature for the existing pressure. Unless the pressure on the liquid is very high (in which case appendix D should be used), the various thermodynamic properties can be considered to be functions of the liquid's temperature only.

Assuming that the properties are a function of temperature only, enthalpy, entropy, internal energy, and specific volume can be read directly from the property tables (appendix A or B for water) in the h_f, s_f, u_f, and υ_f columns respectively, regardless of the pressure.

If the liquid is compressed greatly, appendix D must be used. However, such extreme accuracy is seldom called for.

[14]Without regard to any arbitrary reference point, the Third Law of Thermodynamics says that *the absolute value of entropy of a perfect crystalline solid is zero.*

Example 6.4

What is the enthalpy of water at 30 psia and 240°F?

Although the substance is water, we do not know what phase the water is in. It could be liquid, vapor, or a combination. From appendix B, the saturation (boiling) temperature for 30 psia is 250.33°F. Since the water temperature is less than the boiling temperature, it is a liquid.

As a liquid, the properties are a function of temperature only. From appendix A for 240°F, h = 208.34 BTU/lbm.

Example 6.5

A piston compresses 1 lbm of 300°F saturated water to 1000 psia. What is the specific volume?

From appendix A or D, the specific volume of the water in its original saturated state is .01745 ft³/lbm.

The difference in specific volumes for the saturated and compressed liquids is given by appendix D.

$$(\upsilon - \upsilon_f)\ \text{EE5} = -6.9$$

For ease of tabulation, this entry has been multiplied by EE5. The actual correction must be multiplied by EE−5 to recover the correct value.

$$\upsilon_2 = .01745 - 6.9\ \text{EE}{-5} = .01738\ \text{ft}^3/\text{lbm}$$

C. Properties of Saturated Liquids

Either a saturated liquid's temperature or its pressure must be given to identify its thermodynamic state. Knowing one defines the other, since there is a one-to-one relationship between saturation pressure and saturation temperature. The first two columns of appendix A and appendix B can be used to determine saturation temperatures and pressures.

Since the properties tables are set up specifically for saturated substances, enthalpy, entropy, internal energy, and specific volume can be read directly from the h_f, s_f, u_f, and υ_f columns respectively. Density can be calculated as the reciprocal of the specific volume. The liquid's vapor pressure is the same as the saturation pressure listed in the table.

D. Properties of Saturated Vapors

Properties of saturated vapors can be read directly from the property tables. The vapor's pressure or temperature can be used to define the thermodynamic state. Enthalpy, entropy, internal energy, and specific volume can be read directly as h_g, s_g, u_g, and υ_g respectively.

E. Properties of a Liquid-Vapor Mixture

If a thermodynamic property has a value between the saturated liquid and vapor values (i.e., h is between h_f

and h_g), it will consist of a mixture of liquid and vapor. The quality of such a mixture can be calculated from equation 6.30, where h_f and h_{fg} are read from the property table.

$$x = \frac{h - h_f}{h_{fg}} \qquad 6.30$$

Similar equations can be written for entropy, internal energy, and specific volume.

Once the quality is known, it can be used to calculate the other thermodynamic properties by using equations 6.31 through 6.34.

$$h = h_f + xh_{fg} \qquad 6.31$$

$$s = s_f + xs_{fg} \qquad 6.32$$

$$u = u_f + xu_{fg} \qquad 6.33$$

$$\upsilon = \upsilon_f + x\upsilon_{fg} \qquad 6.34$$

Example 6.6

What is the specific volume of a 200°F steam mixture with a quality of 90%?

Using appendix A and equation 6.34,

$$\upsilon = .01663 + (.90)(33.62)$$
$$= 30.27 \text{ ft}^3\text{/lbm}$$

Example 6.7

What is the final enthalpy of steam which is expanded isentropically through a turbine from 100 psia and 500°F (superheated) to 3 psia?

From the superheat table (appendix C), $s_1 = 1.7085$ BTU/lbm-°R.

Since the expansion is isentropic (see equation 6.17), $s_2 = s_1$.

From appendix B for 3 psia vapor, the entropy of a saturated vapor (s_g) is 1.8863. Since s_2 is less than s_g, the expanded steam is in the liquid vapor region. The quality of the mixture can be found from equation 6.30.

$$x = \frac{s - s_f}{s_{fg}} = \frac{1.7085 - .2008}{1.6855}$$
$$= .895$$

The final steam enthalpy can be found from equation 6.31.

$$h = 109.37 + (.895)(1013.2)$$
$$= 1016.2 \text{ BTU/lbm}$$

F. Properties of a Superheated Vapor

Unless a vapor is highly superheated, its properties should be found from a superheat table, such as appendix C for water vapor. Since temperature and pressure are independent for a superheated vapor, both must be known to define the thermodynamic state.

If the vapor's temperature and pressure do not correspond to the superheat table entries, single or double interpolation will be required. Such interpolation can be avoided by using more complete tables, but where required, interpolation is standard practice.

Example 6.8

What is the enthalpy of water at 200 psia and 900°F?

It is not obvious which phase the water is in. From appendix B, the saturation (boiling) temperature for 200 psia water is 381.79°F. Since the water temperature exceeds the boiling temperature, the water exists as a vapor.

From appendix C, enthalpy can be read directly as 1476.2 BTU/lbm.

G. Properties of an Ideal Gas

A gas can be considered *ideal* if its pressure is very low and the temperature is much higher than its critical temperature. By definition, ideal gases behave according to the various ideal gas laws.

The first of two specific laws that define the behavior of an ideal gas is *Boyle's Law*. This law states that the volume and pressure vary inversely when the temperature is held constant.

$$p_1V_1 = p_2V_2 \qquad 6.35$$

The second specific law is *Charles' Law:* volume and temperature vary proportionally when the pressure is constant.

$$\frac{T_1}{V_1} = \frac{T_2}{V_2} \qquad 6.36$$

Equations 6.35 and 6.36 can be combined into a general law that applies to ideal gases undergoing any process.

$$\frac{p_1V_1}{T_1} = \frac{p_2V_2}{T_2} \qquad 6.37$$

Avogadro's Law states that equal volumes of different gases with the same temperature and pressure contain equal numbers[15] of molecules. For one mole of any gas, Avogadro's Law can be reformulated as the *Equation of State.*

$$\frac{pV}{T} = R^* \qquad 6.38$$

R^* is known as the *universal gas constant.* It is *universal* because the same number can be used with any gas.

[15]It is frequently stated that *Avogadro's* number is approximately 6.023 EE23 molecules per mole. This is correct value for a gram-mole (gmole). This chapter uses pound-moles (pmoles), so the number of molecules would be considerably larger.

THERMO

Due to the different units which can be used with the variables *p*, *V*, and *T*, there are different values of R*.

Table 6.3 Values of the Universal Gas Constant

1545.33 ft-lbf/pmole-°R
0.08206 atm-liter/gmole-°K
1.986 BTU/pmole-°R
1.986 cal/gmole-°K
8.314 joule/gmole-°K
0.730 atm-ft³/pmole-°R

Equation 6.38 can be modified to allow more than one mole of gas. If there are *n* moles, then

$$pV = nR^*T \qquad 6.39$$

The number of moles can be calculated from the substance's mass and molecular weight.

$$n = \frac{m}{M} \qquad 6.40$$

Equations 6.39 and 6.40 can be combined. *R* (no asterisk) is the *specific gas constant.* It is *specific* because it is valid only for a gas with a molecular weight of *M*.

$$pV = \frac{mR^*T}{M} = m(R^*/M)T \qquad 6.41$$

$$= mRT \qquad 6.42$$

Values of the specific gas constant and the molecular weight of various gases are contained in table 6.4.

Example 6.9

What mass of nitrogen is contained in a 2000 ft³ tank if the pressure and temperature are 14.7 psia and 70°F respectively?

From table 6.4, the specific gas constant of nitrogen is 55.2 ft-lbf/lbm-°R. From equation 6.42,

$$m = \frac{pV}{RT} = \frac{(14.7)(144)(2000)}{(55.2)(460 + 70)}$$

$$= 144.7 \text{ lbm}$$

Example 6.10

A 25 ft³ tank contains 10 lbm of an ideal gas with a molecular weight of 44. What is the pressure if the temperature is 70°F?

The specific gas constant is

$$R = \frac{R^*}{M} = \frac{1545.33 \text{ ft-lbf/pmole-°R}}{44 \text{ lbm/pmole}}$$

$$= 35.1 \frac{\text{ft-lbf}}{\text{lbm-°R}}$$

From equation 6.42,

$$p = \frac{mRT}{V} = \frac{(10)(35.1)(460 + 70)}{25}$$

$$= 7441 \text{ lbf/ft}^2$$

Table 6.4 Approximate Properties of Gases

GAS	SYMBOL	MOLECULAR WEIGHT	R	c_p	c_v	k
Acetylene	C_2H_2	26.0	59.4	0.350	0.2737	1.30
Air	---------	29.0	53.3	0.241	0.1724	1.40
Ammonia	NH_3	17.0	91.0	0.523	0.4064	1.32
Argon	A	39.9	38.7	0.124	0.0743	1.67
Carbon dioxide	CO_2	44.0	35.1	0.205	0.1599	1.28
Carbon monoxide	CO	28.0	55.2	0.243	0.1721	1.40
Chlorine	Cl_2	70.9	21.8	0.115	0.0865	1.33
Ethane	C_2H_6	30.07	51.3	0.422	0.357	1.18
Ethylene	C_2H_4	28.0	55.1	0.40	0.3292	1.22
Freon (R-12)	CCl_2F_2	120.9	12.6	---------	---------	1.13
Helium	He	4.0	386.3	1.25	0.754	1.66
Hydrogen	H_2	2.0	766.8	3.42	2.435	1.41
Isobutane	C_4H_{10}	58.12	26.6	0.420	0.387	1.09
Krypton	Kr	82.9	18.6	---------	---------	1.67
Methane	CH_4	16.0	96.4	0.593	0.4692	1.32
Neon	Ne	20.18	76.4	0.248	0.151	1.64
Nitrogen	N_2	28.0	55.2	0.247	0.1761	1.40
Oxygen	O_2	32.0	48.3	0.217	0.1549	1.40
Propane	C_3H_8	44.09	35.0	0.404	0.360	1.12
Steam (see note)	H_2O	18.0	85.8	0.46	0.36	1.28
Sulfur dioxide	SO_2	64.1	24.0	0.154	0.1230	1.26
Xenon	Xe	130.2	11.9	---------	---------	1.67

Note: Values for steam are approximate and may be used for low pressures and high temperatures only.

R is in ft-lbf/lbm-°R. Both c_p and c_v are in BTU/lbm-°F.

THERMO

Values of *h*, *u*, and *υ* for gases are usually read from gas tables, such as appendix F for air. Such tables are valid for gases under low pressure. The thermodynamic properties are considered to be functions of temperature only.

Enthalpy can be related to the equation of state because both contain a *pV* term.

$$h = u + \frac{p\upsilon}{J} = u + \frac{RT}{J} \qquad 6.43$$

Furthermore, density is the reciprocal of specific volume. So, the density of an ideal gas can be derived from the equation of state. If $m = 1$, then

$$p = \left(\frac{1}{\upsilon}\right)RT = \rho RT \qquad 6.44$$

Or,

$$\rho = \frac{p}{RT} \qquad 6.45$$

The specific heats of an ideal gas are related to the gas constants. The following equations make it possible to find one specific heat if the other is known.

$$c_p - c_v = \frac{R}{J} \qquad 6.46$$

$$C_p - C_v = \frac{R^*}{J} \qquad 6.47$$

$$c_p = \frac{Rk}{J(k-1)} \qquad 6.48$$

$$C_p = \frac{R^*k}{J(k-1)} \qquad 6.49$$

$$k = c_p/c_v = C_p/C_v \qquad 6.50$$

Example 6.11

What is the enthalpy of air at 100°F and 50 psia?

Since the pressure is low (less than 300 psia), appendix F can be used. From the T = 560°R line, the enthalpy is read directly as 133.86 BTU/lbm.

Statistical thermodynamics predicts the kinetic behavior of gas molecules. This *kinetic gas theory* results in an equation which gives a distribution of the number of gas molecules versus velocity. The *Maxwell-Boltzmann distribution* is illustrated in figure 6.9. The distribution law itself is given by equation 6.51, where *k* is the Boltzmann constant[16] and *m* is the mass of a molecule. Kinetic gas theory calculations are traditionally done in SI units. However, English units can be used if all constants and units are consistent.

$$\frac{dN}{dv} = \frac{4N}{\sqrt{\pi}}(v^2)\left(\frac{m}{2kT}\right)^{3/2}\exp\left(\frac{-mv^2}{2kT}\right) \qquad 6.51$$

[16]The Boltzmann constant has a value of 1.3803 EE−23 J/molecule-°K. These units are the same as kg-m²/sec²-molecule-°K.

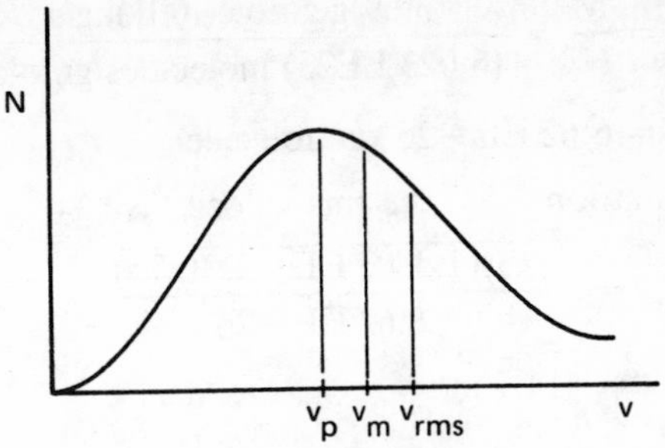

Figure 6.9
Maxwell-Boltzmann Velocity Distribution

The *most probable speed* of the molecules is

$$v_p = \sqrt{\frac{2kT}{m}} \qquad 6.52$$

The *average speed* is

$$v_m = 2\sqrt{\frac{2kT}{\pi m}} \qquad 6.53$$

The *root-mean-square speed* is

$$v_{rms} = \sqrt{\frac{3kT}{m}} = \sqrt{\frac{3RT}{M}} \qquad 6.54$$

The three velocities illustrated by figure 6.9 and defined by equations 6.52, 6.53, and 6.54 are related.

$$\frac{v_m}{v_p} = 1.128 \qquad 6.55$$

$$\frac{v_{rms}}{v_p} = 1.225 \qquad 6.56$$

The thermodynamic property called *temperature* has a molecular interpretation derived from the kinetic gas theory. It can be shown that the root-mean-square velocity is related to the absolute temperature. That is,

$$T = \frac{m}{3k}(v_{rms})^2 \qquad 6.57$$

Since $\frac{1}{2}mv^2$ is the definition of kinetic energy, the mean translational kinetic energy of the molecule is proportional to the mean absolute temperature.

$$\tfrac{1}{2}m(v_{rms})^2 = \tfrac{3}{2}kT \qquad 6.58$$

Example 6.12

What are the kinetic energy and rms velocity for 275°K argon molecules?

From equation 6.58,

$$E_k = (\tfrac{3}{2})(1.3803\ \text{EE}-23)(275) = 5.69\ \text{EE}-21\ \text{J/molecule}$$

The molecular mass of argon is its mass per mole divided by the number of molecules in a mole.

$$m = \frac{M}{N_o} = \frac{(39.9)\text{g/gmole}(.001)\text{kg/g}}{(6.023\ \text{EE23})\ \text{molecules/gmole}}$$

$$= 6.62\ \text{EE}-26\ \text{kg/molecule}$$

From equation 6.54, the rms velocity is

$$v_{rms} = \sqrt{\frac{(3)(1.3803\ \text{EE}-23)(275)}{6.62\ \text{EE}-26}}$$

$$= 414.7\ \text{m/s}$$

H. Properties of Ideal Gas Mixtures

A gas mixture consists of an aggregation of molecules of each gas component, the molecules of any single component being distributed uniformly and moving as if they occupied the space alone. As a consequence, the total pressure exerted by the mixture against the walls of its container is the sum of the *partial pressures* of the various gas components. The partial pressure of the components is the pressure that the gas would have if it occupied the total volume at the same temperature. The values of the partial pressures of the gases in a mixture are:

$$p_A = \frac{m_A R_A T}{V} \qquad 6.59$$

$$p_B = \frac{m_B R_B T}{V} \qquad 6.60$$

$$p_C = \frac{m_C R_C T}{V} \qquad 6.61$$

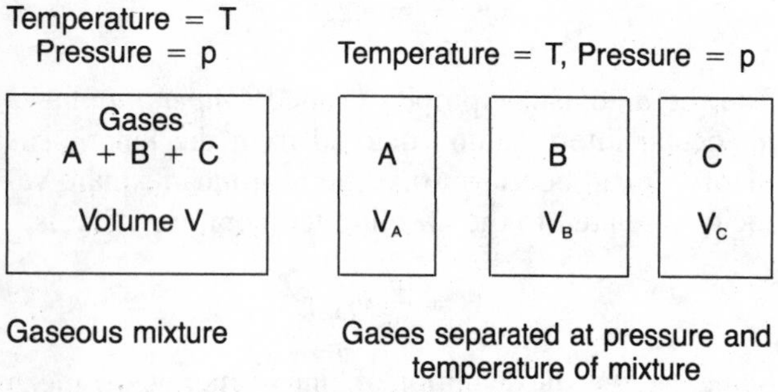

Figure 6.10
A Mixture of Ideal Gases

According to *Dalton's law,* the *total pressure* of the mixture is the sum of the partial pressures.

$$p = p_A + p_B + p_C \qquad 6.62$$

The ideal gas law can be used with one gas component and the entire mixture to calculate the partial pressure of the component.

$$\frac{p_A V}{pV} = \frac{n_A R^* T}{nR^* T} \qquad 6.63$$

The ratio (n_A/n) is known as the *mole fraction.*

$$x_A = \frac{n_A}{n} = \frac{n_A}{n_A + n_B + n_C} \qquad 6.64$$

Once the mole fraction is known for a component, it can be used with equation 6.63 to calculate the partial pressure. Since the V, R*, and T terms cancel,

$$p_A = x_A p \qquad 6.65$$

The mole fraction is not the same as the *gravimetric fraction,* the ratio of component masses.

$$G_A = \frac{m_A}{m} = \frac{m_A}{m_A + m_B + m_C} \qquad 6.66$$

The partial pressures can also be found from the gravimetric fraction. However, the average specific gas constant of the mixture is needed.

$$p_A = G_A\left(\frac{R_A}{R_{mixture}}\right)p \qquad 6.67$$

The *volumetric fraction* is defined as the ratio of a component's partial volume to the overall mixture volume. The partial volume is the volume the gas would occupy at the mixture temperature and pressure.

$$B_A = \frac{V_A}{V} = \frac{V_A}{V_A + V_B + V_C} \qquad 6.68$$

Since $pV = nR^*T$, equation 6.68 can be written as

$$B_A = \frac{n_A R^* T/p}{nR^* T/p} = \frac{n_A}{n} \qquad 6.69$$

Thus, the partial pressure ratio, mole fraction, and volumetric fraction are the same for ideal gas mixtures. The partial volume can be found from the volumetric fraction.

$$V_A = B_A V \qquad 6.70$$

Amagat's Law states that the total volume of a mixture of non-reacting gases is equal to the sum of the partial gas volumes.

$$V = V_A + V_B + V_C \qquad 6.71$$

The assumed absence of intermolecular forces in an ideal gas mixture is responsible for the total pressure being the sum of the partial pressures of the various component gases. For the same reason, mixture internal energy and enthalpy are equal to the sum of the internal energies and enthalpies of the various components.

If the mixing is reversible and adiabatic, then the entropy will also be equal to the sum of the individual entropies. This is an essential element of *Gibb's theorem,* which can be stated as follows: the total property (U, H, or S) of a mixture of ideal gases is the sum of the properties that the individual gases would have if each occupied the total mixture volume alone at the same temperature.

A summary of the composite gas properties is given in table 6.5.

Table 6.5
Summary of the Composite Gas Properties

Gravimetrically Weighted	Volumetrically Weighted
u	
h	
c_p	M
c_v	ρ
R	
s	

Example 6.13

0.14 lbm of octane vapor (M = 114) is mixed with 2 lbm of air in the manifold of the engine. The total pressure in the manifold is 12.5 psia and the temperature is 520°R. Assume octane vapor behaves ideally. (a) What is the total volume of this mixture? (b) What is the partial pressure of the air in the mixture?

(a) The average molecular weight of air is

$$M = \frac{R^*}{R_{air}} = \frac{1545.33}{53.3} = 29.0$$

The number of moles of octane and air are

$$n_{oct} = \frac{0.14}{114} = .001228$$

$$n_{air} = \frac{2}{29} = .068966$$

From equation 6.39,

$$V = \frac{(.001228 + .068966)(1545.33)(520)}{(12.5)(144)}$$

$$= 31.34 \text{ ft}^3$$

(b) The mole fraction of air is

$$x_{air} = \frac{.068966}{.001228 + .068966} = .983$$

$$p_{air} = (.983)(12.5) = 12.29 \text{ psia}$$

I. Properties of Real Gases

Real gases do not meet the basic assumptions set forth for an ideal gas. The molecules of the real gas occupy a definite volume that is not negligible in comparison with the total volume of the gas. This is especially true for gases at low temperatures. Furthermore, real gases are subject to *Van der Waals' forces,* which are attractive forces existing between molecules.

A modification of the perfect gas law accounts for molecular volumes and weakly-attractive intermolecular forces. *Van der Waals' equation of state* is:

$$(p + \frac{a}{V^2})(V - b) = nR^*T \qquad 6.72$$

For an ideal gas, the terms *a* and *b* are zero. When the spacing between molecules is close, such as at low temperatures, the molecules tend to attract one another and reduce the pressure exerted by the gas. The pressure is then corrected by the term (a/V^2). *b* is a constant dependent on the volume occupied by molecules in a dense state. Usually these corrections need to be applied only when the gas is below the critical pressure. Table 6.6 contains some typical correction factors.

Table 6.6 Van der Waals' Factors

Material	a (atm-ft^6/pmole)	b (ft^3/pmole)
air	345.2	.585
CO_2	926	.686
H_2	62.8	.427
O_2	348	.506
steam	1400	.488

The ideal gas equation may also be modified to predict real gas behavior by introducing a *compressibility factor.* The compressibility factor, Z, is a dimensionless constant that is dependent upon the pressure, temperature, and type of gas.

$$p\upsilon = ZRT \qquad 6.73$$

Compressibility factors for each gas can be plotted against pressure and temperature. One diagram may be constructed for several gases if the *principle of corresponding states* is applied. This principle states that *all gases behave alike whenever they have the same reduced variables.* These *reduced variables* are the ratios of pressure, volume, and temperature to their corresponding critical values.

$$p_r = \frac{p}{p_c} \qquad 6.74$$

$$T_r = \frac{T}{T_c} \qquad 6.75$$

$$V_r = \frac{\upsilon}{\upsilon_c} \qquad 6.76$$

The critical properties needed to calculate the reduced properties can be found from table 6.7.

Example 6.14

What is the specific volume of carbon dioxide at 2680 psia and 300°F?

The absolute temperature is (460 + 300) = 760°R. From table 6.7, the critical temperature and pressure of carbon dioxide are 547.8°R and 1071.0 psia. From equations 6.74 and 6.75,

$$T_r = \frac{760}{547.8} = 1.39 \qquad p_r = \frac{2680}{1071} = 2.5$$

THERMO

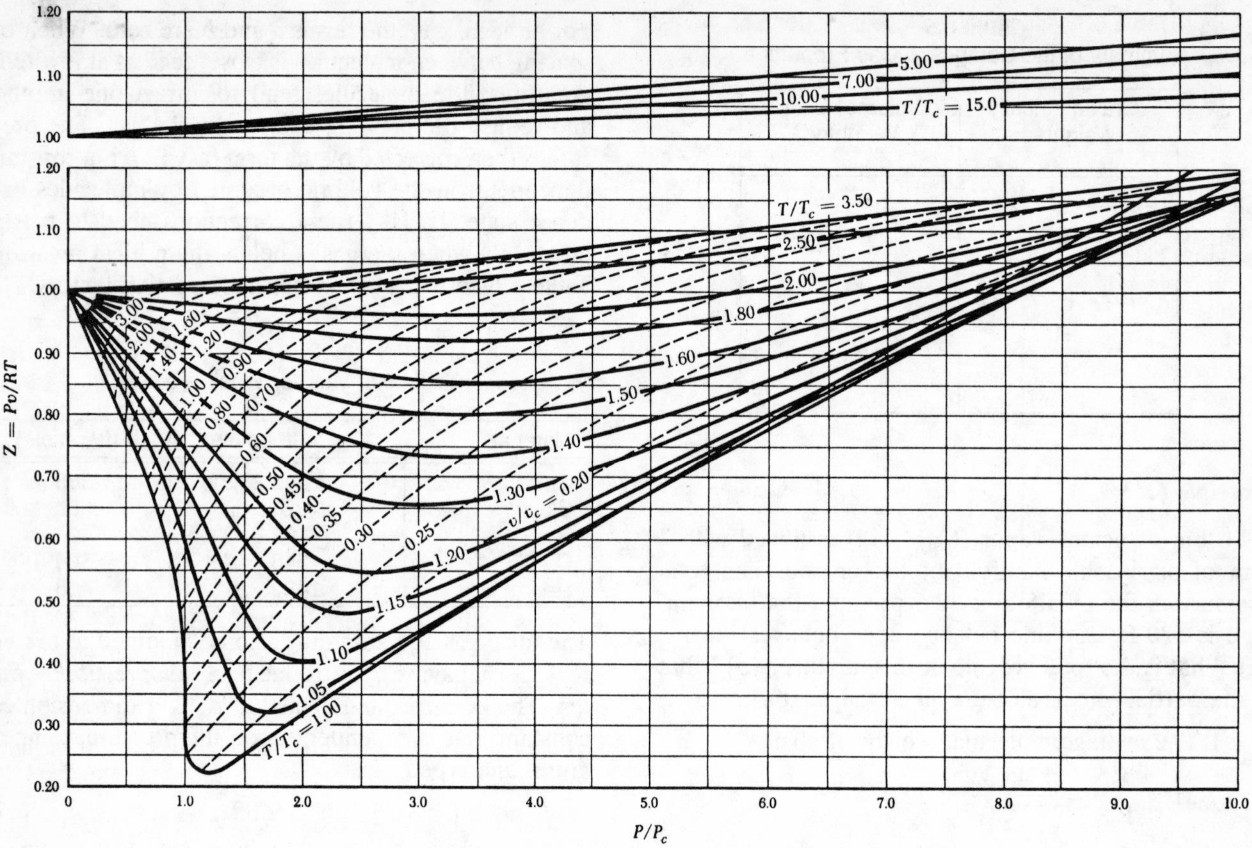

Figure 6.11 Compressibility Factors for Low Pressures

As originally presented by Professor Edward F. Obert and L.C. Nelson in "Generalized P-V-T Properties of Gases", ASME Transactions, 76, 1057 (1954).

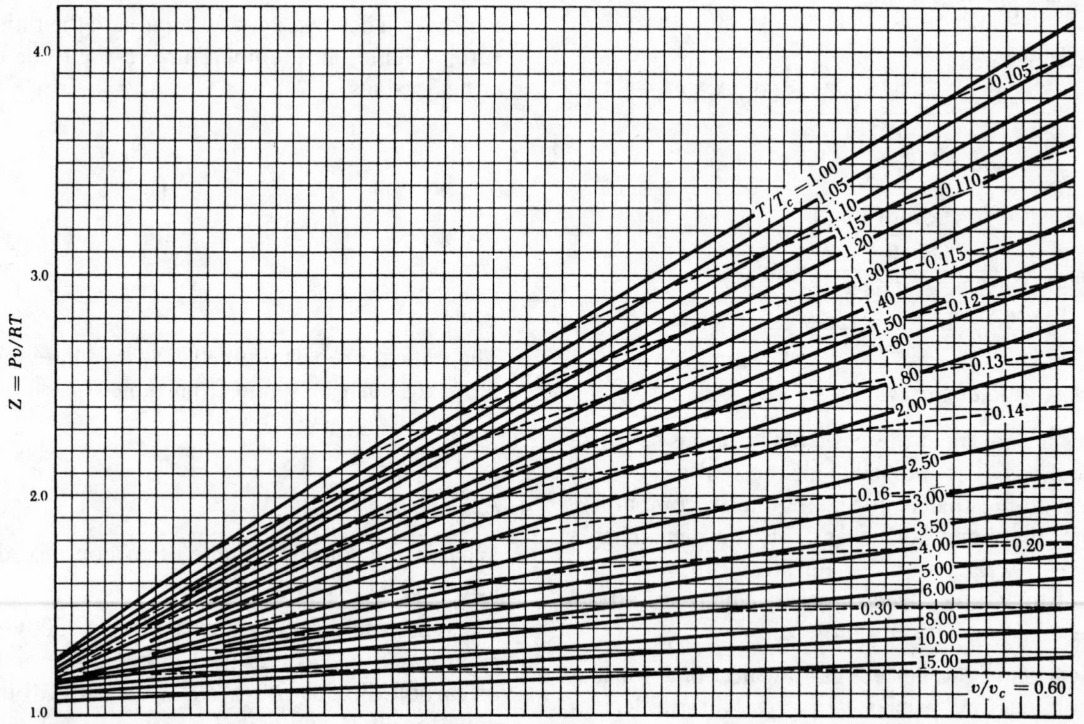

Figure 6.12 Compressibility Factors for High Pressures

Table 6.7 Approximate Critical Properties

Gas	Critical Temperature (°R)	Critical Pressure (psia)
air	235.8	547.0
ammonia	730.1	1639.0
argon	272.2	705.0
carbon dioxide	547.8	1071.0
carbon monoxide	242.2	508.2
chlorine	751.0	1116.0
ethane	549.8	717.0
ethylene	509.5	745.0
helium	10.0	33.8
hydrogen	60.5	188.0
mercury	2109.0	2646.0
neon	79.0	377.8
nitrogen	227.2	492.5
oxygen	278.1	730.9
propane	666.3	617.0
sulfur dioxide	775.0	1141.0
water vapor	1165.4	3206.0

From figure 6.11, Z = .75. From equation 6.73,

$$\upsilon = \frac{(.75)(35.1)(760)}{(2680)(144)} = .0518 \text{ ft}^3\text{/lbm}$$

J. Properties of Atmospheric Air

Dry atmospheric air is a mixture of oxygen, nitrogen, and small amounts of carbon dioxide, water vapor, argon, and other inert gases. If all constituents except oxygen are grouped with the nitrogen, the air composition is as given in table 6.8. It is necessary to supply (1/0.2315) or 4.32 pounds of air to obtain one pound of oxygen. The molecular weight of air is 28.9.

Table 6.8 Composition of Dry Atmospheric Air (Rare inert gases included as N_2)

	% by weight	% by volume
Oxygen (O_2)	23.15	20.9
Nitrogen (N_2)	76.85	79.1

Moist atmospheric air can be considered a mixture of two ideal gases — air and water vapor. As a mixture, everything in the previous section applies. For example, Dalton's law is applicable.

$$p_a = p_{air} + p_w \qquad 6.77$$

Since the study of atmospheric air is so important, it has its own name: *psychrometrics*. Psychrometrics initially seems complicated by the three different definitions of temperature. The terms dry bulb temperature, wet bulb temperature, and dewpoint temperature are not interchangeable.

The *dry bulb temperature* is the temperature that a regular thermometer would measure if exposed to atmospheric air. The *wet bulb temperature* is the temperature of air which has gone through an *adiabatic saturation process*. The *dewpoint temperature* is the drybulb temperature at which water starts to condense out when cooled at a constant pressure.

The wet bulb temperature is always less than the dry bulb temperature unless the air is saturated, in which case the two temperatures are the same. An adiabatic saturation process can be achieved with a *psychrometer*, which is essentially a regular thermometer with its bulb covered with wet cotton or gauze. If the thermometer is moved rapidly through the air (usually by twirling at the end of a string), the water in the gauze will evaporate, with the heat of vaporization coming from the air itself. The thermometer measures the temperature of the air, which has cooled due to the removal of the heat of vaporization.

For every temperature, there is a unique pressure for which water vapor is saturated. This pressure can be found from the second column of appendix A. If the partial pressure of the water vapor in atmospheric air is equal to this saturation pressure, the air is said to be *saturated*. (Actually, the water vapor is saturated, not the air. However, this inconsistency is part of psychrometrics.)

The amount of water vapor in the atmosphere can be measured by two different indexes. The *humidity ratio* (also known as *specific humidity*) is the ratio of water vapor and dry air masses.

$$\omega = \frac{m_w}{m_{air}} \qquad 6.78$$

Since $m = \rho V$, and since $V_w = V_{air}$, the humidity ratio can also be written as

$$\omega = \frac{\rho_w}{\rho_{air}} \qquad 6.79$$

Also, from the ideal gas equation of state, $m = pV/RT$, $V_w = V_{air}$ and $T_w = T_{air}$, so the ratio of masses is

$$\omega = \frac{R_a p_w}{R_w p_a} = \frac{53.3 p_w}{85.8 p_a} = .621\frac{p_w}{p_a} \qquad 6.80$$

The other index of moisture content is the *relative humidity*. The relative humidity is the partial water vapor pressure divided by the saturation pressure.

$$\phi = \frac{p_w}{p_{sat}} \qquad 6.81$$

From the ideal gas equation of state, $\rho = p/RT$, so the relative humidity can be written as

$$\phi = \frac{\rho_w}{\rho_{sat}} \qquad 6.82$$

Combining equations 6.80 and 6.81 yields equation 6.83.

$$\phi = 1.61\omega\left(\frac{p_a}{p_{sat}}\right) \qquad 6.83$$

THERMO

It is possible to develop mathematical relationships for enthalpy, entropy, internal energy, and specific volume for atmospheric air. However, this is almost never done in practice. Psychrometric properties can be read directly from a *psychrometric chart*, such as figure 6.14.

A psychrometric chart is easy to use, despite the multiplicity of scales on it. The thermodynamic state is defined by specifying any two intersecting scales (e.g., dry bulb and wet bulb temperatures, or a temperature and humidity). Once the air state has been located on the chart, all other properties can be read directly.

The psychrometric chart is additionally useful since many air conditioning processes follow along straight paths on the chart. The paths taken by air during typical air conditioning processes are shown in figure 6.13.

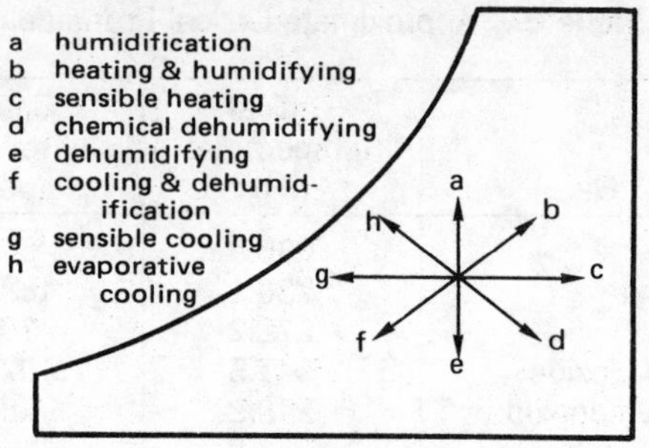

Figure 6.13 Air Conditioning Processes

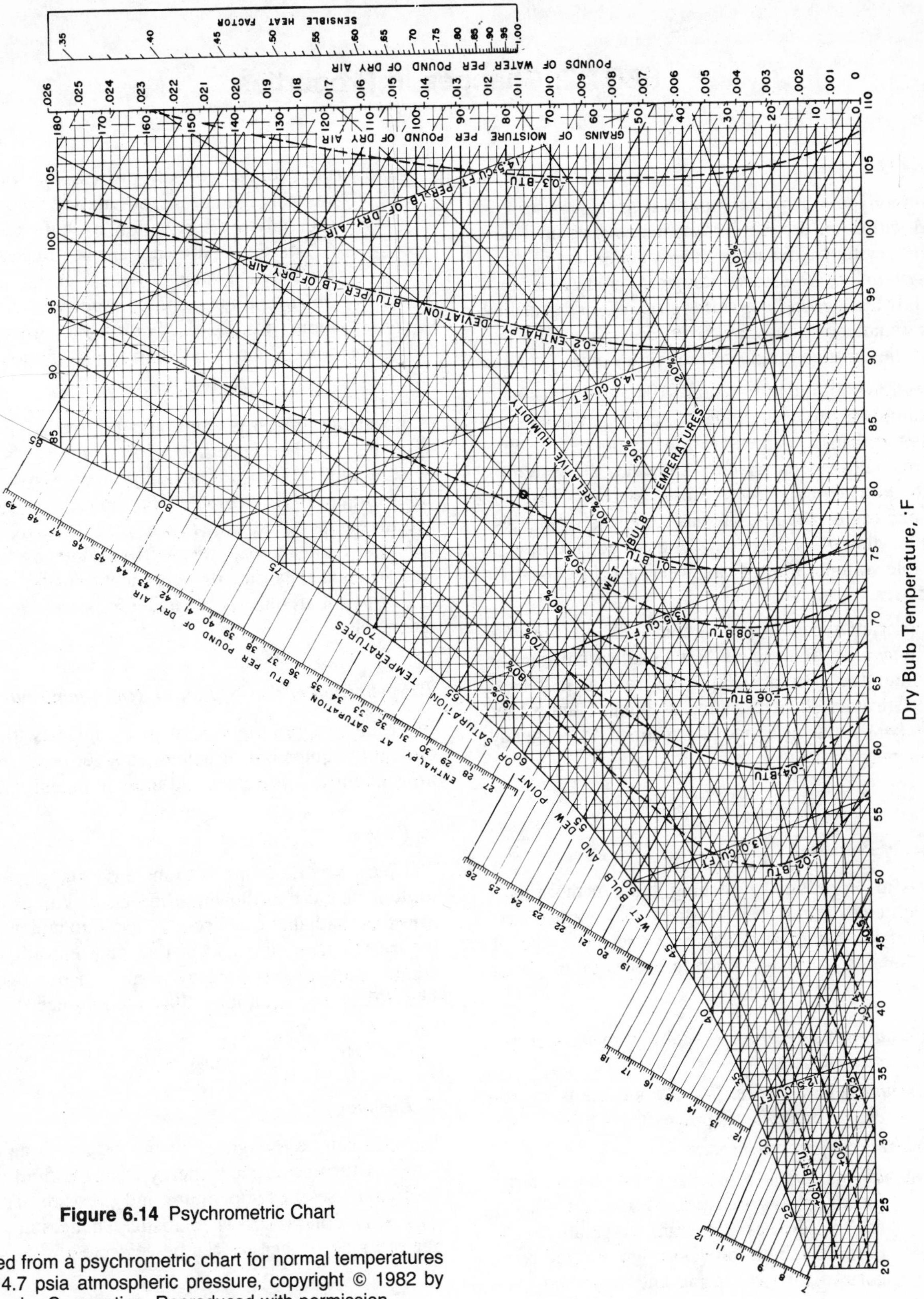

Figure 6.14 Psychrometric Chart

Derived from a psychrometric chart for normal temperatures and 14.7 psia atmospheric pressure, copyright © 1982 by the Carrier Corporation. Reproduced with permission.

PART 2: Changes in Properties

1. Systems

A thermodynamic *system* is defined as the matter enclosed within an arbitrarily but precisely defined volume. Everything external to the system is *surroundings* or *environment.* The surroundings and system are separated by the *system boundaries.* The region defined by the boundaries is known as a *control volume.* The surface of the control volume is known as a *control surface.*

If mass flows through a system, it is an *open system.* Typical open systems are pumps, heat exchangers, and jet engines. Energy can enter or leave an open system. If no mass crosses the system boundary, the system is said to be a *closed system.* Closed systems need not have a constant volume. The gas compressed by a piston in a cylinder is an example of a closed system with a variable volume. Energy can also cross closed system boundaries.

An important type of open system is the *steady flow open system.* To be a steady flow open system, matter must enter the system at the same rate as it leaves. Pumps, turbines, heat exchangers, boilers, and other thermodynamic devices may be assumed to be steady flow open systems.

2. Types of Processes

Changes in properties to working fluids in thermodynamic systems often depend on the type of process experienced by the fluids. This is particularly true of gases. Some of the most common processes are listed here.

- *constant pressure process* — also known as an *isobaric* process
- *constant volume process* — also known as an *isochoric* or *isometric* process
- *constant temperature process*
- *adiabatic process* — a process in which no energy (heat) crosses the system boundary. This is not the same as a constant temperature process. Adiabatic processes are further categorized into isentropic and throttling processes.
  - *isentropic process* — a process in which no change in entropy occurs
  - *throttling process* — a process in which no change in enthalpy occurs
- *polytropic process* — any process for which the working fluid properties may be predicted by the polytropic equation of state (equation 6.84). Generally, polytropic processes are limited to systems of ideal gases. n is the *polytropic exponent,* a property of the equipment, not the gas. For efficient air compressors, n is typically between 1.25 and 1.30.

$$p_1(V_1)^n = p_2(V_2)^n \qquad 6.84$$

A *reversible process* is one that is performed in such a way that, at the conclusion of the process, both the system and the local surroundings may be restored to their initial states. A process that does not meet these requirements is said to be *irreversible.* Some of the factors that render a real process irreversible are friction, unrestrained expansion of gases, heat transfer, and mixing of two materials. A reversible process is implicitly isentropic.

3. Application of the 1st Law to Real Equipment

In order to apply thermodynamic principles to actual operating equipment, it is necessary to make assumptions about the processes and the equipment itself.

A. Pumps

The purpose of a pump is to increase the total energy content of the fluid flowing through it. Pumps can be considered adiabatic devices. If the pump inlet and outlet are the same size and at the same elevations, the kinetic and potential energy terms can be neglected. The SFEE (see equation 6.104) then reduces to equation 6.85.

$$P = \dot{m}(h_2 - h_1) \qquad 6.85$$

B. Turbines

Turbines can be thought of as pumps operating in reverse. A turbine extracts energy from the fluid as the fluid decreases in temperature and pressure. The energy extraction process is adiabatic. Changes in potential and kinetic energies can be neglected.

$$P = \dot{m}(h_2 - h_1) \qquad 6.86$$

C. Heat Exchangers and Feedwater Heaters

A heat exchanger transfers energy from one fluid to another. Since energy cannot be created or destroyed, the total energy content of both input streams must be the same as the total energy content of both output streams. Heat exchangers can be considered adiabatic. No work is done within a heat exchanger. The potential and kinetic energies of the fluids can be neglected. Therefore, from the SFEE,

$$h_2 - h_1 = 0 \tag{6.87}$$

This can be rewritten as equation 6.88.

$$\dot{m}_A h_{A,in} + \dot{m}_B h_{B,in} = \dot{m}_A h_{A,out} + \dot{m}_B h_{B,out} \tag{6.88}$$

D. Condensers

Condensers remove the heat of vaporization from fluids. This heat is transfered to the environment. It is possible to analyze a condenser as a heat exchanger. However, if the total heat removal is known, the SFEE can be written as equation 6.89.

$$Q = \dot{m}(h_2 - h_1) \tag{6.89}$$

Since the heat flow is out of the condenser, both *Q* and $(h_2 - h_1)$ will be negative.

E. Compressors

Compressors can be evaluated using the assumptions made for pumps.

F. Valves

Flow through valves is adiabatic. No work is done on the fluid as it passes through a valve. If the potential energy changes are neglected, the SFEE reduces to equation 6.90.

$$h_1 + \frac{v_1^2}{2g_cJ} = h_2 + \frac{v_2^2}{2g_cJ} \tag{6.90}$$

If the kinetic energy changes are neglected, $h_1 = h_2$. This is the definition of a *throttling* process.

Throttling is a constant temperature process for ideal gases. However, real gases experience a temperature drop upon throttling over normal temperatures. Whether or not a rise or drop in temperature occurs is dependent on the range of pressure and temperature over which the throttling occurs. There is one temperature at which no temperature changes occur upon throttling. This is called the *inversion temperature*. Below the inversion point, gases cool when throttled. Above the inversion point, their temperatures increase.

Table 6.9
Approximate Inversion Temperatures

Gas	Temperature, °R
air	1085 (max)
argon	1301 (max)
carbon dioxide	2700 (max)
helium	45 (1 atm)
	72 (max)
hydrogen	364 (max)
nitrogen	1118 (max)

The *Joule-Thompson coefficient* is defined as the ratio of the change in temperature to the change in pressure when a gas is throttled. The Joule-Thompson coefficient is zero for an ideal gas.

$$\mu = \frac{\partial T}{\partial p} \tag{6.91}$$

G. Nozzles and Orifices

If a high energy fluid is directed through a nozzle or orifice, its velocity will increase at the expense of the energy. This acceleration is adiabatic. No work is performed. The potential energy change can be neglected. Therefore, the SFEE reduces to equation 6.92.

$$h_1 + \frac{v_1^2}{2g_cJ} = h_2 + \frac{v_2^2}{2g_cJ} \tag{6.92}$$

If the initial velocity is neglected, the exit velocity is

$$v_2 = \sqrt{2g_cJ(h_1 - h_2)} \tag{6.93}$$

4. Property Changes in Processes

One or more properties of a system will change when the system takes part in a process. For example, the temperature and pressure of a gas will increase if that gas is compressed in a cylinder by a piston. The remainder of this chapter is devoted to predicting the changes to the various properties.

A definite sign convention is used in calculating property changes. This convention is explained here, and it is consistent with the signs that are automatically generated by the formulas in this chapter.

- *Q* is positive if energy (heat) flows into the system.
- *W* is positive if the system does work on the surroundings.
- *H, U,* and *S* are positive if they increase within the system.

In calculating changes to properties, it should be remembered that most formulas can be written in several forms. All equations can be written in terms of both *lbm* and *pmole* bases. The change in enthalpy for a perfect gas, for example, can be written as either equation 6.94 or 6.95.

$$\Delta h = c_p \Delta T \tag{6.94}$$

$$\Delta H = C_p \Delta T \tag{6.95}$$

It is not practical to list every relationship for property changes. You will have to use your knowledge of the behavior of the system to make substitutions into the various equations.

5. The First Law of Thermodynamics

There is one basic principle that underlies all property changes as a system experiences a process: all energy

THERMO

must be accounted for. Energy that enters a system must either leave the system or be stored in some manner.

A. Closed Systems

This principle can be stated at the *First Law of Thermodynamics:* energy cannot be created or destroyed. The First Law may be written in differential form for *closed systems.*

$$dQ = dU + W \qquad 6.96$$

Most thermodynamic problems can be solved without resorting to differential calculus. The First Law can be written in finite terms.

$$Q = \Delta U + W \qquad 6.97$$

Equation 6.97 is the First Law formulation for closed systems. It says that heat entering a closed system can either increase the temperature (increase U) or be used to perform work (W) on the surroundings. Heat energy entering the system can also leak to the surroundings (a non-adiabatic closed system) but, in that case, the term Q is understood to be the *net* heat entering the system.

If the net heat exchange to the system is a loss, then Q will be negative. ΔU will be negative if the internal energy of the system decreases. W will be negative if the surroundings do work on the system (e.g., a piston compresses gas in a cylinder). These signs are consistent with the convention used in this chapter.

Since Q and U contain units of BTU, the work term must also be expressed in BTU. If work is given in ft-lbf, Joule's constant must be incorporated into the First Law.

$$Q = \Delta U + \frac{W}{J} \qquad 6.98$$

B. Open Systems

The First Law of Thermodynamics can be written for open systems also, but more terms are required to account for the many energy forms that can change. If the mass flow rate is constant, the system is a steady flow system. The applicable First Law formulation is essentially the Bernoulli energy conservation equation extended to non-adiabatic processes. This equation is known as the *Steady Flow Energy Equation (SFEE).*

$$Q = \Delta U + \Delta E_p + \Delta E_k + W_{flow} + W_{shaft} \qquad 6.99$$

The terms in equation 6.99 can be illustrated by figure 6.15. Q is the net heat flow into the system. It may be supplied by furnace flame, electrical heating, nuclear reaction, or any other method. Of course, if the system is known to be adiabatic, Q is zero.

ΔU is the change in the internal energy of the system. It can be evaluated by using one of the many formulas

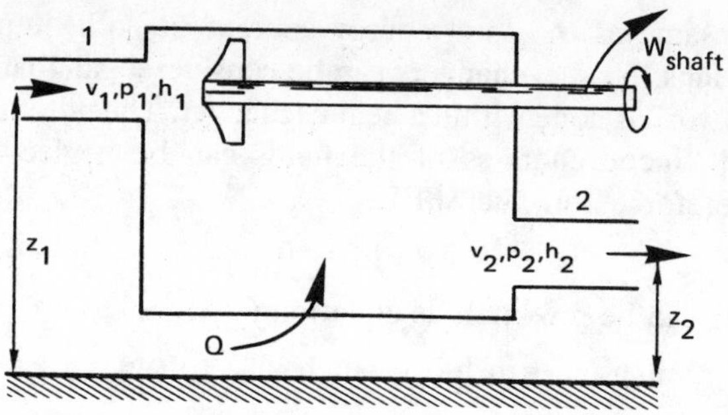

Figure 6.15 A Steady Flow Device

presented in this chapter. However, it is seldom necessary to work with internal energy with a steady flow device.

ΔE_p is the change in *potential energy* of the fluid. This can be found on a per-pound basis from equation 6.100.

$$\Delta E_p = \frac{g(z_2 - z_1)}{Jg_c} \qquad 6.100$$

Similarly, ΔE_k is the change in *kinetic energy.* This can be found on a per-pound basis from equation 6.101.

$$\Delta E_k = \frac{v_2^2 - v_1^2}{2g_c J} \qquad 6.101$$

W_{flow} is the *flow work (pV work)* previously presented in the definition of enthalpy. Looking at figure 6.15, there is a pressure (p_2) at the exit of the steady flow device. This exit pressure opposes the entrance of fluid. Therefore, the flow work term represents the work required to cause the flow into the system against the existing pressure. The flow work can be calculated on a per-pound basis from equation 6.102.

$$W_{flow} = \frac{p_2 \upsilon_2 - p_1 \upsilon_1}{J} \qquad 6.102$$

W_{shaft} is known as *shaft work.* This represents work that the steady flow device does on the surroundings. Its name is derived from the output shaft that almost always transmits the energy out of the system. For example, turbines and internal combustion engines have output shafts that perform useful tasks. Of course, W_{shaft} can also be negative, as it would be in the case of a pump or compressor.

Enthalpy was previously defined as the sum of internal energy and flow energy. Therefore, the internal energy change and flow work terms can be combined in an enthalpy change. That is,

$$\Delta h = u_2 - u_1 + \frac{p_2 \upsilon_2 - p_1 \upsilon_1}{J} \qquad 6.103$$

The complete formulation of the *Steady Flow Energy Equation* is given as equation 6.104. The shaft work term must be in units of ft-lbf/lbm. It does not represent the total work of the device. Similarly, Q is on a per-pound basis.

$$Q = h_2 - h_1 + \frac{v_2^2 - v_1^2}{2g_cJ} + \frac{g(z_2 - z_1)}{g_cJ} + \frac{W_{shaft}}{J} \quad 6.104$$

Both sides of equation 6.104 can be multiplied by $\dot{m}$ to obtain units of BTU/sec. If both sides are multiplied by $\dot{m}J$, the units become ft-lbf/sec.

Example 6.15

4 lbm/sec of steam enter a turbine with velocity of 65 ft/sec and enthalpy of 1350 BTU/lbm. The steam enters the condenser after being expanded to 1075 BTU/lbm and 125 ft/sec. The total heat loss from the turbine casing is 50 BTU/sec. What power is generated by this turbine?

Equation 6.104 can be used to solve this problem. Solving for the shaft work, including the mass flow term, and neglecting the potential energy change,

$$\begin{aligned} P &= -50 + (4)(1350 - 1075) \\ &\quad + (4)\left[\frac{(65)^2 - (125)^2}{(2)(32.2)(778)}\right] \\ &= 1049 \text{ BTU/sec} \end{aligned}$$

1049 BTU/sec corresponds to approximately 1500 horsepower.

A simple application of the first law involves the establishment of a thermal equilibrium between two substances. A thermal equilibrium is reached when all parts of a system are at the same temperature. Thermal equilibrium is achieved naturally (without the addition of external work) whenever two liquids or a solid and liquid with different temperatures are mixed.

A thermal equilibrium problem may be solved by the First Law. The entering energy comes from the heat given off by the cooling substance. Energy is stored in increasing the temperature of the warming substance. Since no work is added, the *W* terms is zero.

$$(\text{Heat loss})_A = (\text{Heat gain})_B \quad 6.105$$

$$Q_A = Q_B \quad 6.106$$

The form of the equation for *Q* depends on the phases of the substances. If both substances are liquid or solid, equation 6.18 can be used.

$$m_A c_A (T_{1,A} - T_{2,A}) = m_B c_B (T_{2,B} - T_{1,B}) \quad 6.107$$

For an open system, the *m* terms should be replaced by $\dot{m}$.

Example 6.16

A 2 pound block of steel (c = .11 BTU/lbm-°F) is removed from a furnace and quenched in a 5 pound aluminum (c = .21 BTU/lbm-°F) tank filled with 12 pounds of water. The water and tank are initially in equilibrium at 75°F, and their temperature rises to 100°F after quenching. What is the initial steel temperature?

From equation 6.105, the heat lost by the steel is equal to the heat gained by the tank and water.

$$m_s c_s (T_{s,1} - T_2) = (m_a c_a + m_w c_w)(T_2 - T_{w,1})$$

$$(2)(.11)(T_{s,1} - 100) = [(5)(.21) + (12)(1)](100 - 75)$$

$$T_{s,1} = 1583°\text{F}$$

6. Availability

The maximum possible work that may be obtained from a system is known as the system's *availability.* Both the first and second laws must be applied to determine this availability.

Refer to figure 6.15. Heat enters the system from an environment which is at a constant temperature T_o. Work is done on the environment at a steady rate *W*. Assuming steady flow, and neglecting kinetic and potential energies, the first law may be written as

$$\dot{m}h_1 + Q = \dot{m}h_2 + W \quad 6.108$$

Entropy is not a part of the first law. However, the leaving entropy can be calculated from the entering entropy and the entropy production.

$$\dot{m}s_2 = \dot{m}s_1 + \frac{Q}{T_o} \quad 6.109$$

Since equations 6.108 and 6.109 both contain *Q*, they may be combined. Since the second law of thermodynamics requires that $s_2 \geqslant s_1$, the combined equation may be written as an inequality.

$$W \leqslant \dot{m}(h_1 - T_o s_1 - h_2 + T_o s_2) \quad 6.110$$

This equation may be simplified by introducing the *steady flow availability function,* Φ. The maximum work output (availability) is

$$W_{max} = \Phi_1 - \Phi_2 \quad 6.111$$

$$\Phi = h - T_o s \quad 6.112$$

If the equality in equation 6.110 holds, the process within the control volume and the energy transfers between the system and environment must both be reversible. Maximum work output will, therefore, be obtained in a reversible process. The difference between the maximum and actual work output is known as the *process irreversibility.*

Example 6.17

What is the maximum useful work which can be produced per pound of saturated steam which enters a steady flow system at 800 psia and leaves in equilibrium with the atmosphere at 70°F and 14.7 psia?

From the saturated steam table, h_1 = 1198.6 BTU/lbm and s_1 = 1.4153 BTU/lbm-°R.

The final properties are obtained from the saturated steam table for water at 70°F. h_2 = 38.04 BTU/lbm and s_2 = .0745 BTU/lbm-°R.

From equation 6.111 using T_o = 530°R, the availability is

$$W_{max} = [1198.6 - 530(1.4153)] - [38.04 - 530(.0745)] = 449.94 \text{ BTU/lbm}$$

7. Property Changes in Ideal Gases

For solids, liquids, and vapors, the changes in properties must be calculated the 'hard way.' That is, the change is found by subtracting the initial property value from the final property value. For ideal gases, however, the changes may be found directly, without knowing the initial and final property values.

Some of the methods available for finding property changes do not depend on the type of process experienced by the gas. For example, the equation of state can be applied to all problems.

$$pV = mRT \quad 6.113$$

Similarly, the changes in enthalpy and internal energy can be found from equation 6.114 and 6.115, regardless of the process.

$$\Delta h = c_p \Delta T \quad 6.114$$

$$\Delta u = c_v \Delta T \quad 6.115$$

Changes in entropy can be calculated from equations 6.116 and 6.117.

$$\Delta s = c_p \ln\left(\frac{T_2}{T_1}\right) - R \ln\left(\frac{p_2}{p_1}\right) \quad 6.116$$

$$= c_v \ln\left(\frac{T_2}{T_1}\right) + R \ln\left(\frac{\upsilon_2}{\upsilon_1}\right) \quad 6.117$$

Equations 6.114 and 6.115 should not be confused with equation 6.19, which gives the energy (heat) transfer required to achieve a change in energy. The heat equation depends on the type of process.

$$\frac{Q}{m} = c_p \Delta T \quad \textit{(constant pressure processes)} \quad 6.118$$

$$\frac{Q}{m} = c_v \Delta T \quad \textit{(constant volume processes)} \quad 6.119$$

Relationships between the variables and properties are given in the following paragraphs for specific processes. It should be remembered that all of these specific equations can be written in terms of pounds and moles. For the sake of compactness, only the *lbm*-basis equations are listed. However, you can convert all equations to *pmole*-basis by substituting V for υ, H for h, R* for R, etc. The sign conventions previously established apply to the following equations.

A. Constant Pressure, Closed Systems

$$p_2 = p_1 \quad 6.120$$

$$T_2 = T_1\left(\frac{\upsilon_2}{\upsilon_1}\right) \quad 6.121$$

$$\upsilon_2 = \upsilon_1\left(\frac{T_2}{T_1}\right) \quad 6.122$$

$$Q = h_2 - h_1 \quad 6.123$$

$$= c_p(T_2 - T_1) \quad 6.124$$

$$= c_v(T_2 - T_1) + p(\upsilon_2 - \upsilon_1) \quad 6.125$$

$$u_2 - u_1 = c_v(T_2 - T_1) \quad 6.126$$

$$= \frac{c_v p(\upsilon_2 - \upsilon_1)}{R} \quad 6.127$$

$$= \frac{p(\upsilon_2 - \upsilon_1)}{k - 1} \quad 6.128$$

$$W = p(\upsilon_2 - \upsilon_1) \quad 6.129$$

$$= R(T_2 - T_1) \quad 6.130$$

$$s_2 - s_1 = c_p \ln\left(\frac{T_2}{T_1}\right) \quad 6.131$$

$$= c_p \ln\left(\frac{\upsilon_2}{\upsilon_1}\right) \quad 6.132$$

$$h_2 - h_1 = Q \quad 6.133$$

$$= c_p(T_2 - T_1) \quad 6.134$$

$$= \frac{kp(\upsilon_2 - \upsilon_1)}{k - 1} \quad 6.135$$

B. Constant Volume, Closed Systems

$$p_2 = p_1\left(\frac{T_2}{T_1}\right) \quad 6.136$$

$$T_2 = T_1\left(\frac{p_2}{p_1}\right) \quad 6.137$$

$$\upsilon_2 = \upsilon_1 \quad 6.138$$

$$Q = u_2 - u_1 \quad 6.139$$

$$= c_v(T_2 - T_1) \quad 6.140$$

$$u_2 - u_1 = Q \quad 6.141$$

$$= c_v(T_2 - T_1) \quad 6.142$$

$$= \frac{c_v \upsilon(p_2 - p_1)}{R} \quad 6.143$$

$$= \frac{\upsilon(p_2 - p_1)}{k - 1} \quad 6.144$$

$$W = 0 \quad 6.145$$

THERMO

$$s_2 - s_1 = c_v \ln\left(\frac{T_2}{T_1}\right) \quad 6.146$$

$$= c_v \ln\left(\frac{p_2}{p_1}\right) \quad 6.147$$

$$h_2 - h_1 = c_p(T_2 - T_1) \quad 6.148$$

$$= \frac{k\upsilon(p_2 - p_1)}{k - 1} \quad 6.149$$

C. Constant Temperature, Closed Systems

$$p_2 = p_1\left(\frac{\upsilon_1}{\upsilon_2}\right) \quad 6.150$$

$$T_2 = T_1 \quad 6.151$$

$$\upsilon_2 = \upsilon_1\left(\frac{p_1}{p_2}\right) \quad 6.152$$

$$Q = W \quad 6.153$$

$$= T(s_2 - s_1) \quad 6.154$$

$$= p_1\upsilon_1 \ln\left(\frac{\upsilon_2}{\upsilon_1}\right) \quad 6.155$$

$$= RT \ln\left(\frac{\upsilon_2}{\upsilon_1}\right) \quad 6.156$$

$$u_2 - u_1 = 0 \quad 6.157$$

$$W = Q \quad 6.158$$

$$= p_1\upsilon_1 \ln\left(\frac{\upsilon_2}{\upsilon_1}\right) \quad 6.159$$

$$= RT \ln\left(\frac{\upsilon_2}{\upsilon_1}\right) \quad 6.160$$

$$= RT \ln\left(\frac{p_1}{p_2}\right) \quad 6.161$$

$$s_2 - s_1 = \frac{Q}{T} \quad 6.162$$

$$= R \ln\left(\frac{\upsilon_2}{\upsilon_1}\right) \quad 6.163$$

$$= R \ln\left(\frac{p_1}{p_2}\right) \quad 6.164$$

$$h_2 - h_1 = 0 \quad 6.165$$

D. Isentropic, Closed Systems (Reversible Adiabatic)

$$p_2 = p_1\left(\frac{\upsilon_1}{\upsilon_2}\right)^k \quad 6.166$$

$$= p_1\left(\frac{T_2}{T_1}\right)^{\frac{k}{k-1}} \quad 6.167$$

$$T_2 = T_1\left(\frac{\upsilon_1}{\upsilon_2}\right)^{k-1} \quad 6.168$$

$$= T_1\left(\frac{p_2}{p_1}\right)^{\frac{k-1}{k}} \quad 6.169$$

$$\upsilon_2 = \upsilon_1\left(\frac{p_1}{p_2}\right)^{\frac{1}{k}} \quad 6.170$$

$$= \upsilon_1\left(\frac{T_1}{T_2}\right)^{\frac{1}{k-1}} \quad 6.171$$

$$Q = 0 \quad 6.172$$

$$u_2 - u_1 = -W \quad 6.173$$

$$= c_v(T_2 - T_1) \quad 6.174$$

$$= \frac{c_v(p_2\upsilon_2 - p_1\upsilon_1)}{R} \quad 6.175$$

$$= \frac{p_2\upsilon_2 - p_1\upsilon_1}{k - 1} \quad 6.176$$

$$W = u_1 - u_2 \quad 6.177$$

$$= c_v(T_1 - T_2) \quad 6.178$$

$$= \frac{p_1\upsilon_1 - p_2\upsilon_2}{k - 1} \quad 6.179$$

$$= \frac{p_1\upsilon_1}{k-1}\left[1 - \left(\frac{p_2}{p_1}\right)^{\frac{k-1}{k}}\right] \quad 6.180$$

$$s_2 - s_1 = 0 \quad 6.181$$

$$h_2 - h_1 = c_p(T_2 - T_1) \quad 6.182$$

$$= \frac{k(p_2\upsilon_2 - p_1\upsilon_1)}{k - 1} \quad 6.183$$

E. Polytropic, Closed Systems

$$p_2 = p_1\left(\frac{\upsilon_1}{\upsilon_2}\right)^n \quad 6.184$$

$$= p_1\left(\frac{T_2}{T_1}\right)^{\frac{n}{n-1}} \quad 6.185$$

$$T_2 = T_1\left(\frac{\upsilon_1}{\upsilon_2}\right)^{n-1} \quad 6.186$$

$$= T_1\left(\frac{p_2}{p_1}\right)^{\frac{n-1}{n}} \quad 6.187$$

$$\upsilon_2 = \upsilon_1\left(\frac{p_1}{p_2}\right)^{\frac{1}{n}} \quad 6.188$$

$$= \upsilon_1\left(\frac{T_1}{T_2}\right)^{\frac{1}{n-1}} \quad 6.189$$

$$Q = \frac{c_v(n - k)(T_2 - T_1)}{n - 1} \quad 6.190$$

THERMO

$$u_2 - u_1 = c_v(T_2 - T_1) \quad 6.191$$

$$= \frac{p_2 v_2 - p_1 v_1}{n - 1} \quad 6.192$$

$$W = \frac{R(T_1 - T_2)}{n - 1} \quad 6.193$$

$$= \frac{p_1 v_1 - p_2 v_2}{n - 1} \quad 6.194$$

$$= \frac{p_1 v_1}{n-1}\left[1 - \left(\frac{p_2}{p_1}\right)^{\frac{n-1}{n}}\right] \quad 6.195$$

$$s_2 - s_1 = \frac{c_v(n - k)}{n - 1}\left[\ln\left(\frac{T_2}{T_1}\right)\right] \quad 6.196$$

$$h_2 - h_1 = c_p(T_2 - T_1) \quad 6.197$$

$$= \frac{n(p_2 v_2 - p_1 v_1)}{n - 1} \quad 6.198$$

F. Isentropic, Steady Flow Systems

p_2, v_2, and T_2 are the same as for isentropic, closed systems.

$$Q = 0 \quad 6.199$$

$$W = h_2 - h_1 \quad 6.200$$

$$= c_p T_1\left[1 - \left(\frac{p_2}{p_1}\right)^{\frac{k-1}{k}}\right] \quad 6.201$$

$$u_2 - u_1 = c_v(T_2 - T_1) \quad 6.202$$

$$h_2 - h_1 = W \quad 6.203$$

$$= c_p(T_2 - T_1) \quad 6.204$$

$$= \frac{k(p_2 v_2 - p_1 v_1)}{k - 1} \quad 6.205$$

$$s_2 - s_1 = 0 \quad 6.206$$

G. Polytropic, Steady Flow Systems

p_2, v_2, and T_2 are the same as for polytropic, closed systems.

$$Q = \frac{c_v(n - k)(T_2 - T_1)}{n - 1} \quad 6.207$$

$$W = h_2 - h_1 \quad 6.208$$

$$= \frac{n c_v(n - k)T_1}{n - 1}\left[1 - \left(\frac{p_2}{p_1}\right)^{\frac{n-1}{n}}\right] \quad 6.209$$

$$u_2 - u_1 = c_v(T_2 - T_1) \quad 6.210$$

$$h_2 - h_1 = W \quad 6.211$$

$$= c_p(T_2 - T_1) \quad 6.212$$

$$= \frac{n(p_2 v_2 - p_1 v_1)}{n - 1} \quad 6.213$$

$$s_2 - s_1 = \frac{c_v(n - k)}{n - 1}\left[\ln\left(\frac{T_2}{T_1}\right)\right] \quad 6.214$$

H. Throttling, Steady Flow Systems

$$p_1 v_1 = p_2 v_2 \quad 6.215$$

$$p_2 < p_1 \quad 6.216$$

$$v_2 > v_1 \quad 6.217$$

$$T_2 = T_1 \quad 6.218$$

$$Q = 0 \quad 6.219$$

$$W = 0 \quad 6.220$$

$$u_2 - u_1 = 0 \quad 6.221$$

$$h_2 - h_1 = 0 \quad 6.222$$

$$s_2 - s_1 = R \ln\left(\frac{p_1}{p_2}\right) \quad 6.223$$

$$= R \ln\left(\frac{v_2}{v_1}\right) \quad 6.224$$

Example 6.18

Two pounds of hydrogen are cooled at constant volume from 760°F to 660°F. What heat is removed?

Equation 6.140 is on a per pound basis. The total heat for *m* pounds is $Q = mc_v(T_2 - T_1)$. c_v is found from table 6.4 to be 2.435 BTU/lbm-°R.

$$Q = (2)(2.435)(660 - 760)$$

$$= -487 \text{ BTU}$$

The minus sign is consistent with the convention that a heat loss is a negative heat flow.

Example 6.19

Four pounds of air are initially at 14.7 psia and 530°R. The air is compressed in a constant temperature, closed process to 100 psia. What heat is removed during the compression?

Either equation 6.155 or 6.156 could be used if the volumes were known. The volumes could be found from the equation of state, but it is not necessary to do so. Combining equations 6.152 and 6.156, and including the mass term,

$$Q = mRT \ln\left(\frac{p_1}{p_2}\right)$$

$$= (4)(53.3)(530) \ln\left(\frac{14.7}{100}\right)$$

$$= -216{,}650 \text{ ft-lbf}$$

Notice that (a) the pressure did not have to be converted to *psf* in the ratio, (b) the units are ft-lbf, not BTU, and (c) the heat flow is out of the gas.

8. Isentropic Flow of High-Velocity Gas

A *high-velocity gas* is defined as one with a velocity in excess of 200 ft/sec. The reasons that high-velocity gases must be handled differently from low-velocity gases are:

(a) The Bernoulli equation assumes a constant density and it cannot account for the conversion of internal energy into kinetic energy in an expanding gas. This conversion results in a lower temperature and a higher velocity than would be predicted by the Bernoulli equation.

(b) Density changes complicate the continuity equation, often resulting in more than one unknown.

(c) Pressure fluctuations and other local disturbances cannot be transmitted in flows with velocities greater than the speed of sound.

Refer to figure 6.16 and consider a compressible, ideal gas with constant heat capacity, c_p, flowing in a duct with varying area. The flow can be assumed to be one-dimensional as long as the cross-section varies slowly along the length. Flow in this duct is governed by the Steady Flow Energy Equation.

$$\frac{v_1^2}{2g_c} + z_1 + Jh_1 = \frac{v_2^2}{2g_c} + z_2 + Jh_2 \qquad 6.225$$

Since the potential energy changes are minimal, this can be rewritten as:

$$\frac{v_2^2 - v_1^2}{2g_c} = J(h_1 - h_2) \qquad 6.226$$

$$v_2 = \sqrt{2g_cJ(h_1 - h_2) + v_1^2} \qquad 6.227$$

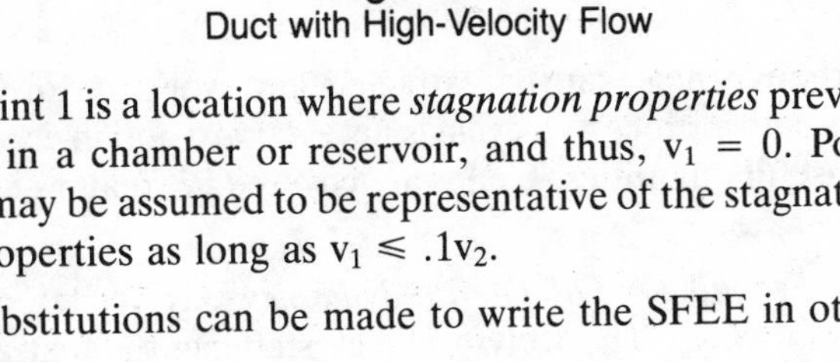

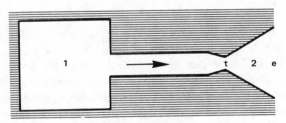

Figure 6.16
Duct with High-Velocity Flow

Point 1 is a location where *stagnation properties* prevail, as in a chamber or reservoir, and thus, $v_1 = 0$. Point 1 may be assumed to be representative of the stagnation properties as long as $v_1 \leq .1v_2$.

Substitutions can be made to write the SFEE in other forms

$$v_2 = \sqrt{2g_cJc_p(T_1 - T_2)} \qquad 6.228$$

$$= \sqrt{\frac{2g_cRk(T_1 - T_2)}{k - 1}} \qquad 6.229$$

$$= \sqrt{\frac{2g_ck}{k - 1}\left[\frac{p_1}{\rho_1} - \frac{p_2}{\rho_2}\right]} \qquad 6.230$$

If the gas flow is isentropic, the change in entropy is zero. As a practical matter, completely isentropic flow does not exist. Some steady state flow processes, however, proceed with little increase in entropy and are considered as isentropic. The reversible effects are accounted for by various correction factors, such as nozzle and discharge coefficients.

In isentropic flow, *total pressure, total temperature,* and *total density* remain constant regardless of the flow area or velocity. This is not to imply that the static properties do not change, for in fact, they do.

From equation 6.229,

$$v_2^2 = \frac{2g_cRk(T_T - T_2)}{k - 1} \qquad 6.231$$

Dividing both sides of equation 6.231 by kg_cRT_2 gives

$$\frac{v_2^2}{kg_cRT_2} = \frac{2\left(\frac{T_T}{T_2} - 1\right)}{k - 1} \qquad 6.232$$

But, kg_cRT is the square of the speed of sound at the conditions at point 2, and the left hand side is M_2^2. Rearranging gives:

$$(T_T/T_2) = \tfrac{1}{2}(k - 1)M_2^2 + 1 \qquad 6.233$$

Similar results can be derived for the ratio of (p_T/p_2) and (ρ_T/ρ_2). The subscript '2' is usually omitted as being understood.

$$(p_T/p) = [\tfrac{1}{2}(k - 1)M^2 + 1]^{\frac{k}{k-1}} \qquad 6.234$$

$$= [T_T/T]^{\frac{k}{k-1}} \qquad 6.235$$

$$(\rho_T/\rho) = [\tfrac{1}{2}(k - 1)M^2 + 1]^{\frac{1}{k-1}} \qquad 6.236$$

$$= [T_T/T]^{\frac{1}{k-1}} \qquad 6.237$$

Example 6.20

Air flows isentropically from a large tank at 70°F through a converging/diverging nozzle and is expanded to supersonic velocities. What is the gas temperature at a point where the Mach number is 2.5? What is the actual velocity where $M = 2.5$?

$$(T_T/T_2) = \tfrac{1}{2}(1.4 - 1)(2.5)^2 + 1 = 2.25$$

$$T_T = 70° + 460° = 530°R$$

$$T_2 = 530/2.25 = 236°R = -224°F$$

$$v_2 = 2.5\sqrt{(1.4)(32.2)(53.3)(236)} = 1883 \text{ fps}$$

In order to design a nozzle capable of expanding a gas to some given velocity, it is necessary to be able to calculate the area perpendicular to the flow. Since the reservoir cross sectional area is an unrelated variable, it is not possible to develop a ratio for (A_T/A) as was done for temperature, pressure, and density. The usual choice for a reference area is that area at which the gas velocity is (or could be) sonic. This area is designated as A*. The mass balance at this critical point is:

$$A^*v^*\rho^* = Av\rho \qquad 6.238$$

Various substitutions in the above formula can be made to arrive at equation 6.239.

$$\frac{A}{A^*} = \frac{1}{M}\left[\frac{\tfrac{1}{2}(k-1)M^2 + 1}{\tfrac{1}{2}(k-1) + 1}\right]^{\frac{k+1}{2(k-1)}} \qquad 6.239$$

Equation 6.239 can be plotted versus *M* with the startling results shown in figure 6.17. Notice that, as long as *M* is less than one, the area must decrease for the velocity to increase. However, if *M* exceeds one, the area must increase for the velocity to increase!

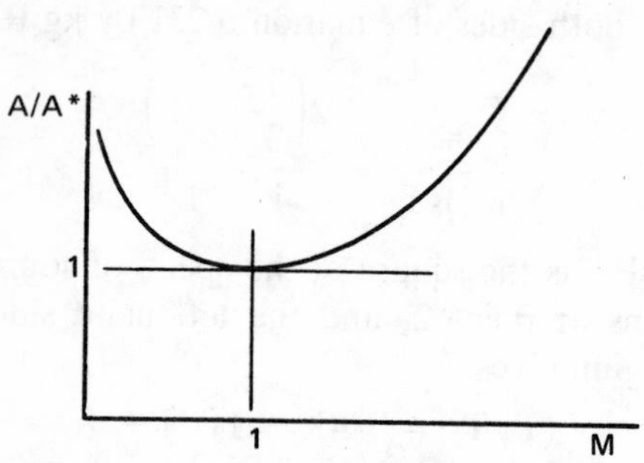

Figure 6.17 A/A* versus Mach Number

In addition, the following equation, derived from the continuity formulas, gives the ratio of the mass flow rate to the critical area. The formula is useful when you know the flow rate and want the critical area.

$$\frac{\dot{m}}{A^*} = \frac{\rho_T\sqrt{kg_cRT_T}}{[\tfrac{1}{2}(k-1)+1]^{\frac{k+1}{2(k-1)}}} \qquad 6.240$$

A final ratio is that of (v/c*). This ratio is useful if you know the velocity and the total properties but need to find *M*. It is, of course, possible to solve for the local temperature and the speed of sound, but this is tedious.

$$\frac{v}{c^*} = \sqrt{\frac{\tfrac{1}{2}(k+1)M^2}{\tfrac{1}{2}(k-1)M^2 + 1}} \qquad 6.241$$

By now it should be apparent that the algebraic burden of solving a gas dynamics problem is tremendous. The easiest way to solve such problems is with the use of gas tables, such as appendix K of this chapter. The notation and symbols in this table, or in any other table that might be available, should be studied carefully in order to avoid confusion.

Example 6.21

A frictionless, adiabatic duct receives 10 lbm/sec of air from a reservoir at 200°F and 30 psia. Somewhere in the duct a velocity of 1400 ft/sec is attained. What is the cross sectional area, temperature, pressure, and Mach number at this point?

In the reservoir,

$$\rho = \frac{p}{RT} = \frac{(30)(144)}{(53.3)(660)} = .1228 \text{ lbm/ft}^3$$

For a point where M = 1, the isentropic flow table gives

$$\frac{T}{T_T} = .8333 \quad \text{or } T^* = (.8333)(660) = 550°R$$

$$\frac{\rho}{\rho_T} = .6339 \quad \text{or } \rho^* = (.6339)(.1228) = .0778$$

The sonic velocity at the throat is

$$c^* = \sqrt{kg_cRT}$$
$$= \sqrt{(1.4)(32.2)(53.3)(550)}$$
$$= 1149.2 \text{ fps}$$

From $\dot{m} = A\rho v$,

$$A^* = \frac{10}{(.0778)(1149.2)} = .1118 \text{ ft}^2$$

Then, $\frac{v}{c^*} = \frac{1400}{1149.2} = 1.218$

Searching the (v/c*) column for 1.218 gives M = 1.28. For this Mach number, read

$$T/T_T = .7532, \; p/p_T = .3708, \; A/A^* = 1.058$$

$$T = .7532(660) = 497.1°R$$

$$p = .3708(30) = 11.12 \text{ psia}$$

$$A = (1.058)(.1118) = .1183 \text{ ft}^2$$

9. Cycles

A *cycle* is a process or series of processes that eventually brings the working fluid back to its original condition. Most cycles repeat over and over. For example, the same water may be alternately vaporized in a boiler and expanded through a turbine. Also, a four-stroke internal combustion engine operates by continually repeating the compression-expansion-exhaust-intake processes.[17]

Although heat can be extracted and work performed by a single process, a cycle is necessary to obtain energy in useful quantities. Cycles are covered in greater detail in chapter 7.

The *Second Law of Thermodynamics* may be stated in several ways. The Kelvin-Planck statement is that it is impossible to operate an engine operating in a cycle that will have no other effect than to extract heat from a reservoir and turn it into an equivalent amount of work. That is, it is impossible to build a cyclic engine that will have a thermal efficiency of 100%.

[17]The definition of a cycle is not dependent on the same working fluid being used repeatedly as it is with a boiler/turbine.

THERMO

The Second Law is not a contradiction of the First Law. The First Law does not preclude the possiblility of converting heat entirely into work—it only denies the possibility of creating or destroying energy. The Second Law does not preclude the conversion of heat into work either. The Second Law says that if some heat is converted entirely into work, then some other energy must be lost to the surroundings.

The *thermal efficiency* of a process is defined as the ratio of useful work output to the supplied input energy.

$$\eta_{th} = \frac{\text{work output}}{\text{energy input}} = \frac{W}{Q_{in}} \qquad 6.242$$

The First Law may be written as

$$Q_{in} = Q_{out} + W \qquad 6.243$$

Combining equations 6.242 and 6.243 produces equation 6.244.

$$\eta_{th} = 1 - \frac{Q_{out}}{Q_{in}} \qquad 6.244$$

If a heat engine could be built with a 100% thermal efficiency, Q_{out} would be zero. Such an engine is precluded by the Second Law. The most efficient engine cycle possible is the *Carnot cycle*. This cycle is discussed in detail in chapter 7.

THERMO

Appendix A

Saturated Steam: Temperatures

Temp. T,°F	Abs. Press. psia p	Specific Volume, ft³/lbm Sat. Liquid v_f	Evap. v_{fg}	Sat. Vapor v_g	Enthalpy, BTU/lbm Sat. Liquid h_f	Evap. h_{fg}	Sat. Vapor h_g	Entropy, BTU/lbm-°R Sat. Liquid s_f	Evap. s_{fg}	Sat. Vapor s_g	Temp. T,°F
32°	0.08854	0.01602	3306	3306	0.00	1075.8	1075.8	0.0000	2.1877	2.1877	32°
35	0.09995	0.01602	2947	2947	3.02	1074.1	1077.1	0.0061	2.1709	2.1770	35
40	0.12170	0.01602	2444	2444	8.05	1071.3	1079.3	0.0162	2.1435	2.1597	40
45	0.14752	0.01602	2036.4	2036.4	13.06	1068.4	1081.5	0.0262	2.1167	2.1429	45
50	0.17811	0.01603	1703.2	1703.2	18.07	1065.6	1083.7	0.0361	2.0903	2.1264	50
60°	0.2563	0.01604	1206.6	1206.7	28.06	1059.9	1088.0	0.0555	2.0393	2.0948	60°
70	0.3631	0.01606	867.8	867.9	38.04	1054.3	1092.3	0.0745	1.9902	2.0647	70
80	0.5069	0.01608	633.1	633.1	48.02	1048.6	1096.6	0.0932	1.9428	2.0360	80
90	0.6982	0.01610	468.0	468.0	57.99	1042.9	1100.9	0.1115	1.8972	2.0087	90
100	0.9492	0.01613	350.3	350.4	67.97	1037.2	1105.2	0.1295	1.8531	1.9826	100
110°	1.2748	0.01617	265.3	265.4	77.94	1031.6	1109.5	0.1471	1.8106	1.9577	110°
120	1.6924	0.01620	203.25	203.27	87.92	1025.8	1113.7	0.1645	1.7694	1.9339	120
130	2.2225	0.01625	157.32	157.34	97.90	1020.0	1117.9	0.1816	1.7296	1.9112	130
140	2.8886	0.01629	122.99	123.01	107.89	1014.1	1122.0	0.1984	1.6910	1.8894	140
150	3.718	0.01634	97.06	97.07	117.89	1008.2	1126.1	0.2149	1.6537	1.8685	150
160°	4.741	0.01639	77.27	77.29	127.89	1002.3	1130.2	0.2311	1.6174	1.8485	160°
170	5.992	0.01645	62.04	62.06	137.90	996.3	1134.2	0.2472	1.5822	1.8293	170
180	7.510	0.01651	50.21	50.23	147.92	990.2	1138.1	0.2630	1.5480	1.8109	180
190	9.339	0.01657	40.94	40.96	157.95	984.1	1142.0	0.2785	1.5147	1.7932	190
200	11.526	0.01663	33.62	33.64	167.99	977.9	1145.9	0.2938	1.4824	1.7762	200
210°	14.123	0.01670	27.80	27.82	178.05	971.6	1149.7	0.3090	1.4508	1.7598	210°
212	14.696	0.01672	26.78	26.80	180.07	970.3	1150.4	0.3120	1.4446	1.7566	212
220	17.186	0.01677	23.13	23.15	188.13	965.2	1153.4	0.3239	1.4201	1.7440	220
230	20.780	0.01684	19.365	19.382	198.23	958.8	1157.0	0.3387	1.3901	1.7288	230
240	24.969	0.01692	16.306	16.323	208.34	952.2	1160.5	0.3531	1.3609	1.7140	240
250°	29.825	0.01700	13.804	13.821	218.48	945.5	1164.0	0.3675	1.3323	1.6998	250°
260	35.429	0.01709	11.746	11.763	228.64	938.7	1167.3	0.3817	1.3043	1.6860	260
270	41.858	0.01717	10.044	10.061	238.84	931.8	1170.6	0.3958	1.2769	1.6727	270
280	49.203	0.01726	8.628	8.645	249.06	924.7	1173.8	0.4096	1.2501	1.6597	280
290	57.556	0.01735	7.444	7.461	259.31	917.5	1176.8	0.4234	1.2238	1.6472	290
300°	67.013	0.01745	6.449	6.466	269.59	910.1	1179.7	0.4369	1.1980	1.6350	300°
310	77.68	0.01755	5.609	5.626	279.92	902.6	1182.5	0.4504	1.1727	1.6231	310
320	89.66	0.01765	4.896	4.914	290.28	894.9	1185.2	0.4637	1.1478	1.6115	320
330	103.06	0.01776	4.289	4.307	300.68	887.0	1187.7	0.4769	1.1233	1.6002	330
340	118.01	0.01787	3.770	3.788	311.13	879.0	1190.1	0.4900	1.0992	1.5891	340
350°	134.63	0.01799	3.324	3.342	321.63	870.7	1192.3	0.5029	1.0754	1.5783	350°
360	153.04	0.01811	2.939	2.957	332.18	862.2	1194.4	0.5158	1.0519	1.5677	360
370	173.37	0.01823	2.606	2.625	342.79	853.5	1196.3	0.5286	1.0287	1.5573	370
380	195.77	0.01836	2.317	2.335	353.45	844.6	1198.1	0.5413	1.0059	1.5471	380
390	220.37	0.01850	2.0651	2.0836	364.17	835.4	1199.6	0.5539	0.9832	1.5371	390
400°	247.31	0.01864	1.8447	1.8633	374.97	826.0	1201.0	0.5664	0.9608	1.5272	400°
410	276.75	0.01878	1.6512	1.6700	385.83	816.3	1202.1	0.5788	0.9386	1.5174	410
420	308.83	0.01894	1.4811	1.5000	396.77	806.3	1203.1	0.5912	0.9166	1.5078	420
430	343.72	0.01910	1.3308	1.3499	407.79	796.0	1203.8	0.6035	0.8947	1.4982	430
440	381.59	0.01926	1.1979	1.2171	418.90	785.4	1204.3	0.6158	0.8730	1.4887	440
450°	422.6	0.0194	1.0799	1.0993	430.1	774.5	1204.6	0.6280	0.8513	1.4793	450°
460	466.9	0.0196	0.9748	0.9944	441.4	763.2	1204.6	0.6402	0.8298	1.4700	460
470	514.7	0.0198	0.8811	0.9009	452.8	751.5	1204.3	0.6523	0.8083	1.4606	470
480	566.1	0.0200	0.7972	0.8172	464.4	739.4	1203.7	0.6645	0.7868	1.4513	480
490	621.4	0.0202	0.7221	0.7423	476.0	726.8	1202.8	0.6766	0.7653	1.4419	490
500°	680.8	0.0204	0.6545	0.6749	487.8	713.9	1201.7	0.6887	0.7438	1.4325	500°
520	812.4	0.0209	0.5385	0.5594	511.9	686.4	1198.2	0.7130	0.7006	1.4136	520
540	962.5	0.0215	0.4434	0.4649	536.6	656.6	1193.2	0.7374	0.6568	1.3942	540
560	1133.1	0.0221	0.3647	0.3868	562.2	624.2	1186.4	0.7621	0.6121	1.3742	560
580	1325.8	0.0228	0.2989	0.3217	588.9	588.4	1177.3	0.7872	0.5659	1.3532	580
600°	1542.9	0.0236	0.2432	0.2668	617.0	548.5	1165.5	0.8131	0.5176	1.3307	600°
620	1786.6	0.0247	0.1955	0.2201	646.7	503.6	1150.3	0.8398	0.4664	1.3062	620
640	2059.7	0.0260	0.1538	0.1798	678.6	452.0	1130.5	0.8679	0.4110	1.2789	640
660	2365.4	0.0278	0.1165	0.1442	714.2	390.2	1104.4	0.8987	0.3485	1.2472	660
680	2708.1	0.0305	0.0810	0.1115	757.3	309.9	1067.2	0.9351	0.2719	1.2071	680
700°	3093.7	0.0369	0.0392	0.0761	823.3	172.1	995.4	0.9905	0.1484	1.1389	700°
705.4	3206.2	0.0503	0	0.0503	902.7	0	902.7	1.0580	0	1.0580	705.4

THERMO

Appendix B
Saturated Steam: Pressures

Abs. Press. psia p	Temp. T,°F	Specific Volume: Sat. Liquid υ_f	Specific Volume: Sat. Vapor υ_g	Enthalpy, BTU/lbm: Sat. Liquid h_f	Enthalpy, BTU/lbm: Evap h_{fg}	Enthalpy, BTU/lbm: Sat. Vapor h_g	Entropy, BTU/lbm-°R: Sat. Liquid s_f	Entropy, BTU/lbm-°R: Evap. s_{fg}	Entropy, BTU/lbm-°R: Sat. Vapor s_g	Internal Energy: Sat. Liquid u_f	Internal Energy: Sat. Vapor u_g	Abs. Press psia p
1.0	101.74	0.01614	333.6	69.70	1036.3	1106.0	0.1326	1.8456	1.9782	69.70	1044.3	1.0
2.0	126.08	0.01623	173.73	93.99	1022.2	1116.2	0.1749	1.7451	1.9200	93.98	1051.9	2.0
3.0	141.48	0.01630	118.71	109.37	1013.2	1122.6	0.2008	1.6855	1.8863	109.36	1056.7	3.0
4.0	152.97	0.01636	90.63	120.86	1006.4	1127.3	0.2198	1.6427	1.8625	120.85	1060.2	4.0
5.0	162.24	0.01640	73.52	130.13	1001.0	1131.1	0.2347	1.6094	1.8441	130.12	1063.1	5.0
6.0	170.06	0.01645	61.98	137.96	996.2	1134.2	0.2472	1.5820	1.8292	137.94	1065.4	6.0
7.0	176.85	0.01649	53.64	144.76	992.1	1136.9	0.2581	1.5586	1.8167	144.74	1067.4	7.0
8.0	182.86	0.01653	47.34	150.79	988.5	1139.3	0.2674	1.5383	1.8057	150.77	1069.2	8.0
9.0	188.28	0.01656	42.40	156.22	985.2	1141.4	0.2759	1.5203	1.7962	156.19	1070.8	9.0
10	193.21	0.01659	38.42	161.17	982.1	1143.3	0.2835	1.5041	1.7876	161.14	1072.2	10
14.696	212.00	0.01672	26.80	180.07	970.3	1150.4	0.3120	1.4446	1.7566	180.02	1077.5	14.696
15	213.03	0.01672	26.29	181.11	969.7	1150.8	0.3135	1.4415	1.7549	181.06	1077.8	15
20	227.96	0.01683	20.089	196.16	960.1	1156.3	0.3356	1.3962	1.7319	196.10	1081.9	20
25	240.07	0.01692	16.303	208.42	952.1	1160.6	0.3533	1.3606	1.7139	208.34	1085.1	25
30	250.33	0.01701	13.746	218.82	945.3	1164.1	0.3680	1.3313	1.6993	218.73	1087.8	30
35	259.28	0.01708	11.898	227.91	939.2	1167.1	0.3807	1.3063	1.6870	227.80	1090.1	35
40	267.25	0.01715	10.498	236.03	933.7	1169.7	0.3919	1.2844	1.6763	235.90	1092.0	40
45	274.44	0.01721	9.401	243.36	928.6	1172.0	0.4019	1.2650	1.6669	243.22	1093.7	45
50	281.01	0.01727	8.515	250.09	924.0	1174.1	0.4110	1.2474	1.6585	249.93	1095.3	50
55	287.07	0.01732	7.787	256.30	919.6	1175.9	0.4193	1.2316	1.6509	256.12	1096.7	55
60	292.71	0.01738	7.175	262.09	915.5	1177.6	0.4270	1.2168	1.6438	261.90	1097.9	60
65	297.97	0.01743	6.655	267.50	911.6	1179.1	0.4342	1.2032	1.6374	267.29	1099.1	65
70	302.92	0.01748	6.206	272.61	907.9	1180.6	0.4409	1.1906	1.6315	272.38	1100.2	70
75	307.60	0.01753	5.816	277.43	904.5	1181.9	0.4472	1.1787	1.6259	277.19	1101.2	75
80	312.03	0.01757	5.472	282.02	901.1	1183.1	0.4531	1.1676	1.6207	281.76	1102.1	80
85	316.25	0.01761	5.168	286.39	897.8	1184.2	0.4587	1.1571	1.6158	286.11	1102.9	85
90	320.27	0.01766	4.896	290.56	894.7	1185.3	0.4641	1.1471	1.6112	290.27	1103.7	90
95	324.12	0.01770	4.652	294.56	891.7	1186.2	0.4692	1.1376	1.6068	294.25	1104.5	95
100	327.81	0.01774	4.432	298.40	888.8	1187.2	0.4740	1.1286	1.6026	298.08	1105.2	100
110	334.77	0.01782	4.049	305.66	883.2	1188.9	0.4832	1.1117	1.5948	305.30	1106.5	110
120	341.25	0.01789	3.728	312.44	877.9	1190.4	0.4916	1.0962	1.5878	312.05	1107.6	120
130	347.32	0.01796	3.455	318.81	872.9	1191.7	0.4995	1.0817	1.5812	318.38	1108.6	130
140	353.02	0.01802	3.220	324.82	868.2	1193.0	0.5069	1.0682	1.5751	324.35	1109.6	140
150	358.42	0.01809	3.015	330.51	863.6	1194.1	0.5138	1.0556	1.5694	330.01	1110.5	150
160	363.53	0.01815	2.834	335.93	859.2	1195.1	0.5204	1.0436	1.5640	335.39	1111.2	160
170	368.41	0.01822	2.675	341.09	854.9	1196.0	0.5266	1.0324	1.5590	340.52	1111.9	170
180	373.06	0.01827	2.532	346.03	850.8	1196.9	0.5325	1.0217	1.5542	345.42	1112.5	180
190	377.51	0.01833	2.404	350.79	846.8	1197.6	0.5381	1.0116	1.5497	350.15	1113.1	190
200	381.79	0.01839	2.288	355.36	843.0	1198.4	0.5435	1.0018	1.5453	354.68	1113.7	200
250	400.95	0.01865	1.8438	376.00	825.1	1201.1	0.5675	0.9588	1.5263	375.14	1115.8	250
300	417.33	0.01890	1.5433	393.84	809.0	1202.8	0.5879	0.9225	1.5104	392.79	1117.1	300
350	431.72	0.01913	1.3260	409.69	794.2	1203.9	0.6056	0.8910	1.4966	408.45	1118.0	350
400	444.59	0.0193	1.1613	424.0	780.5	1204.5	0.6214	0.8630	1.4844	422.6	1118.5	400
450	456.28	0.0195	1.0320	437.2	767.4	1204.6	0.6356	0.8378	1.4734	435.5	1118.7	450
500	467.01	0.0197	0.9278	449.4	755.0	1204.4	0.6487	0.8147	1.4634	447.6	1118.6	500
550	476.94	0.0199	0.8424	460.8	743.1	1203.9	0.6608	0.7934	1.4542	458.8	1118.2	550
600	486.21	0.0201	0.7698	471.6	731.6	1203.2	0.6720	0.7734	1.4454	469.4	1117.7	600
650	494.90	0.0203	0.7083	481.8	720.5	1202.3	0.6826	0.7548	1.4374	479.4	1117.1	650
700	503.10	0.0205	0.6554	491.5	709.7	1201.2	0.6925	0.7371	1.4296	488.8	1116.3	700
750	510.86	0.0207	0.6092	500.8	699.2	1200.0	0.7019	0.7204	1.4223	498.0	1115.4	750
800	518.23	0.0209	0.5687	509.7	688.9	1198.6	0.7108	0.7045	1.4153	506.6	1114.4	800
850	525.26	0.0210	0.5327	518.3	678.8	1197.1	0.7194	0.6891	1.4085	515.0	1113.3	850
900	531.98	0.0212	0.5006	526.6	668.8	1195.4	0.7275	0.6744	1.4020	523.1	1112.1	900
950	538.43	0.0214	0.4717	534.6	659.1	1193.7	0.7355	0.6602	1.3957	530.9	1110.8	950
1000	544.61	0.0216	0.4456	542.4	649.4	1191.8	0.7430	0.6467	1.3897	538.4	1109.4	1000
1100	556.31	0.0220	0.4001	557.4	630.4	1187.8	0.7575	0.6205	1.3780	552.9	1106.4	1100
1200	567.22	0.0223	0.3619	571.7	611.7	1183.4	0.7711	0.5956	1.3667	566.7	1103.0	1200
1300	577.46	0.0227	0.3293	585.4	593.2	1178.6	0.7840	0.5719	1.3559	580.0	1099.4	1300
1400	587.10	0.0231	0.3012	598.7	574.7	1173.4	0.7963	0.5491	1.3454	592.7	1095.4	1400
1500	596.23	0.0235	0.2765	611.6	556.3	1167.9	0.8082	0.5269	1.3351	605.1	1091.2	1500
2000	635.82	0.0257	0.1878	671.7	463.4	1135.1	0.8619	0.4230	1.2849	662.2	1065.6	2000
2500	668.13	0.0287	0.1307	730.6	360.5	1091.1	0.9126	0.3197	1.2322	717.3	1030.6	2500
3000	695.36	0.0346	0.0658	802.5	217.8	1020.3	0.9731	0.1885	1.1615	783.4	972.7	3000
3206.2	705.40	0.0503	0.0503	902.7	0	920.7	1.0580	0	1.0580	872.9	872.9	3206.2

THERMO

Appendix C
Superheated Steam

Abs. Press. psia (Sat. Temp.) (°F)		200°	300°	400°	500°	600°	700°	800°	900°	1000°	1100°	1200°	1400°	1600°
	υ	392.6	452.3	512.0	571.6	631.2	690.8	750.4	809.9	869.5	929.1	988.7	1107.8	1227.0
1	h	1150.4	1195.8	1241.7	1288.3	1335.7	1383.8	1432.8	1482.7	1533.5	1585.2	1637.7	1745.7	1857.5
(101.74)	s	2.0512	2.1153	2.1720	2.2233	2.2702	2.3137	2.3542	2.3923	2.4283	2.4625	2.4952	2.5566	2.6137
	υ	78.16	90.25	102.26	114.22	126.16	138.10	150.03	161.95	173.87	185.79	197.71	221.6	245.4
5	h	1148.8	1195.0	1241.2	1288.0	1335.4	1383.6	1432.7	1482.6	1533.4	1585.1	1637.7	1745.7	1857.4
(162.24)	s	1.8718	1.9370	1.9942	2.0456	2.0927	2.1363	2.1767	2.2148	2.2509	2.2851	2.3178	2.3792	2.4363
	υ	38.85	45.00	51.04	57.05	63.03	69.01	74.98	80.95	86.92	92.88	98.84	110.77	122.69
10	h	1146.6	1193.9	1240.6	1287.5	1335.1	1383.4	1432.5	1482.4	1533.2	1585.0	1637.6	1745.6	1857.3
(193.21)	s	1.7927	1.8595	1.9172	1.9689	2.0160	2.0596	2.1002	2.1383	2.1744	2.2086	2.2413	2.3028	2.3598
	υ	......	30.53	34.68	38.78	42.86	46.94	51.00	55.07	59.13	63.19	67.25	75.37	83.48
14.696	h	......	1192.8	1239.9	1287.1	1334.8	1383.2	1432.3	1482.3	1533.1	1584.8	1637.5	1745.5	1857.3
(212.00)	s	......	1.8160	1.8743	1.9261	1.9734	2.0170	2.0576	2.0958	2.1319	2.1662	2.1989	2.2603	2.3174
	υ	......	22.36	25.43	28.46	31.47	34.47	37.46	40.45	43.44	46.42	49.41	55.37	61.34
20	h	......	1191.6	1239.2	1286.6	1334.4	1382.9	1432.1	1482.1	1533.0	1584.7	1637.4	1745.4	1857.2
(227.96)	s	......	1.7808	1.8396	1.8918	1.9392	1.9829	2.0235	2.0618	2.0978	2.1321	2.1648	2.2263	2.2834
	υ	......	11.040	12.628	14.168	15.688	17.198	18.702	20.20	21.70	23.20	24.69	27.68	30.66
40	h	......	1186.8	1236.5	1284.8	1333.1	1381.9	1431.3	1481.4	1532.4	1584.3	1637.0	1745.1	1857.0
(267.25)	s	......	1.6994	1.7608	1.8140	1.8619	1.9058	1.9467	1.9850	2.0212	2.0555	2.0883	2.1498	2.2069
	υ	......	7.259	8.357	9.403	10.427	11.441	12.449	13.452	14.454	15.453	16.451	18.446	20.44
60	h	......	1181.6	1233.6	1283.0	1331.8	1380.9	1430.5	1480.8	1531.9	1583.8	1636.6	1744.8	1856.7
(292.71)	s	......	1.6492	1.7135	1.7678	1.8162	1.8605	1.9015	1.9400	1.9762	2.0106	2.0434	2.1049	2.1621
	υ	......	.....	6.220	7.020	7.797	8.562	9.322	10.077	10.830	11.582	12.332	13.830	15.325
80	h	......	.....	1230.7	1281.1	1330.5	1379.9	1429.7	1480.1	1531.3	1583.4	1636.2	1744.5	1856.5
(312.03)	s	......	.....	1.6791	1.7346	1.7836	1.8281	1.8694	1.9079	1.9442	1.9787	2.0115	2.0731	2.1303
	υ	......	.....	4.937	5.589	6.218	6.835	7.446	8.052	8.656	9.259	9.860	11.060	12.258
100	h	......	.....	1227.6	1279.1	1329.1	1378.9	1428.9	1479.5	1530.8	1582.9	1635.7	1744.2	1856.2
(327.81)	s	......	.....	1.6518	1.7085	1.7581	1.8029	1.8443	1.8829	1.9193	1.9538	1.9867	2.0484	2.1056
	υ	......	.....	4.081	4.636	5.165	5.683	6.195	6.702	7.207	7.710	8.212	9.214	10.213
120	h	......	.....	1221.4	1277.2	1327.7	1377.8	1428.1	1478.8	1530.2	1582.4	1635.3	1743.9	1856.0
(341.25)	s	......	.....	1.6287	1.6869	1.7370	1.7822	1.8237	1.8625	1.8990	1.9335	1.9664	2.0281	2.0854
	υ	......	.....	3.468	3.954	4.413	4.861	5.301	5.738	6.172	6.604	7.035	7.895	8.752
140	h	......	.....	1221.1	1275.2	1326.4	1376.8	1427.3	1478.2	1529.7	1581.9	1634.9	1743.5	1855.7
(353.02)	s	......	.....	1.6087	1.6683	1.7190	1.7645	1.8063	1.8451	1.8817	1.9163	1.9493	2.0110	2.0683
	υ	......	.....	3.008	3.443	3.849	4.244	4.631	5.015	5.396	5.775	6.152	6.906	7.656
160	h	......	.....	1217.6	1273.1	1325.0	1375.7	1426.4	1477.5	1529.1	1581.4	1634.5	1743.2	1855.7
(363.06)	s	......	.....	1.5908	1.6519	1.7033	1.7491	1.7911	1.8301	1.8667	1.9014	1.9344	1.9962	2.0535
	υ	......	.....	2.649	3.044	3.411	3.764	4.110	4.452	4.792	5.129	5.466	6.136	6.804
180	h	......	.....	1214.0	1271.0	1323.5	1374.7	1425.6	1476.8	1528.6	1581.0	1634.1	1742.9	1855.2
(373.53)	s	......	.....	1.5745	1.6373	1.6894	1.7355	1.7776	1.8167	1.8534	1.8882	1.9212	1.9831	2.0404
	υ	......	.....	2.361	2.726	3.060	3.380	3.693	4.002	4.309	4.613	4.917	5.521	6.123
200	h	......	.....	1210.3	1268.9	1322.1	1373.6	1424.8	1476.2	1528.0	1580.5	1633.7	1742.6	1855.0
(381.79)	s	......	.....	1.5594	1.6240	1.6767	1.7232	1.7655	1.8048	1.8415	1.8763	1.9094	1.9713	2.0287
	υ	......	.....	2.125	2.465	2.772	3.066	3.352	3.634	3.913	4.191	4.467	5.017	5.565
220	h	......	.....	1206.5	1266.7	1320.7	1372.6	1424.0	1475.5	1527.5	1580.0	1633.3	1742.3	1854.7
(389.86)	s	......	.....	1.5453	1.6117	1.6652	1.7120	1.7545	1.7939	1.8308	1.8656	1.8987	1.9607	2.0181
	υ	......	.....	1.9276	2.247	2.533	2.804	3.068	3.327	3.584	3.839	4.093	4.597	5.100
240	h	......	.....	1202.5	1264.5	1319.2	1371.5	1423.2	1474.8	1526.9	1579.6	1632.9	1742.0	1854.5
(397.37)	s	......	.....	1.5319	1.6003	1.6546	1.7017	1.7444	1.7839	1.8209	1.8558	1.8889	1.9510	2.0084
	υ	......	.....	.....	2.063	2.330	2.582	2.827	3.067	3.305	3.541	3.776	4.242	4.707
260	h	......	.....	.....	1262.3	1317.7	1370.4	1422.3	1474.2	1526.3	1579.1	1632.5	1741.7	1854.2
(404.42)	s	......	.....	.....	1.5897	1.6447	1.6922	1.7352	1.7748	1.8118	1.8467	1.8799	1.9420	1.9995
	υ	......	.....	.....	1.9047	2.156	2.392	2.621	2.845	3.066	3.286	3.504	3.938	4.370
280	h	......	.....	.....	1260.0	1316.2	1369.4	1421.5	1473.5	1525.8	1578.6	1632.1	1741.4	1854.0
(411.05)	s	......	.....	.....	1.5796	1.6354	1.6834	1.7265	1.7652	1.8033	1.8383	1.8716	1.9337	1.9912
	υ	......	.....	.....	1.7675	2.005	2.227	2.442	2.652	2.859	3.065	3.269	3.674	4.078
300	h	......	.....	.....	1257.6	1314.7	1368.3	1420.6	1472.8	1525.2	1578.1	1631.7	1741.0	1853.7
(417.33)	s	......	.....	.....	1.5701	1.6268	1.6751	1.7184	1.7582	1.7954	1.8305	1.8638	1.9260	1.9835
	υ	......	.....	.....	1.4923	1.7036	1.8980	2.084	2.266	2.445	2.622	2.798	3.147	3.493
350	h	......	.....	.....	1251.5	1310.9	1365.5	1418.5	1471.1	1523.8	1577.0	1630.7	1740.3	1853.1
(431.72)	s	......	.....	.....	1.5481	1.6070	1.6563	1.7002	1.7403	1.7777	1.8130	1.8463	1.9086	1.9663
	υ	......	.....	.....	1.2851	1.4770	1.6508	1.8161	1.9767	2.134	2.290	2.445	2.751	3.055
400	h	......	.....	.....	1245.1	1306.9	1362.7	1416.4	1469.4	1522.4	1575.8	1629.6	1739.5	1852.5
(444.59)	s	......	.....	.....	1.5281	1.5984	1.6398	1.6842	1.7247	1.7623	1.7977	1.8311	1.8936	1.9513

Appendix C — Continued

Abs. Press. psia (Sat. Temp.) (°F)		Temperature-Degrees Fahrenheit													
		500°	550°	600°	620°	640°	660°	680°	700°	800°	900°	1000°	1200°	1400°	1600°
	υ	1.1231	1.2155	1.3005	1.3332	1.3652	1.3967	1.4278	1.4584	1.6074	1.7516	1.8928	2.170	2.443	2.714
450	h	1238.4	1272.0	1302.8	1314.6	1326.2	1337.5	1348.8	1359.9	1414.3	1467.7	1521.0	1628.6	1738.7	1851.9
(456.28)	s	1.5095	1.5437	1.5735	1.5845	1.5951	1.6054	1.6153	1.6250	1.6699	1.7108	1.7486	1.8177	1.8803	1.9381
	υ	0.9927	1.0800	1.1591	1.1893	1.2188	1.2478	1.2763	1.3044	1.4405	1.5715	1.6996	1.9504	2.197	2.442
500	h	1231.3	1266.8	1298.6	1310.7	1322.6	1334.2	1345.7	1357.0	1412.1	1466.0	1519.6	1627.6	1737.9	1851.3
(467.01)	s	1.4919	1.5280	1.5588	1.5701	1.5810	1.5915	1.6016	1.6115	1.6571	1.6982	1.7363	1.8056	1.8683	1.9262
	υ	0.8852	0.9686	1.0431	1.0714	1.0989	1.1259	1.1523	1.1783	1.3038	1.4241	1.5414	1.7706	1.9957	2.219
550	h	1223.7	1261.2	1294.3	1306.8	1318.9	1330.8	1342.5	1354.0	1409.9	1464.3	1518.2	1626.6	1737.1	1850.6
(476.94)	s	1.4751	1.5131	1.5451	1.5568	1.5680	1.5787	1.5890	1.5991	1.6452	1.6868	1.7250	1.7946	1.8575	1.9155
	υ	0.7947	0.8753	0.9463	0.9729	0.9988	1.0241	1.0489	1.0732	1.1899	1.3013	1.4096	1.6208	1.8279	2.033
600	h	1215.7	1255.5	1289.9	1302.7	1315.2	1327.4	1339.3	1351.1	1407.7	1462.5	1516.7	1625.5	1736.3	1850.0
(486.21)	s	1.4586	1.4990	1.5323	1.5443	1.5558	1.5667	1.5773	1.5875	1.6343	1.6762	1.7147	1.7846	1.8476	1.9056
	υ	.....	0.7277	0.7934	0.8177	0.8411	0.8639	0.8860	0.9077	1.0108	1.1082	1.2024	1.3853	1.5641	1.7405
700	h	.....	1243.2	1280.6	1294.3	1307.5	1320.3	1332.8	1345.0	1403.2	1459.0	1513.9	1623.5	1734.8	1848.8
(503.10)	s	.....	1.4722	1.5084	1.5212	1.5333	1.5449	1.5559	1.5665	1.6147	1.6573	1.6963	1.7666	1.8299	1.8881
	υ	.....	0.6154	0.6779	0.7006	0.7223	0.7433	0.7635	0.7833	0.8763	0.9633	1.0470	1.2088	1.3662	1.5214
800	h	.....	1229.8	1270.7	1285.4	1299.4	1312.9	1325.9	1338.6	1398.6	1455.4	1511.0	1621.4	1733.2	1847.5
(518.23)	s	.....	1.4467	1.4863	1.5000	1.5129	1.5250	1.5366	1.5476	1.5972	1.6407	1.6801	1.7510	1.8146	1.8729
	υ	.....	0.5264	0.5873	0.6089	0.6294	0.6491	0.6680	0.6863	0.7716	0.8506	0.9262	1.0714	1.2124	1.3509
900	h	.....	1215.0	1260.1	1275.9	1290.9	1305.1	1318.8	1332.1	1393.9	1451.8	1508.1	1619.3	1731.6	1846.3
(531.98)	s	.....	1.4216	1.4653	1.4800	1.4938	1.5066	1.5187	1.5303	1.5814	1.6257	1.6656	1.7371	1.8009	1.8595
	υ	.....	0.4533	0.5140	0.5350	0.5546	0.5733	0.5912	0.6084	0.6878	0.7604	0.8294	0.9615	1.0893	1.2146
1000	h	.....	1198.3	1248.8	1265.9	1281.9	1297.0	1311.4	1325.3	1389.2	1448.2	1505.1	1617.3	1730.0	1845.0
(544.61)	s	.....	1.3961	1.4450	1.4610	1.4757	1.4893	1.5021	1.5141	1.5670	1.6121	1.6525	1.7245	1.7886	1.8474
	υ	.....	.....	0.4532	0.4738	0.4929	0.5110	0.5281	0.5445	0.6191	0.6866	0.7503	0.8716	0.9885	1.1031
1100	h	.....	.....	1236.7	1255.3	1272.4	1288.5	1303.7	1318.3	1384.3	1444.5	1502.2	1615.2	1728.4	1843.8
(556.31)	s	.....	.....	1.4251	1.4425	1.4583	1.4728	1.4862	1.4989	1.5535	1.5995	1.6405	1.7130	1.7775	1.8363
	υ	.....	.....	0.4016	0.4222	0.4410	0.4586	0.4752	0.4909	0.5617	0.6250	0.6843	0.7967	0.9046	1.0101
1200	h	.....	.....	1223.5	1243.9	1262.4	1279.6	1295.7	1311.0	1379.3	1440.7	1499.2	1613.1	1726.9	1842.5
(567.22)	s	.....	.....	1.4052	1.4243	1.4413	1.4568	1.4710	1.4843	1.5409	1.5879	1.6293	1.7025	1.7672	1.8263
	υ	.....	.....	0.3174	0.3390	0.3580	0.3753	0.3912	0.4062	0.4714	0.5281	0.5805	0.6789	0.7727	0.8640
1400	h	.....	.....	1193.0	1218.4	1240.4	1260.3	1278.5	1295.5	1369.1	1433.1	1493.2	1608.9	1723.7	1840.0
(587.10)	s	.....	.....	1.3639	1.3877	1.4079	1.4258	1.4419	1.4567	1.5177	1.5666	1.6093	1.6836	1.7489	1.8083
	υ	.....	.....	.....	0.2733	0.2936	0.3112	0.3271	0.3417	0.4034	0.4553	0.5027	0.5905	0.6738	0.7545
1600	h	.....	.....	.....	1187.8	1215.2	1238.7	1259.6	1278.7	1358.4	1425.3	1487.0	1604.6	1720.5	1837.5
(604.90)	s	.....	.....	.....	1.3489	1.3741	1.3952	1.4137	1.4303	1.4964	1.5476	1.5914	1.6669	1.7328	1.7926
	υ	.....	.....	.....	.....	0.2407	0.2597	0.2760	0.2907	0.3502	0.3986	0.4421	0.5218	0.5968	0.6693
1800	h	.....	.....	.....	.....	1185.1	1214.0	1238.5	1260.3	1347.2	1417.4	1480.8	1600.4	1717.3	1835.0
(621.03)	s	.....	.....	.....	.....	1.3377	1.3638	1.3855	1.4044	1.4765	1.5301	1.5752	1.6520	1.7185	1.7786
	υ	.....	.....	.....	.....	0.1936	0.2161	0.2337	0.2489	0.3074	0.3532	0.3935	0.4668	0.5352	0.6011
2000	h	.....	.....	.....	.....	1145.6	1184.9	1214.8	1240.0	1335.5	1409.2	1474.5	1596.1	1714.1	1832.5
(635.82)	s	.....	.....	.....	.....	1.2945	1.3300	1.3564	1.3783	1.4576	1.5139	1.5603	1.6384	1.7055	1.7660
	υ	.....	.....	.....	.....	.....	.....	0.1484	0.1686	0.2294	0.2710	0.3061	0.3678	0.4244	0.4784
2500	h	.....	.....	.....	.....	.....	.....	1132.3	1176.8	1303.6	1387.8	1458.4	1585.3	1706.1	1826.2
(668.13)	s	.....	.....	.....	.....	.....	.....	1.2687	1.3073	1.4127	1.4772	1.5273	1.6088	1.6775	1.7389
	υ	.....	.....	.....	.....	.....	.....	.....	0.0984	0.1760	0.2159	0.2476	0.3018	0.3505	0.3966
3000	h	.....	.....	.....	.....	.....	.....	.....	1060.7	1267.2	1365.0	1441.8	1574.3	1698.0	1819.9
(695.36)	s	.....	.....	.....	.....	.....	.....	.....	1.1966	1.3690	1.4439	1.4984	1.5837	1.6540	1.7163
	υ	.....	.....	.....	.....	.....	.....	.....	.....	0.1583	0.1981	0.2288	0.2806	0.3267	0.3703
3206.2	h	.....	.....	.....	.....	.....	.....	.....	.....	1250.5	1355.2	1434.7	1569.8	1694.6	1817.2
(705.40)	s	.....	.....	.....	.....	.....	.....	.....	.....	1.3508	1.4309	1.4874	1.5742	1.6452	1.7080
	υ	.....	.....	.....	.....	.....	.....	.....	0.0306	0.1364	0.1762	0.2058	0.2546	0.2977	0.3381
3500	h	.....	.....	.....	.....	.....	.....	.....	780.5	1224.9	1340.7	1424.5	1563.3	1689.8	1813.6
	s	.....	.....	.....	.....	.....	.....	.....	0.9515	1.3241	1.4127	1.4723	1.5615	1.6336	1.6968
	υ	.....	.....	.....	.....	.....	.....	.....	0.0287	0.1052	0.1462	0.1743	0.2192	0.2581	0.2943
4000	h	.....	.....	.....	.....	.....	.....	.....	763.8	1174.8	1314.4	1406.8	1552.1	1681.7	1807.2
	s	.....	.....	.....	.....	.....	.....	.....	0.9347	1.2757	1.3827	1.4482	1.5417	1.6154	1.6795
	υ	.....	.....	.....	.....	.....	.....	.....	0.0276	0.0798	0.1226	0.1500	0.1917	0.2273	0.2602
4500	h	.....	.....	.....	.....	.....	.....	.....	753.5	1113.9	1286.5	1388.4	1540.8	1673.5	1800.9
	s	.....	.....	.....	.....	.....	.....	.....	0.9235	1.2204	1.3529	1.4253	1.5235	1.5990	1.6640
	υ	.....	.....	.....	.....	.....	.....	.....	0.0268	0.0593	0.1036	0.1303	0.1696	0.2027	0.2329
5000	h	.....	.....	.....	.....	.....	.....	.....	746.4	1047.1	1256.5	1369.5	1529.5	1665.3	1794.5
	s	.....	.....	.....	.....	.....	.....	.....	0.9152	1.1622	1.3231	1.4034	1.5066	1.5839	1.6499
	υ	.....	.....	.....	.....	.....	.....	.....	0.0262	0.0463	0.0880	0.1143	0.1516	0.1825	0.2106
5500	h	.....	.....	.....	.....	.....	.....	.....	741.3	985.0	1224.1	1349.3	1518.2	1657.0	1788.1
	s	.....	.....	.....	.....	.....	.....	.....	0.9090	1.1093	1.2930	1.3821	1.4908	1.5699	1.6369

Appendix D
Compressed Water

Abs. Press., psia (Sat. Temp.)		Temperature, °F 32	100	200	300	400	500	600	700
	Saturated Liquid P	0.08854	0.9492	11.526	67.013	247.31	680.8	1542.9	3093.7
	υ_f	0.016022	0.016132	0.016634	0.017449	0.018639	0.020432	0.023629	0.03692
	h_f	0	67.97	167.99	269.59	374.97	487.82	617.0	823.3
	s_f	0	0.12948	0.29382	0.43694	0.56638	0.68871	0.8131	0.9905
	$(\upsilon-\upsilon_f)$EE5	−1.1	−1.1	−1.1	−1.1				
200	$(h-h_f)$	+0.61	+0.54	+0.41	+0.23				
(381.79)	$(s-s_f)$EE3	+0.03	−0.05	−0.21	−0.21				
	$(\upsilon-\upsilon_f)$EE5	−2.3	−2.1	−2.2	−2.8	−2.1			
400	$(h-h_f)$	+1.21	+1.09	+0.88	+0.61	+0.16			
(444.59)	$(s-s_f)$EE3	+0.04	−0.16	−0.47	−0.56	−0.40			
	$(\upsilon-\upsilon_f)$EE5	−4.6	−4.0	−4.4	−5.6	−6.5	−1.7		
800	$(h-h_f)$	+2.39	+2.17	+1.78	+1.35	+0.61	−0.05		
(518.23)	$(s-s_f)$EE3	+0.10	−0.40	−0.97	−1.27	−1.48	−0.53		
	$(\upsilon-\upsilon_f)$EE5	−5.7	−5.1	−5.4	−6.9	−8.7	−6.4		
1000	$(h-h_f)$	+2.99	+2.70	+2.21	+1.75	+0.84	−0.14		
(544.61)	$(s-s_f)$EE3	+0.15	−0.53	−1.20	−1.64	−2.00	−1.41		
	$(\upsilon-\upsilon_f)$EE5	−8.4	−7.5	−8.1	−10.4	−14.1	−17.3		
1500	$(h-h_f)$	+4.48	+3.99	+3.36	+2.70	+1.44	−0.29		
(596.23)	$(s-s_f)$EE3	+0.20	−0.86	−1.79	−2.53	−3.32	−3.56		
	$(\upsilon-\upsilon_f)$EE5	−11.0	−9.9	−10.8	−13.8	−19.5	−27.8	−32.6	
2000	$(h-h_f)$	+5.97	+5.31	+4.51	+3.64	+2.03	−0.38	−2.5	
(635.82)	$(s-s_f)$EE3	+0.22	−1.18	−2.39	−3.42	−4.57	−5.58	−4.3	
	$(\upsilon-\upsilon_f)$EE5	−16.3	−14.7	−16.0	−20.7	−30.0	−47.1	−87.9	
3000	$(h-h_f)$	+9.00	+7.88	+6.76	+5.49	+3.33	−0.41	−6.9	
(695.36)	$(s-s_f)$EE3	+0.28	−1.79	−3.56	−5.12	−7.03	−9.42	−12.4	
	$(\upsilon-\upsilon_f)$EE5	−21.5	−19.2	−21.0	−27.5	−40.0	−64.5	−132.2	−821
4000	$(h-h_f)$	+11.88	+10.49	+9.03	+7.41	+4.71	−0.16	−10.0	−59.5
	$(s-s_f)$EE3	+0.29	−2.42	−4.74	−6.77	−9.40	−13.03	−19.3	−55.8
	$(\upsilon-\upsilon_f)$EE5	−26.7	−23.6	−26.0	−34.0	−49.6	−80.5	−169.3	−1017
5000	$(h-h_f)$	+14.75	+13.08	+11.30	+9.36	+6.08	+0.25	−12.1	−76.9
	$(s-s_f)$EE3	+0.22	−3.07	−5.92	−8.40	−11.74	−16.47	−25.3	−75.3

Appendix E
Mollier Diagram

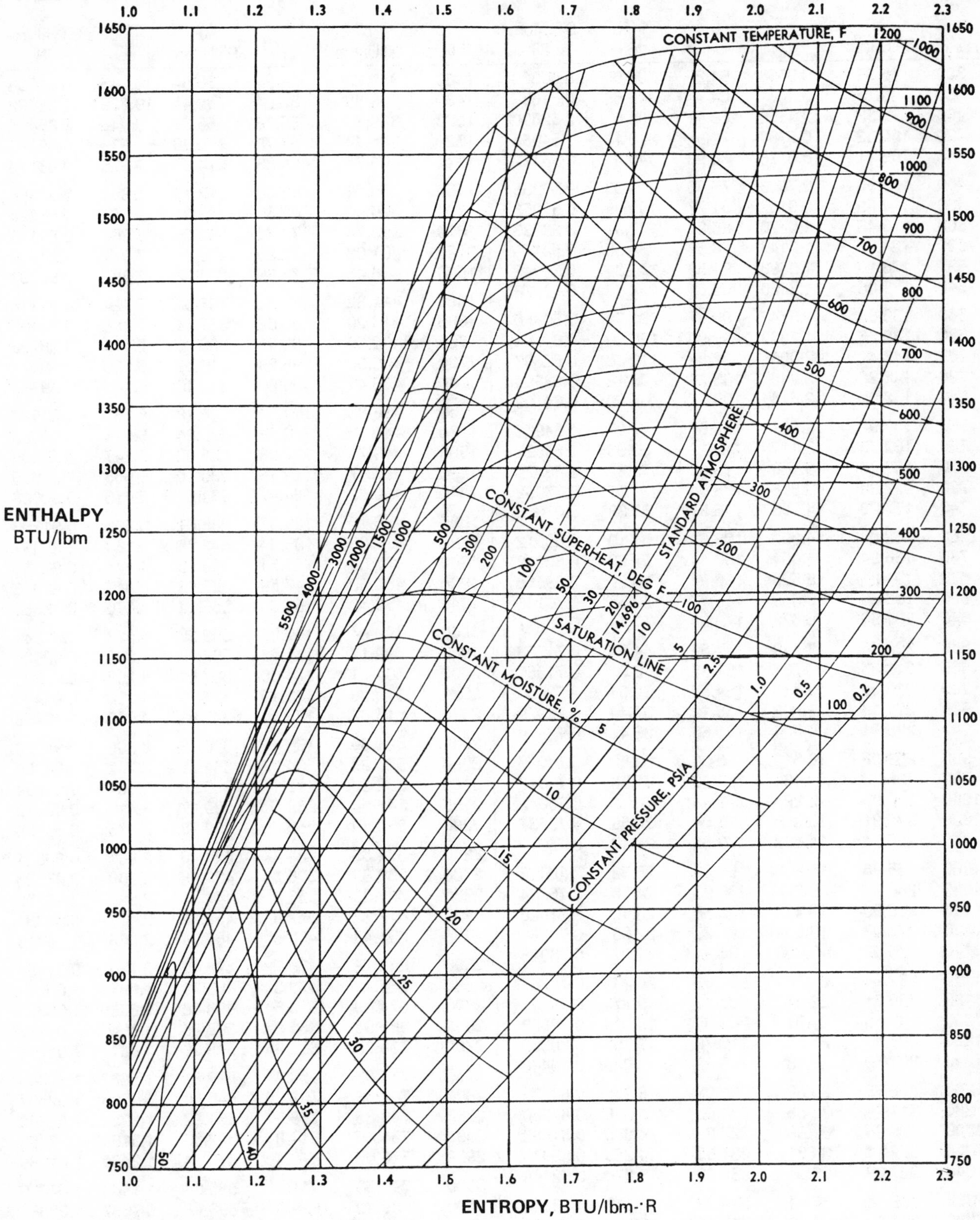

THERMO

Appendix F
Air Table

T°R	h BTU/lbm	p_r	u BTU/lbm	υ_r	φ BTU/lbm-°R
360	85.97	0.3363	61.29	396.6	0.50369
380	90.75	0.4061	64.70	346.6	0.51663
400	95.53	0.4858	68.11	305.0	0.52890
420	100.32	0.5760	71.52	270.1	0.54058
440	105.11	0.6776	74.93	240.6	0.55172
460	109.90	0.7913	78.36	215.33	0.56235
480	114.69	0.9182	81.77	193.65	0.57255
500	119.48	1.0590	85.20	174.90	0.58233
520	124.27	1.2147	88.62	158.58	0.59173
537	128.10	1.3593	91.53	146.34	0.59945
540	129.06	1.3860	92.04	144.32	0.60078
560	133.86	1.5742	95.47	131.78	0.60950
580	138.66	1.7800	98.90	120.70	0.61793
600	143.47	2.005	102.34	110.88	0.62607
620	148.28	2.249	105.78	102.12	0.63395
640	153.09	2.514	109.21	94.30	0.64159
660	157.92	2.801	112.67	87.27	0.64902
680	162.73	3.111	116.12	80.96	0.65621
700	167.56	3.446	119.58	75.25	0.66321
720	172.39	3.806	123.04	70.07	0.67002
740	177.23	4.193	126.51	65.38	0.67665
760	182.08	4.607	129.99	61.10	0.68312
780	186.94	5.051	133.47	57.20	0.68942
800	191.81	5.526	136.97	53.63	0.69558
820	196.69	6.033	140.47	50.35	0.70160
840	201.56	6.573	143.98	47.34	0.70747
860	206.46	7.149	147.50	44.57	0.71323
880	211.35	7.761	151.02	42.01	0.71886
900	216.26	8.411	154.57	39.64	0.72438
920	221.18	9.102	158.12	37.44	0.72979
940	226.11	9.834	161.68	35.41	0.73509
960	231.06	10.610	165.26	33.52	0.74030
980	236.02	11.430	168.83	31.76	0.74540
1000	240.98	12.298	172.43	30.12	0.75042
1020	245.97	13.215	176.04	28.59	0.75536
1040	250.95	14.182	179.66	27.17	0.76019
1060	255.96	15.203	183.29	25.82	0.76496
1080	260.97	16.278	186.93	24.58	0.76964
1100	265.99	17.413	190.58	23.40	0.77426
1120	271.03	18.604	194.25	22.30	0.77880
1140	276.08	19.858	197.94	21.27	0.78326
1160	281.14	21.18	201.63	20.293	0.78767
1180	286.21	22.56	205.33	19.377	0.79201
1200	291.30	24.01	209.05	18.514	0.79628
1220	296.41	25.53	212.78	17.700	0.80050
1240	301.52	27.13	216.53	16.932	0.80466
1260	306.65	28.80	220.28	16.205	0.80876
1280	311.79	30.55	244.05	15.518	0.81280
1300	316.94	32.39	227.83	14.868	0.81680
1320	322.11	34.31	231.63	14.253	0.82075
1340	327.29	36.31	235.43	13.670	0.82464
1360	332.48	38.41	239.25	13.118	0.82848
1380	337.68	40.59	243.08	12.593	0.83229
1400	342.90	42.88	246.93	12.095	0.83604
1420	348.14	45.26	250.79	11.622	0.83975
1440	353.37	47.75	254.66	11.172	0.84341

T°R	h BTU/lbm	p_r	u BTU/lbm	υ_r	φ BTU/lbm-°R
1460	358.63	50.34	258.54	10.743	0.84704
1480	363.89	53.04	262.44	10.336	0.85062
1500	369.17	55.86	266.34	9.948	0.85416
1520	374.47	58.78	270.26	9.578	0.85767
1540	379.77	61.83	274.20	9.226	0.86113
1560	385.08	65.00	278.13	8.890	0.86456
1580	390.40	68.30	282.09	8.569	0.86794
1600	395.74	71.73	286.06	8.263	0.87130
1620	401.09	75.29	290.04	7.971	0.87462
1640	406.45	78.99	294.03	7.691	0.87791
1660	411.82	82.83	298.02	7.424	0.88116
1680	417.20	86.82	302.04	7.168	0.88439
1700	422.59	90.95	306.06	6.924	0.88758
1720	428.00	95.24	310.09	6.690	0.89074
1740	433.41	99.69	314.13	6.465	0.89387
1760	438.83	104.30	318.18	6.251	0.89697
1780	444.26	109.08	322.24	6.045	0.90003
1800	449.71	114.03	326.32	5.847	0.90308
1820	455.17	119.16	330.40	5.658	0.90609
1840	460.63	124.47	334.50	5.476	0.90908
1860	466.12	129.95	338.61	5.302	0.91203
1880	471.60	135.64	342.73	5.134	0.91497
1900	477.09	141.51	346.85	4.974	0.91788
1920	482.60	147.59	350.98	4.819	0.92076
1940	488.12	153.87	355.12	4.670	0.92362
1960	493.64	160.37	359.28	4.527	0.92645
1980	499.17	167.07	363.43	4.390	0.92926
2000	504.71	174.00	367.61	4.258	0.93205
2020	510.26	181.16	371.79	4.130	0.93481
2040	515.82	188.54	375.98	4.008	0.93756
2060	521.39	196.16	380.18	3.890	0.94026
2080	526.97	204.02	384.39	3.777	0.94296
2100	532.55	212.1	388.60	3.667	0.94564
2150	546.54	233.5	399.17	3.410	0.95222
2200	560.59	256.6	409.78	3.176	0.95868
2250	574.69	281.4	420.46	2.961	0.96501
2300	588.82	308.1	431.16	2.765	0.97123
2350	603.00	336.8	441.91	2.585	0.97732
2400	617.22	367.6	452.70	2.419	0.98331
2450	631.48	400.5	463.54	2.266	0.98919
2500	645.78	435.7	474.40	2.125	0.99497
2550	660.12	473.3	485.31	1.9956	1.00064
2600	674.49	513.5	496.26	1.8756	1.00623
2650	688.90	556.3	507.25	1.7646	1.01172
2700	703.35	601.9	518.26	1.6617	1.01712
2750	717.83	650.4	529.31	1.5662	1.02244
2800	732.33	702.0	540.40	1.4775	1.02767
2850	746.88	756.7	551.52	1.3951	1.03282
2900	761.45	814.8	562.66	1.3184	1.03788
2950	776.05	876.4	573.84	1.2469	1.04288
3000	790.68	941.4	585.04	1.1803	1.04779
3500	938.40	1829.3	698.48	0.7087	1.09332
4000	1088.26	3280	814.06	0.4518	1.13334
4500	1239.86	5521	931.39	0.3019	1.16905
5000	1392.87	8837	1050.12	0.20959	1.20129
6000	1702.29	20120	1291.00	0.11047	1.25769
6500	1858.44	28974	1412.87	0.08310	1.28268

Appendix G
Saturated Ammonia

Abstracted from Tables of *Thermodynamic Properties of Ammonia,* U.S. Dept. of Commerce, Bureau of Standards Circular No. 142, 1945.

BY TEMPERATURE

Temp. (°F)	Press. (psia)	Specific Volume		Enthalpy			Entropy			Temp. (°F)
		Sat. liquid	Sat. vapor	Sat. liquid	Evap.	Sat. vapor	Sat. liquid	Evap.	Sat. vapor	
T	p	v_f	v_g	h_f	h_{fg}	h_g	s_f	s_{fg}	s_g	T
−60	5.55	0.02278	44.73	−21.2	610.8	589.6	−0.0517	1.5286	1.4769	−60
−50	7.67	0.02299	33.08	−10.6	604.3	593.7	−0.0256	1.4753	1.4497	−50
−40	10.41	0.02322	24.86	0.0	597.6	597.6	0.0000	1.4242	1.4242	−40
−30	13.90	0.02345	18.97	10.7	590.7	601.4	0.0250	1.3751	1.4001	−30
−20	18.30	0.02369	14.68	21.4	583.6	605.0	0.0497	1.3277	1.3774	−20
−10	23.74	0.02393	11.50	32.1	576.4	608.5	0.0738	1.2820	1.3558	−10
0	30.42	0.02419	9.116	42.9	568.9	611.8	0.0975	1.2377	1.3352	0
5	34.27	0.02432	8.150	48.3	565.0	613.3	0.1092	1.2161	1.3253	5
10	38.51	0.02446	7.304	53.8	561.1	614.9	0.1208	1.1949	1.3157	10
20	48.21	0.02474	5.910	64.7	553.1	617.8	0.1437	1.1532	1.2969	20
30	59.74	0.02503	4.825	75.7	544.8	620.5	0.1663	1.1127	1.2790	30
40	73.32	0.02533	3.971	86.8	536.2	623.0	0.1885	1.0733	1.2618	40
50	89.19	0.02564	3.294	97.9	527.3	625.2	0.2105	1.0348	1.2453	50
60	107.6	0.02597	2.751	109.2	518.1	627.3	0.2322	0.9972	1.2294	60
70	128.8	0.02632	2.312	120.5	508.6	629.1	0.2537	0.9603	1.2140	70
80	153.0	0.02668	1.955	132.0	498.7	630.7	0.2749	0.9242	1.1991	80
86	169.2	0.02691	1.772	138.9	492.6	631.5	0.2875	0.9029	1.1904	86
90	180.6	0.02707	1.661	143.5	488.5	632.0	0.2958	0.8888	1.1846	90
100	211.9	0.02747	1.419	155.2	477.8	633.0	0.3166	0.8539	1.1705	100
110	247.0	0.02790	1.217	167.0	466.7	633.7	0.3372	0.8194	1.1566	110
120	286.4	0.02836	1.047	179.0	455.0	634.0	0.3576	0.7851	1.1427	120

BY PRESSURE

Press. (psia)	Temp. (°F)	Specific Volume		Enthalpy			Entropy			Press. (psia)
		Sat. liquid	Sat. vapor	Sat. liquid	Evap.	Sat. vapor	Sat. liquid	Evap.	Sat. vapor	
p	T	v_f	v_g	h_f	h_{fg}	h_g	s_f	s_{fg}	s_g	p
5	−63.11	0.02271	49.31	−24.5	612.8	588.3	−0.0599	1.5456	1.4857	5
10	−41.34	0.02319	25.81	−1.4	598.5	597.1	−0.0034	1.4310	1.4276	10
15	−27.29	0.02351	17.67	13.6	588.8	602.4	0.0318	1.3620	1.3938	15
20	−16.64	0.02377	13.50	25.0	581.2	606.2	0.0578	1.3122	1.3700	20
30	−0.57	0.02417	9.236	42.3	569.3	611.6	0.0962	1.2402	1.3364	30
40	11.66	0.02451	7.047	55.6	559.8	615.4	0.1246	1.1879	1.3125	40
50	21.67	0.02479	5.710	66.5	551.7	618.2	0.1475	1.1464	1.2939	50
60	30.21	0.02504	4.805	75.9	544.6	620.5	0.1668	1.1119	1.2787	60
80	44.40	0.02546	3.655	91.7	532.3	624.0	0.1982	1.0563	1.2545	80
100	56.05	0.02584	2.952	104.7	521.8	626.5	0.2237	1.0119	1.2356	100
120	66.02	0.02618	2.476	116.0	512.4	628.4	0.2452	0.9749	1.2201	120
140	74.79	0.02649	2.132	126.0	503.9	629.9	0.2638	0.9430	1.2068	140
170	86.29	0.02692	1.764	139.3	492.3	631.6	0.2881	0.9019	1.1900	170
200	96.34	0.02732	1.502	150.9	481.8	632.7	0.3090	0.8666	1.1756	200
230	105.30	0.02770	1.307	161.4	472.0	633.4	0.3275	0.8356	1.1631	230
260	113.42	0.02806	1.155	171.1	462.8	633.9	0.3441	0.8077	1.1518	260

Appendix H
Superheated Ammonia

Abstracted from Tables of *Thermodynamic Properties of Ammonia,* U.S. Dept. of Commerce, Bureau of Standards Circular No. 142, 1945.

Abs. press. psia (Sat. temp.)		Temperature—Degrees Fahrenheit											
		−40	−20	0	20	40	60	80	100	150	200	250	300
5	v	52.36	54.97	57.55	60.12	62.69	65.24	67.79	70.33	76.68	......	......	......
	h	600.3	610.4	620.4	630.4	640.4	650.5	660.6	670.7	696.4	......	......	......
(−63.11)	s	1.5149	1.5385	1.5608	1.5821	1.6026	1.6223	1.6413	1.6598	1.7038	......	......	......
10	v	......	26.58	28.58	29.90	31.20	32.49	33.78	35.07	38.26	41.45	......	......
	h	......	603.2	618.9	629.1	639.3	649.5	659.7	670.1	695.8	722.2	......	......
(−41.34)	s	......	1.4420	1.4773	1.4992	1.5200	1.5400	1.5593	1.5779	1.6222	1.6637	......	......
15	v	......	18.01	18.92	19.82	20.70	21.58	22.44	23.31	25.46	27.59	......	......
	h	......	606.4	617.2	627.8	638.2	648.5	658.9	669.2	695.3	721.7	......	......
(−27.29)	s	......	1.4031	1.4272	1.4497	1.4709	1.4912	1.5108	1.5296	1.5742	1.6158	......	......
20	v	......	......	14.09	14.78	15.45	16.12	16.78	17.43	19.05	20.66	......	......
	h	......	......	615.5	626.4	637.0	647.5	658.0	668.5	694.7	721.2	......	......
(−16.64)	s	......	......	1.3907	1.4138	1.4856	1.4562	1.4760	1.4950	1.5399	1.5817	......	......
25	v	......	......	11.19	11.75	12.30	12.84	13.37	13.90	15.21	16.50	17.79	......
	h	......	......	613.8	625.0	635.8	646.5	657.1	667.7	694.1	720.8	748.0	......
(−7.96)	s	......	......	1.3616	1.3855	1.4077	1.4287	1.4487	1.4679	1.5131	1.5552	1.5948	......
30	v	......	......	......	9.731	10.20	10.65	11.10	11.55	12.65	13.73	14.81	......
	h	......	......	......	623.5	634.6	645.5	656.2	666.9	693.5	720.3	747.5	......
(−0.57)	s	......	......	......	1.3618	1.3845	1.4059	1.4261	1.4456	1.4911	1.5334	1.5733	......
35	v	......	......	......	8.287	8.695	9.093	9.484	9.869	10.82	11.75	12.68	......
	h	......	......	......	622.0	633.4	644.4	655.3	666.1	692.9	719.9	747.2	......
(5.89)	s	......	......	......	1.3413	1.3646	1.3863	1.4069	1.4265	1.4724	1.5148	1.5547	......
40	v	......	......	......	7.203	7.568	7.922	8.268	8.609	9.444	10.27	11.08	11.88
	h	......	......	......	620.4	632.1	643.4	654.4	665.3	692.3	719.4	746.8	774.6
(11.86)	s	......	......	......	1.3231	1.3470	1.3692	1.3900	1.4098	1.4561	1.4987	1.5387	1.5766
50	v	......	......	......	......	5.988	6.280	6.564	6.843	7.521	8.185	8.840	9.489
	h	......	......	......	......	6.295	641.2	652.6	663.7	691.1	718.5	746.1	774.0
(21.67)	s	......	......	......	......	1.3169	1.3399	1.3613	1.3816	1.4286	1.5716	1.5219	1.5500
60	v	......	......	......	......	4.933	5.184	5.428	5.665	6.239	6.798	7.348	7.892
	h	......	......	......	......	626.8	639.0	650.7	662.1	689.9	717.5	745.3	773.3
(30.21)	s	......	......	......	......	1.2913	1.3152	1.3373	1.3581	1.4058	1.4493	1.4898	1.5281
70	v	......	......	......	......	4.177	4.401	4.615	4.822	5.323	5.807	6.287	6.750
	h	......	......	......	......	623.9	636.6	648.7	660.4	688.7	716.6	744.5	772.7
(37.70)	s	......	......	......	......	1.2688	1.2937	1.3166	1.3378	1.3863	1.4302	1.4710	1.5095
80	v	......	......	......	......	......	3.812	4.005	4.190	4.635	5.063	5.487	5.894
	h	......	......	......	......	......	634.3	646.7	658.7	687.5	715.6	743.8	772.1
(44.40)	s	......	......	......	......	......	1.2745	1.2981	1.3199	1.3692	1.4136	1.4547	1.4933
90	v	......	......	......	......	......	3.353	3.529	3.698	4.100	4.484	4.859	5.228
	h	......	......	......	......	......	631.8	644.7	657.0	688.7	714.7	743.0	771.5
(50.47)	s	......	......	......	......	......	1.2571	1.2814	1.3038	1.3863	1.3988	1.4401	1.4789
100	v	......	......	......	......	......	2.985	3.149	3.304	3.672	4.021	4.361	4.695
	h	......	......	......	......	......	629.3	642.6	655.2	685.0	713.7	742.2	770.8
(56.05)	s	......	......	......	......	......	1.2409	1.2661	1.2891	1.3401	1.3854	1.4271	1.4660
120	v	......	......	......	......	......	......	2.576	2.712	3.029	3.326	3.614	3.895
	h	......	......	......	......	......	......	638.3	651.6	682.5	711.8	740.7	769.6
(66.02)	s	......	......	......	......	......	......	1.2386	1.2628	1.3157	1.3620	1.4042	1.4435
140	v	......	......	......	......	......	......	2.166	2.288	2.569	2.830	3.080	3.323
	h	......	......	......	......	......	......	633.8	647.8	679.9	709.9	739.2	768.3
(74.79)	s	......	......	......	......	......	......	1.2140	1.2396	1.2945	1.3418	1.3846	1.4243
160	v	......	......	......	......	......	......	......	1.969	2.224	2.457	2.679	2.895
	h	......	......	......	......	......	......	......	643.9	677.2	707.9	737.6	767.1
(82.64)	s	......	......	......	......	......	......	......	1.2186	1.2757	1.3240	1.3675	1.4076

Appendix H — Continued
Superheated Ammonia

Abstracted from Tables of *Thermodynamic Properties of Ammonia,* U.S. Dept. of Commerce, Bureau of Standards Circular No. 142, 1945.

Abs. press. psia (Sat. temp.)		Temperature—Degrees Fahrenheit											
		−40	−20	0	20	40	60	80	100	150	200	250	300
180	υ	......	......	......	......	......	......	......	1.720	1.995	2.167	2.367	2.561
	h	......	......	......	......	......	......	......	639.9	674.6	705.9	736.1	765.8
(89.78)	s	......	......	......	......	......	......	......	1.1992	1.2586	1.3081	1.3521	1.3926
200	υ	......	......	......	......	......	......	......	......	1.740	1.935	2.118	2.295
	h	......	......	......	......	......	......	......	......	671.8	703.9	734.5	764.5
(96.34)	s	......	......	......	......	......	......	......	......	1.2429	1.2935	1.3382	1.3791
220	υ	......	......	......	......	......	......	......	......	1.564	1.745	1.914	2.076
	h	......	......	......	......	......	......	......	......	669.0	701.9	732.9	763.2
(102.42)	s	......	......	......	......	......	......	......	......	1.2281	1.2801	1.3255	1.3668
240	υ	......	......	......	......	......	......	......	......	1.416	1.587	1.745	1.895
	h	......	......	......	......	......	......	......	......	666.1	699.8	731.3	762.0
(108.09)	s	......	......	......	......	......	......	......	......	1.2145	1.2677	1.3137	1.3554
260	υ	......	......	......	......	......	......	......	......	1.292	1.453	1.601	1.741
	h	......	......	......	......	......	......	......	......	663.1	697.7	729.7	760.7
(113.42)	s	......	......	......	......	......	......	......	......	1.2014	1.2560	1.3027	1.3349
280	υ	......	......	......	......	......	......	......	......	1.184	1.339	1.478	1.610
	h	......	......	......	......	......	......	......	......	660.1	695.6	728.1	759.4
(118.45)	s	......	......	......	......	......	......	......	......	1.1888	1.2449	1.2924	1.3350
300	υ	......	......	......	......	......	......	......	......	1.091	1.239	1.372	1.496
	h	......	......	......	......	......	......	......	......	656.9	693.5	726.5	758.1
(123.21)	s	......	......	......	......	......	......	......	......	1.1767	1.2344	1.2827	1.3257

Appendix I

Saturated Freon-12 (Refrigerant 12)

Freon is a registered trademark of the E. I. du Pont de Nemours & Co., which originally tabulated this data.

BY TEMPERATURE

Temp. (°F)	Press. (psia)	Specific volume		Enthalpy			Entropy			Temp. (°F)
		Sat. liquid	Sat. vapor	Sat. liquid	Evap.	Sat. vapor	Sat. liquid	Evap.	Sat. vapor	
T	p	v_f	v_g	h_f	h_{fg}	h_g	s_f	s_{fg}	s_g	T
−60	5.37	0.01036	6.516	−4.20	75.33	71.13	−0.0102	0.1681	0.1783	−60
−50	7.13	0.01047	5.012	−2.11	74.42	72.31	−0.0050	0.1717	0.1767	−50
−40	9.32	0.0106	3.911	0.00	73.50	73.50	0.00000	0.17517	0.17517	−40
−30	12.02	0.0107	3.088	2.03	72.67	74.70	0.00471	0.16916	0.17387	−30
−20	15.28	0.0108	2.474	4.07	71.80	75.87	0.00940	0.16335	0.17275	−20
−10	19.20	0.0109	2.003	6.14	70.91	77.05	0.01403	0.15772	0.17175	−10
0	23.87	0.0110	1.637	8.25	69.96	78.21	0.01869	0.15222	0.17091	0
5	26.51	0.0111	1.485	9.32	69.47	78.79	0.02097	0.14955	0.17052	5
10	29.35	0.0112	1.351	10.39	68.97	79.36	0.02328	0.14687	0.17015	10
20	35.75	0.0113	1.121	12.55	67.94	80.49	0.02783	0.14166	0.16949	20
30	43.16	0.0115	0.939	14.76	66.85	81.61	0.03233	0.13654	0.16887	30
40	51.68	0.0116	0.792	17.00	65.71	82.71	0.03680	0.13153	0.16833	40
50	61.39	0.0118	0.673	19.27	64.51	83.78	0.04126	0.12659	0.16785	50
60	72.41	0.0119	0.575	21.57	63.25	84.82	0.04568	0.12173	0.16741	60
70	84.82	0.0121	0.493	23.90	61.92	85.82	0.05009	0.11692	0.16701	70
80	98.76	0.0123	0.425	26.28	60.52	86.80	0.05446	0.11215	0.16662	80
86	107.9	0.0124	0.389	27.72	59.65	87.37	0.05708	0.10932	0.16640	86
90	114.3	0.0125	0.368	28.70	59.04	87.74	0.05882	0.10742	0.16624	90
100	131.6	0.0127	0.319	31.16	57.46	88.62	0.06316	0.10268	0.16584	100
110	150.7	0.0129	0.277	33.65	55.78	89.43	0.06749	0.09793	0.16542	110
120	171.8	0.0132	0.240	36.16	53.99	90.15	0.07180	0.09315	0.16495	120

BY PRESSURE

Press. (psia)	Temp. (°F)	Specific volume		Enthalpy			Entropy			Press. (psia)
		Sat. liquid	Sat. vapor	Sat. liquid	Evap.	Sat. vapor	Sat. liquid	Evap.	Sat. vapor	
p	T	v_f	v_g	h_f	h_{fg}	h_g	s_f	s_{fg}	s_g	p
5	−62.5	0.01034	6.953	−4.73	75.56	70.83	−0.0115	0.1943	0.1788	5
10	−37.3	0.0106	3.662	0.54	73.28	73.82	0.00127	0.17360	0.17487	10
15	−20.8	0.0108	2.518	3.91	71.87	75.78	0.00902	0.16381	0.17283	15
20	−8.2	0.0109	1.925	6.53	70.74	77.27	0.01488	0.15672	0.17160	20
30	11.1	0.0112	1.324	10.62	68.86	79.48	0.02410	0.14597	0.17007	30
40	25.9	0.0114	1.009	13.86	67.30	81.16	0.03049	0.13865	0.16914	40
50	38.3	0.0116	0.817	16.58	65.94	82.52	0.03597	0.13244	0.16841	50
60	48.7	0.0117	0.688	18.96	64.69	83.65	0.04065	0.12726	0.16791	60
80	66.3	0.0120	0.521	23.01	62.44	85.45	0.04844	0.11872	0.16716	80
100	80.9	0.0123	0.419	26.49	60.40	86.89	0.05483	0.11176	0.16659	100
120	93.4	0.0126	0.350	29.53	58.52	88.05	0.06030	0.10580	0.16610	120
140	104.5	0.0128	0.298	32.28	56.71	88.99	0.06513	0.10053	0.16566	140
160	114.5	0.0130	0.260	34.78	54.99	89.77	0.06958	0.09564	0.16522	160
180	123.7	0.0133	0.228	37.07	53.31	90.38	0.07337	0.09139	0.16476	180
200	132.1	0.0135	0.202	39.21	51.65	90.86	0.07694	0.08730	0.16424	200
220	139.9	0.0138	0.181	41.22	50.28	91.50	0.08021	0.08354	0.16375	220

Appendix J

Superheated Freon-12 (Refrigerant 12)

Abs. press. psia (Sat. temp.)		Temperature—Degrees Fahrenheit											
		−40	−20	0	20	40	60	80	100	150	200	250	300
	υ	7.363	7.726	8.088	8.450	8.812	9.173	9.533	9.893	10.79	11.69	......	......
5	h	73.72	76.36	79.05	81.78	84.56	87.41	90.30	93.25	100.84	108.75	......	......
(−62.5)	s	0.1859	0.1920	0.1979	0.2038	0.2095	0.2150	0.2205	0.2258	0.2388	0.2513	......	......
	υ	......	3.821	4.006	4.189	4.371	4.556	4.740	4.923	5.379	5.831	6.281	......
10	h	......	76.11	78.81	81.56	84.35	87.19	90.11	93.05	100.66	108.63	116.88	......
(−37.3)	s	......	0.1801	0.1861	0.1919	0.1977	0.2033	0.2087	0.2141	0.2271	0.2396	0.2517	......
	υ	......	2.521	2.646	2.771	2.895	3.019	3.143	3.266	3.571	3.877	4.191	......
15	h	......	75.89	78.59	81.37	84.18	87.03	89.94	92.91	100.53	108.49	116.78	......
(−20.8)	s	......	0.17307	0.17913	0.18499	0.19074	0.19635	0.20185	0.20723	0.22028	0.23282	0.24491	......
	υ	......	......	1.965	2.060	2.155	2.250	2.343	2.437	2.669	2.901	3.130	......
20	h	......	......	78.39	81.14	83.97	86.85	89.78	92.75	100.40	108.38	116.67	......
(−8.2)	s	......	......	0.17407	0.17996	0.18573	0.19138	0.19688	0.20229	0.21537	0.22794	0.24005	......
	υ	......	......	......	1.712	1.793	1.873	1.952	2.031	2.227	2.422	2.615	......
25	h	......	......	......	80.95	83.78	86.67	89.61	92.56	100.26	108.26	116.56	......
(2.2)	s	......	......	......	0.17637	0.18216	0.18783	0.19336	0.19748	0.21190	0.22450	0.23665	......
	υ	......	......	......	1.364	1.430	1.495	1.560	1.624	1.784	1.943	2.099	......
30	h	......	......	......	80.75	83.59	86.49	89.43	92.42	100.12	108.13	116.45	......
(11.1)	s	......	......	......	0.17278	0.17859	0.18429	0.18983	0.19527	0.20843	0.22105	0.23325	......
	υ	......	......	......	1.109	1.237	1.295	1.352	1.409	1.550	1.689	1.827	......
35	h	......	......	......	80.49	83.40	86.30	89.26	92.26	99.98	108.01	116.33	......
(18.9)	s	......	......	......	0.16963	0.17591	0.18162	0.18719	0.19266	0.20584	0.21849	0.23069	......
	υ	......	......	......	......	1.044	1.095	1.144	1.194	1.315	1.435	1.554	......
40	h	......	......	......	......	83.20	86.11	89.09	92.09	99.83	107.88	116.21	......
(25.9)	s	......	......	......	......	0.17322	0.17896	0.18455	0.19004	0.20325	0.21592	0.22813	......
	υ	......	......	......	......	0.821	0.863	0.904	0.944	1.044	1.142	1.239	1.332
50	h	......	......	......	......	82.76	85.72	88.72	91.75	99.54	107.62	116.00	124.69
(38.3)	s	......	......	......	......	0.16895	0.17475	0.19040	0.18591	0.19923	0.21196	0.22419	0.23600
	υ	......	......	......	......	......	0.708	0.743	0.778	0.863	0.946	1.028	1.108
60	h	......	......	......	......	......	85.33	88.35	91.41	99.24	107.36	115.54	124.29
(48.7)	s	......	......	......	......	......	0.17120	0.17689	0.18246	0.19585	0.20865	0.22094	0.23280
	υ	......	......	......	......	......	0.553	0.642	0.673	0.750	0.824	0.896	0.967
70	h	......	......	......	......	......	84.94	87.96	91.05	98.94	107.10	115.54	124.29
(57.9)	s	......	......	......	......	......	0.16765	0.17399	0.17961	0.19310	0.20597	0.21830	0.23020
	υ	......	......	......	......	......	......	0.540	0.568	0.636	0.701	0.764	0.826
80	h	......	......	......	......	......	......	87.56	90.68	98.64	106.84	115.30	124.08
(66.3)	s	......	......	......	......	......	......	0.17108	0.17675	0.19035	0.20328	0.21566	0.22760
	υ	......	......	......	......	......	......	......	0.505	0.568	0.627	0.685	0.742
90	h	......	......	......	......	......	......	......	90.31	98.32	106.56	115.07	123.88
(73.6)	s	......	......	......	......	......	......	......	0.17443	0.18813	0.20111	0.21356	0.22554
	υ	......	......	......	......	......	......	......	0.442	0.499	0.553	0.606	0.657
100	h	......	......	......	......	......	......	......	89.93	97.99	106.29	114.84	123.67
(80.9)	s	......	......	......	......	......	......	......	0.17210	0.18590	0.19894	0.21145	0.22347
	υ	......	......	......	......	......	......	......	0.357	0.407	0.454	0.500	0.543
120	h	......	......	......	......	......	......	......	89.13	97.30	105.70	114.35	123.25
(93.4)	s	......	......	......	......	......	......	......	0.16803	0.18207	0.19529	0.20792	0.22000
	υ	......	......	......	......	......	......	......	......	0.341	0.383	0.423	0.462
140	h	......	......	......	......	......	......	......	......	96.65	105.14	113.85	122.85
(104.5)	s	......	......	......	......	......	......	......	......	0.17868	0.19205	0.20479	0.21701
	υ	......	......	......	......	......	......	......	......	0.318	0.335	0.372	0.408
160	h	......	......	......	......	......	......	......	......	95.82	104.50	113.33	122.39
(114.5)	s	......	......	......	......	......	......	......	......	0.17561	0.18927	0.20213	0.21444
	υ	......	......	......	......	......	......	......	......	0.294	0.287	0.321	0.353
180	h	......	......	......	......	......	......	......	......	94.99	103.85	112.81	121.92
(123.7)	s	......	......	......	......	......	......	......	......	0.17254	0.18648	0.19947	0.21187
	υ	......	......	......	......	......	......	......	......	0.241	0.255	0.288	0.317
200	h	......	......	......	......	......	......	......	......	94.16	103.12	112.20	121.42
(132.1)	s	......	......	......	......	......	......	......	......	0.16970	0.18395	0.19717	0.20970
	υ	......	......	......	......	......	......	......	......	0.188	0.232	0.254	0.282
220	h	......	......	......	......	......	......	......	......	93.32	102.39	111.59	120.91
(139.9)	s	......	......	......	......	......	......	......	......	0.16685	0.18142	0.19387	0.20753

THERMO

Appendix K
Isentropic Flow Factors For k = 1.4

M = Mach number, P = pressure, TP = total pressure, D = density, TD = total density, T = temperature, TT = total temperature, A = area at a point, A* = throat area for M = 1, v = velocity, and c* = speed of sound at throat.

M	P/TP	D/TD	T/TT	A/A*	V/C*
.00	1.0000	1.0000	1.0000	-------	.0000
.01	.9999	1.0000	1.0000	57.8737	.0110
.02	.9997	.9998	.9999	28.9420	.0219
.03	.9994	.9996	.9998	19.3005	.0329
.04	.9989	.9992	.9997	14.4815	.0438
.05	.9983	.9988	.9995	11.5914	.0548
.06	.9975	.9982	.9993	9.6659	.0657
.07	.9966	.9976	.9990	8.2915	.0766
.08	.9955	.9968	.9987	7.2616	.0876
.09	.9944	.9960	.9984	6.4613	.0985
.10	.9930	.9950	.9980	5.8218	.1094
.11	.9916	.9940	.9976	5.2992	.1204
.12	.9900	.9928	.9971	4.8643	.1313
.13	.9883	.9916	.9966	4.4969	.1422
.14	.9864	.9903	.9961	4.1824	.1531
.15	.9844	.9888	.9955	3.9103	.1639
.16	.9823	.9873	.9949	3.6727	.1748
.17	.9800	.9857	.9943	3.4635	.1857
.18	.9776	.9840	.9936	3.2779	.1965
.19	.9751	.9822	.9928	3.1123	.2074
.20	.9725	.9803	.9921	2.9635	.2182
.21	.9697	.9783	.9913	2.8293	.2290
.22	.9668	.9762	.9904	2.7076	.2398
.23	.9638	.9740	.9895	2.5968	.2506
.24	.9607	.9718	.9886	2.4956	.2614
.25	.9575	.9694	.9877	2.4027	.2722
.26	.9541	.9670	.9867	2.3173	.2829
.27	.9506	.9645	.9856	2.2385	.2936
.28	.9470	.9619	.9846	2.1656	.3043
.29	.9433	.9592	.9835	2.0979	.3150
.30	.9395	.9564	.9823	2.0351	.3257
.31	.9355	.9535	.9811	1.9765	.3364
.32	.9315	.9506	.9799	1.9218	.3470
.33	.9274	.9476	.9787	1.8707	.3576
.34	.9231	.9445	.9774	1.8229	.3682
.35	.9188	.9413	.9761	1.7780	.3788
.36	.9143	.9380	.9747	1.7358	.3893
.37	.9098	.9347	.9734	1.6961	.3999
.38	.9052	.9313	.9719	1.6587	.4104
.39	.9004	.9278	.9705	1.6234	.4209
.40	.8956	.9243	.9690	1.5901	.4313
.41	.8907	.9207	.9675	1.5587	.4418
.42	.8857	.9170	.9659	1.5289	.4522
.43	.8807	.9132	.9643	1.5007	.4626
.44	.8755	.9094	.9627	1.4740	.4729
.45	.8703	.9055	.9611	1.4487	.4833
.46	.8650	.9016	.9594	1.4246	.4936
.47	.8596	.8976	.9577	1.4018	.5038
.48	.8541	.8935	.9560	1.3801	.5141
.49	.8486	.8894	.9542	1.3595	.5243
.50	.8430	.8852	.9524	1.3398	.5345

M	P/TP	D/TD	T/TT	A/A*	V/C*
.51	.8374	.8809	.9506	1.3212	.5447
.52	.8317	.8766	.9487	1.3034	.5548
.53	.8259	.8723	.9468	1.2865	.5649
.54	.8201	.8679	.9449	1.2703	.5750
.55	.8142	.8634	.9430	1.2549	.5851
.56	.8082	.8589	.9410	1.2403	.5951
.57	.8022	.8544	.9390	1.2263	.6051
.58	.7962	.8498	.9370	1.2130	.6150
.59	.7901	.8451	.9349	1.2003	.6249
.60	.7840	.8405	.9328	1.1882	.6348
.61	.7778	.8357	.9307	1.1767	.6447
.62	.7716	.8310	.9286	1.1656	.6545
.63	.7654	.8262	.9265	1.1551	.6643
.64	.7591	.8213	.9243	1.1451	.6740
.65	.7528	.8164	.9221	1.1356	.6837
.66	.7465	.8115	.9199	1.1265	.6934
.67	.7401	.8066	.9176	1.1179	.7031
.68	.7338	.8016	.9153	1.1097	.7127
.69	.7274	.7966	.9131	1.1018	.7223
.70	.7209	.7916	.9107	1.0944	.7318
.71	.7145	.7865	.9084	1.0873	.7413
.72	.7080	.7814	.9061	1.0806	.7508
.73	.7016	.7763	.9037	1.0742	.7602
.74	.6951	.7712	.9013	1.0681	.7696
.75	.6886	.7660	.8989	1.0624	.7789
.76	.6821	.7609	.8964	1.0570	.7883
.77	.6756	.7557	.8940	1.0519	.7975
.78	.6691	.7505	.8915	1.0471	.8068
.79	.6625	.7452	.8890	1.0425	.8160
.80	.6560	.7400	.8865	1.0382	.8251
.81	.6495	.7347	.8840	1.0342	.8343
.82	.6430	.7295	.8815	1.0305	.8433
.83	.6365	.7242	.8789	1.0270	.8524
.84	.6300	.7189	.8763	1.0237	.8614
.85	.6235	.7136	.8737	1.0207	.8704
.86	.6170	.7083	.8711	1.0179	.8793
.87	.6106	.7030	.8685	1.0153	.8882
.88	.6041	.6977	.8659	1.0129	.8970
.89	.5977	.6924	.8632	1.0108	.9058
.90	.5913	.6870	.8606	1.0089	.9146
.91	.5849	.6817	.8579	1.0071	.9233
.92	.5785	.6764	.8552	1.0056	.9320
.93	.5721	.6711	.8525	1.0043	.9406
.94	.5658	.6658	.8498	1.0031	.9493
.95	.5595	.6604	.8471	1.0021	.9578
.96	.5532	.6551	.8444	1.0014	.9663
.97	.5469	.6498	.8416	1.0008	.9748
.98	.5407	.6445	.8389	1.0003	.9832
.99	.5345	.6392	.8361	1.0001	.9916
1.00	.5283	.6339	.8333	1.0000	1.0000

ISENTROPIC FLOW AND NORMAL SHOCK PARAMETERS FOR AIR AND OTHER GASES WITH k = 1.4

(X refers to upstream conditions. Y refers to downstream conditions.)
(For example, PX/TPY is the ratio of static pressure before the shock wave to total pressure after the shock wave.)

M	P/TP	D/TD	T/TT	A/A*	V/C*	MY	PY/PX	DY/DX	TY/TX	TPY/TPX	PX/TPY
1.00	.5283	.6339	.8333	1.0000	1.0000	1.0000	.1000 EE+1	.1000 EE+1	.1000 EE+1	1.0000	.5283
1.10	.4684	.5817	.8052	1.0079	1.0812	.9118	.1245 EE+1	.1169 EE+1	.1065 EE+1	.9989	.4689
1.20	.4124	.5311	.7764	1.0304	1.1583	.8422	.1513 EE+1	.1342 EE+1	.1128 EE+1	.9928	.4154
1.30	.3609	.4829	.7474	1.0663	1.2311	.7860	.1805 EE+1	.1516 EE+1	.1191 EE+1	.9794	.3685
1.40	.3142	.4374	.7184	1.1149	1.2999	.7397	.2120 EE+1	.1690 EE+1	.1255 EE+1	.9582	.3280
1.50	.2724	.3950	.6897	1.1762	1.3646	.7011	.2458 EE+1	.1862 EE+1	.1320 EE+1	.9298	.2930
1.60	.2353	.3557	.6614	1.2502	1.4254	.6684	.2820 EE+1	.2032 EE+1	.1388 EE+1	.8952	.2628
1.70	.2026	.3197	.6337	1.3376	1.4825	.6405	.3205 EE+1	.2198 EE+1	.1458 EE+1	.8557	.2368
1.80	.1740	.2868	.6068	1.4390	1.5360	.6165	.3613 EE+1	.2359 EE+1	.1532 EE+1	.8127	.2142
1.90	.1492	.2570	.5807	1.5553	1.5861	.5956	.4045 EE+1	.2516 EE+1	.1608 EE+1	.7674	.1945
2.00	.1278	.2300	.5556	1.6875	1.6330	.5774	.4500 EE+1	.2667 EE+1	.1687 EE+1	.7209	.1773
2.10	.1094	.2058	.5313	1.8369	1.6769	.5613	.4978 EE+1	.2812 EE+1	.1770 EE+1	.6742	.1622
2.20	.9352 EE−1	.1841	.5081	2.0050	1.7179	.5471	.5480 EE+1	.2951 EE+1	.1857 EE+1	.6281	.1489
2.30	.7997 EE−1	.1646	.4859	2.1931	1.7563	.5344	.6005 EE+1	.3085 EE+1	.1947 EE+1	.5833	.1371
2.40	.6840 EE−1	.1472	.4647	2.4031	1.7922	.5231	.6553 EE+1	.3212 EE+1	.2040 EE+1	.5401	.1266
2.50	.5853 EE−1	.1317	.4444	2.6367	1.8257	.5130	.7125 EE+1	.3333 EE+1	.2137 EE+1	.4990	.1173
2.60	.5012 EE−1	.1179	.4252	2.8960	1.8571	.5039	.7720 EE+1	.3449 EE+1	.2238 EE+1	.4601	.1089
2.70	.4295 EE−1	.1056	.4068	3.1830	1.8865	.4956	.8338 EE+1	.3559 EE+1	.2343 EE+1	.4236	.1014
2.80	.3685 EE−1	.9463 EE−1	.3894	3.5001	1.9140	.4882	.8980 EE+1	.3664 EE+1	.2451 EE+1	.3895	.9461 EE−1
2.90	.3165 EE−1	.8489 EE−1	.3729	3.8498	1.9398	.4814	.9645 EE+1	.3763 EE+1	.2563 EE+1	.3577	.8848 EE−1
3.00	.2722 EE−1	.7623 EE−1	.3571	4.2346	1.9640	.4752	.1033 EE+2	.3857 EE+1	.2679 EE+1	.3283	.8291 EE−1
3.10	.2345 EE−1	.6852 EE−1	.3422	4.6573	1.9866	.4695	.1104 EE+2	.3947 EE+1	.2799 EE+1	.3012	.7785 EE−1
3.20	.2023 EE−1	.6165 EE−1	.3281	5.1210	2.0079	.4643	.1178 EE+2	.4031 EE+1	.2922 EE+1	.2762	.7323 EE−1
3.30	.1748 EE−1	.5554 EE−1	.3147	5.6286	2.0278	.4596	.1254 EE+2	.4112 EE+1	.3049 EE+1	.2533	.6900 EE−1
3.40	.1512 EE−1	.5009 EE−1	.3019	6.1837	2.0466	.4552	.1332 EE+2	.4188 EE+1	.3180 EE+1	.2322	.6513 EE−1
3.50	.1311 EE−1	.4523 EE−1	.2899	6.7896	2.0642	.4512	.1412 EE+2	.4261 EE+1	.3315 EE+1	.2129	.6157 EE−1
3.60	.1138 EE−1	.4089 EE−1	.2784	7.4501	2.0808	.4474	.1495 EE+2	.4330 EE+1	.3454 EE+1	.1953	.5829 EE−1
3.70	.9903 EE−2	.3702 EE−1	.2675	8.1691	2.0964	.4439	.1580 EE+2	.4395 EE+1	.3596 EE+1	.1792	.5526 EE−1
3.80	.8629 EE−2	.3355 EE−1	.2572	8.9506	2.1111	.4407	.1668 EE+2	.4457 EE+1	.3743 EE+1	.1645	.5247 EE−1
3.90	.7532 EE−2	.3044 EE−1	.2474	9.7990	2.1250	.4377	.1758 EE+2	.4516 EE+1	.3893 EE+1	.1510	.4987 EE−1
4.00	.6586 EE−2	.2766 EE−1	.2381	10.7187	2.1381	.4350	.1850 EE+2	.4571 EE+1	.4047 EE+1	.1388	.4747 EE−1
4.10	.5769 EE−2	.2516 EE−1	.2293	11.7147	2.1505	.4324	.1944 EE+2	.4624 EE+1	.4205 EE+1	.1276	.4523 EE−1
4.20	.5062 EE−2	.2292 EE−1	.2208	12.7916	2.1622	.4299	.2041 EE+2	.4675 EE+1	.4367 EE+1	.1173	.4314 EE−1
4.30	.4449 EE−2	.2090 EE−1	.2129	13.9549	2.1732	.4277	.2140 EE+2	.4723 EE+1	.4532 EE+1	.1080	.4120 EE−1
4.40	.3918 EE−2	.1909 EE−1	.2053	15.2099	2.1837	.4255	.2242 EE+2	.4768 EE+1	.4702 EE+1	.9948 EE−1	.3938 EE−1
4.50	.3455 EE−2	.1745 EE−1	.1980	16.5622	2.1936	.4236	.2346 EE+2	.4812 EE+1	.4875 EE+1	.9170 EE−1	.3768 EE−1
4.60	.3053 EE−2	.1597 EE−1	.1911	18.0178	2.2030	.4217	.2452 EE+2	.4853 EE+1	.5052 EE+1	.8459 EE−1	.3609 EE−1
4.70	.2701 EE−2	.1464 EE−1	.1846	19.5828	2.2119	.4199	.2560 EE+2	.4893 EE+1	.5233 EE+1	.7809 EE−1	.3459 EE−1
4.80	.2394 EE−2	.1343 EE−1	.1783	21.2637	2.2204	.4183	.2671 EE+2	.4930 EE+1	.5418 EE+1	.7214 EE−1	.3319 EE−1
4.90	.2126 EE−2	.1233 EE−1	.1724	23.0671	2.2284	.4167	.2784 EE+2	.4966 EE+1	.5607 EE+1	.6670 EE−1	.3187 EE−1
5.00	.1890 EE−2	.1134 EE−1	.1667	25.0000	2.2361	.4152	.2900 EE+2	.5000 EE+1	.5800 EE+1	.6172 EE−1	.3062 EE−1
5.10	.1683 EE−2	.1044 EE−1	.1612	27.0696	2.2433	.4138	.3018 EE+2	.5033 EE+1	.5997 EE+1	.5715 EE−1	.2945 EE−1
5.20	.1501 EE−2	.9620 EE−2	.1561	29.2833	2.2503	.4125	.3138 EE+2	.5064 EE+1	.6197 EE+1	.5297 EE−1	.2834 EE−1
5.30	.1341 EE−2	.8875 EE−2	.1511	31.6491	2.2569	.4113	.3260 EE+2	.5093 EE+1	.6401 EE+1	.4913 EE−1	.2730 EE−1
5.40	.1200 EE−2	.8197 EE−2	.1464	34.1748	2.2631	.4101	.3385 EE+2	.5122 EE+1	.6610 EE+1	.4560 EE−1	.2631 EE−1
5.50	.1075 EE−2	.7578 EE−2	.1418	36.8690	2.2691	.4090	.3512 EE+2	.5149 EE+1	.6822 EE+1	.4236 EE−1	.2537 EE−1
5.60	.9643 EE−3	.7012 EE−2	.1375	39.7402	2.2748	.4079	.3642 EE+2	.5175 EE+1	.7038 EE+1	.3938 EE−1	.2448 EE−1
5.70	.8663 EE−3	.6496 EE−2	.1334	42.7974	2.2803	.4069	.3774 EE+2	.5200 EE+1	.7258 EE+1	.3664 EE−1	.2364 EE−1
5.80	.7794 EE−3	.6023 EE−2	.1294	46.0500	2.2855	.4059	.3908 EE+2	.5224 EE+1	.7481 EE+1	.3412 EE−1	.2284 EE−1
5.90	.7021 EE−3	.5590 EE−2	.1256	49.5075	2.2905	.4050	.4044 EE+2	.5246 EE+1	.7709 EE+1	.3179 EE−1	.2208 EE−1
6.00	.6334 EE−3	.5194 EE−2	.1220	53.1798	2.2953	.4042	.4183 EE+2	.5268 EE+1	.7941 EE+1	.2965 EE−1	.2136 EE−1
6.10	.5721 EE−3	.4829 EE−2	.1185	57.0772	2.2998	.4033	.4324 EE+2	.5289 EE+1	.8176 EE+1	.2767 EE−1	.2067 EE−1
6.20	.5173 EE−3	.4495 EE−2	.1151	61.2102	2.3042	.4025	.4468 EE+2	.5309 EE+1	.8415 EE+1	.2584 EE−1	.2002 EE−1
6.30	.4684 EE−3	.4187 EE−2	.1119	65.5899	2.3084	.4018	.4614 EE+2	.5329 EE+1	.8658 EE+1	.2416 EE−1	.1939 EE−1
6.40	.4247 EE−3	.3904 EE−2	.1088	70.2274	2.3124	.4011	.4762 EE+2	.5347 EE+1	.8905 EE+1	.2259 EE−1	.1880 EE−1
6.50	.3855 EE−3	.3643 EE−2	.1058	75.1343	2.3163	.4004	.4912 EE+2	.5365 EE+1	.9156 EE+1	.2115 EE−1	.1823 EE−1
6.60	.3503 EE−3	.3402 EE−2	.1030	80.3227	2.3200	.3997	.5065 EE+2	.5382 EE+1	.9411 EE+1	.1981 EE−1	.1768 EE−1
6.70	.3187 EE−3	.3180 EE−2	.1002	85.8049	2.3235	.3991	.5220 EE+2	.5399 EE+1	.9670 EE+1	.1857 EE−1	.1716 EE−1
6.80	.2902 EE−3	.2974 EE−2	.9758 EE−1	91.5935	2.3269	.3985	.5378 EE+2	.5415 EE+1	.9933 EE+1	.1741 EE−1	.1667 EE−1
6.90	.2646 EE−3	.2785 EE−2	.9504 EE−1	97.7017	2.3302	.3979	.5538 EE+2	.5430 EE+1	.1020 EE+2	.1634 EE−1	.1619 EE−1
7.00	.2416 EE−3	.2609 EE−2	.9259 EE−1	104.1429	2.3333	.3974	.5700 EE+2	.5444 EE+1	.1047 EE+2	.1535 EE−1	.1573 EE−1
7.10	.2207 EE−3	.2446 EE−2	.9024 EE−1	110.9309	2.3364	.3968	.5864 EE+2	.5459 EE+1	.1074 EE+2	.1443 EE−1	.1530 EE−1
7.20	.2019 EE−3	.2295 EE−2	.8797 EE−1	118.0799	2.3393	.3963	.6031 EE+2	.5472 EE+1	.1102 EE+2	.1357 EE−1	.1488 EE−1
7.30	.1848 EE−3	.2155 EE−2	.8578 EE−1	125.6046	2.3421	.3958	.6200 EE+2	.5485 EE+1	.1130 EE+2	.1277 EE−1	.1448 EE−1
7.40	.1694 EE−3	.2025 EE−2	.8367 EE−1	133.5200	2.3448	.3954	.6372 EE+2	.5498 EE+1	.1159 EE+2	.1202 EE−1	.1409 EE−1
7.50	.1554 EE−3	.1904 EE−2	.8163 EE−1	141.8415	2.3474	.3949	.6546 EE+2	.5510 EE+1	.1188 EE+2	.1133 EE−1	.1372 EE−1
7.60	.1427 EE−3	.1792 EE−2	.7967 EE−1	150.5849	2.3499	.3945	.6722 EE+2	.5522 EE+1	.1217 EE+2	.1068 EE−1	.1336 EE−1
7.70	.1312 EE−3	.1687 EE−2	.7777 EE−1	159.7665	2.3523	.3941	.6900 EE+2	.5533 EE+1	.1247 EE+2	.1008 EE−1	.1302 EE−1
7.80	.1207 EE−3	.1589 EE−2	.7594 EE−1	169.4030	2.3546	.3937	.7081 EE+2	.5544 EE+1	.1277 EE+2	.9510 EE−2	.1269 EE−1
7.90	.1111 EE−3	.1498 EE−2	.7417 EE−1	179.5114	2.3569	.3933	.7264 EE+2	.5555 EE+1	.1308 EE+2	.8982 EE−2	.1237 EE−1
8.00	.1024 EE−3	.1414 EE−2	.7246 EE−1	190.1094	2.3591	.3929	.7450 EE+2	.5565 EE+1	.1339 EE+2	.8488 EE−2	.1207 EE−1
8.10	.9449 EE−4	.1334 EE−2	.7081 EE−1	201.2148	2.3612	.3925	.7638 EE+2	.5575 EE+1	.1370 EE+2	.8025 EE−2	.1177 EE−1
8.20	.8723 EE−4	.1260 EE−2	.6921 EE−1	212.8461	2.3632	.3922	.7828 EE+2	.5585 EE+1	.1402 EE+2	.7592 EE−2	.1149 EE−1
8.30	.8060 EE−4	.1191 EE−2	.6767 EE−1	225.0221	2.3652	.3918	.8020 EE+2	.5594 EE+1	.1434 EE+2	.7187 EE−2	.1122 EE−1
8.40	.7454 EE−4	.1126 EE−2	.6617 EE−1	237.7622	2.3671	.3915	.8215 EE+2	.5603 EE+1	.1466 EE+2	.6806 EE−2	.1095 EE−1
8.50	.6898 EE−4	.1066 EE−2	.6472 EE−1	251.0862	2.3689	.3912	.8412 EE+2	.5612 EE+1	.1499 EE+2	.6449 EE−2	.1070 EE−1

THERMO

Note: Use $g = 32.2$ ft/sec^2 unless told otherwise.

CONSISTENT UNITS

1. A 10 lbm object is acted upon by a 4 lbf force. What is the acceleration in ft/min^2?

2. If mud is 70% sand (density of 140 lbm/ft^3) and 30% water by weight, what is the mass density of mud?

3. How much does a 10 lbm earth object weigh on the moon?

4. How much does a 10 slug object weigh on the moon? How much does it weigh on the earth?

5. What force is required to accelerate 11 kg at .8 m/sec^2?

6. What force is acting if a 40 slug object accelerates at 8 ft/sec^2?

7. Convert the Boltzmann gas constant (1.38 EE-23 joules/·K) to English engineering units.

TEMPERATURE

8. Convert (a) 20·C to ·K
(b) 20·C to ·F
(c) 70·F to ·R

IDEAL GAS LAWS

9. The temperature of a closed cylinder containing air at 25·C and 76 cm of mercury is raised to 100·C. What is the new pressure?

10. A quantity of gas occupies 1.2 cubic feet at STP. The gas is allowed to expand to 1.5 cubic feet and 15 psia. What is the new temperature?

11. A gas has a density of .094 lbm/ft^3 at 100·F and 2 atm. What pressure is needed to change the density to .270 ibm/ft^3 at 250·F?

12. A cold tire contains 1000 cubic inches of air at 24 psig and 32·F. What is the pressure in the tire if the temperature and volume are increased to 35·F and 1020 cubic inches respectively?

13. What is the pressure of 4.87 grams of hydrogen that occupy 6.8 liters at 10·C?

14. A spherical helium balloon at STP has a diameter of 60 feet. What is its lifting power?

15. A cylindrical tin can 4 inches in diameter and 4 inches long has 2/3 of its air removed. What is the force on each end?

16. A tank contains 3 cubic feet of 120 psig air at 80·F. How many tires of volume 1.2 ft^3 can be inflated to 28 psig at 80·F?

HEAT

17. How many BTU's are required to raise 20 lbm of copper (c_p = .091 BTU/lbm-·F) from 30·F to 500·F?

18. How many calories are required to raise 20 kg of aluminum (c_p = .208 BTU/lbm-·F) from 80·C to 600·C?

19. How much heat is required to change 30 kg of 20·C water into 100·C steam?

20. How many BTU are required to raise 4 lbm of air from 70·F to 180·F (a) in a constant volume process? (b) in a constant pressure process?

21. 2 lbm of 200·F iron (c_p=.1) are dropped into a gallon of 40·F water. What is the final temperature of the mixture.

22. Heat is supplied to 20 lbm of ice (c_p = .5) at 0·F at the rate of 160 BTU/sec. How long will it take to convert the ice to steam (c_p = .5) at 213·F? (c_p in BTU/lbm-·F.)

23. Water at 100 psia and 80·F is mixed with 100 psia, 80% quality steam. Hot water at 20 psia and 180·F results. If 4000 lbm of hot water are produced each hour, what are the weight flows of the input ingredients?

24. Two liquids enter a mixing chamber and are discharged at 80·F at the rate of 50 gpm. Liquid A enters at 140·F with a specific heat of 10 BTU/gal-·F. Liquid B enters at 65·F with a specific heat of 8.33 BTU/gal-·F. What are the volume flows for the liquids? What is the specific heat of the mixture?

25. How much heat is required to increase the temperature of 10 pounds of air from 90· to 360·F in a constant volume process if the average value of c_p over this range is .245 BTU/lbm-·R?

26. One-half of a molecular weight of a perfect gas exists at a pressure of 100 psia. A heat transfer of 200 BTU at constant volume causes the temperature to change by 100·F. Find the specific heat for constant pressure processes on a molar basis.

THE FIRST LAW

27. 10 BTU are transferred in a process where a piston compresses a spring and in so doing does 1500 ft-lbf of work. Find the change in internal energy of the system.

28. A 5 gram bullet is propelled by a powder charge producing 500 calories of usable energy. What is the bullet muzzle velocity?

29. What is the rise in water temperature of water dropping over a 200 foot waterfall and settling in a basin below?

THERMO

30. A 200 gram glass jar (c_p=.2) contains 2000 grams of 20·C water. A 1/10 hp motor drives a stirrer for 15 minutes. What is the final water temperature neglecting other losses?

31. A steady-flow hydroelectric plant drops water 40 feet. The water enters at 10 fps and exits at 30 fps. What is the electrical output per pound of water?

ENTHALPY

32. What is the enthalpy of
- (a) water at 14.7 psia and 80·F?
- (b) steam at 20 psia and 1000·F?
- (c) saturated steam at 300·F
- (d) saturated steam at 30 psia?
- (e) ammonia at 60 psia and 100·F?
- (f) freon-12 at 45 psia and 170·F?
- (g) air at 400·F?

33. What is the specific enthalpy of 4 pounds of water which is initially at 80·F and to which 240 BTU are added?

ENTROPY

34. What is the entropy of
- (a) saturated water at 30 psia?
- (b) superheated steam at 5 psia and 500·F?
- (c) steam expanded isentropically from 100 psia and 800·F to 5 psia?
- (d) steam expanded from 100 psia and 800·F to 5 psia in a process with an isentropic efficiency of 80%?

AVAILABILITY

35. What is the specific availability of 100 psia steam at 1000·F if the environment is 14.7 psia and 60·F?

36. A 900·F flame provides 400·F saturated steam. What is the maximum useful work that can be obtained from one combustion BTU if the environment is at 80·F and 1 atm?

37. What is the specific availability of 300·F air expanding from 40 psia if the surroundings are at 80·F and 1 atm?

CLOSED IDEAL GAS PROCESSES

38. What is the heat flow when 3 lbm of nitrogen undergo a closed system isothermal process at 300·F from an initial volume of 40 cubic feet to final volume of 22.5 cubic feet?

39. The combustion process occuring after top dead center of a piston engine behaves polytropically with n = -1.2. At the start of the process the volume is .02 ft^3, the pressure is 200 psia, and the temperature is 1600·F. Find the work done if the final volume is .028 ft^3.

40. How many BTU's are needed to isothermally compress 4 lbm of 240·F air to twice the original pressure?

41. A quantity of air at 180·F originally occupies 20 ft^3 at 30 psig. The gas is compressed reversibly and adiabatically to 180 psig. What is the heat flow?

42. One pound of air goes through a reversible cycle containing the following three processes:

(a) From a volume of 2 ft^3 at 20·F, compressed adiabatically to 1 ft^3.

(b) Heat is added at constant pressure until the volume is 2 ft^3.

(c) Air is returned reversibly to the original state.

Find the work, internal energy change, and heat transfer for each process. Express all answers in ft-lbf.

OPEN IDEAL GAS PROCESSES

43. Air is compressed in steady flow to one-tenth of its original volume and from 14.7 psia and 70·F. Find the initial volume, the final volume, the final temperature, and the compression work per pound of air if the isentropic efficiency is 90%.

44. Air expands in steady flow from 60·F and 500 psia to 14.7 psia and -100·F. Find the change in entropy, the work output, and the isentropic efficiency of the process.

45. 3 pmoles of an ideal gas are throttled per second from 9.2 psia to 6.4 psia. What is the entropy change?

46. Ideal oxygen is throttled at 140·F from 10 atmospheres to 5 atmospheres. What is its temperature change?

MECHANICAL ENERGY BALANCES

47. The turbine of a jet engine operates adiabatically and receives a steady flow of gases at 114 psia, 1340·F, and 540 fps. It discharges at 30.6 psia, 820·F, and 1000 fps. Find the work output per pound of gas.

48. A turbine uses 100,000 lbm/hr of steam which enters with an enthalpy of 1400 BTU/lbm and essentially zero entrance velocity. 10,000 horsepower are developed. The exit velocity of the steam is 500 fps. Expansion is adiabatic. What is the exit enthalpy?

49. A steady flow of 275 lbm/min develops 1000 hp with the conditions below. What is the heat transfer in BTU/hr?

initial conditions	final conditions
160 psia	2 psia
400·F	126.1·F
3.008 ft^3/lbm	156 ft^3/lbm
1217.6 BTU/lbm	1012.5 BTU/lbm
70 fps	400 fps
10.2 ft elevation	1.8 ft elevation

50. The entrance enthalpy, velocity, and height of a 5 lbm/sec system are 1000 BTU/lbm, 100 ft/sec, and 100 feet respectively. 50 BTU/sec are lost to the surroundings. The exit enthalpy, velocity, and height are 1020 BTU/lbm, 50 ft/sec, and 0 feet respectively. What is the input horsepower?

KINETIC GAS THEORY

51. What is the rms velocity of oxygen at 70·F?

52. What is the average kinetic energy of a gas molecule at 0·F?

THERMO

53. What is the average velocity of a nitrogen molecule at 20·C?

GAS MIXTURES

54. A mixture of gas is 30% CO, 15% CO_2, and 55% H_2 by volume. Find the gravimetric analysis, and the molecular weight, gas constant, specific heat at constant pressure, and specific heat at constant volume for the mixture.

55. A gravimetric analysis of a gas mixture is 20% He, 40% air, and 40% CO_2. Find the volumetric analysis, average molecular weight, and the composite gas constant.

56. A mixture of 60% N_2, 10% CO_2, and 30% H_2 by volume is heated from 40·F to 250·F at constant pressure. The total initial volume is 5 cubic feet.
   (a) What is the total mixture weight?
   (b) What is the specific heat at constant pressure of the mixture?
   (c) How much energy is added to the mixture?

VAPOR TABLES AND DIAGRAMS

57. What is the enthalpy of
   (a) 80 psia steam with 388·F superheat?
   (b) 100 psia steam with a quality of 45%?
   (c) 300·F water compressed to 1000 psia?
   (d) steam at 50 psia and 800·F?

58. What is the quality of steam at 400 psia if the enthalpy is 600 BTU/lbm?

59. What is the specific volume of 600·F steam at 80 psia?

60. What is the final pressure of steam which is expanded isentropically from 100 psia and 700·F to 1100 BTU/lbm?

61. What is the final entropy of steam which undergoes an ideal throttling process from 500 psia and 900·F to 30 psia?

PSYCHROMETRICS (Optional)

(A) Atmospheric air has a dry-bulb temperature of 85·F and a wet-bulb temperature of 70·F. Find the dew point, the relative humidity, and the specific humidity.

(B) A mixture of air and water vapor exists at 10 psia and 70·F with a dewpoint of 50·F. Find the specific humidity, relative humidity, and the specific volume.

(C) 1000 pounds of air at 95·F dry bulb and 78·F wet bulb are cooled to 55·F and 75% relative humidity. Find the
   (a) total heat removed
   (b) sensible heat removed
   (c) latent heat removed
   (d) final wet bulb temperature
   (e) initial dew point
   (f) final dew point
   (g) weight of moisture condensed

THERMO

7 Vapor, Combustion, Refrigeration and Compression Cycles

Part 1: Vapor Cycles

Nomenclature for Vapor Cycles

h	enthalpy	BTU/lbm
J	Joule's constant (778)	ft-lbf/BTU
p	pressure	psf
Q	heat	BTU/lbm
s	entropy	BTU/°R-lbm
T	temperature	°R
W	work	BTU/lbm
x	quality or bleed fraction	decimal
y	bleed fraction	decimal

Symbols

υ	specific volume	ft³/lbm
η	efficiency	decimal

Subscripts

comp	compressor
ext	external
id	ideal
int	internal
m	mechanical
s	isentropic
th	thermal
turb	turbine

1. Energy Conversions

Multiply	By	To Get
BTU	3.929 EE−4	hp-hrs
BTU	778.3	ft-lbf
BTU	2.930 EE−4	kw-hrs
BTU	1.0 EE−5	therms
BTU/hr	.2161	ft-lbf/sec
BTU/hr	3.929 EE−4	hp
BTU/hr	.2930	watts
ft-lbf	1.285 EE−3	BTU
ft-lbf	3.766 EE−7	kw-hrs
ft-lbf	5.051 EE−7	hp-hrs
ft-lbf/sec	4.6272	BTU/hr
ft-lbf/sec	1.818 EE−3	hp
ft-lbf/sec	1.356 EE−3	kw
hp	2545.0	BTU/hr
hp	550	ft-lbf/sec
hp	.7457	kw
hp-hr	2545.0	BTU
hp-hr	1.976 EE6	ft-lbf
hp-hr	.7457	kw-hrs
kw	1.341	hp
kw	3412.9	BTU/hr
kw	737.6	ft-lbf/sec

2. Symbols for Drawings

There are many different symbols in use. Those that are commonly used are noted here. Should a symbol be needed that is not given, a simple rectangular box labeled inside with the function provided is usually sufficient.

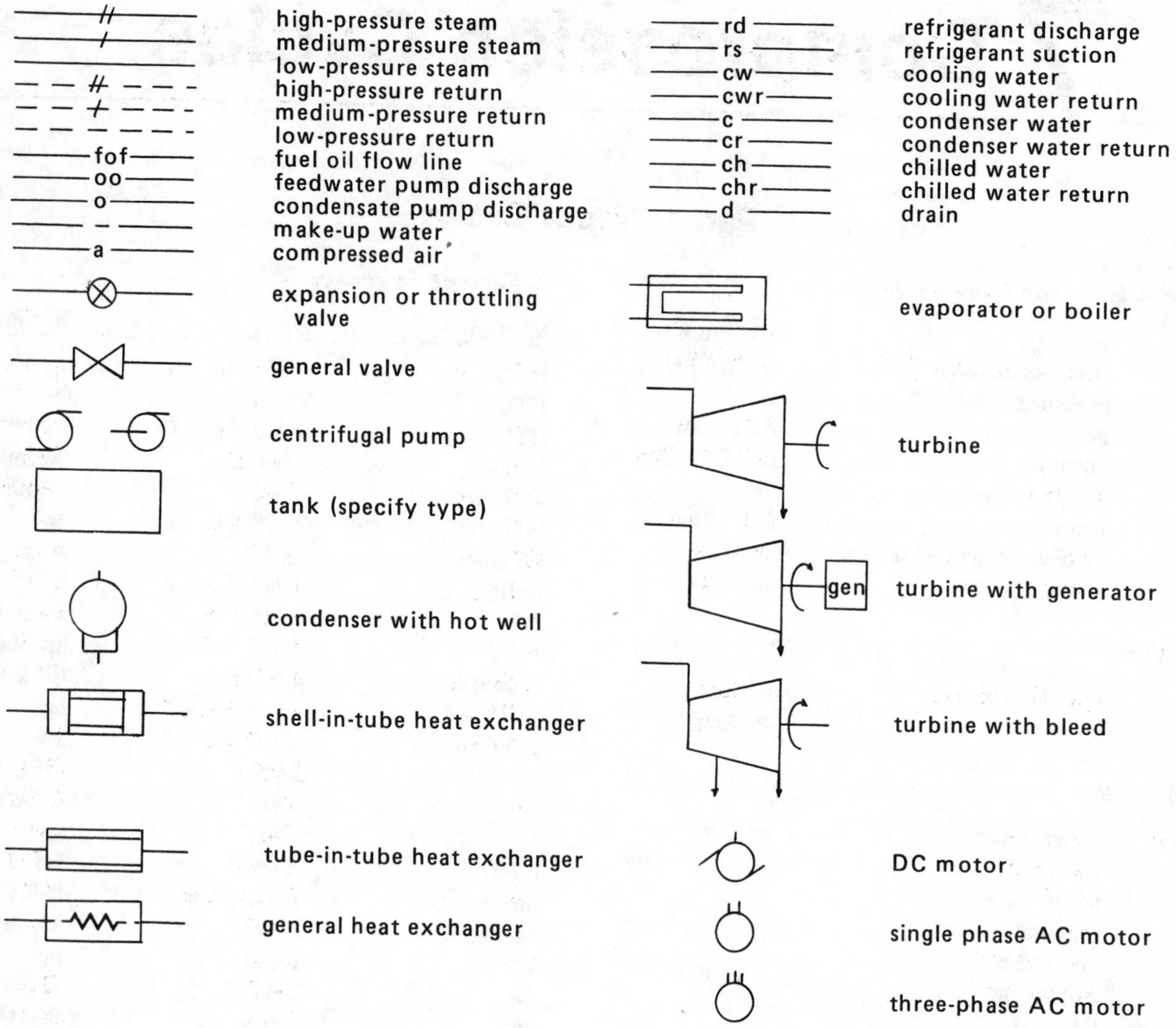

3. Definitions

Air heater: See 'Air preheater'.

Air preheater: A device that heats combustion air by recovering energy from stack gases. A convection preheater uses a conventional heat exchanger; a regenerative preheater uses a rotating drum which is alternately exposed to both gas flows.

Back-pressure turbine: A turbine that exhausts to a second turbine operating in a lower pressure range. Alternative definition: a turbine which exhausts to a pressure greater than atmospheric.

Bleed: A removal of partially expanded steam from a turbine.

Co-generation cycle: A cycle in which steam from an electricity generating process is used for subsequent processes, or vice-versa.

Condensing cycle: A cycle in which steam is condensed and returned to the boiler after expanding through a turbine.

Condition line: The locus of all states of steam during the expansion process, as in line d-e in Figure 7.2.

Deaerator: A heat exchanger in which water is heated to a point where dissolved corrosive gases (primarily oxygen) are liberated.

Economizer: A heat exchanger that heats feedwater by exposing it to stack gases.

Evaporator: A closed heat exchanger that vaporizes untreated water at atmospheric pressure by high-pressure steam. The vaporized water is condensed and stored for use in the boiler.

Extraction heater: A feedwater heater using extracted steam as a heating source. Also see 'feedwater heater'.

Extraction rate: The rate, usually expressed in BTU/hr, at which partially expanded steam is bled off from a turbine.

Extraction turbine: A turbine with one or more bleeds.

Feedwater heater: A device used to heat water from the condenser prior to pumping into the boiler. Open heaters mix extracted steam directly with feedwater. Closed heaters are conventional shell and tube heat exchangers, with the extracted steam typically being routed to the hot well.

Flue gas: See 'Stack gas'.

Heat rate: For an electrical generating system, the heat rate is defined as the total energy input of the process in BTU/hr divided by the energy output in kilowatts. If the output is mechanical, the total energy is divided by the horsepower output.

High-pressure turbine: A turbine exhausting to a pressure greater than atmospheric.

Hot well: The lower portion of a condenser containing water in its liquid form.

Impulse turbine: A turbine with steam directed at moving blades through stationary nozzles.

Intercooling: Cooling of compressed air between gas turbine and compressor stages.

Low-pressure turbine: A turbine exhausting to a pressure lower than atmospheric.

Non-condensing cycle: A cycle in which steam is exhausted to the atmosphere or otherwise utilized without passing through a condenser.

Preheater: See 'Air preheater'.

Reaction turbine: A turbine with steam discharging from moving nozzles.

Regenerator: A heat exchanger in a gas turbine used to preheat compressed air prior to combustion by exposure to exhaust gases.

Reheat factor: The ratio of actual work output in a multi-stage expansion to ideal work assuming a one-stage expansion.

Reheater: A section of the boiler used to reheat steam after partial expansion in a turbine.

Stack gases: Products of a combustion reaction, consisting essentially of nitrogen, carbon dioxide, and water vapor.

Superposed turbine: See 'Back-pressure turbine'.

Surface condenser: A heat exchanger in which the vapor and coolant do not mix.

Throttle: A series of valves designed to vary the amount of steam admitted to a turbine in accordance with varying loads. May also be known as a 'throttle valve'.

Throttling valve: A device used to drop the pressure of a vapor without any significant change in enthalpy.

Topping turbine: See 'Back-pressure turbine'.

Water rate: For electrical generating systems, the water rate is defined as the steam flow rate in pounds per hour divided by the output in kilowatts. If the output is mechanical, the steam flow rate is divided by the horsepower output.

4. Typical Efficiencies

Table 7.1 lists typical efficiencies which may be assumed in a problem in which such assumptions are required.

5. General Vapor Power Cycles

The general vapor power cycle makes use of a boiler, turbine, condenser, and boiler feed pump (compressor) as shown in Figure 7.1.

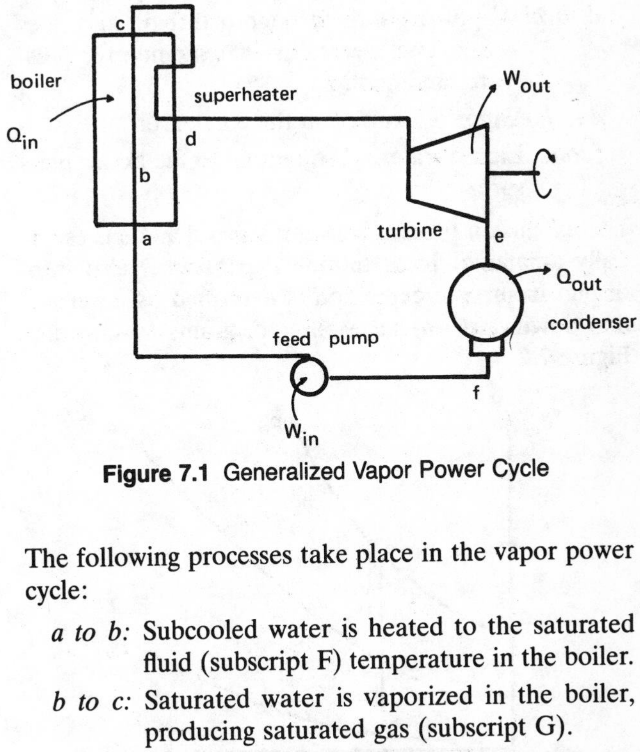

Figure 7.1 Generalized Vapor Power Cycle

The following processes take place in the vapor power cycle:

a to b: Subcooled water is heated to the saturated fluid (subscript F) temperature in the boiler.

b to c: Saturated water is vaporized in the boiler, producing saturated gas (subscript G).

c to d: An optional superheating process increases the steam temperature while maintaining the pressure.

POWER

Table 7.1 Approximate Full-Load Efficiencies

Device	Rating, or type	mechanical	isentropic	volumetric	electrical	thermal
steam turbine	EE3 kw	.98	.62–.66			
	EE4 kw	.98	.72–.78			
	EE5 kw	.98	.80–.82			
electrical generator	EE2 kva				.90	
	EE3 kva				.94	
	2 EE3 kva				.96	
	10 EE3 kva				.97	
	20 EE3 kva				.98	
	75 EE3 kva				.99	
electrical motor (AC)	1 hp				.80	
	10 hp				.85	
	50 hp				.90	
	EE3 hp				.95	
	5 EE3 hp				.96	
boiler	hand-fired					.50–.60
	chain-grate					.60–.70
	pulverized coal					.80–.90
	oil					.85–.90
pump, piston, water	1000 psi	.90		.71		
pump, centrifugal	200 gpm	.60–.70				
	500 gpm	.70–.75				
	1000 gpm	.75–.80				
	3000 gpm	.80–.85				
	10,000 gpm	.85–.87				
compressor, piston		.88–.93	.85–.93			
compressor, turbine	500 hp	.95	.62			
	5000 hp	.98	.76			
	10000 hp	.99	.84			
Otto IC engine	4-stroke	.85		.90		.25–.30
Diesel IC engine		.85		.92		

d to e: Vapor expands through a turbine and does work as it decreases in temperature, pressure, and quality.

e to f: Vapor is liquified in the condenser.

f to a: Liquid water is brought up to the boiler pressure.

The expansion process between states d and e is essentially adiabatic. Ideal turbine expansion is also isentropic. Isentropic expansion is described by a vertical line downward on the Mollier diagram, as shown in Figure 7.2.

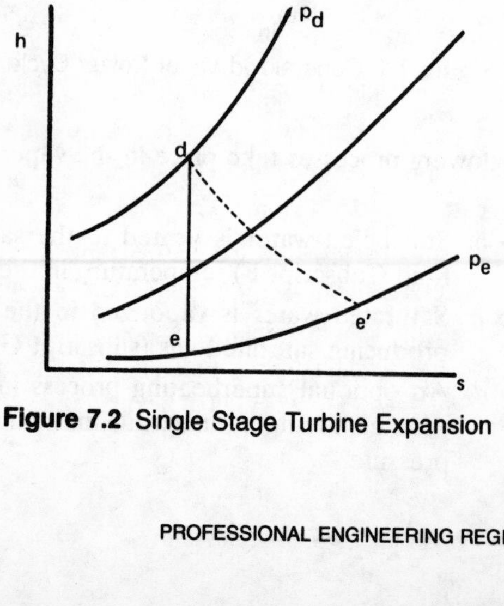

Figure 7.2 Single Stage Turbine Expansion

If the flow is steady, and if changes in potential and kinetic energies are insignificant (as they are in a turbine), and if the expansion is isentropic, then the energy extracted per pound of steam is

$$W_{id} = h_d - h_e \qquad 7.1$$

If the expansion process is not isentropic, then entropy will increase and h'_e will be higher than h_e. The actual energy extracted is

$$W' = h_d - h'_e \qquad 7.2$$

If friction losses are deducted, then the turbine shaft work can be found:

$$W_{turb} = W' - W_{friction} = \eta_{mech} W' \qquad 7.3$$

The isentropic efficiency is defined as

$$\eta_s = W'/W_{id} \qquad 7.4$$

The overall turbine efficiency is

$$\eta_{turb} = W_{turb}/W_{id} = \eta_s \eta_{mech} \qquad 7.5$$

Inasmuch as the frictional losses are usually very small (see table 7.1), the isentropic and turbine efficiencies are essentially identical, and

$$W_{turb} = W' = \eta_s W_{id} = \eta_s (h_d - h_e) \qquad 7.6$$

$$\eta_{turb} = \eta_s \qquad 7.7$$

Example 7.1

Steam is expanded from 700°F and 200 psia through a 87% efficient turbine to 5 psia. What energy is extracted per pound of steam?

Refer to figure 7.2. From the superheated steam tables:

$$h_d = 1373.6 \text{ BTU/lbm}$$

$$s_d = 1.7232 \text{ BTU/lbm-°R}$$

Proceed as if the turbine were 100% efficient. From the saturated steam tables for 5 psia:

$$s_F = .2347$$

$$s_{FG} = 1.6094$$

Since it was assumed that expansion was isentropic (100% efficient), $s_e = s_d = 1.7232$. The quality at point e can be found from:

$$x_e = \frac{s_e - s_F}{s_{FG}} = \frac{1.7232 - .2347}{1.6094} = .92$$

Now that the quality is known, the enthalpy can be found:

$$h_e = h_F + xh_{FG} = 130.13 + (.92)(1001.0) = 1051.1$$

However, this value of h_e was found assuming isentropic expansion through the turbine. Since the turbine is capable of extracting only 87% of the ideal energy, the actual value of h_e is:

$$h'_e = h_d - \eta_{turb}(h_d - h_e)$$

$$= 1373.6 - (.87)(1373.6 - 1051.1)$$

$$= 1093.0 \text{ BTU/lbm}$$

Example 7.2

Repeat example 7.1 using the Mollier diagram.

h_d is read directly from the Mollier diagram at the intersection of 700°F and 200 psia. $h_d \approx 1375$. Greater accuracy is possible with larger Mollier diagrams.

h_e is found by dropping straight down (which keeps entropy constant) to the 5 psia line. h_e is read as approximately 1050. h'_e is calculated as in example 7.1.

If the expansion is multiple stage, or if the turbine has a bleed at an intermediate pressure p_m, the expansion process for each stage or bleed will be illustrated by Figure 7.3.

In the first stage, the ideal and actual outputs per pound of steam are

$$W_{id,1} = h_d - h_m \qquad 7.8$$

$$W'_1 = h_d - h'_m = \eta_{s,1}(W_{id,1}) \qquad 7.9$$

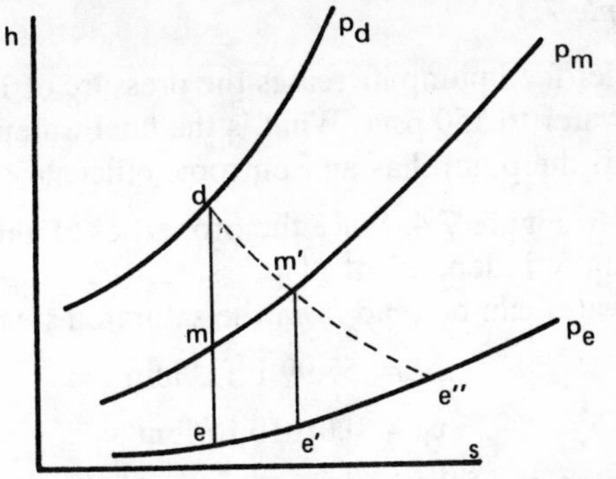

Figure 7.3 Two-stage Turbine Expansion

If y% of the original pound of steam expands to pressure p_e, the additional ideal and actual outputs are:

$$W_{id,2} = y(h'_m - h'_e) \qquad 7.10$$

$$W'_2 = y(h'_m - h''_e) = \eta_{s,2}y(h'_m - h'_e) \qquad 7.11$$

The actual work done per pound of steam is

$$W_{turb} = W'_1 + W'_2 = h_d - h'_m + y(h'_m - h''_e) \qquad 7.12$$

A similar analysis can be made for a pump or compressor. The ideal and actual work inputs are shown by the lines (f to a) and (f to a′) respectively in Figure 7.4.

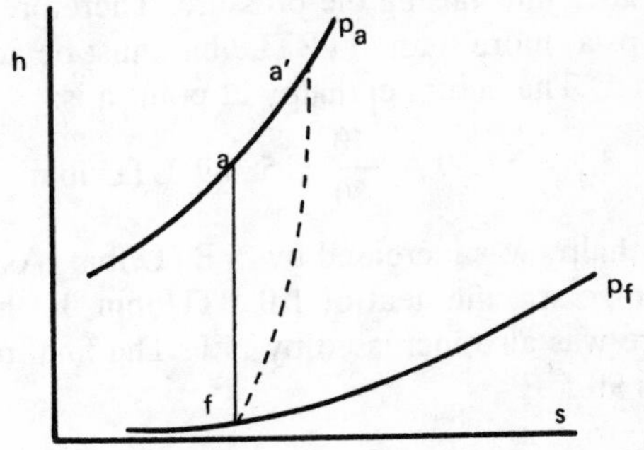

Figure 7.4 One-stage Pump Compression

The pump relationships are:

$$W_{id} = h_a - h_f \qquad 7.13$$

$$W' = h'_a - h_f = W_{id}/\eta_s \approx \upsilon_f(p_a - p_f)/\eta_s J \qquad 7.14$$

$$\eta_s = \frac{h_a - h_f}{h'_a - h_f} \qquad 7.15$$

$$\eta_{pump} = \eta_{mech}\eta_s \qquad 7.16$$

Since W_{id} is so small — in the neighborhood of only a few BTU/lbm — the pump mechanical efficiency is often ignored and η_{pump} is taken as η_s.

The thermal efficiency of an entire power cycle with singe-stage expansion is:

$$\eta_{th} = \frac{Q_{in} - Q_{out}}{Q_{in}} = \frac{W_{turb} - W_{comp}}{Q_{in}}$$

$$= \frac{(h_d - h'_e) - (h'_a - h_f)}{h_d - h'_a} \qquad 7.17$$

POWER

Example 7.3

A boiler feed pump increases the pressure of 14.7 psia, 90°F water to 150 psia. What is the final water temperature if the pump has an isentropic efficiency of 80%?

Refer to Figure 7.4. Since the properties of a liquid are essentially independent of pressure, the properties of 90°F water can be read from the saturated steam table.

$$h_f = 57.99 \text{ BTU/lbm}$$

$$\upsilon_f = 0.01610 \text{ ft}^3/\text{lbm}$$

The enthalpy of the boiler feedwater (point a on Figure 7.4) is equal to the enthalpy at point f plus the energy put into the water by the pump. Assuming the water to be incompressible, the specific volumes at points a and f are the same.

$$h_a = h_f + \frac{\upsilon_f(p_a - p_f)}{J}$$

$$= 57.99 + \frac{(.01610)(150 - 14.7)(144)}{778}$$

$$= 57.99 + .40 = 58.39$$

The above calculation assumed that the pump was capable of isentropic compression. Because of the pump's inefficiency, not all of the .40 BTU/lbm enthalpy increase goes into raising the pressure. Therefore, to get to 150 psia, more than .40 BTU/lbm must be added to the water. The actual enthalpy at point a is:

$$h_a' = 57.99 + \frac{.40}{.80} = 58.49 \text{ BTU/lbm}$$

The enthalpy was increased by .5 BTU/lbm. Assuming an average specific heat of 1.0 BTU/lbm-°F, the temperature was also increased by .5°F. The final temperature is 90.5°F.

6. The Carnot Cycle

The Carnot cycle is an ideal power cycle which is impractical to implement. However, its work output sets the maximum attainable from any heat engine, as evidenced by the isentropic (reversible) processes between states (d and a) and (b and c) in Figure 7.5. The working fluid is irrelevant in a Carnot cycle. As with most other cycles, it is necessary to have property tables for the working fluid if the cycle is to be evaluated.

The processes involved are:

a to b: isothermal expansion of saturated fluid to saturated gas
b to c: isentropic expansion
c to d: isothermal compression
d to a: isentropic compression

The properties at the various states may be found from the following solution methods. The capital letters 'F' and 'G' stand for saturated fluid and saturated gas respectively. They do not correspond to states f and g on any cycle diagram. The Carnot cycle may be evaluated by working around, finding T, p, x, h, and s at each node.

at a: From the property table for T_{high}, read p_a, h_a, and s_a for a saturated fluid.

at b: $T_b = T_a$; $p_b = p_a$; $x = 1$; h_b read from the table as h_G; s_b read as s_G.

at c: Either p_c or T_c will be given. Read p_c from the T_c line on the property table or vice versa; $x_c = (s_b - s_F)/s_{FG}$; $h_c = h_F + x_c(h_{FG})$; $s_c = s_b$.

at d: $T_d = T_c$; $p_d = p_c$; $x_d = (s_a - s_F)/s_{FG}$; $h_d = h_F + x_d(h_{FG})$; $s_d = s_a$.

The turbine and compressor work terms are:

$$W_{turb} = h_b - h_c \qquad 7.18$$

$$W_{comp} = h_a - h_d \qquad 7.19$$

The heat flows into and out of the system are:

$$Q_{in} = T_{high}(s_b - s_a) = h_b - h_a \qquad 7.20$$

$$Q_{out} = T_{low}(s_c - s_d) = T_{low}(s_b - s_a) = h_c - h_d \qquad 7.21$$

The thermal efficiency of the entire cycle is

$$\eta_{th} = \frac{Q_{in} - Q_{out}}{Q_{in}} = \frac{W_{turb} - W_{comp}}{Q_{in}} \qquad 7.22$$

$$= \frac{(h_b - h_c) - (h_a - h_d)}{h_b - h_a} = \frac{T_{high} - T_{low}}{T_{high}} \qquad 7.23$$

If isentropic efficiencies for the pump and turbine are given, proceed as follows: Calculate all properties assuming the efficiencies are 100%. Then, modify h_c and h_a as given below. Use the new values to find the actual thermal efficiency of the cycle.

$$h_c' = h_b - \eta_{turb}(h_b - h_c) \qquad 7.24a$$

$$h_a' = h_d + (h_a - h_d)/\eta_{comp} \qquad 7.24b$$

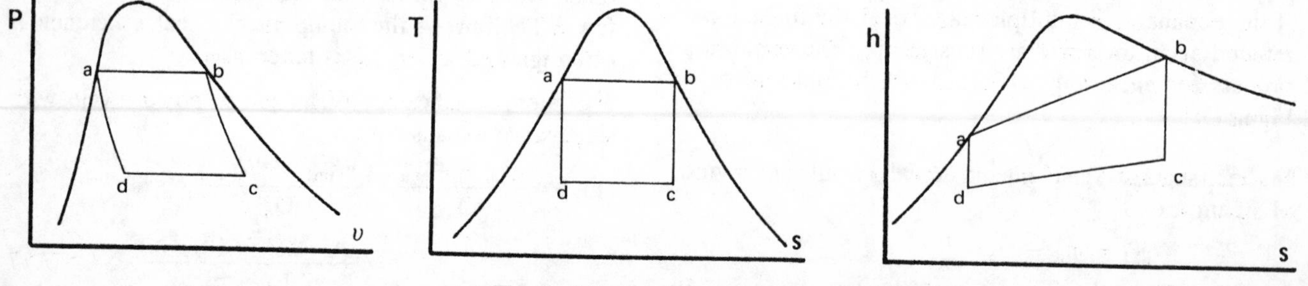

Figure 7.5 The Carnot Cycle

$$W'_{turb} = h_b - h'_c \quad 7.25a$$

$$W'_{comp} = h'_a - h_d \quad 7.25b$$

7. The Basic Rankine Cycle

The basic Rankine cycle is similar to the Carnot cycle except that the compression process occurs in the liquid region. The Rankine cycle is closely approximated in actual steam turbine plants. The efficiency of the Rankine cycle is lower than that of a Carnot cycle operating between the same temperature limits because the mean temperature at which heat is added to the system is lower than T_{high}. The piping diagram and property plots are shown in Figures 7.6 and 7.7.

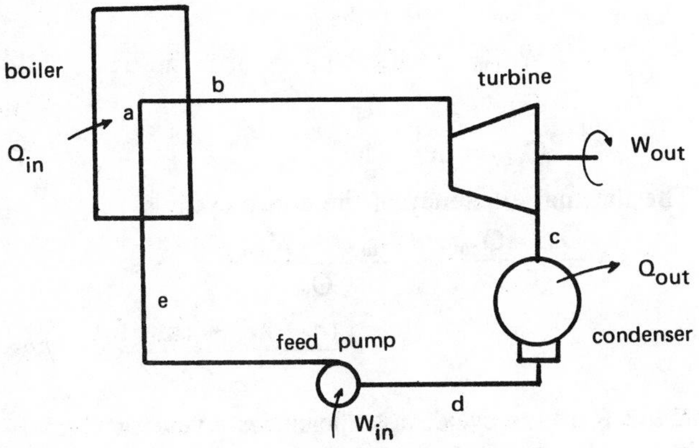

Figure 7.6 Basic Rankine Piping Diagram

The processes used in the basic Rankine cycle are:

a to b: vaporization in the boiler
b to c: adiabatic expansion in the turbine
c to d: condensation
d to e: adiabatic compression to boiler pressure
e to a: heating to fluid saturation temperature

The properties at each point may be found from the following procedure. The capital letters 'F' and 'G' refer to 'saturated fluid' and 'saturated gas' respectively. They do not correspond to locations f and g on any diagram. Usually T_{high} and T_{low} are given. The procedure is to work around the cycle, finding T, p, x, h, and s at each node.

at a: From the property table for T_{high}, read p_a, h_a, and s_a for a saturated fluid.

at b: $T_b = T_a$; $p_b = p_a$; $x = 1$; h_b read from the table as h_G; s_b read as s_G.

at c: Either p_c or T_c will be given. Read p_c from the T_c line on the property table or vice versa; $x_c = (s_b - s_F)/s_{FG}$; $h_c = h_F + x_c(h_{FG})$; $s_c = s_b$.

at d: $T_d = T_c$; $p_d = p_c$; $x_d = 0$; h_d read as h_F; s_d read as s_F; υ_d read as υ_F

at e: $p_e = p_a$; $s_e = s_d$; $h_e = h_d + W_{pump} = h_d + \upsilon_d(p_e - p_d)/J$ (watch units); T_e found as the saturation temperature for a fluid with enthalpy equal to h_e.

The work and heat flow terms are:

$$W_{turb} = h_b - h_c \quad 7.26a$$

$$W_{pump} = h_e - h_d \approx \upsilon_d(p_e - p_d)/J \quad 7.26b$$

$$Q_{in} = h_b - h_e \quad 7.27a$$

$$Q_{out} = h_c - h_d \quad 7.27b$$

The thermal efficiency of the entire cycle is:

$$\eta_{th} = \frac{Q_{in} - Q_{out}}{Q_{in}} = \frac{W_{turb} - W_{pump}}{Q_{in}} = \frac{(h_b - h_c) - (h_e - h_d)}{h_b - h_e} \quad 7.28$$

If isentropic efficiencies are given for the pump and turbine, calculate all properties as though these efficiencies were 100%. Then, use the following relationships to modify h_c and h_e. Use the new values to recalculate the thermal efficiency.

$$h'_c = h_b - \eta_{turb}(h_b - h_c) \quad 7.29$$

$$h'_e = h_d + (h_e - h_d)/\eta_{pump} \quad 7.30$$

$$W'_{turb} = h_b - h'_c \quad 7.31$$

$$W'_{pump} = h'_e - h_d \quad 7.32$$

$$Q'_{in} = h_b - h'_e \quad 7.33$$

8. The Rankine Cycle with Superheat

Superheating occurs when heat is added to the water in excess of that required to produce saturated vapor. Superheat is used to raise the vapor above the critical temperature, raise the mean effective temperature at

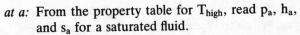

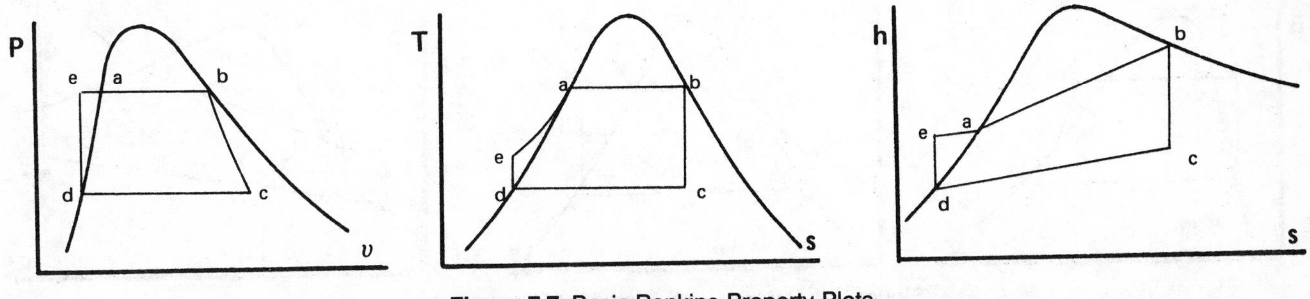

Figure 7.7 Basic Rankine Property Plots

which heat is added, and to keep the expansion primarily in the vapor region to reduce wear on the turbine blades. A maximum practical metallurgical limit on superheat is 1150°F. The piping and property plots are shown in Figures 7.8 and 7.9 respectively.

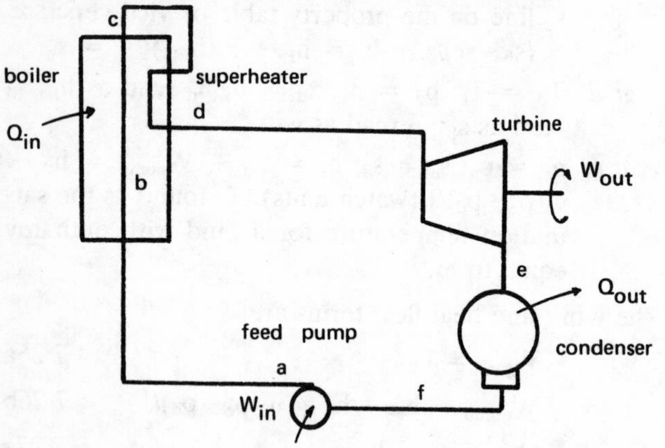

Figure 7.8 Rankine with Superheat Piping Diagram

The processes in the Rankine cycle with added superheat are similar to the basic Rankine cycle:

a to b: heating water to the saturation temperature in the boiler

b to c: vaporization of water in the boiler

c to d: superheating steam in the superheater region of the boiler

d to e: adiabatic expansion in the turbine

e to f: condensation

f to a: adiabatic compression to boiler pressure

The properties at each point may be found from the following procedure. The subscripts 'F' and 'G' refer to 'saturated fluid' and 'saturated gas' respectively. They do not correspond to any points on the property plot diagram.

at a: Point a is covered below. Start the analysis at point b.

at b: From the property table for T_b or p_b, read h_b and s_b for a saturated fluid.

at c: $T_c = T_b$; $p_c = p_b$; $x = 1$; h_c read as h_G; s_c read as s_G.

at d: T_d is usually known; $p_d = p_c$; h_d read from superheat tables with T_d and p_d known. Same for s_d and υ_d.

at e: T_e is usually known. p_e is read from saturated table for T_e. $x_e = (s_e - s_F)/s_{FG}$; $h_e = h_F + x_e(h_{FG})$; $s_e = s_d$. If point e is in the superheated region, the Mollier diagram should be used to find the properties at point e.

at f: $T_f = T_e$; $p_f = p_e$; $x = 0$; h_f, s_f, and υ_f read as h_F, s_F, and υ_F from the saturated table.

at a: $p_a = p_b$; $h_a = h_f + \upsilon_f(p_a - p_f)/J$ (watch units); $s_a = s_f$. T_a is equal to the saturation temperature for a liquid with enthalpy equal to h_a.

The work and heat flow terms are:

$$W_{turb} = h_d - h_e \tag{7.34}$$

$$W_{pump} = \upsilon_f(p_a - p_f) = h_a - h_f \tag{7.35}$$

$$Q_{in} = h_d - h_a \tag{7.36}$$

$$Q_{out} = h_e - h_f \tag{7.37}$$

The thermal efficiency of the entire cycle is

$$\eta_{th} = \frac{Q_{in} - Q_{out}}{Q_{in}} = \frac{W_{turb} - W_{pump}}{Q_{in}}$$

$$= \frac{(h_d - h_a) - (h_e - h_f)}{h_d - h_a} \tag{7.38}$$

If the Rankine cycle has efficiencies given for the turbine and pump, calculate all quantities as though those efficiencies were 100%. Then modify h_e and h_a prior to recalculating the thermal efficiency.

$$h'_e = h_d - \eta_{turb}(h_d - h_e) \tag{7.39}$$

$$h'_a = h_f + (h_a - h_f)/\eta_{pump} \tag{7.40}$$

$$W'_{turb} = h_d - h'_e \tag{7.41}$$

$$W'_{pump} = h'_a - h_f \tag{7.42}$$

$$Q'_{in} = h_d - h'_a \tag{7.43}$$

9. Rankine Cycle with Superheat and Reheat — The Reheat Cycle

Reheat is used to increase the mean effective temperature at which heat is added without producing significant expansion in the liquid-vapor region. The analysis below assumes that $T_d = T_f$ as is usually the case. How-

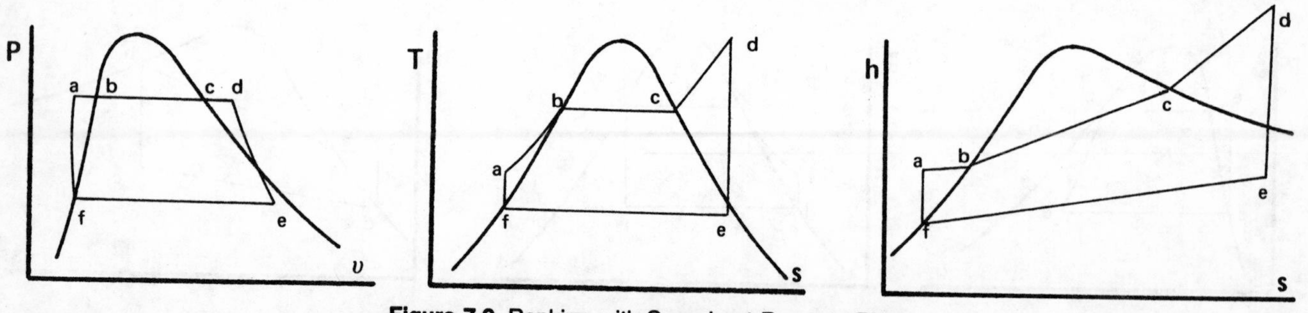

Figure 7.9 Rankine with Superheat Property Plots

ever, it is possible that the two temperatures will be different.

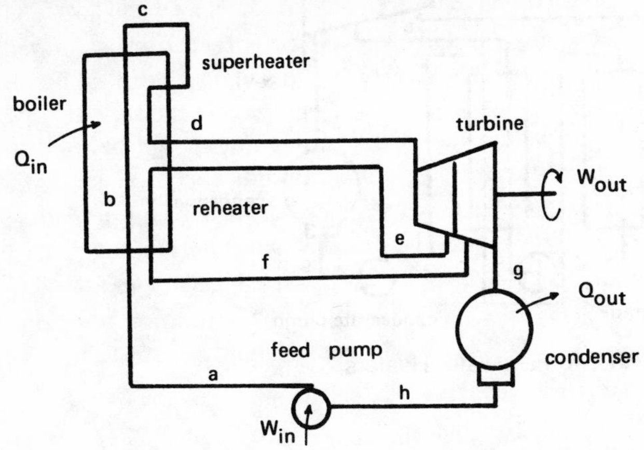

Figure 7.10 Reheat Cycle Piping Diagram

The properties at each point can be found from the following procedure.

at a: Point a is covered below. Start the analysis at point b.

at b: From the vapor table for T_b, read p_b, h_b, and s_b as a saturated fluid.

at c: $T_c = T_b$; $p_c = p_b$; $x_c = 1$; h_c read as h_G; s_c read as s_G.

at d: T_d usually known; $p_d = p_c$; h_d read from superheat tables with T_d and p_d known. Same for s_d and υ_d.

at e: p_e usually known; T_e read from property table for p_e; $s_e = s_d$; $x_e = (s_e - s_F)/s_{FG}$; $h_e = h_F + x_e(h_{FG})$. (Use the Mollier diagram if superheated.)

at f: T_f usually known; $p_f = p_e$; h_f read from superheat tables with T_f and p_f known. Same for s_f and υ_f.

at g: T_g usually known; p_g read from property tables for T_g; $s_g = s_f$; $x_g = (s_g - s_F)/s_{FG}$; $h_g = h_F + x_g(h_{FG})$. (Use the Mollier diagram if superheated.)

at h: $T_h = T_g$; $p_h = p_g$; $x_h = 0$; h_h, s_h, and υ_h read as h_F, s_F, and υ_F.

at a: $p_a = p_b$; $h_a = h_h + \upsilon_h (p_a - p_h)/J$ (watch units); $s_a = s_h$. T_a is the saturation temperature for a liquid with enthalpy equal to h_a.

The work and heat flow terms are:

$$W_{turb} = (h_d - h_e) + (h_f - h_g) \qquad 7.44$$

$$W_{pump} = \upsilon_h (p_a - p_h) = h_a - h_h \qquad 7.45$$

$$Q_{in} = (h_d - h_a) + (h_f - h_e) \qquad 7.46$$

$$Q_{out} = h_g - h_h \qquad 7.47$$

The thermal efficiency of the entire cycle is:

$$\eta_{th} = \frac{(h_d - h_a) + (h_f - h_e) - (h_g - h_h)}{(h_d - h_a) + (h_f - h_e)} \qquad 7.48$$

If the reheat cycle is specified with isentropic efficiencies for the pump and turbine, calculate all of the above quantities as though the efficiencies were 100%. Then, modify h_e, h_g, and h_a prior to recalculating the thermal efficiency.

$$h'_e = h_d - \eta_{turb}(h_d - h_e) \qquad 7.49$$

$$h'_g = h_f - \eta_{turb}(h_f - h_g) \qquad 7.50$$

$$h'_a = h_h + (h_a - h_h)/\eta_{pump} \qquad 7.51$$

$$W'_{turb} = (h_d - h'_e) + (h_f - h'_g) \qquad 7.52$$

$$W'_{pump} = (h'_a - h_h) \qquad 7.53$$

$$Q'_{in} = (h_d - h'_a) + (h_f - h'_e) \qquad 7.54$$

10. The Rankine Cycle with Regeneration — The Regenerative Cycle

If the mean effective temperature at which heat is added can be increased, the overall thermal efficiency of the cycle will be improved. This may be accomplished by raising the temperature at which the condensed fluid enters the boiler.

In the regenerative cycle, portions of the steam in the turbine are withdrawn at various points. Heat is transferred from this bleed steam to the feedwater coming from the condenser. Although only two bleeds are used in the following analysis, 7 or more exchange locations may be used in a large installation. Although it conceptually need not, the regenerative cycle always involves superheating.

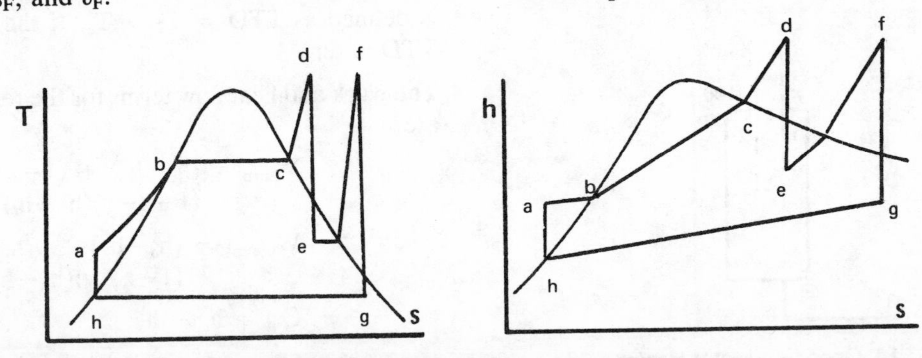

Figure 7.11 Reheat Cycle Property Plot

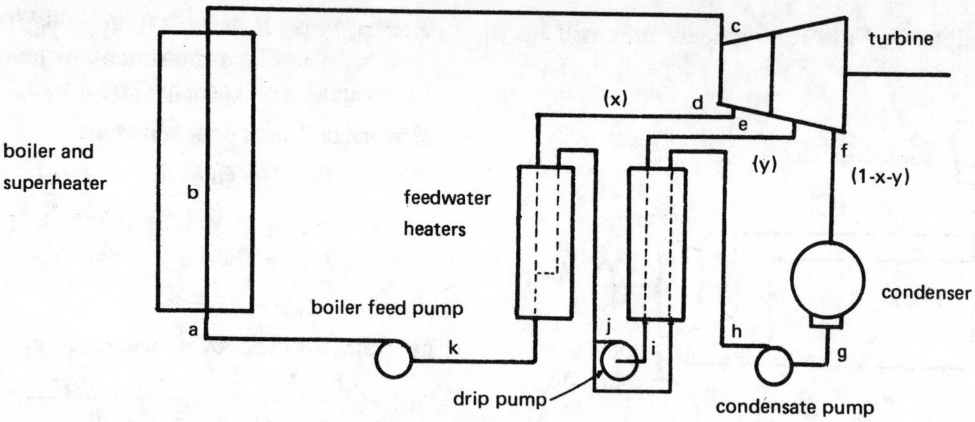

Figure 7.12 Regenerative Cycle with Two Feedwater Heaters

In the analysis below, x is the first bleed fraction and y is the second bleed fraction.

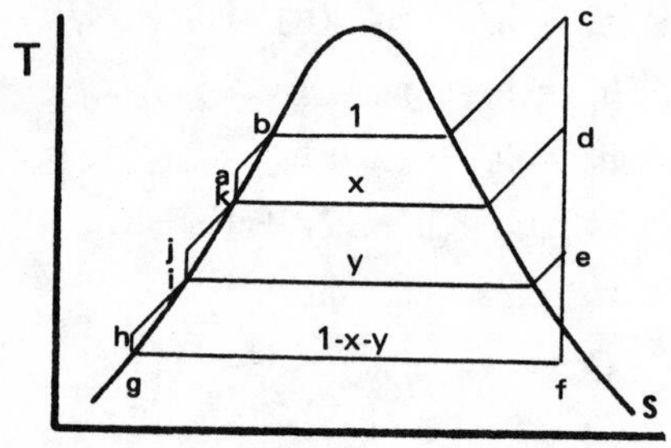

Figure 7.13 Regenerative Cycle Property Plot

The heat exchange may be by direct mixing of the bleed and feedwater in an open heater, or by use of a conventional heat exchanger known as a closed heater. These feedwater heaters are illustrated in Figures 7.14 and 7.15.

In the open heater with x% bleed and (1 − x)% feedwater, the following energy balance may be used to find the bleed fraction if the enthalpies are known. The open heater is assumed to be an adiabatic device.

$$(1 - x)h_2 + xh_1 = h_3 \qquad 7.55$$

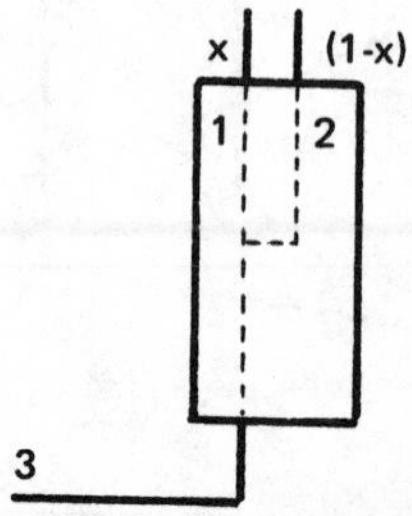

Figure 7.14 Open Feedwater Heater

If the heater is closed, a special drip pump may be used. This drip pump is in addition to the condensate and boiler feed pumps. Various possibilities exist for disposing of the drips, including mixing with the feedwater (as in Figures 7.12 and 7.15) or piping back to the hot well. For a closed heater with y% bleed, the adiabatic energy balance is given by equation 7.56.

$$(1-x-y)h_5 + y(h_4) + W_{\text{drip pump}} = (1-x)h_8 \qquad 7.56$$

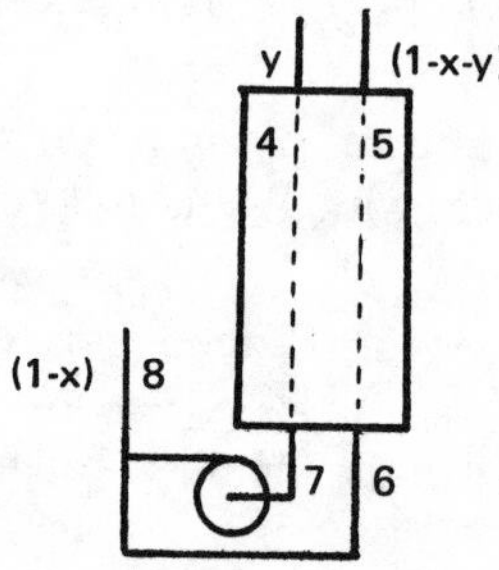

Figure 7.15 Closed Feedwater Heater

Because it is small, the drip pump work may be omitted for first approximations. Equation 7.56 may be solved for y once x is known. For that reason, it is always best to start with the heater nearest the boiler when evaluating the regenerative cycle.

The terminal temperature difference for a closed heater is defined as TTD = $T_7 - T_6$. If the heater is ideal, TTD is zero.

The work and heat flow terms for the regenerative cycle are:

$$W_{turb} = (h_c - h_d) + (1-x)(h_d - h_e) + (1-x-y)(h_e - h_f) \qquad 7.57$$

$$W_{pumps} = (h_a - h_k) + y(h_j - h_i) + (1-x-y)(h_h - h_g) \qquad 7.58$$

$$Q_{in} = h_c - h_a \qquad 7.59$$

$$Q_{out} = (1-x-y)(h_f - h_g) \qquad 7.60$$

The thermal efficiency of the entire cycle is:

$$\eta_{th} = \frac{Q_{in} - Q_{out}}{Q_{in}} = \frac{W_{turb} - W_{pumps}}{Q_{in}}$$

$$= \frac{(h_c - h_a) - (1-x-y)(h_f - h_g)}{h_c - h_a} \qquad 7.61$$

11. The Binary Cycle

The binary cycle utilizes two different fluids, typically mercury and steam, achieving conditions unobtainable with a single working fluid. Operation of the binary cycle is essentially two Rankine cycles, and Rankine procedures should be used to evaluate it.

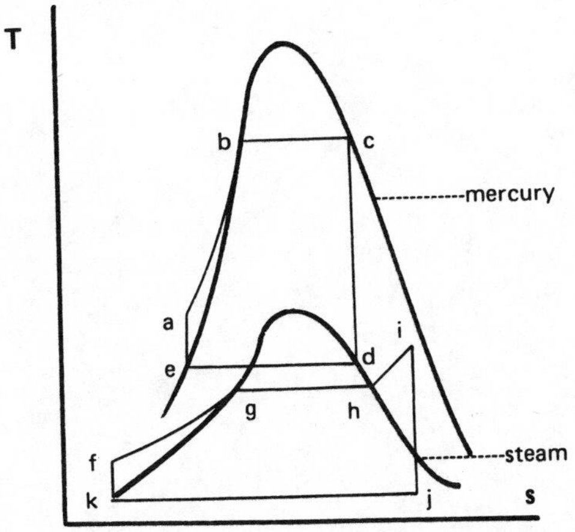

Figure 7.16 Binary Cycle Property Plot

The processes in a binary cycle are:

a to b: Mercury is heated to its saturation temperature

b to c: Mercury is vaporized in the boiler

c to d: Mercury expands adiabatically in the turbine

d to e: Mercury condenses in the combination condenser-boiler

e to a: Mercury is compressed adiabatically

f to g: Water is heated to its saturation temperature

g to h: Water vaporizes in a boiler

h to i: Steam is superheated

i to j: Steam expands adiabatically in the turbine

j to k: Water condenses

k to f: Water is compressed adiabatically

If x pounds of mercury flow for every pound of steam, the thermal efficiency of the binary cycle is:

$$\eta_{th} = \frac{W_{turbines} - W_{pumps}}{Q_{in}}$$

$$= \frac{x(h_c - h_d) + (h_i - h_j) - x(h_a - h_e) - (h_f - h_k)}{x(h_c - h_a) + (h_i - h_f)} \qquad 7.62$$

12. Magnetohydrodynamics

If an ionized gas flows through a magnetic field, an electric field is generated. The electric field can be used as a potential source to generate useful electricity. This concept is employed in a magnetohydrodynamic (MHD) generator.

In a MHD generator, a high-temperature, electrically-conducting plasma is passed through a nozzle and directed through a duct. The plasma passes through a toroidal coil or between two plates which contain the magnetic field. The electric field will be generated perpendicularly to the magnetic field.

Very high temperatures (2000°C to 2500°C) are required to achieve the desired plasma properties. This is achieved through the combustion of fossil fuels, although fission and fusion power plants could also generate the necessary temperatures.

The high-temperature gas is doped with easily-ionized elements, such as potassium or cesium, to obtain the necessary ionization (less than 1%). These alkali metal ions provide a secondary benefit by combining with sul-

POWER

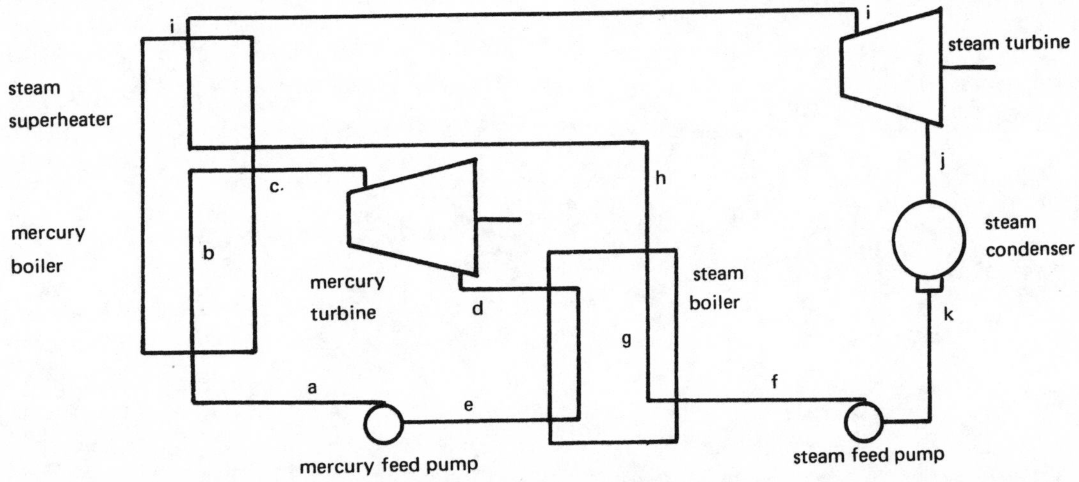

Figure 7.17 Binary Cycle Piping Diagram

fur in the fossil fuel combustion products. Since the sulfur compounds (such as potassium sulfate) can be recovered in traps or electrostatic precipitators, sulfur emissions would be minimal.

The required magnetic field strength is high, in the neighborhood of 50,000 gauss.

The following assumptions are typically made in evaluating simple MHD problems:

(1) The gas velocity is much less than the speed of light.

(2) Collisions determine the fluid behavior. (Electrostatic interaction between the particles is negligible.)

(3) No plasma oscillations occur.

(4) Gas particles travel in straight lines, not spirals.

(5) Capacitance effects are negligible.

Under the above assumptions, Ohm's law can be written as:

$$\mathbf{I} = \sigma (\mathbf{E} + \mathbf{v} \times \mathbf{B}) \qquad 7.63$$

In the above equation, **I** is the current vector, σ is the gas conductivity (reciprocal of resistivity), **E** is the generated electrical field vector, **v** is the plasma velocity vector, and **B** is the magnetic field intensity vector.

Part 2: Combustion Cycles

Nomenclature for Combustion Cycles

c	specific heat	BTU/lbm-°R
h	enthalpy	BTU/lbm
k	adiabatic exponent (c_p/c_v)	–
p	pressure	psf
Q	heat flow	BTU/lbm
R	compression ratio	–
s	entropy	BTU/lbm-°R
T	temperature, or torque	°R, or ft-lbf
V	actual volume	ft^3
w	mass	lbm
W	work	BTU/lbm

Symbols

η	efficiency	–
ρ	density	lbm/ft^3
υ	specific volume	ft^3/lbm

Subscripts

a	air
a/f	air/fuel
c	cut-off
f	fuel
p	constant pressure
th	thermal
v	constant volume

Combustion power cycles differ from vapor power cycles in that combustion products cannot be returned to their initial conditions for reuse. Due to the difficulties of working with mixtures of fuel and air, combustion power cycles are often analyzed as 'air-standard' cycles. The cycle is said to be 'air-standard' if it is analyzed as a closed system using ideal air as a working fluid and with the heat of combustion added without regard to source. Actual engine efficiencies may be 10% to 30% lower than calculated for air-standard analyses. However, if excess air is used in combustion, there may be fair agreement.

13. The Air-Standard Carnot Cycle

The T-s diagram (Figure 7.18) is the same for any Carnot cycle. However, the isothermal expansion and compression do not occur at constant pressure as they do in a vapor power cycle. The Carnot air-standard cycle is not a practical engine cycle, but it does represent an upper maximum on efficiency due to the isentropic processes.

The air-standard Carnot cycle is described by the following processes:

a to b: isentropic compression

b to c: isothermal expansion power stroke with compression ignition and metered fuel injection

c to d: isentropic expansion

d to a: isothermal compression

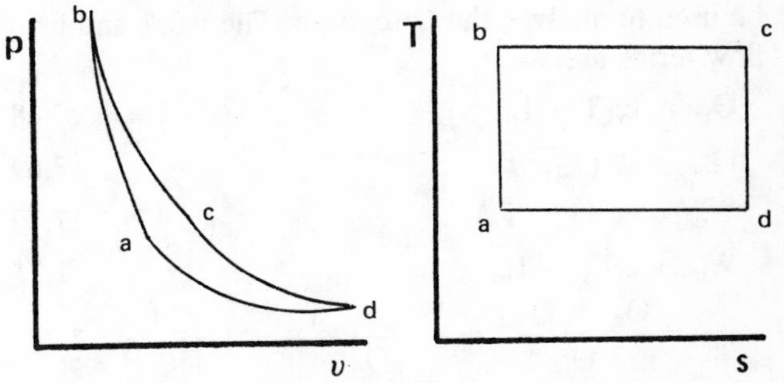

Figure 7.18 Air-Standard Carnot Cycle

The p, V, and T properties may be found from the isentropic and isothermal relationships in chapter 6. The compression ratio is defined as:

$$R = (V_a/V_b) = (V_d/V_c) \qquad 7.64$$

The thermal efficiency of the process is:

$$\eta_{th} = \frac{T_b - T_a}{T_b} = \frac{T_c - T_d}{T_c} = 1 - \frac{1}{R^{k-1}} \qquad 7.65$$

14. The Air-Standard Otto Cycle

The air-standard Otto cycle is less efficient than the Carnot cycle operating within the same temperature limits. However, equal efficiencies will be obtained if the compression ratios are made equal. This will require the Otto engine to operate at higher temperatures than the Carnot cycle. The processes in the cycle are shown in Figure 7.19.

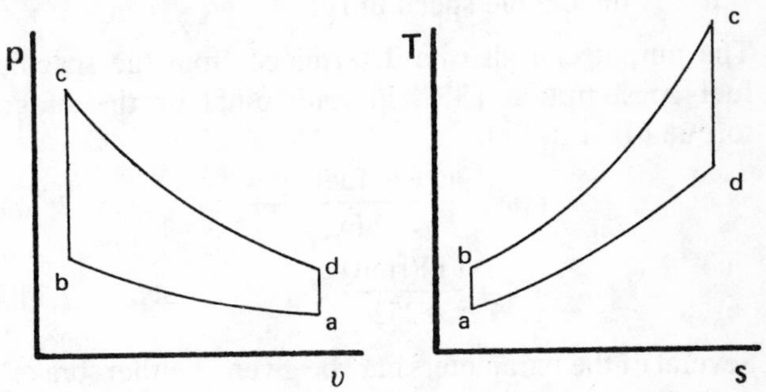

Figure 7.19 The Air-Standard Otto Cycle

The following processes comprise the Otto cycle:

a to b: isentropic compression

b to c: constant volume heat addition

c to d: isentropic expansion

d to a: constant volume heat rejection

It is useful to note the following relationships:

$$R = (V_a/V_b) = (V_d/V_c) \qquad 7.66$$

$$(T_d/T_c) = (T_a/T_b) \qquad 7.67$$

The p, V, and T relationships for the isentropic and constant volume ideal gas processes from chapter 6 can be used to analyze the Otto cycle. The work and heat flow terms are:

$$Q_{in} = c_v(T_c - T_b) \qquad 7.68$$

$$Q_{out} = c_v(T_d - T_a) \qquad 7.69$$

$$W_{out} = c_v(T_c - T_d) \qquad 7.70$$

$$W_{in} = c_v(T_b - T_a) \qquad 7.71$$

$$\eta_{th} = \frac{Q_{in} - Q_{out}}{Q_{in}} = \frac{W_{out} - W_{in}}{Q_{in}} = 1 - \frac{1}{R^{k-1}} \qquad 7.72a$$

$$= 1 - (T_d/T_c) = 1 - (T_a/T_b) \qquad 7.72b$$

Performance of an internal combustion engine operating on the Otto cycle at full throttle can also be analyzed on a macroscopic basic by use of the 'PLAN' formula. It is necessary to know the size of the engine (bore × stroke) in inches, the number of engine cycles per minute, and the operating pressures. Then, the horsepower output may be found from equation 7.73. It is important to use the units given.

$$hp = \frac{PLAN}{33000} \qquad 7.73$$

hp is the engine output in horsepower

P (or MEP) is the mean effective pressure in psig

L is the stroke length in feet

A is the piston area in square inches

N is the number of power strokes per minute, equal to (2n)(#cylinders) ÷ (#strokes per cycle)

n is the engine speed in rpm

The output can also be determined from the specific fuel consumption (SFC in lbm/hp-hr) or the brake torque (T in ft-lbf).

$$hp = \frac{\text{actual fuel flow}}{\text{SFC}} \qquad 7.74a$$

$$= \frac{(T)(\text{rpm})}{5252} \qquad 7.74b$$

Several of the parameters may be given as either 'brake' or 'indicated', such as BHP and IHP, BMEP and IMEP, or BSFC and ISFC. The term 'brake' refers to the actual performance at the output shaft as installed in an operating environment. The term 'indicated' refers to a test of the engine under frictionless conditions. It is important that all terms in equations 7.73 and 7.74 be consistent with regard to 'brake' and 'indicated'.

Various efficiencies can be calculated:

$$\eta_{th} = \frac{2545}{(\text{SFC})(\text{heating value})}$$

$$= \frac{(2545)(\text{hp})}{(\text{fuel consumption})(\text{heating value})} \qquad 7.75$$

$$\eta_{mechanical} = \frac{\text{BHP}}{\text{IHP}} \qquad 7.76$$

$$\eta_{volumetric} = \frac{(\text{air intake in ft}^3/\text{hr})}{(30n)(\text{engine displacement in ft}^3)} \qquad 7.77$$

$$\eta_{ideal} = 1 - \frac{1}{R^{k-1}} \qquad 7.78$$

$$\eta_{relative} = \eta_{th}/\eta_{ideal} \qquad 7.79$$

Since a lower atmospheric density decreases the available oxygen per intake stroke, engine output will decrease with altitude. A numerical approach for determining the output under new conditions of altitude, pressure, or temperature is given in the following steps. It is assumed that engine speed is constant.

step 1: Let 1 and 2 be the lower and higher altitudes respectively.

step 2: Calculate the frictionless horsepower, $IHP_1 = BHP_1/\eta_{mech}$.

step 3: Calculate the friction horsepower, which is assumed to be constant at constant speed: $FHP = IHP_1 - BHP_1$.

step 4: Find the air densities ρ_{a1} and ρ_{a2}.

step 5: Calculate $IHP_2 = IHP_1\,(\rho_{a2}/\rho_{a1})$.

step 6: Calculate the new shaft horsepower: $BHP_2 = IHP_2 - FHP$.

step 7: Calculate the volume flow: $V_{a2} = V_{a1}$.

step 8: Calculate the weight flow of air: $\dot{w}_{a2} = V_{a2}\rho_{a2}$.

step 9: Calculate the weight flow of fuel: $\dot{w}_{f2} = \dot{w}_{a2}/R_{a/f}$.

step 10: Calculate the new fuel consumption: $BSFC_2 = \dot{w}_{f2}/BHP_2$.

15. The Air-Standard Diesel Cycle

The air-standard Diesel cycle is less efficient than the Otto cycle given the same compression ratio and heat addition. However, it is more efficient than the Otto cycle with the same pressure and heat addition. Figure 7.20 illustrates the Diesel cycle.

The processes in the Diesel cycle are:

a to b: isentropic compression

b to c: constant pressure heat addition

c to d: isentropic expansion

d to a: constant volume heat rejection

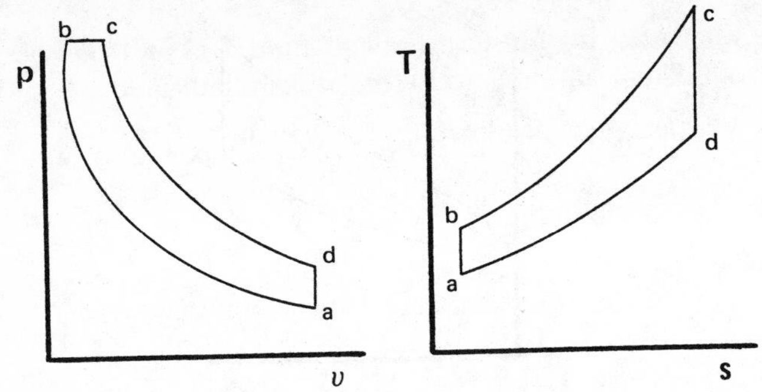

Figure 7.20 The Air-Standard Diesel Cycle

The following definitions are commonly used with the Diesel cycle:

$$R = (V_a/V_b) \quad \text{(compression ratio)} \qquad 7.80$$

$$R_c = (V_c/V_b) = (T_c/T_b) \quad \text{(cut-off ratio)} \qquad 7.81$$

The p, V, and T properties can readily be evaluated by using the ideal gas process equations presented in chapter 6. The work and heat flow terms are:

$$Q_{in} = c_p(T_c - T_b) \qquad 7.82$$

$$Q_{out} = c_v(T_d - T_a) \qquad 7.83$$

$$W_{in} = c_v(T_b - T_a) \qquad 7.84$$

$$W_{out} = c_v(T_c - T_d) + (c_p - c_v)(T_c - T_b) \qquad 7.85$$

$$\eta_{th} = \frac{Q_{in} - Q_{out}}{Q_{in}} = \frac{W_{out} - W_{in}}{Q_{in}}$$

$$= 1 - \frac{T_d - T_a}{k(T_c - T_b)} \qquad 7.86$$

16. The Air-Standard Dual Cycle

The air-standard dual cycle is a compromise between the Diesel and Otto cycles, with an intermediate efficiency. It most closely approximates the operation of an actual internal combustion cycle. The dual cycle is illustrated in Figure 7.21.

The processes of the dual cycle are:

a to b: isentropic compression
b to c: constant volume compression
c to d: constant pressure heat addition
d to e: isentropic expansion
e to a: constant volume heat rejection

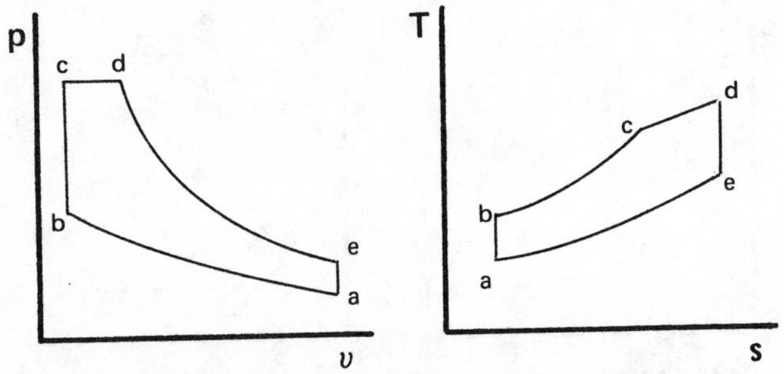

Figure 7.21 The Air-Standard Dual Cycle

The following definitions are used to describe the dual cycle:

$$R = (V_a/V_b) \quad \text{(compression ratio)} \qquad 7.87$$

$$R_c = (V_d/V_c) \quad \text{(cut-off ratio)} \qquad 7.88$$

$$R_p = (p_c/p_b) \quad \text{(pressure ratio)} \qquad 7.89$$

17. Gas Turbine Brayton Cycle (The Air-Standard Joule Cycle)

Strictly speaking, the Brayton cycle is an internal combustion cycle. However, it differs from previous cycles in that each process is carried out in a different location, flow is steady, and air-standard calculations are realistic since a large air/fuel ratio is used to keep combustion temperatures below metallurgical limits.

Figure 7.22 illustrates the physical arrangement of components used to achieve the Brayton cycle. Almost all installations drive the compressor off of the turbine. The actual arrangement differs from the air-standard property plot shown in Figure 7.23 in that the exhaust products exiting at point d are not cooled and returned to the compressor at point a.

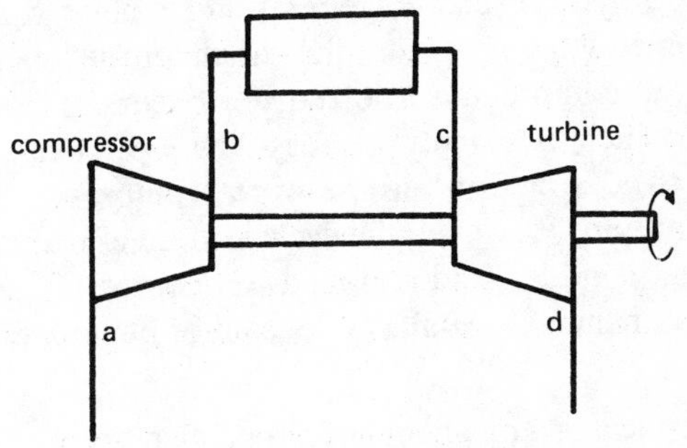

Figure 7.22 Gas Turbine

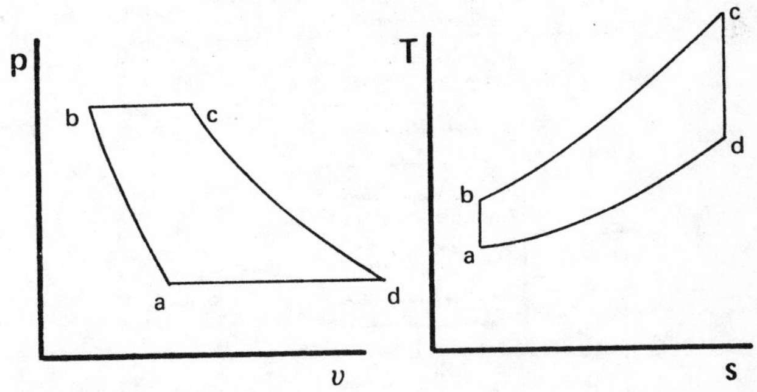

Figure 7.23 Air-Standard Brayton Cycle Property Plots

The processes in the air-standard Brayton cycle are as follows:

a to b: isentropic compression in the compressor
b to c: constant pressure heat addition in the combustor
c to d: isentropic expansion in the turbine
d to a: constant pressure heat rejection to the sink

Relationships from chapter 6 for steady flow systems can be used to evaluate the p, V, and T properties at points a, b, c, and d. Constant values of k, c_p, and c_v may be assumed if air is considered ideal. An air table may also be used if the isentropic efficiencies are known. It is usually assumed that p_b and p_c are equal.

The work and heat flow terms are:

$$Q_{in} = c_p(T_c - T_b) = h_c - h_b \tag{7.90}$$

$$W_{turb} = c_p(T_c - T_d) = h_c - h_d \tag{7.91}$$

$$W_{comp} = c_p(T_b - T_a) = h_b - h_a \tag{7.92}$$

$$\eta_{th} = \frac{Q_{in} - Q_{out}}{Q_{in}} = \frac{W_{turb} - W_{comp}}{Q_{in}}$$

$$= \frac{(h_c - h_b) - (h_d - h_a)}{h_c - h_b} \tag{7.93}$$

If the gas is ideal so that c_p is constant, then

$$\eta_{th} = \frac{(T_c - T_b) - (T_d - T_a)}{T_c - T_b} \tag{7.94}$$

18. The Air-Standard Brayton Cycle with Regeneration

Regeneration is typically used to improve the efficiency of the Brayton cycle. Regeneration involves transferring some of the heat from the exhaust products to the air in the compressor. The transfer occurs in a regenerator which is usually a cross-flow heat exchanger. There is no effect on turbine work, compressor work, or net output. However, the cycle is more efficient since less heat is added and removed. Of course, T_b cannot be greater than T_f. Similarly, T_c cannot be greater than T_e.

The physical arrangement and property plots are shown in Figures 7.24 and 7.25.

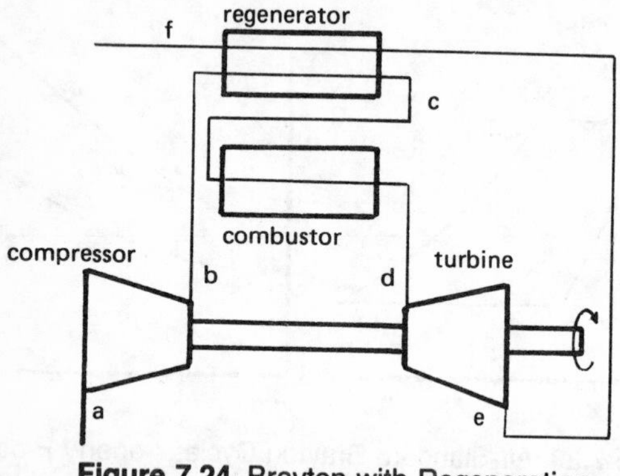

Figure 7.24 Brayton with Regeneration

The processes involved are:

a to b: isentropic compression

b to c: constant pressure heat addition in regenerator

c to d: constant pressure heat addition in combustor

d to e: isentropic expansion

e to f: constant pressure heat removal in regenerator

f to a: constant pressure heat removal in the sink

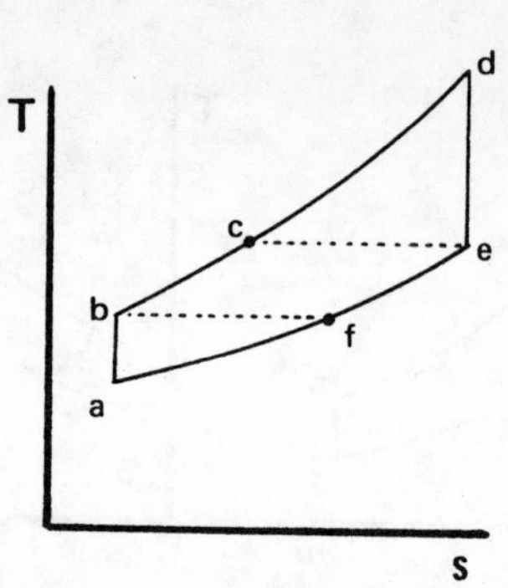

Figure 7.25 Brayton with Regeneration Property Plot

If $T_c = T_e$, the regenerator is said to be 100% efficient. Otherwise, the regenerator efficiency is calculated as:

$$\eta_{regen} = \frac{h_c - h_b}{h_e - h_b} \tag{7.95}$$

The thermal efficiency of the air-standard Brayton cycle with regeneration is:

$$\eta_{th} = \frac{W_{out} - W_{in}}{Q_{in}} = \frac{(h_d - h_e) - (h_b - h_a)}{h_d - h_c} \tag{7.96}$$

If air is considered to be an ideal gas, temperatures may be substituted for enthalpies in equations 7.95 and 7.96.

19. The Ericcson Cycle

The Ericcson cycle offers the best chance of achieving an efficiency approaching that of the Carnot cycle. The processes shown in Figure 7.26 are as follows:

a to b: isothermal compression

b to c: constant pressure heat addition

c to d: isothermal expansion

d to a: constant pressure heat rejection

The efficiency of the Ericcson cycle is equal to the Carnot efficiency if reversible regeneration is used to transfer heat from the (d to a) process to the (b to c) process. Isothermal processes can be approximated by reheating and intercooling. The constant-pressure processes can be approached with counter-flow heat exchangers. Modifications of the Brayton cycle, including regeneration, intercooling, and reheat, approximate the Ericcson cycle.

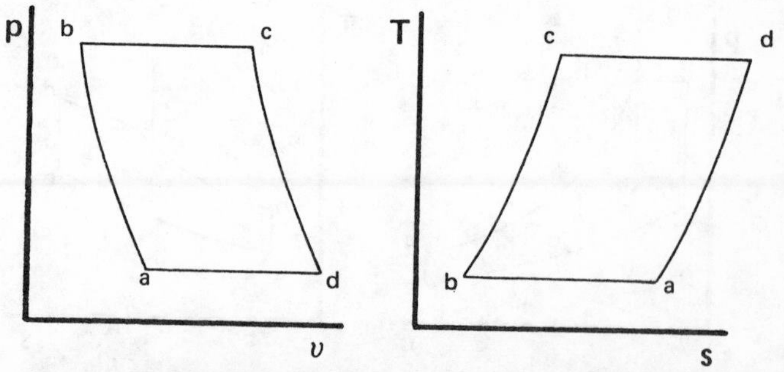

Figure 7.26 The Ericcson Cycle

The work and heat flow terms for the cycle as shown in Figure 7.26 are:

$$W_{out} = W_{c-d} \quad 7.97$$

$$W_{in} = W_{a-b} \quad 7.98$$

$$Q_{in} = Q_{b-c} + Q_{c-d} \quad 7.99$$

If a reversible regenerator is used such that $Q_{b-c} = 0$, then

$$\eta_{th} = \frac{T_{high} - T_{low}}{T_{high}} \quad 7.100$$

20. *Brayton Cycle with Regeneration, Intercooling, and Reheating*

Multiple compression and expansion may be used to improve the efficiency of the Brayton cycle still further. Physical limitations usually preclude more than two stages of intercooling and reheat. This section is written assuming only one stage of each. The physical arrangement is shown in Figure 7.27.

The processes are:

a to b: isentropic compression
b to c: cooling at constant pressure (usually back to T_a)
c to d: isentropic compression
d to f: constant pressure heat addition
f to g: isentropic expansion
g to h: reheating at constant pressure in combustor or reheater (usually back to T_f)
h to i: isentropic expansion
i to a: constant pressure heat rejection

Calculation of the work and heat flow terms and the thermal efficiency is similar to that in the previous cycles, except that there are two W_{turb}, W_{comp}, and Q_{in} terms. If efficiencies for the compressor, turbine, and regenerator are given, then the following relationships are required:

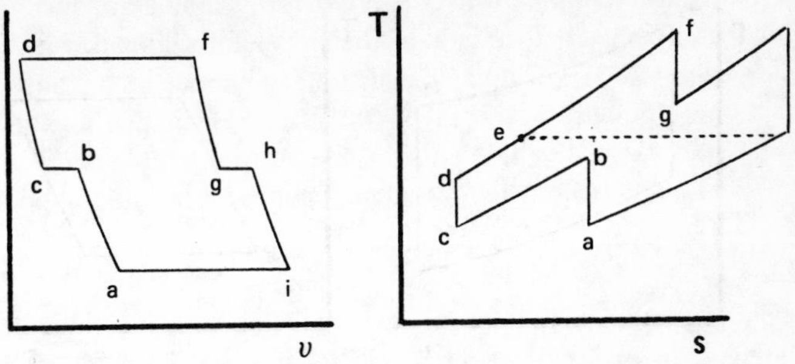

Figure 7.28 Property Plots for Augmented Brayton Cycle

$$p'_b = p_b \quad 7.101a$$

$$h'_b = h_a + (h_b - h_a)/\eta_{comp} \quad 7.101b$$

$$p'_d = p_d \quad 7.102a$$

$$h'_d = h_c + (h_d - h_c)/\eta_{comp} \quad 7.102b$$

$$p'_e = p_e \quad 7.103a$$

$$h'_e = h_d + \eta_{regen}(h_i - h_d) \quad 7.103b$$

$$p'_g = p_g \quad 7.104a$$

$$h'_g = h_f - \eta_{turb}(h_f - h_g) \quad 7.104b$$

$$p'_i = p_i \quad 7.105a$$

$$h'_i = h_h - \eta_{turb}(h_h - h_i) \quad 7.105b$$

21. *The Stirling Cycle*

The Stirling cycle shown in Figure 7.29 can have a thermal efficiency equal to the Carnot efficiency if a reversible regenerator is used to transfer heat from the (c to d) process to the (a to b) process. The isothermal processes are possible with reheating and intercooling, but the steady flow constant volume processes are not. The processes are:

a to b: constant volume heat addition
b to c: isothermal expansion with heat addition
c to d: constant volume heat rejection
d to a: isothermal compression with heat rejection

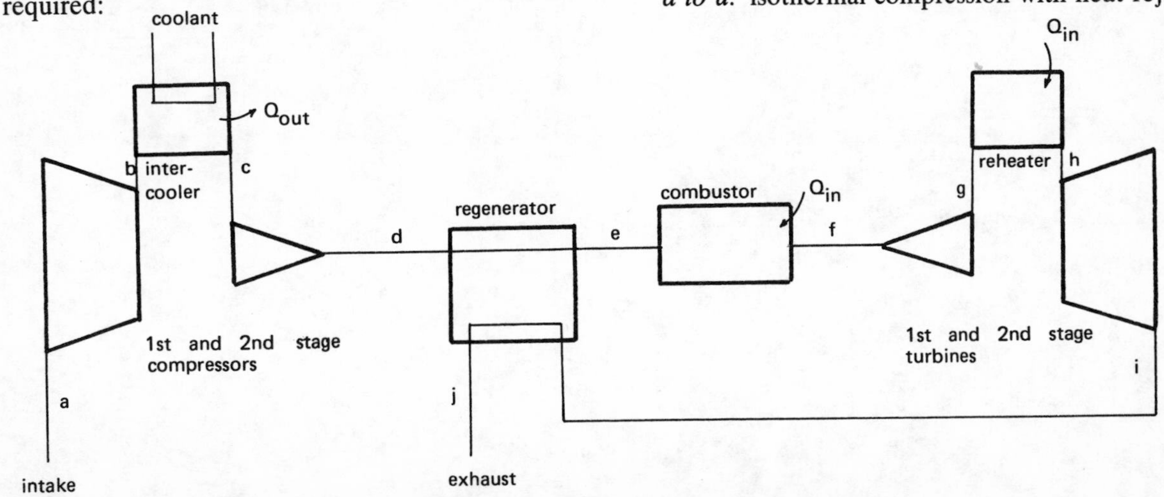

Figure 7.27 Augmented Brayton Cycle

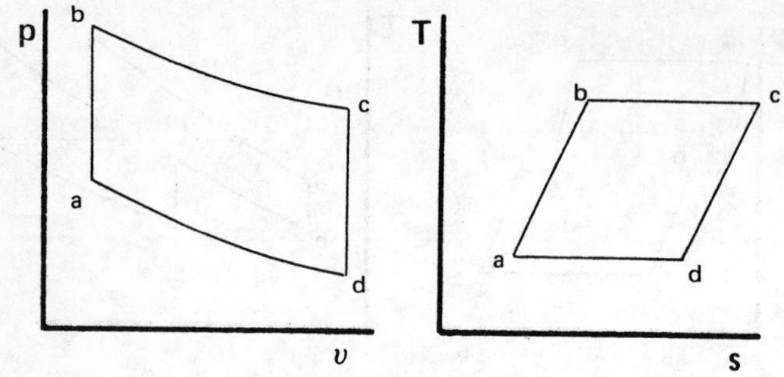

Figure 7.29 The Stirling Cycle

Because both work and heat transfer occur in the (b to c) process, the Stirling cycle must be analyzed with the aid of the ideal gas process relationships in chapter 6. Enthalpy relationships cannot be easily used.

$$W_{out} = W_{b-c} \quad 7.106$$

$$W_{in} = W_{d-a} \quad 7.107$$

$$Q_{in} = Q_{a-b} + Q_{b-c} \quad 7.108$$

If a reversible regenerator is used so that Q_{a-b} is recovered from the (c to d) process, then

$$\eta_{th} = \frac{T_{high} - T_{low}}{T_{high}} \quad 7.109$$

POWER

Part 3: Refrigeration Cycles

Nomenclature for Refrigeration Cycles

COP	coefficient of performance	–
h	enthalpy	BTU/lbm
k	adiabatic exponent (c_p/c_v)	–
m	mass	lbm
p	pressure	psf
Q	heat flow	BTU/lbm
R	compression ratio	–
s	entropy	BTU/lbm-°R
T	temperature	°R
W	work	BTU/lbm

22. General Refrigeration Cycles

Refrigeration cycles are essentially power cycles in reverse. It is necessary to do work on the refrigerant in order to get it to discharge energy to the high-temperature sink. A heat pump also operates on a refrigeration cycle. The difference between a refrigerator and a heat pump is only in the purpose of each. The refrigerator cools a low temperature area as its main purpose and rejects the absorbed heat to a high temperature area. The heat pump rejects the absorbed heat to a high temperature area as its main purpose, having obtained that heat from a low temperature area.

The coefficient of performance (COP) of a device operating on a refrigeration cycle is defined as the ratio of useful heat transferred to the work input. If Q_{in} is the heat absorbed from the low temperature area, the COP for a refrigerator is:

$$COP_{refrig} = \frac{Q_{in}}{Q_{out} - Q_{in}} = \frac{Q_{in}}{W_{net}} \qquad 7.110$$

The COP for a heat pump includes the desired heating effect of the compressor work input.

$$COP_{heat\ pump} = \frac{Q_{in} + W_{in}}{Q_{out} - Q_{in}} = \frac{Q_{in} + W_{in}}{W_{net}} = COP_{refrig} + 1 \qquad 7.111$$

The capacity of a refrigerator is the rate at which heat is removed, expressed in 'tons'. A ton is 12,000 BTU/hr or 200 BTU/min. The ton is derived from the heat flow required to melt a ton of ice in 24 hours. The relationship between horsepower, COP, and the capacity of the refrigerator in tons is given by equation 7.112.

$$COP = \frac{(4.715)(\text{tonnage})}{\text{horsepower}} \qquad 7.112$$

A new term used to evaluate the performance of cooling devices is the energy efficiency ratio (EER) as defined in equation 7.113.

$$EER = \frac{\text{cooling in BTU/hr}}{\text{input power in watts}} \qquad 7.113$$

In general, the required refrigerant flow in lbm/hr can be found from equation 7.114.

$$\dot{m} = \frac{Q_{in}}{h_c - h_b} \qquad 7.114$$

23. The Reversed Carnot Refrigeration Cycle

The reversed Carnot cycle can actually be constructed if a condensable vapor, such as ammonia or Freon, is used. The cycle's main disadvantage is the complex linkwork required for the (a to b) process. The cycle's processes, shown in Figure 7.30, are:

a to b: isentropic expansion behind a piston
b to c: isothermal heat addition (vaporization)
c to d: isentropic compression
d to a: isothermal cooling (condensation)

The solution method is reversed, but identical in concept to the Carnot power cycle. The coefficients of performance are:

$$COP_{refrig} = \frac{T_{low}}{T_{high} - T_{low}} \qquad 7.115$$

$$COP_{heat\ pump} = \frac{T_{high}}{T_{high} - T_{low}} = COP_{refrig} + 1 \qquad 7.116$$

24. The Vapor Compression Cycle

In the vapor compression cycle, an irreversible expansion through a throttling valve replaces the isentropic expansion behind a piston in the reversed Carnot cycle.

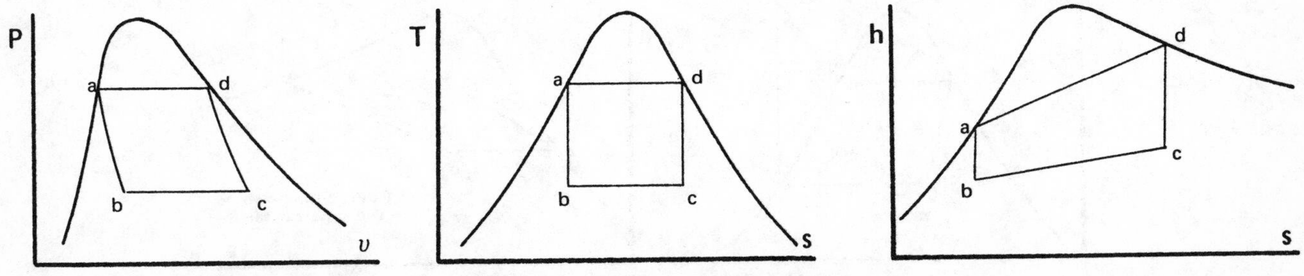

Figure 7.30 The Carnot Refrigeration Cycle

POWER

The processes illustrated in Figure 7.31 are:

a to b: irreversible, isenthalpic expansion
b to c: isothermal vaporization
c to d: isentropic compression
d to a: isothermal condensation

Compression of saturated or superheated vapor is said to be 'dry compression' and is favored over 'wet compression' due to wear in the compressor. For that reason, the refrigerator is usually designed so that the refrigerant leaves the evaporator either saturated or slightly superheated, as shown in Figure 7.32.

If the refrigerant is saturated when leaving the evaporator at point c, the following solution method may be used:

at a: saturated liquid; $h_a = h_b$; $p_a = p_d$
at b: $T_b = T_c$; $h_b \approx h_a$
at c: saturated gas; $T_c = T_b$; $s_c = s_d$
at d: h_d found from the refrigerant table given s_d and either p_d or T_d; $s_d = s_c$

25. The Air Refrigeration Cycle (Reversed Brayton Cycle)

The air refrigeration cycle is not very common due to its high power consumption. However, air as a refrigerant is non-flammable, readily available, and very safe. Therefore, the air refrigeration cycle sees considerable use in aircraft air-conditioning and gas liquefication. The processes shown in Figure 7.33 are:

a to b: isentropic expansion in a turbine
b to c: constant pressure heat addition

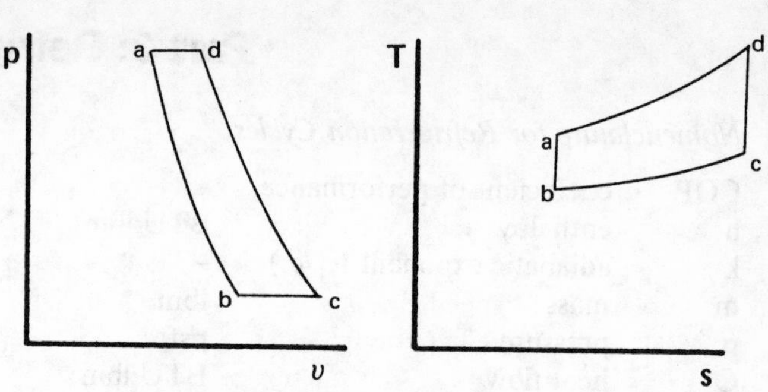

Figure 7.33 The Air Refrigeration Cycle

c to d: isentropic compression
d to a: constant pressure cooling

If air is considered to be an ideal gas, then the ideal gas formulas from chapter 6 can be used to find the properties at each point. In that instance, the COP will be

$$COP = \frac{T_c - T_b}{(T_d - T_a) - (T_c - T_b)} \qquad 7.117a$$

If the (a to b) and (c to d) processes are both isentropic, the following equation may be used to find the coefficient of performance.

$$COP = \frac{1}{\left(R^{\frac{k-1}{k}} - 1\right)} \qquad 7.117b$$

$$R = p_{high}/p_{low} \qquad 7.118$$

If it cannot be assumed that air is an ideal gas, then an air table will have to be used to find the COP. Enthalpy can then be substituted for temperature in equation 7.117a.

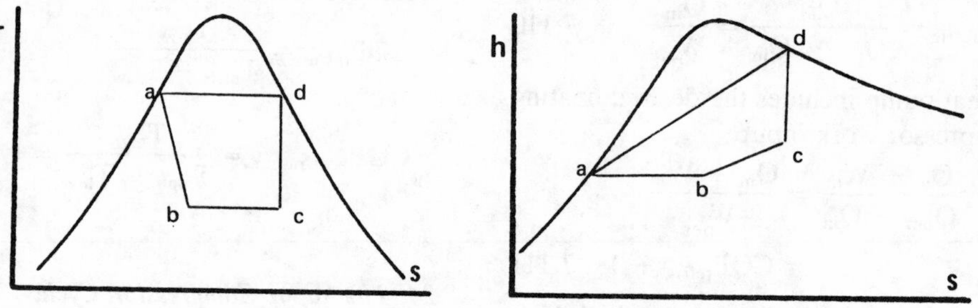

Figure 7.31 Wet Vapor Compression Cycle

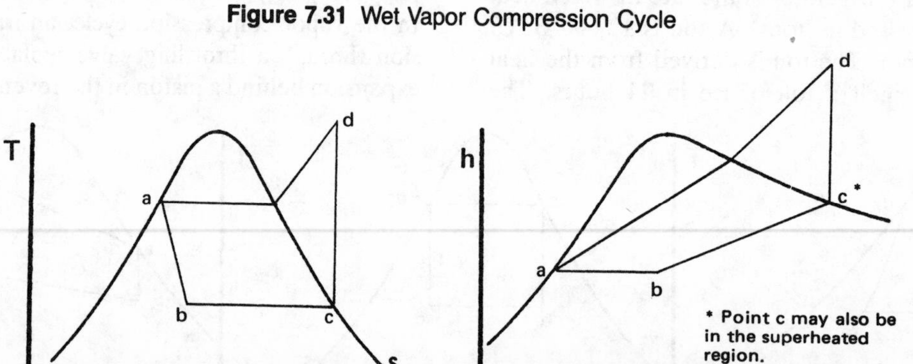

Figure 7.32 Dry Vapor Compression Cycle

Part 4: Compression Cycles

Nomenclature for Compression Cycles

BHP	brake horsepower	hp
c	clearance	decimal
h	enthalpy	BTU/lbm
k	ratio of specific heats	–
n	polytropic exponent	–
p	pressure	psf
Q	heat	BTU/lbm
R	a ratio	–
s	entropy	BTU/lbm-°R
T	temperature	°R
V	volume	ft^3
W	work	BTU/lbm

Symbols

η	efficiency

Subscripts

m	mechanical
s	isentropic
v	volumetric

26. Reciprocating, Single-Stage Compressor with Zero Clearance

The following processes describe the zero-clearance compressor illustrated in Figure 7.34.

a to b: suction at constant pressure (intake valve open)

b to c: compression following polytropic relationships

c to d: discharge at constant delivery pressure (outlet valve open)

d to a: This is not a closed cycle, so the gas properties do not return to their original state.

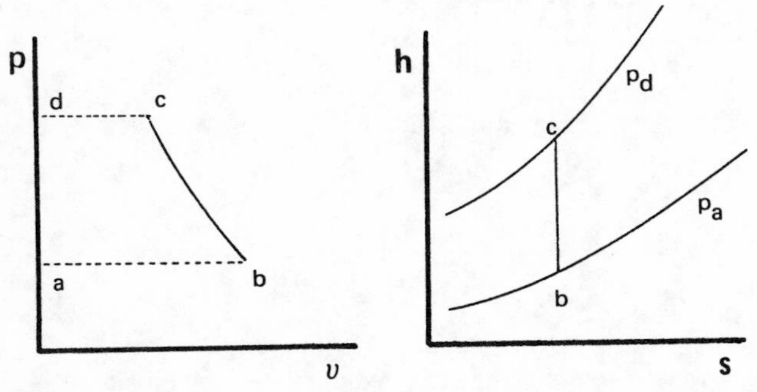

Figure 7.34 Single-Stage Compression with Zero Clearance

The dotted lines on Figure 7.34 are used to indicate the cylinder conditions, not the gas. The work done per cycle depends on the polytropic exponent, since the (b to c) process is assumed to be polytropic. For efficient air compressors, n is between 1.25 and 1.30. All properties may be analyzed using the closed system polytropic process equations given in chapter 6.

In addition to the work of compression in the (b to c) process, the motor must supply power to overcome friction and power to discharge the air in the (c to d) process. The brake work can be found from equation 7.119.

$$\text{brake work} = \frac{W_{compression}\,\eta_v}{\eta_m\,\eta_s} \qquad 7.119$$

27. Reciprocating Single-Stage Compressor with Clearance

If the compressor has clearance (V_d in Figure 7.35), the residual gases expand along with the intake stroke to reduce the capacity per stroke. The dotted line (d to a) in Figure 7.35 is for the residual gases, not the main charge. For convenience, n for the expansion is assumed to be the same as for the compression. For two compressors with the same air flow rate, clearance does not affect the required power.

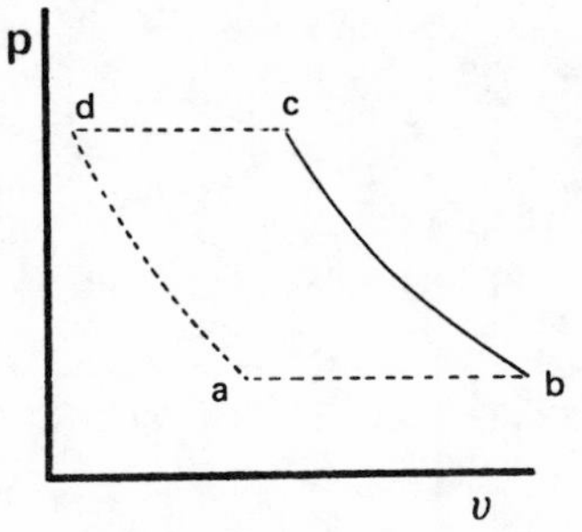

Figure 7.35 Single Stage Compressor with Clearance

Two parameters affecting performance are the compression ratio and the volumetric efficiency. Unlike that for internal combustion engines, the compression ratio for compressors is a ratio of pressures. That is,

$$R = (p_d/p_a) = (p_c/p_b) \qquad 7.120$$

The theoretical formula for volumetric efficiency is:

$$\eta_v = 1 - (R^{1/n} - 1)c \qquad 7.121$$

As with the zero clearance case, the work per cycle may be evaluated with the aid of air tables or the ideal gas process equations in chapter 6.

28. Reciprocating Multi-Stage Compressors with Clearance

In the multi-stage reciprocating compressor, the partially-compressed gas is withdrawn, cooled, and compressed further. Desirable results include a reduction in required power, a decrease in working temperature, and a decrease in valve and ring loading due to the reduced pressure differential. The term 'perfect intercooling' refers to the case when $T_b = T_d$ (see Figure 7.36). The discharge from an intercooler will usually be about 20°F higher than the jacket water temperature.

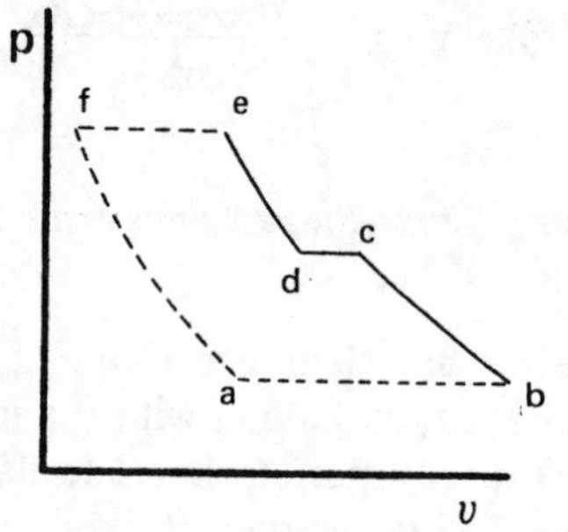

Figure 7.36 Multi-Stage Compressor with Clearance

Analysis of the multi-stage compressor is similar to that of the single-stage compressor.

29. Centrifugal (Dynamic) Compression

The analysis of dynamic compressors is not exceptionally difficult. It can be handled by the open system (steady flow) process relationships in chapter 6. One added consideration is that of kinetic energy, which may have to be included in the calculations. Although this can be done adequately by using the steady flow energy equation (SFEE), it is most expedient to use isentropic gas flow tables.

GENERAL CONCEPTS

1. Steam with negligible velocity enters a horizontal nozzle at 100 psia and 500·F. It expands isentropically to 70 psia. What is the final velocity of the steam?

2. The exhaust velocity of a rocket is 8207 fps. If combustion occurs at 1700·F and the temperature is 274·F at the point where the exit velocity is measured, find the nozzle efficiency. Use 1.9 BTU/lbm-·R as the value of c_p.

3. An adiabatic turbine produces 4000 horsepower by expanding 900·F steam at 300 psia to a saturated 5 psia vapor. What is the steam rate in lbm/hr? What is the isentropic efficiency of the turbine?

CARNOT POWER CYCLE

4. A steam cycle operates between 650·F and 100·F. What is its maximum thermal efficiency?

5. What is the efficiency of a steam Carnot cycle operating between 650·F and 100·F if the turbine and compressor have efficiencies of .9 and .8 respectively?

6. A Carnot cycle operates from two reservoirs - one with boiling water and the other at water's triple point. If 100 BTU are absorbed, find the work done and the rejected heat.

BASIC RANKINE CYCLE

7. A steam Rankine cycle operates between 360·F and 100·F. What is the thermal efficiency assuming isentropic expansion and compression?

8. A steam Rankine cycle operates with 100 psia saturated steam which is reduced to 1 atmosphere through expansion in an 80% efficient turbine. It is 80·F liquid at one atmosphere upon entering the 60% efficient pump. What is the cycle thermal efficiency?

RANKINE CYCLE WITH SUPERHEAT

9. Find the thermal efficiency of a Rankine cycle with superheat at 700·F and 300 psia. Condensing pressure is 1 psia.

10. 500 psia steam is superheated to 1000·F before expanding through a 75% efficient turbine to 5 psia. No sub-cooling occurs and the pump work is negligible compared to the 200,000 kw generated. What quantity of steam is required? What is the heat rejected in the condenser?

11. A superheat steam cycle operates on 200 psia, 400·F steam exhausting at 1.5 inch mercury. What is the thermal efficiency of the cycle?

RANKINE CYCLE WITH SUPERHEAT AND REHEAT

12. A turbine and pump are 88% and 96% efficient respectively. The cycle operates between 70·F and 700·F with superheated 600 psia steam. Steam is reheated when its pressure drops to 200 psia. What is the thermal efficiency of the cycle?

RANKINE CYCLE WITH REGENERATION

13. The cycle in problem 12 is modified to include a bleed of 270·F steam (within the second expansion) for feedwater heating in a closed feedwater heater. The drips are returned to the condensate line. The terminal temperature difference is 6·F. Neglecting drip pump work, what is the thermal efficiency of the cycle?

14. Steam enters a turbine at 1200 psia and 1100·F and expands to 1.5" Hg. A feedwater heater produces 180·F water when supplied with 900·F extraction steam. What is the thermal efficiency of the process?

AIR-STANDARD CARNOT CYCLE

15. An air-standard Carnot cycle operates between 360·F and 100·F. The maximum pressure is 147 psia. Q_{in} is 100 BTU/lbm-cycle. Find the expansion and compression work per pound of air. What is the net work?

AIR-STANDARD OTTO CYCLE

16. The air in an air-standard Otto cycle enters an 11 ft^3 cylinder at 80·F and 14.3 psia. During the compression stroke 160 BTU are added. The compression ratio is 10. What is the temperature at the end of heat addition? What is the thermal efficiency?

17. What is the thermal efficiency of an air-standard Otto cycle operating between 2800·F and 100·F with a compression ratio of 6?

PLAN PROBLEMS

18. Tests of a 6 cylinder, 4-inch bore, 3.125-inch stroke engine at full throttle show BHP to be 79.5 at 3400 rpm. Find the brake torque and mean effective cylinder pressure.

19. A 10"x18" 2-cylinder, 4-stroke engine operates with mean effective pressure of 95 psig at 200 rpm. If the actual developed torque is 600 ft-lbf, what is the friction horsepower?

AIR-STANDARD DIESEL (optional)

20. At 60·F and 14.7 psia, a fully-loaded 1000 BHP diesel runs at 2000 rpm. The A/F ratio is 23 and BSFC is .45 lbm/BHP-hr. The mechanical efficiency is a constant 80%. What are the BHP and BSFC at 5000 feet altitude? (The air density at 5000 feet is .06592 lbm/ft^3.)

21. What is the thermal efficiency of an air-standard diesel cycle operating on 14.2 psia air at 75°F? The temperatures of the air before and after heat addition are 750° and 2900°F respectively.

22. An air-standard diesel cycle with a compression ratio of 16 takes 14.7 psia, 65°F air, compresses it, and adds heat to increase the temperature to 2600°F. What is the thermal efficiency of the cycle?

BRAYTON CYCLE

23. In an air-standard gas turbine, 60°F and 14.7 psia air is compressed through a volume ratio of 5. Air enters at 1500°F and expands to 14.7 psia. If the isentropic efficiency of the compressor and turbine are .83 and .92 respecitively, what is the thermal efficiency of the cycle?

BRAYTON CYCLE WITH REGENERATION (optional)

24. What is the thermal efficiency if a 65% efficient regenerator is added to problem 23?

CARNOT REFRIGERATION CYCLE

25. Ammonia is used in a reversed Carnot cycle with reservoirs at 110°F and 10°F . If 1000 BTU/hr are to be removed, find the COP, work input, and rejected heat.

26. What is the COP for an ammonia Carnot heat pump operating between 700°F and 40°F?

27. A refrigerator cools a continuous brine flow of 100 gpm from 80°F to 20° in a 80°F environment. What is the minimum horsepower requirement?

VAPOR COMPRESSION CYCLE

28. An ammonia compressor is used in a heat pump cycle. Suction pressure is 30 psia. Discharge pressure is 160 psia. Saturated liquid ammonia enters the throttle valve. The refrigeration effect is 500 BTU/lbm ammonia. Find the coefficient of performance as a heat pump.

29. A refrigeration cycle uses Freon-12 as a refrigerant between a 70°F environment and a -30°F heat source. If the vapor leaving the evaporator and liquid leaving the condenser are both saturated, find the volume flow of refrigerant per ton of refrigeration. Assume isentropic compression.

30. Solve problem 29 using ammonia.

AIR REFRIGERATION CYCLE

31. An air refrigeration cycle compresses air from 70°F and 14.7 psia to 60 psia in a 70% efficient compressor. The air is cooled to 25°F in a constant pressure process before entering a turbine with isentropic efficiency of .80. The air then passes through a walk-in meat locker before re-entering the compressor. What is the temperature of the air leaving the compressor and turbine? What is the coefficient of performance of the cycle?

RECIPROCATING COMPRESSORS (optional)

32. A positive displacement compressor with 7% clearance discharges 48 pounds of air per minute at 65 psia. If the polytropic exponent is 1.33, how many pounds of air are compressed each minute?

33. 5000 cfh of 14.5 psia, 70°F air are compressed in a reciprocating, single-stage compressor to 100 psia. Assuming 0% clearance, calculate the required horsepower for
(a) isothermal compression
(b) adiabatic compression

CENTRIFUGAL COMPRESSORS

34. Atmospheric air at 500°F is compressed in a centrifugal compressor to 6 atmospheres with a 65% isentropic efficiency. Find the work done, the final temperature, and the entropy change.

35. 200 lbm/min of 25 psia saturated steam is compressed adiabatically in a centrifugal compressor to 95 psia. If the isentropic efficiency is 70%, what is the required horsepower of the compressor motor? What is the final enthalpy, entropy, temperature, and specific volume of the steam?

POWER

8 Chemistry

Nomenclature

A.W.	atomic weight	grams/gmole
F	force, formality	newtons, -
GEW	gram equivalent weight	grams
I	current	amps
K	constant	various
m	mass, molality	grams, -
M	molarity	–
N	normality	–
p	pressure	N/m^2
r	radius	meters
R	radius	meters
t	time	seconds
T	temperature	°C
v	reaction rate	–
w	combining mass	grams
x	distance, mole fraction	meters, -
[X]	ion concentration of *X*	moles/liter

Subscripts

b	boiling
eq	equilibrium
f	freezing
sp	solubility product

1. Atomic Structure

Chemistry studies the manner in which atoms combine into molecules. An *atom* is the smallest subdivision of an element that can take part in a chemical reaction. A *molecule* is the smallest subdivision of an element or compound that can exist in a free state.

As covered in chapter 19, Niels Bohr suggested that electrons circle[1] a nucleus similar to the way the planets orbit the sun. The electrons in the outer-most orbit are known as the *valence electrons*. These valence electrons are important in bond formation.

The nucleus is constructed of neutrons and protons. The masses of neutrons and protons are essentially the same—one *amu*. The *atomic weight*[2] in *amu* of an atom is approximately equal to the number of neutrons and protons in the nucleus. The *atomic number* of an atom is equal to the positive charge in electrostatic units (*esu*) on the nucleus. This is the same as the number of protons in the nucleus.

Because of the nature of the electron motion around the nucleus, the word *orbit* is not used. Rather, it is said that an electron occupies an *orbital*. The hydrogen orbital, named the *1s* orbital, has a maximum capacity of two electrons. The *1* in the *1s* classification is the *principal quantum number,* **n**. The *s* orbital[3] is the most stable state an electron can occupy.

An atom with two electrons in its *s* orbital will be in a lower energy state than an atom with only one *s* orbital electron. In fact, the $1s^2$ element (helium) is so stable that it will not combine with any other element.

For a principal quantum number of **n** = 2, there are two orbitals—the *s* and *p* orbitals. A total of eight electrons can occupy the *s* and *p* orbitals. The element which has both the *s* and *p* orbitals completely filled is neon. Neon also will not combine with other elements.

It can be generalized that very stable atomic structures result when all of the orbitals corresponding to a particular principal quantum number are filled. The elements which possess this filled structure (i.e., helium, neon, argon, krypton, xenon, and radon) are known as the *inert* or *noble gases* because they do not combine with other elements.

[1]This is essentially correct, but not quite. The electrons travel around the nucleus in random motion, not in circles or ellipses. By moving very rapidly, electrons can be considered to occupy all space around the nucleus.

[2]Although an element can have only one atomic number, it can have several different atomic weights. Many elements possess *isotopes*. The nuclei of isotopes differ from one another only in the number of neutrons.

[3]The *s* stands for the word *sharp*.

CHEM

2. The Periodic Table

The periodic table (table 8.1) is based on the concept that the properties of elements are periodic functions of their atomic weights. Elements are arranged in order of increasing atomic weights from left to right. Adjacent elements in horizontal rows differ decidedly in both physical and chemical properties. However, elements in the same columns have similar properties.

Graduations in properties, both chemical and physical, are most pronounced in the horizontal rows *(periods)*. The most electronegative elements are those at the end of the periods. Elements with low electronegativities are found at the beginning of the periods.

Each vertical group except 0 and 1 has *A* and *B* families. The elements of a *family* resemble each other more than they do those of the other family in the same group. Graduations in properties are definite but less pronounced in vertical groups of families. The trend in any family is toward more metallic properities as the atomic weight increases.

Non-metals (elements at the right of the periodic chart) have high electron affinities, are oxidizing agents, form negative ions, and have negative valences. Metals (at the left of the periodic chart) have low electron affinities, are reducing agents, form positive ions, and have positive valences.

Elements in the periodic table are often categorized into the following groups.

- *alkali metals:* group *1A*
- *alkaline earth metals:* group *2A*
- *noble gases:* group 0
- *lanthanons:* elements 57–71
- *rare earths:* same as lanthanons
- *actinons:* elements 89–103
- *actinide series:* same as actinons
- *transition metals:* all *B* families
- *radioactive elements:* elements 84 upwards
- *halogens:* group *7A*
- *metals:* everything except the non-metals
- *non-metals:* elements 2, 5–10, 14–18, 33–36, 52–54, 85, 86

3. Ionic Bonds

The atomic number of chlorine is 17. That means there are 17 protons in the nucleus of a chlorine atom. There are also 17 electrons in the various orbitals surrounding the nucleus. The atomic structure of chlorine is written as $1s^2 2s^2 2p^6 3s^2 3p^5$.

Notice that there are only seven electrons in the outer shell corresponding to $\mathbf{n} = 3$. Since eight electrons is a more stable configuration, chlorine atoms have a high tendency to absorb neighboring electrons. This tendency is known as *electron affinity.* The electrons which are absorbed by chlorine come from neighboring atoms with low *ionization energies.*

The chlorine atom prior to the absorption of a neighboring electron is neutral. The fact that it needs one electron to complete its outer subshell does not mean that chlorine needs one electron to become neutral. On the contrary, the chlorine atom becomes negatively charged when it absorbs an electron.

Negatively charged ions are known as *anions.*[4] Anions lose electrons at the *anode* during electro-chemical reactions. Anions must lose electrons to become neutral. The loss of electrons is known as *oxidation.*

The charge on an anion is equal to the number of electrons absorbed from neighboring atoms. This charge is known as the *oxidation number.* The oxidation number is equal to the number of electrons that must be lost for charge neutrality. A more common term for the charge on an ion is *valence.*[5] For the chlorine example, the valence is -1.

Sodium has an electronic configuration of $1s^2 2s^2 2p^6 3s^1$. The orbitals corresponding to $\mathbf{n} = 1$ and $\mathbf{n} = 2$ are completely filled. The last electron, which occupies the $3s^1$ orbital, is very easily removed (i.e., sodium has a low ionization energy).

If its outer electron is removed, sodium becomes positively charged. Positively charged ions are known as *cations.*[6] Cations gain electrons and are formed at the cathode in electro-chemical reactions. The gaining of electrons is known as *reduction.* Cations must gain electrons to become neutral.

If a chlorine atom becomes an anion at the expense of a sodium atom (which becomes a cation), the two ions will be attracted to each other by electrostatic force. The electrostatic attraction of the positive sodium to the negative chlorine effectively bonds the two ions together. This type of bonding, in which electrostatic attraction is the characteristic feature, is known as *ionic* or *electrovalent bonding.*

Ionic bonding is characteristic of compounds formed between atoms with high electron affinities and low ionization energies. Compounds containing ionic bonds typically do not have distinct molecules—they consist

[4]Pronounced an′-ion like the girl's name (Ann) followed by *ion.* Not pronounced like *onion.*

[5]Although the valence and oxidation number of an ion are the same in an ionic bond, the oxidation number of an atom in a free-state molecule is zero. Thus, the valence of hydrogen in the molecule H_2 is one. However, its oxidation number is zero. The same is true for O_2, N_2, Cl_2, etc.

[6]Pronounced cat′-ion, like the animal (cat) followed by *ion.* Not pronounced like *cashew.*

Table 8.1 The Periodic Table of Elements (Long Form)

Table 9.1: THE PERIODIC TABLE OF ELEMENTS (Long Form)

The number of electrons in filled shells is shown in the column at the extreme left; the remaining electrons for each element are shown immediately below the symbol for each element. Atomic numbers are enclosed in brackets. Atomic weights (rounded, based on Carbon-12) are shown above the symbols. Atomic weight values in parentheses are those of the isotopes of longest half-life for certain radioactive elements whose atomic weights cannot be precisely quoted without knowledge of origin of the element.

METALS — TRANSITION METALS — NON–METALS

periods	I A	II A	III B	IV B	V B	VI B	VII B	VIII	VIII	VIII	I B	II B	III A	IV A	V A	VI A	VII A	O
1 0	1.0079 H[1] 1																1.0079 H[1] 1	4.0026 He[2] 2
2 2	6.941 Li[3] 1	9.0122 Be[4] 2											10.81 B[5] 3	12.011 C[6] 4	14.007 N[7] 5	16.000 O[8] 6	19.000 F[9] 7	21.179 Ne[10] 8
3 2, 8	22.99 Na[11] 1	24.305 Mg[12] 2											26.982 Al[13] 3	28.086 Si[14] 4	30.974 P[15] 5	32.06 S[16] 6	35.453 Cl[17] 7	39.948 Ar[18] 8
4 2, 8	39.098 K[19] 8, 1	40.08 Ca[20] 8, 2	44.956 Sc[21] 9, 2	47.90 Ti[22] 10, 2	50.941 V[23] 11, 2	51.996 Cr[24] 13, 1	54.938 Mn[25] 13, 2	55.847 Fe[26] 14, 2	58.933 Co[27] 15, 2	58.70 Ni[28] 16, 2	63.546 Cu[29] 18, 1	65.38 Zn[30] 18, 2	69.72 Ga[31] 18, 3	72.59 Ge[32] 18, 4	74.922 As[33] 18, 5	78.96 Se[34] 18, 6	79.904 Br[35] 18, 7	83.80 Kr[36] 18, 8
5 2,8,18	85.468 Rb[37] 8, 1	87.62 Sr[38] 8, 2	88.906 Y[39] 9, 2	91.22 Zr[40] 10, 2	92.906 Nb[41] 12, 1	95.94 Mo[42] 13, 1	(97) Tc[43] 14, 1	101.07 Ru[44] 15, 1	102.906 Rh[45] 16, 1	106.4 Pd[46] 18	107.868 Ag[47] 18, 1	112.41 Cd[48] 18, 2	114.82 In[49] 18, 3	118.69 Sn[50] 18, 4	121.75 Sb[51] 18, 5	127.60 Te[52] 18, 6	126.905 I[53] 18, 7	131.30 Xe[54] 18, 8
6 2,8,18	132.905 Cs[55] 18,8,1	137.33 Ba[56] 18,8,2	* [57-71]	178.49 Hf[72] 32,10,2	180.948 Ta[73] 32,11,2	183.85 W[74] 32,12,2	186.207 Re[75] 32,13,2	190.2 Os[76] 32,14,2	192.22 Ir[77] 32,15,2	Pt[78] 32,17,1	196.967 Au[79] 32,18,1	200.59 Hg[80] 32,18,2	204.37 Tl[81] 32,18,3	207.2 Pb[82] 32,18,4	208.980 Bi[83] 32,18,5	(209) Po[84] 32,18,6	(210) At[85] 32,18,7	(222) Rn[86] 32,18,8
7 2,8,18,32	(223) Fr[87] 18,8,1	226.025 Ra[88] 18,8,2	† [89-103]	Rf[104] 32,10,2	Ha[105] 32,11,2	[106] 32,12,2	[107]	[108]										

* LANTHANIDE SERIES	138.906 La[57] 18,9,2	140.12 Ce[58] 20,8,2	140.908 Pr[59] 21,8,2	144.24 Nd[60] 22,8,2	(145) Pm[61] 23,8,2	150.4 Sm[62] 24,8,2	151.96 Eu[63] 25,8,2	157.25 Gd[64] 25,9,2	158.925 Tb[65] 27,8,2	162.50 Dy[66] 28,8,2	164.930 Ho[67] 29,8,2	167.26 Er[68] 30,8,2	168.934 Tm[69] 31,8,2	173.04 Yb[70] 32,8,2	174.97 Lu[71] 32,9,2
† ACTINIDE SERIES	(227) Ac[89] 18,9,2	232.038 Th[90] 18,10,2	231.036 Pa[91] 20,9,2	238.029 U[92] 21,9,2	237.048 Np[93] 23,8,2	(244) Pu[94] 24,8,2	(243) Am[95] 25,8,2	(247) Cm[96] 25,9,2	(247) Bk[97] 26,9,2	(251) Cf[98] 28,8,2	(254) Es[99] 29,8,2	(257) Fm[100] 30,8,2	(258) Md[101] 31,8,2	(255) No[102] 32,8,2	(260) Lr[103] 32,9,2

CHEM

of a group of charged ions in a lattice. The formulas (e.g., NaCl) are indications of the *relative* number of ions in the compound.

Ionic compounds are usually hard crystalline solids. They have high melting points and low vapor pressures. Ionic solids become electrically conductive when dissociated.

In an ionic bond, the anion and cation are drawn together until the coulomb attraction is exactly balanced by the repulsion of the electrons in the outer shells. The *bonding energy* is the sum of these forces integrated from infinity to the *equilibrium distance*. This is illustrated by figure 8.1.

$$\text{bonding energy} = \int_{\infty}^{x_{eq}} (F_{\text{coulomb}} + F_{\text{repulsive}})dx \qquad 8.1$$

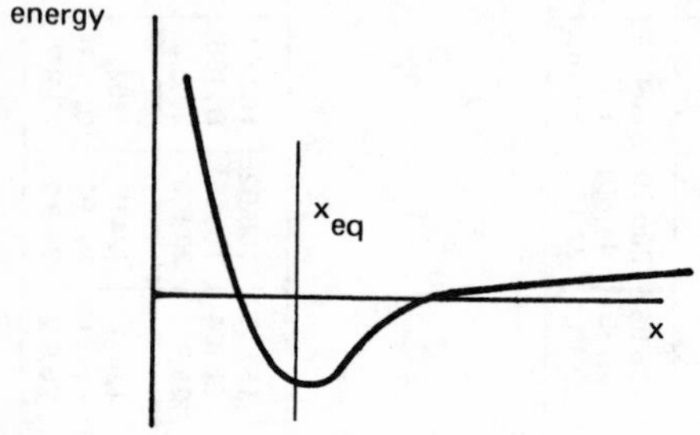

Figure 8.1 Equilibrium Spacing in Ionic Solids

The equilibrium spacing is dependent on several factors:

- *temperature*—As the temperature increases, the equilibrium spacing increases.
- *ionic charge*—Since the electron repulsive forces diminish as an ion loses electrons, the equilibrium spacing will decrease as the ion becomes more positive.
- *coordination number*—The equilibrium spacing increases as the number of adjacent ions in the molecule increases. This is due to the increase in repulsive electronic forces.
- *bond type*—The stronger the bond, the smaller will be the equilibrium spacing. The double covalent bond between the two carbon atoms in ethylene (C_2H_4) is smaller than the single covalent bond between the two carbon atoms in ethane (C_2H_6).

4. Covalent Bonds

Several common gases in their free states form diatomic molecules. Examples are hydrogen (H_2), oxygen (O_2),[7] and nitrogen (N_2). Since two atoms of the same element will have the same electron affinity and ionization energy, it is apparent that the bond formed is not ionic in nature. The electrons in these diatomic molecules are shared equally in order to fill the outer subshells.

This type of bonding, in which sharing of electrons is the characteristic feature, is known as *covalent bonding*. Covalent bonds are typical of the bonds formed in organic compounds.

The valence of an atom that forms a covalent bond is equal to the number of shared electron *pairs*. For example, the chemical reaction in which a hydrogen molecule forms is

$$H\cdot + H\cdot \rightarrow H\text{—}H \qquad 8.2$$

The line between the two hydrogen symbols on the right-hand side of the chemical equation means *a pair of shared electrons*. The valence of hydrogen in this covalent bond is one—the number of shared electron pairs.

Example 8.1

Write the chemical reaction for the formation of a nitrogen gas molecule.

Nitrogen has an atomic number of seven and an electronic configuration of $1s^2 2s^2 2p^3$. There are five electrons in the **n** = 2 shell. In order to form a full set of eight electrons, three more electrons would have to be obtained from another nitrogen atom. These three electrons are shared.

$$:\dot{\underset{\cdot}{N}}\cdot + :\dot{\underset{\cdot}{N}}\cdot \rightarrow :N{\equiv}N:$$

Since there are three shared electron pairs, the valence of nitrogen is three.

If the atoms forming a covalent molecule are not both the same element, the electrons will not be shared equally. For example, the bond between hydrogen and chlorine in HCl is both covalent and ionic in nature. Thus, there is no sharp dividing line between ionic and covalent bonds for most compounds.

The bond between hydrogen and chlorine is predominantly (81%) covalent in nature. This implies that the electron which hydrogen provides spends approximately 19% of its time surrounding the chlorine. The attraction of an atom for an electron in a covalent molecule is known as *electronegativity*. Chlorine has a higher electronegativity than hydrogen. This is why the chlorine atom can capture and keep the shared electron 19% of the time.

[7]Oxygen also forms an O_3 molecule known as *ozone*. Different molecular forms of the same element are known as *allotropes*. Allotropes are not the same as isotopes. In the case of ozone, all oxygen atoms have the same atomic weight as the oxygen atoms in the O_2 molecule.

CHEM

5. Coordination Numbers

The *coordination number* is defined as the number of closest neighboring atoms in a molecule. For example, the coordination number of carbon[8] in a methane (CH_4) molecule is four.

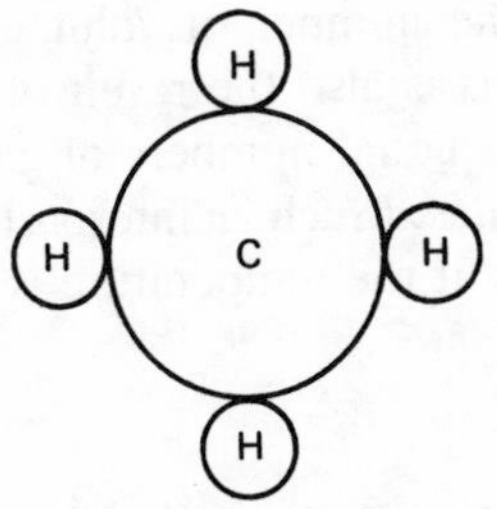

Figure 8.2 A Methane Molecule

The coordination number depends on both the valence of the interacting atoms and the ratio of their ionic radii. If an atom has a valence of one, its coordination number can only be one. This is because it takes only one donor atom to supply one electron. However, if the valence is two, the coordination number can be either one (as for oxygen in an O_2 molecule) or two (as for oxygen in an H_2O molecule).

The effect of the ionic radii is geometric. The size of an atom that can fit into the void formed between other atoms is a function of their relative sizes. This is illustrated by figure 8.3. Specific coordination numbers for various radius ratios are given in table 8.2.

Table 8.2
Radius Ratios Versus Coordination Numbers

r/R	Maximum Coordination Number
.155 – .224	3
.225 – .413	4
.414 – .731	6
.732 – .999	8
1.0 up	12

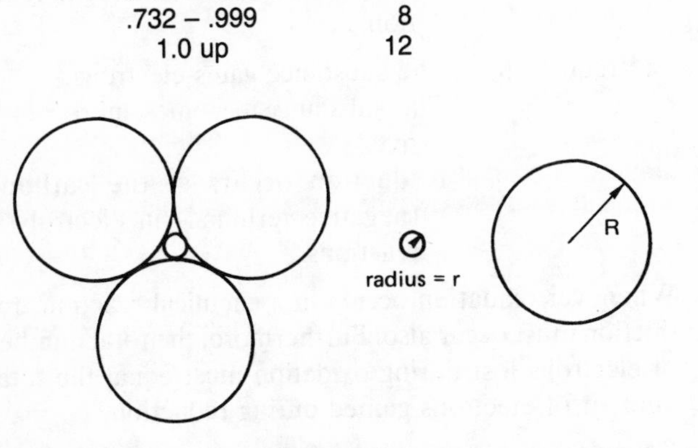

Figure 8.3
An Ion with Coordination Number of 3

[8]However, the coordination number of hydrogen in a methane molecule is one.

6. Formation of Compounds

The sum of the valences must be zero if a neutral compound is to form. For example, H_2O is a valid compound since the two hydrogen atoms have a total positive valence of $2 \times 1 = +2$. The oxygen ion has a valence of -2. These valences sum to zero.

On the other hand, $NaCO_3$ is not a valid compound formula. The sodium (Na) has a valence of $+1$. However, the carbonate radical[9] has a valence of -2. The correct sodium carbonate molecule is Na_2CO_3.

Table 8.3
Valences of Ions and Radicals

Name	Symbol	Valence
acetate	$C_2H_3O_2$	−1
aluminum	Al	+3
ammonium	NH_4	+1
barium	Ba	+2
boron	B	+3
borate	BO_3	−3
bromine	Br	−1
calcium	Ca	+2
carbon	C	+4, −4
carbonate	CO_3	−2
chlorate	ClO_3	−1
chlorine	Cl	−1
chlorite	ClO_2	−1
chromate	CrO_4	−2
chromium	Cr	+2, +3, +6
copper	Cu	+1, +2
dichromate	Cr_2O_7	−2
fluorine	F	−1
gold	Au	+1, +3
hydrogen	H	+1
hydroxide	OH	−1
hypochlorite	ClO	−1
iron	Fe	+2, +3
lead	Pb	+2, +4
lithium	Li	+1
magnesium	Mg	+2
mercury	Hg	+1, +2
nickel	Ni	+2, +3
nitrate	NO_3	−1
nitrite	NO_2	−1
nitrogen	N	−3, +1, +2, +3, +4, +5
oxygen	O	−2
permanganate	MnO_4	−1
phosphate	PO_4	−3
phosphorous	P	−3, +3, +5
potassium	K	+1
silicon	Si	+4, −4
silver	Ag	+1
sodium	Na	+1
sulfate	SO_4	−2
sulfite	SO_3	−2
sulfur	S	−2, +4, +6
tin	Sn	+2, +4
zinc	Zn	+2

[9]A *radical* is a charged group of atoms that combines as a single ion.

In order to evaluate whether or not a compound formula is valid, it is necessary to know the valences of the interacting atoms. Although some atoms have more than one possible valence, most do not. The valences of some common ions and radicals are given in table 8.3.

7. Chemical Reactions

During chemical reactions, chemical bonds between atoms are broken and new bonds are formed. *Reactants* are either converted to simpler products or are synthesized into more complex products. There are five common types of chemical reactions.

- *direct combination or synthesis:* This is the simplest type of reaction where two elements or compounds combine directly to form a compound.

$$2H_2 + O_2 \rightarrow 2H_2O \qquad 8.3$$

$$SO_2 + H_2O \rightarrow H_2SO_3 \qquad 8.4$$

- *decomposition:* Bonds uniting a compound are disrupted by heat or another energy source to yield simpler compounds or elements.

$$2HgO \rightarrow 2Hg + O_2 \qquad 8.5$$

$$H_2CO_3 \rightarrow H_2O + CO_2 \qquad 8.6$$

- *single displacements:* This type of reaction is identified by one element and one compound as the reactants.

$$2Na + 2H_2O \rightarrow 2NaOH + H_2 \qquad 8.7$$

$$2KI + Cl_2 \rightarrow 2KCl + I_2 \qquad 8.8$$

- *double decomposition:* These are reactions characterized by having two compounds as reactants and forming two new compounds.

$$AgNO_3 + NaCl \rightarrow AgCl + NaNO_3 \qquad 8.9$$

$$H_2SO_4 + ZnS \rightarrow H_2S + ZnSO_4 \qquad 8.10$$

- *oxidation-reduction (Redox):* These reactions involve oxidation of one substance and reduction of another. In the example, calcium loses electrons and is oxidized; oxygen gains electrons and is reduced.

$$2Ca + O_2 \rightarrow 2CaO \qquad 8.11$$

The coefficients of the chemical symbols represent the number of molecules taking part in the reaction. Since matter cannot be destroyed in a chemical reaction, the number of atoms of each element must be equal on both sides of the arrow. This is the principle used in balancing chemical equations.

The coefficients can also represent the number of *moles*[10] taking part in a reaction. *Avogadro's law* states that the number of atoms in a mole of any substance is the same (6.023 EE23 molecules/gmole). Therefore, it is only necessary to multiply all of the coefficients in a chemical reaction equation by 6.023 EE23 to show that the coefficients represent moles.

[10]A *mole* of a substance has a mass equal to its atomic or molecular weight. Mass may be measured in grams (the *gmole*), kilograms (the *kgmole*), or in pounds (the *pmole*). Avogadro's number is valid only when the mass of a mole is measured in grams.

If the reactants and products are gases, the coefficients also represent the number of volumes taking part in the reaction. This is also the result of *Avogadro's law* which states that equal numbers of gas molecules occupy equal volumes. (Such an interpretation is valid for water vapor only if the temperature is high.)

Example 8.2

Balance the reaction equation between aluminum and sulfuric acid.

$$Al + H_2SO_4 \rightarrow Al_2(SO_4)_3 + H_2$$

step 1: Since there are two aluminums on the right, multiply *Al* by 2.

$$2Al + H_2SO_4 \rightarrow Al_2(SO_4)_3 + H_2$$

step 2: Since there are three sulfate radicals (SO_4) on the right, multiply H_2SO_4 by 3.

$$2Al + 3H_2SO_4 \rightarrow Al_2(SO_4)_3 + H_2$$

step 3: Now there are six hydrogens on the left, so multiply H_2 by 3 to balance the equation.

$$2Al + 3H_2SO_4 \rightarrow Al_2(SO_4)_3 + 3H_2$$

8. Redox Reactions

Redox is short for *reduction-oxidation*. Redox reactions are characterized by the transfer of electrons from one element or compound to another.

- oxidation – The substance loses electrons.
  The substance becomes less negative.
  Oxidation occurs at the anode (positive terminal) in electrolytic reactions.
- reduction – The substance gains electrons.
  The substance becomes more negative.
  Reduction occurs at the cathode (negative terminal) in electrolytic reactions.

Whenever oxidation occurs in a chemical reaction, reduction must occur also. Furthermore, the total number of electrons lost during oxidation must equal the total number of electrons gained during reduction.

Consider, for example, the formation of sodium chloride from sodium and chlorine.

$$2Na + Cl_2 \rightarrow 2NaCl \qquad 8.12$$

This reaction is a combination of the oxidation of sodium and the reduction of chlorine.

$$Na \rightarrow Na^+ + e^- \qquad 8.13$$

$$Cl + e^- \rightarrow Cl^- \qquad 8.14$$

The requirement that all electrons released during oxidation be used in a reduction reaction helps balance complex chemical reactions.

Example 8.3

How many $AgNO_3$ molecules are formed per NO molecule in the reaction of silver with nitric acid?

The unbalanced reaction is given with constants c_i to represent the unknown coefficients.

$$c_1Ag + c_2HNO_3 \rightarrow c_3AgNO_3 + c_4NO + c_5H_2O$$

The oxidation number of Ag as a reactant is zero. The oxidation number of Ag in $AgNO_3$ is +1. Therefore, silver has become less negative (more positive) and has been oxidized through the loss of one electron.

$$Ag \rightarrow Ag^+ + e^-$$

The *N* in HNO_3 has an oxidation number of +5. The *N* in NO has an oxidation number of +2. The nitrogen has become more negative (less positive) and has been reduced through the gain of three electons.

$$N^{+++++} + 3e^- \rightarrow N^{++}$$

Since three electrons are required for every NO molecule formed, it is necessary that $c_3 = 3c_4$.

9. Stoichiometry

Stoichiometry is the study of the proportions in which elements combine into compounds.

Stoichiometric problems are known as *weight and proportion* problems because their solutions use simple ratios to determine the weight of reactants required to produce given amounts of products. The procedure for solving these problems is essentially the same regardless of the reaction.

step 1: Write and balance the chemical equation.

step 2: Calculate the molecular weight of each compound or element in the equation.

step 3: Multiply the molecular weights by their respective coefficients and write the products under the formulas.

step 4: Write the given weight data under the molecular weights calculated from step 3.

step 5: Fill in missing information by using simple ratios.

Example 8.4

Caustic soda (NaOH) is made from sodium carbonate (Na_2CO_3) and slaked lime ($Ca(OH)_2$). How many pounds of caustic soda can be made from 2000 pounds of sodium carbonate?

The balanced chemical equation is

	Na_2CO_3 +	$Ca(OH)_2$ →	2NaOH +	$CaCO_3$
MW's:	106	74	2 × 40	100
given:	2000		X	

The ratio used is

$$\frac{NaOH}{Na_2CO_3} = \frac{80}{106} = \frac{X}{2000}$$

$$X = 1509 \text{ pounds}$$

10. Empirical Formula Development

A relationship between the atomic weights, combining masses, and the chemical formula can be developed for *binary compounds* — compounds constructed from two elements.

For example, suppose elements *A* and *B* combine to form compound A_mB_n according to the reaction

$$mA + nB \rightarrow A_mB_n \qquad 8.15$$

Then, if w_A and w_B are the masses of *A* and *B* which combine together,

$$\frac{w_A}{w_B} = \left(\frac{m}{n}\right)\left(\frac{A.W._A}{A.W._B}\right) \qquad 8.16$$

Solving for *m* and *n* gives the formula. Solving for w_A and w_B gives the combining masses. If, in solving for *m* and *n*, one of the values is assumed to be one, the other will often come out to be some simple fraction. In such a case, multiply both the assumed value and the derived value by the smallest integer which clears the fraction. For example, $A_1B_{1.5}$ would not be allowed, since only a whole atom can combine. The cleared formula would be A_2B_3.

Example 8.5

How many grams of carbon are required to form 18.7 grams of methane?

The chemical reaction is

$$C + 2H_2 \rightarrow CH_4$$

In this example $m = 1$, $n = 4$, $A.W._A = 12$, $A.W._B = 1$.

w_A is unknown, and $w_B = 18.7 - w_A$.

$$\frac{w_A}{18.7 - w_A} = \left(\frac{1}{4}\right)\left(\frac{12}{1}\right)$$

$$w_A = 14.0 \text{ grams}$$

CHEM

A *tertiary compound* (3 elements) formula can be analyzed in the following manner:

step 1: Convert all combining weights into percentages by dividing each by the sum of the combining weights.

step 2: Divide the percentages by the atomic weight of each of the respective elements. Call the result Y_i.

step 3: Divide all Y_i by the smallest Y_i. Call the result X_i. One of the X_i (corresponding to the element with the smallest Y_i) should be 1.

step 4: If all of the X_i are integers, the formula is $A_{x1}B_{x2}C_{x3}$. Otherwise, clear the fraction as before.

Example 8.6

An alcohol is analyzed and the following gravimetric analysis is recorded: carbon 37.5%, hydrogen 12.5%, oxygen 50%. What is the alcohol?

step 1: The percentages are given.

step 2: Divide the gravimetric analysis by the atomic weight.

$$\text{C: } \frac{37.5}{12} = 3.125$$

$$\text{H: } \frac{12.5}{1} = 12.5$$

$$\text{O: } \frac{50}{16} = 3.125$$

step 3: The smallest Y is 3.125. Dividing all by this gives,

$$\text{C: } \frac{3.125}{3.125} = 1$$

$$\text{H: } \frac{12.5}{3.125} = 4$$

$$\text{O: } \frac{3.125}{3.125} = 1$$

step 4: The basic formula is CH_4O. The alcohol is CH_3OH (methyl alcohol).

In example 8.6, there was insufficient information in the chemical analysis to determine whether the formula was CH_4O or CH_3OH. (It might have been known that the compound analyzed behaved as though it possessed a hydroxyl (OH) radical.) Similarly, chemical properties would have to be used to distinguish between ethyl alcohol (C_2H_5OH) and dimethyl ether ($(CH_3)_2O$). Different arrangements of the same atoms are known as *isomers*.

11. Equivalent Weights

The *equivalent weight* of a molecule is its molecular weight divided by the change in charge which is experienced by the molecule in a chemical reaction.

Example 8.7

What are the equivalent weights of the following compounds?

(a) *Al* in the reaction

$$Al^{+++} + 3e^- \rightarrow Al$$

(b) H_2SO_4 in the reaction

$$H_2SO_4 + H_2O \rightarrow 2H^+ + SO_4^{--} + H_2O$$

(c) NaOH in the reaction

$$NaOH + H_2O \rightarrow Na^+ + OH^- + H_2O$$

(a) The atomic weight of aluminum is 27. Since the change in charge is three, the equivalent weight is $27 \div 3 = 9$.

(b) The molecular weight of sulfuric acid is 98.1. Since it changes from neutral to ions with 2 charges each, the equivalent weight is $98.1 \div 2 = 49.05$.

(c) Sodium hydroxide has a molecular weight of 40. The molecule went from neutral to a singly-charged state. So, the equivalent weight is $40 \div 1 = 40$.

12. Solutions of Solids in Liquids

Various methods exist for measuring the strengths of solutions.

- *N* *normality*—the number of gram equivalent weights of solute per liter of solution. A solution is *normal* if there is exactly one gram equivalent weight per liter.
- *M* *molarity*—the number of gram moles of solute per liter of solution. A *molar* solution contains one gram mole per liter of solution.
- *F* *formality*—the number of gram formula weights per liter of solution.
- *m* *molality*—the number of gram moles of solute per 1000 grams of solvent. A *molal* solution contains 1 gmole per 1000 grams.
- *x* *mole fraction*—the number of moles of solute divided by the number of moles of solvent and all solutes.
- *mg/l* *milligrams per liter*—the number of milligrams per liter. (Same as *ppm* for solutions of water.)
- *ppm* *parts per million*—the number of pounds (or grams) of solute per million pounds (or grams) of solution. (Same as *mg/l* for solutions of water.)
- *meq/l* the number of milligram equivalent weights of solute per liter of solution. Calculated by multiplying normality by 1000.

CHEM

Example 8.8

A solution is made by dissolving .353 grams of $Al_2(SO_4)_3$ in 730 grams of water. Assuming 100% ionization, what is the normality, molarity, and concentration in *mg/l*?

The molecular weight of $Al_2(SO_4)_3$ is

$$2(26.98) + 3[32.06 + 4(16)] = 342.14$$

The equivalent weight is

$$342.14/6 = 57.02$$

The number of gram equivalent weights used is

$$.353/57.02 = 6.19\ \text{EE}-3.$$

The number of liters of solution (same as the solvent volume if the small amount of solute is neglected) is .73.

$$N = \frac{6.19\ \text{EE}-3}{.73} = 8.48\ \text{EE}-3$$

The number of moles of solute used is

$$.353/342.14 = 1.03\ \text{EE}-3.$$

$$M = \frac{1.03\ \text{EE-3}}{.73} = 1.41\ \text{EE}-3$$

The number of milligrams is .353/.001 = 353.

$$mg/l = \frac{353}{.73} = 483.6$$

Raoult's Law gives the drop in a solvent's vapor pressure as a solute is added. Raoult's law applies to non-volatile liquids which are not electrolytes.

$$\Delta p_{vapor} = (p_{vapor,\ pure})\begin{pmatrix}\text{mole fraction}\\ \text{of solute}\end{pmatrix} \qquad 8.17$$

$$p_{vapor,\ solution} = (p_{vapor,\ pure})\begin{pmatrix}\text{mole fraction}\\ \text{of solvent}\end{pmatrix} \qquad 8.18$$

13. Boiling and Freezing Point Changes

The addition of a solute to a solvent will change the boiling and freezing points. If m is the molality, then the rise in the boiling point or the lowering of the freezing point is given by the following equations. K_b and K_f are from table 8.4.

$$\Delta T_b = mK_b \qquad 8.19$$

$$\Delta T_f = mK_f \qquad 8.20$$

These formulas are good only for non-electrolytic solvents. Otherwise, the actual change will be two to three times larger.

14. Solutions of Gases in Liquids

Henry's law states that the amount of gas dissolved in a liquid is proportional to the partial pressure of the gas. This applies separately to each gas to which the liquid is exposed.

Table 8.4 Boiling and Freezing Point Constants

solvent	boiling point, °C	K_b	freezing point, °C	K_f
water	100.0	.512	0.0	1.86
acetic acid	118.5	3.07	16.7	3.9
benzene	80.2	2.53	5.5	5.12
camphor	208.3	5.95	178.4	40.0
chloroform	60.2	3.63	63.5	4.68
ethyl alcohol (ethanol)	78.3	1.22		
ethyl ether	34.4	2.02		
methyl alcohol (methanol)	64.7	0.83		
naphthalene	218.0	5.65	80.2	6.9
nitrobenzene	210.9	5.24	5.7	8.1
phenol	181.2	3.56	42.0	7.27

Example 8.9

At 20°C and 760 mm Hg, one liter of water will absorb .043 grams of oxygen or .19 grams of nitrogen. Find the masses of oxygen and nitrogen in one liter of water exposed to air at 20°C and 760 mm Hg total pressure.

Partial pressure is volumetrically weighted. Air is 20.9% oxygen by volume and the remainder is assumed to be nitrogen.

$$\text{oxygen} = (.209)(.043) = .009\ \text{g/l}$$

$$\text{nitrogen} = (.791)(.19) = .150\ \text{g/l}$$

15. Acids and Bases

An *acid* is any substance (compound) that dissociates in water into H^+ (or H_3O^+). A *base* dissociates in water and gives up OH^-. A measure of the strength of the acid or base is the number of these hydrogen or hydroxide ions in a liter of solution. Since these are large numbers, a logarithmic scale is used.

$$\text{pH} = -\log_{10}[H^+] \qquad 8.21$$

$$\text{pOH} = -\log_{10}[OH^-] \qquad 8.22$$

The relationship between pH and pOH is

$$\text{pH} + \text{pOH} = 14 \qquad 8.23$$

[X] is the *ion concentration* in moles per liter. The number of moles can be calculated from Avogadro's law by dividing the actual number of ions by 6.023 EE23. Alternately,

$$[X] = (\text{fraction ionized})(\text{molarity}) \qquad 8.24$$

Example 8.10

Calculate the concentrations of H^+, OH^-, pH, and pOH in a 4.2% ionized .010*M* ammonia solution prepared from ammonium hydroxide (NH_4OH).

$$[OH^-] = (.042)(.010) = 4.2\ \text{EE}-4\ \text{moles/liter}$$

$$\text{pOH} = -\log(4.2\ \text{EE}-4) = 3.38$$

$$pH = 14 - 3.38 = 10.62$$

$$[H^+] = 10^{-10.62} = 2.4\ \text{EE}-11\ \text{moles/liter}$$

Since $H^+ + OH^- \rightarrow H_2O$, acids and bases neutralize each other by forming water. The volumes required for complete neutralization are

$$V_{base}N_{base} = V_{acid}N_{acid} \qquad 8.25$$

If the concentrations are expressed as molarities,

$$V_{base}M_{base}\Delta_{base\ charge} = V_{acid}M_{acid}\Delta_{acid\ charge} \qquad 8.26$$

Both equations assume 100% ionization of the solute.

Compounds which form during the complete or partial neutralization of acids are known as *salts*. Such neutralization occurs through a replacement of H^+ atoms with a metal or electropositive radical. Salts which result from complete neutralization of an acid by a base (e.g., Na_2SO_4) are known as *normal salts*. Compounds containing H^+ ions (e.g., $NaHSO_4$) are known as *acid salts*.

16. Reversible Reactions

Some reactions are capable of going in either direction. Such reactions are called *reversible reactions*. They are characterized by the concurrent presence of all reactants and all products. For example, the formation of ammonia is reversible.

$$N_2 + 3H_2 \longleftrightarrow 2NH_3 + 24{,}500\ \text{calories} \qquad 8.27$$

Le Chatelier's principle specifies the direction a reaction will go when a property or condition is changed. That principle says that when a reversible reaction has reached an equilibrium, and when the reactants and products are stressed by changing the pressure, temperature or concentration, a new equilibrium will be formed in the direction which reduces the stress.

Consider the formation of ammonia. When the reaction proceeds in the forward direction, heat is given off (an *exothermic* reaction). Now, if the system is stressed by increasing the temperature, the reaction will proceed in the reverse direction because that direction will absorb heat and reduce the temperature.

For reactions that involve gases, the coefficients of the molecules can be taken as volumes. Thus, in the nitrogen-hydrogen reaction, four volumes combine to form two volumes. If the equilibrium system is stressed by increasing the pressure, the forward reaction will occur because this direction reduces the volume and pressure.

If the concentration of any substance is increased, the reaction proceeds in a direction away from the substance with the increased concentration.

If a *catalyst*[11] is introduced, the equilibrium will not be changed. However, the reaction speed is increased, and equilibriums are reached more quickly. This reaction speed is known as the *rate of reaction* or *reaction velocity*.

The *law of mass action* says that the reaction speed is proportional to the concentrations of reactants. Given the reaction in equation 8.28, the *rates of reaction* can be calculated as the products of the reactants' concentrations.

$$A + B \longleftrightarrow C + D \qquad 8.28$$

$$v_{forward} = C_1[A][B] \qquad 8.29$$

$$v_{reverse} = C_2[C][D] \qquad 8.30$$

The constants C_i are needed to obtain the proper units. For reversible reactions, the *equilibrium constant* is defined by equation 8.31. This equilibrium constant is essentially independent of pressure but is dependent on temperature.

$$K_{eq} = \frac{[C][D]}{[A][B]} \qquad 8.31$$

If the reaction is given by equation 8.32, the equilibrium constant is given by equation 8.33.

$$aA + bB \longleftrightarrow cC + dD \qquad 8.32$$

$$K_{eq} = \frac{[C]^c[D]^d}{[A]^a[B]^b} \qquad 8.33$$

If any of A, B, C, or D are solids, then their concentrations are omitted in the calculation of the equilibrium constant.

Example 8.11

Acetic acid dissociates according to the following equation. Calculate the equilibrium constant using the ion concentrations given.

$$HC_2H_3O_2 + H_2O \longleftrightarrow H_3O^+ + C_2H_3O_2^-$$

$[HC_2H_3O_2] = .09866$ moles/liter

$[H_2O] = 55.5555$ moles/liter

$[H_3O^+] = .00134$ moles/liter

$[C_2H_3O_2^-] = .00134$ moles/liter

From equation 8.31,

$$K_{eq} = \frac{(.00134)(.00134)}{(.09866)(55.5555)} = 3.3\ \text{EE}-7$$

[11]A *catalyst* is a substance which increases the reaction rate without taking part chemically in the reaction.

CHEM

For weak aqueous solutions, the concentration of water is very large and essentially constant. Therefore, it can be omitted. The new constant is known as the *ionization constant* to distinguish it from the equilibrium constant which does include the water concentration.

$$K_{ionization} = (K_{eq})[H_2O] \qquad 8.34$$

Pure water is a very weak electrolyte and it ionizes only slightly by itself.

$$2H_2O \longleftrightarrow H_3O^+ + OH^- \qquad 8.35$$

At equilibrium, the ion concentrations are

$$[H_3O^+] = EE-7$$

$$[OH^-] = EE-7$$

The ionization constant for pure water is

$$K_{ionization} = [H_3O^+][OH^-] = (EE-7)(EE-7) = (EE-14) \qquad 8.36$$

Taking logs of both sides of equation 8.36 will result in equation 8.23.

Example 8.12

A .1*M* acetic acid solution is 1.34% ionized. Find the

(a) hydrogen ion concentration
(b) acetate ion concentration
(c) un-ionized acid concentration
(d) ionization constant

(a) From equation 8.24, $[H_3O^+] = (.0134)(.1) = .00134$ moles/liter

(b) Since every hydrogen ion has a corresponding acetate ion (see example 8.10), the acetate ion concentration is also .00134 moles per liter.

(c) The concentration of un-ionized acid is

$$[HC_2H_3O_2] = (\text{fraction not ionized})(\text{molarity}) = (1 - .0134)(.1) = .09866 \text{ moles/liter}$$

(d) The ionization constant is

$$K_{ionization} = \frac{(.00134)(.00134)}{(.09866)} = 1.82\ EE-5$$

A faster way to find the ionization constant is to use the *mass action equation*.

$$K_{ionization} = \frac{(\text{molarity})(\text{fraction ionized})^2}{(1 - \text{fraction ionized})} \qquad 8.37$$

Example 8.13

Find the concentration of the hydrogen ion for a .2*M* acetic acid solution if $K_{ionization}$ is 1.8 EE−5.

Let *X* be the fraction ionized, and use equation 8.37.

$$1.8\ EE-5 = \frac{(.2)(X)^2}{1 - X}$$

If *X* is small, then $(1 - X) \approx 1$. Then $X = 9.49\ EE-3$.

From equation 8.24, the concentration is

$$[H_3O^+] = (9.49\ EE-3)(.2) = 1.9\ EE-3$$

The *common ion effect* is a form of Le Chatelier's law. This is a statement to the effect that if a salt containing a common ion is added to a weak acid or base solution, ionization will be repressed. This is a consequence of the need to have an unchanged ionization constant.

Example 8.14

What is the hydrogen ion concentration of a solution with .1 gmole of 80% ionized ammonium acetate in one liter of .1*M* acetic acid?

$$1.8\ EE-5 = \frac{[H_3O^+][C_2H_3O_2^-]}{[HC_2H_3O_2]}$$

Let $X = [H_3O^+]$. Then $[HC_2H_3O_2] = .1 - X \approx .1$

But $[C_2H_3O_2^-] = X + (.8)(.1) \approx (.8)(.1) = .08$

So, from the ionization constant equation,

$$1.8\ EE-5 = \frac{(X)(.08)}{.1}$$

$$X = 2.2\ EE-5$$

17. Solubility Product

When an ionic solid (such as silver chloride) is dissolved, it *dissociates* or *ionizes*.

$$AgCl \longleftrightarrow Ag^+ + Cl^- \qquad 8.38$$

When the equation for the equilibrium constant is written, the terms for solid components are omitted. When the equation for the ionization constant is written, the term for water concentration is also omitted. Thus, when an ionic solid is placed in water, the ionization constant will consist only of the ion concentrations. This ionization constant is known as the *solubility product*.

$$K_{sp} = [Ag^+][Cl^-] \qquad 8.39$$

Table 8.5
Solubility Products

solute	T	K_{sp}
aluminum hydroxide	15°C	4 EE−13
	18°C	1.1 EE−15
	25°C	3.7 EE−15
calcium carbonate	15°C	.99 EE−8
	25°C	.87 EE−8
calcium fluoride	18°C	3.4 EE−11
	26°C	4.0 EE−11
ferric hydroxide	18°C	1.1 EE−36
magnesium hydroxide	18°C	1.2 EE−11

CHEM

As with ionization constants, the solubility product is essentially constant for slightly soluble solutes. Any time that the product of terms exceeds the standard value of the solubility product, solute will *precipitate* until the product of the remaining ion concentrations attains the standard value. If the product is less than the standard value, the solution is not *saturated.*

Example 8.15

What is the solubility product of lead sulfate ($PbSO_4$) if its solubility is 38 mg/l?

$$PbSO_4 \longleftrightarrow Pb^{++} + SO_4^{--} \text{ (in water)}$$

The molecular weight of $PbSO_4$ is

$$\text{M.W.} = [207.19 + 32.06 + 4(16)] = 303.25$$

The number of moles of $PbSO_4$ in a liter of saturated solution is

$$\frac{.038}{303.25} = 1.25 \text{ EE}-4$$

1.25 EE−4 is also the number of moles of Pb^{++} and SO_4^{--} that will form in the solution. Therefore,

$$K_{sp} = [Pb^{++}][SO_4^{--}] = (1.25 \text{ EE}-4)^2$$

$$= 1.56 \text{ EE}-8$$

The method used in example 8.15 to find the solubility product works well with *chromates* (CRO_4^{--}), *halides* (F^-, CL^-, Br^-, I^-), *sulfates* (SO_4^{--}), and *iodates* (IO_3^-). However, *sulfides* (S^{--}), *carbonates* (CO_3^{--}) and *phosphates* (PO_4^{---}) and the salts of transition elements, such as iron, hydrolize and must be handled differently.

18. Thermochemical Reactions

The *enthalpy of formation* is the energy absorbed or released by an element or compound taking part in a reaction. The enthalpy of formation is assigned a value of zero for elements in their free states.

Reactions which give off energy during the formation of compounds are known as *exothermic reactions.* Many exothermic reactions begin spontaneously. On the other hand, *endothermic reactions* require the application of heat or other energy to start the reaction.

Table 8.6 contains enthalpies of formation for several elements and compounds. The enthalpy of formation depends on the phase of the compound. Compounds in table 8.6 are solid unless indicated to be gases (*g*) or liquids (*l*).

The enthalpy of formation of a compound formed in a chemical reaction is found by summing the enthalpies of reaction over all products and subtracting the sum of enthalpies of reaction over all reactants.

Table 8.6
Standard Enthalpy of Formation for Various Compounds in kcal/mole at 25°C

element	name	Δh_f
Al (g)	aluminum	75.0
Al_2O_3		−399.09
C	graphite	0.00
C	diamond	0.45
C (g)		171.70
C_2 (g)		234.7
CO (g)		−26.42
CO_2 (g)		−94.05
CH (g)		142.1
CH_2 (g)		95
CH_3 (g)		32.0
CH_4 (g)		−17.90
C_2H_2 (g)		54.19
C_2H_4 (g)		12.50
C_2H_6 (g)		−20.24
CCl_4 (g)		−25.5
$CHCl_4$ (g)		−24
CH_2Cl_2 (g)		−21
CH_3Cl (g)		−19.6
CS_2 (g)		27.55
COS (g)		−32.80
$(CH_3)_2S$ (g)		−8.98
CH_3OH (g)	methanol	−48.08
C_2H_5OH (g)	ethanol	−56.63
$(CH_3)_2O$ (g)	dimethyl ether	−44.3
C_3H_6 (g)	cyclopropane	9.0
C_6H_{12} (g)	cyclohexane	−29.98
C_6C_{10} (g)	cyclohexene	−1.39
C_6H_6 (g)	benzene	19.82
Fe	iron	0.00
Fe (g)		99.5
Fe_2O_3		−196.8
Fe_3O_4		−267.8
H_2 (g)	hydrogen	0.00
H_2O (g)		−57.80
H_2O (l)		−68.32
H_2O_2 (g)		−31.83
H_2S (g)		−4.82
N_2 (g)	nitrogen	0.00
NO (g)		21.60
NO_2 (g)		8.09
NO_3 (g)		13
NH_3 (g)		−11.04
O_2 (g)	oxygen	0.00
O_3 (g)		34.0
S	sulfur	0.00
SO (g)		1.4
SO_2 (g)		−70.96
SO_3 (g)		−94.45

Example 8.16

What is the heat of stoichiometric combustion of gaseous methane and oxygen at 25°C?

Balance the chemical equation and obtain enthalpies of reaction for each element and compound.

$$CH_4 + 2O_2 \rightarrow 2H_2O + CO_2$$

$$(-17.90) + 2(0) \rightarrow 2(-57.80) + (-94.05)$$

Subtract the reactants' enthalpy sum from the products' enthalpy sum.

$$2(-57.80)-94.05-(-17.90) = -191.75\frac{\text{kcal}}{\text{mole-methane}}$$

The negative sign indicates the reaction is exothermic.

19. Combustion

A. Introduction

Combustion reactions involving oxygen, combustible fuels, and combustible elements occur according to fixed stoichiometric principles. These stoichiometric principles are the same as for any other chemical reaction. Table 8.7 lists some of the more common chemical reactions.

Table 8.7
Common Combustion Reactions

Substance	Molecular Symbol	Reaction Equation
carbon (to CO)	C	$2C + O_2 = 2CO$
carbon (to CO_2)	C	$2C + 2O_2 = 2CO_2$
sulfur (to SO_2)	S	$S + O_2 = SO_2$
sulfur (to SO_3)	S	$2S + 3O_2 = 2SO_3$
carbon monoxide	CO	$2CO + O_2 = 2CO_2$
methane	CH_4	$CH_4 + 2O_2 = CO_2 + 2H_2O$
acetylene	C_2H_2	$2C_2H_2 + 5O_2 = 4CO_2 + 2H_2O$
ethylene	C_2H_4	$C_2H_4 + 3O_2 = 2CO_2 + 2H_2O$
ethane	C_2H_6	$2C_2H_6 + 7O_2 = 4CO_2 + 6H_2O$
hydrogen	H_2	$2H_2 + O_2 = 2H_2O$
hydrogen sulfide	H_2S	$2H_2S + 3O_2 = 2H_2O + 2SO_2$
propane	C_3H_8	$C_3H_8 + 5O_2 = 3CO_2 + 4H_2O$
n-butane	C_4H_{10}	$2C_4H_{10} + 13O_2 = 8CO_2 + 10H_2O$

B. Fuels

Hydrocarbon fuels are generally classified by phase—solid, liquid, or gas. Solid fuels include wood, coal, coke,[12] and other various waste products from industrial and agricultural operations.

Coals consist of volatile matter, fixed carbon, moisture, ash, and sulfur. *Volatile matter* is driven off as a vapor when the coal is heated. *Fixed carbon* is the combustible residue left after the volatile matter is burned off. Moisture is present in coal as free water and water of hydration. Oxides of sulfur contribute to air pollution and cause corrosion of combustion equipment. However, sulfur contributes to coal heat content and should be considered in the calculation of heating value.

Fuel components may be found from a proximate or ultimate anaysis. A *proximate* analysis gives the percentage by weight of moisture, volatile matter, fixed carbon, and ash (sulfur combined or given separately). The *ultimate analysis* gives percent by weight of the individual elements of carbon, hydrogen, sulfur, nitrogen, and moisture.

Moisture may be broken into hydrogen and oxygen in an ultimate analysis of coal, in which case all oxygen is assumed locked up as water. For every percent oxygen, ⅛ percent hydrogen is considered locked up as water. The remaining hydrogen is combustible.

Liquid fuels are commonly hydrocarbon products refined from crude petroleum oil. They include liquid petroleum gases (LPG), gasoline, kerosene, jet fuel, diesel fuels, and light heating oils. Heating oils are classed according to their viscosity. Number one (No. 1) oil is very light. No. 6 (also known as Bunker-C oil) is the heaviest. It is frequently used in industrial applications.

Most applications using gaseous fuels are limited to natural gas and LPG. Natural gas is a mixture of methane (55% to 95%), higher hydrocarbons (primarily ethane), and noncombustible gases. Typical heating values range from 950 to 1100 BTU/ft^3 at standard industrial STP.[13] Liquid petroleum gases are available as butane, propane, and a mixture of the two.

C. Stoichiometric Combustion

Atmospheric air is a mixture of oxygen, nitrogen, and small amounts of carbon dioxide, water vapor, argon, and other inert gases. If all constituents except oxygen are grouped with the nitrogen, the air composition is as given in table 8.8. It is necessary to supply $1 \div .2315 = 4.32$ pounds of air to obtain one pound of oxygen.

Table 8.8
Percent Oxygen and Nitrogen in Air
(Rare inert gases included as N_2)

	% by weight	% by volume
oxygen (O_2)	23.15	20.9
nitrogen (N_2)	76.85	79.1

Stoichiometric air, or *theoretical air,* is the exact quantity of air necessary to provide the required amount of oxygen for complete combustion of a fuel. Stoichiometric air requirements are usually stated on the basis of pounds of air for solid and liquid fuels, and on the basis of cubic feet of air for gaseous fuels. Stoichiometric air can be found from a standard analysis of the chemical combustion equation.

CHEM

[12]*Coke* is an industrial fuel derived from coal. Coke is produced by driving off the volatiles from coal in an oxygen-deficient atmosphere.

[13]Standard industrial STP is one atmosphere and 60°F.

Complete combustion occurs when all fuel is burned. *Excess air* is usually required to accomplish this. Excess air is expressed as a percentage by weight of the theoretical air.

D. Heat of Combustion

The *heating value* of a fuel can be determined experimentally in a *calorimeter* or can be estimated from its chemical analysis. The analysis percentage (gravimetric or volumetric) is multiplied by the heating value per unit (pound or ft^3) for each combustible element, and all values are summed. The correct percentage of free hydrogen is

$$\%H_{2,\text{free}} = \%H_2 - \frac{\%O_2}{8} \qquad 8.40$$

The *higher* or *gross heating value* of a fuel includes the heat of vaporization of water vapor formed during combustion. The *lower* or *net heating value* assumes that all the products of combustion remain gaseous.

Table 8.9
Approximate Heats of Combustion

		Heat of Combustion BTU per cu. ft.		BTU per lbm	
Substance	Formula	Gross (high)	Net (low)	Gross (high)	Net (low)
Carbon	C			14,093	14,093
Hydrogen	H_2	325	275	60,958	51,623
Carbon Monoxide	CO	322	322	4,347	4,347
Methane	CH_4	1,013	913	23,879	21,520
Ethane	C_2H_6	1,792	1,641	22,320	20,432
Propane	C_3H_8	2,590	2,385	21,661	19,944
n-Butane	C_4H_{10}	3,370	3,113	21,308	19,680
Isobutane	C_4H_{10}	3,363	3,105	21,257	19,629
Acetylene	C_2H_2	1,499	1,448	21,500	20,776
Methyl alcohol	CH_3OH	868	768	10,259	9,078
Ethyl alcohol	C_2H_5OH	1,600	1,451	13,161	11,929
Sulfur	S			3,983	3,983
Hydrogen Sulfide	H_2S	647	596	7,100	6,545

Example 8.17

What is the higher heating value of a coal sample with an ultimate analysis of 93.9% carbon, 2.1% hydrogen, 2.3% oxygen, .3% nitrogen, and 1.4% ash?

The ash, oxygen, and nitrogen do not contribute to heating value. Some of the hydrogen is locked up in the form of water. From equation 8.40, the free hydrogen available for combustion is

$$2.1\% - \frac{2.3\%}{8} = 1.8\%$$

From table 8.9, the heating values of carbon and hydrogen are 14,093 and 60,958 BTU/lbm respectively. The total heating value per pound of coal is

$$\begin{aligned} \text{HHV} &= (.939)(14{,}093) + (.018)(60{,}958) \\ &= 14{,}330 \text{ BTU/lbm} \end{aligned}$$

E. Flue Gas Analysis

The composition of a flue gas[14] can be obtained with an *Orsat apparatus*. This apparatus determines the volumetric percentages of CO_2, CO, O_2, and N_2 in the flue gases. The analysis is a dry analysis — the percentage of water vapor is not determined.

20. Electrolysis

Electrolysis is the passing of an electric current through an electrolytic[15] solution. The solution may be an aqueous solution of some soluble compound, or it may be a molten solution of a solid substance. Electrolysis is accomplished by placing both the positive terminal (*anode*) and the negative terminal (*cathode*) of a voltage source in the solution. Negative ions will be attracted to the anode where they will be *oxidized*. Positive ions will be attracted to the cathode where they will be *reduced*.

Faraday's laws of electrolysis can be used to predict the duration and magnitude of a direct current needed.

- *Law 1:* The mass of a substance generated by electrolysis is proportional to the amount of electricity used.
- *Law 2:* For any constant amount of electricity, the mass of substance is proportional to its equivalent weight.
- *Law 3:* One *Faraday* (96,500 coulombs or 96,500 amp-seconds) will produce one gram equivalent weight.

The number of grams of a substance produced at an electrode can be found from equation 8.41 or 8.42.

$$\#\text{ grams} = \frac{(\text{I})(\text{t})(\text{A.W.})}{(96{,}500)(\text{valence})} \qquad 8.41$$

$$= (\#\text{ Faradays})(\text{GEW}) \qquad 8.42$$

The number of gmoles produced is

$$\#\text{ gmoles} = \frac{\#\text{ grams produced}}{\text{A.W.}} \qquad 8.43$$

$$= (\#\text{ Faradays})/(\text{change in charge}) \qquad 8.44$$

Example 8.18

What current is required to produce two grams of metallic copper from a copper sulfate solution in 1½ hours?

[14]*Flue gases* may also be called *stack gases* or *combustion gases*.

[15]An electrolyte is a solution that partially or completely dissociates into positive and negative ions. These ions move under the influence of an electric potential. This movement constitutes current.

CHEM

The electrolysis reaction is

$$Cu^{++} + 2e^- \rightarrow Cu$$

Since the change in charge on the copper ion is 2, the equivalent weight is

$$GEW = \frac{63.6}{2} = 31.8$$

From equation 8.41,

$$2 = \frac{(I)(1.5)(3600)(31.8)}{96{,}500}$$

The current is I = 1.12 amps.

21. Corrosion

Corrosion is a chemical or physical reaction with parts of the environment. Conditions within the aggregate or crystalline structures can amplify or inhibit corrosion. Three categories of corrosion are galvanic action, stress corrosion, and fretting corrosion.

A. Galvanic Action

Galvanic action results from the electro-chemical variation in the potential of metallic ions. If two metals of different potentials are placed in an electrolytic medium, the one with the higher potential will act as an anode and will corrode. The metal with the lower potential, being the cathode, will be unchanged. Metals are often classified according to their position in the *galvanic series.* Several metals are listed in table 8.10 in the order of their anodic-cathodic characteristics.

Table 8.10
The Galvanic Series
(Anodic to Cathodic)

magnesium
zinc
aluminum
cadmium
steel
cast iron
lead
tin
copper
nickel
titanium
gold

B. Stress Corrosion

When subjected to sustained surface tensile stresses in the presence of corrosive environments, certain metals exhibit *stress corrosion* cracking. This cracking can lead to failure at stresses well below normal working stresses.

Stress corrosion occurs because the more highly-stressed fibers or grains are slightly more anodic than neighboring fibers with lower stresses. The cracks are intergranular and propagate through grain boundaries until failure occurs. One extreme type of stress corrosion is called *exfoliation,* in which open end-grains separate into layers.

C. Fretting Corrosion

Fretting corrosion occurs when two highly loaded members have a common surface at which differential movement takes place. The phenomenon is a combination of wear and chemical corrosion. Metals which depend on a surface oxide for protection, such as aluminum, are especially susceptible.

22. Qualities of Supply Water

A. Acidity and Alkalinity

Acidity is a measure of acids in solution. Acidity in surface water is caused by formation of *carbonic acid* from carbon dioxide in the air. Carbonic acid is very aggressive and must be neutralized to eliminate water pipe corrosion.

If the pH of water is greater than 4.5, carbonic acid ionizes to form bicarbonate (equation 8.46). If the pH is greater than 8.3, carbonate ions form which cause water hardness by combining with calcium (equation 8.47).

$$CO_2 + H_2O \rightarrow H_2CO_3 \quad 8.45$$

$$H_2CO_3 + H_2O \rightarrow HCO_3^- + H_3O^+ \ (pH > 4.5) \quad 8.46$$

$$HCO_3^- + H_2O \rightarrow CO_3^{--} + H_3O^+ \ (pH > 8.3) \quad 8.47$$

Measurement of acidity is done by *titration* with a standard basic measuring solution. Acidity in water is typically given in terms of the concentration of $CaCO_3$ (in mg/l) that would neutralize the acid. Alkalinity is also measured in terms of mg/l of $CaCO_3$ by using an acidic titrant.

Alkalinity and acidity of a titrated sample is determined from color changes in indicators[16] added to the titrant. Some common indicators are listed in table 8.11.

B. Hardness

Water hardness is caused by doubly-charged (but not singly- or triply-charged) positive metallic ions such as calcium, magnesium, and iron. Calcium and magnesium are found in the greatest quantities. Hardness reacts with soap to reduce its effectiveness and to form scum on the surface of water and ring around the bathtub.

[16]The simplest colorimetric indicator is litmus paper. Litmus paper turns red in acidic solutions. It turns blue in basic solutions.

CHEM

Table 8.11
Colorimetric Indicators

This titrant	turns from (basic)	to (less basic)	at pH
phenolphthalein	pink	clear	10.0 to 8.3
methyl orange	orange	pink	4.4 to 4.0
eriochrome black T	wine red	blue	10 to 7
bromocresol green/with methyl red	gray	pink	4.8 to 4.6
thymol blue	yellow	red	2.8 to 1.2
methyl red	yellow	red	6.3 to 4.2
bromthymol blue	blue	yellow	7.6 to 6.0
cresol red	red	yellow	8.8 to 7.2
alizarin yellow	red	yellow	12.0 to 10.1

Water containing *bicarbonate* ions (HCO_3^-) can be heated to precipitate a carbonate molecule. This hardness is known as *temporary hardness* or *carbonate hardness.*

$$Ca^{++} + 2HCO_3^- + \text{heat} \rightarrow CaCO_3\downarrow + CO_2 + H_2O \quad 8.48$$

$$Mg^{++} + 2HCO_3^- + \text{heat} \rightarrow MgCO_3\downarrow + CO_2 + H_2O \quad 8.49$$

(The symbol $\downarrow$ means that the compound precipitates until the water is saturated.)

Remaining hardness due to sulfates, chlorides, and nitrates is known as *permanent hardness* or *non-carbonate hardness* because it cannot be removed by heating. Permanent hardness may be calculated numerically by causing precipitation, drying, and then weighing the precipitate.

$$Ca^{++} + SO_4^{--} + Na_2CO_3 \rightarrow 2Na^+ + SO_4^{--} + CaCO_3\downarrow \quad 8.50$$

$$Mg^{++} + 2Cl^- + 2NaOH \rightarrow 2Na^+ + 2Cl^- + Mg(OH)_2\downarrow \quad 8.51$$

23. Water Softening

A. Lime and Soda Ash Softening

Water softening can be accomplished with lime and soda ash to precipitate calcium and magnesium ions from the solution. Lime treatment has added benefits of disinfection, iron removal, and clarification. Practical limits of *precipitation softening* are 30 mg/l of $CaCO_3$ and 10 mg/l of $Mg(OH)_2$ (as $CaCO_3$) because of intrinsic solubilities. Water treated by this method usually leaves the softening apparatus with a hardness of between 50 and 80 mg/l as $CaCO_3$.

Lime (CaO) is available as granular *quicklime* (90% CaO, 10% MgO) or *hydrated lime* (68% CaO, 32% water). Both forms are *slaked* prior to use, which means that water is added to form a lime slurry in an exothermic reaction.

$$CaO + H_2O \rightarrow Ca(OH)_2 + \text{heat} \quad 8.52$$

Soda ash is usually available as 98% pure sodium carbonate (Na_2CO_3).

In the *first stage treatment*[17], lime added to water reacts with free carbon dioxide to form calcium carbonate precipitate.

$$CO_2 + Ca(OH)_2 \rightarrow CaCO_3\downarrow + H_2O \quad 8.53$$

Next, the lime reacts with calcium bicarbonate.

$$Ca(HCO_3)_2 + Ca(OH)_2 \rightarrow 2CaCO_3\downarrow + 2H_2O \quad 8.54$$

Magnesium hardness is next removed.

$$Mg(HCO_3)_2 + Ca(OH)_2 \rightarrow CaCO_3\downarrow + 2H_2O + MgCO_3 \quad 8.55$$

To remove the soluble $MgCO_3$, the pH must be above 10.8. This is accomplished by adding an excess of CaO or $Ca(OH)_2$.

$$MgCO_3 + Ca(OH)_2 \rightarrow CaCO_3\downarrow + Mg(OH)_2\downarrow \quad 8.56$$

B. Ion Exchange Softening

In the *ion exchange process,* (also known as *zeolite process* or *base exchange method*) water is passed through a filter bed of exchange material. This exchange material is known as *zeolite.* Ions in the insoluble exchange material are displaced by ions in the water. When the exchange material is spent, it is regenerated with a rejuvenating solution such as sodium chloride (salt).

The processed water will have a zero hardness. However, since there is no need for water with zero hardness, some water is usually bypassed around the unit.

There are three types of ion exchange materials. *Greensand (glauconite)* is a natural substance which is mined and treated with manganese dioxide. *Siliceous-gel zeolite* is an artificial solid used in small volume deionizer columns. *Polystyrene resins* are also synthetic. Polystyrene resins currently dominate the softening field.

During operation, the calcium and magnesium ions are removed according to the following reaction in which *R* is the zeolite anion.

[17]First stage softening removes the carbonate hardness. A second softening process is necessary if there is *non-carbonate hardness* resulting from sulfates and chlorides.

CHEM

$$\left\{\begin{matrix} Ca \\ Mg \end{matrix}\right\} \left\{\begin{matrix} (HCO_3)_2 \\ SO_4 \\ Cl_2 \end{matrix}\right\} + Na_2R \rightarrow Na_2 \left\{\begin{matrix} (HCO_3)_2 \\ SO_4 \\ Cl_2 \end{matrix}\right\} + \left\{\begin{matrix} Ca \\ Mg \end{matrix}\right\} R \qquad 8.57$$

The resulting sodium compounds are soluble.

Example 8.19

A municipal plant processes water with a total initial hardness of 200 mg/l. The designed discharge hardness is 50 mg/l. If an ion exchange unit is used, what is the bypass factor?

The water passing through the ion exchange unit is reduced to zero hardness. If x is the water fraction bypassed around the zeolite bed,

$$(1-x)0 + x(200) = 50$$

$$x = .25$$

24. *Wastewater Quality Characteristics*

A. *Dissolved Oxygen*

Aquatic life requires oxygen. The biological decomposition of organic solids is also dependent on oxygen. If the *dissolved oxygen* content of water is less than saturated, it is likely that the water is organically polluted.

B. *Biological Oxygen Demand*

Biological organisms remove oxygen from water when they oxidize organic waste matter. Therefore, the oxygen depletion is an indication of the organic waste content. The more oxygen that is removed over a given period, the greater will be the amount of organic food present in the water.

Biological oxygen demand (BOD) is determined by adding a measured amount of wastewater (which supplies the organic material) to a measured amount of dilution water (which reduces toxicity and supplies dissolved oxygen). An oxygen use curve similar to figure 8.4 will result. The standard BOD test typically calls for a 5-day incubation period at 20°C.

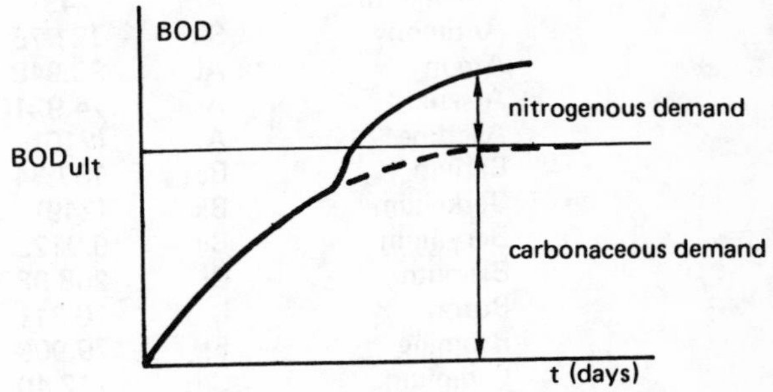

Figure 8.4 BOD Time Curve

C. *Chemical Oxygen Demand*

Unlike BOD which is a measure of oxygen removed by biological organisms, chemical oxygen demand (COD) is a measure of the total oxidizable substances. COD is a good measure of effluent strength when chemical contamination is present.

D. *Chlorine Demand*

Chlorination destroys bacteria, hydrogen sulfide, and other noxious substances by oxidation. For example, hydrogen sulfide is oxidized according to equation 8.58.

$$H_2S + 4H_2O + 4Cl_2 \rightarrow H_2SO_4 + 8HCl \qquad 8.58$$

The *chlorine demand* is the amount of chlorine required to give a 2 mg/l residual after fifteen minutes of contact time. Fifteen minutes is the recommended contact and mixing time prior to discharge since this period will kill nearly all pathogenic bacteria in the water.

The chlorine dose applied to waste water is frequently determined by the water's breakpoint. *Breakpoint chlorination* implies that chlorine is added to the water until free chlorine residuals begin to appear.

Appendix A
Atomic Weights of Elements Referred to Carbon (12)

Element	Symbol	Atomic Weight	Element	Symbol	Atomic Weight
Actinium	Ac	(227)	Mercury	Hg	200.59
Aluminum	Al	26.9815	Molybdenum	Mo	95.94
Americium	Am	(243)	Neodymium	Nd	144.24
Antimony	Sb	121.75	Neon	Ne	20.183
Argon	Ar	39.948	Neptunium	Np	(237)
Arsenic	As	74.9216	Nickel	Ni	58.71
Astatine	At	(210)	Niobium	Nb	92.906
Barium	Ba	137.34	Nitrogen	N	14.0067
Berkelium	Bk	(249)	Osmium	Os	190.2
Beryllium	Be	9.0122	Oxygen	O	15.9994
Bismuth	Bi	208.980	Palladium	Pd	106.4
Boron	B	10.811	Phosphorus	P	30.9738
Bromine	Br	79.909	Platinum	Pt	195.09
Cadmium	Cd	112.40	Plutonium	Pu	(242)
Calcium	Ca	40.08	Polonium	Po	(210)
Californium	Cf	(251)	Potassium	K	39.102
Carbon	C	12.01115	Praseodymium	Pr	140.907
Cerium	Ce	140.12	Promethium	Pm	(145)
Cesium	Cs	132.905	Protactinium	Pa	(231)
Chlorine	Cl	35.453	Radium	Ra	(226)
Chromium	Cr	51.996	Radon	Rn	(222)
Cobalt	Co	58.9332	Rhenium	Re	186.2
Copper	Cu	63.54	Rhodium	Rh	102.905
Curium	Cm	(247)	Rubidium	Rb	85.47
Dysprosium	Dy	162.50	Ruthenium	Ru	101.07
Einsteinium	Es	(254)	Samarium	Sm	150.35
Erbium	Er	167.26	Scandium	Sc	44.956
Europium	Eu	151.96	Selenium	Se	78.96
Fermium	Fm	(253)	Silicon	Si	28.086
Fluorine	F	18.9984	Silver	Ag	107.870
Francium	Fr	(223)	Sodium	Na	22.9898
Gadolinium	Gd	157.25	Strontium	Sr	87.62
Gallium	Ga	69.72	Sulfur	S	32.064
Germanium	Ge	72.59	Tantalum	Ta	180.948
Gold	Au	196.967	Technetium	Tc	(99)
Hafnium	Hf	178.49	Tellurium	Te	127.60
Helium	He	4.0026	Terbium	Tb	158.924
Holmium	Ho	164.930	Thallium	Tl	204.37
Hydrogen	H	1.00797	Thorium	Th	232.038
Indium	In	114.82	Thulium	Tm	168.934
Iodine	I	126.9044	Tin	Sn	118.69
Iridium	Ir	192.2	Titanium	Ti	47.90
Iron	Fe	55.847	Tungsten	W	183.85
Krypton	Kr	83.80	Uranium	U	238.03
Lanthanum	La	138.91	Vanadium	V	50.942
Lead	Pb	207.19	Xenon	Xe	131.30
Lithium	Li	6.939	Ytterbium	Yb	173.04
Lutetium	Lu	174.97	Yttrium	Y	88.905
Magnesium	Mg	24.312	Zinc	Zn	65.37
Manganese	Mn	54.9380	Zirconium	Zr	91.22
Mendelevium	Md	(256)			

CHEM

Appendix B
Common Acids

Acetic	CH_3COOH
Acrylic	C_2H_3COOH
Benzene Sulfonic	$C_6H_5SO_3H$
Benzoic	C_6H_5COOH
Butyric	C_3H_7COOH
Carbolic	C_6H_5OH
Carbonic	H_2CO_3
Chloric	$HClO_3$
Formic	$HCOOH$
Hydrobromic	HBr
Hydrochloric	HCl
Hydrosulfuric	H_2S
Nitric	HNO_3
Oleic	$C_{17}H_{33}COOH$
Oxalic	$H_2C_2O_4$
Perchloric	$HClO_4$
Phenol	C_6H_5OH
Phosphoric	H_3PO_4
Propionic	C_2H_5COOH
Stearic	$C_{17}H_{35}COOH$
Sulfuric	H_2SO_4
Sulfurous	H_2SO_3
Valeric	C_4H_9COOH

CHEM

NOTE: All moles are gram-moles. Scientific STP is 0°C and one atmosphere.

DEFINITIONS

1. Define the following terms:
   (a) halogen
   (b) alkali earth metals
   (c) catalyst
   (d) base
   (e) hydroxyl ion
   (f) ozone
   (g) hydrolysis
   (h) sublimation
   (i) distillation
   (j) water hardness
   (k) coke
   (l) Orsat method
   (m) stoichiometric air
   (n) Dulong and Petit law
   (o) Biological Oxygen Demand (BOD)
   (p) titration
   (q) bubble point
   (r) hydration
   (s) precipitation
   (t) electrolyte
   (u) salt
   (v) solution
   (w) suspension
   (x) emulsion
   (y) scale (as a noun)
   (z) allotrope
   (aa) gram atom
   (ab) alcohol
   (ac) alkane
   (ad) alkene
   (ae) alkyne
   (af) aromatic hydrocarbon
   (ag) phenol
   (ah) ether
   (ai) aldehyde
   (aj) ketone
   (ak) carboxylic acid
   (al) halide
   (am) anhydride
   (an) amide
   (ao) ester
   (ap) amino acid
   (aq) carbohydrate
   (ar) endothermic reaction
   (as) exothermic reaction
   (at) metalloid
   (au) flocculation
   (av) anaerobic
   (aw) isomer
   (ax) aqueous solution
   (ay) buffer
   (az) Brownian movement

IDEAL GAS LAW

2. What is the universal gas constant in units of liters, moles, and °K?

3. 287.5 ml of vapor at 100°C and 753 mm Hg has a mass of .725 gram. What is the molecular weight of the vapor?

4. A gas sample occupies 300 ml at 20°C and 700 mm Hg. What volume will the sample occupy at 5°C and 740 mm Hg?

5. The density of helium is .178 grams/liter at STP. What is the density at 25°C and 740 mm Hg?

6. At what temperature will one liter of oxygen weigh one gram if the pressure is 760 mm Hg?

7. At STP, the density of chlorine is 3.22 grams/liter. What weight of this gas is contained in a flask of 100 ml at 24°C and 750 mm Hg?

ATOMS, MOLECULES, AND MOLES

8. If a sample of calcium nitrate contains 20 grams of nitrogen, how many grams of calcium are contained in the sample?

9. What is the percentage of calcium oxide (CaO) in calcium carbonate ($CaCO_3$)?

10. How many moles of ammonia are there in 125 liters at STP?

11. What is the STP volume of 49 grams of chlorine?

12. How many molecules are there in
    (a) 57 grams of oxygen?
    (b) 15 liters of carbon dioxide at STP?
    (c) 9 liters of hydrogen at 25°C and 770 mm Hg?

13. .1225 grams of a gas occupy 110 ml when measured over water at 22°C and 743 mm Hg. What is the molecular weight of the gas?

EQUIVALENT WEIGHTS

14. Assume complete neutralization and calculate the equivalent weights for each of the following:
    (a) HBr
    (b) H_2SO_3
    (c) H_3PO_4
    (d) LiOH

15. How many equivalent weights of $H_4P_2O_7$ should be used to prepare 400 grams of $Na_2H_2P_2O_7$?

CHEM

16. What is the equivalent weight of the element in each of the following reactions?
(a) $Na^+ + e^- \rightarrow Na$
(b) $Ag \rightarrow Ag^+ + e^-$
(c) $Cu^{++} + 2e^- \rightarrow Cu$

PERIODIC TABLE

17. For the hypothetical element 118, predict the following:
(a) the atomic weight
(b) the melting point
(c) the boiling point

18. Describe the characteristics of the compound formed when chlorine combines with the hypothetical element 119.

19. Which of the following are metalloids?
(a) magnesium
(b) aluminum
(c) xenon
(d) silicon
(e) phosphorous

BALANCING CHEMICAL EQUATIONS

20. Complete and balance the following chemical reactions. What kind of reaction is each?

(a) $CH_4 + Cl_2 \rightarrow C + ?$
(b) $AgNO_3 + HCl \rightarrow HNO_3 + ?$
(c) $AsCL_3 + H_2S \rightarrow As_2S_3 + HCl$
(d) $Cu_2O + Cu_2S \rightarrow Cu + SO_2$
(e) $B_2O_3 + Mg \rightarrow MgO + B$
(f) $BaSO_4 + C \rightarrow BaS + CO$
(g) $Li_2O + P_2O_5 \rightarrow Li_3PO_4$
(h) $H_2SO_4 + Ba(OH)_2 \rightarrow H_2O + ?$
(i) $2HNO_3 + CaO \rightarrow Ca(NO_3)_2 + ?$
(j) $H_3PO_4 + MgCO_3 \rightarrow Mg_3(PO_4)_2 + H_2O + ?$

WEIGHTS AND VOLUMES

21. What weight of Na_2SO_4 (83.4 % pure) can be formed from 250 pounds of 94.5% pure salt?

$$2NaCl + H_2SO_4 \rightarrow Na_2SO_4 + 2HCl$$

22. How many gallons of ocean water must be processed to produce one pound of bromine if bromine constitutes 65 ppm by weight in sea water? Assume 100% recovery and a specific gravity of one for sea water.

23. Consider the following reaction:

$$H_2S + 2NH_4OH \rightarrow (NH_4)_2S + 2H_2O$$

(a) How many moles of water are produced from 9 grams of hydrogen sulfide?
(b) How many STP liters of hydrogen sulfide are needed to produce 8 grams of water?

24. Consider the following reaction:

$$H_2SO_4 + 2NaOH \rightarrow Na_2SO_4 + 2H_2O$$

(a) How many moles of sulfuric acid are needed to produce 4 moles of water?
(b) How many grams of sodium sulfate are produced from 18 grams of sulfuric acid?
(c) How many grams of water are produced from 100 grams of reactants?
(d) How many moles of products result from 9 grams of reactants?

EMPIRICAL FORMULA DEVELOPMENT

25. A compound is analyzed. What is the simplest molecular formula if the gravimetric analysis is as follows?
(a) 79.9% Cu, 20.1% O
(b) 46.56% Fe, 53.44% S
(c) 63.53% Fe, 36.47% S
(d) 85.62% C, 14.38% H
(e) 40% C, 6.7% H, 53.3% O
(f) 15.89% Ca, 2.4% H, 24.6% P, 57.11% O, 7.14% H_2O

COMPOUND ANALYSIS

26. A silver coin weighing 5.82 grams is completely dissolved in nitric acid. Sodium chloride is added to precipitate the silver as silver chloride. The precipitate weighs 7.2 grams. What was the coin's purity?

27. A sample of copper weighing 10 grams is heated in sulfuric acid. After the copper is dissolved, water is added. The solution is evaporated to dryness and 30 grams of $CuSO_4 \cdot 5H_2O$ are obtained. What was the copper purity?

COMBUSTION

28. Assuming that the furnace efficiency is 50%, how many kg of water can be heated from 15°C to 95°C by burning 200 standard liters of methane? The heating value of methane is 24,000 BTU/lbm.

29. 15 pounds of propane (C_3H_8) are burned in air each hour stoichiometrically. How many cubic feet of dry CO_2 are formed after cooling to 70°F and 14.7 psia?

30. How many pounds of nitrogen pass through a furnace that is burning 4000 cfh of methane? Assume 30% excess air (15 psia, 100°F) is needed for complete combustion.

31. How many pounds of air are required to burn one pound of a fuel which is 84% carbon, 15.3% hydrogen, .3% sulfur, and .4% nitrogen? Assume complete combustion.

FLUE GAS

32. Propane (C_3H_8) is burned with 20% excess air. What percent of CO_2 by weight will be in the flue gas?

CHEM

OXIDATION–REDUCTION (optional)

33. For the following reactions, indicate which element is oxidized, which is reduced, and the number of electrons lost or gained by each element.

(a) $H_2 + Cl_2 \rightarrow 2HCl$
(b) $Zn + H_2SO_4 \rightarrow H_2 + ZnSO_4$
(c) $2KBr + CL_2 \rightarrow Br_2 + 2KCl$

34. Balance the following reactions :

(a) $Zn + N_2 \rightarrow Zn_3N_2$
(b) $FeCl_3 + SnCl_2 \rightarrow FeCl_2 + SnCl_4$

35. Balance and complete the following reactions:

(a) $FeCl_2 + Cl_2 \rightarrow$
(b) $H_2SO_4 + KMnO_4 \rightarrow$
(c) $Al + H_2SO_4 \rightarrow$
(d) $Fe + Ag_2SO_4 \rightarrow$

THERMOCHEMISTRY

36. What is the heating value in kcal/liter for complete combustion of a fuel with the following volumetric analysis? 6% CO_2, 22% CO, 12% H_2, 60% N_2

37. What is the approximate flame temperature of carbon monoxide burned with 25% excess air?

SOLUTIONS

38. Find the molarity and molality of 100 mls of a solution containing 13.5 grams of sucrose if the solution has a density of 1.05 g/ml.

39. A mixture of 10 grams of acetic acid in 125 grams of water is produced. What is the molality? What are the mole fractions?

40. It is desired to have 60 liters of .5N solution. How many kilograms of wet NaOH (12% water by weight) are required?

41. The specific gravity of a 10% (by weight) calcium chloride solution is 1.0835. Water weighs 62.43 pounds per cubic foot and there are 7.48 gallons per cubic foot. Calculate the number of pounds of calcium chloride needed to make 55 gallons of 10% (by weight) solution.

42. How many grams of solute are contained in
(a) 2.5 liters of 2M H_2SO_4?
(b) 5 liters of .525M $Ba(OH)_2$?
(c) 350 mls of .5N $Al_2(SO_4)_3$?
(d) 25 mls of 1.9 N $KMnO_4$?

pH

43. What is the pH if $[H_3O^+]$ is 3 EE-3 g-ion/liter?

44. What is $[H_3O^+]$ and $[OH^-]$ if pH is 6.5?

45. What is the pH of a .5N hydrochloric acid that is 93% ionized?

46. What is pOH if $[H_3O^+]$ is 5 EE-6 g-ion/liter?

IONIZATION CONSTANTS

47. What is the pH of a solution containing 0.01 M acetic acid and 0.01 M $NaC_2H_3O_2$ (100% ionized)? The ionization constant for acetic acid is 1.8 EE-5.

48. A .005M acid solution has a pH of 5. What percentage of this solution is ionized?

49. What is the ionization constant of a 1.0M ammonia solution that is 0.4% ionized?

50. What is $[OH^-]$ of a 0.05M ammonia solution?

BOILING AND FREEZING POINTS

51. How many kilograms of methyl alcohol must be added to 20 liters of water to depress the freezing point 12°C?

52. How many degrees (°C) will the freezing point of 15 liters of water be lowered by adding 5 kg of glycol?

53. When 1.15 grams of naphthalene are dissolved in 100 grams of benzene, the resulting solution has a freezing point of 4.95°C. What is the molecular weight of naphthalene?

54. How many degrees will the boiling point of 2 quarts of water be raised by the addition of 5 grams of salt?

NEUTRALIZATION

55. 65.8 ml of 3 N HCl are used to titrate 50 ml of Na_2CO_3 solution with a specific gravity of 1.25. What is the gravimetric fraction of Na_2CO_3 in the solution?

$$CO_3^{-2} + 2H^+ \rightarrow CO_2 + H_2O$$

56. 20 ml of H_2SO_4 solution is titrated with 38.3 ml of .103 F NaOH. What is the formality of the acid?

57. In order to standardize 25 ml of HCl solution, 24.3 ml of .1035 F NaOH is required. What is the formality of the acid?

ELECTROCHEMISTRY

58. 2 grams of silver are deposited from $AgNO_2$ with some quantity of electricity. How much tin will the same amount of electricity deposity from $SnCl_2$?

59. How long will it take to deposit 100 grams of aluminum (at the cathode only) from an electrolytic cell containing Al_2O_3 if the current is 125 amps?

60. A current of 2.5 amps is passed through a solution of sulfuric acid for one hour. The hydrogen and oxygen liberated at the electrodes are collected and dried. What is the volume of the mixed gases at 25°C and 780 mm Hg?

GAS MIXTURES

61. Ethylene bromide ($C_2H_4Br_2$) and 1,2-dibromopropane ($C_3H_6Br_2$) form ideal solutions when the two are mixed. At 85°C, the vapor pressures of the two pure liquids are 173 mm Hg and 127 mm Hg respectively. If 10 grams of ethylene bromide are dissolved in 80 grams of 1,2-dibromopropane, what are the partial pressures of each?

62. The vapor pressure of 20°C water is approximately 17.54 mm Hg. If 57 grams of sucrose are dissolved in 500 grams of 20°C water, the vapor pressure is lowered by 0.092 mm Hg. What is the molecular weight of sucrose?

SOLUBILITY PRODUCT (optional)

63. The solubility of $BaCO_3$ is 1.4 EE-3 g/100 ml. What is the solubility product constant?

64. The solubility product constant of lead fluoride is 3.2 EE-8. What are the concentrations of Pb^{++} and F^- in a saturated solution?

65. The solubility product constant of silver chromate is 1.1 EE–12. Determine the minimum concentration of CrO_4^{--} required to precipitate Ag_2CrO_4 in 2 ml of 0.5 $AgNO_3$ solution. Assume complete ionization of the silver nitrate.

9 Statics

PART 1: Determinate Structures

Nomenclature

a	horizontal distance from point of maximum sag	ft
A	area	ft^2
c	parameter in catenary equations	ft
d	distance	ft
E	modulus of elasticity	lb/ft^2
F	vertical force	lb
H	horizontal component of tension	lb
I	moment of inertia	ft^4
J	polar moment of inertia	ft^4
k	radius of gyration	ft
L	length, cable length from point of maximum sag	ft
M	moment, or mass	ft-lb, slugs
p	pressure	lb/ft^2
P	load, or product of inertia	lb, ft^4
S	maximum cable sag	ft
T	tension, or temperature	lb, °F
w	load per unit weight	lb/ft

Symbols

α	coefficient of linear thermal expansion	1/°F
δ	deflection	ft

Subscripts

c	centroidal, or concrete
i	the ith component
R	resultant
s	steel

1. Concentrated Forces and Moments

Forces are vector quantities having magnitude, direction, and location in 3-dimensional space. The direction of a force **F** is given by its *direction cosines,* which are cosines of the true angles made by the force vector with the *x, y,* and *z* axes. The components of the force are given by equations 9.1, 9.2 and 9.3.

$$\mathbf{F_x} = \mathbf{F}(\cos\theta_x) \quad 9.1$$

$$\mathbf{F_y} = \mathbf{F}(\cos\theta_y) \quad 9.2$$

$$\mathbf{F_z} = \mathbf{F}(\cos\theta_z) \quad 9.3$$

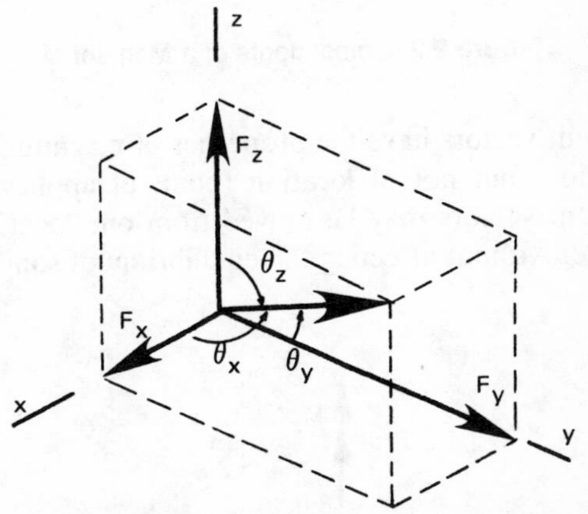

Figure 9.1 Components of a Force **F**

A force which would cause an object to rotate is said to contribute a *moment* to the object. The magnitude of a moment can be found by multiplying the magnitude of the force times the appropriate moment arm. That is, **M** = **F**·d.

The moment arm is a perpendicular distance from the force's line of application to some arbitrary reference point. This reference point should be chosen to elimi-

STATICS

nate one or more unknowns. This can be done by choosing the reference as a point at which unknown reactions are applied.

Moments can also be treated as vector quantities, and they are shown as double-headed arrows. Using the right-hand rule as shown below, the direction cosines are again used to give the *x*, *y*, and *z* components of a moment vector.

$$\mathbf{M_x} = \mathbf{M}(\cos\theta_x) \qquad 9.4$$

$$\mathbf{M_y} = \mathbf{M}(\cos\theta_y) \qquad 9.5$$

$$\mathbf{M_z} = \mathbf{M}(\cos\theta_z) \qquad 9.6$$

$$|\mathbf{M}| = \sqrt{\mathbf{M_x^2} + \mathbf{M_y^2} + \mathbf{M_z^2}} \qquad 9.7$$

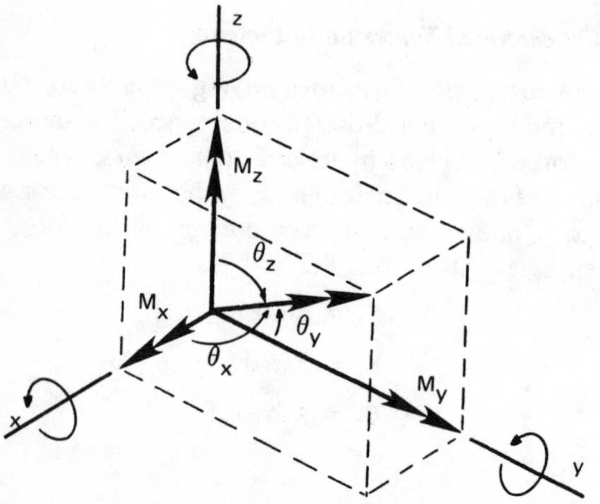

Figure 9.2 Components of a Moment **M**

Moment vectors have the properties of magnitude and direction, but not of location (point of application). Moment vectors may be moved from one location to another without affecting the equilibrium of solid bodies.

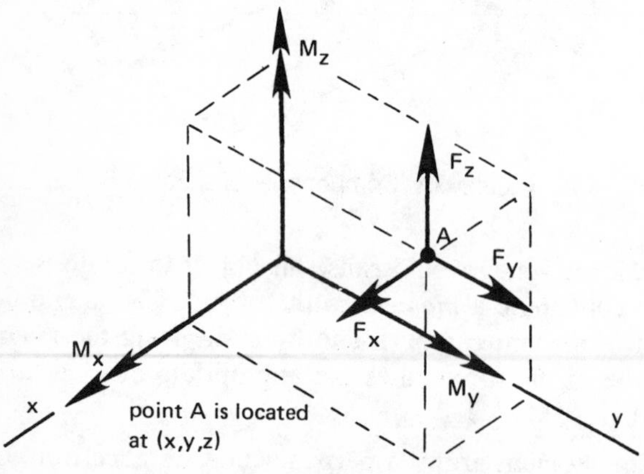

Figure 9.3 Coordinates of a Point A

If a force is not parallel to an axis, it produces a moment around that axis. The moment is evaluated by finding the components of the force and their respective distances to the axis. In figure 9.3, a force acts through point *A* located at (x,y,z) and produces moments given by equations 9.8, 9.9, and 9.10.

$$\mathbf{M_x} = \mathbf{yF_z} - \mathbf{zF_y} \qquad 9.8$$

$$\mathbf{M_y} = \mathbf{zF_x} - \mathbf{xF_z} \qquad 9.9$$

$$\mathbf{M_z} = \mathbf{xF_y} - \mathbf{yF_x} \qquad 9.10$$

Any two equal, opposite, and parallel forces constitute a *couple*. A couple is statically equivalent to a single moment vector. In figure 9.4, the two forces $\mathbf{F_1}$ and $\mathbf{F_2}$ of equal magnitude produce a moment vector $\mathbf{M_z}$ of magnitude **Fy**. The two forces can be replaced by this moment vector which can then be moved to any location on the object.

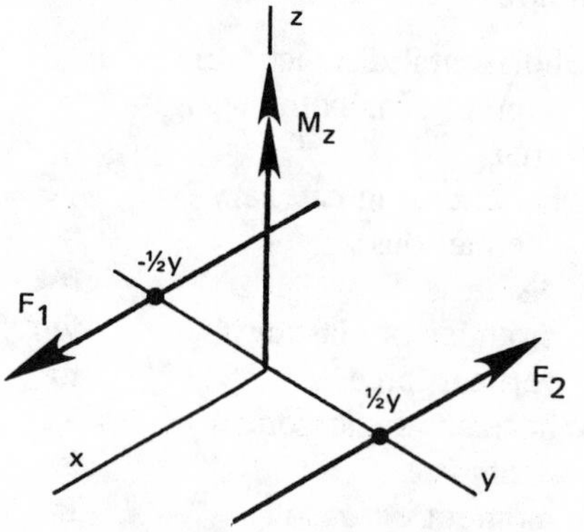

Figure 9.4 A Couple

2. *Distributed Loads*

If an object is loaded by its own weight or by another type of continuous loading, it is said to be subjected to a *distributed* load. Provided that the load per unit length, *w*, is acting in the same direction everywhere, the statically equivalent concentrated load can be found from equation 9.11 by integrating over the line of application.

$$\mathbf{F_R} = \int \mathbf{w\ dx} \qquad 9.11$$

The location of the resultant is given by equation 9.12.

$$\bar{x} = \frac{\int (\mathbf{wx})\ \mathbf{dx}}{\mathbf{F_R}} \qquad 9.12$$

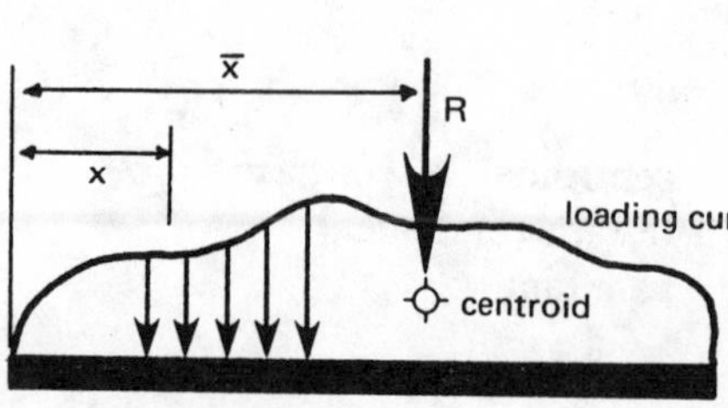

Figure 9.5 A Distributed Load and Resultant

In the case of a straight beam under *transverse*[1] loading, the magnitude of **F** equals the area under the loading curve. The location of the resultant **F** coincides with the centroid of that area. If the distributed load is uniform so that w is constant along the beam, then

$$F = wL \quad 9.13$$

$$\bar{x} = \tfrac{1}{2}L \quad 9.14$$

If the distribution is triangular and increases to a maximum of w pounds per unit length as x increases, then

$$F = \tfrac{1}{2}wL \quad 9.15$$

$$\bar{x} = \tfrac{2}{3}L \quad 9.16$$

Example 9.1

Find the magnitude and location of the resultant of the distributed loads on each span of the beam shown.

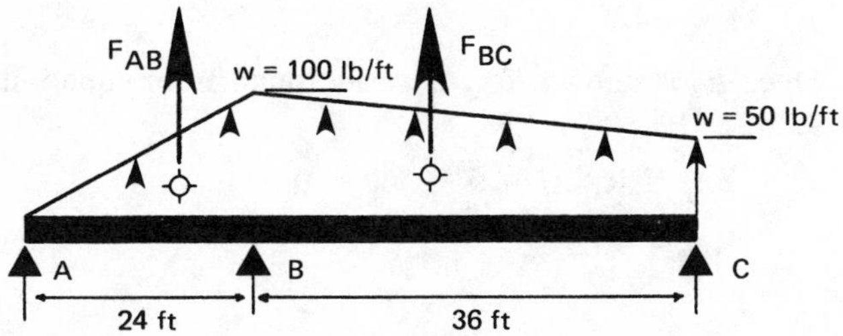

For span A-B: The area under the loading curve is ½(100)(24) = 1200 pounds. The centroid of the loading triangle is (⅔)(24) = 16 feet from point A. Therefore, the triangular load on the span A-B can be replaced (for the purposes of statics) with a concentrated load of 1200 pounds located 16 feet from the left end.

For span B-C: The area under the loading curve is

$$(50)(36) + \tfrac{1}{2}(50)(36) = 2700 \text{ lbs}$$

The centroid of the trapezoid is

$$\frac{36[(2)(100) + 50]}{3(100+50)} = 20 \text{ ft from pt } C$$

Therefore, the distributed load on span B-C can be replaced (for the purposes of statics) with a concentrated load of 2700 pounds located 20 feet to the left of point C.

3. Pressure Loads

Hydrostatic pressure is an example of a pressure load that is distributed over an area. The pressure is denoted as p pounds per unit area of surface. It is normal to the surface at every point. If the surface is plane, the statically equivalent concentrated load can be found by integrating over the area. The resultant is numerically equal to the average pressure times the area. The point of application will be the centroid of the area over which the integration was performed.

[1]Loading is said to be *transverse* if its line of action is perpendicular to the length of the beam.

4. Resolution of Forces and Moments

Any system (collection) of forces and moments is statically equivalent to a single resultant force vector plus a single resultant moment vector in 3-dimensional space. Either or both of these resultants may be zero.

The x-component of the resultant force is the sum of all of the x-components of the individual forces, and similarly for the y- and z-components of the resultant force.

$$\mathbf{F}_{Rx} = \Sigma \mathbf{F}_i(\cos\theta_{x,i}) \quad 9.17$$

$$\mathbf{F}_{Ry} = \Sigma \mathbf{F}_i(\cos\theta_{y,i}) \quad 9.18$$

$$\mathbf{F}_{Rz} = \Sigma \mathbf{F}_i(\cos\theta_{z,i}) \quad 9.19$$

The determination of the resultant moment vector is more complex. The resultant moment vector includes the moments of all system forces around the reference axes plus the components of all system moments.

$$\mathbf{M}_{Rx} = \Sigma(y_i\mathbf{F}_{z,i} - z_i\mathbf{F}_{y,i}) + \Sigma\mathbf{M}_i(\cos\theta_{x,i}) \quad 9.20$$

$$\mathbf{M}_{Ry} = \Sigma(z_i\mathbf{F}_{x,i} - x_i\mathbf{F}_{z,i}) + \Sigma\mathbf{M}_i(\cos\theta_{y,i}) \quad 9.21$$

$$\mathbf{M}_{Rz} = \Sigma(x_i\mathbf{F}_{y,i} - y_i\mathbf{F}_{x,i}) + \Sigma\mathbf{M}_i(\cos\theta_{z,i}) \quad 9.22$$

5. Conditions of Equilibrium

An object which is not moving is said to be static. All forces on a static object are in equilibrium. For an object to be in equilibrium, it is necessary that the resultant force vector and the resultant moment vectors both be equal to zero.

$$\mathbf{F}_R = \sqrt{\mathbf{F}_{Rx}^2 + \mathbf{F}_{Ry}^2 + \mathbf{F}_{Rz}^2} = 0 \quad 9.23$$

$$\mathbf{M}_R = \sqrt{\mathbf{M}_{Rx}^2 + \mathbf{M}_{Ry}^2 + \mathbf{M}_{Rz}^2} = 0 \quad 9.24$$

Since the square of any quantity cannot be negative, equations 9.25 through 9.30 follow directly from equations 9.23 and 9.24.

$$\mathbf{F}_{Rx} = 0 \quad 9.25$$

$$\mathbf{F}_{Ry} = 0 \quad 9.26$$

$$\mathbf{F}_{Rz} = 0 \quad 9.27$$

$$\mathbf{M}_{Rx} = 0 \quad 9.28$$

$$\mathbf{M}_{Ry} = 0 \quad 9.29$$

$$\mathbf{M}_{Rz} = 0 \quad 9.30$$

6. Free-Body Diagrams

A free-body diagram is a representation of an object in equilibrium, showing all external forces, moments, and support reactions. Since the object is in equilibrium, the resultant of all forces and moments on the free-body is zero.

If any part of the object is removed and replaced by the forces and moments which are exerted on the cut surface, a free-body of the remaining structure is obtained, and the conditions of equilibrium will be satisfied by the new free-body also.

By dividing the object into a sufficient number of free-bodies, the internal forces and moments can be found at all points of interest, providing that the conditions of equilibrium are sufficient to give a static solution.

7. Reactions

A typical first step in solving statics problems is to determine the supporting reaction forces. The manner in which the structure is supported will determine the type, location, and direction of the reactions. Conventional symbols may be used to define the type of reactions which occur at each point of support. Some examples are shown in table 9.1.

Example 9.2

Find the reactions $\mathbf{R}_1$ and $\mathbf{R}_2$.

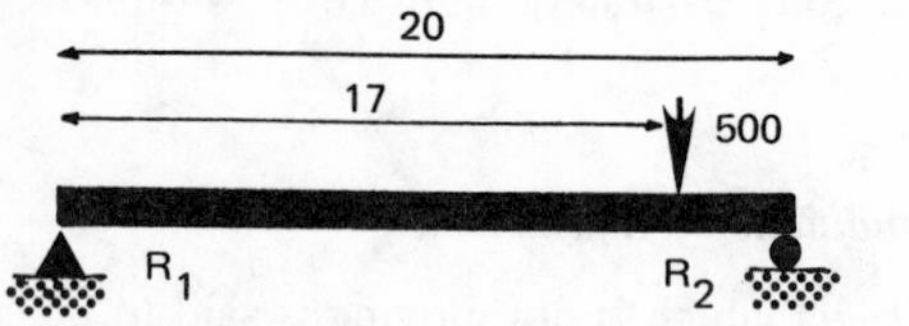

Since the left support is a simple support, its reaction can have any direction. $\mathbf{R}_1$ can, therefore, be written in terms of its x and y components, $\mathbf{R}_{1,x}$ and $\mathbf{R}_{1,y}$ respectively. $\mathbf{R}_2$ is a roller support which cannot sustain an x component.

From equation 9.25, choosing forces to the right as positive, $\mathbf{R}_{1,x}$ is found to be zero.

$$\Sigma F_x = \mathbf{R}_{1,x} = 0$$

Equation 9.26 can be used to obtain a relationship between the y components of force. Forces acting upward are considered positive.

$$\Sigma F_y = \mathbf{R}_{1,y} + \mathbf{R}_2 - 500 = 0$$

Since both $\mathbf{R}_{1,y}$ and $\mathbf{R}_2$ are unknown, a second equation is needed. Equation 9.30 is used. The reference point is chosen as the left end to make the moment arm for $\mathbf{R}_{1,y}$ equal to zero. This eliminates $\mathbf{R}_{1,y}$ as an unknown, allowing $\mathbf{R}_2$ to be found directly. Clockwise moments are considered positive.

$$\Sigma M_{\text{left end}} = (500)(17) - (\mathbf{R}_2)(20) = 0$$

$$\mathbf{R}_2 = 425$$

Once $\mathbf{R}_2$ is known, $\mathbf{R}_{1,y}$ is easily found from equation 9.26.

$$\Sigma F_y = \mathbf{R}_{1,y} + 425 - 500 = 0$$

$$\mathbf{R}_{1,y} = 75$$

8. Influence Lines For Reactions

An influence line ('influence graph') is an x-y plot of the magnitude of a reaction (any reaction on the object) as it would vary as the load is placed at different points on the object. The x-axis corresponds to the location along the object (as along the length of a beam); the y-axis corresponds to the magnitude of the reaction. For uniformity, the load is taken as 1 unit. Therefore, for an actual load of $\mathbf{P}$ units, the actual reaction would be given by equation 9.31.

$$\text{actual reaction} = \mathbf{P}\begin{bmatrix}\text{influence graph} \\ \text{ordinate}\end{bmatrix} \qquad 9.31$$

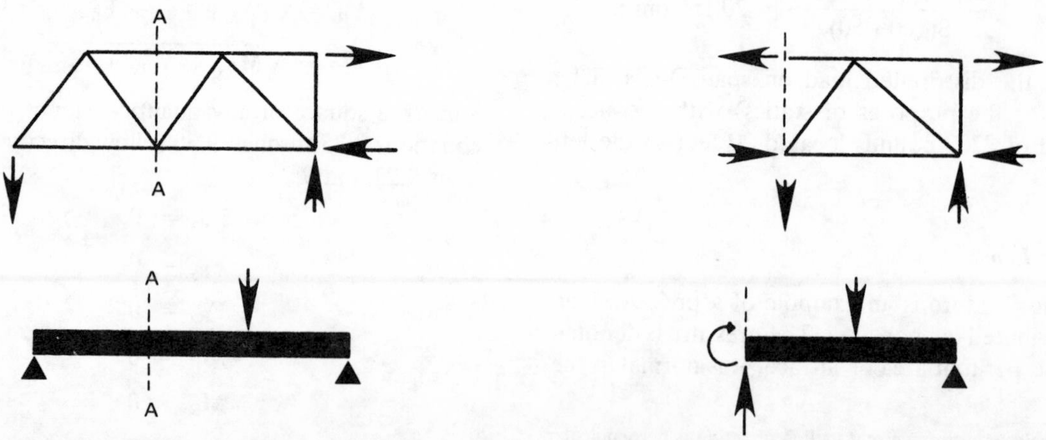

Figure 9.6 Original and Cut Free-bodies

Table 9.1
Common Support Symbols

Type of Support	Symbol	Characteristics
Built-in		Moments and forces in any direction
Simple		Load in any direction; no moment
Roller		Load normal to surface only; no moment
Cable		Load in cable direction; no moment
Guide	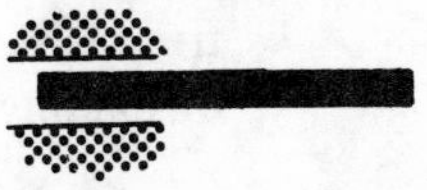	No load or moment in guide direction
Hinge		Load in any direction; no moment

Example 9.3

Draw the influence graphs for the left and right reactions for the beam shown.

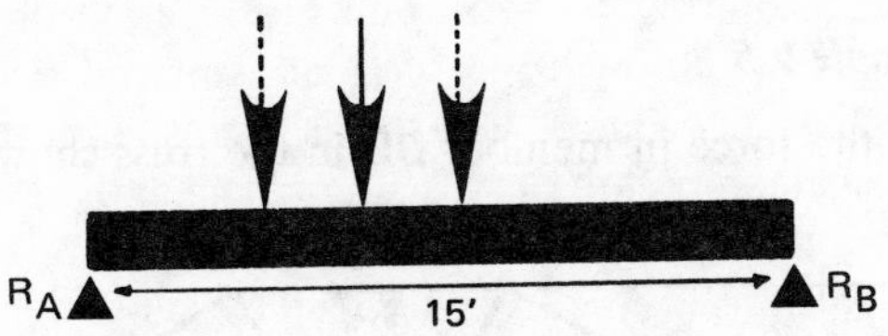

If a unit load was at the left end, reaction $\mathbf{R_A}$ would be equal to 1. If the unit load was at the right end, it would be supported entirely by $\mathbf{R_B}$, so $\mathbf{R_A}$ would be zero. The influence line for $\mathbf{R_A}$ is, then

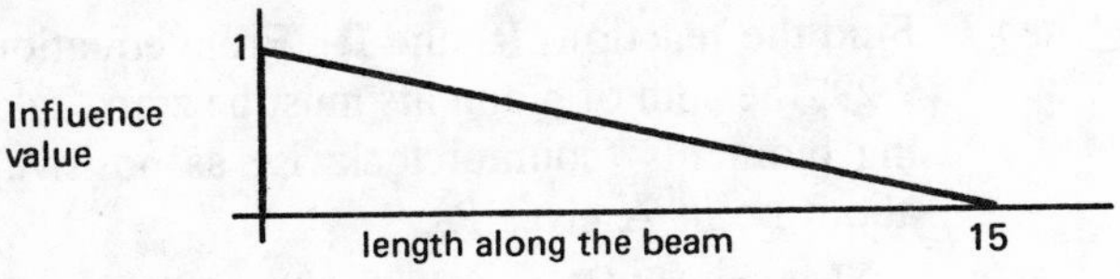

The influence line for $\mathbf{R_B}$ is found similarly.

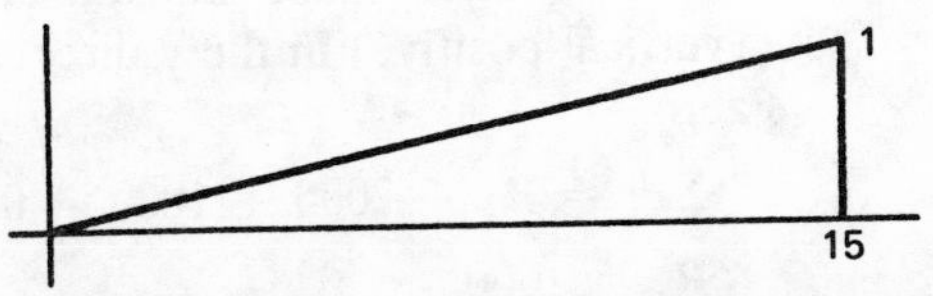

9. Axial Members

A member which is in equilibrium when acted upon by forces at each end and by no other forces or moments is an axial member. For equilibrium to exist, the resultant forces at the ends must be equal, opposite, and colinear. In an actual truss, this type of loading can be approached through the use of frictionless bearings or pins at the ends of the axial members. In simple truss analysis, the members are assumed to be axial members regardless of the end conditions.

A typical inclined axial member is illustrated in figure 9.7. For that member to be in equilibrium, the following equations must hold:

$$\mathbf{F_{Rx}} = \mathbf{F_{Bx}} - \mathbf{F_{Ax}} = 0 \qquad 9.32$$

$$\mathbf{F_{Ry}} = \mathbf{F_{By}} - \mathbf{F_{Ay}} = 0 \qquad 9.33$$

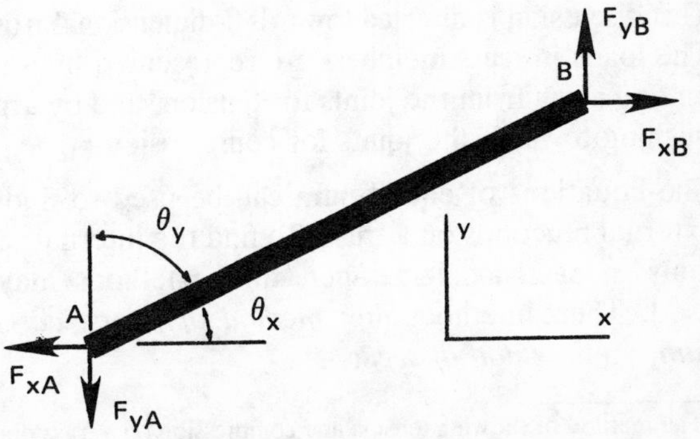

Figure 9.7 An Axial Member

The resultant force, $\mathbf{F_R}$, may be derived from the components by trigonometry and direction cosines.

$$\mathbf{F_{Rx}} = \mathbf{F_R}\cos\theta_x \qquad 9.34$$

$$\mathbf{F_{Ry}} = \mathbf{F_R}\cos\theta_y = \mathbf{F_R}\sin\theta_x \qquad 9.35$$

If, however, the geometry of the axial member is known, similar triangles may be used to find the resultant and/or the components. This is illustrated in the following example.

Example 9.4

A 12-foot long axial member carrying an internal load of 180 pounds is inclined as shown. What are the *x*- and *y*-components of the load?

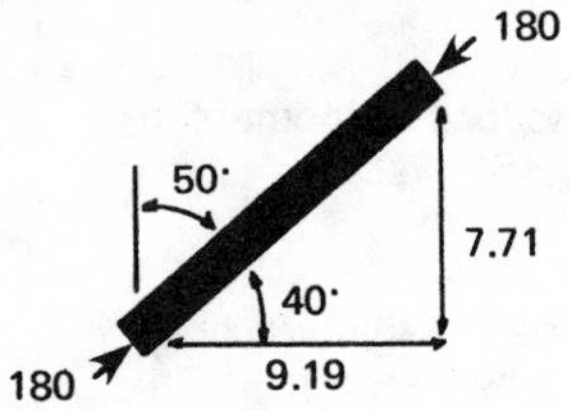

Method 1: Direction Cosines

$$\mathbf{F_x} = 180(\cos 40°) = 137.9$$

$$\mathbf{F_y} = 180(\cos 50°) = 115.7$$

Method 2: Similar Triangles

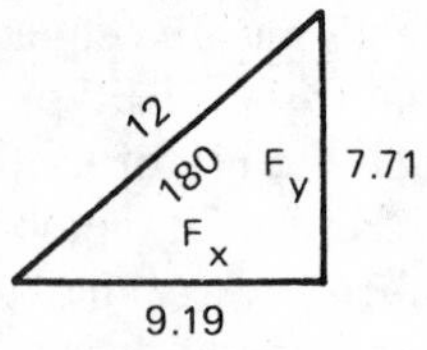

$$\mathbf{F_x} = \left(\frac{9.19}{12}\right)(180) = 137.9$$

$$\mathbf{F_y} = \left(\frac{7.71}{12}\right)(180) = 115.7$$

10. Trusses

This discussion is directed towards 2-dimensional trusses. The loads in truss members are represented by arrows pulling away from the joints for tension, and by arrows pushing towards the joints for compression.[2]

The equations of equilibrium can be used to find the external reactions on a truss. To find the internal resultants in each axial member, three methods may be used. These methods are: *method of joints, cut-and-sum,* and *method of sections.*

[2]The method of showing tension and compression on a truss drawing appears incorrect. This is because the arrows show the forces on the joints, not the forces in the axial members.

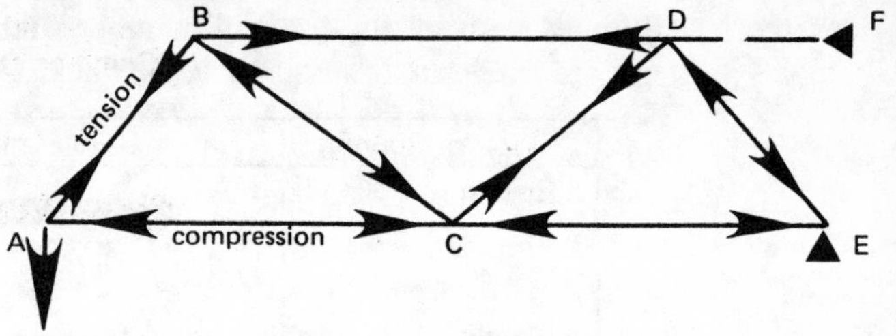

Figure 9.8 Truss Notation

All joints in a truss will be *determinate* and all member loads can be found if equation 9.36 holds.

$$\#\ \text{truss members} = 2(\#\ \text{joints}) - 3 \qquad 9.36$$

If the left-hand side is greater than the right-hand side, indeterminate methods must be used to solve the truss. If the left-hand side is less than the right-hand side, the truss is not rigid and will collapse under certain types of loading.

A. Method of Joints

The *method of joints* is a direct application of equations 9.25 and 9.26. The sums of forces in the *x*- and *y*-directions are taken at consecutive joints in the truss. At each joint, there may be up to two unknown axial forces, each of which may have two components. Since there are two equations of equilibrium, a joint with two unknown forces will be determinate.

Example 9.5

Find the force in member **BD** in the truss shown.

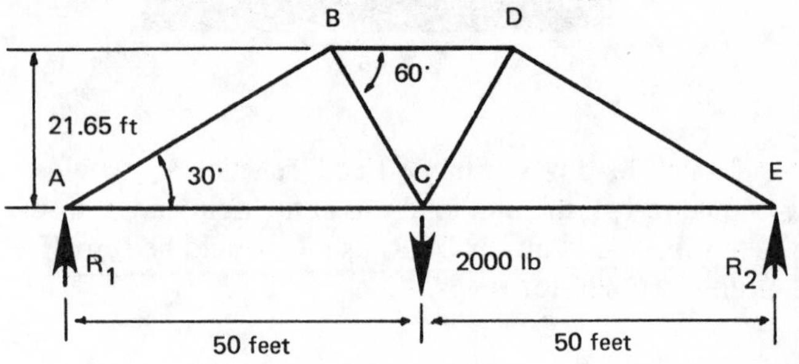

step 1: Find the reactions $\mathbf{R_1}$ and $\mathbf{R_2}$. From equation 9.29, the sum of moments must be zero. Taking moments (counterclockwise as positive) about point A gives $\mathbf{R_2}$.

$$\Sigma\mathbf{M_A} = 100(\mathbf{R_2}) - 2000(50) = 0$$

$$\mathbf{R_2} = 1000$$

From equation 9.26, the sum of the forces (vertical positive) in the y direction must be zero.

$$\Sigma\mathbf{F_y} = \mathbf{R_1} - 2000 + 1000 = 0$$

$$\mathbf{R_1} = 1000$$

step 2: Although we want the force in member **BD**, there are three unknowns at joints B and D. Therefore, start with joint A where there are only two unknowns (**AB** and **AC**). The free-body of joint A is shown below. The direction of $\mathbf{R}_1$ is known. However, the directions of the member forces are usually not known and need to be found by inspection or assumption. If an incorrect direction is assumed, the force will show up with a negative sign in later calculations.

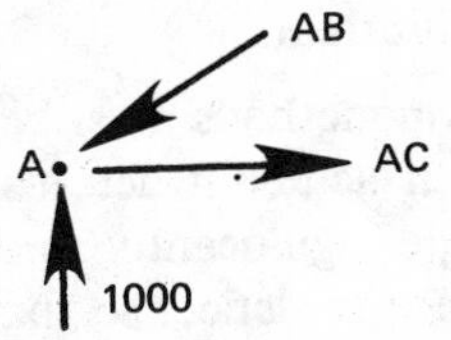

step 3: Resolve all inclined forces on joint A into horizontal and vertical components using trigonometry or similar triangles. $\mathbf{R}_1$ and **AC** are already parallel to the *y* and *x* axes, respectively. Only **AB** needs to be resolved into components. By observation, it is clear that $\mathbf{AB}_y = 1000$. If this wasn't true, equation 9.26 would not hold.

$$\mathbf{AB}_y = \mathbf{AB}(\sin 30°)$$

$$1000 = \mathbf{AB}(.5) \text{ or } \mathbf{AB} = 2000$$

$$\mathbf{AB}_x = \mathbf{AB}(\cos 30°) = 1732$$

step 4: Draw the free-body diagram of joint B. Notice that the direction of force **AB** is towards the joint, just as it was for joint A. The direction of load **BC** is chosen to counteract the vertical component of load **AB.** The direction of load **BD** is chosen to counteract the horizontal components of loads **AB** and **BC**.

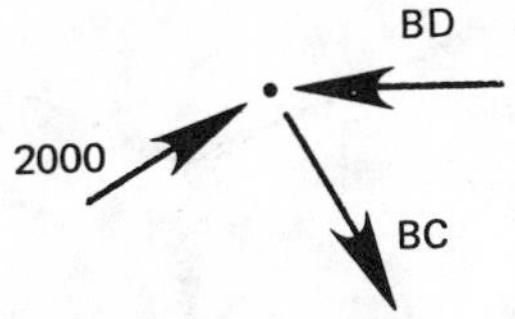

step 5: Resolve all inclined forces into horizontal and vertical components.

$$\mathbf{AB}_x = 1732$$

$$\mathbf{AB}_y = 1000$$

$$\mathbf{BC}_x = \mathbf{BC}(\sin 30°) = .5\mathbf{BC}$$

$$\mathbf{BC}_y = \mathbf{BC}(\cos 30°) = .866\mathbf{BC}$$

step 6: Write the equations of equilibrium for joint B.

$$\Sigma \mathbf{F}_x = 1732 + .5\mathbf{BC} - \mathbf{BD} = 0$$

$$\Sigma \mathbf{F}_y = 1000 - .866\mathbf{BC} = 0$$

BC from the second equation is found to be 1155. Substituting 1155 into the first equilibrium condition equation gives

$$1732 + .5(1155) - \mathbf{BD} = 0$$

$$\mathbf{BD} = 2310$$

Since **BD** turned out to be positive, its direction was chosen correctly. The direction of the arrow indicates that the member is compressing the pin joint. Consequently, the pin is compressing the member; and member **BD** is in compression.

B. Cut-and-Sum Method

The *cut-and-sum method* may be used if a load in an inclined member in the middle of a truss is wanted. The method is strictly an application of the equilibrium condition requiring the sum of forces in the vertical direction to be zero. The method is illustrated in the following example.

Example 9.6

Find the force in member **BC** for the truss shown in example 9.5.

step 1: Find the external reactions. This is the same step as in example 9.5. $\mathbf{R}_1 = \mathbf{R}_2 = 1000$.

step 2: Cut the truss through, making sure that the cut goes through only one member with a vertical component. In this case, that member is **BC.**

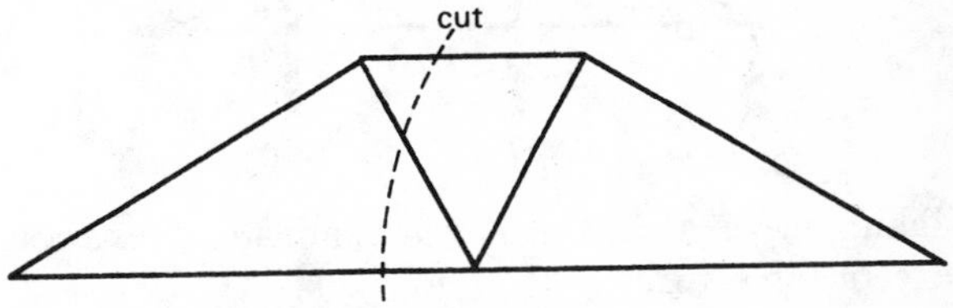

step 3: Draw the freebody of either part of the remaining truss.

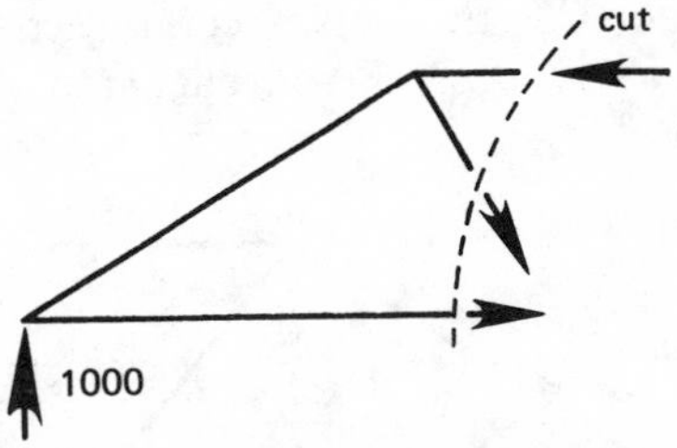

step 4: Resolve the unknown inclined force into vertical and horizontal components.

$$\mathbf{BC}_x = .5(\mathbf{BC})$$

$$\mathbf{BC}_y = .866(\mathbf{BC})$$

step 5: Sum forces in the y direction for the entire freebody.

$$\Sigma F_y = 1000 - .866(\mathbf{BC}) = 0$$

$$\mathbf{BC} = 1155$$

C. Method of Sections

The cut-and-sum method will not work if it is not possible to cut the truss without going through two members with vertical components.

The *method of sections* is a direct approach for finding member loads at any point in a truss. In this method, the truss is cut at an appropriate section, and the conditions of equilibrium are applied to the resulting freebody. This is illustrated in the following example.

Example 9.7

For the truss shown, find the load in members **CE** and **CD.**

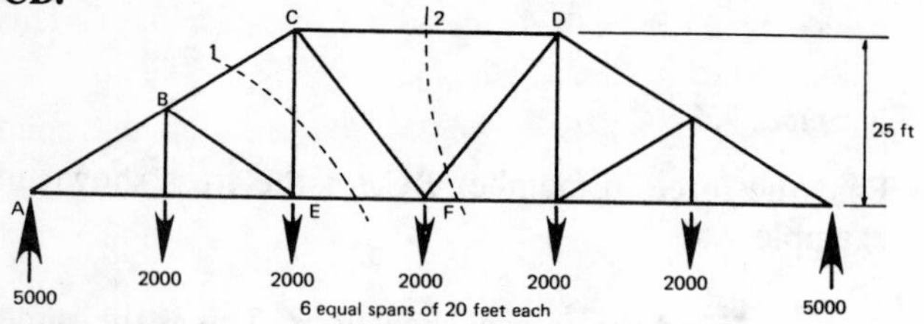

For member force **CE**, the truss is cut at section 1.

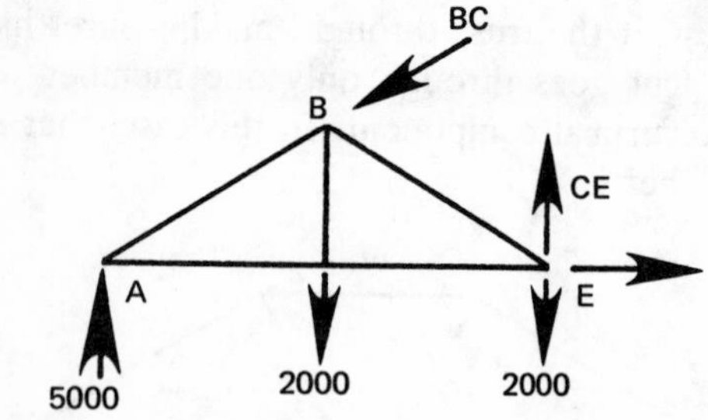

Taking moments about A will eliminate all unknowns except force **CE.**

$$\Sigma M_A = \mathbf{CE}(40) - 2000(20) - 2000(40) = 0$$

$$\mathbf{CE} = 3000$$

For member **CD,** the truss is cut at section 2. Taking moments about point F will eliminate all unknowns except **CD.**

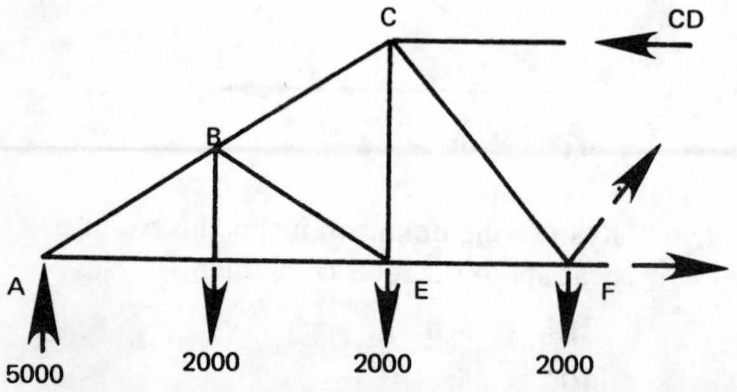

$$\Sigma M_F = \mathbf{CD}(25) + 2000(20) + 2000(40) - 5000(60) = 0$$

$$\mathbf{CD} = 7200$$

11. Superposition of Loadings

For any group of forces and moments which satisfy the conditions of equilibrium, the resultant force and moment vectors are zero. The resultant of two zero vectors is another zero vector. Therefore, any number of such equilibrium systems can be combined without disturbing the equilibrium.

Superposition methods must be used with discretion in working with actual structures since some structures change shape significantly under load. If the actual structure were to deflect so that the points of application of loads were quite different than in the undeflected structure, then superposition would not be applicable.

In simple truss analysis, the change of shape under load is neglected when finding the member loads. Superposition, therefore, can be assumed to apply.

12. Cables

A. Cables Under Concentrated Loads

An ideal cable is assumed to be completely flexible. It therefore acts as an axial member in tension between any two points of concentrated load application.

The method of joints and sections used in truss analysis apply equally well to cables under concentrated loads. However, no compression members will be found. As in truss analysis, if the cable loads are unknown, some information concerning the geometry of the cable must be known in order to solve for the axial tension in the segments.

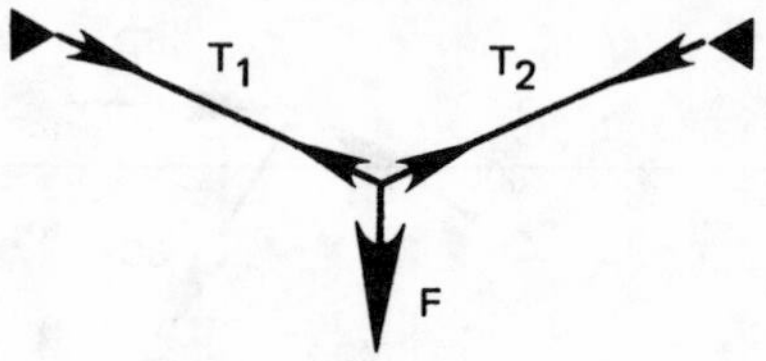

Figure 9.9 Cable Under Transverse Loading

Example 9.8

Find the tension T_2 between points B and C.

step 1: Take moments about point A to find T_3.

$$\Sigma M_A = aF_1 + bF_2 + dT_3\cos\theta_3 - cT_3\sin\theta_3 = 0$$

step 2: Sum forces in the x direction at point A to find T_1.

$$\Sigma F_x = T_1\cos\theta_1 - T_3\cos\theta_3 = 0$$

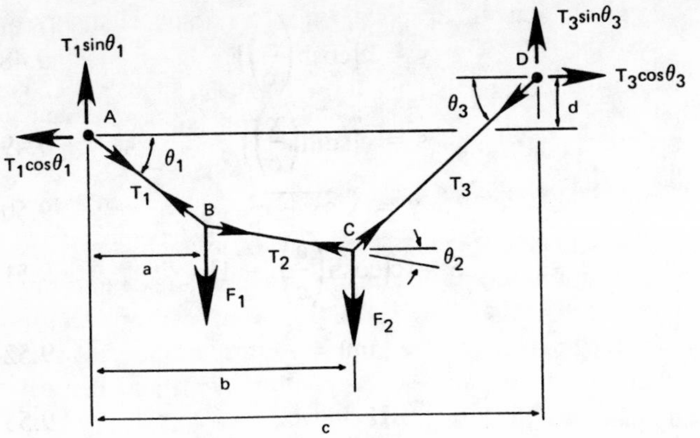

step 3: Sum forces in the x direction at point B to find T_2.

$$\Sigma F_x = T_2\cos\theta_2 - T_1\cos\theta_1 = 0$$

B. Cables Under Distributed Loads

An idealized tension cable under a distributed load is similar to a linkage made up of a very large number of axial members. The cable is an axial member in the sense that the internal tension acts in a direction which is along the centerline of the cable everywhere.

Figure 9.10 illustrates a cable under a unidirectionally distributed load. A free-body diagram of segment B-C of the cable is also shown. **F** is the vertical resultant of the distributed load on the segment.

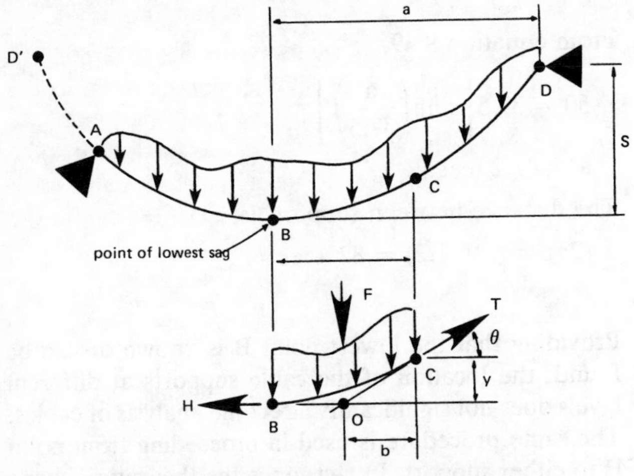

Figure 9.10 Cable Under Distributed Load

T is the cable tension at point C, and **H** is the cable tension at the point of lowest sag. From the conditions of equilibrium for free-body B-C, it is apparent that the three forces **H**, **F**, and **T** must be concurrent at point O. Taking moments about point C, the following equations are obtained.

$$\Sigma M_C = Fb - Hy = 0 \qquad 9.37$$

$$H = \frac{Fb}{y} \qquad 9.38$$

But $\tan\theta = \left(\frac{y}{b}\right)$. So,

$$H = \frac{F}{\tan\theta} \qquad 9.39$$

From the summation of forces in the vertical and horizontal directions,

$$T\cos\theta = H \qquad 9.40$$

$$T\sin\theta = F \qquad 9.41$$

$$T = \sqrt{H^2 + F^2} \qquad 9.42$$

The shape of the cable is function of the relative amount of sag at point B and the relative distribution (not the absolute magnitude) of the applied running load.

C. Parabolic Cables

If the distribution load per unit length, w, is constant with respect to a horizontal line (as is the load from a bridge floor) the cable will be parabolic in shape. This is illustrated in figure 9.11.

The horizontal component of tension can be found from equation 9.38 using $F = wa$, $b = ½a$, and $y = S$.

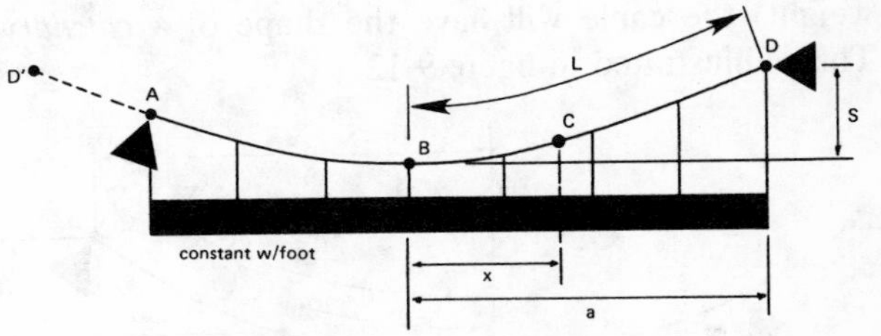

Figure 9.11 Parabolic Cable

$$H = \frac{Fb}{y} = \frac{wa^2}{2S} \qquad 9.43$$

$$T = \sqrt{H^2 + F^2} = \sqrt{(wa^2/2S)^2 + (wx)^2} \qquad 9.44$$

$$= w\sqrt{x^2 + (a^2/2S)^2} \qquad 9.45$$

The shape of the cable is given by equation 9.46.

$$y = \frac{wx^2}{2H} \qquad 9.46$$

The approximate length of the cable from the lowest point to the support is given by equation 9.47.

$$L \approx \left[1 + {}^2\!/\!_3\left(\frac{S}{a}\right)^2 - {}^2\!/\!_5\left(\frac{S}{a}\right)^4\right](a) \qquad 9.47$$

Example 9.9

A pedestrian bridge has two suspension cables and a flexible floor. The floor weighs 28 pounds per foot. The

STATICS

span of the bridge is 100 feet between the two end supports. When the bridge is empty, the tension at point A is 1500 pounds. What is the cable sag, S, at the center? What is the approximate cable length?

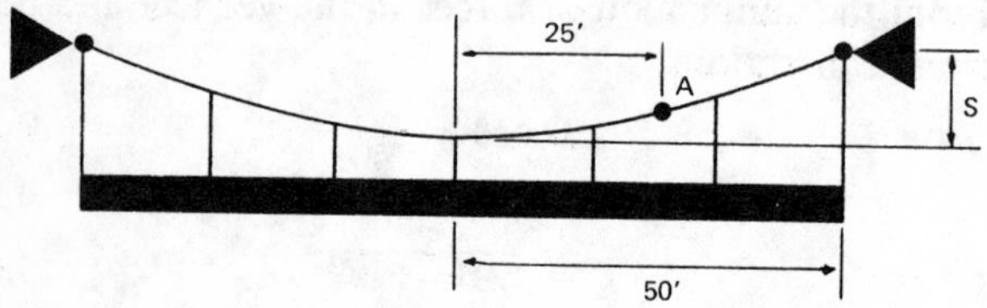

The floor weight per cable is 28/2 = 14 lb/ft. From equation 9.45,

$$1500 = 14\sqrt{[25]^2 + [(50)^2/2S]^2}$$

$$S = 12 \text{ feet}$$

From equation 9.47,

$$L = 50[1 + (2/3)(12/50)^2 - (2/5)(12/50)^4]$$

$$= 51.9 \text{ ft}$$

D. The Catenary

If the distributed load, *w*, is constant along the length of the cable (as in the case of a cable loaded by its own weight) the cable will have the shape of a *catenary*. This is illustrated in figure 9.12.

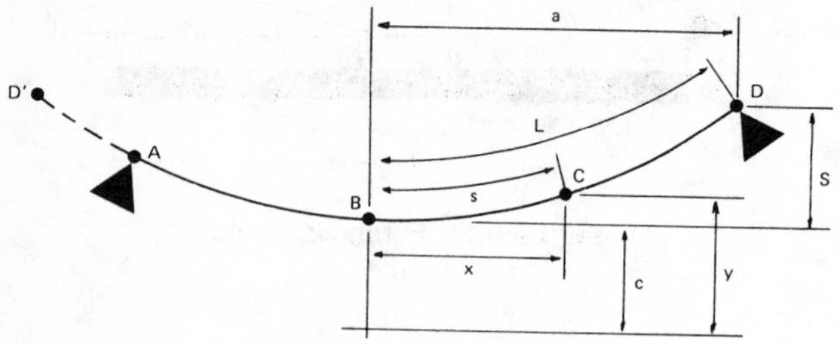

Figure 9.12 A Catenary

As shown in figure 9.12, *y* is measured from a reference plane located a distance *c* below the lowest point of the cable, point B. The location of this reference plane is a parameter of the cable which must be determined before equations 9.48 through 9.53 are used. The value of *c* does not correspond to any physical distance, nor does the reference plane correspond to the ground.

The equations of the catenary are presented below. Some judgment is usually necessary to determine which equations should be used and in which order they should be used. In order to define the cable shape, it is necessary to have some initial information which can be entered into the equations. For example, if *a* and *S* are given, equation 9.51 can be solved by trial and error to obtain *c*. Once *c* is known, the cable geometry and forces are defined by the remaining equations.

$$y = c\left[\cosh\left(\frac{x}{c}\right)\right] \qquad 9.48$$

$$s = c\left[\sinh\left(\frac{x}{c}\right)\right] \qquad 9.49$$

$$y = \sqrt{s^2 + c^2} \qquad 9.50$$

$$S = c\left[\cosh\left(\frac{a}{c}\right) - 1\right] \qquad 9.51$$

$$\tan\theta = \frac{s}{c} \qquad 9.52$$

$$\mathbf{H} = wc \qquad 9.53$$

$$\mathbf{F} = ws \qquad 9.54$$

$$\mathbf{T} = wy \qquad 9.55$$

$$\tan\theta = \frac{ws}{\mathbf{H}} \qquad 9.56$$

$$\cos\theta = \frac{\mathbf{H}}{\mathbf{T}} \qquad 9.57$$

Example 9.10

A cable 100 feet long is loaded by its own weight. The sag is 25 feet and the supports are on the same level. What is the distance between the supports?

From equation 9.50 at point D:

$$c + 25 = \sqrt{(50)^2 + c^2}$$

$$c = 37.5$$

From equation 9.49,

$$50 = 37.5\left[\sinh\left(\frac{a}{37.5}\right)\right]$$

$$a = 41.2 \text{ feet}$$

The distance between supports is:

$$(2a) = (2)(41.2) = 82.4 \text{ feet}$$

Providing that the lowest point B is known or can be found, the location of the cable supports at different levels does not significantly affect the analysis of cables. The same procedure is used in proceeding from point B to either support. In fact, once the theoretical shape of the cable has been determined, the supports can be relocated anywhere along the cable line without affecting the equilibrium of the supported segment.

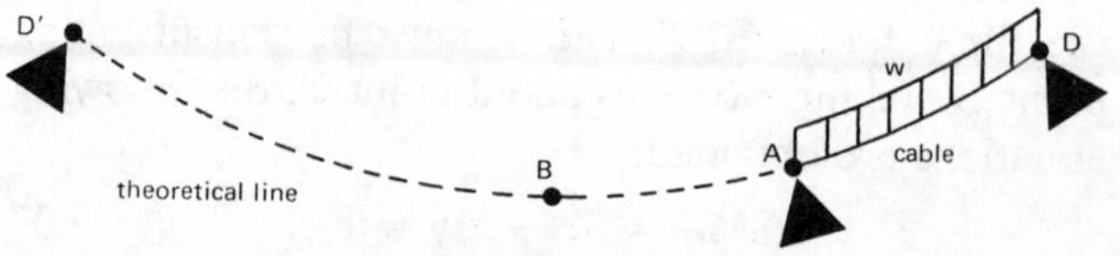

Figure 9.13 Non-Symmetrical Segment of Symmetrical Cable

13. 3-Dimensional Structures

The static analysis of 3-dimensional structures usually requires the following steps:

step 1: Determine the components of all loads and reactions. This is usually accomplished by finding the *x, y,* and *z* coordinates of all points and then using direction cosines.

step 2: Draw three free-bodies of the structure — one each for the *x, y,* and *z* components of loads and reactions.

step 3: Solve for unknowns using $\Sigma \mathbf{F} = 0$ and $\Sigma \mathbf{M} = 0$.

Example 9.11

Beam ABC is supported by the two cables as shown. Find the cable tensions $\mathbf{T}_1$ and $\mathbf{T}_2$.

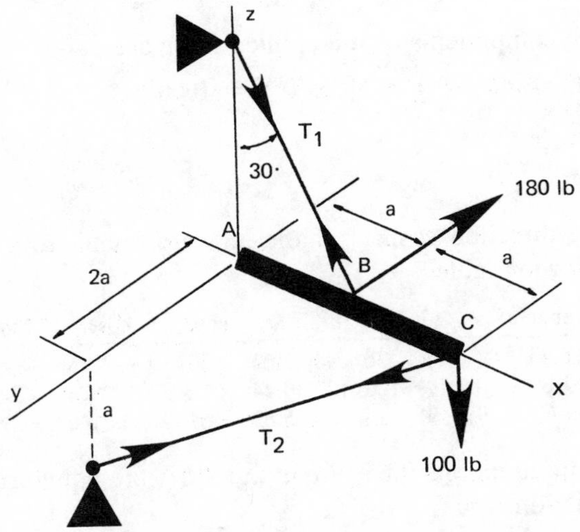

step 1: For the 180 pound load:

$F_x = 0$

$F_y = -180$

$F_z = 0$

For the 100 pound load:

$F_x = 0$

$F_y = 0$

$F_z = -100$

For cable 1:

$\cos\theta_x = \cos 120° = -.5$

$\cos\theta_y = 0$

$\cos\theta_z = \cos 30° = .866$

$\mathbf{T}_{1x} = -.5\mathbf{T}_1$

$\mathbf{T}_{1y} = 0$

$\mathbf{T}_{1z} = .866\mathbf{T}_1$

For cable 2:

The length of the cable is

$L = \sqrt{(2a)^2 + (2a)^2 + (-a)^2} = 3a$

$\cos\theta_x = \frac{-2a}{3a} = -.667$

$\cos\theta_y = \frac{2a}{3a} = .667$

$\cos\theta_z = \frac{-a}{3a} = -.333$

$\mathbf{T}_{2x} = -.667\mathbf{T}_2$

$\mathbf{T}_{2y} = .667\mathbf{T}_2$

$\mathbf{T}_{2z} = -.333\mathbf{T}_2$

step 2:

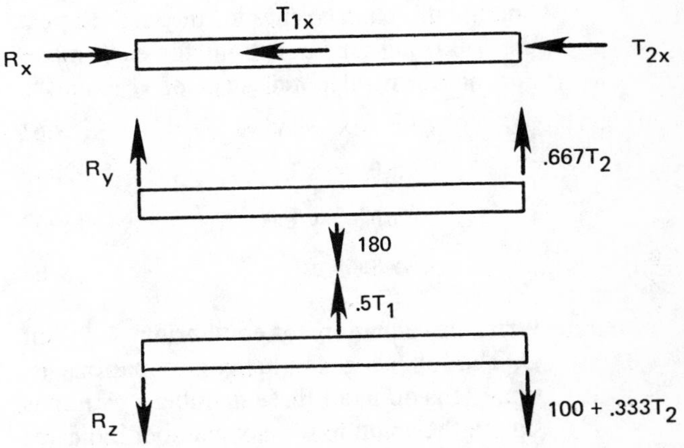

step 3: Summing moments about point A for the *y* case gives $\mathbf{T}_2$.

$\Sigma \mathbf{M}_{Ay} = .667\mathbf{T}_2(2a) - 180(a) = 0$

$\mathbf{T}_2 = 135$

Summing moments about point A for the z case gives $\mathbf{T}_1$.

$\Sigma \mathbf{M}_{Az} = .866\,\mathbf{T}_1(a) - .333(135)(2a) - 100(2a) = 0$

$\mathbf{T}_1 = 335$

14. General Tripod Solution

The procedure given in the preceding section will work quite well with a tripod consisting of 3 axial members with a load in any direction applied at the apex. However, the tripod problem occurs frequently enough to develop a specialized procedure for solution.

step 1: Use the direction cosines of the force **F** to find its components.

$$\mathbf{F}_x = \mathbf{F}(\cos\theta_x) \quad 9.58$$

$$\mathbf{F}_y = \mathbf{F}(\cos\theta_y) \quad 9.59$$

$$\mathbf{F}_z = \mathbf{F}(\cos\theta_z) \quad 9.60$$

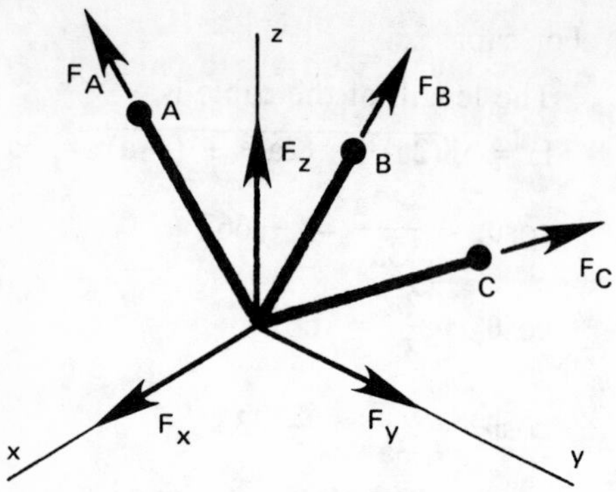

Figure 9.14 A General Tripod.

step 2: Using the x, y, and z coordinates of points A, B, and C (taking the origin at the apex) find the direction cosines for the legs. Repeat the following four equations for each member, observing algebraic signs of x, y, and z.

$$L^2 = x^2 + y^2 + z^2 \qquad 9.61$$

$$\cos\theta_x = x/L \qquad 9.62$$

$$\cos\theta_y = y/L \qquad 9.63$$

$$\cos\theta_z = z/L \qquad 9.64$$

step 3: Write the equations of equilibrium for joint O. The following simultaneous equations assume tension in all three members. A minus sign in the solution for any member indicates compression instead of tension.

$$F_A\cos\theta_{xA} + F_B\cos\theta_{xB} + F_C\cos\theta_{xC} + F_x = 0 \qquad 9.65$$

$$F_A\cos\theta_{yA} + F_B\cos\theta_{yB} + F_C\cos\theta_{yC} + F_y = 0 \qquad 9.66$$

$$F_A\cos\theta_{zA} + F_B\cos\theta_{zB} + F_C\cos\theta_{zC} + F_z = 0 \qquad 9.67$$

Example 9.12

Find the load on each leg of the tripod shown.

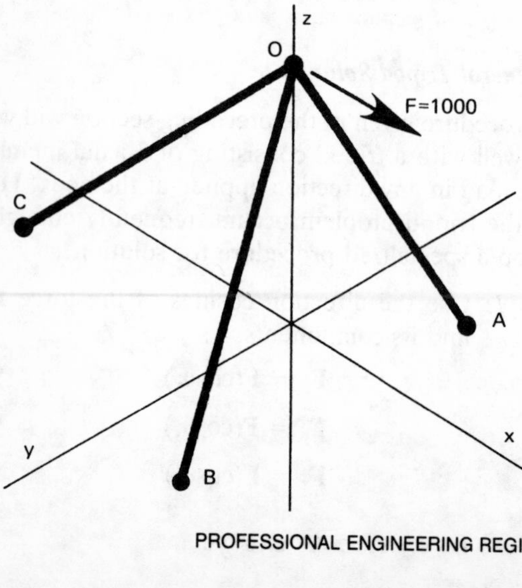

Point	x	y	z
O	0	0	6
A	2	−2	0
B	0	3	−4
C	−3	3	2

Since the origin is not at point (0,0,0), it is necessary to transfer the origin to the apex. This is done by the following equations. Only the z values are actually affected.

$$x' = x - x_0$$
$$y' = y - y_0$$
$$z' = z - z_0$$

The new coordinates with the origin at the apex are:

Point	x	y	z
O	0	0	0
A	2	−2	−6
B	0	3	−10
C	−3	3	−4

The components of the applied force are:

$$\mathbf{F_x} = \mathbf{F}(\cos\theta_x) = \mathbf{F}[\cos(0°)] = 1000$$

$$\mathbf{F_y} = 0$$

$$\mathbf{F_z} = 0$$

The direction cosines of the legs are found from the following table.

Member	x^2	y^2	z^2	L^2	L	$\cos\theta_x$	$\cos\theta_y$	$\cos\theta_z$
O-A	4	4	36	44	6.63	.3015	−.3015	−.9046
O-B	0	9	100	109	10.44	0	.2874	−.9579
O-C	9	9	16	34	5.83	−.5146	.5146	−.6861

From equations 9.65, 9.66, and 9.67, the equilibrium equations are:

$$.3015\mathbf{F_A} + \quad - .5146\mathbf{F_C} + 1000 = 0$$

$$-.3015\mathbf{F_A} + .2874\mathbf{F_B} + .5146\mathbf{F_C} + \quad 0 = 0$$

$$-.9046\mathbf{F_A} - .9579\mathbf{F_B} - .6861\mathbf{F_C} + \quad 0 = 0$$

The solution to this set of simultaneous equations is:

$$\mathbf{F_A} = +1531 \text{ (tension)}$$

$$\mathbf{F_B} = -3480 \text{ (compression)}$$

$$\mathbf{F_C} = +2841 \text{ (tension)}$$

15. Properties of Areas

A. Centroids

The location of the *centroid* of a 2-dimensional area which is defined mathematically as $y = f(x)$ can be found from equations 9.68 and 9.69. This is illustrated in example 9.13.

$$\bar{x} = \frac{\int x\,dA}{A} \qquad 9.68$$

$$\bar{y} = \frac{\int y\,dA}{A} \qquad 9.69$$

$$A = \int f(x)\,dx \qquad 9.70$$

$$dA = f(x)\,dx = f(y)\,dy \qquad 9.71$$

Example 9.13

Find the x component of the centroid of the area bounded by the x and y axes, $x = 2$, and $y = e^{2x}$.

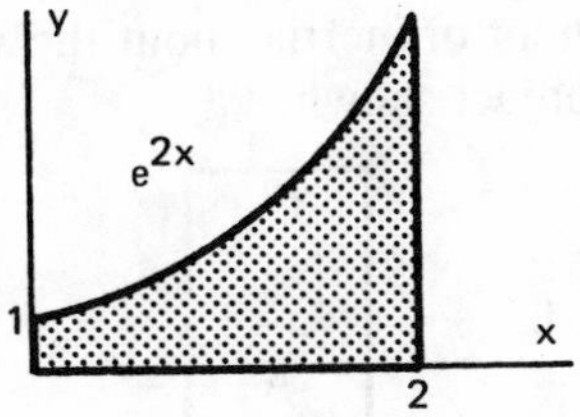

step 1: Find the area. $A = \int_0^2 e^{2x}\,dx = [{}^1\!/\!{}_2 e^{2x}]_0^2$

$= 27.3 - .5 = 26.8$

step 2: Put dA in terms of dx. $dA = f(x)\,dx = e^{2x}\,dx$

step 3: Use equation 9.68 to find $\bar{x}$.

$$\bar{x} = \frac{1}{26.8}\int_0^2 xe^{2x}\,dx = \frac{1}{26.8}[{}^1\!/\!{}_2 xe^{2x} - {}^1\!/\!{}_4 e^{2x}]_0^2$$

$$= 1.54$$

With very few exceptions, most areas for which the centroidal location is needed will be either rectangular or triangular. The locations of the centroids for these and other common shapes are given in this chapter's appendix.

The centroid of a complex 2-dimensional area which can be divided into the simple shapes in appendix A can be found from equations 9.72 and 9.73.

$$\bar{x}_{composite} = \frac{\Sigma(A_i\bar{x}_i)}{\Sigma A_i} \qquad 9.72$$

$$\bar{y}_{composite} = \frac{\Sigma(A_i\bar{y}_i)}{\Sigma A_i} \qquad 9.73$$

Example 9.14

Find the y-coordinate of the centroid for the object shown.

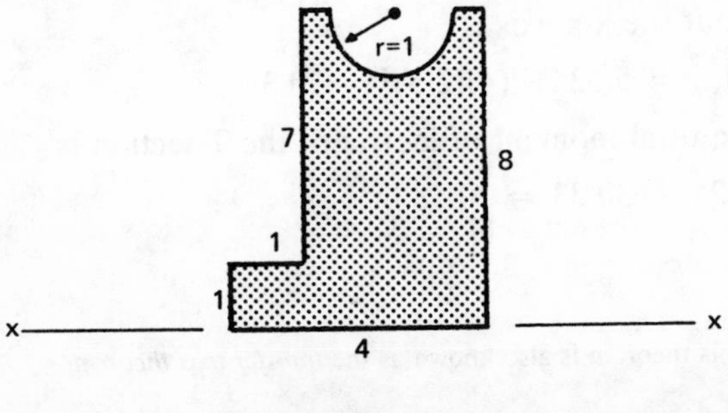

The object is divided into three parts: a 1 × 4 rectangle, a 3 × 7 rectangle, and a half-circle of radius 1. Then, the areas and distances from the *x-x* axis to the individual centroids are found.

$$A_1 = (1)(4) = 4$$

$$A_2 = (3)(7) = 21$$

$$A_3 = (-\tfrac{1}{2})\pi(1)^2 = -1.57$$

$$\bar{y}_1 = \tfrac{1}{2}$$

$$\bar{y}_2 = 4\tfrac{1}{2}$$

$$\bar{y}_3 = 8 - .424 = 7.576$$

Using equation 9.73,

$$\bar{y} = \frac{(4)(\tfrac{1}{2}) + (21)(4\tfrac{1}{2}) - (1.57)(7.576)}{4 + 21 - 1.57}$$

$$= 3.61$$

B. Moment of Inertia

The moment of inertia, I, of a 2-dimensional area is a parameter which is often needed in mechanics of materials problems. It has no simple geometric interpretation, and its units (length to the fourth power) add to the mystery of this quantity. However, it is convenient to think of the moment of inertia as a resistance to bending.

If the moment of inertia is a resistance to bending, it is apparent that this quantity must always be positive. Since bending of an object (e.g., a beam) may be in any direction, the resistance to bending must depend on the direction of bending. Therefore, a reference axis or direction must be included when specifying the moment of inertia.

In this chapter, I_x is used to represent a moment of inertia with respect to the x axis. Similarly, I_y is with respect to the y axis. I_x and I_y are not components of the 'resultant' moment of inertia. The moment of inertia taken with respect to a line passing through the area's centroid is known as the *centroidal moment of inertia,* I_c. The centroidal moment of inertia is the smallest possible moment of inertia for the shape.

The moments of inertia of a function which can be expressed mathematically as $y = f(x)$ are given by equations 9.74 and 9.75.

$$I_x = \int y^2\,dA \qquad 9.74$$

$$I_y = \int x^2\,dA \qquad 9.75$$

In general, however, moments of inertia will be found from appendix A.

Example 9.15

Find I_y for the area bounded by the y axis, $y = 8$, and $y^2 = 8x$.

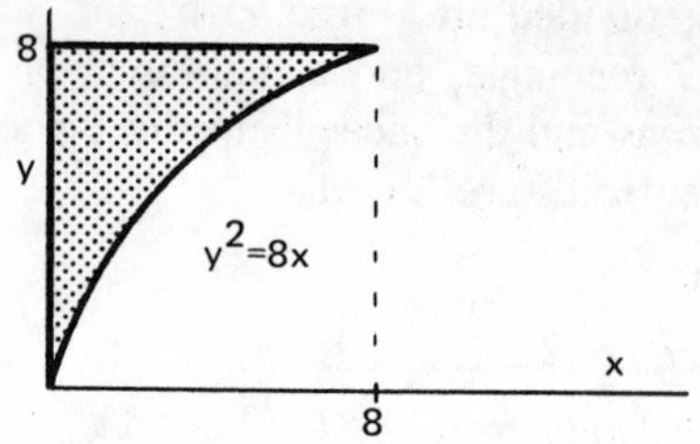

$$I_y = \int_0^8 x^2\, dA = \int_0^8 x^2(8-y)\, dx$$

But $y = \sqrt{8x}$.

$$I_y = \int_0^8 (8x^2 - \sqrt{8}x^{5/2})\, dx$$

$$= 195.04 \text{ inches}^4$$

Example 9.16

What is the centroidal moment of inertia of the area shown?

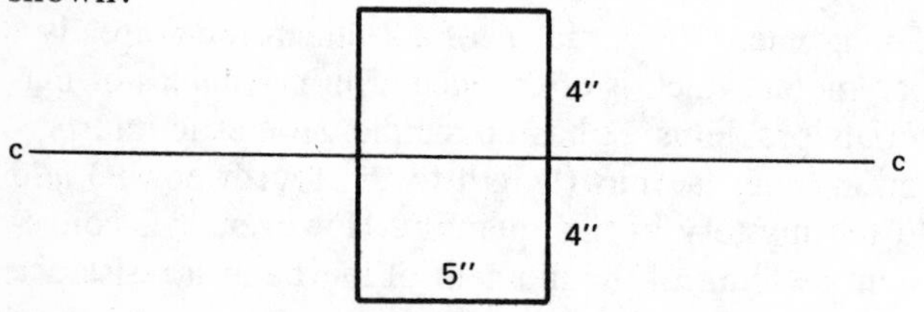

From appendix A, $I_c = \dfrac{(5)(8)^3}{12} = 213.3 \text{ inches}^4$

The *polar moment of inertia* of a 2-dimensional area can be thought of as a measure of the area's resistance to torsion (twisting). Although the polar moment of inertia can be evaluated mathematically by equation 9.76, it is more expedient to use equation 9.77 if I_x and I_y are known.

$$J_z = \int(x^2 + y^2)\, dA \qquad 9.76$$

$$J_z = I_x + I_y \qquad 9.77$$

The *radius of gyration, k,* is a distance at which the entire area can be assumed to exist. This distance is measured from the axis about which the moment of inertia was taken.

$$I = k^2A \qquad 9.78$$

$$k = \sqrt{I/A} \qquad 9.79$$

Example 9.17

What is the radius of gyration for the section shown in example 9.16? What is the significance of this value?

$$A = (5)(4+4) = 40$$

From equation 9.79, $k = \sqrt{213.3/40} = 2.31$ inches

2.31" is the distance from the axis c-c that an infinitely long strip (with area of 40 square inches) would have to be located to have a moment of inertia of 213.3 inches^4.

The *parallel axis theorem*[4] is usually needed to evaluate the moment of inertia of a composite object made up of several simple 2-dimensional shapes. The parallel axis theorem relates the moment of inertia of an area taken with respect to any axis to the centroidal moment of inertia. In equation 9.80, *A* is the 2-dimensional object's area, and *d* is the distance between the centroidal and new axes.

$$I_{\text{any parallel axis}} = I_c + Ad^2 \qquad 9.80$$

Example 9.18

Find the moment of inertia about the *x* axis for the 2-dimensional object shown.

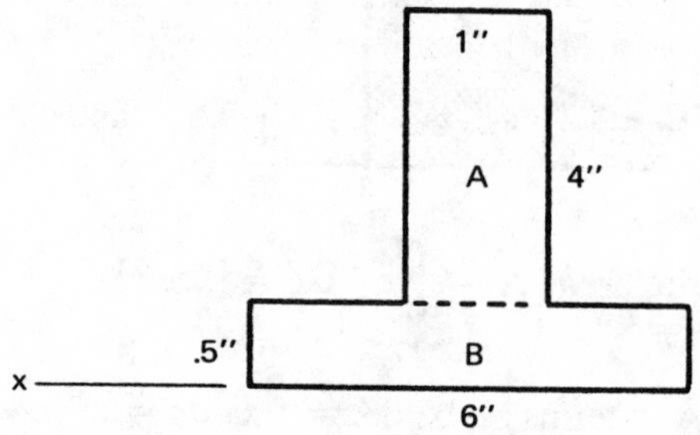

The T-section is divided into two parts: A and B. The moment of inertia of section B can be readily evaluated by using appendix A.

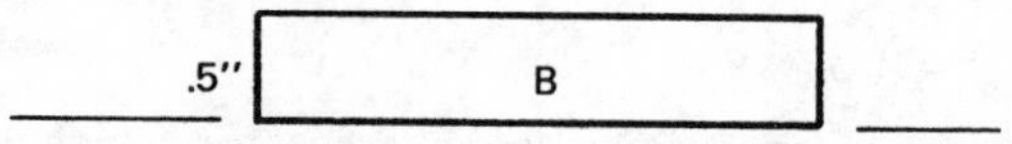

$$I_{x\text{-}x} = \frac{(6)(.5)^3}{3} = .25$$

The moment of inertia of the stem about its own centroidal axis is

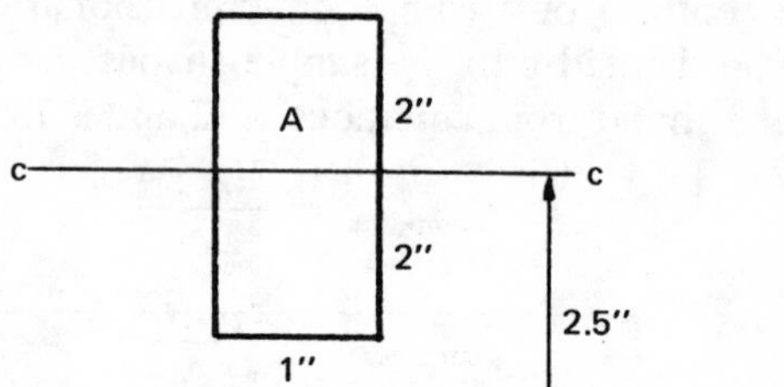

$$I_{c\text{-}c} = \frac{(1)(4)^3}{12} = 5.33$$

Using equation 9.80, the moment of inertia of the stem about the x-x axis is

$$I_{x\text{-}x} = 5.33 + (4)(2.5)^2 = 30.33$$

The total moment of inertia of the T-section is

$$.25 + 30.33 = 30.58 \text{ in}^4$$

[4]This theorem is also known as the *transfer axis theorem.*

C. Product of Inertia

The *product of inertia* of a two-dimensional object is found by multiplying each differential element of area times its *x* and *y* coordinates, and then integrating over the entire area.

$$P_{xy} = \int xy\, dA \qquad 9.81$$

The product of inertia is zero when either axis is an axis of symmetry. The product of inertia may be negative.

16. Rotation of Axes

Suppose the various properties of an area are known for one set of axes, *x* and *y*. If the axes are rotated through an angle without rotating the area itself, the new properties can be found from the old properties.

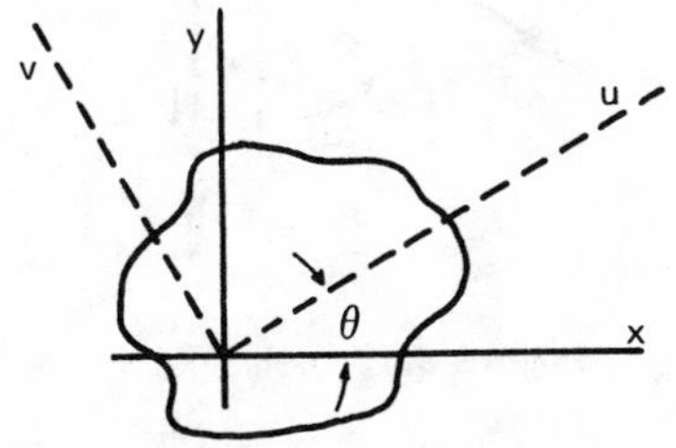

Figure 9.15 Rotation of Axes

$$I_u = I_x\cos^2\theta - 2P_{xy}\sin\theta\cos\theta + I_y\sin^2\theta \qquad 9.82$$

$$= \tfrac{1}{2}(I_x + I_y) + \tfrac{1}{2}(I_x - I_y)\cos 2\theta - P_{xy}\sin 2\theta \qquad 9.83$$

$$I_v = I_x\sin^2\theta + 2P_{xy}\sin\theta\cos\theta + I_y\cos^2\theta \qquad 9.84$$

$$= \tfrac{1}{2}(I_x + I_y) - \tfrac{1}{2}(I_x - I_y)\cos 2\theta + P_{xy}\sin 2\theta \qquad 9.85$$

$$P_{uv} = I_x\sin\theta\cos\theta + P_{xy}(\cos^2\theta - \sin^2\theta) - I_y\sin\theta\cos\theta \qquad 9.86$$

$$= \tfrac{1}{2}(I_x - I_y)\sin 2\theta + P_{xy}\cos 2\theta \qquad 9.87$$

Since the polar moment of inertia about a fixed axis is constant, the sum of the two area moments of inertia is also constant.

$$I_x + I_y = I_u + I_v \qquad 9.88$$

There is one angle that will maximize the moment of inertia, I_u. This angle can be found from calculus by setting $dI_u/d\theta = 0$. The resulting equation defines two angles, one of which maximizes I_u, the other of which minimizes I_u.

$$\tan 2\theta = -\frac{2P_{xy}}{I_x - I_y} \qquad 9.89$$

The two angles which satisfy equation 9.89 are 90° apart. These are known as the *principal axes*. The moments of inertia about these two axes are known as the *principal moments of inertia*. These principal moments are given by equation 9.90.

$$I_{max,min} = \tfrac{1}{2}(I_x + I_y) \pm \sqrt{\tfrac{1}{2}(I_x - I_y)^2 + P_{xy}^2} \qquad 9.90$$

17. Properties of Masses

A. Center of Gravity

The *center of gravity* in three-dimensional objects is analogous to centroids in two-dimensional areas. The center of gravity can be located mathematically if the object can be described by a mathematical function.

$$\bar{x} = \frac{\int x\, dM}{M} \qquad 9.91$$

$$\bar{y} = \frac{\int y\, dM}{M} \qquad 9.92$$

$$\bar{z} = \frac{\int z\, dM}{M} \qquad 9.93$$

The location of the center of gravity is often obvious for simple objects. It is always located on an axis of symmetry. If the object is complex or composite, the overall center of gravity can be found from the individual centers of gravity of the constituent objects.

$$\bar{x}_{composite} = \frac{\Sigma(M_i\bar{x}_i)}{\Sigma M_i} \qquad 9.94$$

$$\bar{y}_{composite} = \frac{\Sigma(M_i\bar{y}_i)}{\Sigma M_i} \qquad 9.95$$

$$\bar{z}_{composite} = \frac{\Sigma(M_i\bar{z}_i)}{\Sigma M_i} \qquad 9.96$$

B. Mass Moment of Inertia

The mass moment of inertia may be thought of as a measure of resistance to rotational motion. Although it can be found mathematically from equations 9.97, 9.98, and 9.99, it is more expedient to use appendix B to evaluate simple objects.

$$I_x = \int (y^2 + z^2)\, dM \qquad 9.97$$

$$I_y = \int (x^2 + z^2)\, dM \qquad 9.98$$

$$I_z = \int (x^2 + y^2)\, dM \qquad 9.99$$

The *centroidal mass moment of inertia* is found by evaluating the moment of inertia about an axis passing through the object's center of gravity. Once this centroidal mass moment of inertia is known, the parallel axis theorem can be used to find the moment of inertia about any parallel axis.

$$I_{any\ parallel\ axis} = I_c + Md^2 \qquad 9.100$$

The radius of gyration of a three-dimensional object is defined by equation 9.101.

$$k = \sqrt{I/M} \qquad 9.101$$

$$I = k^2M \qquad 9.102$$

18. Friction

Friction is a force which resists motion or attempted motion. It always acts parallel to the contacting surfaces. The frictional force exerted on a stationary object is known as *static friction* or *coulomb friction.* If the object is moving, the friction is known as *dynamic friction*. Dynamic friction is less than static friction in most situations.

The actual magnitude of the frictional force depends on the *normal force* and the *coefficient of friction, f,* between the object and the surface. For an object resting on a horizontal surface, the normal force is the weight, *w*.

$$F_f = fN = fw \qquad 9.103$$

If the object is resting on an inclined surface, the normal force will be

$$N = w \cos\theta \qquad 9.104$$

The frictional force is again equal to the product of the normal force and the coefficient of friction.

$$F_f = fN = fw \cos\theta \qquad 9.105$$

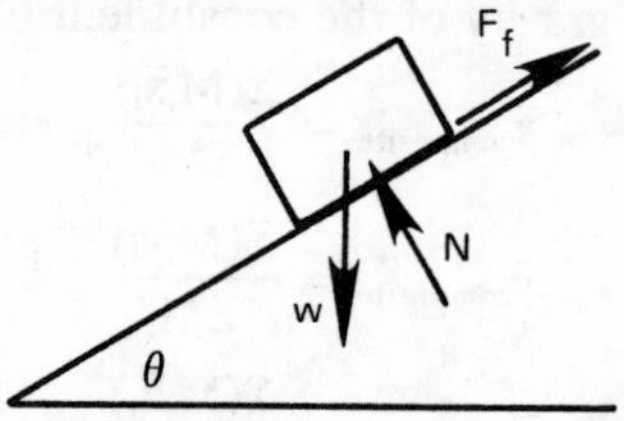

Figure 9.16 Frictional Force

The object shown in figure 9.16 will not slip down the plane until the angle reaches a critical angle known as the *angle of repose.* This angle is given by equation 9.106.

$$\tan\theta = f \qquad 9.106$$

Typical values of the coefficient of friction are given in table 9.2.

Friction also exists between a belt, rope, or band wrapped around a drum, pulley, or sheave. If T_1 is the tight side tension, T_2 is the slack side tension, and ϕ is the contact angle in *radians,* the relationship governing *belt friction* is given by equation 9.107.

$$\frac{T_1}{T_2} = e^{f\phi} \qquad 9.107$$

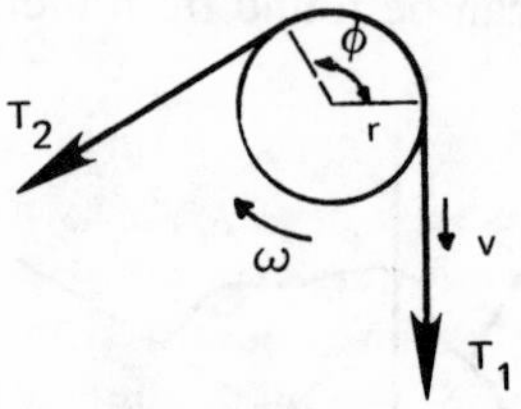

Figure 9.17 Belt Friction

The transmitted torque in ft-lb is

$$\text{torque} = (T_1 - T_2)r \qquad 9.108$$

The power in ft-lb/sec transmitted by the belt running at speed *v* in ft/sec is

$$\text{power} = (T_1 - T_2)v \qquad 9.109$$

Table 9.2 Coefficients of Friction

Material	Condition	Dynamic	Static
cast iron on cast iron	dry	.15	1.10
plastic on steel	dry	.35	
grooved rubber on pavement	dry	.40	.55
bronze on steel	oiled		.09
steel on graphite	dry		.21
steel on steel	dry	.42	.78
steel on steel	oiled	.08	.10
steel on asbestos-faced steel	dry		.15
steel on asbestos-faced steel	oiled		.12
press fits (shaft in hole)	oiled		.10–.15

STATICS

PART 2: Indeterminate Structures

A structure that is statically indeterminate (redundant) is one for which the equations of statics are not sufficient to determine all reactions, moments, and internal stress distributions. Additional formulas involving deflection relationships are required to completely determine these unknowns.

The *degree of redundancy* is equal to the number of reactions or members that would have to be removed in order to make the structure statically determinate. For example, a 2-span beam on three simple supports is redundant to the first degree. The degree of redundancy of a truss can be calculated from equation 9.110.

$$\text{degree of redundancy} = \#\text{ members} - 2(\#\text{ joints}) + 3 \qquad 9.110$$

The *method of consistent deformation* may be used to evaluate simple structures consisting of two or three members in tension or compression. Although this method cannot be proceduralized, it is very simple to learn and to apply. The method makes use of geometry to develop relationships between the deflections (deformations) between different members or locations on the structure.

Example 9.19

A pile is constructed of concrete with a steel jacket. What is the stress in the steel and concrete if a load **P** is applied? Assume the end caps are rigid and the steel-concrete bond is perfect.

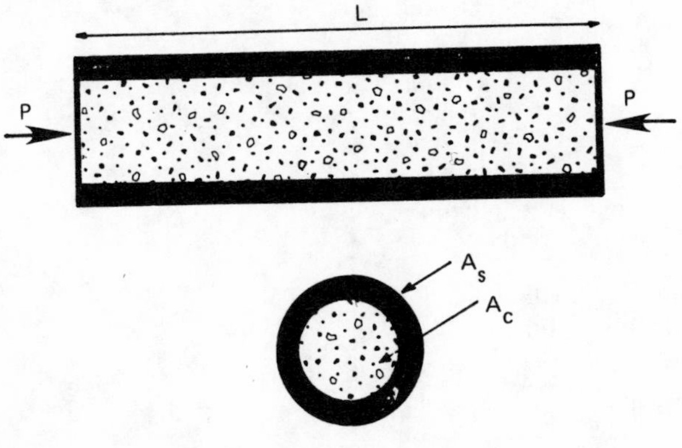

Let $\mathbf{P_c}$ and $\mathbf{P_s}$ be the loads carried by the concrete and steel respectively. Then,

$$\mathbf{P_c} + \mathbf{P_s} = \mathbf{P} \qquad 9.111$$

The deformation of the steel is

$$\delta_s = \frac{P_sL}{A_sE_s} \qquad 9.112$$

Similarly, the deflection of the concrete is

$$\delta_c = \frac{P_cL}{A_cE_c} \qquad 9.113$$

But $\delta_c = \delta_s$ since the bonding is perfect. Therefore,

$$\frac{P_c}{A_cE_c} - \frac{P_s}{A_sE_s} = 0 \qquad 9.114$$

Equations 9.111 and 9.114 are solved simultaneously to determine $\mathbf{P_c}$ and $\mathbf{P_s}$. The respective stresses are:

$$\sigma_s = P_s/A_s \qquad 9.115$$

$$\sigma_c = P_c/A_c \qquad 9.116$$

Example 9.20

A uniform bar is clamped at both ends and the axial load applied near one of the supports. What are the reactions?

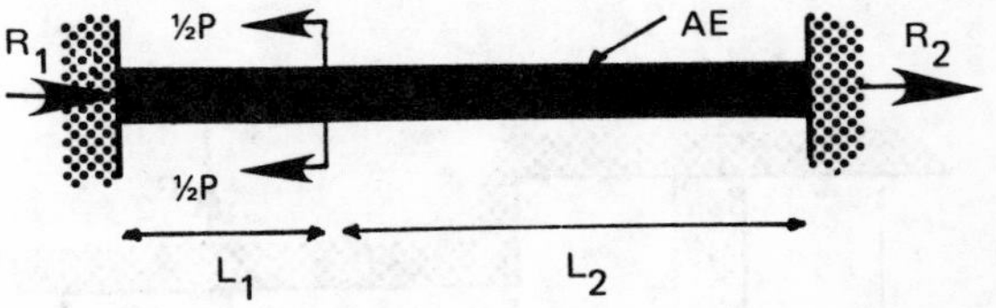

Clearly, the first required equation is

$$\mathbf{R_1} + \mathbf{R_2} = \mathbf{P} \qquad 9.117$$

The shortening of section 1 due to the reaction $\mathbf{R_1}$ is

$$\delta_1 = \frac{-R_1L_1}{AE} \qquad 9.118$$

The elongation of section 2 due to the reaction $\mathbf{R_2}$ is

$$\delta_2 = \frac{R_2L_2}{AE} \qquad 9.119$$

However, the bar is continuous, so $\delta_1 = -\delta_2$. Therefore,

$$R_1L_1 = R_2L_2 \qquad 9.120$$

Equations 9.117 and 9.120 are solved simultaneously to find $\mathbf{R_1}$ and $\mathbf{R_2}$.

Example 9.21

The non-uniform bar shown is clamped at both ends. What are the reactions of both ends if a temperature change of ΔT is experienced?

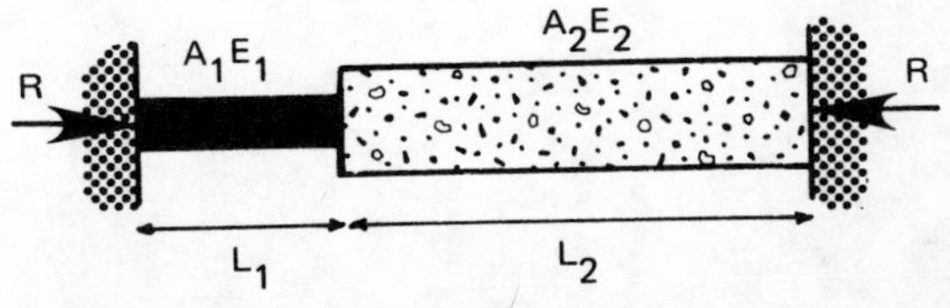

The thermal deformations of sections 1 and 2 can be calculated directly.

$$\delta_1 = \alpha_1 L_1 \Delta T \quad 9.121$$

$$\delta_2 = \alpha_2 L_2 \Delta T \quad 9.122$$

The total deformation is $\delta = \delta_1 + \delta_2$. However, the deformation can also be calculated from the principles of mechanics of material.

$$\delta = \frac{\mathbf{R}L_1}{A_1E_1} + \frac{\mathbf{R}L_2}{A_2E_2} \quad 9.123$$

Combining equations 9.121, 9.122, and 9.123 gives

$$(\alpha_1 L_1 + \alpha_2 L_2)\Delta T = \left(\frac{L_1}{A_1E_1} + \frac{L_2}{A_2E_2}\right)\mathbf{R} \quad 9.124$$

Equation 9.124 can be solved directly for **R**.

Example 9.22

The beam shown is supported by dissimilar tension members. What are the reactions in the tension members? Assume the horizontal bar is rigid and remains horizontal.

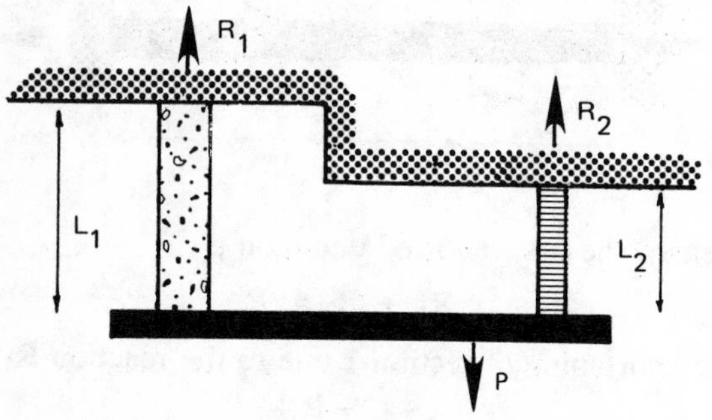

The required equilibrium condition is

$$\mathbf{R}_1 + \mathbf{R}_2 = \mathbf{P} \quad 9.125$$

The elongations of the two tension members are

$$\delta_1 = \frac{\mathbf{R}_1 L_1}{A_1E_1} \quad 9.126$$

$$\delta_2 = \frac{\mathbf{R}_2 L_2}{A_2E_2} \quad 9.127$$

If the horizontal bar remains horizontal, then $\delta_1 = \delta_2$. Therefore,

$$\frac{\mathbf{R}_1 L_1}{A_1E_1} = \frac{\mathbf{R}_2 L_2}{A_2E_2} \quad 9.128$$

Equations 9.125 and 9.128 are solved simultaneously to find $\mathbf{R}_1$ and $\mathbf{R}_2$.

Example 9.23

Find the forces in the three tension members.

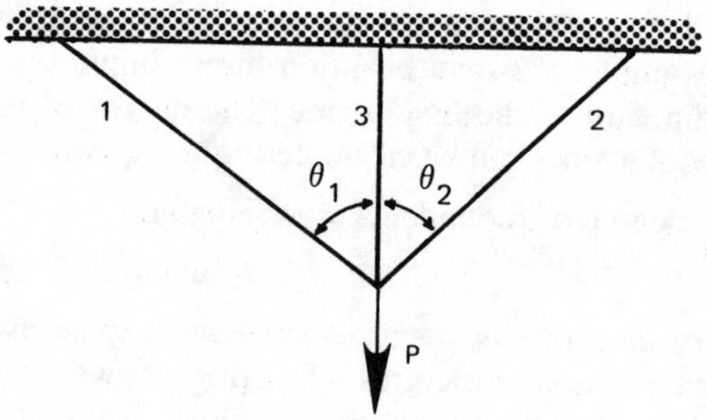

The equilibrium requirement is

$$\mathbf{P}_{1y} + \mathbf{P}_3 + \mathbf{P}_{2y} = \mathbf{P} \quad 9.129$$

$$\mathbf{P}_1\cos\theta_1 + \mathbf{P}_3 + \mathbf{P}_2\cos\theta_2 = \mathbf{P} \quad 9.130$$

The vertical elongations of all three tension members are the same at the junction.

$$\frac{\mathbf{P}_1 L_1}{A_1E_1}(\cos\theta_1) = \frac{\mathbf{P}_3 L_3}{A_3E_3} = \frac{\mathbf{P}_2 L_2}{A_2E_2}(\cos\theta_2) \quad 9.131$$

Equations 9.130 and 9.131 may be solved simultaneously to find $\mathbf{P}_1$, $\mathbf{P}_2$, and $\mathbf{P}_3$. It may be necessary to work with the *x*-components of the deflections in order to find a third equation.

Appendix A
Area Moments of Inertia

SHAPE	DIMENSIONS	CENTROID (x_c, y_c)	AREA MOMENT OF INERTIA
Rectangle	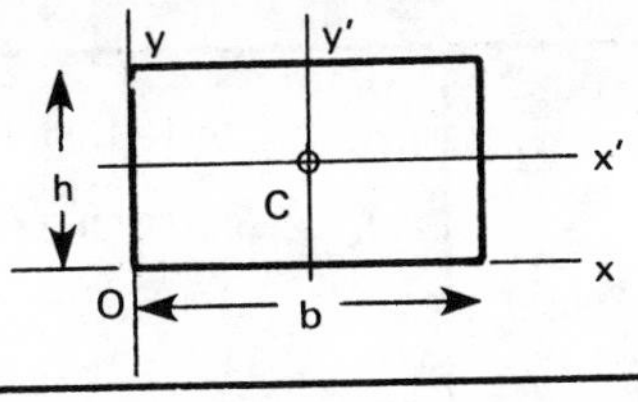	$(\frac{1}{2}b,\ \frac{1}{2}h)$	$I_{x'} = (1/12)bh^3$ $I_{y'} = (1/12)hb^3$ $I_x = (1/3)bh^3$ $I_y = (1/3)hb^3$ $J_C = (1/12)bh(b^2 + h^2)$
Triangle	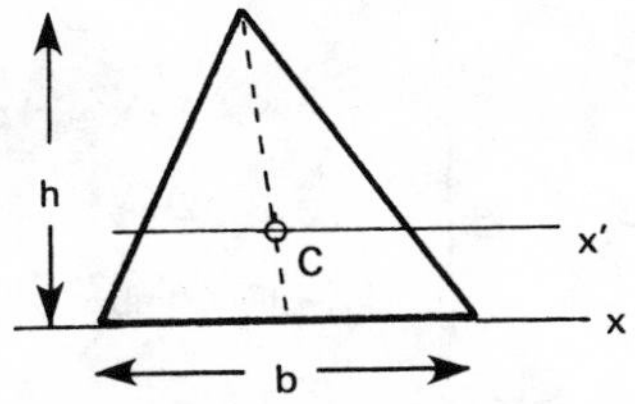	$y_c = (h/3)$	$I_{x'} = (1/36)bh^3$ $I_x = (1/12)bh^3$
Trapezoid	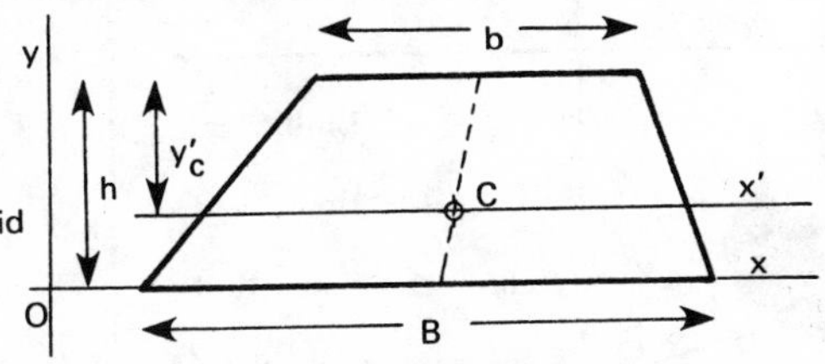	$y'_c = \dfrac{h(2B + b)}{3(B + b)}$ Note that this is measured from the top surface.	$I_{x'} = \dfrac{h^3(B^2+4Bb+b^2)}{36(B+b)}$ $I_x = \dfrac{h^3(B+3b)}{12}$
Quarter-Circle, of radius r	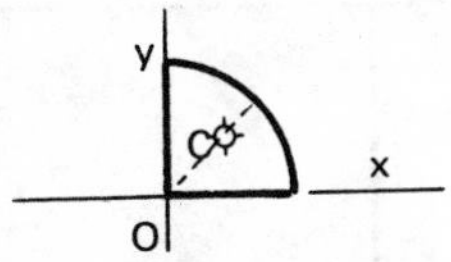	$((4r/3\pi),\ (4r/3\pi))$	$I_x = I_y = (1/16)\pi r^4$ $J_O = (1/8)\pi r^4$
Half Circle, of radius r	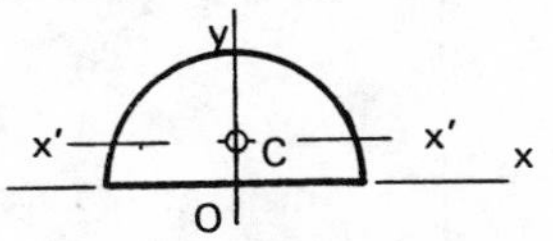	$(0,\ (4r/3\pi))$	$I_x = I_y = (1/8)\pi r^4$ $J_O = \frac{1}{4}\pi r^4$ $\quad I_{x'} = .11r^4$
Circle, of radius r	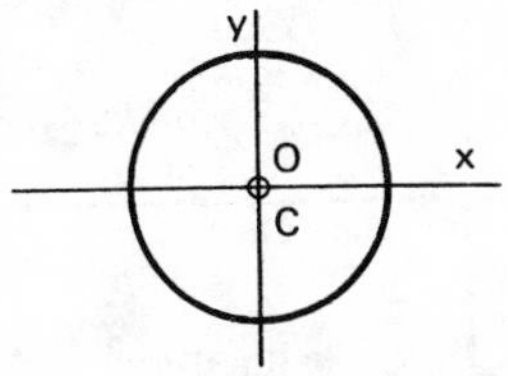	$(0,0)$	$I_x = I_y = \frac{1}{4}\pi r^4$ $J_O = \frac{1}{2}\pi r^4$
Parabolic Area	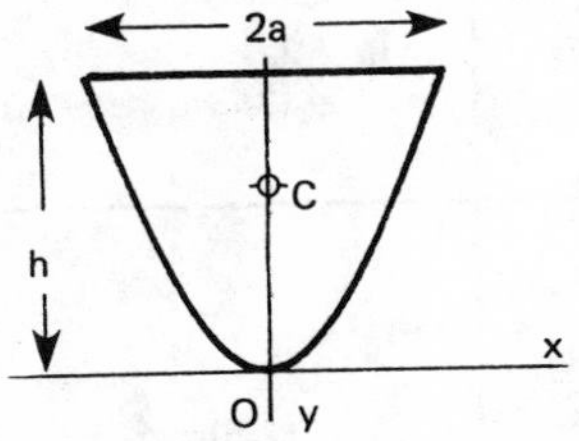	$(0,\ (3h/5))$	$I_x = 4h^3a/7$ $I_y = 4ha^3/15$
Parabolic Spandrel	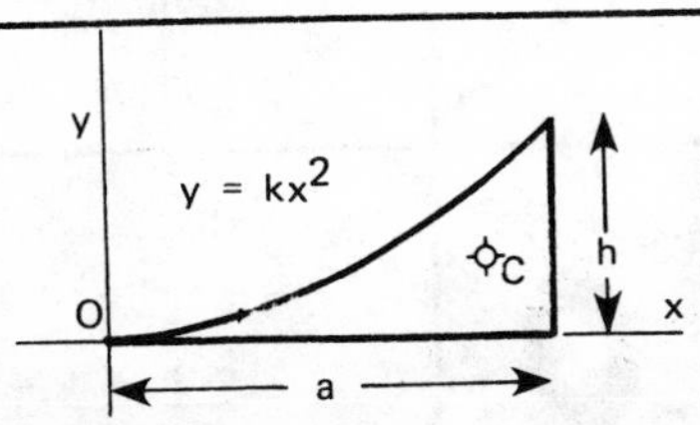	$((3a/4),(3h/10))$	$I_x = ah^3/21$ $I_y = 3ha^3/15$

Appendix B
Mass Moments of Inertia

(m is in slugs, Lengths are in feet)

Slender rod	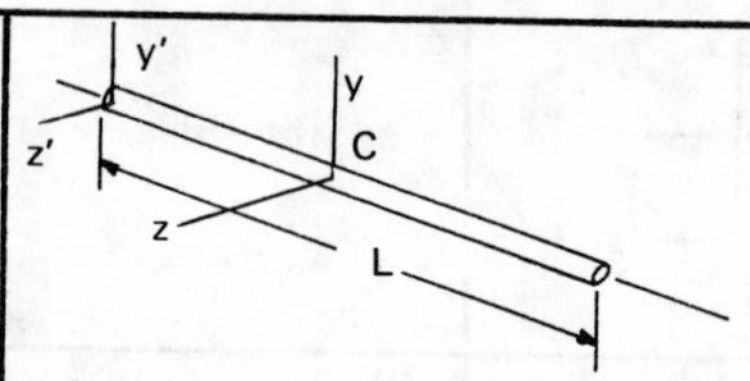	$I_y = I_z = (1/12)mL^2$ $I_{y'} = I_{z'} = (1/3)mL^2$
Thin rectangular plate	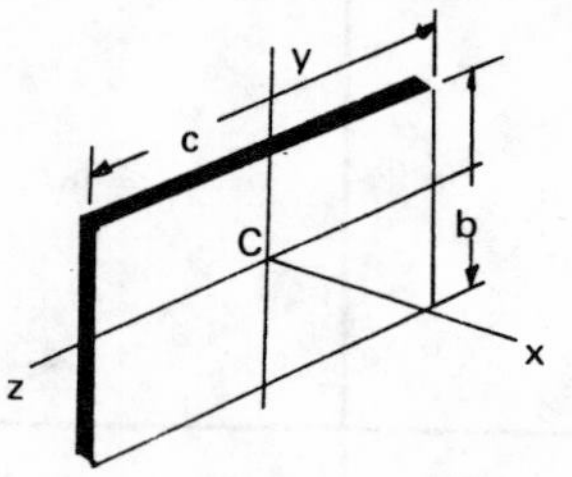	$I_x = (1/12)(b^2+c^2)\ m$ $I_y = (1/12)mc^2$ $I_z = (1/12)mb^2$
Rectangular Parallelepiped	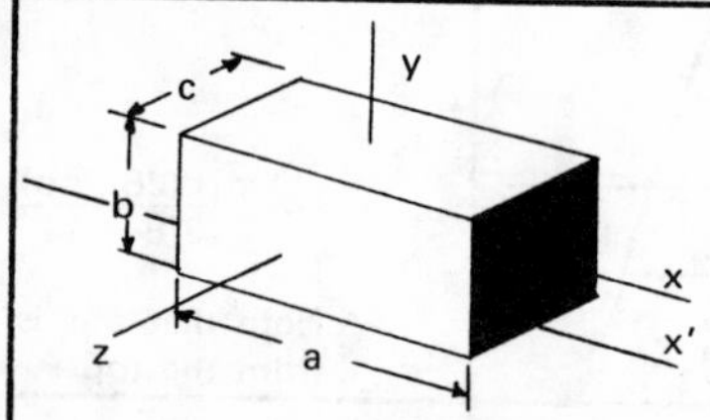	$I_x = (1/12)m(b^2+c^2)$ $I_y = (1/12)m(c^2+a^2)$ $I_z = (1/12)m(a^2+b^2)$ $I_{x'} = (1/12)m(4b^2+c^2)$
Thin disk, radius r	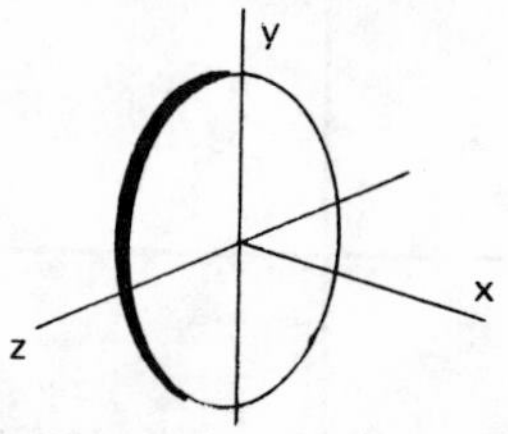	$I_x = \frac{1}{2}mr^2$ $I_y = I_z = \frac{1}{4}mr^2$
Circular cylinder, radius r	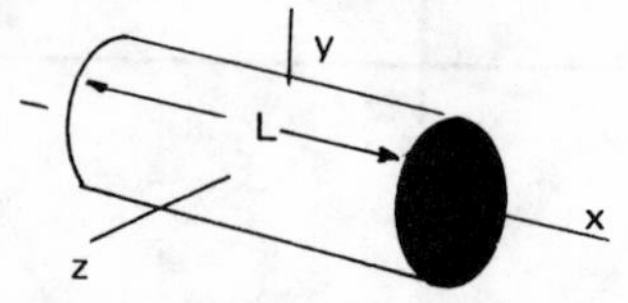	$I_x = \frac{1}{2}mr^2$ $I_y = I_z$ $= (1/12)m(3r^2+L^2)$
Circular cone, base radius r	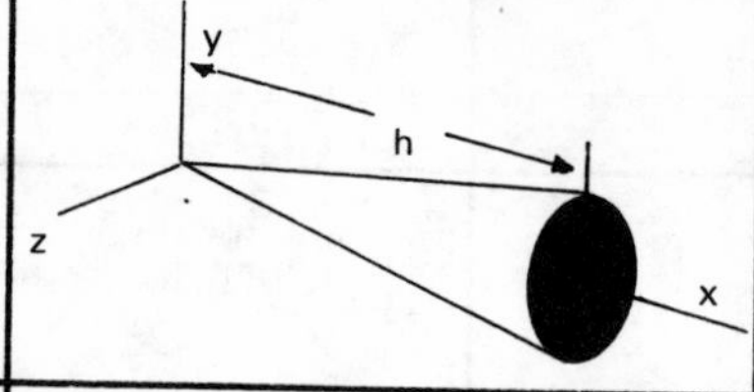	$I_x = (3/10)mr^2$ $I_y = I_z$ $= (3/5)m(\frac{1}{4}r^2+h^2)$
Sphere, radius r	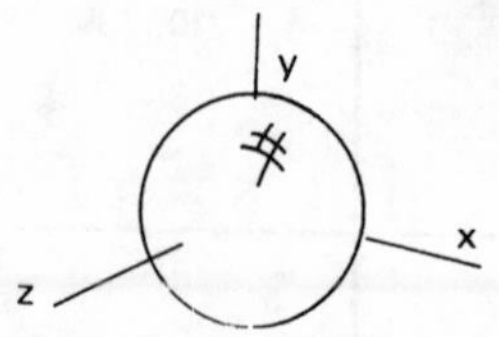	$I_x = I_y = I_z$ $= (2/5)mr^2$
Hollow circular cylinder	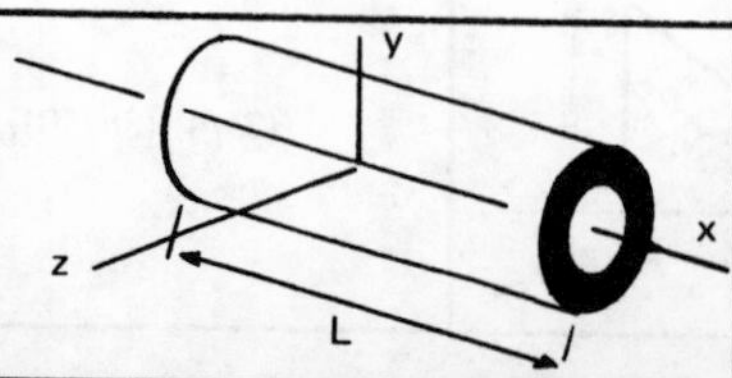	$I_x = \frac{1}{2}m(r_o^2 + r_i^2)$ $= \frac{\pi\rho L}{2}(r_o^4 - r_i^4)$

FORCES, MOMENTS, AND RESULTANTS

1. What is the required resisting force P if the 50 pound block rests on a frictionless plane?

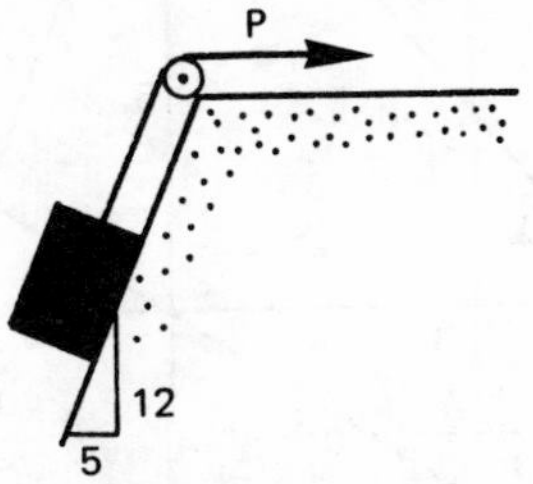

2. Neglecting the beam thickness, find the moment about point A. Find the resultant and its location with respect to point A.

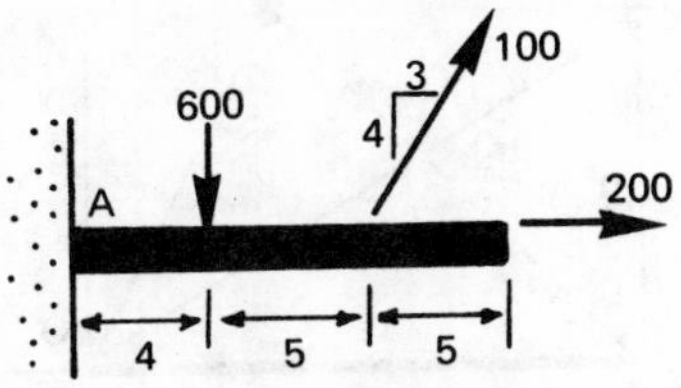

3. Find the x and y components of the resultant, as well as the moment about the center of the disc.

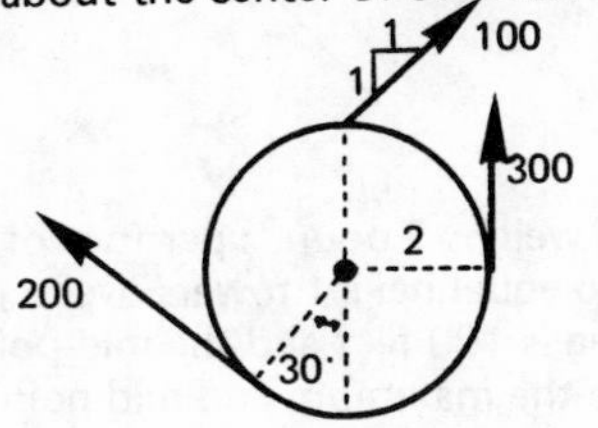

REACTIONS

4. Find the forces in the guy wire and the pole if AC = 19′ and CB = 24′.

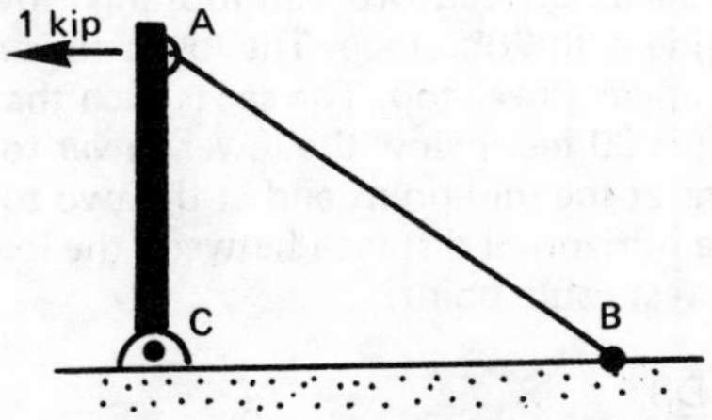

5. What are the reactions on the plate at points A and B?

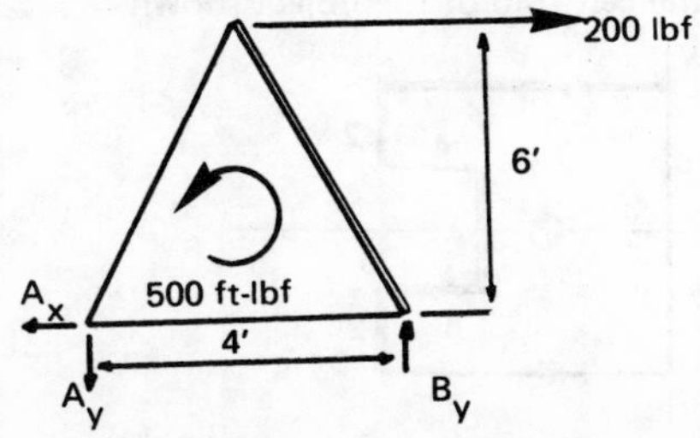

6. What are the reactions at point A and at the wall?

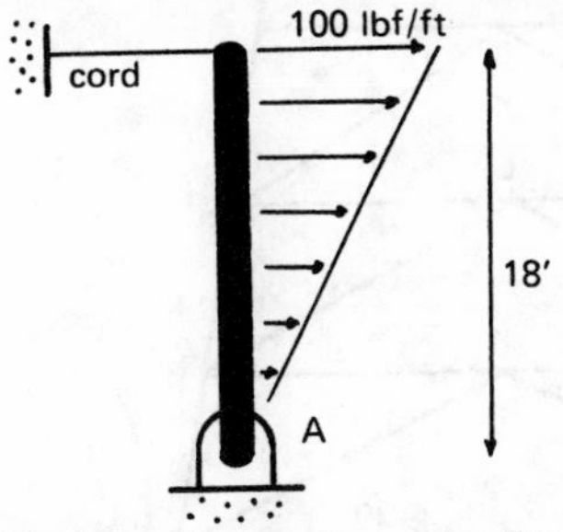

7. What are the reactions at point B?

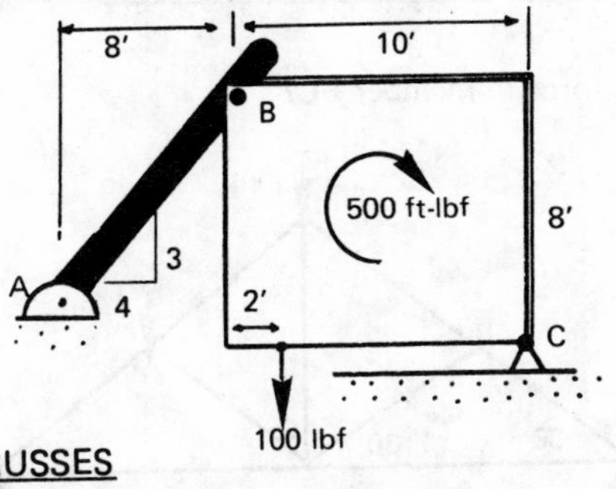

TRUSSES

8a. What is the force in member DE?

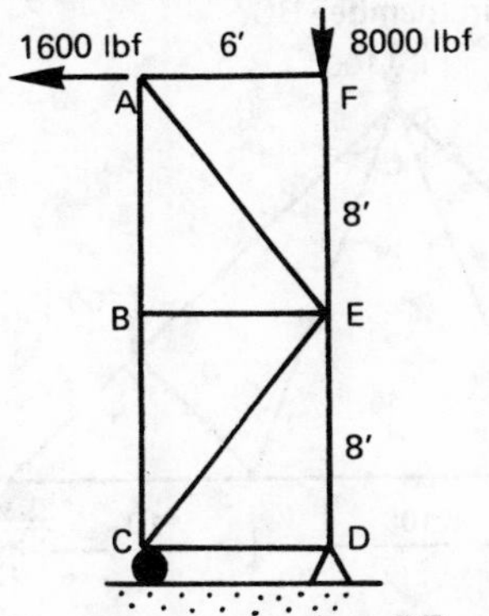

8b. Find the forces in members DE and HJ.

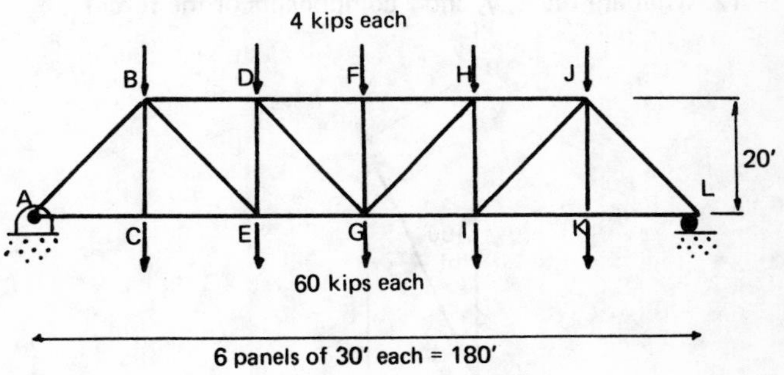

9. What is the force in member MG?

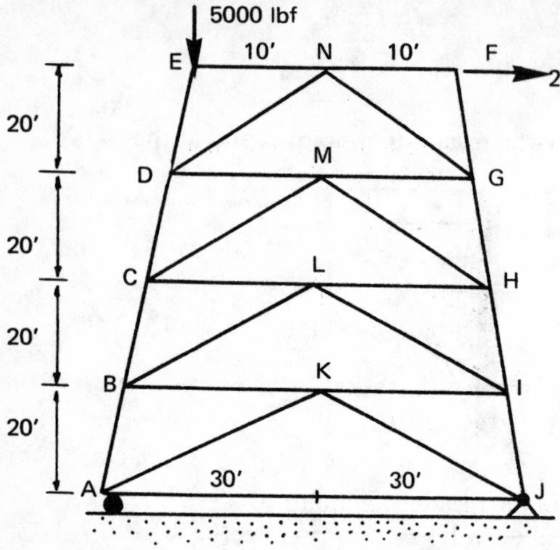

10. What is the force in member BC?

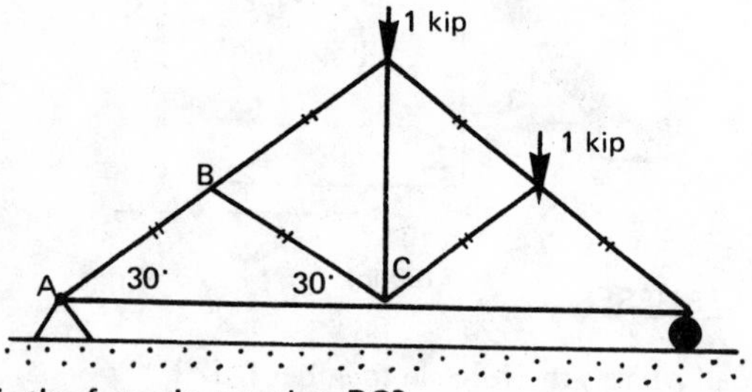

11. What is the force in member BC?

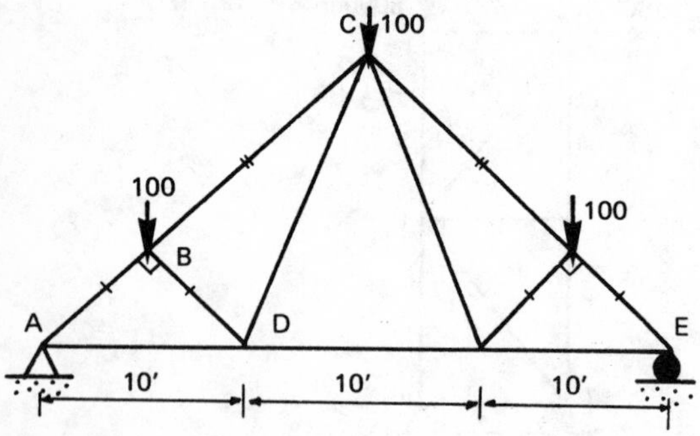

3-D FRAMEWORKS

12. What are the x, y, and z components of the force?

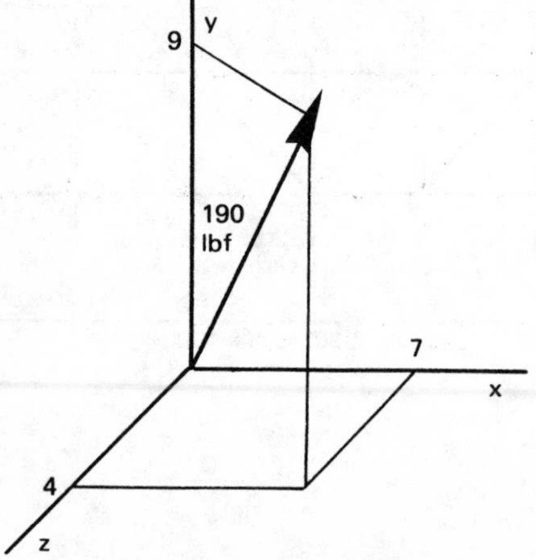

13. What are the forces in the legs?

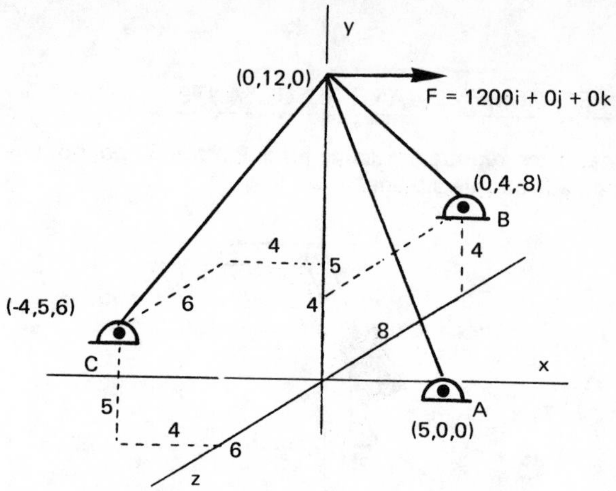

14. What are the x, y, and z components of the forces at A, B, and C?

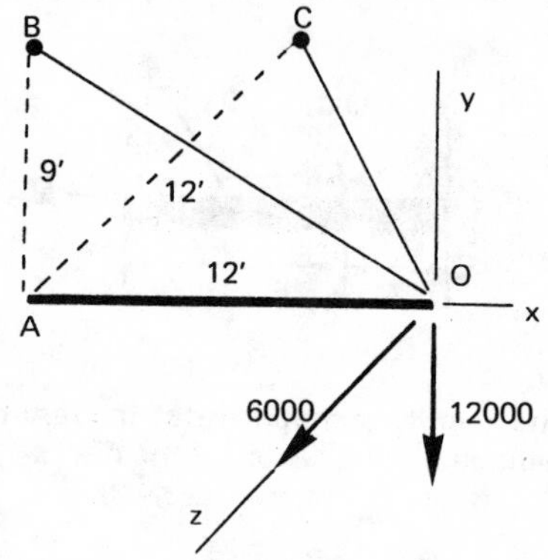

CABLES

15. A power line weighs 2 pounds per foot of length. It is supported by two equal height towers over a level forest. The tower spacing is 100 feet and the mid-point sag is 10 feet. What are the maximum and mid-point tensions?

16. What is the sag for the cable described in problem number 15 if the maximum tension is 500 pounds?

17. An armor shielded power line weighing 10 pounds per foot is installed between two 100 foot high towers on a mountainside with 20% slope. The lower tower top is 120' below the upper tower top. The sag is such that the lowest cable point is 30 feet below the lower tower top. What are the tensions at the mid-point and at the two towers? What is the horizontal distance between the lower tower and the lowest cable point?

CENTROIDS

18. Locate the centroid of the object shown.

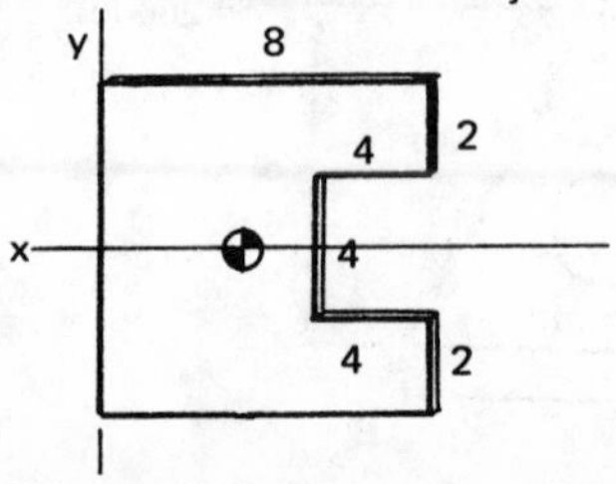

STATICS

19. Locate the centroid of the object shown.

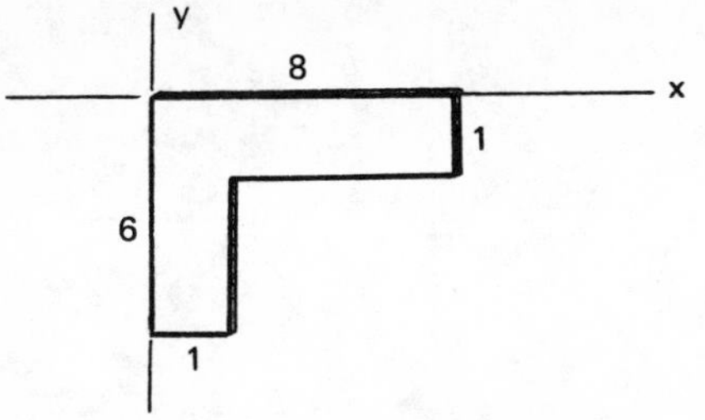

20. Replace the distributed load with three concentrated loads, and indicate the points of application.

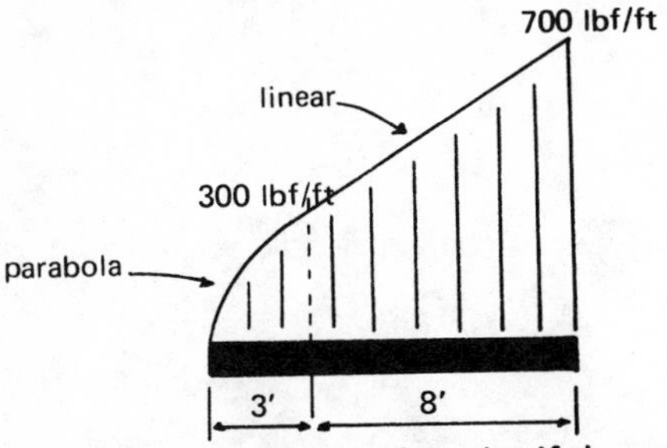

21. Locate the center of gravity if the weights of the sphere, rod, and cylinder at 64.4, 32.2, and 64.4 pounds respectively.

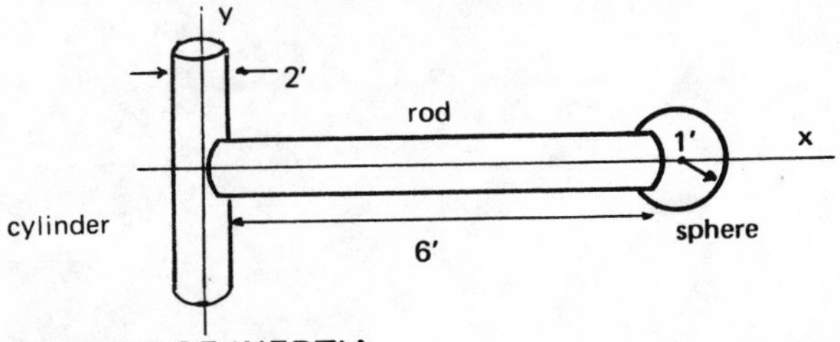

MOMENTS OF INERTIA

22. Find the moment of inertia about the BB axis if I_{AA} is 90 slug-ft^2 and the weight is 64.4 pounds.

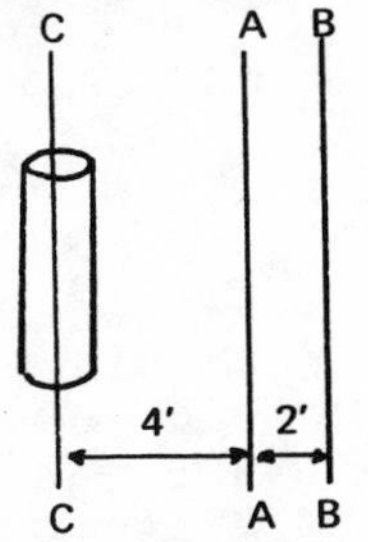

23. Find the centroidal moment of inertia about an axis parallel to the x axis.

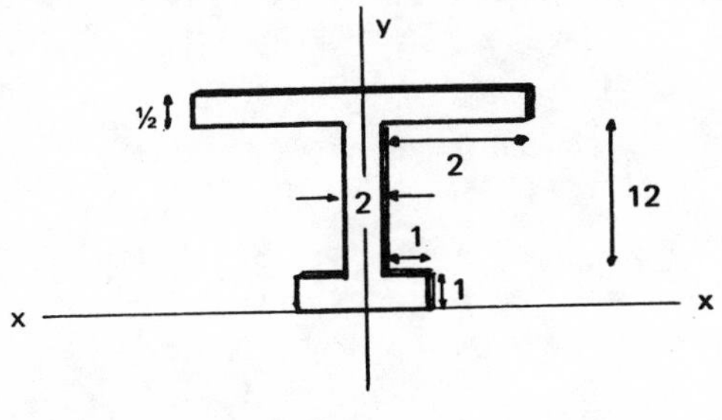

24. Find the moment of inertia about the horizontal centroidal axis.

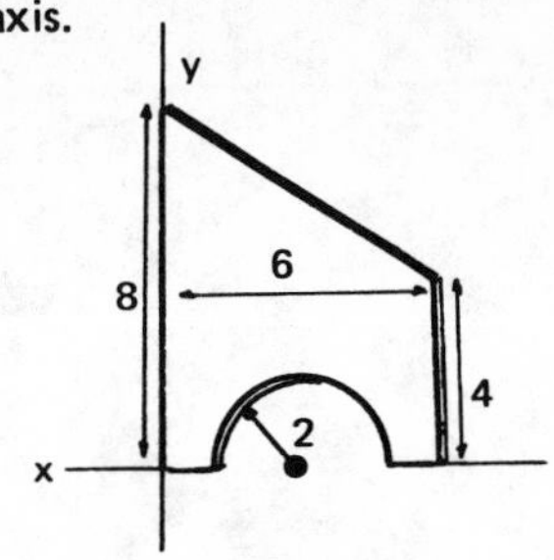

25. Find the radius of gyration for the object shown in problem 24.

CONTINUOUS BEAMS (optional)

26. What are the reactions R_1, R_2, and R_3?

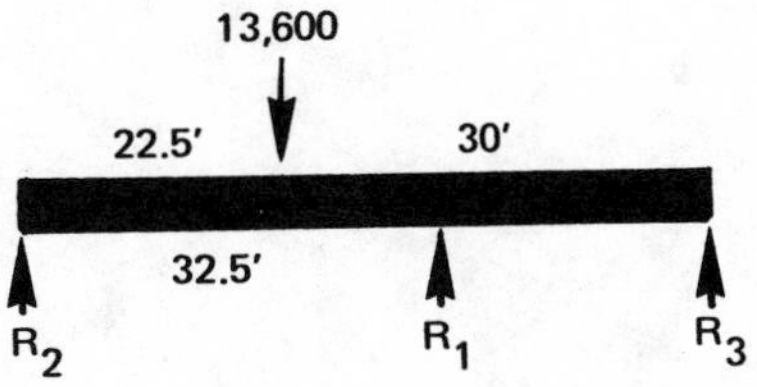

10 Materials Science

PART 1: Engineering Materials

Nomenclature

a	cubic lattice dimension	Å
BHN	Brinell hardness number	–
C	defect concentration	1/cm^3
d	ionic plane spacing	Å
D	diffusivity (diffusion constant)	cm^2/sec
g	local gravitational acceleration	ft/sec^2
g_c	gravitational constant (32.2)	lbm-ft/lbf-sec^2
h	height	ft
J	diffusion rate	1/cm^2-sec
m	mass	lbm
n	integer	–
S_{ut}	ultimate tensile strength	ksi

Symbols

λ	wavelength	Å
θ	angle	degrees

1. Metals

A. Iron and Steel

Iron is obtained from its oxides Fe_2O_3 and Fe_3O_4. These oxides contain 67% and 70% iron by weight respectively. The percentage of iron in ore is approximately 50%, having been reduced by the *gangue*[1] accompanying the oxides.

The process used to reduce iron ore to *iron* requires coke and limestone. The limestone lowers the melting temperature of the gangue so it can be drawn off as molten slag. The process takes place in a *blast furnace,* as shown in figure 10.1.

The top of the furnace is provided with a pair of conical bells for loading the charge while limiting the escape of gases. High temperature air is injected through openings called *tuyeres*[2] in the lower portion of the furnace. The air oxidizes the coke to produce heat and carbon monoxide. The carbon monoxide rises to the top of the furnace and, at a temperature of about 600°F, reduces the Fe_2O_3 and Fe_3O_4 to FeO.

$$C + O_2 \rightarrow CO_2 \qquad 10.1$$

$$CO_2 + C \rightarrow 2CO \qquad 10.2$$

$$2C + O_2 \rightarrow 2CO \qquad 10.3$$

$$3Fe_2O_3 + CO \rightarrow 2Fe_3O_4 + CO_2 \qquad 10.4$$

$$Fe_3O_4 + CO \rightarrow 3FeO + CO_2 \qquad 10.5$$

As the process continues, the FeO drops down to a region where the temperature ranges from 1300° to 1500°F. The FeO is reduced by surrounding carbon to iron. This iron drops into an area where the temperature ranges from 1500° to 2500°F. The iron becomes saturated with carbon in the form of both carbides and free carbon. The absorbed carbon lowers the melting point of the iron so that it runs as a liquid to the bottom of the furnace.

$$FeO + CO \rightarrow Fe + CO_2 \qquad 10.6$$

The *slag* melts at the same temperature as the iron and floats on the liquid iron. This allows the slag and iron to be drawn off into separate receptacles — the iron to molds where it is cast into *pigs,* and the slag to a slag heap (disposal).

Hot air is provided by stoves which adjoin the blast furnace. Two or more stoves are provided for the blast furnace. Each stove is alternately heated by the hot furnace gases and then heats outside air to 1400°F before being injected into the furnace.

Pig iron, which contains approximately 4% carbon, is brittle. Carbon must be removed before the iron can be used. The pigs also contain other undesirable elements such as sulfur, phosphorous, silicon, and manganese.

[1]Pronounced *găng*. Gangue is the earth and stone mixed with the iron oxides.

[2]Pronounced *twee'-yair.*

MATL SCI

Four processes are available for the further refinement of the iron: the Bessemer, open hearth, electric furnace, and wrought iron processes.

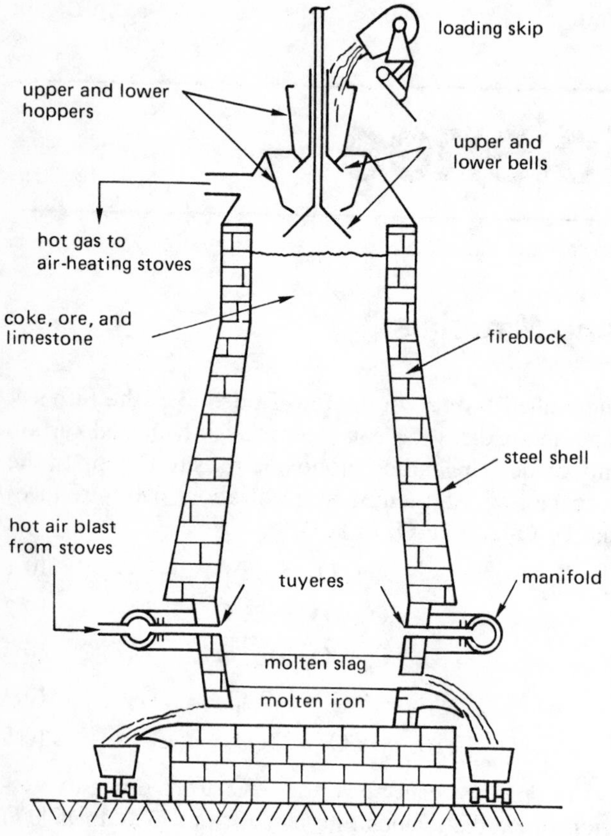

Figure 10.1 Blast Furnace

The *Bessemer process* uses a pear-shaped steel crucible lined with refractory materials. The crucible, which is provided with air inlets in its base, is filled with molten pig iron at about 2200°F. High pressure air is blown through the iron, oxidizing the carbon and other impurities and causing a temperature rise to 3500°F.

The burning impurities are consumed in about 20 minutes after which measured amounts of carbon, manganese and other alloying agents are added to obtain the desired grade of steel. The Bessemer process is a crude process since its speed leads to poor control of constituents. The product is used to make cheaper grades of sheet, wire, pipe, and screw stock.

The *open hearth process* (figure 10.2) is used to manufacture most of the steel used in the United States. The *charge* is contained in a shallow receptacle and is composed of a mixture of pig iron, scrap iron, iron oxide, and limestone. Heat is provided by burning CO above the iron. Since the combustion products contain very little oxygen, oxidation of the impurities is accomplished by the iron oxide in the charge. The function of the limestone is to provide a protective slag over the melt. Since this is a slow process (8 to 12 hours), continuous monitoring of the steel composition is possible.

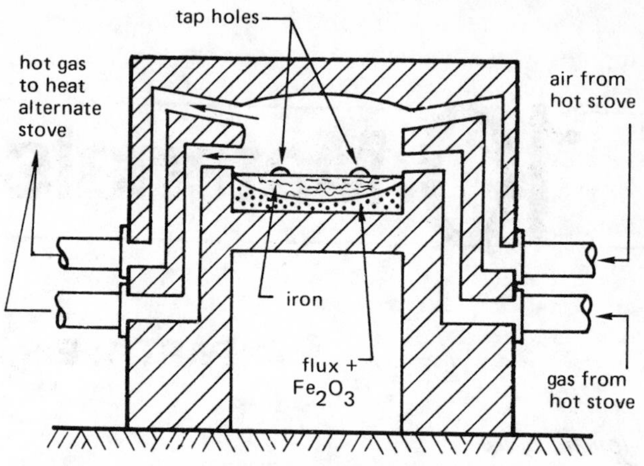

Figure 10.2 Open Hearth Furnace

Electric furnaces employing electric arc or induction heating are used for the production of tool and alloy steels. Since no air or gaseous fuels are required, the impurities introduced by them are eliminated. The furnace walls are lined with either silica (acidic) or magnesite (basic) brick, the selection being based on the iron being refined. This process also depends on the use of iron oxide as the reducing agent.

Wrought iron is steel with small amounts of slag in the form of fibrous inclusions. Wrought iron is produced in a *reverberatory furnace* similar to the open hearth furnace. The molten iron floats on a layer of iron oxide which provides the oxygen for removal of carbon and sulfur. The limestone flux combines with silicon and phosphorus to form slag. Spongy masses of iron and slag are collected on the ends of steel rods inserted into the pool of molten metal. These masses are then removed and hammered or pressed to squeeze out most of the slag. The remaining product consists of slag coated iron particles welded together by the forging process.

B. Aluminum

Most of the aluminum in use today is recovered from *bauxite* ore. This ore is a mixture of hydroxides of aluminum mixed with iron, silicon, and titanium oxides. The ore is crushed and ground to fine powder. It is then treated with a hot solution of sodium hydroxide producing sodium aluminate and water. This solution is drawn off into a separate tank, leaving the remaining ore constituents as a solid deposit.

$$Al(OH)_3 + NaOH \rightarrow NaAl(OH)_4 \qquad 10.7$$

As the solution cools, aluminum hydroxide precipitates out leaving a sodium hydroxide solution. The aluminum hydroxide is then baked to form aluminum oxide Al_2O_3. Final reduction is accomplished through an electrolytic process which uses molten *cryolite* (Na_3AlF_6) as the electrolyte and large carbon blocks as anodes. The steel process tank is lined with carbon and acts as the cathode. The aluminum oxide dissolves in the cryolite and is separated into molten aluminum and oxygen by the applied electric current. The aluminum collects in the bottom of the tank. Carbon dioxide is released at the anodes. The cryolite does not enter into the reaction and can be reused.

$$Al_2O_3 \rightarrow 2Al^{+++} + 3O^{--} \qquad 10.8$$

$$Al^{+++} + 3e^- \rightarrow Al \qquad 10.9$$

$$C + 2O^{--} \rightarrow CO_2 + 4e^- \qquad 10.10$$

C. Copper

Copper occurs in the free state as well as in ores containing its oxides, sulfides, and carbonates. *Native copper* is recovered by the simple process of heating finely crushed ore. Molten copper flows to the bottom of the furnace. Oxides and carbonates of copper are reduced in a reverberatory furnace by combining with the hot unburned gases used in the process.

Sulfides are heated in air, and the sulfur is replaced by oxygen. The product, which contains both copper and iron oxides, is reduced in a reverberatory furnace. This removes the oxygen but leaves some sulfur and iron. The final treatment takes place in a furnace similar to the Bessemer converter in which air is injected into the molten copper, removing the sulfur. The iron oxide combines with a silica furnace liner.

Electrolysis is required to remove remaining impurities. Thick sheets of impure copper are immersed, together with thin sheets of pure copper, in an electrolyte of copper sulfate. The pure copper acts as the cathode. A direct current causes copper from the impure material to migrate to the pure sheets. Impurities drop to the bottom of the tank.

2. Natural Polymers

Many of the natural organic materials (e.g., rubber and asphalt) are polymers. Natural rubber is a polymer of *isoprene*. *Vulcanization* is accomplished by heating raw rubber with small amounts of sulfur. The process raises the tensile strength of the material from around 300 psi to 3,000 psi. The addition of carbon black as a reinforcing filler raises this value to 4,500 psi and provides tear resistance and toughness.

3. Synthetic Polymers

Polymers are extremely large molecules built up of long chains of repeating units. The basic repeating unit is called the *monomer* or *mer*. The *degree* of the polymer is the average number of units in the compound, typically 75 to 750 mers per molecule.

A compound having only two, three, or four monomers may be called a *di-*, *tri-*, or *tetramer* respectively. In general, compounds with degrees of less than ten are called *telenomers* or *oligomers*.

The stiffness and hardness of polymers vary with their degree. Polymers with low degrees of polymerization are liquids or oils. With increasing degree, they go through waxy to hard resin stages. High degree polymers have hardness or elastic qualities which make them useful for engineering applications.

Example 10.1

A polyvinyl chloride molecule is found to contain 860 carbon atoms, 1290 hydrogen atoms, and 430 chlorine atoms. What is the degree of polymerization? (The vinyl chloride mer is C_2H_3Cl with a molecular weight of 62.5.)

The molecular weight of the polymer is

$$860(12) + 1290(1) + 430(35.5) = 26{,}875$$

The degree of polymerization is

$$\frac{26{,}875}{62.5} = 430$$

A polymer will contain molecules with different chain lengths. Therefore, the degree of polymerization will vary from molecule to molecule. A determination of the polymer's degree requires statistical analysis.

Example 10.2

What is the degree of polymerization of a polyvinyl acetate sample having the following molecular weight analysis?

range of molecular weights	fraction
5,000 — 15,000	.30
15,000 — 25,000	.47
25,000 — 35,000	.23

Using the midpoint of each range, the average molecular weight is

$$(.30)(10{,}000) + (.47)(20{,}000) + (.23)(30{,}000) = 19{,}300$$

The vinyl acetate mer is $C_4H_6O_2$ with a molecular weight of 86. The degree of polymerization is

$$\frac{19{,}300}{86} = 224.4$$

Polymers are named by adding the prefix *poly* to the name of the basic mer. For example, C_2H_4 is the chem-

MATL SCI

Table 10.1 Names of Some Mers

name	repeating unit	combined formula
ethylene	CH_2CH_2	C_2H_4
propylene	$CH_2(HCCH_3)$	C_3H_6
styrene	$CH_2CH(C_6H_5)$	C_8H_8
vinyl acetate	$CH_2CH(C_2H_3O_2)$	$C_4H_6O_2$
vinyl chloride	CH_2CHCl	C_2H_3Cl
isobutylene	$CH_2C(CH_3)_2$	C_4H_8
methyl methacrylate	$CH_2C(CH_3)(COOCH_3)$	$C_5H_8O_2$
acrylonitrile	CH_2CHCN	C_3H_3N
epoxide (ethoxylene)	CH_2CH_2O	C_2H_4O
amide (nylon)	$CONH_2$ or CONH	$CONH_2$ or CONH

ical formula for ethylene. Chains of C_2H_4 are called polyethylene. Some important repeating units are listed in table 10.1.

Two processes are used to form polymers: addition polymerization and condensation polymerization. With *addition polymerization,* mers combine sequentially using single covalent bonds to form chains. For example, the formation of polyethylene is given by equation 10.11.

$$2(CH_2 = CH_2) \rightarrow —CH_2—CH_2—CH_2—CH_2— \qquad 10.11$$

Substances called *initiators* are used in addition polymerization. Initiators break down under heat or light and provide free radicals. These free radicals act as chain carriers as well as openers of double bonds. The reaction produces unpaired electrons, and the process continues until the initiator is used up or until another free radical reacts to form a termination molecule. The latter occurrence is called *saturation.*

A typical initiator is benzoil peroxide which, upon decomposition, produces a free radical.

$$(C_6H_5COO_2)_2 \rightarrow 2C_6H_5— + 2CO_2 \qquad 10.12$$

Propagation starts with the free radical reacting with the monomer (vinyl chloride in equations 10.13 and 10.14). The radical opens the monomer's double bond and forms a new free radical. The new free radical reacts with an additional monomer, continuing the reaction.

$$C_6H_5— + CH_2 = CHCl \rightarrow C_6H_5—CH_2—CHCl— \qquad 10.13$$

$$C_6H_5—CH_2—CHCl— + CH_2 = CHCl \rightarrow C_6H_5—CH_2—CHCl—CH_2—CHCl— \qquad 10.14$$

Condensation polymerization opens bonds in two molecules to form a larger molecule. This formation is often accompanied by the release of small molecules such as H_2O, CO_2, and N_2. The repeating units derived from the condensation process are not the same as the monomers from which they are formed since portions of the original monomer form small molecules and are eliminated.

Phenolic plastics are produced from the condensation of formaldehyde and ammonia into a crystalline solid.

$$6HCHO + 4NH_3 \rightarrow (CH_2)_6N_4 + 6H_2O \qquad 10.15$$

Some types of polymers form multiple chains linked by cross branches. These are called *framework–bonded polymers.* Some framework structures are resistant to heat and will not melt or flow. These are called *thermosetting resins.* Conversely, most chain polymers can be softened and formed under the application of heat and pressure. These are called *thermoplastic resins.*

Thermosetting resins are represented by the phenolics (bakelites) and polyesters. The thermoplastic resins include cellulose derivatives, polystyrene, vinyl polymers, polyethylene, nylon, epoxies and the methacrylics.

Table 10.2
Approximate Properties of Woods

variety	flexural strength (psi)	tensile strength (psi)*	compression strength (psi)**	modulus of elasticity (psi × EE−6)
oak (red)	14,400	820	6,920	1.81
mahogany	11,460	740	6,780	1.50
douglas fir	12,200	340	7,430	1.95
cedar	7,700	220	5,020	1.12
fir	9,800	300	5,480	1.49
spruce	10,200	710	5,610	1.57
yellow pine	12,800	470	7,080	1.80
redwood	10,000	240	6,150	1.34

*Perpendicular to grain **Parallel to grain

MATL SCI

4. Wood

Wood is classified broadly as *softwood* or *hardwood*. Softwoods contain tubelike fibers oriented with the longitudinal axis. Hardwoods contain complex structures (storage cells) in addition to longitudinal fibers. Fibers in hardwoods are much smaller and shorter than those in softwoods.

The mechanical properties of woods are influenced by moisture content and grain orientation. Strengths of dry woods are approximately twice that of wet or green woods. The mechanical properties of several dry wood varieties are listed in table 10.2.

5. Inorganic Materials

A. Ceramics

Ceramics are compounds of metals and non-metals. For example, MgO, SiO_2, and ZnS are ceramics. Ceramics are typically good insulators, although some (e.g., Fe_3O_4) possess semiconductor properties. Other ceramics, such as $BaTiO_3$, are *piezoelectric* (i.e., generate a voltage when compressed). Ferrimagnetic materials *(ferrites, spinels,* or *ferrispinels)* are ceramics. Some common or important ceramics are listed in table 10.3.

Table 10.3
Common or Important Ceramics

compound	mineral name	use
Al_2O_3	corundum	abrasives
$Al_2Si_2O_5(OH)_4$	kaolinite clay	porcelain paste
$BaTiO_3$		piezoelectricity
BN		refractory
CaF_2	fluorite	flux
Fe_3O_4 or $FeFe_2O_4$	magnetite	thermistors
$KAlSi_3O_8$	feldspar	—
MgO	periclase	refractory
$MgSiO_4$	forsterite	refractory
$MnFe_2O_4$		ferrimagnetism
$MgCr_2O_4$		piezoelectricity
$MgFe_2O_4$		antiferromagnetism
$NiFe_2O_4$		ferrimagnetism
NaCl	salt	—
$PbZrO_3$		piezoelectricity
SiC		refractory
SiO_2	quartz*	—
TiC		refractory
UO_2		nuclear fuel
ZrN		refractory
$ZnFe_2O_4$		ferrimagnetism

*SiO_2 has several temperature-dependent polymorphs, including coesite, cristobalite, and tridymite.

B. Concrete

Concrete is a mixture of mineral aggregates locked into a solid structure by a binding material. The concrete is produced by adding water to the aggregate and binder, and then casting the mixture in place. The semi-fluid mixture hardens to concrete by chemical action.

The binding material is known as *cement*. There are two types of cement: bituminous and non-bituminous. Asphalt and tar are the most common bituminous cements. These cements harden upon cooling. They do not require water.

The most common non-bituminous cements are alumina cement and portland cement. Since portland cement is the most widely used, the terms *cement* and *concrete* are frequently used interchangeably. It is assumed in this chapter that the binding agent is portland cement.

Portland cement is manufactured from lime, silica, and alumina in the appropriate proportions. After being ground and blended, the mixture is kilned at approximately 2700°F. The resulting *clinkers* are ground and mixed with a small amount of plaster of Paris.

There are five common types of portland cement, although special cements are available on special order. The type of cement used is dependent on the application intended for it.

type 1: *normal portland cement:* This is a general-purpose cement used whenever sulfate hazards are absent and when the heat of hydration will not produce objectionable rises in temperature. Typical applications are sidewalks, pavement, beams, columns, and culverts.

type 2: *modified portland cement:* This cement has a moderate sulfate resistance. It is generally used in hot weather in the construction of large concrete structures. Its heat rate and total heat generation are lower than for normal portland cement.

type 3: *high-early-strength portland cement:* This type of cement develops its strength quickly. It is suitable for use when the structure must be put into early use or when long-term protection against cold temperatures is not feasible. Its shrinkage rate, however, is higher than for types 1 and 2, and extensive cracking may result.

type 4: *low-heat portland cement:* This is used in massive concrete structures such as gravity dams. Low-heat cement is required to minimize the curing heat. The ultimate strength also develops more slowly than for the other types.

type 5: *sulfate-resistant portland cement:* This type of cement is used when exposure to sulfates is expected. This typically occurs in some western states having highly alkaline soils.

MATL SCI

Cement itself is very fine. However, the majority of concrete consists of sand and rock particles that have been added to increase weight and bulk. These sand and rock particles are known as *aggregate.*

Table 10.4
Relative Strengths of Concrete

concrete type	compressive strength, % of normal strength 3 days	28 days	3 months
1	100	100	100
2	80	85	100
3	190	130	115
4	50	65	90
5	65	65	85

Sand that will pass through a #4 sieve (openings less than .25") is known as *fine aggregate*. Any particles that are larger than this are known as *coarse aggregate.* Coarse aggregate is produced from crushed stone.

Proportions of a concrete mixture are designated as a ratio of cement, fine aggregate, and coarse aggregate, in that order. For example, 1:2:3 means that one part of cement, two parts of fine aggregate, and three parts of coarse aggregate are to be combined. The ratio may be either in terms of weight or volume. Weight ratios are more common.

The amount of water used is called out in terms of gallons of water per sack of cement. This is known as the *water-cement ratio.*

Anything added to concrete to improve workability, hardening, or strength characteristics is known as an *admixture.* Hydrated lime, diatomaceous silica, fly ash, and bentonite are added to concrete which has too little fine aggregate. These admixtures separate the coarse aggregate and reduce the friction of the mixture. Calcium chloride may be added as a curing accelerator. It is also an antifreeze, but its use as such is not recommended in pre-stressed concrete or concrete with aluminum embedments.

Sulfonated soaps and oils, as well as natural resins, increase air entrainment. This increases the durability of the concrete while decreasing the strength and weight.

The amount of concrete that can be made from mixing known proportions of ingredients can be found from the *absolute* or *solid volume method.* This method assumes that there will be no voids in the placed concrete. Therefore, the amount of concrete will be the sum of the solid volumes of the cement, sand, coarse aggregate, and water.

To use the absolute volume method, it is necessary to know the solid densities of the constituents. In the absence of any other information, the following data can be used for solid densities: cement, 195 pcf; fine aggregate, 165 pcf; coarse aggregate, 165 pcf; water, 62.4 pcf.[3]

A sack of cement weighs 94 pounds. 7.48 gallons of water make a cubic foot. There are 239.7 gallons per ton of water.

If the mix proportions are volumetric, it will be necessary to multiply the ratio values by the bulk densities to get the weights of the constituents. Then, weight ratios can be calculated and the absolute volume method applied directly.

Example 10.3

A mix was designed as 1:1.9:2.8 by weight. The water-cement ratio was chosen as 7 gallons of water per sack of cement. (a) What is the concrete yield in cubic feet per sack of cement? (b) How much of each constituent is needed to make 45 cubic yards of concrete?

(a) The solution can be tabulated.

material	ratio	weight per sack cement	solid density	absolute volume
cement	1.0	94	195	94/195 = .48
sand	1.9	179	165	179/165 = 1.08
coarse	2.8	263	165	263/165 = 1.60
water				7/7.48 = .94
				4.10

The yield is 4.1 cubic feet of concrete per sack cement.

(b) The number of one-sack batches is

$$\frac{(45)\ \text{yd}^3\ (27)\ \text{ft}^3/\text{yd}^3}{(4.1)\ \text{ft}^3/\text{sack}} = 296.3\ \text{sacks}$$

So, order 297 sacks of cement.

$$\frac{(297)(1.9)(94)}{2000} = 26.5\ \text{tons of sand}$$

$$\frac{(297)(2.8)(94)}{2000} = 39.1\ \text{tons of coarse aggregate}$$

$$(297)(7) = 2079\ \text{gallons of water}$$

Calculation of solid densities can be complicated by air entrainment and/or the water content of the aggregate. The following guidelines can be used:

- The absolute volume yield is increased by the addition of air. This can be accounted for by dividing the solid yield by (1 – %air).
- Any water content in the aggregate above the *saturated surface dry (SSD)* water content must be subtracted from the water requirements.
- Any porosity (water affinity) below the *SSD* water content must be added to the water requirement.
- The densities used in the calculation of the yield should be the *SSD* densities.

[3]The abbreviation *pcf* is commonly used in concrete work for *pounds per cubic foot.*

MATL SCI

Example 10.4

50 cubic feet of 1:2½:4 (by weight) concrete are to be produced. The constituents have the following properties:

constituent	*SSD* density	moisture (from *SSD*)
fresh cement	197	—
sand	164	5% excess
coarse aggregate	168	2% dry

What are the required order quantities if 5.5 gallons of water are to be used per sack and the mixture is to have 6% entrained air?

As in example 10.3, the solution can be tabulated.

constituent	ratio	weight per sack cement	*SSD* density	absolute volume
cement	1.0	94	197	.477
sand	2.5	235	164	1.433
coarse	4.0	376	168	2.238
water			5.5/7.48	.735
				4.883

The yield with 6% air is

$$\left(\frac{4.883}{1-.06}\right) = 5.19 \text{ ft}^3\text{/sack.}$$

The number of one sack batches is

$$\frac{50}{5.19} = 9.63$$

The required sand weight (ordered as is, not *SSD*) is

$$\frac{(9.63)(1.05)(94)(2.5)}{2000} = 1.19 \text{ tons}$$

The required coarse aggregate weight (ordered as is, not *SSD*) is

$$\frac{(9.63)(.98)(94)(4)}{2000} = 1.77 \text{ tons}$$

The excess water contained in the sand is

$$(1.19)\left(\frac{.05}{1.05}\right)(239.7) = 13.58 \text{ gallons}$$

The water needed to bring the coarse aggregate to *SSD* conditions is

$$(1.77)\left(\frac{.02}{.98}\right)(239.7) = 8.66 \text{ gallons}$$

The total water needed is

$$(5.5)(9.63) + 8.66 - 13.58 = 48.0 \text{ gallons}$$

Typical ultimate compressive strengths for concrete vary from 2000 psi to 8000 psi after curing 28 days.[4] The ultimate tensile strength is very low, seldom greater than 400 psi.

The ultimate compressive strength is dependent on the water/cement ratio. It is fairly independent of the mix proportion. Provided that the mix is workable, *Abram's strength law* states that the compressive strength varies directly with the cement/water ratio.

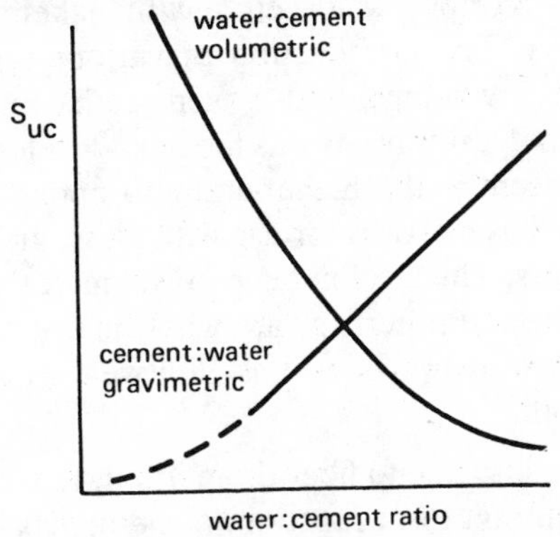

Figure 10.3 Concrete Compressive Strengths

The results of a compressive stress-strain test are shown in figure 10.4. The slope of the stress-strain line varies with the applied stress. There are three ways of evaluating the *modulus of elasticity.* These methods give the initial *tangent modulus,* actual tangent modulus, and *secant modulus.* The secant modulus is most frequently used.

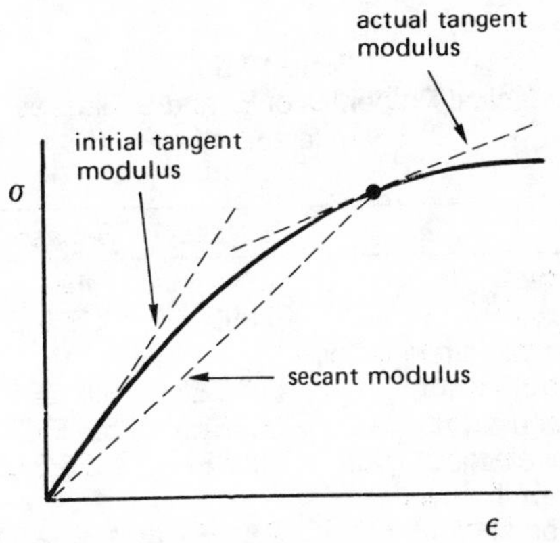

Figure 10.4 Concrete Compressive Test

The modulus of elasticity is greatly dependent on age, quality, and proportions. It can vary from 1 EE6 to 5 EE6 psi at 28 days.

Concrete density can vary from 100 pcf to 160 pcf depending on the mixture ratios and the specific gravities of the constituents. Generally, the range will be 140 pcf to 160 pcf. Steel reinforced concrete will be between 3% and 5% higher in density than similar plain concrete. An average of 150 pcf can be used in most calculations.

[4]Since the ultimate strength increases with time, all values should specify the concrete age. If no age is given, a standard 28-day age is assumed.

6. Composite Materials

A. Fiber/Resin Mixtures

The initial processes leading to the widespread use of fiber-reinforced polymers were the impregnation of natural fibers (cotton, wool, etc.) with bakelite and phenolic resins. The introduction of various types of glass (rovings,[5] windings, and woven cloth) for reinforcement of polyester resins was the next development step. More recently, the higher-strength epoxy resins have been used as matrices for use with glass, graphite, and boron fibers. The resultant composite materials are highly *anisotropic*. Properties vary with the orientation of monofilament layups as well as with weave orientation in the cloth.

In directions transverse to fiber orientation, tensile and compressive strength is a function of the matrix material. Loads parallel to the fibers are carried by the reinforcement, while flexural strength is limited by the shear bond between the filaments and the matrix.

Typical reinforcing materials for fiberglass are *E*- and *S*-glass.[6] *S*-glass is a silica-alumina-magnesia compound developed for improved tensile properties and is used mainly in non-woven, monodirectional and wound configurations. *E*-glass is a lime-alumina-borosilicate compound used primarily in woven fabrics.

Table 10.5
Typical Properties of E- and S-Glasses
(room temperature)

property	E-glass	S-glass
specific gravity	2.54	2.48
density (lbm/in³)	0.092	0.090
ultimate tensile strength (psi)		
• monofilament	5.0 EE5	6.6 EE5
• 12-end roving	3.7 EE5	5.5 EE5
modulus of elasticity (psi)	10.5 EE6	12.5 EE6
coefficient of thermal expansion (1/°F)	2.8 EE−6	1.6–2.2 EE−6
specific heat (BTU/lbm-°F)	0.192	0.176

Graphite fibers are used where high stiffness and low coefficients of thermal expansion are needed. These properties must be balanced against the disadvantages of brittleness and high cost. Ultimate tensile strengths for graphite vary inversely with modulus of elasticity. Graphite fibers range in tensile strength from 180 *ksi* for yarn configurations to 350 *ksi* for *tow* (loose, untwisted fibers), while modulus of elasticity varies from 60 EE6 to 20 EE6 psi.

Boron is being produced in fibrous form for use as a reinforcing material. Its modulus of elasticity is approximately 60 EE6 psi, while tensile strengths can reach 500,000 psi.

B. Steel-Reinforced Concrete

Since concrete is essentially incapable of resisting tension, steel reinforcing bars are used. Figure 10.5 shows three types of reinforcing in concrete beams.

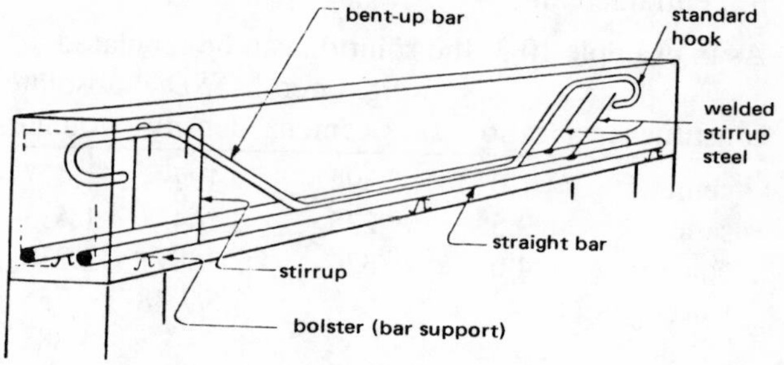

Figure 10.5 Beam Reinforcement

Straight and bent-up bars resist flexural tension in the central part of the beam. Since the bending moment is smaller near the ends of the beam, less reinforcing is necessary there. Some of the bar is bent up in order to resist diagonal shear near the beam ends. In continuous beams, the horizontal upper parts of the bent-up bars are continued on to the next span.

Since the bent-up bars cannot resist all of the diagonal tension, *stirrups* are used. These pass underneath the bottom steel for anchoring or are welded to the bottom steel. The stirrups may be placed at any convenient angle, as shown in figure 10.5.

The horizontal steel is supported on *bolsters (chairs)* of which there are a variety of designs and heights.

Columns are reinforced with longitudinal steel. This steel runs the full length of the column and is placed around the periphery or at the corners of a rectangular column. The longitudinal steel is wrapped with individual circumferential bars *(tied columns)* or with a continuous spiral bar *(spiral columns)*.

Reinforcing steel bars are available in a variety of sizes. Steel is also available as wire for spiral wrapping and wire mesh for shrinkage and expansion reinforcement. The steel comes in several grades depending on its yield strength.

Steel is specified by its yield strength. Grade 40 is the most common, although 50, 60, 75, and 80 are also available. However, not all sizes may be available in every grade.

[5]A *roving* consists of a number of parallel strands of glass fiber. The strands are side by side, forming a flat ribbon. However, the strands are not interwoven or twisted together.

[6]Other types are: A-glass (common soda-lime glass used for windows, bottles, and jars), C-glass (glass developed for greater chemical and corrosion resistance), D-glass (glass possessing a low dielectric constant), and M-glass (glass containing BeO to increase the elastic modulus).

Table 10.6
Standard Reinforcing Bars

size	weight lb/ft	diameter (in)	area (in²)	perimeter (in)
#2	.167	.250	.05	.786
#3	.376	.375	.11	1.178
#4	.668	.500	.20	1.571
#5	1.043	.625	.31	1.963
#6	1.502	.750	.44	2.356
#7	2.044	.875	.60	2.749
#8	2.670	1.000	.79	3.142
#9	3.400	1.128	1.00	3.544
#10	4.303	1.270	1.27	3.990
#11	5.313	1.410	1.56	4.430
#14S	7.65	1.693	2.25	5.32
#18S	13.60	2.257	4.00	7.09

Although steel can have a yield strength of 40,000 psi or above, its working stress is limited to less than this. Generally, a factor of safety of between 1.8 to 2.5 is used to keep the stress in the 20,000 to 24,000 psi range.

MATL SCI

PART 2: Materials Testing

The most important test used to predict material properties is the tensile test. This test will be covered in chapter 11.

1. Hardness Tests

Hardness tests measure the capacity of a surface to resist deformation. Through correlation, it is possible to predict the ultimate tensile strength of a metallic material through hardness tests. Hardness tests are also used to verify heat treatments.

The Brinell hardness number (BHN) is determined by pressing a hardened steel ball into the surface of the specimen. The diameter of the resulting depression is a measure of the material hardness. The standard ball is 10 mm in diameter and the loads are 3000 *kg* and 500 *kg* for hard and soft materials respectively.

The approximate ultimate tensile strength of steel can be calculated from the Brinell hardness number.

$$S_{ut} \approx (500)(BHN) \qquad 10.16$$

The *Rockwell hardness test* is similar to the Brinell test. The depth of penetration of a steel ball or a diamond spheroconical penetrator is measured. The machine applies an initial load (10 *kg*) which sets the penetrator below surface anomalies. Then a load is applied, and the hardness is shown on a dial. Although a number of Rockwell scales exist, the most familiar scales are the *B* and *C* scales. The *B* scale is used for mild steel and high-strength aluminum, while the *C* scale is used for hard steels having ultimate tensile strengths up to 300 ksi.

2. Toughness Testing

Toughness is the capability of a metal to absorb (through yielding) highly localized and rapidly applied stresses. In the *Charpy test,* a beam specimen is provided with a 45° *V* notch. This beam is centered on simple supports with the notch down.

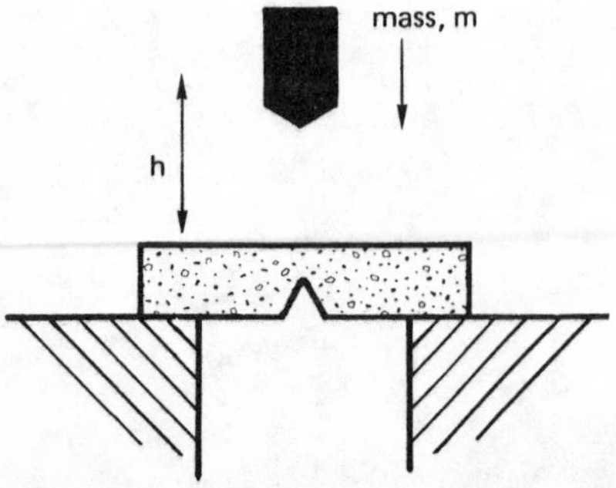

Figure 10.6 The Charpy Test

A falling-weight striker hits the center of the beam. By measuring the height at which the striker is released, the kinetic energy expended at impact can be determined. The objective is to determine the energy required for failure.

$$\text{Energy} = \frac{mgh}{g_c} \qquad 10.17$$

At 70°F failure impact energy ranges from 45 ft-lbs for the carbon steels to 110 ft-lbs for the chromium-manganese steels. As temperature is reduced, the toughness decreases. The *transition temperature* is taken as the point at which an impact of 15 ft-lbs will cause failure.

Table 10.7
Typical Transition Temperatures for Steel

type of steel	Transition Temperature, °F
carbon steel	30°
high-strength, low-alloy steel	0° to 30°
heat treated, high-strength, carbon steel	−25°
heat treated, construction alloy steel	−40° to −80°

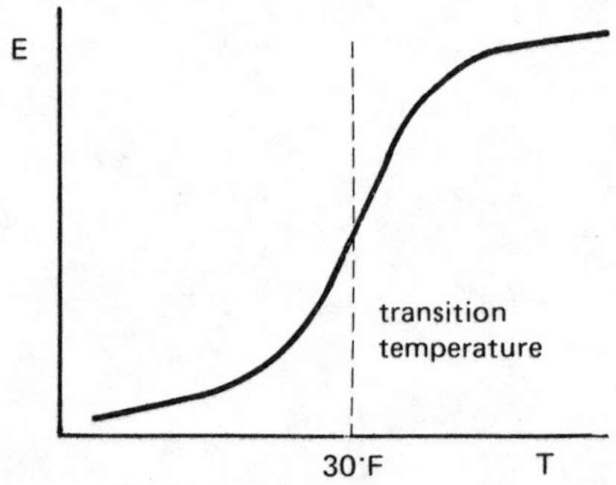

Figure 10.7
Failure Energy versus Temperature for Low-Carbon Steel

Not all materials exhibit the energy-temperature curve shape shown in figure 10.7. Figure 10.8 illustrates the curves for high-strength steels and FCC (face-centered cubic) metals such as aluminum.

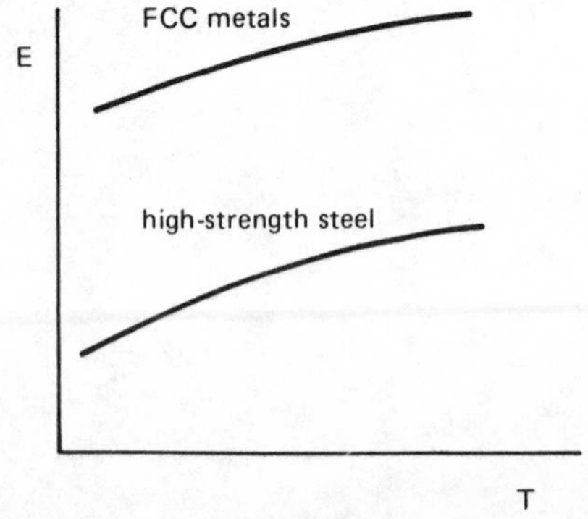

Figure 10.8
Failure Energy Versus Temperature for Other Metals

Another toughness test is the *Izod test.* This is illustrated in figure 10.9. Equation 10.17 is also applicable to the Izod test.

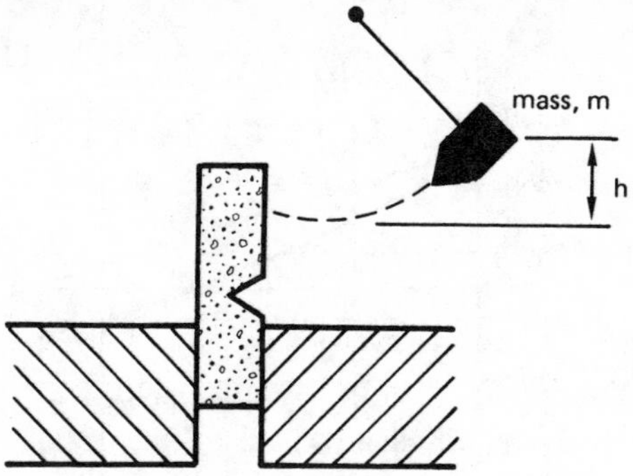

Figure 10.9 The Izod Test

3. Creep Testing

During a creep test, a low tensile load of constant magnitude is applied to a sample. The strain is measured as a function of time.

The *creep strength* is the stress which results in a given creep rate, usually .001% or .0001% per hour. The *rupture strength* is the stress which results in failure after a given amount of time, usually 100, 1000, or 10,000 hours.

If the creep is plotted as a function of time (figure 10.10), three different sections on the curve will be apparent. During the first stage, the creep rate ($d\epsilon/dt$) increases since strain hardening occurs at a greater rate than annealing. During the second stage, the creep rate is constant. Strain hardening and annealing occur at the same rate during this second stage. The sample begins to neck down during the third stage and rupture occurs.

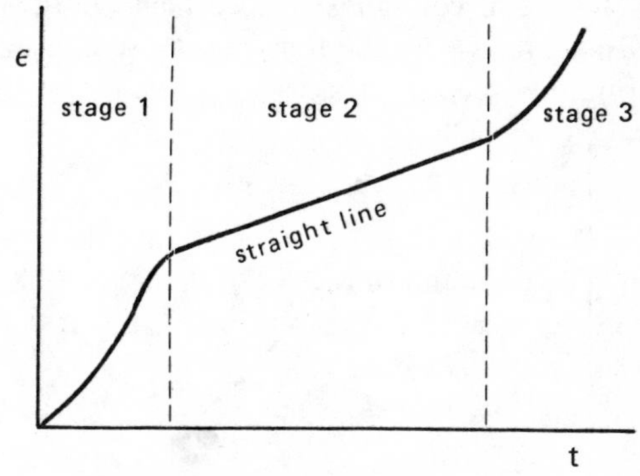

Figure 10.10 Creep versus Time

The *creep rate* is very temperature dependent. The slope of the line during the second stage increases with temperature.

PART 3: Thermal Treatments of Metals

1. Equilibrium Conditions

Figure 10.11 illustrates the graph of temperature versus time as a pure metal cools from liquid to solid state. At a particular point, the temperature remains constant. This temperature is known as the *freezing point* of the liquid. (The metal continues to lose heat energy—its *heat of fusion*. However, the temperature remains constant during the freezing process.)

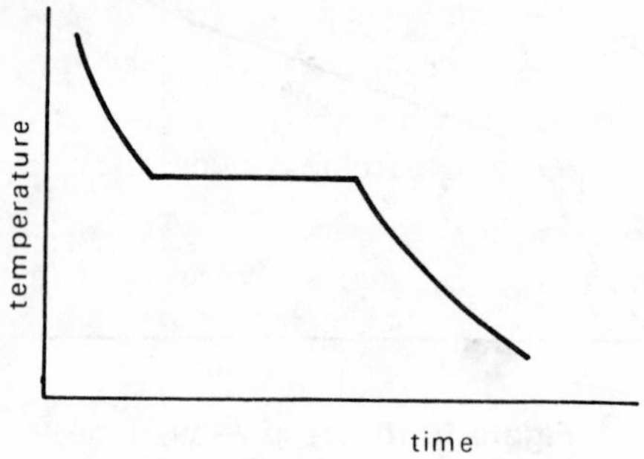

Figure 10.11
Temperature versus Time for a Pure Metal

If the same experiment is performed with an alloy of two metals, the transition temperature will vary with the proportions of the constituent. The locus of constant temperature points is a line above which the mixture is entirely liquid and below which it is entirely solid.

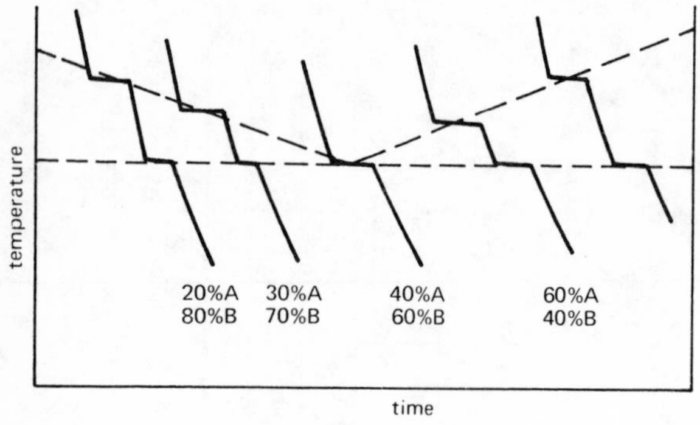

Figure 10.12
Temperature versus Time for a Binary Alloy

Figure 10.13 illustrates the variations of liquid and solid equilibrium temperatures as a function of the relative concentration of the constituents in a completely miscible alloy. The *liquidus line* represents the limit above which no solid can exist while the *solidus line* is the limit below which no liquid occurs. The area between these curves denotes a mixture of solid and liquid phase materials.

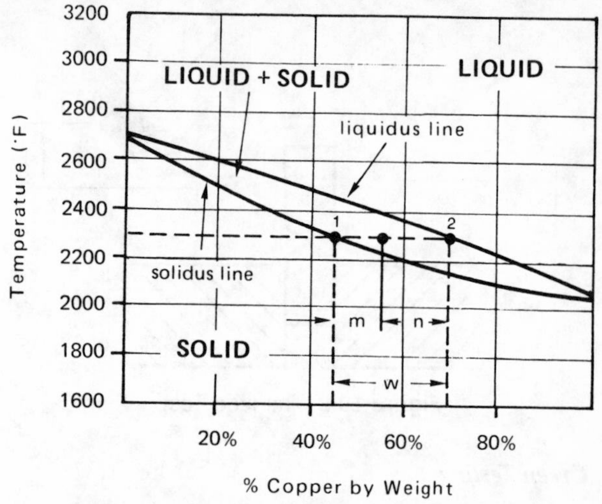

Figure 10.13 Copper-Nickel Phase Diagram

Example 10.5

Consider a mixture of 55% copper and 45% nickel at 2300°F. What are the percentages of solid and liquid materials?

The solid portion of the mixture will have the composition at point *1* (48% copper) while the liquid will be of composition *2* (67.8% copper). To find the relative amounts of solid and liquid in the mixture, the *lever rule* is used. The amounts of each phase are:

$$\%\text{ solid} = \frac{n}{w} = \frac{67.8 - 55}{67.8 - 48} = .647\ (64.7\%)$$

$$\%\text{ liquid} = 100 - 64.7 = 35.3\%$$

The elements of most alloys are not completely miscible in all states and phases. For instance, the elements of a two-component alloy may be perfectly soluble in the liquid state but only partially soluble in the solid state. Moreover, a number of different semi-solid and solid phases can occur, depending on the composition and temperature of the alloy.

In figure 10.14, the constituents are only perfectly miscible at point *C*, called the *eutectic point.*[7] The material in area *ABC* consists of a mixture of crystals of **A** and a liquid solution of **A** and **B**. The eutectic materials will not solidify until the line *B-D* is reached, the lowest point at which liquid eutectic can exist. Below the eutectic line, the material will be in the form of a mixture of solid **A** and eutectic crystals. To the right of the eutectic composition line, the inverse relationship will oc-

[7]The *eutectic point* should not be confused with the *eutectoid point*. A eutectic composition is associated with a liquid-to-solid reaction. A eutectoid composition is associated with a solid-to-solid reaction.

MATL SCI

cur with a mixture of **B** crystals and liquid or solid **B** and eutectic crystals depending on the temperature.

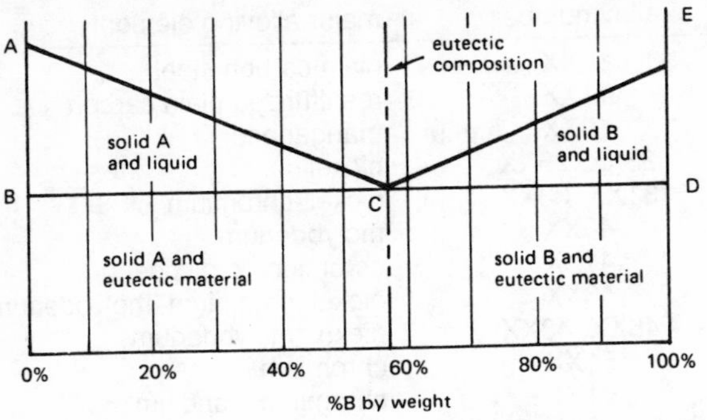

Figure 10.14 Binary Equilibrium Diagram

The *iron-carbon equilibrium* diagram is more complicated. Figure 10.15 is a simplified version.

The following alphabetical list of definitions may be helpful in understanding figure 10.15.

Allotropic changes: Reversible changes which occur in an iron-carbon mixture at the critical points. Compositions remain constant but properties such as atomic structure, magnetism, and electrical resistance change.

Alpha iron: A BCC (body-centered cubic) structure which exists beneath the A_3 line only. Maximum carbon solubility is 0.03%, the lowest of all forms. Alpha iron is stable from −460° to 1670°F, quite soft, and strongly magnetic up to 1418°F.

Austenite: A solid solution of carbon in gamma iron: It is non-magnetic, decomposes on slow cooling, and does not normally exist below 1333°F. It can be partially preserved through extreme cooling rates.

Bainite: A structure induced through an interrupted quenching process called *austempering*.

Beta iron: A non-magnetic form of alpha iron which exists between 1418° and 1670°F.

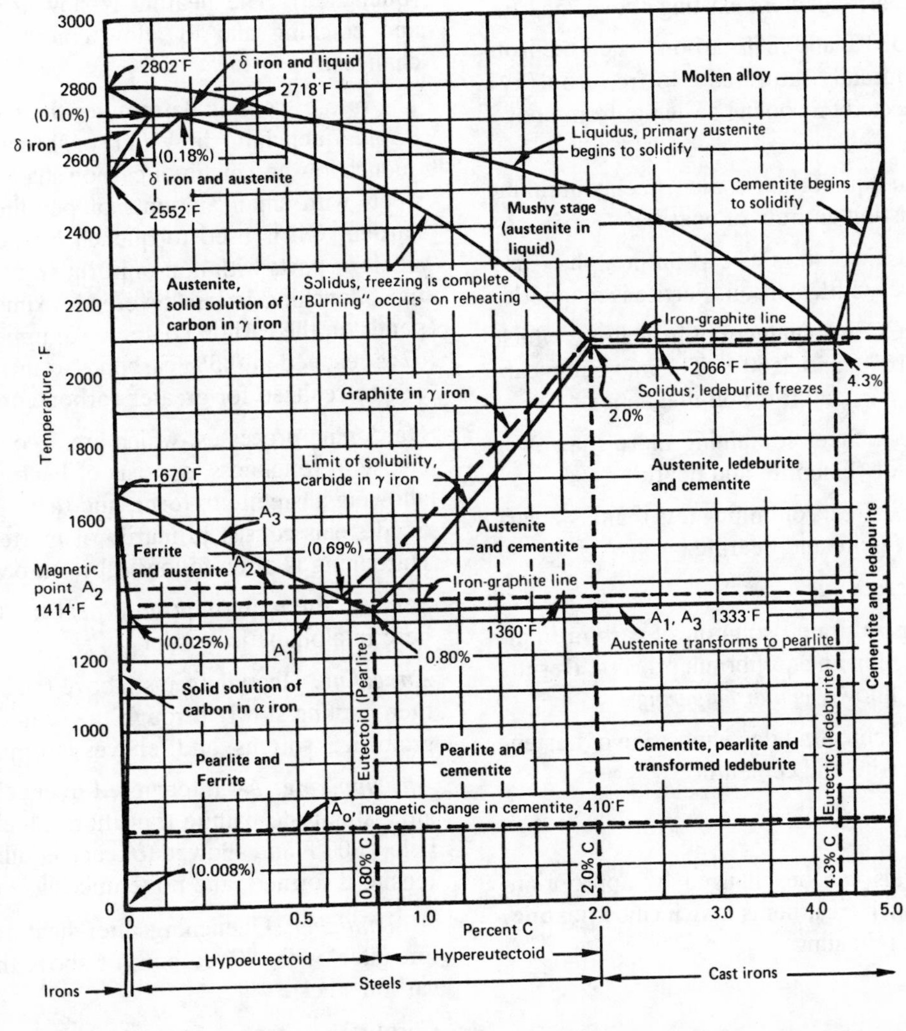

Figure 10.15
Simplified Iron-Carbon Diagram

Cementite: The chemical compound Fe_3C. It is the hardest of all forms of iron and is quite brittle. Cementite undergoes a magnetic change at the A_0 line.

Critical points: The temperatures at which allotropic changes take place. *Pure iron* has three critical points:

2802°F — liquid iron to delta (BCC) iron
2552°F — delta iron to gamma (FCC) iron
1670°F — gamma iron to alpha (BCC) iron

The critical lines for an iron-carbon mixture are shown in figure 10.15:

- $\mathbf{A_0}$: About 410°F, where cementite becomes non-magnetic.
- $\mathbf{A_1}$: The eutectoid temperature, 1333°F.
- $\mathbf{A_2}$: About 1410°F, below which the proper compositions become magnetic. Also known as the *Curie point.*
- $\mathbf{A_3}$: A temperature dependent on composition. Above $\mathbf{A_3}$, steel is austenite. Below, it is ferrite.

Delta iron: A BCC arrangement existing above 2552°F.

Eutectoid point: 1333°F and .8% carbon, when the iron and carbon are mutually saturated. No free iron (hypoeutectoid) is rejected on cooling austenite from above $\mathbf{A_1}$.

Eutectoid steel: An equilibrium-cooled steel containing .8% carbon, consisting entirely of pearlite.

Ferrite: A solid BCC solution of carbon in alpha iron, generally having very little carbon. Ferrite is magnetic.

Gamma iron: A FCC arrangement of iron atoms which is stable between 1670° and 2550°F. The maximum carbon solubility is 1.7%, and it is non-magnetic.

Hypereutectoid steel: Steel containing more than .8% carbon, consisting of cementite and pearlite.

Hypoeutectoid steel: Steel containing less than .8% carbon, consisting of ferrite and pearlite.

Iron carbide: See *cementite.*

Martensite: A supersaturated solution of carbon in alpha iron, not formed in an equilibrium process. Results from an operation known as *martempering.*

Pearlite: A result of eutectoid decomposition of austenite into layers of ferrite and cementite.

2. Heat Treatments

Properties of metals can be changed by appropriate heat treatments. Such treatments may include heating, rapid cooling, and reheating.

A. Steel

As low-carbon steel is heated, the grain size remains the same until the $\mathbf{A_1}$ line is reached. Between line $\mathbf{A_1}$ and line $\mathbf{A_3}$, the grain size of austenite in solution decreases. This characteristic is used in heat treatments wherein steel is heated and then cooled rapidly (quenched). The heating is slow to prevent warping and cracking and to allow completion of all phase changes.

Table 10.8
Alloying Ingredients in Steel
(XX is the carbon content, 0.XX%)

alloy number	major alloying elements
10XX	plain carbon steel
11XX	resulfurized plain carbon
13XX	manganese
23XX, 25XX	nickel
31XX, 33XX	nickel, chromium
40XX	molybdenum
41XX	chromium, molybdenum
43XX	nickel, chromium, molybdenum
46XX, 48XX	nickel, molybdenum
51XX	chromium
61XX	chromium, vandium
81XX, 86XX, 87XX	nickel, chromium, molybdenum
92XX	silicon

The rate of cooling determines the hardness obtained. Rapid quenching in water or brine is necessary to quench low and medium-carbon steels since low-carbon steels with small amounts of pearlite are difficult to harden. Oil is used to quench high-carbon and alloy steels or parts with non-uniform cross-sections because the quenching is less severe. Maximum hardness depends on the carbon content. An upper limit of R_C 66-67 is reached at 0.5% carbon. No increase in hardness can be realized for greater carbon content.

Hardening processes, which are accompanied by a decrease in toughness, consists of heating above line $\mathbf{A_3}$, allowing austenite to form, and then quenching. Hardened steels consist primarily of martensite or bainite. Hardening is often followed by tempering.

Some of the more important heat treatments are listed here in alphabetical order.

Annealing: Heating to just above the critical point and then cooling slowly through the critical range. Refines grain size, softens, and relieves internal stresses.

Austempering: An interrupted quenching process with an austenite-to-bainite transition. Steel is quenched to below 800°F and allowed to reach equilibrium. No martensite is formed, and no tempering is required.

Austenitizing: Quenching after heating above the $\mathbf{A_3}$ line (steel with .8% carbon) or above the $\mathbf{A_1}$ line (high-carbon steel).

Carburizing: heating for up to 24 hours at about 1650°F in contact with a carbonaceous material, followed by rapid cooling. Also known as *cementation.*

Cyaniding: Heating at 1700°F in a cyanide-rich atmosphere to produce a hardened surface.

Flame hardening: Supplying flame heat at the surface in quantities greater than can be conducted into the interior, followed by drastic quenching.

Induction hardening: Using high-frequency electric currents to heat the metal surface, followed by drastic quenching.

Nitriding: Heating for up to 100 hours in an ammonia atmosphere at about 1000°F with slow cooling.

Normalizing: Similar to annealing, but more rapid due to air cooling. Heating is 200°F above critical point.

Tempering: Hypo-eutectoid steels are tempered by reheating below the critical temperature after hardening. Tempering changes martensite into pearlite. The higher the temperature, the softer and tougher the steel becomes. Tempering is also known as *drawing* or *toughening*.

Time-temperature transformation (TTT) diagrams have been developed as an aid in designing heat treatment processes. Figure 10.16 is a *TTT* diagram for high-carbon (.95%) steel.

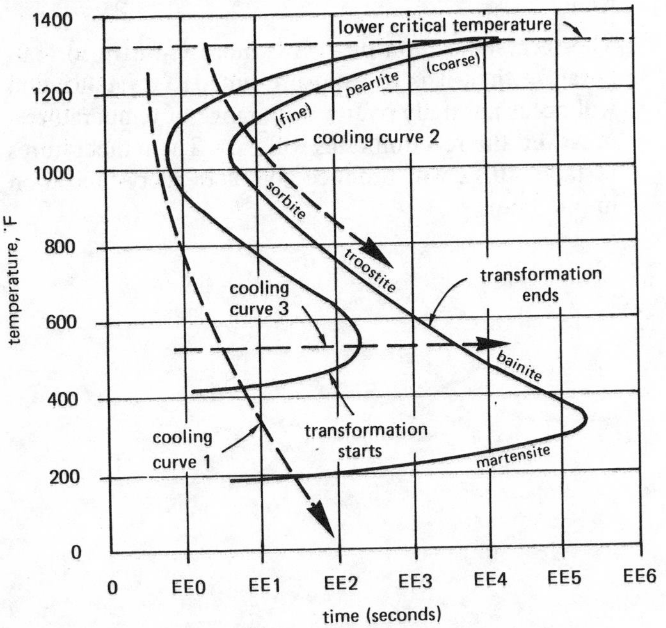

Figure 10.16 TTT Diagram for High-Carbon Steel

Curve 1 represents extremely rapid cooling. The transformation begins at 420°F and continues for 8 to 10 seconds, changing all of the austenite to martensite. Curve 2 is a slow quench which converts all of the austenite to fine pearlite. An intermediate curve passing through transition twice would produce combined material (e.g., sorbite and martensite).

If the temperature decreases rapidly along curve 1 to 520°F and is then held constant along curve 3, the structure would be bainite. This is the principle of *austempering*. Performing the same procedure at 350 – 400°F is *martempering*. Martempering produces a tough and soft steel.

B. Aluminum

The primary method of heat treating aluminum alloys is *precipitation* or *age hardening*. This involves the formation of a new crystalline structure through the application of controlled quenching and tempering procedures. Precipitation hardening disperses hard particles throughout a ductile matrix. These particles serve to disrupt the long dislocation planes of the matrix, increasing the strength and stiffness of the alloy. The strength limit is raised to the rupture strength of either the particles or the surrounding matrix.

Solution heat treatment culminates in rapid quenching. Quenching speeds must be consistent with the size of the specimen. Massive specimens may require slower processes using oil or boiling water. Post-treatment cooling for precipitation hardening is relatively unimportant.

Aluminum alloys are identified by a number and a letter (e.g., 2014-T4). The number indicates the chemical composition of the alloy, as determined from table 10.9. The letter indicates the condition of the alloy, as determined from table 10.10.

Table 10.9
Alloying Ingredients in Aluminum

alloy number	major alloying element
1XXX	commercially pure aluminum (99+%)
2XXX	copper
3XXX	manganese
4XXX	silicon
5XXX	magnesium
6XXX	magnesium & silicon
7XXX	zinc
8XXX	other

Table 10.10
Aluminum Alloy Conditions

letter	alloy condition
F	as fabricated
O	soft (after annealing)
H	strain hardened (cold worked)
T	heat treated

The letters *H* and *T* are followed by numbers. These numbers provide additional detail about the type of hardening process used to achieve the material properties. Table 10.11 lists the types of tempers associated with the *T* condition letter.

Table 10.11
Aluminum Tempers

temper	description
T2	annealed (castings only)
T3	solution heat-treated, followed by cold working
T4	solution heat-treated, followed by natural aging
T5	artificial aging
T6	solution heat-treated, followed by artificial aging
T7	solution heat-treated, followed by stabilizing by overaging heat treating
T8	solution heat treated, followed by cold working and subsequent artificial aging.

3. Recrystallization

Recrystallization involves heating a metal specimen to a specific temperature (the *recrystallization temperature*) and holding it there for a long time.[8] This results in the formation and growth of strain-free grains within grains already formed. The grain structure that results is essentially the structure that existed prior to any cold-working (e.g., deep drawing or bending). The new structure is softer and more ductile than the original structure.

Recrystallization can be used with all metals to relieve stresses induced during cold-working. Table 10.12 lists approximate recrystallization temperatures for metals and alloys.

[8]Recrystallization is a specific form of annealing.

Table 10.12
Approximate Recrystallization Temperatures

material	recrystallization temperature, °F
copper (99.999% pure)	250
(5% zinc)	600
(5% aluminum)	550
(2% beryllium)	700
aluminum (99.999% pure)	175
(99.0+% pure)	550
(alloys)	600
nickel (99.99% pure)	700
(99.4% pure)	1100
(monel metal)	1100
iron (pure)	750
(low-carbon steel)	1000
magnesium (99.99% pure)	150
(alloys)	450
zinc	50
tin	25
lead	25

The recrystallization process is more sensitive to temperature than it is to exposure time. Recrystallization will occur naturally over a wide range of temperatures; however, the reaction rates will vary. The temperatures in table 10.12 will produce complete recrystallization in one hour.

MATL SCI

PART 4: Crystalline Structure Theory

1. Crystalline Lattices

Stable aggregations of atoms produce energy states which are lower than for free atoms. These stable states are attained spontaneously, without the application of external energy. For example, the formation of sodium chloride (NaCl) proceeds spontaneously upon contact. The resultant compound is a stable crystalline solid.

Sodium and chlorine ions in NaCl form a 3-dimensional lattice. Each ion of one sign has six neighbors of the opposite sign. An electrostatic attraction results which is greater than the repulsion of the next twelve neighbors of like sign. If the ions were merely positive and negative point charges, the system would collapse into itself. However, the inner electron shells of both ionized atoms provide a repulsion force which just balances the net attractive forces between the positive and negative ions. This is the simplest crystal lattice in which a pattern is repeated in all coordinate axis directions.

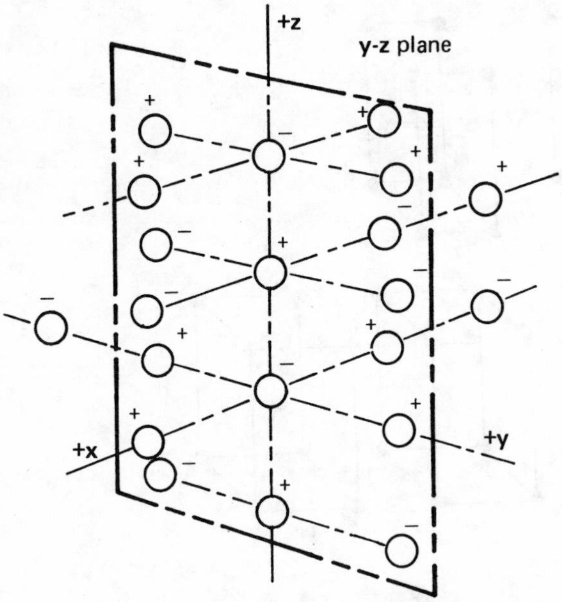

Figure 10.17 Basic Ionic Cubic Structure

Most metallic crystals form in one of three lattice structures *(cells)*: the body-centered cubic *(BCC)*, the face-centered cubic *(FCC)*, and the close-packed hexagonal *(HCP)* cells. Figure 10.18 illustrates the crystalline forms in which the atoms can be arranged.

The following metals are *FCC:* aluminum, copper, gold, gamma iron (between 906°C and 1401°C), lead, nickel, platinum, and silver.

The following metals are *BCC:* chromium, alpha iron (below 906°C), delta iron (above 1401°C), lithium, molybdenum, potassium, sodium, tantalum, and alpha tungsten.

The following metals are *HCP:* beryllium, cadmium, magnesium, alpha titanium, and zinc.

In the study of crystalline lattices, it is convenient to assume that the atoms have definite sizes. These sizes are functions of the ionic radii, approximated by representing the ions as hard spheres of radius *r*. On the basis of this assumption, figures 10.19 through 10.21 show three types of cubic lattices and give formulas for the center-to-center distances between the lattice atoms.

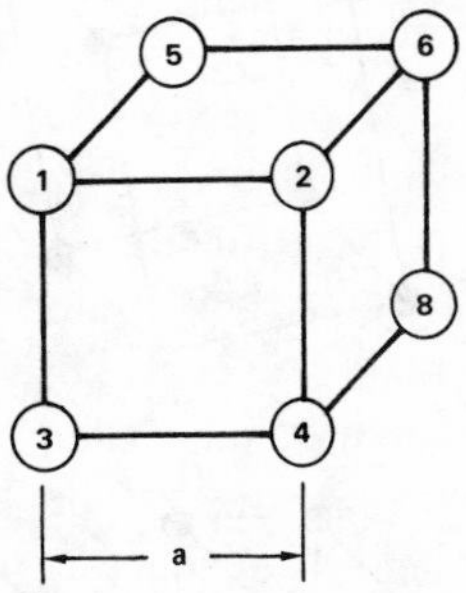

Figure 10.19
Simple Cubic Lattice Dimensions
(assuming touching hard spheres of radius *r*)

distance between atoms	in terms of *r*	in terms of **a**
1 and 2	2r	**a**
1 and 4	$2\sqrt{2}r$	$\sqrt{2}$**a**
1 and 8	$2\sqrt{3}r$	$\sqrt{3}$**a**

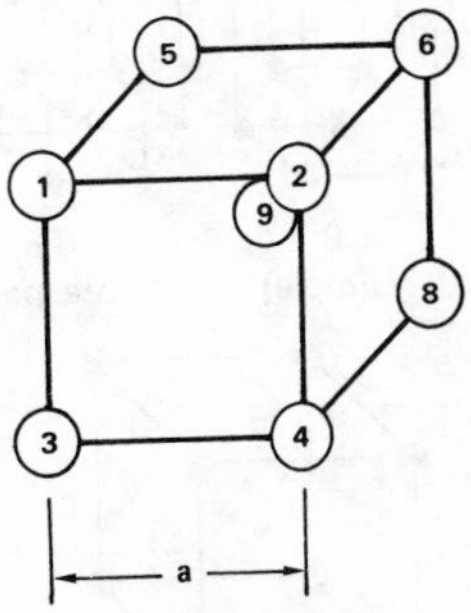

Figure 10.20
BCC Lattice Dimensions
(assuming touching hard spheres of radius *r*)

distance between atoms	in terms of *r*	in terms of **a**
1 and 2	$4r/\sqrt{3}$	**a**
1 and 4	$4\sqrt{(2/3)}r$	$\sqrt{2}$**a**
1 and 9	2r	$\frac{1}{2}\sqrt{3}$**a**
1 and 8	4r	$\sqrt{3}$**a**

MATL SCI

Figure 10.18 Crystalline Lattice Structures

The 14 basic point-lattices are illustrated by a unit cell of each: (1) simple triclinic, (2) simple monoclinic, (3) base-centered monoclinic, (4) simple orthorhombic, (5) base-centered orthorhombic, (6) body-centered orthorhombic, (7) face-centered orthorhombic, (8) hexagonal, (9) rhombohedral, (10) simple tetragonal, (11) body-centered tetragonal, (12) simple cubic, (13) body-centered cubic, (14) face-centered cubic.

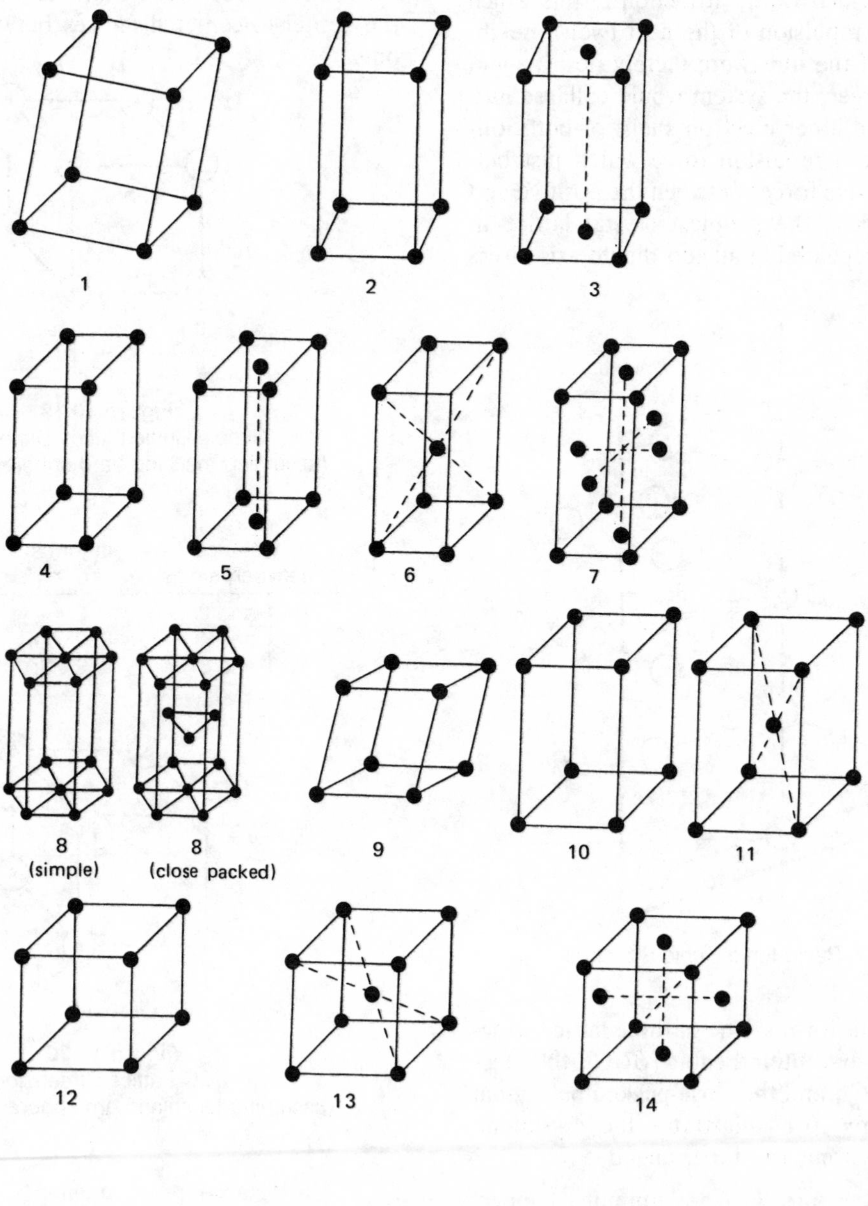

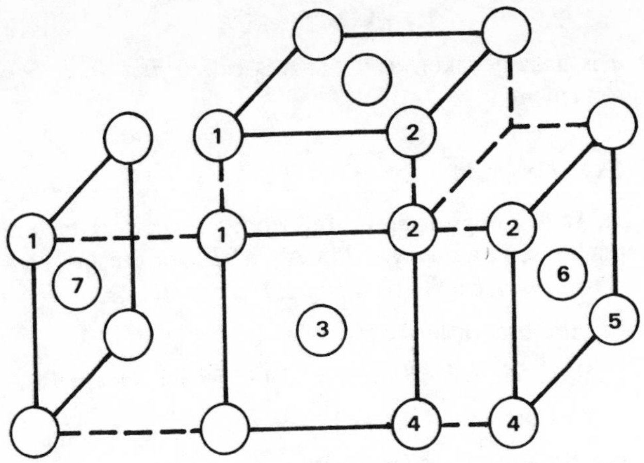

Figure 10.21
FCC Lattice Dimensions
(assuming touching hard spheres of radius *r*)

distance between atoms	in terms of *r*	in terms of **a**
1 and 2	$2\sqrt{2}r$	**a**
1 and 3	2r	$\frac{1}{2}\sqrt{2}$**a**
1 and 4	4r	$\sqrt{2}$**a**
1 and 5	$2\sqrt{6}r$	$\sqrt{3}$**a**
1 and 6	3.7r	1.3 **a**
3 and 6	2r	$\frac{1}{2}\sqrt{2}$**a**
7 and 6	$2\sqrt{2}r$	**a**

Because of sharing, the number of atoms in a cell is not the total number of atoms appearing in figures 10.19, 10.20, or 10.21. For example, there are nine atoms shown in figure 10.20 (*BCC*). However, atoms *1* through *8* are shared by other cells. Only atom *9* is completely enclosed by the cell. Since each corner atom is shared by eight other cells, the number of atoms in a *BCC* cell is $1 + (\frac{1}{8})(8) = 2$.

The *coordination number* of an atom in a cell is the number of closest (touching) atoms. The *packing factor* for a cell is the volume of the atoms divided by the total cell volume.

Table 10.13
Lattice Parameters

type of cell	number of atoms	packing factor	coordination number
simple cubic	1	.52	6
BCC	2	.68	8
FCC	4	.74	12
simple hexagonal	2	.52	8
HCP	6	.74	12

Since their packing factors are low (i.e., the formation is wasteful of space), the simple cubic and hexagonal lattices seldom form in nature.

2. Lattice Directions

Crystallography uses a system to designate directions of lines and planes in the lattice network. Consider a cubic cell that has cell dimensions **a**, **b**, and **c** in the *x*-, *y*-, and *z*-axes. If a point is located at (*u***a**, *v***b**, *w***c**) relative to a chosen origin, the line from the origin to the point is written as [*uvw*]. *u*, *v*, and *w* are fractions or multiples of the unit cell distances **a**, **b**, and **c**. Minus signs are placed *over* the integers to indicate negative coordinate directions.

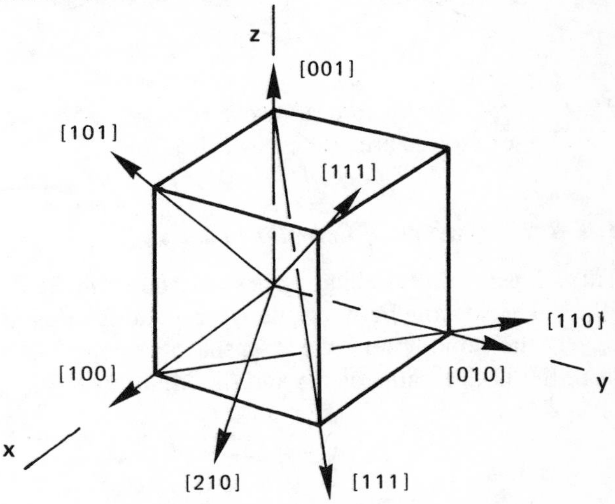

Figure 10.22 Sample Lattice Directions

The *Miller indexing system* is used to specify crystalline planes. The following procedure is used to obtain the Miller plane indices.

step 1: Note the orientation of the coordinate axes in case a non-standard orientation is used.

step 2: Find the *x*-, *y*-, and *z*-intercepts of the plane on the axes. For planes running parallel to an axis, the intercept is taken as infinity. If one intercept has a value of zero, shift the origin over by one cell dimension.

step 3: Express the intercepts as fractions of the unit cell dimensions: (x/**a**, y/**b**, z/**c**).

step 4: Take reciprocals of the fractions from step 3.

step 5: Clear all fractions. The resulting values are designated *h*, *k*, and *l* respectively.

step 6: Place the indices in parentheses without commas. Minus signs are placed over the index.

Example 10.6

What are the Miller indices of the plane shown?

The unit cell dimensions are **a**, **b**, and **c**. The intercepts along the respective axes are **a**/2, *b*, and ∞. Dividing these intercepts by **a**, **b**, and **c** produces quotients of

MATL SCI

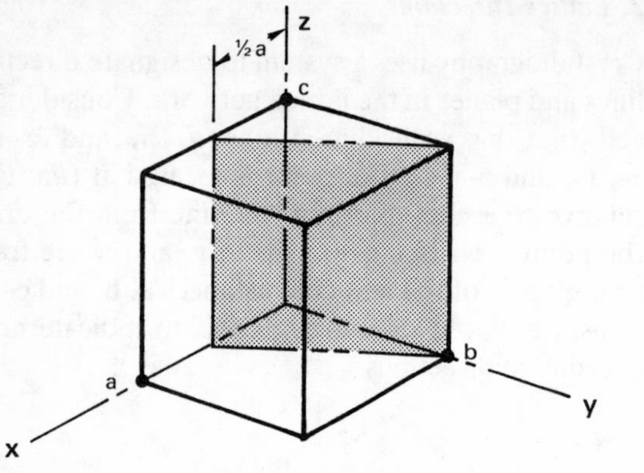

½, 1, and ∞. The reciprocals are $h=2$, $k=1$, and $l=0$. The Miller indices are written as (210).

3. *X-Ray Diffraction of Crystalline Planes*

Three layers of crystalline planes are shown in figure 10.23. The spacing between the layers is designated as d_{hkl}. In the cubic system, the layer spacing can be found from the lattice parameter **a** and the Miller indices.

$$d_{hkl} = \frac{\mathbf{a}}{\sqrt{h^2+k^2+l^2}} \qquad 10.18$$

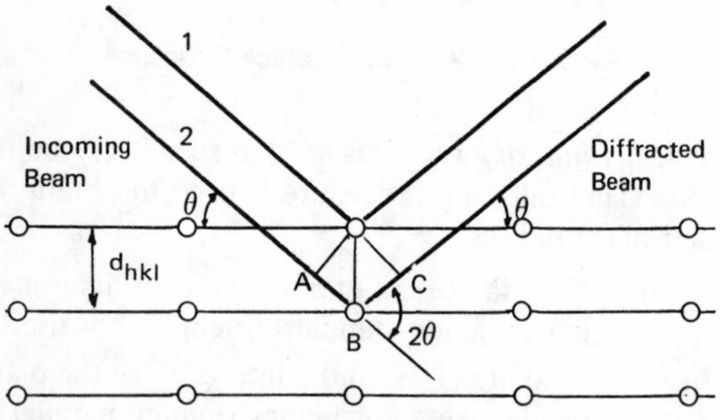

Figure 10.23
X-Ray Diffraction of Crystalline Planes

If a crystalline solid is hit by a monochromatic X-ray beam, reinforced diffracted beams will occur only in specific directions. The conditions governing these directions are defined by *Bragg's law,* equation 10.21.

If the reinforcement[9] is to be in-phase, then the combined distance (AB + BC) must be an integral number of wavelengths.

$$n\lambda = AB + BC \qquad 10.19$$

However, the distance (AB + BC) can be calculated from the plane spacing and the incident ray angle.

$$AB + BC = 2d_{hkl}\sin\theta \qquad 10.20$$

Combining equations 10.19 and 10.20 produces equation 10.21.

[9]Reinforcement is also called *in-phase return, superposition,* and *diffraction peak.*

$$n\lambda = 2d_{hkl}\sin\theta \qquad 10.21$$

n is usually taken as 1. This is called *first-order reinforcement.*

Example 10.7

1.541 Å X-rays are reflected from a quartz crystal with planes separated by 4.255 Å. What are the first- and second-order reflection angles?

For the first-order angle, $n = 1$.

$(1)(1.541\ \text{EE}-8\ \text{cm}) = (2)(4.255\ \text{EE}-8\ \text{cm})(\sin\theta)$

$\theta = \arcsin(.1811) = 10.43°$

For the second-order angle, $n = 2$.

$(2)(1.541\ \text{EE}-8) = (2)(4.255\ \text{EE}-8)(\sin\theta)$

$\theta = 21.23°$

If the wavelength of the X-rays is fixed, then the angle θ must be varied to obtain reinforcement. More than one angle will cause reinforcement. One value of sinθ, however, will be minimum.

The multiples of $\sin^2\theta_{min}$ can be used to predict the type of lattice for cubic structures.

Table 10.14
Multiples of $\sin^2\theta_{min}$ for Cubic Lattices

lattice type	multiples of $\sin^2\theta_{min}$
simple cubic	1,2,3,4,5,6,7,8,9,10
BCC	2,4,6,8,10,12,14,16,18
FCC	3,4,8,11,12,16,19

4. *Crystalline Imperfections*

Real crystals possess a variety of imperfections which affect their behavior. The imperfections can be categorized into point, line, and surface imperfections. The more important *point defects* are shown in figure 10.24.

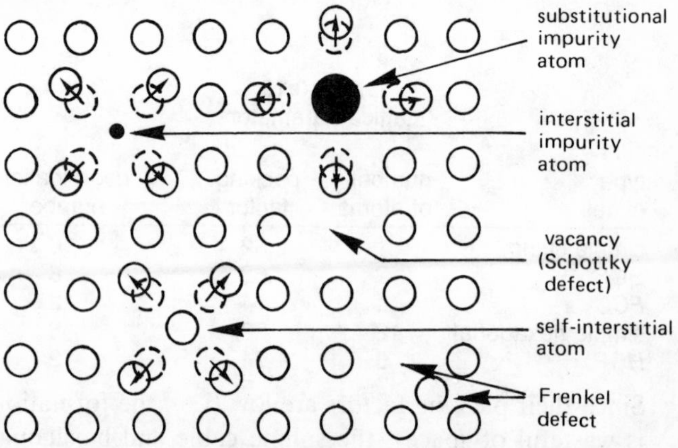

Figure 10.24 Point Defects

- *Schottky defect:* A lattice vacancy.
- *Frenkel defect:* A lattice vacancy with the missing atom transferring to an interstitial point.
- *Substitutional atoms:* The substitution of a foreign atom for an atom in a crystal.
- *Interstitial atoms:* Atoms which occupy interstitial spaces.
- *Self interstitial atoms:* A self diffusion or spontaneous rearrangement of atoms in a crystal.

All of the defects shown in figure 10.24 can move (*diffuse*) from one position to another. This movement requires *activation energy.* Such energy can come from heat, strain, (i.e., bending or forming), or other environmental sources.

Fick's laws govern diffusion. The first law predicts the number of defects moving across a unit surface per unit time. This movement, called the flux J, is proportional to the defect concentration gradient dC/dx in the direction of movement.

$$J = -D \frac{dC}{dx} \qquad 10.22$$

D is a diffusion coefficient which is dependent on the material and temperature.

As defects migrate in the x direction, the concentration changes. Fick's second law of diffusion relates the change in concentration with time to the instantaneous value of concentration.

$$\frac{dC}{dt} = D \frac{d^2C}{dx^2} \qquad 10.23$$

Point defects move individually and independently through the lattice network. Line defects, however, have extension in one particular direction. Line defects move as a unit. The two common line defects are *edge* and *screw dislocations*. These are illustrated in figures 10.25 and 10.26.

If a circuit is made around a crystal lattice with a dislocation, the **x**- and **y**-vectors will not close unless an additional *Burgers' vector,* **b**, is included. The Burgers' vector is an important parameter in the calculation of the strain energy produced by the dislocations.

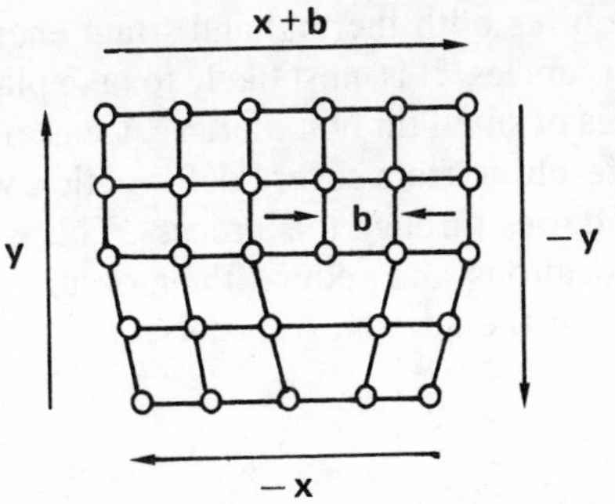

Figure 10.25 Edge Dislocation

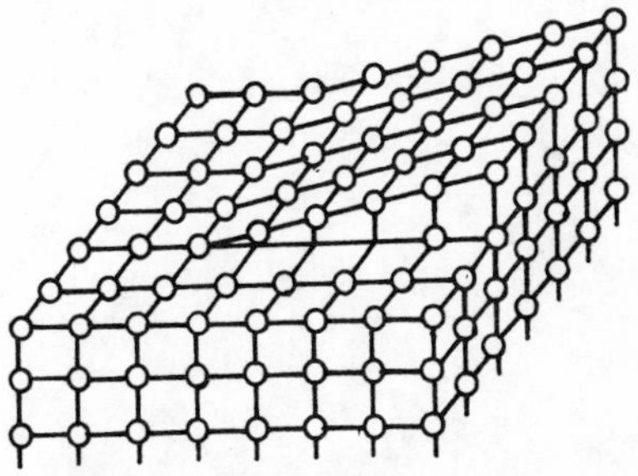

Figure 10.26 Screw Dislocation

Grain boundaries are the interfaces between two or more crystals of different orientation. In special cases, called *twin boundaries,* the crystals are perfect up to the interface. More often, however, the boundary is occupied by a corridor of transitional structure or amorphous material. Another special case is a boundary consisting of a series of edge dislocations. This is called a *tilt* or *narrow angle boundary.*

Slip is a phenomenon caused by shearing whereby two portions of a crystal are displaced with respect to each other. The location of the displacement is along a crystal plane called the *slip plane* in the slip direction.

Climb is motion of an edge dislocation. It can occur parallel to the slip plane or in a direction normal to the slip plane as shown in figure 10.27.

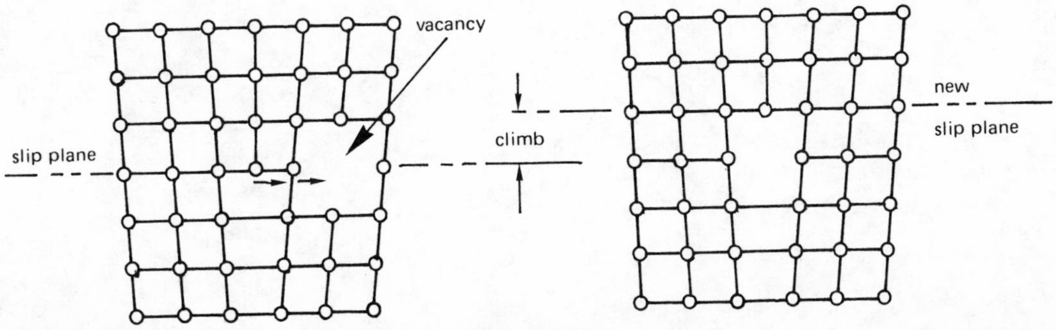

Figure 10.27
Movement of Edge Dislocation by Climb

MATL SCI

Climb requires both thermal and strain energy as well as local vacancies. It is most likely to take place in grain boundaries of small tilt boundaries. Areas of dense dislocation resulting from severe deformation will line up into boundaries through this process. This will concentrate dislocations and reduce their occurrence in the remainder of the adjoining crystals.

The *dislocation density* is the number of dislocations which occur per unit area. It is typically in the order of EE6 to EE9 per square inch. This density can be increased to as much as EE13 dislocations per square inch through severe deformation and plastic strain. A high dislocation density represents a considerable increase in stored strain energy.

CRYSTAL LATTICES

1. A BCC cell is shown below. How far apart are the atoms (a) 1 and 2? (b) 1 and 3? Express your answers in terms of the lattice dimension, a.

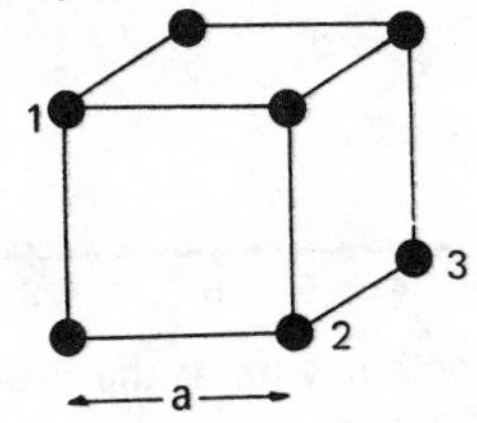

2. A BCC cell of iron is shown. What are the indices of the crystallographic directions shown?

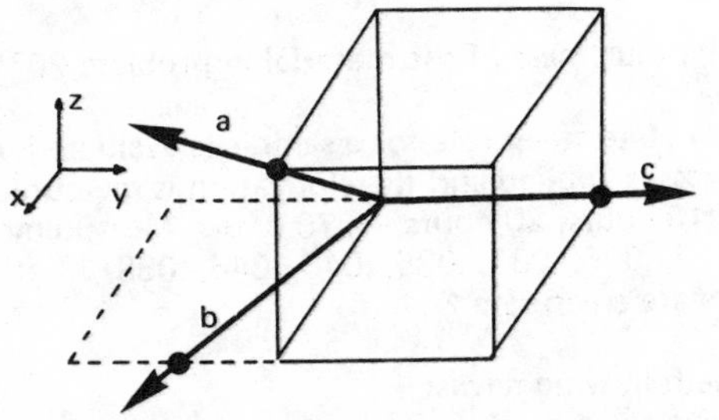

3. A FCC cell of aluminum is shown. What are the Miller indices of the planes shown?

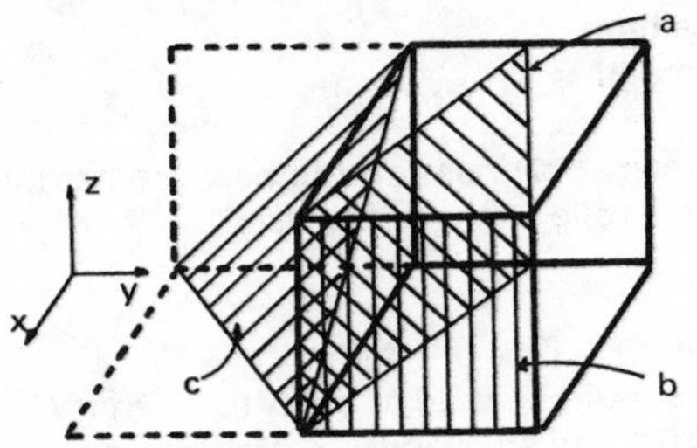

4. How many atoms are there in a
   (a) BCC cell?
   (b) FCC cell?
   (c) simple hexagonal cell?

5. An unknown crystalline substance has diffraction peaks at the following angles, using a radiation with 1.54 A wavelength: 13.7°, 16.0°, 22.9°, 27.1°, 28.3°, 32.6°. What is the structure of the material? What is its lattice parameter?

6. X-rays (.58 A wavelength) are used in a crystal diffraction analysis. If the diffraction angle is 9.5°, what is the unit cell size?

7. Oxygen ions (valence = -2) have an ionic radius of 1.32 A. What is the smallest cation that can have
   (a) a coordination number of 6?
   (b) a coordination number of 8?

8. What is the packing factor of the following cells? Assume touching spheres of radius 1.5 A.
   (a) simple cubic
   (b) FCC
   (c) BCC

LATTICE DEFECTS

9. What is Fick's law of diffusion?

10. Substitutional silicon in BCC iron at room temperature has a diffusion constant of .4 EE-8 cm^2/sec. How many silicon atoms will pass through a 2 square cm area each second if the silicon concentration decreases linearly 50% every 5 cm in the direction of diffusion?

11. Identify and sketch the following imperfections:
   (a) vacancy
   (b) interstitiality
   (c) Schottky defect
   (d) Frenkel defect
   (e) edge dislocation
   (f) screw dislocation
   (g) mixed dislocation
   (h) grain boundary

12. Define and sketch the Burgers circuit and the Burgers vector for the following dislocations:
   (a) edge dislocation
   (b) screw dislocation
   (c) mixed dislocation

13. Define the following terms:
   (a) creep
   (b) dislocation climb
   (c) hot working
   (d) cold working

MATERIAL TREATMENTS

14. Explain the principle of recrystallization. How is it affected by grain size, temperature, and time? What is the direction of grain growth?

15. For steel, define the following terms:
   (a) normalizing
   (b) tempering
   (c) austempering
   (d) ausforming
   (e) martempering
   (f) precipitation hardening
   (g) age hardening
   (h) strain hardening

16. What methods are used for hardening aluminum?

EQUILIBRIUM DIAGRAMS

17. Refer to the equilibrium diagram below.
   (a) For a 4%-96% alloy at temperature T_1, what are the compositions of solids α and β?
   (b) For a 1%-99% alloy at temperature T_2, how much liquid and how much solid are present?

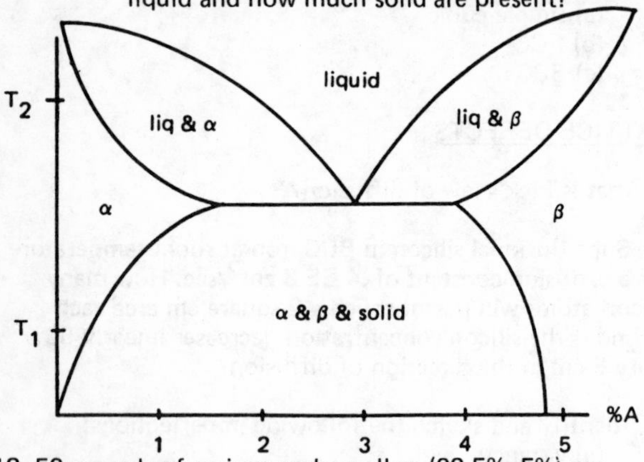

18. 50 pounds of an iron-carbon alloy (99.5%-.5%) are cooled from 1650°F to room temperature. How many pounds of pearlite are formed?

19. Approximately how much carbon do 10 pounds of AISI 1040 steel contain?

MAGNETISM

20. Define the following terms:
   (a) ferromagnetism
   (b) paramagnetism
   (c) diamagnetism
   (d) ferrimagnetism
   (e) anti-ferromagnetism

21. What is the curie temperature for iron?

CORROSION

22. Define the following terms:
   (a) corrosion
   (b) stress corrosion
   (c) corrosion fatigue
   (d) fretting corrosion
   (e) cavitation
   (f) electrode potential
   (g) galvanic cell

23. What are the standard electrode potentials for the following reactions? (Use standard hydrogen reference).
   (a) $F^{+++} + e \rightarrow Fe^{++}$

   (b) $Al \rightarrow Al^{+++} + 3e$

24. Explain how sacrificial zinc anodes help to protect steel.

MATERIALS TESTING

25. The engineering stress and strain for a copper sample (poisson's ratio = .3) were 20,000 psi and .0200 in/in respectively. What were the true stress and true strain?

26. An engineering stress-strain curve is shown below. Find the .5% parallel offset yield strength, the elastic modulus, the ultimate strength, the fracture strength, and the % elongation at fracture.

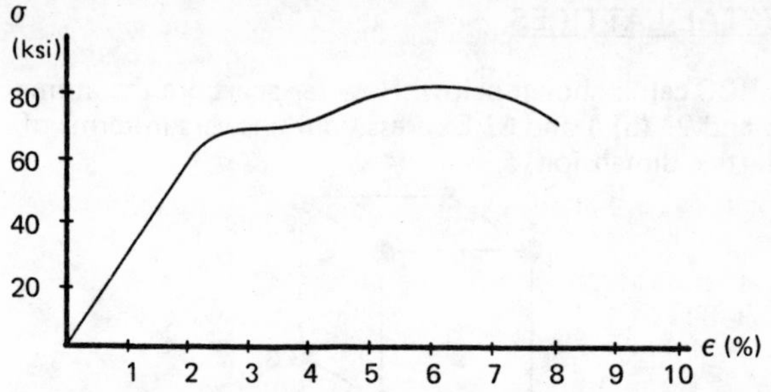

27. For the above problem, what is the shear modulus? Assume poisson's ratio is .3.

28. A 4 in^2 sample necks down to 3.42 in^2 before breaking in a tensile test. What is the sample's ductility?

29. What is the toughness of the material in problem 26?

30. The steady state creep rate for a sample is evaluated. A 15,000 psi stress is applied and the elongation is measured at t=5 hours, 10 hours, 20 hours,⋯, 70 hours. Measurements were .018, .022, .026, .031, .035, .040, .046, .058. What is the steady state creep rate?

31. Define the following terms:
   (a) fatigue
   (b) S-N curve
   (c) fatigue life
   (d) endurance limit
   (e) fatigue strength

32. What is the mathematical relationship between hardness (BHN) and ultimate tensile strength for steel?

POLYMERS

33. How much HCl should be used as an initiator in PVC if the efficiency is 20% and an average molecular weight of 7000 grams/gmole is desired? The final polymer has the following structure:

$$H-\underset{\underset{H}{|}}{\overset{\overset{H}{|}}{C}}-\underset{\underset{Cl}{|}}{\overset{\overset{H}{|}}{C}}-\underset{\underset{H}{|}}{\overset{\overset{H}{|}}{C}}-\underset{\underset{Cl}{|}}{\overset{\overset{H}{|}}{C}}-\cdots-\underset{\underset{H}{|}}{\overset{\overset{H}{|}}{C}}-\underset{\underset{Cl}{|}}{\overset{\overset{H}{|}}{C}}-\underset{\underset{H}{|}}{\overset{\overset{H}{|}}{C}}-\underset{\underset{Cl}{|}}{\overset{\overset{H}{|}}{C}}-Cl$$

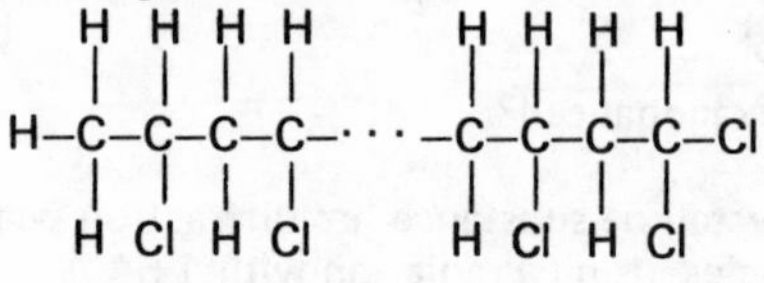

34. A .2% solution (by weight) of hydrogen peroxide is added to ethylene to stabilize the polymer. What is the average degree of polymerization if the hydrogen peroxide is completely utilized? Assume that hydrogen peroxide breaks down according to

$$H_2O_2 \rightarrow 2(OH)^- + \cdots$$

Assume that the stabilized polymer has the following structure:

$$OH-\underset{\underset{H}{|}}{\overset{\overset{H}{|}}{C}}-\underset{\underset{H}{|}}{\overset{\overset{H}{|}}{C}}-\underset{\underset{H}{|}}{\overset{\overset{H}{|}}{C}}-\cdots-\underset{\underset{H}{|}}{\overset{\overset{H}{|}}{C}}-\underset{\underset{H}{|}}{\overset{\overset{H}{|}}{C}}-\underset{\underset{H}{|}}{\overset{\overset{H}{|}}{C}}-OH$$

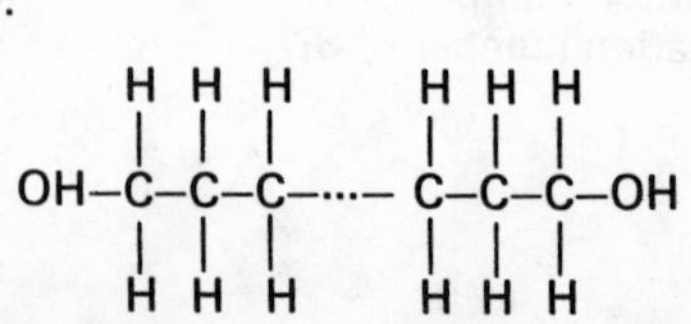

35. Distinguish between mers, monomers, and polymers.

36. Distinguish between bi-, tri-, and tetrafunctional monomers.

37. Distinguish between thermoplastic and thermosetting resins. What are the mechanical and chemical characteristics of each?

38. What is an initiator?

39. Distinguish between saturated and unsaturated polymers.

40. List and rate the mechanical properties of 5 common plastics.

41. What is the degree of polymerization? How is it calculated?

42. Distinguish between addition and condensation polymerization.

43. Distinguish between linear (chain) and framework bonding.

CONCRETE

44. A concrete mixture was designed as 1:1.9:2.8 by weight. The densities of the cement, sand, and coarse aggregate are 195, 165, and 165 pounds per cubic foot respectively. The water-cement ratio was chosen as 7 gallons of water per 94-pound sack of cement. (a) What is the concrete yield in cubic feet per sack of cement? (b) How much of each constituent is needed to make 45 cubic yards of concrete?

45. Draw an accurately scaled graph of concrete compressive strength versus

(a) curing time

(b) original water content in gallons per sack

CERAMICS

46. What are the chemical, mechanical, and electrical properties of ceramics?

47. What are the elemental compound formulas for the following ceramics?

(a) clay

(b) furnace refractory brick

(c) glass

(d) corundum

(e) fluorite

11 Mechanics of Materials

PART 1: Strength of Materials

Nomenclature

A	area	in^2
b	width	in
c	distance from neutral axis to extreme fiber	in
C	end restraint coefficient, or correction	–
D	diameter	in
e	eccentricity	in
E	modulus of elasticity	psi
F	force, or load	lbf
FS	factor of safety	–
g	local gravitational acceleration	ft/sec^2
g_c	gravitational constant (32.2)	$\frac{lbm\text{-}ft}{lbf\text{-}sec^2}$
G	shear modulus	psi
I	moment of inertia	in^4
J	polar moment of inertia	in^4
k	radius of gyration, or spring constant	in, lbf/in
K	stress concentration factor	–
L	length	in
m	mass	lbm
M	moment	in-lbf
n	ratio, rotational speed, or number	–, rpm, –
N	number of cycles	–
p	pressure	psi
Q	statical moment	in^3
r	radius	in
S	strength, or axial load	psi, lbf
t	thickness	in
T	temperature, or torque	°F, in-lbf
u	virtual truss load	lbf
U	energy	in-lbf
V	shear, or volume	lbf, in^3
w	load per unit length, or width	lbf/in, in
W	work	in-lbf
x	distance, or displacement	in
y	deflection, or distance	in
Z	section modulus	in^3

Symbols

δ	elongation or displacement	in
θ	angle	degrees
ϕ	angle	radians
σ	normal stress	psi
α	coefficient of linear thermal expansion	1/°F
β	coefficient of volumetric thermal expansion	1/°F
γ	coefficient of area thermal expansion	1/°F
τ	shear stress	psi
ε	strain	–
μ	Poisson's ratio	–

Subscripts

a	allowable
b	bending
br	bearing
c	centroidal, or compressive
e	endurance, euler, or equivalent
ext	external
h	hoop
i	inside
l	long
o	original, or outside
p	pull
s	shear
t	transformed, tension, or temperature
th	thermal
T	torsion
u	ultimate
y	yield

1. Properties of Structural Materials

A. The Tensile Test

Many material properties can be derived from the standard tensile test. In a tensile test, a material sample is loaded axially in tension, and the elongation is measured as the load is increased. A graphical representation of typical test data for steel is shown in figure 11.1 in which the elongation, δ, is plotted against the applied load, *F*.

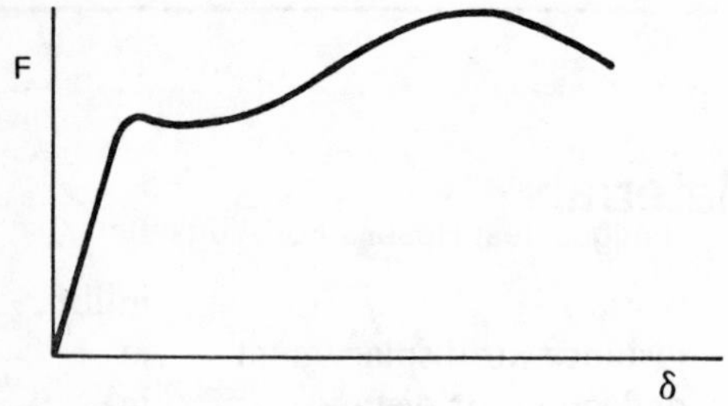

Figure 11.1
Typical Tensile Test Results for Steel

Since this graph is only applicable to an object with the same length and area as the test sample, the data are converted to *stresses* and *strains* by use of equations 11.1 and 11.2. σ is known as the *normal stress,* and ε is known as the *strain.* Strain is the percentage elongation of the sample.

$$\sigma = \frac{F}{A} \qquad 11.1$$

$$\varepsilon = \frac{\delta}{L} \qquad 11.2$$

The stress-strain data can also be graphed, and the shape of the resulting curve will be the same as figure 11.1 with the scales changed.

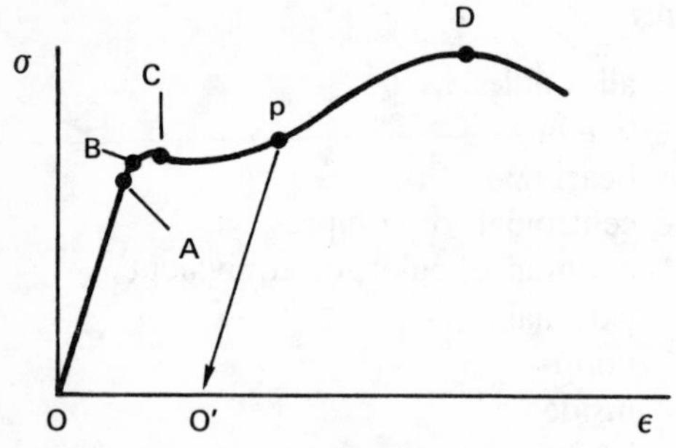

Figure 11.2
A Typical Stress-Strain Curve for Steel

The line O-A in figure 11.2 is a straight line. The relationship between the stress and the strain is given by *Hooke's law,* equation 11.3. *E* is the *modulus of elasticity (Young's modulus)* and is the slope of the line segment O-A. The stress at point *A* is known as the *proportionality limit.* The modulus of elasticity for steel is approximately 3 EE 7 psi.

$$\sigma = E\varepsilon \qquad 11.3$$

Slightly above the proportionality limit is the *elastic limit* (point *B*). As long as the stress is kept below the elastic limit, there will be no permanent strain when the applied stress is removed. The strain is said to be *elastic* and the stress is said to be in the *elastic region.*

If the elastic limit stress is exceeded before the load is removed, recovery will be along a line parallel to the straight line portion of the curve, as shown in the line segment p-O′. The strain that results (line O-O′) is permanent and is known as *plastic strain.*

The *yield point* (point *C*) is very close to the elastic limit. For all practical purposes, the *yield stress*, S_y, can be taken as the stress which accompanies the beginning of plastic strain. Since permanent deformation is to be avoided, the yield stress is used in calculating safe stresses in ductile materials such as steel. A36 structural steel has a minimum yield strength of 36,000 psi.

$$\sigma_a = \frac{S_y}{FS} \qquad 11.4$$

Some materials, such as aluminum, do not have a well-defined yield point. This is illustrated in figure 11.3 In such cases, the yield point is taken as the stress which will cause a .2% parallel offset.

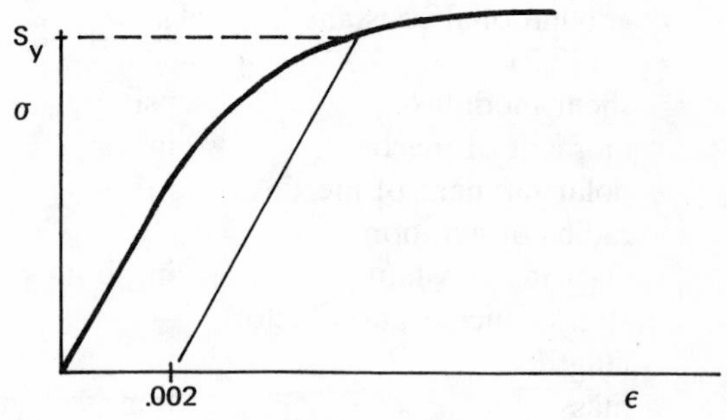

Figure 11.3
A Typical Stress-Strain Curve for Aluminum

The *ultimate tensile strength,* point *D* in figure 11.2, is the maximum load carrying ability of the material. However, since stresses near the ultimate strength are accompanied by large plastic strains, this parameter should not be used for the design of ductile materials such as steel and aluminum.

As the sample is elongated during a tensile test, it will also be decreasing in thickness (width, or diameter). The ratio of the lateral strain to the axial strain is known as *Poisson's ratio,* μ. μ is typically taken as .3 for steel and .33 for aluminum.

$$\mu = \frac{\varepsilon_{lateral}}{\varepsilon_{axial}} = \frac{\dfrac{\Delta D}{D_o}}{\dfrac{\Delta L}{L_o}} \qquad 11.5$$

B. Fatigue Tests

A part may fail after repeated stress loading even if the stress never exceeds the ultimate fracture strength of the material. This type of failure is known as *fatigue failure.*

The behavior of a material under repeated loadings can be evaluated in a fatigue test. A sample is loaded repeatedly to a known stress, and the number of applications of that stress is counted until the sample fails. This procedure is repeated for different stress levels. The results of many of these tests can be graphed, as is done in figure 11.4.

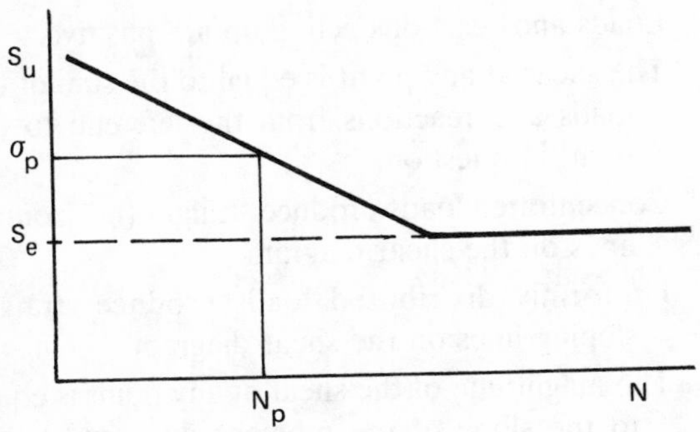

Figure 11.4
Results of Many Fatigue Tests for Steel

For any given stress level, say σ_p in figure 11.4, the corresponding number of applications of the stress which will cause failure is known as the *fatigue life.* That is, the fatigue life is just the number of cycles of stress required to cause failure. If the material is to fail after only one application of stress, then the required stress must equal or exceed the ultimate strength of the material.

Below a certain stress level, called the *endurance limit* or *endurance strength,* the part will be able to withstand an infinite number of stress applications without experiencing failure. Therefore, if a dynamically loaded part is to have an infinite life, the applied stress must be kept below the endurance limit.

Some materials, such as aluminum, do not have a well-defined endurance limit. In such cases, the endurance limit is taken as the stress that will cause failure at EE8 or 5 EE8 applications of the stress.

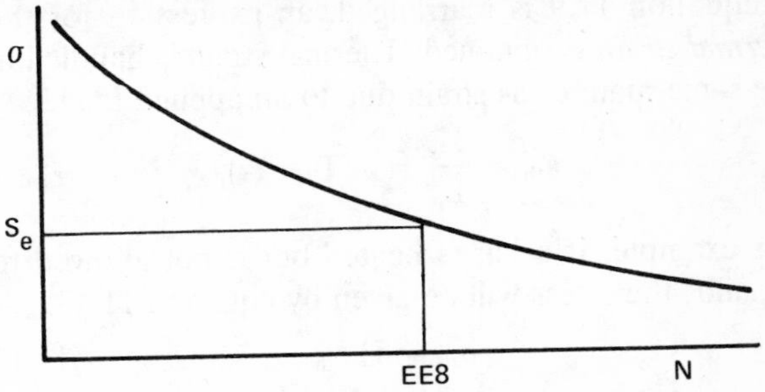

Figure 11.5
Fatigue Test Results For Aluminum

C. Estimates of Material Properties

Although the properties of a material will depend on its classification (ASTM, AISC, etc.), average values are given in table 11.1.

2. Deformation Under Loading

Equation 11.2 can be rearranged to give the elongation of an axially loaded member in compression or tension.

$$\delta = L\varepsilon = \frac{L\sigma}{E} = \frac{LF}{AE} \qquad 11.6$$

A tension load is taken as positive and a compressive load is taken as negative. The actual length of a member under loading is given by equation 11.7 where the algebraic sign of the deformation must be observed.

$$L_{actual} = L_o + \delta \qquad 11.7$$

The energy stored in a loaded member is equal to the work required to deform it. Below the proportionality limit, this energy is given by equation 11.8.

$$U = \tfrac{1}{2}F\delta = \tfrac{1}{2}\left(\frac{F^2L}{AE}\right) \qquad 11.8$$

3. Thermal Deformation

If the temperature of an object is changed, the object will experience length, area, and volume changes. These changes can be predicted by equations 11.9, 11.10, and 11.11.

Table 11.1
Average Material Properties

material	E (psi)	G (psi)	μ	ρ (pcf)	α (1/°F)
steel (hard)	30 EE6	11.5 EE6	.30	489	6.5 EE−6
steel (soft)	29 EE6	11.5 EE6	.30	489	6.5 EE−6
aluminum alloy	10 EE6	3.9 EE6	.33	173	12.8 EE−6
magnesium alloy	6.5 EE6	2.4 EE6	.35	112	14.5 EE−6
titanium alloy	15.4 EE6	6.0 EE6	.34	282	4.9 EE−6
cast iron	20 EE6	8 EE6	.27	442	5.6 EE−6

$$\Delta L = \alpha L_o(T_2-T_1) \qquad 11.9$$

$$\Delta A = \gamma A_o(T_2-T_1) \approx 2\alpha A_o(T_2-T_1) \qquad 11.10$$

$$\Delta V = \beta V_o(T_2-T_1) \approx 3\alpha V_o(T_2-T_1) \qquad 11.11$$

If equation 11.9 is rearranged, an expression for the *thermal strain* is obtained. Thermal strain is handled in the same manner as strain due to an applied load.

$$\epsilon_{th} = \frac{\Delta L}{L_o} = \alpha(T_2-T_1) \qquad 11.12$$

For example, if a bar is heated but is not allowed to expand, the stress will be given by equation 11.13.

$$\sigma_{th} = E\,\epsilon_{th} \qquad 11.13$$

4. Shear and Moment Diagrams

It was illustrated in chapter 9 that for an object in equilibrium the sums of forces and moments are equal to zero everywhere. For example, the sum of moments about point *A* for the beam shown in figure 11.6 is zero.

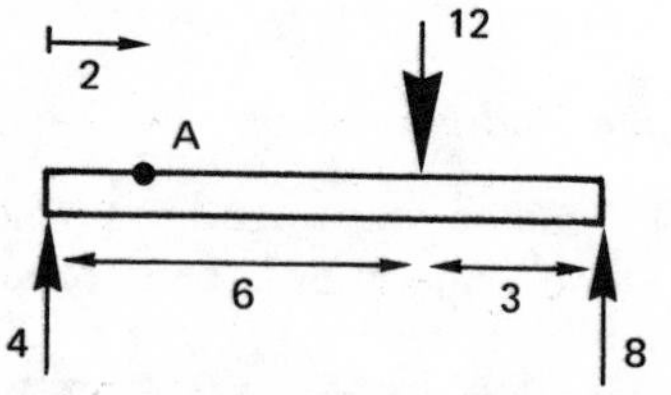

Figure 11.6 A Beam in Equilibrium

Nevertheless, the beam shown in figure 11.6 will bend under the influence of the forces. This bending is evidence of the stress experienced by the beam. Since the sum of moments about any point is zero, the moment used to find stresses and deflection is taken from the point in question to one end of the beam only. This is called the *one-way moment.* The absolute value of the moment will not depend on the end used. This can be illustrated by the beam shown in figure 11.6.

$$\Sigma M_{A\ (\text{to right end})} = -8(7) + 4(12) = -8$$

$$\Sigma M_{A\ (\text{to left end})} = 4(2) = +8$$

The moment obtained will depend on the location chosen. A graphical representation of the one-way moment at every point along a beam is known as a *moment diagram.* The following guidelines should be observed in constructing moment diagrams.

- Moments should be taken from the left end to the point in question. If the beam is cantilever, place the built-in end at the right.
- Clockwise moments are positive. The left-hand rule should be used to determine positive moments.
- Concentrated loads produce linearly increasing lines on the moment diagram.
- Uniformly distributed loads produce parabolic lines on the moment diagram.
- The maximum moment will occur when the shear (V) is zero.
- The moment at any point is equal to the area under the shear diagram up to that point. That is,

$$M = \int V\,dx \qquad 11.14$$

- The moment is zero at a free end or hinge.

Similarly, the sum of forces in the *y* direction on a beam in equilibrium is zero. However, the shearing stress at a point along the beam will depend on the sum of forces and reactions from the point in question to one end only.

A *shear diagram* is drawn to graphically represent the shear at any point along a beam. The following guidelines should be observed in constructing a shear diagram.

- Loads and reactions acting up are positive.
- The shear at any point is equal to the sum of the loads and reactions from the left end to the point in question.
- Concentrated loads produce straight (horizontal) lines on the shear diagram.
- Uniformly distributed loads produce straight sloping lines on the shear diagram.
- The magnitude of the shear at any point is equal to the slope of the moment diagram at that point.

$$V = \frac{dM}{dx} \qquad 11.15$$

Example 11.1

Draw the shear and moment diagrams for the following beam.

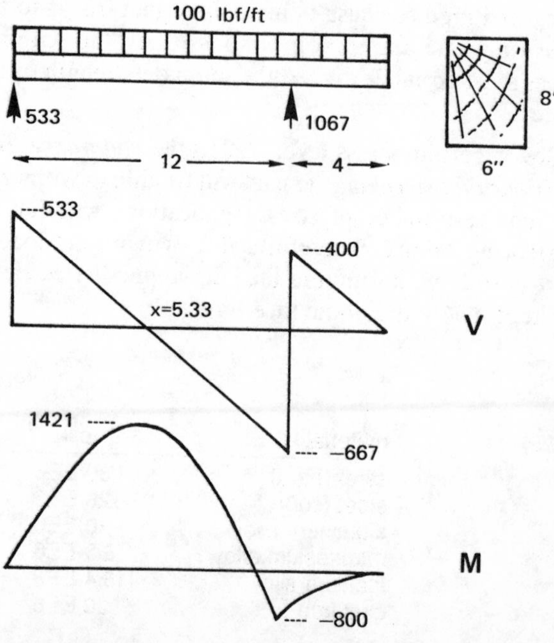

5. Stresses in Beams

A. Normal Stress

Normal stress is the type of stress experienced by a member which is axially loaded. The normal stress is the load divided by the area.

$$\sigma = \frac{F}{A} \qquad 11.16$$

Normal stress also occurs when a beam bends, as shown in figure 11.7. The lower part of the beam experiences normal tensile stress (which causes lengthening). The upper part of the beam experiences a normal compressive stress (which causes shortening). There is no stress along a horizontal plane passing through the centroid of the cross section. This plane is known as the *neutral plane* or *neutral axis*.

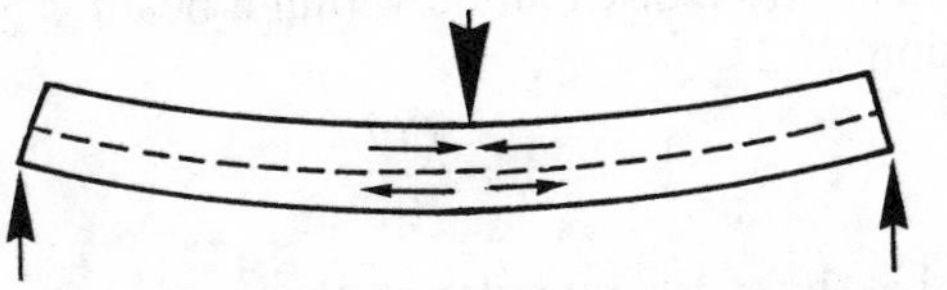

Figure 11.7 Normal Stress Due to Bending

Although it is a normal stress, the stress produced by the bending is usually called *bending stress* or *flexure stress*. Bending stress varies with position within the beam. It is zero at the neutral axis, but increases linearly with distance from the neutral axis.

$$\sigma_b = \frac{-My}{I_c} \qquad 11.17$$

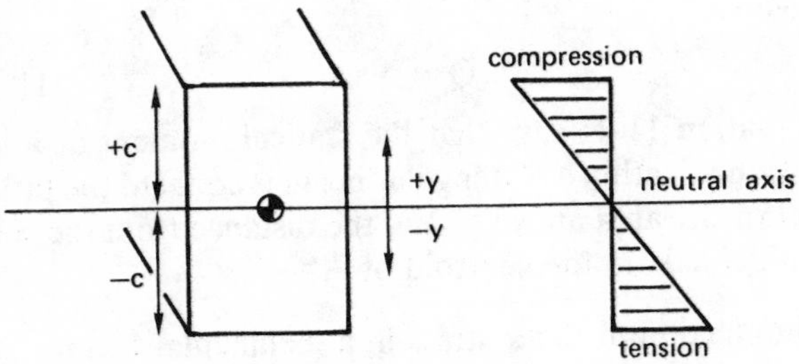

Figure 11.8
Bending Stress Distribution in a Beam

The moment, *M*, used in equation 11.17 is the *one-way moment* previously discussed. I_c is the centroidal moment of inertia of the beam's cross sectional area. The negative sign in equation 11.17 is typically omitted. However, it is required to be consistent with the convention that compression is negative.

Since the maximum stress will govern the design, *y* may be set equal to *c* to obtain the maximum stress. *c* is the distance from the neutral axis to the *extreme fiber*.

$$\sigma_{b,max} = \frac{Mc}{I_c} \qquad 11.18$$

For any given structural shape, *c* and I_c are fixed. Therefore, these two terms can be combined into the *section modulus, Z*.

$$\sigma_{b,max} = \frac{M}{Z} \qquad 11.19$$

$$Z = \frac{I_c}{c} \qquad 11.20$$

For most beams, the section modulus *Z* is constant along the length of the beam. Equation 11.19 shows that the maximum stress along the length of a beam is proportional to the moment at that point. The location of the maximum bending moment is called the *dangerous section*. The dangerous section can be found directly from a moment or shear diagram of the beam.

If an axial member is loaded eccentrically, it will experience axial stress (equation 11.16) as well as bending stress (equation 11.17). This is illustrated by figure 11.9 in which a load is not applied to the centroid of a column's cross sectional area.

Because the beam bends and supports a compressive load, the stress produced is a sum of bending and normal stress.

$$\sigma_{max,min} = \frac{F}{A} \pm \frac{Mc}{I_c} = \frac{F}{A} \pm \frac{Fec}{I_c} \qquad 11.21$$

If a cross section is loaded with an eccentric compressive load, part of the section can be in tension. This is illustrated in example 11.3. There will be no stress sign reversal, however, as long as the load is applied within a diamond-shaped area formed from the middle-thirds of the centroidal axes. This area is known as the *kern* or *kernal*. It is particularly important to keep eccentric compressive loads within the kern on concrete and masonry piers since these materials do not tolerate tension.

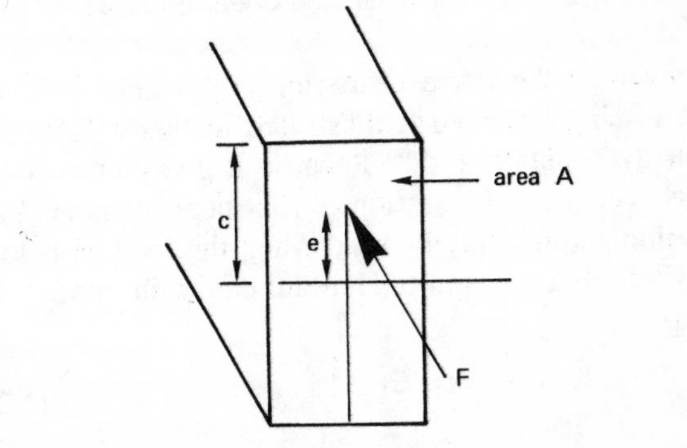

Figure 11.9
Eccentric Loading of an Axial Member

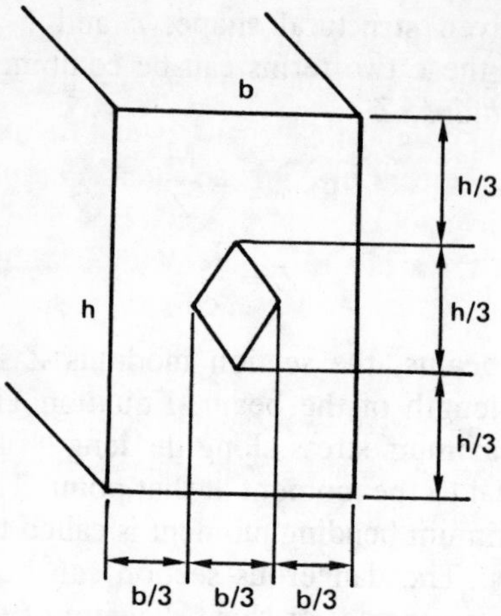

Figure 11.10 The Kern

The *elastic strain energy* stored in a beam experiencing a moment (bending) is

$$U = \tfrac{1}{2}\int \frac{M^2}{EI}\,dx \qquad 11.22$$

B. Shear Stress

Normal stress is produced when a load is absorbed by an area normal to it. On the other hand, *shear stress* is produced by a load being carried by an area parallel to the load. This is illustrated in figure 11.11.

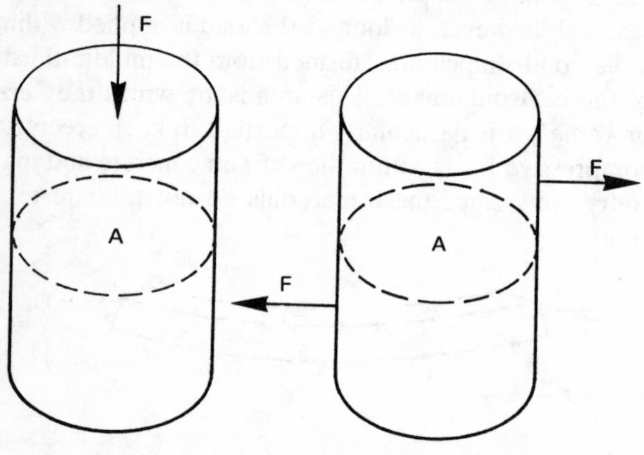

Figure 11.11 Normal and Shear Stresses

The average shear stress experienced by a pin, bolt, or rivet in single shear (as illustrated in figure 11.11) is given by equation 11.23. Because it gives an average value over the cross section of the shear member, this equation should only be used when the loading is low or when there is multiple redundancy in the shear group.

$$\tau = \frac{F}{A} \qquad 11.23$$

The actual shear stress in a beam is dependent on the location within the beam, just as was the bending stress. Shear stress is zero at the outer edges of a beam and maximum at the neutral axis. This is illustrated in figure 11.12.

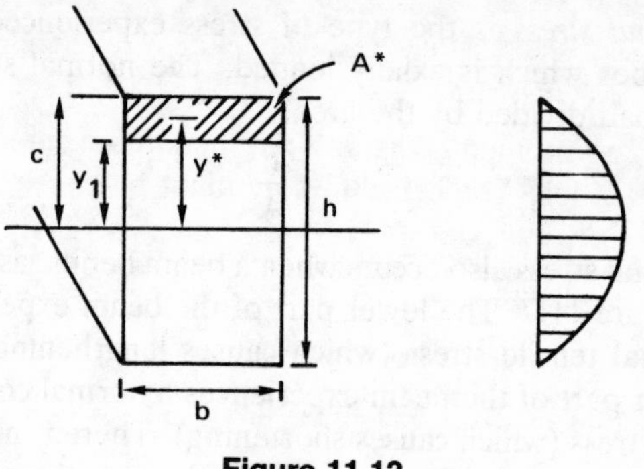

Figure 11.12
Shear Stress Distribution in a Rectangular Beam

The shear stress distribution within a beam is given by equation 11.24.

$$\tau = \frac{QV}{Ib} \qquad 11.24$$

V is the shear (in pounds) at the section where the shear stress is wanted. *V* can be found from a shear diagram. *I* is the beam's centroidal moment of inertia. *b* is the width of the beam at the depth y_1 within the beam where the shear stress is wanted. *Q* is the *statical moment*[1], as defined by equation 11.25.

$$Q = \int_{y_1}^{c} y\,dA \qquad 11.25$$

For rectangular beams, dA = b dA. Equation 11.25 can be simplified to equation 11.26 for rectangular beams.

$$Q = y^*A^* \qquad 11.26$$

Equation 11.26 says that the statical moment at a location y_1 within a rectangular beam is equal to the product of the area above y_1 and the distance from the centroidal axis to the centroid of A*.

The maximum shear stress in a rectangular beam is

$$\tau_{max} = \frac{3V}{2A} = \frac{3V}{2bh} \qquad 11.27$$

For a round beam of radius *r* and area *A*, the maximum shear stress is

$$\tau_{max} = \frac{4V}{3A} = \frac{4V}{3\pi r^2} \qquad 11.28$$

The shear, *V*, used in equations 11.27 and 11.28 is the *one-way shear.*

Example 11.2

What are the maximum shear and bending stresses for the beam shown in example 11.1?

[1]The statical moment is also known as the *first moment of the area.*

From the shear diagram, the maximum shear is −667 pounds. From equation 11.27, the maximum shear stress is

$$\tau_{max} = \frac{(3)(-667)}{(2)(6)(8)} = -20.8 \text{ psi}$$

From the moment diagram, the maximum moment is +1420 ft-lbs. The centroidal moment of inertia is

$$I_c = \frac{(6)(8)^3}{12} = 256 \text{ in}^4$$

From equation 11.18, the maximum bending stress is

$$\sigma_{b,max} = \frac{(1420)(12)(4)}{256} = 266.3 \text{ psi}$$

Example 11.3

The chain hook shown carries a load of 500 pounds. What are the minimum and maximum stresses in the vertical portion of the hook?

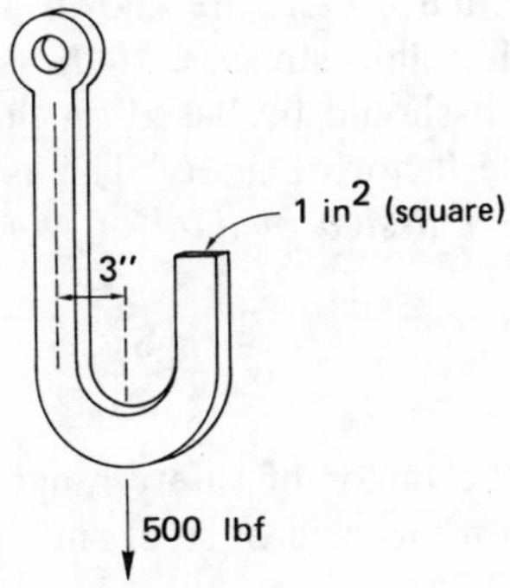

The hook is eccentrically loaded because the load and supporting force are not in line. The centroidal moment of inertia of the 1″ × 1″ section is

$$I_c = \frac{bh^3}{12} = \frac{(1)(1)^3}{12} = .0833 \text{ in}^4$$

From equation 11.21,

$$\sigma_{max,min} = \frac{500}{1} \pm \frac{(500)(3)(.5)}{.0833}$$
$$= 500 \pm 9{,}000$$
$$= +9{,}500 \text{ and } -8{,}500$$

The 500 psi direct stress is tensile. However, the flexural compressive stress of 9,000 psi counteracts this tensile stress, resulting in 8,500 psi compressive stress at the outer face of the hook. The stress is 9,500 psi tension at the inner face.

6. Stresses in Composite Structures

A *composite structure* is one in which two or more different materials are used, both of which carry a part of the load. Unless the various materials used all have the same modulus of elasticity, the stress analysis will be dependent on the assumptions made.

Some simple composite structures can be analyzed using the assumption of *consistent deformations*. This is illustrated in examples 11.4 and 11.5. The technique used to analyze structures for which the strains are consistent is known as the *transformation method*.

step 1: Determine the modulus of elasticity for each of the materials used in the structure.

step 2: For each of the materials used, calculate the ratio

$$n = \frac{E}{E_{weakest}} \qquad 11.29$$

$E_{weakest}$ is the smallest modulus of elasticity of any of the materials used in the composite structure.

step 3: For all of the materials except the weakest, multiply the actual material stress area by *n*. Consider this expanded (*transformed*) area to have the same composition as the weakest material.

step 4: If the structure is a tension or compression member, the distribution or placement of the transformed areas is not important. Just assume that the transformed areas carry the axial load. For beams in bending, the transformed area can add to the width of the beam, but it cannot change the depth of the beam or the thickness of the reinforcing.

step 5: For compression or tension numbers, calculate the stresses in the weakest and stronger materials.

$$\sigma_{weakest} = \frac{F}{A_t} \qquad 11.30$$

$$\sigma_{stronger} = \frac{nF}{A_t} \qquad 11.31$$

step 6: For beams in bending, proceed through step 9. Find the centroid of the transformed beam.

step 7: Find the centroidal moment of inertia of the transformed beam, I_{ct}.

step 8: Find V_{max} and M_{max} by inspection or from the shear and moment diagrams.

step 9: Calculate the stresses in the weakest and stronger materials.

$$\sigma_{weakest} = \frac{Mc_{weakest}}{I_{ct}} \qquad 11.32$$

$$\sigma_{stronger} = \frac{nMc_{stronger}}{I_{ct}} \qquad 11.33$$

Example 11.4

Find the stress in the steel inner cylinder and the copper tube which surrounds it if a uniform compressive load of 100 kips is applied axially. The copper and steel are well bonded. Use E_{steel} = 3 EE7 psi and E_{copper} = 1.75 EE7 psi.

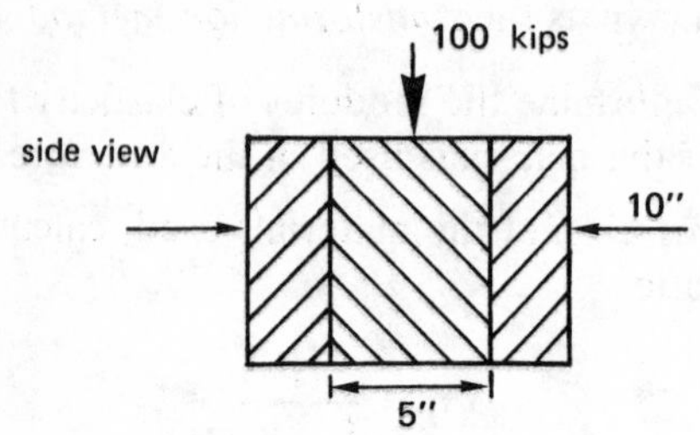

$$n = \frac{3 \text{ EE7}}{1.75 \text{ EE7}} = 1.714$$

The actual steel area is $\frac{1}{4}\pi(5)^2 = 19.63 \text{ in}^2$.
The actual copper area is $\frac{1}{4}\pi[(10)^2-(5)^2] = 58.9 \text{ in}^2$.
The transformed area is $A_t = 58.9 + 1.714(19.63) = 92.55 \text{ in}^2$.

$$\sigma_{copper} = \frac{100{,}000}{92.55} = 1080.5 \text{ psi}$$

$$\sigma_{steel} = (1.714)(1080.5) = 1852.0 \text{ psi}$$

Example 11.5

Find the maximum bending stress in the steel-reinforced wood beam shown below at a point where the moment is 40,000 ft-lb. Use E_{steel} = 3 EE7 psi and E_{wood} = 1.5 EE6 psi.

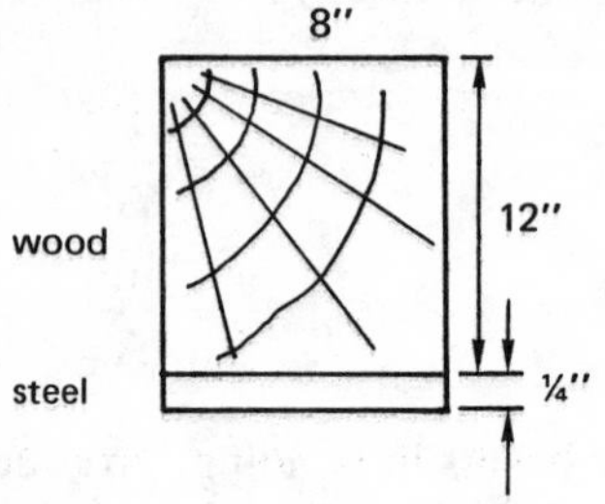

$$n = \frac{3 \text{ EE7}}{1.5 \text{ EE6}} = 20$$

The actual steel area is (.25)(8) = 2.

The area of the steel is expanded to 20(2) = 40. Since the depth of beam and reinforcement cannot be increased, the width must increase. The 160″ dimension is arrived at by dividing the area of 40 square inches by the thickness of ¼″.

The centroid is located at $\bar{y}$ = 4.45 inches from the x-x axis. The centroidal moment of inertia of the transformed section is $I_c = 2211.5 \text{ in}^4$. Then, from equations 11.32 and 11.33,

$$\sigma_{max,wood} = \frac{(40{,}000)(12)(7.8)}{(2211.5)} = 1692 \text{ psi}$$

$$\sigma_{max,steel} = \frac{(20)(40{,}000)(12)(4.45)}{2211.5} = 19{,}320 \text{ psi}$$

7. Allowable Stresses

Once the actual stresses are known, they must be compared to allowable stresses. If the allowable stress is calculated, it should be based on the yield stress and a reasonable factor of safety. This is known as the *allowable stress design method* or *working stress design method.*

$$\sigma_a = \frac{S_y}{FS} \qquad 11.34$$

For steel, the factor of safety ranges from 1.5 to 2.5 depending on the type of steel and application.

The allowable stress method is being replaced in structural work by the *load factor design method,* also known as the *ultimate strength method* and *plastic design method*. In this method, the applied loads are multiplied by a load factor. The product must be less than the structural member's ultimate strength, usually determined from a table.

8. Beam Deflections

A. Double Integration Method

The deflection and slope of a loaded beam are related to the applied moment and shear by equations 11.35 through 11.38.

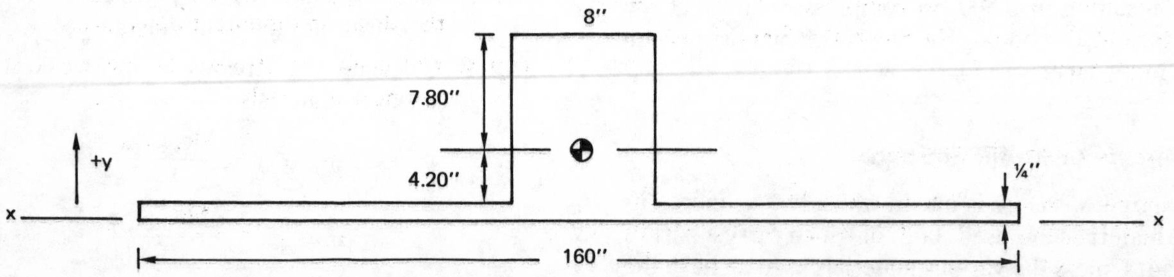

$$y = \text{deflection} \qquad 11.35$$

$$\frac{dy}{dx} = \text{slope} \qquad 11.36$$

$$\frac{d^2y}{dx^2} = \frac{M}{EI} \qquad 11.37$$

$$\frac{d^3y}{dx^3} = \frac{V}{EI} \qquad 11.38$$

Thus, if the moment function, M(x), is known for a section of beam, the deflection at any point can be found from equation 11.39.

$$y = \frac{1}{EI}\int\int M(x)\,dx \qquad 11.39$$

In order to find the deflection, constants need to be introduced during the integration process. These constants can be found from table 11.2.

Table 11.2
Beam Boundary Conditions

end condition	y	y′	y″	V	M
simple support	0				
built-in support	0	0			
free end			0	0	0
hinge					0

Example 11.6

Find the tip deflection of the beam shown. *EI* is 5 EE10 lbf-in^2 everywhere on the beam.

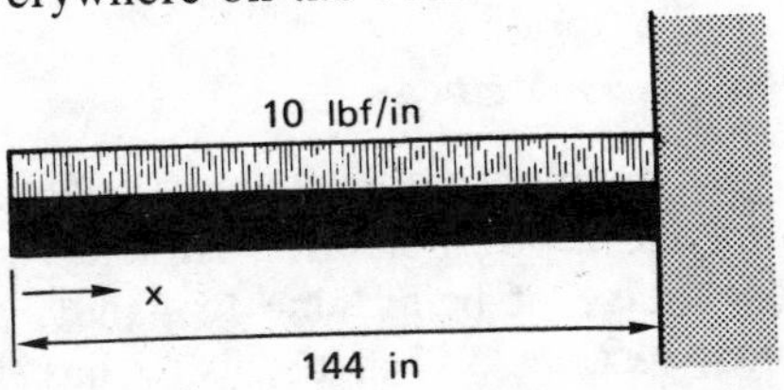

The moment at any point *x* from the left end of the beam is

$$M(x) = (-10)(x)(\tfrac{1}{2}x) = -5x^2$$

This is negative by the left-hand rule convention. From equation 11.37,

$$y'' = \frac{M}{EI}$$

So,

$$EIy'' = -5x^2$$

$$EIy' = \int -5x^2\,dx = -\tfrac{5}{3}x^3 + C_1$$

Since $y' = 0$ at a built-in support (table 11.2) and $x = 144$ inches at the built-in support,

$$0 = -\tfrac{5}{3}(144)^3 + C_1$$

or $C_1 = 4.98$ EE6

$$EIy = \int(-\tfrac{5}{3}x^3 + 4.98\text{ EE6})\,dx$$

$$= -\tfrac{5}{12}x^4 + (4.98\text{ EE6})x + C_2$$

Again, $y = 0$ at $x = 144$, so $C_2 = -5.38$ EE8.

Therefore, the deflection as a function of *x* is

$$y = \left(\frac{1}{EI}\right)\left[(-\tfrac{5}{12})x^4 + (4.98\text{ EE6})x - 5.38\text{ EE8}\right]$$

At the tip, $x = 0$, so the deflection is

$$y_{tip} = \frac{-5.38\text{ EE8}}{5\text{ EE 10}} = -.0108\text{ inches}$$

B. Moment Area Method

The moment area method is a semi-graphical technique which is applicable whenever slopes of deflection beams are not too great. This method is based on the following two theorems.

Theorem I: The angle between tangents at any two points on the elastic line of a beam is equal to the area of the moment diagram between the two points divided by *EI*. That is,

$$\theta = \int\frac{M(x)\,dx}{EI} \qquad 11.40$$

Theorem II: One point's deflection away from the tangent of another point is equal to the statical moment of the bending moment between those two points divided by *EI*. That is,

$$y = \int\frac{xM(x)\,dx}{EI} \qquad 11.41$$

The application of these two theorems is aided by the following two comments.

- If EI is constant, the statical moment $\int xM(x)\,dx$ may be calculated as the product of the total moment diagram area times the distance from the point whose deflection is wanted to the centroid of the moment diagram.
- If the moment diagram has positive and negative parts (areas above and below the zero line) the statical moment should be taken as the sum of two products, one for each part of the moment diagram.

Example 11.7

Find the deflection, *y*, and the angle, θ, at the free end of the cantilever beam shown below.

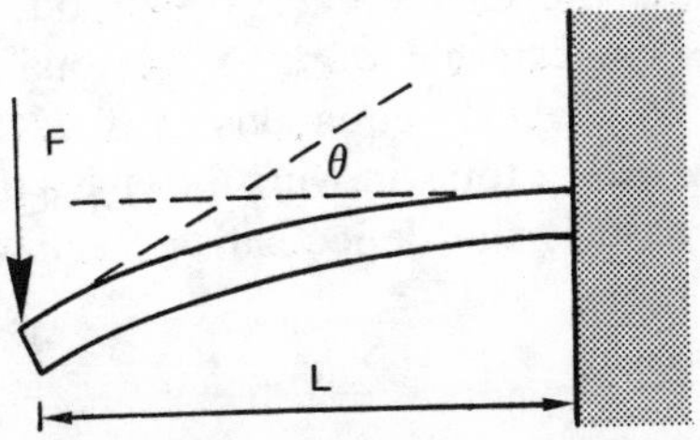

The deflection angle, θ, is the angle between the tangents at the free and built-in ends (Theorem **I**). The moment diagram is

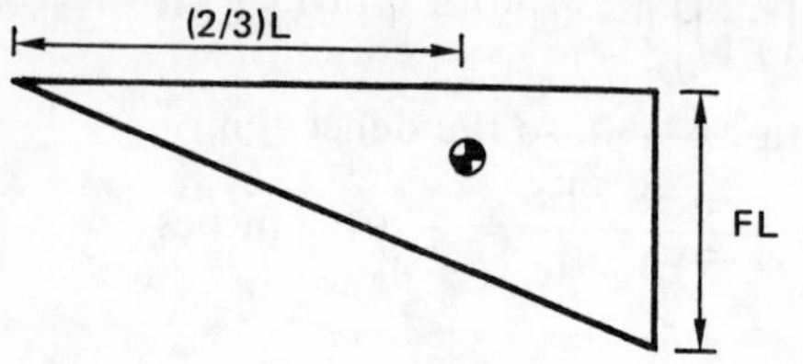

The area of the moment diagram is

$$\tfrac{1}{2}(FL)(L) = \tfrac{1}{2}FL^2$$

From Theorem **I**,

$$\theta = \frac{FL^2}{2EI}$$

From Theorem **II**,

$$y = \frac{FL^2}{2EI}(\tfrac{2}{3}L) = \frac{FL^3}{3EI}$$

Example 11.8

Find the deflection of the free end of the cantilever beam shown.

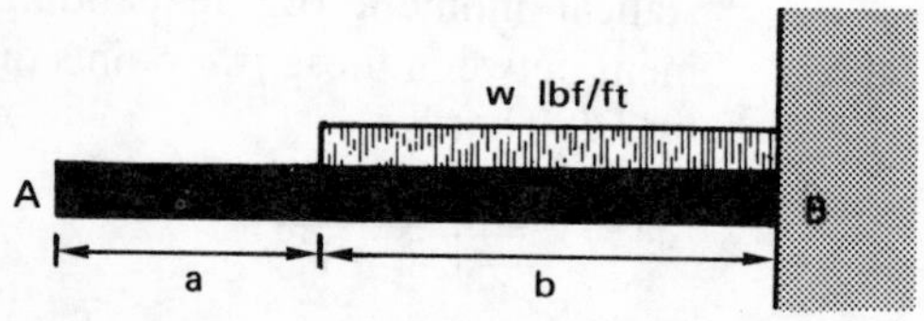

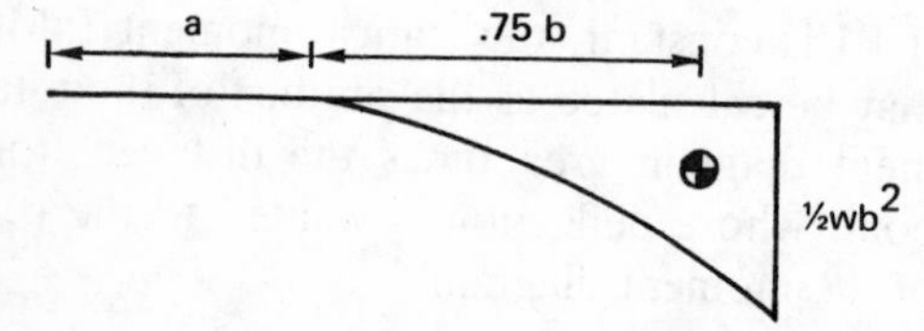

The distance from point *A* (where the deflection is wanted) to the centroid is (a + .75b). The area of the moment diagram is ($wb^3/6$). From Theorem **II**,

$$y = \frac{wb^3}{6EI}(a + .75b)$$

C. Strain Energy Method

The deflection at a point of load application may be found by the strain energy method. This method equates the external work to the total internal strain energy as given by equations 11.8, 11.22, and 11.73. Since work is a force moving through a distance (which in this case is the deflection) we can write equation 11.42.

$$\tfrac{1}{2}Fy = \Sigma U \qquad 11.42$$

Example 11.9

Find the deflection at the tip of the stepped beam shown below.

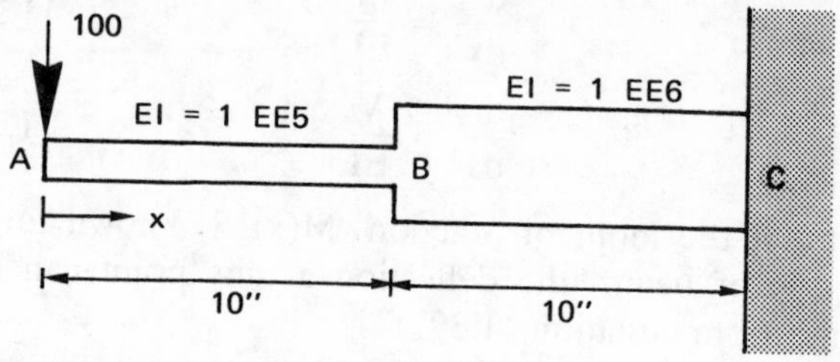

In section A-B: M = 100x in-lbf

From equation 11.22,

$$U = \tfrac{1}{2}\int_0^{10} \frac{(100x)^2}{1\ \text{EE5}}\,dx = 16.67 \text{ in-lbf}$$

In section B-C: M = 100x

$$U = \tfrac{1}{2}\int_{10}^{20} \frac{(100x)^2}{1\ \text{EE6}}\,dx = 11.67 \text{ in-lbf}$$

Equating the internal work (U) and the external work,

$$16.67 + 11.67 = \tfrac{1}{2}(100)y$$

$$y = .567 \text{ in}$$

D. Conjugate Beam Method

The *conjugate beam method* changes a deflection problem into one of drawing moment diagrams. The method has the advantage of being able to handle beams of varying cross sections and materials. It has the disadvantage of not easily being able to handle beams with two built-in ends. The following steps constitute the conjugate beam method.

step 1: Draw the moment diagram for the beam as it is actually loaded.

step 2: Construct the *M/EI* diagram by dividing the value of *M* at every point along the beam by the product of *EI* at that point. If the beam is of constant cross section, *EI* will be constant and the *M/EI* diagram will have the same shape as the moment diagram. However, if the beam cross section varies with *x*, then *I* will change. In this case, the *M/EI* diagram will not look the same as the moment diagram.

step 3: Draw a conjugate beam of the same length as the original beam. The material and cross sectional area of this conjugate beam are not relevant.

(a) If the actual beam is simply supported at its ends, the conjugate beam will be simply supported at its ends.

(b) If the actual beam is simply supported away from its ends, the conjugate beam has hinges at the support points.

(c) If the actual beam has free ends, the conjugate beam has built-in ends.

(d) If the actual beam has built-in ends, the conjugate beam has free ends.

step 4: Load the conjugate beam with the M/EI diagram. Find the conjugate reactions by methods of statics. Use the superscript * to indicate conjugate parameters.

step 5: Find the conjugate moment at the point where the deflection is wanted. The deflection is numerically equal to the moment as calculated from the conjugate beam forces.

Example 11.10

Find the deflections at the two load points. EI has a constant value of 2.356 EE7 lb-in^2.

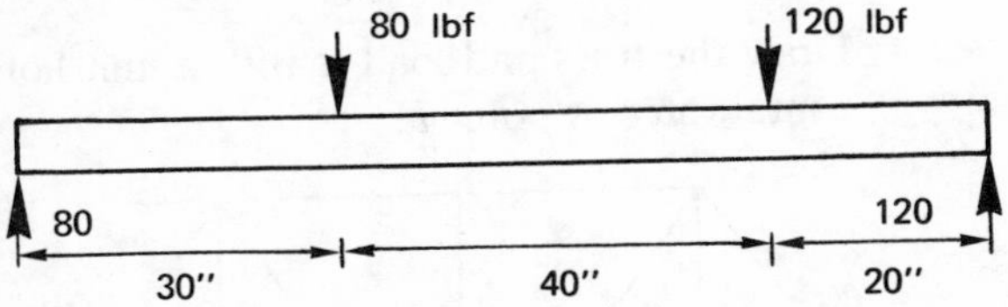

step 1: The moment diagram for the actual beam is:

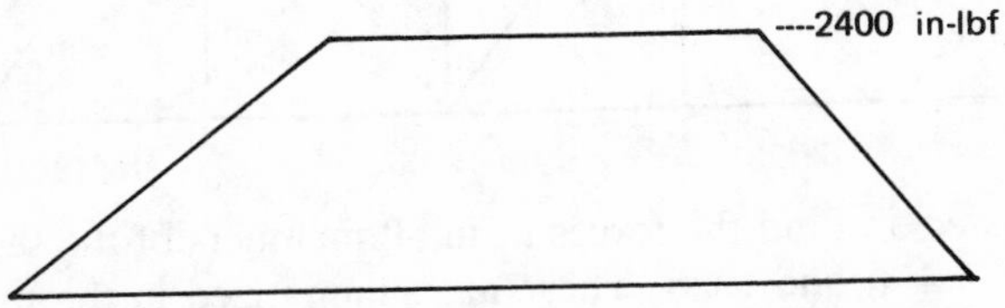

steps 2, 3, and 4: Since the cross section is constant, the conjugate load has the same shape as the original moment diagram. The peak load on the conjugate beam is

$$\frac{2400 \text{ in-lb}}{2.356 \text{ EE7 lb-in}^2} = 1.019 \text{ EE}-4 \text{ (1/in)}$$

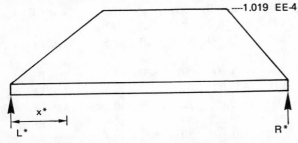

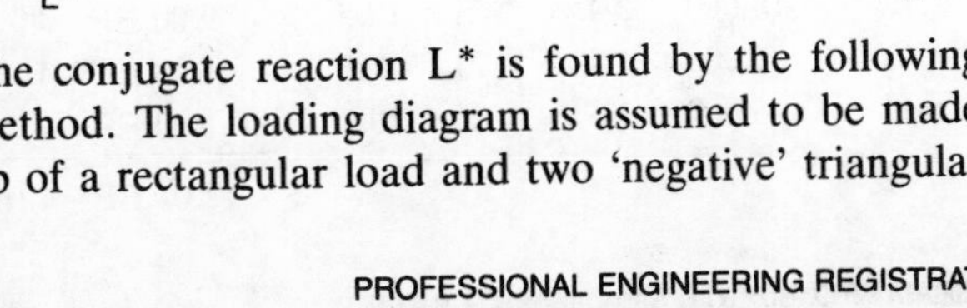

The conjugate reaction L* is found by the following method. The loading diagram is assumed to be made up of a rectangular load and two 'negative' triangular loads. The area of the rectangular load (which has a centroid at $x^* = 45$) is (90)(1.019 EE−4) = 9.171 EE−3.

Similarly, the area of the left triangle (which has a centroid at $x^* = 10$) is ½(30)(1.019 EE−4) = 1.529 EE−3. The area of the right triangle (which has a centroid at $x^* = 83.33$) is ½(20)(1.019 EE−4) = 1.019 EE−3.

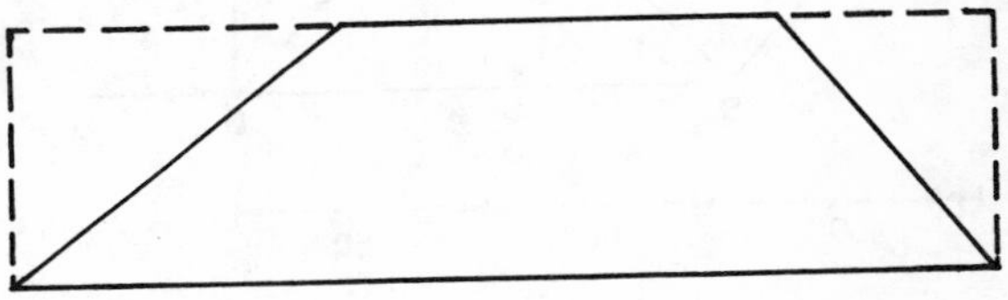

$$\Sigma M^*_{(L^*)} = 90R^* + (1.019 \text{ EE}-3)(83.3) + (1.529 \text{ EE}-3)(10) - (9.171 \text{ EE}-3)(45) = 0$$

$$R^* = 3.472 \text{ EE}-3 \text{ (1/in)}$$

Then, $L^* = (9.171 - 1.019 - 1.529 - 3.472)\text{EE}-3 = 3.151 \text{ EE}-3$ (1/in)

step 5: The conjugate moment at $x^* = 30$ is

$$M^* = (3.151 \text{ EE}-3)(30) + (1.529 \text{ EE}-3)(30-10) - (9.171 \text{ EE}-3)(30/90)(15) = 6.266 \text{ EE}-2 \text{ in}$$

The conjugate moment at the right-most load is

$$M^* = (3.472 \text{ EE}-3)(20) + (1.019 \text{ EE}-3)(13.3) - (9.171 \text{ EE}-3)(20/90)(10) = 6.266 \text{ EE}-2 \text{ in}$$

E. Table Look-Up Method

Appendix A is an extensive listing of the most commonly needed beam formulas. The use of these formulas is recommended whenever they can be applied singly or as part of a superposition solution.

F. Method of Superposition

If the deflection at a point is due to the combined action of two or more loads, the deflections at that point due to the individual loads may be added to find the total deflection.

9. Truss Deflections

A. Strain-Energy Method

The deflection of a truss at the point of a single load application may be found by the *strain-energy method* if all member forces are known. This method is illustrated by example 11.11.

Example 11.11

Find the vertical deflection of point A under the external load of 707 pounds. AE = 10 EE5 pounds for all members. The internal forces have been determined.

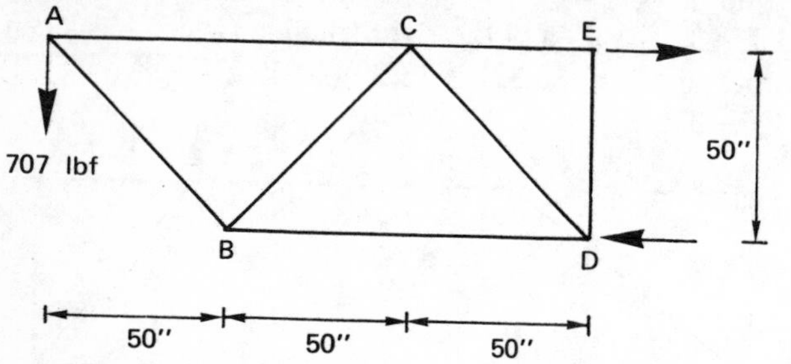

The length of member AB is $\sqrt{(50)^2 + (50)^2} = 70.7$ inches. From equation 11.8, the internal strain energy in member AB is

$$U = \frac{(-1000)^2(70.7)}{2(10\ EE5)} = 35.4 \text{ in-lbf}$$

Similarly, the energy in all members can be determined.

Member	L	F	U
AB	70.7	−1000	+35.4
BC	70.7	+1000	+35.4
AC	100	+707	+25.0
BD	100	−1414	+100.0
CD	70.7	−1000	+35.4
CE	50	+2121	+112.5
DE	50	+707	+ 12.5
			356.2

The external work is $W_{ext} = \frac{1}{2}(707)y$, so

$$(\tfrac{1}{2})(707)y = 356.2$$

$$y = 1 \text{ in}$$

B. *Virtual Work Method (Hardy Cross Method)*

An extension of the strain-energy method results in an easy procedure for computing the deflection of *any* point on a truss.

step 1: Draw the truss twice.

step 2: On the first truss, place all of the actual loads.

step 3: Find the forces, S, in all of the members due to the actual applied loads.

step 4: On the second truss, place a dummy one pound load in the direction of the desired displacement.

step 5: Find the forces, u, in all members due to the one pound dummy load.

step 6: Find the desired displacement from equation 11.43.

$$\delta = \Sigma\frac{SuL}{AE} \qquad 11.43$$

In equation 11.43, the summation is over all truss members which have non-zero forces in *both* trusses.

Example 11.12

What is the horizontal deflection of point F on the truss shown? Use E = 3 EE7 psi. Joint A is restrained horizontally.

steps 1 and 2: Use the truss as drawn.

step 3: The forces in all of the truss members are summarized in step 5.

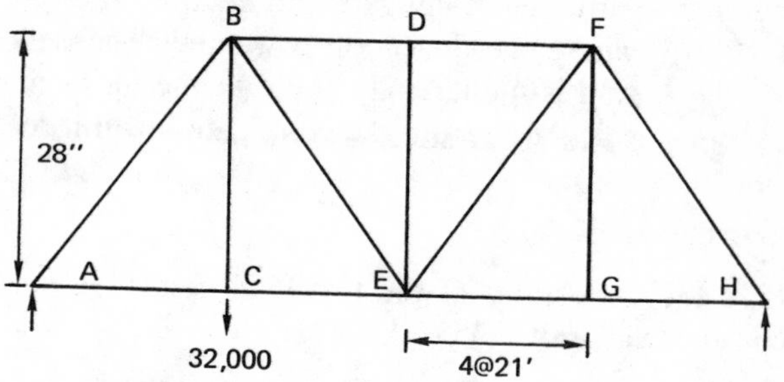

step 4: Draw the truss and load it with a unit horizontal force at point F.

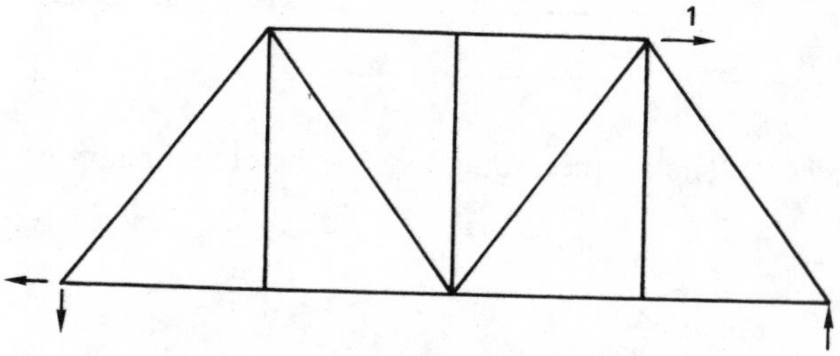

step 5: Find the forces u, in all members of the second truss. These are summarized in the following table. Notice the sign convention: + for tension and − for compression.

member	S(lbs)	u	L(ft)	A(in²)	$\frac{SuL}{AE}$(ft)
AB	−30,000	5/12	35	17.5	−8.33 EE−4
CB	32,000	0	28	14	0
EB	−10,000	−5/12	35	17.5	2.78 EE−4
ED	0	0	28	14	0
EF	10,000	5/12	35	17.5	2.78 EE−4
GF	0	0	28	14	0
HF	−10,000	−5/12	35	17.5	2.78 EE−4
BD	−12,000	1/2	21	10.5	−4.00 EE−4
DF	−12,000	1/2	21	10.5	−4.00 EE−4
AC	18,000	3/4	21	10.5	9.00 EE−4
CE	18,000	3/4	21	10.5	9.00 EE−4
EG	6,000	1/4	21	10.5	1.00 EE−4
GH	6,000	1/4	21	10.5	1.00 EE−4
					12.01 EE−4 (ft)

Since 12.01 EE−4 is positive, the deflection is in the direction of the dummy unit load. In this case, the deflection is to the right.

10. Combined Stresses

Most practical cases of combined stresses have normal stresses on two perpendicular planes and a known shear stress acting parallel to these two planes. Based on knowledge of these stresses, the shear and normal stresses on all other planes can be found from conditions of equilibrium.

Under any condition of stress at a point, a plane can be found where the shear stress is zero. The normal stresses on this plane are known as the *principal stresses*. The principal stresses are the maximum and minimum stresses at the point in question.

The normal and shear stresses on a plane inclined an angle θ from the horizontal are given by equations 11.44 and 11.45.

$$\sigma_\theta = \tfrac{1}{2}(\sigma_x+\sigma_y) + \tfrac{1}{2}(\sigma_x-\sigma_y)\cos 2\theta + \tau\sin 2\theta \qquad 11.44$$

$$\tau_\theta = -\tfrac{1}{2}(\sigma_x-\sigma_y)\sin 2\theta + \tau\cos 2\theta \qquad 11.45$$

The maximum and minimum values of σ_θ and τ_θ (as θ is varied) are the principal stresses. These are given by equations 11.46 and 11.47.

$$\sigma(\text{max,min}) = \tfrac{1}{2}(\sigma_x+\sigma_y) \pm \tau(\text{max}) \qquad 11.46$$

$$\tau(\text{max,min}) = \pm\tfrac{1}{2}\sqrt{(\sigma_x-\sigma_y)^2 + (2\tau)^2} \qquad 11.47$$

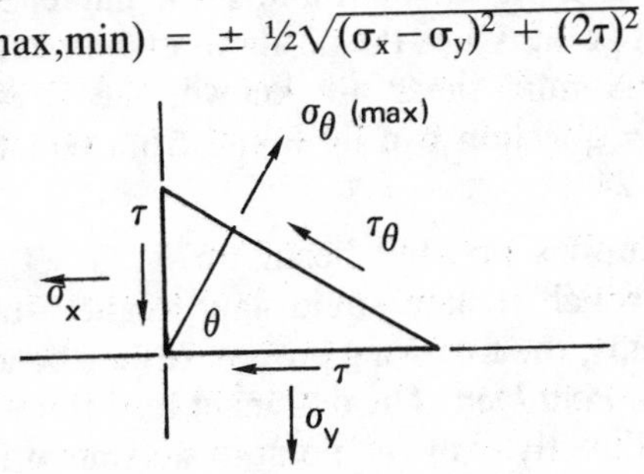

Figure 11.13 Plane of Principal Stress

The angles of the planes on which the normal stresses are minimum and maximum are given by equation 11.48. θ is measured from the *x* axis, clockwise if negative, and counter-clockwise if positive. Equation 11.48 will yield two angles. These angles must be used in equation 11.44 to determine which angle corresponds to the minimum normal stress and which angle corresponds to the maximum normal stress.

$$\theta = \tfrac{1}{2}\arctan\left(\frac{2\tau}{\sigma_x-\sigma_y}\right) \qquad 11.48$$

The angles of the planes on which the shear stress is minimum and maximum are given by equation 11.49. The same angle sign convention used for equation 11.48 applies to equation 11.49.

$$\theta = \tfrac{1}{2}\arctan\left(\frac{\sigma_x-\sigma_y}{-2\tau}\right) \qquad 11.49$$

Proper sign convention must be adhered to when using equations 11.44 through 11.49. Normal tensile stresses are positive; normal compressive stresses are negative. Shear stresses are positive as shown in figure 11.14

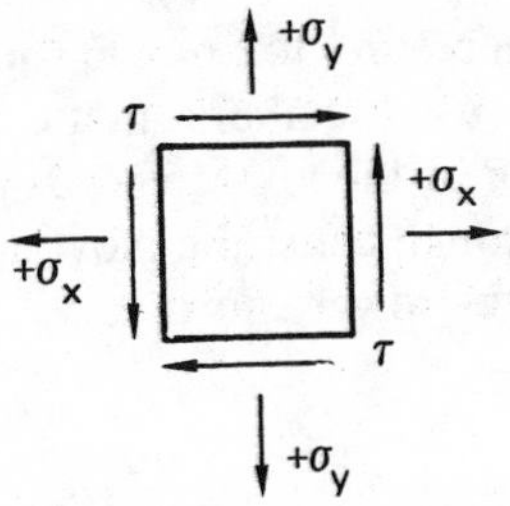

Figure 11.14 Sign Convention

Example 11.13

Find the maximum shear stress and the maximum normal stress on the object shown.

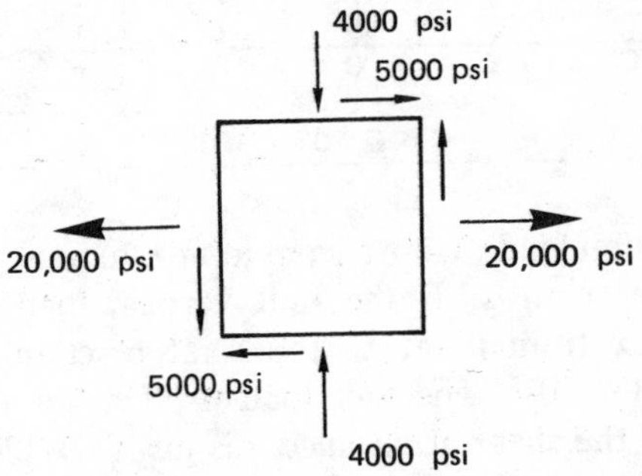

By the sign convention of figure 11.14, the 4000 psi is negative. From equation 11.47, the maximum shear stress is

$$\tau_{max} = \tfrac{1}{2}\sqrt{[20{,}000 - (-4000)]^2 + [(2)(5000)]^2}$$

$$= 13{,}000 \text{ psi (tension)}$$

From equation 11.46, the maximum normal stress is

$$\sigma_{max} = \tfrac{1}{2}[20{,}000 + (-4000)] + 13{,}000$$

$$= 21{,}000 \text{ psi}$$

11. Dynamic Loading

If a load is applied suddenly to a structure, the transient response may create stresses greater than would normally be calculated from the concepts of statics and mechanics of materials alone. Although a dynamic analysis of the structure is appropriate, the procedure is extremely lengthy and complicated. Therefore, arbitrary dynamic factors are applied to the static stress. For example, if the load is applied quickly compared to the natural period of the structure, a dynamic factor of 2 may be used. This assumes the load is applied as a ramp function.

12. Influence Diagrams

Shear, moment, and reaction *influence diagrams (influence lines)* can be drawn for any point on a truss. This is a necessary step in the evaluation of stresses induced by moving loads. It is important to realize, however, that the influence line applies to only one point on the truss.

To begin, it is necessary to know if the loads are transmitted to the truss members at the lower chords (a *through truss*) or at the upper chords (a *deck truss*). If the truss is a through truss, the moving load is assumed to move along the lower chords.

Example 11.14

Draw the influence diagram for vertical shear in panel *DF* of the through truss shown.

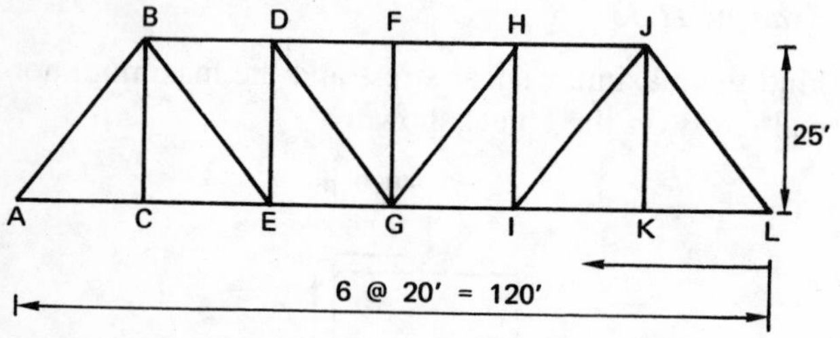

Allow a unit load to move from joint *L* to joint *G* along the lower chords. If the unit vertical load is at a distance *x* from point *L*, the right reaction will be $+[1 - (x/120)]$. The unit load itself has a value of (-1), so the shear at distance *x* is just $(-x/120)$.

Next, allow a unit load to move from joint *A* to joint *E* along the lower chords. If the unit load is a distance *x* from point *L*, the left reaction will be $(x/120)$, and the shear at distance *x* will be $[(x/120) - 1]$.

These two lines can be graphed.

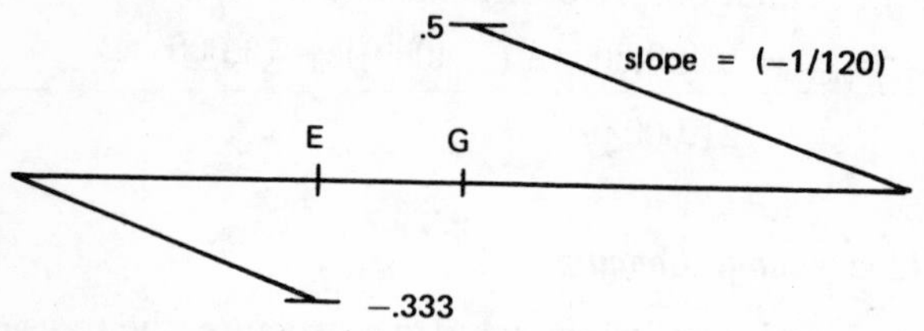

The influence line is completed by connecting the two lines as shown below. Therefore, the maximum shear in panel *DF* will occur when a load is at point *G* on the truss.

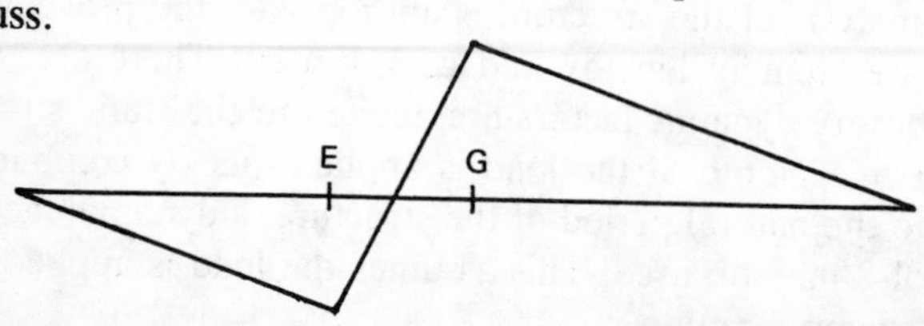

Example 11.15

Draw the moment influence diagram for panel *DF* on the truss shown in example 11.14.

The left reaction is $(x/120)$ where *x* is the distance from the unit load to the right end. If the unit load is to the right of point *G*, the moment can be found by summing moments from point *G* to the left. The moment is $(x/120)(60) = .5x$.

If the unit load is to the left of point *E*, the moment will be

$$\left(\frac{x}{120}\right)(60) - (1)(x-60) = 60 - .5x$$

These two lines can be graphed. The moment for a unit load between points *E* and *G* is obtained by connecting the two end points of the lines derived above.

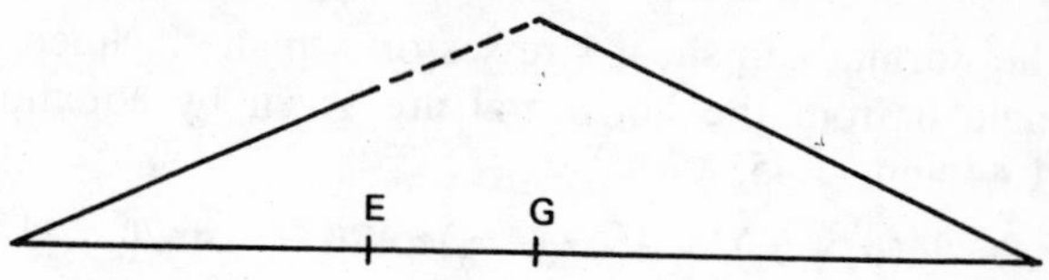

13. Moving Loads on Beams

If a beam supports a single moving load, the maximum bending and shearing stresses at any point can be found by drawing the moment and shear influence diagrams for that point. Once the positions of maximum moment and maximum shear are known, the stresses at the point in question can be found from equations 11.18 and 11.24.

If a simply-supported beam carries a set of moving loads (which remain equidistant as they travel across the beam), the following procedure may be used to find the *dominant load*. The dominant load is the one which occurs directly over the point of maximum moment.

step 1: Calculate and locate the resultant of the load group.

step 2: Assume one of the loads is dominant. Place the group on the beam such that the distance from one support to the assumed dominant load is equal to the distance from the other support to the resultant of the load group.

step 3: Check to see that all loads are on the span and that the shear changes sign under the assumed dominant load. If the shear does not change sign under the assumed dominant load, the maximum moment may occur when only some of the load group is on the beam. If it does change sign, calculate the bending moment under the assumed dominant load.

step 4: Repeat steps 2 and 3 assuming the other loads are dominant.

step 5: Find the maximum shear by placing the load group such that the resultant is a minimum distance from a support.

14. Columns

The *Euler load* is the theoretical maximum load that an initially straight column can support without buckling. For columns with pinned ends, this load is given by equation 11.50.

$$F_e = \frac{\pi^2 EI}{L^2} = \frac{(k\pi)^2 EA}{L^2} \qquad 11.50$$

The corresponding column stress is

$$\sigma_e = \frac{F_e}{A} = \frac{\pi^2 E}{(L/k)^2} \qquad 11.51$$

Equations 11.50 and 11.51 assume that the column is long so that the Euler stress is reached before the yield stress is reached. If the column is short, then the yield stress of the material may be less than the Euler stress. In that case, short-column curves based on test data are used to predict the allowable column stress.

The value of L/k at the point of intersection of the short column and Euler curves is known as the critical *slenderness ratio.* The critical slenderness ratio becomes smaller as the compressive yield stress increases. The region in which the short column formulas apply is determined by tests for each particular type of column and material. Typical critical slenderness ratios range from 80 to 120.

In general, the Euler allowable stress formula may be used if the stress obtained from equation 11.51 does not exceed one-half of the compressive yield stress.

Example 11.16

As S-type, 4 × 9.5 A36 steel I-beam 8.5 feet long is used as a column. What is the working stress for a safety factor of 3? Use E = 2.9 EE7 psi. The yield stress for A36 steel is 36,000 psi. The required properties of the I beam are $A = 2.79 \text{ in}^2$, $I = .903 \text{ in}^4$, and $k = .569$ in.

From equation 11.51, the Euler stress is

$$\sigma_e = \frac{\pi^2(2.9 \text{ EE7})}{[(8.5)(12)/.569]^2} = 8907 \text{ psi}$$

Since 8907 is less than one-half of 36,000, the Euler formulas are valid. The allowable working stress is

$$\sigma_a = \frac{8907}{3} = 2969 \text{ psi}$$

An ultimate load for any column can be found using the *secant formula.* The secant formula is particularly suited for use when the column is intermediate in length.

$$\sigma = \frac{F}{A} = \frac{S_y}{\left(1 + \frac{ec}{k^2}\right)\sec\theta} \qquad 11.52$$

$$\theta = \tfrac{1}{2}\left(\frac{L}{k}\right)\sqrt{F/AE} \qquad 11.53$$

The formula is solved by trial and error for F/A with the given eccentricity, e. If the value of e is not known, the eccentricity ratio (ec/k^2) is taken as .25. Substituting this value and E = 2.9 EE7 for steel and 1.0 EE7 for aluminum respectively, the following formulas result which converge quickly to the known L/k ratio when assumed values of F/A are substituted.

$$\theta = \arccos\left[\frac{1.25}{S_y}\left(\frac{F}{A}\right)\right] \qquad 11.54$$

$$\frac{L}{k} = 2\theta\sqrt{EA/F} \qquad 11.55$$

$$\left(\frac{L}{k}\right)_{\text{steel}} = \frac{10{,}770(\theta)}{\sqrt{F/A}} \qquad 11.56$$

$$\left(\frac{L}{k}\right)_{\text{aluminum}} = \frac{6{,}325(\theta)}{\sqrt{F/A}} \qquad 11.57$$

Example 11.17

A steel column has a compressive yield strength of 90,000 psi and an L/k ratio of 75. Find the working stress assuming a factor of safety of 2.5.

From equation 11.51, the Euler stress is

$$\sigma_e = \frac{\pi^2(2.9 \text{ EE7})}{(75)^2} = 50{,}880 \text{ psi}$$

Since this exceeds one-half of the compressive yield strength, an intermediate column formula should be used. Use the secant formula. Assume that the maximum stress F/A is 30,000 psi. Substituting into equations 11.54 and 11.56,

$$\theta = \arccos\left[\frac{1.25(30{,}000)}{90{,}000}\right] = 1.141 \text{ radians}$$

$$\frac{L}{k} = \frac{(10{,}770)(1.141)}{\sqrt{30{,}000}} = 70.9$$

Since F/A will be lower when L/k is higher, try F/A = 28,000. If this is done, θ = 1.1714 radians and L/K = 75.4 (close enough). Therefore, 28,000 psi is the ultimate strength of the column. Applying a factor of safety,

$$\sigma_a = \frac{28{,}000}{2.5} = 11{,}200 \text{ psi}$$

All of the preceding column formulas are for columns with frictionless round or pinned ends. For other end

MECH

conditions, the *effective length* L' should be used in place of L.

$$L' = CL \qquad 11.58$$

C is the *end restraint coefficient* which varies from .5 to 2. For practical columns, C smaller than .7 should not be used since infinite stiffness of the support structure is not normally achievable.

Table 11.3
End-Restraint Coefficients

end conditions	C
both ends pinned	1
both ends built in	.5
one end pinned, one end built in	.7
one end built in, one end free	2

PART 2: Application to Design

1. Springs

Springs are assumed to be perfectly elastic within their working range. Hooke's law can be used to predict the amount of compression experienced when a load is placed on a spring.

$$F = kx \qquad 11.59$$

k is the spring constant. It has units of pounds per unit length.

When a spring is compressed, it stores energy. This energy can be recovered by restoring the spring's original length. It is assumed that no energy is lost through friction or hysteresis when a spring returns to its original length. The energy storage in a spring is given by equation 11.60. This energy is the same as the work required to compress the spring.

$$W = \Delta U = \tfrac{1}{2}kx^2 \qquad 11.60$$

If a weight is dropped from height *h* onto a spring, the compression can be found by equating the change in potential energy to the energy storage.

$$m\left(\frac{g}{g_c}\right)(h + x) = \tfrac{1}{2}kx^2 \qquad 11.61$$

2. Thin-Walled Cylinders

A cylinder can be considered *thin-walled* if its wall thickness-to-radius ratio is less than .1. The circumferential *hoop stress* can be easily derived from the free-body diagram of a cylinder half. This hoop stress is

$$\sigma_h = \frac{pr}{t} \qquad 11.62$$

Since the cylinder is assumed to be thin-walled, the radius used in equation 11.62 is taken as the inside radius.

If the cylinder is part of a tank, the axial force on the end plates produces an axial stress. The axial force is equal to the tank pressure times the end plate area. The stress produced is at right angles to the hoop stress. Accordingly, it is called *longitudinal stress* or *long stress.*

$$\sigma_l = \frac{pr}{2t} \qquad 11.63$$

Equation 11.63 also gives the stress in a spherical tank. In a spherical tank, the hoop and long stresses are the same.

The hoop and long stresses are principal stresses. They do not combine into a larger stress.

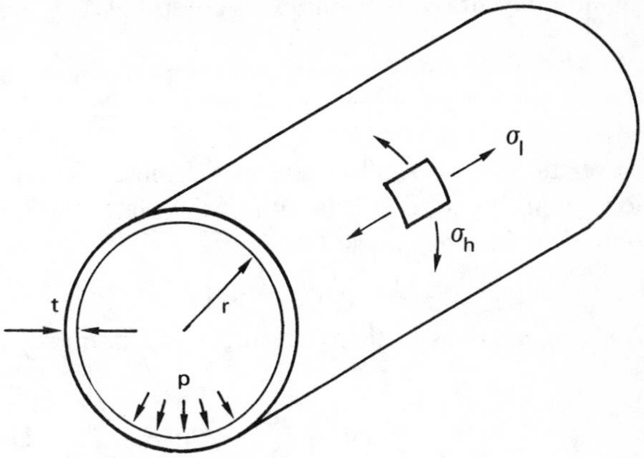

Figure 11.15 Hoop and Long Stresses

3. Rivet and Bolt Connections

A tension splice using rivets or bolts can fail in one of three ways: bearing failure, shear failure, or tension failure. All three failure mechanisms must be checked to determine the maximum load the splice can carry.

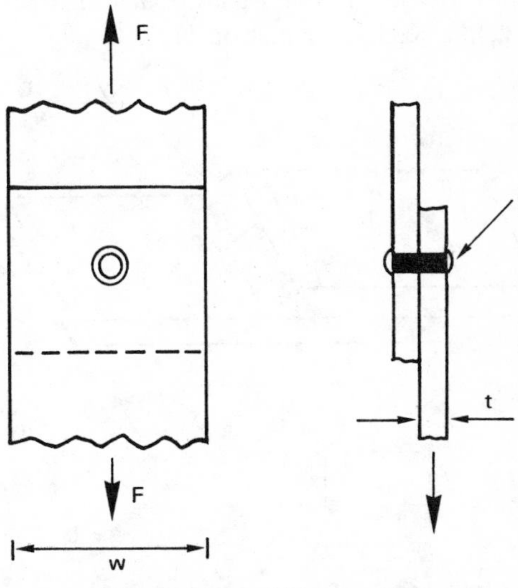

Figure 11.16 A Tension Splice

The plate can fail in bearing. For one connector, the *bearing stress* in the plate is

$$\sigma_{br} = \frac{F}{Dt} \qquad 11.64$$

The number of rivets required to keep the actual bearing stress below the allowable bearing stress is

$$n_{br} = \frac{\sigma_{br}}{\text{allowable bearing stress}} \qquad 11.65$$

The rivet can fail in shear. The shear stress in the rivet is

MECH

$$\sigma_s = \frac{F}{\frac{1}{4}\pi D^2} \quad 11.66$$

The number of rivets required, as determined by shear, is

$$n_s = \frac{\sigma_s}{\text{allowable shear stress}} \quad 11.67$$

The plate can also fail in tension. If there are *n* rivet holes in a line across the width of the plate, the minimum area in tension will be

$$A_t = t(w - nD) \quad 11.68$$

The tensile stress in the plate at the minimum section is

$$\sigma_t = \frac{F}{A_t} \quad 11.69$$

The maximum number of rivets across the plate width must be chosen to keep the tensile stress less than the allowable stress.

4. *Fillet Welds*

The most common weld type is the *fillet weld* shown in figure 11.17. Such welds are commonly used to connect one plate to another. The applied load is assumed to be carried by the *effective weld throat* which is related to the weld size, *y*, by equation 11.70.

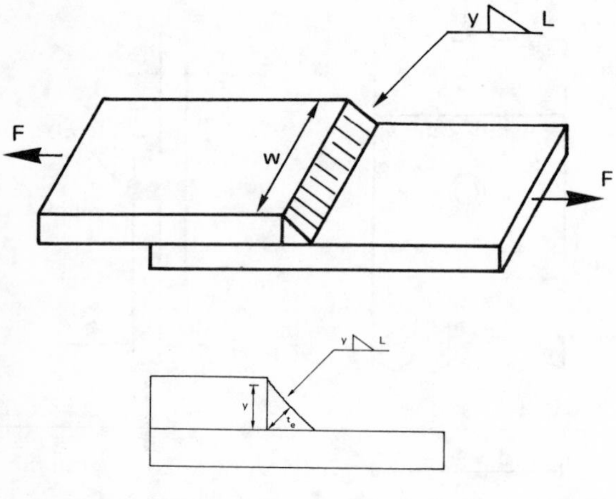

Figure 11.17 Fillet Lap Weld and Symbol

The effective weld throat size is

$$t_e = (.707)y \quad 11.70$$

Weld sizes (*y*) of 3/16″, 1/4″, and 5/16″ are desirable because they can be made in a single pass. However, fillet welds from 3/16″ to 1/2″ in 1/16″ increments are available. The increment is 1/8″ for larger welds.

Neglecting any effects due to eccentricity, the stress in the fillet lap weld shown in figure 11.17 is

$$\sigma = \frac{F}{wt_e} \quad 11.71$$

5. *Shaft Design*

Shear stress occurs when a shaft is placed in torsion. The shear stress at the outer surface of a bar of radius *r* which is torsionally loaded by a torque *T* is

$$\tau = \frac{Tr}{J} \quad 11.72$$

The total strain energy due to torsion is

$$U = \frac{T^2L}{2GJ} \quad 11.73$$

J is the shaft's polar moment of inertia, as defined in chapter 9. For a solid round shaft, *J* is

$$J = \frac{\pi r^4}{2} = \frac{\pi D^4}{32} \quad 11.74$$

For a hollow round shaft, the polar moment of inertia is

$$J = \frac{\pi}{2}[r_o^4 - r_i^4] \quad 11.75$$

If a shaft of length *L* carries a torque *T*, the angle of twist (in radians) will be

$$\phi = \frac{TL}{GJ} \quad 11.76$$

G is the *shear modulus,* approximately equal to 11.5 EE6 psi for steel. The shear modulus can be calculated from the modulus of elasticity by using equation 11.77.

$$G = \frac{E}{2(1+\mu)} \quad 11.77$$

The torque, *T*, carried by a shaft spinning at *n* revolutions per minute is related to the transmitted horsepower.

$$T = \frac{(63{,}025)(\text{horsepower})}{n} \quad 11.78$$

6. *Eccentric Connector Analysis*

An eccentric torsion connection is illustrated in figure 11.18. This type of connection gets its name from the tendency that the load has to rotate the bracket. This rotation must be resisted by the shear stress in the connectors.

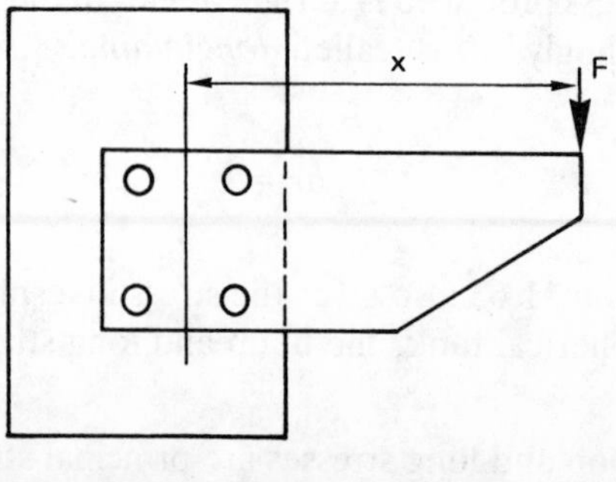

Figure 11.18 Torsion Resistance

An extension of equation 11.72 can be used to evaluate the maximum stresses in the connector group. To use equation 11.72, the following changes in definition must be made.

- The torque, T, is replaced by the moment on the bracket. This moment is the product of the eccentric load, F, and the distance from the load to the centroid of the fastener group, x.
- r is taken as the distance from the centroid of the fastener group to the critical fastener. The critical fastener is the one for which the vector sum of the vertical and torsional shear stresses is the greatest.
- J is based on the parallel-axis theorem. As bolts and rivets have little resistance to twisting in their holes, their polar moments of inertia (J_i) are omitted.

$$J = \Sigma(r_i^2 A_i) \qquad 11.79$$

r_i is the distance from the fastener group centroid to the ith fastener, which has an area of A_i.

- The vertical shear stress in the critical fastener must be added in a vector sum to the torsional shear stress. This vertical shear stress is

$$\text{vertical shear stress} = \frac{PA_{\text{critical}}}{\Sigma A_i} \qquad 11.80$$

Example 11.18

For the bracket shown, find the load on the most critical fastener. All fasteners have a nominal ½″ diameter.

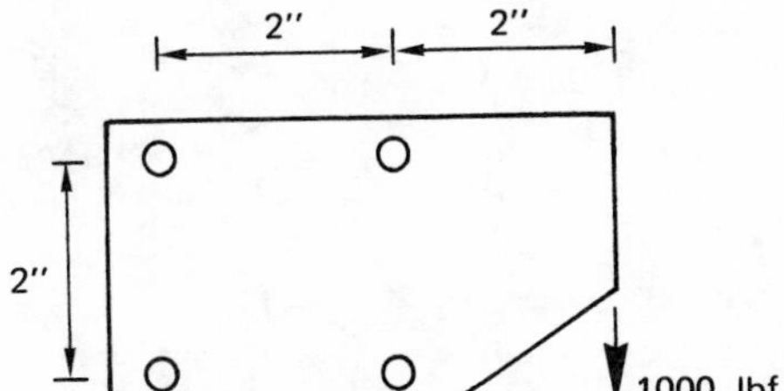

Since the fastener group is symmetrical, the group centroid is centered within the 4 fasteners. This makes the eccentricity of the load equal to 3 inches. Each fastener is located r from the centroid, where

$$r = \sqrt{(1)^2 + (1)^2} = 1.414$$

The area of each fastener is

$$A_i = \tfrac{1}{4}\pi(.5)^2 = .1963$$

Using the parallel axis theorem for polar moments of inertia,

$$J = 4[.1963(1.414)^2] = 1.570 \text{ in}^2$$

The torsional stress on each fastener is

$$\tau_T = \frac{(1000)(3)(1.414)}{(1.570)} = 2702 \text{ psi}$$

This torsional shear stress is directed perpendicularly to a line connecting each fastener with the centroid.

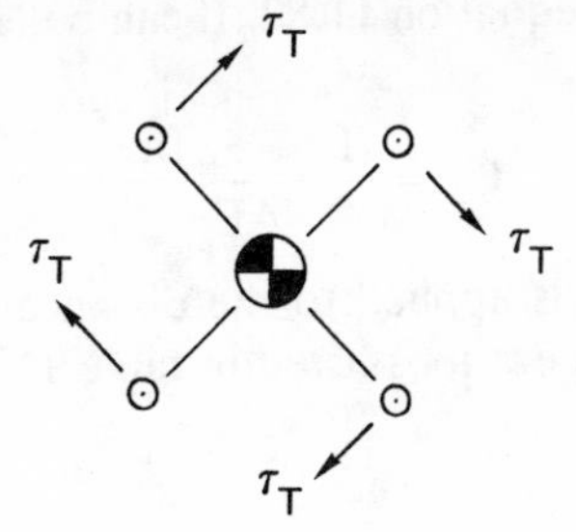

τ_T can be divided into horizontal stresses of $\tau_{T,x}$ and vertical stresses of $\tau_{T,y}$. Both of these components are equal to 1911 psi. In addition, each fastener carries a vertical shear load equal to (1000/4) = 250 pounds. The vertical shear stress due to this load is (250/.1963) = 1274.

The two right fasteners have vertical downward components of τ_T which add to the vertical downward stress of 1274. Thus, both of the two right fasteners are critical. The total stress in each of these fasteners is

$$\tau = \sqrt{(1911)^2 + (1911 + 1274)^2} = 3714 \text{ psi}$$

7. Surveyor's Tape Corrections

The standard surveyor's tape consists of a flat steel ribbon with a length very close to 100 feet. Such tapes are standardized at a particular temperature and with a specific tension and type of support. Since the tape may not be used in the conditions under which it was standardized, corrections are needed.

A. Temperature Correction

If the tape is not used at the standardized temperature, the change in length will be

$$C_t = \alpha(T - T_{std})L \qquad 11.81$$

α for steel has an approximate value of 6.5 EE−6 1/°F, although low-coefficient tapes containing nickel can reduce this expansion 75%. The correction given by equation 11.81 can be positive or negative, depending on the values of T and T_{std}. The correction is applied to the distance according to the algebraic operations listed in table 11.4.

Table 11.4
Corrections for Surveyor's Tapes

measuring a distance	setting out points
add C_t	subtract C_t
add C_p	subtract C_p

B. Tension Correction

The correction due to non-standard pull (tension) can be found from equation 11.82. It can be either positive or negative.

$$C_p = \frac{(F - F_{std})L}{AE} \qquad 11.82$$

The correction is applied to the distance according to the algebraic operations listed in table 11.4

Example 11.19

A steel surveyor's tape is standardized at 68°F. It is used at 50°F to place two monuments exactly 79 feet apart. What will be the tape reading used to place the monuments?

From equation 11.81,

$$C_t = (6.5\ \text{EE}-6)(50-68)(79) = -9.2\ \text{EE}-3$$

From table 11.4,

$$\text{tape reading} = 79.0000 - (-9.2\ \text{EE}-3) = 79.0092$$

(The tape cannot be read to the degree of precision indicated by this answer.)

8. Stress Concentration Factors

Stress concentration factors[2] are correction factors used to account for non-uniform stress distributions within objects. Non-uniform distributions result from non-uniform shapes. Examples of non-uniform shapes requiring stress concentration factors are stepped shafts, plates with holes, shafts with keyways, etc.

The actual stress experienced is the product of the stress concentration factor and the ideal stress. Values of K are always greater than 1.0, and they typically range between 1.2 and 2.5 for most designs. The exact values must be determined graphically from published results of extensive experimentation.

$$\sigma' = K\sigma \qquad 11.83$$

[2]Stress concentration factors are also known as *stress risers.*

Appendix A
Beam Formulas

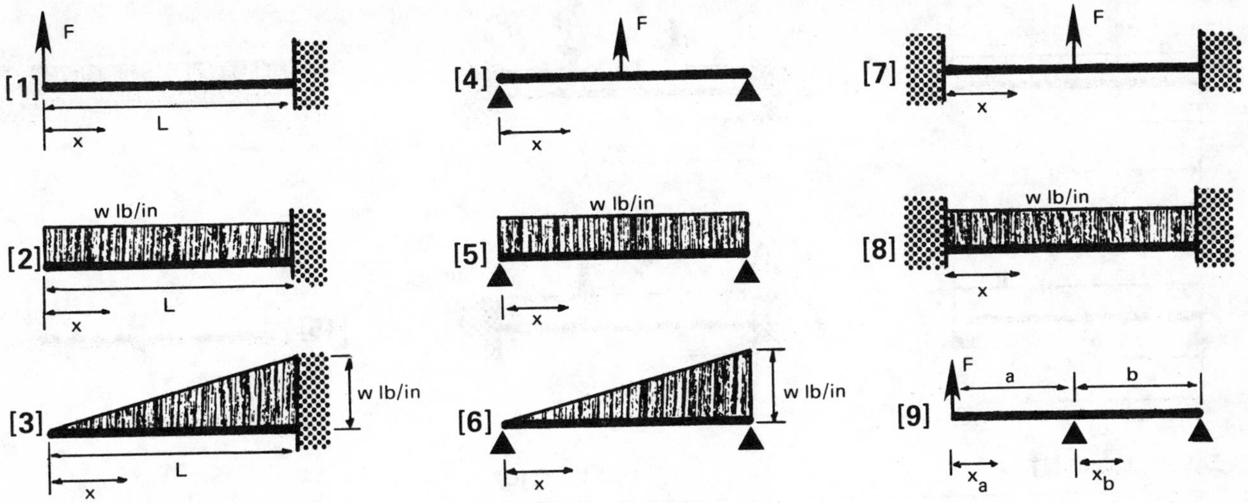

CASE	MOMENT	DEFLECTION
1	$M = Fx$ $M_{max} = FL$	$y = (F/6EI)(2L^3 - 3L^2x + x^3)$ $y_{max} = FL^3/3EI$
2	$M = ½wx^2$ $M_{max} = ½wL^2$	$y = (w/24EI)(3L^4 - 4L^3x + x^4)$ $y_{max} = wL^4/8EI$
3	$M = wx^3/6L$ $M_{max} = wL^2/6$	$y = (w/120EIL)(4L^5 - 5L^4x + x^5)$ $y_{max} = wL^4/30EI$
4	$M = -½Fx$ $M_{max} = -¼FL$	$y = (Fx/48EI)(3L^2 - 4x^2)$ $y_{max} = FL^3/48EI$
5	$M = (½wx)(x - L)$ $M_{max} = -wL^2/8$	$y = (wx/24EI)(L^3 - 2Lx^2 + x^3)$ $y_{max} = 5wL^4/384EI$
6	$M = (-wx/6L)(L^2 - x^2)$ $M_{max} = -.064wL^2$ at $x = .5774L$	$y = (wx/360EIL)(7L^4 - 10L^2x^2 + 3x^4)$ $y_{max} = .00652wL^4/EI$ at $x = .5193L$
7	$M = ½F[(¼L) - x]$ $M_{max} = FL/8$ at $x = 0$ $M_{max} = -FL/8$ at $x = ½L$	$y = (Fx^2/48EI)(3L - 4x)$ $y_{max} = FL^3/192EI$
8	$M = (½wL^2)[(1/6) - (x/L) + (x/L)^2]$ $M_{max} = wL^2/12$ at $x = 0$ and $x = L$ $M = -wL^2/24$ at $x = ½L$	$y = (wx^2/24EI)(L - x)^2$ $y_{max} = wL^4/384EI$
9	$M_a = Fx_a$ $M_b = (Fa/b)(b - x_b)$ $M_{max} = Fa$ at $x_a = a$	$y_a = (F/3EI)[(a^2+ab)(a-x_a) + (x_a/2)(x_a^2 - a^2)]$ $y_b = (Fax_b/6EI)[3x_b - (x_b^2/b) - 2b]$ $y_{tip} = (Fa^2/3EI)(a + b)$ (max up) $y_{max} = (0.06415)Fab^2/EI$ at $x_b = .4226b$ (max down)

Appendix A, continued

[10]
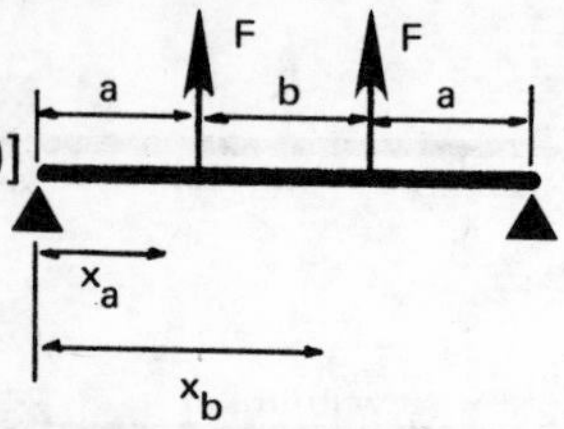

[12]
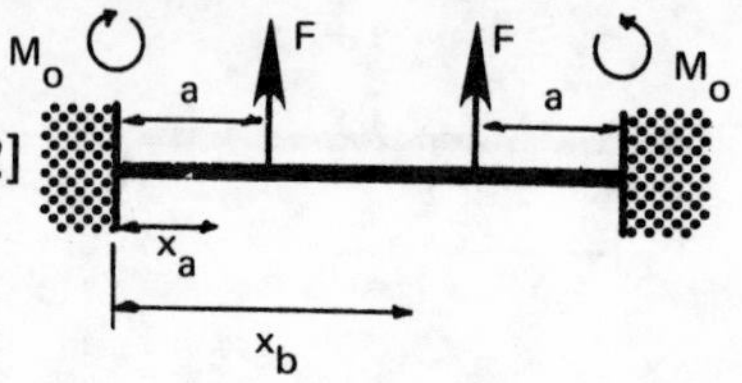

[14]

[11]
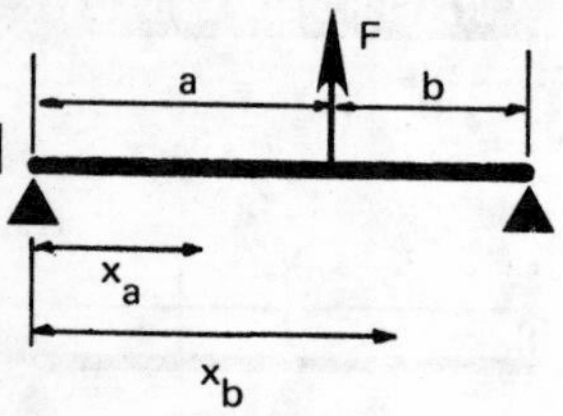

[13]
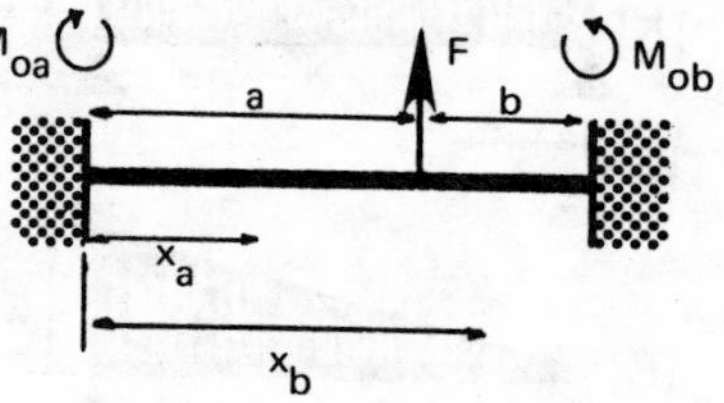

[15]
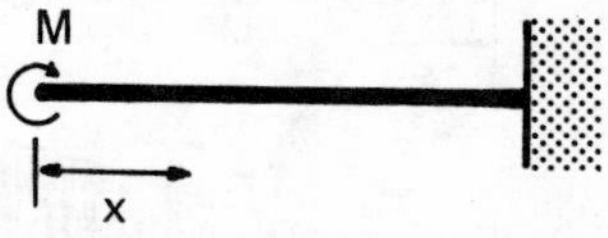

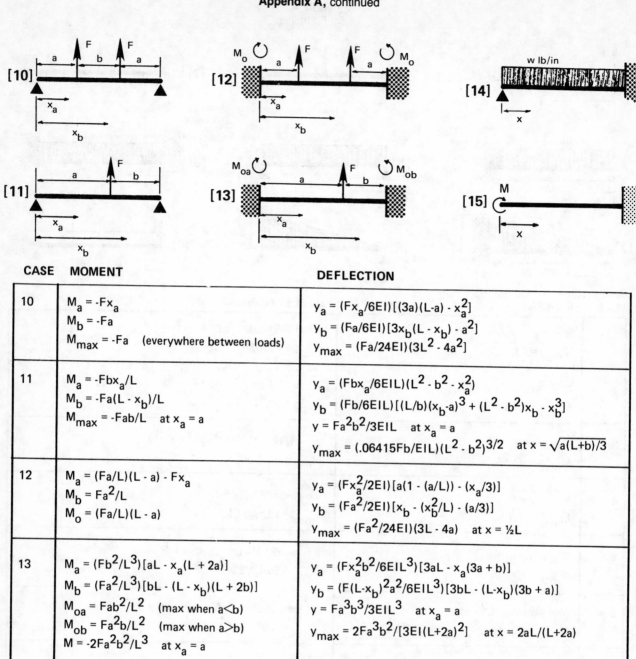

CASE	MOMENT	DEFLECTION
10	$M_a = -Fx_a$ $M_b = -Fa$ $M_{max} = -Fa$ (everywhere between loads)	$y_a = (Fx_a/6EI)[(3a)(L-a) - x_a^2]$ $y_b = (Fa/6EI)[3x_b(L - x_b) - a^2]$ $y_{max} = (Fa/24EI)(3L^2 - 4a^2]$
11	$M_a = -Fbx_a/L$ $M_b = -Fa(L - x_b)/L$ $M_{max} = -Fab/L$ at $x_a = a$	$y_a = (Fbx_a/6EIL)(L^2 - b^2 - x_a^2)$ $y_b = (Fb/6EIL)[(L/b)(x_b-a)^3 + (L^2 - b^2)x_b - x_b^3]$ $y = Fa^2b^2/3EIL$ at $x_a = a$ $y_{max} = (.06415Fb/EIL)(L^2 - b^2)^{3/2}$ at $x = \sqrt{a(L+b)/3}$
12	$M_a = (Fa/L)(L - a) - Fx_a$ $M_b = Fa^2/L$ $M_o = (Fa/L)(L - a)$	$y_a = (Fx_a^2/2EI)[a(1 - (a/L)) - (x_a/3)]$ $y_b = (Fa^2/2EI)[x_b - (x_b^2/L) - (a/3)]$ $y_{max} = (Fa^2/24EI)(3L - 4a)$ at $x = ½L$
13	$M_a = (Fb^2/L^3)[aL - x_a(L + 2a)]$ $M_b = (Fa^2/L^3)[bL - (L - x_b)(L + 2b)]$ $M_{oa} = Fab^2/L^2$ (max when $a<b$) $M_{ob} = Fa^2b/L^2$ (max when $a>b$) $M = -2Fa^2b^2/L^3$ at $x_a = a$	$y_a = (Fx_a^2b^2/6EIL^3)[3aL - x_a(3a + b)]$ $y_b = (F(L-x_b)^2a^2/6EIL^3)[3bL - (L-x_b)(3b + a)]$ $y = Fa^3b^3/3EIL^3$ at $x_a = a$ $y_{max} = 2Fa^3b^2/[3EI(L+2a)^2]$ at $x = 2aL/(L+2a)$
14	$M = (3wLx/8) - ½wx^2$ $M_{max} = wL^2/8$ at $x = L$	$y = (wx/48EI)[L^3 - 3Lx^2 + 2x^3]$ $y_{max} = wL^4/185EI$ at $x = .4215L$
15	$M = M$ everywhere	$y = Mx^2/2EI$ $y_{max} = ML^2/2EI$ at free end

Appendix B
Mechanical Properties of Representative Metals

The following mechanical properties are not guaranteed since they are averages for various sizes, product forms, and methods of manufacture. Thus, this data is not for design use, but is intended only as a basis for comparing alloys and tempers.

KEY TO HEAT TREATMENT ABBREVIATIONS: acrt, air cooled to room temperature; anat, annealed at; ct, cooled to; ht, heated to; oqf, oil quench from; rcf, rapid cool from; ta, tempered at.

Material designation, composition, typical use, and source if applicable	Condition, heat treatment	S_{tu} (ksi)	S_{ty} (ksi)
IRON BASED			
Armco ingot iron, for fresh and salt water piping	normalized (ht 1700°F, acrt)	44	24
AISI 1020, plain carbon steel, for general machine parts and screws, and carburized parts	hot rolled	65	43
	cold worked	78	66
AISI 1030, plain carbon steel, for gears, shafts, levers, seamless tubing, and carburized parts	cold drawn	87	73.9
AISI 1040, plain carbon steel, for high-strength parts, shafts, gears, studs, connecting rods, axles, and crane hooks	hot rolled	91	58
	cold worked	100	88
	hardened (wqf 1525°F, ta 1000°F)	113	86
AISI 1095, plain carbon steel, for hand-tools, music wire springs, leaf springs, knives, saws, agricultural tools such as plows and disks	annealed (ht 1450°F, ct 1200°F, acrt)	100	53
	hot rolled	142	84
	hardened (oqf 1475°F, ta 700°F)	180	118
AISI 1330, manganese steel, for axles, drive shafts	annealed	97	83
	cold drawn	113	93
	hardened (wqf 1525°F, ta 1000°F)	122	100
AISI 4130, chromium molybdenum steel, for high-strength aircraft structures	annealed (ht 1500°F, ct 1230°F, acrt)	81	52
	hardened (wqf 1575°F, ta 900°F)	161	137
AISI 4340, nickel-chromium-molybdenum steel, for large-scale, heavy duty high-strength structures	annealed	119	99
	as-rolled	192	147
	hardened (wqf/oqf 1500°F, ta 800°F)	220	200
AISI 2315, nickel steel, for carburized parts	as rolled	85	56
	cold drawn	95	75
AISI 2330, nickel steel	as rolled	98	65
	cold drawn	110	90
	annealed at 1450°F	80	50
	normalized at 1675°F	95	61
AISI 3115, nickel chromium steel for carburized parts	cold drawn	95	70
	as rolled	75	60
	annealed at 1500°F	71	62
STAINLESS STEELS			
AISI 302 stainless steel, most widely used, same as 18:8.	annealed	90	35
	cold drawn	105	60
AISI 303 austenitic stainless steel, good machineability	annealed, rqf 1950°F	90	35
	cold worked	110	75
AISI 304 austenitic stainless steel, good machineability and weldability	annealed, rqf 1950°F	85	30
	cold worked	110	75
AISI 309 stainless steel, good weldability, high strength at high temperatures, used in furnaces and ovens	annealed	90	35
	cold drawn	110	65
AISI 316 stainlesss steel, excellent corrosion resistance	annealed, rqf 2000°F	85	35
	cold drawn	105	60
AISI 410, magnetic, martensitic, can be quenched and tempered to give varying strength	annealed	60	32
	cold drawn	180	150
	oil quenched and drawn at 1100°F	110	91
AISI 430, magnetic, ferritic, used for auto and architectural trim, and for equipment in food and chemical industries.	annealed	60	35
	cold drawn	100	
AISI 502, magnetic, ferritic, low cost, widely used in oil refineries	annealed	60	25

Appendix B, continued

Material, designation, composition, typical use, and source if applicable	Condition, heat treatment	S_{tu}(ksi)	S_{ty}(ksi)
ALUMINUM BASED			
2011, for screw machine parts, excellent machineability, but not weldable and corrosion sensitive	T3	55	43
	T8	59	45
2014, for aircraft structures; weldable	T3	63	40
	T4, T451	61	37
	T6, T651	68	60
2017, for screw machine parts	T4, T451	62	40
2018, for engine cylinders, heads, pistons	T61	61	46
2024, for truck wheels, screw machine parts, aircraft structures	T3	65	45
	T4, T351	64	42
	T361	72	57
2025, for forgings	T6	58	37
2117, for rivets	T4	43	24
2219, high temperatures applications (up to 600·F), excellent weldability and machineability.	T31, T351	52	36
	T37	57	46
	T42	52	27
3003, for pressure vessels and storage tanks, poor machineability but good weldability. Excellent corrosion resistance.	0	16	6
	H12	19	18
	H14	22	21
	H16	26	25
3004, same characteristics as 3003	0	26	10
	H32	31	25
	H34	35	29
	H36	38	33
4032 pistons	T6	55	46
5083, unfired pressure vessels, cryogenics, towers, drilling rigs	0	42	21
	H116, H117, H321	46	33
5154, salt water services, welded structures, storage tanks	0	35	17
	H32	39	30
	H34	42	33
5454, same characteristics as 5154	0	36	17
	H32	40	30
	H34	44	35
5456, same characteristics as 5154	0	45	23
	H111	47	33
	H321, H116, H117	51	37
6061 corrosion resistant and good weldability. Used in railroad cars.	T4	33	19
	T6	42	37
7178	0	33	15
	T6	88	78

CAST IRON (Note redefinition of columns)		S_{tu}(ksi)	S_{us}(ksi)	S_{uc}(ksi)
Gray cast iron	class 20	20	32.5	80
	class 25	25	34	100
	class 30	30	41	110
	class 35	35	49	125
	class 40	40	52	135
	class 50	50	64	160
	class 60	60	60	150

NOTE: Unless told otherwise, use the following properties: **Steel:** E = 30 EE6 psi
G = 11.5 EE6 psi
α = 6.5 EE–6 1/°F
μ = .3
Aluminum: E = 10 EE6 psi
Copper: E = 17.5 EE6 psi

SIMPLE STRESSES AND ELONGATIONS

1. A steel support must connect two 30,000 pound tensile loads separated by 200 inches. The maximum allowable stress is 10,000 psi and the maximum allowable elongation is .02″. What is the required area?

2. What is the decrease in height of an 8″ round by 16″ high concrete cylinder (E = 2.5 EE6 psi) when the unit deformation is .0012? What is the stress?

3. Find the required diameter of a steel member if the tensile design load is 7000 pounds. Assume a safety factor of 5 based on an ultimate strength of 60,000 psi.

4. The ultimate shear strength of a 5/8″ thick steel plate is 42,000 psi. What force is necessary to punch a 3/4″ diameter round hole?

5. What is the actual hole diameter made by a steel punch (¾″ diameter) which is subjected to a 40,000 pound load?

6. An elevator weighs 1000 pounds and is supported by a 5/16″ cable, 1500 feet long. When the elevator carries a 1500 pound load, the cable elongates 6″ more. What is the modulus of elasticity of the cable?

THERMAL STRESS

7. At 90°F, the stress in a steel rod is 2000 psi (C). What is the stress at 0°F?

8. What is the tensile load if a ½″x4″x12′ steel tie rod experiences an 80°F temperature decrease from the no-stress temperature?

9. The distance between two points is exactly 100 feet. This measurement is read correctly on a steel surveyor's tape at 70°F and with a 10 pound tension. If the tape measures another distance as 100 feet at 100°F and with a 20 pound tension, what is the actual distance? The tape is 3/8″ wide and 1/32″ thick.

SHEAR AND MOMENT DIAGRAMS

10. A beam 25 feet long is simply supported at the left end and 5 feet from the right end. A uniform load of 2 kips/ft extends over a 10′ length starting from the left end. There is also a concentrated 10 kip load at the right end. Draw the shear and moment diagram.

11. A 19 foot long beam supports 1500 pounds/foot on two supports 14 feet apart. The right end of the beam extends 2 feet past the support. Draw the shear and moment diagrams.

12. Draw the shear and moment diagrams for the beam shown below. What is the location of the maximum moment? What is the maximum moment?

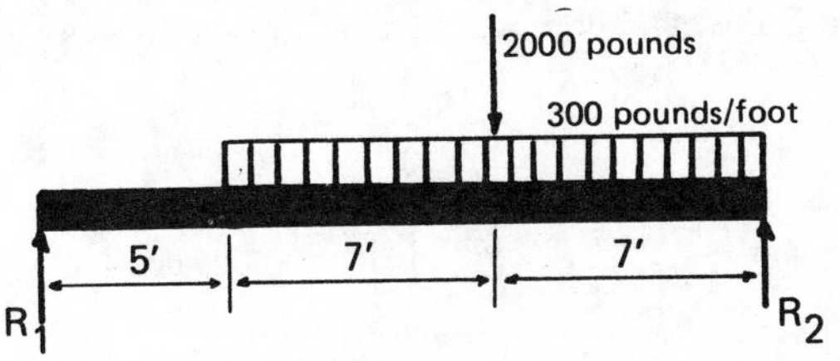

13. For the beam shown below, draw the shear and moment diagrams. What is the maximum moment?

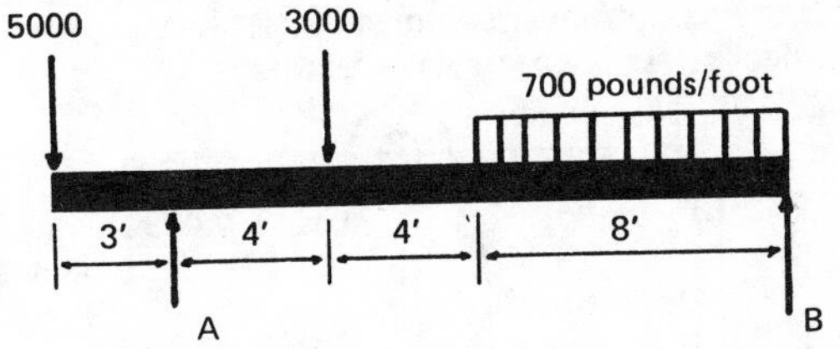

BEAM STRESSES

14. The beam in problem 11 is 10″ wide and 20″ high. What are the bending and shear stresses at mid-span?

15. The beam in problem 12 is 5″ wide and 10″ deep. What are the maximum bending and shearing stresses?

16. A 14 foot long simple beam is uniformly loaded with 200 pounds per foot over its entire length. If the beam is 3.625″ wide and 7.625″ deep, what is the maximum bending stress?

17. The beam in problem 13 is constructed of steel with an allowable stress of 24,000 psi. What is the required section modulus?

18. A 10″ deep beam with I = 78.5 in^4 is loaded as shown. What is the maximum value of w if the allowable stress is 22,000 psi?

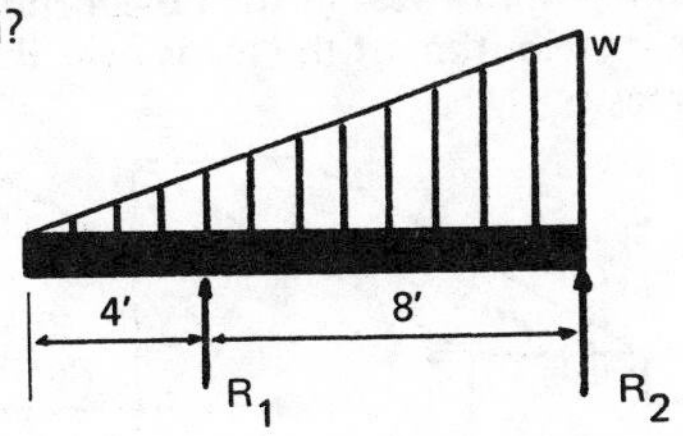

19. A 4″ wide, 8″ deep wood beam is simply supported and loaded as shown. What is the maximum shearing stress?

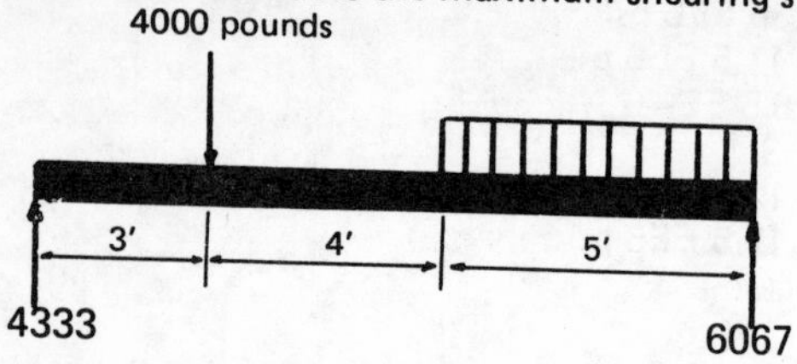

ECCENTRIC LOADING

20. What is the maximum tensile stress at section A-A? The material is 1″x1″ steel.

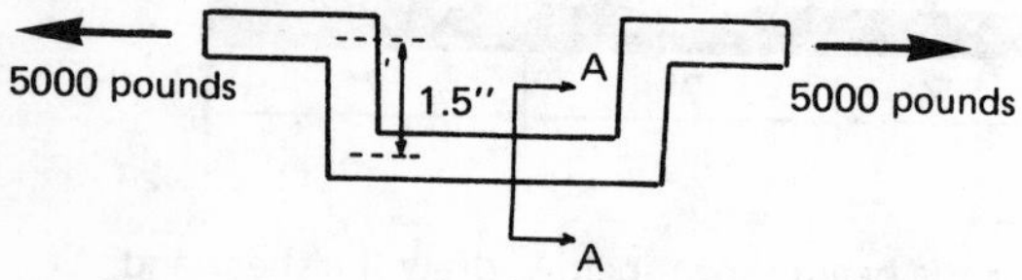

21. What is the maximum stress in the offset link shown below? Neglect stress concentration factors.

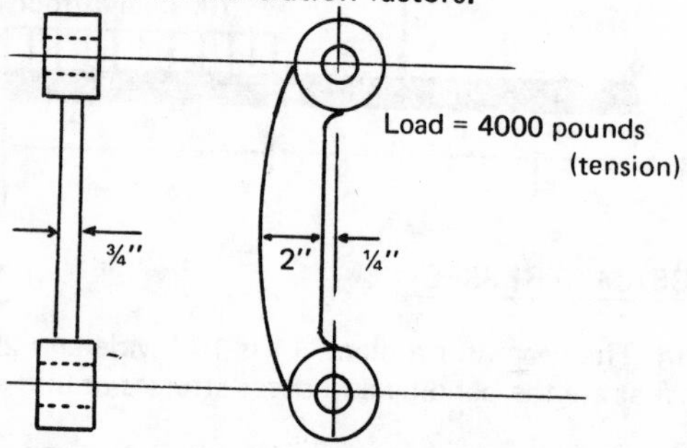

22. What is the maximum stress in the structure shown below?

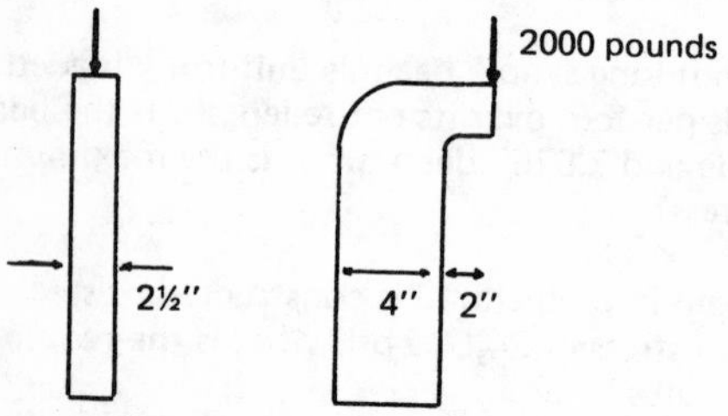

COMBINED STRESSES (optional)

23. What is the maximum stress in the object shown below? Assume uniform distribution of the loads over their perpendicular areas.

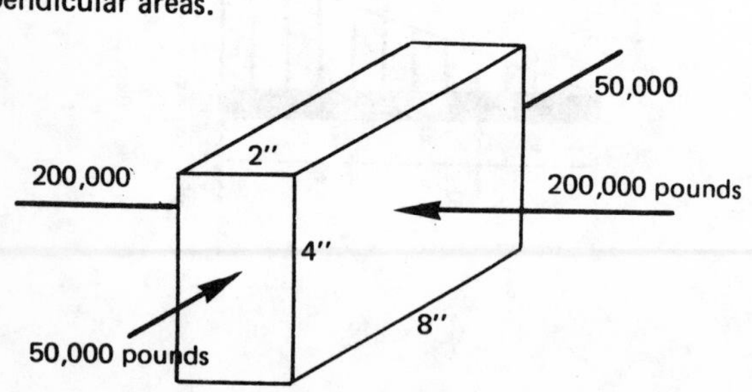

24. What is the maximum stress in the 2″ diameter, 3′ long steel shaft?

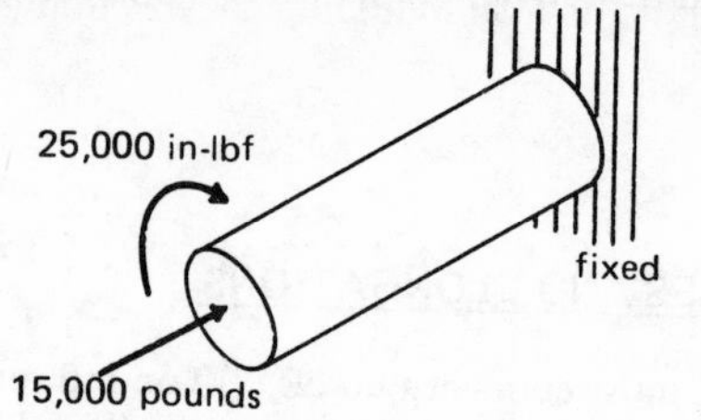

BEAM DEFLECTIONS

25. A beam of length L has a deflection formula which is

$$48EIy = w(2x^4 - 5Lx^3 + 3L^2x^2)$$

Where does the maximum deflection occur?

26. What is the mid-span deflection for the steel beam shown below if the cross sectional moment of inertia is 200 inches^4?

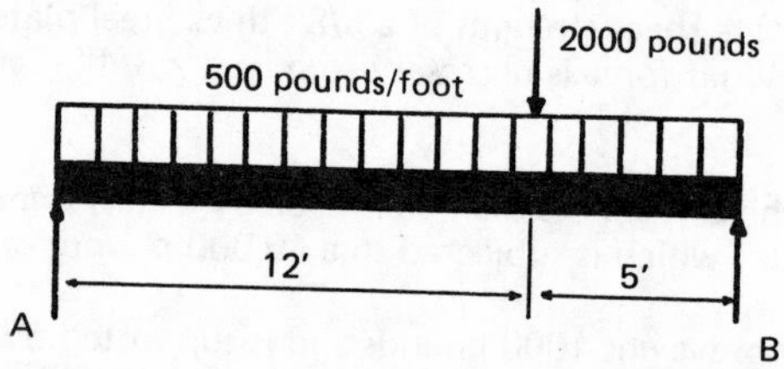

27. A 12″x4″ beam will deflect .2″ under a 4000 pound concentrated load applied mid-span along its 10 foot length when the 12″ side is vertical and the 4″ side is horizontal. What will be the deflection if the beam is rotated 90 degrees and the load stays in the original plane?

28. What is the tip deflection of a cantilever beam which carries a distributed load of 8000 pounds/foot over the first 4 feet (from the support) of its 10 feet length? Neglect the weight of the steel beam, which is a T-beam with 5″ width, 12″ total height, and 2″ web thickness.

29. What uniform load will cause a simple beam which is 10 feet long to deflect .3 inches if it is supported (in addition to the end supports) by a spring at the beam's mid-point? The spring has a spring constant of 30,000 pounds per inch. Assume the beam is steel, 10″ deep, rectangular, and with a centroidal moment of inertia of 100 inches^4.

TORSION

30. A 2.5″ diameter steel shaft is 2 feet long. Its maximum shear stress is 10,000 psi. What is the applied torque? What is the angular deflection in degrees?

31. What diameter shaft is required to transmit 200 horsepower at 1850 rpm if the maximum shear stress is 10,000 psi and the maximum twist is 1 degree per foot of length? The shaft is round steel.

32. Find the maximum shearing stress in a steel shaft of 2″ diameter which transmits 200 horsepower at 875 rpm.

33. A 2" solid steel shaft is driven by a 4" diameter gear and transmits power at 100 rpm. If the allowable shear stress is 12 ksi, what is the maximum horsepower that can be transmitted?

COLUMNS

34. A 24 foot long, 4"x4" white oak timber (E = 1.5 EE6 psi) is used to support a sign. On end is pinned solid, the other end is free. Find the critical buckling sign weight.

35. What increase in temperature will buckle a slender steel rod of 1" diameter and 8 foot length which is fixed at both ends and which carries no initial load?

36. A compressive load of 5000 pounds is applied to a slender steel tube hinged at both ends. The outside diameter is 3" and the length is 10 feet. The allowable compressive stress is 22,000 psi, and a factor of safety of 2 is desired. What is the required tube wall thickness?

THIN-WALLED CYLINDERS

37. A 5" diameter cylinder with a 1/8" wall is pressurized to 1900 psig. What is the
   (a) hoop stress?
   (b) long stress?
   (c) combined stress?

38. What is the circumferential stress in the walls if the inside diameter is 10" and the wall thickness is 1"?

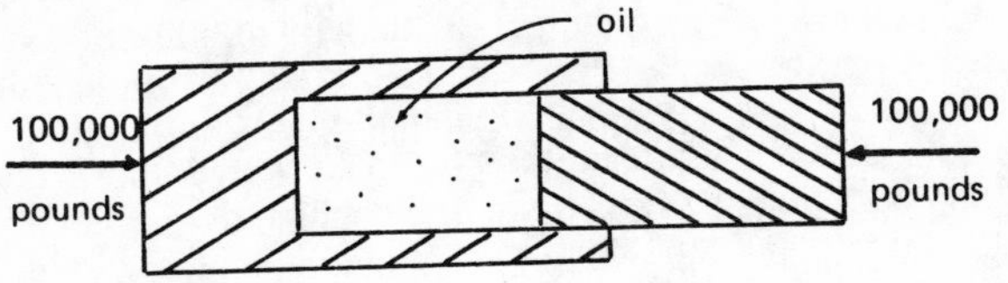

SPRINGS

39. What is the deflection of a helical coil spring carrying an 18 pound load if the spring constant is 12 pounds/foot?

COMPOSITE STRUCTURES

40. The total cross section of a 3-layer composite bar is 6"x¾". The center layer is steel, 6"x½". The two outer layers are copper. A tensile load of 100 kips is applied to each end. Find the strains and the stresses in both materials.

41. Find the stress in the steel cylinder and the copper tube which surrounds it if a uniform load of 100 kips is applied axially. The diameter of the steel cylinder is 5". The outside diameter of the copper tube, which is is intimate contact with the steel cylinder, is 10".

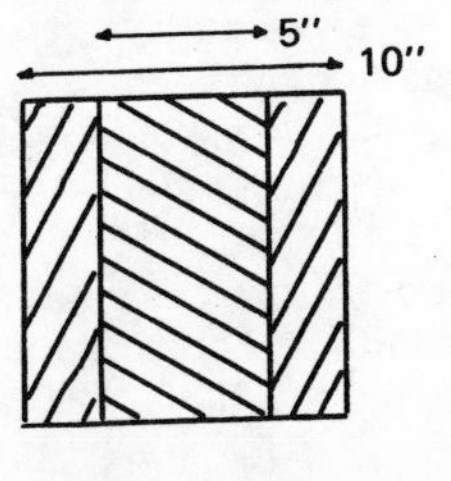

INDETERMINATE REACTIONS

42. Bar AB is rigid and remains horizontal. It is supported by two steel rods on the outside, each with a cross-sectional area of .2 square inches. The central rod is copper with an area of .6 square inches. All rods are 6 feet long. What is the force in each bar?

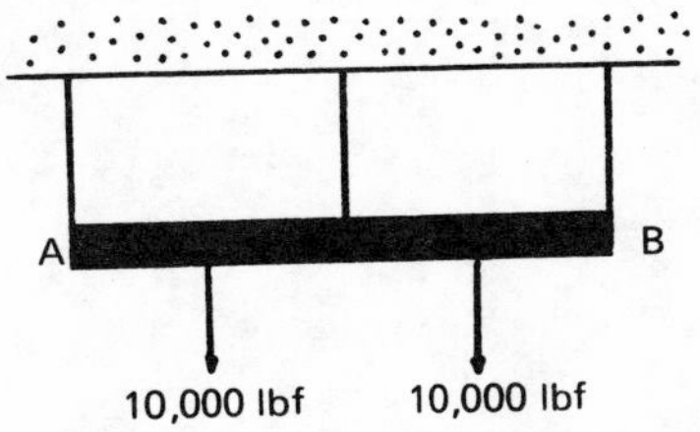

43. What is the deflection at point C? BC is steel with a cross-sectional area of 4 square inches. CD is copper with a cross-sectional area of 1 square inch. Neglect the column effect of CD. AC is rigid.

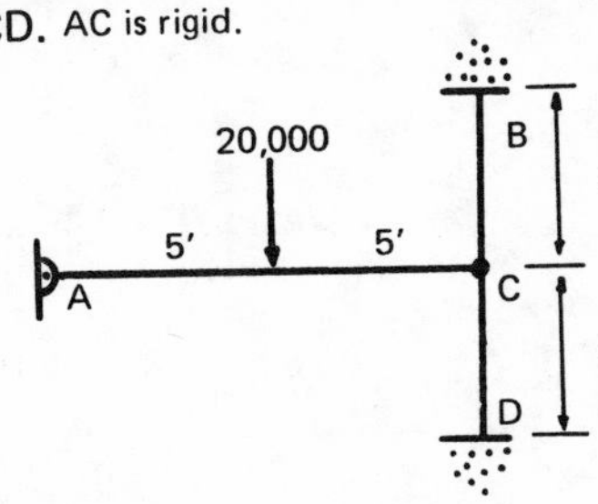

TORSION

$\theta = \frac{TL}{GJ}$

$\tau = \frac{Tr}{J}$

AXIAL

$\delta = \frac{PL}{EA}$

$\sigma = \frac{P}{A}$

12 Dynamics

Nomenclature

a	acceleration	ft/sec^2
e	coefficient of restitution	—
E	energy	ft-lbf
f	coefficient of friction	—
F	force	lbf
g	gravitational acceleration	ft/sec^2
g_c	gravitational constant (32.2)	lbm-ft/lbf-sec^2
G	gravitational constant (3.44 EE−8)	ft^4/lbf-sec^4
h	height	ft
I	moment of inertia	lbm-ft^2
k	spring constant	lbf/ft
m	mass	lbm
N	normal force	lbf
P	power	ft-lbf/sec
r	radius or separation	ft
R	total plane reaction	lbf
s	distance	ft
t	time	sec
v	velocity	ft/sec
w	weight	lbf
W	work	ft-lbf

Symbols

α	angular acceleration	rad/sec^2
τ	torque	ft-lbf
θ	angle	radians
ϕ	angle	degrees
ω	angular velocity	rad/sec

Subscripts

0	initial
c	centrifugal, coriolis, or centroidal
d	dynamic
f	friction
i	inertial
k	kinetic
n	normal
o	with respect to center
p	potential
r	radial
t	tangential, total

1. Introduction

Dynamics is the study of bodies which are not in equilibrium. The subject is subdivided into kinematics and kinetics. *Kinematics* is the study of motion without consideration of the forces causing motion. Kinematics deals with the relationships between position, velocity, acceleration, and time.

Kinetics is concerned with motion and the forces causing motion. Kinetics deals with the variables of force, mass, and acceleration. The study of kinetics requires the application of Newton's laws.

Bodies in motion are called *particles* if rotation is absent or insignificant. Particles do not possess rotational kinetic energy. All parts of a particle possess the same instantaneous displacement, velocity, and acceleration. *Rigid bodies* are objects whose parts exhibit different displacements, velocities, and accelerations.

2. Newton's Laws

Much of this chapter is based on *Newton's laws of motion.* These laws can be stated in many forms. They are presented here, along with *Newton's law of gravitation,* in common wording.

First law: A particle will remain in a state of rest or will continue to move with constant velocity unless an unbalanced external force acts on it.

Second law: The acceleration of a particle is directly proportional to the force acting on it and is inversely proportional to the particle mass.[1] The direction of acceleration is the same as the force direction.

[1]The pound *(lbm)* is used as the unit of mass in this chapter. This choice makes it necessary to include g_c in equations 12.1, 12.2, and others that follow. g_c can be omitted if a consistent set of units is used (e.g., pounds-force, slugs, and ft/sec^2).

$$\mathbf{F} = \frac{d\left(\dfrac{mv}{g_c}\right)}{dt} \qquad 12.1$$

If the mass is constant with respect to time, equation 12.1 can be rewritten as equation 12.2.

$$\mathbf{F} = \left(\frac{m}{g_c}\right)\frac{dv}{dt} = \left(\frac{m}{g_c}\right)a \qquad 12.2$$

Third law: There is an equal and opposite reacting force for every acting force.

$$\mathbf{F}_{reacting} = -\mathbf{F}_{acting} \qquad 12.3$$

Law of Universal Gravitation: The attractive force between two particles is directly proportional to the product of masses and inversely proportional to the square of distances between their centroids.

$$F = \frac{Gm_1m_2}{(g_cr)^2} \qquad 12.4$$

3. Particle Motion

A. Basic Variables

A *linear system* is one in which particles move only in straight lines. The relationships between force, position, velocity, and acceleration for a linear system are given by equations 12.5 through 12.8.

$$a = \frac{dv}{dt} = \frac{d^2s}{dt^2} \qquad 12.5$$

$$v = \frac{ds}{dt} = \int a\,dt \qquad 12.6$$

$$s = \int v\,dt = \iint a\,dt \qquad 12.7$$

$$\mathbf{F} = \left(\frac{m}{g_c}\right)a \qquad 12.8$$

A *rotational system* is one in which particles move in circular paths. The relationships between torque, angular position, angular velocity, and angular acceleration are given by equations 12.9 through 12.12.

$$\alpha = \frac{d\omega}{dt} = \frac{d^2\theta}{dt^2} \qquad 12.9$$

$$\omega = \frac{d\theta}{dt} = \int \alpha\,dt \qquad 12.10$$

$$\theta = \int \omega\,dt = \iint \alpha\,dt \qquad 12.11$$

$$\boldsymbol{\tau} = \left(\frac{I}{g_c}\right)\alpha \qquad 12.12$$

Example 12.1

The velocity of a 40 lbm particle as a function of time is

$$v(t) = 8t - 6t^2 \text{ (ft/sec)}$$

The velocity and position are both zero at $t = 0$. What are the acceleration and position functions?

$$a(t) = \frac{dv(t)}{dt} = 8 - 12t \text{ (ft/sec}^2\text{)}$$

$$s(t) = \int v(t)\,dt = 4t^2 - 2t^3 \text{ (ft)}$$

Example 12.2

What is the force acting at $t = 6$ on the particle described in example 12.1?

The acceleration at $t = 6$ is

$$a(6) = 8 - (12)(6) = -64 \text{ ft/sec}^2$$

From equation 12.8, the force is

$$F = \left(\frac{40}{32.2}\right)(-64) = -79.5 \text{ lbf}$$

It is necessary to distinguish between position, displacement, and distance traveled. The *position* of a particle is an actual location. Position is determined from the *position function,* s(t). The *displacement* of a particle is the difference in positions at two different times. Displacement is found by subtracting values of the position function.

$$\Delta s = s(t_2) - s(t_1) \qquad 12.13$$

The *distance traveled* includes distance covered during all direction reversals. It can be found by adding displacements during periods in which the velocity sign does not change.

Example 12.3

What distance is traveled by the particle described in example 12.1 during the period $t = 0$ to $t = 6$?

The velocity changes from positive to negative at $t = 4/3$. The initial position is $s = 0$ at $t = 0$. At $t = 4/3$, the position is

$$s(4/3) = 4(4/3)^2 - 2(4/3)^3 = 2.37 \text{ ft}$$

The displacement is given by equation 12.13.

$$\Delta s = 2.37 - 0 = 2.37 \text{ ft}$$

The position at $t = 6$ is

$$s(6) = 4(6)^2 - 2(6)^3 = -288 \text{ ft}$$

The displacement between $t = 4/3$ and $t = 6$ is

$$s = s(6) - s(4/3)$$
$$= -288 - 2.37 = -290.37 \text{ ft}$$

The total distance traveled is

$$2.37 + 290.37 = 292.74 \text{ ft}$$

If the acceleration is constant, the *a* term can be taken out of the integrals in equations 12.6 and 12.7.

$$v(t) = a\int dt = at + v_0 \qquad 12.14$$

$$s(t) = v\int dt = \tfrac{1}{2}at^2 + v_0t + s_0 \qquad 12.15$$

Table 12.1 summarizes the equations required to solve uniform acceleration problems. The table can be used for rotational problems by substituting the analogous variables of α, ω, and θ for *a, v,* and *s* respectively.

Table 12.1
Uniform Acceleration Formulas

to find	given these	use this equation
t	a v_o v	$t = \frac{v - v_o}{a}$
t	a v_o s	$t = \frac{\sqrt{2as + v_o^2} - v_o}{a}$
t	v_o v s	$t = \frac{2s}{v_o + v}$
a	t v_o v	$a = \frac{v - v_o}{t}$
a	t v_o s	$a = \frac{2s - 2v_o t}{t^2}$
a	v_o v s	$a = \frac{v^2 - v_o^2}{2s}$
v_o	t a v	$v_o = v - at$
v_o	t a s	$v_o = \frac{s}{t} - \frac{1}{2}at$
v_o	a v s	$v_o = \sqrt{v^2 - 2as}$
v	t a v_o	$v = v_o + at$
v	a v_o s	$v = \sqrt{v_o^2 + 2as}$
s	t a v_o	$s = v_o t + \frac{1}{2}at^2$
s	a v_o v	$s = \frac{v^2 - v_o^2}{2a}$
s	t v_o v	$s = \frac{1}{2}t(v_o + v)$

Example 12.4

A locomotive traveling at 80 *mph* locks its wheels and skids 580 feet before stopping. If the deceleration is constant, how long (in seconds) will it take for the locomotive to come to a standstill?

Convert the 80 *mph* to ft/sec.

$$\frac{(80)\ \text{mi/hr}\ (5280)\ \text{ft/mi}}{(3600)\ \text{sec/hr}} = 117.3\ \text{ft/sec}$$

t is unknown. v_0 = 117.3 ft/sec. *v* = 0. *s* = 580. From table 12.1,

$$t = \frac{2s}{v_0 + v} = \frac{(2)(580)}{117.3 + 0}$$

$$= 9.89\ \text{sec}$$

B. Relationships Between Linear and Rotational Motion

If a particle travels in a circular path with instantaneous radius *r*, the particle's *tangential velocity* can be calculated from the angular velocity.

$$v_t = \omega r \qquad 12.16$$

$$v_{t,x} = v_t \cos\phi \qquad 12.17$$

$$v_{t,y} = v_t \sin\phi \qquad 12.18$$

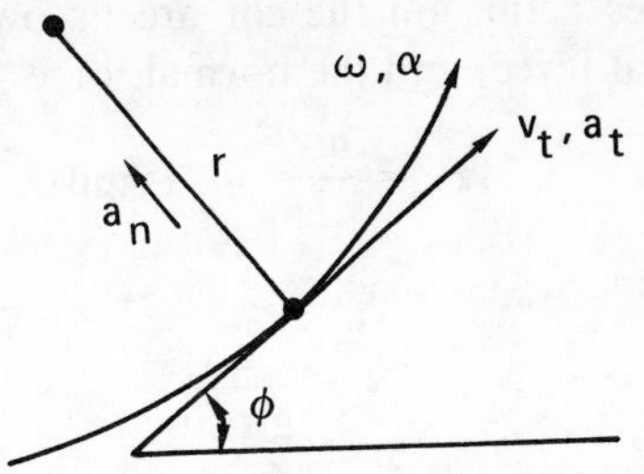

Figure 12.1 Tangential and Normal Variables

If the particle is accelerating as it travels around the curve, its angular velocity will be changing. The relationship between the tangential and angular accelerations is given by equation 12.19.

$$a_t = \alpha r \qquad 12.19$$

A particle traveling in a circular path will tend to continue traveling along its tangent. If the particle is restrained (e.g., a rock being twirled on a string), it will continue to travel in the circular path. This restraint will be due to an applied force (e.g., tension in the string).

From equation 12.2, an acceleration acts whenever a mass experiences a force. The force that keeps the particle in circular motion is directed towards the center of rotation. Therefore, the acceleration is also directed towards the center. This acceleration is known as the *normal acceleration.*

$$a_n = \frac{v_t^2}{r} = r\omega^2 = v_t\omega \qquad 12.20$$

This normal acceleration produces the apparent *centrifugal*[2] *force.*

$$F_c = \frac{ma_n}{g_c} = \frac{mv_t^2}{g_c r} \qquad 12.21$$

Example 12.5

A 4000 pound car travels at 40 *mph* around a banked curve with radius of 500 feet. What should the banking angle be such that tire friction is not needed to prevent the car from sliding?

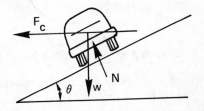

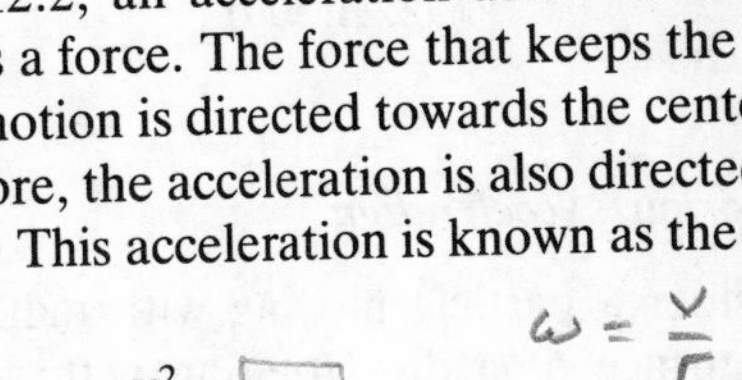

[2]The centrifugal force is an *apparent force* on the object. Centrifugal force is directed outward, away from the center of rotation. It is most commonly experienced by riders in cars. However, the real force on the particle is *towards* the center of rotation. This real force is known as *centripetal force.* The centrifugal and centripetal forces are equal in magnitude but opposite in sign.

The forces acting on the car are its own weight, the centripetal force, and the normal force.

$$F_c = \frac{mv_t^2}{g_c r} = N \sin\theta \qquad 12.22$$

Solving for **N**,

$$N = \frac{mv_t^2}{g_c r \sin\theta} \qquad 12.23$$

$$w = \frac{mg}{g_c} = N \cos\theta \qquad 12.24$$

Solving for **N**,

$$N = \frac{mg}{g_c \cos\theta} \qquad 12.25$$

Equating both expressions for **N**,

$$\tan\theta = \frac{v_t^2}{gr} \qquad 12.26$$

The car velocity is

$$\frac{(40)\ \text{mi/hr}\ (5280)\ \text{ft/mi}}{(3600)\ \text{sec/hr}} = 58.67\ \text{ft/sec}$$

The required banking angle is

$$\theta = \arctan\left[\frac{(58.67)^2}{(32.2)(500)}\right] = 12.07°$$

C. Coriolis Acceleration

Consider a particle moving with radial velocity v_r on the surface of a disk rotating with velocity ω. From equation 12.20, the normal acceleration towards the center is $r\omega^2$. The particle's tangential velocity increases as the particle moves outward. This increase is said to be produced by the *coriolis acceleration* acting tangentially.

$$a_c = 2v_r\omega \qquad 12.27$$

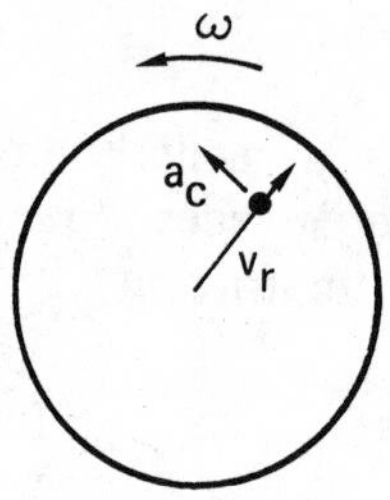

Figure 12.2
Coriolis Acceleration with a Rotating Disk

Motion on the surface of a moving sphere is more complex due to the changing velocity v_r. Consider an airplane flying with constant airspeed *v* from the equator to the north pole while the earth rotates below it. Three accelerations act on the aircraft. The first is the normal acceleration.

$$a_n = r\omega^2 = R(\cos\phi)\omega^2 \qquad 12.28$$

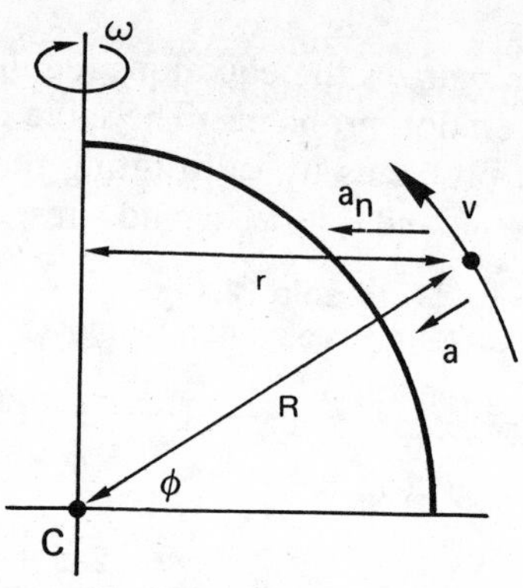

Figure 12.3
Coriolis Acceleration with a Rotating Sphere

The second acceleration is the coriolis acceleration which depends on the latitude ϕ. It acts perpendicularly to figure 12.3.

$$a_c = 2\omega v_x = 2\omega v(\sin\phi) \qquad 12.29$$

The third acceleration (also a normal acceleration) is directed toward the earth's center. Its magnitude is

$$a = \frac{v^2}{R} \qquad 12.30$$

Example 12.6

A slider moves with a constant velocity of 20 ft/sec along a rod rotating at 5 radians per second. What is the slider's acceleration when it is 4 feet from the center of rotation?

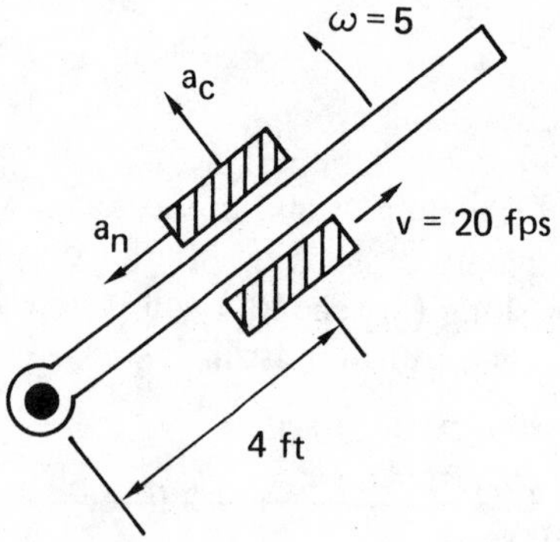

The normal acceleration is

$$a_n = (4)(5)^2 = 100\ \text{ft/sec}^2$$

The coriolis acceleration is

$$a_c = (2)(20)(5) = 200\ \text{ft/sec}^2$$

The resultant acceleration experienced by the slider is

$$a = \sqrt{(100)^2 + (200)^2} = 223.6\ \text{ft/sec}^2$$

D. Projectile Motion

Consider a projectile launched at an angle of ϕ from the horizontal with initial velocity v_0. Neglecting air resistance, the trajectory is a parabola with coordinates

(x,y) at time t. The impact velocity is equal to the launch velocity. The maximum range is achieved when $\phi = 45°$.

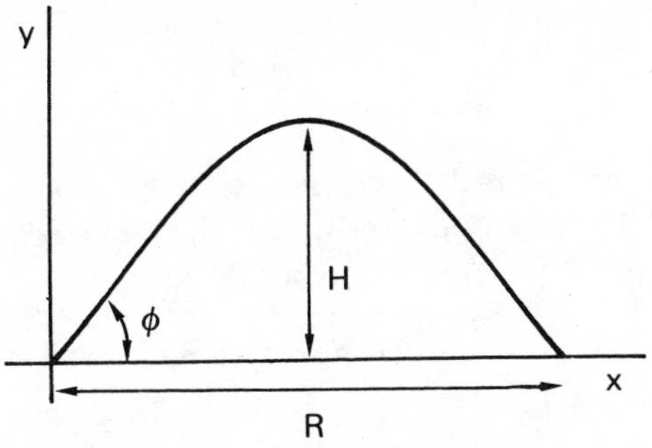

Figure 12.4 Projectile Motion

After a projectile is launched, it is acted upon by a constant gravitational acceleration directed downward. A projectile's altitude above the launch plane varies with time. This altitude is a composite function of the vertical velocity component and gravity.

$$y(t) = (v_0 \sin\phi)t - \tfrac{1}{2}gt^2 \qquad 12.31$$

Other relationships are easily derived from the laws of uniform acceleration. Special nomenclature is used with equations 12.32 through 12.42.

g	acceleration due to gravity
H	maximum altitude
R	total range
t	time after launch
T	total flight time
v_0	initial launch velocity
ϕ	launch angle

$$x = (v_0 \cos\phi)t \qquad 12.32$$

$$y = (v_0 \sin\phi)t - \tfrac{1}{2}gt^2 \qquad 12.33$$

$$v = \sqrt{v_0^2 - 2gy} \qquad 12.34$$

$$v_x = v_0 \cos\phi \qquad 12.35$$

$$v_y = v_0 \sin\phi - gt \qquad 12.36$$

$$H = \frac{v_0^2 \sin^2\phi}{2g} \qquad 12.37$$

$$R = \frac{v_0^2 \sin 2\phi}{g} \qquad 12.38$$

$$T = \frac{2v_0 \sin\phi}{g} \qquad 12.39$$

If the projection is horizontal from height H, then $\phi = 0$.

$$x = v_0 t \qquad 12.40$$

$$y = H - \tfrac{1}{2}gt^2 \qquad 12.41$$

$$T = \sqrt{2y/g} \qquad 12.42$$

Example 12.7

A bomber flies horizontally at 275 mph at an altitude of 9000 feet. At what angle α should the bombs be dropped? Neglect air friction.

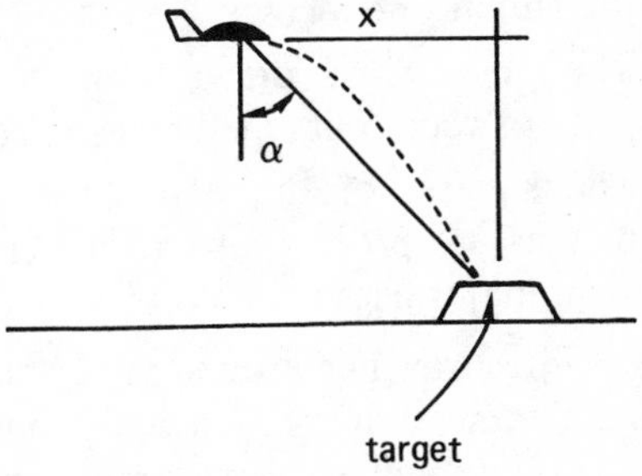

This is a case of horizontal projection. The falling time depends only on the height of the bomber.

$$T = \sqrt{(2)(9000)/32.2} = 23.64 \text{ sec}$$

The range is

$$x = 23.64(275)\left(\frac{5280}{3600}\right) = 9535 \text{ ft}$$

Therefore, the angle is

$$\alpha = \arctan\left(\frac{9535}{9000}\right) = 46.7°$$

Example 12.8

A projectile is launched at 600 ft/sec with a 30° inclination from the horizontal. The launch point is 500 feet above the plane of impact. Neglecting friction, find the range and maximum altitude.

The maximum altitude above the launch elevation can be found from equation 12.37. The total altitude above the impact plane is

$$H_t = 500 + \frac{(600)^2(.5)^2}{(2)(32.2)} = 1897.5 \text{ ft}$$

From equation 12.39, the time to maximum altitude is

$$t_{max} = \tfrac{1}{2}T = \frac{(600)(.5)}{32.2} = 9.32 \text{ sec}$$

The time to fall from the maximum altitude to the impact plane can be found from table 12.1.

$$t_{fall} = \sqrt{(2)(1897.5)/32.2} = 10.86 \text{ sec}$$

The total flight time is

$$T = t_{max} + t_{fall}$$
$$= 9.32 + 10.86 = 20.18 \text{ sec}$$

The x-component of velocity is

$$v_x = 600(\cos 30°) = 519.6 \text{ ft/sec}$$

The maximum range is

$$R = (519.6)(20.18) = 10{,}485 \text{ ft}$$

E. Satellite Motion

Kepler's laws of planetary motion can be used to describe the movement of satellites.

law 1: Each planet moves in an elliptical path with the sun at one focus.

law 2: The radius vector drawn from the sun to the planet sweeps out equal areas in equal times.

law 3: The squares of the periodic times of the planets are proportional to the cubes of the semi-major axes of the orbits.

For light satellites traveling around the earth, it is possible to make some simplifying assumptions. It is assumed that (1) the satellite's mass is much smaller than the earth's, and (2) the satellite is unaffected by objects other than the earth.

The force exerted on the satellite is given by *Newton's law of universal gravitation.*

$$F = \frac{Gm_{earth}m_{satellite}}{(rg_c)^2} \qquad 12.43$$

G is a gravitational constant with a value of 3.44 EE−8 $ft^4/lbf\text{-}sec^4$. The product (Gm_{earth}) has a value of 4.55 EE17 $ft^4\text{-}lbm/lbf\text{-}sec^4$.

4. Work, Energy, and Power

There are three general categories of external[3] forces: applied, gravitational, and friction. Positive work is performed when a force acts in the direction of motion. Negative work is performed when a force opposes motion.

Applied and gravitational forces can do either positive or negative work depending on the direction of motion. Friction always opposes motion. Therefore, friction can only do negative work.

The work performed by a variable force in the direction of motion is calculated by the scalar products in equations 12.44 and 12.45.

$$W = \int \mathbf{F} \cdot d\mathbf{s} \text{ (linear systems)} \qquad 12.44$$

$$= \int \boldsymbol{\tau} \cdot d\boldsymbol{\theta} \text{ (rotational systems)} \qquad 12.45$$

If the force or torque is constant, the integral can be eliminated.

$$W = \mathbf{F} \cdot \Delta\mathbf{s} \text{ (linear systems)} \qquad 12.46$$

$$= \boldsymbol{\tau} \cdot \Delta\boldsymbol{\theta} \text{ (rotational systems)} \qquad 12.47$$

Energy is the capacity to do work. Since energy cannot be created or destroyed, any work performed on a conservative system goes into increasing the system's energy. This is known as the *Work-Energy Principle.*

$$\Delta E = W \qquad 12.48$$

[3]Internal forces, such as inertia, can do no work on a moving object.

Potential energy is possessed by a mass m due to its relative height h in a gravitational field. This energy is equal to the work that would raise the mass a distance h. Conversely, it is the energy released when the mass falls a distance h.

$$E_p = \frac{mgh}{g_c} \qquad 12.49$$

Kinetic energy is the work necessary to bring a moving object to rest. Conversely, it is the work required to accelerate a stationary object to velocity v.

$$E_k = \frac{mv^2}{2g_c} \text{ (linear systems)} \qquad 12.50$$

$$= \frac{I\omega^2}{2g_c} \text{ (rotational systems)} \qquad 12.51$$

Example 12.9

A 2 lbm projectile is launched straight up with an initial velocity of 700 ft/sec. Neglect air friction and calculate the

(a) kinetic energy immediately after launch
(b) kinetic energy at maximum height
(c) potential energy at maximum height
(d) total energy at an elevation where the velocity is 300 ft/sec
(e) maximum height

(a) From equation 12.50

$$E_k = \tfrac{1}{2}\left(\frac{2}{32.2}\right)(700)^2 = 15{,}217 \text{ ft-lbf}$$

(b) At the maximum height, the velocity is zero. Therefore, $E_k = 0$.

(c) At the maximum height, all of the kinetic energy has been converted to potential energy. Therefore, E_p = 15,217 ft-lbf.

(d) Although some of the kinetic energy has been transformed into potential energy, the total energy is still 15,217 ft-lbf.

(e) Since all of the kinetic energy has been converted to potential energy, the maximum height can be found from equation 12.49.

$$15{,}217 = \left(\frac{2}{32.2}\right)(32.2)(h)$$

$$h = 7608.5 \text{ ft}$$

Example 12.10

A lawn mower engine is started by pulling a cord wrapped around a pulley. The pulley radius is 3 inches. The cord is wrapped around the pulley 5 times. If a constant tension of 20 lbf is maintained in the cord during starting, what work is done?

The torque on the engine is

$$\tau = \mathbf{F}r = (20)(3/12) = 5 \text{ ft-lbf}$$

The cord rotates the engine $(5)(2\pi) = 31.4$ radians. From equation 12.47, the work done is

$$W = (5)(31.4) = 157 \text{ ft-lbf}$$

Power is the amount of work done per unit time.

$$P = \frac{W}{\Delta t} \qquad 12.52$$

Power can be calculated from force and velocity.

$$P = \mathbf{F}v \text{ (linear systems)} \qquad 12.53$$

$$= \tau\omega \text{ (angular systems)} \qquad 12.54$$

Although the general mechanical work unit is the foot-pound, other units are used. Some useful conversions are listed in table 12.2.

Table 12.2
Useful Power Conversions

horsepower	$= \frac{\text{ft-lbf/sec}}{550}$	$= \frac{\text{ft-lbf/min}}{33{,}000}$	= 1.34 (kw)
kilowatts	$= \frac{\text{ft-lbf/sec}}{737}$	$= \frac{\text{ft-lbf/min}}{44{,}220}$	= .746 (hp)
BTU/sec	$= \frac{\text{ft-lbf/sec}}{778}$	$= \frac{\text{ft-lbf/min}}{46{,}680}$	= .707 (hp)

Example 12.11

A 5-ton ore car traveling at 4 feet per second passes point A, rolls down an incline, and is stopped by a spring bumper which compresses 2 feet. A constant frictional force of 50 pounds acts on the ore car. What spring modulus is required?

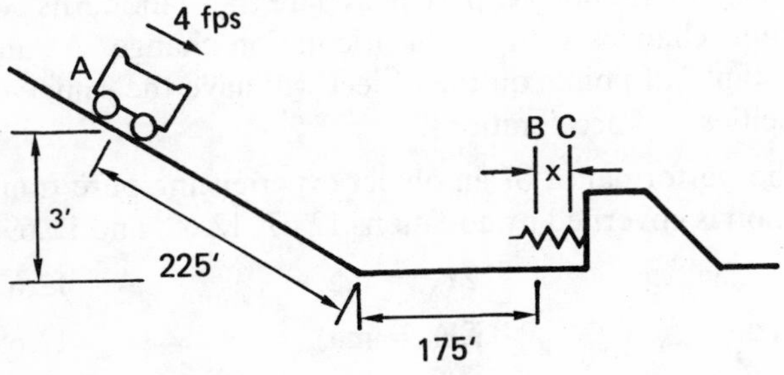

The car's mass is

$$m = (5)(2000) = 10{,}000 \text{ lbm}$$

The car's total energy at A is

$$E_t = E_k + E_p = \tfrac{1}{2}\left(\frac{10{,}000}{32.2}\right)(4)^2 + \left(\frac{10{,}000}{32.2}\right)(32.2)(3) = 32{,}484 \text{ ft-lbf}$$

Since the frictional force does negative work, the total energy at *B* is

$$E_{t,B} = 32{,}484 - (50)(225 + 175) = 12{,}484 \text{ ft-lbf}$$

At point *C*, the maximum compression point, the energy has gone into compressing the spring and performing a small amount of frictional work.

$$12{,}484 = \tfrac{1}{2}kx^2 + 50x$$

Since $x = 2$, $k = 6192$ lbf/ft

Example 12.12

A 200 lbm crate is pushed 25 feet across a warehouse floor. The crate's velocity is constant. What work is done if the coefficient of sliding friction between the crate and floor is .3?

The frictional force is

$$F_f = \frac{(\mu)(m)(g)}{g_c} = \frac{(.3)(200)(32.2)}{(32.2)} = 60 \text{ lbf}$$

From equation 12.46, the work done is

$$W = (60)(25) = 1500 \text{ ft-lbf}$$

5. Impulse and Momentum

The vector *impulse* is defined by equations 12.55 and 12.56.

$$\text{impulse} = \mathbf{F}\Delta t \text{ (linear systems)} \qquad 12.55$$

$$= \tau\Delta t \text{ (angular systems)} \qquad 12.56$$

The vector *momentum* is defined by equations 12.57 and 12.58.

$$p = \frac{mv}{g_c} \text{ (linear systems)} \qquad 12.57$$

$$= \frac{I\omega}{g_c} \text{ (angular systems)} \qquad 12.58$$

The *impulse-momentum equations* (equations 12.59 and 12.60) relate an applied impulse to a change in momentum.

$$\mathbf{F}\Delta t = \frac{m\Delta v}{g_c} \text{ (linear systems)} \qquad 12.59$$

$$\tau\Delta t = \frac{I\Delta\omega}{g_c} \text{ (angular systems)} \qquad 12.60$$

Example 12.13

A 1.62 ounce marble attains a velocity of 170 mph in a hunting slingshot. Contact with the sling is 1/25th of a second. What is the average force on the marble during contact?

Solving equation 12.59 for F,

$$F = \frac{\dfrac{1.62}{(16)(32.2)}(170)\left(\dfrac{5280}{3600}\right)}{1/25} = 19.6 \text{ lbf}$$

6. Impact Problems

According to Newton's second law, momentum is conserved unless the object is acted upon by an external force. In an impact, there are no external forces, only internal forces. Thus, momentum is conserved even though energy may be lost.

Direct impact occurs when the velocities of the two colliding bodies are perpendicular to the contacting surfaces. *Central impact* occurs when the force of impact is along the line of connecting centroids. The disposition of the velocities can be found from the *coefficient of restitution, e,* which corrects for frictional and other losses. It is used even when the particles are smooth and frictionless. However, it should be used only with velocity components along a mutual line.

$$e = \frac{\text{relative separation velocity}}{\text{relative approach velocity}} = \frac{-(v_2' - v_1')}{(v_2 - v_1)} \qquad 12.61$$

A collision is said to be perfectly *elastic* if $e = 1$. A collision is said to be *inelastic* ($e < 1$) if kinetic energy is lost, and is *perfectly inelastic* ($e = 0$) if the objects stick together. In the case of rebounding objects from a stationary surface,

$$e = \frac{\tan(\phi_{\text{rebound}})}{\tan(\phi_{\text{incident}})} \qquad 12.62$$

Since equation 12.61 generally will have two unknowns, either the *conservation of momentum* equation (equation 12.63) or the *conservation of kinetic energy* equation (equation 12.64) must also be used.[4] (Equation 12.64 is valid only if $e = 1$.)

$$m_1v_1 + m_2v_2 = m_1v_1' + m_2v_2' \qquad 12.63$$

$$m_1v_1^2 + m_2v_2^2 = m_1(v_1')^2 + m_2(v_2')^2 \qquad 12.64$$

If the impact is *oblique,* the coefficient of restitution should be used to find the *x*-components of the resultant velocities. Then equations 12.65 and 12.66 can be used.

$$v_{1y} = v_{1y}' \qquad 12.65$$

$$v_{2y} = v_{2y}' \qquad 12.66$$

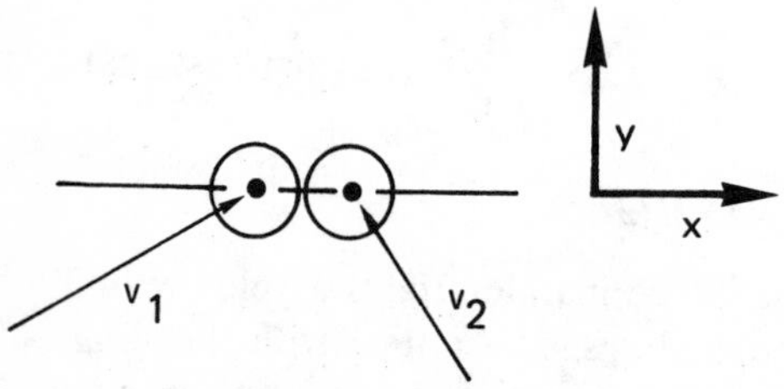

Figure 12.5 Oblique Impact

[4]Equations 12.63 and 12.64 contain mass variables in all four terms. The dimensional constant g_c should be included if momentum and kinetic energy units are desired. However, the constant will be applied to all four terms in each equation. For that reason, g_c can be omitted without compromising the equations.

Example 12.14

A golf ball dropped 8 feet onto a hard surface rebounds 5½ feet. What is the coefficient of restitution?

The impact velocity of the ball is found by equating the initial potential energy to the incident kinetic energy. This yields

$$v_1 = \sqrt{2gh} = \sqrt{(2)(32.2)(8)} = 22.7 \text{ ft/sec}$$

Similarly, the rebound velocity is

$$v_1' = \sqrt{(2)(32.2)(5.5)} = 18.8 \text{ ft/sec}$$

From equation 12.61 with $v_2 = v_2' = 0$,

$$e = \frac{-(0 - 18.8)}{(0-(-22.7))} = .828$$

Example 12.15

A 1.25 lbm projectile is fired from a cannon at 1200 ft/sec. The mass of the cannon is 350 lbm. What is the cannon's recoil velocity?

Although frictional forces will impede the cannon soon after it begins moving, its initial momentum is due to an impulse internal to the cannon/projectile system. Therefore, the principle of momentum conservation can be used to find the recoil velocity immediately after separation.

From equation 12.63,

$$(350)(0) + (1.25)(0) = (350)(v) + (1.25)(1200)$$

$$v = 4.29 \text{ ft/sec}$$

Since the mass variable appears in all four terms in equations 12.63, it is not necessary to divide each *lbm* by g_c.

7. Motion of Rigid Bodies

A. Translation

When a rigid body experiences pure translation,[5] its position changes without an orientation change. At any instant, all points on the object will have the same velocities and accelerations.

The performance of an object experiencing pure transition is governed by equations 12.67, 12.68, and 12.69.

$$\Sigma F_x = ma_x \qquad 12.67$$

$$\Sigma F_y = ma_y \qquad 12.68$$

$$\Sigma \tau = I\alpha \qquad 12.69$$

A moving object is not in static equilibrium. If internal (inertial) forces are added to the static equilibrium equation, the object will be placed in *dynamic equilib-*

[5]The word *translation* means movement.

rium. This procedure, known as *D'Alembert's principle,* is illustrated in example 12.16.

Example 12.16

A 5000 lbm truck skids with a deceleration of −15 ft/sec². What are the horizontal and vertical reactions at the wheels? What is the coefficient of sliding friction?

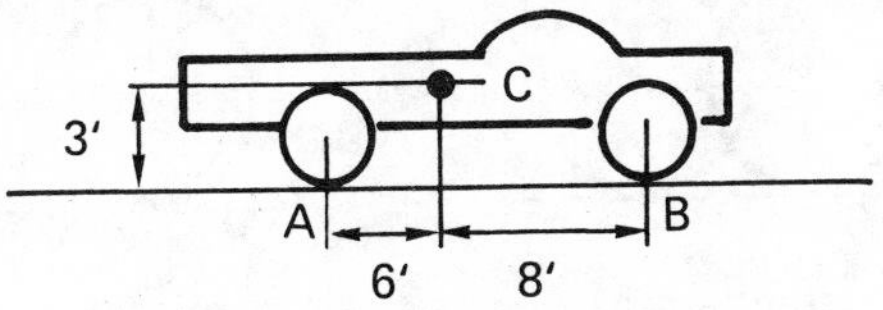

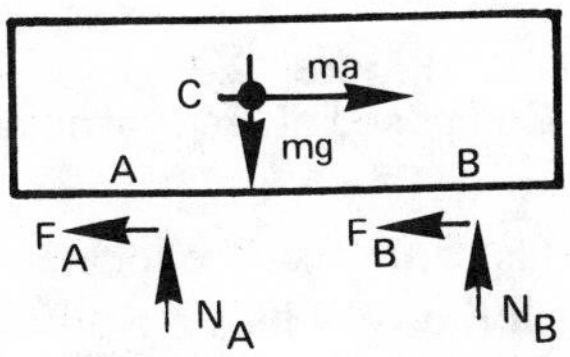

The freebody diagram of the truck in dynamic equilibrium is shown. The equations of dynamic equilibrium are:

$$\Sigma \mathbf{F}_x = 0: \left(\frac{m}{g_c}\right)a - \mathbf{F}_A - \mathbf{F}_B = 0$$

$$\frac{5000}{32.2}(15) - 5000f = 0$$

$$f = .466$$

$$\Sigma \mathbf{M}_A = 0: 14\mathbf{N}_B - 6(5000) - 3\left(\frac{5000}{32.2}\right)15 = 0$$

$$\mathbf{N}_B = 2642 \text{ lbf}$$

$$\Sigma \mathbf{F}_y = 0: \mathbf{N}_A + \mathbf{N}_B - m\left(\frac{g}{g_c}\right) = 0$$

$$\mathbf{N}_A + 2642 - 5000 = 0$$

$$\mathbf{N}_A = 2358 \text{ lbf}$$

Since **N** is the total reaction at both wheels,

$$\mathbf{F}_A = (.5)(.466)(2358) = 549 \text{ lbf}$$

$$\mathbf{F}_B = (.5)(.466)(2642) = 616 \text{ lbf}$$

B. Rotation About A Fixed Axis

Rotation is a motion produced from a turning moment or torque. The relationship between turning moment and angular acceleration is

$$\mathbf{T} = \left(\frac{I}{g_c}\right)\alpha \qquad 12.70$$

The rotation will usually be about the centroid axis. If it is not, the correct moment of inertia should be calculated from the *transfer formula,* equation 12.71. *d* is the distance between the centroidal axis and axis of rotation.

$$I = I_c + md^2 \qquad 12.71$$

The *tangential inertial force* on a rotating object acts normal to the centrifugal force. *r* in equation 12.72 is the object's distance from the rotational axis.

$$\mathbf{F}_i = \frac{ma_t}{g_c} = \frac{mr\alpha}{g_c} \qquad 12.72$$

Example 12.17

A turntable starts from rest and accelerates uniformly at α = 1.5 radian/sec². How many revolutions will be made before the rotational speed of 33⅓ *rpm* is attained?

The 33⅓ *rpm* must be converted to radians per second. Since there are 2π radians per revolution,

$$\omega = \frac{(33^1/3)(2\pi)}{60} = 3.49 \text{ rad/sec}$$

α, ω_0, and ω are known. θ is the unknown. (This is analogous to knowing *a*, v_0, and *v*, and needing *s*.) From table 12.1,

$$\theta = \frac{\omega^2 - \omega_0^2}{2\alpha} = \frac{(3.49)^2 - (0)^2}{(2)(1.5)}$$

$$= 4.06 \text{ rad}$$

Since there are 2π radians per revolution, the turntable is up to speed in

$$\frac{4.06}{2\pi} = .65 \text{ revolutions}$$

Example 12.18

A 4 foot diameter pulley with centroidal moment of inertia of 1610 lbm-ft² is subjected to tight-side and loose-side tensions of 200 lbf and 100 lbf respectively. A frictional moment of 15 ft-lbf is acting. What is the angular acceleration?

The net torque is

$$\mathbf{T} = (2)(200-100) - 15$$

$$= 185 \text{ ft-lbf}$$

From equation 12.69, the angular acceleration is

$$\alpha = \frac{(32.2)(185)}{1610} = 3.7 \text{ radians/sec}^2$$

Example 12.19

A 25 lbm suitcase is placed 6 feet from the center of a revolving table. The table has a centroidal moment of inertia of 9660 lbm-ft². A torque of 180 ft-lbf is applied to the table which is initially at rest. The coefficient of friction between the suitcase and table is .4. After how many seconds will the suitcase begin to slip off the table?

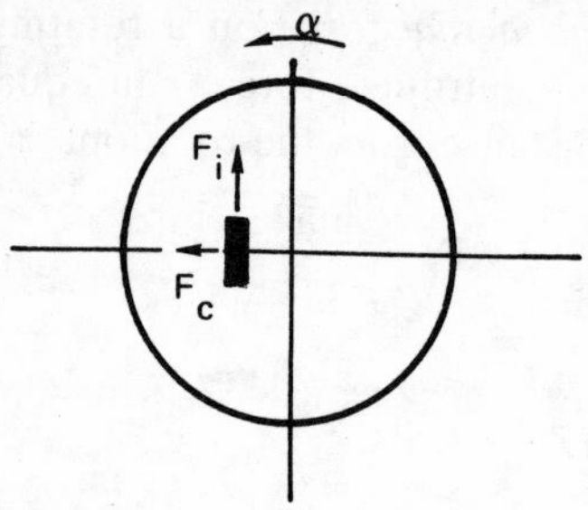

The total moment of inertia is

$$I = 9660 + (25)(6)^2 = 10{,}560 \text{ lbm-ft}^2$$

From equation 12.69, the angular acceleration is

$$\alpha = \frac{(32.2)(180)}{(10{,}560)} = .549 \text{ rad/sec}$$

The tangential inertial force on the suitcase is

$$\mathbf{F}_i = \frac{(25)(6)(.549)}{32.2} = 2.56 \text{ lbf}$$

The angular velocity of the table increases with time. At time t, the velocity is $\omega = (.549)t$. The centrifugal force at time t is

$$\mathbf{F}_c = \frac{(25)(6)[(.549)(t)]^2}{32.2} = 1.4t^2$$

The total plane reaction on the suitcase is

$$\mathbf{R} = \sqrt{(2.56)^2 + (1.4t^2)^2}$$

The frictional force is

$$\mathbf{F}_f = (25)(.4) = 10 \text{ lbf}$$

Using D'Alembert's principle, the applied frictional and centrifugal forces balance the inertial force.

$$10 = \mathbf{R}$$

Solving for t,

$$t = 2.62 \text{ sec}$$

C. General Plane Motion

Plane motion can be illustrated in two dimensions. Examples are rolling wheels, gear sets, and linkages. Plane motion can always be considered as the sum of a translation and a rotation. This is illustrated in figure 12.6.

Analysis of an object in plane motion can sometimes be simplified by working with the object's *instantaneous center.* The instantaneous center is a point at which the moving object could be pinned without changing the angular velocities of the particles in the body. Thus, as far as the angular veocities are concerned, the body seems to rotate about the instantaneous center.

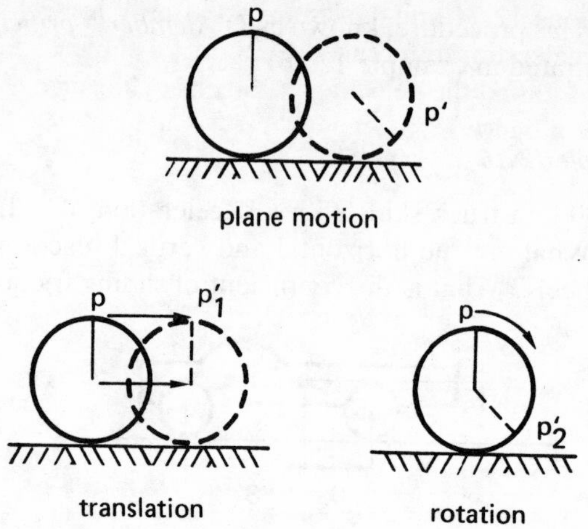

Figure 12.6
Plane Motion Composed of Translation and Rotation

The instantaneous center can be located by finding two particles for which the velocities are different and known in direction. Draw perpendiculars to the velocities through the points. The two perpendiculars will intersect at the instantaneous center. The instantaneous center of a rolling wheel is always the contact point with the supporting surface.

Once the instantaneous center has been located, the velocity of a particle located a distance l from the instantaneous center is given by equation 12.73. ω is the actual angular velocity of the object.

$$v = l\omega \qquad 12.73$$

Use of the instantaneous center reduces problems to simple geometry. However, it is important to use the correct reference point.

The velocity of any point on a wheel rolling (figure 12.7) with the translational velocity v_o can be found from its instantaneous center. Assume that the wheel is pinned at point C and rotates with the actual angular velocity. The angular velocities and accelerations with respect to point C can be found from geometry.

$$v = l\omega = \frac{lv_o}{r} \qquad 12.74$$

$$a_t = l\alpha = \frac{la_o}{r} \qquad 12.75$$

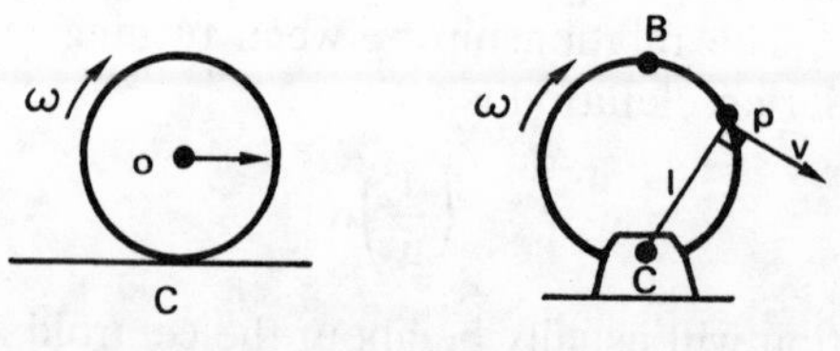

Figure 12.7
Rolling Wheel Pinned at its Instantaneous Center

Equations 12.74 and 12.75 are valid only for velocity and acceleration referenced to point *C*. Table 12.3 can be used to find the velocities with respect to other points shown in figure 12.7.

Table 12.3
Relative Velocities on a Rolling Wheel

point	reference point o	C	B
v_o	0	$v_o \rightarrow$	$\leftarrow v_o$
v_C	$\leftarrow v_o$	0	$\leftarrow 2v_o$
v_B	$v_o \rightarrow$	$2v_o \rightarrow$	0

Example 12.20

A car with 35 inch diameter tires is traveling at a constant 35 *mph*. What is the velocity of point *p* with respect to the ground?

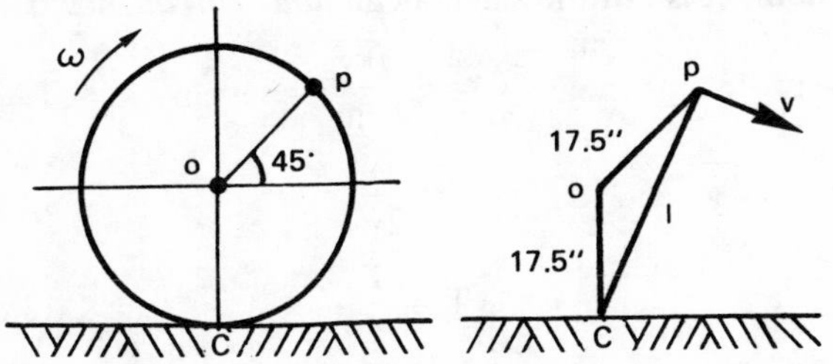

The translational velocity of point *o* is

$$v_o = 35\left(\frac{5280}{3600}\right) = 51.3 \text{ ft/sec}$$

The angular velocity of the wheel is

$$\omega = \frac{v_o}{r} = \frac{51.3}{\frac{35}{(2)(12)}} = 35.2 \text{ rad/sec}$$

(This angular velocity could also have been found by dividing v_o by the tire circumference and then multiplying by 2π.)

The law of cosines can be used to find the distance *l*.

$$l^2 = (17.5)^2 + (17.5)^2 - (2)(17.5)(17.5)\cos 135°$$

$$l = 32.3 \text{ inches}$$

The velocity of point *p* with respect to point *G* is found from equation 12.74.

$$v = l\omega = \left(\frac{32.3}{12}\right)(35.2)$$

$$= 94.7 \text{ ft/sec}$$

The instantaneous center, point *C*, of a slider rod assembly can be found from the perpendiculars of the velocity vectors.

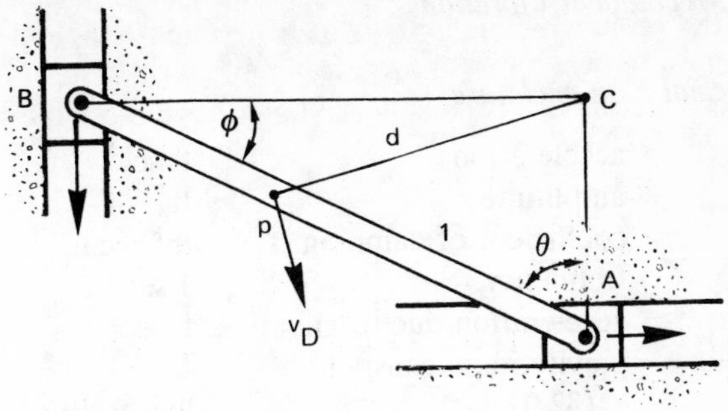

Figure 12.8 Slider Rod in Plane Motion

If the velocity with respect to point *C* of one end of the slider is known, say v_A, then v_B can be found from geometry. Since the slider can be assumed to rotate about point *C* with angular velocity ω,

$$\omega = \frac{v_A}{AC} = \frac{v_A}{l\cos\theta} = \frac{v_B}{BC} = \frac{v_B}{l\cos\phi} \qquad 12.76$$

Since $\cos\phi = \sin\theta$,

$$v_B = v_A(\tan\theta) \qquad 12.77$$

If the velocity with respect to point *C* of any other point is required, it can be found from

$$v_p = d\omega \qquad 12.78$$

D. Relative Plane Motion

The motion of one moving point relative to another moving point should be evaluated using vector subtraction. The following procedure can be used to determine the velocity of one point (*B*) with respect to another point (*A*) as illustrated in figure 12.9.

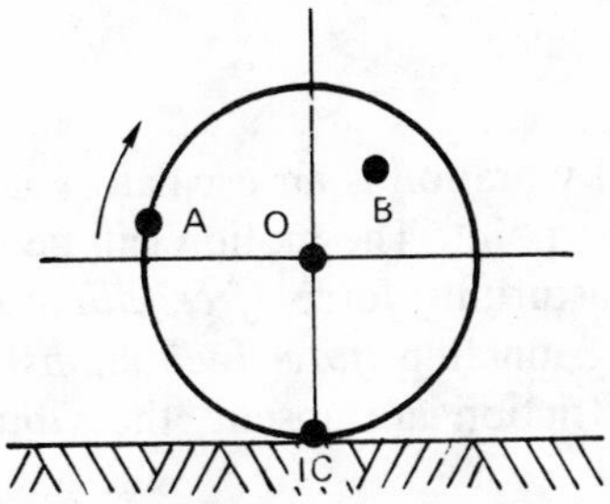

Figure 12.9 Relative Motion of Two Points

step 1: Find the velocity of *B* with respect to any point. For rolling wheels or rotating pulleys, choose the center as the reference point to simplify the problem. For more complex problems, choose the instantaneous center as the reference point.

step 2: Find the velocity of point *A* with respect to the same reference point.

step 3: Subtract the v_A vector from v_B. This can be done by reversing the direction of v_A and adding it heads-to-tails to v_B.

8. Mechanical Vibrations

Special Nomenclature

a	acceleration	ft/sec^2
A	amplitude	ft
C	coefficient of damping	lbf-sec/ft
f	frequency	1/sec
g	acceleration due to gravity	ft/sec^2
g_c	gravitational constant (32.2)	$lbm\text{-}ft/lbf\text{-}sec^2$
h	distance from centroid to suspension point	ft
I	moment of inertia	$lbm\text{-}ft^2$
k	spring constant	lbf/ft
L	pendulum length	ft
m	mass	lbm
n	damping ratio	–
P	magnitude of forcing function	lbf
t	time	sec
T	period	sec
v	velocity	ft/sec
w	weight	lbf
x	displacement	ft

Symbols

β	magnification factor	–
ω	frequency	rad/sec
δ	deflection	ft
ζ	log decrement	–

Subscripts

d	damped
eq	equivalent
st	static

A mechanical vibration is an oscillatory motion about an equilibrium point. The motion can be the result of a one-time disturbing force *(free vibration)* or a repeated forcing function *(forced vibrations)*. If external and internal friction are absent, the vibration is undamped.

A. Undamped, Free Vibrations

The common example of free vibration *(simple harmonic motion)* is a mass hanging from an ideal spring. After the mass is displaced and released, it will oscillate up and down. Since there is no friction (i.e., the vibration is undamped), the oscillations will continue forever.

After the initial disturbing force is removed, the object is acted upon by only the restoring force ($-kx$) and the inertial force (ma). Acceleration and the restoring force

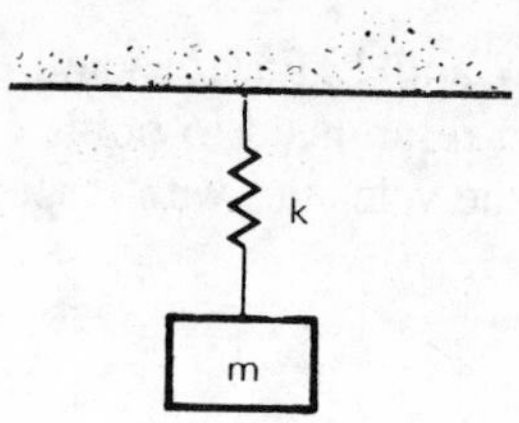

Figure 12.10 Spring and Mass

are proportional to the displacement from the equilibrium point, and they are opposite to the displacement. From D'Alembert's principle,

$$-kx = \frac{m}{g_c}\left(\frac{d^2x}{dt^2}\right) \qquad 12.79$$

The solution to this second-order differential equation is

$$x(t) = x_0\cos\omega t + \left(\frac{v_0}{\omega}\right)\sin\omega t \qquad 12.80$$

ω is known as the *natural frequency* of the vibrating system. It is not the same as the *linear frequency, f.*

$$\omega = \sqrt{\frac{kg_c}{m}} \qquad 12.81$$

$$f = \frac{\omega}{2\pi} \qquad 12.82$$

$$T = \frac{1}{f} \qquad 12.83$$

An alternate form of the solution is

$$x(t) = A\cos(\omega t - \alpha) \qquad 12.84$$

$$A = \sqrt{(x_0)^2 + (v_o/\omega)^2} \qquad 12.85$$

$$\alpha = \arctan\left(\frac{v_o}{\omega x_0}\right) \qquad 12.86$$

The maximum values are

$$x_{max} = A \qquad 12.87$$

$$v_{max} = A\omega \qquad 12.88$$

$$a_{max} = A(\omega)^2 \qquad 12.89$$

Equation 12.81 can be written in terms of the weight of the mass suspended from the spring.

$$w = \frac{mg}{g_c} \qquad 12.90$$

$$\omega = \sqrt{\frac{kg}{w}} \qquad 12.91$$

However, *w/k* is the static deflection[6] experienced by the spring when the mass is initially attached to it.

$$\omega = \sqrt{g/\delta_{st}} \qquad 12.92$$

[6]This initial deflection is known as the *static deflection*. It is not the same as the displacement experienced when the mass is acted upon by a disturbing force.

Springs in parallel share the applied load. The composite spring constant for parallel springs is

$$k_{eq} = k_1 + k_2 + \cdots \qquad 12.93$$

The composite spring constant for springs in series can be found from equation 12.94.

$$\frac{1}{k_{eq}} = \frac{1}{k_1} + \frac{1}{k_2} + \cdots \qquad 12.94$$

Example 12.21

A 120 lbm block is supported by three springs as shown. The initial displacement is 2 inches from the equilibrium position. No external forces act on the block after being released. What are the maximum velocity and acceleration?

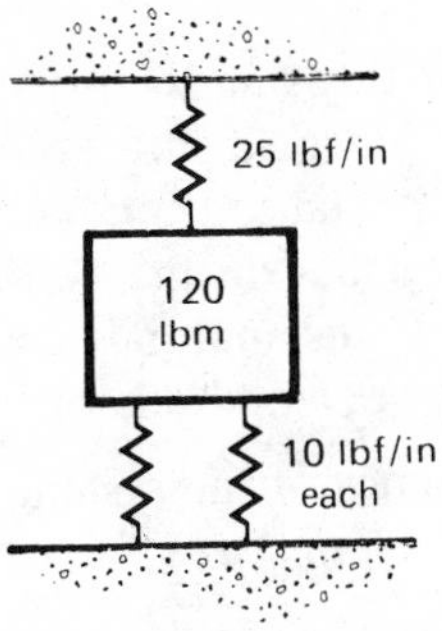

The equivalent spring constant is

$$k_{eq} = 25 + 10 + 10 = 45 \text{ lbf/in}$$

The static deflection is

$$\delta_{st} = \frac{120}{45} = 2.67 \text{ in}$$

The angular velocity is given by equation 12.92.

$$\omega = \sqrt{\frac{32.2}{(2.67/12)}} = 12.0 \text{ rad/sec}$$

The initial conditions are

$$v_0 = 0$$

$$x_0 = \frac{2}{12} = .167 \text{ ft}$$

From equations 12.88 and 12.89,

$$v_{max} = (.167)(12.0) = 2.0 \text{ ft/sec}$$

$$a_{max} = (.167)(12.0)^2 = 24.0 \text{ ft/sec}^2$$

The movement of pendulums which oscillate in small arcs can be described by the equations of simple harmonic motion. A *simple pendulum* (e.g., a mass on a string of length *L*) oscillates according to equation 12.95.

$$T = 2\pi\sqrt{\frac{L}{g}} \qquad 12.95$$

The period of a *compound pendulum* depends on the moment of inertia taken about the suspension point.

$$T = 2\pi\sqrt{\frac{I}{mgh}} \qquad 12.96$$

B. Damped, Free Vibrations

If damping is present such that the resisting force is proportional to the velocity, the magnitude of the frictional force is

$$F_d = C\frac{dx}{dt} \qquad 12.97$$

C is the *coefficient of viscous damping*. The differential equation of motion is

$$\frac{m}{g_c}\left(\frac{d^2x}{dt^2}\right) = -kx - C\left(\frac{dx}{dt}\right) \qquad 12.98$$

The general solution is

$$x(t) = e^{-nt}(c_1\cos\omega_d t + c_2\sin\omega_d t) \qquad 12.99$$

$$n = \frac{Cg_c}{2m} \qquad 12.100$$

ω_d is the *damped frequency*. It is not the same as the natural frequency, ω.

$$\omega_d = \sqrt{(\omega)^2 - (n)^2} \qquad 12.101$$

The *damping ratio* is

$$\text{damping ratio} = \frac{n}{\omega} = \frac{C}{C_{crit}} \qquad 12.102$$

case 1: If $n<\omega$, the oscillations will be *underdamped*. Motion will be oscillatory with diminishing amplitude.

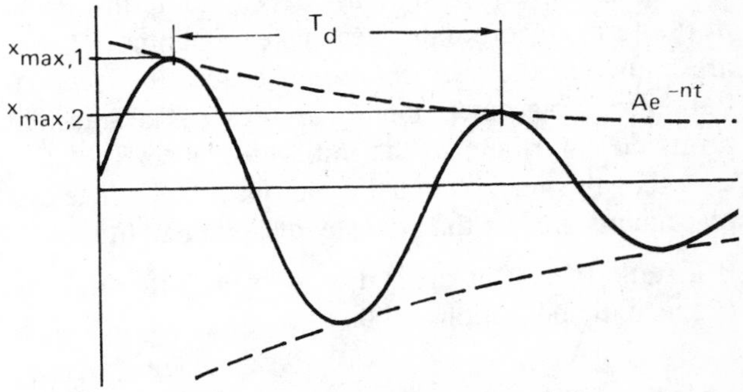

Figure 12.11 Underdamped Vibration

The logarithmic decrement can be used to find *n*.

$$e^{j\zeta} = \frac{x_{max,ith\ cycle}}{x_{max,(i+j)th\ cycle}} \qquad 12.103$$

$$\zeta = \frac{2\pi n}{\omega_d} \approx \frac{2\pi n}{\omega} \qquad 12.104$$

case 2: If $n > \omega$, the motion will be *overdamped.* There will be no oscillation, only a gradual return to the equilibrium position.

case 3: *Critical damping,* also known as *dead-beat motion,* occurs if $n = \omega$. There is no overshoot and the return is the fastest of the three cases. The *critical damping coefficient* is

$$C_{crit} = \frac{2m\omega}{g_c} \qquad 12.105$$

C. Damped, Forced Vibrations[7]

If a system is subjected to a periodic force, as in figure 12.12, vibrations will occur at regular intervals.

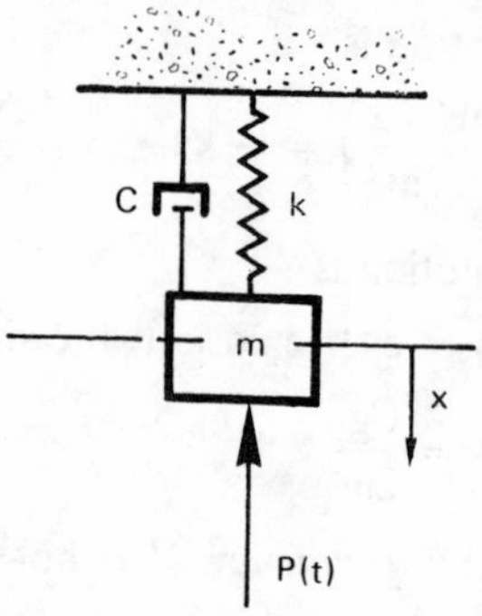

Figure 12.12 Forced Response

If the forcing function is sinusoidal (equation 12.106), the differential equation of motion is given by equation 12.107.

$$P(t) = \mathbf{P}\sin\omega_f t \qquad 12.106$$

$$m\left(\frac{d^2x}{dt^2}\right) = -kx - C\left(\frac{dx}{dt}\right) + \mathbf{P}\cos\omega_f t \qquad 12.107$$

The solution to equation 12.107 has several terms. Some of the terms incorporate decaying exponentials. These are known as the *transient terms* because their contribution to displacement rapidly decreases. The transient terms do contribute to the initial displacement, however. For this reason, initial cycles may experience displacements greater than the steady state amplitudes.

If a static force **P** is applied to a spring with stiffness k, the static deflection will be

$$\delta_{st} = \frac{\mathbf{P}}{k} \qquad 12.108$$

The amplitude of forced vibrations can be found from the pseudo-static deflection and the *magnification factor,* β.

$$A = \beta\left(\frac{\mathbf{P}}{k}\right) \qquad 12.109$$

$$\beta = \frac{1}{\sqrt{\left[1-\left(\frac{\omega_f}{\omega}\right)^2\right]^2 + \left[\frac{g_c C\omega_f}{m(\omega)^2}\right]^2}} \qquad 12.110$$

If the forcing frequency ω_f is equal to the natural frequency ω, the magnification factor will be very large (theoretically infinite). This condition is known as *resonance.*

Example 12.22

A 250 lbm motor turns at 1000 rpm and is supported on a resilient pad having a stiffness of 3000 pounds per inch. Due to an unbalanced condition, a periodic force of 20 pounds is applied in the vertical direction once each revolution. If the motor is constrained to move vertically, what is the amplitude of vibration?

The natural frequency of the system is

$$\omega = \sqrt{k/m} = \sqrt{\frac{(3000)(12)}{(250/32.2)}}$$

$$= 68.1 \text{ rad/sec}$$

The forcing frequency is

$$\omega_f = \frac{(1000)(2\pi)}{60} = 104.7 \text{ rad/sec}$$

The static deflection that would be caused by the unbalanced load is

$$\delta_{st} = \frac{20}{3000} = .00667 \text{ in}$$

The magnification factor (assuming $C = 0$) is found from equation 12.110.

$$\beta = \frac{1}{1-\left(\frac{104.7}{68.1}\right)^2} = -.733$$

Therefore, the total amplitude is

$$A = (-.733)(.00667) = -.0049 \text{ in}$$

The minus sign indicates that the vibrations are out of phase.

[7]These results can also be used with undamped, forced vibrations by setting $C = 0$.

BASIC CONCEPTS

1. A particle moves horizontally according to $x = 2t^2-8t+3$. When t=2, what are the position, velocity, and acceleration? What are the linear displacement and total distance traveled between t=1 and t=3?

2. A point travels in a circle according to $\omega = 6t^2-10t$. At t=2, the direction of motion is clockwise. What is the angular velocity at t=2? What is the displacement between t=1 and t=3? What is the total angle turned through between t=1 and t=3? Assume $\theta(0) = 0$.

3. A balloon is 200 feet above the ground and is rising at a constant 15 fps. An automobile passes under it traveling along a straight road at 45 mph. How fast is the distance between them changing one second later?

UNIFORM ACCELERATION

4. Object A weighs 10 pounds and rests on a frictionless plane with a 36.87° slope. Object B weighs 20 pounds. What is the velocity of B 3 seconds after release?

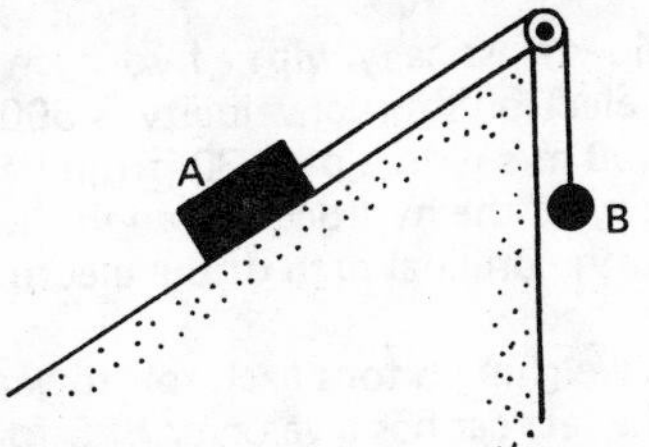

5. A jet plane acquires a speed of 180 mph in 60 seconds. What is its acceleration?

6. What is the acceleration of a train which increases its speed from 5 fps to 20 fps in 2 minutes?

7. A car traveling at 60 mph applies its brakes and stops in 5 seconds. What is its acceleration and distance traveled before stopping?

8. What is the acceleration experienced by a 7.8 kg mass acted upon by a 12 Newton force?

ROTATION

9. What is the linear speed of a point on the edge of a 14" diameter disc turning at 40 rpm?

10. What angular acceleration is required to increase an electric motor's speed from 1200 rpm to 3000 rpm in 10 seconds?

11. An apparatus for determining the speed of a bullet consists of 2 paper disks, mounted 5 feet apart on a single horizontal shaft which is turning at 1750 rpm. A bullet pierces both disks at radius 6" and an angle of 18° exists between each hole. What is the bullet velocity?

12. Disks B and C are in contact and rotate without slipping. A and B are splined together and rotate counter-clockwise. Angular velocity and acceleration of disk C are 2 rad/sec and 6 rad/sec^2 respectively. What is the velocity and acceleration of point D?

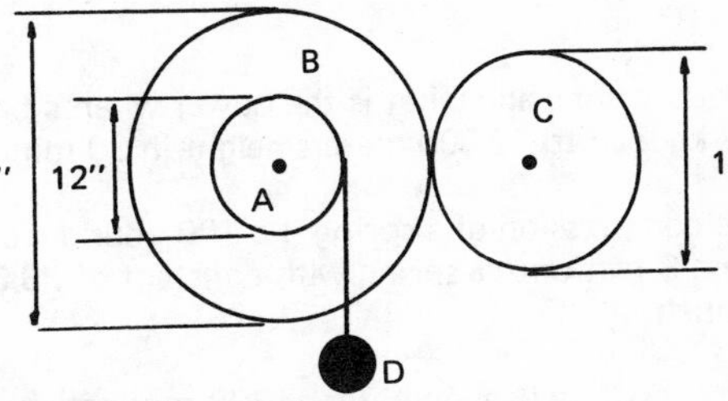

CENTRIFUGAL FORCE

13. Calculate the superelevation in % necessary on a curve with a 6000 foot radius so that 60 mph cars will not have to rely on friction to stay on the roadway.

14. What is the angle between the pole and the wire if the radius of the path is 4 feet? The 8.05 pound object is rotating at 20 fps.

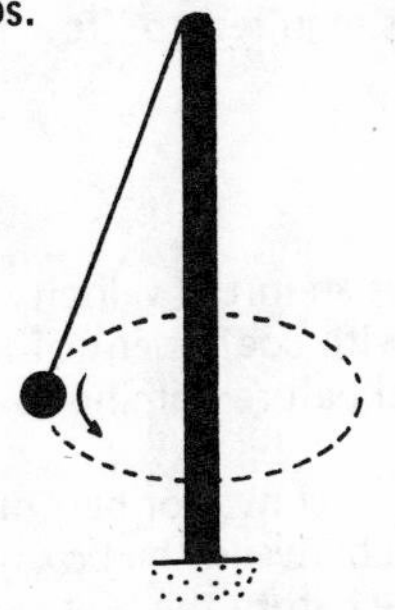

15. A 5 kg mass is tied to a 50 cm string and whirled at 5 revolutions per second horizontally to the ground. Find the
(a) centripetal acceleration
(b) centrifugal force
(c) centripetal force
(d) angular momentum

PROJECTILE MOTION

16. A projectile is fired at 45° from the horizontal with an initial velocity of 2700 fps. Find the maximum altitude and range neglecting air friction.

17. A baseball is hit at 60 fps and 36.87° from the horizontal. It strikes a goal post 72 feet away. Find the velocity components and the elevation above the origin at impact.

18. A bomb is dropped from a plane which is climbing at 30° and 600 fps while traveling at 12,000 feet altitude. What is the bomb's maximum altitude? How long will it take for the bomb to reach the ground from the release point?

WORK, ENERGY, AND POWER

19. A 1000 foot long cable weighs 2 pounds per foot and is suspended from a winding drum down into a vertical shaft. What work must be done to rewind the cable?

20. A solid cast iron sphere of 10" diameter travels without friction at 30 fps horizontally. What is its kinetic energy?

21. How many cubic meters of water can be pumped to a 40 meter height in one hour by a 5 kw pump? Assume 85% efficiency.

22. What work is done and what is the power when a balloon carries a 5.2 kg load to 12,000 meters height in 30 minutes?

23. Find the compression of a spring if a 100 pound weight is dropped from 8 feet onto a spring with constant of 33.33 pounds per inch.

24. A punch press flywheel operates at 300 rpm with a moment of inertia of 15 slug-ft^2. Find the speed in rpm to which the wheel will be reduced after a sudden punching requiring 4500 ft-lbf of work.

25. A force of 2500 N making a 40° angle (upward) from the horizontal pushes a box 6 meters across the floor. What work is done?

26. What power (in kw) is required to lift a 1500 kg mass 80 meters in 14 seconds?

FRICTION

27. A 100 pound body has an initial velocity of 12.88 fps while moving on a plane with coefficient of friction of .2. What distance will it travel before coming to rest?

28. A box is dropped onto a conveyor belt moving at 10 fps. The coefficient of friction between the box and the belt is .333. How long will it take before the box stops slipping on the belt?

29. A motorcyle and rider weigh 400 pounds. They travel horizontally around the inside of a hollow, right-angle cylinder of 100 feet inside diameter. What is the coefficient of friction that will allow a speed of 40 mph?

30. When a force (P) acts on a 10 pound body initially at rest, a speed of 12 fps is attained in 36 feet. What is the force (P) if the coefficient of friction is .25 and the acceleration is constant?

31. A 130 pound block slides up a 22.62° incline with coefficient of friction of .1 and initial velocity of 30 fps. How far will the block slide up the incline before coming to rest?

32. A 100 pound block is acted upon by a 100 pound force while resting on a surface with coefficient of friction of .2. A 50 pound block sits on top of the 100 pound block. What is the minimum coefficient of friction between the 50 and 100 pound blocks for there to be no slipping?

IMPULSE AND MOMENTUM

33. Sand drops at the rate of 560 lb/min onto a conveyor belt moving with a velocity of 3.2 fps. What force is required to keep the belt moving?

34. What is the impulse imparted to a .4 pound baseball that approaches the batter at 90 fps and leaves at 130 fps?

35. At what velocity will a 500 kg gun mounted on wheels recoil if a 1.2 kg projectile is propelled to 650 meters per second?

36. A 60 gram bullet (700 meters per second) embeds itself in a 4.5 kg wood block. What will be the block's initial velocity?

37. A nozzle discharges 40 gpm at 60 fps. Find the total force required to hold a flat plate in front of the nozzle. Assume that there is no splashback.

38. In a water turbine, 100 gallons per second impinge on a stationary blade. The water is turned through an angle of 160 degrees and exits at 57 fps. The water impinges at 60 fps. Find the force exerted by the blade on the stream. What is the angle from the horizontal of the force?

IMPACTS

39. An electron collides elastically with a hydrogen atom initially at rest. The electron's initial velocity is 500 m/s. Its final velocity is 490 m/s with a path 30° from its original path. Find the velocity of the hydrogen atom in the x direction if it recoils 1.2° from the original path of the electron.

40. Two freight cars weighing 5 tons each roll toward each other and couple. The left car has a velocity of 5 fps and the right car has a velocity of 4 fps prior to the impact. What is the velocity of the two cars coupled together after the impact?

41. Two billard balls collide as shown below. What are the velocities after impact? Use e=.8.

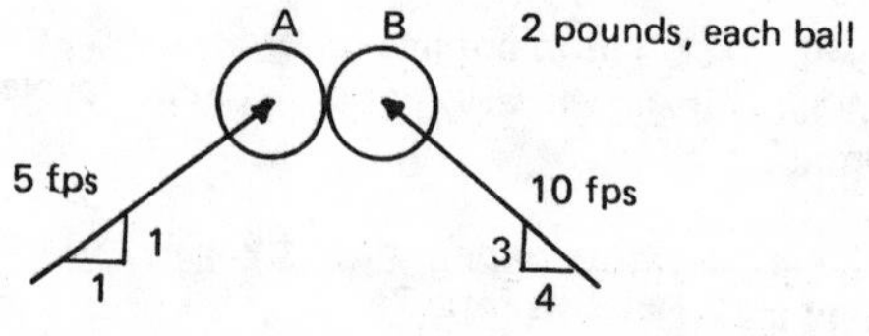

42. A 10 pound pendulum is released from rest and strikes a 50 pound block with e=.7. The block slides on a frictionless surface. Find the (a) velocity of the pendulum at impact, (b) the tension in the cord at impact, (c) the block's velocity immediately after impact, and (d) the required spring constant to stop the block with less than 6 inches deflection.

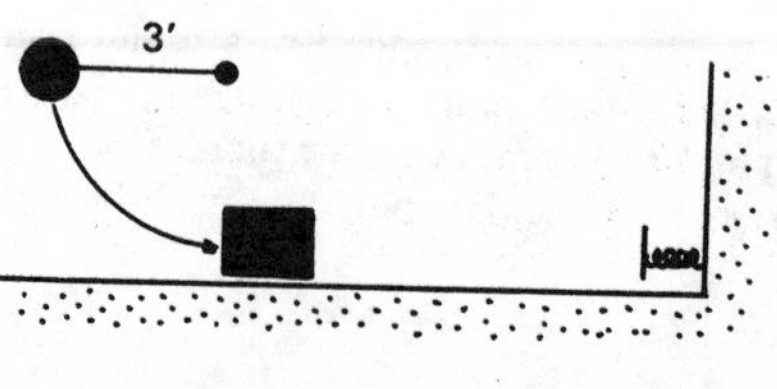

RIGID BODY MOTION

43. A solid sphere rolls without slipping down a 30° incline, starting from rest. What is its speed after two seconds?

44. A constant force of 20 pounds is applied to a 100 pound door which is supported on rollers at A and B. What is the acceleration of the door? What are the reactions at A and B? Where would the force have to be applied if the reactions at A and B were to be equal?

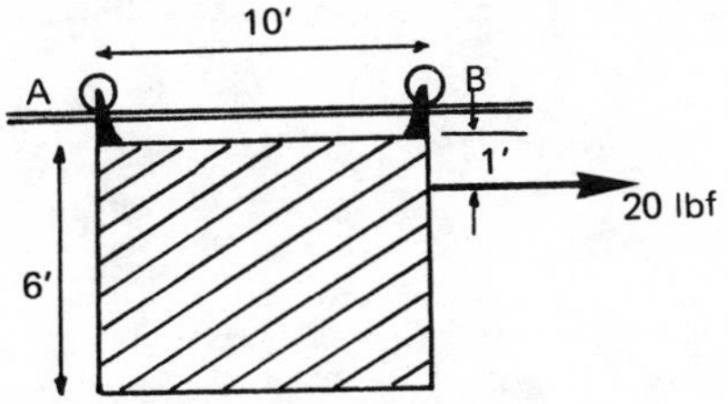

ROLLING WHEELS

45. Find the velocity of points A and B with respect to point O if the wheel rolls without slipping. The axle to which the wheel is attached moves at 10 fps to the right.

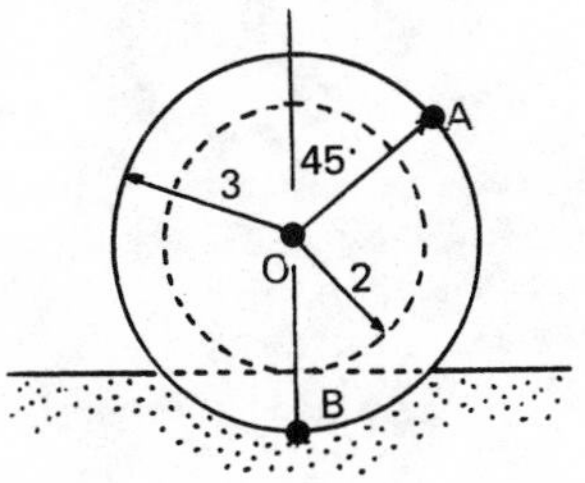

46. What is the velocity of point B with respect to point A in problem 45?

SIMPLE HARMONIC MOTION

47. An 870 gram mass hangs vertically from a spring with a constant of 150,000 dynes/cm. The mass is pulled 5 cm from its neutral hanging point and released. What is
   (a) the frequency of oscillation?
   (b) the period?
   (c) the force on the object when the displacement is 5 cm?
   (d) the maximum acceleration?
   (e) the velocity when the displacement is 3 cm?
   (f) the maximum velocity?

48. A simple pendulum has a mass and length of .2 kg and 1 meter respectively. It is displaced through a 10° angle and then is released. What is
   (a) the maximum force acting on the pendulum?
   (b) the maximum angular acceleration?
   (c) the maximum velocity?

49. The period of oscillation of a clock balance wheel is .2 seconds. The wheel is constructed with its 24 gram mass concentrated around a rim with .8 cm radius. What is the
   (a) wheel's moment of inertia?
   (b) torsional constant of the attached spring?

50. A thin disk is pivoted at its outermost edge and is allowed to oscillate in simple harmonic motion. Find the period of oscillation. (HINT: Consider the disk as a compound pendulum.)

51. What will be the frequency of oscillation of the water in the U-tube? The total length of fluid in the column is 30".

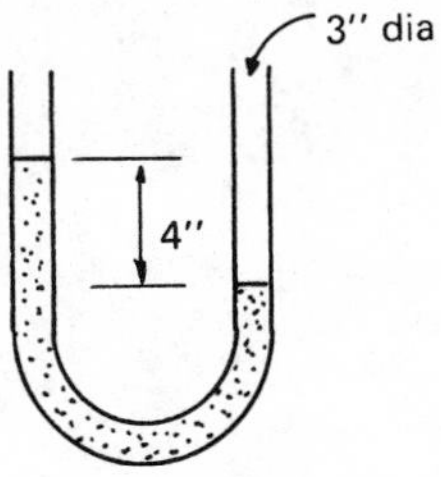

DYNAMICS

13 Direct Current Electricity

Nomenclature

a	acceleration	m/sec^2
$\mathbf{a_r}$	unit radial vector	–
A	area	m^2
B	magnetic flux density	wb/m^2 (tesla)
d	distance	meters
D	electrostatic flux density	lines/m^2
E	electric field strength	N/coul (volts/m)
E_p	potential energy	joules
F	force	newtons
H	magnetic field strength	N/wb, or amp/m
I	current	amps
L	length	meters
m	mass	kg
M	magnetic strength	unit poles
MMF	magnetomotive force	amp-turns
n	number of coils or turns	–
P	power	watts
q	charge	coulombs
r	distance	meters
R	resistance	ohms
s	distance traveled	meters
t	time	seconds
T	temperature	°C
v	velocity	m/sec
V	electrostatic potential	volts
W	work	joules
y	deflection	meters

Symbols

α	thermal coefficient of resistance	1/°C
ϕ	flux	lines, or webers
μ	permeability	henries/meter
λ	flux linkage	lines
ρ	resistivity	various
ρ_L	linear charge density	coulombs/meter
ρ_A	area charge density	coulombs/m^2
ϵ	permittivity	coul^2/n-m^2

Ω	magnetic potential	amps
$\mathcal{R}$	reluctance	rels

Subscripts

c	coil
m	multiplier
o	original, or at 20°C
r	ratio, or radial
s	shunt
v	vacuum

1. Electrostatics

The smallest amount of charge is possessed by an electron. This quantity of charge is known as one *electrostatic unit (esu)*. Since an *esu* is very small, a *coulomb* is defined as 6.24 EE18 *esu*. A point object with charge q_1 in coulombs will set up a radial *electric field* around itself, as shown in figure 13.1.

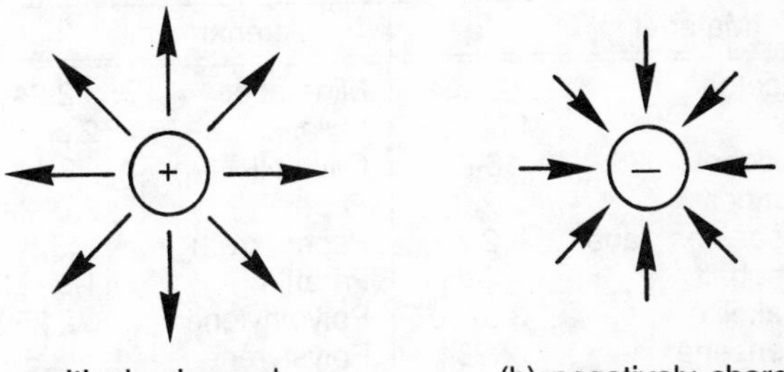

(a) positively charged   (b) negatively charged

Figure 13.1
Electric Field Around a Point Charge

This electric field is known as a *force field* since any other charged object will experience a force if it is near. The direction of the force will be along a *flux line*. Flux lines, also known as *displacement*, are imaginary lines which give the field direction. Flux lines leave perpendicularly to an object's surface. They are directed from positive to negative. The flux, with symbol ϕ and units of lines, is numerically equal to the object's charge in coulombs.

$$\phi_1 = q_1 \qquad\qquad 13.1$$

The number of flux lines per unit area perpendicular to the flux is known as the *flux density* or *displacement density*. Flux density, with units of lines per square meter, is proportional to the field strength.

$$D_1 = \frac{\phi_1}{A} \qquad 13.2$$

Since $\phi = q$,

$$D_1 = \frac{q_1}{A} \qquad 13.3$$

Gauss' law states that all of the flux must pass through a sphere whose surface area depends on the distance from the point charge.

$$D_1 = \frac{q_1}{4\pi r^2} \qquad 13.4$$

The *electric field strength* at any radius *r* is proportional to the flux density. A proportionality constant dependent on the surroundings is used to obtain the proper units.

$$\mathbf{E}_1 = \frac{\mathbf{a}_r D_1}{\varepsilon} = \left(\frac{q_1}{4\pi\varepsilon \mathbf{r}^2}\right)\mathbf{a}_r \qquad 13.5$$

Since D is a scalar and $\mathbf{E}$ is a vector, a unit radial vector, $\mathbf{a}_r$, is used in equation 13.5. The proportionality constant, ε, is known as the *permittivity* or *capacity* of the surrounding medium. It is most easily determined from equation 13.6 and table 13.1. ε_v is the permittivity of free space, equal[1] to 8.85 EE − 12 coul^2/n-m^2, and ε_r is the *dielectric constant (capacity ratio)*.

$$\varepsilon = \varepsilon_v \varepsilon_r \qquad 13.6$$

Table 13.1
Approximate Dielectric Constants
(20°C and 1 atmosphere)

Material	ε_r	Material	ε_r
Acetone	21.3	Mineral oil	2.24
Air	1.00059	Mylar	2.8–3.5
Alcohol	16–31	Olive oil	3.11
Amber	2.9	Paper	2.0–2.6
Asbestos paper	2.7	Paper (kraft)	3.5
Asphalt	2.7	Paraffin	1.9–2.5
Bakelite	3.5–10	Polyethylene	2.25
Benzene	2.284	Polystyrene	2.6
Carbon dioxide	1.001	Porcelain	5.7–6.8
Carbon tetrachloride	2.238	Quartz	5
		Rock	≈ 5
Castor oil	4.7	Rubber	2.3–5.0
Diamond	16.5	Shellac	2.7–3.7
Glass	5–10	Silicon oil	2.2–2.7
Glycerine	56.2	Slate	6.6–7.4
Hydrogen	1.003	Sulfur	3.6–4.2
Lucite	3.4	Teflon	2.0–2.2
Marble	8.3	Vacuum	1.000
Methanol	22	Water	80.37
Mica	2.5–8	Wood	2.5–7.7

[1]The units coul^2/n-m^2 are the same as farads/meter.

Equation 13.5 can be used to find the field strength around a point charge. Other configurations and their field strengths are given in table 13.2.

The force that is experienced by a point object with charge q_2 in an electric field of strength $\mathbf{E}$ is given by equation 13.7.

$$\mathbf{F} = q_2\mathbf{E} \qquad 13.7$$

If the electric field is also due to a point charge, $\mathbf{E}$ is given by equation 13.5. The force is given by equation 13.8, known as *Coulomb's law*.

$$\mathbf{F} = \frac{q_1 q_2}{4\pi\varepsilon r^2}\mathbf{a}_r \qquad 13.8$$

Any consistent set of units can be used with equation 13.8. The sign convention is such that repulsion is positive, and attraction is negative. The force direction can usually be found from observation as the direction the object would move if released. The method of vector superposition can be used with systems of multiple point charges.

Example 13.1

Find the force on point charge *A* in a vacuum.

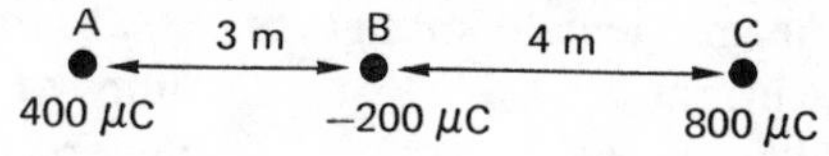

Since all charges are in line, the net force on *A* can be found as the sum of the individual forces computed from Coulomb's law.

$$
\begin{aligned}
\mathbf{F}_{A\text{-}B,C} &= \mathbf{F}_{A\text{-}B} + \mathbf{F}_{A\text{-}C} \\
&= \frac{q_A q_B}{4\pi\varepsilon(r_{A\text{-}B})^2} + \frac{q_A q_C}{4\pi\varepsilon(r_{A\text{-}C})^2} \\
&= \frac{(400\ \text{EE}-6)(-200\ \text{EE}-6)}{4\pi(8.85\ \text{EE}-12)(3)^2} \\
&\quad + \frac{(400\ \text{EE}-6)(800\ \text{EE}-6)}{4\pi(8.85\ \text{EE}-12)(7)^2} \\
&= -80 + 58.7 \\
&= -21.3\ \text{newtons}
\end{aligned}
$$

Since the sign is negative, the force on *A* will be attractive, towards *B* and *C*.

For non-linear systems, electrostatic forces must be broken into *x* and *y* components. The procedure is illustrated by example 13.2, which makes use of the following definitions:

$\mathbf{F}_{A\text{-}B}$ is the force on object *A* due to object *B*, expressed as a vector in terms of unit rectangular vectors.

$\mathbf{R}_{A\text{-}B}$ is the vector from object *B* to object *A*.

$\mathbf{a}_{r,A\text{-}B}$ is a unit radial vector from *B* to *A*.

Example 13.2

Completely determine the force on object A due to objects B and C.

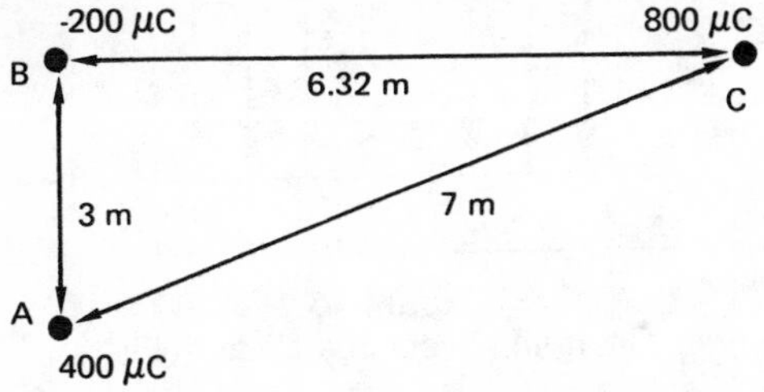

Vector superposition can be used by adding the x and y components of the forces computed individually. It is convenient to put the origin of the coordinate system at point A. The force on A due to B is -80 N (attractive), the numerical value having been calculated in example 13.1.

$$\mathbf{R}_{A\text{-}B} = (x_A - x_B)\mathbf{i} + (y_A - y_B)\mathbf{j}$$

$$= (0-0)\mathbf{i} + (0-3)\mathbf{j} = -3\mathbf{j}$$

$$|\mathbf{R}_{A\text{-}B}| = \sqrt{(x_A - x_B)^2 + (y_A - y_B)^2}$$

$$= \sqrt{(0-0)^2 + (0-3)^2} = 3$$

$$\mathbf{a}_{r,A\text{-}B} = \mathbf{R}_{A\text{-}B}/|\mathbf{R}_{A\text{-}B}| = -3\mathbf{j}/3 = -\mathbf{j}$$

$$\mathbf{F}_{A\text{-}B} = (-80)\mathbf{a}_{r,A\text{-}B} = +80\mathbf{j}$$

$\mathbf{F}_{A\text{-}B}$ is positive because object A is being attracted toward B, in the positive y direction. The force on A due to C is $+58.7$ N, also found in example 13.1. Proceeding similarly,

$$\mathbf{R}_{A\text{-}C} = (0-6.32)\mathbf{i} + (0-3)\mathbf{j} = -6.32\mathbf{i} - 3\mathbf{j}$$

$$|\mathbf{R}_{A\text{-}C}| = \sqrt{(0-6.32)^2 + (0-3)^2} = 7$$

$$\mathbf{a}_{r,A\text{-}C} = (-6.32\mathbf{i} - 3\mathbf{j})/7 = -.90\mathbf{i} - .43\mathbf{j}$$

Table 13.2
Electrostatic Field Configurations

Point Charge 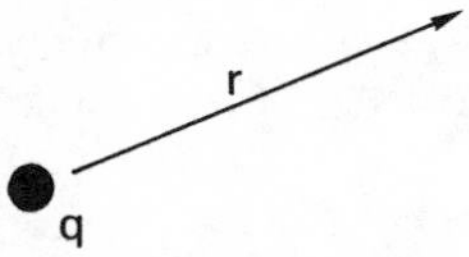 $E = q/4\pi\epsilon r^2$	**Infinite Line Distribution Inside an Infinite Cylinder** 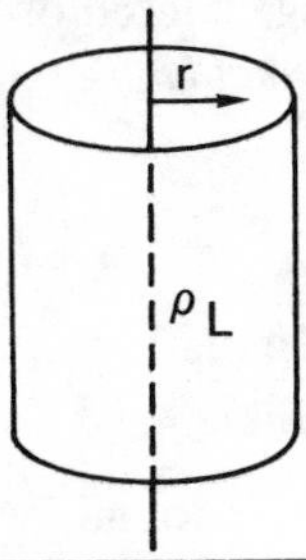 $E = \rho_L/2\pi\epsilon r$
Infinite Line Distribution 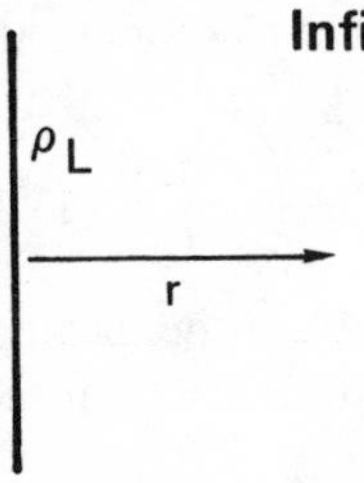 $E = \rho_L/2\pi\epsilon r$	**Sphere** 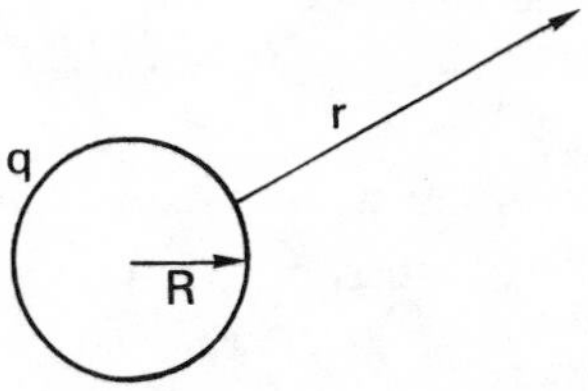 $E = q/4\pi\epsilon(R+r)^2$
Infinite Sheet Distribution 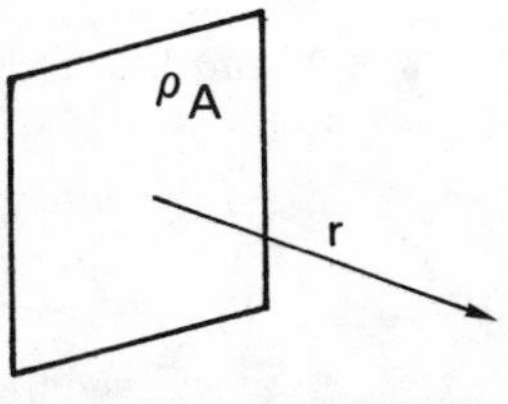 $E = \rho_A/2\epsilon$	**Infinite Coaxial Capacitor** 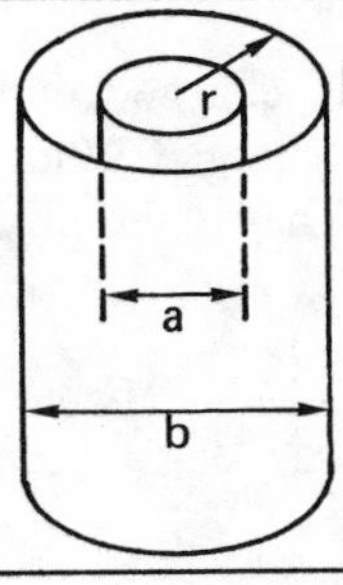 $\frac{a\rho_A}{2r\epsilon} = \frac{\rho_L}{2\pi r\epsilon}$ $\frac{1}{2}a < r < \frac{1}{2}b$ (ρ is for the inner cylinder)
Infinite Parallel Plates 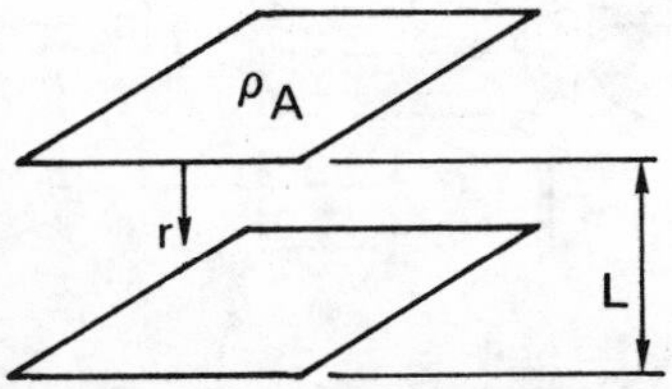 $E = \rho_A/\epsilon = V/L$ $(r < L)$	**Concentric Spheres** 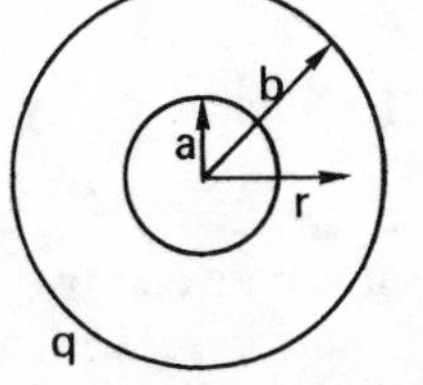 $E = q/4\pi\epsilon r^2$ $(a < r < b)$

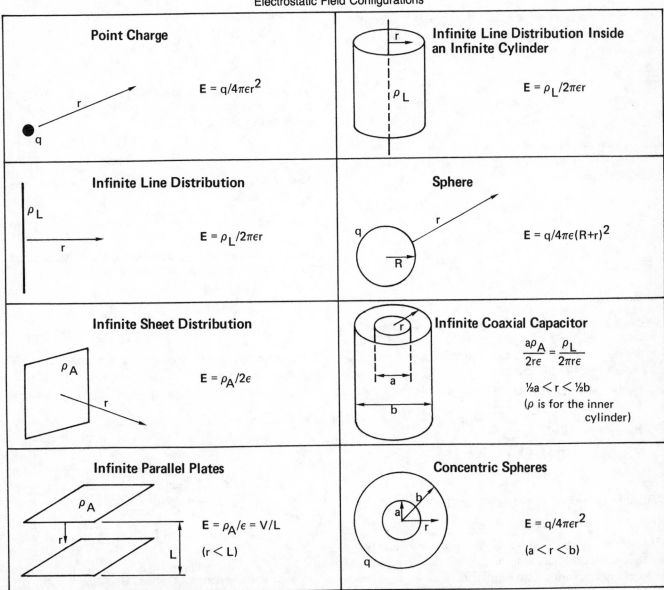

$$F_{A\text{-}C} = 58.7(-.90\mathbf{i} - .43\mathbf{j}) = -52.8\mathbf{i} - 25.2\mathbf{j}$$

Combining x and y components from $F_{A\text{-}B}$ and $F_{A\text{-}C}$ gives

$$F_A = (-52.8)\mathbf{i} + (80-25.2)\mathbf{j} = -52.8\mathbf{i} + 54.8\mathbf{j}$$

$$|F_A| = \sqrt{(-52.8)^2+(54.8)^2} = 76.1 \text{ N}$$

The *work* that must be done to move a charge B radially from distance r_1 to r_2 in a field due to charged particle A is given by equation 13.9. W is positive if the charge system does work on the surroundings. W is negative if the surroundings do work on the charge system.

$$W = \int_{r_1}^{r_2} \mathbf{F} \cdot d\mathbf{r} = \int_{r_1}^{r_2} \frac{q_A q_B}{4\pi\varepsilon r^2} \cdot d\mathbf{r} = \frac{-q_A q_B}{4\pi\varepsilon}\left[\frac{1}{r_2} - \frac{1}{r_1}\right] \qquad 13.9$$

If r_1 is infinite, the work required to bring charged object B to distance r_0 is known as the *potential energy* of the system.

$$E_p = W_{\infty - r_0} = \int_{\infty}^{r_0} \mathbf{F} \cdot d\mathbf{r} = \int_{\infty}^{r_0} \frac{q_A q_B}{4\pi\varepsilon r^2} \cdot d\mathbf{r} = \frac{-q_A q_B}{4\pi\varepsilon r_0} \qquad 13.10$$

The *potential* of an object in volts can be found by dividing the potential energy of the system by the object's charge.

$$V_A = \frac{E_p}{q_A} \qquad 13.11$$

$$V_B = \frac{E_p}{q_B} \qquad 13.12$$

Example 13.3

What work is required to bring a 400 μC charge from infinity to within 3 meters of a 600 μC charge? What is the potential energy of the system? What is the potential of the 400 μC charge?

$$W = \frac{-(400\text{ EE}-6)(600\text{ EE}-6)}{4\pi(8.85\text{ EE}-12)}\left[\frac{1}{3} - \frac{1}{\infty}\right] = -719 \text{ J}$$

$$E_p = W = 719 \text{ J}$$

$$V_{400\ \mu C} = 719/(400\text{ EE}-6) = 1.8\text{ EE}6 \text{ V}$$

The energy in J/m^3 stored in an electric field is

$$E_{stored} = \tfrac{1}{2}\varepsilon E^2 \qquad 13.13$$

If the electrostatic field is *uniform*, as it is between two plates connected across a voltage, V, the force, work, and energy equations are simplified.

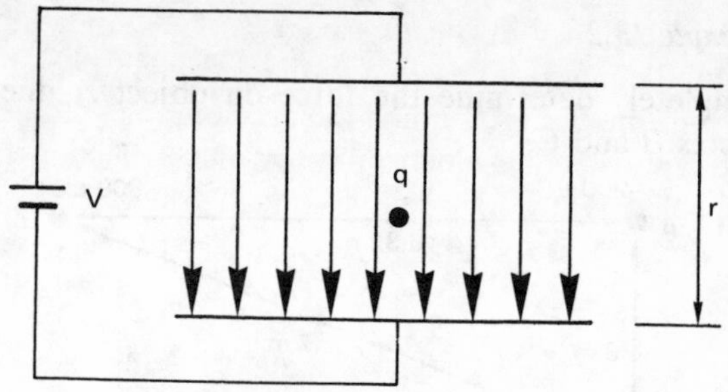

Figure 13.2
Charged Object in a Uniform Field

The electric field intensity, $\mathbf{E}$, is given by equation 13.14.

$$\mathbf{E} = \left(\frac{V}{r}\right)\mathbf{a}_r \qquad 13.14$$

The force on the object in the field is

$$\mathbf{F} = \mathbf{E}q \qquad 13.15$$

The work done in moving the object a distance d parallel to the field is

$$W = \mathbf{F}d = \mathbf{E}qd = \left(\frac{Vqd}{r}\right)\mathbf{a}_r \qquad 13.16$$

An object of mass m and charge q moving freely a distance d under influence of a constant force F will be subjected to an acceleration of

$$a = \frac{F}{m} = \frac{Eq}{m} = \frac{Vq}{rm} \qquad 13.17$$

After being exposed to this force for t seconds, the velocity will be

$$v = v_0 + at = v_0 + \left(\frac{Eqt}{m}\right) = v_0 + \left(\frac{Vqt}{rm}\right) \qquad 13.18$$

The velocity can also be determined from the distance traveled while under the influence of the constant force, as given by equation 13.19.

$$v = \sqrt{2ad + v_0^2} = \sqrt{2Eqd/m + v_0^2} = \sqrt{2Vqd/rm + v_0^2} \qquad 13.19$$

If the object is an electron with $v_0 = 0$ and the acceleration is across a plate so that $d = r$, then

$$v = 5.93\text{ EE}5\sqrt{V} \qquad 13.20$$

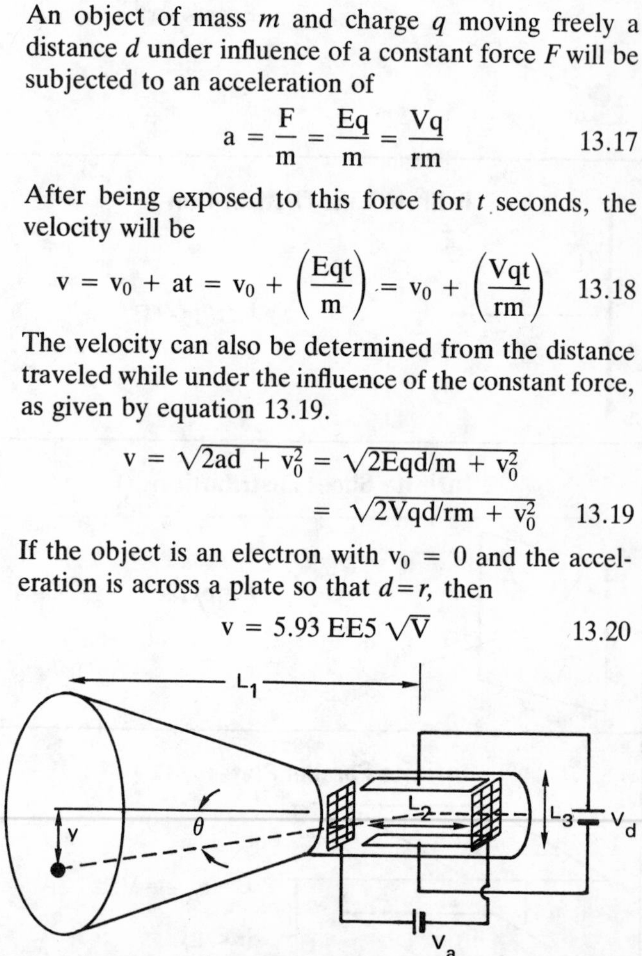

Figure 13.3
Deflection in a Cathode Ray Tube

The deflection y in a *cathode ray tube* will depend on the tube dimensions as well as the accelerating and deflecting voltages. Assuming the accelerating voltage V_a is zero so that the electron enters the deflection field with a constant horizontal velocity v_0, the deflection is

$$y = L_1 \tan\theta \qquad 13.21$$

$$\theta = \arctan\left(\frac{V_d q L_2}{L_3 m (v_0)^2}\right) \qquad 13.22$$

L_1 is measured from the screen to the center of the deflection plates.

If the electron is accelerated from a standstill by V_a while it is being deflected,

$$y = (L_1 + {}^1\!/_2 L_2)\tan\theta \qquad 13.23$$

$$\theta = \arctan\left(\frac{L_2 V_d}{L_3 V_a}\right) \qquad 13.24$$

2. Magnetism

Magnetism is attributed to electron spin. A magnetic field can only exist in the presence of two opposite poles, unlike an electric field which can be produced by a single charged object. Figure 13.4 shows two common magnetic field configurations. It also illustrates the North-to-South flux direction convention.

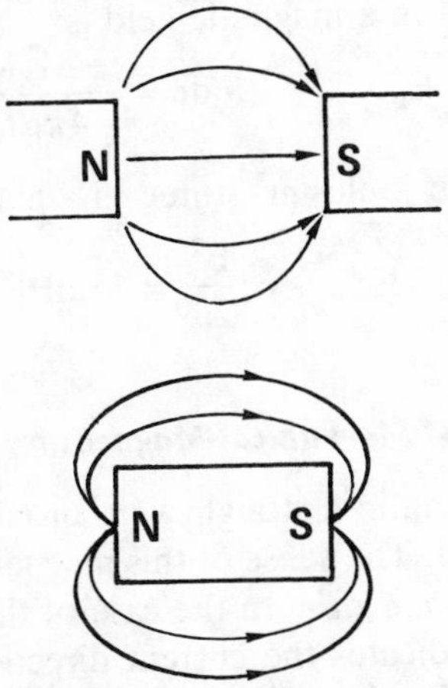

Figure 13.4 Magnetic Fields

Magnetic field strength depends on the influenced material and the surrounding medium. The different types of magnetism are:

Ferromagnetism: A spontaneous alignment of electron spins producing strong magnetic behavior in alpha-iron, cobalt, nickel, and gadolinium. The spin alignment and its ferromagnetism are destroyed by random molecular vibrations when the material is heated above its *Curie temperature,* which is about 750°C for iron. Iron is paramagnetic above its Curie temperature. Relative permeability is much greater than 1 in ferromagnetic materials.

Ferrimagnetism: The form of magnetism present in soft *ferrites* such as $MnFe_2O_4$ or $ZnFe_2O_4$.

Anti-Ferromagnetism: The weakly-attractive effect produced in salts of the transition elements such as MnO.

Paramagnetism: The weakly-attractive effect produced in most alkalai and transition metals. Paramagnetic materials have relative permeabilities slightly greater than 1.

Diamagnetism: The weakly-repulsive effect present in most non-metals and organic materials. Relative permeability is slightly less than 1.

The graph of a material's magnetic response to an applied magnetic field is known as a *hysteresis loop,* shown in figure 13.5. If a material without any internal magnetism is exposed to an external magnetic field, the material will become slightly magnetic. This internal magnetism will increase from zero to a maximum point (the *saturation point*) as shown by the line starting at the origin in figure 13.5.

If the external field is removed, some internal magnetism will remain. This remaining magnetism is known as *residual magnetism.* In order to totally remove any internal magnetism, a reverse external field with strength equal to the *coercive strength* will have to be applied.

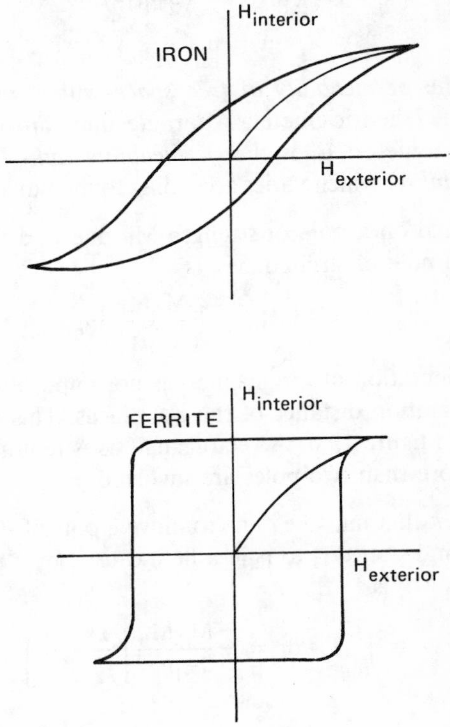

Figure 13.5
Typical Hysteresis Loops

The analysis of magnetic effects requires variables similar in name and utilization to that of electrostatics, as shown in table 13.3.

Table 13.3
Comparison of Magnetism and Electrostatics

Variable	Electrostatic Symbol and Units	Magnetic Symbol and Units
Total strength	q (coulombs)	M (unit-poles)
Flux	ϕ (lines)	ϕ (webers)
Flux density	D (lines/m²)	B (wb/m², or tesla)
Field strength	**E** (newton/coul, or volt/meter)	**H** (newton/weber, or amp/meter)
Space resistance	ε (farads/meter)	μ (henries/meter)
Potential	V (volts)	Ω (amps)

The amount of *flux*, ϕ, is numerically equal to the *magnetic pole strength.*

$$\phi = M \qquad 13.25$$

The *flux density* can be found by dividing the flux by an area perpendicular to it. The flux density at a distance *r* from a pole of strength *M* is

$$B = \frac{\phi}{A} = \frac{M}{4\pi r^2} \qquad 13.26$$

The magnetic *field strength* can be found by dividing the flux density by a proportionality constant which is dependent on the environment and the flux density itself.

$$\mathbf{H} = \left(\frac{B}{\mu}\right)\mathbf{a}_r = \left(\frac{M}{4\pi\mu r^2}\right)\mathbf{a}_r \qquad 13.27$$

$$\mu = \mu_v \mu_r \qquad 13.28$$

μ_v is the *permeability of free space* with a value of 4π EE−7 henries/meter. Alternate units are webers/amp-turn-meter. μ_r is the *permeability ratio (relative permeability)* which varies according to the flux density.

The force on a pole of strength M_B due to a field set up by a pole of strength M_A is

$$\mathbf{F} = M_B\mathbf{H}_A = \left(\frac{M_A M_B}{4\pi\mu r^2}\right)\mathbf{a}_r \qquad 13.29$$

The orientation of the magnets is not important, only the separation distance of the pole faces. This is illustrated in figure 13.6. A vector analysis is required any time more than two poles are involved.

The *work* that must be done to move a pole of strength M_B from distance r_1 to r_2 in a field caused by a pole of strength M_A is

$$W = \int_{r_1}^{r_2} F \cdot dr = \frac{-M_A M_B}{4\pi\mu}\left[\frac{1}{r_2} - \frac{1}{r_1}\right] \qquad 13.30$$

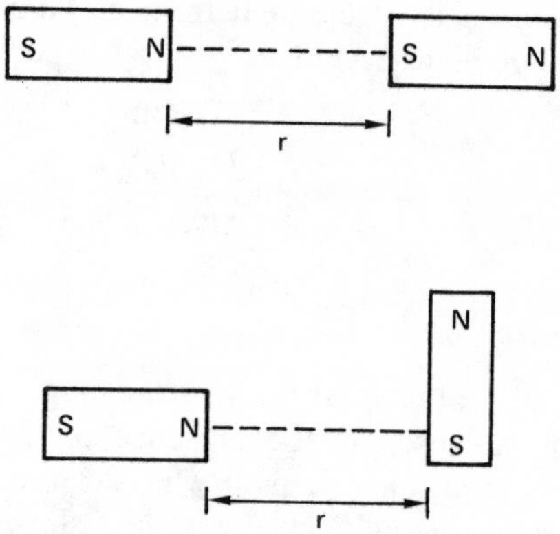

Figure 13.6 Pole-Face Orientation

The *potential energy* possessed by a pole by virtue of its placement in a magnetic field is

$$E_p = \int_{\infty}^{r_0} F \cdot dr = \frac{-M_A M_B}{4\pi\mu r_0} \qquad 13.31$$

The energy in joules/m³ stored in a magnetic field is

$$E_{stored} = \frac{B^2}{2\mu} = \tfrac{1}{2}\mu|\mathbf{H}|^2 \qquad 13.32$$

3. Electro- and Mechanical-Magnetism

Current flowing in a straight wire or coil will induce a magnetic field. The sense of this magnetic field is given by the *right-hand rule.* In the case of the straight wire, the thumb indicates the current direction and the fingers curl in the field direction. For the wire coil, the fingers indicate current flow and the thumb indicates the field direction. Assuming a constant permeability, field strength, flux, and current are all proportional. That is,

$$\frac{\mathbf{H}_1}{\mathbf{H}_2} = \frac{\phi_1}{\phi_2} = \frac{I_1}{I_2} \qquad 13.33$$

Table 13.4
Approximate Relative Permeabilities

Material	Flux Density (wb/m²) .2	.4	.6	.8	1.0	1.2	1.4	1.6
cast iron	400	300	200	120	90	60	50	50
cast steel	750	1050	1200	1225	1175	950		
silicon sheet steel	3000	5300	6000	5300	4000	2400		

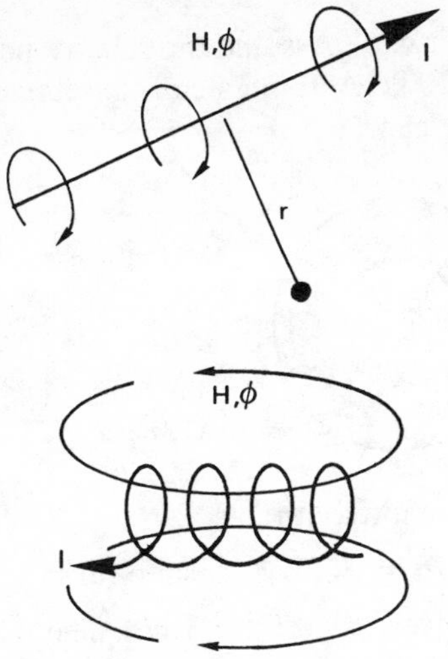

Figure 13.7 Magnetic Field Directions

The magnetic field at a distance r from the center of a long thin conductor is given by equations 13.34 and 13.35. μ is the permeability of the material surrounding the wire.

$$B = \frac{\mu I}{2\pi r} \qquad 13.34$$

$$\mathbf{H} = \frac{B}{\mu} = \frac{I}{2\pi r} \qquad 13.35$$

Example 13.4

A long wire with radius of .003 meters carries 4 amps. Find the magnetic flux density and the field strength .01 meter from the wire surface.

$$B = \frac{(4\pi \text{ EE}-7)(4)}{2\pi(.013)} = 6.2 \text{ EE}-5 \text{ tesla}$$

$$\mathbf{H} = \frac{4}{2\pi(.013)} = 49.0 \text{ amps/meter}$$

The flux density, field strength, and flux inside a torus with $r<<R$ is given by equations 13.36, 13.37, and 13.38. μ is the permeability of the toroidal core material.

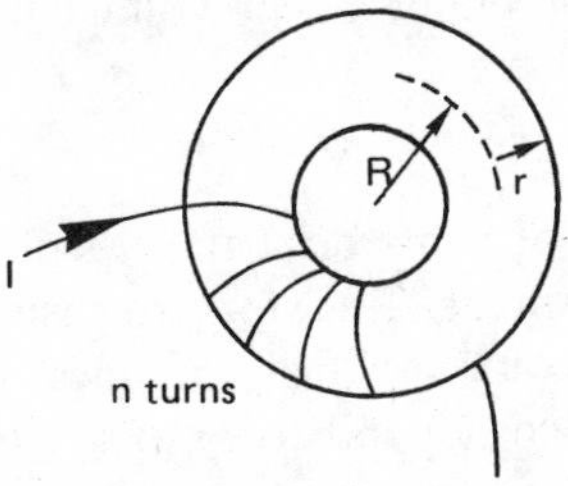

Figure 13.8 Toroidal Coil

$$B = \frac{\mu n I}{2\pi R} \qquad 13.36$$

$$\mathbf{H} = \frac{nI}{2\pi R} \qquad 13.37$$

$$\phi = BA = \frac{\mu r^2 n I}{2R} \qquad 13.38$$

A voltage will be induced in a conductor moving through a magnetic field, as illustrated in figure 13.9. It is assumed that the conductor and field are orthogonal, otherwise a sine term must be introduced.

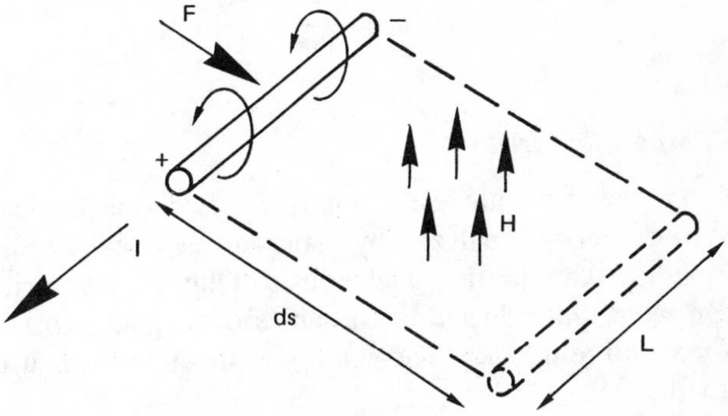

Figure 13.9
Conductor Moving in a Magnetic Field

The conversion of linear motion to current is given by *Faraday's law.* The induced voltage is

$$V = n\frac{d\phi}{dt} = \frac{d\lambda}{dt} \qquad 13.39$$

λ is known as the *flux linkage,* defined by equation 13.40.

$$d\lambda = n\, d\phi \qquad 13.40$$

If only one conductor is involved ($n=1$), then

$$d\phi = B\, dA = BL\, ds \qquad 13.41$$

$$V = n\frac{d\phi}{dt} = BL\frac{ds}{dt} = BLv \qquad 13.42$$

The power dissipated in the conductor is

$$P = IV = IBLv \qquad 13.43$$

Since power is the time rate of work,

$$P = \frac{dW}{dt} = \mathbf{F}\frac{ds}{dt} = \mathbf{F}v \qquad 13.44$$

Setting equations 13.43 and 13.44 equal, the force on the conductor is

$$\mathbf{F} = IBL \qquad 13.45$$

Equation 13.45 is a special case of *Ampere's law,* which allows for a non-orthogonal conductor/flux configuration. If θ is the angle between the current and the flux, then the force per unit length is

$$\frac{F}{L} = IB \sin\theta \qquad 13.46$$

A similar result can be obtained for a particle of charge q being moved through a magnetic field. If the velocity is perpendicular to the field, the initial force on the particle is

$$F = qvB \qquad 13.47$$

The acceleration is specified by Newton's second law. The particle will travel in a circle with radius and angular velocity given by equations 13.48 and 13.49.

$$r = \frac{mv}{qB} \qquad 13.48$$

$$\omega = \frac{qB}{m} \qquad 13.49$$

4. *Magnetic Circuits*

A magnetic circuit in a composite electro-mechanical system can be analyzed by using an electrical circuit analogy. An equation analogous to Ohm's law is written which introduces the variables of magnetomotive force and reluctance. The analogy is illustrated in figure 13.10.

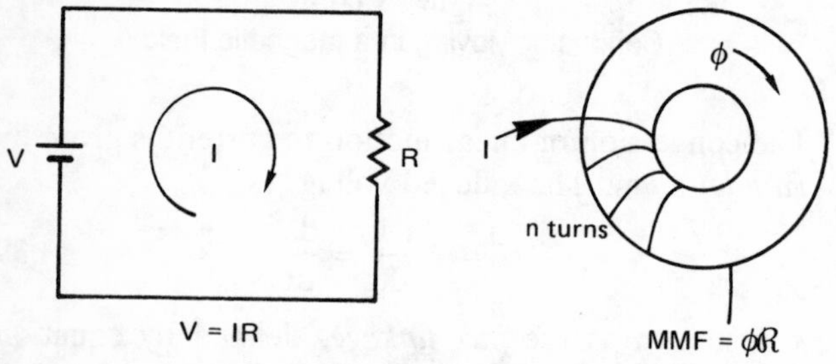

Figure 13.10
The Magnetic Circuit Analogy

In a magnetic circuit, flux will be proportional to the *magnetomotive force (MMF)*. Flux is proportional to the current flowing in the coil and the number of turns, n.

$$MMF = In \quad \text{(amp-turns)} \qquad 13.50$$

Reluctance is analogous to electrical resistance. Its units are rels, dimensionally the same as amp-turns/weber. Reluctance is calculated from the path area, length, and permeability. Of course, the value of permeability used must be consistent with the flux density. Several iterations may be required if either the reluctance or the flux density are unknown. Reluctance is found from equation 13.51.

$$\mathcal{R} = \frac{L}{\mu A} \qquad 13.51$$

The magnetic circuit can be evaluated by using the following relationship:

$$MMF = \phi\mathcal{R} \qquad 13.52$$

Example 13.5

What current is required to establish a flux of 10 mWb across the air gap? Assume the relative permeability of cast iron is 2000. Assume a constant flux density in the core and air gap.

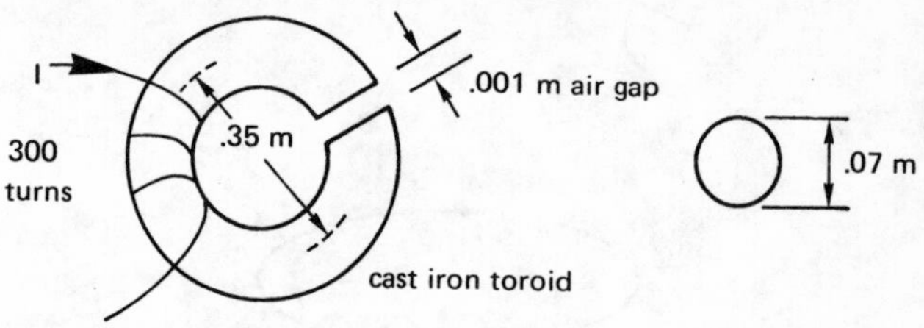

The magnetomotive force is

$$MMF = In = 300I \quad \text{amp-turns}$$

The flux is given as 10 EE−3 Wb. The mean cast iron path length is

$$(.35)(\pi) - .001 = 1.0986 \text{ m}$$

The path cross sectional area is

$$(\pi/4)(.07)^2 = .00385 \text{ m}^2$$

The total path reluctance is

$$\mathcal{R} = \Sigma\frac{L}{\mu A} = \frac{1.0986}{(2000)(4\pi \text{ EE}-7)(.00385)} + \frac{.001}{(4\pi \text{ EE}-7)(.00385)}$$

$$= 3.2 \text{ EE5 rels}$$

From equation 13.52,

$$300I = (3.2 \text{ EE5})(10 \text{ EE}-3)$$

$$I = 10.67 \text{ amps}$$

5. *Electrical Resistance*

A *current* is said to exist whenever charges move. By convention, the current moves in a direction opposite to the flow of negative particles (electrons). Current, in amps, is defined as the number of coulombs passing a point per second.

$$I = \frac{dq}{dt} \qquad 13.53$$

The current in a simple electrical circuit, as shown in figure 13.10, is related to the voltage and resistance by *Ohm's law.*

$$V = IR \qquad 13.54$$

The property of a material to impede current flow is known as *resistance.* Resistors are usually constructed from carbon compositions, ceramics, oxides, or wire coils. Resistance depends on the *resistivity* of the material and the cross sectional current area.

Referring to figure 13.11, the resistance of a solid through which a current flows is

$$R = \frac{\rho L}{A} \qquad 13.55$$

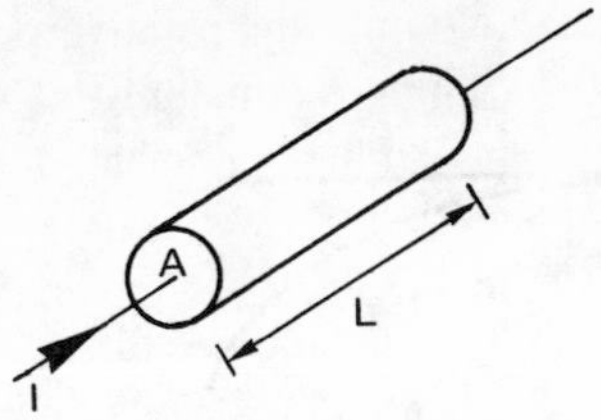

Figure 13.11 A Resistor

In the case of a circular conductor, area is usually measured in units of *circular mils*, abbreviated *CM*. One circular mil is the area of a .001″ diameter circle. The following relationships can be used to evaluate areas of conductors.

$$\text{Area (CM)} = \left(\frac{\text{D, inches}}{.001}\right)^2 \qquad 13.56$$

$$\text{Area (sq. in)} = (\text{\# of circular mils})(7.843\ \text{EE}-7) \qquad 13.57$$

Resistivity is not constant. It depends on the operating temperature. For most conducting materials, resistivity increases as temperature increases.

$$\rho = \rho_o(1 + \alpha\Delta T) \qquad 13.58$$

$$R = R_o(1 + \alpha\Delta T) \qquad 13.59$$

Values of the *thermal coefficient of resistance*, α, are given in table 13.5.

Table 13.5
Approximate Temperature Coefficients of Resistance

Material	α per °C	Material	α per °C
Aluminum,	0.00403	Mercury	0.00089
hard-drawn	0.0038	Monel metal	0.0020
Antimony	0.0036	Palladium	0.0033
Cadmium	0.004300	Platinum	0.003
Cobalt	0.00551	Rhodium	0.00457
Constantan	0.00002	Silver,	0.0038
Copper, IACS	0.00393	annealed	0.004050
hard-drawn	0.00382	Steel, cold-	
Gold	0.0034	rolled	0.0042
Iridium	0.00398	Tin	0.0042
Iron	0.0065	Titanium	0.00469
Lead	0.0039	Tungsten	0.00510
Magnesium	0.0040	Zinc	0.0037
		Zirconium	0.00438

The reciprocal of resistivity is *conductivity*. The ratio of a metal's conductivity to the conductivity of standard copper is known as the *percent conductivity*.

$$\%\ \text{conductivity} = \frac{\rho_{\text{std. copper}}}{\rho_{\text{metal}}} \times 100 \qquad 13.60$$

The resistivity of standard copper depends on the units chosen. Some values are

1.7241 EE − 6 ohm-cm²/cm

.67879 EE − 6 ohm-in²/in

10.371 ohm-CM/ft

Table 13.6
Approximate Percent Conductivites

Material	Percent conductivity	Material	Percent conductivity
Aluminum,	65	Mercury	1.8
hard-drawn	60.5	Monel metal	3.87
Antimony	4.11	Nichrome	1.724
Cadmium	22.7	Palladium	15.97
Cobalt	27.4	Platinum	16.5
Constantan	4.01	Rhodium	36.4
Copper, IACS,	100.0	Silver,	106.4
hard-drawn	97.5	annealed	105.1
Chromium	12.2	Steel, cold-	
Gold	73.4	rolled	17.4
Iridium	32.5	Tin	15.0
Iron	17.75	Titanium	1.95
Lead	8.35	Tungsten	31.4
Magnesium	39.9	Zinc	29.1
		Zirconium	3.84

Example 13.6

What is the resistance of a 2 foot long, .03 inch diameter circular wire which has a resistivity of 11 ohms-CM/ft?

From equation 13.56,

$$A = (.03/.001)^2 = 900\ \text{CM}$$

From equation 13.55,

$$R = \frac{(11)(2)}{900} = .0244\ \text{ohms}$$

There are two situations when the resistance of a substance can be zero. The first case is when the temperature of the substance is absolute zero. The quantum theory predicts that the conductivity of a *pure metal* without lattice imperfections is infinite at absolute zero.

The second case is when a real metal with lattice imperfections exhibits infinite conductivity (zero resistance) at a temperature a few degrees above absolute zero. This latter case is known as *superconductivity*.

Superconductivity is explained with the aid of quantum mechanics. Although the mechanism is not fully understood, it appears that electrons initially set in motion by a voltage source attract other electrons in the metallic lattice. This attractive force exceeds the electrostatic repulsive force. All of the electrons remain in a state of motion because that is the lowest energy state available. The motion of the electrons constitute a current. The current continues as long as the temperature is maintained, even after the voltage source is removed.

Superconductivity does not appear in all metals. And, the temperature at which it appears is different for metals which exhibit the phenomenon. The *transition temperature* below which superconductivity occurs is listed in table 13.7.

Table 13.7
Superconductivity Transition Temperatures

metal	temperature, °K
aluminum	1.2
lead	7.26
mercury	≈4
Nb_3Sn alloy	≈18
niobium	9.22
tantalum	4.39
tin	3.69

6. Power

The power dissipated across two terminals with resistance *R* and voltage drop *V* can be calculated from equation 13.61.

$$P = IV = I^2R = \frac{V^2}{R} \qquad 13.61$$

The *decibel* is a unit used to express power, voltage, and current ratios.

$$db = 10 \log_{10}\left(\frac{P_2}{P_1}\right) \qquad 13.62$$

$$= 20 \log_{10}\left(\frac{V_2}{V_1}\right) \qquad 13.63$$

$$= 20 \log_{10}\left(\frac{I_2}{I_1}\right) \qquad 13.64$$

An alternate use of the decibel is to measure a single power source with reference to some given power level. For example, sound power measurements are made with reference to a source of EE − 12 watts.

$$db = 10 \log_{10}\left(\frac{\text{sound power}}{\text{EE} - 12}\right) \qquad 13.65$$

When the power ratio is less than one, it is common to invert the fraction and express the result as a decibel loss (negative). For cascaded stages in series, the total decibel gain is the sum of the individual stage decibels.

Strictly defined, decibels can be used to express voltage and current ratios only when the two points being measured have the same apparent impedances. This applies to equations 13.63 and 13.64. However, common practice allows the use of these equations when the impedances are not the same.

7. Power Sources

Power to an electrical network can be supplied by two different types of sources. An ideal *voltage source* supplies power at a constant voltage, regardless of the current drawn. True voltage sources cannot maintain a constant voltage when currents become large, as indicated by figure 13.12.

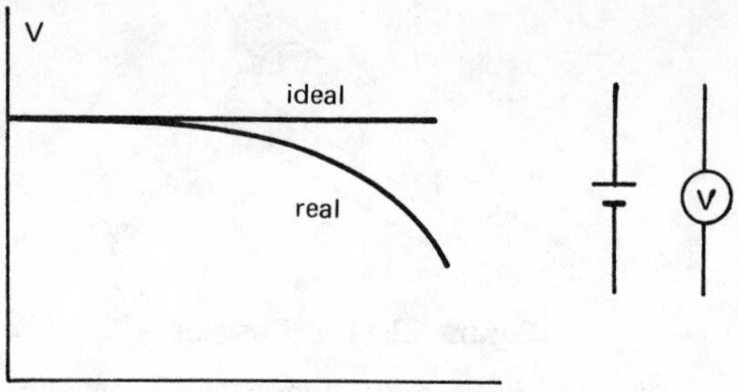

Figure 13.12 Ideal and Real Voltage Sources

The *regulation* of a voltage source is defined by equation 13.66. Regulation may be either positive or negative.

$$\text{Regulation} = \frac{\text{no load voltage} - \text{full load voltage}}{\text{full load voltage}} \qquad 13.66$$

An ideal *current source* supplies constant current regardless of the load attached to it. That is, the current through the source is independent of the voltage between its terminals.

Ideal voltage and current sources are known as *independent sources.* Their outputs are independent of current and voltage respectively. Real sources, however, are known as *dependent sources.*

8. Resistive Network Analysis

In series or single-loop circuits, illustrated by figure 13.13, the following statements are true:

- Current is constant throughout the circuit.
- The total resistance is the sum of the individual resistances.
- The applied voltage is equal to the sum of the voltage drops across the components.

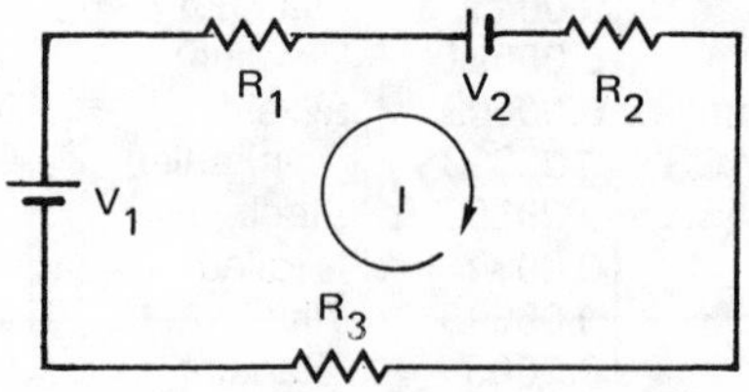

Figure 13.13 Series Circuits

$$\Sigma V_i = RI \qquad 13.67$$

$$R = R_1 + R_2 + R_3 \qquad 13.68$$

$$V_1 + V_2 = (R_1 + R_2 + R_3)I \qquad 13.69$$

In parallel circuits with one active source, illustrated by figure 13.14, the following statements are true:

- The total current is equal to the sum of the branch currents.
- The reciprocal of the total resistance is equal to the sum of reciprocals of the individual resistances.
- The voltage drop is constant across all branches.

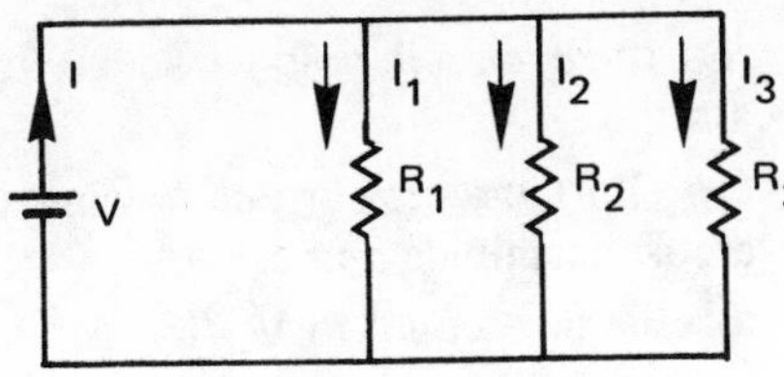

Figure 13.14 Parallel Circuits

$$V = I_1R_1 = I_2R_2 = I_3R_3 \qquad 13.70$$

$$I = I_1 + I_2 + I_3 \qquad 13.71$$

$$\frac{1}{R} = \frac{1}{R_1} + \frac{1}{R_2} + \frac{1}{R_3} \qquad 13.72$$

Several theorems and methods exist for determining voltage and current unknowns in more complex resistive circuits. Typically, such circuits are multi-loop networks with more than one voltage source.

A. *Kirchoff's Laws*

Kirchoff's laws can be used to analyze multi-loop circuits. *Kirchoff's current law* says there is as much current flowing toward a connection (node) as is flowing away. *Kirchoff's voltage law* says that the algebraic sum of voltage drops around any closed path within a circuit equals the sum of the voltage sources.

$$\Sigma I_{in} = \Sigma I_{out} \qquad 13.73$$

$$\Sigma V = \Sigma(IR) \qquad 13.74$$

B. *Loop Current Method*

Kirchoff's laws are incorporated into all network analysis methods. The *Maxwell loop current method* is a direct extension of Kirchoff's voltage law. The *loop method*, which requires writing two simultaneous voltage drop equations for a three-loop system, is illustrated by example 13.7.

step 1: Choose two of the three loops.

step 2: Assume current directions for the chosen loops.

step 3: Write Kirchoff's voltage law for the chosen loops.

step 4: Simultaneously solve for the unknown currents. If a direction was incorrectly chosen, it will have a negative value.

Example 13.7

Find the current in the .5 ohm resistor.

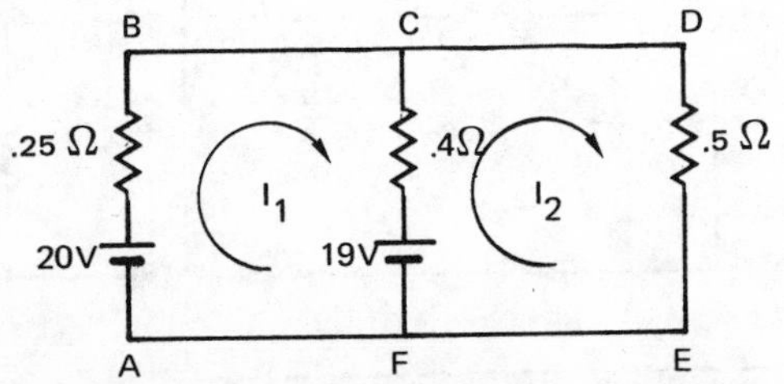

This is a three-loop network — ABCF, FCDE, and ABCDEF. Two simultaneous equations are required. If the first loop is taken as ABCF and the second is FCDE, the assumed currents are I_1 and I_2. The directions chosen are arbitrary.

For the left loop, Kirchoff's voltage law is

$$\Sigma V = \Sigma(IR)$$

$$20 - 19 = .25I_1 + .4(I_1 - I_2)$$

Notice that the battery polarities determine the signs on the left-hand side of the equation. The polarity of the 19 volt battery is counter to the assumed current flow direction. Also, notice that the voltage drops are always positive. It is the current directions which determine the sign.

For the loop FCDE, the voltage drop equation is

$$19 = .4(I_2 - I_1) + .5(I_2)$$

The simultaneous solution to these two equations is

$$I_1 = 20 \text{ amps}$$

$$I_2 = 30 \text{ amps}$$

C. *Node-Voltage Method*

The *nodal (node-voltage) method* incorporates Kirchoff's current law. The nodal method also provides current information, but its primary use is in finding voltage potentials at various points in the circuit.

step 1: Choose one node as the reference voltage node. Usually this will be the circuit ground — a node to which at least one negative battery terminal is connected.

step 2: Label the voltage potential at all other nodes as unknowns.

step 3: Write Kirchoff's current law for all unknown nodes.

step 4: Write all currents in terms of voltage drops.

step 5: Write all voltage drops in terms of the node voltages.

Example 13.8

Find I_1 and voltage potential for node *A*.

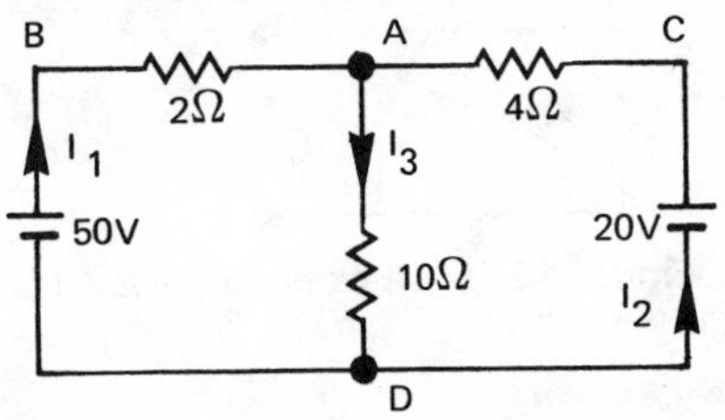

Node *D* is chosen as the reference. Thus, $V_D = 0$. *A* is the only true node since *B* and *C* are not true junctions. Kirchoff's current law is then written in terms of voltage drops.

DC ELEC

$$I_1 + I_2 = I_3$$

$$I_1 = \frac{V_{BA}}{2} \quad I_2 = \frac{V_{CA}}{4} \quad I_3 = \frac{V_{AD}}{10}$$

$$\frac{V_{BA}}{2} + \frac{V_{CA}}{4} = \frac{V_{AD}}{10}$$

V_{BA}, V_{CA}, and V_{AD} are voltage drops. The current law in terms of node voltages is:

$$\frac{(V_B - V_A)}{2} + \frac{(V_C - V_A)}{4} = \frac{(V_A - V_D)}{10}$$

However, $V_D = 0$ (the reference node), $V_B = 50$, and $V_C = 20$, so

$$\frac{(50 - V_A)}{2} + \frac{(20 - V_A)}{4} = \frac{V_A}{10}$$

$$V_A = 35.3 \text{ volts}$$

$$I_1 = \frac{(V_B - V_A)}{2} = \frac{(50 - 35.3)}{2} = 7.35 \text{ amps}$$

D. Superposition

The superposition theorem states that the response of a linear circuit element fed by two or more independent sources is equal to the sum of the responses to each source, with all remaining independent sources set to zero (voltage sources shorted, current sources opened).

E. Thevenin's Theorem

The external behavior of a linear, two-terminal network with dependent and independent sources can be represented by a Thevenin equivalent circuit consisting of a voltage source in series with a resistor.

- V_{ab} the open circuit voltage across terminals *a* and *b*
- I the short circuit current flowing in a shunt across terminals *a* and *b*
- R a resistance equal to V_{ab}/I

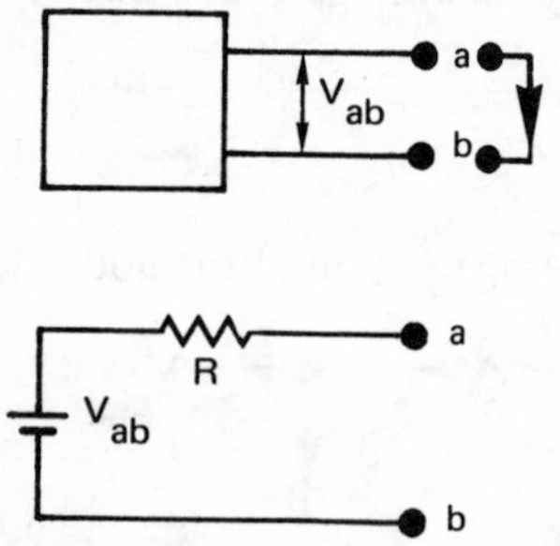

Figure 13.15 Thevenin Equivalent

F. Norton's Theorem

The external behavior of a linear, two-terminal network with dependent and independent sources can be represented by a Norton equivalent circuit consisting of a current source and a resistor in parallel.

- V_{ab} the open circuit voltage across terminals *a* and *b*
- I the short circuit current flowing in a shunt across terminals *a* and *b*
- R a resistance equal to V_{ab}/I

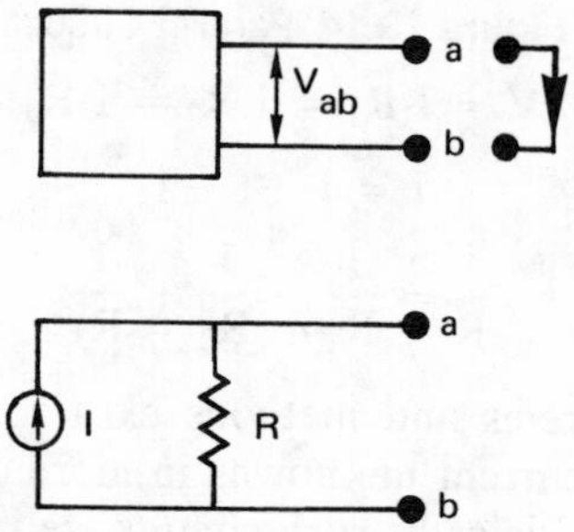

Figure 13.16 Norton Equivalent

Example 13.9

Find I_2 in example 13.8 using Thevenin's theorem.

The circuit is redrawn to isolate the load resistor through which I_2 flows.

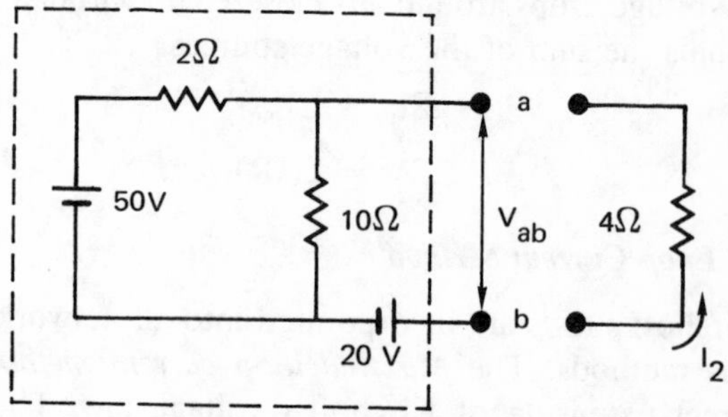

The voltage across terminals *a* and *b* is dependent on both batteries. The 20 volt battery appears totally across the 10 ohm resistor. The 50 volt battery appears across the 2 ohm and 10 ohm resistors.

$$V_{ab} = 50\left(\frac{10}{10 + 2}\right) - 20 = 21.67$$

If terminals *a* and *b* are shorted, the short circuit current can be found by superposition to be 13 amps. The Thevenin resistance is

$$R = \frac{21.67}{13} = 1.67$$

The Thevenin equivalent circuit is

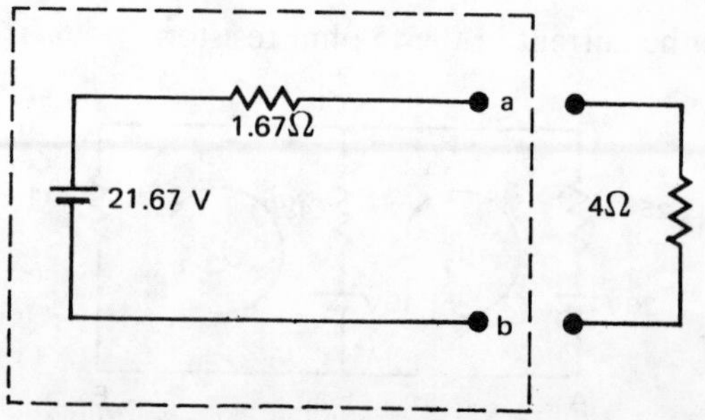

The current is

$$I_2 = \frac{21.67}{4 + 1.67} = 3.82 \text{ amps}$$

G. Reciprocity Theorem

In a network of linear resistances, the ratio of applied voltage to current measured at any point is identical to the ratio obtained if the source and the meter are exchanged. This ratio is known as the *transfer impedance.*

$$Z_{transfer} = \frac{V_1}{I_1} = \frac{V_2}{I_2} \qquad 13.75$$

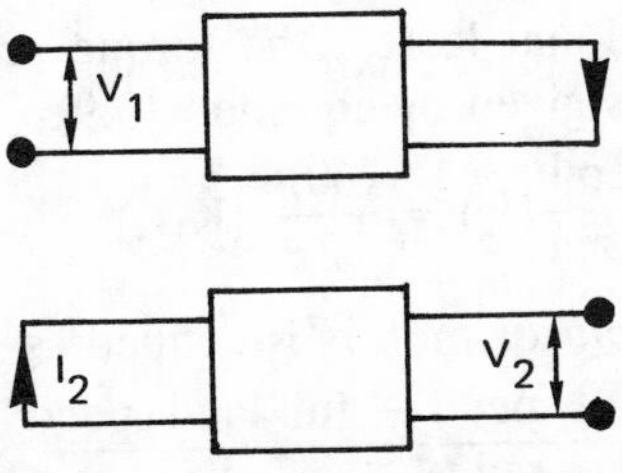

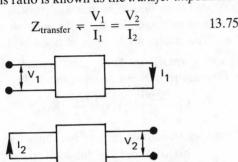

Figure 13.17 Reciprocal Measurements

H. Delta/Wye Conversions

The equivalent circuits of three-phase networks as viewed from the terminals can be found from the following relationships.

Delta to Wye:

$$R^* = R_a + R_b + R_c \qquad 13.76$$

$$R_1 = \frac{R_a R_c}{R^*} \qquad 13.77$$

$$R_2 = \frac{R_a R_b}{R^*} \qquad 13.78$$

$$R_3 = \frac{R_b R_c}{R^*} \qquad 13.79$$

Wye to Delta:

$$R^* = R_1R_2 + R_1R_3 + R_2R_3 \qquad 13.80$$

$$R_a = \frac{R^*}{R_3} \qquad 13.81$$

$$R_b = \frac{R^*}{R_1} \qquad 13.82$$

$$R_c = \frac{R^*}{R_2} \qquad 13.83$$

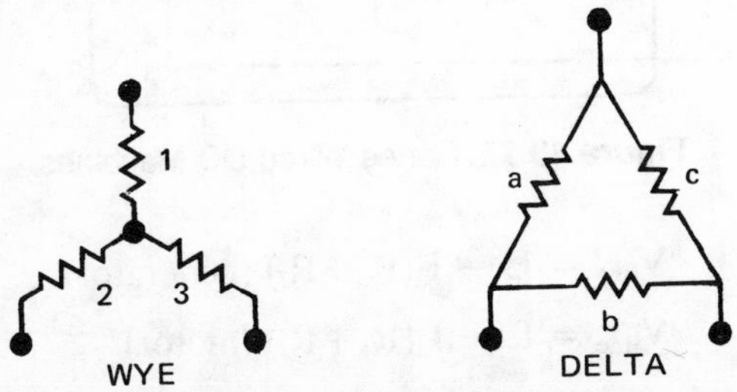

Figure 13.18 Wye and Delta Networks

9. Basic Meter Circuits

The *d'Arsonval movement* is essentially a current measuring device. The current in a circuit flows through a coil, creating a magnetic field. This field interacts with the field from a permanent magnet, producing a torque which moves the indicator needle. The meter can be represented by the simple equivalent circuit shown in figure 13.19a.

Since the meter movement is limited, a current, I_c^*, exists which will cause maximum needle movement. To measure a greater current, a *shunt resistor* (figure 13.19b) must be used. If I_{max} is to cause full-scale movement, then the shunt resistance is given by equation 13.84.

$$R_s = \frac{I_c^* R_c}{I_{max} - I_c^*} \qquad 13.84$$

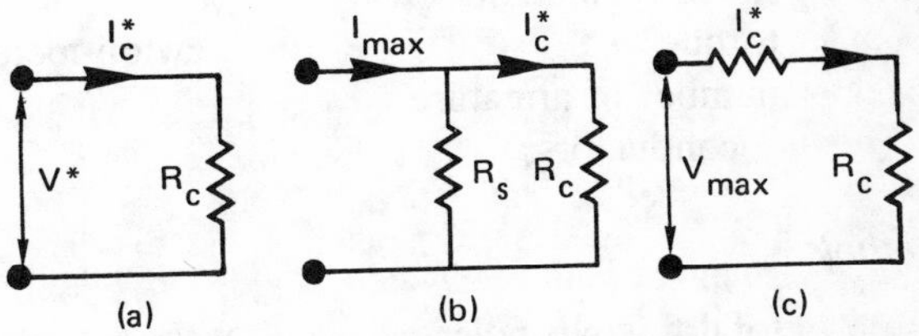

Figure 13.19 Equivalent Meter Circuits

The meter reading must be multiplied by $\left(\frac{R_s + R_c}{R_c}\right)$ to obtain the actual current being measured when a shunt resistor is used.

If the meter is to be used to measure a voltage greater than V* (which causes maximum needle movement), then a *multiplier resistor* will be required, as in figure 13.19c. If V_{max} is to produce full-scale movement, then the multiplier resistor must be

$$R_m = \frac{V_{max} - I_c^* R_c}{I_c^*} \qquad 13.85$$

The voltage reading must be multiplied by $\left(\frac{R_m + R_c}{R_c}\right)$ to obtain the actual applied voltage.

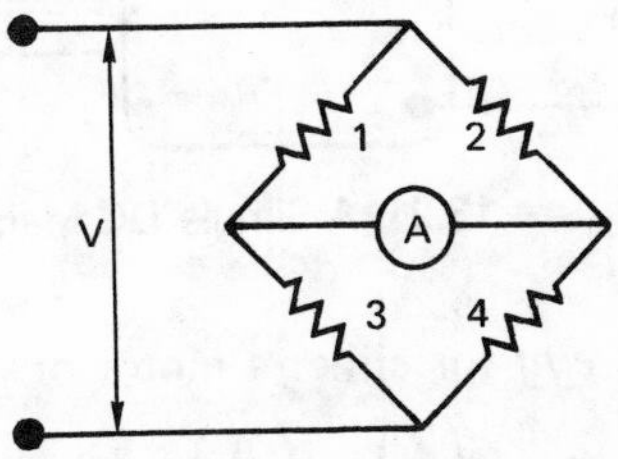

Figure 13.20 Wheatstone Bridge

A zero-indicating *Wheatstone bridge* (figure 13.20) can be evaluated with the following relationships. The reciprocity theorem can be used if the voltage source and ammeter are reversed.

$$I_2 = I_4 \quad 13.86$$

$$I_1 = I_3 \quad 13.87$$

$$V_1 + V_3 = V_2 + V_4 \quad 13.88$$

10. DC Machinery

Special Nomenclature

a	number of parallel armature paths	–
E	induced *emf*	volts
I	current	amps
K	a constant	–
n	speed	rpm
p	number of poles	–
R	resistance	ohms
SR	speed regulation	%
T	torque	newton-meters
Z	number of armature conductors	–

Symbols

Φ	total flux per pole	webers

Subscripts

a	armature
f	field

A. Introduction

A DC machine (generator or motor) is shown in figure 13.21. When the shaft turns, a voltage known as *induced emf* will be produced in the armature. For a generator, this induced *emf* is the desired product of the work input. For a motor, the induced emf opposes the input current and is called *back emf.*

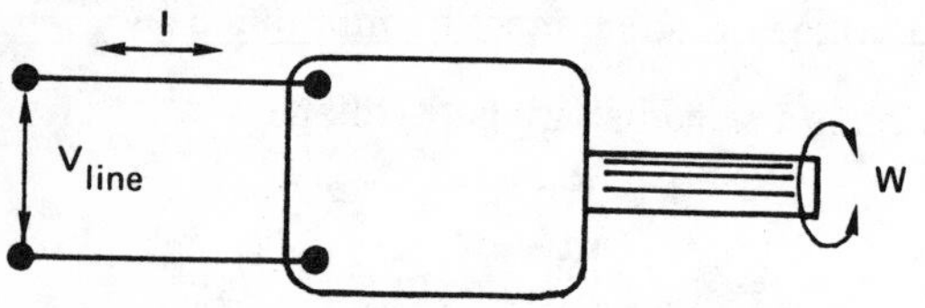

Figure 13.21 A Simple DC Machine

The induced *emf* for either a motor or generator is

$$E = \left(\frac{Z}{a}\right) p\Phi \left(\frac{n}{60}\right) = K_1 \Phi n \quad 13.89$$

$$K_1 = \frac{Zp}{(60)a} \quad 13.90$$

Flux is frequently given in units other than webers. Table 13.8 can be used to convert other units to webers.

Table 13.8
Flux Conversions

multiply	by	to obtain
lines	EE−8	webers
maxwells	EE−8	webers
gauss	(pole face area) EE−8	webers
tesla	(pole face area)	webers

If the machine is *simplex lap wound,* then *p* and *a* are equal. *p* and *a* are numerically equal to the number of brushes. If the machine has a *wave-wound* armature, then *a* is two and the minimum number of brushes is two. The minimum number of poles is also two.

Ignoring rotational losses, the torque produced by a DC machine is given by equation 13.91.

$$T = \frac{Zp\Phi}{2\pi a}(I_a) = \frac{(60)\Phi}{2\pi}(K_1 I_a) = K_2 \Phi I_a \quad 13.91$$

Speed regulation for motors is defined as

$$SR = \frac{\text{no load speed} - \text{full load speed}}{\text{full load speed}} \times 100 \quad 13.92$$

B. Series Wired Machines

A series wired motor and generator are shown in figure 13.22. The only components in the equivalent electrical circuit are the field resistance and the armature resistance. A brush resistance may also be specified, but it is usually combined with the armature resistance.

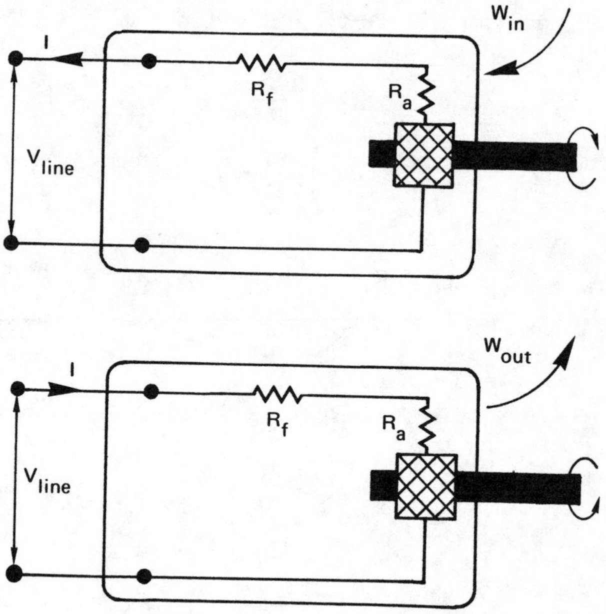

Figure 13.22 Series Wired DC Machines

$$V_{line} = E - I_a(R_a + R_f) \text{ (generator)} \quad 13.93$$

$$V_{line} = E + I_a(R_a + R_f) \text{ (motor)} \quad 13.94$$

In a series motor, a load (torque) reduction will cause

a decrease in armature current. However, since the field and armature currents are identical, the flux will also be reduced. The speed will increase to maintain equation 13.94. Therefore, a series motor is not a constant speed device. A load should never be removed from a DC motor. Gears, not belts, should be used to connect DC motors to their loads.

In a series motor, *E* is initially zero when the motor starts from rest. Therefore, I_a will be dangerously high to keep equation 13.94 valid. Thus, reduced voltages are required when starting, and R_f is often a rheostat or switchable resistor bank. At higher speeds, the back *emf* counteracts the applied voltage. *Stall speed* is reached when equation 13.95 is valid.

$$E = I_a(R_a + R_f) = \tfrac{1}{2}V_{line} \qquad 13.95$$

C. Shunt Wired Machines

Shunt fields are used to produce an essentially constant speed motor since the field current is constant. A shunt DC generator is also known as a *self-excited machine* since the armature-produced *emf* causes the field current to flow.

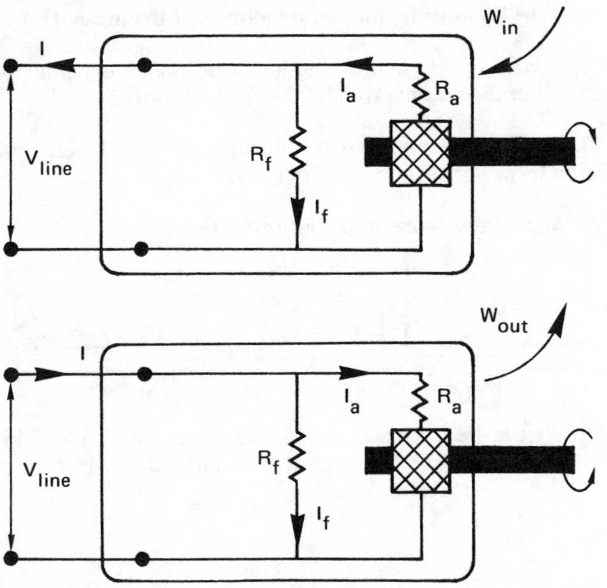

Figure 13.23 Shunt Wired DC Machines

$$V_{line} = E - I_aR_a = I_fR_f \text{ (generator)} \qquad 13.96$$

$$I = I_a - I_f \quad \text{(generator)} \qquad 13.97$$

$$V_{line} = E + I_aR_a = I_fR_f \text{ (motor)} \qquad 13.98$$

$$I = I_f + I_a \quad \text{(motor)} \qquad 13.99$$

In both the series- and shunt-wired machines, care must be used in the analysis of problems giving *loaded* and *no-load* data. For a generator, the term *no-load* implies $I = 0$. For a motor, the term *no-load* implies $W = 0$, and consequently, I_a must be nearly zero.

Example 13.10

A DC generator with two poles runs at 1800 rpm. There are 100 conductors between the brushes. The average magnetic flux density in the air gap between the pole faces and the armature is 1.2 T. The pole faces have an area of .03 m². What is the no-load terminal voltage?

The flux per pole is

$$\Phi = \phi A = (1.2)(.03) = .036 \text{ Wb}$$

From equation 13.89, assuming $a = 1$,

$$E = \left(\frac{100}{1}\right)(2)(.036)\left(\frac{1800}{60}\right)$$

$$= 216 \text{ V}$$

DC ELEC

ENERGY CONVERSIONS

1. How much energy is required to operate an 80% efficient 5 horsepower motor used 1 hour every other day and a 500 watt lamp used for 4 hours a day, both over a 30-day month?

2. A pump supplies 2500 gpm of water against a 120 foot head. What electric power is required if the pump is 70% efficient?

3. An electric lift truck is powered by a DC motor connected to a 110 volt battery. The truck exerts a 200 pound pull while moving at 7 mph. If the overall efficiency of the motor and drive mechanism is 75%, what is the current drain on the battery?

4. An 800 watt electric iron operates on 110 volts. How many electrons pass through the iron per second?

ELECTROSTATICS

Assume all point charges are in a vacuum unless the problem states otherwise. All distances are in meters.

5. A -1 uC point charge is located at (0,0). A -2 uC point charge is located at (0,1). What is the magnitude and direction of the force between them? What is the potential of the -2 uC charge?

6. Charge one of +4 uC is located at (2,0,0). Charge two of -5 uC is located at (0,2,0). What is the force between them? What is the potential at a point with coordinates (3,5,2)?

7. What is the electric field intensity at (2,0) due to a ½uC point charge located at (0,0)?

8. What is the electric field intensity at (0,3) due to a +2 uC charge at (0,0) and a -5uC charge at (4,0)?

9. What is the force on a +5 EE-11 coulomb charge that is 5 cm from each of two +2 EE-9 coulomb charges that are separated by 8 cm?

UNIFORM ELECTRIC FIELDS

10. What is the electric field strength between 2 plates separated by .005 meter which are connected across 100 volts? What energy density is stored in the electric field?

11. A dielectric (ϵ_r = 3) .005 meter thick separates 2 oppositely charged plates. An electric field of EE6 volts/meter exists between the plates. What is the applied voltage?

12. A vacuum tube consists of a glass cylinder .5 meter long with electrodes at each end connected across 1200 volts. What is the force on and potential of an electron adjacent to the negative terminal?

VACUUM TUBES

13. A source at zero potential emits electrons with negligible velocity. An open grid at +18V is located .003 meter from the source. At what velocity will the electrons pass through the grid?

14. In problem 13, what will be the velocity of electrons striking a plate at +15V potential located .012 meter from the grid?

15. An electron with velocity of 2 EE7 meter/second enters a 4 cm long deflecting field in a cathode ray tube. The field strength is 20,000 newtons/coulomb. A fluorescent screen is located 12 cm from the edge of the deflecting field. What will be the deflection ahgle? What will be the deflection?

MAGNETISM

16. What is the flux density (in tesla) in a .02 square meter area if the perpendicular flux is 34 mwb?

17. An armco iron magnet (.01 meter diameter, .1 meter long) has an internal magnetic field of 1.2 wb/m^2. What is the pole strength? What is the magnetic moment of the magnet?

18. A magnet with pole strength of 4 unit-poles experiences a force of 3 newtons. What is the field intensity?

19. Two 8 unit-pole magnets are separated by .10 meter. What is the force between the two magnets?

20. What is the force on the left-most pole?

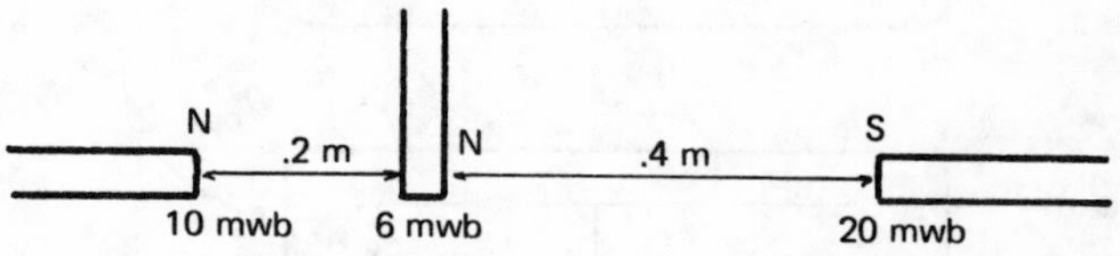

21. What is the force and its direction, and what is the field intensity and its direction, on the 15 unit-pole magnet?

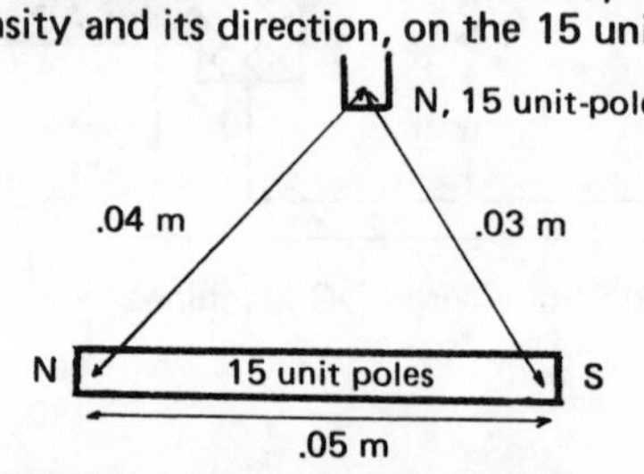

ELECTROMAGNETIC INDUCTION

22. A .3 meter wire is moved through a uniform orthogonal magnetic field (B = 1.1 wb/m^2). What force is required if a 6 amp current is induced?

23. When the current in a wire is 5 amps, the flux 3 meters from the wire is 30 mwb. If the current is decreased to 4 amps, what will be the flux?

24. A coil of 100 turns is cut by a magnetic field which increases at the rate of 40 webers/minute. What is the induced voltage?

25. What voltage is induced in a .10 meter long straight conductor traveling at 1 meter/second through an orthogonal 4 wb/m^2 magnetic field?

MAGNETIC CIRCUITS

26. What is the magnetomotive force of a 400 turn coil carrying .9 amps?

27. A flux of 3.6 EE-4 webers is to be established in the center leg of the medium silicon steel core (μ = 8400) shown below. If the coil has 300 turns, what current is required? All dimensions are in centimeters.

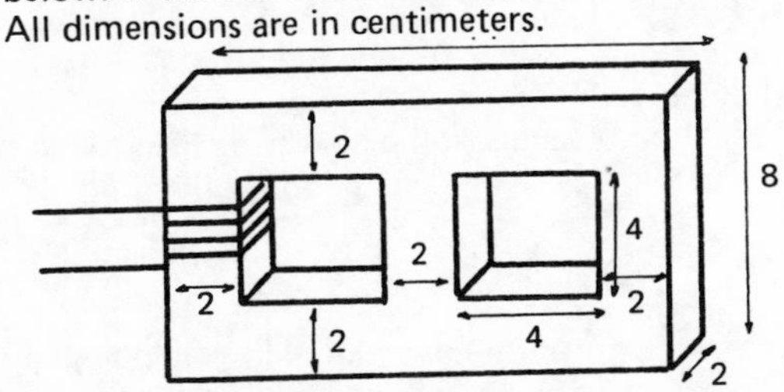

28. How many turns are required to support a 600 kiloline flux across the air gap? A 5 amp current is to be used. Assume the permeability is a constant 800. All dimensions are in centimeters.

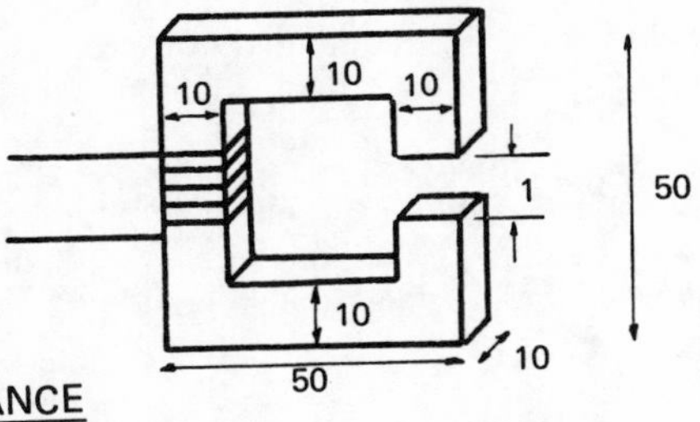

RESISTANCE

29. A standard copper wire is .064" in diameter. What is the resistance of 500 feet of wire
(a) at 20°C?
(b) at 80°C?

30. The resistance of a wire is 400 ohms at 200°C and 30 ohms at 100°C. What is the average value of α?

31. A 240V motor requiring 2000 watts is located 1 km from a power source. What minimum copper wire diameter is to be used if the power loss is to be kept less than 5%?

32. What is the inferred zero resistance temperature of a resistor with resistance of 11 ohms at 5°C and 12 ohms at 31°C?

POWER

33. What power is dissipated when a ½ amp current flows through a 5 ohm resistor?

34. What is the power loss in decibels of a 400 watt signal is received as 100 mW?

35. What power is supplied by each of the voltage sources?

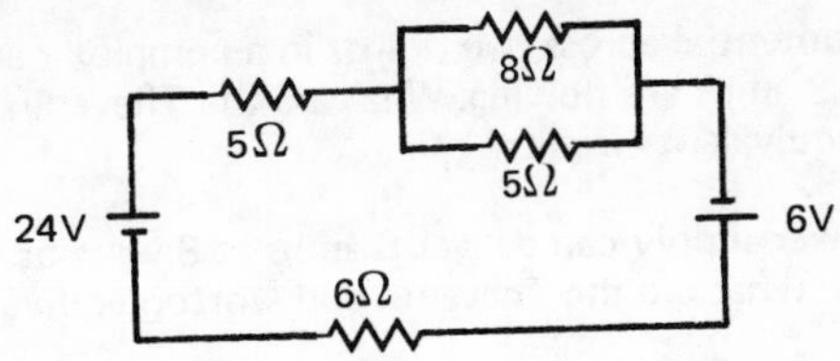

36. A 100 volt DC generator with a 1 ohm internal resistance is connected through a 5 ohm rheostat to an opposing 66 volt battery with 2 ohms internal resistance. What is the power dissipated in the circuit? What energy is stored during one hour of charging?

37. Three cascaded amplifier states have amplifications of 100 db, 25 db, and 9 db. What is the overall amplification?

38. What is the cost (at $.05/kw-hr) of operating a 70% efficient, 20 horsepower pump for 20 days, 8 hours per day?

NETWORK ANALYSIS

39. A battery consists of 6 cells, each having an EMF of 1.1 volts and resistance of 4 ohms, connected in two parallel rows of 3 cells in series. This battery is connected to a group of two resistances in parallel of 30 ohms and 6 ohms respectively. What is the current?

40. What current is flowing in the center leg?

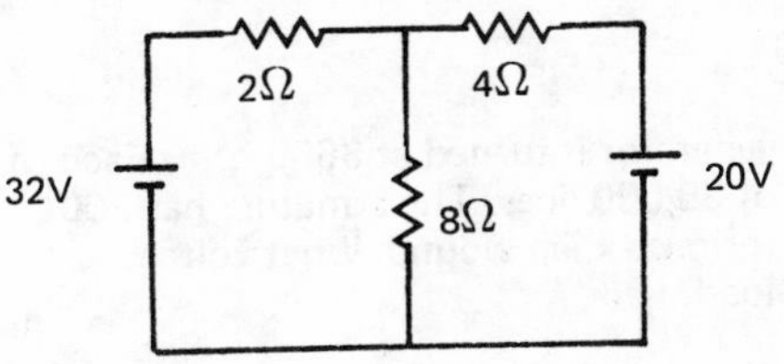

41. What current is flowing in the 6 ohm resistor?

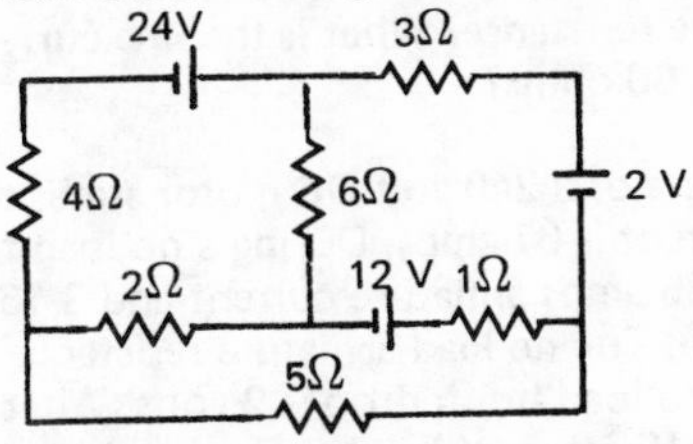

42. What current is flowing in the 4 ohm resistor?

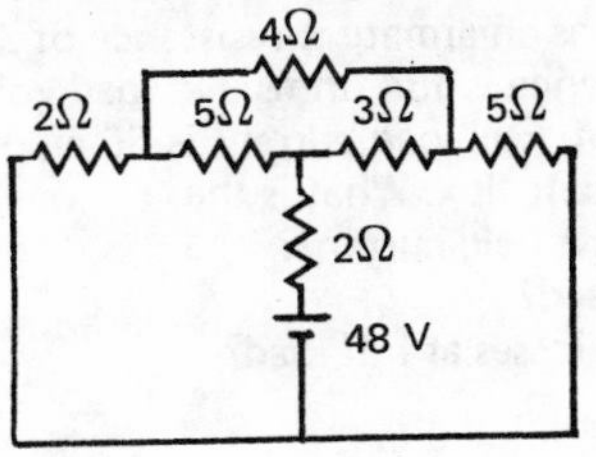

DC ELEC

EQUIVALENT CIRCUITS

43. The potential across two points in a complex network is 50 volts. 8 amps are flowing. What are the Thevenin and Norton equivalent circuits?

44. A power supply can deliver 5 amps at 8 volts or 10 amps at 6 volts. What are the Thevenin and Norton equivalent circuits?

45. A delta resistor network has $R_a = 30$ ohms, $R_b = 10$ ohms, and $R_c = 20$ ohms. What is the equivalent wye network?

METER CIRCUITS

46. An electrical milliammeter having an internal resistance of 500 ohms reads full scale when one milliamp flows. If the meter is to be used as an ammeter reading .1 amp full scale, what size resistor should be used in parallel?

47. If the meter described in problem 46 is to be used as a voltmeter reading 100 volts full scale, what size resistor should be used in series?

48. A DC 0-50 millivoltmeter is used to measure 5 amps current by means of a shunt. The meter was tested and found to have a terminal resistance of 4.17 ohms. How many ohms resistance must be connected in parallel with the meter so that it will indicate a full scale deflection at 5 amps? How many amps will pass through the shunt resistor at full scale? By what quantity must the meter reading be multiplied to get the true current?

DC MACHINES

49. A four-pole generator is turned at 3600 rpm. Each of its poles has a flux of 30,000 lines. The armature has 200 conductors and is simplex lap wound. What voltage is generated at no load?

50. What is the back EMF of a 10 horsepower shunt motor operating on 110 volt terminals if 90 amps flow through a .05 ohm armature resistance? What is the line current if the field resistance is 60 ohms?

51. The nameplate of a 240 volt DC motor states that the full load line current is 67 amps. During a no load test at rated voltage, 3.35 amps armature current and 3.16 field current are drawn. The no load armature resistance is .207 ohms. The no load brush drop is 2 volts. What are the horsepower and efficiency at full load?

52. A DC generator has an armature resistance of .31 ohms. The shunt field resistance is 134 ohms. No load voltage is 121 volts at 1775 rpm. Full load current is 30 amps at 110 volts. Assume a constant flux. What is the

(a) no load shunt field current?
(b) full load speed?
(c) stray power losses at full load?

14 Alternating Current Electricity

Nomenclature

a	turns ratio	–
A	area	$meter^2$
B	magnetic flux density, or susceptance	$webers/meter^2$, mhos
C	capacitance	farads
CF	crest factor	–
E	back EMF	volts
f	frequency	hertz
FF	form factor	–
G	conductance	mhos
I	current	amps
k	coefficient of coupling	–
K	proportionality constant	various
L	inductance	henrys
M	mutual inductance	henrys
n	rotational speed	rpm
N	number of coils or turns	–
p	number of poles	–
P	power	watts
q	charge	coulombs
Q	quality factor	–
r	vector length	various
R	resistance	ohms
s	slip	–
S	sinusoidal variable	–
t	time	seconds
T	period, or torque	seconds, newton-meter
V	voltage	volts
X	reactance	ohms
Y	admittance	mhos
Z	impedance	ohms

Symbols

θ	angle	degrees
φ	magnetic flux, or phase difference angle	lines, degrees
δ	torque angle	degrees
λ	flux linkage	lines
ω	angular velocity	radians/sec
ρ	energy	joules
ε	permittivity	$coul^2/N\text{-}m^2$

Subscripts

*	resonant
ave	average
C	capacitive
eff	effective
f	forced
L	inductive, or load
m	maximum
n	natural
o	original
p	primary
r	rotor, or ratio
rms	effective
s	secondary, synchronous, or stator
t	terminal
v	vacuum

1. Production of Alternating Potential

A potential of alternating polarity is the natural product of an alternator (AC generator). Unlike a DC generator, an alternator does not use slip rings or other methods to rectify the produced potential. Figure 14.1 shows how several coils of wire can be rotated in a *coil dynamo* to produce a continuously varying potential.

Assuming the magnetic field is uniform, the maximum flux linked by a coil of area *A* will be

$$\lambda_{max} = NAB \qquad 14.1$$

Since the coil is rotating, the *flux linkage* is a function of the projected coil area.

$$\lambda(t) = NAB\cos\omega t = \lambda_{max}\cos\omega t \qquad 14.2$$

Faraday's law, (equation 14.3) predicts the induced voltage at any instant. Only the coil sides cut the mag-

AC ELEC

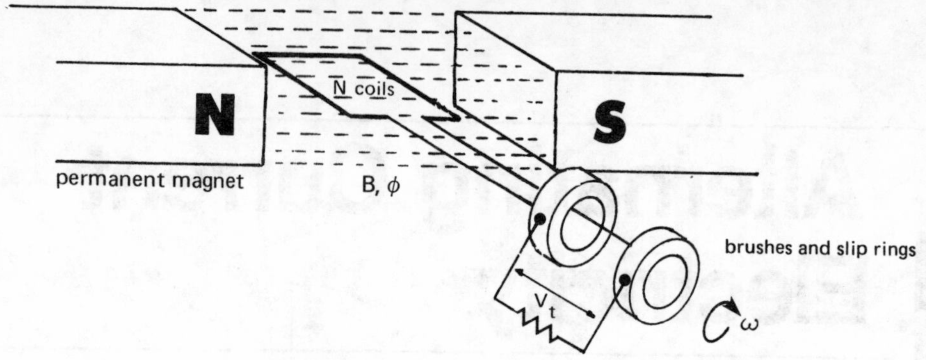

Figure 14.1 Generation of an AC Potential

netic field. The coil ends do not contribute to the generation of potential.

$$V(t) = -\frac{d\lambda}{dt} = \lambda_{max}\omega\ \sin\omega t = V_{max}\sin\omega t \qquad 14.3$$

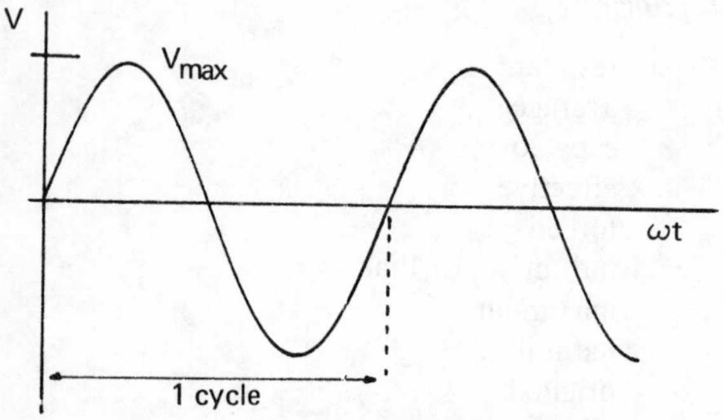

Figure 14.2 An Alternating Potential

The two-pole alternator produces one complete potential cycle per revolution. A *p*-pole machine produces ½p cycles per revolution. Since the armature turns at a constant speed, *n*, the generated *frequency* is

$$f = \left(\frac{n}{60}\right)\left(\frac{p}{2}\right) \qquad 14.4$$

The *period* of the waveform is the reciprocal of the frequency.

$$T = \frac{1}{f} \qquad 14.5$$

The linear frequency, *f*, is related to the angular frequency, ω, by equation 14.6. The angular frequency of the potential should not be confused with the angular frequency of the armature.

$$\omega_{potential} = 2\pi f \qquad 14.6$$

Example 14.1

At what speed must a four-pole alternator be turned to produce a 60 hz alternating potential? What is the angular velocity of the armature? What is the angular velocity of the potential?

From equation 14.4,

$$n = \frac{(120)(60)}{4} = 1800 \text{ rpm}$$

The angular velocity of the armature is

$$\omega_{armature} = \frac{2\pi n}{60} = 2\pi(1800/60)$$
$$= 188.5 \text{ radians/sec}$$

From equation 14.6,

$$\omega_{potential} = 2\pi f = 2\pi(60)$$
$$= 377 \text{ radians/sec}$$

The *torque* opposing rotation, which must be overcome by the mechanical power source, is

$$T = NBAI\ \cos\omega t \qquad 14.7$$

An improvement on the coil dynamo of figure 14.1 is shown in figure 14.3(a). The field is produced by an electromagnet, not by permanent magnets. There are also numerous coils in the continuous armature winding. This design, however, has several deficiencies.

- The armature coils must be well insulated to prevent shorting.
- Structural bracing must be used to counteract the centrifugal force. This results in a large rotational moment of inertia.
- It is difficult to make efficient high-voltage connections through slip rings.

Practical alternators reverse the location of the field and induction coils, as shown in figure 14.3(b). The DC magnetization current is applied to the field windings through brushes attached to slip rings. The armature with its field coils is known as a *rotor*.

A voltage is induced in series-connected conductors placed in slots a small distance from the rotor. These conductors are known as the *stator*. The close proximity of the rotor and stator minimizes the magnetization current needed in the rotor.

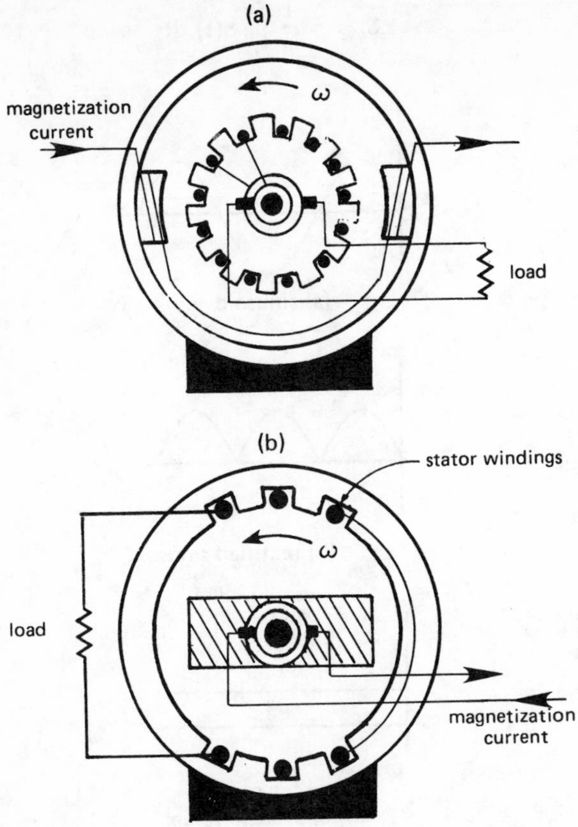

Figure 14.3 Improved Alternator Designs

An ideal source of AC potential is represented in circuits by any one of the symbols shown in figure 14.4.

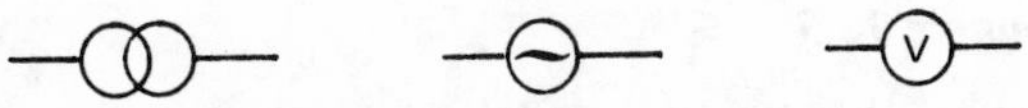

Figure 14.4 AC Sources

2. General Sinusoids

The term *alternating waveform* can be used to describe any symmetrical waveform of time-varying polarity. Included in the category of alternating waveforms are square waves, sawtooth waves, and triangular waves, as well as sinusoidal waves. Nevertheless, the great majority of problems encountered involve only sinusoidal waves. Only sinusoidal waves will be considered in the remainder of this chapter.

Instantaneous values of sinusoidal variables may be specified by one of two basic trigonometric relationships. This is illustrated in figure 14.5. The choice of sine-based or cosine-based calculations is a matter of preference. However, it is imperative that the same choice be used throughout a problem.

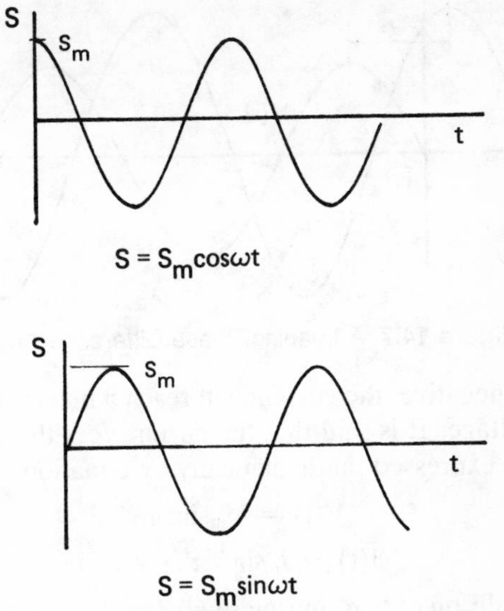

Figure 14.5 Sine and Cosine Sinusoids

The sine form of sinusoidal variables will be used and implied throughout this chapter. If it is necessary to convert from or to the sine form, equation 14.8 or 14.9 may be used.

$$\cos\omega t = \sin(\omega t + \tfrac{1}{2}\pi) = -\sin(\omega t - \tfrac{1}{2}\pi) \quad 14.8$$

$$\sin\omega t = \cos(\omega t - \tfrac{1}{2}\pi) = -\cos(\omega t + \tfrac{1}{2}\pi) \quad 14.9$$

If the value of S(t) is not zero at $t=0$, a *phase angle correction* must be introduced.

$$S(t) = S_m \sin(\omega t + \theta) \quad 14.10$$

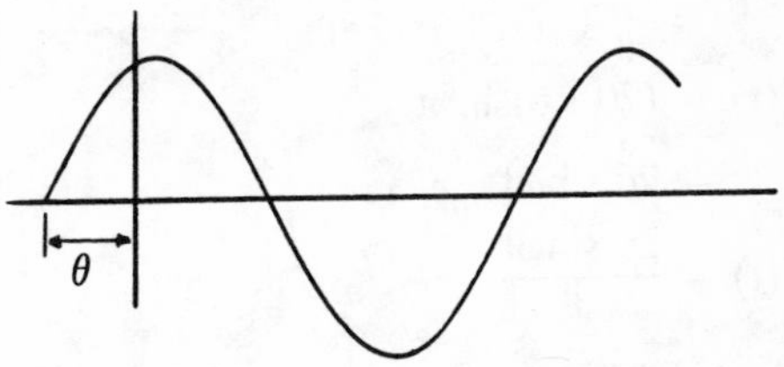

Figure 14.6 Phase Angle Correction

The phase angle correction is not the same as the *phase difference angle* between two sinusoids. A simultaneous plot is often used to compare current and voltage. One of the sinusoids (usually voltage) is chosen as the reference and is given a zero phase angle. The difference in phases between the two sinusoids is called the phase difference angle.

If ϕ is positive, the current will reach a maximum before the reference voltage. The current is said to *lead* the voltage. This is expressed mathematically by equation 14.12.

$$V(t) = V_m \sin\omega t \quad 14.11$$

$$I(t) = I_m \sin(\omega t + \phi) \quad 14.12$$

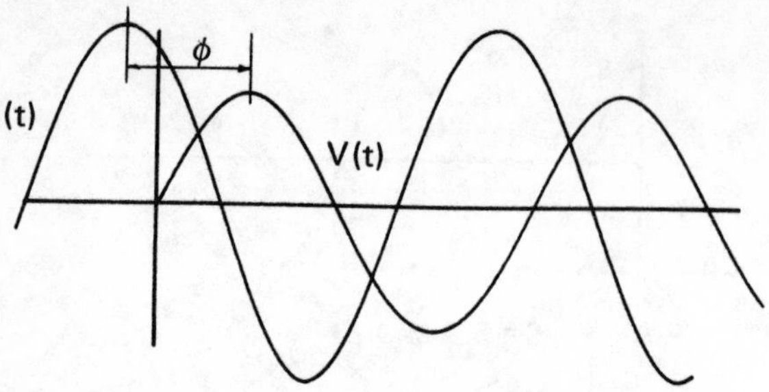

Figure 14.7 A Leading Phase Difference Angle

If ϕ is negative, the current will reach a maximum after the voltage. It is said that the current *lags* the voltage. This is expressed mathematically by equation 14.14.

$$V(t) = V_m \sin\omega t \qquad 14.13$$

$$I(t) = I_m \sin(\omega t - \phi) \qquad 14.14$$

The addition of two sinusoids with identical frequencies and phase angles can be done with equation 14.15.

$$S_1(t) + S_2(t) = (S_{m,1} + S_{m,2})\sin\omega t \qquad 14.15$$

Example 14.2

What is the current in the resistor?

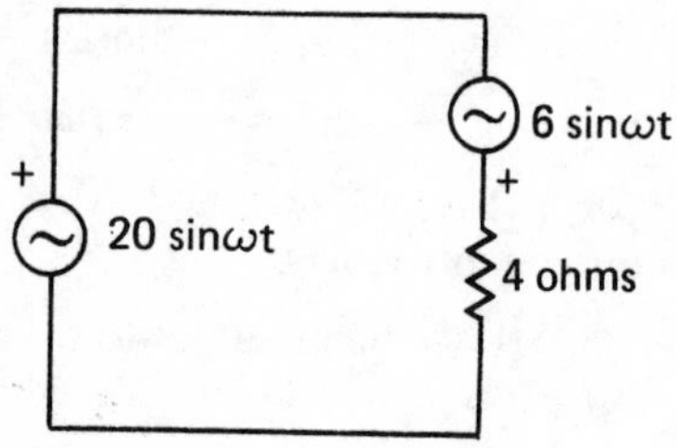

$$V_{total}(t) = (20+6)\sin\omega t$$

$$= 26 \sin\omega t$$

$$I(t) = \frac{26 \sin\omega t}{4}$$

$$= 6.5 \sin\omega t$$

3. Forms and Factors

The average and effective values of a sinusoid can be derived from the trigonometric sinusoidal equations. The *average value* of any periodic variable is

$$S_{ave} = \frac{1}{2\pi}\int_0^{2\pi} S(t)\,dt \qquad 14.16$$

For waveforms that are symmetric about the *x* axis, equation 14.16 is zero. Therefore, the average is taken over only half a cycle. This is equivalent to taking the average over the waveform after it has been rectified. A rectified AC current with maximum I_m is equivalent to a DC current of I_{ave} in terms of electrolytic action (e.g., capacitor charging, plating, and ion formation).

$$S_{ave} = \frac{1}{\pi}\int_0^{\pi} S(t)\,dt \qquad 14.17$$

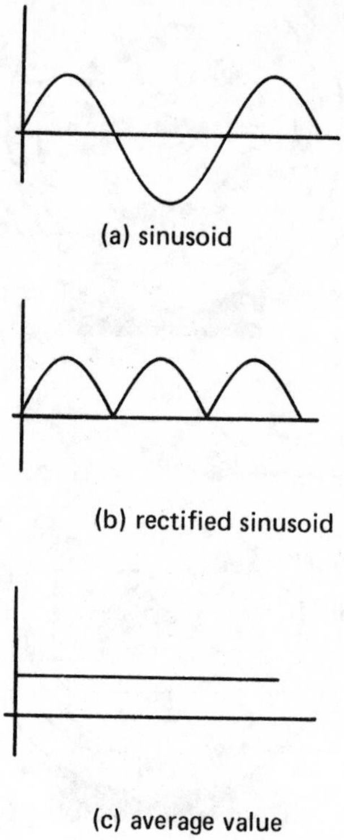

Figure 14.8 The Average of a Rectified Sinusoid

If S(t) is a sinusoid, the average value is

$$S_{ave} = \frac{2S_m}{\pi} \qquad 14.18$$

Example 14.3

A plating tank with an effective fluid resistance of 100 ohms is connected to a full-wave diode rectifier. The plating requires .005 faradays. How much time should the plating take?

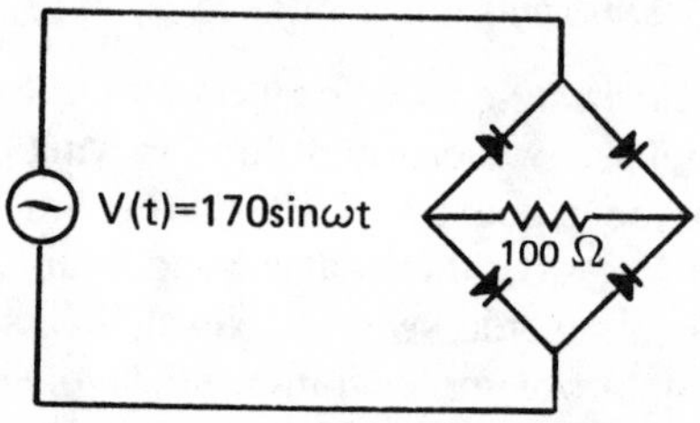

The average value of the rectified sinusoid is

$$V_{ave} = \frac{2(170)}{\pi} = 108.2 \text{ volts}$$

The average current will be

$$I = \frac{V}{R} = \frac{108.2}{100} = 1.082 \text{ amps}$$

Since one faraday is equal to 96,500 amp-seconds, the plating time is

$$t = \frac{\#\ \text{faradays}}{I} = \frac{(96{,}500)(.005)}{1.082}$$

$$= 445.9 \text{ seconds}$$

The *effective value* of a periodic function is given by equation 14.19. Because of the structure of the formula, this value is also known as the *root-mean-square value* or *rms value*.

$$S_{rms} = \sqrt{\frac{1}{2\pi}\int_0^{2\pi} S^2(t)\,dt} \qquad 14.19$$

An AC current with maximum I_m is equivalent to a DC current I_{rms} in terms of power dissipation and heat generation.

If the periodic function is a sinusoid, then the effective value is

$$S_{eff} = \frac{S_m}{\sqrt{2}} \qquad 14.20$$

The *form factor* is defined as

$$FF = \frac{S_{eff}}{S_{ave}} \qquad 14.21$$

The *crest factor* (also known as the *peak factor* and *amplitude factor*) is defined as

$$CF = \frac{S_m}{S_{eff}} \qquad 14.22$$

The average value, effective value, form factor, and crest factor for various common waveforms are summarized in table 14.1.

Example 14.4

What power is dissipated in the resistor?

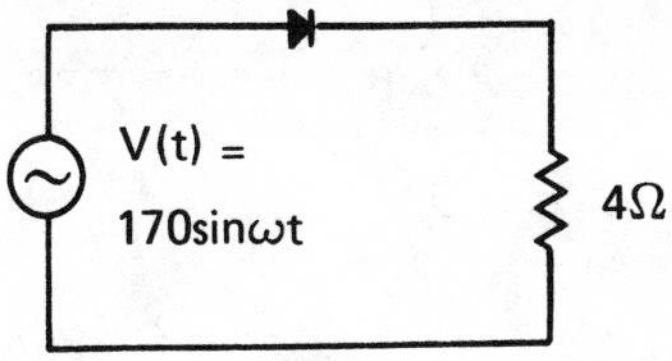

From equation 14.20, the effective voltage is $170/\sqrt{2} = 120$ volts. Since the diode will pass only one half of the sinusoidal cycle, the power is

$$P = \tfrac{1}{2}\left(\frac{V^2}{R}\right) = \frac{(.5)(120)^2}{4}$$

$$= 1800 \text{ watts}$$

4. AC Circuit Elements

Unlike DC circuits which usually contain only resistors, AC circuits can contain three different circuit elements. These are the resistor, inductor, and capacitor. The term *impedance* (with units of ohms) is used to describe the current-modifying characteristics of AC circuit elements. Impedance is given the symbol **Z**. It is defined as the quantity which, when divided into the voltage, gives the current.

AC circuit elements have the ability to cause a phase angle difference between the applied voltage and the resulting line current. That is, the current and voltage will not ordinarily reach maximum simultaneously.

Table 14.1 Forms and Factors

Waveform	S_{ave}	S_{rms}	FF	CF
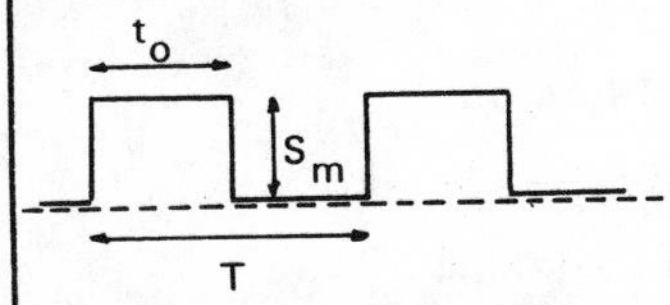	$S_m t_o/T$	$S_m\sqrt{t_o/T}$	$\sqrt{T/t_o}$	$\sqrt{T/t_o}$
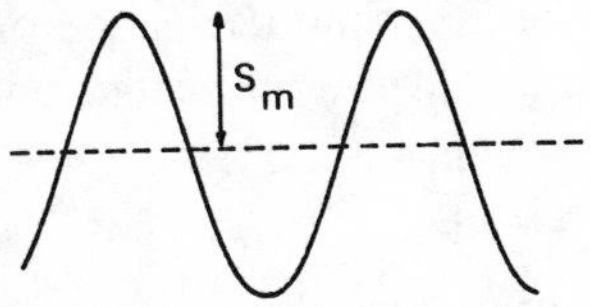	$.636S_m$	$.707S_m$	1.11	1.41
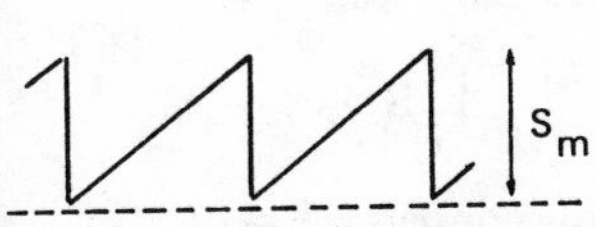	$.5S_m$	$.577S_m$	1.15	1.73

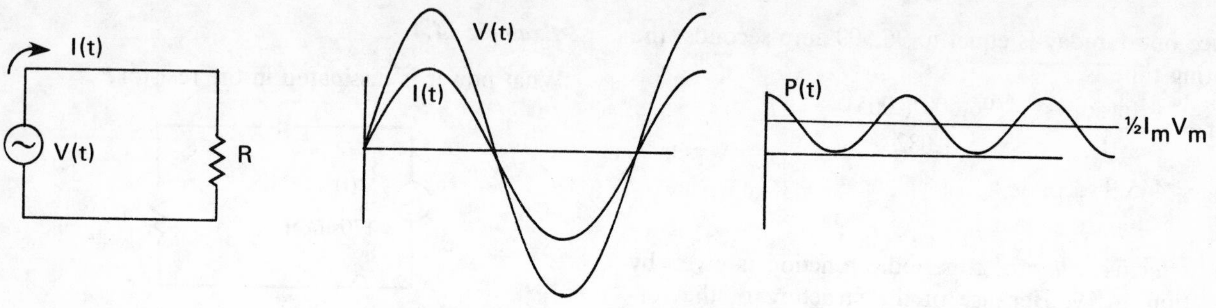

Figure 14.9 AC Circuit with Pure Resistance

Therefore, each AC circuit element is also assigned an angle, ϕ, known as the *impedance angle.* This angle is the phase difference angle produced when a sinusoidal voltage is applied across the element alone. In many calculations, the impedance is written in phasor form as $Z\angle\phi$.

A. Resistors

A pure resistor in an AC circuit is shown in figure 14.9. The applied voltage and resistance are known — the current is unknown. Writing Ohm's law in terms of instantaneous values results in equation 14.23.

$$V_m \sin\omega t = I(t)R \qquad 14.23$$

The current, therefore, is

$$I(t) = \left(\frac{V_m}{R}\right)\sin\omega t = I_m \sin\omega t \qquad 14.24$$

Notice that R must be divided into V_m to obtain I_m. Therefore, the impedance of a resistor is just the resistance. Furthermore, the current and voltage are in-phase. Therefore, the impedance of a resistor in phasor form is $R\angle 0°$.

The power dissipated in a purely resistive circuit can be obtained in a manner similar to that used for DC circuits.

$$P(t) = I(t)V(t) = (I_m \sin\omega t)(V_m \sin\omega t) \qquad 14.25$$

$$= I_m V_m \sin^2\omega t \qquad 14.26$$

Trigonometric identities can be used to derive equation 14.27 from equation 14.26.

$$P(t) = \tfrac{1}{2} I_m V_m - \tfrac{1}{2} I_m V_m \cos 2\omega t \qquad 14.27$$

Since the second term in equation 14.27 averages to zero, the *average power* dissipated in a purely resistive circuit is

$$P_{ave} = 1/2 I_m V_m = \frac{V_m^2}{2R} \qquad 14.28$$

If the current and voltage are given in effective values instead of maximum values, then

$$P_{ave} = I_{eff} V_{eff} \qquad 14.29$$

B. Capacitors

A *capacitor* is constructed of two conducting surfaces separated by a dielectric. If the conducting surfaces are connected across the terminals of a power source, charge will build up and create an electric field between the surfaces. The amount of charge built up is proportional to the applied voltage. The constant of proportionality is known as the *capacitance, C,* with units of *farads.* Capacitance is the ability to store charge. A capacitance of one farad exists when a change of one volt per second produces a one amp current.

$$q = CV \qquad 14.30$$

Capacitance of parallel plates may be found from equation 14.31. *A* is the plate area, and *r* is the plate separation.

$$C = \frac{\varepsilon A}{r} \qquad 14.31$$

Example 14.5

What is the capacitance of two parallel plates (.04 meter square) separated by a .1 cm thick insulator with a dielectric constant of 3.4? What charge is held by the capacitor if it is connected across 200 volts?

$$C = \frac{\varepsilon A}{r} = \frac{\varepsilon_v \varepsilon_r A}{r} = \frac{(8.85\ \text{EE}-12)(3.4)(.04)^2}{.001}$$

$$= 4.8\ \text{EE}-11 \text{ farads}$$

$$q = CV = (4.8\ \text{EE}-11)(200)$$

$$= 9.6\ \text{EE}-9 \text{ coulombs}$$

Since the current is the first time derivative of charge, the relationship between current and capacitance is

$$I(t) = \frac{dq}{dt} = C\frac{dV}{dt} \qquad 14.32$$

Equation 14.32 can be integrated to obtain a relationship for voltage across the capacitor:

$$V(t) = \left(\frac{1}{C}\right)\int I(t)\,dt + V_o \qquad 14.33$$

Substituting the trigonometric form of V(t) into equation 14.33,

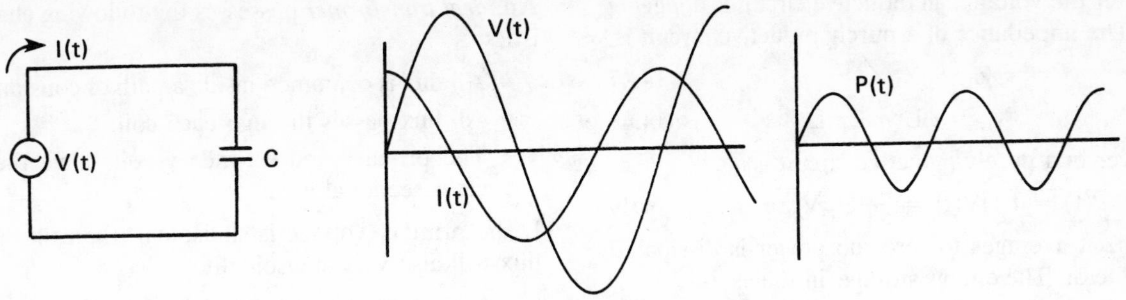

Figure 14.10 AC Circuit with Pure Capacitance

$$I(t) = C\frac{dV}{dt} = C\frac{d}{dt}(V_m \sin\omega t) = \frac{V_m}{\left(\frac{1}{\omega C}\right)}\cos\omega t \qquad 14.34$$

$$= \frac{V_m}{\left(\frac{1}{\omega C}\right)}\sin(\omega t + 90°) = I_m \sin(\omega t + 90°) \qquad 14.35$$

The quantity $\left(\frac{1}{\omega C}\right)$ is known as the *capacitive reactance* with symbol X_C and units of ohms. Since the current reaches a maximum value prior to the voltage, a capacitive circuit is a *leading circuit.* The impedance of a purely capacitive circuit is

$$Z_C = X_C \angle -90° \qquad 14.36$$

$$X_C = \frac{1}{\omega C} = \frac{1}{2\pi fC} \qquad 14.37$$

The power in a purely capacitive circuit is

$$P(t) = I(t)V(t) = (I_m \cos\omega t)(V_m \sin\omega t) \qquad 14.38$$

$$= I_m V_m (\sin\omega t)(\cos\omega t) \qquad 14.39$$

But $(\sin\omega t)(\cos\omega t) = ½\sin 2\omega t$. So,

$$P(t) = ½ I_m V_m \sin 2\omega t \qquad 14.40$$

Since $\sin 2\omega t$ averages to zero, no power is dissipated in a capacitor. However, energy is stored in a capacitor during its charging process. The *energy* in joules given instantaneous values of *V* and *q* is

$$\rho_C = ½CV^2 = \frac{q^2}{2C} \qquad 14.41$$

For capacitors in series,

$$\frac{1}{C} = \frac{1}{C_1} + \frac{1}{C_2} + \frac{1}{C_3} + \cdots \qquad 14.42$$

For capacitors in parallel,

$$C = C_1 + C_2 + C_3 + \cdots \qquad 14.43$$

C. Inductors

An *inductor* is a coil of wire connected across a voltage source. The magnetic field produced in the coil opposes current changes in the coil. From Faraday's law, the induced voltage is proportional to the change in flux linkage, which in turn is proportional to the current change. The proportionality constant, *L*, is the *inductance* expressed in *henrys*.

$$V(t) = L\frac{dI(t)}{dt} \qquad 14.44$$

Solving equation 14.44 for I(t),

$$I(t) = \frac{I}{L}\int V(t)\,dt = \frac{1}{L}\int V_m \sin(\omega t)\,dt \qquad 14.45$$

$$= \frac{-V_m}{\omega L}\cos\omega t$$

$$= \frac{V_m}{\omega L}\sin(\omega t - 90°) = I_m \sin(\omega t - 90°) \qquad 14.46$$

ωL is the *inductive reactance* with symbol X_L and units of ohms. Since current in an inductance reaches a max-

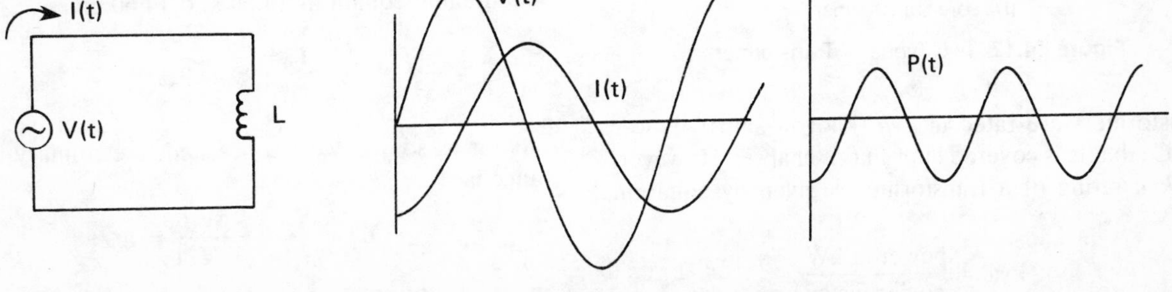

Figure 14.11 AC Circuit with Pure Inductance

imum after the voltage, an inductive circuit is a *lagging* circuit. The impedance of a purely inductive circuit is

$$\mathbf{Z}_L = X_L\angle 90^\circ \qquad 14.47$$

$$X_L = \omega L = 2\pi f L \qquad 14.48$$

The power in a purely inductive circuit is

$$P(t) = I(t)V(t) = -\tfrac{1}{2}I_m V_m \sin 2\omega t \qquad 14.49$$

Since sin2ωt averages to zero, no power is dissipated in an inductor. The energy storage in joules is

$$\rho_L = \tfrac{1}{2}LI^2 \qquad 14.50$$

For inductors in series,

$$L = L_1 + L_2 + L_3 + \cdots \qquad 14.51$$

For inductors in parallel,

$$\frac{1}{L} = \frac{1}{L_1} + \frac{1}{L_2} + \frac{1}{L_3} + \cdots \qquad 14.52$$

5. *Transformers*

Transformers are used to change voltages, to match impedances, and to isolate circuits. They consist of windings of wire on a magnetically permeable core. The primary current produces a magnetic flux in the core. That flux then induces a current in the secondary coil. Two typical transformer designs are shown in figure 14.12.

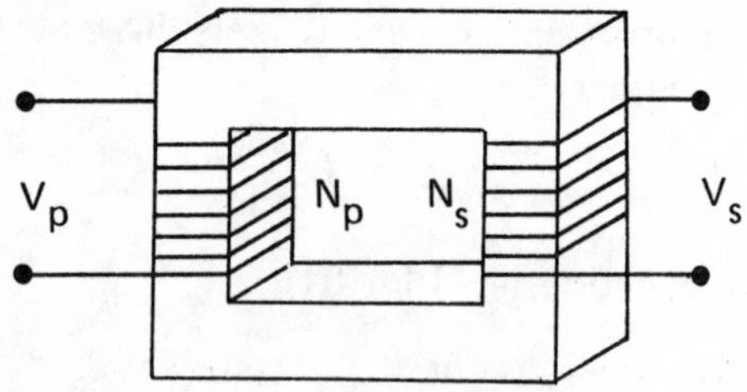

(a) shell transformer

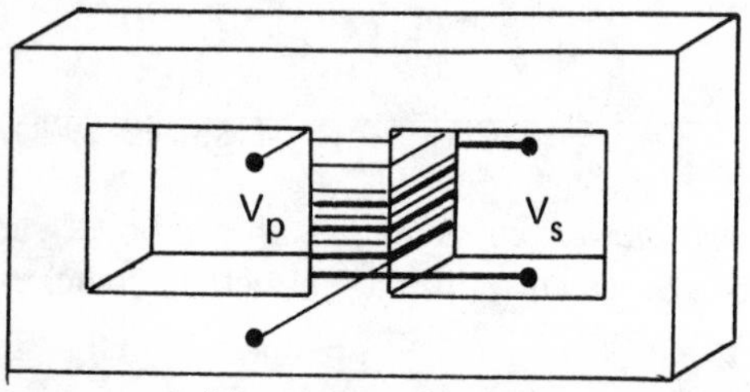

(b) core transformer

Figure 14.12 Two Types of Transformers

Transformers are rated in *kva* (kilovolt-amps). Power in AC circuits is covered later in this chapter. However, the *kva* rating of a transformer is given by equation 14.53.

$$\text{kva} = \frac{\text{power in kW}}{\text{power factor}} \qquad 14.53$$

An *ideal transformer* possesses the following characteristics:

- All flux is contained inside a path of constant area.
- All flux passes through each coil.
- The primary and secondary coils have the same cross sectional area.

If the primary voltage is sinusoidal, then the induced flux will also vary sinusoidally.

$$\phi(t) = \phi_m \sin\omega t \qquad 14.54$$

The induced secondary voltage is given by *Faraday's law.*

$$V_s(t) = -\frac{d\lambda}{dt} = -N_s\frac{d\phi}{dt} = -N_s\omega\phi_m\cos\omega t \qquad 14.55$$

$V_s(t)$ is the maximum secondary voltage. If the effective secondary voltage is needed, equation 14.55 can be divided by $\sqrt{2}$. If $2\pi f$ is substituted for ω, the effective secondary voltage is

$$V_{s,\text{eff}}(t) = -N_s\left(\frac{2\pi}{\sqrt{2}}\right)f\phi_m\cos\omega t$$

$$= -4.44N_s f\phi_m\cos\omega t \qquad 14.56$$

An analogous equation for $V_p(t)$ can be developed. The ratio of these voltage relationships is known as the *turns ratio* or *ratio of transformation, a.*

$$a = \frac{\mathbf{V}_p}{\mathbf{V}_s} = \frac{N_p}{N_s} \qquad 14.57$$

If *a* is greater than 1, the transformer decreases voltage and is known as a *step-down transformer.* If *a* is less than 1, the transformer increases voltage and is known as a *step-up transformer.* Since the power into an ideal transformer equals the power out,

$$\mathbf{V}_p\mathbf{I}_p = \mathbf{V}_s\mathbf{I}_s \qquad 14.58$$

Combining equations 14.57 and 14.58 gives a current relationship.

$$\frac{\mathbf{I}_s}{\mathbf{I}_p} = a \qquad 14.59$$

A transformer can be used to match impedances. Consider figure 14.13. The secondary current is

$$\mathbf{I}_s = \frac{\mathbf{V}_s}{\mathbf{Z}_s} \qquad 14.60$$

Combining equations 14.59 and 14.60 gives

$$\mathbf{I}_p = \frac{\mathbf{I}_s}{a} = \frac{\mathbf{V}_s}{\mathbf{Z}_s a} \qquad 14.61$$

As far as source $\mathbf{V}_p$ is concerned, the primary impedance is

$$\mathbf{Z}_p = \frac{\mathbf{V}_p}{\mathbf{I}_p} = \frac{a\mathbf{V}_s}{\mathbf{V}_s/\mathbf{Z}_s a} = a^2\mathbf{Z}_s \qquad 14.62$$

Equation 14.62 can be used to match impedance.

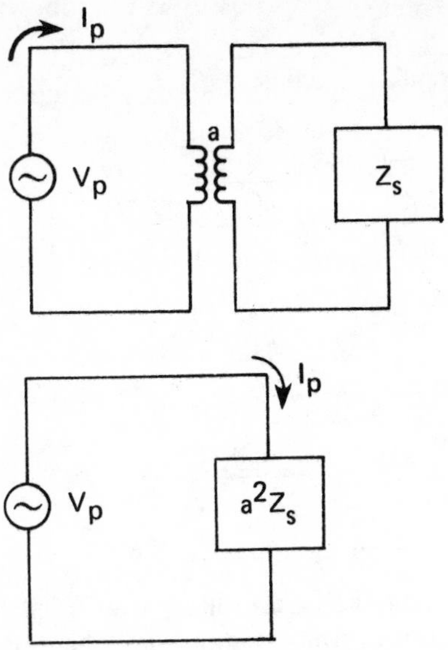

Figure 14.13 Impedance Matching

Example 14.6

An ideal step-down transformer has 200 primary turns and 50 secondary turns. 440 volts are applied across the primary. The secondary load resistance is 5 ohms. What is the secondary current? What is the impedance seen by the primary source?

$$a = N_p/N_s = 200/50 = 4$$

$$\mathbf{V}_s = \mathbf{V}_p/a = 440/4 = 110 \text{ volts}$$

$$\mathbf{I}_s = \mathbf{V}_s/R_s = 110/5 = 22 \text{ amps}$$

$$R_p = a^2R_s = (4)^2 5 = 80 \text{ ohms}$$

Real transformers possess winding resistance and flux leakage. Flux which links both coils is known as *mutual flux. Primary leakage flux* links only the primary coil. *Secondary leakage flux* links only the secondary coil.

According to Faraday's law, each leakage flux induces a voltage loss in its corresponding coil. Each induced voltage loss can be considered to have been caused by an equivalent reactance. For example, the *primary leakage reactance* is defined as

$$X_p = \omega L_p = \frac{V_{p,loss}}{I_p} = \frac{4.44 f \phi_{m,loss} N_p}{I_p} \qquad 14.63$$

The equivalent transformer parameters are shown in figure 14.14.

The *coefficient of coupling* for a real transformer is given by equation 14.64. *M* is the *mutual inductance* and *L* is the *leakage inductance.* The coefficient of coupling is equal to one for an ideal transformer.

$$k = \frac{M}{\sqrt{L_p L_s}} = \frac{X_M}{\sqrt{X_p X_s}} \qquad 14.64$$

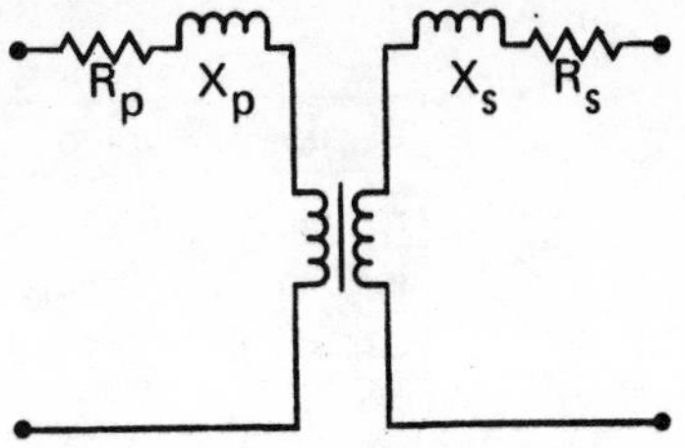

Figure 14.14 A Real Transformer

6. Admittance

Admittance is the reciprocal of impedance. Its units are *mhos,* the reciprocal of ohms.

$$\mathbf{Y} = \frac{1}{\mathbf{Z}} = \frac{\mathbf{I}}{\mathbf{V}} \qquad 14.65$$

Since admittance is a complex quantity, it can be written in phasor form or as a combination of its real and imaginary parts. The real part of **Y** is known as conductance, *G*. The imaginary part of **Y** is known as susceptance, *B*. (Susceptance is negative in an inductive circuit.) Both *G* and *B* have units of mhos.

$$\mathbf{Y} = Y\angle\phi_Y = G + jB \qquad 14.66$$

Complex conjugates can be used to write **Y** in terms of resistance and reactance.

$$\mathbf{Y} = G + jB = \frac{1}{R + jX} = \frac{R - jX}{R^2 + X^2} \qquad 14.67$$

Admittance is primarily used to analyze parallel circuits. Admittances in parallel can be added together to obtain the composite circuit admittance.

7. Complex Numbers and Circuit Elements

It was shown in chapter 1 how the imaginary exponentials $e^{j\phi}$ and $e^{-j\phi}$ could be combined into $\sin\phi$ and $\cos\phi$ terms. The angle ϕ in $e^{j\phi}$ does not need to be constant. If ϕ is a function of time, as in $\phi = \omega t$, then a sinusoidal voltage and current can be written as equations 14.68 and 14.69.

$$V_m \sin\omega t = V_m e^{j\omega t} = V_m\angle\omega t \qquad 14.68$$

$$I_m \sin(\omega t + \phi) = I_m e^{(j\omega t + \phi)}$$

$$= I_m\angle(\omega t + \phi) \qquad 14.69$$

Thus, the trigonometric expressions have been rewritten as varying exponentials and phasors. In the analysis of AC circuits, it is sufficient to let t=0 since the magnitudes of **I** and **V** are not changed and the phase relationship is unaltered.

$$V_m \sin\omega t = V_m\angle 0° \qquad 14.70$$

$$I_m \sin(\omega t + \phi) = I_m\angle\phi \qquad 14.71$$

The phasor impedance forms for the ideal resistor, inductor, and capacitor have already been developed in this chapter. The various methods of expressing these circuit elements are summarized in table 14.2.

Table 14.2 AC Impedance Forms

element	value	reactance	rectangular	phasor
resistor	*R* ohms	0	$R+j0$	$R\angle 0°$
inductor	*L* henrys	ωL ohms	$0+j(\omega L)$	$\omega L\angle +90°$
capacitor	*C* farads	$(1/\omega C)$ ohms	$0-j(1/\omega C)$	$(1/\omega C)\angle -90°$

AC ELEC

8. Application to AC Circuits

All of the circuit analysis methods and theorems presented in the DC chapter can be used to analyze AC circuits as long as complex arithmetic is used. Among others, the following laws and techniques can be used: Ohm's law, Kirchoff's laws, parallel circuit voltage uniformity, series circuit current uniformity, loop current methods, node voltage methods, voltage and current division laws, and equivalent circuits.

The following examples illustrate the application of complex number arithmetic to various typical problems.

Example 14.7

In the circuit shown, what is

- I(t) in trigonometric form?
- the voltage across the inductor?
- the current through the capacitor?

Assume the voltage source is specified by a maximum value.

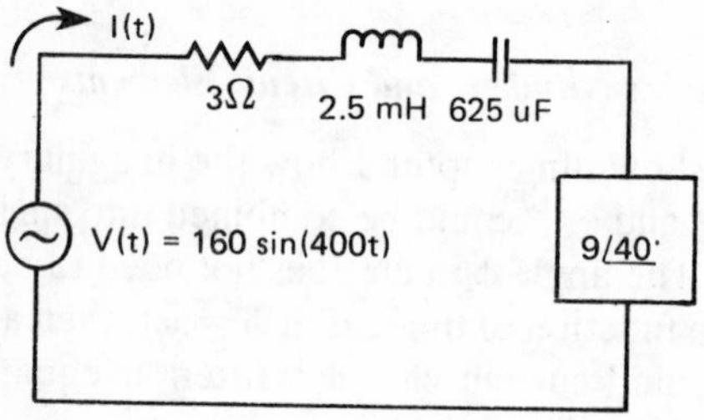

Ohm's law can be used to find the current as long as the impedance sum is found in complex arithmetic. Since it is necessary to add the impedances (and adding complex numbers is easiest in rectangular form), the first step is to rewrite the voltage as a phasor and rewrite the circuit elements in rectangular form.

voltage source: $160\angle 0°$

$$\begin{aligned}
&\textit{resistor:} && = 3 + j0\\
&\textit{capacitor: } X_C = 1/\omega C = 1/(400)(625\text{ EE}-6) && = 0 - j4\\
&\textit{inductor: } X_L = \omega L = (400)(2.5\text{ EE}-3) && = 0 + j1\\
&\textit{black box: } 9(\cos 40° + j\sin 40°) && = 6.89 + j5.79\\
\hline
&\mathbf{Z} && = 9.89 + j2.79
\end{aligned}$$

Now convert **Z** to a phasor.

$$|\mathbf{Z}| = \sqrt{(9.89)^2 + (2.79)^2} = 10.28$$

$$\phi = \arctan\left(\frac{2.79}{9.89}\right) = 15.75°$$

$$\mathbf{Z} = 10.28\angle 15.75°$$

$$\mathbf{I}(t) = \frac{\mathbf{V}(t)}{\mathbf{Z}} = \frac{160\angle 0°}{10.28\angle 15.75°} = 15.56\angle -15.75°$$

$$= 15.56\sin(400t - 15.75°)$$

Note that (400t) is in radians and 15.75° is in degrees. This mixing of units is common. The value of **I**(t) calculated above is constant throughout the entire circuit.

The voltage across the inductor is

$$\begin{aligned}
\mathbf{V}_L(t) &= \mathbf{I}(t)\mathbf{Z}_L = (15.56\angle -15.75°)(1\angle 90°)\\
&= 15.56\angle 74.25° = 15.56\sin(400t + 74.25°)
\end{aligned}$$

Since the current angle is negative, the circuit is lagging.

Example 14.8

In the parallel circuit shown find

- **V**(t).
- **I**(t) expressed as a sinusoid with phase difference measured with respect to a zero phase voltage.
- the current through the capacitor, measured with respect to a zero phase voltage.

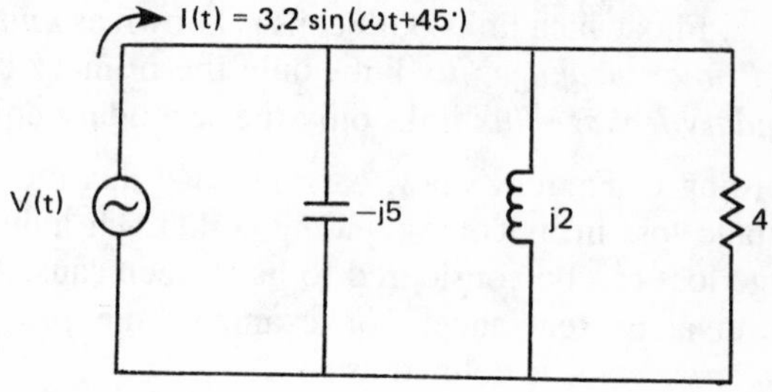

Adding parallel admittances,

$$\mathbf{Y} = \frac{1}{\mathbf{Z}} = \Sigma\frac{1}{\mathbf{Z}_i}$$

$$= \frac{1}{\mathbf{Z}_C} + \frac{1}{\mathbf{Z}_L} + \frac{1}{\mathbf{Z}_R}$$

$$= \frac{1}{5\angle -90°} + \frac{1}{2\angle 90°} + \frac{1}{4\angle 0°}$$

$$= .2\angle 90° + .5\angle -90° + .25\angle 0°$$

To perform this addition, it is necessary to convert the phasors to rectangular form.

$$\mathbf{Y} = (0+j.2)+(0-j.5)+(.25+j0) = .25-j.3$$

Then,

$$\mathbf{Z} = \frac{1}{\mathbf{Y}} = \frac{1}{.25-j.3} = \frac{1}{.39\angle-50.2} = 2.56\angle 50.2°$$

So,

$$\mathbf{V}(t) = \mathbf{I}(t)\mathbf{Z} = (3.2\angle 45°)(2.56\angle 50.2°)$$
$$= 8.2\angle 95.2° = 8.2\sin(\omega t+95.2°)$$

If the time scale is changed so that **V** is zero at t=0 (a zero phase voltage), then

$$\mathbf{V}(t) = 8.2\sin(\omega t+95.2°-95.2°) = 8.2\sin\omega t$$

and

$$\mathbf{I}(t) = 3.2\sin(\omega t+45°-95.2°) = 3.2\sin(\omega t-50.2°)$$

The current flowing through the capacitor is

$$\mathbf{I}_C(t) = \frac{\mathbf{V}(t)}{\mathbf{Z}_C} = \frac{8.2\angle 0°}{5\angle-90°} = 1.64\angle 90°$$

9. Resonance

Resonance occurs in a circuit when the phase angle difference is zero. This is equivalent to saying that the circuit is purely resistive in its response to AC voltage. Three types of resonance are common:

- series resonance
- parallel resonance
- parallel C-series RL resonance

Resonant circuits are frequency selective since they pass more (or less) current at certain frequencies than at others. A series *LCR* circuit will pass the selected frequency (as measured across *C*) if *R* is high and the source resistance is low. If *R* is low and the source resistance is high, then the circuit will reject the chosen frequency. A parallel resonant circuit placed in series with the load will reject the chosen frequency.

A. Series Resonance

The overall impedance in rectangular form of a series *RLC* circuit is $R + j(X_L - X_C)$. If an *RLC* circuit is to behave resistively, then

$$X_L = X_C \qquad 14.72$$

$$\omega^* L = \frac{1}{\omega^* C} \qquad 14.73$$

$$\omega^* = \sqrt{1/LC} \qquad 14.74$$

ω^* is known as the *resonant frequency.* The power consumed by a resonant series RLC circuit is

$$P^*_{ave} = I^2 R = \frac{V^2}{R} \qquad 14.75$$

With series RLC circuits at resonance,

- impedance is minimum
- impedance equals resistance
- the phase angle difference is zero
- current is maximum
- power is maximum

The *quality factor* for coils and resonant circuits is a dimensionless ratio which compares the energy storage in the inductor to the energy loss in the resistor. It is assumed that the capacitor contributes little to the quality factor. The quality factor is

$$Q = 2\pi\left(\frac{\text{maximum energy stored per cycle}}{\text{energy dissipated per cycle}}\right) \qquad 14.76$$

For a series RLC circuit,

$$Q = \frac{\omega^* L}{R} = \frac{1}{\omega^* CR} = \frac{\omega^*}{\omega_2 - \omega_1} \qquad 14.77$$

ω_1 and ω_2 are known as the *70% points* or the *half-power points.* If **V** is fixed and ω is changed to ω_1 and ω_2, then

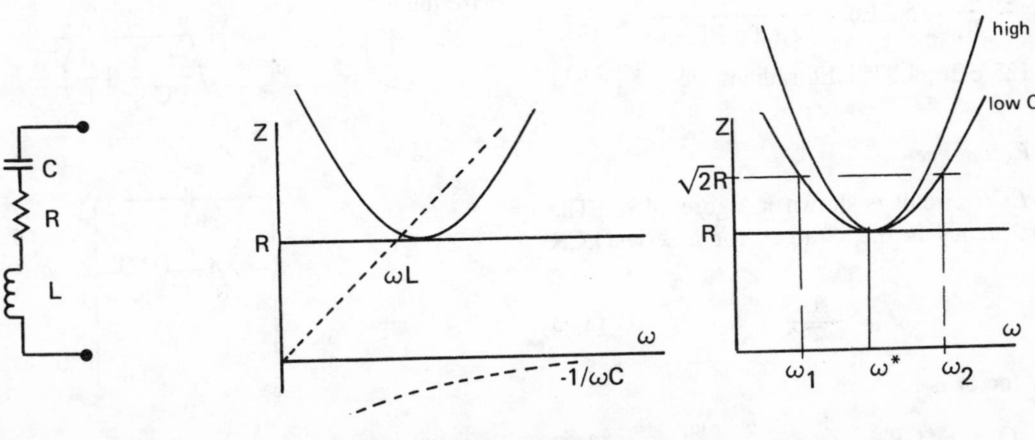

Figure 14.15 Series Resonance

$$Z = \sqrt{2}R \qquad 14.78$$

$$I = V/Z = V/\sqrt{2}R = .707I^* \qquad 14.79$$

$$P = I^2R = (.707I^*)^2R = {}^1\!/_2P^* \qquad 14.80$$

$$\phi_I = 45° \qquad 14.81$$

$(\omega_2 - \omega_1)$ is known as the *circuit bandwidth.* Bandwidth is a measure of circuit selectivity. The relationships between the half-power points, bandwidth, and Q are given by equations 14.82 and 14.83.

$$\omega_1,\omega_2 = \omega^*\sqrt{1+\left(\frac{1}{2Q}\right)^2} \mp \frac{\omega^*}{2Q} = \omega^* \mp \frac{R}{2L} \qquad 14.82$$

$$\omega_2 - \omega_1 = \frac{\omega^*}{Q} \qquad 14.83$$

Example 14.9

For the circuit shown, find the

- resonant frequency.
- current at resonance.
- voltage across each component at resonance.
- half-power frequencies.

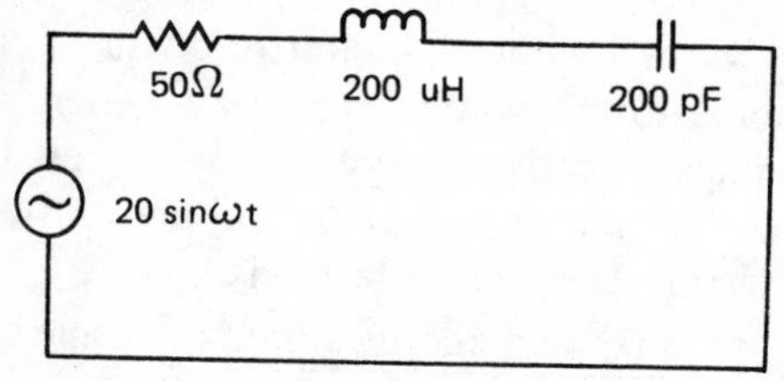

$$\omega^* = \frac{1}{\sqrt{LC}} = \frac{1}{\sqrt{(200\text{ EE}-6)(200\text{ EE}-12)}}$$

$$= 5\text{ EE6 radians/sec}$$

$$I^* = \frac{V}{R} = \frac{20\sin\omega t}{50} = .4\sin\omega t\text{ amps}$$

$$V_R^* = I^*R = (.4)(50) = 20\text{ volts}$$

$$V_L^* = I^*X_L = (.4)(5\text{ EE6})(200\text{ EE}-6) = 400\text{ volts}$$

$$V_C^* = I^*X_C = (.4)/(5\text{ EE6})(200\text{ EE}-12) = 400\text{ volts}$$

$$\omega_1,\omega_2 = \omega^* \mp \frac{R}{2L} = 5\text{ EE6} \mp \frac{50}{(2)(200\text{ EE}-6)}$$

$$= 5.125\text{ EE6},\ 4.875\text{ EE6 radians/sec}$$

B. Parallel Resonance

A *parallel LCR* circuit is shown in figure 14.16. The resonant frequency is the same as for a series LCR circuit.

$$\omega^* = \frac{1}{\sqrt{LC}} \qquad 14.84$$

The quality factor is

$$Q = \omega^*CR = \frac{R}{\omega^*L} = \frac{\omega_2-\omega_1}{\omega^*} \qquad 14.85$$

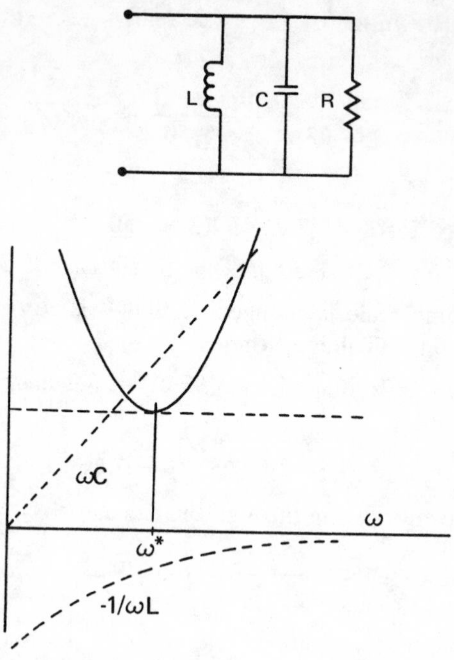

Figure 14.16 Parallel Resonance

At resonance in a parallel *LCR* circuit,

- impedance is maximum
- impedance is equal to R
- the phase difference angle is zero
- current is minimum
- power is minimum

Equations 14.82 and 14.83 can be used to find the 70% points.

C. Parallel C — Series RL Resonance

Practical resonant circuits generally consist of an inductor and variable capacitor in parallel. Since the inductor will possess some resistance, the equivalent circuit is shown in figure 14.17. With this form of resonance, the current will be minimized. The resonant frequency is

$$\omega^* = \sqrt{\frac{1}{LC} - \left(\frac{R}{L}\right)^2} \qquad 14.86$$

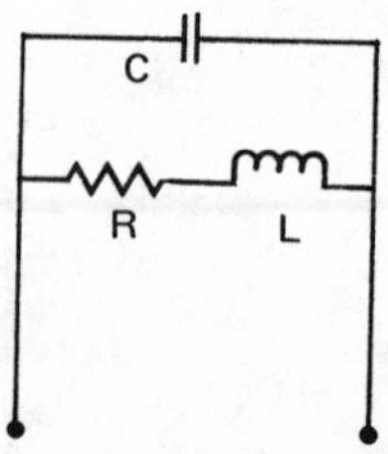

Figure 14.17 Parallel C — Series RL Resonance

10. Complex Power

Figure 14.18 illustrates an AC voltage source supplying power to a parallel *RL* circuit. The equivalent black box circuit is also shown. All currents have been evaluated. Since the inductor does not dissipate energy, the only energy dissipated in the circuit is due to the current in the resistor. The maximum value of the instantaneous power in the resistor is

$$P_{max,real} = \tfrac{1}{2} I_R V_R = (.5)(18\angle 0°)(180\angle 0°) = 1620\angle 0° \text{ W}$$

This 1620 watts is energy that is really dissipated in the form of heat or useful work. For that reason, it is called *real power*. (Other names are *average power* and *true power*.) The units of real power are watts and kilowatts.

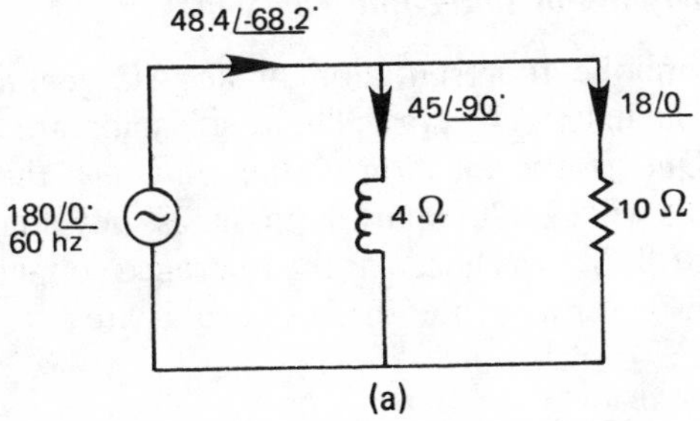

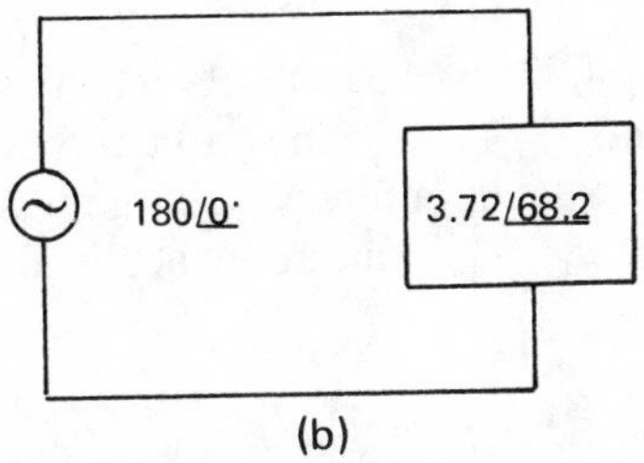

Figure 14.18 A Parallel RL Circuit

If power is graphed in the real-imaginary coordinate system, real power could be represented by the point (1620,0) as shown in figure 14.19(a). Alternately, it could be represented by a vector $(1620 + j0)$ in the polar coordinate system, as in figure 14.19(b).

Similarly, the maximum instantaneous power stored per cycle in the inductor is

$$P_{max,reactive} = -\tfrac{1}{2} I_L V_L = -(.5)(45\angle -90°)(180\angle 0°) = -4050\angle -90° \text{ vars (volt-amps reactive)}$$

This could be represented as $-4050\angle -90°$ in the polar coordinate system shown in figure 14.19(c). Since this power is not dissipated — only periodically stored and released — it is distinguished from real power by the name *reactive power*. It is also given different units from real power. Reactive power has the units of volt-amps reactive (vars). Reactive power results from an actual current draw. This current supplies the magnetization energy in motors and the charge on capacitors.

It is interesting to evaluate the power based on the supply voltage and the total current drain. From the data given, it appears as though the maximum instantaneous circuit power is

$$P_{max,apparent} = -\tfrac{1}{2} V_m I_m = -(.5)(180\angle 0°)(48.4\angle -68.2°) = -4360\angle -68.2° \text{ va}$$

Although the actual power dissipated is only 1620 watts, it appears (based on the information available at the source terminals) that the power has a magnitude of 4360. For this reason, $\tfrac{1}{2} I_m V_m$ is known as the *apparent* power. Its units are volt-amps (va). In this example, the apparent power could be represented as $-4360\angle -68.2°$ in the polar coordinate system, as shown in figure 14.19(d).

The three power vectors close on themselves to make the power triangle shown in figure 14.20.

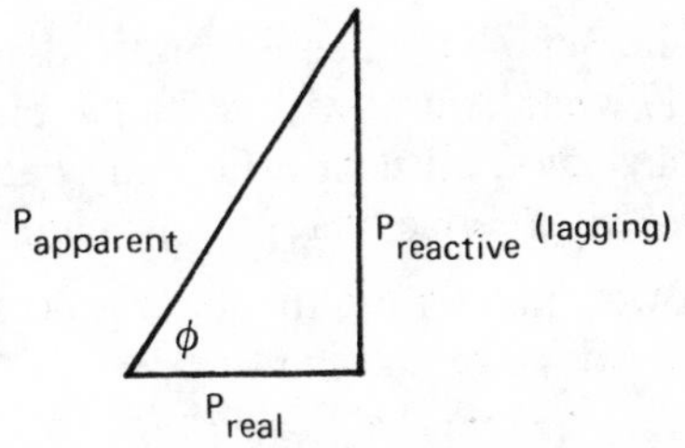

Figure 14.20 The Power Triangle

The following points should be noted.

- The overall current angle (−68.2°) and the triangle angle φ are the same. This angle is known as the *impedance angle* or *power angle*.
- The powers and angle are all related by trigonometry.

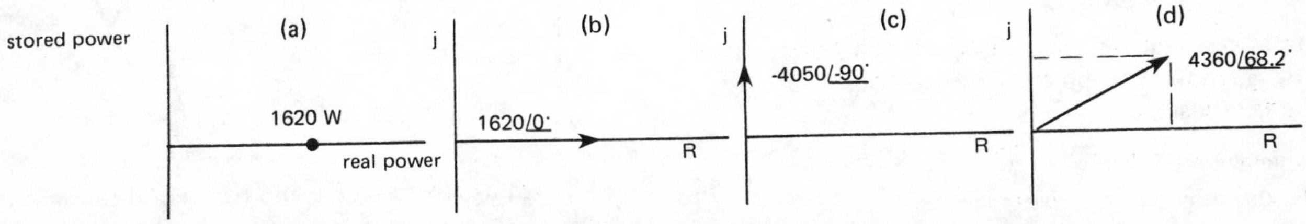

Figure 14.19 Complex Power

$$P^2_{\text{apparent}} = P^2_{\text{real}} + P^2_{\text{reactive}} \quad 14.87$$

$$P_{\text{real}} = P_{\text{apparent}}\cos\phi \quad 14.88$$

$$P_{\text{reactive}} = P_{\text{apparent}}\sin\phi \quad 14.89$$

- $\cos\phi$ is known as the *power factor.* It is usually given in percent. Since the cosines of positive and negative angles are both positive, the power factor is said to be *lagging* if the circuit is inductive, and *leading* if the circuit is capacitive.
- $\sin\phi$ is known as the *reactive factor.*

Inasmuch as apparent power is paid for but only real power is dissipated, a cost reduction is possible if the power angle can be changed without changing the real power. This is accomplished by changing the circuit reactance to modify the reactive power. The act of changing the reactive power is known as *power factor correction.*

The required change in reactive power to change the power angle from ϕ_1 to ϕ_2 is

$$\Delta P_{\text{reactive}} = P_{\text{real}}(\tan\phi_1 - \tan\phi_2) \quad 14.90$$

If this reactive power change is accomplished by adding a capacitor across the voltage source, the capacitor can be sized from equation 14.91. V_{line} is the maximum value of the sinusoid. If V_{line} is an RMS value, divide equation 14.91 by 2.

$$C = \frac{\Delta P_{\text{reactive}}}{\pi f (V_{\text{line}})^2} \quad \text{(farads)} \quad 14.91$$

Example 14.10

A 60 hz, 5 horsepower induction motor draws 53 amps (rms) at 117 volts (rms), with a 78.5% electrical/mechanical conversion efficiency. Find the parallel capacitance required to reduce the power factor to 92%.

The real power drawn from the line can be found from the real work done by the motor:

$$P_{\text{real}} = \frac{W}{\eta} = \frac{(5)\text{ hp }(746)\text{ watts/hp}}{.785} = 4750\text{ W}$$

The apparent power is found from the observed voltage and current.

$$P_{\text{apparent}} = IV = (53)(117) = 6200$$

Therefore, the reactive power and power angle are

$$P_{\text{reactive}} = \sqrt{(6200)^2 - (4750)^2} = 3985\text{ vars}$$

$$\phi = \arccos\left(\frac{4750}{6200}\right) = 40°$$

After the capacitor is installed, the new power factor is 92%, so the new power angle is arccos (.92) = 23°. The new reactive power is

$$P_{\text{reactive,new}} = (\tan 23°)(4750) = 2016\text{ vars}$$

The difference in reactive power is

$$\Delta P_{\text{reactive}} = 3985 - 2016 = 1969\text{ vars}$$

The required capacitance is found from equation 14.91. Because *V* and *I* are rms values, the factor 2 is required in the denominator.

$$C = \frac{1969}{2\pi(60)(117)^2} = 3.82\text{ EE}-4\text{ farads}$$

11. Three-Phase Circuits

Three-phase energy distribution systems are more efficient than multiple single-phase systems, both in the number of wires used and in the size of the wiring. In addition, a three-phase motor will operate at constant power, unlike the instantaneous power from a single-phase line which varies sinusoidally.

A. Generation of Three-Phase Potential

The symbolic representation of an AC generator is shown in figure 14.21(a). Three sinusoids are generated. Due to the location of the windings, the three sinusoids will be 120° apart in phase, as shown in figure 14.21(b). If $V_{aa'}$ is chosen as the reference voltage, then the phasor forms of the three sinusoids are

$$\mathbf{V}_{aa'} = V\angle 0° \quad 14.92$$

$$\mathbf{V}_{bb'} = V\angle -120° \quad 14.93$$

$$\mathbf{V}_{cc'} = V\angle -240° \quad 14.94$$

Equations 14.92, 14.93, and 14.94 are a *positive sequence.* That is, $\mathbf{V}_{aa'}$ reaches a peak before $\mathbf{V}_{bb'}$, and $\mathbf{V}_{bb'}$ reaches a peak before $\mathbf{V}_{cc'}$. If the order is reversed (e.g., $\mathbf{V}_{cc'}$, $\mathbf{V}_{bb'}$, $\mathbf{V}_{aa'}$), the sequence is negative.

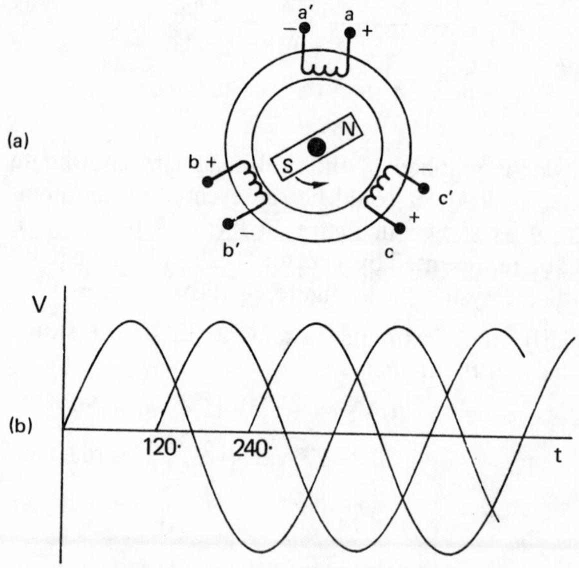

Figure 14.21 A Three-Phase Generator

Although a six-wire line could be used to transmit the three voltages, it is more efficient to interconnect the windings. Savings in wire and other benefits can be gained by so doing. Two possible interconnection methods are commonly used — the *delta (mesh)* connection and the *wye (star)* connection.

Figure 14.22 illustrates delta-interconnected phases. The voltage seen by an impedance **Z** connected across two lines (the system or line voltage) is equal to the generated voltage, **V**. Any of the coils can be selected as the reference, as long as the sequence is maintained.

$$\mathbf{V}_{CA} = \mathbf{V}_{aa'} = V\angle 0° \qquad 14.95$$

$$\mathbf{V}_{AB} = \mathbf{V}_{bb'} = V\angle -120° \qquad 14.96$$

$$\mathbf{V}_{BC} = \mathbf{V}_{cc'} = V\angle -240° \qquad 14.97$$

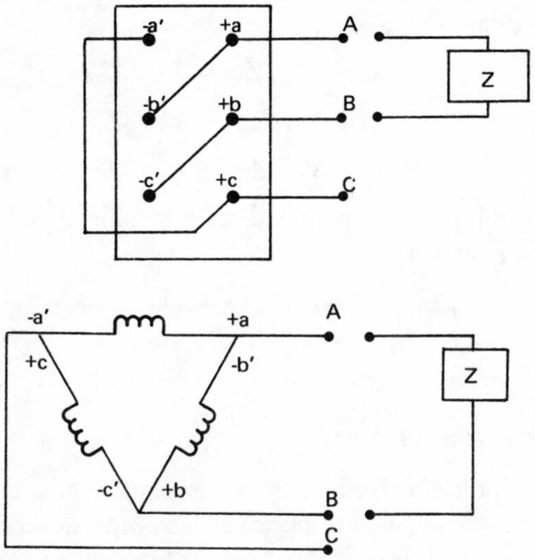

Figure 14.22 Delta-Connected Sources

Wye-connected sources are shown in figure 14.23. Wire *G* is a ground or neutral wire which carries current only if the load becomes unbalanced.

$$\mathbf{V}_{AG} = V\angle 0° \qquad 14.98$$

$$\mathbf{V}_{BG} = V\angle -120° \qquad 14.99$$

$$\mathbf{V}_{CG} = V\angle -240° \qquad 14.100$$

The *system* or *line voltages* for sequence *ABC* are

$$\mathbf{V}_{AB} = \sqrt{3}V\angle 30° \qquad 14.101$$

$$\mathbf{V}_{BC} = \sqrt{3}V\angle -90° \qquad 14.102$$

$$\mathbf{V}_{CA} = \sqrt{3}V\angle -210° \qquad 14.103$$

Although the system voltage depends on whether the generator coils are delta or wye connected, each connection results in three equal sinusoidal voltages, each 120° out of phase with the others. In the remainder of this chapter, the distinction between source connections will not be made, with reference being made only to the line voltage. It will be assumed that the system voltages have angles of 0°, −120°, and −240°.

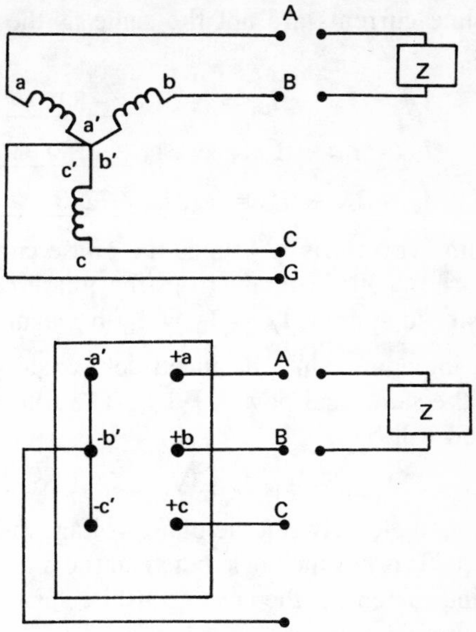

Figure 14.23 Wye-Connected Sources

B. Three-Phase Loads

Three impedances are required to fully load a three-phase source. If all three impedances are equal in magnitude and angle, the system is said to be *balanced*. These three impedances may themselves be either delta- or wye-connected.

Three-phase circuits are only slightly more difficult to evaluate than single-phase circuits. The first requirement is to develop a clearly-labeled schematic. A delta-wired load is illustrated in figure 14.24.

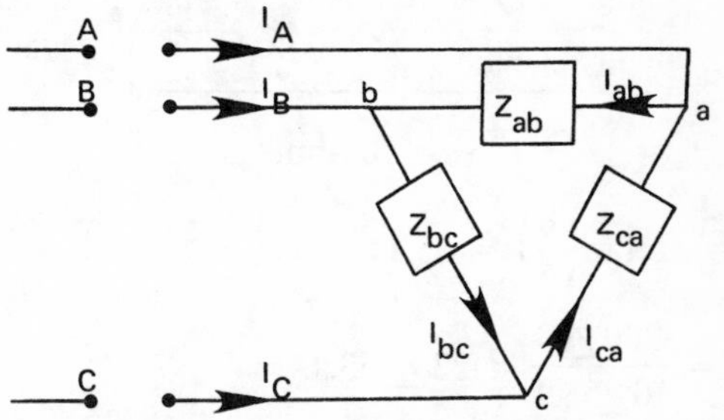

Figure 14.24 Delta-Wired Loads

Using phasor notation, the phase currents for a balanced system are given below. **V** is the line voltage, the same as the phase voltage.

$$\mathbf{I}_{ab} = \mathbf{V}_{AB}/\mathbf{Z}_{ab} = \frac{V\angle 0°}{Z\angle \phi_z} = \left(\frac{V}{Z}\right)\angle -\phi_z \qquad 14.104$$

$$\mathbf{I}_{bc} = \mathbf{V}_{BC}/\mathbf{Z}_{bc} = \left(\frac{V}{Z}\right)\angle -120° - \phi_z \qquad 14.105$$

$$\mathbf{I}_{ca} = \mathbf{V}_{CA}/\mathbf{Z}_{ca} = \left(\frac{V}{Z}\right)\angle -240° - \phi_z \qquad 14.106$$

The line currents are not the same as the phase currents.

$$\mathbf{I}_A = \mathbf{I}_{ab} - \mathbf{I}_{ca} = \sqrt{3}\mathbf{I}_{ab}\angle{-30° - \phi_z} \qquad 14.107$$

$$\mathbf{I}_B = \mathbf{I}_{bc} - \mathbf{I}_{ab} = \sqrt{3}\mathbf{I}_{bc}\angle{-150 - \phi_z} \qquad 14.108$$

$$\mathbf{I}_c = \mathbf{I}_{ca} - \mathbf{I}_{bc} = \sqrt{3}\mathbf{I}_{ca}\angle{-270 - \phi_z} \qquad 14.109$$

The line current is $\sqrt{3}$ times the phase current and is displaced $-30°$ in phase from the phase current. For a balanced system, $\mathbf{I}_A = \mathbf{I}_B = \mathbf{I}_C$ in magnitude.

Each impedance in a balanced delta system will dissipate the same real power, P_{phase}. The total power dissipated will be

$$P_{total} = 3P_{phase} = 3V_{phase}I_{phase}\cos\phi \qquad 14.110$$

ϕ is the angle between the phase voltage and the phase current. It is not the angle between the line voltage and the line current. If the power must be derived from line measurements, then

$$P_{total} = \sqrt{3}V_{line}I_{line}\cos\phi \qquad 14.111$$

Example 14.11

Given the balanced delta load shown, find the *ba* phase current, the *B* line current, the *ba* phase power, and the total circuit power. Assume $\mathbf{V}_{AB}$ is $220\angle{0°}$ *rms*, $\mathbf{V}_{BC}$ is $220\angle{-120°}$, and V_{CA} is $220\angle{-240°}$.

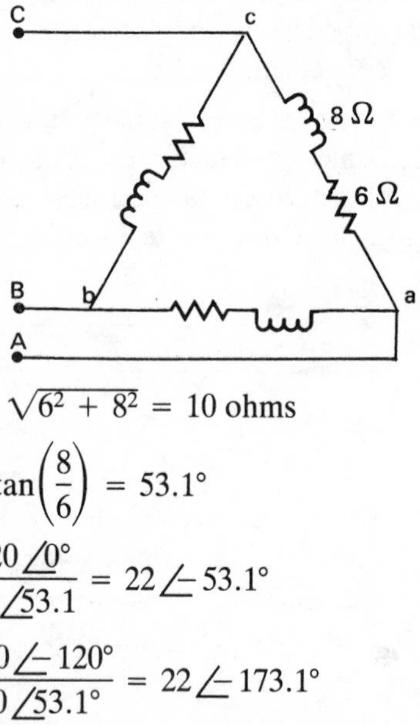

$$Z_{phase} = \sqrt{6^2 + 8^2} = 10 \text{ ohms}$$

$$\phi = \arctan\left(\frac{8}{6}\right) = 53.1°$$

$$\mathbf{I}_{ab} = \frac{220\angle{0°}}{10\angle{53.1}} = 22\angle{-53.1°}$$

$$\mathbf{I}_{bc} = \frac{220\angle{-120°}}{10\angle{53.1°}} = 22\angle{-173.1°}$$

$$\mathbf{I}_B = \mathbf{I}_{bc} - \mathbf{I}_{ab} = 38.1\angle{-203.1°}$$

$$P_{phase} = I^2_{phase}R = (22)^2 6 = 2904 \text{ W}$$

$$P_{total} = 3P_{phase} = 8712 \text{ W}$$

Three equal impedances connected in a wye configuration are shown in figure 14.25.

Using phasor notation, the phase currents are given in equations 14.112, 14.113, and 14.114. V is the line voltage. Notice that the line and phase currents are equal. However, the phase voltage is found by dividing the line voltage by $\sqrt{3}$.

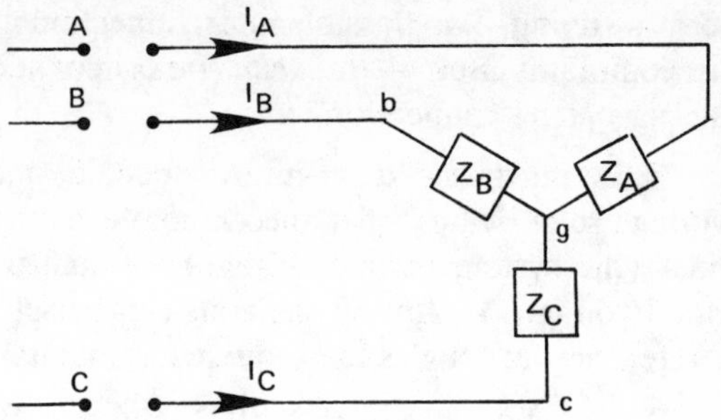

Figure 14.25 Wye-Connected Loads

$$\mathbf{I}_A = \mathbf{V}_{AG}/\mathbf{Z}_A = \mathbf{V}/\sqrt{3}\mathbf{Z}_A \qquad 14.112$$

$$\mathbf{I}_B = \mathbf{V}_{BG}/\mathbf{Z}_B = \mathbf{V}/\sqrt{3}\mathbf{Z}_B \qquad 14.113$$

$$\mathbf{I}_C = \mathbf{V}_{CG}/\mathbf{Z}_C = \mathbf{V}/\sqrt{3}\mathbf{Z}_C \qquad 14.114$$

The total power dissipated in a balanced wye-connected system is

$$P_{total} = 3P_{phase} = 3V_{phase}I_{phase}\cos\phi \qquad 14.115$$

$$= \sqrt{3}V_{line}I_{line}\cos\phi \qquad 14.116$$

12. Transient Analysis

The response in some circuits is neither steady-state nor sinusoidal. Such response usually indicates a charging or discharging energy source. One case is that of the series *RL* circuit shown in figure 14.26. At some long time after the switch is closed, a current equal to (V/R) will be flowing. However, at t = 0, the current will be zero. How the current goes from zero to (V/R) is found from a transient analysis.

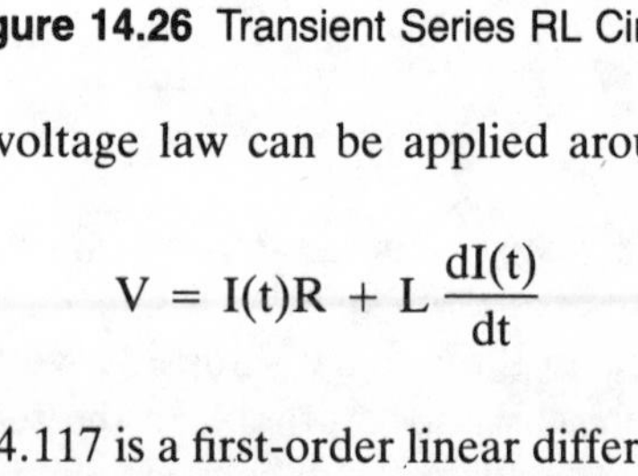

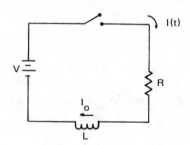

Figure 14.26 Transient Series RL Circuit

Kirchoff's voltage law can be applied around the circuit:

$$V = I(t)R + L\frac{dI(t)}{dt} \qquad 14.117$$

Equation 14.117 is a first-order linear differential equation with the solution

$$I(t) = \frac{V}{R}\left[1 - e^{-Rt/L}\right] + I_o \qquad 14.118$$

I_o is the initial current, if any, flowing in the circuit prior to the switch closing. Equation 14.118 is plotted in figure 14.26(b). Notice that the time axis is divided into two scales—seconds and time constants. The *time constant*[1] for a series *LR* circuit is the value of *t* which makes the instantaneous current equal to 63.3% of its steady state value.

$$.633 = 1 - e^{-Rt/L} \qquad 14.119$$

$$e^{-Rt/L} = .367 \qquad 14.120$$

$$\text{time constant} = \frac{L}{R} \quad \text{(LR circuit)} \qquad 14.121$$

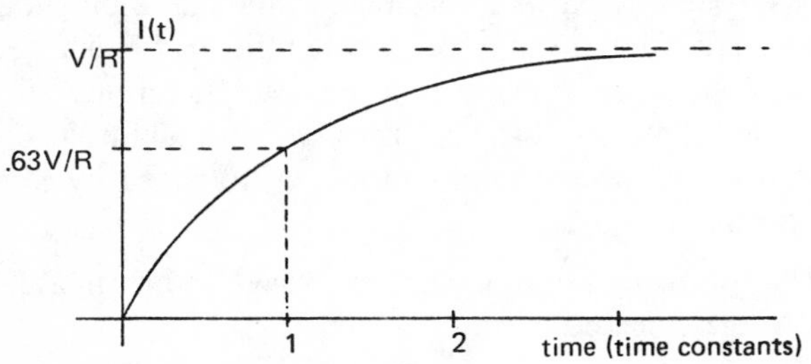

[1]The time constant for a series *RC* circuit is RC.

Care must be taken in generalizing about circuit response versus the time constant. Some parameters have reached 63% of their steady state values after one time constant. Others are at 37% of their final value. However, transient variables have essentially reached their steady state values after five time constants.

Solutions to differential equations resulting from transient circuits with DC excitation are readily available. These are listed in table 14.3. Complex circuits or circuits with AC excitation are best solved with Laplace transforms.

Example 14.12

A series *RC* circuit is connected to an AC source at t=0. What is the current as a function of time? Assume the capacitor is initially uncharged.

Because this is a forced circuit, the total current is the sum of the natural (transient) and forced (steady-state) currents.

The impedance is found first.

$$X_C = \frac{1}{\omega C} = \frac{1}{(377)(300\ \text{EE}-6)} = 8.84 \text{ ohms}$$

Table 14.3
DC Transient Solutions

TYPE OF CIRCUIT	*ENERGIZING*	*DISCHARGING*
Series RC: time constant = RC seconds 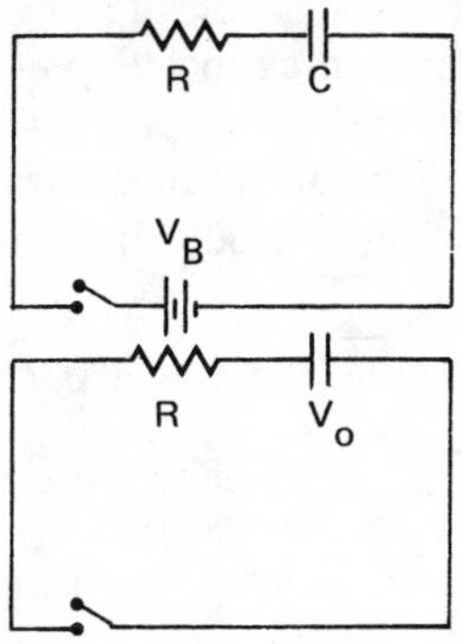	$V_R(t) = V_B \exp(-t/RC)$ $V_C(t) = V_B(1 - \exp(-t/RC))$ $I(t) = (V_B/R)\exp(-t/RC)$ $V_B = V_R + V_C$ $V_B = I(t)R + (1/C)\int I(t)dt$ $V_B = R(dq/dt) + (q/C)$ $q_C(t) = CV_B(1 - \exp(-t/RC))$	$V_R(t) = -V_o \exp(-t/RC)$ $V_C(t) = V_o \exp(-t/RC)$ $I(t) = (-V_o/R)\exp(-t/RC)$ $0 = V_R + V_C$ $q_C(t) = CV_o(\exp(-t/RC))$
Series RL: time constant = L/R seconds 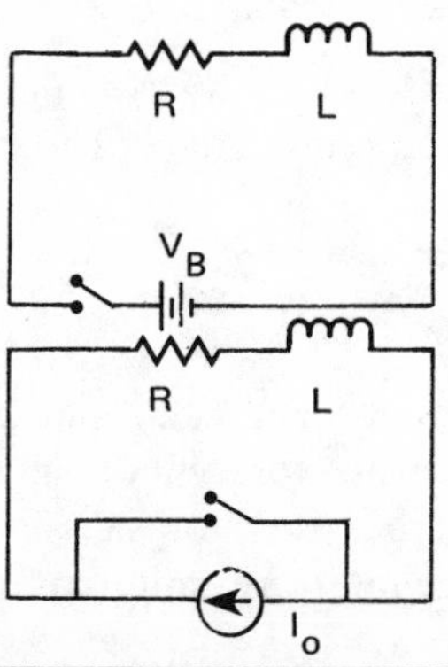	$V_R(t) = V_B(1 - \exp(-Rt/L))$ $V_L(t) = V_B \exp(-Rt/L)$ $I(t) = (V_B/R)(1 - \exp(-Rt/L))$ $V_B = V_R + V_L$ $V_B = RI(t) + L(dI(t)/dt)$	$V_R(t) = I_o R(\exp(-Rt/L))$ $V_L(t) = -I_o R(\exp(-Rt/L))$ $I(t) = I_o \exp(-Rt/L))$ $0 = V_R + V_L$

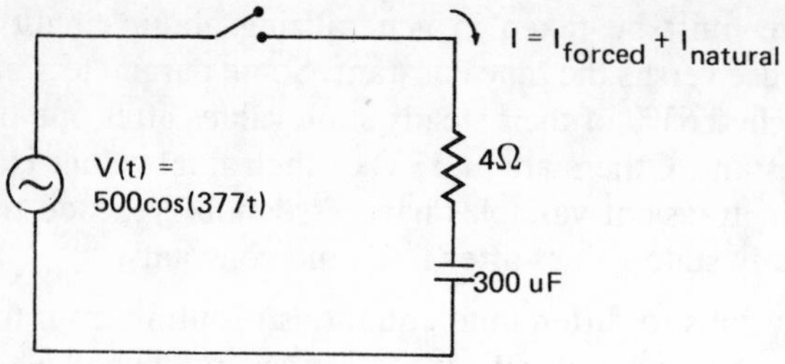

$$Z = \sqrt{(4)^2 + (8.84)^2} = 9.7 \text{ ohms}$$

$$\phi = \arctan\left(\frac{-8.84}{4}\right) = -65.65°$$

The forced response is

$$I_f(t) = \frac{\mathbf{V}(t)}{\mathbf{Z}} = \frac{500\angle 0°}{9.7\angle -65.65°} = 51.5\angle 65.65° \text{ amps}$$

At t=0, the forced current is 51.5 × cos(65.65°) = 21.2 amps.

At t=0, the applied voltage is 500 volts. This voltage appears across the resistor only, because voltage cannot change instantaneously across the capacitor. Therefore, the current through the resistor (which is also the current through the capacitor) is

$$I(0) = \frac{V}{R} = \frac{500}{4} = 125 \text{ amps}$$

However,

$$I = I_f + I_n$$

So,

$$I_n(0) = I(0) - I_f(0) = 125 - 21.2 = 103.8 \text{ amps}$$

The reciprocal of the time constant for the natural response is

$$\frac{1}{RC} = \frac{1}{(4)(300 \text{ EE}-6)} = 833$$

Therefore, the total response is

$$I(t) = 51.5\cos(377t + 65.65°) + 103.8e^{-833t}$$

13. AC Machinery

AC generators have been covered earlier in this chapter. Therefore, this section will be devoted to AC motors. Two types of AC motors are in widespread use — synchronous motors and induction motors.

A. Synchronous Motors

Synchronous motors are alternators operating in reverse. They are almost always three-phase. Alternating current is supplied to the stator windings. DC current is supplied to the rotor by slip rings and brushes, or by magnetic induction (transformer action) and rotor-mounted diodes. In *cylindrical-rotor motors,* the field windings are imbedded in a smooth-surfaced rotor. However, *salient-pole rotors* resemble electromagnets with windings in a direction normal to the shaft.

A desirable feature of the synchronous motor is its ability to draw a leading current. Thus, a synchronous motor can be used for power factor correction. The motor is then called a *synchronous capacitor.*

The speed at which a synchronous motor runs is known as the *synchronous speed.*

$$n_s = \frac{(120)f}{p} \qquad 14.122$$

Synchronous motors revolve at this constant synchronous speed regardless of load, although momentary speed fluctuations may occur when the load is changed. *Stalling* occurs when a motor's counter-torque is exceeded. Starting torque is zero. It is usually necessary to bring the synchronous motor up to speed by some external means.

The following relationships are based on one phase of a wye-connected motor.

$$V_{phase} = \frac{V_{line}}{\sqrt{3}} \qquad 14.123$$

$$T_{phase} = \tfrac{1}{3} T_{total} \qquad 14.124$$

$$P_{phase} = \tfrac{1}{3} P_{total} \qquad 14.125$$

The back EMF and torque generated are

$$E = K_1 n\phi = K_2 I_f \qquad 14.126$$

$$T = K_3 \phi_r \phi_s \sin\delta \qquad 14.127$$

ϕ is the air gap flux. ϕ_r and ϕ_s are the internal rotor and stator fluxes. δ is known as the *torque angle,* the phase difference angle between the applied voltage and the generated back EMF. Torque is maximum when the torque angle is 90°, a condition producing a unity power factor.

If I, E, and V are rms value, the power per phase is

$$P_{phase} = IV\cos\theta = \frac{VE}{X_s}\sin\delta \text{ (watts)} \qquad 14.128$$

$$= T\omega \quad \left(\frac{\text{ft-lbf}}{\text{sec}}\right) \qquad 14.129$$

The vector voltage relationship is

$$\mathbf{V} = \mathbf{E} + jIX_s \qquad 14.130$$

The torque angle, δ, is found by solving equation 14.130 for **E** and using its phase angle. The developed torque per phase is

$$T = \frac{P}{\omega} = \frac{VE}{\omega_s X_s}\sin\delta \qquad 14.131$$

The relationships in equations 14.128 and 14.131 involving the torque angle are for cylindrical-rotor motors only. They cannot be used for salient-pole machines. Salient-pole machines are more common than cylindrical-rotor machines.

AC ELEC

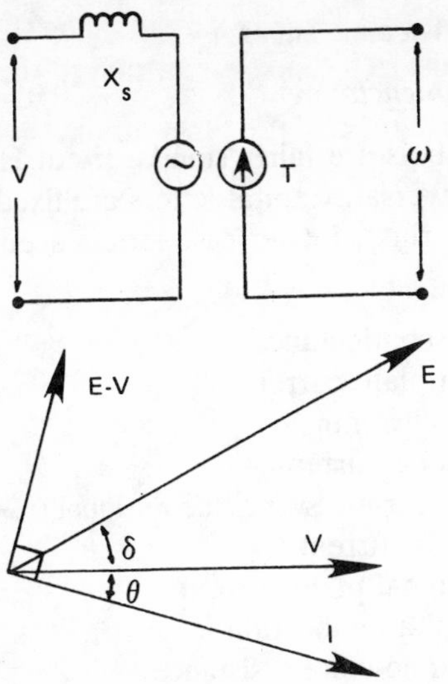

Figure 14.27 Synchronous Motor Equivalent Circuit

Example 14.13

A wye-connected three-phase motor is tested on 220 *rms* volts, 60 hz. The motor, which has a synchronous reactance of 3 ohms, is rated 10 *kva* at synchronous speed. The motor has 6 poles. At unity power factor, find the

- synchronous speed
- phase voltage
- phase current
- voltage drop across the synchronous reactance
- generated back EMF
- torque angle

The synchronous speed is

$$n_s = \frac{(120)f}{p} = \frac{(120)(60)}{6} = 1200 \text{ rpm}$$

The phase voltage is

$$220/\sqrt{3} = 127 \text{ volts } rms$$

The rated phase current is

$$I = \left(\frac{P}{3}\right)/V\cos\theta = \frac{10{,}000}{(3)(127)(1)}$$
$$= 26.25 \text{ amps}$$

The voltage drop across each winding is

$$IX = (26.23)(3) = 78.69 \text{ volts}$$

The back EMF is

$$\mathbf{E} = \mathbf{V} - jIX = 127 - j(78.69)$$
$$= 149.7\angle{-31.8°}$$

The torque angle is $-31.8°$.

B. Induction Motors

Induction motors are almost always three-phase devices. They are common, rugged, and inexpensive. Induction motors are available in almost every size. The motor's construction is unique because the rotor receives power through induction — there are no brushes. The motor is essentially a rotating transformer secondary with a stationary stator. There are no windings at all in a *squirrel-cage rotor.* However, a wound rotor may be incorporated to obtain speed control or high torque.

The squirrel-cage rotor consists of axial conductors with shorted ends. This is illustrated in figure 14.28. The stator is identical to that of a synchronous motor.

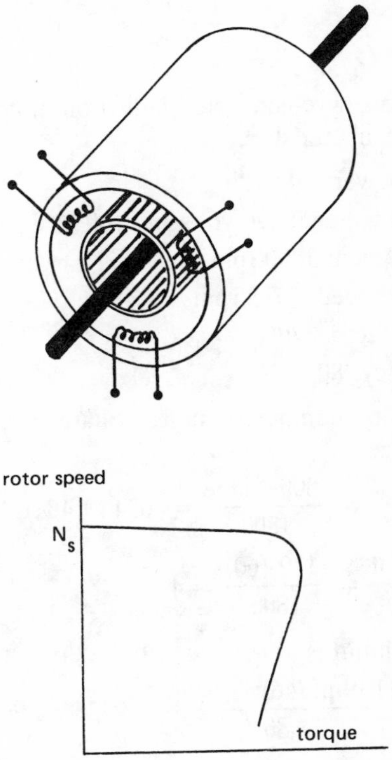

Figure 14.28 Squirrel-Cage Rotor

As the starter field moves past the rotor conductors, an EMF is induced by transformer action. Because the rotor windings have reactance, the rotor field lags the induced EMF by an angle θ. The stator field rotates at synchronous speed.

$$n_s = \frac{(120)f}{p} \qquad 14.132$$

In order for there to be a change in flux linkage, the rotor must turn slower than the synchronous speed. The difference in speed is small but essential. The *slip,* defined in equation 14.133, is typically 2% to 5%.

$$s = \frac{n_s - n_r}{n_s} \qquad 14.133$$

The frequency of the rotor field (relative to itself) is

$$f_r = sf \qquad 14.134$$

Induction motors have the following performance characteristics.

- Speed is essentially constant.
- Maximum current is drawn during starting, when slip is 1.
- Total torque is three times the phase torque.
- Torque is directly proportional to slip.
- Torque is inversely proportional to the rotor winding resistance.
- Starting torque is directly proportional to rotor winding resistance and line voltage.

Example 14.14

A three-phase wye-connected induction motor has the following characteristics:

- horsepower: 10
- system voltage: 220 volts *rms*, 60 hz
- rated speed: 1800 rpm
- loaded speed: 1738 rpm
- power factor: 70%
- efficiency: 80%

Find the slip, number of poles, phase current, and phase voltage.

$$s = \frac{n_s - n_r}{n_s} = \frac{1800 - 1738}{1800} = .034\ (3.4\%)$$

$$p = \frac{(120)f}{n_s} = \frac{(120)(60)}{1800} = 4$$

The power input is

$$P_{input} = \frac{(10)\text{hp}(746)\text{watts/hp}}{.80} = 9325 \text{ watts}$$

For a wye-connected motor,

$$P_{in} = V_{line} I_{line} \sqrt{3} \cos\phi$$

$$I = I_{phase} = I_{line} = \frac{9325}{(220)\sqrt{3}(.70)} = 34.96 \text{ amps}$$

The voltage across each phase is

$$220/\sqrt{3} = 127 \text{ volts}$$

14. Triode Vacuum Tubes

Special Nomenclature

These symbols are fairly standardized. However, they are not universal. Capital letters are fixed values (usually DC or rms). Lower case letters are instantaneous values.

g_m	transconductance	mhos
i_b	total plate current	amps
i_c	grid current	amps
i_k	cathode current	amps
i_p	plate current swing due to signal ($i_b - I_b$)	amps
I_a	heater current	amps
I_b	no-signal plate current	amps
K	signal amplification	–
r_p	dynamic plate resistance	ohms
R_L	load resistance	ohms
v_b	plate potential due to bias and signal ($V_b + v_p$)	volts
v_c	grid potential	volts
v_g	input signal potential	volts
v_o	output signal ($i_p R_L$)	volts
v_p	plate potential swing due to signal	volts
V_a	heater voltage	volts
V_b	no-signal plate potential due to bias battery	volts
V_{bb}	plate supply bias battery voltage	volts
V_c	no-signal grid potential due to bias battery	volts
V_{cc}	grid bias battery voltage	volts
μ	amplification factor	–

A triode vacuum tube consists of a filament, grid, and plate enclosed in an evacuated metal or glass tube. The filament is a thin wire which is heated by the application of a current supplied from battery V_a. Upon heating, electrons boil off of the filament which is gradually consumed. This cloud of electrons is called the *space charge* and surrounds the filament until the plate is positively charged by battery V_{bb}. This attraction is called the *Edison effect.*

For any fixed filament voltage, increasing the plate voltage will increase the plate current, I_b, up to the *satu-*

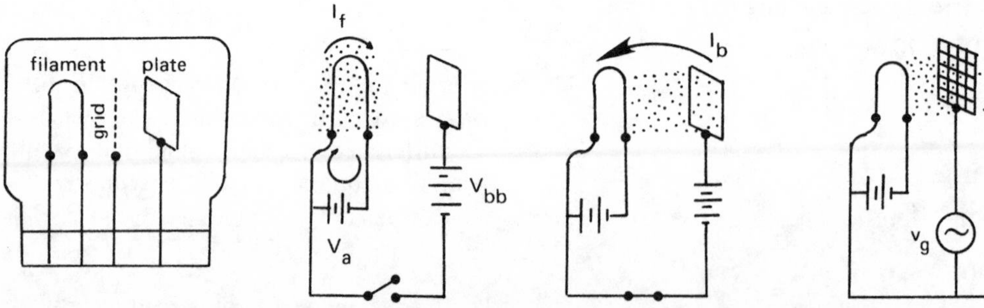

Figure 14.29 Triode Operation

ration point. Increasing the plate potential further will not increase plate current. Increases are not linear due to inter-electron repulsion. Even at the saturation point, not all electrons that boil off are absorbed by the plate. Some are repelled back to the filament.

Usually the filament and plate voltage are held constant. The plate current is controlled by the grid voltage applied to a wire screen *grid.* This grid is connected to the variable signal potential, v_g.

If the grid is at a positive potential, some of the electrons that would normally be repelled back into the filament are compelled to travel toward the grid and plate. Of course, some electrons actually hit the grid and are lost.

Due to the nearness of the grid to the filament, it only takes a small change in grid potential to affect plate current. It is this *leverage* that allows a vacuum tube to amplify low-power signals.

The grid is usually negative. Therefore, no grid current flows at any time. However, the degree of negativity is changed by the signal. Since the effect on the plate current by the grid is non-linear, no simple relationship can be developed for the variables. However, since input signals are usually very small and in a limited range, differentials can be used to show that the change in plate potential depends on the grid voltage and the plate current changes.

$$dv_b = \frac{\partial v_b}{\partial v_c} dv_c + \frac{\partial v_b}{\partial i_b} di_b \qquad 14.135$$

If the grid voltage, v_c, is held constant, the first term in equation 14.135 drops out.

$$dv_b = \frac{\partial v_b}{\partial i_b} di_b = r_p \, di_b \qquad 14.136$$

The coefficient of di_b is an equivalent tube parameter known as the *dynamic plate resistance,* r_p, in ohms. It is not a measurable resistance. Dynamic plate resistance is essentially constant. It can be found from the plate characteristic curves.

$$r_p = \tan\phi_1 \qquad 14.137$$

If plate current, i_b, is constant, the last term in equation 14.135 drops out.

$$dv_b = \frac{\partial v_b}{\partial v_c} dv_c = (-\mu) dv_c \qquad 14.138$$

The coefficient of dv_c is known as the *amplification factor,* μ. The negative sign is needed because of the phase reversal between input signal and response. The amplification factor can be found from the constant-current characteristic curves.

$$\mu = \tan\phi_2 \qquad 14.139$$

The *transconductance* (also known as the *mutual conductance*) with units of mhos is defined as

$$\frac{\partial i_b}{\partial v_c} = g_m \qquad 14.140$$

Transconductance can be found from the tube's transfer characteristics.

$$g_m = \tan\phi_3 \qquad 14.141$$

In most tube problems, these parameters can be approximated by using large swings (Δ) instead of infinitesimals (∂). Two of the parameters can be used to find the third.

$$\mu = (g_m)(r_p) \qquad 14.142$$

Equation 14.135 can be rewritten as equation 14.143.

$$\Delta v_b = -\mu \Delta v_c + r_p \Delta i_b \qquad 14.143$$

Example 14.15

When the plate and grid of a vacuum tube are at 220 volts and -5 volts respectively, the plate current is 5 *ma*. When the grid voltage is reduced to -3 volts, the plate current doubles. To reduce the plate current back to its original value, the plate voltage must be reduced to 150 volts. Find the tube parameters.

The amplification factor is

$$\mu = \frac{-\partial v_b}{\partial v_c} = \frac{-(v_{b2} - v_{b1})}{(v_{c2} - v_{c1})} = \frac{-(150-220)}{(-3-(-5))}$$

$$= -35$$

AC ELEC

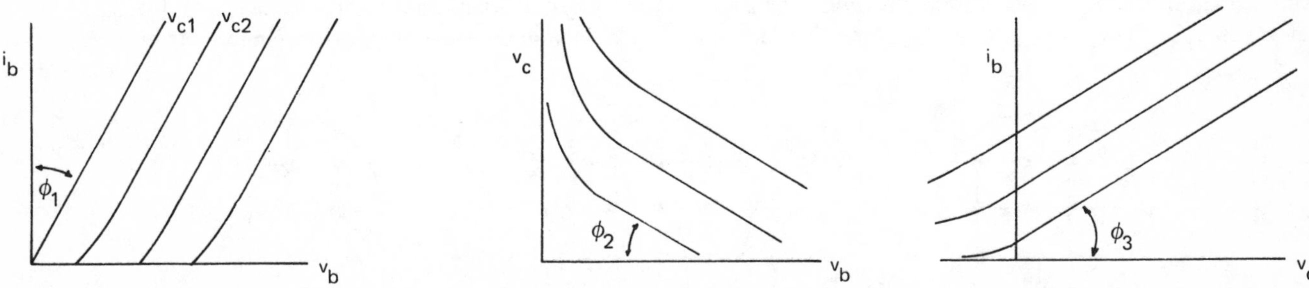

(a) plate (b) constant current (c) transfer

Figure 14.30 Triode Characteristics

The mutual conductance is

$$g_m = \frac{\partial i_b}{\partial v_c} = \frac{i_{b2}-i_{b1}}{v_{c2}-v_{c1}} = \frac{(10-5)\ \text{EE}-3}{(-3-(-5))}$$

$$= 2.5\ \text{EE}-3\ \text{mhos}$$

The plate resistance is

$$r_p = \frac{\mu}{g_m} = \frac{35}{2.5\ \text{EE}-3}$$

$$= 1.4\ \text{EE4 ohms}$$

15. Triode Amplifier Circuits

Usually the heater is omitted in schematic drawings. The schematic representation of a triode is shown in figure 14.31(a). Notice that the symbol *k* is used to indicate the *cathode* (filament).

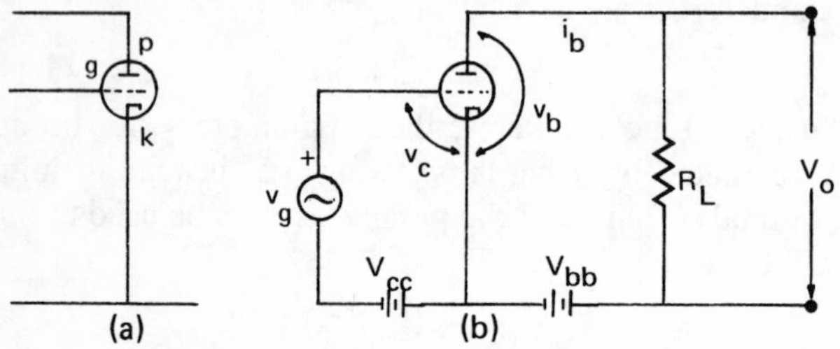

Figure 14.31 Grounded-Cathode Circuit

The most common application of a triode is an amplification circuit in the *grounded-* (or *common-*) *cathode* configuration. This circuit is shown in figure 14.31(b). V_{cc} must be larger than v_g to keep the grid negative. If the grid is kept negative by V_{cc}, then $i_c = 0$. However, changes in v_c due to the application of v_g will change the repulsive power of the grid, changing the flow of electrons to the plate.

Due to the leverage of the grid, massive changes in i_p take place when v_g changes. This results in a varying voltage across R_L. Since the voltage change across R_L is many times greater than the signal swing, the circuit amplifies. This amplification is shown in figure 14.32.

In figure 14.32, the straight diagonal line is known as the *load line.* The load line is defined by points $(V_{bb},0)$ and $(0, V_{bb}/R_L)$ on the plate curves. The intersection of the load line at $v_c = -V_{cc}$ defines the *quiescent point* at which $v_g = 0$, $v_c = V_{cc}$, $v_b = V_{bb}$, and $i_b = I_b$.

When a small signal, v_g, is introduced, v_c, v_b, and i_b change.

$$v_c = V_c + v_g \quad 14.144$$

$$v_b = V_b + v_p \quad 14.145$$

$$i_b = I_b + i_p \quad 14.146$$

The output voltage seen across the load resistance is

$$v_o = \frac{-\mu v_g R_L}{r_p + R_L} \quad 14.147$$

The *amplification (gain)* of the circuit is

$$K = \frac{v_o}{v_g} \quad 14.148$$

Gain can be increased if R_L is increased, but a larger V_{bb} will be required. Choice of V_{bb} will depend on the tube and economic factors.

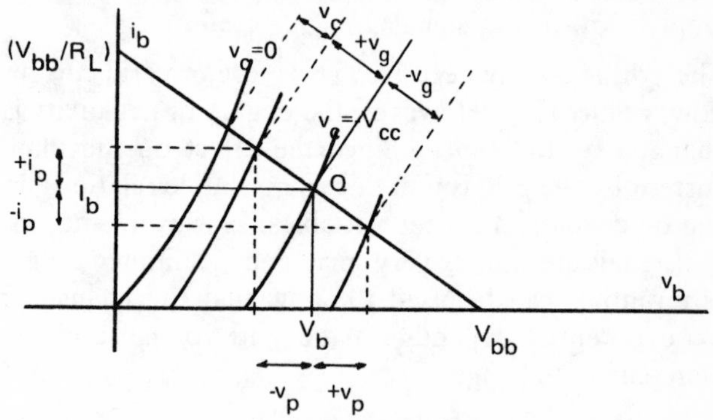

Figure 14.32 Amplifier Operation

16. Semiconductors

A. Introduction

Silicon and germanium, which form the basis for semiconductors, are slightly more conductive than insulators but are not nearly as conductive as metals. Both silicon and germanium possess four valence electrons. Their crystal lattice structure uses covalent bonding to fill the outer shell of eight (the octet). As in other chemical structure diagrams, each line in figure 14.33 represents a shared electron pair.

At normal temperatures, the lattice is not perfect. An occasional free electron or missing electron (*hole*) ex-

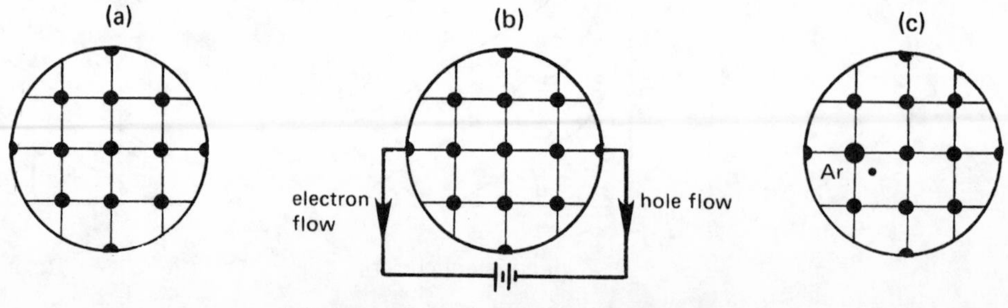

Figure 14.33 Semiconductor Structure

ists. The formation of holes and free electrons increases as temperature increases. This formation is known as *thermal carrier generation.* In a semiconductor, both electron and hole movement contribute significantly to current when a voltage is applied. This is illustrated in figure 14.33(b).

The addition of minute quantities (e.g., 1 in EE8) of impurities with either three or five valence electrons increases semiconductor conductivity. Intentional impurities are called *dopes* and their addition is known as *doping*. Typical dopes are phosphorus, arsenic, and antimony.

When a 5-valence dope is used in a germanium lattice, an electron which requires little ionization energy is made available for conduction. Thus, arsenic, for example, donates a negative charge carrier. Arsenic is therefore a *donor,* and the resultant semiconductor is called an *n-type* since negative electrons are the *majority carriers.* Some holes will still be present, of course, and these are the *minority carriers.*

Aluminum, boron, indium, and gallium have only three valence electrons. A *vacancy* is created when one of these elements is used as a dope. Due to the incomplete structure, the vacancy is easily shifted from one atom to another. These dopes are called *acceptors*. Semiconductors manufactured with acceptor dopes are called *p-types* since positive holes are the *majority carriers.* Electrons are the *minority carriers.*

B. Semiconductor Diodes

Special Nomenclature

I	total junction current	amps
I_i	injection or recombination current	amps
I_s	saturation or thermal current	amps
k	Boltzmann constant (1.38 EE−23)	joule/°K
q	electron charge (1.6 EE−19)	coulomb
T	temperature	°K
V	applied voltage	volts

A diode consists of a p-type semiconductor in contact with an n-type semiconductor. The boundary between the two is called a *junction*. The p-type is the *anode,* and the n-type is the *cathode.* Due to diffusion, holes and electrons cross the junction. The ions left behind set up an electric field which stops further diffusion and establishes an equilibrium. The material in the vicinity of the junction is called the *depletion region* since holes and free electrons have been depleted.

Even in equilibrium, however, an occasional hole or electron manages to overcome the electric field. This charge carrier combines with its counterpart on the other side of the depletion region. This carrier movement is known as *recombination current.* Absorbed energy in the form of heat causes holes and electrons to move across the electric field. This carrier movement is called *thermal current* or *saturation current.* At equilibrium and without an applied voltage, the junction current is

$$I = I_i - I_s = 0 \qquad 14.149$$

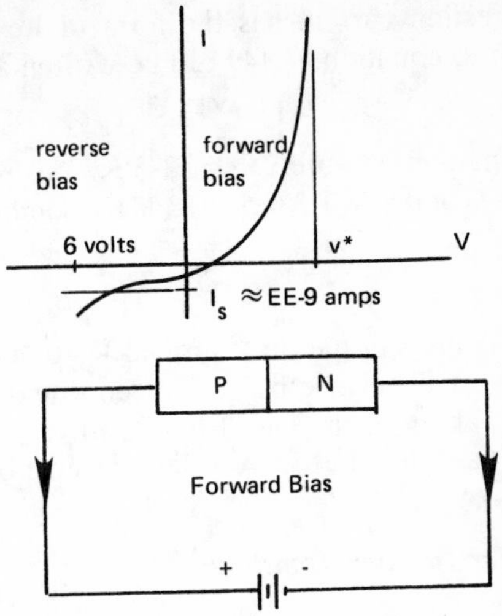

Figure 14.34 Semiconductor Diode Characteristics

This equilibrium is destroyed when a battery is connected across the diode, as shown in figure 14.34. Assume the battery is connected as shown, with the positive terminal to the p-type anode. Thermal current is not affected as long as the temperature is constant. However, holes are repelled from the left side and electrons are repelled from the right side. This neutralizes some ions and reduces the electric field which kept the recombination current small. A sizeable increase in recombination current results. Connecting the battery in this fashion is known as *forward bias*.

If the battery is reversed, the connection is known as *reverse bias.* The electric field will be increased and recombination current flow will be reduced almost to zero. Thermal current will remain the same as long as the temperature is constant. Therefore, a net junction current (about EE−9 amps) will exist in an opposite direction from the forward bias current. This reverse bias current has three causes.

- thermal current due to minority carriers
- surface leakage which is dependent on the surface resistance
- avalanche breakdown when the reverse bias exceeds about 6 volts. Each minority carrier collides with and forcibly removes several other electrons.

A typical static characteristic curve is shown in figure 14.34. V* is the voltage required to overcome the electric field. V* is about .6-.8 volts for silicon and about .2 volts for germanium.

AC ELEC

The injection current has the form of $I_s\exp(qV/kT)$. Therefore, equation 14.149 can be written as

$$I = I_s[e^{qV/kT} - 1] \qquad 14.150$$

At room temperature, T = 293°K. Also, q = 1.6 EE−19 C and k = 1.38 EE−23 J/°K. Combining these terms,

$$I \approx I_s[e^{40V} - 1] \qquad 14.151$$

Due to flatness of the curve around $V = 0$, any value of I with a small reverse bias (between 0 and −1 volts) can be taken as I_S (e.g., EE−5 amps). Equation 14.151 assumes that the diode is reverse-biased if V is negative.

C. Bipolar Junction Transistors

Special Nomenclature

i_B	total base current	amps
i_C	total collector current	amps
i_E	total emitter current	amps
i_g	input signal current	amps
I_B	no signal (quiescent) base current	amps
I_C	no signal (quiescent) collector current	amps
I_{CBO}	collector cut-off current	amps
I_E	no signal (quiescent) emitter current	amps
R_E	equivalent emitter resistance	ohms
R_{in}	input resistance as seen by the base/emitter terminals	ohms
T	temperature	°K
v_g	input signal voltage	volts
V	bias voltage	volts

Symbols

α	a ratio (I_C/I_E)	—
β	current gain, or current transfer ratio $[\alpha/(1-\alpha)]$	—

A *transistor*[2] consists of a sandwich of two p-type semiconductors with one thin n-type semiconductor (*pnp*), or two n-type semiconductors with one thin p-type semiconductor (*npn*). Usually silicon *npn* transistors are used in high-temperature applications. Germanium *pnp* transistors offer lower collector-emitter saturation voltages and usually lower cost.

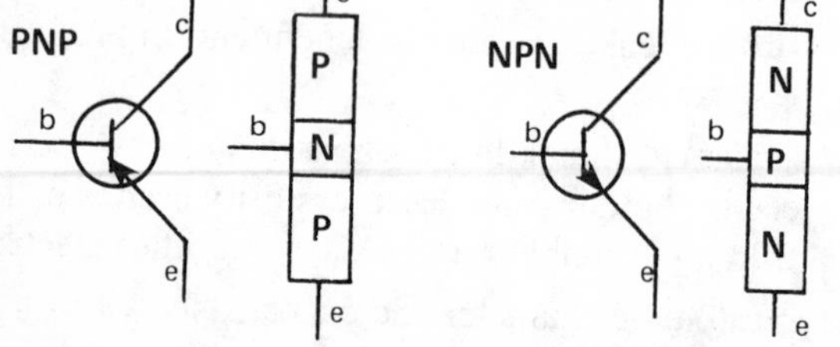

Figure 14.35 pnp and npn Transistors

[2]This type of transistor is known as a *bipolar junction transistor,* BJT.

The center semiconductor (n-type in a *pnp* transistor, or the p-type in an *npn* transistor) is known as the *base.* The outer semiconductor pieces are known as the *emitter* and *collector.* The base, emitter, and collector are labeled on a circuit diagram with the letters *B, E,* and *C* respectively. The arrow in the transistor symbols always points towards the n-type material.

Normally, the emitter-base junction is forward-biased as in figure 14.36(a). The majority carrier holes are emitted from the emitter and injected into the base region. A small electron flow also occurs from the base to the emitter. Some holes in the base combine with electrons from the battery and are lost. However, many holes make it across to the collector-base junction.

Normally, the collector-base junction is reverse-biased as in figure 14.36(b). This reverse biasing prohibits hole movement from the collector to the base. However, holes are attracted to the collector where they combine with electrons from the battery.

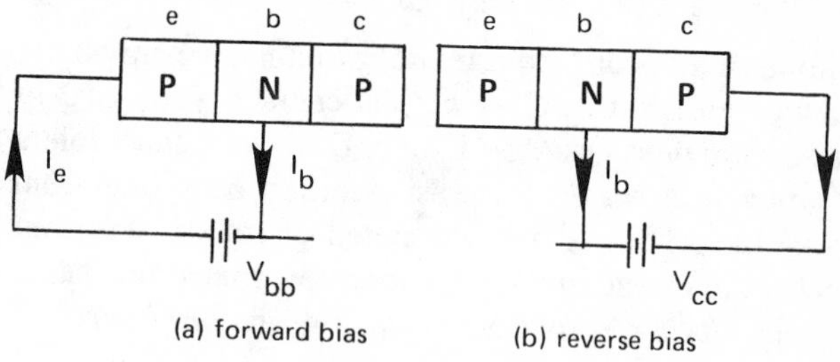

Figure 14.36 Forward and Reverse Biasing

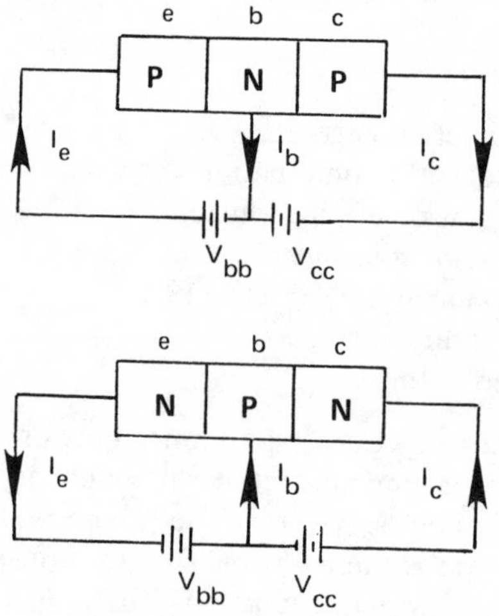

Figure 14.37 Fully Biased Transistors

Since hole-electron combination in the base is undesirable, the base is kept very thin. Usually I_B is two or three orders of magnitude smaller than I_C or I_E.

$$I_E = I_C + I_B \qquad 14.152$$

Thermal current is also present. It can be included in equation 14.152.

$$I_C = \alpha I_E + I_{CBO} \qquad 14.153$$

I_{CBO} is the *collector cut-off current.* This is the thermal current at the collector-base junction. It is very small at room temperature. However, it doubles every 10°C, making germanium transistors useless around 100°C. Silicon remains useful up to around 200°C.

17. Transistor Amplifier Circuits

Since transistors are often used in common-emitter configuration, only the *common-emitter amplifier* shown in figure 14.39 is analyzed in this chapter. Typical characteristics for a common-emitter *npn* transistor are shown in figure 14.38.

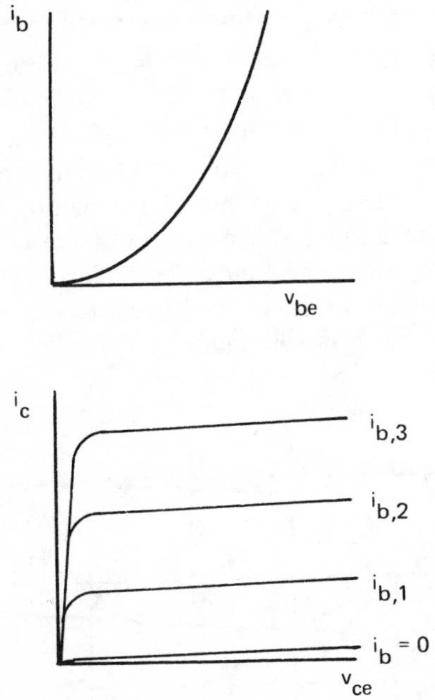

Figure 14.38 Common-Emitter Characteristics

The analysis of a simple common-emitter amplifier is illustrated in the following procedure. Some steps may be out of order, depending on the given and required information.

step 1: The load line is defined by any two of the three points shown in figure 14.40.

point 1: $v_{CE} = V_{cc}$ and $i_C = 0$

point 2: $v_{CE} = 0$ and $i_C = V_{cc}/R_L$

point 3: knowledge of I_C and v^*_{CE}
or I_C and I_B
or v^*_{CE} and I_B

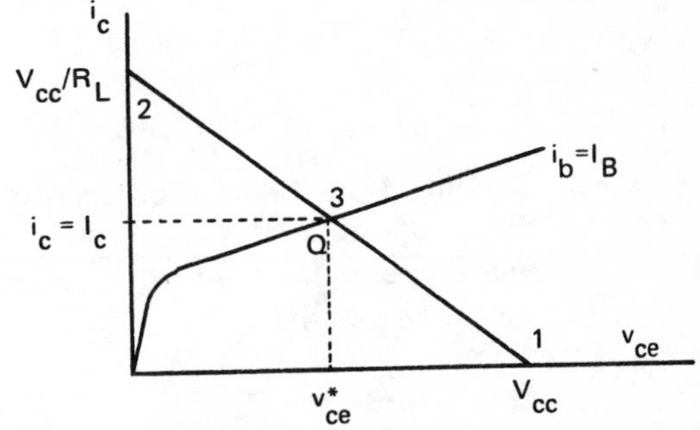

Figure 14.40 The Load Line

step 2: Assuming no signal, determine I_B or R_1 by using Kirchoff's voltage law around the input loop.

$$V_{bb} = I_B R_1 + v_{BE} \qquad 14.154$$

The base-emitter junction is forward biased and v_{BE} is less than one volt.

$$V_{bb} \approx I_B R_1 \qquad 14.155$$

step 3: Assuming no signal, use Kirchoff's voltage law around the output loop to get the load-line equation. Solve for v_{CE}, R_L, or I_C.

$$V_{cc} = v^*_{CE} + I_C R_L \qquad 14.156$$

step 4: When the signal is applied, calculate the total base current.

$$i_B = I_B + i_g = I_B + \frac{v_g}{R_1} \qquad 14.157$$

step 5: On the characteristic curve, find the signal i_C from i_B.

step 6: Calculate the *current* gain.

$$\beta = \frac{i_C - I_C}{i_g} = \frac{R_1(i_C - I_C)}{v_g} \qquad 14.158$$

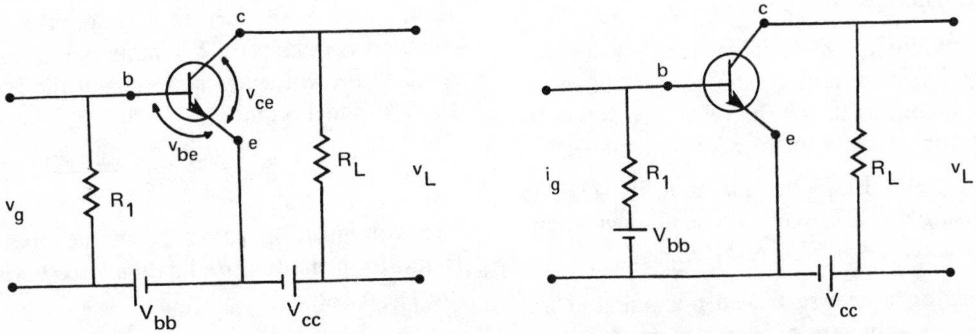

Figure 14.39 Common Emitter Amplifiers

step 7: Calculate the incremental voltage across the load resistance.

$$\Delta V_L = (i_C - I_C)R_L \qquad 14.159$$

step 8: Calculate the amplifier input impedance.

8.1: $$I_E = \frac{I_B}{1-\alpha} \qquad 14.160$$

8.2: $$R_E \approx \frac{25{,}000}{I_E} \qquad 14.161$$

8.3: $$R_{in} = R_E(\beta+1) = \frac{V_{BE}}{I_B} \qquad 14.162$$

8.4: Combine R_{in} with any other resistors in the input circuit to calculate the *input impedance.* In figure 14.39, the resistances are in parallel.

$$Z_{in} = \frac{R_1 R_{in}}{R_1 + R_{in}} \qquad 14.163$$

step 9: Calculate the *output impedance.* This is a parallel combination of R_L and R_C. R_C is a transistor characteristic, not a circuit element. However, since R_C is very large, $Z_{out} \approx R_L$.

step 10: Find the *resistance gain.*

$$A_R = \frac{Z_{out}}{Z_{in}} \qquad 14.164$$

step 11: Find the *voltage gain.*

$$A_V = \frac{\Delta V_L}{v_g} = A_R\beta = \frac{\beta R_L}{(\beta+1)R_E} \approx \frac{R_L}{R_E} \approx \frac{R_L I_E}{25{,}000} \qquad 14.165$$

step 12: Find the *power gain:*

$$A_P = \beta^2 A_R = \beta A_V \qquad 14.166$$

step 13: Find the collector power dissipation. Check it against the maximum allowable for the transistor. (Use maximum values, not effective values.)

$$P_C = i_C v_{CE} \qquad 14.167$$

18. Field Effect Transistors

Field effect transistors, unlike bipolar junction transistors, draw negligible current at the gate. (The *gate, drain,* and *source* terminals for the FET correspond to the base, collector, and emitter terminals for the BJT.)

There are two types of FETS, the *junction FET (JFET)* and the *insulated gate FET (IGFET)* also known as the *metal-oxide-semiconductor FET* or *MOSFET.*

The JFET relies on a reverse biased p-n junction for its operation. The mechanism of control is the *depletion region* of the junction which widens with increasing

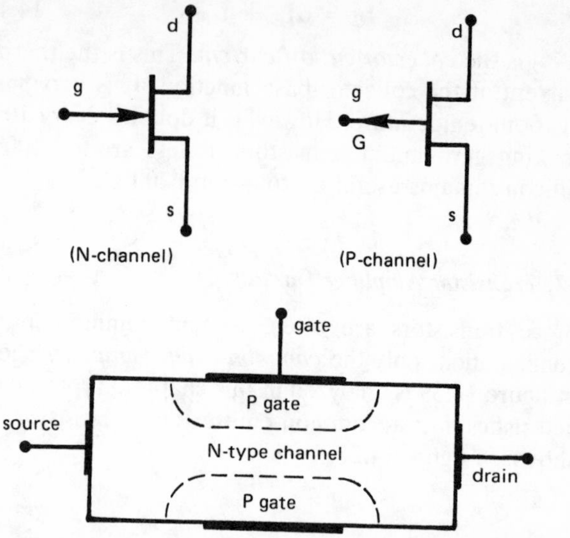

Figure 14.41 Junction Field Effect Transistor (JFET)

negative junction bias. The thickness of the depletion region controls the resistance of the channel between the drain and source. When the depletion regions of the two gates meet at the middle of the channel, the channel is said to be *pinched off.* The FET then behaves similarly to the BJT. Typical characteristics for the JFET are shown in figure 14.42.

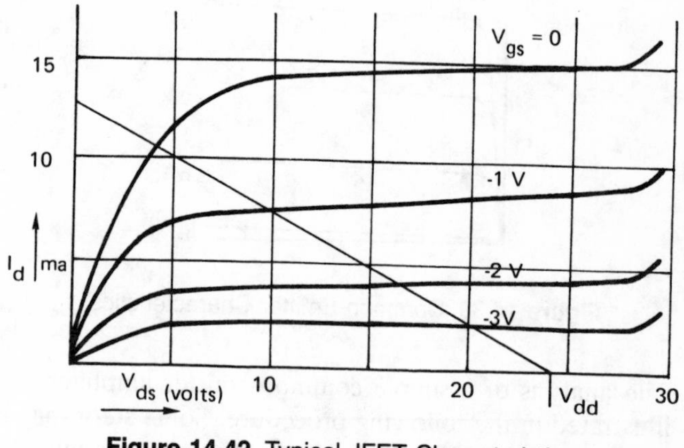

Figure 14.42 Typical JFET Characteristics

A *self-biasing* circuit for a JFET is shown in figure 14.43. The gate current through the reverse-biased junction is negligible. When an operating point is chosen and the voltage V_{dd} is known, the load line can be used to find R_d and R_s.

$$R_d + R_s = \frac{V_{dd} - V_{ds}}{I_s} \qquad 14.168$$

The voltage drop across R_s at the operating point is found from the required gate-source voltage at the operating point.

$$I_s R_s \Big|_{\text{operating point}} = -V_{gs}\Big|_{\text{operating point}} \qquad 14.169$$

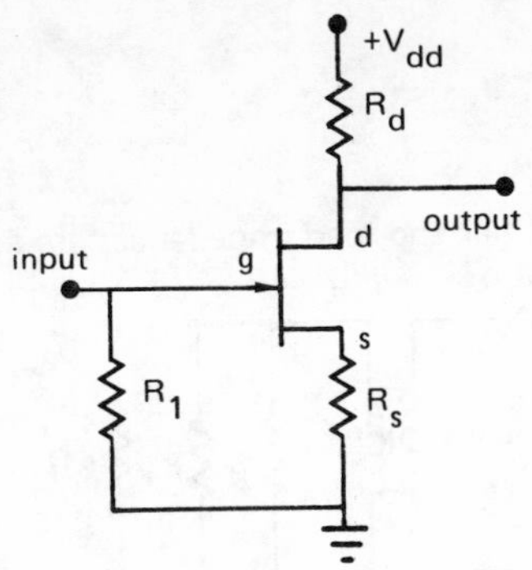

Figure 14.43 Self-Biasing JFET Circuit

Example 14.16

A JFET with characteristics shown in figure 14.42 is to be operated as a small-signal amplifier. The available supply voltage is 24 volts. A bias source current of 5 mA is acceptable at a bias voltage of V_{ds} = 15 volts. Design the biasing circuit.

From equation 14.168,

$$R_d + R_s = \frac{24-15}{.005} = 1800 \text{ ohms}$$

Since the gate draws negligible current, $I_d = I_s$.

Reading from figure 14.42 for $V_{ds} = 15$ and $I_d = .005$, $V_{gs} \approx -1.75$ volts.

From equation 14.169,

$$R_s = \frac{-(-1.75)}{.005} = 350 \text{ ohms}$$

Since $R_d + R_s = 1800$,

$$R_d = 1800 - 350 = 1450 \text{ ohms}$$

The construction of an insulated gate field effect transistor (IGFET) is illustrated by figure 14.44.

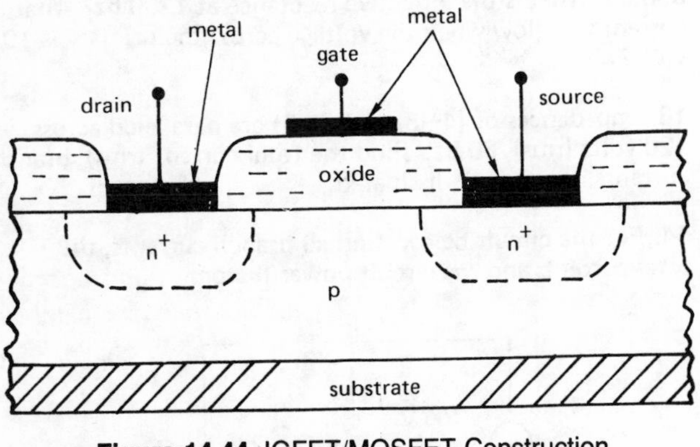

Figure 14.44 IGFET/MOSFET Construction

IGFET/MOSFET's are usually operated with a positive V_{gs} bias. This biasing requires a separate network. A typical biasing circuit for an IGFET/MOSFET is illustrated by figure 14.45. As no gate current flows, the resistor R_g is used to increase the input impedance. R_g is normally very large.

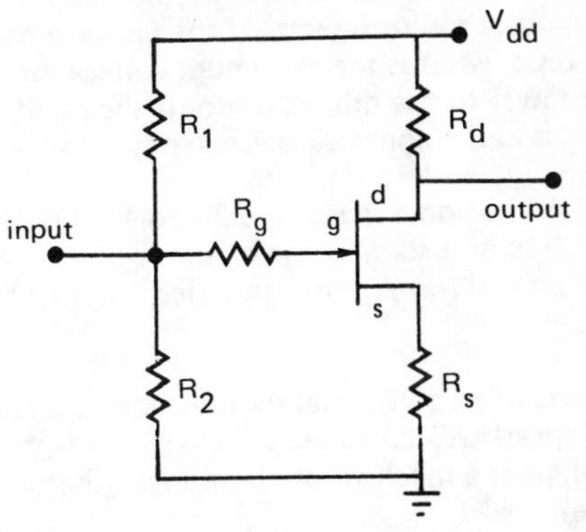

Figure 14.45 IGFET/MOSFET Biasing Circuit

Design of the biasing circuit is straightforward. R_1 and R_2 form a voltage divider. Since the gate current is negligible, the voltage divider is unloaded. Therefore, the gate voltage is

$$V_g = V_{dd}\left[\frac{R_2}{R_1 + R_2}\right] \qquad 14.170$$

Using the load line equation (equation 14.168),

$$V_{gs} = V_g - I_s R_s \qquad 14.171$$

Typical characteristic curves for an IGFET/MOSFET are shown in figure 14.46.

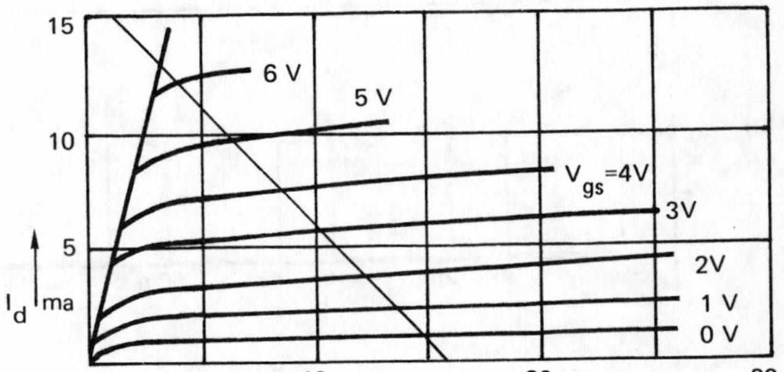

Figure 14.46 IGFET/MOSFET Characteristics

AC ELEC

A.C. GENERATION

1. A 4-pole armature rotating at 1800 rpm has 12 coils of 20 turns each. The effective armature length is 16.7 cm. The effective armature radius is 5 cm. If the uniform magnetic field is one tesla, what is the maximum voltage induced? What is the horsepower of the driving motor if the rated output is one kilowatt and the conversion efficiency is 90%?

2. What is the rotational speed of a 24-pole alternator that produces a 60 hz sinusoid? If the effective EMF is 2200 volts and each phase coil has 20 turns in series, what is the maximum flux?

3. The outputs of two identical six-pole alternators rotating at the same speed are connected in series. If the phase difference angle is a mechanical 20 degrees, what is the total voltage?

FORMS AND FACTORS

4. Find the average value, rms value, crest factor, and form factor for the waveforms shown below.

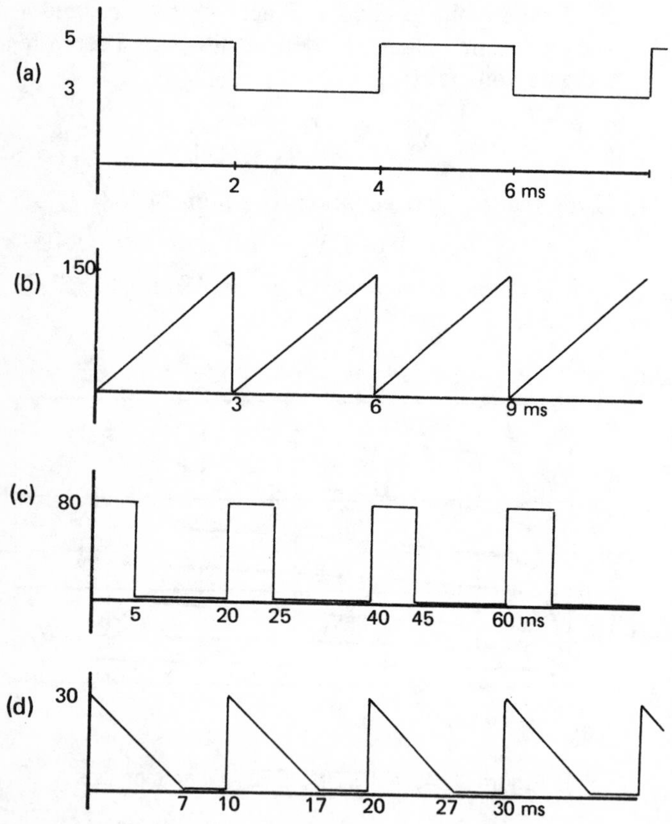

TRANSFORMERS

5. A step-down transformer consists of 200 primary turns and 40 secondary turns. The primary voltage is 550 volts. If an impedance of 4.2 ohms is the secondary load, what is the (a) secondary voltage (b) primary current (c) secondary current?

6. If the amplifier and load impedances are matched, what is the turns ratio?

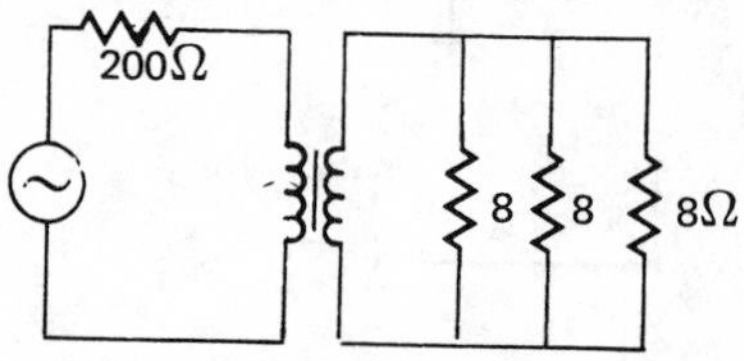

7. A 100 volt source with an internal resistance of 8 ohms is connected to a 200/100 turn step-down transformer. An impedance of (6+j8) is connected to the secondary. Find the primary current and secondary power.

COMPLEX AND PHASOR CALCULATIONS

8. Express each of the following in rectangular form.
   (a) $15\angle -180^\circ$
   (b) $10\angle 37^\circ$
   (c) $50\angle 120^\circ$
   (d) $21\angle -90^\circ$

9. Express each of the following in phasor notation.
   (a) (6+j7)
   (b) (50-j60)
   (c) (-75+j45)
   (d) (90-j180)

10. Calculate the following and express in phasor notation.
   (a) (7+j5)/6
   (b) (13+j17)/(15-j10)
   (c) $.020\angle 90^\circ \div .034\angle 56^\circ$

A.C. CIRCUIT ANALYSIS

11. Two capacitors are connected in series across a 12 volt line. One has a capacitance of 3 uf and the other has a capacitance of 6 uf. Calculate the charge and potential across each.

12. The primary of an r-f transformer has an inductance of 350 uh. What is the inductive reactance at 1200 hz? What current will flow when the voltage across the primary is 10 volts?

13. Impedances of (4+j5) and (5-j3) are paralleled across 120 volts (rms), 60 hz. Find the total current (rms), branch currents, and power dissipated.

14. For the circuit below, find all branch currents, the total current, and the circuit power factor.

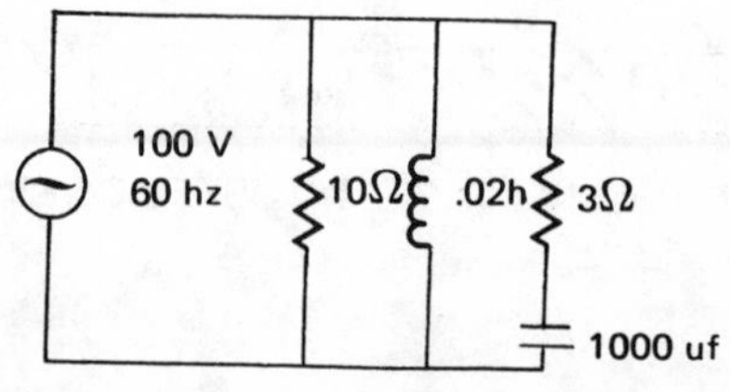

15. What is the voltage across the capacitor?

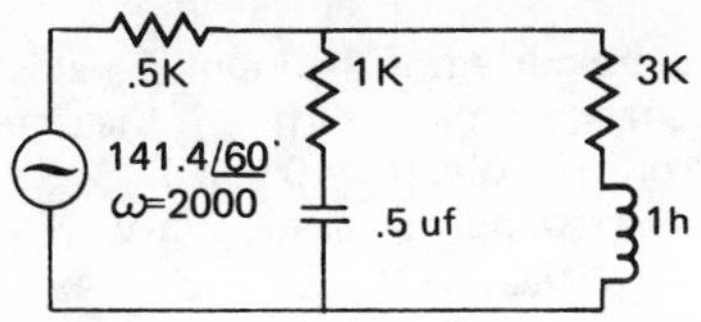

16. What is the voltage across the inductance?

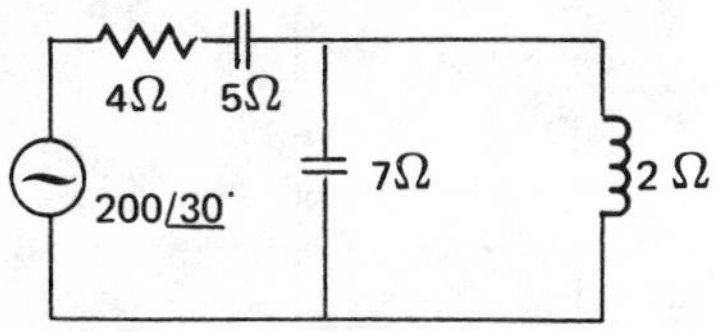

RESONANCE

17. A 120-volt (rms) 60 hz circuit consists of a capacitive reactance of 46 ohms and a resistance of 11 ohms in series. Find the impedance of the circuit, the current, the size of inductor which should be added in series to maximize current, and the combined voltage drop across the capacitor and inductor at resonance.

18. What capacitance is required for 60 hz series resonance with a 20 ohm coil? What is the circuit current if the applied voltage is 220 volts?

19. A coil (4 ohms reactance, 4 ohms resistance) is in series with a capacitor. The circuit is connected across a 110 volt (rms) 60 hz source. What is the capacitance required for resonance? What are the quality factor and bandwidth?

20. A 100 uh coil has an internal resistance of 15 ohms. What parallel capacitance will cause resonance at 100 khz?

21. What are the 70% points of a 200 hz resonant circuit which consists of a 10 ohm resistance, a 1 mH inductor, and a 2 nF capacitor in parallel?

COMPLEX POWER

22. A resistance of 6 ohms and an unknown impedance coil in series take 12 amps from a 120 volt (rms) 60 hz line. If the total power taken from the line is 1152 watts, what is the coil resistance? What is the coil inductance?

23. A small industrial plant operates with an average load of 400 kw, consisting of small induction motors and some electric heating. The average power factor is 60% lagging. A 150 kw synchronous motor driving a compressor is added to the plant. Neglecting motor losses, calculate the power factor at which this motor must operate in order to raise the power factor of the plant to 80%.

24. Three 2000 horsepower synchronous motors of 92% efficiency and 80% power factor are to be added to an existing 10,000 kva load with lagging 82% power factor. If the motors are operating fully loaded, what will be the new kva and power factor?

25. A .001 henry coil and an unknown resistance draw 50 va from a 100 volt (rms) 4000 hz source. What is the load resistance?

THREE PHASE CIRCUITS

26. A three phase 208 volt (rms) system supplies heating elements connected in wye. What is the resistance of each element if the total balanced load is 3 kw?

27. A three phase system has a balanced load of 5000 kw at 84% power factor. If the line voltage is 4160 volts (rms), what is the line current? The system is delta connected.

28. A 120 volt (rms) three phase system has a balanced load consisting of three 10 ohm resistances. What power is dissipated if the connection is (a) wye (b) delta?

29. A balanced delta load consists of three 20∠25˚ impedances. The 60 hz line voltage is 208 volts (rms). Find the phase current, the line current, the phase voltage, the power consumed by each phase, and the total power.

30. A balanced wye load consists of three (3+j4) impedances. The system voltage is 110 volts (rms). Find the phase voltage, the line current, the phase current, and the total power.

AC ELEC

TRANSIENTS

31. The switch in the circuit below is closed at t=0 and opened again at t = .05 seconds. How much energy is stored in the capacitor?

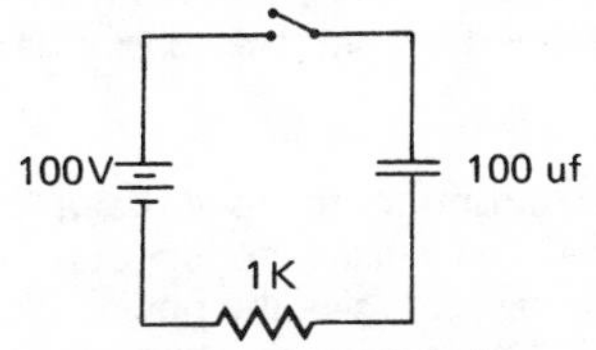

32. Find the current through the 80 ohm resistor (a) immediately after the switch is closed, and (b) 2 seconds after the switch is closed. The switch has been open for a long time.

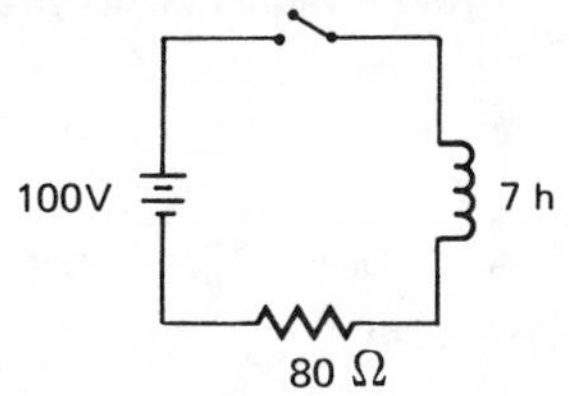

33. The switch moves from a to b at t=0. What is the current through the 15 ohm resistor after one second?

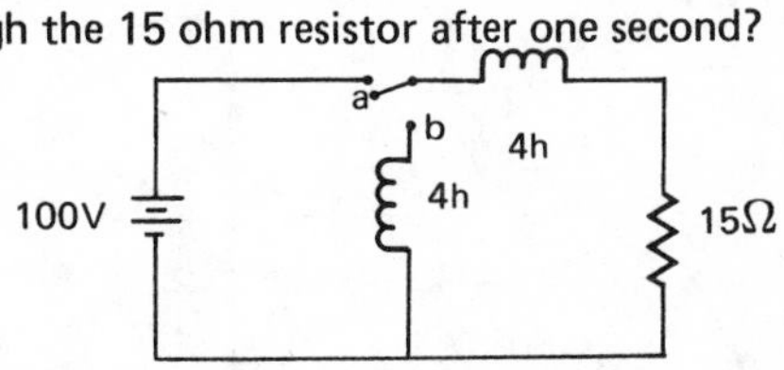

34. At t=0 the switch connects to position 1. At t=5 EE-4 seconds, the switch is moved to position 2. Assume there is no current anywhere in the circuit prior to t=0. What is the current flowing in the circuit just prior to the switch moving to position 2? What is the current after the switch is moved to position 2?

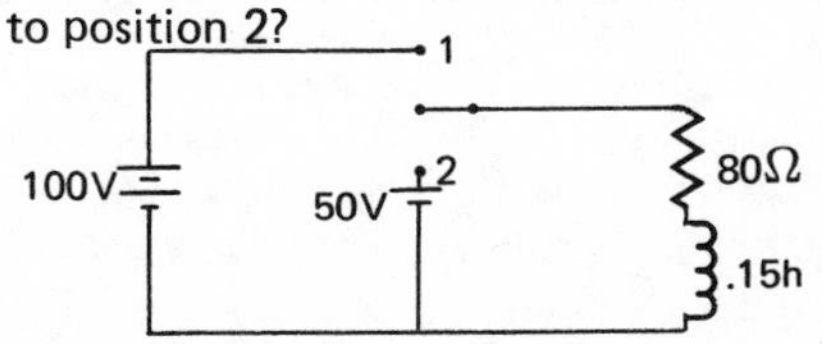

35. The capacitor has an initial charge of 600 uC with polarity as shown. What is the current after the switch is closed?

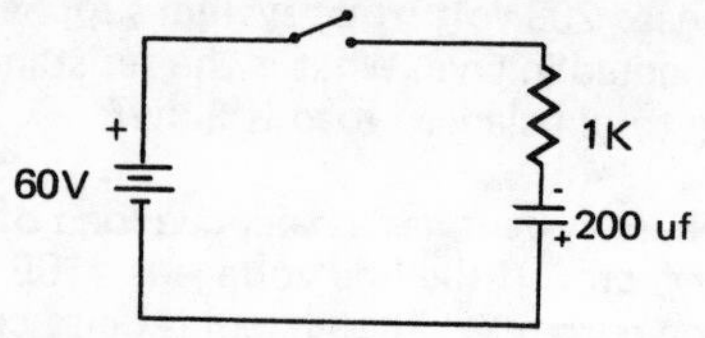

36. If the capacitor is initially uncharged, what is the current after the switch is closed? Assume ϕ = 90° when the switch is closed.

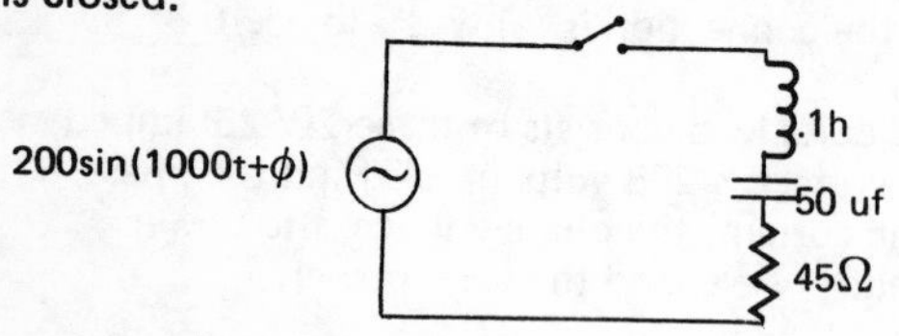

A.C. MOTORS

37. Calculate the full load current drawn by a 440 volt (rms) 20 horsepower induction motor having a full load efficiency of 86% and a full load power factor of 76%.

38. A 200 horsepower, three phase, four pole, 60 hz, 440 volt (rms) squirrel cage induction motor operates at full load with an efficiency of 85%, power factor of 91%, and 3% slip. Find the speed in rpm, the torque developed, and the line current.

39. A factory's induction motor load draws 550 kw at 82% power factor. What size synchronous motor is required to carry 250 horsepower and raise the power factor to 95%? The line voltage is 220 volts (rms).

40. The nameplate of an induction motor lists 960 rpm as the full load speed. For what frequency do you think the motor was designed?

VACUUM TUBES

41. Estimate u, g_m, and r_p for the 12AX7 triode whose characteristics are given below.

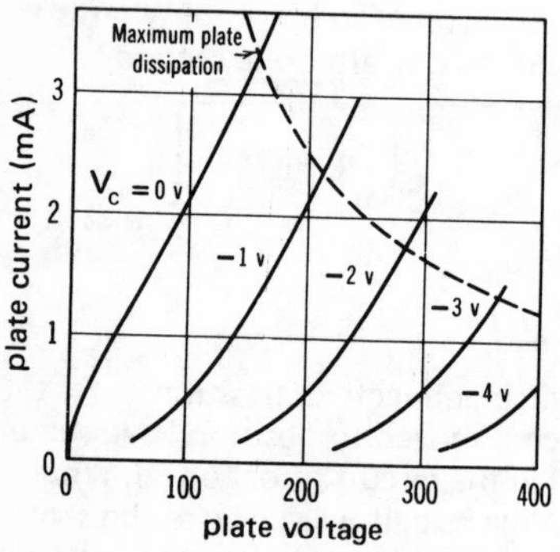

42. The 12AX7 triode is used in the amplifier circuit shown below.

(a) What is the plate current if the input signal is +9 volts?
(b) What is the change in plate current if the signal changes from +8 volts to +10 volts?
(c) What is the corresponding variation in V_L?
(d) What is the amplification?

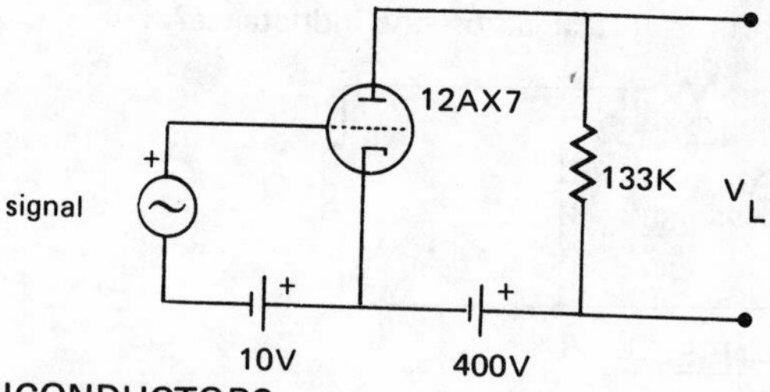

SEMICONDUCTORS

43. At 20°C, a diode flows 70 uA when the reverse bias is -1.5 volts. What is the saturation current? What is the current which flows when a forward bias of +.2 volts is applied

(a) at 20°C?
(b) at 40°C?

15 Peripheral Sciences

PART 1: Illumination

Nomenclature

A	area	$meter^2$
c	velocity of light (3 EE8)	meter/sec
E	energy, or illuminance	joules, or foot-candles
f	frequency	hz
F	luminous flux	lumens
h	Planck's constant (6.626 EE-34)	joule-sec
I	intensity	candles
L	brightness	foot-lamberts
P	power	watts
r	radius	meters
s	distance	meters
v	velocity	meter/sec

Symbols

θ	angle	degrees
λ	wavelength	meters
η	efficiency	—
ω	solid angle	steradian

Subscripts

L luminous

Visible light consists of electromagnetic waves with lengths (wavelengths) in the vicinity of 5 EE-7 meters. By definition, light is only that part of the electromagnetic spectrum which affects the eye. Visible light is bounded on either side of the electromagnetic spectrum by infrared and ultraviolet waves having longer and smaller wavelengths respectively. Visible light's placement in the electromagnetic spectrum is shown in figure 15.1.

For convenience, optical wavelengths are often specified in microns (EE-6 meters), millimicrons (EE-9 meters), and angstroms (EE-10 meters). Thus, the center of the visible spectrum, which has a wavelength of 5.550 EE-7 meter for an average observer, is said to have a wavelength of 5550 Å.

The lower and upper limits of the visible spectrum are approximately 4300 Å and 6900 Å. The visual sensations at these wavelengths are intense violet and red respectively. The intermediate color sensations are given in table 15.1.

TABLE 15.1 Color versus Wavelength

violet	< 4500 Å
blue	4500–5000 Å
green	5000–5700 Å
yellow	5700–5900 Å
orange	5900–6100 Å
red	> 6100 Å

By careful analysis of experimental evidence, the speed of the electromagnetic waves which constitute light has been determined to be 2.998 EE8 meters per second. Figure 15.2 shows the electrical field strength of a light wave as a function of time or distance from the source.

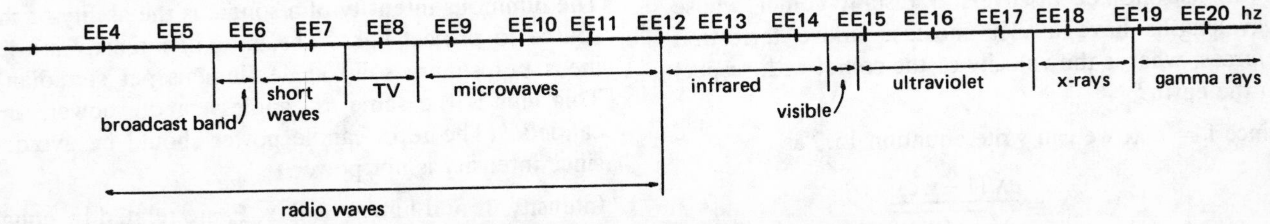

FIGURE 15.1 The Electromagnetic Spectrum

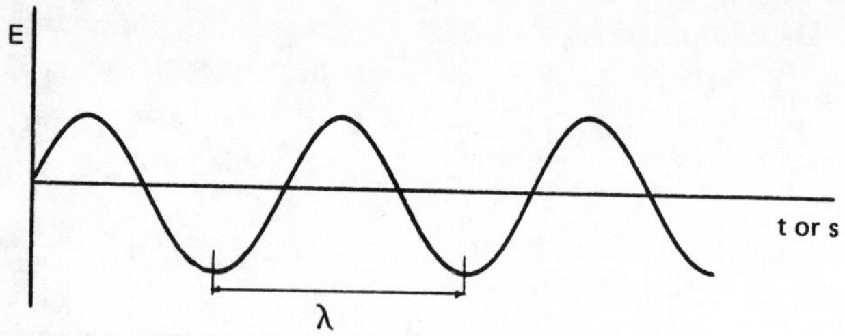

FIGURE 15.2 An Electromagnetic Wave

The distance traveled in one second by the wave is

$$s = \lambda f t = \lambda f \qquad 15.1$$

Since the distance traveled in one second is also the velocity of light,

$$c = \lambda f \qquad 15.2$$

Example 15.1

What is the length of a quarter-wave FM antenna which operates in the 100 Mhz region?

$$\text{length} = \frac{\lambda}{4} = \frac{c}{4f} = \frac{3 \text{ EE8}}{(4)(\text{EE8})} = .75 \text{ m}$$

Einstein's theory of relativity asserts that the velocity of light is independent of the reference frame chosen or the velocity of the observer. However, the frequency and wavelength of the observed light do depend on the relative motion between the source and observer. The product of frequency and wavelength, as given in equation 15.2, will always be the same—however, the frequency and wavelength will vary. Such changes in the frequency of the observed light are known as Doppler shifts.

If the actual emitted frequency is f and the relative separation velocity is v, the observed Doppler frequency will be given by equation 15.3.

$$f' = \frac{f(1-v/c)}{\sqrt{1-(v/c)^2}} \qquad 15.3$$

If two objects, such as the earth and a distant star, are separating, f' will be less than f. As a consequence of this, λ' will be greater than λ.

Example 15.2

If an astronomer discovers a distant comet whose wavelengths have a .3% increase above terrestrial sources, what is the velocity of the comet with respect to the earth?

Since $f = c/\lambda$, we can write equation 15.3 as

$$\lambda = \frac{\lambda'(1-v/c)}{\sqrt{1-(v/c)^2}} \qquad 15.4$$

In this example, $\lambda' = 1.003\lambda$ so

$$\frac{1}{1.003} = \frac{(1-v/3 \text{ EE8})}{\sqrt{1-(v/3 \text{ EE8})^2}} \text{ so } v = 9.1 \text{ EE5 meters/sec}$$

Because the velocity is positive, the comet is receding.

Maxwell's electromagnetic theory provides an accurate explanation of light propagation. However, radiant energy in its interaction with matter (either at the source or the observer) behaves as though it consists of discrete quanta of energy. Each quantum of energy is defined by equation 15.5. h is known as Planck's constant, with a value of 6.626 EE-34 joule-sec. A quantum of light is called a photon.

$$\Delta E = hf \qquad 15.5$$

The most important light source is the photosphere of the sun, a plasma envelope at a temperature of about 6300° K. Artificial light sources include the mercury arc (5000° K), the carbon arc (4000° K), the incandescent lamp (600° K), and the fluorescent lamp (room temperature).

All of these sources emit radiation, some of which is not in the visible region. Therefore, the source power is not an indication of the lighting effect. The rate at which visible light is emitted is called the luminous flux, with units of lumens. A lumen is approximately equal to 1.47 EE-3 watts of visible light with a wavelength of 5550 Å.

For an artificial light source, the ratio of lumens output to watts input is known as the luminous efficiency. Typical values of luminous efficiency are given in table 15.2.

$$\eta_L = \frac{\text{luminous flux (lumens)}}{\text{power input (watts)}} = \frac{F}{P} \qquad 15.6$$

TABLE 15.2 Typical Luminous Efficiencies (lumens/watt)

25W	tungsten lamp	10.4
100W	tungsten lamp	16.3
1000W	tungsten lamp	21.5
1000W	carbon arc	60
100W	mercury arc (in quartz)	35
1000W	mercury arc (in quartz)	65
40W	fluorescent lamp	58

The luminous intensity of a source is the ability of the source to project light. The units of intensity are lumens per square solid angle (lumens per steradian). This unit is the same as candles, candle power, and candela. (The term candle power should be avoided since intensity is not power.)

Intensity, I, and luminous flux, F, are related by equation 15.7. An ideal omnidirectional one-candle source projects 1 lumen on each square foot of a sphere with radius of 1 foot.

$$F = 4\pi I \qquad 15.7$$

Strictly speaking, the intensity of an ideal point source illuminating a surface also depends on the orientation of the plane. If the plane is rotated θ degrees from the normal, the intensity is given by equation 15.8. This is illustrated in figure 15.3.

$$I = \frac{F}{\omega} = \frac{Fr^2}{A\cos\theta} \quad \text{(candles)} \qquad 15.8$$

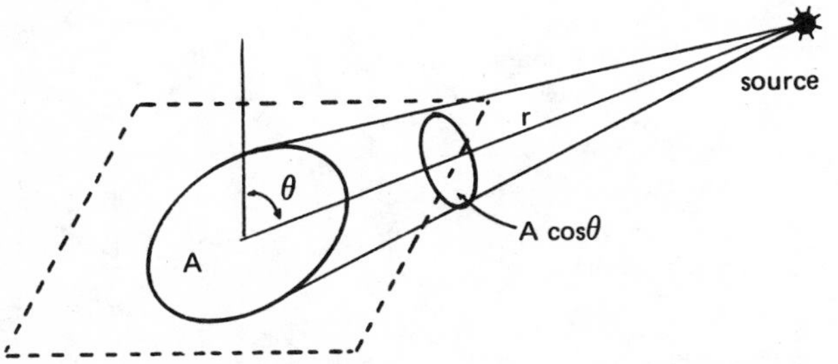

FIGURE 15.3 Luminous Intensity

Example 15.3

A 100 watt fluorescent lamp has a luminous flux of 4400 lumens. What is the luminous efficiency of the lamp?

$$\eta_L = \frac{4400}{100} = 44 \text{ lumens/watt}$$

Example 15.4

If 2 lumens pass through the circular hole shown below, what is the

(a) solid angle subtended by the hole?
(b) intensity of the source in the direction of the source?
(c) total lumens emitted by the source?

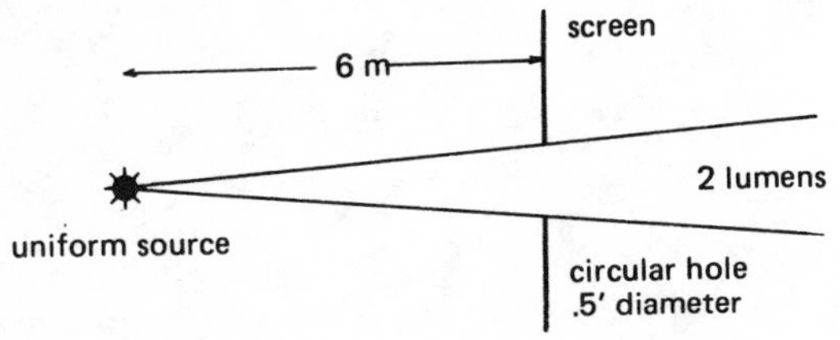

(a) $\omega = A/r^2 = \frac{(\frac{1}{4}\pi)(.5)^2}{(6)^2} = .0055$ steradians

(b) $I = F/\omega = 2/.0055 = 366.7$ candles

(c) $F = 4\pi I = 4\pi(366.7) = 4608$ lumens

The illumination of a surface, known as the illuminance (E) is the rate at which the surface receives visible radiation. That is,

$$E = F/A \qquad 15.9$$

For an omnidirectional source and a surface at a distance r, the illuminance is

$$E = F/4\pi r^2 \qquad 15.10$$

If F is in lumens and r is in feet, the illuminance has the units of lumens/ft². An older unit of illuminance is the foot-candle, which is the illuminance produced on a surface 1 foot from a 1 candle source. Foot-candles and lumens/ft² are numerically the same.

Combining equations 15.10 and 15.7 results in the following equation:

$$E = F/4\pi r^2 = 4\pi I/4\pi r^2 = I/r^2 \qquad 15.11$$

The brightness of an object is the density of luminous flux issuing or projected from a light source. A unit of brightness is the foot-lambert, which is numerically equal to lumens per square foot.

$$L = F/A \quad \text{(foot-lambert)} \qquad 15.12$$

If the area in equation 15.12 is expressed in square centimeters, brightness will have the units of lamberts.

PHYSICS

Table 15.3 Recommended Illuminance (foot-candles)

social activities	5–15
typical assembly	100
fine assembly	500–1000
hospital operating table	2500
library reading table	30–70
office	100–200
roadway	1–2
evening sports	30–100

Example 15.5

A 700 square foot work bench uniformly receives 3500 lumens from a suspended luminaire. What is the average illumination on the work bench?

$$E = F/A = \frac{3500}{700} = 5 \text{ ft-candles}$$

Example 15.6

A tungsten filament emitting 1600 lumens has a surface area of .35 in². What is the brightness in (a) lamberts? (b) foot-lamberts?

$$L_a = \frac{F}{(6.45)A_{inches}^2} = \frac{1600}{(6.45)(.35)} = 709 \text{ lamberts}$$

$$L_b = \frac{F}{A_{foot}^2} = \frac{1600}{(.35/144)} = 6.58 \text{ EE5 ft-lamberts}$$

Example 15.7

A lamp rated at 2000 lumens is positioned 20 feet above a walkway. What is the illumination on the walkway? Assume the lamp is an ideal hemispherical radiator.

$$E = F/2\pi r^2 = 2000/2\pi(20)^2 = .8 \text{ ft-candle}$$

The following hydraulic analogy may help to explain the various terms used in the illuminating industry.

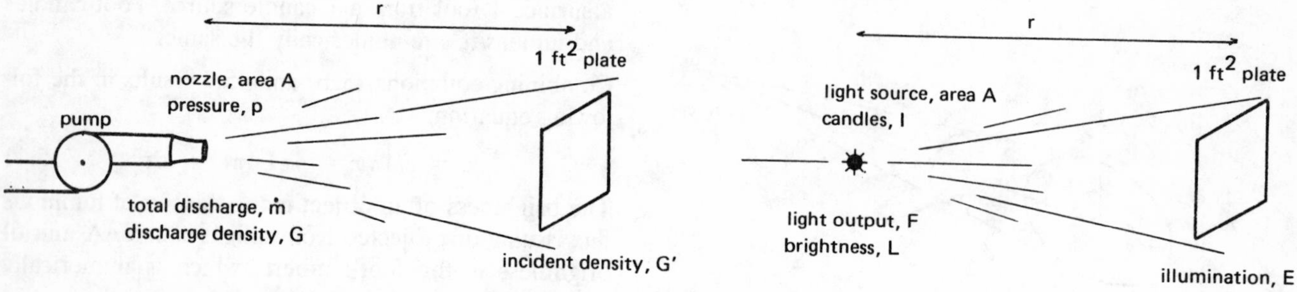

FIGURE 15.4 Hydraulic Analogy

TABLE 15.4 Analogous Variables

photometric variable	units	abbreviation	symbol	analogous hydraulic variable	units
luminous flux	lumen	lm	F	discharge, $\dot{m}$	lbm/sec
luminous intensity	candela, candle power	cp	I	discharge pressure, p	lbf/ft^2
illumination (luminous flux density)	ft-candle, lumens/ft²	ft-c	E	incident fluid density, G′	$\frac{lbm}{sec\text{-}ft^2}$
brightness	ft-lamberts, lumens/ft²	—	L	discharge rate per unit area, G	$\frac{lbm}{sec\text{-}ft^2}$

PART 2: Optics

Nomenclature

A	prism angle	degrees
c	speed of light (3 EE8)	meters/sec
d	slit or grating separation	meters
E	energy	joules
f	frequency	hz
h	Planck's constant (6.626 EE-34)	joule-sec
L	length	meters
m	an integer (1, 2, 3, etc)	—
M	magnification	—
n	index of refraction	—
r	radius	meters
t	film thickness	meters
x	distance	meters
y	distance	meters

Symbols

ø	angle	degrees
λ	wavelength	meters

Subscripts

d	deviated
i	incident
r	reflected

Light is an electromagnetic wave when it travels through a vacuum. When light makes contact with matter (air, glass, water, etc.), it causes electrons to change their energy states. Some energy is absorbed and transformed into heat. Some is re-emitted when the electrons drop to a lower energy level. The re-emitted energy may or may not have the same wavelength as the incident energy.

A polished metal surface, for example, will absorb about 10% of the incident energy. The remaining 90% is reflected back into the original medium. If the metal surface is sufficiently smooth, the reflected light is said to be regularly reflected. If the surface is rough, the reflection is diffuse. This is illustrated in figure 15.5.

The energy that is absorbed is said to be refracted. In the case of an opaque material, the refracted energy is absorbed within a very thin layer. If the material is optically transparent, as in figure 15.5(c), the refracted light is able to pass through without being totally absorbed.

Reflected light has the following characteristics:

1. The reflected light is in the same plane as the incident light ray and a line normal to the surface.
2. The reflection angle is the same as the incident angle.
3. If a flat mirror receiving light from a stationary source is rotated θ degrees, the reflected beam will rotate 2θ degrees from its original position.

The method of determining the image position of an object seen in a plane mirror is illustrated in figure 15.6. The image appears to be at point B. The image is called a virtual image because no rays of light actually pass through or originate from point B—they only appear to do so. Furthermore, if a screen was to be placed at point B, no image would be projected upon it. (Conversely, the image is real if it can be projected onto a screen or photographic plate.)

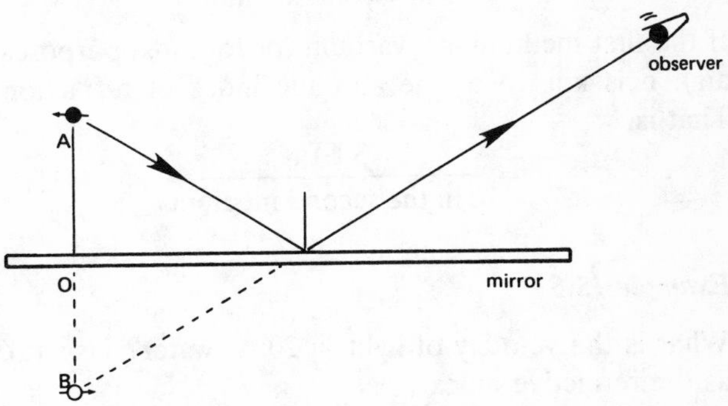

FIGURE 15.6 Image in a Plane Mirror

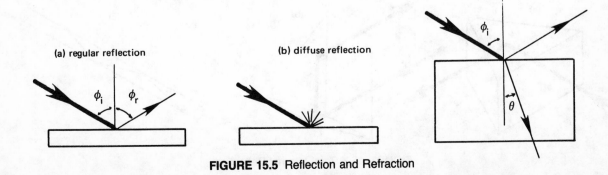

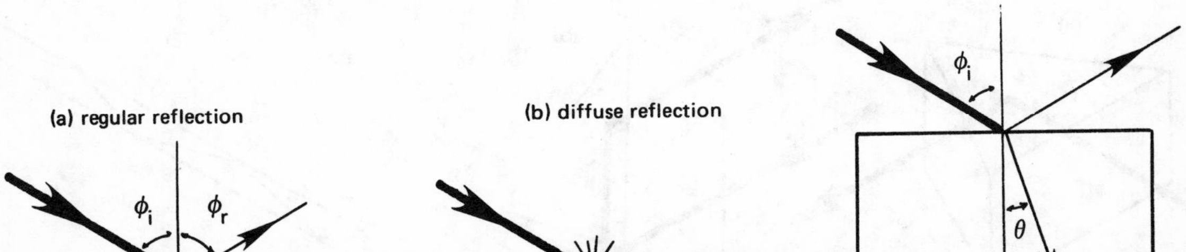

FIGURE 15.5 Reflection and Refraction

TABLE 15.5 Absolute Refractive Indices

substance	typical critical angle	wavelength, angstroms 4047	4861	5893	6563	7682
air, 20° C, 1 atm	88.6°			1.0003		
water, 20° C	48.7°		1.3371	1.3330	1.3311	1.3289
borosilicate crown glass	41.1°	1.5382	1.5301	1.5243	1.5219	1.5191
dense flint glass	37.3°	1.6901	1.6691	1.6555	1.6501	1.6441
diamond	24.4°			2.417		
fused quartz	43.2°			1.46		
salt (NaCl)	40.8°			1.53		

Distances OA and OB are equal, as are the image and object sizes. Of course, the image and object are reversed from right to left.

If the incident angle is sufficiently large, the light will be totally reflected back into the medium from which it originates. The critical angle is defined by equation 15.13. Total reflection occurs when the incident angle exceeds the critical angle.

$$\text{Ø}_{ci} = \arcsin\left(\frac{1}{n}\right) \qquad 15.13$$

n is the relative index of refraction. The refractive index depends on the material doing the reflecting as well as the light wavelength. It can be calculated from equation 15.14 as the ratio of speeds of light in the two media.

$$n = \frac{\text{c in first medium}}{\text{c in second medium}} \qquad 15.14$$

If the first medium is a vacuum (or for most purposes, air), n is known as the absolute index of refraction. That is,

$$n = \frac{3\text{ EE8}}{\text{c in the second medium}} \qquad 15.15$$

Example 15.8

What is the velocity of light in 20° C water? Use 1.33 as the refractive index.

$$c_{water} = c_{vacuum}/n = (3\text{ EE8})/1.33 = 2.26\text{ EE8 m/s}$$

This total reflecting characteristic can be used advantageously. Total reflecting prisms are used instead of silvered mirrors when precise reflection is required. Images can be reversed from left to right or from top to bottom, and the path of light can be changed up to 180° by the reflecting prisms shown in figure 15.7(a) and 15.7(b). Fiber optics (light pipes) also rely on total internal reflection for their light transmitting characteristics.

A silvered spherical surface may also be used as a mirror, although the image thus produced will not usually be the same size as the original object. Two types of spherical mirrors, concave and convex, are illustrated in figure 15.8.

The following statements apply to spherical mirrors:

1. The principal axis is a line through the center of curvature (C) and the vertex of the mirror (V).
2. The focal distance is the distance from the vertex to the focus (F). It is one half of the radius of curvature.

$$VF = VC/2 \qquad 15.16$$

3. The type and size of the image at B depend on the placement of the actual object at A. The various combinations of variables are listed in table 15.6.

The mirror equation numerically relates distances VA and VB to the focal length, VF. As in table 15.6,

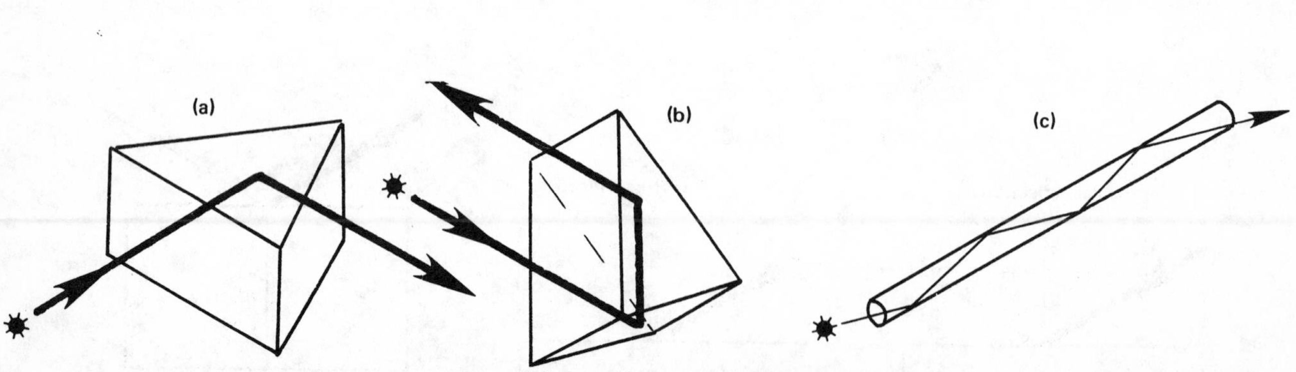

FIGURE 15.7 Total Internal Reflection

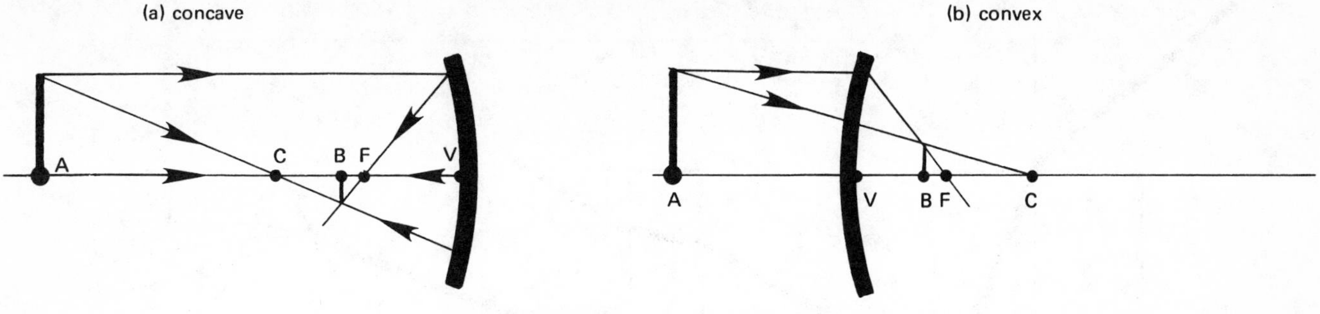

FIGURE 15.8 Spherical Mirrors

distances to the right of the mirror are negative. Thus, VF and VC are negative for the convex mirror.

$$\frac{1}{VA} + \frac{1}{VB} = \frac{1}{VF} = \frac{2}{VC} \qquad 15.17$$

The magnification produced by the mirror is

$$M = -\frac{VB}{VA} \qquad 15.18$$

The actual position of the image can be found from equation 15.19.

$$VB = \frac{(VA)(VF)}{(VA)-(VF)} \qquad 15.19$$

Combining equations 15.18 and 15.19 produces an alternative expression for the magnification:

$$M = \frac{(VF)}{(VF)-(VA)} \qquad 15.20$$

Example 15.9

An object is placed 50 cm from a convex mirror with a focal length of −20 cm. What is the image distance and the magnification?

$$VB = \frac{(VA)(VF)}{(VA)-(VF)} = \frac{(50)(-20)}{(50)-(-20)} = -14.3 \text{ cm}$$

$$M = -\frac{VB}{VA} = \frac{-(-14.3)}{50} = .29$$

The virtual image is located 14.3 cm behind the mirror. It is upright, but smaller than the object.

If the spherical mirror is constructed from a larger part of a sphere's surface, the reflected light rays will not all pass through the same focal point. This inability to accurately focus all incoming rays to the same point is known as spherical aberration. (Chromatic aberration is the inability of lenses to focus different wavelengths to the same point.) Spherical aberration can be eliminated by using parabolic mirrors if all of the incident rays are parallel.

TABLE 15.6 Spherical Mirror Images
(Distance is positive to the left of V)

mirror type	object position	image position	image type	image size	image orientation
concave	VA = ∞	VB = VF	real	zero	—
	VA > VC	VC > VB > VF	real	smaller	inverted
	VA = VC	VB = VC	real	smaller	inverted
	VC > VA > VF	∞ > VB > VC	real	larger	inverted
	VA = VF	VB = ∞	—	—	—
	VF > VA >0	0 > VB> ∞	virtual	larger	erect
	VA = 0	VB = 0	—	same	erect
convex	VA = ∞	VB = VF	virtual	zero	—
	VA > 0	0 > VB > VF	virtual	smaller	erect
	VA = 0	VB = 0	virtual	same	erect

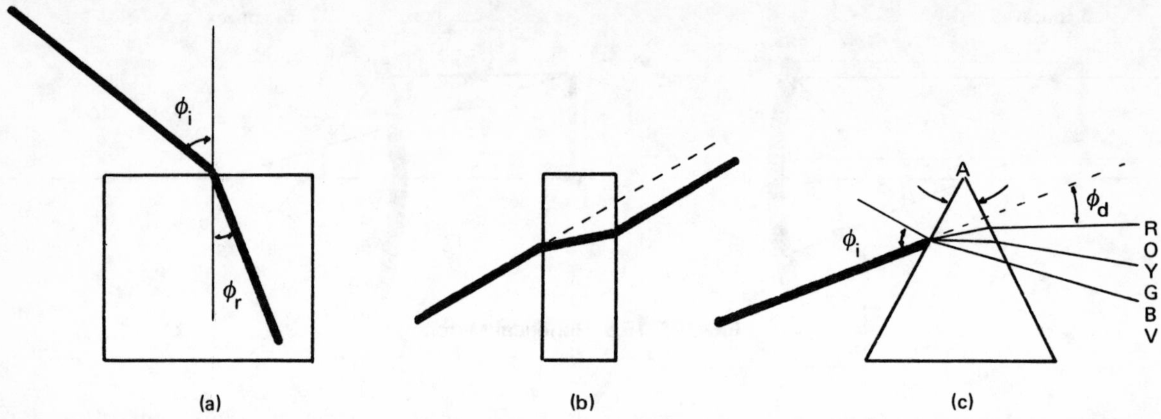

FIGURE 15.9 Refraction

Refraction

Refraction occurs whenever light passes from one transparent medium into another. When light passes from a vacuum (or air) into a solid medium, it is bent towards the normal, as shown in figure 15.9. The refraction angle can be found from Snell's law:

$$\frac{\sin\phi_i}{\sin\phi_r} = n \qquad 15.21$$

The velocity of light in a vacuum is independent of the light's wavelength. However, the light velocity in a material medium is wavelength dependent. It is this principle that creates the rainbow effects when white light passes through a prism. The prism is said to disperse the white light into its spectrum.

A light consisting of multiple wavelengths that can be dispersed by a prism is known as polychromatic light. Light with only one wavelength is known as monochromatic light. The angle of deviation of a monochromatic light depends on the incident angle, as shown in figure 15.9(c). The minimum angle of deviation is found from equation 15.22.

$$n = \frac{\sin\frac{1}{2}(A + \phi_{d,min})}{\sin(\frac{1}{2}A)} \qquad 15.22$$

If light passes through a transparent medium with parallel surfaces, the emergent light will be parallel to the incident light. This is illustrated in figure 15.9(b). Another effect of refraction is to make submerged objects appear closer to the surface than they actually are. The apparent depth can be found from equation 15.23.

$$\text{apparent depth} = \frac{\text{actual depth}}{n} \qquad 15.23$$

Lenses use refraction to form images of objects that send light to them. Lenses are classified as converging or diverging dependent on what they do to parallel light rays. Regardless of overall curvature, a converging lens is thicker at its center; a diverging lens is thicker at its edges. Parallel light rays will be converged to the focus, F, of a converging lens. Parallel rays will appear to have originated from the virtual focus of a diverging lens.

Regardless of the source direction of the incident rays, the image will always appear on the focal plane. This is illustrated in figure 15.11.

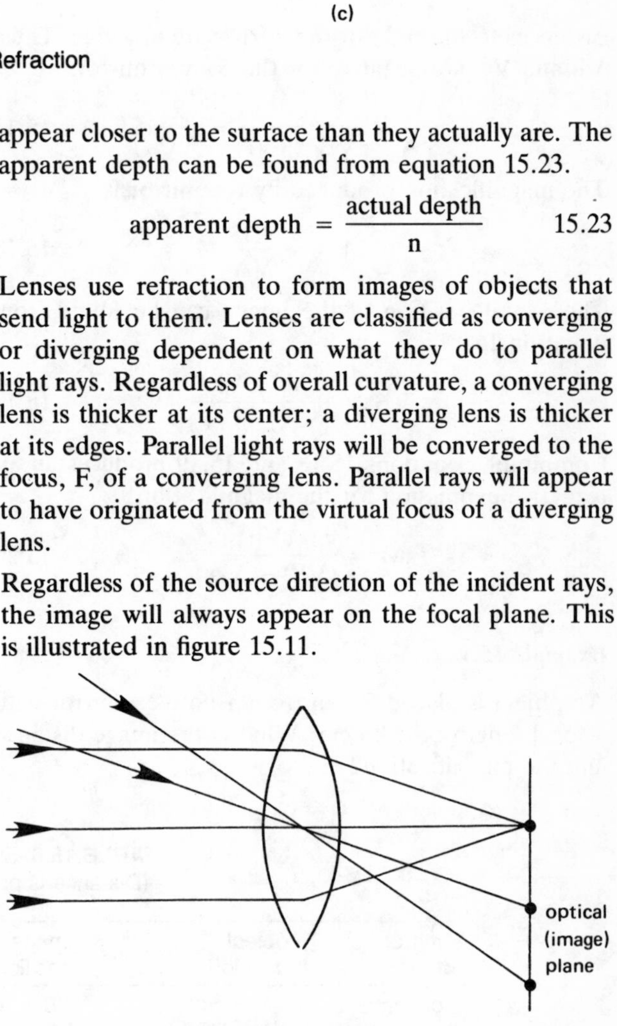

FIGURE 15.11 Optical Plane

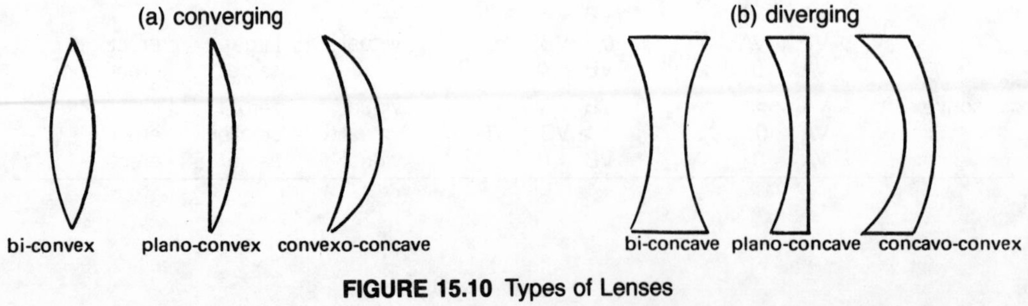

FIGURE 15.10 Types of Lenses

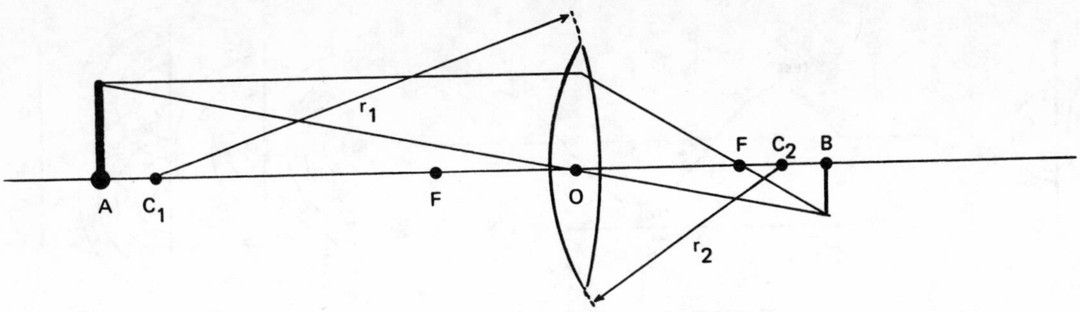

FIGURE 15.12 Spherical Lenses

A thin lens has a thickness which is small compared to its diameter. Only thin lenses are covered in this chapter. Because they are easiest to produce, most thin lenses have spherical surfaces. Thus, these lenses are classified as spherical lenses. The two surfaces of a lens, however, need not have the same radii of curvature.

The principal axis is a line which passes through the centers of curvature, C_1 and C_2. The foci are also along the principal axis. The foci are equidistant from the optical center, O, of the lens. The optical center of a thin lens is the midpoint of the thickness along the principal axis. All light rays, regardless of the direction from which they come, pass through the optical center without deviation. This is illustrated in figure 15.12.

The size, location, and type of image can be found from table 15.7.

The lens equation enables us to actually calculate the image position.

$$\frac{1}{OA} + \frac{1}{OB} = \frac{1}{OF} \qquad 15.24$$

As in table 15.7, OA is positive to the left of the lens, and OB is positive to the right of the lens. OF is negative for a diverging lens. The magnification is

$$M = -\frac{OB}{OA} \qquad 15.25$$

The focal length itself can be obtained from the lens-makers' equation, equation 15.26. The radius r is negative for concave surfaces.

$$\frac{1}{OF} = (n-1)\left(\frac{1}{r_1} + \frac{1}{r_2}\right)$$

$$= (n-1)\left(\frac{1}{OC_1} + \frac{1}{OC_2}\right) \qquad 15.26$$

If OF is in meters, (1/OF) is the power of the lens, as measured in units of diopters.

PHYSICS

Example 15.10

A bi-convex lens (n = 1.52) has radii of curvature of 14 cm and 20 cm. What is the focal length? What is the magnification if a 2 cm object is placed 28 cm from the optical center?

$$\frac{1}{OF} = (1.52-1)\left(\frac{1}{14} + \frac{1}{20}\right) = .063$$

$$OF = 15.84 \text{ cm}$$

$$\frac{1}{OB} = \frac{1}{15.84} - \frac{1}{28} = .0274$$

$$OB = 36.47$$

The magnification is $M = -\frac{36.47}{28.00} = -1.3$

The image is larger than the object. Because M is negative, the image is inverted.

Table 15.7 Spherical Lens Images
(OA to the left of O is positive, to the right of O is negative.)
(OB to the right of O is positive, to the left of O is negative.)

type of lens	object position	image position	image type	image size	image orientation
converging	OA = ∞	OB = OF	real	zero	—
	2(OF) < OA < ∞	OF < OB < 2(OF)	real	smaller	inverted
	OA = 2(OF)	OB = OF	real	same	inverted
	OF < OA < 2(OF)	2(OF) < OB < ∞	real	larger	inverted
	OA = OF	OB = ∞	—	—	—
	OA < OF	−∞ < OB < O	virtual	larger	erect
diverging	OA = ∞	OB = −OF	virtual	zero	—
	O < OA < ∞	−OF < OB < O	virtual	smaller	erect

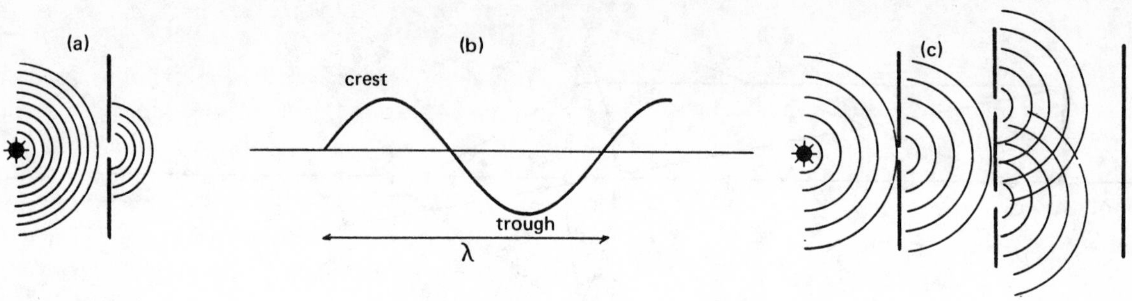

FIGURE 15.12 Huygen's Principle

With a few exceptions, the effect of combinations of lenses is difficult to evaluate. One of those exceptions is the case when lenses are placed in contact so that their principal axes coincide. In that case, the focal length of the lens combination can be predicted from equation 15.27.

$$1/(OF)_{combined} = 1/(OF_1) + 1/(OF_2) \qquad 15.27$$

If the two lenses are separated by distance d, it is simplest to find the image produced by the first (closest) lens, and then to use that image as the object of the second lens. The magnification effect of two lenses will be the product of the individual magnifications.

$$M_{combined} = M_1M_2 \qquad 15.28$$

The focal length of a lens combination is

$$\frac{1}{(OF)_{combined}} = \frac{1}{(OF_1)} + \frac{1}{(OF_2)} - \frac{d}{(OF_1)(OF_2)} \qquad 15.29$$

Because different wavelengths will also be refracted differently in passing through a homogeneous lens, not all light rays will focus at the same point on the principal axis. Lenses which exhibit this tendency are said to possess chromatic aberrations. Special lens combinations (e.g., flint glass and crown glass) may be used to create an achromatic lens.

Interference

Most light sources emit diverging light at random, and the waves which constitute this light have arbitrary phase relationships. However, a source of parallel rays can be obtained by allowing light to pass through a narrow slit. According to Huygen's principle, each point in a wave front can be considered as a source of waves. Therefore, the slit will act as a light source, as shown in figure 15.12(a). In this figure, the circular lines represent the wave crests, and the spaces between the lines represent the wave troughs.

Interference can be obtained by allowing the parallel rays to pass through two parallel slits. Huygen's principle again makes these two slits sources of waves. These secondary sources are in-phase because they derive from the same source.

The effect on a distant screen is shown in figure 15.13. The screen will exhibit regions of darkness and light. Wherever two wave crests or two wave troughs coincide, they reinforce and form a bright spot. The two sources are said to be in-phase at such a bright spot. If a trough coincides with a crest, the two sources are 180° out-of-phase, and the result is a dark spot on the screen.

For in-phase reinforcement, the difference in path lengths $(L_1 - L_2)$ must be a whole number of wavelengths. For out-of-phase cancellation, the difference is an odd number of half-wavelengths. This is described mathematically by the following equations:

$$L - L_2 = m\lambda \quad \text{in-phase} \qquad 15.30$$

$$y = m\lambda x/d \quad \text{in-phase} \qquad 15.31$$

$$y = (m + \tfrac{1}{2})\lambda x/d \quad \text{out-of-phase} \qquad 15.32$$

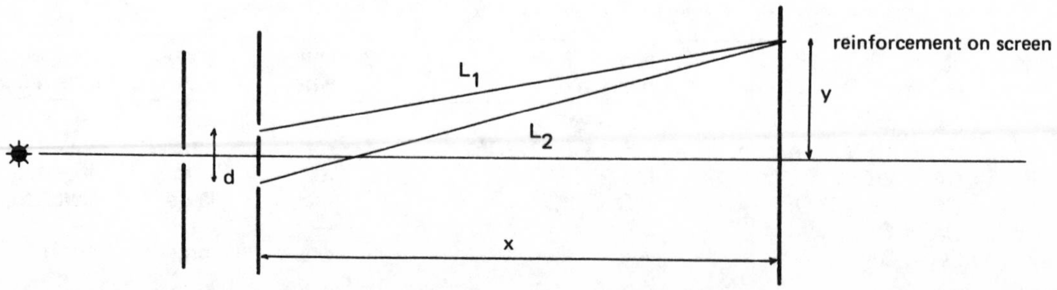

FIGURE 15.13 Interference From Two Sources

Example 15.11

A screen is placed 2.4 meters from two sodium gas sources (λ = 5893 Å) producing in-phase light. The separation of the two sources is .0005 meter. What is the distance (y) from the central image to the first reinforcement?

$$y = \frac{m\lambda x}{d} = \frac{(1)(5893\ \text{EE-10})(2.4)}{.0005} = .0028\ \text{meter}$$

Interference also occurs when light passes through thin films (soap bubbles, layers of oil on water, etc.). If white light is used, each wavelength will produce its own interference effect, resulting in a rainbow effect. In figure 15.14, ray D is composed of a partial reflection of ray B and the remainder of ray A. Because of the 180° reversal upon reflection at point G, cancellation along ray D requires that path FGH be an integral number of wavelengths. Since FGH ≈ 2t, equations 15.33 and 15.34 give the requirements for reinforcement and cancellation.

$$(m + \tfrac{1}{2})\lambda_{film} = 2t \quad \text{(in-phase)} \qquad 15.33$$

$$m\lambda_{film} = 2t \quad \text{(out-of-phase)} \qquad 15.34$$

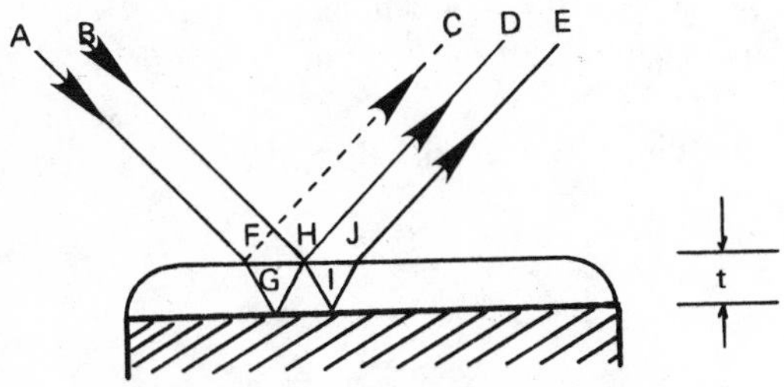

FIGURE 15.14 Thin Film Interference

Of course, the wavelengths used are the film wavelengths, not the air wavelengths.

$$\lambda_{film} = \lambda_{air}/n$$

Interference from thin air films (Newton's rings) can be observed by viewing the image produced when a plano-convex lens is placed convex side down on a reflecting surface.

Diffraction

Diffracted light is light whose path has been changed by passing around corners or through narrow slits. This diffracted light also produces interference and reinforcement effects. In the case of single slit diffraction (figure 15.15), each edge of the slit can be assumed to be a source of light waves (Huygen's principle) with the following equation giving the requirements for the mth dark band.

$$m\lambda = d(\sin\theta_m) \quad \text{out-of-phase} \qquad 15.35$$

Similar results can be obtained for the diffraction grating with slit spacing d (use equation 15.36) and crystal x-ray diffraction (see chapter 14).

$$m\lambda = d(\sin\theta_m) \qquad 15.36$$

$$m_{max} = d/\lambda \qquad 15.37$$

The wavelength may be found from equation 15.36 if m, d, and θ_m are known, as would be the case when a diffraction grating spectrometer was in use.

PHYSICS

Polarized Light

Light is a transverse wave. Most sources produce waves which oscillate in random planes. However, polarized light (light with waves in only one plane) can be obtained by several methods:

1. By reflecting light from a glass surface at a special incident angle.
2. By sending light through special minerals (e.g., tourmaline) which only pass light in one plane.
3. By scattering light from small particles.

To obtain linearly polarized light by reflection, the incident angle is given by equation 15.38.

$$\theta = \arctan(n) \qquad 15.38$$

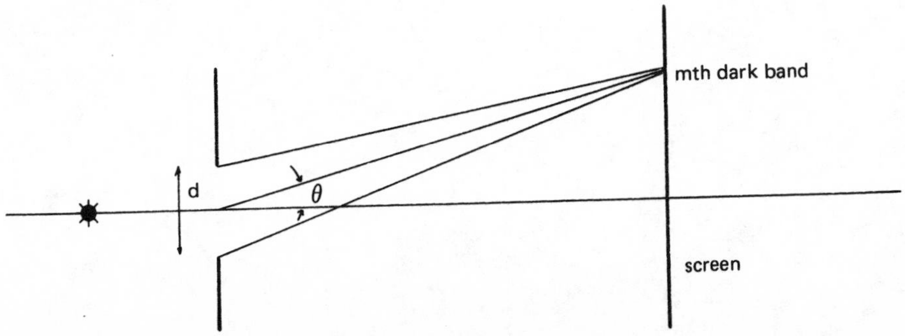

FIGURE 15.15 Single Slit Diffraction

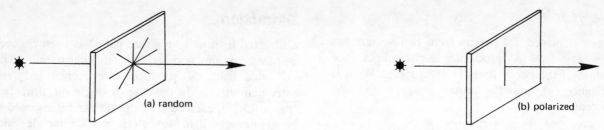

(a) random
(b) polarized

FIGURE 15.16 Planes of Transverse Waves

Lasers

Light rays which are all in-phase are said to be coherent. A device which produces coherent radiation in or near the visible spectrum is known as an optical MASER (microwave amplification by stimulated emission of radiation) or a LASER (light amplification by stimulated emission of radiation).

One method of producing coherent radiation is to use a transparent tube with mirrored ends (figure 15.17(a)). If monochromatic light is produced in the tube from a gas discharge or a flash coil, it will be reflected back and forth by the end mirrors. If the tube length is an integral number of wavelengths, standing waves will be produced and the radiation will be coherent. The tube is then known as a resonant cavity. The cavity can be solid (e.g., ruby crystal) or it may be filled with a gas (CO_2, helium-neon) or a liquid.

Atoms in the gas or crystal are usually in the ground (lowest) energy state, E_1. Atoms at higher energy states (E_2, E_3, etc) will emit energy when they drop to the ground state. The frequency of this emitted energy may be found from the quantum equation:

$$E_2 - E_1 = hf \qquad 15.39$$

In a laser, the normal proportion of high and low energy atoms must be reversed (population inversion). For there to be emitted light energy, the number of atoms at energy E_2 must exceed the number at energy E_1.

Laser operation is also enhanced by the stimulation of high energy atoms to release their energy. When a photon of energy already in motion hits a high energy atom, it stimulates the atom to release its energy. This energy release, in the form of a photon, is in phase with the original photon. Of course, spontaneous decay also occurs, in which case there is no phase relationship. Therefore, it is essential to have a high density of photons in motion.

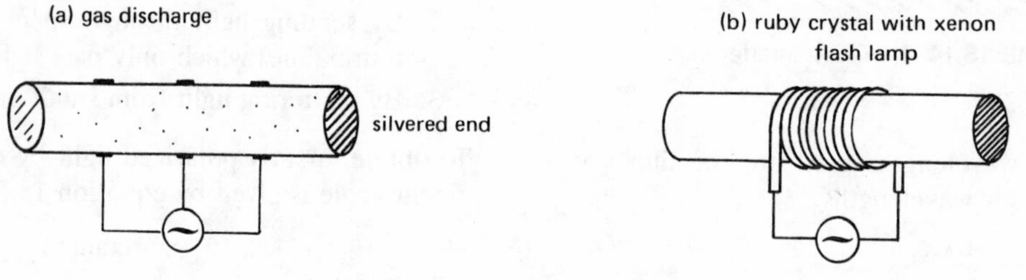
(a) gas discharge
silvered end
(b) ruby crystal with xenon flash lamp

FIGURE 15.17 Laser Construction

PART 3: Waves and Sound

Nomenclature

a	speed of sound	meters/sec, or ft/sec
E	bulk modulus	pascals, or lbf/ft^2
f	frequency	hz
F	tension in wire	newtons, or pounds
g	acceleration due to gravity	meters/second^2, or ft/sec^2
I	sound intensity	db
k	ratio of specific heats	
L	length	meters, or feet
m	an integer (1, 2, 3, etc)	—
M	total wire mass	kilograms, or slugs
R	specific gas constant	meter/°F, or ft/°R
T	temperature	°K, or °R
v	wave velocity	meters/sec, ft/sec

Symbols

λ	wavelength	meters
ρ	density	kg/(meter-sec)^2, or lbm/ft^3

Subscripts

o	observer
s	source

Sound and light are both wave phenomena. However, sound has two distinct differences from electromagnetic waves such as light. First, sound is transmitted in the form of a longitudinal wave. These waves are actual alternating compressions and rarefactions of the medium through which they travel. Electromagnetic waves, on the other hand, are transverse waves such as are seen in a stretched cord or wire.

Second, longitudinal waves require a medium through which to travel—they cannot travel through a vacuum. For ease of illustration, a longitudinal wave may be represented by a transverse wave. This is illustrated in figure 15.18. Of course, the usual wavelength, period, and frequency relationships are valid:

$$v = f\lambda \qquad 15.40$$

$$f = \frac{1}{\text{period}} \qquad 15.41$$

The velocity of a longitudinal wave depends on the compressibility of the supporting medium.

$$v_{long} = \sqrt{gE/\rho} \qquad 15.42$$

For gases, this velocity is found from equation 15.43.

$$a = \sqrt{kgRT} \qquad 15.43$$

The speed of a transverse wave in a wire depends on the wire tension and the wire mass.

$$v_{trans} = \sqrt{FL/M} \qquad 15.44$$

The fundamental (first harmonic) wavelength generated in a string of length L is $\lambda = 2L$. The wavelength of the mth overtone (same as the (m+1)th harmonic) is

$$\lambda_{\text{mth overtone}} = \frac{2L}{(m+1)} \qquad 15.45$$

TABLE 15.8 Longitudinal Wave Velocities (Speed of Sound)

material	a(meters/sec)	a(feet/sec)
air	330 @ 0°C	1130 @ 70°F
carbon dioxide	260 @ 0°C	870 @ 70°F
hydrogen	970 @ 0°C	3310 @ 70°F
pure water	1490 @ 20°C	4880 @ 70°F
steel	5140	16860
aluminum	4990	16370

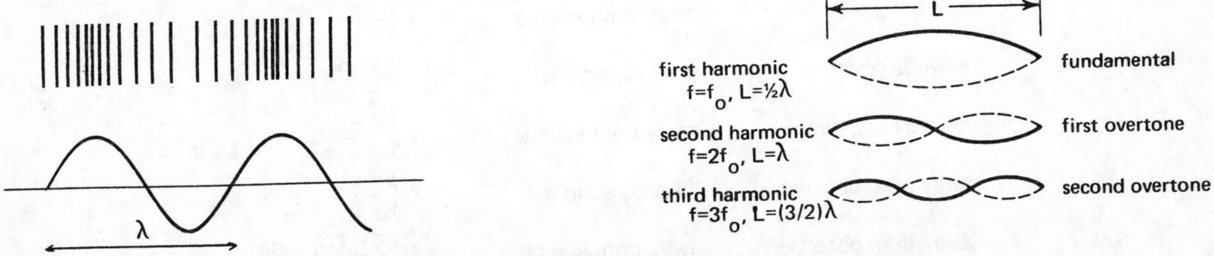

FIGURE 15.18 Fundamental and Harmonic Wavelengths

The fundamental and harmonic frequencies produced by a taut wire are given by equation 15.46. As in equation 15.45, m = 0 for the fundamental frequency.

$$f = \frac{(m+1)v_{trans}}{2L} \qquad 15.46$$

Sympathetic vibration is the same as resonance. Resonance occurs when a body (such as an air column or stretched wire) which is capable of standing waves at a definite frequency receives impulses at that exact frequency.

The gas in a hollow cylinder closed at one end will be resonant if its length is any quarter multiple of the wavelength. Thus, for resonance at the fundamental frequency, the length of the tube with one closed end will be

$$L = \tfrac{1}{4}\lambda \qquad 15.47$$

If the tube (with one closed end) has a fixed length, the fundamental and harmonic wavelengths can be found from the following equation.

$$\lambda = \frac{4L}{2m+1} \qquad 15.48$$

If the cylinder is open at both ends, the resonant length for the fundamental frequency is

$$L = \tfrac{1}{2}\lambda \qquad 15.49$$

The fundamental and harmonic wavelengths for tubes with two open ends can be found from the following equation.

$$\lambda = \frac{2L}{m+1} \qquad 15.50$$

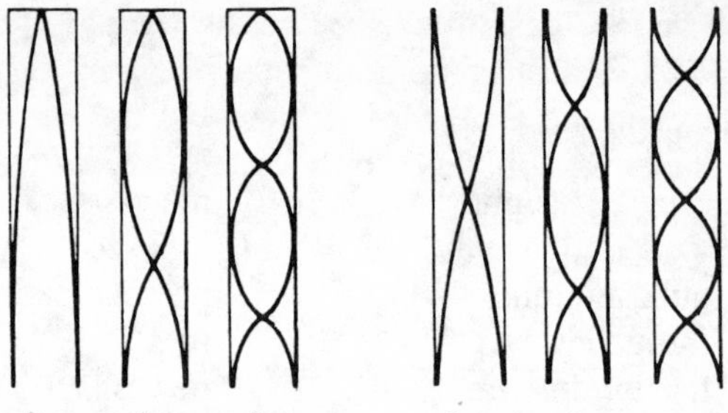

FIGURE 15.19 Resonant Gas Columns

If two sources produce sound with slightly different frequencies, the combined sound will beat. The beating effect is due to successive reinforcement and interference. Beats are only produced when the two frequencies are very close. The beat frequency is the difference between the two combining frequencies.

$$f_{beats} = f_1 - f_2 \qquad 15.51$$

If the source, observer, or both are moving, the observed frequency of a sound emitted by the source will be different from the emitted frequency. The observed frequency can be predicted from table 15.9.

Example 15.12

What frequency is observed by a stationary pedestrian when a police siren (1200 hz) recedes at 45 mph? (Note that 45 mph = 66 ft/sec).

$$f' = f\left(\frac{a}{a+v_s}\right) = 1200\left(\frac{1130}{1130+66}\right) = 1134 \text{ hz}$$

TABLE 15.9 The Doppler Effect
(f is the actual source frequency; a is the local speed of sound.)

Source Velocity (v_s)	Observer Velocity (v_o)	Apparent Frequency (f')
0	0	f
towards observer	0	$f\left(\frac{a}{a-v_s}\right)$
away from observer	0	$f\left(\frac{a}{a+v_s}\right)$
0	towards source	$f\left(\frac{a+v_o}{a}\right)$
0	away from source	$f\left(\frac{a-v_o}{a}\right)$
towards observer	towards source	$f\left(\frac{a+v_o}{a-v_s}\right)$
towards observer	away from source	$f\left(\frac{a-v_o}{a-v_s}\right)$
away from observer	towards source	$f\left(\frac{a+v_o}{a+v_s}\right)$
away from observer	away from source	$f\left(\frac{a-v_o}{a+v_s}\right)$

PHYSICS

Sound can be described in scientific terms. However, sound is also a physiological phenomenon. Longitudinal waves in the air can be detected by the ear if their frequencies are in the 20 to 20,000 hz region. The average human ear is most sensitive to frequencies around 3000 hz.

Even if sound is in the proper frequency range, the intensity must be large enough to be detected. For example, the threshold of audibility at 3000 hz is about EE-17 watts/cm^2. If the sound is too intense (the intensity is above the threshold of pain), the sound is felt as a painful sensation.

Both the threshold of audibility and the threshold of pain are frequency dependent, although the threshold of pain does not vary much from EE-4 watts/cm^2 (120 dB). The current 8-hour OSHA industrial exposure limit is 90 dB.

Since sound is both a physical and physiological phenomenon, it is not unexpected that qualitative terms have been developed by the non-scientific community to describe it. The relationships between the qualitative and scientific terms are given below.

pitch: A qualitative sensation dependent on the sound frequency.

intensity: A quantitative term used to describe the amount of energy passing through a unit area each second. Typical units are watts per square meter. Intensity is proportional to the square of the vibrational amplitude.

timbre: A qualitative sensation dependent on frequency, intensity, and the number of overtones present in the sound.

loudness: A qualitative sensation which can be described mathematically. If I_o is the zero intensity level (EE-12 watts/m_2), the loudness is

$$\text{loudness} = 10 \log_{10} (I/I_o) \qquad \text{(dB)} \qquad 15.52$$

quality: See 'timbre.'

overtone: Any tone which is produced by a sound source in addition to the main (fundamental) tone. The overtones of stringed and wind instruments are octaves of the fundamental tone.

octave: A sound whose frequency is an even multiple of the fundamental frequency. Alternatively, the wavelength of an octave tone is a whole fraction (½, ⅓, ¼ etc.) of the fundamental wavelength.

PHYSICS

TABLE 15.10 Sound Loudness

Source	Loudness (dB)	Qualitative
jet engine, thunder	120	painful
jack hammer	110	deafening
sheet metal shop	90	very loud
street noise	70	loud
office noise, normal speech	50	moderate
quiet conversation	30	quiet
whisper	20	faint
anechoic room	10	very faint
none	0	silence

PART 4: Heat Transfer

Nomenclature

a	a constant	—
A	area	ft^2
b	a constant	—
C	a constant	—
h	film coefficient	BTU/hr-ft^2-°F
k	thermal conductivity	BTU-ft/hr-ft^2-°F
L	thickness	feet
N_{nu}	Nusselt number	—
N_{pr}	Prandtl number	—
N_{re}	Reynolds number	—
q	heat transfer	BTU/hr
T	temperature	°F

Symbols

σ	Stefan-Boltzmann constant (.1713 EE-8)	BTU/hr-ft^2-$°R^4$
ϵ	emissivity	—

Subscripts

A	at one end
B	at the other end
m	logarithmic

Heat is energy in motion. If energy is transferred through a material by molecular vibration, the mechanism is known as conduction. If energy is transferred from one point to another by a moving fluid, the transfer mechanism is known as convection. (Natural convection transfers heat by relying on density changes to cause fluid motion. Forced convection involves the use of fans or pumps to move the fluid.) If no medium is involved at all, the mechanism is radiation.

Steady state conduction through a constant area wall is described by Fourier's law:

$$q = \frac{kA\Delta T}{L} \qquad 15.53$$

If the wall is a series composite, as shown in figure 15.20, equation 15.54 must be used:

$$q = \frac{A\Delta T}{\Sigma\left(\frac{L_i}{k_i}\right)} \qquad 15.54$$

Typical values of k are given in table 15.11. The thermal conductivity is also widely given in other units, so use equation 15.53 with compatible variables.

Thin fluid films (air, water, oil, etc.) also contribute to resistance to heat flow. The magnitude of the resistance is given by a film coefficient, h. The film coefficient may be found from the fluid properties by using the Nusselt equation (equation 15.55). Each of the dimensionless numbers in the Nusselt equation is made up of other variables, and it is possible to extract the film coefficient.

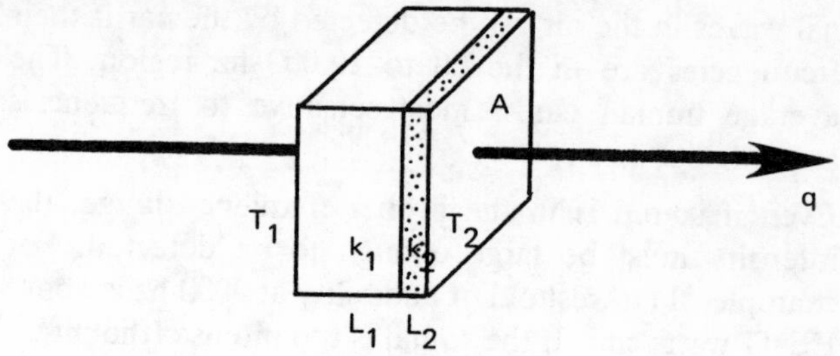

FIGURE 15.20 Series Composite Wall Structure

TABLE 15.11 Average Thermal Conductivities, BTU-ft/hr-ft^2-°F

copper	224
aluminum	117
steel	27
ice	1.3
water	.32
air	.014

$$N_{nu} = C(N_{re})^a(N_{pr})^b \qquad 15.55$$

Use of the Nusselt equation can be time consuming. Many approximations have been developed for evaluating the film coefficient. Typical values and ranges are listed in table 15.12.

TABLE 15.12 Typical Film Coefficients, BTU/hr-ft^2-°F

still air	1.65
air, 15 mph	6.0
water (no change of state)	150–2000
condensing steam	1000–2000
evaporating water	800–2000

If films are present on walls through which heat is being conducted, equation 15.54 must be modified to include the film coefficients:

$$q = \frac{A\Delta T}{\Sigma\left(\frac{L_i}{k_i}\right) + \Sigma\left(\frac{1}{h_j}\right)} \qquad 15.56$$

If the film coefficient is the primary thermal resistance (as it is in heat exchangers and other thin-wall applications), the heat transfer is

$$q = hA\Delta T \qquad 15.57$$

In a heat exchanger, ΔT is not uniform along the length of the tubes. For that reason, the logarithmic mean temperature difference, ΔT_m, is used in equation 15.57 instead of ΔT. In equation 15.58, ΔT_A and ΔT_B are the actual temperature differences at the two ends of the heat exchanger.

$$T_m = \frac{\Delta T_A - \Delta T_B}{\ln\left(\frac{\Delta T_A}{\Delta T_B}\right)} \qquad 15.58$$

The energy radiated by a hot body at temperature T (degrees Rankine) is given by the Stefan-Boltzmann law, also known as the fourth power law.

$$q = \sigma \epsilon A T^4 \qquad 15.59$$

σ is the Stefan-Boltzmann constant, equal to .1713 EE-8 BTU/hr-ft^2-°R^4. ϵ is the emissivity of the object. $\epsilon = 1$ for black bodies.

The net heat transfer between two bodies at temperatures T_1 and T_2 (°R) is

$$q = \sigma A_1 F_{1-2}[(T_1)^4 - (T_2)^4] \qquad 15.60$$

F_{1-2} is a factor which depends on the shapes, emissivities, and orientations of the two objects. If object 1 is small and is completely enclosed by object 2, $F_{1-2} = \epsilon_1$.

PHYSICS

ILLUMINATION

1. A 20 candela standard lamp and a test lamp are placed 200 centimeters apart. It is found that the two lamps produce equal illumination on a screen placed between them when the screen is 80 centimeters from the standard lamp. What is the intensity of the test lamp?

2. What is the illuminance on a flat object placed 5 meters from a 100 watt tungsten filament lamp? Assume the lamp is an ideal spherical radiator.

3. What is the radiation pressure 20 meters away from a 400 watt light bulb? Assume the bulb radiates spherically.

4. What is the speed of a cosmic object whose 4350 Å light spectrum reaches earth with a 6580 Å wavelength?

5. What is the energy of a photon whose wavelength is 22 cm?

OPTICS

6. The frosted glass screen of a projection microscope is placed 3 meters behind a 1 meter focal length objective lens. What is the magnification if the object is 1.5 meters from the lens?

7. Light of wavelength 5 EE-5 cm is diffracted by a grating possessing 2000 lines/cm. What is the distance on a screen 3 meters from the grating between the zeroeth and first-order image?

8. A gas discharge emits 5890 Å light. What angle will exist in the first order spectrum if a diffraction grating with 10,000 lines/cm is used?

9. A converging lens with 5 cm focal length is located at x=0. Where will be the image if the object is at
   (a) x=3?
   (b) x=7?
   (c) x=12?

10. What is the refractive index for a glass in which the speed of light is 124,000 miles per second?

11. What is the maximum refractive angle for light hitting a glass surface (n=1.70)?

12. A prism with a 60° vertex angle has a 45° minimum deviation angle. What is the refractive index?

WAVES AND SOUND

13. A tuning fork (256 Hz) resonates with a 4 foot long tube of hydrogen. Find the velocity of sound in the hydrogen if one end of the tube is closed.

14. A tuning fork (512 Hz) beats at 2 Hz with a vibrating string. Tightening the string eliminates the beats. What was the relative increase in string tension?

15. Two speakers emit a pure tone which travels in the surrounding medium at 1100 fps. 10 feet from the left speaker and 8 feet from the right speaker is a location where the sound is minimum. What is the sound frequency?

16. A string has a mass of 30 grams and is 30 cm long. If the string is tuned to 262 Hz, what is its tension?

17. What is the frequency of the high C located 2 octaves above middle C (262 Hz)?

18. What is the wavelength of a 20 Hz sound in air at 70°F?

19. What is the velocity of sound in
   (a) aluminum?
   (b) air at 120°F?
   (c) water at 50°F?

20. A low piccolo note has a frequency of 528 Hz. What are the frequencies of the 2nd and 3rd harmonics of this note?

21. What is the speed of a wave in a 90 cm long piano wire which produces a 198 Hz tone when hit?

22. Referenced to EE-12 watts/m^2, what is the sound intensity in decibels of a source with sound power of 5 EE-10 watts/m^2?

23. A plucked steel wire 4 feet long emits a tone with a fundamental frequency of 150 Hz. If the wire weighs .038 pounds, find the wire tension.

ENERGY AND HEAT TRANSFER

24. A 4 cm thick insulator (k = 2 EE-4 cal-cm/sec-cm^2-°C) has an area of 1000 square cm. If the temperatures on its two sides are 170°C and 50°C, what is the heat transfer by conduction? (Neglect films)

25. Two ideal radiators in the form of parallel plates of equal areas are maintained at 530°R and 1000°R respectively. What is the approximate heat transfer between them on a unit area basis? Use $F_{1\text{-}2}$=1.0.

26. A small tungsten filament (emissivity of .35, surface area of .5 in^2) is placed within a spherical enclosure whose temperature is 70°F. What power input is needed if the filament is to be kept at 5000°R?

27. The peak electric field 10 miles north of a vertical antenna is EE-3 volt/cm. What are the direction and magnitude of the magnetic field?

28. What are the variables in the
   (a) Reynold's number?
   (b) Prandtl number?
   (c) Nusselt number?
   (d) Biot number?

16 Modeling of Engineering Systems

PART 1: Mechanical Systems

Nomenclature

a	acceleration
B	coefficient of damping
F	force
k	spring constant (stiffness)
m	mass
t	time
x	position
v	velocity

1. Mechanical Elements

A *mechanical system* consists of interconnected masses, springs, dampers,[1] and energy sources. System modeling is used to predict the performance of a mechanical system without actually observing the system in operation.

A *lumped mass* is a rigid body which behaves like a particle. All parts of the mass experience identical velocities, accelerations, and displacements. The lumped mass is *ideal* if Newtonian physics applies (i.e., there are no relativistic changes in mass). The performance of ideal lumped masses is predicted by *Newton's second law,* equation 16.1.

$$F(t) = m\,a(t) \qquad 16.1$$

Various symbols are used to diagram masses in system models. Typical symbols are illustrated in figure 16.1.

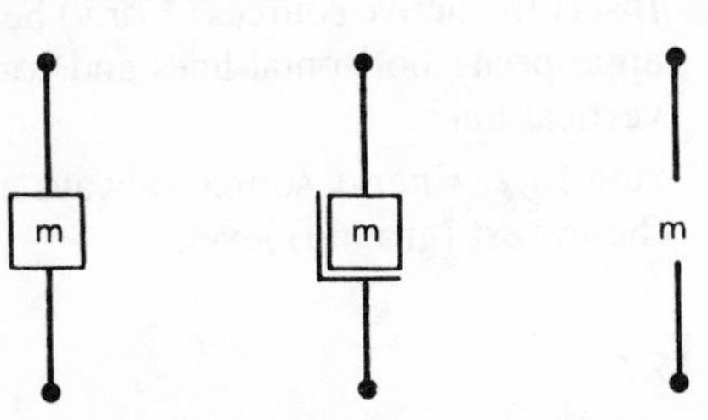

Figure 16.1 Ideal Lumped Mass Symbols

[1]Dampers are also known as *dashpots.* An automobile shock absorber is an example of a damping device.

An ideal *spring* is massless, has a constant *stiffness,* and is immune to set, creep, and fatigue. The performance of a spring is predicted by *Hooke's law,* equation 16.2.

$$F(t) = k[x_1(t) - x_2(t)] \qquad 16.2$$

x_1 and x_2 in equation 16.2 are displacements from the initial equilibrium positions of the two spring ends. If both ends are displaced the same distance, there will be no net extension or compression. Consequently, the spring will not experience a force.

Typical symbols used to diagram springs are illustrated in figure 16.2.

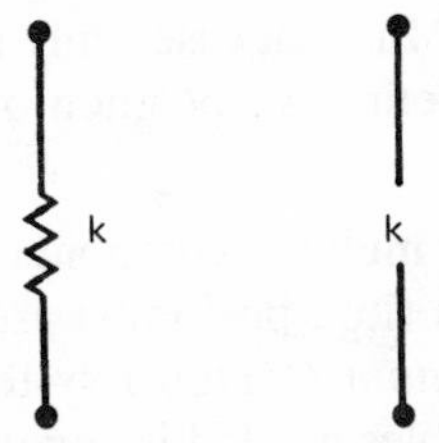

Figure 16.2 Ideal Spring Symbols

An ideal *damper* is massless, has no stiffness, and behaves linearly according to equation 16.3. *B* is the damping coefficient. The damping force is proportional to the difference in velocities of the two damper ends.

$$F(t) = B[v_1(t) - v_2(t)] \qquad 16.3$$

Typical symbols used to diagram dampers are illustrated in figure 16.3.

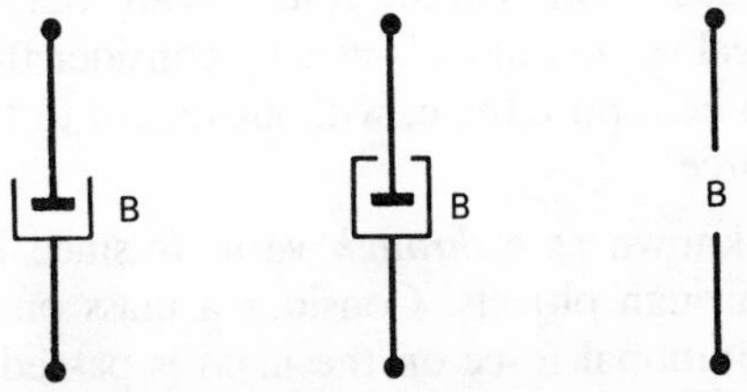

Figure 16.3 Ideal Damper Symbols

2. Response Variables

A dependent variable which predicts the performance of a mechanical system is known as a *response variable.*[2] Position, velocity, and acceleration vary with time and are the dependent response variables. Time is usually the independent variable.

Springs and dashpots are *two-port devices*. They have two ends, each of which can have a different value of the response variable. For example, the velocity of both ends of a damper need not be the same. Even though masses do not have ends in the traditional sense of the word, masses are also considered to be two-port devices. Rules for assigning values of response variables to these ends are covered later in this chapter.

The goal of system modeling is to determine the *response* of a system. This means finding the position, velocity, and acceleration of each part of the system as a function of time.

It is not necessary to work separate problems to find x(t), v(t), and a(t) functions. If x(t) is known, it can be consecutively differentiated to find v(t) and a(t). If any one of the three response functions is known, the other two can be derived.

If a system can be completely defined by a single response variable (e.g., x), it is known as a *single-degree-of-freedom (SDOF) system*. Examples of SDOF systems are a mass on a spring, a swinging pendulum, and a rotating pulley. In each case, only one variable (x or θ) is needed to define the position of the primary system object.

Systems in which multiple components can take on independent values (i.e., positions) are known as *multiple-degree-of-freedom (MDOF)* systems. The number of response variables needed is equal to the number of independent system objects. This number is known as the *degree of freedom* of the system.

3. Energy Sources

Masses, springs, and dampers are known as *passive devices* because they dissipate and absorb energy. Energy is required to start a mechanical system oscillating. Energy sources are known as *active components.*

An energy source need not be an actual component such as a battery, fuel cell, or wound spring. Anything that produces motion in the system is an energy source. Mechanical system models usually consider the energy source to be a pure force, without regard to the origin of that force.

Force is known as a *through variable* since it can be passed through objects. Consider a mass on a string. The gravitational force on the mass is passed through the string to the support. Through variables have the same magnitude at both ends of an element.

The velocity attained by a mass acted upon by a force will depend on the magnitude of the force. Another type of energy source, a *velocity source,* produces a specific velocity regardless of the system mass. Velocity is known as an *across variable* since it must be measured with respect to another reference frame. Across variables have different magnitudes at the two ends of an element.

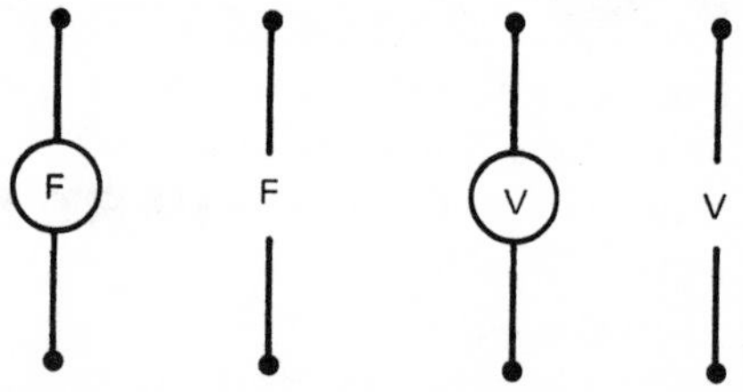

Figure 16.4 Symbols for Energy Sources

4. System Diagrams

The system elements are interconnected to produce the system diagram. This diagram is then used to write the differential equations which describe the response of the system.

step 1: Choose a response variable. Although v is a common choice, x or a can also be used.

step 2: Identify all parts of the mechanical system which have unique values of the chosen response variable. It is not necessary to know the actual values of the response variable. It is only necessary to know which parts have different values.

step 3: On paper, draw horizontal lines and associate these lines with the different values of the response variable.

step 4: Insert the passive elements (m, B, or k) between the appropriate horizontal lines. Connect the elements to the horizontal lines with vertical lines.

rule 16.1: Masses always connect to the lowest (ground) level.

step 5: Insert the active sources (F or v) between the appropriate horizontal lines and connect with vertical lines.

rule 16.2: Energy sources always connect to the lowest (ground) level.

Example 16.1

A mass is connected through a damper to a solid wall as shown. The mass slides without friction on its support. A force is applied to the mass. What is the system diagram?

[2]Response variables are also known as *state variables*.

SYSTEMS

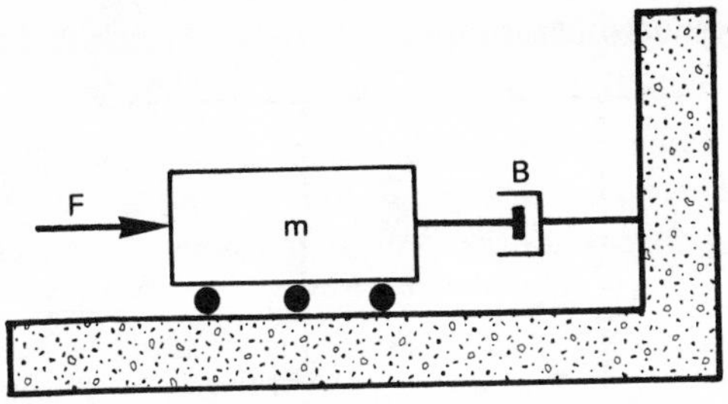

step 1: Choose velocity as the response variable.

step 2: All parts of the mass move with the same velocity. Call this velocity v_1. The plunger also moves with velocity v_1 since it is attached to the mass. The body of the damper is rigidly attached to the wall, which remains stationary. Thus, there are two unique velocities in this system: v_1 and $v = 0$.

step 3: Two horizontal lines are drawn. The top line is associated with v_1. The lower line is associated with $v = 0$.

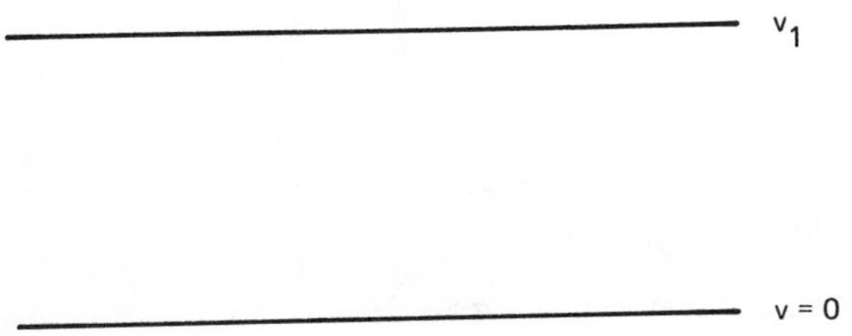

step 4: One end of the damper travels at v_1; the other is stationary. Therefore, insert the damper symbol so that its two ends connect to the v_1 and $v = 0$ lines. The mass moves at v_1, so one of its vertical lines should connect to v_1. By rule 16.1 the other line connects to $v = 0$.

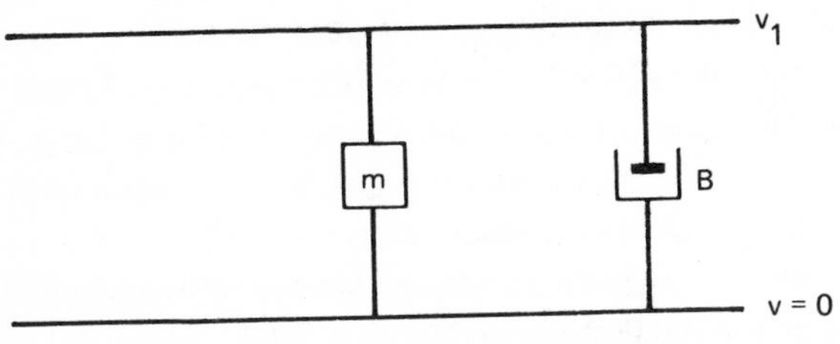

step 5: The force contacts the mass which moves at v_1. Therefore, one end of the force symbol connects to v_1. By rule 16.2, the other line connects to $v = 0$.

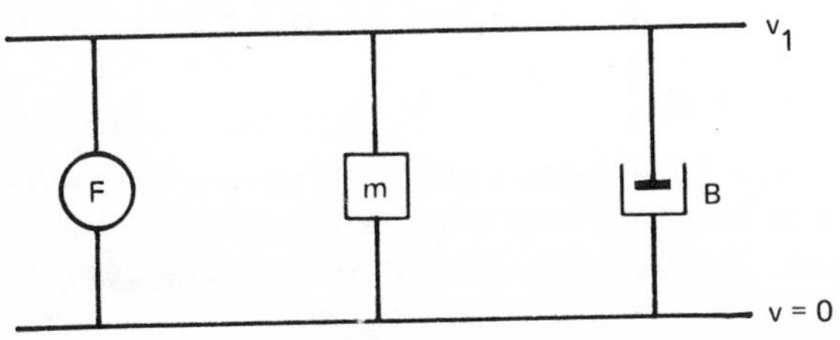

Example 16.2

A force is applied through a damper to a mass. The mass rolls on frictionless bearings. What is the system diagram?

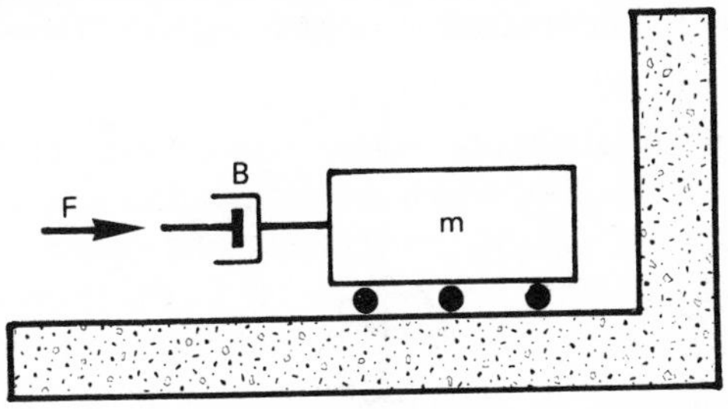

step 1: Choose velocity as the response variable.

step 2: The damper plunger moves. Call its velocity v_1. The mass moves. Call the mass's velocity v_2. The support does not move ($v = 0$).

step 3: Draw three lines corresponding to v_1, v_2, and $v = 0$.

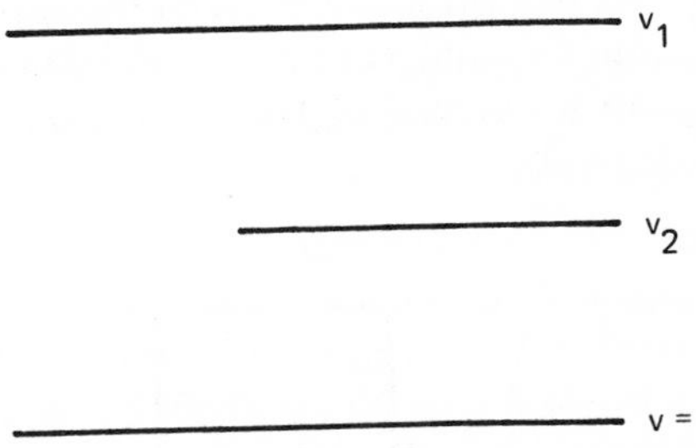

step 4: One end of the damper moves at v_1; the other moves at v_2. So, connect the damper between lines v_1 and v_2. The mass moves at v_2 and connects to $v = 0$ by rule 16.1.

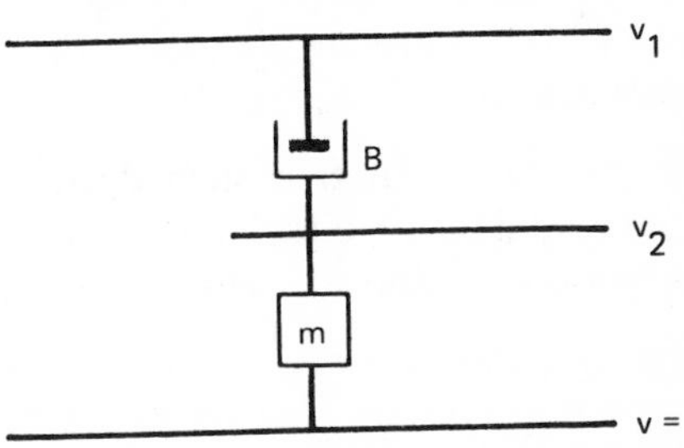

step 5: The force contacts the damper plunger moving at v_1 and connects to $v = 0$ by rule 16.2.

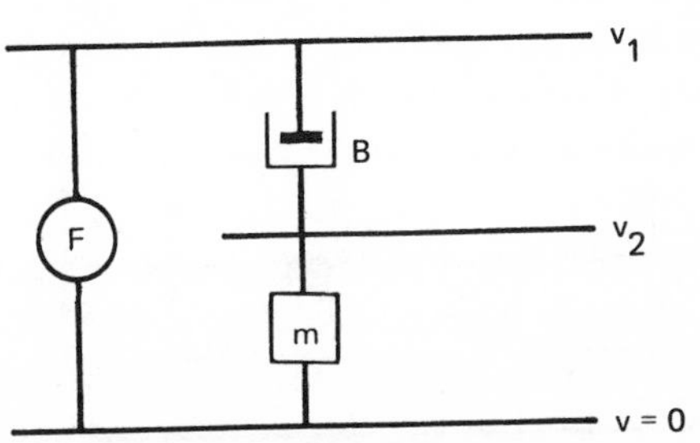

Example 16.3

A high-performance shock absorber is constructed with an integral coil spring as shown. What is the system diagram if a force is applied to one end?

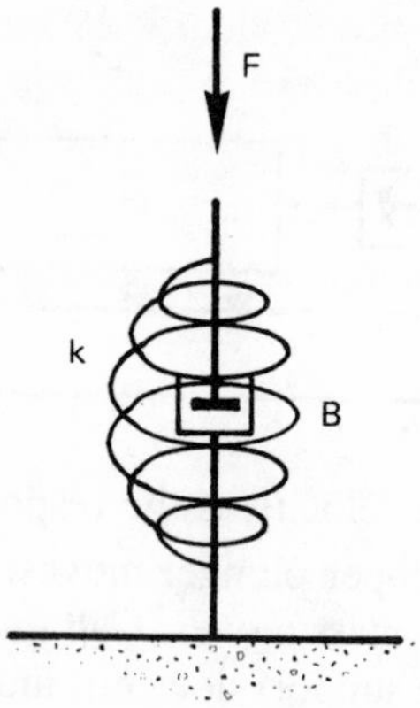

step 1: Choose velocity as the response variable.

step 2: The plunger and top end of the spring move with the same velocity, v_1. The damper body and lower end of the spring do not move ($v = 0$).

steps 3, 4, and 5:

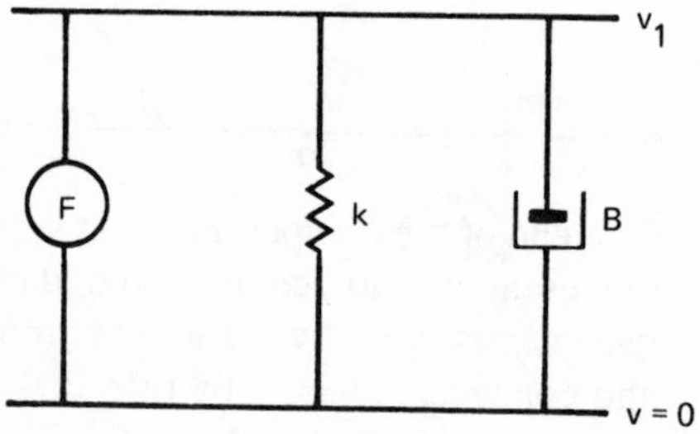

5. Line Diagrams

Line diagrams are similar to system diagrams. However, the horizontal lines are replaced by nodes and the element symbols are removed, leaving only their variables.

Example 16.4

Draw the system and line diagrams for the spring and dashpot shown.

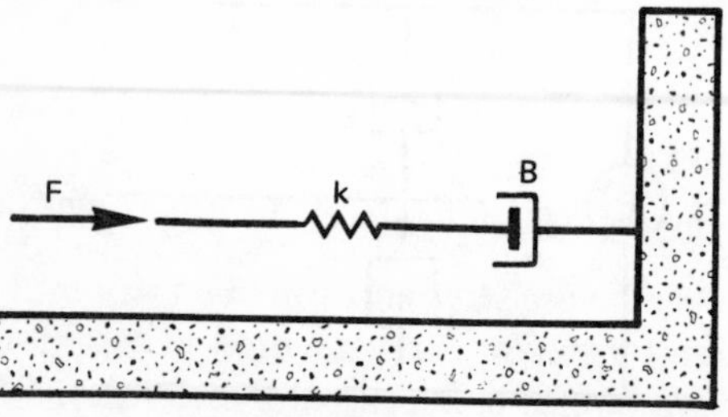

The system diagram is

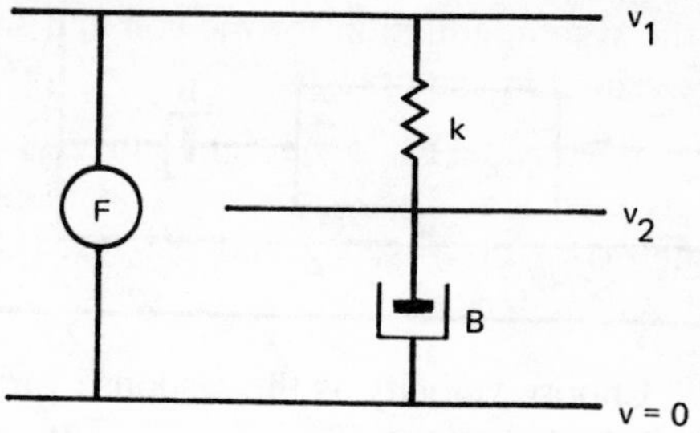

The line diagram is

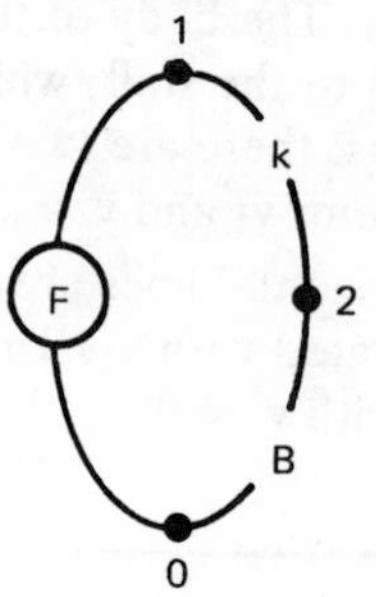

6. System Equations

Once the system diagram is drawn, the *system equations* can be derived. The system equations describe the response variables as functions of the independent variable.

The system equations derived from the system diagram do not explicitly give the response variables. Rather, the system equations are differential equations which must be solved before the response functions can be used. Solution procedures for differential equations (e.g., traditional methods, analog computers, and Laplace transforms) are covered in other chapters of this book.

The principle used to derive the system equations is analogous to Kirchoff's current law and other conservation laws. These conservation laws say ". . . what goes in must come out . . ." In the case of system modeling, the quantity being conserved is force. The total force being supplied by the energy source must equal the sum of the forces leaving through all parallel legs.

rule 16.3: The force passing through a leg consisting of series elements can be determined from the conditions across the first element in that leg.

rule 16.4: When writing a difference in a response variable (e.g., $v_2 - v_1$, $x_2 - x_1$, etc.), the first subscript is the same as the node number for which the equation is being written.

Example 16.5

What is the system equation for the coil and shock absorber described in example 16.3?

The force is considered to be a quantity flowing up through its own leg. This force splits, with part of it (F_B) going down the first leg and the rest (F_k) going down the second leg.

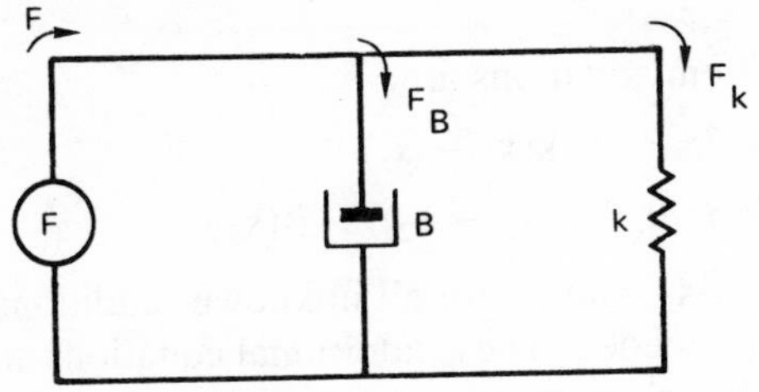

Since force is conserved, the system equation is

$$F(t) = F_B(t) + F_k(t)$$
$$= Bv(t) + kx(t)$$

However, v(t) is the first derivative of x(t). Therefore, this is a first-order differential equation.

$$F(t) = Bx' + kx$$

Example 16.6

The coupling of a railroad car is modeled as the mechanical system shown. Assume all elements are linear. What are the system equations which describe the positions x_1 and x_2 as functions of time?

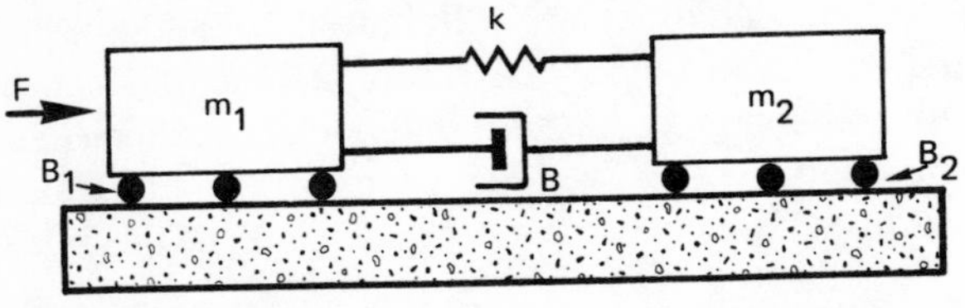

The system diagram is

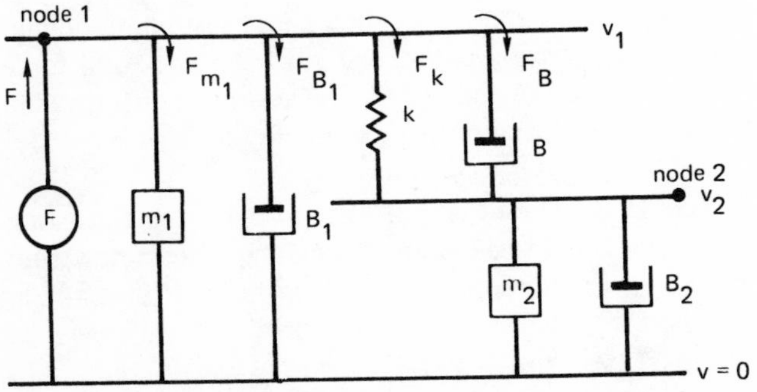

The force that enters node *1* is equal to the sum of forces leaving node *1*. The system equation is

$$F = F_{m_1} + F_{B_1} + F_k + F_B$$

It is correct to write $F_{B_1} = B_1v_1$ since the lower end of the damper has a zero velocity. However, it would be incorrect to write F_B in terms of v_1 only, since the two ends of damper *B* move with velocities v_1 and v_2.

Using rule 16.3, the force F_{B_1} can be written in terms of Δv. But, it may not be clear if Δv means $(v_1 - v_2)$ or $(v_2 - v_1)$. Since this system equation is being written for node *1*, rule 16.4 requires that the difference be $(v_1 - v_2)$.

The system equation in terms of the response variables is

$$F = m_1a_1 + B_1v_1 + B(v_1 - v_2) + k(x_1 - x_2)$$

This can be written as a differential equation.

$$F = m_1x_1'' + B_1x_1' + B(x_1' - x_2') + k(x_1 - x_2)$$

This differential equation contains two variables — x_1 and x_2. A second differential equation is required. Once it is found, the two system differential equations can be solved simultaneously.

The second system equation can be found from node *2*. The same principle applies to node *2* as was used to write the node *1* system equation — the forces into and out of node *2* are equal. There is no force going into node *2* since there is no force source across the terminals on the right side of the system diagram.

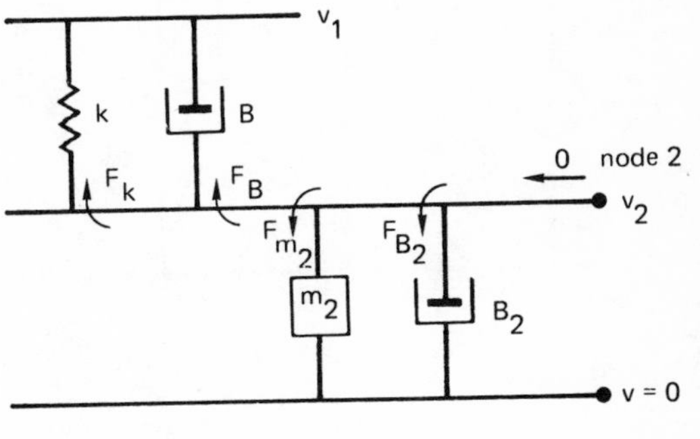

The node *2* system equation is

$$0 = F_{B_2} + F_{m_2} + F_k + F_B$$
$$= B_2v_2 + m_2a_2 + B\Delta v + k\Delta x$$

Using rule 16.4, Δv and Δx become $(v_2 - v_1)$ and $(x_2 - x_1)$.

$$0 = B_2v_2 + m_2a_2 + B(v_2 - v_1) + k(x_2 - x_1)$$
$$= B_2x_2' + m_2x_2'' + B(x_2' - x_1') + k(x_2 - x_1)$$

7. *Energy Transformation*

Levers can be used to transform one force into another. This transformation is represented in system diagrams by the symbol for an electrical transformer.

Example 16.7

What are the system equations for the mechanical system shown?

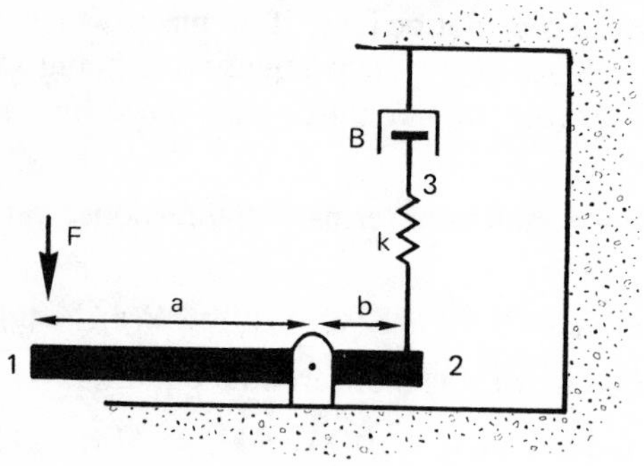

The lever transforms the force, displacement, velocity, and acceleration at point *1* into force, displacement, velocity, and acceleration at point *2*. The system diagram is

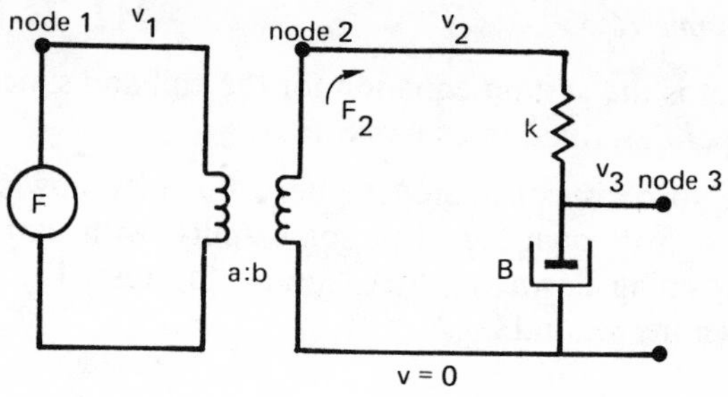

The system equations are:

node 2: $F_2 = k(x_2 - x_3)$

node 3: $0 = k(x_3 - x_2) + B(x_3')$

Since F_2, x_2, and x_3 are all unknown, additional equations are needed. These additional equations are based on the lever's ratio of transformation.

$$x_2 = \left(\frac{b}{a}\right)x_1$$

$$F_2 = \left(\frac{a}{b}\right)F_1$$

SYSTEMS

PART 2: Rotational Systems

Nomenclature

a	ratio of transformation
B	coefficient of damping
J	rotational moment of inertia
k	rotational spring stiffness
N	number of teeth

Symbols

α	angular acceleration
θ	angular rotation
τ	torque
ω	angular velocity

1. Rotational Elements

Rotational elements are the rotational mass (flywheel), rotational spring (torsional spring), and rotational damper (fluid coupling).

When a *rotational mass* is acted upon by a torque, it behaves according to equation 16.4.

$$\tau = J\alpha \qquad 16.4$$

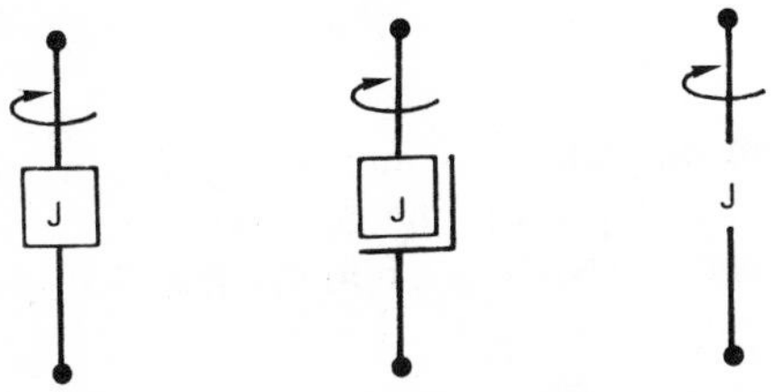

Figure 16.5 Symbols for Rotational Mass

The torque needed to twist a *rotational spring* is given by equation 16.5. *k* is the spring's *stiffness*.

$$\tau = k\Delta\theta \qquad 16.5$$

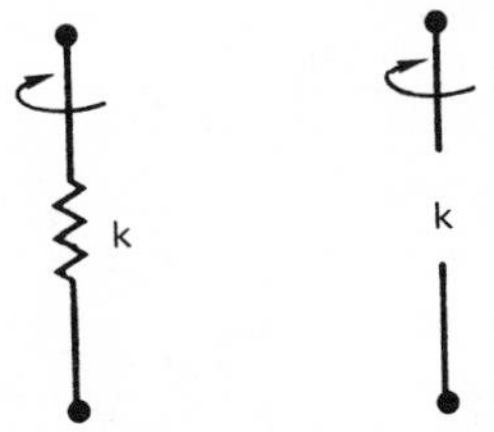

Figure 16.6 Symbols for Rotational Spring

The performance of a *rotational damper* is governed by equation 16.6.

$$\tau = B(\omega_2 - \omega_1) \qquad 16.6$$

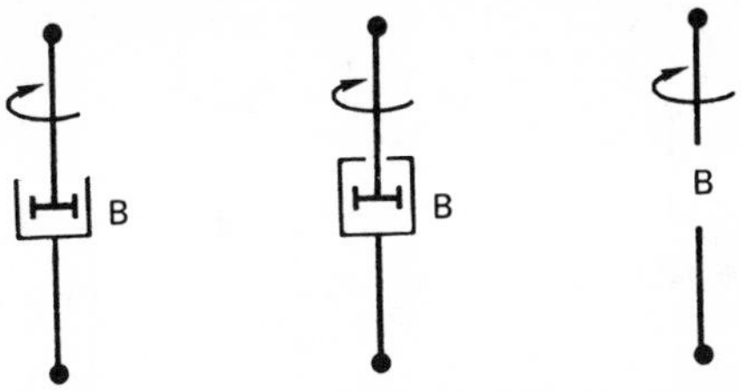

Figure 16.7 Symbols for Rotational Damper

Gearsets can be used to transform one torque into another. The ratio of speeds and torques depends on the number of teeth possessed by each gear.

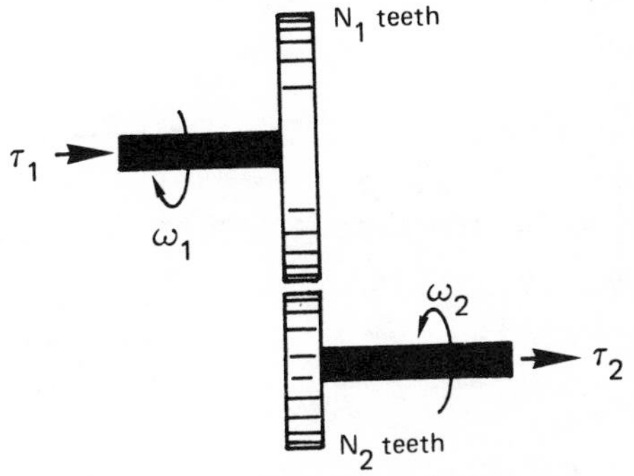

Figure 16.8 Rotational Transformer

The transformation ratio is

$$a = \frac{N_1}{N_2} \qquad 16.7$$

The transformed torques and speeds can be found from equation 16.8. The minus signs imply a direction change.

$$a = -\frac{\tau_1}{\tau_2} = -\frac{\omega_2}{\omega_1} \qquad 16.8$$

2. Response Variables

The rotational response variables are angular displacement (θ), angular velocity (ω), and angular acceleration (α). These response variables are usually expressed in terms of *radians,* although *revolutions* can also be used.

As with translational mechanical systems, any of these three response variables can be used to write the system equations. The three variables are related. If one is known, the others can be found by integration or differentiation.

3. Energy Sources

Energy can be provided to rotational systems by constant-torque sources or constant-velocity sources. Symbols for these energy sources are shown in figure 16.9.

SYSTEMS

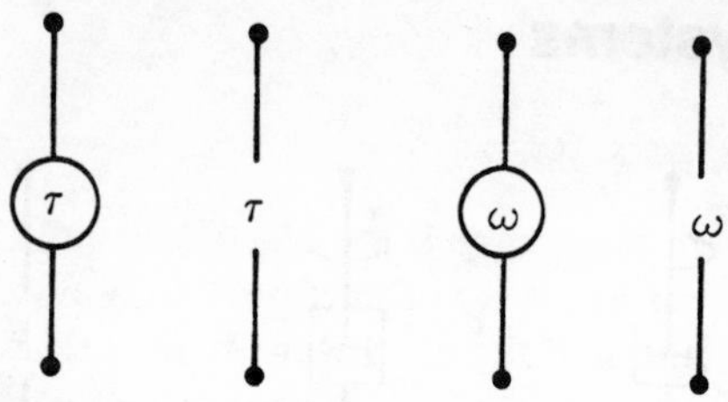

Figure 16.9 Symbols for Rotational Energy Sources

Example 16.8

Two rotating flywheels are connected by a flexible shaft with known stiffness. The second flywheel is acted upon by a viscous force proportional to the velocity. Draw the system diagram and write the differential equations of motion.

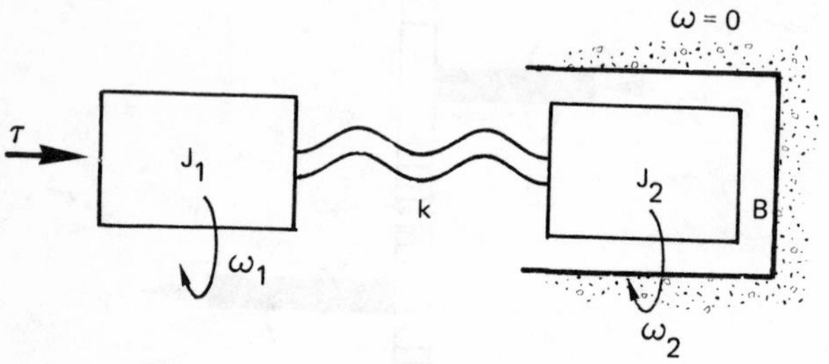

step 1: Choose ω as the response variable.

step 2: There are three different rotational speeds: ω_1, ω_2, and $\omega = 0$.

steps 3, 4, and 5:

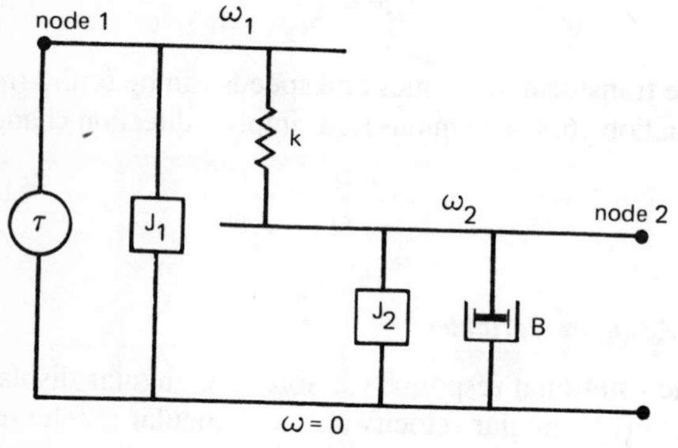

The system equations in terms of the response variables are

node 1: $\tau = J_1\alpha_1 + k(\theta_1 - \theta_2)$

node 2: $0 = B\omega_2 + J_2\alpha_2 + k(\theta_2 - \theta_1)$

These simultaneous equations can be written as differential equations.

node 1: $\tau = J_1\theta_1'' + k(\theta_1 - \theta_2)$

node 2: $0 = B\theta_2' + J_2\theta_2'' + k(\theta_2 - \theta_1)$

Example 16.9

A motor drives a flywheel through a set of reduction gears. The flywheel is connected to the driven gear by a flexible shaft. What are the system equations?

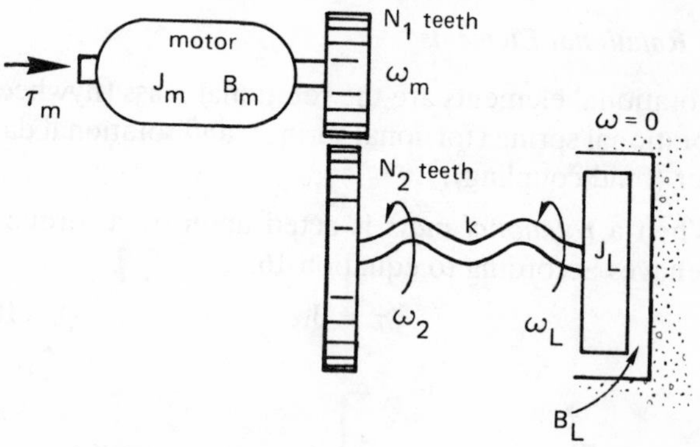

The system diagram uses the symbol for an electrical transformer to represent the gear set. The transformation ratio is $a = (N_1/N_2)$, a number less than *one*.

node m: $\tau_m = J_m\alpha_m + B_m\omega_m + a\tau_2$

node 2: $\tau_2 = k(\theta_2 - \theta_L)$

node L: $0 = B_L\omega_L + J_L\alpha_L + k(\theta_L - \theta_2)$

Since ω_m, ω_2, ω_L, and τ_2 are all unknown, a fourth equation is necessary. This equation is

$a\omega_m = \omega_2$

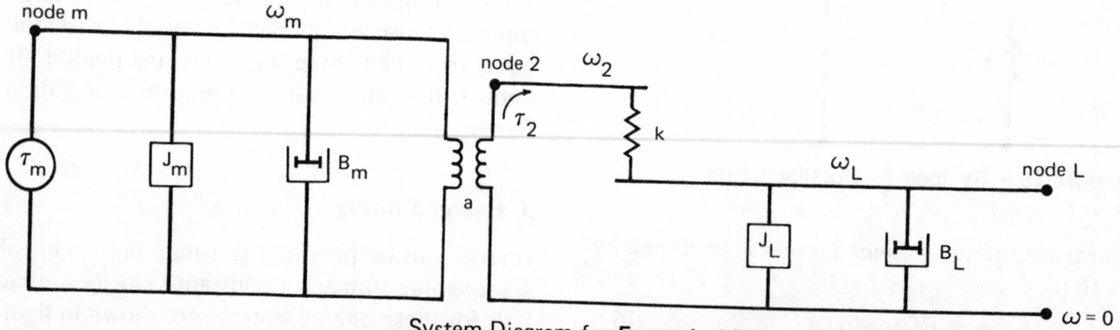

System Diagram for **Example 16.9**

PART 3: Electrical Systems

Nomenclature

C	capacitance
I	current
L	inductance
R	resistance
t	time
V	voltage

1. Electrical Elements

The passive elements in an electrical circuit are the resistor, capacitor, and inductor. These elements were described in detail in chapter 14. Their governing equations are briefly presented here.

resistor: $$I = \frac{1}{R}(V_1 - V_2) \qquad 16.9$$

capacitor: $$I = C\frac{d(V_1 - V_2)}{dt} \qquad 16.10$$

inductance: $$I = \frac{1}{L}\int (V_1 - V_2)dt \qquad 16.11$$

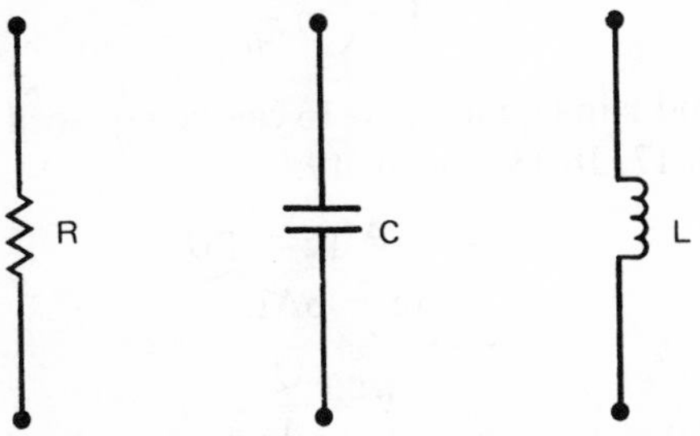

Figure 16.10 Passive Electrical Symbols

2. Response Variables

Voltage is the response variable chosen when working with electrical systems. The integral and derivative of voltage are not commonly encountered.

3. Energy Sources

Voltage and current sources can both be used with electrical systems. Current is a *through variable* since it passes through the circuit elements. Voltage is an *across variable* since it must be measured across two terminals.

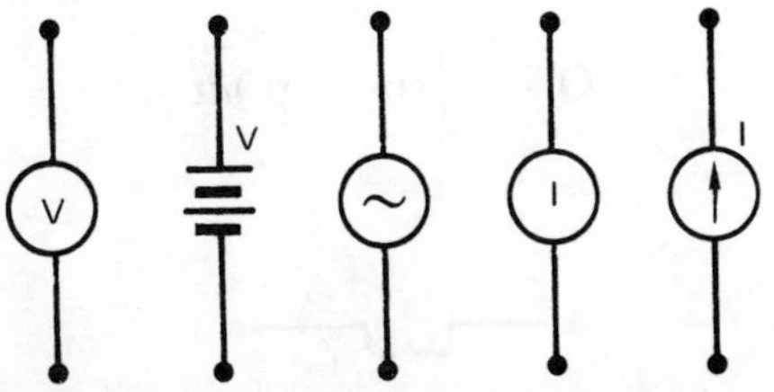

Figure 16.11 Electrical Energy Source Symbols

4. System Diagrams

The system diagram and the electrical circuit are identical. Therefore, it is not necessary to draw a different system diagram if the electrical circuit is shown.

Example 16.10

What are the system equations for the electrical circuit shown?

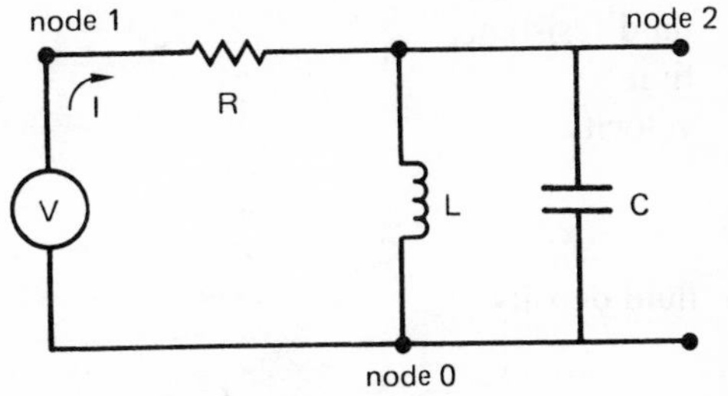

Nodes *1* and *2* have different potentials (voltages).

node 1: $$I = \frac{1}{R}(V_1 - V_2)$$

node 2: $$0 = C\frac{dV_2}{dt} + \frac{1}{L}\int V_2\,dt + \frac{1}{R}(V_2 - V_1)$$

Example 16.11

What are the system equations for the electrical circuit shown?

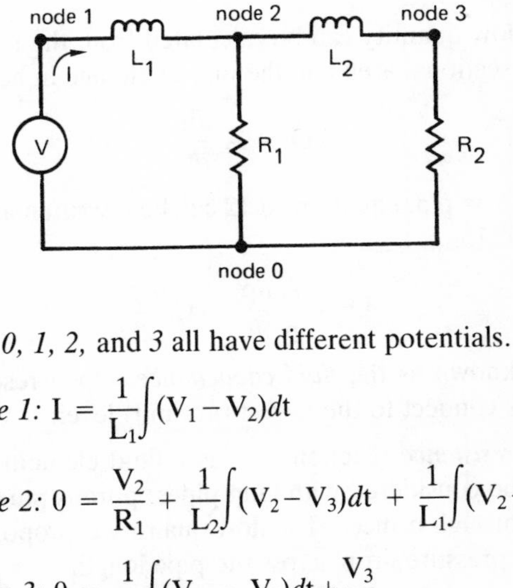

Nodes *0, 1, 2,* and *3* all have different potentials.

node 1: $$I = \frac{1}{L_1}\int (V_1 - V_2)dt$$

node 2: $$0 = \frac{V_2}{R_1} + \frac{1}{L_2}\int (V_2 - V_3)dt + \frac{1}{L_1}\int (V_2 - V_1)dt$$

node 3: $$0 = \frac{1}{L_2}\int (V_3 - V_2)dt + \frac{V_3}{R_2}$$

PART 4: Fluid Systems

Nomenclature

A	area
C_f	fluid capacitance
F	force
h	height
I	fluid inertance
L	length
m	mass
p	pressure
Q	flow quantity
R_f	fluid resistance
t	time
v	velocity

Symbols

ρ	fluid density

Just as electrical capacitors store charge, *reservoirs (fluid capacitors)* store fluid. Figure 16.12 shows a reservoir with constant cross sectional area.

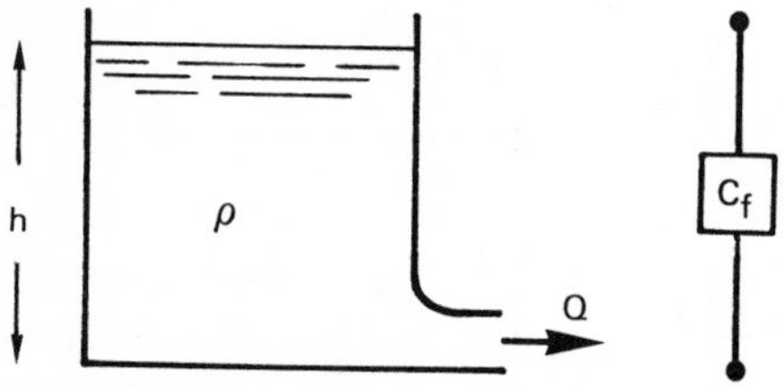

Figure 16.12 Reservoir and Symbol

The flow quantity can be calculated from the reservoir cross sectional area and the rate of change in height.

$$Q = A\frac{dh}{dt} \qquad 16.12$$

Since $h = p/\rho$, equation 16.12 can be rewritten as equation 16.13.

$$Q = \frac{A}{\rho}\frac{dp}{dt} = C_f\frac{dp}{dt} \qquad 16.13$$

C_f is known as the *fluid capacitance.* Open reservoirs always connect to the lowest (ground) level.

Flow resistance (friction loss) is a fluid element which must be considered with long pipes, porous plugs, and flow through orifices. The flow quantity is proportional to the pressure drop across the pipe length.

$$Q = \frac{1}{R_f}(p_2 - p_1) \qquad 16.14$$

R_f is the *fluid resistance coefficient.* Its form depends on the type of flow, quantity of flow, and physical configuration. In most cases the pressure drop is proportional to the *square* of the flow quantity. Therefore, equation 16.14 is an approximation which is valid over limited ranges of the flow quantity. For that reason, R_f is usually found by direct experimentation.

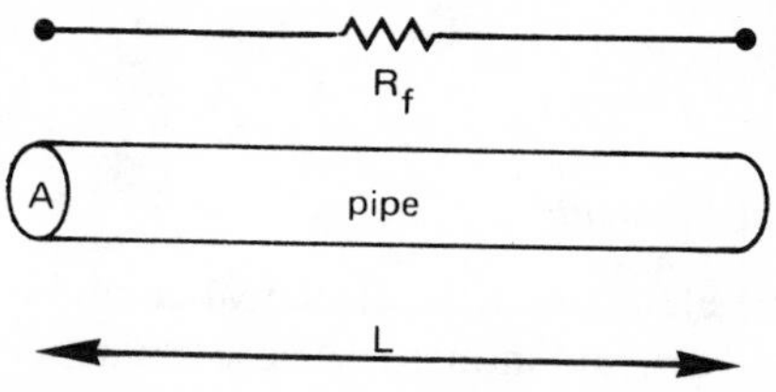

Figure 16.13 Fluid Resistance Symbol

Fluid inertance accounts for the inertia of the fluid flow. The impulse-momentum principle (equation 16.15) is the basis for fluid inertance.

$$F\,dt = m\,dv \qquad 16.15$$

Equation 16.15 can be reorganized by dividing both sides by *dt*.

$$F = m\frac{dv}{dt} \qquad 16.16$$

F, m, and *v* in equation 16.16 can be replaced by equations 16.17, 16.18, and 16.19.

$$F = A(p_2 - p_1) \qquad 16.17$$

$$m = \rho A L \qquad 16.18$$

$$v = \frac{Q}{A} \qquad 16.19$$

Combining equations 16.16 through 16.19 produces equation 16.20.

$$A(p_2 - p_1) = \rho A L\left[\frac{1}{A}\left(\frac{dQ}{dt}\right)\right] \qquad 16.20$$

$$p_2 - p_1 = \frac{\rho L}{A}\left(\frac{dQ}{dt}\right) = I\left(\frac{dQ}{dt}\right) \qquad 16.21$$

I is known as the *fluid inertance.* It is also known as the *fluid inductance.*

Integrating both sides of equation 16.21 results in an expression for the flow quantity.

$$\int (p_2 - p_1)\,dt = I\,dQ \qquad 16.22$$

$$Q = \frac{1}{I}\int (p_2 - p_1)dt \qquad 16.23$$

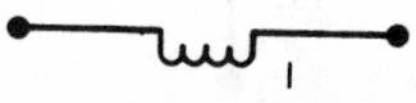

Figure 16.14 Fluid Inertance Symbol

Example 16.12

A pump is used to keep a liquid flowing through a filter pack as shown. What is the system equation? (Neglect the pipe friction.)

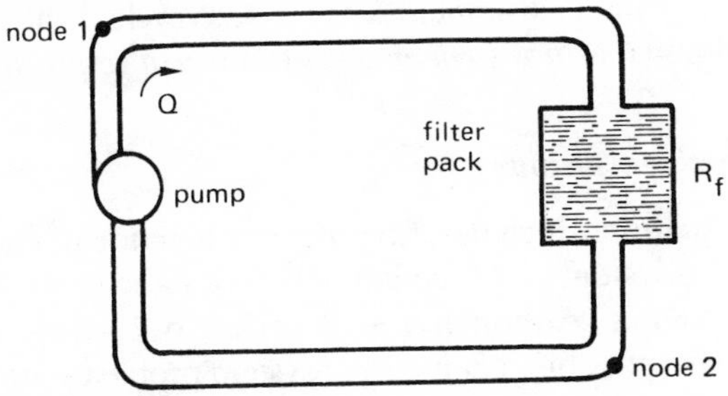

step 1: Pressure is the response variable.

step 2: Neglecting pressure drop in the pipe, there are two different pressures in the system: p_1 before the filter pack and p_2 after the filter pack.

steps 3, 4, and 5:

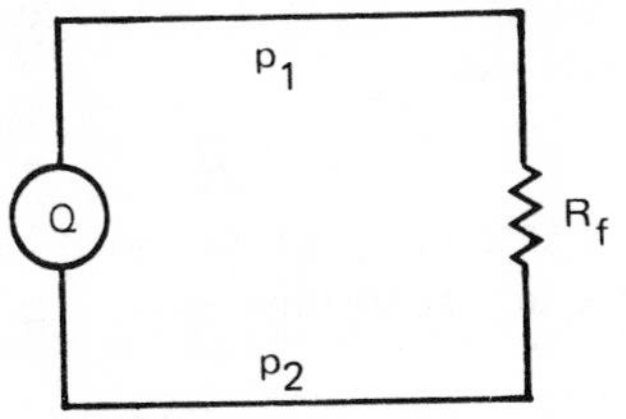

The system equation is

$$Q = \frac{1}{R_f}(p_2 - p_1)$$

Example 16.13

The reservoir shown discharges through a long pipe. The reservoir is not refilled. What is the system equation?

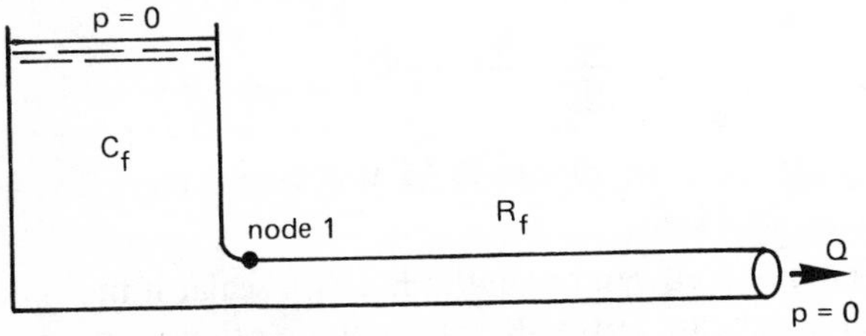

Although there is flow, there is no energy source (e.g., pump) in the system. The system diagram is

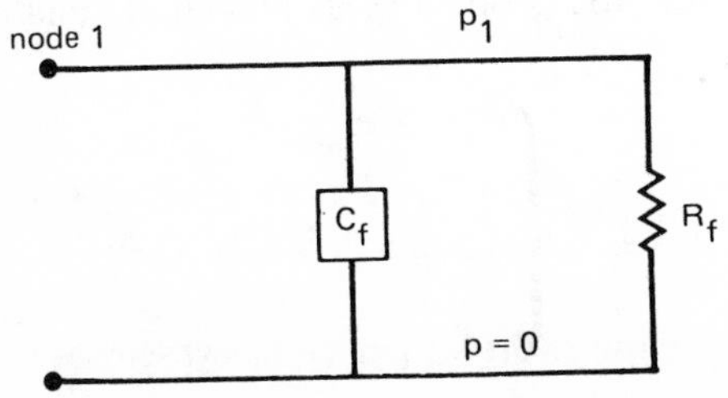

The system equation is

$$0 = \frac{p_1}{R_f} + C_f\left(\frac{dp_1}{dt}\right)$$

Example 16.14

A pump is used to transfer liquid from one reservoir to another as shown. What are the system equations? (Neglect pipe friction.)

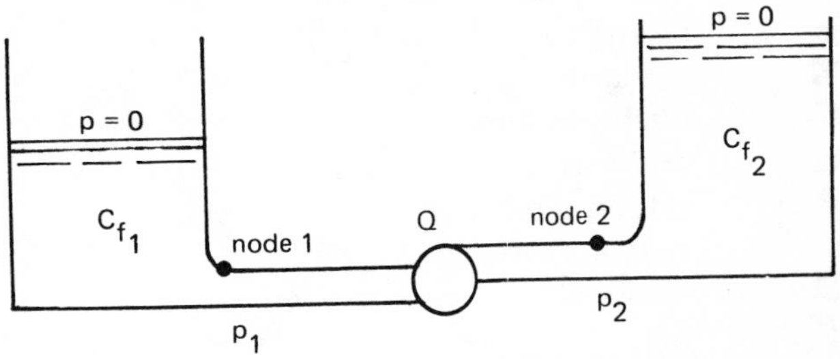

Since open reservoirs always connect to the lowest level, the system diagram is

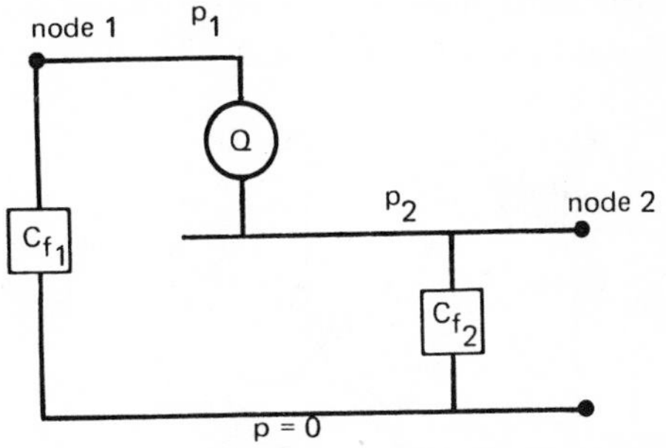

The system equations are:

node 1: $Q = C_{f1}\left(\frac{dp_1}{dt}\right)$

node 2: $Q = C_{f2}\left(\frac{dp_2}{dt}\right)$

PART 5: Analysis of Engineering Systems

Nomenclature

a	imaginary part
B	bandwidth
E	error
F	forcing function
G	forward transfer function
h	height of pulse
H	reverse transfer function
j	square root of −1
K	scalar
Q	quality factor
r	real part
R	response function
s	LaPlace variable
S	sensitivity
t	time

Symbols

ω	angular frequency

Subscripts

f	feedback
n	natural
o	output
s	signal

1. Natural, Forced, and Total Response

If motion or change is induced in an engineering system, the system equation can be solved to determine the response function. This function gives the condition (e.g., position, voltage, pressure, etc.) as a function of time. The response function can have constant terms, sine and cosine terms, exponential terms, or multiplicative combinations and sinusoids and exponentials.

Natural response is induced when energy is applied to an engineering system and is subsequently removed. The system is left alone and is allowed to do what it would do naturally, without the application of further disturbing forces.

Natural response is characterized by pure sine and cosine terms in the system equation if friction is absent. Frictionless systems oscillate continuously without decay. If friction is present, the natural terms will be products of sinusoids and exponentials.

If a system is acted upon by a force which repeats at regular intervals, the system will move in accordance with that force. This is known as *forced response.* It is characterized by sinusoidal terms having the same frequency as the forcing function.

The natural and forced responses are present simultaneously in forced systems. The sum of the two responses is known as the *total response.* Since the natural effects usually disappear after a few cycles, they are also known as *transient effects* or *transient response.*

2. Forcing Functions

An equation which describes the introduction of energy into the system as a function of time is known as a *forcing function.* Although a wide variety of forcing functions are possible, engineering system problems are too complicated for manual calculations if not limited to the simpler types.

A *homogenous forcing function* is a constant zero force. A zero forcing function does not preclude initial disturbance. For example, a spring/mass system which is displaced, released, and allowed to oscillate freely is an example of homogenous forcing function. The forcing function is homogenous if it ceases to act after the system begins to move.

$$F(t) = 0 \qquad 16.24$$

A *unit step* is a forcing function which has zero magnitude up to a particular instant (say t_1) and a magnitude of *one* therafter.

$$F(t) = \begin{cases} 0 & t < t_1 \\ 1 & t \geq t_1 \end{cases} \qquad 16.25$$

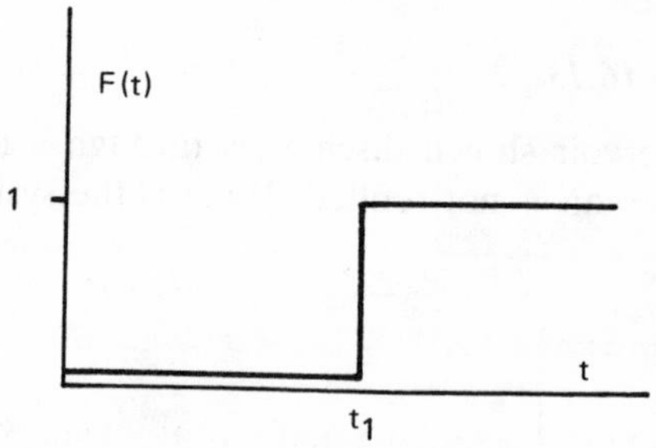

Figure 16.15 Unit Step

The unit step can be multiplied by a scalar if the actual force appears at t_1 with a magnitude other than *one.*

A *unit pulse* is a limited duration force whose total area under the curve is *one.* A necessary condition of a unit pulse is that the product of its magnitude and duration is *one.*

$$F(t) = \begin{cases} 0 & t < t_1 \\ \dfrac{1}{\Delta t} & t_1 \leq t < t_1 + \Delta t \\ 0 & t \geq t_1 + \Delta t \end{cases} \qquad 16.26$$

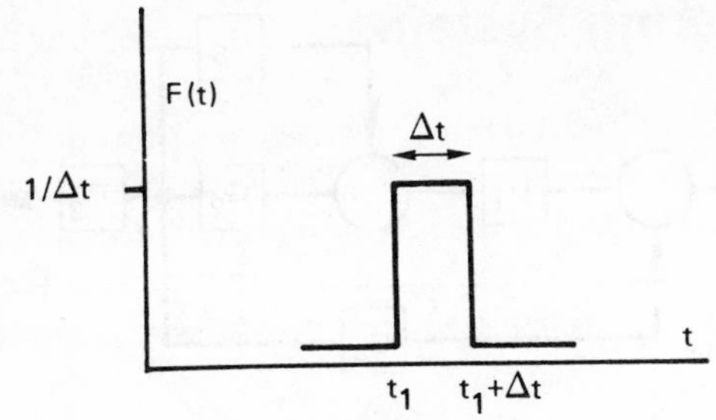

Figure 16.16 Unit Pulse

The unit pulse can be multiplied by a scalar if the actual pulse has a magnitude greater than 1/Δt.

The terms *impulse* and *pulse* are often used interchangeably. However, *impulse* is more appropriate for very short impacts with large magnitudes. Hitting a bell with a hammer is illustrative of an impulse.

The most common forcing functions used in the analysis of engineering systems are the sine and cosine functions. Sinusoids, when combined with Fourier analysis, can be used to approximate all other forcing functions.

$$F(t) = \sin\omega t \qquad 16.27$$

3. Transfer Functions

A system model can be thought of as a *black box*. The input to the black box is the forcing function. The output is the performance of the system, known as the *system response function*, R(t). This analogy is particularly valid with electrical systems. The input voltage is applied across the input terminals. The output voltage is measured across the output terminals.

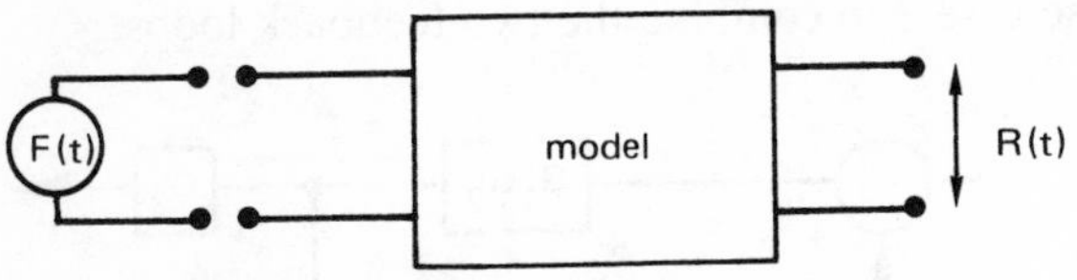

Figure 16.17 A Two-Port Black Box

The ratio of the system response (output) to the forcing function (input) is known as the *transfer function,*[3] T(t). The transfer function can be a simple scalar. However, it can also be a phasor (vector) and possess both magnitude and phase change capability.

$$T(t) = \frac{R(t)}{F(t)} \qquad 16.28$$

Transfer functions are generally written in terms of the *s* variable.[4] This is accomplished, if T(t) is known, by taking the Laplace transform of the transfer function. The result is the *transform of the transfer function*[5].

$$T(s) = \mathcal{L}[T(t)] \qquad 16.29$$

[3]The transfer function is also known as the *rational function, system function, closed-loop function,* and *control ratio.*

Transforming T(t) into T(*s*) is more than a simple change of variables. The variable *s* can be thought of as an operator which performs differentiation. Specifically, x′ could be written in terms of *s* as *s*x. Similarly, x″ could be written as s^2x. The operation of integration can be represented by the reciprocal of *s*.

Example 16.15

Write the following system differential equation in terms of the *s* operator.

$$I(t) = \frac{1}{L}\int (V_1 - V_2)dt + C\frac{d(V_2 - V_1)}{dt}$$

If all derivatives are replaced by *s* and the integral is replaced by 1/*s*, the transformed system equation is

$$I(s) = \frac{V_1}{sL} - \frac{V_2}{sL} + CsV_2 - CsV_1$$

Example 16.16

Determine the transformed transfer function for the electrical system shown.

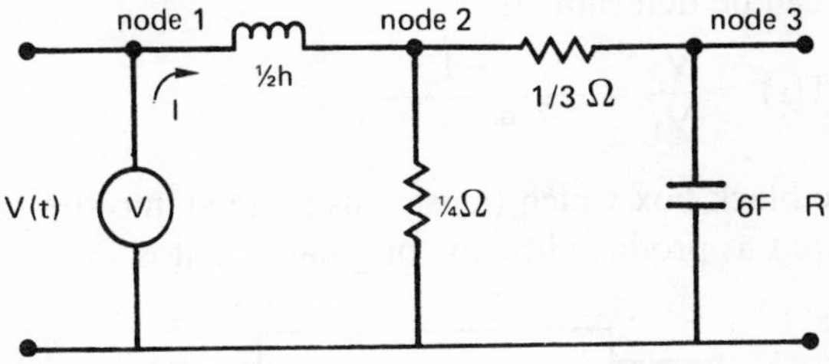

node 1:

$$I(t) = \frac{1}{L}\int (V_1 - V_2)dt$$

Substituting ½ for *L* and using the *s* operator,

$$I(s) = \frac{2(V_1 - V_2)}{s} \qquad \text{Equation I}$$

node 2:

$$0 = \frac{1}{L}\int (V_2 - V_1)dt + \frac{1}{R_2}(V_2) + \frac{1}{R_3}(V_2 - V_3)$$

$$0 = \frac{2(V_2 - V_1)}{s} + 4V_2 + 3(V_2 - V_3)$$

Equation II

[4]A review of the material on Laplace transforms in chapter 1 is recommended.

[5]T(*s*) is frequently just called the *transfer function.* In actual practice, T(t) is almost never encountered. It is acceptable, therefore, for T(*s*) to share the name of T(t).

node 3:

$$0 = C\frac{dV_3}{dt} + \frac{1}{R}(V_3 - V_2)$$

$$0 = 6sV_3 + 3(V_3 - V_2) \qquad \text{Equation III}$$

The transfer function is the ratio of the output to the input. It does not depend on the intermediate voltage V_2.

$$T(t) = \frac{V_3(t)}{V_1(t)}$$

Since $V_2(t)$ does not appear in T(t), V_2 must be eliminated from equations **II** and **III.** (Equation **I** cannot be used unless the current is known. It generally is unknown.)

From equation **II**,

$$V_2 = \frac{-2V_1 - 3sV_3}{2 + 7s}$$

From equation **III**,

$$V_2 = V_3(1 + 2s)$$

Setting these two expressions for V_2 equal, the ratio V_3/V_1 can be determined.

$$T(s) = \frac{V_3}{V_1} = \frac{-1}{7s^2 + 7s + 1}$$

The black box which transforms a signal into the same output as produced by the original circuit is

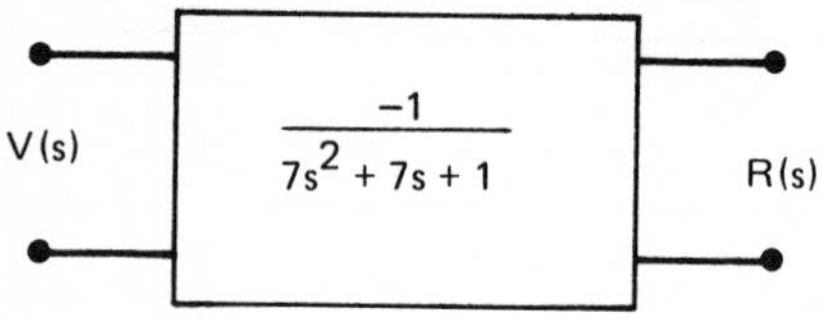

4. *Block Diagram Algebra*

Example 16.16 illustrated how an engineering system can be modeled by a single operational black box. Several such boxes can be grouped together to obtain the desired response.

Complex systems of several box diagrams can be simplified by using the equivalent structures shown in table 16.1.

Example 16.17

A complex block system is constructed from five blocks and two summing points as shown.

Simplify the system and determine its overall effect on an input.

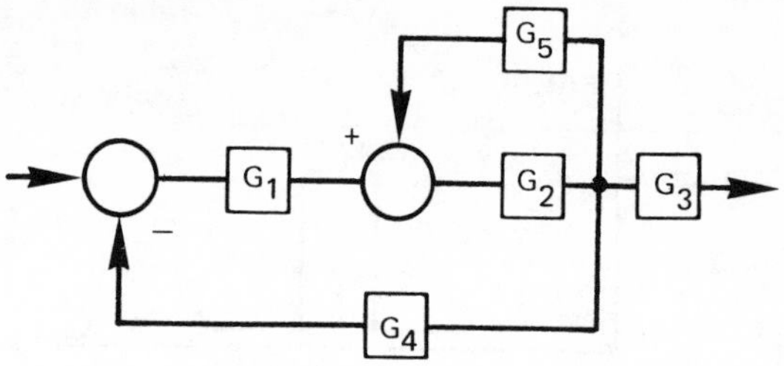

Use case *5* from table 16.1 to move the second summing point back to the first summing point.

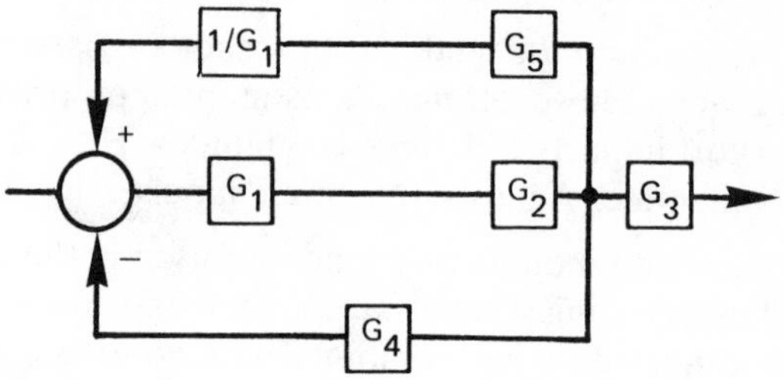

Use case *1* to combine boxes in series.

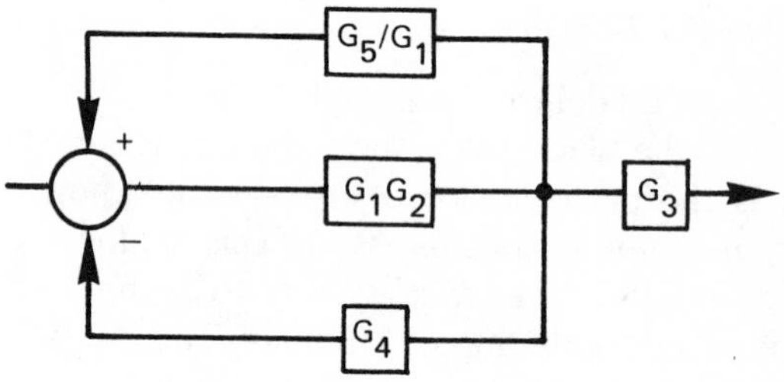

Use case *2* to combine the two feedback loops.

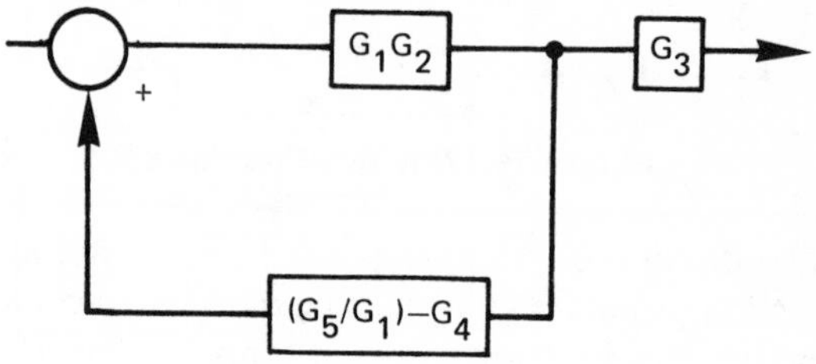

Use case *8* to move the pick-off point outside of the G_3 box.

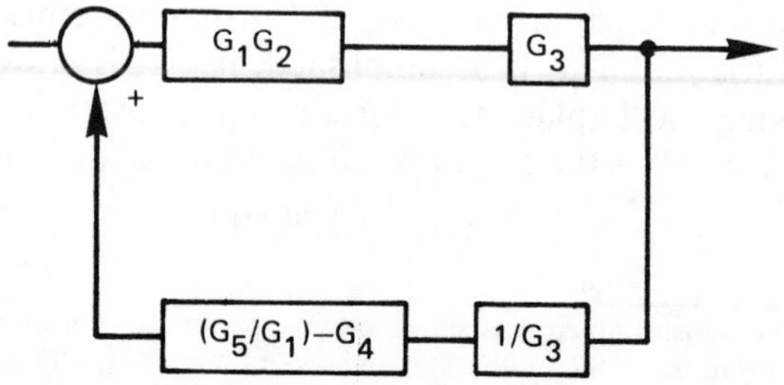

Table 16.1 Equivalent Block Diagrams

case	original structure	equivalent structure
1	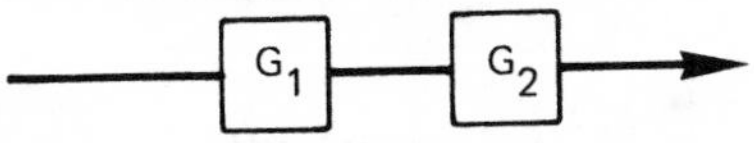	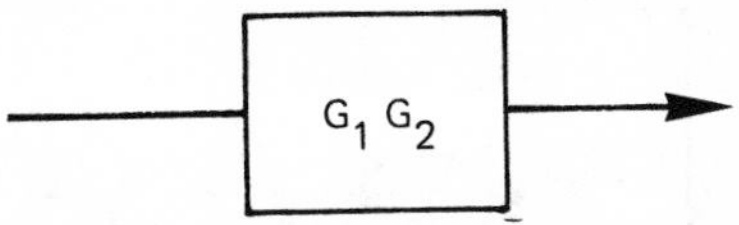
2	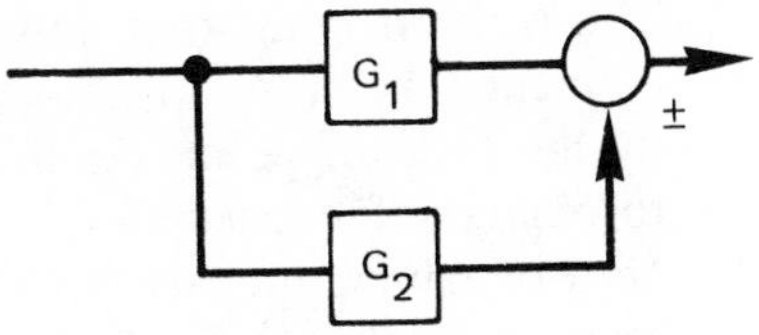	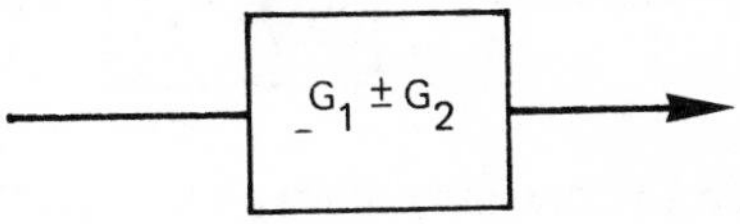
3	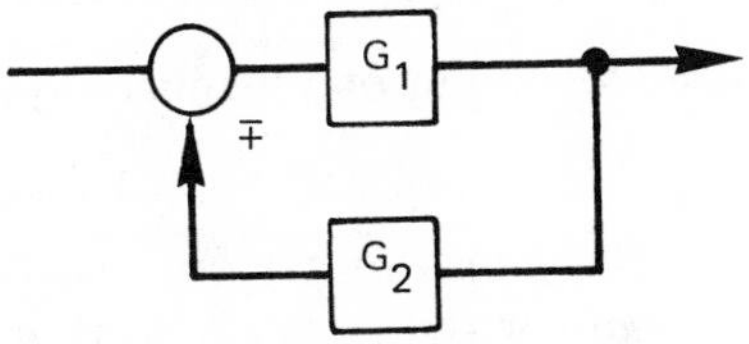	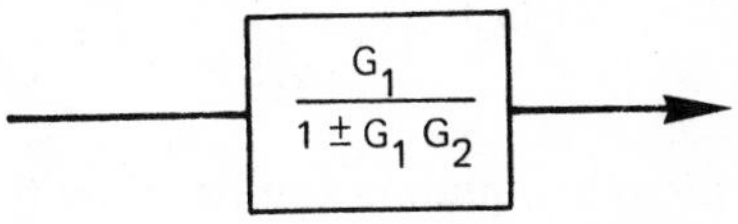
4	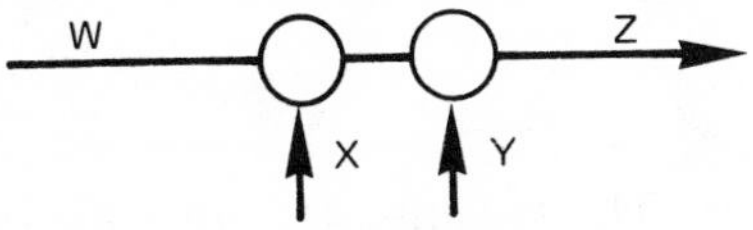	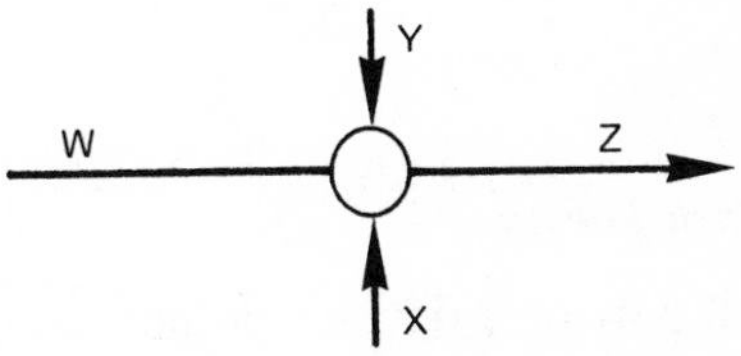
5	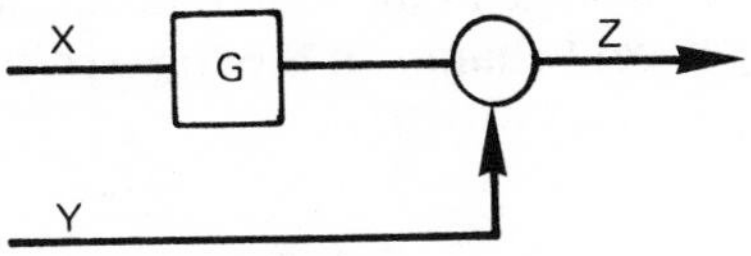	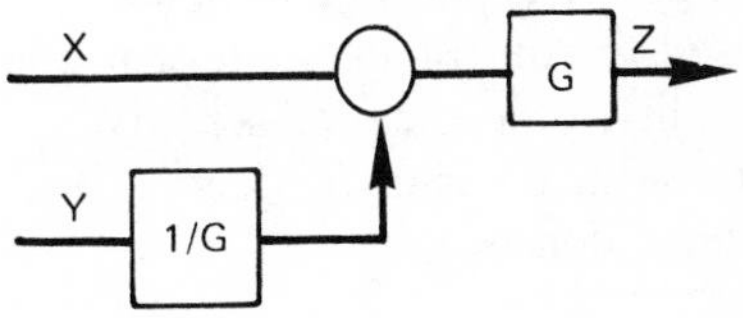
6	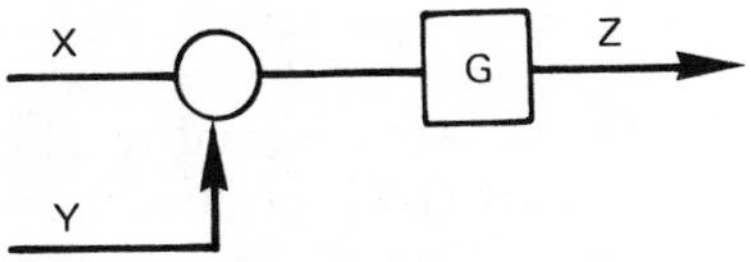	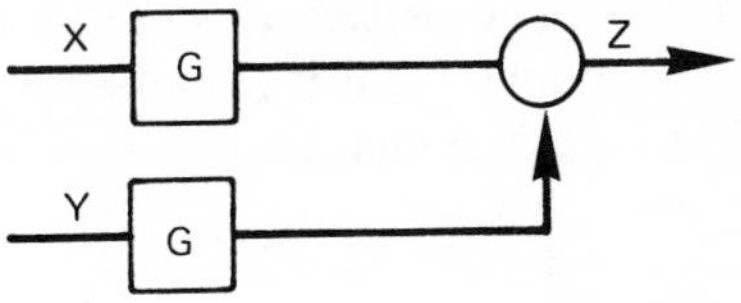
7	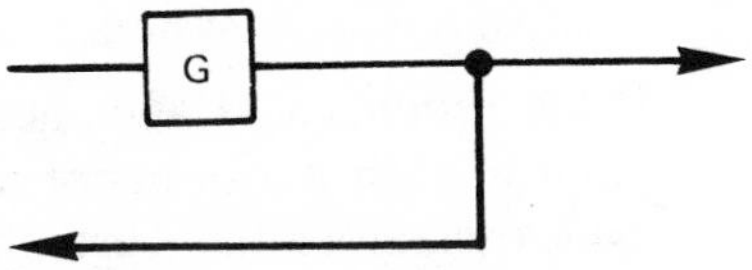	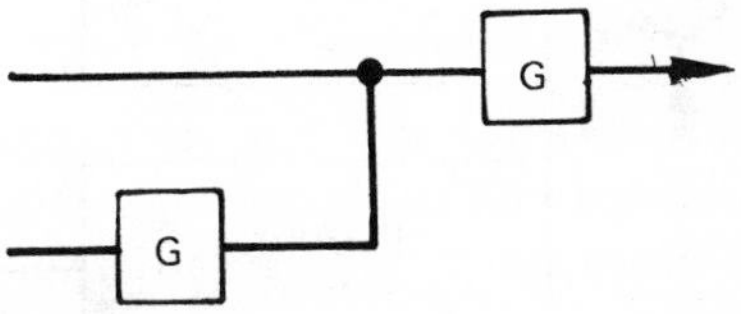
8	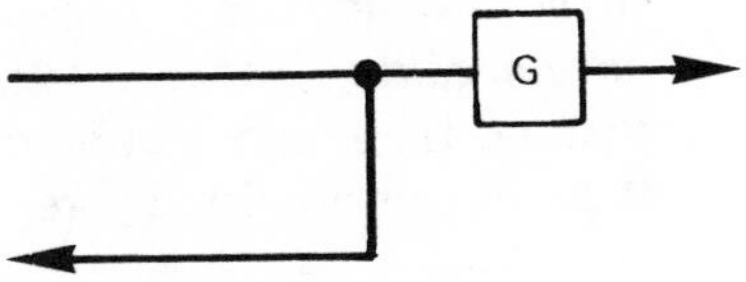	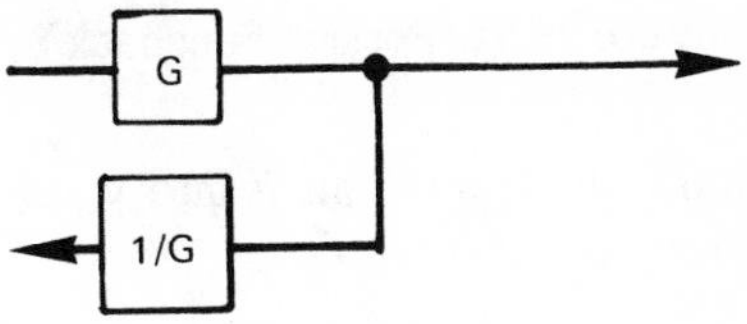

Use case *1* to combine the boxes in series.

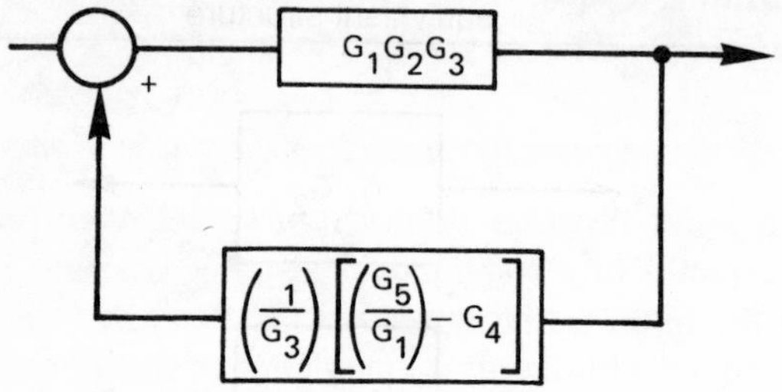

Use case *3* to determine the system gain.

$$G = \frac{G_1G_2G_3}{1 - (G_1G_2G_3)\left(\frac{1}{G_3}\right)\left[\left(\frac{G_5}{G_1}\right) - G_4\right]}$$

The system gain equation can be simplified.

$$G = \frac{G_1G_2G_3}{1 - G_2G_5 + G_1G_2G_4}$$

5. *Feedback Theory*

If a small part of a device's output is returned to the input, the device is part of a *feedback system.* Feedback can be used to control a system or to improve the system's stability, efficiency, or sensitivity. Feedback can also decrease such negative effects as distortion and frequency dependence.

The basic feedback system consists of a *dynamic unit,* a *feedback element,* a *pick-off point,* and a *summing point.* The summing point is assumed to perform positive addition unless a minus sign is present.

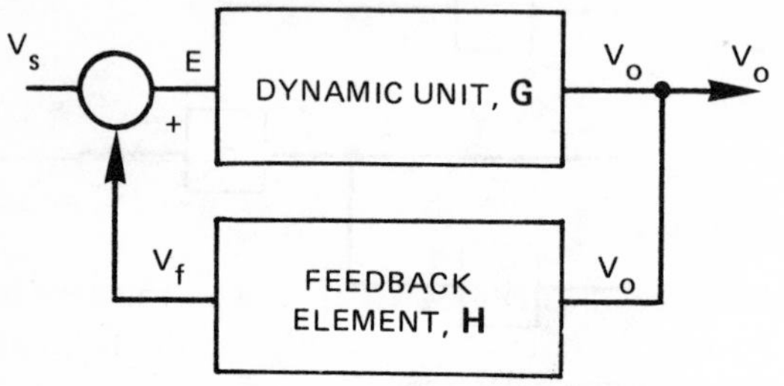

Figure 16.18 Positive Feedback System

The dynamic unit transforms E into V_o according to the *forward transfer function, G.*

$$V_o = GE \qquad 16.30$$

For amplifiers, the forward transfer function is known as the *direct* or *forward gain.* G can be a simple scalar if the dynamic unit merely scales its input. However, E, V_o, and G can also be phasors if the dynamic unit shifts the signal phase.

The difference between the signal and the feedback is known as the *error.* Whether addition or subtraction occurs in equation 16.31 depends on the summing point. Addition is used with additive summing points, subtraction with subtractive.

$$E = V_s \pm V_f \qquad 16.31$$

E/V_s is known as the *error ratio.* V_f/V_s is known as the *primary feedback ratio.*

The pick-off point transmits V_o back to the feedback element. The output of the dynamic unit is not reduced by the pick-off point. As the picked-off signal travels through the feedback loop, it is acted upon by the *feedback* or *reverse transfer function, H. H* can be a simple scalar or a phasor. For typical control systems, $H = 1$ (i.e., $HV_o = V_o$) and there is no transformation at all.

The output of a feedback system is

$$V_o = GV_s + GHV_o \qquad 16.32$$

The *loop transfer function* (also known as the *control ratio* or *system function*) is the ratio of the output to the signal.

$$G_{loop} = \frac{V_o}{V_s} = \frac{G}{1 - GH} \qquad 16.33$$

The sensitivity, S, of a feedback system is the percent change in the loop transfer function divided by the percent change in the forward transfer function.

$$S = \frac{\dfrac{dG_{loop}}{G_{loop}}}{\dfrac{dG}{G}} = \frac{1}{1 - GH} \qquad 16.34$$

The network is said to have *positive feedback* if the product GH is positive and less than *one.* In that case, the denominator in equation 16.33 will be less than *one,* making G_{loop} larger than G. Thus, a characteristic of a positive feedback system is an increase in the gain.

If the product GH approaches *one,* G_{loop} approaches infinity. Such performance may be undesirable. However, oscillators make use of this ability to produce an output in the absence of an input.

If the product GH is negative, the network is said to have *negative feedback.* Although the G_{loop} will be less than G, there may be other desirable effects. Some of these effects are listed here.

- reduction in sensitivity to temperature
- reduction in sensitivity to changes in the operating characteristics of the circuit components
- reduction in sensitivity to frequency of the signal
- reduction in sensitivity to noise and other variations in the signal

- improvement in the circuit's input and output impedances[6]

Example 16.18

Three inverting amplifier stages with gains of −100 each are cascaded as shown. Feedback is provided by a resistor network. What is the overall gain with feedback?

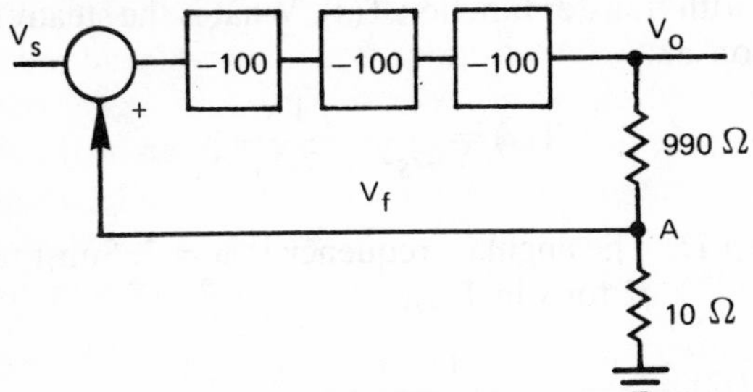

The two resistors form a voltage divider. The fraction of V_o fed back to the summing point is

$$H = \frac{10}{10 + 990} = .01$$

The forward transfer function is

$$G = (-100)^3 = -1{,}000{,}000$$

Without feedback, the output would be $(-1{,}000{,}000)V_s$. The minus sign indicates that the output is 180° out of phase with the input.

Since the summing point adds the feedback to the signal, equation 16.33 should be used to calculate the overall gain.

$$G_{loop} = \frac{-1{,}000{,}000}{1 - (-1{,}000{,}000)(.01)} \approx -100$$

Example 16.19

A closed-loop gain of −100 and stability of 99% is required from an inverting amplifier. The forward gain of the amplifier varies by 20%. What values of G and H satisfy the design requirements?

From equation 16.33,

$$G_{loop} = \frac{G}{1 - GH} = -100$$

The sensitivity requirement limits dG_{loop}/G_{loop} to .01 when dG/G is .2. From equation 16.34,

$$.01 = \left[\frac{1}{1 - GH}\right](.2)$$

[6]For circuits to be directly connected in series without affecting the performance of each, all input impedances must be infinite and all output impedances must be zero.

Solving for GH as an unknown results in $GH = -19$. Substituting GH into the equation for G_{loop} gives the value $G = -2000$. H is found to be .0095.

6. Predicting System Response From Transfer Functions

The transfer function alone is not sufficient to predict the response of a system. It is also necessary to know what the input to the system is. (The transfer function was derived without prior knowledge of the input.) The response of the system will depend on whether the input is homogenous, a step function, sinusoid, etc.

The *transformed transfer function* is the ratio of the transformed response function to the transformed forcing function.

$$T(s) = \frac{\mathcal{L}[R(t)]}{\mathcal{L}[F(t)]} = \frac{R(s)}{F(s)} \qquad 16.35$$

Assuming that $T(s)$ and $\mathcal{L}[F(t)]$ are known, the response function can be found by performing an inverse transformation.

$$R(t) = \mathcal{L}^{-1}[R(s)] = \mathcal{L}^{-1}[\mathcal{L}[F(t)]T(s)] \qquad 16.36$$

If the extreme (initial and final) values of R(t) are wanted, they can be found from the *initial value theorem* and *final value theorem* for Laplace transforms.

$$R(0^+) = \lim_{s\to\infty} [sR(s)] \qquad 16.37$$

$$R(\infty) = \lim_{s\to 0} [sR(s)] \qquad 16.38$$

Example 16.20

What is the transformed forcing function for a step of height 8 occurring at $t = 0$?

The Laplace transform of a unit step is $(1/s)$. The Laplace transform of a step with height 8 is

$$F(s) = \mathcal{L}[F(t)] = \frac{8}{s} \ .$$

Example 16.21

A mechanical system is acted upon by a constant force of 8 pounds starting at $t = 0$. What is the response if the transfer function is $T(s)$?

$$T(s) = \frac{6}{(s+2)(s+4)}$$

$F(s)$ was found in example 16.20 to be $(8/s)$. From equation 16.36,

$$R(t) = \mathcal{L}^{-1}\left[\left(\frac{8}{s}\right)\left(\frac{6}{(s+2)(s+4)}\right)\right]$$

$$= \mathcal{L}^{-1}\left[\frac{48}{s(s+2)(s+4)}\right]$$

SYSTEMS

R(t) is found by taking the inverse transform. Using a table of Laplace transforms and recognizing that the product of linear terms in the denominator translates into a sum of exponential terms gives R(t).

$$R(t) = 6 - 12e^{-2t} + 6e^{-4t}$$

The last two terms in R(t) are decaying exponentials. They represent the *transient natural response.* The 6 in R(t) does not vary with time. It is the *steady state response.* The sum of the transient and steady state responses is the *total response.*

Example 16.22

What is the final value of R(t)?

$$R(s) = \frac{1}{s(s+1)}$$

The limit is taken to zero for final values.

$$R(\infty) = \lim_{s\to 0}\,[s\,R(s)] = \lim_{s\to 0}\left[\frac{s}{s(s+1)}\right]$$

$$= \lim_{s\to 0}\left[\frac{1}{s+1}\right] = 1$$

Abbreviated methods are available if only the steady state response is wanted.

unit step: The steady state response for a unit step can be obtained by substituting 0 for s everywhere in T(s). If the step has a magnitude h, the steady state response must be multiplied by h.

unit pulse: The steady state response for a unit pulse is the Laplace inverse of the transfer function. That is, a pulse has no long-term effect on an engineering system.

sinusoids: The steady state response can be found by the following procedure:

step 1: Substitute $j\omega$ for s in T(s).
step 2: Convert T($j\omega$) into phasor form.
step 3: Convert the input sinusoid to phasor form.
step 4: Multiply T($j\omega$) by the input phasor.

Example 16.23

Determine the steady state response when the system described in example 16.21 is acted upon by a step of height 8 at $t = 0$.

Substitute 0 for s in T(s) and multiply by 8.

$$R(t)_{\text{steady state}} = 8\left[\frac{6}{(0+2)(0+4)}\right] = 6$$

Example 16.24

$4[\sin(2t + \frac{\pi}{4})]$ is applied as a sinusoidal input to a system with transfer function T(s). What is the steady state response?

$$T(s) = \frac{-1}{7s^2 + 7s + 1}$$

step 1: The angular frequency is $\omega = 2$. Substituting $j2$ for s in T(s),

$$T(j2) = \frac{-1}{7(j2)^2 + 7(j2) + 1}$$

Simplify this expression by recognizing that $j^2 = -1$.

$$T(j2) = \frac{-1}{-28 + 14j + 1} = \frac{-1}{14j - 27}$$

step 2: $(14j - 27)$ in phasor form is $30.4\angle 152.6°$. The negative inverse of this is

$$T(j2) = \frac{-1}{30.4\angle 152.6°} = -.033\angle -152.6°$$

This system has a magnitude gain of $-.033$ and a phase shift of $-152.6°$.

step 3: The input phasor is $4\angle 45°$.

step 4: The steady state phasor is

$$V(t) = (-.033\angle -152.6°)(4\angle 45°)$$
$$= -.132\angle -107.6°$$

7. Poles and Zeros

A *pole* of the transfer function is a value of s which makes T(s) infinite. Specifically, a pole is a value of s which makes the denominator of T(s) zero. A *zero* of the transfer function makes the numerator of T(s) zero. Poles and zeros can be real or complex quantities. Poles and zeros can be repeated within a given transfer function — they need not be unique.

A rectangular coordinate system based on the real-imaginary axes is known as an *s-plane.* If poles and zeros are plotted on the s-plane, the result is a *pole-zero diagram.*

Poles are represented on the pole-zero diagram as ×'s. Zero's are represented as O's.

SYSTEMS

Example 16.25

Draw the pole-zero diagram for the transfer function T(*s*).

$$T(s) = \frac{(5)(s + 3)}{(s + 2)(s^2 + 2s + 2)}$$

The numerator can be zero only if $s = -3$. This is the only zero of the transfer function.

The denominator can be zero if $s = -2$ or if $s = -1 \pm j$. These three values are the poles of the transfer function.

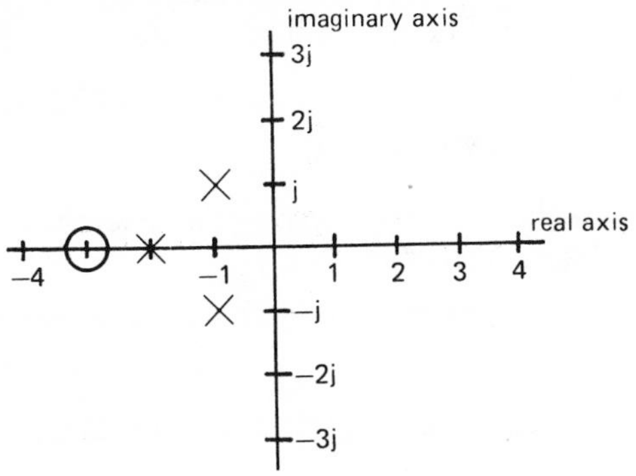

The pole-zero diagram can be used to partially find the transfer function T(*s*). The pole-zero diagram effectively gives the factors of the numerator and denominator. It does not give any scalars *(scale factors)* in the numerator.

Example 16.26

A pole-zero diagram has a pole at $s = -2$ and a zero at $s = -7$. What is the transfer function?

$$T(s) = \frac{K(s + 7)}{(s + 2)}$$

The scalar *K* must be determined separately. This analysis assumes that the forcing function F(*s*) has not been included in the pole-zero diagram.

8. Frequency Response to Sinusoidal Forcing Functions

The gain and phase angle response of a system will change as ω is varied. The *gain characteristic* is the relationship between gain and frequency. The *phase characteristic* is the relationship between phase and frequency.

A. Gain Characteristic

A plot of gain versus frequency can be obtained from the following procedure.

step 1: Factor the denominator of the transfer function, T(*s*) to determine its roots. Poles with imaginary parts correspond to values of *j*ω (in radians/sec) where the output peaks in magnitude.

step 2: Calculate the system gain for all values of the roots.

step 3: Set the input frequency to zero and determine the system gain.

step 4: Assume an infinite input frequency and calculate the system gain.

step 5: Choose several points around and close to each peak. Determine the system gain for these points.

step 6: Determine the *half-power points*. These are the frequencies for which the gain is .707 of the maximum values determined in step 2.

step 7: Determine the *bandwidth*. This is the difference between the upper and lower half-power points.

step 8: Calculate the *quality factor*.

$$Q = \frac{\text{frequency at peak}}{\text{bandwidth}} \qquad 16.39$$

Figure 16.19 illustrates the gain characteristics of transfer functions with the form given by equation 16.40. (The coefficient of the s^2 term must be 1.)

$$T(s) = \frac{as + b}{s^2 + Bs + \omega_n^2} \qquad 16.40$$

The zero defined by *a* and *b* is not significant. ω_n is much larger than *B*, so that the pole is close to the imaginary axis. *B* will correspond to the bandwidth (in radians/sec). ω_n will be the natural or *resonant frequency* of the system, the frequency at which the gain peaks.

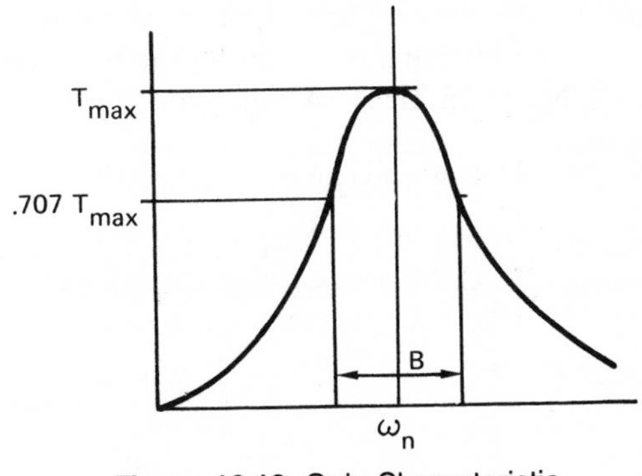

Figure 16.19 Gain Characteristic

Example 16.27

Predict the resonant frequency and bandwidth for the transfer function T(*s*).

$$T(s) = \frac{s + 19}{s^2 + 7s + 1000}$$

SYSTEMS

The coefficient of the s^2 term is 1, and the form is the same as equation 16.40. Therefore, the bandwidth is 7 radians/sec, and the resonant frequency is $\sqrt{1000} = 31.6$ radians/sec.

B. Phase Characteristic

Figure 16.20 illustrates the phase characteristic for a transfer function with the form given by equation 16.40. Values of the angle versus the frequency are determined simultaneously with the derivation of the gain characteristic.

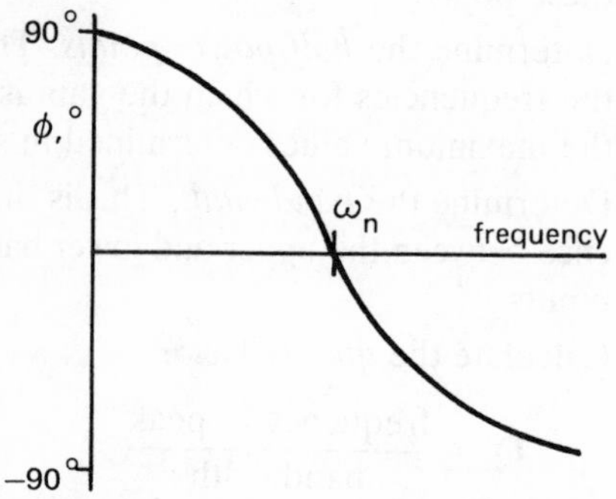

Figure 16.20 Phase Characteristic

9. Predicting System Response from Pole-Zero Diagrams

Poles on the pole-zero diagram can be used to predict the usual response of engineering systems. Zeros are not used.

- *pure oscillation:* Sinusoidal oscillation will occur if a pole-pair[7] falls on the imaginary axis as in figure 16.21(a). A pole with a value of $\pm ja$ will produce oscillation with a natural frequency of $\omega = a$ radians/sec.
- *exponential decay:* Pure exponential decay is indicated when a pole falls on the real axis as in figure 16.21(b). A pole with a value of $-r$ will produce an exponential with time constant $(1/r)$.
- *damped oscillation:* Decaying sinusoids result from pole-pairs in the second and third quadrants of the s-plane. A pole-pair having the value $r \pm ja$ will produce oscillation with natural frequency of

$$\omega_n = \sqrt{r^2 + a^2} \qquad 16.41$$

The closer the poles are to the real axis, the greater will be the damping effect. The closer the poles are to the imaginary axis, the greater will be the oscillatory effect.

[7] Poles off the real axis always occur in conjugate pairs.

Example 16.28

What is the natural response of a system with the pole-zero diagram shown?

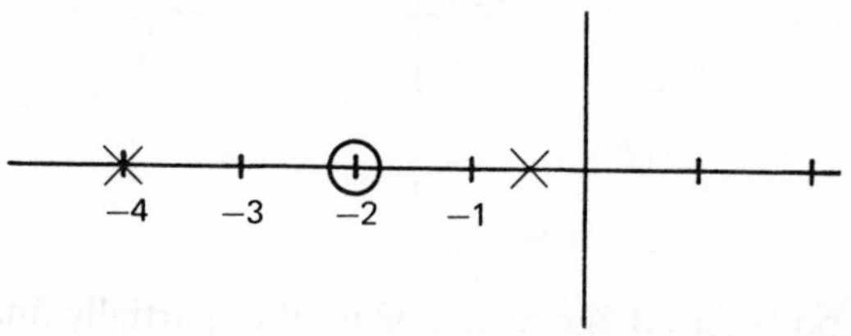

The poles are at $r = -\frac{1}{2}$ and $r = -4$. The response is

$$R(t) = C_1e^{-.5t} + C_1e^{-4t}$$

C_1 and C_2 can be found by methods not covered in this book.

Since the response of an engineering system depends on the type of input (step, pulse, etc.), a pole-zero diagram based on T(s) alone cannot be used to predict the system response. The system response must be determined from the product of the transfer function and the forcing function. This is equivalent to plotting T(s) and F(s) simultaneously on the pole-zero diagram.

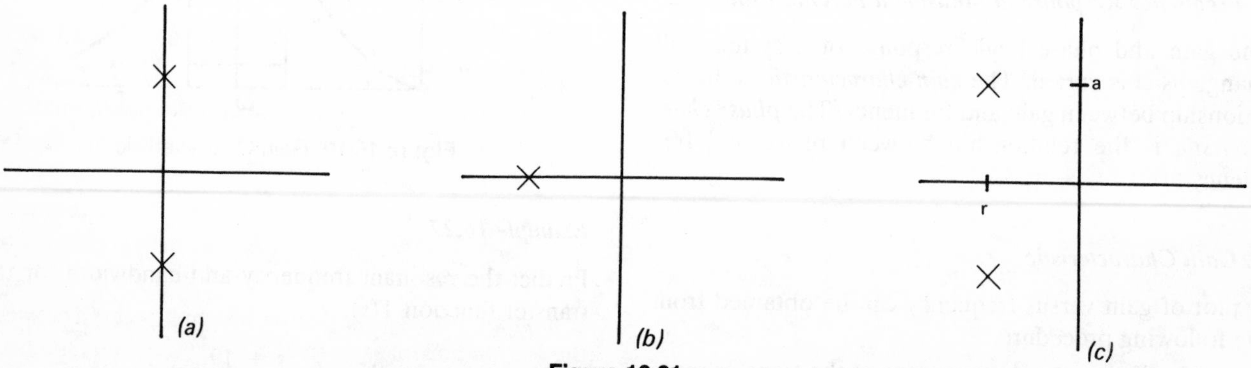

Figure 16.21
Using the Pole-Zero Diagram to Predict Natural Response

Example 16.29

What is the system response if $T(s) = \dfrac{(s+2)}{(s+3)}$ and the input is a unit step?

The response is found from equation 16.35.

$$R(s) = F(s)T(s) = \frac{(s+2)}{s(s+3)}$$

The pole-zero diagram is

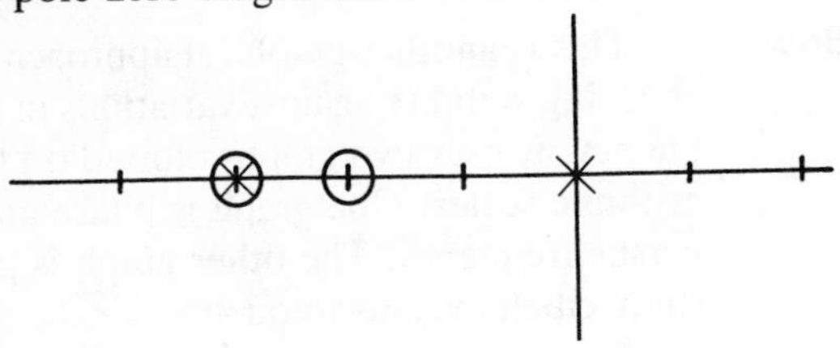

The pole at $r = 0$ contributes the exponential C_1e^{-0t} to the total response. Since this is equal to C_1, the total response is

$$R(t) = C_1 + C_2e^{-3t}$$

10. Root-Locus Diagrams

A root-locus diagram is a pole-zero diagram in which one system parameter is varied. The locus of points defined by the poles can be used to predict critical operating points (e.g., instability). The gain factor (scalar multiplier) of a system is frequently the varied parameter.

The root-locus diagram gets its name from the necessity of finding roots for the denominator of the transfer function.

Example 16.30

Draw the root-locus diagram for the feedback system shown. *K* is a scalar constant which can be varied.

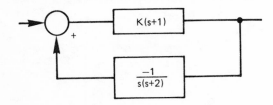

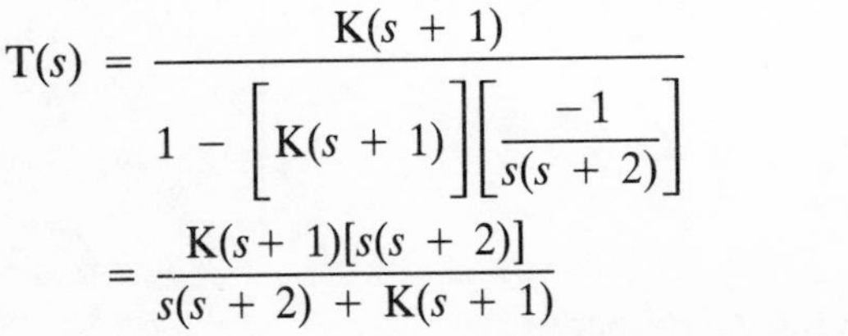

The transfer function is

$$T(s) = \frac{K(s+1)}{1 - \left[K(s+1)\right]\left[\dfrac{-1}{s(s+2)}\right]}$$

$$= \frac{K(s+1)[s(s+2)]}{s(s+2) + K(s+1)}$$

The poles are found by solving for the roots of the denominator.

$$s_1, s_2 = -\tfrac{1}{2}(2+K) \pm \sqrt{1 + \tfrac{1}{4}K^2}$$

Since the second term can be either added or subtracted, there are two roots for each value of *K* chosen. Allowing *K* to vary from zero to infinity produces the following root-locus diagram.

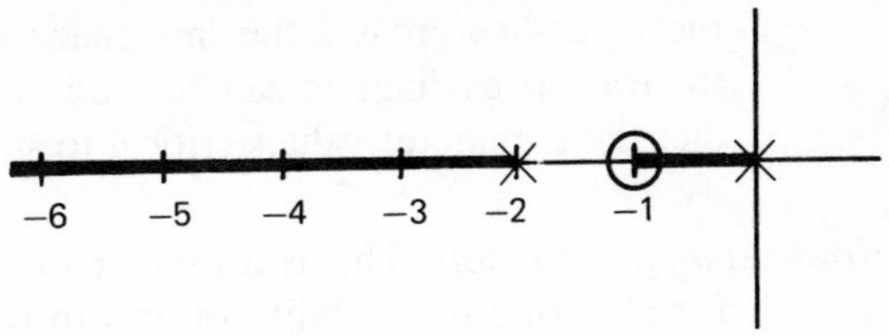

The root-locus diagram shows two distinct parts known as *branches*. The first branch extends from the pole at the origin to the zero corresponding to $s = -1$. The second branch extends from the pole corresponding to $s = -2$ to infinity.

Root-locus diagrams do not always result in loci confined to the real axis. They frequently leave the real axis at *break-away points* and continue on with constant or varying slopes.

11. Stability

A pole with a value of $-r$ on the real axis corresponds to an exponential response of e^{-rt}. Similarly, a pole with a value of $+r$ on the real axis corresponds to an exponential response of e^{rt}. However, e^{rt} increases without limit. For that reason, such a pole is said to be unstable.

Since any pole in the first and fourth quadrants of the *s*-plane will correspond to a positive exponential, a stable system must have poles limited to the left half of the *s*-plane (i.e., quadrants two and three).

Passive systems are always stable. A passive system does not contain an energy source. The system may experience an initial disturbance, but once movement begins, the energy source is removed. In the absence of an energy source, exponential growth cannot occur.

Active systems contain energy sources. Not all active sources, however, are unstable.

- If any coefficient is negative in the denominator of the transfer function for an active system, the system is unstable.
- If all coefficients are positive in the denominator of the transfer function for an active system, the system can be stable or unstable.

If a single pole exists on the imaginary axis (between the left and right parts of the *s*-plane), the response is stable. However, a double pole on the imaginary axis (resulting from an s^2 term in the denominator of the transfer function) usually corresponds to a linearly increasing term in the response. Such a system is unstable.

A number of tabular and graphical methods exist for evaluating the stability of engineering systems.

- *root-locus diagrams:* If a parameter is varied until the locus-line crosses the imaginary axis, the root-locus diagram can be used to predict the parameter values critical to stability.
- *Routh stability criterion:* This is a tabular method for checking the stability of an nth order characteristic equation.
- *Hurwitz stability criterion:* This method calculates determinants from the coefficients of the characteristic equation to determine stability.
- *Nyquist analysis:* This is a graphical method of checking stability. It is particularly useful when time delays are present in a system or when frequency response data is available on a system.
- *Bode plots:* This is another graphical approach for checking stability against variations in frequency. It uses two graphs plotted on logarithmic scales. One graph is phase angle versus frequency. The other graph is gain (in decibels) versus frequency.

Appendix A
Summary of Ideal System Elements

name	common symbol	value	governing equation
Through-Type Energy Storers			
translational spring		k	$F = kx$
rotational spring		k	$\tau = k\theta$
inductance		L	$I = \int V$
fluid inertance		I	$Q = \frac{1}{I}\int p$
Across-Type Energy Storers			
transactional mass	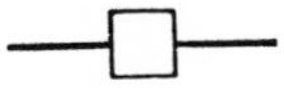	m	$F = ma$
rotational inertia	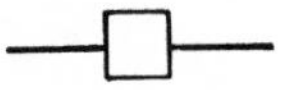	J	$\tau = J\alpha$
electrical capacitance	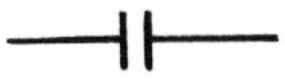	C	$I = C\frac{dV}{dt}$
fluid capacitance	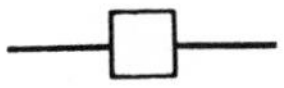	C_f	$Q = C_f\frac{dp}{dt}$
Energy Dissipators			
translational damper	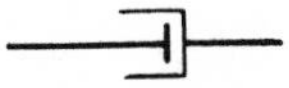	B	$F = Bv$
rotational damper		B	$\tau = B\omega$
electrical resistance		R	$I = \frac{V}{R}$
fluid resistance		R_f	$Q = \frac{p}{R_f}$

Practice problems for *Modeling of Engineering Systems* have been combined in this edition with problems from chapter 17, *General Systems Modeling*.

17 General Systems Modeling

PART 1: Introduction

Systems modeling attempts to develop mathematical models capable of predicting the outcomes of real-world activities. A mathematical model can be built around a set of fixed rules such that any given input always results in a specific output. This type of model is known as a *deterministic model.* If an input can produce a variety of outputs whose selection is determined by rules of probability, the model is known as a *probabilistic* or *stochastic model.*

Since models must be validated by comparison with real-world responses, measures of validity are needed.

1. Accuracy

An experiment is said to be *accurate* if it is unaffected by experimental error. In this case, *error* is not synonymous with *mistake*, but rather includes all variations not within the experimenter's control.

For example, suppose a gun is aimed at a point on a target and five shots are fired. The mean distance from the point of impact to the sight-in point is a measure of the alignment accuracy between the barrel and sights. The difference between the actual value and the experimental value is known as *bias.*

2. Precision

Precision is not synonymous with accuracy. Precision is concerned with the repeatability of the experimental results. If an experiment is repeated with identical results, the experiment is said to be precise.

In the previous example, the average distance of each impact from the centroid of the impact group is a measure of the precision of the experiment. Thus, it is possible to have a highly precise experiment with a large bias.

Most of the techniques which are applied to experiments in order to improve the accuracy of the experimental results (e.g., repeating the experiment, refining the experimental methods, or reducing variability), actually increase the precision.

Sometimes the word *reliability* is used with regards to the precision of an experiment. Thus, a reliable estimate is used in the same sense as a precise estimate.

3. Stability

Stability and *insensitivity* are synonymous terms. A stable experiment will be insensitive to minor changes in the experiment parameters. For example, suppose the centroid of a bullet group is 2.1 inches from the sight-in point at 65°F and 2.3 inches away at 80°F. The sensitivity of the experiment to temperature changes would be $(2.3 - 2.1)/(80 - 65) = .0133$ inches/°F.

PART 2: Dimensional Analysis

Nomenclature

c_p	specific heat	BTU/lbm-°F
C_i	a constant	–
D	diameter	ft
F	force	lbf
g_c	gravitational constant (32.2)	lbm-ft/sec^2-lbf
$\bar{h}$	average film coefficient	BTU/hr-ft^2-°F
J	Joule's constant (778)	ft-lbf/BTU
k	the number of pi-groups (m – n)	–
L	length	ft
m	number of relevant independent variables	–
M	mass	lbm
n	number of independent dimensional quantities	–
N_{Nu}	Nusselt number	–
N_{Pe}	Peclet number	–
N_{Re}	Reynolds number	–
v	velocity	ft/sec
x_i	ith independent variable	various
y	the dependent variable	various

Symbols

ρ	density	lbm/ft^3
θ	time	sec
π_i	ith dimensionless group	–
μ	viscosity	lbm/ft-sec

MODELS

Dimensional analysis is a means of obtaining an equation for some phenomenon without understanding the inner mechanism of the phenomenon. The most serious limitation to this method is the need to know beforehand which variables influence the phenomenon. Once these are known or assumed, dimensional analysis can be applied by a routine procedure.

The first step is to select a system of primary dimensions. Usually the MLθT system (mass, length, time, and temperature) is used, although this choice may require the use of g_c and J in the final results. The dimensional formulas and symbols for variables most frequently encountered are given in table 17.1.

The second step is to write a functional relationship between the dependent variable and the independent variables, x_i.

$$y = \mathbf{f}(x_1, x_2, \cdots, x_m) \qquad 17.1$$

This function can be expressed as an exponentiated series.

$$y = C_1 x_1^{a_1} x_2^{b_1} x_3^{c_1} \cdots x_m^{z_1} + C_2 x_1^{a_2} x_2^{b_2} x_3^{c_2} \cdots x_m^{z_2} + \cdots \qquad 17.2$$

The C_i, a_i, b_i, $\cdots$ z_i in equation 17.2 are unknown constants.

The key to solving the above equation is that each term on the right-hand side must have the same dimensions as *y*. Simultaneous equations are used to determine some of the a_i, b_i, c_i, and z_i. Experimental data is required to determine the C_i and remaining exponents. In most analyses, it is assumed that the C_i = 0 for *i* = 2 and up.

Example 17.1

A sphere submerged in a fluid rolls down an incline. Find an equation for the velocity, *v*.

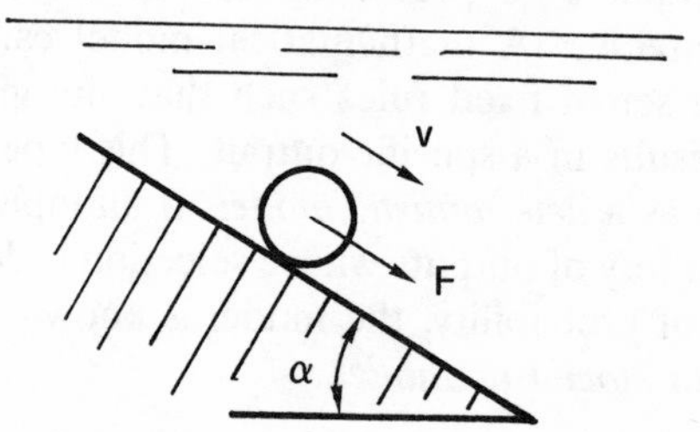

It is assumed that the velocity depends on the force *F* due to the inclination, the diameter of the sphere *D*, the density of the fluid ρ, and the viscosity of the fluid μ.

$$v = \mathbf{f}(F, D, \rho, \mu)$$

This equation can be written in terms of the dimensions of the variables.

$$\frac{L}{\theta} = C\left(\frac{ML}{\theta^2}\right)^a (L)^b \left(\frac{M}{L^3}\right)^c \left(\frac{M}{L\theta}\right)^d$$

Since *L* on the left-hand side has an implied exponent of one, the necessary equation is

$$1 = a + b - 3c - d \qquad (L)$$

Similarly, the other necessary equations are

$$-1 = -2a - d \qquad (\theta)$$

$$0 = a + c + d \qquad (M)$$

Solving simultaneously yields

$$b = -1$$

$$c = a - 1$$

$$d = 1 - 2a$$

or

$$v = C\left(\frac{\mu}{D\rho}\right)\left(\frac{F\rho}{\mu^2}\right)^a$$

C and *a* would have to be determined experimentally.

Table 17.1
Units and Dimensions of Typical Variables

QUANTITY	SYMBOL	DIMENSIONS $ML\theta T$ System	DIMENSIONS $ML\theta TFQ$ System	Units in Engineering System
Length	L or x	L	L	ft
Time	θ	θ	θ	sec or hour
Mass	M	M	M	lbm
Force	F	ML/θ^2	F	lbf
Temperature	T	T	T	°F
Heat	Q	ML^2/θ^2	Q	BTU
Velocity	V	L/θ	L/θ	ft/sec
Acceleration	a or g	L/θ^2	L/θ^2	ft/sec^2
Dimensional conversion factor	g_c	none	ML/θ^2F	32.2 lbm-ft/sec^2-lbf
Energy conversion factor	J	none	FL/Q	778 ft-lbf/BTU
Work	W	ML^2/θ^2	FL	ft-lbf
Pressure	p	M/θ^2L	F/L^2	lbf/ft^2
Density	ρ	M/L^3	M/L^3	lbm/ft^3
Internal energy and enthalpy	u, h	L^2/θ^2	Q/M	BTU/lbm
Specific heat	c	L^2/θ^2T	Q/MT	BTU/lbm-°F
Dynamic viscosity	u_f	$M/L\theta$	$F\theta/L^2$	lbf-sec/ft^2
Absolute viscosity	u	$M/L\theta$	$M/L\theta$	lbm/ft-sec
Kinematic viscosity	$\nu = u/\rho$	L^2/θ	L^2/θ	ft^2/sec
Thermal conductivity	k	ML/θ^3T	$Q/LT\theta$	BTU/hr-ft-°F
Coefficient of expansion	β	$1/T$	$1/T$	1/°F
Surface tension	σ	M/θ^2	F/L	lbf/ft
Stress	σ or τ	$M/L\theta^2$	F/L^2	lbf/ft^2
Film coefficient	h	M/θ^3T	$Q/\theta L^2T$	BTU/hr-ft^2-°F
Mass flow rate	m	M/θ	M/θ	lbm/sec

Since the above method requires working with m different variables and n different independent dimensional quantities (such as M, L, T, and θ), an easier method is desirable. One simplification is to combine the m variables into dimensionless groups, called *pi-groups.*

If these dimensionless groups are represented by π_1, π_2, π_3, · · ·, π_k, the equation expressing the relationship between the variables is given by the *Buckingham π-Theorem.*

$$f(\pi_1, \pi_2, \pi_3, \cdots, \pi_k) = 0 \qquad 17.3$$

$$k = m - n \qquad 17.4$$

The dimensionless pi-groups are usually found from the m variables according to an intuitive process. A formalized method is possible as long as the following conditions are met:

- The dependent variable and independent variables chosen contain all of the variables affecting the phenomenon. Extraneous variables may be included at the expense of obtaining extra pi-groups.
- The pi-groups must include all of the original x_i at least once.
- The dimensions must all be independent.

The formal procedure is to select n variables (x_i) out of the total m as repeating variables to appear in all k pi-groups. These variables are used in turn with the remaining variables in each successive pi-group. Each of the repeating variables must have different dimensions and the repeating variables must collectively contain all of the dimensions. This procedure is illustrated in example 17.2.

Example 17.2

It is desired to determine a relationship giving the heat transfer to air flowing across a heated tube. The following variables affect the heat flow:

Variable	Symbol	Dimensional Equation
tube diameter	D	L
fluid conductivity	k	ML/θ^3T
fluid velocity	v	L/θ
fluid density	ρ	M/L^3
fluid viscosity	μ	$M/L\theta$
fluid specific heat	c_p	L^2/θ^2T
film coefficient	$\bar{h}$	M/θ^3T

There are $m=7$ variables and $n=4$ primary dimensions (L,M,θ, and T). Accordingly, there are $k = 7-4 = 3$ dimensionless groups that are required to correlate the data. The four repeating variables are chosen such that all dimensions are represented. Then, the π_i are written as a function of these repeating variables in turn with the remaining variables.

MODELS

The repeating variables should not include any of the unknown quantities. For example, $\bar{h}$ should not be chosen as a repeating variable since it is directly related to the unknown heat flow. In addition, important material properties, such as c_p and k, are also often omitted. Trial and error is required to include all four primary dimensions.

Using trial and error, omitting $\bar{h}$ as a repeating variable, and representing all four primary dimensions, the repeating variables are chosen arbitrarily as D, k, v, and ρ.

Then, the pi-groups are

$$\pi_1 = D^{a_1} k^{a_2} v^{a_3} \rho^{a_4} \mu$$
$$\pi_2 = D^{a_5} k^{a_6} v^{a_7} \rho^{a_8} c_p$$
$$\pi_3 = D^{a_9} k^{a_{10}} v^{a_{11}} \rho^{a_{12}} \bar{h}$$

Since the π_i are dimensionless, we write for π_1:

$$0 = a_1 + a_2 + a_3 - 3a_4 - 1 \qquad \text{(L)}$$
$$0 = a_2 + a_4 + 1 \qquad \text{(M)}$$
$$0 = -3a_2 - a_3 - 1 \qquad (\theta)$$
$$0 = -a_2 \qquad \text{(T)}$$

So,

$$a_2 = 0 \quad a_3 = -1 \quad a_4 = -1 \quad a_1 = -1$$
$$\pi_1 = \frac{\mu}{Dv\rho}$$

π_1 is the reciprocal of the Reynolds number. Proceeding similarly with π_2,

$$0 = a_5 + a_6 + a_7 - 3a_8 + 2 \qquad \text{(L)}$$
$$0 = a_6 + a_8 \qquad \text{(M)}$$
$$0 = -3a_6 - a_7 - 2 \qquad (\theta)$$
$$0 = -a_6 - 1 \qquad \text{(T)}$$

So,

$$a_6 = -1 \quad a_7 = 1 \quad a_8 = 1 \quad a_5 = 1$$
$$\pi_2 = \frac{Dv\rho c_p}{k}$$

π_2 is the *Peclet number* (product of the Reynolds number and the Prandtl number).

π_3 is found to be $\frac{D\bar{h}}{k}$, which is the *Nusselt number.*

Therefore, the seven original variables have been combined into three dimensionless groups, making data correlation much easier. The implicit equation for heat transfer is

$$f_1(\pi_1,\pi_2,\pi_3) = f_1(N_{Re},N_{Nu},N_{Pe}) = 0$$

Rearrangement of the pi-groups is needed to isolate the dependent variable (in this case $\bar{h}$).

$$N_{Nu} = f_2(N_{Re}, N_{Pe}) = C(N_{Re})^{e_1}(N_{Pe})^{e_2}$$

C, e_1, and e_2 are found experimentally.

The selection of the repeating and non-repeating variables is the key step. The choice of the repeating variables determines which dimensionless groups are obtained. The theoretical maximum number of valid dimensionless groups is

$$\frac{m!}{(n+1)!(m-n-1)!} \qquad 17.5$$

However, not all dimensionless groups obtained are equally useful to researchers. For example, the Peclet number was obtained in the above example. However, researchers would have chosen D, k, ρ, and μ as repeating variables in order to obtain the Prandtl number as a dimensionless group. This choice of repeating variables is a matter of intuition.

PART 3: Critical Path Techniques

1. Introduction

Critical path techniques are used to graphically represent the multiple relationships between stages in a complicated project. The graphical network shows the *precedence relationships* between the various activities. The graphical network can be used to control and monitor the progress, cost, and resources of a project. A critical path technique will also identify the most critical activities in the project.

Definitions

Activity: Any subdivision of a project whose execution requires time and other resources.

Critical path: A path connecting all activities which have minimum or zero slack times. The critical path is the longest path through the network.

Duration: The time required to perform an activity. All durations are *normal* durations unless otherwise referred to as *crash* durations.

Event: The beginning or completion of an activity.

Event time: Actual time at which an event occurs.

Float: Same as slack time.

Slack time: The maximum time that an activity can be delayed without causing the project to fall behind schedule. Slack time is always minimum or zero along the critical path.

Critical path techniques use *directed graphs* to represent a project. These graphs are made up of *arcs* (arrows) and *nodes* (junctions). The placement of the arcs and nodes completely specifies the precedences of the project. Durations and precedences are usually given in a precedence table or matrix.

2. Critical Path Method

One technique is known as the *Critical Path Method,* CPM. This deterministic method is applicable when all activity durations are known in advance. CPM is usually represented as an *Activity-on-Node* model since arcs are used to specify precedence and the nodes actually represent the activities. Events are not present on the graph, other than as the heads and tails of the arcs. Two *dummy nodes* taking zero time may be used to specify the start and finish of the project.

Example 17.3

Given the project listed in the precedence table below, construct the precedence matrix and draw an activity-on-node network.

Activity	Duration (days)	Predecessors
A, Start	0	—
B	7	A
C	6	A
D	3	B
E	9	B, C
F	1	D, E
G	4	C
H, Finish	0	F, G

The precedence matrix is

predecessor \ successor	A	B	C	D	E	F	G	H
A		■	■					
B				■	■			
C					■		■	
D						■		
E						■		
F								■
G								■
H								

The activity-on-node network is

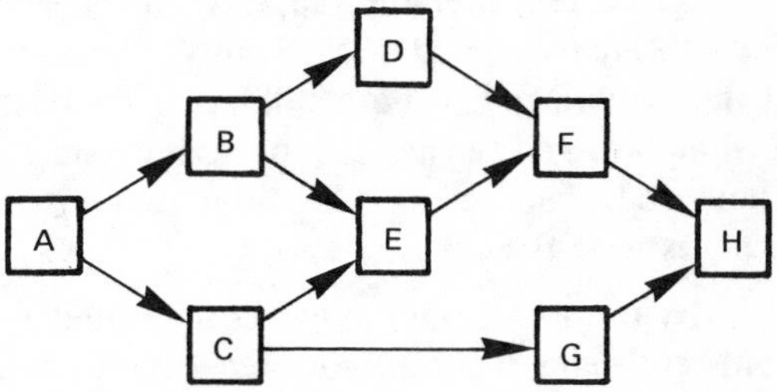

3. Solving a CPM Problem

The solution of a critical path problem results in a knowledge of the earliest and latest times that an activity can be started and finished. It also identifies the critical path and generates the slack times for each activity.

To facilitate the solution method, each node should be replaced by a square which has been quartered. The compartments have the meanings indicated by the key.

ES	**EF**
LS	**LF**

key

ES: Earliest Start
EF: Earliest Finish
LS: Latest Start
LF: Latest Finish

The following procedure may be used to solve a CPM problem.

step 1: Place the project start time or date in the **ES** and **EF** positions of the start activity. The start time is zero for relative calculations.

step 2: Consider any unmarked activity, all of whose predecessors have been marked in the **EF** and **ES** positions. (Go to step 4 if there is none.) Mark in its **ES** position the largest number marked in the **EF** position of those predecessors.

step 3: Add the activity time to the **ES** time and write this in the **EF** box. Go to step 2.

step 4: Place the value of the latest finish date in the **LS** and **LF** boxes of the finish node.

step 5: Consider unmarked predecessors whose successors have all been marked. Their **LF** is the smallest **LS** of the successors. Go to step 7 if there are no unmarked predecessors.

step 6: The **LS** for the new node is **LF** minus its activity time. Go to step 5.

step 7: The slack for each node is (**LS** − **ES**) or (**LF** − **EF**).

step 8: The critical path encompasses nodes for which the slack equals (**LS** − **ES**) from the start node. There may be more than one critical path.

4. Activity-on-Branch Networks

Much of the current literature applies critical path calculations to *Activity-on-Branch* graphs. The arcs represent the activities, which are still labeled with letters of the alphabet, and the nodes represent events, which are numbered. The activity durations may appear in parentheses near the activity letter.

The activity-on-branch method is complicated by the frequent requirement for *dummy activities* to maintain precedence. Consider the following part of a precedence table:

Activity	Predecessors
L	–
M	–
N	L,M
P	M

Notice that activity *P* depends on the completion of only *M*. It can be represented as:

However, *N* depends on the completion of both *L* and *M*. It would be incorrect to draw the network as below since the activity *N* appears twice.

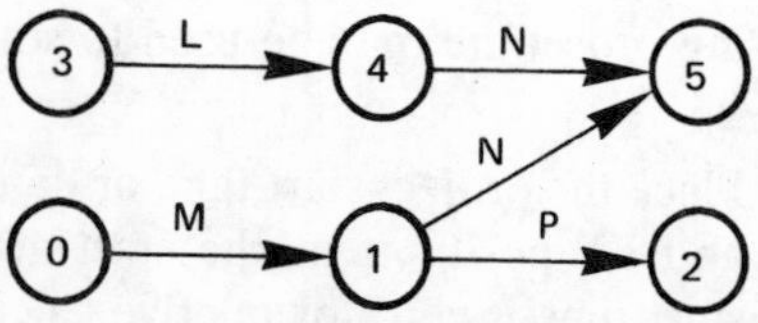

Incorrect

To represent the project, the dummy activity X must be used.

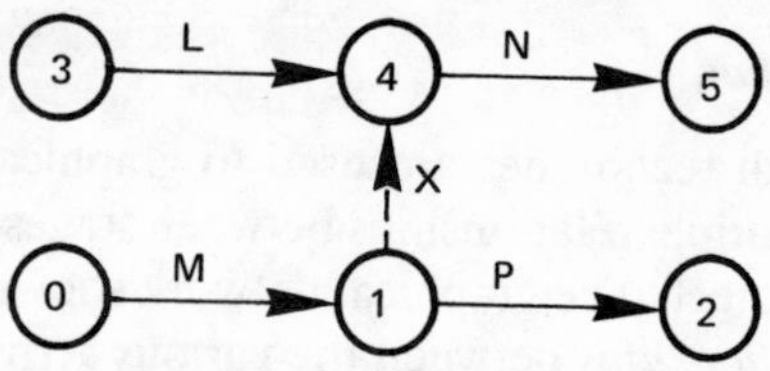

Also, if two activities have the same starting and ending event, a dummy node is required to give one activity a uniquely identifiable completion event.

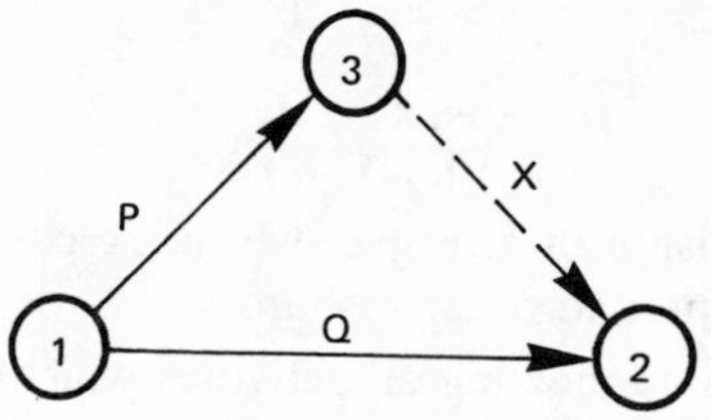

instead of

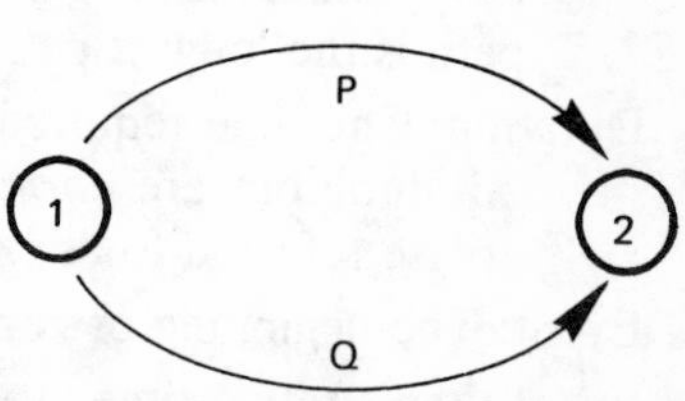

The solution method for an activity-on-branch problem is essentially the same as for the activity-on-node problem, requiring forward and reverse passes to determine earliest and latest dates.

5. Stochastic Critical Path Models

Stochastic models differ from deterministic models only in the way in which the activity durations are found. Whereas durations are known explicitly for the deterministic model, the time for a stochastic activity is distributed as a random variable.

This stochastic nature complicates the problem greatly since the actual distribution is often unknown. Such problems are solved as a deterministic model using the mean of an assumed duration distribution as the activity duration.

The most common stochastic critical path model is PERT, which stands for *Program Evaluation and Review Technique*. In PERT, all duration variables are assumed to come from a *beta distribution,* with mean and standard deviation given by equations 17.6 and 17.7 respectively.

$$t_{mean} = (1/6)(t_{minimum} + 4t_{most\ likely} + t_{maximum}) \qquad 17.6$$

$$\sigma = (1/6)(t_{maximum} - t_{minimum}) \qquad 17.7$$

The project *completion time* for large projects is assumed to be normally distributed with mean equal to the critical path length and overall variance equal to the sum of the variances along the critical path.

Example 17.4

Complete the network for the previous example and find the critical path. Assume the desired completion date is in 19 days.

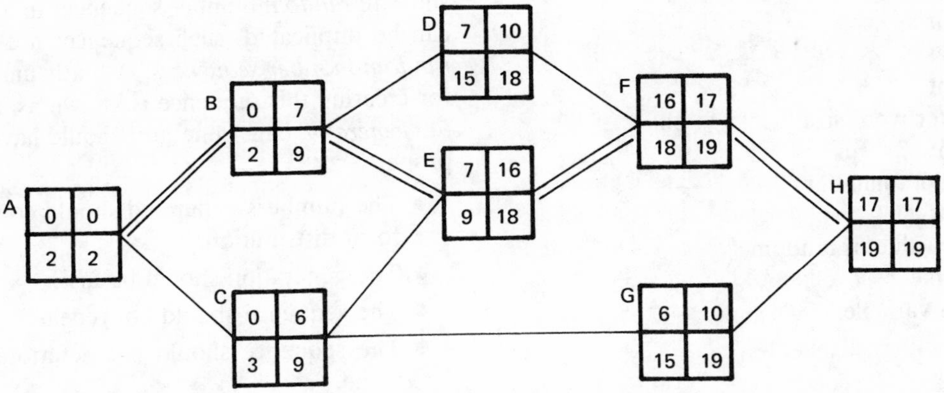

Example 17.5

Represent the project in the previous example as an activity-on-branch network.

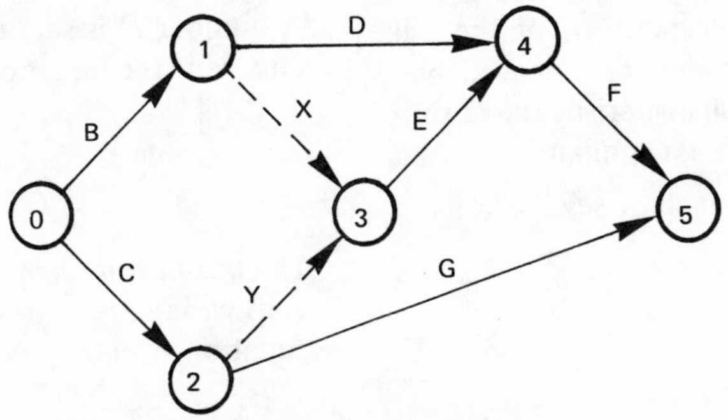

Event	Event description
0	start project
1	finish B, start D
2	finish C, start G
3	finish B and C, start E
4	finish D and E, start F
5	finish F and G

MODELS

PART 4: Simulation

Nomenclature

a	constant	–
c	constant	–
D	constant	–
k	constant	–
M	constant	–
n	computer word size	bits
N	constant	–
r_i	ith random number	–
T	constant	–
x	specific value of random variable	–
X	random variable	–

1. Introduction

Simulation is essentially a technique of performing sampling experiments on a model of the system. The sampling is performed on a model of the system if experimenting with the real system would be inconvenient, expensive, or time consuming. Simulation is also used when it is inappropriate to experiment on the real system or where analytical techniques are not available. Although it is not a necessary characteristic, most simulation models are carried out on a computer.

The technique of simulation involves several related topics. These are

- time control
- random number generation
- stochastic variate generation
- variance reduction

These topics are all handled automatically in some of the special purpose simulation languages such as SIMSCRIPT, GASP, DYNAMO, GPSS, SOL, and SIMULA.

MODELS

2. Time Control

Time control can either be *Fixed-Interval Incrementing* (also known as *Uniform Time Flow*) or *Next-Event Incrementing* (also known as *Variable Time Flow*). With Fixed-Interval Incrementing, the model's master clock is advanced by one unit and the system model is updated by determining what has happened during the elapsed time.

With Next-Event Incrementing, the master clock is incremented by a variable amount of time. In this case, the computer actually proceeds by keeping track of when the simulated events occur and jumping ahead to the first of these events.

3. Generating Random Numbers

A truly random process cannot be repeated. However, since all *random* number sequences in a simulation test can be duplicated, such sequences are called *pseudorandom number sequences.* A mathematical procedure for creating this sequence is known as a *random number generator.* This generator should have the following properties:

- The numbers generated should come from a uniform distribution.
- The generation should be fast.
- The sequence should not repeat.
- The sequence should not deteriorate to a single value.

Historically, the *midsquare, midproduct,* and *Fibonacci methods* have been used to generate random numbers. These, however, do not always yield satisfactory results. The *congruential method* is frequently used in simulation programs.

A number *D* is said to be *congruent* with *N* with modulus *M* if the equation 17.8 is an integer.

$$\frac{(D-N)}{M} \qquad 17.8$$

The *mixed congruential method* calculates the next random variate, r_{i+1}, from the current variate, r_i, by using equation 17.9.

$$r_{i+1} = (ar_i + c)(\textbf{mod}\ T) \qquad 17.9$$

This means that $(ar_i + c)$ is to be divided by *T* and r_{i+1} set equal to the remainder. The starting value, r_0, is known as the *seed* and must be supplied by the user. If c = 0, the method is known as a *multiplicative congruential method.* If a = 1, it is an *additive congruential method.*

The constants *a, c,* and *T* are usually chosen according to established rules in order to ensure the desired properties of the generator.

4. Variance Reduction Techniques

Methods of increasing the precision of the sample estimation without increasing computer time are called *variance reduction techniques.* Two such techniques, *stratified sampling* and *complementary random numbers,* are complicated. They are used when extreme accuracy is required.

PART 5: Monte Carlo Technique

Monte Carlo simulation evaluates the interactions of random variables whose distributions are known but whose combined effects are too complex to be specified mathematically. *Crude Monte Carlo* simulation is essentially random sampling from distributions to obtain values of each interacting variable. The following steps are similar in all Monte Carlo problems.

step 1: Establish the probability distribution for each variable in the study. This does not need to be a mathematical formula—a histogram is sufficient.

step 2: Form the cumulative distribution *function* for each variable.

step 3: Multiply the cumulative variable by 100 to obtain a cumulative axis which runs from 0 to 100 instead of 0 to 1.

step 4: Generate random numbers between 0 and 100. Either a random number table or a pseudo-random number generator may be used.

step 5: For each random number generated, locate the corresponding variable value. If the original distribution was continuous, take the midpoint of the range as the variable value.

Example 17.6

A small bank has one teller. Use the crude Monte Carlo method with ten cycles to find the average time the teller spends with a customer.

step 1: Assume that only three service times are possible and that the distribution is

time	probability
¼ minute	.55
½ minute	.40
1 minute	.05

steps 2, 3: The cumulative and converted distributions are:

time	cumulative probability	time	100[cumulative probability]	range
¼	.55	¼	55	1–55
½	.95	½	95	56–95
1	1.00	1	100	96–00

step 4: Select ten random numbers. From table 17.2 starting in the 4th row (chosen arbitrarily), the numbers are: 01, 90, 25, 29, 09, 37, 67, 07, 15, 38.

step 5: The first random number was 01. This corresponds to a service time of ¼ minute. The random number 90 corresponds to a service time of ½ minute, etc. The times of the first 10 customers are: ¼, ½, ¼, ¼, ¼, ¼, ½, ¼, ¼, ¼.

The average of the first ten customers is

$$\bar{t} = \tfrac{1}{10}\left[\tfrac14+\tfrac12+\tfrac14+\tfrac14+\tfrac14+\tfrac14+\tfrac12+\tfrac14+\tfrac14+\tfrac14\right]$$
$$= .3 \text{ minutes}$$

The actual average service time could have been found as the expected value. However, more iterations would be required to obtain this value by the simulation method.

$$\bar{t} = (.55)(.25)+(.4)(.5)+(.05)(1) = .388 \text{ minutes}$$

In example 17.6, many of the random numbers were located nearer to 0 than to 100. Since the number formed by subtracting the random table number from 100 is also a random number, it can be used with the *Complementary Monte Carlo technique* or *Complementary Random Number technique* to improve the accuracy of the estimation. Another *variance-reduction* technique, *stratified sampling,* is not covered in this chapter.

Example 17.7

Repeat example 17.6 using the Complementary Monte Carlo technique.

The ten complementary random numbers are: 99, 10, 75, 71, 91, 63, 33, 93, 85, and 62. The ten new service times are: 1, ¼, ½, ½, ½, ½, ¼, ½, ½, and ½.

Combining these ten new service times with the original ten times gives an average of .4 minutes, which is closer to the true mean of .388.

A common application of the Monte Carlo Technique is finding the average line size, average waiting time, or percent idleness of servers in queueing problems.

Example 17.8

A hamburger stand has two windows. The cumulative arrival rate and service time distributions are shown below. Use the crude Monte Carlo technique to determine the average idle time of both servers during the first fifteen minutes.

Service Time

time	100[cumulative probability]
.25 minute	55
.50 minute	95
1.00 minute	100

MODELS

Arrival Rate

number of arrivals	100 [cumulative probability]
0 per minute	10
1 per minute	25
2 per minute	60
3 per minute	90
4 per minute	100

Fifteen random numbers are chosen from the last column of table 17.2 to simulate the arrivals in each of the first fifteen minutes.

minute	1	2	3	4	5	6	7	8	9	10	11	12	13	14	15
random #	49	16	36	76	68	91	97	85	56	84	39	78	78	01	41
#arrivals	2	1	2	3	3	4	4	3	2	3	2	3	3	0	2

The service times of the 37 customers arriving in the first fifteen minutes can now be simulated. Random numbers are taken from the last line of table 17.2 working backwards.

customer:	1	2	3	4	5	6	7	8	9	10	11	12	13	14
random #:	11	46	62	94	98	52	62	01	25	87	33	12	15	20
time:	.25	.25	.5	.5	1.0	.25	.5	.25	.25	.5	.25	.25	.25	.25

customer:	15	16	17	18	19	20	21	22	23	24	25	26	27	28
random #:	74	07	27	38	96	06	04	10	95	30	87	54	37	94
time:	.5	.25	.25	.25	1.0	.25	.25	.25	.5	.25	.5	.25	.25	.5

customer:	29	30	31	32	33	34	35	36	37
random #:	19	54	69	88	81	85	54	43	89
time:	.25	.25	.5	.5	.5	.5	.25	.25	.5

A bar chart can be used to analyze the data given:

server 1: 8.0 total minutes idle out of 15; % idle = 53.3%

server 2: 8.25 total minutes idle out of 15; % idle = 55.0%

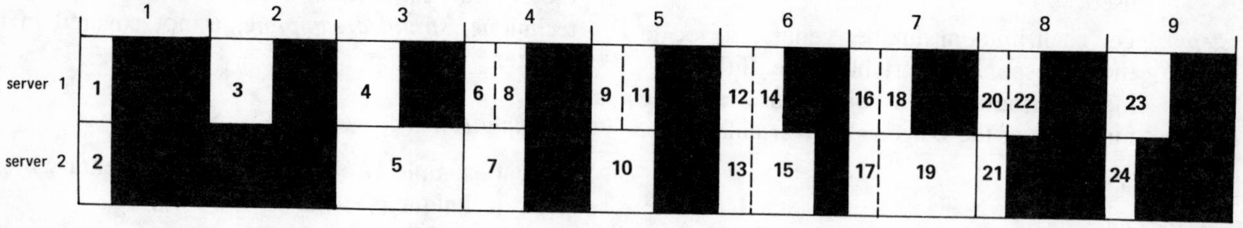

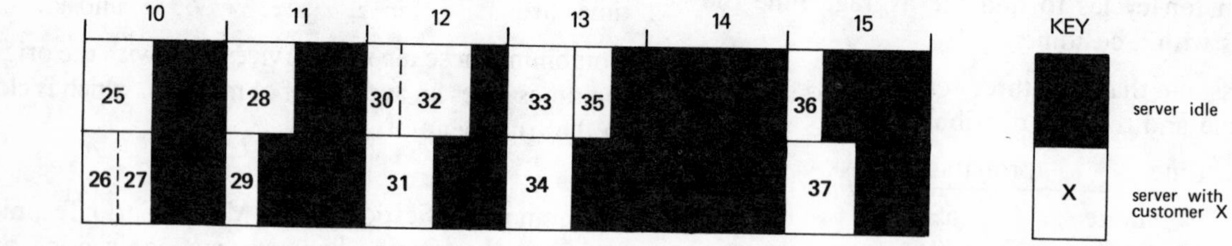

Table 17.2
Random Numbers

24 30 31 51 05	78 49 36 98 90	72 00 96 00 95	04 89 66 55 81	90 85 86 50 49
44 86 03 60 77	28 78 13 22 21	87 85 56 36 03	87 76 20 60 37	22 74 25 93 16
03 88 00 04 27	49 64 37 73 12	37 87 02 20 69	04 57 71 87 22	12 30 11 70 36
01 90 25 29 09	37 67 07 15 38	42 76 98 80 99	15 62 03 43 21	55 01 00 48 76
33 59 35 28 04	65 30 46 11 71	42 66 36 90 42	01 10 28 04 05	14 05 51 76 68
21 00 63 72 63	42 84 04 29 56	68 10 27 35 61	08 02 53 31 56	29 16 50 86 91
45 95 38 13 66	95 96 79 94 32	99 07 72 80 39	81 52 21 05 30	54 06 45 47 97
15 21 91 01 45	85 96 30 50 55	85 23 93 62 09	02 01 84 23 43	65 30 56 83 85
51 35 66 77 35	06 81 79 42 82	53 31 20 70 72	09 88 71 11 65	04 55 09 43 56
44 21 01 55 59	96 36 35 81 04	21 14 26 88 15	13 69 15 64 42	78 04 33 78 84
88 89 50 65 70	42 00 74 76 98	56 25 56 46 26	69 66 18 00 56	95 71 92 83 39
22 20 03 96 98	39 08 19 60 57	47 74 75 66 29	60 58 56 70 70	94 97 71 32 78
03 70 36 94 06	91 88 15 02 40	68 29 72 33 28	20 52 13 19 62	77 41 40 61 78
20 05 67 36 12	45 63 75 83 00	28 85 87 23 08	63 32 92 13 19	80 59 83 60 01
41 90 97 88 48	68 19 80 12 49	82 11 69 95 34	45 91 69 80 32	33 31 14 24 41
03 95 76 92 17	28 51 11 03 52	79 99 86 35 39	53 01 08 65 94	44 82 41 73 21
98 54 84 08 20	82 64 92 78 02	35 21 80 23 60	63 38 04 83 11	42 14 10 80 00
75 91 68 53 18	28 62 16 17 40	81 79 92 18 81	05 16 77 34 39	71 77 27 96 73
75 70 99 63 05	69 11 73 13 18	79 49 91 11 91	92 05 13 13 83	16 55 92 22 35
57 54 55 00 64	51 58 24 50 98	39 88 23 98 23	64 37 66 67 39	19 40 56 02 72
93 76 70 85 75	48 14 18 01 80	03 94 39 99 70	05 44 34 09 87	57 08 27 08 46
21 58 15 83 11	42 60 20 32 22	23 41 08 48 43	98 75 55 75 16	99 20 15 40 64
23 46 76 61 80	51 57 71 15 52	49 73 96 54 47	25 98 56 40 05	92 85 54 08 56
59 99 44 59 61	32 30 46 96 06	61 18 63 55 55	32 14 47 55 22	28 85 21 09 07
63 56 90 90 16	42 98 22 72 10	56 14 75 87 05	15 20 16 70 84	38 00 65 63 87
98 29 88 23 13	25 31 07 15 12	08 66 42 25 80	39 89 62 48 07	38 49 77 71 62
88 57 33 42 05	05 44 82 80 89	17 02 84 29 08	13 00 12 48 70	19 88 89 06 73
71 96 24 18 00	39 96 50 55 91	16 36 90 47 34	71 02 33 66 65	28 57 84 32 67
67 01 18 42 70	81 70 88 05 14	85 92 80 50 99	52 86 98 62 53	39 25 49 42 26
76 94 39 50 27	08 93 30 67 22	17 49 37 79 55	20 57 06 46 65	04 00 00 14 60
82 29 53 23 62	40 26 61 70 16	21 99 03 61 07	83 44 83 58 63	01 05 67 40 29
15 83 00 51 53	56 32 34 96 45	26 56 97 09 26	57 95 25 40 18	00 48 56 84 34
15 28 02 28 37	94 63 03 85 66	36 62 49 41 35	55 19 47 08 82	19 00 57 72 68
41 16 97 23 09	24 65 21 07 86	21 03 22 25 73	03 56 57 38 26	53 38 23 70 69
03 85 95 60 94	00 61 45 85 17	18 43 27 19 70	64 35 52 72 52	23 19 18 04 37
62 86 24 89 32	98 42 50 80 06	04 29 37 73 15	24 31 46 96 08	83 34 85 00 95
64 75 12 92 92	79 80 06 28 22	17 07 82 96 22	16 79 70 82 94	69 48 44 74 35
81 60 43 63 75	31 70 24 95 08	42 64 30 94 80	13 69 27 39 34	33 78 43 08 84
24 17 28 20 40	66 56 67 37 59	67 62 37 13 29	79 31 13 94 71	60 67 19 40 79
13 85 39 33 43	94 59 47 13 09	83 70 59 22 34	89 83 80 09 42	36 54 01 72 71
85 59 73 81 43	02 22 69 48 96	77 65 59 90 10	27 60 03 27 97	41 17 00 53 40
93 35 48 05 41	92 81 86 24 56	48 71 70 27 05	91 29 87 37 45	71 93 61 25 98
45 38 98 83 31	18 10 37 73 80	44 96 55 89 12	21 50 73 07 48	23 58 94 77 71
64 94 41 17 70	27 77 01 53 20	85 66 02 89 43	54 85 81 88 69	54 19 94 37 54
87 30 95 10 04	06 96 38 27 07	74 20 15 12 33	87 25 01 62 52	98 94 62 46 11

MODELS

PART 6: Reliability

Nomenclature

f(t)	probability density function	–
F(t)	cumulative density function	–
k	minimum number for operation	–
MTBF	mean time before failure	time
n	number of items in the system	–
R*	system reliability	–
$\mathbf{R}_i(t)$	ith item reliability	–
t	time	time
x	number of failures	–
X	binary ith item performance variable	–
Y	arbitrary event	–
z(t)	hazard function	1/time

Symbols

λ	constant failure or hazard rate	1/time
φ	binary system performance variable	–

1. Item Reliability

Reliability as a function of time, **R**(t), is the probability that an item will continue to operate satisfactorily up to time *t*. Although other distributions are possible, reliability is often described by the *negative exponential distribution*. Specifically, it is assumed that an item's reliability is

$$\mathbf{R}(t) = 1 - \mathbf{F}(t) = e^{-\lambda t} = e^{-t/\mathrm{MTBF}} \qquad 17.10$$

This infers that the probability of *x* failures in a period of time is given by the Poisson distribution.

$$p\{x\} = \frac{e^{-\lambda}\lambda^x}{x!} \qquad 17.11$$

The negative exponential distribution is appropriate whenever an item fails only by random causes but never experiences deterioration during its life. This implies that the *expected future life* of an item is independent of the previous duration of operation.

Example 17.9

An item exhibits an exponential time to failure distribution with MTBF of 1000 hours. What is the maximum operating time such that the reliability does not drop below .99?

$$.99 = e^{-t/1000}$$

$$t = 10.05 \text{ hours}$$

Hazard function is defined as the conditional probability of failure in the next time interval given that no failure has occurred thus far. For the exponential distribution, the hazard function is

$$\mathbf{z}(t) = \lambda \qquad 17.12$$

In general,

$$\mathbf{z}(t) = \frac{\mathbf{f}(t)}{\mathbf{R}(t)} = \frac{\frac{d\mathbf{F}(t)}{dt}}{1-\mathbf{F}(t)} \qquad 17.13$$

The exponential distribution is summarized by equations 17.14 through 17.17.

$$\mathbf{f}(t) = \lambda e^{-\lambda t} \qquad 17.14$$

$$\mathbf{F}(t) = 1 - e^{-\lambda t} \qquad 17.15$$

$$\mathbf{R}(t) = 1 - \mathbf{F}(t) = e^{-\lambda t} \qquad 17.16$$

$$\mathbf{z}(t) = \lambda e^{-\lambda t}/e^{-\lambda t} = \lambda \qquad 17.17$$

2. System Reliability

The binary variable X_i is defined as 1 if item *i* operates satisfactorily and 0 otherwise. Similarly, the binary variable φ is 1 only if the system operates satisfactorily. Then, φ will be a function of the X_i.

A. Serial Systems

The *performance function* for a system of *n* serial items is

$$\phi = X_1X_2X_3 \cdots X_n = \min\{X_i\} \qquad 17.18$$

Equation 17.18 implies that the system will fail if any of the individual items fail. The system reliability is

$$R^* = R_1R_2R_3 \cdots R_n \qquad 17.19$$

Example 17.10

A block diagram of a system with item reliabilities is shown. What is the performance function and the system reliability?

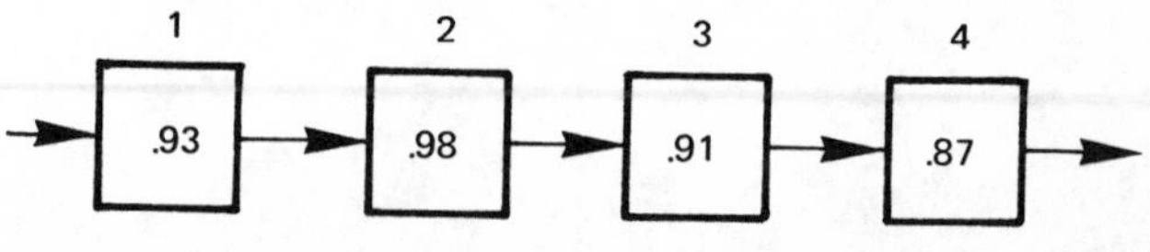

$$\phi = X_1X_2X_3X_4$$

$$R^* = (.93)(.98)(.91)(.87) = .72$$

MODELS

B. Parallel Systems

A parallel system with n items will fail only if all n items fail. This property is called *redundancy* and such a system is said to be redundant. Using redundancy, a highly reliable system can be produced from components with relatively low individual reliabilities.

The performance function of a redundant system is

$$\phi = 1 - (1-X_1)(1-X_2)(1-X_3)\cdots(1-X_n) = \max\{X_i\} \quad 17.20$$

The reliability is

$$R^* = 1-(1-R_1)(1-R_2)(1-R_3)\cdots(1-R_n) \quad 17.21$$

Example 17.11

What is the reliability of the system shown?

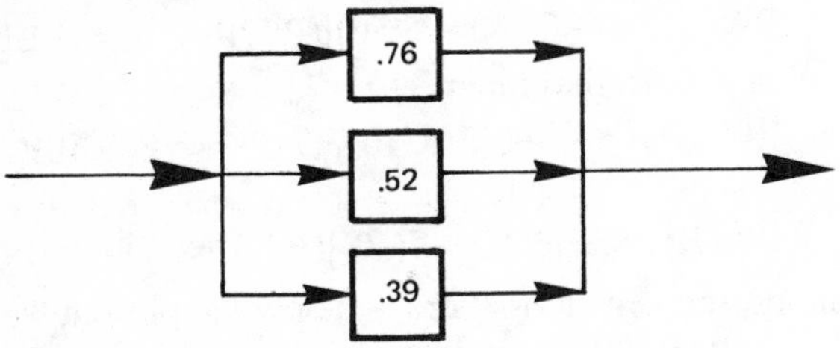

$$R^* = 1 - (1-.76)(1-.52)(1-.39) = .93$$

C. k-out-of-n Systems

If the system operates with any k of its elements operational, it is said to be a k-out-of-n system. The performance function is

$$\phi = \begin{cases} 1 \text{ if } \Sigma X_i \geq k \\ 0 \text{ if } \Sigma X_i < k \end{cases} \quad 17.22$$

The evaluation of the system reliability is quite difficult unless all elements are identical and have identical reliabilities, **R**. In that case, the system reliability follows the binomial distribution.

$$R^* = \sum_{j=k}^{n} \binom{n}{j} R^j (1-R)^{n-j} \quad 17.23$$

D. General System Reliability

A general system can be represented by a graphical network. Each path through the network from the starting node to the finishing node represents a possible operating path. For the 5-path network below, even if BD and AC are cut, the system will operate by way of path ABCD.

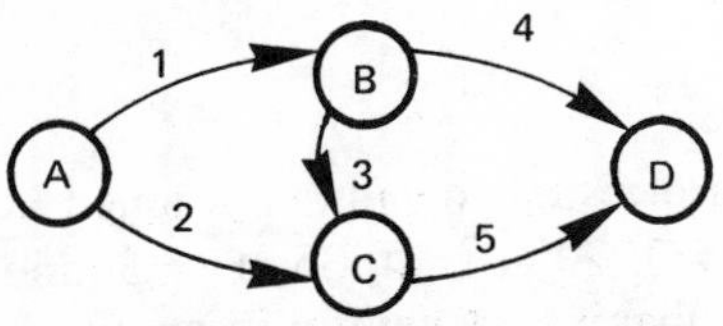

The reliability of the system will be the sum of the serial reliabilities, summed over all possible paths in the system. However, the concepts of minimal paths and minimal cuts are required to facilitate the evaluation of the system reliability.

A *minimal path* is a set of components that, if operational, will ensure the system functioning. In the previous example, components [1 with 4] are a minimal path, as are [2 with 5] and [1 with 3 with 5]. A *minimal cut* is a set of components that, if non-functional, inhibits the system from functioning. Minimal cuts in the previous example are [1 with 2], [4 with 5], [1 with 5], and [2 with 3 with 4].

Since it is usually easier to determine all minimal paths, a method of finding the exact system reliability from the set of minimal paths is needed. In general, the probability of a union of n events contains $(2^n - 1)$ terms and is given by

$$\begin{aligned} p\{Y_1 \text{ or } Y_2 \text{ or } \cdots Y_n\} = {} & p\{Y_1\} + p\{Y_2\} + p\{Y_3\} \\ & + \cdots + p\{Y_n\} \\ & - p\{Y_1 \text{ and } Y_2\} - p\{Y_1 \text{ and } Y_3\} \\ & - \cdots - p\{Y_1 \text{ and } Y_n\} \\ & - p\{Y_1 \text{ and } Y_2 \text{ and } Y_3\} - p\{Y_1 \text{ and } Y_2 \text{ and } Y_4\} \\ & - \cdots - p\{Y_i \text{ and } Y_j \text{ and } Y_k\} \text{ all } i \neq j \neq k \\ & + \{-1\}^{n-1} p\{Y_1 \text{ and } Y_2 \text{ and } Y_3 \\ & \text{and} \cdots \text{and } Y_n\} \end{aligned} \quad 17.24$$

Returning to the 5-path example,

$$Y_1 = [1 \text{ with } 4]$$

$$Y_2 = [2 \text{ with } 5]$$

$$Y_3 = [1 \text{ with } 3 \text{ with } 5]$$

Then,

$$\begin{aligned} p\{\phi = 1\} &= p\{Y_1 \text{ or } Y_2 \text{ or } Y_3\} \\ &= p\{X_1X_4 = 1\} + p\{X_2X_5 = 1\} \\ &\quad + p\{X_1X_3X_5 = 1\} - p\{X_1X_2X_4X_5 = 1\} \\ &\quad - p\{X_1X_3X_4X_5 = 1\} - p\{X_1X_2X_3X_5 = 1\} \\ &\quad + p\{X_1X_2X_3X_4X_5 = 1\} \end{aligned}$$

In terms of the individual item reliabilities, this is

$$\begin{aligned} R^* = {} & R_1R_4 + R_2R_5 + R_1R_3R_5 \\ & - R_1R_2R_4R_5 - R_1R_3R_4R_5 - R_1R_2R_3R_5 \\ & + R_1R_2R_3R_4R_5 \end{aligned}$$

This method requires considerable computation, and an upper bound on R^* would be sufficient. Such an upper bound is close to R^* since the product of individual reliabilities is small. The upper bound is given by

$$p\{\phi = 1\} \leq p\{Y_1\} + p\{Y_2\} + \cdots + p\{Y_n\} \quad 17.25$$

In the 5-path example given,

$$R^* \leq p\{X_1X_4 = 1\} + p\{X_2X_5 = 1\} + p\{X_1X_3X_5 = 1\}$$

PART 7: Replacement

Nomenclature

C_1	item replacement cost with group replacement
C_2	item replacement cost after individual failure
F(t)	number of units failing in the interval ending at t
K(T)	total cost of operating from t = 0 to t = T
MTBF	mean time before failure
n	number of units in original system
p{t}	probability of failing in the interval ending at t
S(t)	number of survivors at the end of time t
t	time
v{t}	conditional probability of failure in the interval (t−1) to t given non-failure before (t−1)

1. Introduction

Replacement and renewal models determine the most economical time to replace existing equipment. Replacement processes fall into two categories depending on the life pattern of the equipment, which either deteriorates gradually (becomes obsolete or less efficient) or fails suddenly.

In the case of gradual deterioration, the solution consists of balancing the cost of new equipment against the cost of maintenance or decreased efficiency of the old equipment. Several models are available for cases with specialized assumptions, but no general solution methods exist.

In the case of sudden failure, of which light bulbs are examples, the solution consists of finding a replacement frequency which minimizes the costs of the required new items, the labor for replacement, and the expected cost of failure. The solution is made difficult by the probabilistic nature of the life spans.

2. Deterioration Models

The replacement criterion with deterioration models is the present worth of all future costs associated with each policy. Solution is by trial and error, calculating the present worth of each policy and incrementing the replacement period by one time period for each iteration.

Example 17.12

Item *A* is currently in use. Its maintenance cost is \$400 this year, increasing each year by \$30. Item *A* can be replaced by item *B* at a current cost of \$3500. However, the cost of *B* is increasing by \$50 each year. Item *B* has no maintenance costs. Disregarding income taxes, find the optimum replacement year. Use 10% as the interest rate.

Calculate the present worth of the various policies.

policy 1: Replacement at t = 5 (starting the 6th year)

$$\text{PW(A)} = -400(\text{P/A},10\%,5) - 30(\text{P/G},10\%,5) = -1722$$

$$\text{PW(B)} = -[3500 + 5(50)](\text{P/F},10\%,5) = -2328$$

policy 2: Replacement at t = 6

$$\text{PW(A)} = -400(\text{P/A},10\%,6) - 30(\text{P/G},10\%,6) = -2033$$

$$\text{PW(B)} = -[3500 + 6(50)](\text{P/F},10\%,6) = -2145$$

policy 3: Replacement at t = 7

$$\text{PW(A)} = -400(\text{P/A},10\%,7) - 30(\text{P/G},10\%,7) = -2330$$

$$\text{PW(B)} = -[3500 + 7(50)](\text{P/F},10\%,7) = -1975$$

The present worth of *B* drops below the present worth of *A* at t = 6. Replacement should take place at that time.

3. Failure Models

The time between installation and failure is not constant for members in the general equipment population. Therefore, in order to solve a failure model, it is necessary to have the distribution of individual item lives (*mortality curve*). The conditional probability of failure in a small time interval, say from t to $(t + \delta t)$, is calculated from the mortality curve. This probability is *conditional* since it is conditioned on non-failure up to time t.

The conditional probability may decrease with time (e.g., *infant mortality*), remain constant (as with an exponential reliability distribution and failure from random causes), or increase with time (as with items that deteriorate with use). If the conditional probability decreases or remains constant over time, operating items should never be replaced prior to failure.

It is usually assumed that all failures occur at the end of a period. The problem is to find the period which minimizes the total cost.

Example 17.13

100 items are tested to failure. 2 failed at t = 1, 5 at t = 2, 7 at t = 3, 20 at t = 4, 35 at t = 5, and 31 at t = 6. Find the probability of failure in any period, the conditional probability of failure, and the mean time before failure.

MODELS

The MTBF is

$$\text{MTBF} = \frac{(2)(1)+(5)(2)+(7)(3)+(20)(4)+(35)(5)+(31)(6)}{100}$$

$$= 4.74$$

elapsed time t	failures F(t)	survivors S(t)	probability of failure p{t} = .01F(t)	conditional probability of failure v{t} = F(t)/S(t − 1)
0	0	100	---	---
1	2	98	.02	.02
2	5	93	.05	.051
3	7	86	.07	.075
4	20	66	.20	.233
5	35	31	.35	.530
6	31	0	.31	1.00

4. Replacement Policy

The expression for the number of units failing in time *t* is

$$\mathbf{F}(t) = n\Big[p\{t\} + \sum_{i=1}^{t-1} p\{i\}p\{t-i\} + \sum_{j=2}^{t-1}\Big[\sum_{i=1}^{j-1} p\{i\}p\{j-i\}\Big]p\{t-j\} + \cdots + \Big] \qquad 17.26$$

The term $np\{t\}$ gives the number of failures in time *t* from the original group.

The term $n\Sigma p\{i\}p\{t-i\}$ gives the number of failures in time *t* from the set of items which replaced the original items.

The third probability term times *n* gives the number of failures in time *t* from the set of items which replaced the first replacement set.

It can be shown that **F**(t) with replacement will converge to a steady state limiting rate of

$$\overline{\mathbf{F}(t)} = \frac{n}{\text{MTBF}} \qquad 17.27$$

The optimum policy is to replace all items in the group, including items just recently installed, when the total cost per period is minimized. That is, we want to find *T* such that **K**(T)/T is minimized.

$$\mathbf{K}(T) = nC_1 + C_2 \sum_{t=0}^{T-1} \mathbf{F}(t) \qquad 17.28$$

Discounting is usually not included in the total cost formula since the time periods are considered short. If the equipment has an unusually long life, discounting would be required.

There are some cases where group replacement is always more expensive than just replacing the failures as they occur. Group replacement will be the most economical policy if equation 17.29 holds.

$$C_2[\overline{\mathbf{F}(t)}] > \left.\frac{\mathbf{K}(T)}{T}\right|_{\text{minimum}} \qquad 17.29$$

If the opposite inequality holds, group replacement may still be the optimum policy. Further analysis is then required.

MODELS

PART 8: Decision Theory

Nomenclature

a_i	the ith alternative
A	the set of alternatives
C	cost of experimentation
D_r	the rth possible decision procedure
E(PI)	the expected loss with perfect information
$E(L(a_i))$	the expected loss associated with the ith alternative
i	index over action alternatives
j	index over actual states of nature
k	index over outcome of experiments
$L(a_i,\theta_j)$	the loss incurred when the ith alternative is chosen when nature is in the jth state
m	number of possible outcomes
n	number of alternatives
p{j,k}	the probability of the jth state of nature occurring given a prediction of the kth state based on experimentation
$p\{\theta_j\}$	prior probability of the jth state of nature
x	experimental outcome variable

Symbols

$\boldsymbol{\Theta}$	set of possible states of nature
θ_j	the jth state of nature
θ_k'	the kth possible experimental outcome or prediction of the state of nature

1. Introduction

When decisions have to be made, there are usually a finite number of alternatives to be selected from. The set of alternatives is known as the *action space.* Each alternative is associated with a *benefit* or *efficiency* which is received if the alternative is selected. Usually these efficiencies are expressed in dollars. If the benefit is negative, the term *loss* is used in place of efficiency.

Example 17.14

A prospector stakes a claim to a gold ore vein. The prospector classifies claims into three categories according to the outcome of digging—rich vein, trace vein, and no vein. The prospector is tired of digging a lot of holes for nothing and wants to make a wise decision.

The prospector has three alternatives: work the land, lease the land, or sell the land. The profit associated with each alternative and each claim type is given in the following matrix.

	rich vein	trace vein	no vein
work	\$100,000	0	−15,000
lease	20,000	5,000	2,000
sell	4,000	4,000	4,000

2. General Problem Formulation

In general, it is necessary to choose from a set of alternatives, **A**.

$$\mathbf{A}: (a_1,a_2,a_3,\cdots,a_n)$$

The state of nature set is $\boldsymbol{\Theta}$.

$$\boldsymbol{\Theta}: (\theta_1,\theta_2,\theta_3,\cdots,\theta_m)$$

In example 17.14, n = m = 3, and

$$a_1 = \text{work the land}$$
$$a_2 = \text{lease the land}$$
$$a_3 = \text{sell the land}$$
$$\theta_1 = \text{rich vein}$$
$$\theta_2 = \text{trace vein}$$
$$\theta_3 = \text{no vein}$$

Decision theory commonly deals in *loss functions,* as opposed to efficiency functions. The difference is in sign only. The loss incurred depends only on the alternative chosen and the state of nature. The loss function is given by $\mathbf{L}(a_i,\theta_j)$. For example, $\mathbf{L}(a_2,\theta_3) = -2000$ in example 17.14

The entire loss matrix for example 17.14 is given below in thousands of dollars.

	θ_1	θ_2	θ_3
a_1	−100	0	15
a_2	− 20	−5	− 2
a_3	− 4	−4	− 4

If the loss function depends on a random variable, the expected loss should be used.

Example 17.15

In the past, the prospector has found 5 trace veins. The cost of exploration has been \$5000 per vein. The trace veins have yielded ore worth \$3000, \$4000, \$5000, \$6000, and \$7000 respectively. What is the loss function for trace veins?

The cost per trace vein is \$5000.

The average ore value per trace vein is

$$\frac{1000}{5}(3 + 4 + 5 + 6 + 7) = 5000$$

The expected loss per trace vein is

$$\text{ore value} - \text{cost} = \$5000 - \$5000 = 0$$

3. *Decisions Made Without Experimentation*

A. *Minimax Criterion*

One decision procedure is the *Minimax Criterion*, which chooses the alternative minimizing the maximum loss. The criterion is very conservative and is seldom used in practice since it assumes that nature is a conscious, malevolent opponent out to inflict maximum damage on the decision maker.

Example 17.16

What is the minimax decision for example 17.14?

The maximum losses associated with each alternative are

alternative	maximum loss
a_1	15
a_2	− 2
a_3	− 4

Inasmuch as the minimum of the values is −4, the minimax criterion would direct the prospector to sell the claim.

B. *Bayes Principle*

Bayes Principle requires knowledge of the probability distribution of the possible states of nature. Specifically, let $p\{\theta_j\}$ be the prior probability of the jth state. Then, the expected loss associated with the ith alternative is

$$E(L(a_i)) = p\{\theta_1\}L(a_i,\theta_1) + p\{\theta_2\}L(a_i,\theta_2) + \cdots + p\{\theta_m\}L(a_i,\theta_m) \qquad 17.30$$

Bayes Principle tells the decision maker to choose the alternative which minimizes the expected loss.

Example 17.17

Out of a total of 10 claims worked during the prospector's career, 4 have had no veins, 5 have had trace veins, and 1 has had a rich vein. What is the superior alternative under Bayes Principle?

Here $p(\theta_3) = \frac{4}{10} = .4 \quad p(\theta_2) = \frac{5}{10} = .5 \quad p(\theta_1) = \frac{1}{10} = .1$

The expected loss functions are

$$E(L(a_1)) = .10(-100) + .50(0) + .40(15) = -4.0$$
$$E(L(a_2)) = .10(-20) + .50(-5) + .40(-2) = -5.3$$
$$E(L(a_3)) = .10(-4) + .50(-4) + .40(-4) = -4.0$$

Alternative 2 should be chosen under Bayes Principle since it minimizes the expected loss. The prospector should lease the claim.

4. *Decision Making with Experimentation*

If the decision maker is allowed to experiment (at a cost) with the problem, valuable data may be obtained which may help in making a decision.

If the prospector in the previous examples had perfect information about his claim, he would know what to do. If he knew there was a rich vein, the best alternative would be to work the land at a loss of −100. If the vein was known to be a trace vein, the best alternative would be to lease the land at a loss of −5. If the claim was without value, the best alternative would be to sell the land at a loss of −4. Since the prior probabilities are known, the expected loss with perfect information is

$$E(PI) = -100(.10) - 5(.5) - 4(.4) = -14.1$$

The Bayes solution in example 17.17 without any data produced a minimum expected loss of −5.3. Therefore, the potential maximum savings from experimentation is 14.1−5.3 = 8.8. Since this value assumes that the experiment will yield perfect information, which it will not, the prospector should not pay more than 8.8 in an attempt to increase his expected return.

MODELS

PART 9: Decision Trees

A tree diagram is a graphical method of enumerating all of the possible outcomes of a sequence of actions. The total number of outcomes is given by the Fundamental Principle of Counting in equation 17.31.

$$N = \Pi n_i \qquad 17.31$$

n_i is the number of elements in the ith set. The value calculated assumes that no outcomes are restricted and that all outcomes are distinctly different.

If a tree is used to model a decision process, it is called a *decision tree.* Usually decision trees have probabilities, losses, and rewards associated with the various possible outcomes.

Example 17.18

Find the possible values of product $P = A \times B$ where A can take on the values (7,8) and B can take on the values (1,3,5). The tree diagram is

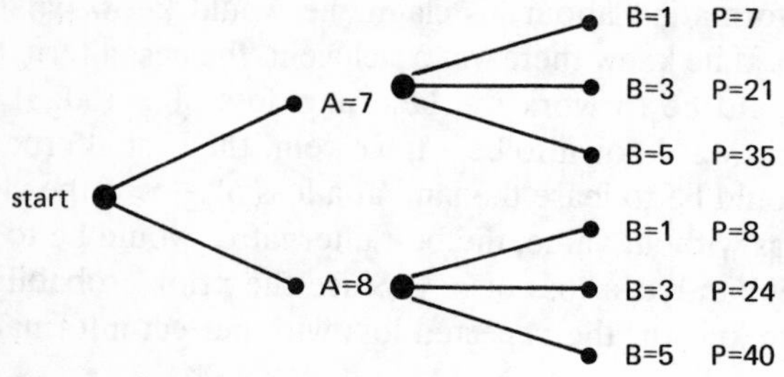

Therefore, P = (7,8,21,24,35,40)

The number of distinctly different values of P is (2)(3) = 6.

Example 17.19

How many members are in the product set $P = C \times D$ where C = (6,8) and D = (3,4)?

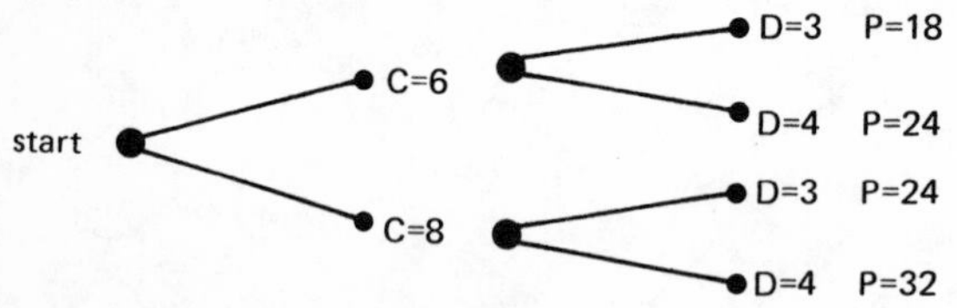

There are 3 distinctly different values of P, so N = 3.

Example 17.20

Michael and Elizabeth want to play a checkers tournament. The first person to win two out of three games will win the tournament. What is the tree diagram for the possible tournament outcomes?

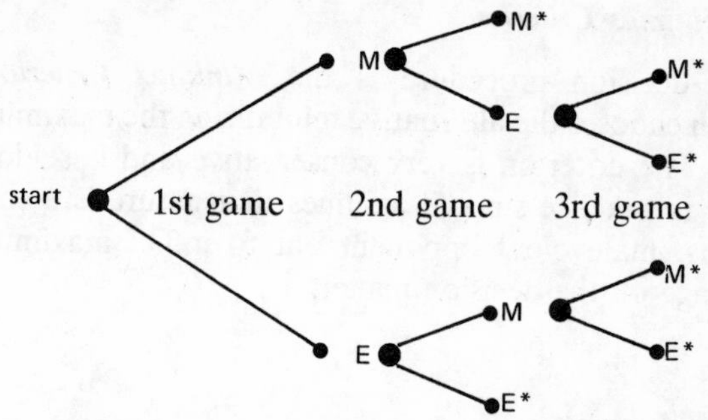

In example 17.20, the tree was divided into segments called *generations* or *stages.* At the end of any given stage, the current status of the tournament could be described by the value of the *state variable.*

For example, suppose Elizabeth had just won the second game. *E* would be the value of the state variable in the second stage. Equivalently, it could be said that the tournament was in the *E* state in the second stage.

Example 17.21

A manufacturer can introduce a new product this month or next month. A new-product announcement can be released this month regardless of when the product is released. The current market is 1000 units. If the manufacturer waits until next month, the market will be 1200 units. A news release will increase sales by 10%.

Draw the tree, list the set of all possible outcomes, and choose the optimum strategy by maximizing the expected sales.

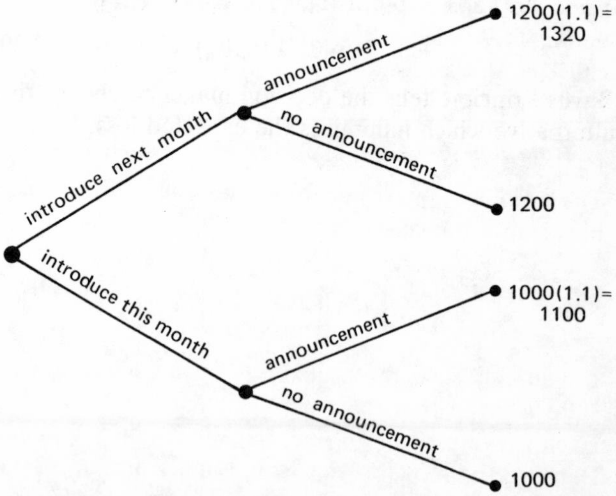

The set of possible sales outcomes is (1000, 1100, 1200, 1320). The optimum strategy is to introduce next month with a news release.

PART 10: Markov Processes

1. Introduction

A Markov process is a random process in which the state variable may have one of a limited number of values at any given time. Since the process is random, it is not possible to predict with certainty which value the random variable will take in the next time increment.

Example 17.22

A parrot in an oasis has three trees from which to choose. On any given day, the parrot can be in any tree, even the previous day's tree, but it may be in only one tree per day.

In example 17.22, the process variable T_j is a random variable representing the tree in which the parrot is in at time *j*. One way to describe the future or history of the parrot's wanderings is by the sequence $(T_0, T_1, T_2, \ldots, T_n)$.

In general, such a random variable is known as a *state variable.* All possible values of the state variable comprise the *state space.* Changes of state are called *transitions.* If the state actually changes, it is a *real transition;* otherwise, it is a *virtual transition.*

At any given time, the probability of moving to one of the other states depends on the current state and the current time. Thus, the T_j are not always identically distributed nor are they independent.

The process is known as a *Markov process* if the current value of the state variable, T_i, depends only on the previous value, T_{i-1}, and affects only the upcoming value, T_{i+1}. Such a linking of the values is sometimes called a *Markov Chain.*

2. Transition and State Probabilities

The usual method of recording the transition probabilities is in a transition matrix, **P**.

$$\mathbf{P} = \begin{vmatrix} p_{11} & p_{12} & p_{13} \\ p_{21} & p_{22} & p_{23} \\ p_{31} & p_{32} & p_{33} \end{vmatrix}$$

In general, p_{jk} is the probability of making a move to state *k* given that the variable is currently in state *j*. Each row must sum to one, but the columns may not. Because the move from *j* to *k* is made in one step or jump, the above matrix is also called the *one-step transition* matrix, $\mathbf{P}^{(1)}$. If the matrix is the same from time increment to time increment, it is said to be a *stationary matrix.*

Example 17.23

Once the parrot gets to tree 1, it will not move directly to tree 3. Moving to tree 2 or staying in tree 1 are equally likely. If the parrot gets to tree 2, it will not move directly to tree 1, but may stay in tree 2 or move to tree 3 with equal likelihood. If the parrot gets to tree 3, it will not stay in tree 3, but will move to tree 1 three times as often as to tree 2.

The one-step transition matrix is

$$\mathbf{P} = \begin{vmatrix} .50 & .50 & .00 \\ .00 & .50 & .50 \\ .75 & .25 & .00 \end{vmatrix}$$

A *two-step transition matrix,* $\mathbf{P}^{(2)}$, contains values of p_{jk} which are probabilities of moving from state *j* to *k* in exactly two steps. The matrix $\mathbf{P}^{(2)}$ can be found by squaring **P** using matrix multiplication.

Example 17.24

Given that the parrot is in tree 1 now, what is the probability that it will be in tree 2 after two moves?

method 1: Enumeration

There are three paths the parrot can take to get from tree 1 to tree 2 in two moves.

$$T_0=1,\ T_1=1,\ T_2=2;\ \text{probability} = (.5)(.5) = .25$$
$$T_0=1,\ T_1=2,\ T_2=2;\ \text{probability} = (.5)(.5) = .25$$
$$T_0=1,\ T_1=3,\ T_2=2;\ \text{probability} = (0)(.25) = \underline{.00}$$
$$p_{12}^{(2)} = .50$$

method 2: Using $\mathbf{P}^{(2)}$

Squaring the one-step transition matrix gives

$$\mathbf{P}^{(2)} = (\mathbf{P})^2 = \begin{vmatrix} .250 & .500 & .250 \\ .375 & .375 & .250 \\ .375 & .500 & .125 \end{vmatrix}$$

The two-step transition probability of $p_{12}^{(2)}$ is again found as .500.

The *n-step transition matrix* is found by raising the one-step matrix to the *n*th power. This is a specific case of the *Chapman-Kolmogorov equation.*

$$\mathbf{P}^{(n)} = (\mathbf{P})^n \qquad 17.32$$

If the value of T_0 is known, the probability of moving from state *j* to *k* in *n* steps can be found directly from the n-step transition matrix. However, often T is unknown—given by its own probability distribution $\mathbf{P}_i^{(0)}$ where $\mathbf{P}_i^{(0)}$ is the probability of the state variable having the value *i* at time 0. Then, the probability of being in state *k* in *n* steps regardless of the initial state is given by equation 17.33.

MODELS

$$p_k^{(n)} = p_1^{(0)}p_{1k}^{(n)} + p_2^{(0)}p_{2k}^{(n)} + p_3^{(0)}p_{3k}^{(n)} \qquad 17.33$$

Notice that the state probabilities are the probabilities of *being* in a state after some number of steps, whereas the transition probabilities are the probabilities of moving *between* two states.

3. Steady-State Probabilities

The transition probabilities stabilize as the number of steps increase. That is, the starting position has increasingly less influence on the transition probabilities as n increases. After five steps, the transition matrix for the example problem is

$$\mathbf{P}^{(5)} = \begin{vmatrix} .336 & .445 & .219 \\ .328 & .445 & .227 \\ .340 & .441 & .219 \end{vmatrix}$$

It can be seen that the three column values appear to be approaching the values of .33, .44, and .22 respectively, independent of the initial state.

The actual *steady-state transition probabilities* (π_1, π_2, π_3) can be found by solving the simultaneous equations specified by the matrix equation 17.34 and the normalizing equation 17.35.

$$[\pi_1, \pi_2, \pi_3] = [\pi_1, \pi_2, \pi_3]\mathbf{P} \qquad 17.34$$

$$\pi_1 + \pi_2 + \pi_3 = 1 \qquad 17.35$$

Example 17.25

Find the steady-state probabilities for the parrot example.

We need to solve the following four equations simultaneously:

$$\pi_1 = .5\pi_1 \qquad + .75\pi_3$$

$$\pi_2 = .5\pi_1 + .5\pi_2 + .25\pi_3$$

$$\pi_3 = \qquad .5\pi_2$$

$$1 = \pi_1 + \pi_2 + \pi_3$$

The solution is $\pi_1^* = .333\cdots$
$\pi_2^* = .444\cdots$
$\pi_3^* = .222\cdots$

The reciprocals of the steady-state probabilities are the *average times between reassignments* to the ith state. For example, an average of three steps is required to return to state 1.

4. Special Cases

If the transition from state j to k is possible, we say that state k is *reachable* from state j. If j is also reachable from k, we say that the two states *communicate*. If all states in the process belong to a single communicating class, the process is said to be *irreducible*. Otherwise, it is *reducible*. A state is an *absorbing state* if a transition to another state is impossible, as in $p_{jj} = 1$.

A *closed set of states* is a set such that no state outside the set is reachable from a state within the set. If there are no absorbing states in the closed set, it is called a *minimal closed set*. States in a minimal closed set are called *recurrent states*. States outside of a closed set are called *transient states*.

If a recurrent state must be reached with a regular frequency due to the structure of the process, it is said to be *periodic;* otherwise, it is *aperiodic*.

A state that is not transient, not periodic, and has a finite mean recurrent time is called an *ergodic state*.

It is possible that the simultaneous equation method for finding steady-state probabilities will fail since it is possible that these probabilities may not exist. However, no difficulties will arise if all states are ergodic.

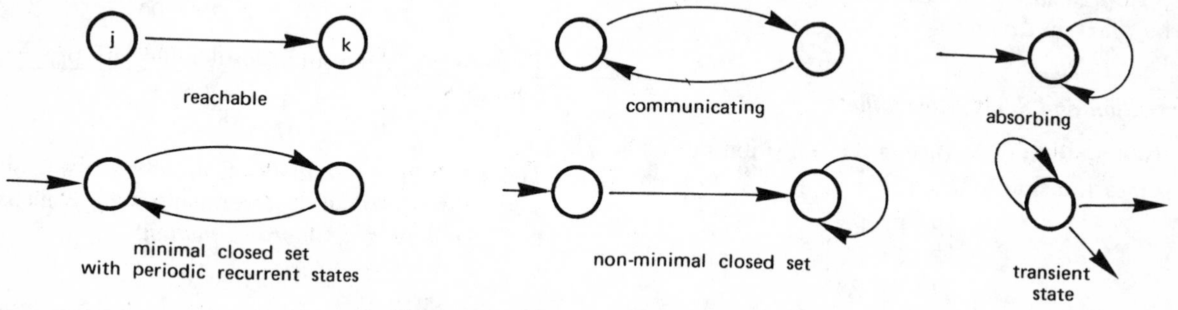

Figure 17.1 Types of Markov States

PART 11: Queueing Models

Nomenclature

L	expected system length
L_q	expected queue length
p{n}	probability of *n* units in the system
s	number of parallel servers
W	expected time in system
W_q	expected time in queue

Symbols

λ	mean arrival rate
ρ	traffic intensity = λ/μ
μ	mean service rate per server

1. Introduction

Queue means waiting line. Queueing theory can predict the length of a waiting line, the average time a customer can expect to spend in the queue, and the probability that *n* customers will be in the queue.

Many queueing models have been developed, all with variations in the primary problem characteristics. These characteristics are

- the distribution of arrivals
- the distribution of service times
- the number of servers
- the size of the calling population
- the order in which the customers are served

Most of the models are too complicated to be presented here. However, two models are given after a brief listing of the general relationships. The relationships given here are for *steady-state operation*. This means that the service facility has been open and in operation for some time.

2. General Relationships

$$L = \lambda W \qquad 17.36$$

$$L_q = \lambda W_q \qquad 17.37$$

$$W = W_q + \frac{1}{\mu} \qquad 17.38$$

$$\lambda < \mu s \qquad 17.39$$

$$\text{average service time} = 1/\mu \qquad 17.40$$

$$\text{average time between arrivals} = 1/\lambda \qquad 17.41$$

3. The M/M/1 System

It is assumed that the following are true in the M/M/1 model:

- There is only one server (s = 1).
- The calling population is infinite.
- The service times are exponentially distributed with mean μ. That is, the probability of a customer's remaining service time exceeding *h* (after already spending time with the server) is

$$p\{t>h\} = e^{-\mu h} \qquad 17.42$$

Notice that equation 17.42 is independent of the time already spent with the server.

- The arrival rate is distributed as Poisson with mean λ. The probability of *x* customers arriving in the next period is

$$p\{x\} = \frac{e^{-\lambda}\lambda^x}{x!} \qquad 17.43$$

The following relationships are valid for the M/M/1 system:

$$p\{0\} = 1 - \rho \qquad 17.44$$

$$p\{n\} = p\{0\}(\rho)^n \qquad 17.45$$

$$W = \frac{1}{\mu - \lambda} = W_q + \frac{1}{\mu} = L/\lambda \qquad 17.46$$

$$W_q = \frac{\rho}{\mu - \lambda} = L_q/\lambda \qquad 17.47$$

$$L = \frac{\lambda}{\mu - \lambda} = L_q + \rho \qquad 17.48$$

$$L_q = \frac{\rho\lambda}{\mu - \lambda} \qquad 17.49$$

Example 17.26

Given an M/M/1 system with μ = 20 customers per hour and λ = 12 per hour, find the steady-state value of W, W_q, L, and L_q. What is the probability that there will be 5 customers in the system?

$$\rho = \frac{12}{20} = .6$$

$$W = \frac{1}{20 - 12} = .125 \text{ hours}$$

$$W_q = \frac{.6}{20 - 12} = .075 \text{ hours}$$

$$L = \frac{12}{20 - 12} = 1.5 \text{ customers}$$

$$L_q = \frac{(.6)(12)}{20 - 12} = .9 \text{ customers}$$

$$p\{0\} = 1 - .6 = .4$$

$$p\{5\} = .4(.6)^5 = .031$$

4. The M/M/s System

The same assumptions are used for the M/M/s system as were used for the M/M/1 system except that there

MODELS

are s servers instead of only 1. Each server has a mean service rate μ. Each server draws from a single line so that the first person in line goes to the first available server. Each server does not have its own line.

However, if customers are allowed to change the lines they are in so that they go to any available server, this model may also be used to predict the performance of a multiple server system where each server has its own line.

$$W = W_q + \frac{1}{\mu} \qquad 17.50$$

$$W_q = L_q/\lambda \qquad 17.51$$

$$L_q = \frac{p\{0\}(\lambda/\mu)^s\rho}{s!(1-\rho)^2} \qquad 17.52$$

$$L = L_q + \rho \qquad 17.53$$

$$p\{0\} = \frac{1}{\dfrac{(\rho)^s}{s![1-(\rho/s)]} + \sum_{j=0}^{s-1}\dfrac{(\rho)^j}{j!}} \qquad 17.54$$

$$p\{n\} = \frac{p\{0\}(\rho)^n}{n!} \quad (n < s) \qquad 17.55$$

$$p\{n\} = \frac{p\{0\}(\rho)^n}{s!s^{n-s}} \quad (n > s) \qquad 17.56$$

MODELS

PART 12: Mathematical Programming

1. Introduction

Mathematical programming is a modeling procedure applicable to problems for which the goal and resource limitations can be described mathematically. If the goal function and all resource constraints are linear (polynomials of degree 1 only) the procedure is known as *linear programming*.

If the variables can take on only integer values, a procedure known as *integer programming* is required. If the polynomials are of any degree or contain other functions, a procedure known as *dynamic programming* is required.

2. Formulation of a Linear Programming Problem

All linear programming problems have a similar format. Each has an *objective function* which is to be optimized. Usually the objective function is to be maximized, as in the case of a profit function. If the objective is to minimize some function, such as cost, the problem may be turned into a maximization problem by maximizing the negative of the original function.

Example 17.27

A cattle rancher buys three types of cattle food. The rancher wants to minimize the cost of feeding his cattle. Write the objective function for this problem on a per animal basis.

food type	cost per pound
1	1.5
2	2.5
3	3.5

Let x_i be the number of pounds of food *i* purchased per animal. Then, the objective function to be minimized is

$$Z = 1.5x_1 + 2.5x_2 + 3.5x_3$$

Each linear programming problem also has a set of limitation functions called constraints. Constraints are used to set the bounds for the objective function.

Example 17.28

The rancher is concerned with meeting published nutritional information on minimum daily requirements (MDR) given in milligrams per animal. The composition of each food type is known and the contributions for each vitamin in mg/pound are

		food type		
vitamin	MDR (mg)	1	2	3
A	100	1	7	13
B	200	3	9	15
C	300	5	11	17

It is also physically impossible for an animal to eat more than the following amounts per day.

food type	maximum feeding
1	50 lbs
2	40
3	30

The constraints on this problem are

$$x_1 + 7x_2 + 13x_3 \geqslant 100$$

$$3x_1 + 9x_2 + 15x_3 \geqslant 200$$

$$5x_1 + 11x_2 + 17x_3 \geqslant 300$$

$$x_1 \leqslant 50$$

$$x_2 \leqslant 40$$

$$x_3 \leqslant 30$$

$$x_1 \geqslant 0$$

$$x_2 \geqslant 0$$

$$x_3 \geqslant 0$$

Linear programming problems are generally solved by computer. Some simple problems may be solved by hand with a procedure known as the *simplex method*. Specialized methods allowing easy manual solutions are available for certain classes of problems, primarily the *transportation* and *assignment problems*.

Once a solution is found, it is possible to determine the effect on the objective function of changing one of the program parameters. This is known as *sensitivity analysis* and is very important in instances where the accuracy of collected data is unknown.

3. Solution to 2-Dimensional Problems

If a linear programming problem can be formulated in terms of only two variables, x_1 and x_2, it can be solved graphically by the following procedure:

step 1: Graph all of the constraints and determine the *feasible region*. Usually this will result in a *convex hull*.

step 2: Evaluate the objective function, Z, at each corner of the hull.

MODELS

step 3: The values of x_1 and x_2 which optimize Z are the coordinates of the corner at which Z is optimized.

Example 17.29

Solve the following linear programming problem graphically.

$$\text{Max } Z = 2x_1 + x_2$$

$$\begin{aligned} \text{such that } x_1 + 4x_2 &\leq 24 \\ x_1 + 2x_2 &\leq 14 \\ 2x_1 - x_2 &\leq 8 \\ x_1 - x_2 &\leq 3 \\ x_1 &\geq 0 \\ x_2 &\geq 0 \end{aligned}$$

The region enclosed by the constraints is shown.

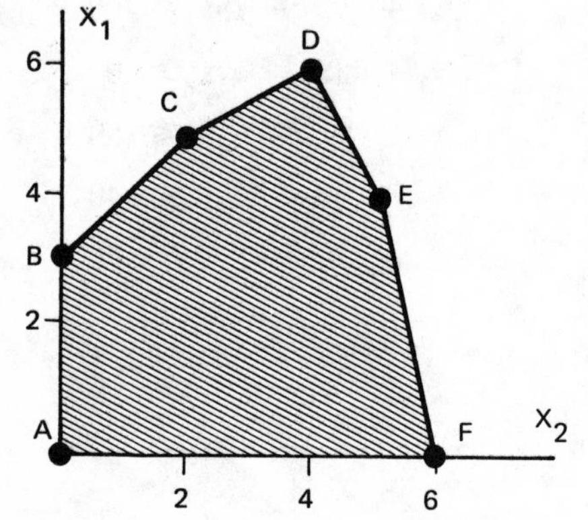

The coordinates and Z value for each corner are

corner	coordinates (x_2,x_1)	Z
A	(0,0)	0
B	(0,3)	6
C	(2,5)	12
D	(4,6)	14
E	(5,4)	13
F	(6,0)	6

Z is maximized when $x_1 = 6$ and $x_2 = 4$.

MODELS

FEEDBACK

1. Simplify the following block diagram and determine the system gain.

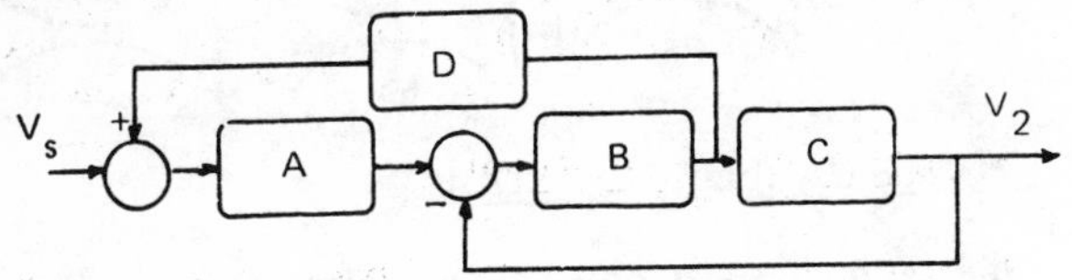

2. Simplify the following block diagram and determine the system gain.

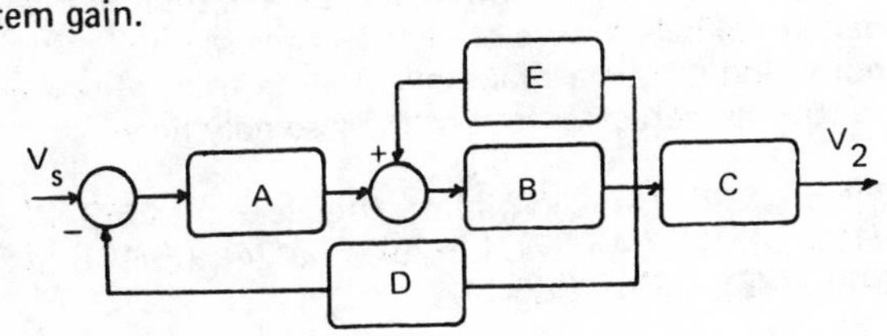

SYSTEM RESPONSE

3. For each of the systems shown below,
   (a) draw the circuit diagram using idealized elements
   (b) write the differential equation or system of differential equations which describe the motion of the system.

3.1

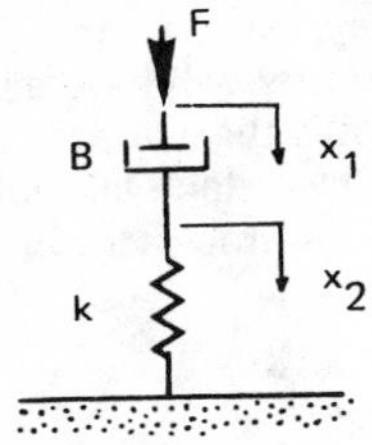

3.2

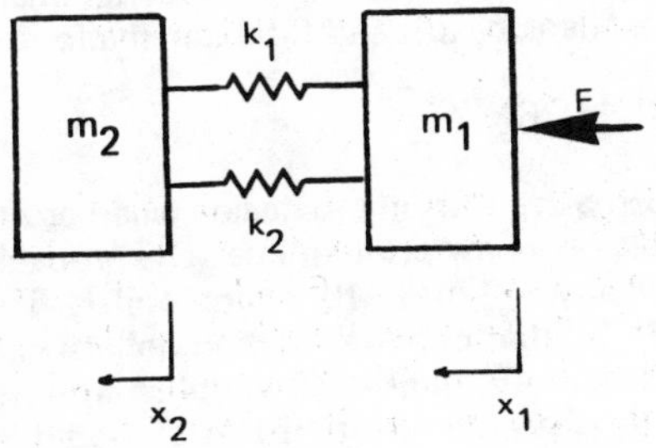

3.3

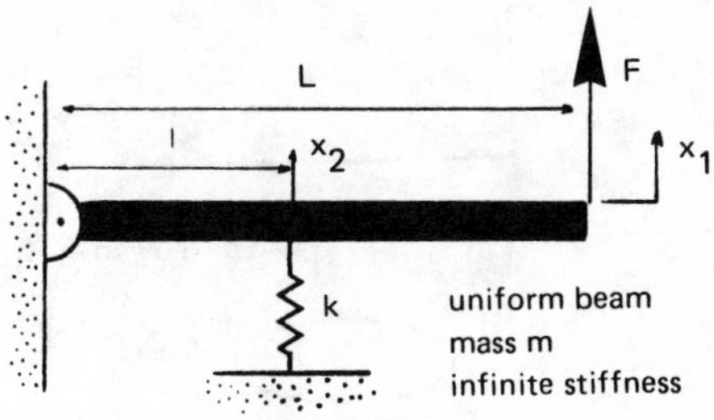

3.4

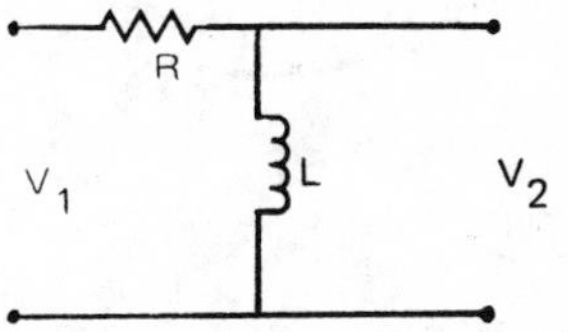

3.5

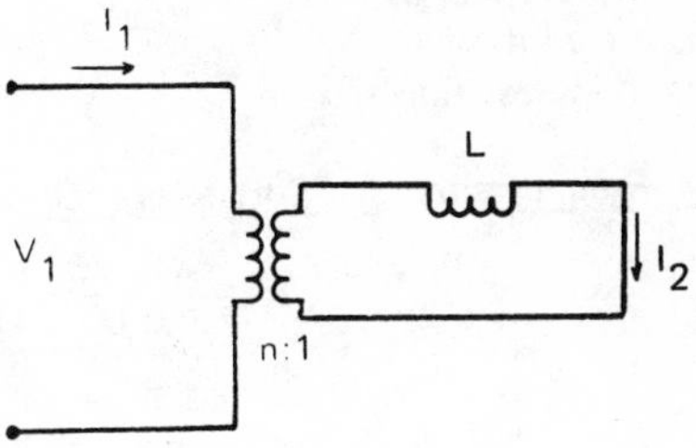

3.6

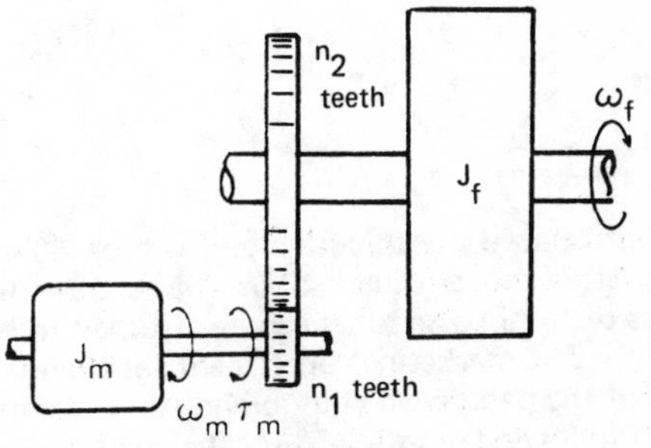

3.7

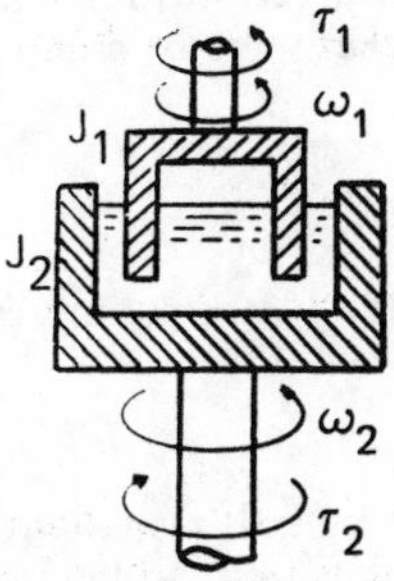

MODELS

3.8

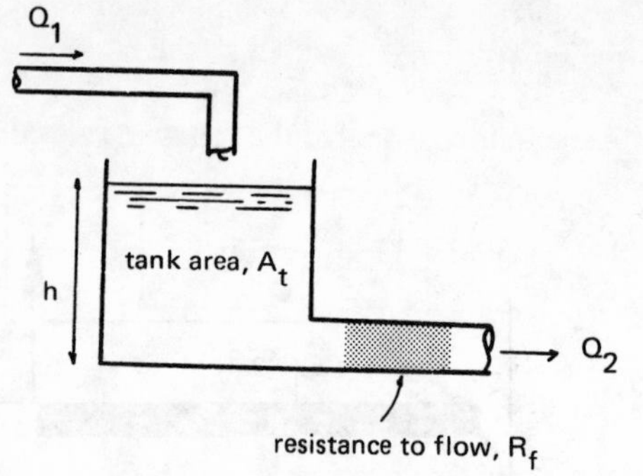

3.9

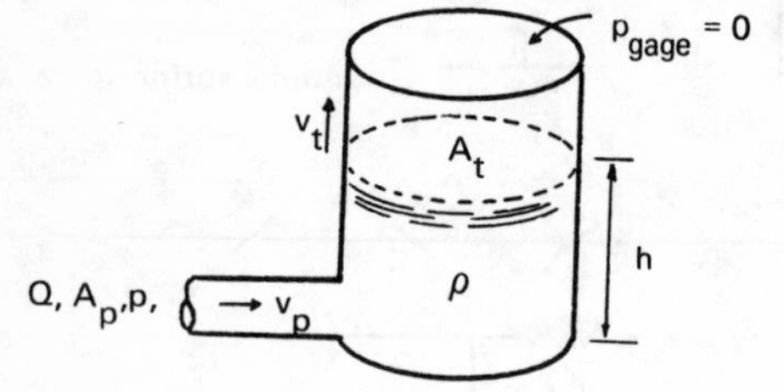

CRITICAL PATHS

4. The activities which constitute a project are listed below. The project starts at t=0.
   (a) Draw the network.
   (b) Indicate the critical path.
   (c) What is the earliest finish?
   (d) What is the latest finish?

Activity	*Predecessors*	*Successors*	*Duration*
start		A	0
A	start	B,C,D	7
B	A	G	6
C	A	E,F	5
D	A	G	2
E	C	H	13
F	C	H,I	4
G	D, B	I	18
H	E,F	finish	7
I	F,G	finish	5
finish	H,I		0

DECISION TREES

5. When introduced, toothpaste brand A took 60% of the market, sharing the remaining 40% with various competitors. All brands of toothpaste tubes can be assumed to be replaced each month. The marketing department for brand A assumes that 75% of the people who buy brand A one month will repeat the following month. 45% of the customers who switch to a competing brand will return to brand A the next month. Find the expected market share of brand A at the end of the third month.

MARKOV PROCESSES

6. Repeat problem 5 using Markov processes analysis.

RELIABILITY

7. A sample of 50 light bulbs is tested until failure. The results in (quantity failing/hours to failure) are: (2/50), (5/100), (18, 150), (22/200), (2/250), and (1/300). Find the MTBF and the conditional failure distribution.

8. What is the reliability of the system shown below? Bank B is a one-out-of-two system; Bank C is a one-out-of-three system; Bank D is a two-out-of-three system.

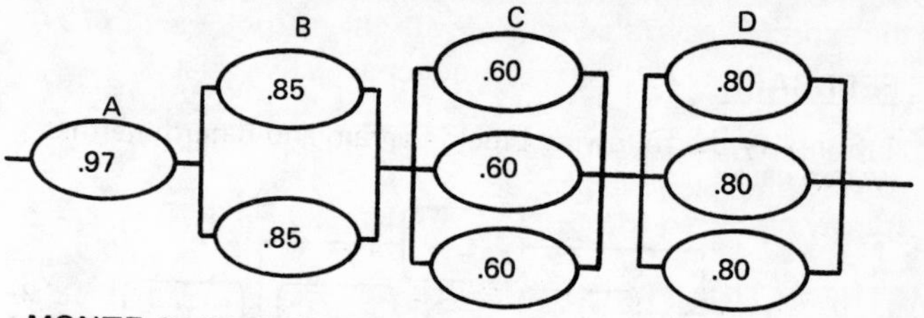

MONTE CARLO TECHNIQUE

9. A product is assembled on a 2-person assembly line. After operator A finishes the first part of the assembly, the product is placed on a belt conveyor which takes it to operator B in .10 minutes. When operations A and B were time studied, the data listed below were obtained. Using Monte Carlo methods, find (a) the average inventory in front of operator B and (b) the average output of the assembly line.

Operation A: (time in minutes/frequency) (.25/3), (.3/2), (.35/10), (.4/22), (.45/27), (.5/26), (.55/24), (.6/16), (.65/15), (.7/6), (.75/5), (.8/5), (.85/2).

Operation B: (.25/2), (.3/10), (.35/17), (.4/20), (.45/18), (.5/15), (.55/10), (.6/6), (.65/5), (.7/5), (.75/4), (.8/3), (.85/2).

QUEUEING MODELS

10. Repeat problem 9 using queueing theory. State your assumptions.

11. A secretary receives an average of 4 small jobs per hour. On the average, each job requires six minutes of time. (a) What proportion of time is the secretary idle? (b) What is the probability that the secretary will have a backlog of three unfinished jobs? (c) What is the average backlog? (d) What is the probability that a job will take more than one-half hour to complete (excluding the time spent in the in-basket)? State your assumptions.

DIMENSIONAL ANALYSIS

12. Develop an expression for the thrust of a screw propeller completely immersed in fluid by using dimensional analysis. You can assume that the following variables are involved: thrust (F), propeller diameter (D), boat velocity (V), propeller revolutions per second (N), gravitational acceleration (g), fluid mass density (ρ), and fluid kinematic viscosity (υ).

DECISION THEORY

13. A new company is trying to decide whether or not it should gamble on a new product line. The probability is .25 that the company will make 15 million dollars. The probability is .30 that only one million dollars will be made. The probability is .45 that 6 million dollars will be lost. (a) What is the expected profit? (b) What should be company do? State your assumptions.

14. The company in problem 13 has a chance to buy a marketing survey for $600,000. The 'track record' of such surveys is given below. What should the company do? What is the expected payoff associated with the course of action?

	product actually was a		
	winner	average	loser
survey predicted a winner	.65	.25	.10
survey predicted average	.25	.45	.15
survey predicted a loser	.10	.30	.75

MATHEMATICAL PROGRAMMING

15. A chemical company has a standing order for 100 barrels of fish fertilizer each month. There are two different formulas, and the customer cannot tell the difference. Enough raw materials are available to produce only 55 barrels of formula B each month. Each barrel of formula A requires 2 tons of fish; each barrel of formula B requires 1 ton of fish. 180 tons of fish are available each month. The company makes a profit of $7 per barrel of A and $2 per barrel of B. How many barrels of each should it make each month?

16. Use the simplex method to solve the following system of equations:

Maximize: $Z = 12x_1 + 18x_2$
Such that $2x_1 + x_2 \leq 4$
$x_1 + 2x_2 \leq 4$
$x_1 \geq 0$
$x_2 \geq 0$

17. Solve the following linear programming problem by any method.

Maximize: $Z = 5x_1 + 2x_2 + 3x_3 - x_4 + x_5$
Such that $x_1 + 2x_2 + 2x_3 + x_4 \leq 0$
$3x_1 + 4x_2 + x_3 + x_5 \leq 0$
$x_1, x_2 \geq 0$

LAPLACE TRANSFORMS

18. Solve the following differential equations using Laplace transforms. u_t is a unit step at time t. p_t is a unit pulse at time t.

(a) $y'' + 3y' + 2y = 0$ $\quad y(0) = 1, y'(0) = 0$
(b) $y'' + 4y' + 4y = 0$ $\quad y(0) = 1, y'(0) = 1$
(c) $y'' + 2y' + 2y = \cos(t)$ $\quad y(0) = 1, y'(0) = 0$
(d) $y'' + 2y' + 2y = \exp(-t)$ $\quad y(0) = 0, y'(0) = 1$
(e) $y'' + y = f(t)$ $\quad y(0) = 0, y'(0) = 1$
$f(t) = 1$ for t less than 3
$= 0$ for t greater than 3
(f) $y'' + 2y' + y = f(t)$ $\quad y(0) = 1, y'(0) = 0$
$f(t) = 1$ for t less than 1
$= 0$ for t greater than 1
(g) $y'' + 3y' + 2y = u_2(t)$ $\quad y(0) = 0, y'(0) = 1$
(h) $y'' + 2y' + 2y = p_{(4)}$ $\quad y(0) = 1, y'(0) = 0$
(i) $y'' + y = 2p_{(1)}$ $\quad y(0) = 1, y'(0) = 0$

FOURIER SERIES

19. Write the fourier series for the following waveforms:

(a)
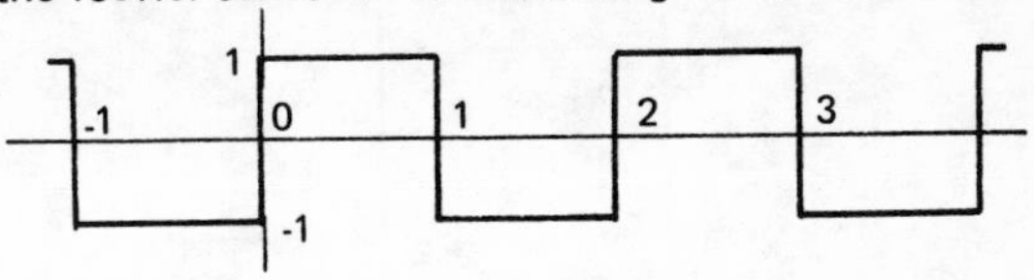

(b)
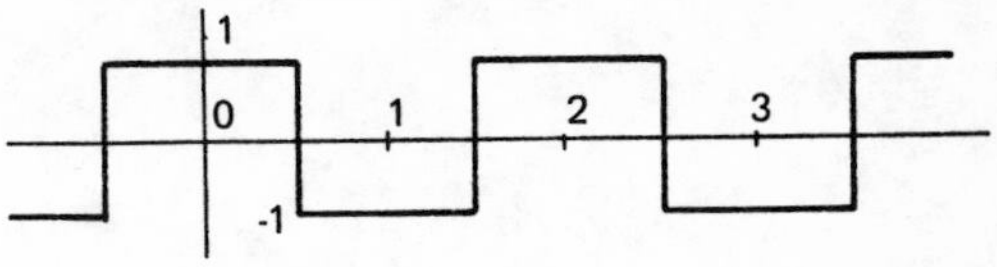

(c)
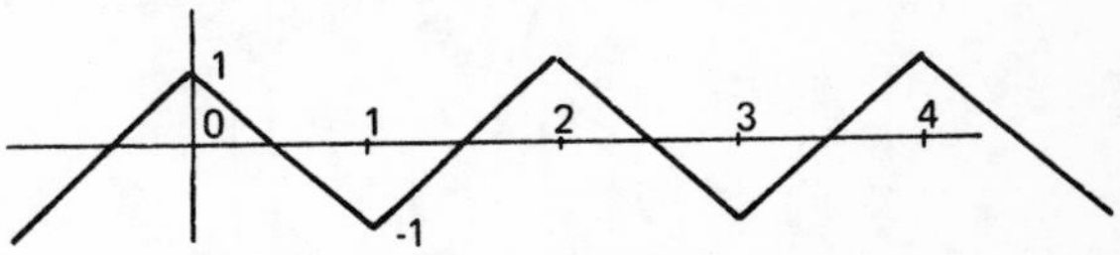

(d)
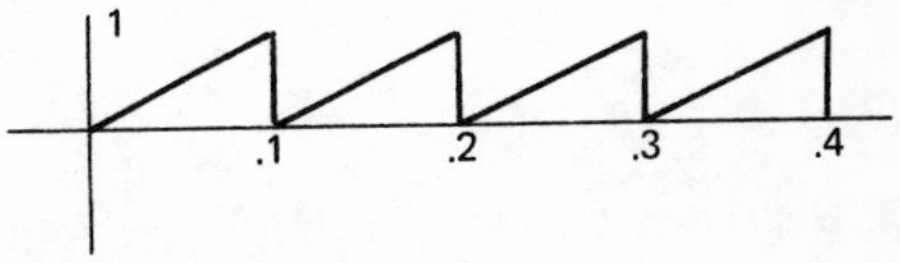

MODELS

18 Computer Science

PART 1: Fundamentals of Data Processing

1. Computer Evolution

The first totally electronic computer was *ENIAC,* built jointly by the Moore School of Electrical Engineering at the University of Pennsylvania and the Ballistics Research Laboratories in Aberdeen, Maryland. *ENIAC* was an acronym for Electronic Numerical Integrator and Computer. Built between 1943 and 1946 for the purpose of solving ballistics problems, its 18,000 vacuum tubes and hard-wired programs were obsolete upon completion of construction.

In 1946 the Moore School began constructing *EDVAC* (Electronic Discrete Variable Automatic Computer) which utilized stored programs. Since it was not completed until 1952, a similar computer completed in 1949 in England was the first stored-program computer.

In 1947 work was begun at M.I.T. on *WHIRLWIND I,* which had electrostatic storage and magnetic core. The core memories later used by IBM and others were based on *WHIRLWIND I* designs.

UNIVAC I (Universal Automatic Computer), built in 1951 by two of the *EDVAC* producers, John Mauchly and J. Presper Eckert as the Eckert-Mauchly Computer Corporation, was the first commercial computer in the United States. For years it was the best system available due to self-checking features and magnetic tape memories.

IBM Corporation was formed in 1924 to produce punched card equipment. Its first commercial data processing electronic computer, the *Model 702,* was introduced in 1955 but was slower than *UNIVAC I.* In 1964 IBM announced its *System 360,* first of the commercial computers to use integrated circuits extensively.

2. Numbering Systems

A. The Binary System

Although the input and output of computers are usually expressed in decimal numbers, all modern computers operate on *binary digits (bits).* Only two digits, *zero* and *one*, are allowed in the binary number system.[1] Thus, all binary numbers are composed of strings of zeros and ones.

In general, a base-*b* number can be converted to a base-10 number by performing the following calculation with base-10 arithmetic.

$$(a_n a_{n-1} \cdots a_2 a_1 a_0)_b = a_n b^n + \cdots + a_2 b^2 + a_1 b^1 + a_0 b^0 \qquad 18.1$$

Since the left-most bit in a binary number contributes the greatest to the number's magnitude, it is known as the *most significant bit (MSB).* Conversely, the right-most bit is known as the *least significant bit (LSB).*

Example 18.1

Convert $(1011)_2$ to base 10.

$$(1)2^3 + (0)2^2 + (1)2^1 + (1)2^0 = 11$$

As with digits from other numbering systems, bits can be added, subtracted, multiplied, and divided. Only the characters 0 and 1 are allowed in the results. The following truth table can be used to predict the results of an addition operation.

Table 18.1
Binary Truth Table for Addition

A	B	A+B
0	0	0
0	1	1
1	0	1
1	1	0 carry 1

[1]Alternatively, the binary states may be called *on* and *off, high* and *low,* or *positive* and *negative.*

COMP SCI

Conversions from base-10 numbers to base-b numbers can be accomplished by the *remainder method*. This method consists of successive divisions by the base b. The base-b number is found by taking the remainders in reverse order.

Example 18.2

Use the remainder method to convert $(75)_{10}$ to base-2.

$75 \div 2 = 37$ remainder 1
$37 \div 2 = 18$ remainder 1
$18 \div 2 = 9$ remainder 0
$9 \div 2 = 4$ remainder 1
$4 \div 2 = 2$ remainder 0
$2 \div 2 = 1$ remainder 0
$1 \div 2 = 0$ remainder 1

The binary representation of $(75)_{10}$ is $(1001011)_2$.

B. The Octal System

Since binary numbers are long and difficult to work with, the *octal* (base-8) *system* is frequently used when working with large amounts of computer output. Only the digits 0 through 7 are allowed in the octal system.

The rules for addition in the octal system are the same as for the decimal system except that the numbers eight and nine do not exist.

$$7+1 = 6+2 = 5+3 = 10$$
$$7+2 = 6+3 = 5+4 = 11$$
$$7+3 = 6+4 = 5+5 = 12$$

Example 18.3

Perform the following operations:

(a) $(2)_8 + (5)_8$
(b) $(7)_8 + (6)_8$
(c) Convert $(75)_{10}$ to base-8.
(d) Convert $(13)_8$ to base-10.

(a) The sum of 2 and 5 in base-10 is 7. 7 is one of the digits in the octal system. Therefore, the answer is $(7)_8$.

(b) The sum of 7 and 6 in base-10 is 13. 13 must be converted to base-8.

$13 \div 8 = 1$ remainder 5
$1 \div 8 = 0$ remainder 1

The answer is $(15)_8$.

(c) $75 \div 8 = 9$ remainder 3
$9 \div 8 = 1$ remainder 1
$1 \div 8 = 0$ remainder 1

The answer is $(113)_8$.

(d) Use the positional numbering method (equation 18.1).

$$(1)(8)^1 + 3(8)^0 = (11)_{10}$$

The octal system is closely related to the binary system since $(2)^3 = 8$. Conversions between the binary and octal systems can be accomplished by starting at the LSB (right-most bit) and grouping the bits in three's. Each group of three bits corresponds to an octal digit.

Example 18.4

Convert $(5431)_8$ to base-2.

$(5)_8 = (101)_2$
$(4)_8 = (100)_2$
$(3)_8 = (011)_2$
$(1)_8 = (001)_2$

The answer is $(101100011001)_2$.

Example 18.5

Convert $(1001011)_2$ to base-8.

First, start at the LSB and group the bits in three's.

1 001 011

Next, convert the groups of three's to octal.

$(1)_2 = (1)_8$
$(001)_2 = (1)_8$
$(011)_2 = (3)_8$

The answer is $(113)_8$.

C. The Hexadecimal System

The *hexadecimal (base-16) system* is a shorthand method of representing the value of four binary digits at a time. To represent the numbers 10 through 15, the capital letters A through F are used. The progression of hexadecimal numbers is illustrated in table 18.2.

Conversions from hexadecimal to base-10 can be accomplished by use of equation 18.1. Conversions to base-16 are more difficult. Care must be taken to convert base-10 digits to the appropriate hexadecimal symbols. This is illustrated in examples 18.6 and 18.7.

Example 18.6

Convert $(4D3)_{16}$ to base-10.

Hex D is $(13)_{10}$.

$$4(16)^2 + 13(16)^1 + 3(16)^0 = (1235)_{10}$$

COMP SCI

Table 18.2
Decimal and Hexadecimal Equivalents

Decimal	Hexadecimal	Decimal	Hexadecimal
0	0	16	10
1	1	17	11
2	2	18	12
3	3	19	13
4	4	20	14
5	5	21	15
6	6	22	16
7	7	23	17
8	8	24	18
9	9	25	19
10	A	26	1A
11	B	27	1B
12	C	28	1C
13	D	29	1D
14	E	30	1E
15	F	31	1F

Example 18.7

Convert $(1475)_{10}$ to base-16.

Use the remainder method.

$1475 \div 16 = 92$ remainder 3

$92 \div 16 = 5$ remainder 12

$5 \div 16 = 0$ remainder 5

Since $(12)_{10}$ is $(C)_{16}$, the answer is $(5C3)_{16}$.

Conversions between the hexadecimal and binary systems are not difficult. It takes four bits to represent the entire range of a single hexadecimal digit. Conversions can be accomplished by grouping the bits into four's, starting with the LSB.

Example 18.8

Convert $(1011111101111001)_2$ to hexadecimal.

First, group the bits in four's.

1011 1111 0111 1001

Next convert the groups to their hexadecimal equivalents.

$(1011)_2 = (B)_{16}$

$(1111)_2 = (F)_{16}$

$(0111)_2 = (7)_{16}$

$(1001)_2 = (9)_{16}$

The answer is $(BF79)_{16}$.

3. Conversions Involving Fractions

Converting base-*b* fractions to base-10 is similar to converting whole numbers and is accomplished by equation 18.2.

$$(.a_1a_2\cdots a_m)_b = a_1(b^{-1}) + a_2(b^{-2}) + \cdots + a_m(b^{-m}) \quad 18.2$$

Example 18.9

Convert $(27.52)_8$ to base-10.

$$2(8)^1 + 7(8)^0 + 5(8)^{-1} + 2(8)^{-2} = (23.656)_{10}$$

Converting a base-10 fraction to base-*b* is somewhat more involved. The procedure involves multiplication of the base-10 fraction and subsequent fractional parts by the base. The base-*b* fraction is formed from the integer parts of the products. This is illustrated by examples 18.10 and 18.11.

Example 18.10

Convert $(.14)_{10}$ to base-8.

$.14 \times 8 = 1.12$

$.12 \times 8 = 0.96$

$.96 \times 8 = 7.68$

$.68 \times 8 = 5.24$

$.24 \times 8 =$ etc.

The answer is taken from the integer parts of the products. This answer is $(.1075\cdots)_8$.

Example 18.11

Convert $(.8)_{10}$ to base-16.

$.8 \times 16 = 12.8$

$.8 \times 16 = 12.8$

$.8 \times 16 =$ etc.

Since $(12)_{10}$ is $(C)_{16}$, the answer is $(.CCCCC\cdots)_{16}$.

4. Computer Arithmetic

Additive computers perform all arithmetic operations by adding. For example, subtraction is accomplished by adding a negative number. There are also *subtractive computers* which do all operations by subtraction.

A. Complements in Base-10

Many computer operations are performed with *complements*. The complement of a number *A* depends on the computer it is formed on. Each computer has a maximum number, *n*, of digits per integer number stored. For machines working in base-10, the complements defined by equations 18.3 and 18.4 are applicable. All of the complements are expressed in base-10.

$$\text{10's complement:} \quad A_{10}^* = 10^n - A \quad 18.3$$

$$\text{9's complement:} \quad A_9^* = A_{10}^* - 1 \quad 18.4$$

The application of complements to computer arithmetic depends on equations 18.5 and 18.6.

$$(A^*)^* = A \quad 18.5$$

$$B - A = B + A^* \quad 18.6$$

COMP SCI

Example 18.12

Find the difference (18 − 6) by simulating the operation on a four-digit machine. Use base-10 complements.

The 10's complement of 6 is

$$A^*_{10} = (10)^4 - 6 = 10{,}000 - 6 = 9994$$

Using equation 18.6,

$$18 - 6 = 9994 + 18 = 10012$$

However, the machine has a maximum capacity of four digits. Therefore, the leading *one* is dropped, leaving 0012 as the correct answer.

B. Complements in Base-2

The *two's complement* of a binary number *A* can be found from equation 18.7. *n* is the maximum number of bits in an integer.

$$A^*_2 = 2^n - A \qquad 18.7$$

The *one's complement* is found from equation 18.8. The one's complement is most easily formed by substituting ones and zeros for all of the zeros and ones in *A* respectively.

$$A^*_1 = A^*_2 - 1 \qquad 18.8$$

Example 18.13

What is the difference $(01101)_2 - (01010)_2$ in a five-bit machine?

Use two's complements.

$$(2)^5 = (32)_{10} = (100000)_2$$

The two's complement of $(01010)_2$ is

$$A^*_2 = (100000)_2 - (01010)_2 = (10110)_2$$

Using equation 18.6,

$$(01101)_2 - (01010)_2 = (01101)_2 + (10110)_2 = (100011)_2$$

Since the machine has a capacity of only 5 bits, the left-most bit is dropped, leaving $(00011)_2$ as the correct difference.

COMP SCI

The one's complement (equation 18.8) is easier to form than the two's complement. It can be combined with a technique known as *end-around carry* to perform subtraction. End-around carry is the adding of the overflow bit to the sum of *A* and its one's complement.

Example 18.14

Calculate the difference $(01101)_2 - (01010)_2$ in a five-bit machine using the one's complement and end-around carry.

A^*_1 of $(01010)_2$ is found by switching all ones to zeros and all zeros to ones. Thus, $A^*_1 = (10101)_2$.

The one's complement is then added.

$$(01101)_2 + (10101)_2 = (100010)_2$$

The leading bit is known as the *overflow bit.* It is removed and added to give the correct difference.

$$(00010)_2 + (1)_2 = (00011)_2$$

5. Coding Systems

A. BCD Codes

Since a computer can only handle binary numbers, all symbolic data such as alphabetic and special characters must be represented by a binary numeric code. The actual code used is computer-dependent.

A code used in the past is *Binary Coded Decimal (BCD).*[2] Each alphanumeric character is represented by six bits. The two left-most bits are called *zone bits* and the remaining bits are *numeric bits.*

Table 18.3
The Standard BCD 6-Bit Codes

Character	BCD Code	Character	BCD Code
0	001010	I	111001
1	000001	J	100001
2	000010	K	100010
3	000011	L	100011
4	000100	M	100100
5	000101	N	100101
6	000110	O	100110
7	000111	P	100111
8	001000	Q	101000
9	001001	R	101001
		S	010010
A	110001	T	010011
B	110010	U	010100
C	110011	V	010101
D	110100	W	010110
E	110101	X	010111
F	110110	Y	011000
G	110111	Z	011001
H	111000		

B. ASCII Codes

The American Standard Code for Information Interchange (ASCII)[3] code is a seven-bit code developed for transmitting and processing data. It is commonly used in machines employing eight-bit microprocessors.[4] Table 18.4 gives the ASCII character codes in base-10. The 7-bit ASCII form can be found by converting the base-10 number to base-2. The meanings of the control codes are given in table 18.5.

[2]The need for additional symbols and minor changes to the original Hollerith codes have stopped BCD coding from achieving extensive usage in modern computers.

[3]Pronounced ask′-key.

[4]The use of the remaining left-most bit has not been standardized among microcomputer manufacturers.

Table 18.4
The ASCII Codes

ASCII Code	Character	ASCII Code	Character	ASCII Code	Character
000	NUL	043	+	086	V
001	SOH	044	,	087	W
002	STX	045	-	088	X
003	ETX	046	.	089	Y
004	EOT	047	/	090	Z
005	ENQ	048	0	091	[
006	ACK	049	1	092	\
007	BEL	050	2	093	]
008	BS	051	3	094	∧
009	HT	052	4	095	_
010	LF	053	5	096	\
011	VT	054	6	097	a
012	FF	055	7	098	b
013	CR	056	8	099	c
014	SO	057	9	100	d
015	SI	058	:	101	e
016	DLE	059	;	102	f
017	DC1	060	<	103	g
018	DC2	061	=	104	h
019	DC3	062	>	105	i
020	DC4	063	?	106	j
021	NAK	064	@	107	k
022	SYN	065	A	108	l
023	ETB	066	B	109	m
024	CAN	067	C	110	n
025	EM	068	D	111	o
026	SUB	069	E	112	p
027	ESC	070	F	113	q
028	FS	071	G	114	r
029	GS	072	H	115	s
030	RS	073	I	116	t
031	US	074	J	117	u
032	SP	075	K	118	v
033	!	076	L	119	w
034	”	077	M	120	x
035	#	078	N	121	y
036	$	079	O	122	z
037	%	080	P	123	{
038	&	081	Q	124	|
039	’	082	R	125	}
040	(	083	S	126	~
041	)	084	T	127	DEL
042	*	085	U		

Table 18.5
The ASCII Control Codes

NUL —Null
SOH —Start of Heading
STX —Start of Text
ETX —End of Text
EOT —End of Transmission
ENQ —Enquiry
ACK —Acknowledge
BEL —Bell
BS —Backspace
HT —Horizontal Tabulation
LF —Line Feed
VT —Vertical Tabulation
FF —Form Feed
CR —Carriage Return
SO —Shift Out
SI —Shift In
DLE —Data Line Escape
DC —Device Control
NAK —Negative Acknowledge
SYN —Synchronous Idle
ETB —End of Transmission Block
CAN —Cancel
EM —End of Medium
SUB —Substitute
ESC —Escape
FS —File Separator
GS —Group Separator
RS —Record Separator
US —Unit Separator
SP —Space
DEL —Delete

Table 18.6
The Standard EBCDIC Codes

Character	Binary Code	Character	Binary Code	Character	Binary Code
Blank	0100 0000	d	1000 0100	G	1100 0111
¢	0100 1010	e	1000 0101	H	1100 1000
.	0100 1011	f	1000 0110	I	1100 1001
<	0100 1100	g	1000 0111	J	1101 0001
(	0100 1101	h	1000 1000	K	1101 0010
+	0100 1110	i	1000 1001	L	1101 0011
|	0100 1111	j	1001 0001	M	1101 0100
&	0101 0000	k	1001 0010	N	1101 0101
!	0101 1010	l	1001 0011	O	1101 0110
$	0101 1011	m	1001 0100	P	1101 0111
*	0101 1100	n	1001 0101	Q	1101 1000
)	0101 1101	o	1001 0110	R	1101 1001
;	0101 1110	p	1001 0111	S	1110 0010
¬	0101 1111	q	1001 1000	T	1110 0011
–	0110 0000	r	1001 1001	U	1110 0100
/	0110 0001	s	1010 0010	V	1110 0101
,	0110 1011	t	1010 0011	W	1110 0110
%	0110 1100	u	1010 0100	X	1110 0111
_	0110 1101	v	1010 0101	Y	1110 1000
>	0110 1110	w	1010 0110	Z	1110 1001
?	0110 1111	x	1010 0111	0	1111 0000
:	0111 1010	y	1010 1000	1	1111 0001
#	0111 1011	z	1010 1001	2	1111 0010
@	0111 1100			3	1111 0011
'	0111 1101	A	1100 0001	4	1111 0100
=	0111 1110	B	1100 0010	5	1111 0101
"	0111 1111	C	1100 0011	6	1111 0110
a	1000 0001	D	1100 0100	7	1111 0111
b	1000 0010	E	1100 0101	8	1111 1000
c	1000 0011	F	1000 0110	9	1111 1001

C. EBCDIC Codes

The Extended Binary Coded Decimal Interchange Code (EBCDIC)[5] is now in widespread use. Its 8-bit coding system allows a maximum of 256 different characters to be represented, including special characters and upper and lower case letters.

D. Packed Decimal

Base-10 digits can be converted to *packed decimal format* by dividing their EBCDIC codes into two strings of four bits each. The two strings are then converted to hexadecimal.

Example 18.15

Represent the number $(7)_{10}$ in packed decimal format.

$(7)_{10}$ in EBCDIC is $(11110111)_2$. $(1111)_2 = (F)_{16}$. $(0111)_2 = (7)_{16}$. Therefore, the packed decimal representation is F7.

E. Hollerith Code

The standard for data entry in card form is the 80-column card, also called the *Hollerith card* after its inventor, Herman Hollerith. The 80-column card is divided horizontally into 12 rows. The first row is the 12-row. Following are the 11-row, the 0-row, and rows 1 through 9. Rows 12, 11, and 0 are known as the *zone rows.* The top and bottom edges of the card are known as the *12-edge* and the *9-edge* respectively.

Figure 18.1 80-Column Card

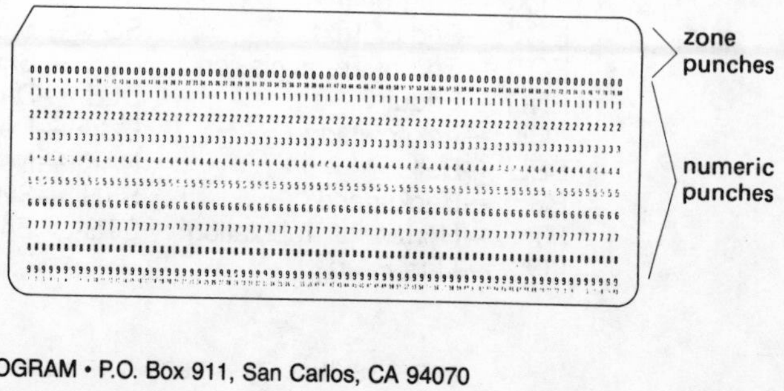

[5]Pronounced ep′-sih-dik.

COMP SCI

Table 18.7
The Hollerith Code

Character	Hollerith Code	Character	Hollerith Code
0	0	W	0-6
1	1	X	0-7
2	2	Y	0-8
3	3	Z	0-9
4	4		
5	5	&	12
6	6	¢	12-2-8
7	7	.	12-3-8
8	8	<	12-4-8
9	9	(	12-5-8
		+	12-6-8
A	12-1	|	12-7-8
B	12-2	-	11
C	12-3	!	11-2-8
D	12-4	$	11-3-8
E	12-5	*	11-4-8
F	12-6	)	11-5-8
G	12-7	;	11-6-8
H	12-8	¬	11-7-8
I	12-9	/	0-1
J	11-1	,	0-3-8
K	11-2	%	0-4-8
L	11-3	_	0-5-8
M	11-4	>	0-6-8
N	11-5	?	0-7-8
O	11-6	:	2-8
P	11-7	#	3-8
Q	11-8	@	4-8
R	11-9	'	5-8
S	0-2	=	6-8
T	0-3	"	7-8
U	0-4	Blank	
V	0-5		

6. *Storage of Integer and Real Numbers*

The IBM 360 and 370 series computers, to which this section is exclusively devoted, normally use a four-byte word, each byte containing eight bits. (There are actually nine bits per byte, but the ninth is a *parity bit* and is omitted in this discussion.) Storage of FORTRAN variables is in the four-byte word form whenever *implicit typing* is used or whenever *explicit* type statements appear in the program.

REAL **[list]**
REAL*4 **[list]**
} identical

INTEGER **[list]**
INTEGER*4 **[list]**
} identical

Double precision (long floating point) variables can be obtained by the use of the REAL*8**[list]** declaration. If the program never encounters integer variables greater than 32,767 ($2^{15} - 1$) and core space is limited, the INTEGER*2 declaration can be used. Execution time will increase (compared to single precision execution time) as it does with double precision variables.

Four-byte integers are stored as 31-bit binary code. The left-most (32nd) bit is used as a sign bit—*zero* for +, *one* for −. There are 2^{31} = 2,147,483,648 possible base-10 integers, of which *zero* is one. Therefore, the largest four-byte integer is 2,147,483,647. Negative numbers are represented with a *one* sign bit in two's complement form.

Real numbers are represented by a fractional part and an exponent. Four-byte floating point numbers are contained in 32 bits, the left-most of which is the sign bit. The next seven bits in both short and long floating point numbers contain the *characteristic* or *exponent.* These seven bits do not include a sign bit. Rather, the shift direction is controlled by the value contained in the seven bits.

The entire range which can be contained in the seven bits is 0 (no bits on) to 127 (all bits on). The middle is 64, which indicates a *zero* exponent. Any number higher than 64 causes a shift to the right of the hexadecimal decimal point (positive exponent) by the difference from 64. Totals of the seven bits less than 64 cause left-shifting (negative exponent). This method is known as *excess-64 scaling.*

The remaining bytes (three for single precision and seven for double precision) contain the fractional part. Using this method, the limits on floating point numbers are approximately EE−78 and EE+75.

All calculations and shifts are done in hexadecimal, as shown in the following example.

Example 18.16

What base-10 number is contained in the four bytes represented by the double precision hexadecimal number 3C333000?

step 1: Convert the two left-most bytes into binary.

$$(3C)_{16} = (0011 \quad 1100)_2$$

step 2: The left-most *zero* bit indicates a positive number. The remaining seven bits total 60. A shift of four hexadecimal decimal places is required.

$$(0111100)_2 = (60)_{10}$$

step 3: Shift the decimal point of the seven right-most bytes four spaces.

.333000 becomes .0000333000

step 4: Convert $(.0000333)_{16}$ to base-10.

$$3(1/16)^5 + 3(1/16)^6 + 3(1/16)^7 = (.000003051)_{10}$$

Notice that the hexadecimal decimal point was moved four spaces. The binary decimal place was moved sixteen spaces.

COMP SCI

The total number of storage locations in a computer can be specified either in bytes or words. This number is always a multiple of 1024, which is written as $1K$. Thus a $64K$ memory would contain 64 times 1024 or 65,536 storage locations.[6]

Vacuum tubes were originally used as the bits in computer memories. Later, the memory *core* received its name from ferrite cores used to record data. The direction of the magnetic current in a core is the *polarity* and can be changed by applying an electric current to wires running through the core's center.

Each core has four wires passing through its center. The x and y wires are used to change the core's polarity. When this change is desired, half of the *critical current* is applied to both the x and y wires. At their junction, the two half-currents superimpose into a critical value, changing the polarity. The *sense* and *inhibit* wires are used to detect and stop polarity reversals respectively.

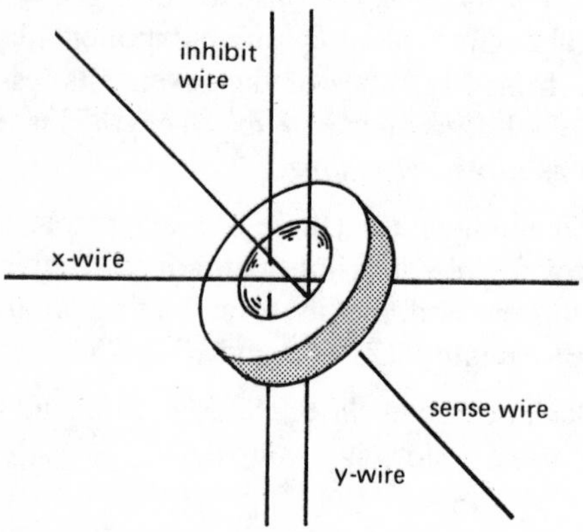

Figure 18.2 A Magnetic Core

Modern computers do not use ferrite cores. Rather, their memories are semiconductor based. This affords considerable size reduction. Semiconductor storage is designated as RAM (Random Access Memory), ROM (Read-Only Memory), PROM (Programmable Read-Only Memory) and EPROM (Erasable Programmable Read-Only Memory). EPROMS can be erased by exposure to ultraviolet light.

Memory is said to be *volatile* if the contents are lost when power is turned off. *Non-volatile* memory retains its contents in the absence of power.

7. Computer Hardware

Most electronic computers available today consist of a control unit, an arithmetic/logical unit, memory, and input/output devices. The term *Central Processing Unit (CPU)* is used to denote the combination of the control unit and the arithmetic/logical unit.

The operation of the computer is governed by the *control unit.* Its purpose is to determine the instruction to be executed next, locate and fetch the instruction, and to interpret the instruction. Execution of the instruction occurs with the aid of the arithmetic unit.

The *arithmetic unit* performs the arithmetic operations directed by the control unit. Once the operation is completed, the results are automatically returned to a memory location.

The primary or central *memory* consists of storage locations, each of which has an address. The contents of each storage location may change, but the address may not. Each memory location has the same number of bits. A group of eight bits together forms a *byte*. Each byte can contain one alphanumeric character.

A computer *word* is the smallest number of bytes that can be used in arithmetic operations. One-byte words are used in microprocessor-based machines. Two-byte words are common in mini-computers although four-byte words are most common in large computers. Most computers also allow a user to specify a double-precision word which doubles the number of bytes normally used.

The fourth part of a computer involves the devices by which information is fed to and received from the computer. These devices are called *input/output (I/O)* devices or *peripherals.* Card readers and printers are common peripherals. However, secondary storage devices such as magnetic tape and disk units are also included.

Peripherals are connected to the computer through *channels.* Selector channels transfer data to and from one I/O device at a time. Although more than one device can be attached to a channel, only one can operate at a time. When the transfer is completed, the channel signals *(interrupts)* the computer.

If the channel is a multiplexer channel, more than one I/O device can appear to operate at a time. If the transfer of data occurs byte-by-byte, the operation is said to be in *byte mode.* Fast I/O devices, such as magnetic tape units, operate in *burst mode,* monopolizing the channel until the transfer is completed.

8. External Storage Devices

Magnetic tape is a *sequential-access* storage medium since it must be written to and read from sequentially. Two recording methods are used. In *Non-Return-To-Zero (NRZI)* recording, only *one*-bits are recorded while a lack of pulse indicates a *zero* bit. In *phase recording,* both *one* and *zero* bits are recorded.

A tape is usually divided into nine *tracks* which run the length of the tape. The width of the tape is divided into *frames* or *characters*. One track is used to record a *par-*

[6]Computer manufacturers use the abbreviations *KB* and *KW* to mean kilo-bytes and kilo-words respectively.

COMP SCI

ity bit. The remaining eight tracks are used to record the data.

If the tape is recorded with *odd-parity,* the frame will always contain an odd number of *one* bits. When the tape is read, each frame is checked to ensure the validity of the data.

Data is transferred to a tape in blocks, with *interblock gaps* of .6 to .75 inches. Typical densities are 200, 556, 800, and 1600 frames per inch. Typical speeds are 75, 112.5, and 200 inches per second. Reflective spots for photoelectric detection are used to indicate the beginning and end of magnetic tapes. These are called *load-points* and *end-of-file markers* respectively.

Magnetic drums are random-access memories since the data can be retrieved in any order. The drum is magnetized over its entire surface and turns at a fixed speed under read/write heads. The band covered by each head is known as a *track* or *channel.* NRZI recording is used. Very large drums may have fewer heads than tracks if the heads are moveable.

Average access time for a magnetic drum with fixed heads is equal to one-half of the time required per revolution, called the *average rotational delay.* If the heads are shared between tracks, the access time also includes the average movement time.

Magnetic disks are also random access storage devices. Disks turn at constant speed in the 2000-3000 rpm range and are accessed by read/write heads on moveable arms. Each disk is divided into concentric *tracks* and pie-shaped *sectors.*

Hard disks can be fixed (not removable) or removable. Removable disks require a case to prevent surface contamination. The combination of case and one or more platters is called a *disk pack. Winchester*[7] *technology* consists of one or more 5¼″, 8″, or 14″ diameter hard disks in a non-removable drive.

Floppy disks (also known as *diskettes*) are removable storage devices. Although their capacity is comparatively low (less than a million characters), they are popular with small computer systems. Floppy disks come in two sizes — 5¼″ and 8″ diameter. The capacity of a floppy disk is dependent on its size, the recording density, and the number of sides used. Single, double, and quad density recording methods are used.

Hard disks and some floppy disks are *soft-sectored.* This means that the sequential sectors in any given track are indicated by magnetic markers. *Hard-sectoring* means that the sectors are indicated by small holes located around the hub of the floppy disk.

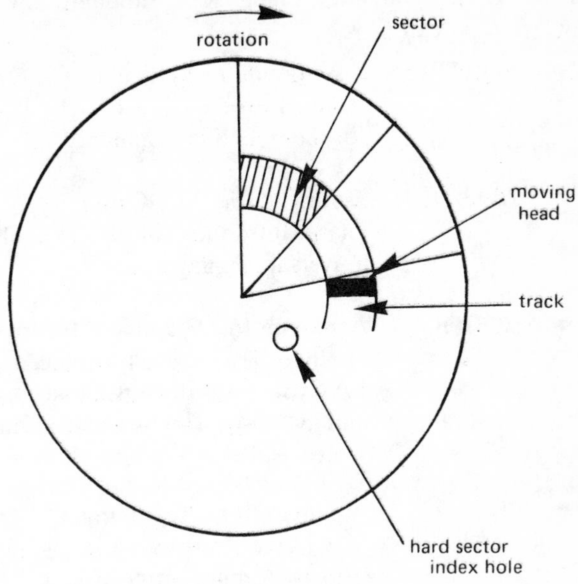

Figure 18.3 Tracks and Sectors

9. Computer Software

Software is a name for the programs that are executed by the computer hardware. It may also relate to the *operating system (OS)* controlling the computer's operation. Software is divided into three types: machine language, assembly language, and high-level languages.

Machine language is compatible with the microprocessor. An instruction in machine language contains two parts: the operation to be performed and the operands expressed as storage locations. Each machine instruction must be expressed as a series of zeros and ones. Most coding, however, is done in octal or hex with translation to binary coding being provided by the computer.

Assembly language is one step higher in sophistication. Mnemonic codes are used to specify the operation to be performed. The hexadecimal machine instruction **1A** for adding two registers together on the IBM 370 system becomes **AR** in assembly language. The operands are referred to by their variable names, not their addresses. Assembly language is translated into machine language by an internal program called an *assembler.*

High-level languages are easier to use than assembly languages because the instructions are more like English. High-level languages must be either compiled or interpreted into machine code. An interpreter checks each instruction and converts it into machine code as the instruction is entered. A compiler performs the checking and conversion functions on all instructions at one time. Error messages are sent by both inter-

[7]IBM introduced an advanced fixed disk drive in 1973. According to legend, this drive was developed under the code name *Winchester.* Although IBM no longer uses the term *Winchester* to describe its products, the name has become synonymous with non-removable hard disk systems.

COMP SCI

preters and compilers if syntax or other error types are encountered.

There are many high-level languages, although only a few are used extensively.

- FORTRAN — Acronym for FORmula TRANslation. Introduced in 1957 by IBM for the scientific community.
- ALGOL — Acronym for ALGOrithmic Language. Introduced in 1958 as a fully universal language.
- COBOL — Acronym for COmmon Business Oriented Language. Introduced in 1958 to serve the business community. Uses English words and sentences.
- PL/1 — Acronym for Programming Language One. Introduced in 1966 by IBM as a multi-purpose language designed for both the scientific and business communities.
- BASIC — Acronym for Beginner's All-purpose Symbolic Instruction Code. The instructions and algebraic equations are English-like and similar to FORTRAN.
- APL — Acronym for A Programming Language. It has operators that carry out functions requiring dozens of statements in other languages.
- Pascal — This language was developed in 1968 and named after the 18th century French mathematician Blaise Pascal. It is essentially machine independent.
- Forth — This language was originally designed for process control.
- Ada — A language developed in the early 1980's for use in U.S. Department of Defense programs.
- Special-Purpose Languages
  * SIMULA, GPSS, SIMSCRIPT, and GASP are simulation languages.
  * LISP and SNOBOL are string and list processors.
  * RPG is a report program generator.

Programs submitted to a computer are called *source programs* or *source codes*. Translation of source programs into *object code* consisting of machine language is done by the compiler or interpreter (as opposed to an *assembler* for source code written in assembly language). Most high-level instructions will result in several machine language instructions.

10. Program Processing

Programs and data are processed on a computer in three main ways. These are batch mode, real-time mode, and time-sharing. Originally, the computer programmer was also the computer operator. Now, however, most programmers submit programs to a computer, which then groups (batches) the programs into efficient categories. This is called *batch processing* since all jobs of a particular type are batched together.

For computers which handle large numbers of programs, a *multiprocessing*[8] *(multiplexed)* capability is necessary. Under the supervision of the operating system, the computer may begin or continue working on a second program while it is waiting for an input or output request to be completed.

The operating system is notified by a hardware interrupt when the I/O request has been completed. An *interrupt* is a signal that stops the execution of the current program and then transfers control to another memory location. Other interrupts signal error conditions. These are typically division by zero, overflow, underflow, and syntax errors.

Teleprocessing is the addressing of a computer by a remote station. The user's terminal and the computer can be separated by many miles. The transmission between computer and terminal uses telephone lines, microwaves, or coaxial cables. Since these media transmit analog signals and the computer operates in a digital mode, a special conversion device called a *modem* (modulator-demodulator) or *data set* is required.

If the transmission is only in one direction, the telecommunication line is *simplex*. The line is *half-duplex* if data can be transmitted in both directions, but only in one direction at a time. With *duplex* transmission, (also known as *full-duplex*), data can be transmitted in both directions simultaneously.

Data transmitted on subvoice-grade lines can reach a 200 bit-per-second (*bps*) transfer rate. Voice-grade lines allow a transfer of 5000 bps. Higher rates, 50,000 to 500,000 bps, are possible on wide-band lines. Transmission speed is also reported in units of *baud,* where

[8]A distinction is made between multi-user and multi-tasking. A *multi-user system* allows different users to share the time on a CPU. A *multi-tasking system* requires multiple users to be executing distinctly different programs.

a baud is a change in the line-condition per second. If *zeros* and *ones* are the only possible characters, baud is the same as *bps*.

If one character is transmitted at a time, the transmission is *asynchronous* or *start-stop*. Each character is preceded and followed by special signals. *Asynchronous equipment* is cheaper than *synchronous equipment* which transmits a block of data continuously without pause. Synchronous transmission is preceded by and interwoven with special syncronizing characters.

Errors in transmission can occur at any rate of 1 in 10,000 or higher. One way of checking the transmisison is to have the receiver send the data back to the sender. This process is known as *loop checking* or *echo checking*.

11. Microcomputers

Microcomputers have essentially the same structure as large mainframes. They have a CPU, an arithmetic/logical unit, memory, and I/O devices. The primary differences involve word size, computational speed, capabilities, and cost.

Typical word size for a microcomputer is eight or sixteen bits. Since the word size limits the memory address, most microcomputers have a maximum addressable memory size of 64K. The memory is typically of the semiconductor type.

12. Microprocessors

A microprocessor is the processing unit in a computer system. Since *large-scale integration (LSI)* technology is used, most microprocessors can be completely contained in a single chip.

The most popular microprocessors are built around eight-bit words. These include Intel's 8080 family, Motorola's 6800 family, and Zilog's Z-80 family. Each of these three microprocessors uses sixteen address lines, allowing 65,536 words of memory to be directly addressed.

Microprocessors communicate with the support chips through *buses*.[9] The *address bus* directs memory and input/output device transfers. The *data bus* actually carries the eight-bit byte data transfers. The *control bus* communicates control and status information.

[9]The spelling *buss* is also used.

COMP SCI

PART 2: FORTRAN Programming

The FORTRAN language currently exists in several versions. Although differences exist between compilers, these are relatively minor. However, some of the instructions listed in this chapter may not be compatible with all compilers.

This section is not intended as instruction in FORTRAN programming, but rather serves as a documentation of the language.

1. Structural Elements

Symbols are limited to the upper-case alphabet, digits 0 through 9, the blank, and the following special characters:

+ = − * / ( ) , . $

Statements written with these characters are generally prepared in an 80-column format. Statements are executed sequentially regardless of the statement numbers.

position	use
1	The letter *C* is used for a *comment.* Comments are not executed.
2–5	The statement number, if used, is placed in positions 2 through 5. Statement numbers can be any integer from 1 through 9999.
6	Any character except *zero* can be placed in position 6 to indicate a continuation from the previous statement.
7–72	The FORTRAN statement is placed here.
73–80	These positions are available for any use and are ignored by the compiler. Usually the final debugged program is numbered sequentially in these positions.

FORTRAN compilers pack the characters. Therefore, blanks may be inserted at any place in most statements. For example, the following statements are compiled the same way:

```
IF (AGE.LT.YEARS) GO TO 10
IF(AGE.LT.YEARS)GOTO10
```

2. Data

Numerical data can either be real or integer. *Integers* are usually limited to nine digits. Unsigned integers and integers preceded by a plus sign are the same. Commas are not allowed in integer constants. For example, ninety thousand would be written as 90000, not 90,000.

Real numbers are distinguished from integers by a decimal point and may contain a fractional part. Scientific notation is indicated by the single letter *E*. Real numbers are limited to one decimal point and usually seven digits.

value	FORTRAN Notation
2 million	2. E6
.00074	7.4 E−4
2.	2.

3. Variables

Variable names can be formed from up to six alphanumeric characters. The first character must be a letter. Variable names starting with the letters *I,J,K,L,M,* or *N* are assumed by the compiler to be integers unless defined otherwise by an *explicit typing statement.* All other variable names represent real variables.

The type convention can be overridden in an explicit typing statement. This is done by defining the desired variable type in the first part of the program with an INTEGER or REAL statement. For example, the statements

```
INTEGER TIME,CLOCK
REAL INSTANT
```

would establish TIME and CLOCK as integer variables and INSTANT as a real variable. The order of such declarations is unimportant. Variables following the standard type convention (implicit typing) do not have to be declared.

Subscripted variables with up to seven dimensions are allowed. They must always be defined in size by the DIMENSION statement. For example, the statements

```
DIMENSION SAMPLE(5)
REAL DIMENSION INCOME(2,7)
```

would establish a 1×5 real *array* called SAMPLE and a 2×7 real array called INCOME. INCOME would have been an integer array without the REAL declaration.

Elements of arrays are addressed by placing the subscripts in parentheses.

```
SAMPLE(2)
INCOME(1,6)
```

The subscripts can also be variables. SAMPLE(K) would be permitted as long as *K* was defined, was between 1 and 5, and was an integer.

Variables and arrays once defined and declared are not automatically initialized. If it is necessary to initialize a storage location prior to use, the DATA statement can be used. Consider the following statements:

COMP SCI

```
REAL X,Y,Z
DIMENSION ONEDIM(5)
DIMENSION TWODIM(2,3)
DATA X,Y,Z/3*0.0/(ONEDIM(I),I=1,5)/5*0.0/
1 ((TWODIM(I,J),I=1,2),J=1,3)/1.,2.,3.,4.,5.,6./
```

Variables X, Y, and Z will all be initialized to 0.0. The entries in ONEDIM will have the values (0,0,0,0,0). The TWODIM array will be initialized to

$$\begin{pmatrix} 1.0 & 2.0 & 3.0 \\ 4.0 & 5.0 & 6.0 \end{pmatrix}$$

After being initialized with a DATA statement, variables can have their values changed by arithmetic operations.

4. *Arithmetic Operations*

FORTRAN provides for the usual arithmetic operations. These are listed in table 18.8.

Table 18.8 FORTRAN Operators

symbol	meaning
=	replacement
+	addition
−	subtraction
*	multiplication
/	division
**	exponentiation
( )	preferred operation

The *equals* symbol is used to replace one quantity with another. For example, the following statement is algebraically incorrect. However, it is a valid FORTRAN statement.

```
Z = Z + 1
```

Each statement is scanned from left-to-right (except that a right-to-left scan is made for exponentiation). Operations are performed in the following order:

exponentiation first
multiplication and division second
addition and subtraction last

Parentheses can be used to modify this hierarchy.

Each operation must be explicitly and unambiguously stated. Thus, AB is not a substitute for $A*B$. Two operations in a row, as in $(A+-B)$ are also unacceptable. Some FORTRAN compilers allow mixed-mode arithmetic. Others, ANSI FORTRAN among them, require all variables in an expression to be either integer or real. Where mixed-mode arithmetic is permitted, care must be taken in the conversion of real data to the integer mode.

Integer variables used to hold the results of a mixed-mode calculation will have their values truncated. This is illustrated in the following example.

Example 18.17

Evaluate J in the expression

J = (6.0 + 3.0)*3.0/6.0 + 5.0 − 6.0**2.0

The expressions within parentheses are evaluated in the first pass.

J = 9.0*3.0/6.0 + 5.0 − 6.0**2.0

The exponentiation is performed in the second pass.

J = 9.0*3.0/6.0 + 5.0 − 36.0

The multiplication and division are performed in the third pass.

J = 4.5 + 5.0 − 36.0

The addition and subtraction are performed in the fourth pass.

J = −26.5

However, J is an integer variable, so the real number −26.5 is truncated and converted to integer. The final result is −26.

5. *Program Loops*

Loops can be constructed from IF and GO TO statements. However, the DO statement is a convenient method of creating loops. The general form of the DO statement is

$$\text{DO } s\ i = j,\ k,\ l$$

where s is a statement number.
i is the integer loop variable.
j is the initial value assigned to i.
k is an inclusive upper bound on i, which must exceed j.
l is the increment for i, with a default value of 1 if omitted.

The DO statement causes the execution of the statements immediately following it through statement s until i equals k or greater.[10] A loop can be *nested* by placing it within another loop. The loop variable may be used to index arrays.

When i equals or exceeds k, the statement following s is executed. However, the loop may be exited at any time before i reaches k if the logic of the loop provides for it.

[10]A peculiarity of FORTRAN *DO* statements is that they are executed at least once, regardless of the values of j and k.

COMP SCI

6. Input/Output Statements

The READ, WRITE, and FORMAT statements are FORTRAN's main I/O statements. Forms of the READ and WRITE statements are

READ (u_1,s) **[list]**
WRITE (u_2,s) **[list]**

where u_1 is the unit number designation for the desired input device, usually 5 for the card reader.

u_2 is the unit number designation for the desired output device, usually 6 for the line printer.

s is the statement number of an associated FORMAT statement.

[list] is a list of variables separated by commas whose values are being read or written.

The **[list]** can also include an implicit DO loop. The following example reads six values, the first five into the array PLACE and the last into SHOW.

```
READ (5,85) (PLACE(J), J=1,5) SHOW
```

The purpose of the FORMAT statement is to define the location, size, and type of the data being read. The form of the FORMAT statement is

s FORMAT **[field list]**

As before, s is the statement number. **[field list]** consists of specifications, set apart by commas, defining the I/O fields. **[list]** can be shorter than **[field list]**.

The format code for integer values is

nIw

n is an optional repeat counter which indicates the number of consecutive variables with the same format. w is the number of character positions.

The format codes for real values are

$nFw.d$ or $nEw.d$

w is again the number of character positons allocated, including the space required for the decimal point. d is the implied number of spaces to the right of the decimal point. In the case of input data, decimal points in any position take precedence over the value of d.

The F format will print a total of $(w-1)$ digits or blanks representing the number. The E format will print a total of $(w-2)$ digits or blanks and give the data in a standard scientific notation with an exponent.

Other formats which may be used in the FORMAT statement are

X horizontal blanks
/ skipping lines
H alphanumeric data
D double precision real
T position (column) indicator
Z hexadecimal
P decimal point modification
L logical data
' ' literal data

The usual output device is a line printer with 133 print positions. The first print position is used for *carriage control.* The data (control character) in the first output position will control the printer advance according to the rules in table 18.9.

Table 18.9
FORTRAN Printer Control Characters

control character	meaning
blank	advance one line
0	advance two lines
1	skip to line one on next page
+	do not advance (overprint)

Carriage control is usually accomplished by the use of literal data. Consider the following statements.

```
      INTEGER K
      K = 193
      WRITE (6,100) K
      WRITE (6,101) K
  100 FORMAT (' ', I3)
  101 FORMAT (I3)
```

The above program would print the number *193* on the next line of the current page and the number *93* on the first line of the next page.

Data can be written to or read from an array by including the array subscripts in the I/O statement.

```
      DIMENSION CLASS (2,5)
      READ (5, 15) ((CLASS(I,J), I=1,2), J=1,5)
   15 FORMAT (10F3.0)
```

7. Control Statements

The STOP statement is used to indicate the logical end of the program. The format is

s STOP

STOP should not be the last statement. When it is reached, program execution is terminated. The value of s is printed out or made available to the next program step. The use of STOP is rarely recommended.

The END statement is required as the last statement. It tells the compiler that there are no more lines in the program to be compiled. A program cannot be compiled or executed without an END statement.

The PAUSE statement will cause execution to temporarily stop. Its format is

s PAUSE

COMP SCI

When the PAUSE statement is reached, the number *s* is transmitted to the computer operator. This gives the operator a chance to set various control switches on the console (the choice of switches being dependent on the value of *s* and the program logic), prior to pushing the START button. The PAUSE statement is used only if the programmer is operating the computer.

The CALL statement is used to transfer execution to a *subroutine*. CALL EXIT will terminate execution and turn control over to the operating system. The CALL EXIT and STOP statements have similar effects. The RETURN statement ends execution of a program-called subroutine and passes execution to the main program. The CONTINUE statement does nothing. It can be used with a statement number as the last line of a DO loop.

The GO TO **[s]** statement transfers control to statement *s*.

The arithmetic IF statement is written

IF **[e]**s_1, s_2, s_3

[e] is any numerical variable or arithmetic expression, and the s_i are statement numbers. The transfer occurs according to the following table.

[e]	statement executed
[e] < 0	s_1
[e] = 0	s_2
[e] > 0	s_3

The logical IF statement has the form

IF **[le] [statement]**

[le] is a logical expression and **[statement]** is any executable statement except DO or IF. Only if **[le]** is true will **[statement]** be executed. Otherwise, the next instruction will be executed.

The logical expression **[le]** is a relational expression using one of several operators.

Table 18.10
FORTRAN Logical Operations

operator	meaning
.LT.	less than
.LE.	less than or equal to
.EQ.	equal to
.NE.	not equal to
.GT.	greater than
.GE.	greater than or equal to

Logical expressions can also incorporate the connectors .AND., .OR., and .NOT..

Example 18.18

IF (A.GT.25.6) A = 27.0

Meaning: If *A* is greater than 25.6, set *A* equal to 27.0.

IF (Z.EQ.(T−4.0).OR.Z.EQ.0.) GO TO 17

Meaning: If *Z* is equal to (T−4.0) or if *Z* is equal to *zero*, go to statement 17.

8. Library Functions

The following single-precision library functions are available. Most are accessed by placing the argument in parentheses after the function name. Placing the letter *D* before the function name will cause the calculation to be performed in double precision. Arguments for trigonometric functions are expressed in radians.

Table 18.11
Some FORTRAN Library Functions

function	use
EXP	e^x
ALOG	natural logarithm
ALOG10	common logarithm
SIN	sine
COS	cosine
TAN	tangent
SINH	hyperbolic sine
SQRT	square root
ASIN	arcsine
MOD	remaindering modulus (integer)
AMOD	remaindering modulus (real)
ABS	absolute value (real)
IABS	absolute value (integer)
FLOAT	convert integer to real
FIX	convert real to integer

9. User Functions

A user-defined function can be created with the FUNCTION statement. Such functions are governed by the following rules.

- The function is defined as a variable in the main program even though it is a function.
- When used in the main program, the function is followed by its arguments in parentheses.
- In the function itself, the function name is type-declared and defined by the word FUNCTION.
- The arguments (parameters) need not have the same names in the main program and function.
- Only the function has a RETURN statement.
- Both the main program and the function have END statements.
- The arguments (parameters) must agree in number, order, type, and length.

These construction rules are illustrated by example 18.19.

COMP SCI

Example 18.19

```
REAL HEIGHT, WIDTH, AREA, MULT
HEIGHT = 2.5
WIDTH = 7.5
AREA = MULT(HEIGHT,WIDTH)
END

REAL FUNCTION MULT(HEIGHT,WIDTH)
REAL HEIGHT, WIDTH
MULT = HEIGHT*WIDTH
RETURN
END
```

10. Subroutines

A subroutine is a user-defined sub-program. It is more versatile than a user-defined function as it is not limited to mathematical calculations. Subroutines are governed by the following rules.

- The subroutine is activated by the CALL statement.
- The subroutine has no type.
- The subroutine does not take on a value. It performs operations on the arguments (parameters) which are passed back to the main program.
- A subroutine has a RETURN statement.
- Both the main program and the subroutine have END statements.
- The arguments (parameters) need not have the same names in the main program and subroutine.
- The arguments (parameters) must agree in number, order, type, and length.
- It is possible to return to any part of the main program. It is not necessary to return to the statement immediately below the CALL statement.

These rules are illustrated by example 18.20.

Example 18.20

```
    REAL HEIGHT, WIDTH, AREA
    CALL GET(HEIGHT, WIDTH)
    AREA = HEIGHT*WIDTH
    END

    SUBROUTINE GET(A,B)
    REAL A,B
    READ (5,100) A,B
100 FORMAT(2F3.1)
    RETURN
    END
```

Variables in functions and subroutines are completely independent of the main program. Subroutine and main program variables which have the same names will not have the same values. A linkage between the main program and the subroutine can be established, however, with the COMMON statement.

The COMMON statement assigns storage locations in memory to be shared by the main program and all of its subroutines. Even the COMMON statement, however, allows different names. It is the order of the common variables which fixes their position in upper memory.

Example 18.21

What are the values of X and Y in the subroutine?

```
COMMON X, Y          main program
X = 2.0
Y = 10.0

COMMON Y, X          subroutine
```

Since Y is the first common subroutine variable which corresponds to X in the main program, $Y = 2.0$. Similarly, $X = 10.0$.

If variables are to be shared with only some of the subroutines, the *named* COMMON statement is required. Whereas there can be only one regular COMMON statement, there can be multiple named COMMON statements.

```
COMMON/PLACE/CAT, COW, DOG main program
COMMON/PLACE/HORSE, PIG, EXPENSE
                                subroutine
```

COMP SCI

PART 3: Analog Computers

Nomenclature

C	capacitance	farads
F(t)	forcing function	various
i	current	amps
K	gain	–
R	resistance	ohms
t	time	seconds
v	voltage	volts
x	dependent variable	various

Subscripts

B	point B
f	feedback
i	in
L	load
o	out
p	potentiometer

1. Introduction

Electrical circuits which model mechanical, fluid, and thermal systems are known as *analog models*. Voltages in analog models correspond (are analogous) to unknown variables in the systems being investigated. Thus, voltage may be used to model displacement in a mechanical system.

Analog models require many electrical components. A collection of amplifiers, potentiometers, dials, and other components can be housed in a single unit and connected together with patch cords. Such a collection is known as an *analog computer*. An analog computer can be configured to solve any number of differential equations.

2. Components of an Analog Computer

A. Potentiometer

Potentiometers can be used to reduce a voltage level. Since the output voltage is less than the input voltage, potentiometers are sometimes called *attenuators*. Figure 18.4 illustrates the symbol for a voltage being reduced by a potentiometer. *K* is the scalar multiplier setting of the potentiometer.

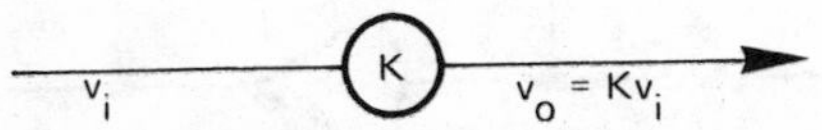

Figure 18.4 Symbol for Potentiometer

Potentiometers must be set to their value under load. Changing the load will change the setting. Figure 18.5 illustrates the electrical schematic of a potentiometer in a circuit.

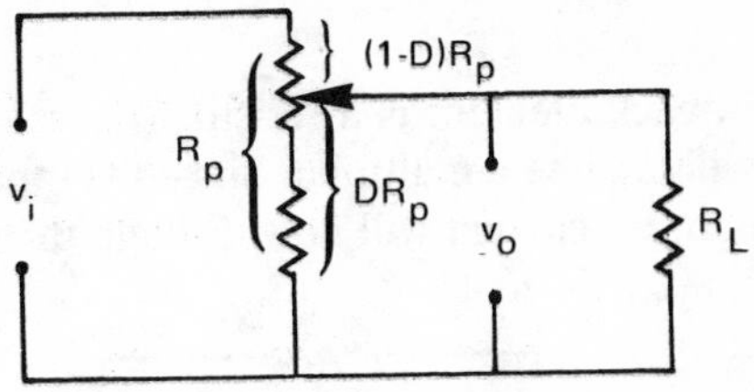

Figure 18.5 Potentiometer Circuit

The high end is the input and the low end is grounded. The slider is the output. R_L is usually an amplifier input resistor. If the unloaded pot setting is *D*, the loaded pot setting is

$$K = \frac{1}{\frac{1}{D} + \frac{(1-D)R_p}{R_L}} \qquad 18.9$$

B. High-Gain Amplifier

The inverting high-gain D.C. amplifier is illustrated in figure 18.6. It has a high gain — at least EE5, but probably around EE8. This gain is constant for all frequencies down to and including $\omega = 0$. The amplifier's input impedance is very high. The high input impedance limits the input current.[11]

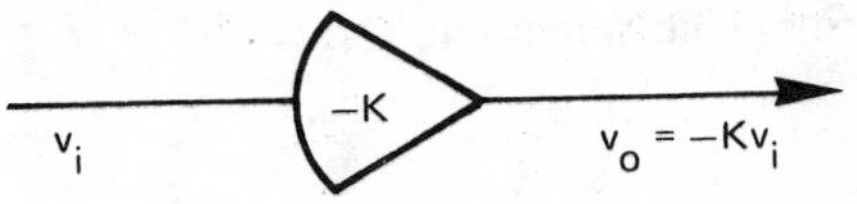

Figure 18.6 The High-Gain D.C. Amplifier

Most semiconductor devices work in the ± 10–13 volt range. Keeping v_o less than 10 volts with a gain of EE8 means that v_i will be very small. (v_i = 10/EE8 = .1 μV.) Since this is such a small voltage, the input is at a *virtual ground*.

C. Operational Amplifier

The operational amplifier (op-amp) is constructed by including a feedback element as shown in figure 18.7. The op-amp can be made to act like a summer, multiplier, inverter, integrator, or differentiator depending on the feedback element chosen.

[11]It is usually assumed that the input impedance is infinite. Therefore, the input current is zero. It is also usually assumed that the amplifier's output impedance is zero.

COMP SCI

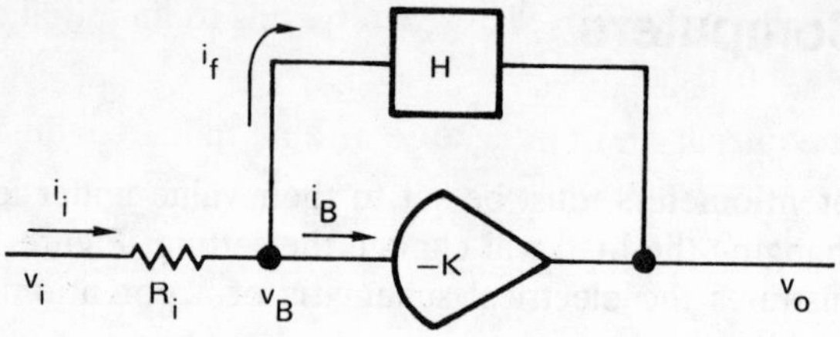

Figure 18.7 The Operational Amplifier

If the feedback element is a resistor (figure 18.8), the op-amp will act like a multipler. Since the input impedance is high, no current will flow through the high-gain amplifier. So, $i_i = i_f$.

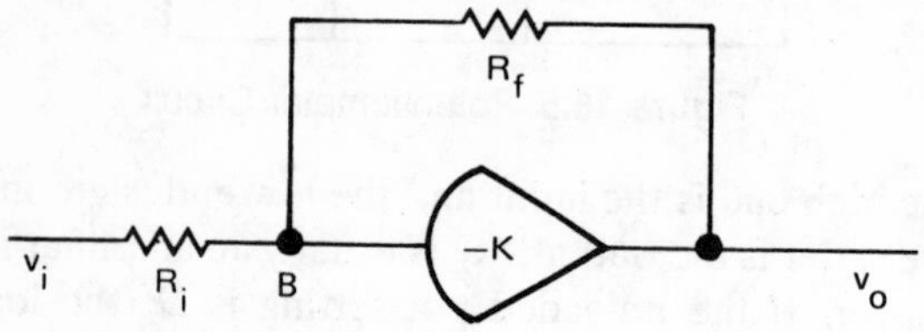

Figure 18.8 Op-Amp Used as a Multiplier

The currents i_i and i_f can be written in terms of the node voltages.

$$i_i = i_f \tag{18.10}$$

$$\frac{v_i - v_B}{R_i} = \frac{v_B - v_o}{R_f} \tag{18.11}$$

Since $v_o = (-K)v_B$, equation 18.11 can be written as

$$\frac{v_i}{R_i} + \frac{v_o}{KR_i} = \frac{-v_o}{KR_f} - \frac{v_o}{R_f} \tag{18.12}$$

Since K is a very large number, its appearance in the denominators of equation 18.12 effectively removes those terms. The remaining terms are

$$\frac{v_i}{R_i} = -\frac{v_o}{R_f} \tag{18.13}$$

Solving for the output voltage,

$$v_o = -\left(\frac{R_f}{R_i}\right)v_i \tag{18.14}$$

The quantity (R_f/R_i) is known as the *gain*. If R_f is greater than R_i, the op-amp multiplies by a number greater than one. This makes up for the potentiometer which can only scale down an input.

COMP SCI

D. Inverter

If R_f and R_i in equation 18.14 are equal, the output voltage will be the negative of the input voltage. This inversion results directly from the inversion performed by the high-gain amplifier.[12]

[12]Op-amps have both inverting and non-inverting output terminals. This chapter assumes that the inverting terminals are used.

Several symbols are used to represent an inverter.

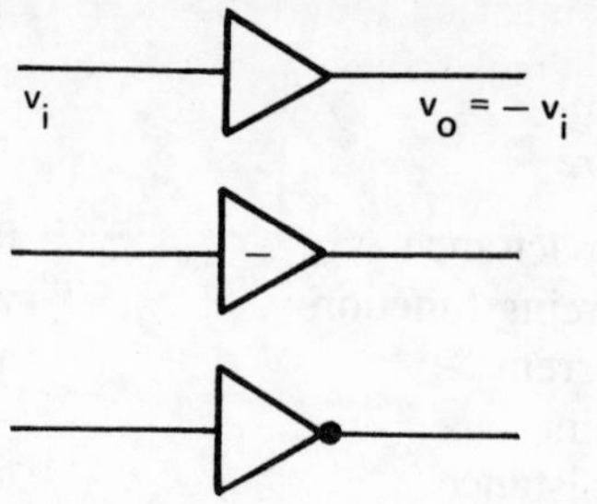

Figure 18.9 Inverter Symbols

E. Adder

If multiple inputs are used as in figure 18.10, the op-amp will act like an *adder*. The analysis is similar to that performed for the single-input op-amp.

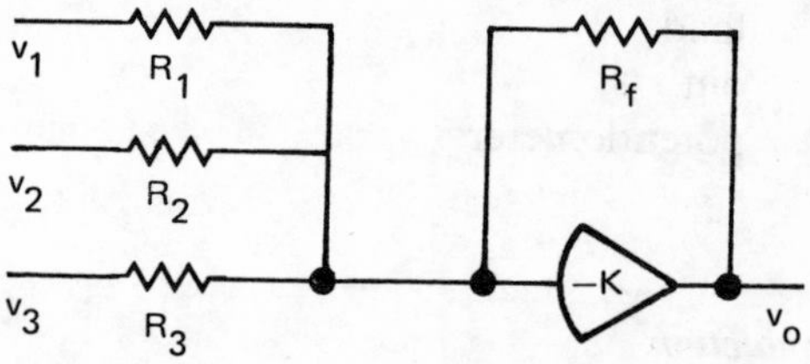

Figure 18.10 Op-Amp Used as an Adder

Using superposition, the output voltage is

$$v_o = -\left[\left(\frac{R_f}{R_1}\right)v_1 + \left(\frac{R_f}{R_2}\right)v_2 + \left(\frac{R_f}{R_3}\right)v_3\right] \tag{18.15}$$

If R_1, R_2, R_3, and R_f are all equal, then the device is called an adder. The symbol for an adder is shown in figure 18.11. If the resistors are not equal, then the output will be a weighted sum of the inputs.

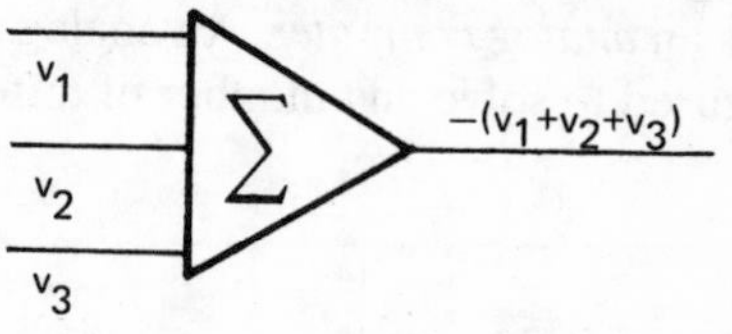

Figure 18.11 Adder Symbol

F. Integrator

If the feedback element is a capacitor, the op-amp will perform the function of integration.

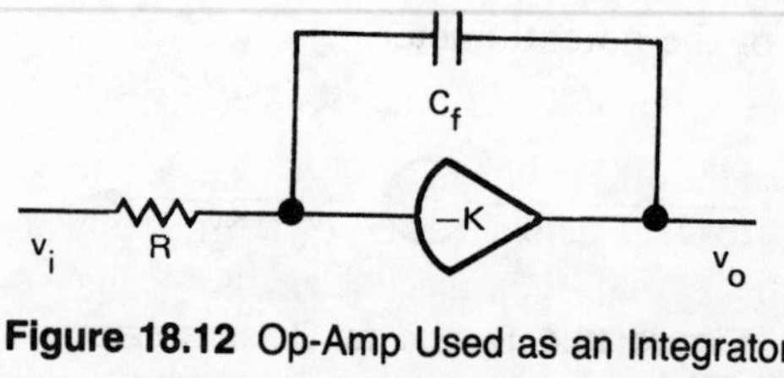

Figure 18.12 Op-Amp Used as an Integrator

As before, the input impedance is large so that $i_i = i_f$. However, i_f must be found from the rate of change of charge on the capacitor.

$$i_f = \frac{dq}{dt} = C_f\left(\frac{dv}{dt}\right) = C_f \frac{d}{dt}(v_B - v_o) \qquad 18.16$$

Since $i_i = i_f$,

$$\frac{v_i - v_B}{R_i} = C_f \frac{d}{dt}(v_B - v_o) \qquad 18.17$$

Since v_B is virtually zero (virtual ground),

$$\frac{dv_o}{dt} = -\left(\frac{1}{R_iC_f}\right)v_i \qquad 18.18$$

Integrating both sides yields equation 18.19. Thus, v_o is a scaled integral of v_i. The product of R_iC_f can be chosen as 1.0.

$$v_o = -\left(\frac{1}{R_iC_f}\right)\int v_i\, dt \qquad 18.19$$

The symbols for an *integrator* are shown in figure 18.13.

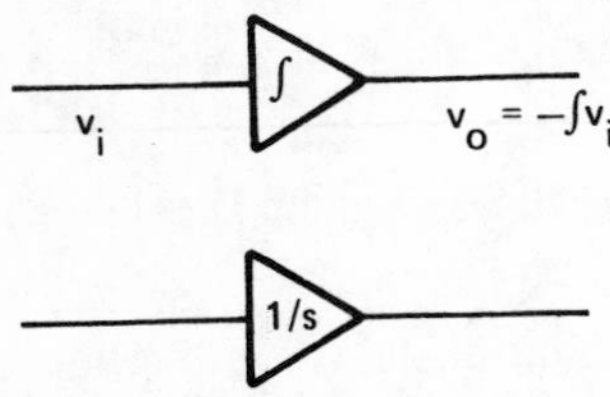

Figure 18.13 Integrator Symbols

The integrator requires a provision for initial conditions. This can be done with the circuit shown in figure 18.14. The potentiometer can be set to apply any voltage to the capacitor. The *reset switch* is closed momentarily to return the model to the initial condition.

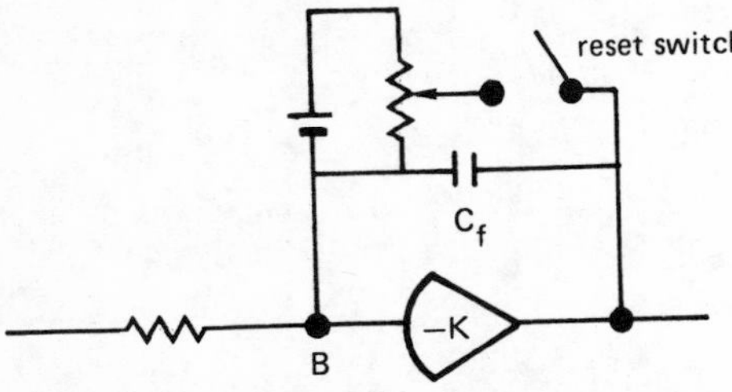

Figure 18.14 Provision for Initial Condition

Consider an integrator that integrates $-x'$ to x using a scale of one volt to one inch. If the initial condition is $x_o = 10''$, then the integrator output must be initially 10 volts.

The potentiometer in figure 18.14 is set to supply 10 volts across the capacitor. When the reset button is momentarily closed, the capacitor charges to 10 volts. Since point *B* is effectively at zero potential, the output is initially at 10 volts. This corresponds to an initial condition of 10 inches.[13]

The output of an integrator with an initial condition is given by equation 18.20.

$$v_{out} = -\left[\frac{1}{R_iC_f}\int_0^t v_i(t) - v_o\right] \qquad 18.20$$

G. Differentiator

If the capacitor and input resistor are reversed as in figure 18.15, the op-amp performs differentiation.

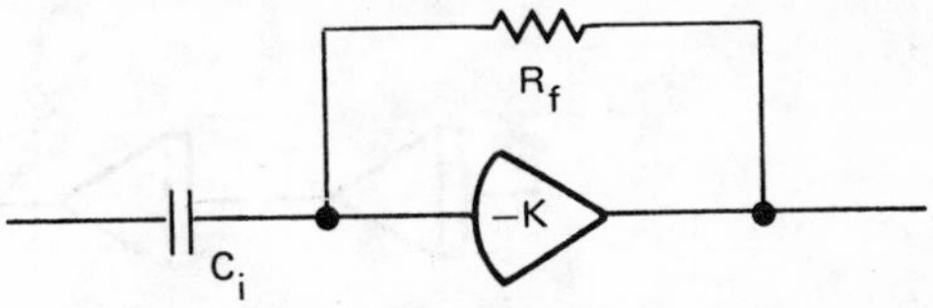

Figure 18.15 Op-Amp Used as a Differentiator

The output is

$$v_o = -R_fC_i\left(\frac{d}{dt}\right)v_i \qquad 18.21$$

Since the derivative of a function corresponds to the function's slope, differentiator output can be very large when the input is random noise or spikes. As the operating voltage must be kept below 10 volts, differentiators are seldom used in analog computers.

3. Indirect Programming

Linear differential equations with constant coefficients can be analyzed with an analog computer. The technique of arranging the analog computer components is known as *indirect programming*. The procedure given here assumes that all variables are *normalized*. To be normalized, the coefficients *a*, *b*, and *c* in equation 18.22 must all be less than 1.0.

$$ax'' + bx' + cx = F(t) \qquad 18.22$$

step 1: Divide the differential equation by the coefficient of the highest-order derivative.

step 2: Solve for the negative of the highest-order derivative. (This accounts for the inversion performed by the final summing amplifier.)

step 3: Assume a source line for the negative of the highest-order derivative. Run this line through successive integrators to obtain all needed derivatives.

step 4: Feed all of the needed derivatives through potentiometers set to the coefficient of those terms.

[13]If the integrator had integrated x′ to −x, then the initial voltage of C_f would have been −10 volts.

COMP SCI

step 5: Feed the scaled forcing function and all of the scaled derivatives into the final summer (see step 2).

step 6: Add initial conditions.

Example 18.22

Draw the analog computer circuit which solves the differential equation

$$2x'' + .32x' + 1.28x = .6$$

step 1: $x'' + .16x' + .64x = .3$

step 2: $-x'' = .16x' + .64x - .3$

step 3:

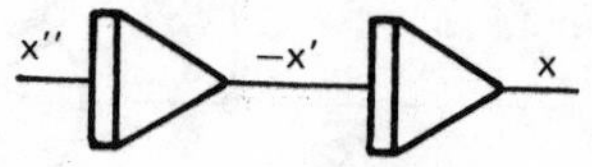

step 4:

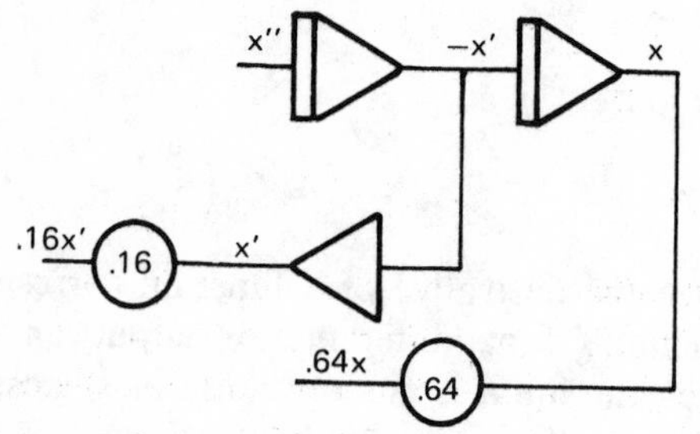

step 5:

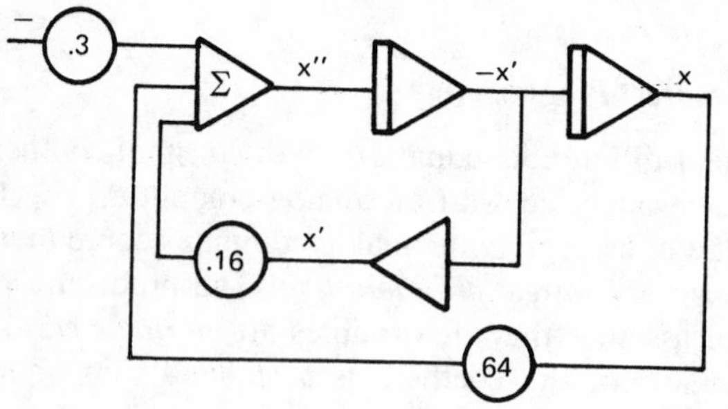

Simultaneous differential equations are easily handled by analog computation. This is illustrated in example 18.23.

Example 18.23

Draw the analog computer circuit which solves the two simultaneous differential equations.

$$-x_1' = .8x_1 - .8x_2 - .3$$

$$-x_2' = x_2 - .4x_1$$

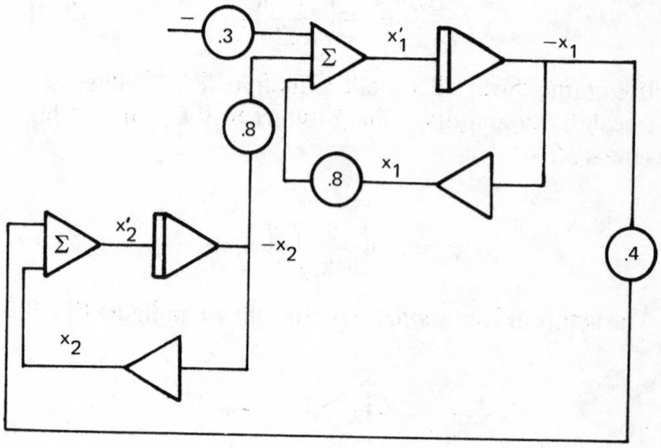

Appendix A
Powers of 2

n	2^n
0	1
1	2
2	4
3	8
4	16
5	32
6	64
7	128
8	256
9	512
10	1,024
11	2,048
12	4,096
13	8,192
14	16,384
15	32,768
16	65,536
17	131,072
18	262,144
19	524,288
20	1,048,576
21	2,097,152
22	4,194,304
23	8,388,608
24	16,777,216
25	33,554,432
26	67,108,864
27	134,217,728

Appendix B
Powers of 8

n	8^n
0	1
1	8
2	64
3	512
4	4,096
5	32,768
6	262,144
7	2,097,152
8	16,777,216
9	134,217,728
10	1,073,741,824

COMP SCI

Appendix C
Powers of 16

n	16^n
0	1
1	16
2	256
3	4,096
4	65,536
5	1,048,576
6	16,777,216
7	268,435,456
8	4,294,967,296
9	68,719,476,736

Appendix D
Hexadecimal-Decimal Conversion Table

0	1	2	3	4	5	6	7	8	9	A	B	C	D	E	F
00 – 0000	0001	0002	0003	0004	0005	0006	0007	0008	0009	0010	0011	0012	0013	0014	0015
01 – 0016	0017	0018	0019	0020	0021	0022	0023	0024	0025	0026	0027	0028	0029	0030	0031
02 – 0032	0033	0034	0035	0036	0037	0038	0039	0040	0041	0042	0043	0044	0045	0046	0047
03 – 0048	0049	0050	0051	0052	0053	0054	0055	0056	0057	0058	0059	0060	0061	0062	0063
04 – 0064	0065	0066	0067	0068	0069	0070	0071	0072	0073	0074	0075	0076	0077	0078	0079
05 – 0080	0081	0082	0083	0084	0085	0086	0087	0088	0089	0090	0091	0092	0093	0094	0095
06 – 0096	0097	0098	0099	0100	0101	0102	0103	0104	0105	0106	0107	0108	0109	0110	0111
07 – 0112	0113	0114	0115	0116	0117	0118	0119	0120	0121	0122	0123	0124	0125	0126	0127
08 – 0128	0129	0130	0131	0132	0133	0134	0135	0136	0137	0138	0139	0140	0141	0142	0143
09 – 0144	0145	0146	0147	0148	0149	0150	0151	0152	0153	0154	0155	0156	0157	0158	0159
0A – 0160	0161	0162	0163	0164	0165	0166	0167	0168	0169	0170	0171	0172	0173	0174	1075
0B – 0176	0177	0178	0179	0180	0181	0182	0183	0184	0185	0186	0187	0188	0189	0190	0191
0C – 0192	0193	0194	0195	0196	0197	0198	0199	0200	0201	0202	0203	0204	0205	0206	0207
0D – 0208	0209	0210	0211	0212	0213	0214	0215	0216	0217	0218	0219	0220	0221	0222	0223
0E – 0224	0225	0226	0227	0228	0229	0230	0231	0232	0233	0234	0235	0236	0237	0238	0239
0F – 0240	0241	0242	0243	0244	0245	0246	0247	0248	0249	0250	0251	0252	0253	0254	0255
10 – 0256	0257	0258	0259	0260	0261	0262	0263	0264	0265	0266	0267	0268	0269	0270	0271
11 – 0272	0273	0274	0275	0276	0277	0278	0279	0280	0281	0282	0283	0284	0285	0286	0287
12 – 0288	0289	0290	0291	0292	0293	0294	0295	0296	0297	0298	0299	0300	0301	0302	0303
13 – 0304	0305	0306	0307	0308	0309	0310	0311	0312	0313	0314	0315	0316	0317	0318	0319
14 – 0320	0321	0322	0323	0324	0325	0326	0327	0328	0329	0330	0331	0332	0333	0334	0335
15 – 0336	0337	0338	0339	0340	0341	0342	0343	0344	0345	0346	0347	0348	0349	0350	0351
16 – 0352	0353	0354	0355	0356	0357	0358	0359	0360	0361	0362	0363	0364	0365	0366	0367
17 – 0368	0369	0370	0371	0372	0373	0374	0375	0376	0377	0378	0379	0380	0381	0382	0383
18 – 0384	0385	0386	0387	0388	0389	0390	0391	0392	0393	0394	0395	0396	0397	0398	0399
19 – 0400	0401	0402	0403	0404	0405	0406	0407	0408	0409	0410	0411	0412	0413	0414	0415
1A – 0416	0417	0418	0419	0420	0421	0422	0423	0424	0425	0426	0427	0428	0429	0430	0431
1B – 0432	0433	0434	0435	0436	0437	0438	0439	0440	0441	0442	0443	0444	0445	0446	0447
1C – 0448	0449	0450	0451	0452	0453	0454	0455	0456	0457	0458	0459	0460	0461	0462	0463
1D – 0464	0465	0466	0467	0468	0469	0470	0471	0472	0473	0474	0475	0476	0477	0478	0479
1E – 0480	0481	0482	0483	0484	0485	0486	0487	0488	0489	0490	0491	0492	0493	0494	0495
1F – 0496	0497	0498	0499	0500	0501	0502	0503	0504	0505	0506	0507	0508	0509	0510	0511
20 – 0512	0513	0514	0515	0516	0517	0518	0519	0520	0521	0522	0523	0524	0525	0526	0527
21 – 0528	0529	0530	0531	0532	0533	0534	0535	0536	0537	0528	0539	0540	0541	0542	0543
22 – 0544	0545	0546	0547	0548	0549	0550	0551	0552	0553	0554	0555	0556	0557	0558	0559
23 – 0560	0561	0562	0563	0564	0565	0566	0567	0568	0569	0570	0571	0572	0573	0574	0575
24 – 0576	0577	0578	0579	0580	0581	0582	0583	0584	0585	0586	0587	0588	0589	0590	0591
25 – 0592	0593	0594	0595	0596	0597	0598	0599	0600	0601	0602	0603	0604	0605	0606	0607
26 – 0608	0609	0610	0611	0612	0613	0614	0615	0616	0617	0618	0619	0620	0621	0622	0623
27 – 0624	0625	0626	0627	0628	0629	0630	0631	0632	0633	0634	0635	0636	0637	0638	0639
28 – 0640	0641	0642	0643	0644	0645	0646	0647	0648	0649	0650	0651	0652	0653	0654	0655
29 – 0656	0657	0658	0659	0660	0661	0662	0663	0664	0665	0666	0667	0668	0669	0670	0671
2A – 0672	0673	0674	0675	0676	0677	0678	0679	0680	0681	0682	0683	0684	0685	0686	0687
2B – 0688	0689	0690	0691	0692	0693	0694	0695	0696	0697	0698	0699	0700	0701	0702	0703
2C – 0704	0705	0706	0707	0708	0709	0710	0711	0712	0713	0714	0715	0716	0717	0718	0719
2D – 0720	0721	0722	0723	0724	0725	0726	0727	0728	0729	0730	0731	0732	0733	0734	0735
2E – 0736	0737	0738	0739	0740	0741	0742	0743	0744	0745	0746	0747	0748	0749	0750	0751
2F – 0752	0753	0754	0755	0756	0757	0758	0759	0760	0761	0762	0763	0764	0765	0766	0767
30 – 0768	0769	0770	0771	0772	0773	0774	0775	0776	0777	0778	0779	0780	0781	0782	0783
31 – 0784	0785	0786	0787	0788	0789	0790	0791	0792	0793	0794	0795	0796	0797	0798	0799
32 – 0800	0801	0802	0803	0804	0805	0806	0807	0808	0809	0810	0811	0812	0813	0814	0815
33 – 0816	0817	0818	0819	0820	0821	0822	0823	0824	0825	0826	0827	0828	0829	0830	0831
34 – 0832	0833	0834	0835	0836	0837	0838	0839	0840	0841	0842	0843	0844	0845	0846	0847
35 – 0848	0849	0850	0851	0852	0853	0854	0855	0856	0857	0858	0859	0860	0861	0862	0863
36 – 0864	0865	0866	0867	0868	0869	0870	0871	0872	0873	0874	0875	0876	0877	0878	0879
37 – 0880	0881	0882	0883	0884	0885	0886	0887	0888	0889	0890	0891	0892	0893	0894	0895

Appendix D (continued)
Hexadecimal-Decimal Conversion Table

	0	1	2	3	4	5	6	7	8	9	A	B	C	D	E	F
38 –	0896	0897	0898	0899	0900	0901	0902	0903	0904	0905	0906	0907	0908	0909	0910	0911
39 –	0912	0913	0914	0915	0916	0917	0918	0919	0920	0921	0922	0923	0924	0925	0926	0927
3A –	0928	0929	0930	0931	0932	0933	0934	0935	0936	0937	0938	0939	0940	0941	0942	0943
3B –	0944	0945	0946	0947	0948	0949	0950	0951	0952	0953	0954	0955	0956	0957	0958	0959
3C –	0960	0961	0962	0963	0964	0965	0966	0967	0968	0969	0970	0971	0972	0973	0974	0975
3D –	0976	0977	0978	0979	0980	0981	0982	0983	0984	0985	0986	0987	0988	0989	0990	0991
3E –	0992	0993	0994	0995	0996	0997	0998	0999	1000	1001	1002	1003	1004	1005	1006	1007
3F –	1008	1009	1010	1011	1012	1013	1014	1015	1016	1017	1018	1019	1020	1021	1022	1023
40 –	1024	1025	1026	1027	1028	1029	1030	1031	1032	1033	1034	1035	1036	1037	1038	1039
41 –	1040	1041	1042	1043	1044	1045	1046	1047	1048	1049	1050	1051	1052	1053	1054	1055
42 –	1056	1057	1058	1959	1060	1061	1062	1063	1064	1065	1066	1067	1068	1069	1070	1071
43 –	1072	1073	1074	1075	1076	1077	1078	1079	1080	1081	1082	1083	1084	1085	1086	1087
44 –	1088	1089	1090	1091	1092	1093	1094	1095	1096	1097	1098	1099	1100	1101	1102	1103
45 –	1104	1105	1106	1107	1108	1109	1110	1111	1112	1113	1114	1115	1116	1117	1118	1119
46 –	1120	1121	1122	1123	1124	1125	1126	1127	1128	1129	1130	1131	1132	1133	1134	1135
47 –	1136	1137	1138	1139	1140	1141	1142	1143	1144	1145	1146	1147	1148	1149	1150	1151
48 –	1152	1153	1154	1155	1156	1157	1158	1159	1160	1161	1162	1163	1164	1165	1166	1167
49 –	1168	1169	1170	1171	1172	1173	1174	1175	1176	1177	1178	1179	1180	1181	1182	1183
4A –	1184	1185	1186	1187	1188	1189	1190	1191	1192	1193	1194	1195	1196	1197	1198	1199
4B –	1200	1201	1202	1203	1204	1205	1206	1207	1208	1209	1210	1211	1212	1213	1214	1215
4C –	1216	1217	1218	1219	1220	1221	1222	1223	1224	1225	1226	1227	1228	1229	1230	1231
4D –	1232	1233	1234	1235	1236	1237	1238	1239	1240	1241	1242	1243	1244	1245	1246	1247
4E –	1248	1249	1250	1251	1252	1253	1254	1255	1256	1257	1258	1259	1260	1261	1262	1263
4F –	1264	1265	1266	1267	1268	1269	1270	1271	1272	1273	1274	1275	1276	1277	1278	1279
50 –	1280	1281	1282	1283	1284	1285	1286	1287	1288	1289	1290	1291	1292	1293	1294	1295
51 –	1296	1297	1298	1299	1300	1301	1302	1303	1304	1305	1306	1307	1308	1309	1310	1311
52 –	1312	1313	1314	1315	1316	1317	1318	1319	1320	1321	1322	1323	1324	1325	1326	1327
53 –	1328	1329	1330	1331	1332	1333	1334	1335	1336	1337	1338	1339	1340	1341	1342	1343
54 –	1344	1345	1346	1347	1348	1349	1350	1351	1352	1353	1354	1355	1356	1357	1358	1359
55 –	1360	1361	1362	1363	1364	1365	1366	1367	1368	1369	1370	1371	1372	1373	1374	1375
56 –	1376	1377	1378	1379	1380	1381	1382	1383	1384	1385	1386	1387	1388	1389	1390	1391
57 –	1392	1393	1394	1395	1396	1397	1398	1399	1400	1401	1402	1403	1404	1405	1406	1407
58 –	1408	1409	1410	1411	1412	1413	1414	1415	1416	1417	1418	1419	1420	1421	1422	1423
59 –	1424	1425	1426	1427	1428	1429	1430	1431	1432	1433	1434	1435	1436	1437	1438	1439
5A –	1440	1441	1442	1443	1444	1445	1446	1447	1448	1449	1450	1451	1452	1453	1454	1455
5B –	1456	1457	1458	1459	1460	1461	1462	1463	1464	1465	1466	1467	1468	1469	1470	1471
5C –	1472	1473	1474	1475	1476	1477	1478	1479	1480	1481	1482	1483	1484	1485	1486	1487
5D –	1488	1489	1490	1491	1492	1493	1494	1495	1496	1497	1498	1499	1500	1501	1502	1503
5E –	1504	1505	1506	1507	1508	1509	1510	1511	1512	1513	1514	1515	1516	1517	1518	1519
5F –	1520	1521	1522	1523	1524	1525	1526	1527	1528	1529	1530	1531	1532	1533	1534	1535
60 –	1536	1537	1538	1539	1540	1541	1542	1543	1544	1545	1546	1547	1548	1549	1550	1551
61 –	1552	1553	1554	1555	1556	1557	1558	1559	1560	1561	1562	1563	1564	1565	1566	1567
62 –	1568	1569	1570	1571	1572	1573	1574	1575	1576	1577	1578	1579	1580	1581	1582	1583
63 –	1584	1585	1586	1587	1588	1589	1590	1591	1592	1593	1594	1595	1596	1597	1598	1599
64 –	1600	1601	1602	1603	1604	1605	1606	1607	1608	1609	1610	1611	1612	1613	1614	1615
65 –	1616	1617	1618	1619	1620	1621	1622	1623	1624	1625	1626	1627	1628	1629	1630	1631
66 –	1632	1633	1634	1635	1636	1637	1638	1639	1640	1641	1642	1643	1644	1645	1646	1647
67 –	1648	1649	1650	1651	1652	1653	1654	1655	1656	1657	1658	1659	1660	1661	1662	1663
68 –	1664	1665	1666	1667	1668	1669	1670	1671	1672	1673	1674	1675	1676	1677	1678	1679
69 –	1680	1681	1682	1683	1684	1685	1686	1687	1688	1689	1690	1691	1692	1693	1694	1695
6A –	1696	1697	1698	1699	1700	1701	1702	1703	1704	1705	1706	1707	1708	1709	1710	1711
6B –	1712	1713	1714	1715	1716	1717	1718	1719	1720	1721	1722	1723	1724	1725	1726	1727
6C –	1728	1729	1730	1731	1732	1733	1734	1735	1736	1737	1738	1739	1740	1741	1742	1743
6D –	1744	1745	1746	1747	1748	1749	1750	1751	1752	1753	1754	1755	1756	1757	1758	1759
6E –	1760	1761	1762	1763	1764	1765	1766	1767	1768	1769	1770	1771	1772	1773	1774	1775
6F –	1776	1777	1778	1779	1780	1781	1782	1783	1784	1785	1786	1787	1788	1789	1790	1791

COMP SCI

CALCULATIONS IN OTHER BASES

1. Perform the following binary operations. Check your work by converting to decimal.
   (a) 101 + 011 (d) 0100 - 1100 (g) 111 x 11
   (b) 101 + 110 (e) 1110 - 1000 (h) 100 x 11
   (c) 101 + 100 (f) 010 - 101 (i) 1011 x 1101

2. Perform the following octal operations. Use sevens complement for subtraction. Check your work by converting to decimal.
   (a) 466 + 457 (d) 71 - 27 (g) 77 x 66
   (b) 1007 + 6661 (e) 1143 - 367 (h) 325 x 36
   (c) 321 + 465 (f) 646 - 677 (i) 3251 x 16.1

3. Perform the following hexadecimal operations. Check your work by converting to decimal.
   (a) BA + C (d) FF - E (g) 4A x 3E
   (b) BB + A (e) 74 - 4A (h) FE x EF
   (c) BE + 10 + 1A (f) FB - BF (i) 17 x 7A

CONVERSIONS BETWEEN BASES

4. Convert the following numbers to decimal (base-10).
   (a) $(674)_8$ (b) $(101101)_2$ (c) $(734.262)_8$
   (d) $(1011.11)_2$

5. Convert the following numbers to octal (base-8).
   (a) $(75)_{10}$ (b) $(.375)_{10}$ (c) $(121.875)_{10}$
   (d) $(1011100.01110)_2$

6. Convert the following numbers to binary (base-2).
   (a) $(83)_{10}$ (b) $(100.3)_{10}$ (c) $(.97)_{10}$
   (d) $(321.422)_8$

FLOWCHARTING

7. Flowchart the following procedure: If A is greater than B and A is greater than C, then add B to C and go to PLACE1. If B is greater than A and B is greater than C, then add A to C and go to PLACE2. If B is greater than A but less than C, then subtract A from B and go to PLACE1.

8. Flowchart the following procedure: If A is greater than 10 and if A is less than 14, subtract X from A. If A is less than 10 or if A is greater than 14, exit the program.

FORTRAN PROGRAMMING

9. If J=3, K=4, L=6, and M=9, what is I = (J*K)+(L-M)?

10. If J=15, K=4, L=9, and M=19, what is the value of the logical variable X?
    (a) X = J.LT.L.OR.K.GE.(M-J)
    (b) X = J.GT.L.OR.K.LT.(M-J)

11. A, B, and C are positive, unequal integers. Write a FORTRAN program to find and print out the smallest.

12. Write a FORTRAN program to perform the procedure described in problem 7.

13. Write a FORTRAN program to perform the procedure described in problem 8.

CHARACTER STORAGE

14. Using an even parity check, what is the value of the parity bit for the following stored values?
    (a) 3 (b) 6 (c) J
    (d) T (e) X

15. If 1000 cards are blocked into 1200 character records on 800 bpi tape, approximately how many inches of tape will be required?

ANALOG COMPUTERS

16. Draw the analog computer simulation for the equation below. Do not normalize.

$$7y''' + 14y'' + 42y' + 2y = 49x$$

17. What is the analog computer simulation for the system shown below?

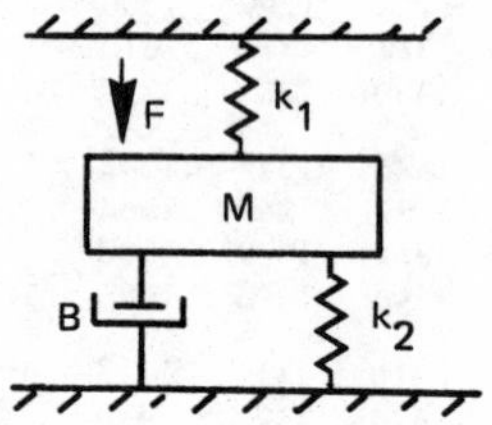

18. Draw an analog circuit to simulate V_2.

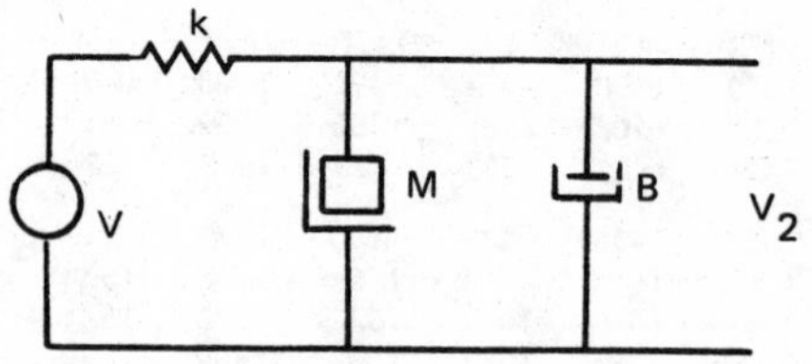

19. Draw an analog circuit to simulate V_4.

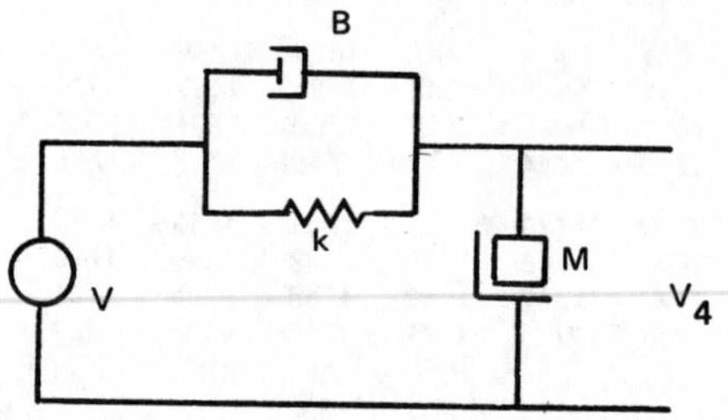

OPERATIONAL AMPLIFIERS

20. What is the differential equation that is solved by the following circuit?

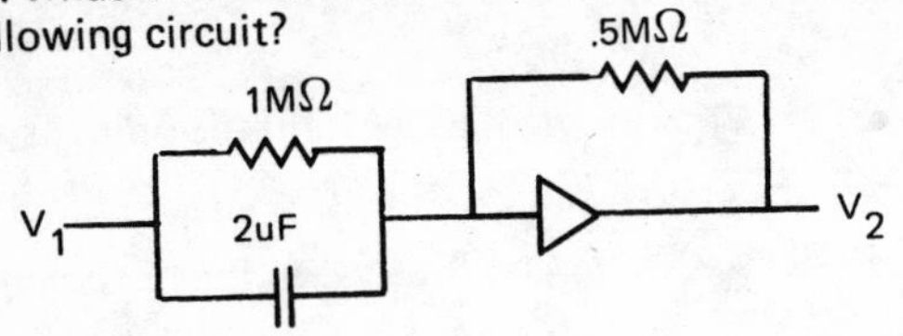

21. What is the differential equation that writes V_2 in terms of V_1?

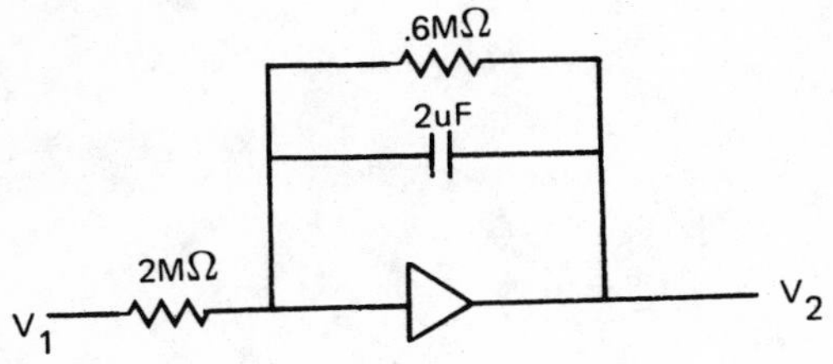

19 Atomic Theory and Nuclear Engineering

Nomenclature

a	acceleration, ellipse parameter	m/sec^2, cm
A	activity, area	1/sec, cm^2
A.W.	atomic weight	g/gmole
b	ellipse parameter	cm
B	magnetic field strength	tesla
c	speed of light	cm/sec, m/sec
C	constant	1/Å
C.C.	current concentration	-
D	absorbed dose	rad
D′	rate of dose absorption	rad/sec
e	eccentricity	-
E	energy	J
f	frequency, thermal utilization	Hz, -
$\bar{f}$	wave number	1/Å
h	Planck's constant (6.626 EE−34)	J-sec
H	equivalent dose	rem
H′	rate of equivalent dose absorption	rem/sec
I	particle intensity	1/cm^2-sec
J	particle density	1/cm^2-sec
k	constant equal to $\left(\frac{1}{4\pi\varepsilon_o}\right)$, relativistic multiplier	N-cm^2/coul^2, -
L	length	m
m	mass	g, kg
n	principal quantum number	-
n_1	starting orbit number	-
n_2	ending orbit number	-
N	number of molecules, number of molecules per unit volume	-, 1/cm^3
N_o	Avogadro's number (6.023 EE23)	1/gmole
p	momentum, resonance escape probability	cm-g/sec, -
q	charge, charge on electron	coul
Q	quality factor	-
r	orbit radius	cm
R	Rydberg constant	1/m
t	time	sec
$\bar{t}$	average life expectancy	sec
T	period, temperature	sec, °K
v	velocity	cm/sec, m/sec
V	voltage	volts
x	distance traveled, thickness	cm
X	exposure	coul/kg
X′	exposure rate	R/sec
Z	atomic number	-

Symbols

α	constant	-
β	constant	-
ε	error, fast fission factor	various, -
ε_o	permittivity of free space	coul^2/N-cm^2
η	multiplication constant	-
λ	wavelength, decay constant	Å, 1/sec
μ	attenuation coefficient	1/cm, or cm^2/g
ρ	density	g/cm^3
σ	cross section	barns
Σ	macroscopic cross section	barns
ψ	amplitude factor	cm
ω	angular frequency	rad/sec

Subscripts

a	absorbed
e	electron
H	hydrogen
k	kinetic
l	linear
m	mass
n	nth orbit, nucleus
o	original, at rest
p	momentum, probable
r	reduced, radial
s	scattered
t	total
v	at velocity v

w	wave
x	position
ϕ	azimuthal
∞	infinite

1. Introduction

In the late 1800's, radioactive substances were known to decay into both positive and negative particles. It was concluded that matter, being neutral, consisted of an equal number of positive and negative charges somehow bound together. This *plum pudding* theory described an atom as a piece of dough through which ran equal but opposite currents of positive and negative charges. The theory was discredited when it was unable to explain the results obtained in Rutherford's nucleus experiment.

2. Rutherford's Nucleus Experiment

In 1911, Rutherford[1] used collimated alpha particles from radium to bombard gold foil. Most of the incident beam passed through the foil. However, a small number of particles was deflected. Some particles were deflected through very large angles approaching 180°. From the observed results, Rutherford made several conclusions.

- The atom is mostly empty space.
- The positive charge is concentrated in a small region.
- The negative charges exist as a cloud around the nucleus.

By neglecting the interaction of the alpha particles and electrons, and by assuming that the nucleus remained stationary, Rutherford was able to calculate the *effective electrostatic radius* of the atom.

For the alpha particles which were reflected 180° (i.e., back to the source), the kinetic energy would have to be exactly balanced by the *coulomb repulsion* when the alpha particles were at a standstill.

$$\tfrac{1}{2}mv^2 = \frac{(2q)(Zq)}{4\pi\,\varepsilon_o r} \qquad 19.1$$

In equation 19.1, q is the charge on an electron, Z is the atomic number of the gold foil, and r is the effective electrostatic radius. The *2* in the numerator is the number of positive charges in the alpha particle.

The effective radius calculated was EE−14 meter, smaller than the EE−10 meter derived from a knowledge of density and Avogadro's number. This discovery of a short-range attractive force between atomic particles rivaled in importance the discovery of the nucleus.

[1]Ernest Rutherford, 1871–1937, England.

3. Atomic Spectra

Visible light from an incandescent solid can be split into color bands by a prism. The multiplicity of colors is dependent on the makeup of the incident light beam since only the colors originally present in the beam can be separated out. Light consisting of various colors is called *polychromatic light.*

Light from an electric discharge through a gas or from an electric arc can also be split by a prism or grating spectrometer, although the results will be solid lines, not color bands. Light which contains only one such line is called *monochromatic light.* The separation and placement of these resultant *spectra* or *spectral lines* is unique for each element, with each spectral line corresponding to a different discharge frequency.

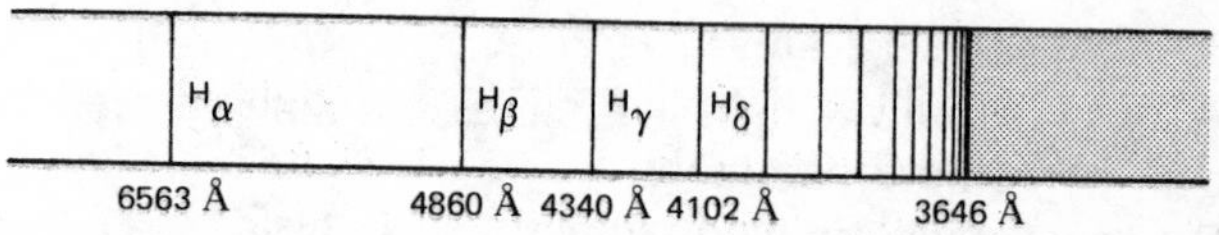

Figure 19.1 Balmer Spectra for Hydrogen

The spectra from visible light produced by a hydrogen gas discharge displays intriguing regularity. This series of spectra is named the *Balmer*[2] *series* after a high school teacher who showed, in a mathematical exercise in 1884, that the wavelength in angstroms of the nine known visible spectra could be derived from equation 19.2

$$\lambda = \frac{3645.6(n_2)^2}{(n_2)^2 - 4} \qquad n_2 = 3, 4, 5, \cdots \qquad 19.2$$

A few years later, Rydberg[3] developed a similar equation for the reciprocal of wavelength, called the *wave number.*

$$\bar{f} = \frac{1}{\lambda} = C - \frac{R_H}{(n_2 + \alpha)^2} \qquad 19.3$$

C and α depend on the element. R_H is the *Rydberg number* for hydrogen, 1.09678 EE7 1/m. Specifically for hydrogen, then

$$\bar{f} = R_H\left[\frac{1}{(2)^2} - \frac{1}{(n_2)^2}\right] \qquad 19.4$$

In 1908, Ritz[4] noted that the constant C in equation 19.3 could be written in the same format as Rydberg's second term.

[2]Johann Jakob Balmer, 1825–1898, Switzerland.
[3]Johannes Robert Rydberg, 1854–1919, Sweden.
[4]Walter Ritz, 1878–1909, Switzerland.

$$\bar{f} = R\left[\frac{1}{(n_1+\beta)^2} - \frac{1}{(n_2+\alpha)^2}\right] \quad 19.5$$

R, α, and β depend on the element, although they are not fixed as single values for a given element. Multiple values characteristic of the element may be used to give the *principal, sharp, diffuse,* and *fundamental* (also known as Bergman) series, abbreviated as *p, s, d,* and *f*. Specifically for hydrogen:

$$\bar{f} = R_H\left[\frac{1}{(n_1)^2} - \frac{1}{(n_2)^2}\right] \quad \begin{array}{l} n_1 = 2 \text{ for visible} \\ n_2 = 3,4,5,\cdots \end{array} \quad 19.6$$

Although formulas 19.2 through 19.6 are correct, there is no theory to back them up. It was the goal of Bohr to develop an atomic structure theory which would explain the characteristic spectra of the elements.

4. Bohr's Atom

In 1913, Bohr[5] treated the electron as a charged particle and hypothesized that the hydrogen atom was a positive nucleus with a revolving electron in orbit. To be stable, the centripetal force would have to equal the coulomb attraction. (The constant *k* in equation 19.7 incorporates all of the constants in the denominator of the formula for electrostatic attraction.)

$$\frac{mv^2}{r} = \frac{kq^2}{r^2} \quad 19.7$$

However, the assumption of a revolving electron led immediately to a problem. A revolving electron, as a charged particle, would be continually accelerating, and it would continuously radiate energy. This energy loss would result in a rapid decay spiral into the nucleus. To circumvent this contradiction to classical physics, Bohr's *first postulate* was that of the existence of stable, non-radiating orbits between which the electron was forbidden to remain permanently.

To jump from one stable orbit (energy level) to the next would require a change in the electron's energy. The minimum of such energy was taken from *Planck's*[6] *quantum hypothesis*. The use of quantized energy was Bohr's *second postulate.* The energy required for an orbit change and the frequency of that energy are given by equation 19.8.

$$E_2 - E_1 = hf \quad 19.8$$

To use the quantum theory, Bohr had to define the permitted stable orbits in terms of frequencies. A *third postulate,* that classical mechanics doesn't hold for the electron between orbits, was made. In a giant stroke of genius, Bohr assumed as his *fourth postulate* that angular momentum was also quantized with the fundamental unit of h/2π. The electron could only have multiples of this unit. Since the angular momentum of an object in a circular path is *mvr,* Bohr wrote

$$mvr = \frac{nh}{2\pi} \quad 19.9$$

n is the *principal quantum number.* If **n** is one, the electron is in the lowest orbit or *ground state.*

Bohr was then able to calculate the radii of the orbits by rearranging equation 19.9 to obtain an expression for velocity. This velocity was then substituted into equation 19.7 to give a formula for radius.

$$v_n = \frac{nh}{2\pi mr} = \frac{2\pi kq^2}{nh} \quad 19.10$$

$$r_n = \frac{(nh/\pi q)^2}{4km} \quad 19.11$$

If **n** is set equal to 1, the radius is .529 Å, which compares favorably with the value obtained from other experiments and calculations. The velocity obtained is 2.2 EE8 cm/sec.

To explain atomic spectra, Bohr reasoned that the *binding energy* holding the electron to the nucleus was the difference between the kinetic energy of the electron and the coulomb attraction between the two.

$$E = {}^1\!/\!_2 mv^2 - \frac{kq^2}{r} \quad 19.12$$

Obviously, the attraction term is larger than the kinetic term. Otherwise, the electron would escape. From equation 19.7,

$${}^1\!/\!_2 mv^2 = \frac{{}^1\!/\!_2 kq^2}{r} \quad 19.13$$

Combining equations 19.12 and 19.13,

$$E_n = \frac{-kq^2}{2r_n} = -2mk^2\left[\frac{\pi q^2}{nh}\right]^2 \quad 19.14$$

This is the *ionization energy* (or *binding energy*, depending on the sign) required to remove the electron completely from the nucleus. It is approximately 13.6 eV for hydrogen in the ground state.

The change in energy for an electron jumping from the n_1 to the n_2 orbit would be

$$E_2 - E_1 = {}^1\!/\!_2 kq^2\left[\frac{1}{r_1} - \frac{1}{r_2}\right] \quad 19.15$$

Equation 19.11 gives the radius in terms of the principal quantum number. This can be combined with equation 19.15 by recognizing that $c = \lambda f = f/\bar{f}$.

$$\bar{f} = \frac{E_2 - E_1}{hc} = R_\infty\left[\frac{1}{n_1^2} - \frac{1}{n_2^2}\right] \quad 19.16$$

$$R_\infty = \frac{2\pi^2 mq^4 k^2}{ch^3} \quad 19.17$$

[5]Niels Henrik David Bohr, 1885–1962, Denmark.
[6]Max Karl Ernst Planck, 1858–1947, Germany.

ATOMIC

Equation 19.16 is known as *Bohr's equation*. It was able to explain the different spectral frequencies in terms of energy changes during electron orbit changes. Electrons dropping down to a lower orbit would release energy and cause a bright band in the emission spectra. Electrons moving up would absorb energy and cause dark bands. The fuzzy area to the right of the spectral record (see figure 19.1) is the result of free electrons with non-quantized energies traveling in hyperbolic paths and dropping into the highest orbit.

Equation 19.16 is identical to that derived by Rydberg (equation 19.4) with n_1 as 2. Bohr was thus able to theoretically calculate the value of the hydrogen Rydberg number and thereby verify his theory. Actually, a small error was introduced by the assumption that the nucleus is stationary. This error can be eliminated by replacing *m* in equation 19.17 with the reduced mass.

$$m_r = \frac{m_n m_e}{m_n + m_e} \qquad 19.18$$

$$R_H = \frac{R_\infty}{1 + \dfrac{m_e}{m_n}} \qquad 19.19$$

The series obtained for the various values of the quantum numbers in equation 19.16 are given in table 19.1, with each series being named after its discoverer.

Table 19.1 Hydrogen Spectra Series

n_1 name	n_2	type
1 Lyman	2 and up	ultraviolet
2 Balmer	3 and up	visible
3 Paschen	4 and up	infrared
4 Brackett	5 and up	infrared
5 Pfund	6 and up	infrared

Bohr's theory was considered successful because it

- explained the spectra of hydrogen
- derived the Rydberg constant
- calculated a satisfactory value for the hydrogen atom radius
- accounted quantitatively for ionization and excitation energy

Bohr's theory, unfortunately, could not be used to explain the spectra for more complex elements, with the exception of singularly ionized helium. The theory was also unable to account for the differences in line intensities, which are related to the probabilities of transition.

5. Extensions to the Bohr Atom

Refined analyses of spectra have shown that spectral lines have *fine structure*. That is, they are not simple, but consist of a number of closely-spaced lines. There is actually a number of energy levels for each quantum number. Sommerfeld[7] was able to partially explain fine structure by postulating the existence of elliptical orbits as well as circular orbits and by recognizing relativistic variations in electron mass.

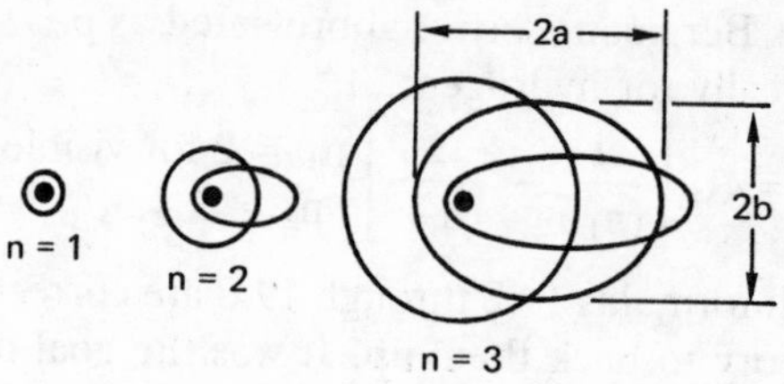

Figure 19.2 Sommerfeld's Elliptical Orbits

Equation 19.9 can be written as:

$$p_\phi = \frac{n_\phi h}{2\pi} \qquad 19.20$$

p_ϕ is the constant angular momentum and $\mathbf{n}_\phi$ is the *azimuthal quantum number* with integer values of 1,2,3, etc. (zero excluded).

By advanced methods, the radial momentum (which is also quantized over the entire orbit) is

$$p_r = 2\pi p_\phi \left[\frac{1}{\sqrt{1-e^2}} - 1\right] = \mathbf{n}_r h \qquad 19.21$$

The orbit *eccentricity* is defined by equation 19.22.

$$e = \sqrt{1 - (b/a)^2} \qquad 19.22$$

$\mathbf{n}_r$ is the *radial quantum number,* which takes on values 0,1,2,3, etc. It follows that

$$1 - e^2 = (b/a)^2 = (\mathbf{n}_\phi/\mathbf{n})^2 = \left[\frac{\mathbf{n}_\phi}{\mathbf{n}_\phi + \mathbf{n}_r}\right] \qquad 19.23$$

$$\mathbf{n} = \mathbf{n}_\phi + \mathbf{n}_r \qquad 19.24$$

$$a = \frac{\mathbf{nhm}}{2\pi q} \qquad 19.25$$

$$b = \frac{a\mathbf{n}_\phi}{\mathbf{n}} \qquad 19.26$$

Equation 19.27 takes into consideration the variation in the mass of the electron as its velocity changes around the elliptical orbit.

$$E_{\mathbf{n},\mathbf{n}\phi} = 2m\left(\frac{\pi q^2}{\mathbf{n}h}\right)\left[1 + \frac{\alpha^2}{\mathbf{n}}\left(\frac{1}{\mathbf{n}_\phi} - \frac{3}{4\mathbf{n}}\right)\right] \qquad 19.27$$

$$\alpha = \frac{2\pi q^2}{hc} = \frac{1}{137} \qquad 19.28$$

α is known as the *fine structure constant*. Fine structure was adequately explained by equation 19.27 with changes in $\mathbf{n}_\phi$ artificially limited to ± 1. Further complications to Bohr's orbit model arose in trying to ex-

[7]Arnold Johannes Wilhelm Sommerfeld, 1868–1951, Germany.

ATOMIC

plain the splitting of spectra produced in a magnetic field. The use of another empirical *magnetic quantum number*, $\mathbf{n}_m$, was only partially successful in explaining the observation.

6. Wave Mechanics

By 1928, the *wave mechanics* or *quantum mechanics* theory was able to deal theoretically with all of the previously mentioned difficulties. This theory, which does not depend on planetary orbits, derives 4 quantum numbers, as opposed to postulating them as required in the Bohr extensions. For each electron, there are four quantum numbers, as given in table 19.2.

Table 19.2 Quantum Numbers

symbol	name	allowed values
n	principal quantum number	1,2,3,. . .
l	orbital quantum number	0,1,2,3,. . .(**n**−1)
$\mathbf{m}_l$	magnetic quantum number	−**l**,. . .0,. . .+**l**
$\mathbf{m}_s$	electron spin number	−½, +½

According to the *Pauli*[8] *exclusion principle,* no two electrons may have the same set of quantum numbers. Thus, the maximum number of electrons with any given principal quantum number is $2\mathbf{n}^2$. The maximum number of electrons with the same value of **l** is $2(2\mathbf{l}+1)$.

The four quantum numbers are sufficient to entirely explain the periodic chart and the structure of the elements. All electrons with the same principal quantum numbers are said to belong to the same *shell.* Shells are named with the letters *K,L,M,N,O,P,* and *Q*.

Table 19.3 Shell Names and Numbers

n	shell name	max. # of electrons
1	*K*	2
2	*L*	8
3	*M*	18
4	*N*	32
5	*O*	50
6	*P*	72
7	*Q*	98

If **n** = 1, then **l** = $\mathbf{m}_l$ = 0. $\mathbf{m}_s$ may be either +½ or −½. So there is a maximum of 2 electrons in the *K* shell.

If **n** = 2, the maximum number of electrons is 8, with 2 electrons associated with **l** = 0 and 6 with **l** = 1. Thus, the electrons in the *L* shell are divided into two subgroups or subshells. These subgroups are designated by the letters *s,p,d,f,g,h,* and *i*. The letters *s,p,d,* and *f* correspond to the spectral line types of *sharp, principal, diffuse,* and *fundamental.*

[8]Wolfgang Pauli, 1900–1958, Austria, Switzerland.

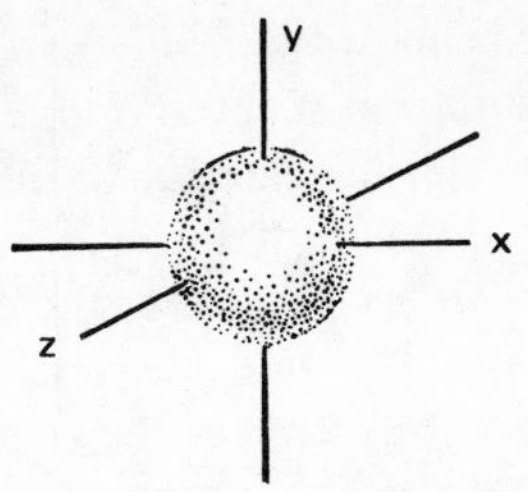

Figure 19.3 An *s* Subgroup

Table 19.4 Subgroup Names

l	subgroup name	max. no. of electrons
0	*s*	2
1	*p*	6
2	*d*	10
3	*f*	14
4	*g*	18
5	*h*	22
6	*i*	26

The shell, subgroup, and quantity of electrons in the subgroup can be designated by a coefficient, subgroup name, and superscript respectively. Thus, a full *K* shell (helium) would be designated as $1s^2$. Three electrons in the *p* subgroup of the *L* shell would be written as $2p^3$. The electronic configurations for some of the elements are given in table 19.5.

Table 19.5
Electronic Configurations of the First 11 Elements

atomic number	element	electronic configuration
1	H	1s
2	He	$1s^2$
3	Li	$1s^22s$
4	Be	$1s^22s^2$
5	B	$1s^22s^22p$
6	C	$1s^22s^22p^2$
7	N	$1s^22s^22p^3$
8	O	$1s^22s^22p^4$
9	F	$1s^22s^22p^5$
10	Ne	$1s^22s^22p^6$
11	Na	$1s^22s^22p^63s$

The *p* subgroup can be further divided into three *orbitals*, corresponding to $\mathbf{m}_l$ = −1 (p_y), $\mathbf{m}_l$ = 0 (p_z), and $\mathbf{m}_l$ = +1 (p_x). Each orbital contains two electrons ($\mathbf{m}_s$ = ±½).

Electrons do not fill up the subgroups in increasing numerical order as atomic number increases. Rather they fill the subgroups which will give the atomic configuration the lowest energy. In order of increasing energy, the subshells are: 1s,2s,2p,3s,3p,4s,3d,4p,5s,4d,5p, 6s,4f,5d,6p,7s,5f,6d. The most stable of all ele-

ATOMIC

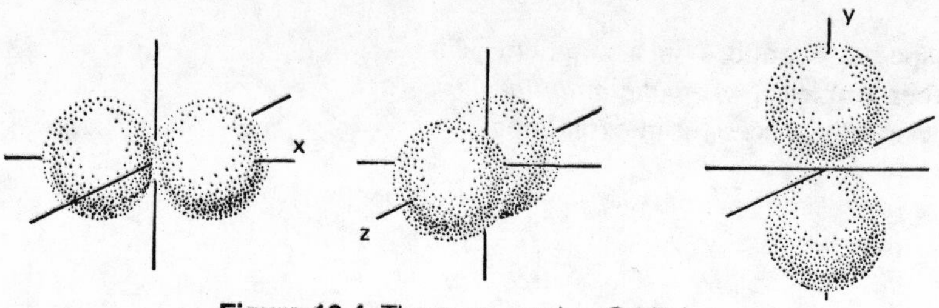

Figure 19.4 The p_x, p_y, and p_z Orbitals

ments are those in which the shells are completely filled. These elements are the *noble* or *inert* gases.

7. The Wave Theory

In 1924 de Broglie[9] suggested a dual particle/wave character for the electron. It was hypothesized and later verified by diffraction grating analysis that the momentum of an electron could be written in terms of wavelength.

$$mv = \frac{h}{\lambda} = \frac{hf}{c} \qquad 19.29$$

The wave theory can be expressed mathematically by equation 19.30, where $\Psi(x,y,z,t)$ is the *amplitude function* of the particle:

$$\frac{\partial^2\Psi}{\partial x^2} + \frac{\partial^2\Psi}{\partial y^2} + \frac{\partial^2\Psi}{\partial z^2} = \frac{1}{v^2}\left(\frac{\partial^2\Psi}{\partial t^2}\right) \qquad 19.30$$

For standing waves (applicable to stable atoms), the wave function can be written as equation 19.31.

$$\Psi = \psi e^{2\pi ift} \qquad 19.31$$

ψ is the *wave amplitude* which is time independent, and $i = \sqrt{-1}$. Also, the wave theory makes use of the quantum theory.

$$E = hf \qquad 19.32$$

$$p = \frac{h}{\lambda} \qquad 19.33$$

Although no method exists for measuring it, the *de Broglie wave velocity* is given by equation 19.34.

$$v_w = \lambda f = \frac{E_{total}}{p} = \frac{c^2}{v} = \frac{hf}{p} \qquad 19.34$$

The velocity of the particle, v, is always less than the velocity of light, c. Therefore, the wave velocity, v_w, is always greater than the velocity of light.

The wavelength of a particle with zero rest mass (e.g., a photon) is given by equation 19.35. If E_{total} is in electron-volts, and λ is in centimeters, then the product hc has a value of 1.24 EE−4.

$$\lambda = \frac{hc}{E_{total}} \qquad 19.35$$

[9]Louis Victor de Broglie, 1892–, France.

The *de Broglie wavelength* of an electron follows from the requirement that the electron wave be a standing wave with average radius r (Bohr's radius) around the nucleus. Since there must be an integral number of wavelengths,

$$2\pi r = n\lambda \qquad 19.36$$

Combining equation 19.36 with 19.34 yields equation 19.16. The assumption of a wave form eliminates the problem of energy transmission from a revolving charged particle.

Example 19.1

An electron moves with an effective velocity of 2 EE9 cm/sec. What is its de Broglie wavelength, wave velocity, particle momentum, and total particle energy?

The wave velocity is

$$v_w = \frac{c^2}{v} = \frac{(3\text{ EE}10)^2}{2\text{ EE }9}$$

$$= 4.5\text{ EE}11\text{ cm/sec}$$

The total relativistic energy of the electron is

$$E_{total} = \frac{(5.486\text{ EE}-4)\text{amu }(9.315\text{ EE}8)\text{ eV/amu}}{\sqrt{1-\left(\frac{2\text{ EE}9}{3\text{ EE}10}\right)^2}}$$

$$= 5.122\text{ EE}5\text{ eV}$$

The momentum is

$$p = \frac{E_{total}}{v_w} = \frac{(5.122\text{ EE}5)\text{eV}}{(4.5\text{ EE}11)\text{ cm/sec}}$$

$$= 1.138\text{ EE}-6\ \frac{\text{eV-sec}}{\text{cm}}$$

And, thus

$$\lambda = \frac{h}{p} = \frac{(4.136\text{ EE}-15)\text{eV-sec}}{(1.138\text{ EE}-6)\text{eV-sec/cm}}$$

$$= 3.634\text{ EE}-9\text{ cm}$$

8. Heisenberg's Uncertainty Principle

Heisenberg[10] stated that the product of the error in the determination of the position (x) of any particle and its momentum (p_x) is in the order of magnitude of Planck's

[10]Werner Karl Heisenberg, 1901–1976, Germany.

ATOMIC

constant. The more accurate the former, the less accurate will be the latter. A similar statement was made for the uncertainty in determining the time at which an event occurs (t) and the change in total energy ($E_2 - E_1$). The physical significance of these uncertainty principles is that the act of measurement actually disturbs the particle.

Heisenberg's uncertainty principles can be written as equations 19.37 and 19.38.

$$\varepsilon_x \varepsilon_p \sim h \qquad 19.37$$

$$\varepsilon_t \varepsilon_{(E_2 - E_1)} \sim h \qquad 19.38$$

Example 19.2

An electron is located with velocity of 2.97 EE2 m/sec, known to an accuracy of 1 part in 10,000. With what fundamental accuracy can the position be located?

The momentum is

$$p = mv = (9.1\ \text{EE}-31)\text{kg}(2.97\ \text{EE}2)\text{m/sec}$$
$$= 2.703\ \text{EE}-28\ \text{kg-m/sec}$$

Since velocity is known to one part in 10,000, the uncertainty in momentum is

$$\varepsilon_p = \frac{1}{10{,}000}(2.703\ \text{EE}-28)$$
$$= 2.703\ \text{EE}-32\ \text{kg-m/sec}$$

The minimum uncertainty in position is found from equation 19.37,

$$\varepsilon_x \sim \frac{h}{\varepsilon_p} = \frac{(6.626\ \text{EE}-34)\text{Joule-sec}}{(2.703\ \text{EE}-32)\text{kg-m/sec}}$$
$$\sim 2.45\ \text{EE}-2 \frac{\text{Joule-sec}^2}{\text{kg-m}}$$

Since a joule is equivalent to units of $\left(\frac{\text{kg-m}^2}{\text{sec}^2}\right)$,

$$\varepsilon_x \sim 2.45\ \text{EE}-2\ \text{m}$$

9. Additional Nuclear Particles

An *elementary particle* is one which cannot be broken down into smaller particles. The electron, proton, and neutron were originally thought to be the only elementary particles. Now, approximately 200 elementary particles have been documented — and, there is consid-

Table 19.6
The Major Elementary Particles

Antiparticles have the same mass and spin as the particles but their charges and strangeness numbers are opposite in sign.

family name	particle name	symbol	mass, electron masses	spin	strangeness	charge	antiparticle	no. of particles	average lifetime, seconds	typical decay products
	photon	γ (gamma ray)	0	1	0	0	same particle	1	infinite	—
electron family	electron	e^-	1	½	—	−e	e^+	2	infinite	—
	electron's neutrino	ν_e	0	½	—	0	$\bar{\nu}_e$	2	infinite	—
muon family	muon	μ^-	206.77	½	—	−e	μ^+	2	2.212×10^{-6}	$e^- + \bar{\nu}_e + \nu_\mu$
	muon's neutrino	ν_μ	0(?)	½	—	0	$\bar{\nu}_\mu$	2	infinite	—
mesons	pion	π^+ π^0	273.2 264.2	0 0	0 0	+e 0	π^- π^0	3	2.55×10^{-8} 1.9×10^{-16}	$\mu^+ + \nu_\mu$ $\gamma + \gamma$
	kaon	K^+ K^0	966.6 974	0 0	+1 +1	+e 0	$\bar{K}^+$ $\bar{K}^0$	4	1.22×10^{-8} 1.00×10^{-10} and 6×10^{-8}	$\pi^+ + \pi^0$ $\pi^+ + \pi^-$
baryons	nucleon	p^+ (proton) n^0 (neutron)	1836.12 1838.65	½ ½	0 0	+e 0	$\bar{p}^+$ $\bar{n}^0$	4	infinite 1013	$p + e^- + \bar{\nu}_e$
	lambda particle	Λ^0	2182.8	½	−1	0	$\bar{\Lambda}^0$	2	2.51×10^{-10}	$p + \pi^-$
	sigma particle	Σ^+ Σ^- Σ^0	2327.7 2340.5 2332	½ ½ ½	−1 −1 −1	+e −e 0	$\bar{\Sigma}^+$ $\bar{\Sigma}^-$ $\bar{\Sigma}^0$	6	8.1×10^{-11} 1.6×10^{-10} about 10^{-20}	$n + \pi^+$ $n + \pi^-$ $\Lambda^0 + \gamma$
	xi particle	Ξ^- Ξ^0	2508 2570	½ ½	−2 −2	−e 0	$\bar{\Xi}^-$ $\bar{\Xi}^0$	4	1.3×10^{-10} about 10^{-10}	$\Lambda^0 + \pi^-$ $\Lambda^0 + \pi^0$

erable evidence that the electron, proton, and neutron are not elementary.[11]

Elementary particles can be categorized into families according to their properties. These families are illustrated in figure 19.5. Typical properties of some elementary particles are given in table 19.6.

- Hadrons (strongly interacting)
  - Baryons
    - Hyperons Λ, Σ, Ξ, Ω (strange)
    - Nucleons *N*, *P*, *N* (nonstrange)
  - Mesons
    - (strange) *K*
    - (nonstrange) π
- Leptons (weakly interacting)
  - Muon family μ, ν_μ
  - Electron family *e*, ν_e
- Photon γ (electromagnetic interaction)

Figure 19.5 Families of Elementary Particles

10. Nuclear Transformations

Nuclear reaction equations are similar to chemical reaction equations. Each particle is assigned a symbol, a superscript equal to its mass in *amu*,[12] and a subscript equal to its atomic number or charge. In nuclear reactions, both the superscripts and subscripts must balance. In the examples below, note the shorthand methods of indicating the incident and product particles.

$$\text{Longhand: } {}_7N^{14} + {}_1H^1 \rightarrow {}_6C^{11} + {}_2He^4 \qquad 19.39$$

$$\text{Shorthand: } {}_7N^{14} + {}_1H^1 \overset{\alpha}{\rightarrow} {}_6C^{11} \qquad 19.40$$

$$\text{Shorthand: } N^{14}(p,\alpha)C^{11} \qquad 19.41$$

Extremely accurate measurements of the product masses may show a decrease from the sum of the reacting particles. The decrease in mass *(mass defect)* is due to the release of *binding energy*. This energy release is given by Einstein's equation. The energy is in joules if *m* is in kilograms and *c* is in m/sec.

$$\Delta E = (\Delta m)c^2 \qquad 19.42$$

If the products have more mass than the reactants, the deficiency must be made up by the kinetic energy of the reactants.

Table 19.7 lists the approximate masses of some basic particles, atoms, and rays. The masses are given in both the carbon-12 and oxygen-16 scales. The oxygen-16 scale has been used extensively in the past. However, the carbon-12 scale is now used exclusively by physicists. Data from the two systems should never be mixed.

Table 19.7
Masses of Atoms, Particles, and Rays

names	symbols	approximate rest mass, amu C-12 scale	approximate rest mass, amu O-16 scale
gamma ray, short wavelength photon	γ	0.000000	0.000000
electron, beta particle	${}_{-1}e^0$, β	.0005486	.0005488
positron, positive electron	${}_{+1}e^0$	.0005486	.0005488
proton, hydrogen atom nucleus	${}_1H^1$, p	1.007277	1.007593
hydrogen atom	H	1.007825	1.008141
neutron	${}_0n^1$, n	1.008665	1.008981
deuteron, deuterium nucleus	${}_1H^2$	2.01355	2.01418
deuterium atom	D	2.01410	2.01473
tritium atom	T	3.01605	3.01700
alpha particle, helium nucleus	${}_2He^4$, α	4.00150	4.00276
helium atom	He	4.00260	4.00386
carbon-12	C	12.00000	12.00376
oxygen-16	O	15.99491	16.00000

Example 19.3

What is the binding energy of the deuterium nucleus?

The deuterium nucleus consists of a proton and a neutron. The mass defect is the difference between the total nucleus mass and the mass of its parts. The particle masses are found in table 19.7.

$$\Delta m = 2.01355 - 1.007277 - 1.008665$$
$$= -.002392 \text{ amu}$$

The negative sign implies that energy would have to be added to break the deuterium nucleus into a proton and a neutron. The binding energy is

$$E_b = (.002392) \text{ amu } (1.66053 \text{ EE}-27) \text{ kg/amu } (3 \text{ EE}8)^2 \text{ (m/sec)}^2$$
$$= 3.575 \text{ EE}-13 \text{ J}$$

11. Nuclear Fission

Fission can occur only in heavy nuclei, such as those listed in table 19.8. The absorption of a neutron with energy above the threshold energy causes the target nucleus to split into two smaller fission fragments. These fragments each emit one or two 2 MeV neutrons. The fission fragments are still radioactive and continue to decay several times. Although the exact fission frag-

[11]Physicists now believe that all particles which interact strongly with other particles are constructed of various types of quarks.

[12]atomic mass units. One *amu* is approximately equal to 1.66000 EE−27 kg on the oxygen scale and 1.66053 EE−27 kg on the carbon scale.

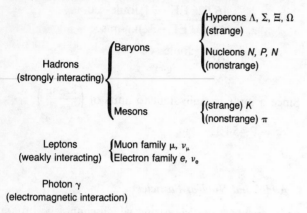

ments vary considerably, a typical fission reaction is given below:

$$_{92}U^{235} + {}_0n^1 \rightarrow {}_{92}U^{236} \rightarrow$$

$$_{54}Xe^{140} + {}_{38}Sr^{94} + 2{}_0n^1 + \gamma + 200 \text{ MeV} \qquad 19.43$$

The energy release in fission reactions involving uranium, plutonium, and thorium is essentially the same — around 200 MeV per fission. This energy is divided up as indicated below. All but the neutrino energy (6%) is useful.

- kinetic energy of the fragments 84%
- kinetic energy of the neutrons 2.5%
- prompt gamma rays 2.5%
- decay of fission fragments 11%

Fission is a chain reaction because only 1 neutron is required to produce 2 or 3 other neutrons. Because the fission fragments also decay and emit delayed neutrons, the average number of neutrons obtained in uranium fission is 2.43. In plutonium fission the value is 2.95.

Thermal neutrons will fission fuel having odd atomic masses such as U-235 and Pu-239. However, fast neutrons are required to fission the stable isotopes with even atomic masses such as U-238 and Pu-240. Fission can also occur when fuel is exposed to high-energy gamma or x-rays. Such fission is known as *photofission*.

Table 19.8 Fission Threshold Energies (MeV)

	neutrons	deuterons	alpha	x,γ rays
U-235	.025 eV	8	21	5.31 ± .25
U-238	1.0 ± .1	8	21	5.8 ± .15
Pu-239	.025 eV			5.31 ± .27
Th-232	1.1 + .05	8	21	5.4 ± .22
Pa-231	.45			
Np-237	.25			
U-234	.3			
U-233	.025 eV			
Pu-241	.025 eV			

The *four-factor formula* determines whether or not fission can sustain itself in a large reactor with various opportunities for neutron loss. This formula calculates *infinite-k* which must be greater than one for sustained fission.

$$k_\infty = \eta\, \varepsilon\, p\, f \qquad 19.44$$

Not all thermal neutrons cause fission. Some are absorbed in other non-fission reactions. The *multiplication constant* of the reactor core, η, is the number of neutrons left over to carry on the reaction. ε is known as the *fast fission factor.* It is the ratio of the total number of neutrons produced by U-235 and U-238 fissions to the number of neutrons produced by U-235 fissions alone.

A minority fraction $(1-p)$ of the moderated neutrons undergo *radiative capture*. This transforms U-238 into U-239 and eventually into Pu-239. p is known as the *resonance escape probability.* Of the remaining neutrons, a small fraction $(1-f)$ are absorbed by the moderator and structure. f is the *thermal utilization,* the neutron fraction entering the fuel.

12. Breeder Reactors

Fissionable material is produced in breeder reactors by placing a blanket of fertile material (U-238 or Th-232) around a core of concentrated fissionable material (Pu-239 or U-233). Core neutrons are captured by the blanket to breed more fissionable fuel according to the following reactions:

$$_{92}U^{238} + \underset{\text{fast}}{{}_0n^1} \rightarrow {}_{92}U^{239} \xrightarrow[\beta-]{(23 \text{ min})} {}_{93}Np^{239} \xrightarrow[\beta-]{(2.3 \text{ days})} {}_{94}Pu^{239} \qquad 19.45$$

$$_{90}Th^{232} + \underset{\text{fast}}{{}_0n^1} \rightarrow {}_{90}Th^{233} \xrightarrow[\beta-]{(23 \text{ min})} {}_{91}Pa^{233} \xrightarrow[\beta-]{(27 \text{ days})} {}_{92}U^{233} \qquad 19.46$$

The *doubling time* of the reactor is the time (usually expressed in days) to double the number of plutonium atoms.

13. Fusion

If two or more light atoms have sufficient energy (available only at high temperatures), they may *fuse* together to form one or more heavier atoms. Because of the high temperatures required, the reaction is called a *thermonuclear reaction.* During fusion, mass is lost and converted to energy according to equation 19.39.

Cold stars (2 EE6 °K) produce energy by the *proton-proton cycle.* The proton-proton cycle consists of three related reactions.

$$_1H^1 + {}_1H^1 \rightarrow {}_1H^2 + e^+ + .4 \text{ MeV} \qquad 19.47$$

$$_1H^1 + {}_1H^2 \rightarrow {}_2He^3 + 5.5 \text{ MeV} \qquad 19.48$$

$$_2He^3 + {}_2He^3 \rightarrow {}_2He^4 + 2{}_1H^1 + 12.9 \text{ MeV} \qquad 19.49$$

Since the first 2 reactions must each occur twice for each $_2He^4$ produced, the total energy is more than is apparent. The energy released per helium atom production after neutrino energy is subtracted is approximately 26.2 MeV. Hotter stars fuse from the *carbon cycle,* which produces about 24.7 MeV per cycle.

The most promising commercial cycles are simpler. Two possibilities are listed here.

$$_1H^2 + {}_1H^2 \rightarrow {}_1H^3 + {}_1H^1 + 4.0 \text{ MeV} \qquad 19.50$$

$$_1H^2 + {}_1H^2 \rightarrow {}_2He^3 + {}_0n^1 + 3.3 \text{ MeV} \qquad 19.51$$

Both reactions have approximately the same probability of occurring. However, the high temperatures (EE6 °K) required and the confinement problems make commercial fusion difficult to achieve. Recent advances have replaced the thermal energy requirements with kinetic energy from high-velocity particles. Very short-duration fusion reactions have been obtained by this method.

14. Radioactivity

Spontaneous disintegration of some neutral nuclei was observed in the early 1900's. Neutral disintegration occurs because the element has excess or insufficient neutrons. These neutrons have zero electrostatic effect, but they provide positive nuclear attraction to balance proton repulsion in the nucleus.

If the nucleus has excess neutrons, one or more may spontaneously transform into protons with emission of electrons to retain charge neutrality. In addition to the beta particle, gamma radiation may be given off to change the binding energy. Decay with the emission of electrons is known as *− beta decay.*

$$ {}_0n^1 \rightarrow {}_{+1}p^1 + {}_{-1}e^0 \qquad 19.52 $$

If the nucleus has too few neutrons, a proton becomes a neutron with a positron emission. This is known as *+ beta decay:*

$$ {}_{+1}p^1 \rightarrow {}_0n^1 + {}_{+1}e^0 \qquad 19.53 $$

Alpha decay decreases both the number of protons and neutrons by 2. Such an equilateral drop in nucleus mass may also place the atom in the stable region.

$$ 2{}_0n^1 + 2{}_{+1}p^1 \rightarrow {}_2He^4 \qquad 19.54 $$

Radioactive decay may be described mathematically through the use of a *decay constant,* λ, which is essentially independent of the environment. The number of disintegrations per second is known as the *activity, A,* typically measured in *curies.* (One curie is equal to 3.7 EE10 disintegrations per second.) The *half-life* is the time required to reduce the number of atoms from the original number, N_o, to ½N_o. The half-life is not the same as the *mean life expectancy,* $\bar{t}$.

Table 19.9
Approximate Half-Lives (years)

C-14	5730
Pu-239	2.411 EE4
Ra-226	1620
Th-232	1.4 EE10
U-234	2.44 EE5
U-235	7.08 EE8
U-238	4.468 EE9

$$ m_t = m_o e^{-\lambda t} \qquad 19.55 $$

$$ A_t = \lambda N_t = A_o e^{-\lambda t} \qquad 19.56 $$

$$ N_t = N_o e^{-\lambda t} \qquad 19.57 $$

$$ t_{1/2} = \frac{.693}{\lambda} \qquad 19.58 $$

$$ \bar{t} = \frac{1}{\lambda} \qquad 19.59 $$

Example 19.4

60 kilograms of Pu-239 are to be sealed in a storage unit deep within a mountain. If the plutonium is not retrieved until after 2000 years have passed, what will its activity be?

From table 19.9, the half-life of Pu-239 is 2.4 EE4 years. From equation 19.58, the decay constant is

$$ \lambda = \frac{.693}{t_{1/2}} = \frac{.693}{2.4 \text{ EE4 years}} $$

$$ = 2.888 \text{ EE}-5 \text{ (1/yr)} $$

The mass of Pu-239 after 2000 years is given by equation 19.55.

$$ m_{2000} = (60)e^{(-2.888 \text{ EE-5})(2000)} $$

$$ = 56.63 \text{ kg} $$

The number of moles of Pu-239 at t = 2000 is

$$ n = \frac{56.63}{239}(1000) = 237.0 $$

Using Avogadro's number, the number of atoms at t = 2000 is

$$ N_{2000} = (237.0) \text{ moles } (6.023 \text{ EE23}) \frac{\text{atoms}}{\text{mole}} $$

$$ = 1.427 \text{ EE26} $$

Using equation 19.56, the activity at t = 2000 is

$$ A_{2000} = \frac{(2.888 \text{ EE}-5) \text{ 1/yr } (1.427 \text{ EE26}) \text{ atoms}}{(3.15 \text{ EE7}) \text{ sec/year}} $$

$$ = 1.308 \text{ EE14 dps} $$

15. Radiocarbon Dating

Neutrons from cosmic rays combine with atmospheric nitrogen according to the following reaction:

$$ {}_7N^{14} + {}_0n^1 \rightarrow {}_6C^{14} + {}_1H^1 \qquad 19.60 $$

Some of the resulting radioactive carbon eventually forms radioactive carbon dioxide, $^{14}CO_2$. The concentration of radioactive carbon is assumed to be stable at 1.29 EE − 12 parts with an equilibrium activity of 15.3 disintegrations per minute per gram of carbon. This concentration is repeated in the tissues of all living animals and plants, but it decreases upon death since decaying radioactive carbon is not replaced by breathing. The decay reaction is

$$ {}_6C^{14} \rightarrow {}_{-1}e^0 + {}_7N^{14} \qquad 19.61 $$

This reaction has a half-life of 5730 years and a decay constant of 1.209 EE − 4 (1/years) or 3.835 EE − 12 (1/sec).

The age of an object can be found from the current concentration (C.C.) of radioactive carbon:

$$\frac{\text{C.C.}}{1.29\ \text{EE}-12} = e^{-\lambda t} \qquad 19.62$$

Example 19.5

A 470 gram sample of burned wood recovered from an excavation has an activity of .5 EE−3 microcuries. What is the approximate age of the sample?

The activity is

$$A = (.5\ \text{EE}-9)\ \text{curies}\ (3.7\ \text{EE}10\ \text{dps/curie})$$
$$= 18.5\ \text{dps}$$

The number of C^{14} atoms is given by equation 19.56.

$$N_t = A_t/\lambda = \frac{18.5}{3.835\ \text{EE}-12}$$
$$= 4.82\ \text{EE}12\ \text{atoms}$$

The approximate mass of the C^{14} is

$$\frac{(4.82\ \text{EE}12)\text{atoms}\ (14)\text{g/gmole}}{(6.023\ \text{EE}23)\text{atoms/gmole}}$$
$$= 1.12\ \text{EE}-10\ \text{g}$$

The mass of the C^{12} in the firewood is

$$470 - (1.12\ \text{EE}-10) \approx 470$$

The number of C^{12} atoms is

$$\frac{470}{12}(6.023\ \text{EE}23) = 2.359\ \text{EE}25$$

Thus, the current concentration of C^{14} atoms is

$$\text{C.C.} = \frac{4.82\ \text{EE}12}{4.82\ \text{EE}12 + 2.359\ \text{EE}25} \approx \frac{4.82\ \text{EE}12}{2.359\ \text{EE}25}$$
$$= 2.04\ \text{EE}-13$$

From equation 19.62,

$$\frac{2.04\ \text{EE}-13}{1.29\ \text{EE}-12} = .1581 = e^{-3.835\ \text{EE}-12\ t}$$

$$t = 4.81\ \text{EE}11\ \text{seconds} = 15{,}480\ \text{years}$$

16. Relativity

Einstein's[13] two basic postulates were:

- The laws of physics are the same in all frames of reference with constant linear velocity.
- The observed speed of light is independent of any motion the observer may have.

The laws pertaining to relativistic motion are given below. The subscripts $_o$ and $_v$ represent stationary and movement at velocity v respectively. The term k is defined as

$$k = \frac{1}{\sqrt{1 - \left(\frac{v}{c}\right)^2}} \qquad 19.63$$

The increased mass at velocity v is

$$m_v = km_o = m_o + \Delta m \qquad 19.64$$

The *length contraction* is

$$L_v = L_o/k = L_o - \Delta L \qquad 19.65$$

The *time dilation* experienced by the stationary observer for every t_o measured by the moving observer is

$$t_v = kt_o = t_o + \Delta t \qquad 19.66$$

If f_o is the frequency observed by the stationary observer, the frequency observed by the moving observer is given by the relativistic *Doppler effect* equation:

$$f_v = f_o k\left(1 - \frac{v}{c}\right) \qquad 19.67$$

If objects 1 and 2 are approaching each other with velocities v_1 and v_2 measured with respect to some stationary point, v^* is the *relativistic velocity sum* measured with respect to either of the objects. Observe proper direction sign conventions when using equation 19.68.

$$v^* = \frac{v_1 - v_2}{1 - \frac{v_1 v_2}{c^2}} \qquad 19.68$$

The velocity as a function of acceleration and time is

$$v = \frac{at}{\sqrt{1 + \left(\frac{at}{c}\right)^2}} \qquad 19.69$$

The relativistic kinetic energy is

$$E_k = (m_v - m_o)c^2$$
$$\approx \tfrac{1}{2}m_o v^2 + (\tfrac{3}{8})m_o\left(\frac{v^4}{c^2}\right) + \cdots \qquad 19.70$$

Example 19.6

A 1000 meter long interstellar craft travels at .7*c* relative to a planet of diameter 10,000 km. (a) What is the apparent length of the craft as viewed from the craft itself? (b) What is the apparent length of the craft as seen from the planet? (c) What is the apparent diameter of the planet as seen from the craft?

(a) Since the observer moves with the craft, the relative velocity v is zero. Thus, $k=1$ and the craft appears to be 1000 meters long.

(b) At $v = .7c$, $k = 1.4$. Thus, the apparent length is 1000/1.4 = 714 meters.

(c) Since motion is relative, the apparent diameter is 10,000/1.4 = 7140 kilometers.

[13] Albert Einstein, 1879–1955, Germany (USA)

ATOMIC

Example 19.7

A beam of pions with a stationary half-life of 1.8 EE – 8 seconds attains a velocity of .6*c*. How far will the beam travel before half of the pions have decayed?

At $v = .6c$, $k = 1.25$. From the standpoint of the stationary observer, the half-life is increased to 1.25(1.8 EE – 8) = 2.25 EE – 8 seconds. The apparent distance traveled is

$$x = (2.25\text{ EE}-8)\text{ sec }(.6)(3\text{ EE8})\text{ m/sec}$$
$$= 4.05\text{ meters}$$

Example 19.8

A spacecraft travels at velocity .4*c* towards a star emitting photons (velocity = 1.0*c*). What is the photon velocity as measured by the spacecraft?

This is an example of a relativistic velocity sum.

$$v^* = \frac{1.0c + .4c}{1 + (1.0)(.4)} = 1.0c$$

Thus, the speed of light is constant regardless of the observer velocity.

17. Acceleration of Charged Particles

Equations 19.71 through 19.77 may be used with any system of consistent units. See Appendix B, chapter 3 for a listing of consistent electromagnetic units.

A. Cyclotrons

Charged particles move in circular paths when exposed to magnetic fields. A *cyclotron* consists of a large electromagnet which produces the uniform magnetic field necessary to keep the charged particles traveling in circles. The charged particles are also accelerated by an electric field. The polarity of the electric field is switched every half revolution to keep the particles accelerating and spiraling outward. For this reason, cyclotrons are also known as *magnetic resonance accelerators*.

Operation of a cyclotron is governed by the following equations.

$$r = \text{path radius} = \frac{mv}{Bq} \qquad 19.71$$

$$\omega = \text{angular velocity} = \frac{v}{r} = \frac{Bq}{m} \qquad 19.72$$

$$v = \text{linear velocity} = \frac{qBr}{m} \qquad 19.73$$

$$T = \frac{\text{time per}}{\text{revolution}} = \frac{2\pi r}{v} = \frac{2\pi m}{qB} = \frac{2\pi}{\omega} \qquad 19.74$$

$$E_k = \frac{\text{kinetic}}{\text{energy}} = {}^1\!/\!_2 mv^2 = \frac{(qBr)^2}{2m} = {}^1\!/\!_2 m(r\omega)^2 \qquad 19.75$$

$$V = \frac{\text{particle}}{\text{voltage}} = \frac{q(Br)^2}{2m} \qquad 19.76$$

$$f = \frac{\text{required}}{\text{frequency}} = \frac{1}{T} \qquad 19.77$$

Since most cyclotrons operate at a fixed frequency, the relativistic variation in mass of the particles sets an upper limit for the energy which may be imparted. Those limits are approximately 10 MeV for protons, 20 MeV for deuterons, and 40 MeV for alpha particles.

However, if a cyclotron is capable of varying its switching frequency, the previous equations can be used if the relativistic mass, m_v, is used in place of *m*.

Example 19.9

A cyclotron oscillates at 1.3 EE7 Hz with a dee radius of 20 inches. What energy can be imparted to deuterons?

From equation 19.74, the required magnetic flux density is

$$B = \frac{2\pi(1.3\text{ EE7})\text{ Hz }(2.01355)\text{ amu }(1.66053\text{ EE}-27)\text{ kg/amu}}{(1.602\text{ EE}-19)\text{ coul}}$$
$$= 1.705\text{ wb/m}^2$$

From equation 19.75, the imparted energy is

$$E_k = \frac{(1.602\text{ EE}-19)^2(1.705)^2(20)^2(.0254)^2(\text{m/ft})^2}{2(2.01355)(1.66053\text{ EE}-27)}$$
$$= 2.88\text{ EE}-12\text{ J}$$
$$= \frac{2.88\text{ EE}-12\text{ J}}{1.602\text{ EE}-19\text{ eV/J}}$$
$$= 1.8\text{ EE7 eV}$$

B. Linear Accelerators

Linear accelerators accelerate charged particles in a straight line. These equations follow directly from the relativistic equations.

$$E_{total} = E_{rest\ mass} + E_{kinetic} = m_o k(c^2) \qquad 19.78$$

$$E_{rest\ mass} = m_o(c^2) \qquad 19.79$$

$$E_{kinetic} = m_o(k-1)(c^2) \qquad 19.80$$

$$m_v = km_o \qquad 19.81$$

$$v = c\sqrt{1-(m_o/m_v)^2} = c\sqrt{1-(1/k)^2} \qquad 19.82$$

Example 19.10

The Stanford Linear Accelerator is capable of delivering 25 BeV electrons. What velocity does this correspond to?

$$25\text{ BeV} = (2.5\text{ EE10})\text{ eV }(1.602\text{ EE}-19)\text{ J/eV}$$
$$= 4.00\text{ EE}-9\text{ J}$$

From equation 19.63,

$$k = \frac{4.00\text{ EE}-9\text{ J}}{(9.11\text{ EE}-31)\text{ kg }(3.00\text{ EE8})^2(\text{m/s})^2}$$
$$= 4.88\text{ EE4}$$

From equation 19.82,

$$v = c\sqrt{1-(1/4.88\ EE4)^2} = .9999999998c$$

18. Interactions of Particles with Matter

Whether or not a moving particle interacts with a nearby atom depends on the effective size of the atom. The effective size of atoms is measured in square centimeters, analogous to the size of a target at which the particles are beamed. Since the effective area is small, *cross sections* (short for *cross sectional areas*) are usually expressed in *barns*, where one barn equals EE−24 cm^2. The symbol for the cross section is σ.

Neutrons, to consider a specific type of particle, can interact with atoms in two ways—they can be absorbed or scattered. The *total cross section,* therefore, may be written as

$$\sigma_t = \sigma_a + \sigma_s \quad 19.83$$

It is useful to think of cross sections as being *interaction probabilities* per unit path length. Thus, the probability of a neutron being absorbed by an atom is

$$p\{\text{absorption}\} = \frac{\sigma_a}{\sigma_t} \quad 19.84$$

The probability that a neutron will interact after traveling distance x through a substance is

$$p\{\text{interacting}\} = \sigma_t N x \quad 19.85$$

N is the number of atoms per unit volume.

$$N = \frac{\rho N_o}{A.W.} \quad 19.86$$

If a neutron beam has an incident density, J, and an area, A, the average number of interactions with the target per second is

$$\text{interaction rate} = JA\sigma_t Nx \quad 19.87$$

The product $N\sigma$ occurs so often in nuclear calculations that it is given the special symbol Σ and the special name *macroscopic cross section.* Thus, equation 19.87 can be rewritten as

$$\text{interaction rate} = JA\Sigma x \quad 19.88$$

Example 19.11

A carbon-12 graphite target .05 cm thick and .5 cm^2 in area is struck by a 0.1 cm^2 neutron beam with intensity of 5 EE8 neutrons/cm^2-sec. If the density is 1.6 g/cm^3 and σ_t = 2.6 barns for graphite, what is the interaction rate and probability that there will be an interaction?

$$N = \frac{(1.6)(6.023\ EE23)}{12}$$

$$= 8.031\ EE22\ \text{atoms/cm}^3$$

The probability of interaction is

$$(2.6\ EE-24)\ \text{cm}^2(8.031\ EE22)\ \text{atoms/cm}^3(.05)\ \text{cm}$$

$$= 1.04\ EE-2\ \text{(about 1\%)}$$

The interaction rate is

$$(5\ EE8)\ \text{neutrons/cm}^2\text{-sec}\ (.1)\ \text{cm}^2\ (1.04\ EE-2)$$

$$= 5.2\ EE5\ \text{neutrons/sec}$$

Particles may be absorbed or scattered in several different ways.

- Scattering (subscript $_s$)
  - *Inelastic scattering* (subscript $_{is}$) occurs when the nucleus is struck by a neutron, gains energy, and then decays by emitting a gamma ray. This reaction is written as (n,n′) since the reflected neutron has a different energy than the incident neutron.
  - *Elastic scattering* (subscript $_{es}$) occurs when kinetic energy is conserved in the collision. The struck nucleus remains at its original energy level. This is written as (n,n).
- Absorption (subscript $_a$)
  - *Radiative capture* (subscript $_c$ or $_\gamma$) occurs when the neutron is captured by the nucleus which then emits gamma radiation. This is written as (n,γ).
  - *Fission* (subscript $_f$) occurs when the neutron splits the atom into two smaller fragments.
  - *Proton decay* (subscript $_p$) is written as (n,p).
  - *Alpha decay* (subscript $_\alpha$) is written as (n,α).

Equation 19.83 can now be written:

$$\sigma_t = \sigma_{is} + \sigma_{es} + \sigma_c + \sigma_f + \sigma_p + \sigma_\alpha \quad 19.89$$

It is customary to describe neutrons according to their energies. The classification is given in table 19.10.

Table 19.10
Classification of Neutrons

energy (eV)	name
.001	cold
.025	thermal
1.	slow (resonant)
100.	slow
EE4	intermediate
EE6	fast
EE8	ultrafast
EE10	relativistic

Most reactors use *thermal neutrons.* The (n,p), (n,α), and (n,n′) reactions are rare at the thermal level, and equation 19.89 can be written as

$$\sigma_t = \sigma_{es} + \sigma_c + \sigma_f \quad 19.90$$

The term *thermal* implies that the neutrons have experienced sufficient collisions to bring them down to

ATOMIC

the same energy level as that of the surrounding atoms, usually assumed to be 20°C. Of course, not all neutrons travel at the exact same velocity. In fact, a Maxwellian velocity distribution occurs, similar to that assumed in the kinetic theory of gases. The most probable particle velocity is

$$v_p = 1.28\text{ EE4}\sqrt{T}\ \text{(cm/sec)} \qquad 19.91$$

Since the velocity is very low, the kinetic energy is

$$E_k = \tfrac{1}{2}mv^2 \qquad 19.92$$

Example 19.12

What is the kinetic energy of thermal neutrons?

$T = 20°C = 293°K$

$v_p = 1.28\text{ EE4}\sqrt{293} = 2.19\text{ EE5 cm/sec}$

$E_k = (.5)(1.008665)\text{amu}(1.66053\text{ EE}-27)\text{kg/amu}$

$\times\ (2.19\text{ EE3})^2(\text{m/sec})^2$

$= 4.017\text{ EE}-21\text{ J}$

19. Particle Attenuation

The thickness of a material required to reduce the intensity of a radiating source depends on the particle, particle energy, type of beam, and shielding material. *Alpha radiation* with its double charge is a short-range radiation because of residual path ionization and electrostatic interaction. *Beta particles* are only singly charged and penetrate to greater, although still short, distances. *Gamma radiation* has no charge and produces no ionization.

Gamma radiation poses the major health threat. It is attenuated by:

- *Compton*[14] *scattering* — elastic collision is the major attenuating mechanism
- *Photoelectric effect* — gamma energy causes an electron from a nearby atom to be ejected
- *Pair production* — gamma radiation with energy greater than 1.02 MeV transforms itself into a positron and an electron in the vicinity of a nucleus

For gamma rays, then

$$\sigma_t = \sigma_{\text{Compton}} + \sigma_{\text{photoelectric}} + \sigma_{\text{pair}}$$

$$\approx \sigma_{\text{Compton}} \qquad 19.93$$

Traditionally, the macroscopic cross section has been called the *attenuation coefficient* or *linear absorption coefficient.*

$$\mu_l = \Sigma_t = \sigma_t N \qquad 19.94$$

[14]Arthur Holly Compton, 1892–1962, USA

Approximate values of μ_l for 2 MeV gamma radiation are given in table 19.11.

Table 19.11 Linear Absorption Coefficients (1/cm)

water	0.049
concrete	0.09
iron	0.33
lead	0.52
aluminum	0.12
uranium	0.92

A second quantity, the *mass attenuation coefficient* (or *mass absorption coefficient*) with units of cm²/g is

$$\mu_m = \mu_l/\rho \qquad 19.95$$

μ_l and μ_m may be used to determine the intensity of a monoenergetic narrow-beam gamma source passing through a shield of thickness x.

$$I = I_o e^{-\mu_l x} = I_o e^{-\mu_m \rho x} \qquad 19.96$$

The *half-value thickness* is the thickness of shielding which will reduce the intensity to 50% of its original value. The half-value thickness and *half-value mass* are given by equations 19.97 and 19.98.

$$x_{1/2} = \frac{.693}{\mu_l} \qquad 19.97$$

$$m_{1/2} = \frac{.693}{\mu_m} \qquad 19.98$$

The thickness of shielding required for any other percentage of attenuation is

$$x_{\%} = \frac{\ln(\text{remaining fraction})}{\mu_l} \qquad 19.99$$

20. Radiological Effects

The previous calculations cannot be used to determine the shielding requirements for humans for two reasons. First, the source is usually not a collimated beam, and second, the attenuated beam is not lost, just scattered into a wider beam. These difficulties are handled by a *build-up factor, B.* Equation 19.96 is modified by that factor.

$$I = I_o B e^{-\mu_l x} = I_o B e^{-\mu_m \rho x} \qquad 19.100$$

The term *exposure* is used as a measure of gamma or x-ray field at the surface of an exposed object. Since this radiation produces an ionization of the air surrounding the object, the exposure is defined as

$$X = \frac{\#\text{ of ions produced}}{\text{mass of air}} \left(\frac{\text{coulombs}}{\text{kg}}\right) \qquad 19.101$$

The unit of exposure is the *roentgen,* equal to 2.58 EE−4 coul/kg. This is the exposure which produces one statcoulomb of ions in one cubic centimeter of air.

ATOMIC

The *exposure rate* is

$$X' = \frac{dX}{dt} = \frac{\text{\# of ions produced per unit time}}{(\text{mass of air})(2.58\ \text{EE}-4)} \left(\frac{\text{R}}{\text{sec}}\right) \qquad 19.102$$

Exposure and exposure rate are measures of ionization surrounding a person. These measures do not determine biological damage, which is dependent on the amount of energy deposited in the body. The change in kinetic energy of radiation particles as they pass through a body is δE_k. This energy is imparted to the body. The *absorbed dose* is

$$D = \delta E_k/\text{mass of object} \qquad 19.103$$

The unit of *dose* is the *rad* (radiation absorbed dose) with 1 rad equal to 100 ergs/g (same as .01 J/kg).

Some types of radiation are more damaging than others. To account for this, the *relative biological effectiveness* (RBE) is used to increase the actual dose. RBE is determined experimentally and is too difficult to use in normal calculations. A rounded up version of RBE is called the *quality factor, Q*. Thus, the *equivalent dose* is

$$H = QD \qquad 19.104$$

The unit of equivalent dose is the *rem* (radiation effective man). Values of *Q* are given in table 19.12.

Table 19.12
Approximate Quality Factors

x-rays and γ rays	1.0
β particles, < .03 MeV	1.7
β particles, > .03 MeV	1.0
thermal neutrons	5.0
fast neutrons	10.0
α particles	20.0
protons	10.0

A *maximum permissible dose* (MPD) of 1.25 rem per quarter is sometimes quoted. This is essentially the same as 100 mrem per week and 2.5 mrem per hour if distributed over 40 hours.

For population individuals not working in a nuclear environment, the suggested MPD as specified by the National Council on Radiation Protection (NCRP) is ½ rem per year, not to exceed 5 rems before age 30 (approximately 170 mrem/year).

Example 19.13

1 MeV gamma radiation with an intensity of 2 EE5 rays/cm^2-sec deposits energy at the rate of EE$-$2 ergs/g-sec. What is the equivalent absorbed tissue dose rate?

$$D' = \frac{\text{EE}-2\ \text{ergs/g-sec}}{100\ \text{ergs/g-rad}} = \text{EE}-4\ \text{rad/sec}$$

Since Q = 1 for gamma radiation,

$$H' = DQ = \text{EE}-4\ \text{rem/sec}$$

ATOMIC

Appendix A
Common Nuclear Conversions

Multiply	By	To Obtain
amu	9.31 EE8	eV
amu	1.492 EE10	J
amu	1.66 EE − 27	kg
BeV	EE9	eV
erg	6.242 EE11	eV
erg	EE − 7	J
erg	1.113 EE − 24	kg
eV	1.074 EE − 9	amu
eV	EE − 9	BeV
eV	1.602 EE − 19	J
eV	1.783 EE − 36	kg
eV	EE − 6	MeV
GeV	EE12	eV
grams	EE − 3	kg
J	6.705 EE9	amu
J	6.242 EE18	eV
J	1.113 EE − 17	kg
kg	6.025 EE26	amu
kg	5.610 EE35	eV
kg	8.987 EE16	J
MeV	EE6	eV

ATOMIC

ATOMIC STRUCTURE

1. What is the de Broglie wavelength of 90 eV electrons?

2. A Balmer series line has a wavelength of 6563A. What is the energy (in eV) of
   (a) a quantum of this radiation?
   (b) a photon of this radiation?

3. What is the energy of emitted radiation when an electron jumps from the second Bohr orbit to the first? The wavelength is 1216 A (the first Lyman line).

4. What is the de Broglie wavelength and wave velocity of an electron moving at 2 EE7 meter/sec?

5. What is the antiparticle of an antineutron?

6. A stationary proton annihilates a stationary antiproton, producing three pions. If the energy is equally divided between the pions, what is the energy of each?

HEISENBERG UNCERTAINTY

7. What is the momentum uncertainty of an electron which passes between capacitor plates with .0005 meter spacing?

8. How closely can we pinpoint the position of an electron moving at 280 ±.1 meter/sec?

NUCLEAR TRANSFORMATIONS

9. What energy is released when four 1.008145 amu particles fuse into one 4.00387 amu particle?

10. What energy must be put into the reaction given below?

$$_2He^4 + {}_7N^{14} \rightarrow {}_8O^{17} + {}_1H^1$$

He	4.00387 amu
N	14.0075 amu
O	17.0045 amu
H	1.00815 amu

RADIOACTIVE DECAY

11. Na^{24} is radioactive with a half-life of 15 hours. If there is an initial sample of 48 grams, how long will it take to reduce the sample to 9 grams?

12. A radioactive substance has a half-life of 4 months. What is the probability that a nucleus will decay
   (a) in the first 4 months?
   (b) after the first 4 months but within 8 months?
   (c) anytime within the first year?

13. What fraction of radium ($t_{1/2}$ = 1620 years) will have decayed after 4000 years?

14. The activity of a radioactive isotope is 200 counts per second 1 minute after formation, 100 counts per second 2 minutes after formation, and 50 counts per second 3 minutes after formation. What is the isotope's half-life ?

RELATIVITY

15. What is the energy (including rest mass energy) of the following particles?
   (a) proton moving at .4c
   (b) electron moving at .8c
   (c) 5000 Å photon moving at 1.0c

16. Consider a frame of reference in which an observer detects two electrons with velocities -.7c and +.8c moving in opposite directions along the x axis. What is the relative velocity of separation of an observer moving with one of the electrons?

17. A proton (rest mass 938 MeV) is accelerated to 985 MeV. What is the mass increase?

18. A pion (rest energy 140 MeV) is given 35 MeV of kinetic energy. If its average life is 1.9 EE-16 seconds if at rest, what is its moving average life?

PARTICLE ACCELERATION

19. What diameter synchroton is required if the magnetic field is 8000 gauss and the energy imparted to protons is 30 BeV?

20. The magnetic field inside a cyclotron is 18,000 gauss. What will be the frequency of oscillation when the protrons reach a velocity of .6c? What is the increase in proton mass?

21. What is the velocity of an electron which has 64 eV of kinetic energy?

20 Postscripts

This chapter collects comments, revisions, and commentary which cannot be incorporated into the body of the text until the next edition. New postscript sections are added as needed when the ENGINEER-IN-TRAINING REVIEW MANUAL is reprinted.

The Greek Alphabet

The Greek Alphabet

A	α	alpha	N	ν	nu
B	β	beta	Ξ	ξ	xi
Γ	γ	gamma	O	o	omicron
Δ	δ	delta	Π	π	pi
E	ϵ	epsilon	P	ρ	rho
Z	ζ	zeta	Σ	σ	sigma
H	η	eta	T	τ	tau
Θ	θ	theta	Υ	υ	upsilon
I	ι	iota	Φ	ϕ	phi
K	κ	kappa	X	χ	chi
Λ	λ	lambda	Ψ	ψ	psi
M	μ	mu	Ω	ω	omega

Cotangent

The definition of *cotangent* was omitted on page 1-12 in the series of equations 1.87 through 1.91.

$$\cot\theta = x/y$$

Cauchy-Schwarz Theorem

When equation 1.144 on page 1-15 is solved for cosθ, it is known as the *Cauchy-Schwarz theorem.*

$$\cos\theta = \frac{x_1x_2 + y_1y_2}{|\mathbf{V}_1|\,|\mathbf{V}_2|}$$

Taylor's Formula

Taylor's formula can be used to approximate a function in the vicinity of a particular point. The approximation consists of a series of terms, each composed of a derivative of the original function and a polynomial. Taylor's formula can be used only if the original function is continuous. The expression $f^{(n)}(x)$ is used to denote the *n*th derivative of the function f(x).

$$f(x) = f(a) + \frac{f'(a)}{1!}(x-a) + \frac{f''(a)}{2!}(x-a)^2 + \cdots + \frac{f^{(n)}(a)}{n!}(x-a)^n + R_{n+1}(x)$$

The term $R_{n+1}(x)$ is the remainder after n+1 terms. It is the difference between the true and approximate values.

Tensors

A scalar has magnitude only. A vector has magnitude and a definite direction. A *tensor* has magnitude in a specific direction, but the direction is not unique. An example of a tensor is stress. From the combined stress equation (equation 11.44 on page 11–13), stress at a point in a solid depends on the direction of the plane passing through that point. Tensors are frequently associated with *anisotropic materials* which have different properties in different directions. Other examples are dielectric constant and magnetic susceptibility.

A vector in three-dimensional space is completely defined by three quantities: F_x, F_y, and F_z. A tensor in three-dimensional space requires nine quantities for complete definition. These nine values are given in matrix form. The tensor definition for stress at a point is

$$\begin{pmatrix} \sigma_{xx} & \sigma_{xy} & \sigma_{xz} \\ \sigma_{yx} & \sigma_{yy} & \sigma_{yz} \\ \sigma_{zx} & \sigma_{zy} & \sigma_{zz} \end{pmatrix}$$

Highlights of the ECONOMIC RECOVERY ACT of 1981

The Economic Recovery Act of 1981 substantially changed the laws relating to personal and corporate income taxes. Some sections in chapter 2 (primarily those dealing with depreciation and advanced tax topics) are affected.

A. Depreciation

Property placed into service in 1981 or after must use the *Accelerated Cost Recovery System (ACRS)*. Other methods (straight-line, declining balance, etc.) cannot be used except in special cases.

Property placed into service in 1980 or before must continue to be depreciated according to the method originally chosen (e.g., straight-line, declining balance, or sum-of-years-digits). *ACRS* cannot be used.

Under *ACRS,* the cost recovery amount in the *j*th year of an asset's cost recovery period is calculated by multiplying the initial cost by a factor found from a table.

$$D_j = (\text{initial cost})(\text{factor})$$

The initial cost used is not reduced by the asset's salvage value for either the regular or alternate *ACRS* calculations. The factor found from the table depends on the asset's cost recovery period.

Table of Recovery Factors

	recovery period		
year	3 yrs	5 yrs	10 yrs
1	.25	.15	.08
2	.38	.22	.14
3	.37	.21	.12
4		.21	.10
5		.21	.10
6			.10
7			.09
8			.09
9			.09
10			.09

The ACRS groups tangible property into four categories depending on the cost recovery period. Tangible property is divided into *real property* and *personal property.* Real property is real estate (i.e., land and its structures). Land is never eligible for cost recovery. Personal property consists of all assets (other than real assets) used for the production of business income. Personal property is not for personal use.

3-Year Personal Property

The cost of cars, light trucks, special tools, all R&D equipment, and other assets with short lives should be recovered over 3 years. The full amount applies regardless of when the asset is placed into service during the first year.

If desired, the cost of the asset can be recovered over a longer period. Straight-line depreciation can be used with either a 5- or 12-year life. This is known as the *alternate ACRS calculation.* No other lives or depreciation methods are allowed. If this alternate method is used, only one-half of the cost recovery amount can be claimed in the first year. The remaining one-half can be claimed in the year following the end of the recovery period.

5-Year Personal Property

Machinery, single-purpose agricultural structures, and equipment not recovered over 3 years should be recovered over 5 years. The full percentage applies regardless of when the asset is placed into service during the first year.

The alternate method uses 12- and 25-year periods with straight-line calculations. No other lives or depreciation methods can be used. As with 3-year property, the half-year rule applies to amounts calculated with this alternate method.

10-Year Personal and Real Property

Selected heavy equipment and machinery (e.g., rail cars), theme park structures, and some public utility property should be recovered over 10 years. The full percentage applies regardless of when the asset is placed into service during the first year.

If desired, the cost of the asset can be recovered over 35 or 45 years using straight-line calculations for the alternate method. No other lives or depreciation methods can be used. The half-year rule applies to amounts calculated for the first year.

15-Year Real Property

Real property, including residential real estate, should be recovered over 15 years. Unlike the previous asset categories, however, the factors used depend on the month in which the asset was placed into service.

If desired, the asset can be recovered over 35 or 45 years using straight-line calculations. The half-year rule limiting first-year deductions does not apply to 15-year assets.

MISC

Recovery Factors for 15-Year Property

Year	Month Placed in Service											
	1	2	3	4	5	6	7	8	9	10	11	12
1	.12	.11	.10	.09	.08	.07	.06	.05	.04	.03	.02	.01
2	.10	.10	.11	.11	.11	.11	.11	.11	.11	.11	.11	.12
3	.09	.09	.09	.09	.10	.10	.10	.10	.10	.10	.10	.10
4	.08	.08	.08	.08	.08	.08	.09	.09	.09	.09	.09	.09
5	.07	.07	.07	.07	.07	.07	.08	.08	.08	.08	.08	.08
6	.06	.06	.06	.06	.07	.07	.07	.07	.07	.07	.07	.07
7	.06	.06	.06	.06	.06	.06	.06	.06	.06	.06	.06	.06
8	.06	.06	.06	.06	.06	.06	.06	.06	.06	.06	.06	.06
9	.06	.06	.06	.06	.05	.06	.05	.05	.05	.06	.06	.06
10	.05	.06	.05	.06	.05	.05	.05	.05	.05	.05	.06	.05
11	.05	.05	.05	.05	.05	.05	.05	.05	.05	.05	.05	.05
12	.05	.05	.05	.05	.05	.05	.05	.05	.05	.05	.05	.05
13	.05	.05	.05	.05	.05	.05	.05	.05	.05	.05	.05	.05
14	.05	.05	.05	.05	.05	.05	.05	.05	.05	.05	.05	.05
15	.05	.05	.05	.05	.05	.05	.05	.05	.05	.05	.05	.05
16	—	—	.01	.01	.02	.02	.03	.03	.04	.04	.04	.05

B. Additional First-Year Depreciation

The additional first-year depreciation of 20% is no longer allowed.

C. Immediate Write-Off

The cost of small assets and part of the cost of large assets can be written off entirely in the year of purchase. The limits on the amounts that can be written off each year are given in the following table.

Limits on Immediate Write-Offs

year	amount
1982	$ 5,000
1983	5,000
1984	7,500
1985	7,500
1986 and on	10,000

D. Salvage Value Reduction

Since salvage value is not included in the calculation of basis for depreciation under *ACRS* guidelines, the salvage value reduction is not applicable.

E. Tax Credits

Investment tax credits for the purchase of assets are allowed in the year of purchase. These credits are a fraction of the assets' cost.

$$\text{tax credit} = \begin{pmatrix}\text{initial}\\\text{cost}\end{pmatrix}\begin{pmatrix}\text{decimal}\\\text{amount}\end{pmatrix}$$

The decimal amounts are taken from the accompanying table. Only assets with lives of 3-years or greater qualify for tax credits. The decimal amounts for 1 and 2 years assume the asset is disposed of prior to the 3 years.

Tax Credit Amounts

life (years)	decimal amount
1	.02
2	.04
3	.06
4 or more	.10

F. Capital Gains

The maximum capital gains tax is 20% for individuals in the highest (50%) tax bracket.

G. Gain on the Sale of a Depreciated Asset

If an asset is sold for more than its current book value, the difference in selling price and book value is taxable income. The gain is taxed at capital gains rates. Excluded from this preferential treatment is non-residential real property depreciated under regular *ACRS* provisions. However, non-residential real property depreciated under the straight-line alternate method qualifies for the capital gains rate.

Minimum Attractive Rate of Return

A company may not know what effective interest rate to use in an economic analysis. In such a case, the com-

MISC

pany can establish a minimum acceptable return on its investment. This *minimum attractive rate of return (MARR)* should be used as the effective interest rate i in economic analyses.

Mortgage Payments

Equation 2.38 (page 2-16) calculates the number of payments necessary to pay off a loan. This equation can be solved with effort for the total periodic payment (PT) or the initial value of the loan (LV). It is easier, however, to use the (A/P,i%,n) factor to find the payment and loan value.

$$PT = (LV)(A/P,i\%,n)$$

If the loan is repaid in yearly installments, then i is the effective annual rate. If the loan is paid off monthly, then i should be replaced by the effective rate per month, ϕ from equation 2.24. For monthly payments, n is the number of months in the payback period.

Common Compounds

The chemical formula of a compound may not be apparent from the compound name. Important compounds in this category are listed in the accompanying table.

Osmosis

When two liquids are separated from each other by a membrane, solvent will flow spontaneously from the diluted to the concentrated liquid until the concentrations of the two liquids are equal. This process is known as *osmosis*. The solute may also move through the membrane. However, this process is known as *diffusion*. If the membrane pores are too large to allow solute molecules to pass, it is said to be a *semipermeable membrane*.

Linear Dynamo

A straight wire moving through an electric field (as shown in figure 13.9 on page 13-7) is sometimes referred to as a *linear dynamo*.

Digital Logic Families

Integrated circuits combine active and passive elements and their interconnections into a single chip. Such circuits are constructed on a silicon base using the production processes of epitaxial growth, masked impurity diffusion, oxide growth and etching, and photolithography. Resistors, capacitors, diodes, and transistors can be produced on a single silicon base by these processes.

Names and Formulas of Compounds

trade name	chemical name	chemical formula
acetone	acetone	$(CH_3)_2 \cdot CO$
acetylene	acetylene	C_2H_2
ammonia	ammonia	NH_3
ammonium	ammonium hydroxide	NH_4OH
aniline	aniline	$C_6H_5 \cdot NH_2$
bauxite	hydrated aluminum oxides	$Al_2O_3 \cdot 2\ H_2O$
bleach	calcium hypochlorite	$CaCl(OCl)$
borax	sodium tetraborate	$Na_2B_4O_7 \cdot 10\ H_2O$
carbide	calcium carbide	CaC_2
carbolic acid	phenol	C_6H_5OH
carbon dioxide	carbon dioxide	CO_2
carborundum	silicon carbide	SiC
caustic potash	potassium hydroxide	KOH
caustic soda	sodium hydroxide	$NaOH$
chalk	calcium carbonate	$CaCO_3$
cinnabar	mercuric sulfide	HgS
ether	di-ethyl ether	$(C_2H_5)_2O$
glycerine	glycerine	$C_3H_5(OH)_3$
graphite	crystaline carbon	C

trade name	chemical name	chemical formula
gypsum	calcium sulfate	$CaSO_4 \cdot 2\ H_2O$
iron chloride	ferrous chloride	$FeCl_2 \cdot 4\ H_2O$
laughing gas	nitrous oxide	N_2O
limestone	calcium carbonate	$CaCO_3$
magnesia	magnesium oxide	MgO
marsh gas	methane	CH_4
potash	potassium carbonate	K_2CO_3
prussic acid	hydrogen cyanide	HCN
pyrolusite	manganese dioxide	MnO_2
quicklime	calcium oxide	CaO
salammoniac	ammonium chloride	$NH_4\ Cl$
slaked lime	calcium hydroxide	$Ca(OH)_2$
soda ash	hydrated sodium carb.	$Na_2CO_3 \cdot 10H_2O$
soot	amorphous carbon	C
stannous chloride	stannous chloride	$SnCl_2 \cdot 2H_2O$
table salt	sodium chloride	$NaCl$
trilene	trichlorethylene	C_2HCl_3
urea	urea	$CO(NH_2)_2$
zinc blende	zinc sulfide	ZnS

The manner in which circuit elements are combined determines the family to which the complete circuit belongs. For example, the *RTL* family uses resistors and transistors to obtain the desired functions. Different families possess different operating characteristics. These characteristics are summarized in the accompanying table.

Some common integrated circuit and logic family terms are defined after the table.

Comparison of Major Digital IC Families
(Representative values given are intended for comparison purposes only.)

	RTL	DTL	HTL	TTL (std)	ECL	MOS	CMOS
maximum fan-out	5	8	10	10	25	20	50 and up
typical power (mW)	12	8	55	10	40	1	.01
noise immunity	nominal	good	excellent	very good	good	nominal	very good
typical propagation delay (ns)	12	30	90	10	2	100	50
maximum clock rate (MHz)	8	12	4	15	60	2	10

Complementary construction: The use of both *n*-channel and *p*-channel transistors in the same integrated circuit.

CML: Current Mode Logic. Same as ECL.

CMOS: Complementary Metallic Oxide Semiconductor logic. A logic family with high packing density and reliability. Capable of higher speeds than MOS logic.

CTL: Complementary Transistor Logic. See CMOS.

DCTL: Direct-Coupled Transistor Logic. An unsophisticated connecting of transistors without diodes or resistors. Results in low power dissipation, but has poor immunity to noise and low logic levels.

DIP: Dual In-Line Package. A packaging method for integrated circuits which brings all leads out to the sides of a rectangular case.

DTL: Diode Transistor Logic. One of the earliest logic families. Compatible with TTL. Requires two voltages.

ECL: Emitter-Coupled Logic. The fastest of all logic families. Transistors do not operate in saturated condition. Requires two voltages.

Fan-out: The number of devices which can be connected to the integrated circuit's output.

HTL: High-Threshold Logic. Similar to DTL in construction, but with higher voltages and diode ratings to produce greater noise immunity.

Hybrid construction: The interconnection of separately-produced components on a single ceramic base.

Logic Level: The voltage taken on by a circuit in its *on* and *off* states.

LSI: Large Scale Integration. More than 100 logic gates on a single chip.

Monolithic construction: Formation of all components on the same semiconductor substrate.

MOS logic: Metallic Oxide Semiconductor logic. Circuitry produced from MOSFETs. Results in simple designs requiring few resistors. High packing density.

MSI: Medium Scale Integration. Between 30 and 100 logic gates on a single chip.

NMOS: *n*-Channel Metallic Oxide Semiconductor. See also CMOS, MOS, and PMOS. More popular than PMOS.

PMOS: *p*-Channel Metallic Oxide Semiconductor. See also CMOS, MOS, and NMOS.

RCTL: Resistor-Capacitor Transistor Logic. A low-cost compromise between speed and power resulting in slow speed.

RTL: Resistor-Transistor Logic. The earliest logic family for integrated circuits. No longer in widespread use. Similar to DCTL with the addition of resistors to limit current hogging.

TTL or T^2L: Transistor-Transistor Logic. The fastest of all saturated transistor logic families. Uses transistors with multiple (parallel) emitters. Available in several variations including: standard, Schotty high-speed, high-power high-speed, Schotty low-power, and low-power low-speed.

VLSI: Very Large Scale Integration.

VTL: Variable Threshold Logic. A variation of DTL providing adjustable immunity to noise.

Questions on Vacuum Tubes

Questions on vacuum tube technology have not reappeared on the E-I-T examination. However, this subject should be studied as background for the design of transistor amplifier circuits.

Spectrum Analysis by FFT

Spectrum analysis (also known as *frequency analysis, signature analysis,* and *time-series analysis*) produces a graph of some property versus frequency. Sound power and electrical noise (i.e., harmonics) are commonly evaluated using spectrum analysis.

The *fast Fourier transform (FFT)* is an efficient method of correlating an input signal with sinusoids whose frequencies are integer multiples of the frequency corresponding to the duration of the signal. A FFT analyzer performs the Fourier series calculations internally, producing magnitude and phase angle data for each frequency range chosen.

If the signal being evaluated is random, the FFT analysis is performed numerous times in fast succession to obtain statistical stability.

The Simplex Method for Linear Programming

There have been two instances in recent E-I-T examinations where linear programming problems were presented. Although these problems were two-dimensional, questions were asked about the variables entering and leaving the basis. Consequently, the graphical solution method could not be used. It is suggested that you study the *simplex method* as a means of solving similar problems. Knowledge of simple *sensitivity analysis* is also required.

Subjects past this point were added as the ENGINEER-IN-TRAINING REVIEW MANUAL went to press. These subjects have not been added to the index.

More Inside Information on the Exam

There are three different versions of each examination distributed concurrently at an examination site. This is done to discourage cheating.

NCEE rescores the answer sheets of examinees who fail by only one or two points.

The minimum raw score needed to pass the October 1980 examination was 69 points out of 140. Although this is approximately 50% correct, scores of examinees were scaled upward. Examinees who achieved 69 points were reported as having achieved 70% on the examination.

Passing Rates for April 1981

The percentages of examinees who passed the April 1981 E-I-T examination have been released by NCEE. (Compare to the percentages reported in the *Introduction* for the October 1980 examination.)

April 1981 Passing Rates

category of examinee	per cent passing
ABET accredited, 4-year engineering degrees	76
ABET accredited, 4-year technology degrees	41
non-accredited, 4-year engineering degrees	65
non-accredited, 4-year technology degrees	27
non-graduates	34
total, all states	72

Ellipse Aspect Ratio

The *aspect ratio* of an ellipse with major length $2a$ and minor length $2b$ is calculated as $a \div b$.

Distance Between a Point and a Plane

The distance between a point (x_1, y_1, z_1) and a plane $(Ax + By + Cz + D = 0)$ is

$$d = \frac{Ax_1 + By_1 + Cz_1 + D}{\sqrt{A^2 + B^2 + C^2}}$$

Hypergeometric Distribution

Sampling from a finite population without replacement can be handled by the *hypergeometric distribution.* If a population of size M contains K items with a given characteristic (e.g., are defective), then the probability of finding x with that characteristic when n are sampled is

$$p\{x\} = \frac{\binom{K}{x}\binom{M-K}{n-x}}{\binom{M}{n}}$$

Triple Point

The *triple point* of a substance is the combination of pressure and temperature at which the solid, liquid, and gas phases can coexist. For water, the triple point is .0099°C and 4.58 mm Hg. (Also, see figure 6.6.)

MISC

Molecular and Formula Weights

The molecular and formula weights are the same for most compounds. They can differ when the exact molecular formula is not known and when a molecule is hydrated.

An ultimate analysis will not determine the molecular formula — it will determine only the relative proportions of each element. For example, an ultimate analysis of hydrogen peroxide will show the compound to have one oxygen atom for each hydrogen atom. In this case, the formula weight would be assumed to be 17 (for HO) although the actual molecular weight is 34 (for H_2O_2).

If a molecule is hydrated (e.g., $FeSO_4 \cdot 7H_2O$), the mass of the *water of hydration* (also known as the *water of crystallization*) is not included when the formula weight is determined.

Dessicants

A *dessicant* is a substance which has a high affinity for water. Dessicants are used as drying agents. Common dessicants are silica gel (produced from silicic acid), activated alumina, and calcium sulfate (anhydrous). Almost all perchlorates (ClO_4 radical) are *hygroscopic* (i.e., absorb moisture from the air). However, $Mg(ClO_4)_2$ and $Ba(ClO_4)_2$ are also powerful oxidizing agents and create an explosion hazard. Calcium chloride may also be used, although it is not a powerful dessicant.

Cast Iron

Cast iron is available in several forms. *White cast iron* has been cooled quickly from a molten state so that graphite is not produced from the cementite. *Malleable cast iron* is produced by reheating white cast iron to 1600° F, followed by slow or intermediate cooling. The most common type of cast iron is *gray cast iron,* available in both pearlitic and ferritic forms. Since gray cast iron has low strength, magnesium or cerium can be added to the molten alloy. The resulting *nodular cast iron* has superior strength and ductility.

Stainless Steel Types

The following table is self-explanatory.

Characteristics of Stainless Steel

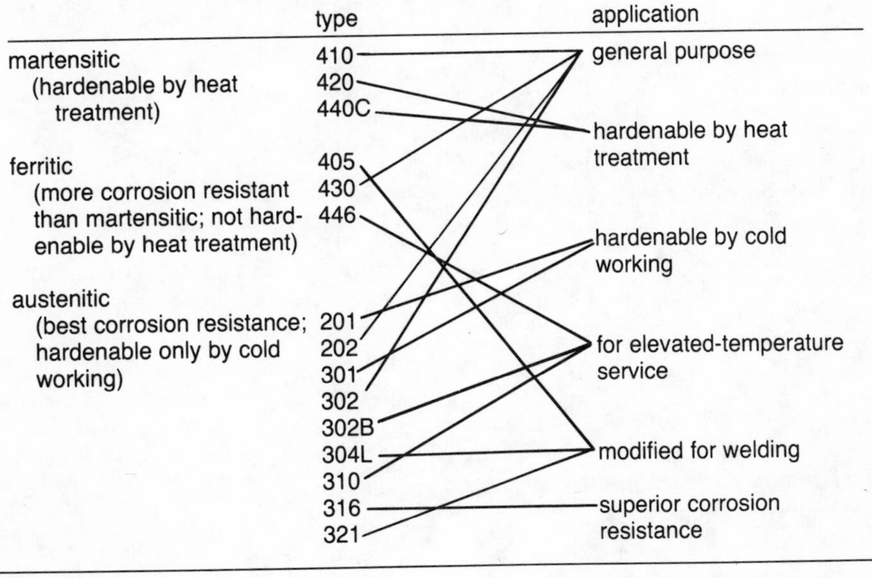

The Baudot Code

The Baudot Code has served as a means of Telex (teletypewriter) communications for several decades. Baudot is a five-bit code with a maximum of 64 characters. Because of its slow speed (50 bps) and the growing popularity of ASCII for TWX (110 bps, asynchronous) and TTY (110 to 9600 bps, synchronous), it is expected that Baudot will be used in the future only for special applications (e.g., teletypewriter communications between deaf persons.)

MISC

Update: January, 1984

There have been four E-I-T examinations since the 6th edition of the *ENGINEER-IN-TRAINING REVIEW MANUAL* was written. Almost all topics on these exams were covered in this book, and subjects that were not generally required specialized knowledge not commonly covered in "breadth" undergraduate courses.

Several afternoon electronics problems have required detailed knowledge of BJT (bipolar junction transistor) and MOSFET (metallic oxide semiconductor field effect transistor) operation. Specifically, it has been necessary to work with small signal models to calculate mid-band gain in FET, BJT, and FET-BJT amplifiers. Also, a few morning problems have been based on being able to recognize (but not evaluate) different types of circuits (e.g., rectifiers, doublers, filters, etc.)

Three-dimensional analysis and the ability to work force problems in unit-vector notation should be given additional emphasis in your preparation.

The E-I-T examination continues to be difficult, although examinees report the morning session to be much easier than the afternoon session.

This page is reserved for future use.

This page is reserved for future use.

This page is reserved for future use.

This page is reserved for future use.

Index

A

B

C

D

E

J

K

L

T

U

V

W

X

Y

Z

DON'T GAMBLE!

These books will extract Every Last Point from the examination for you

Engineering Law, Design Liability, and Professional Ethics

The most difficult problems are essay questions about management, ethics, professional responsibility, and law. Since these questions can ask for definitions of terms you're not likely to know, it is virtually impossible to fake it by rambling on. And yet, these problems are simple if you have the right resources. If you don't feel comfortable with such terms as comparative negligence, discovery proceedings, *and* strict liability in tort, *you should bring* **Engineering Law, Ethics, and Liability** *with you to the examination.*

None of this material is in your review manual. *And, nothing from your review manual has been duplicated here.*

8½" × 11", soft cover, 88 pages, $15.80 (includes postage).

(ISBN 0-932276-37-7)

Expanded Interest Tables

There's nothing worse than knowing how to solve a problem but not having the necessary data. Engineering Economics problems are like that. You might know how to do a problem, but where do you get interest factors for non-integer interest rates? **Expanded Interest Tables** *will prove indispensible for such problems. It has pages for interest rates starting at ¼% and going to 25% in ¼% increments. Factors are given for up to 100 years. There's no other book like it. So, if you want to be prepared for an Engineering Economy problem with 11.75% interest, you need* **Expanded Interest Tables.**

8½" × 11", soft cover, 106 pages, $15.80, including postage.

(ISBN 0-932276-35-0)

Phone and mail orders are accepted.

PROFESSIONAL PUBLICATIONS, INC.
Post Office Box 199
Department 77
San Carlos, CA 94070
(415) 593-9119

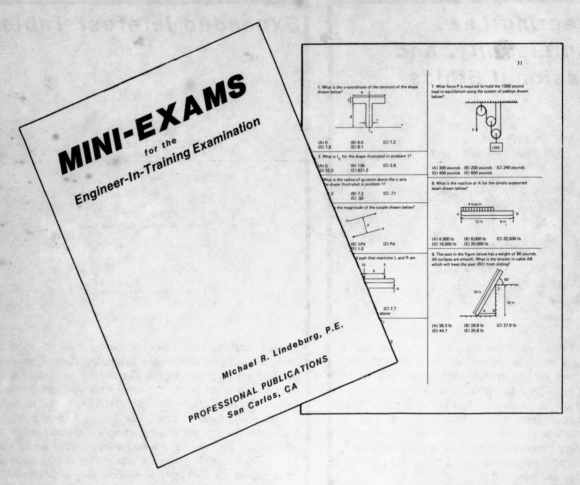